Multivariable

THOMAS'
CALCULUS

FOURTEENTH EDITION

CALCULUS

FOURTEENTH EDITION

Based on the original work by

GEORGE B. THOMAS, JR.
Massachusetts Institute of Technology

as revised by

JOEL HASS
University of California, Davis

CHRISTOPHER HEIL
Georgia Institute of Technology

MAURICE D. WEIR
Naval Postgraduate School

 Pearson

Director, Portfolio Management: Deirdre Lynch
Executive Editor: Jeff Weidenaar
Editorial Assistant: Jennifer Snyder
Content Producer: Rachel S. Reeve
Managing Producer: Scott Disanno
Producer: Stephanie Green
TestGen Content Manager: Mary Durnwald
Manager: Content Development, Math: Kristina Evans
Product Marketing Manager: Claire Kozar
Field Marketing Manager: Evan St. Cyr
Marketing Assistants: Jennifer Myers, Erin Rush
Senior Author Support/Technology Specialist: Joe Vetere
Rights and Permissions Project Manager: Gina M. Cheselka
Manufacturing Buyer: Carol Melville, LSC Communications
Program Design Lead: Barbara T. Atkinson
Associate Director of Design: Blair Brown
Text and Cover Design, Production Coordination, Composition: Cenveo® Publisher Services
Illustrations: Network Graphics, Cenveo Publisher Services

Cover Image: Te Rewa Rewa Bridge, Getty Images/Kanwal Sandhu

Library of Congress Cataloging-in-Publication Data

Names: Hass, Joel. | Heil, Christopher, 1960- | Weir, Maurice D.
Title: Thomas' calculus : multivariable : based on the original work by
    George B. Thomas, Jr., Massachusetts Institute of Technology.
Other titles: Multivariable Thomas' calculus
Description: Fourteenth edition / as revised by Joel Hass,
    University of California, Davis ; Christopher Heil,
    Georgia Institute of Technology; Maurice D. Weir,
    Naval Postgraduate School | Boston: Pearson, [2018] | Includes index.
Identifiers: LCCN 2016031131| ISBN 9780134606088 (pbk.) | ISBN 0134606086 (pbk.)
Subjects: LCSH: Calculus--Textbooks.
Classification: LCC QA303.2 .W4525 2018 | DDC 515--dc23
LC record available at https://lccn.loc.gov/2016031131

Student Edition
ISBN-13: 978-0-13-460608-8
ISBN-10: 0-13-460608-6

6 2023

Contents

Preface

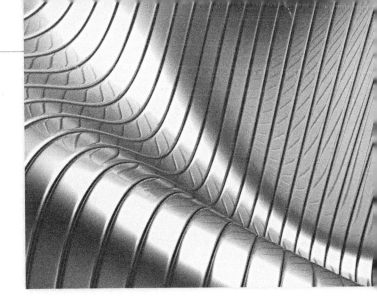

Thomas' Calculus: Early Transcendentals, Fourteenth Edition, provides a modern introduction to calculus that focuses on developing conceptual understanding of the underlying mathematical ideas. This text supports a calculus sequence typically taken by students in STEM fields over several semesters. Intuitive and precise explanations, thoughtfully chosen examples, superior figures, and time-tested exercise sets are the foundation of this text. We continue to improve this text in keeping with shifts in both the preparation and the goals of today's students, and in the applications of calculus to a changing world.

Many of today's students have been exposed to calculus in high school. For some, this translates into a successful experience with calculus in college. For others, however, the result is an overconfidence in their computational abilities coupled with underlying gaps in algebra and trigonometry mastery, as well as poor conceptual understanding. In this text, we seek to meet the needs of the increasingly varied population in the calculus sequence. We have taken care to provide enough review material (in the text and appendices), detailed solutions, and a variety of examples and exercises, to support a complete understanding of calculus for students at varying levels. Additionally, the MyMathLab course that accompanies the text provides adaptive support to meet the needs of all students. Within the text, we present the material in a way that supports the development of mathematical maturity, going beyond memorizing formulas and routine procedures, and we show students how to generalize key concepts once they are introduced. References are made throughout, tying new concepts to related ones that were studied earlier. After studying calculus from *Thomas*, students will have developed problem-solving and reasoning abilities that will serve them well in many important aspects of their lives. Mastering this beautiful and creative subject, with its many practical applications across so many fields, is its own reward. But the real gifts of studying calculus are acquiring the ability to think logically and precisely; understanding what is defined, what is assumed, and what is deduced; and learning how to generalize conceptually. We intend this book to encourage and support those goals.

New to This Edition

We welcome to this edition a new coauthor, Christopher Heil from the Georgia Institute of Technology. He has been involved in teaching calculus, linear algebra, analysis, and abstract algebra at Georgia Tech since 1993. He is an experienced author and served as a consultant on the previous edition of this text. His research is in harmonic analysis, including time-frequency analysis, wavelets, and operator theory.

This is a substantial revision. Every word, symbol, and figure was revisited to ensure clarity, consistency, and conciseness. Additionally, we made the following text-wide updates:

- Updated graphics to bring out clear visualization and mathematical correctness.

- Added examples (in response to user feedback) to overcome conceptual obstacles. See Example 3 in Section 9.1.

- Added new types of homework exercises throughout, including many with a geometric nature. The new exercises are not just more of the same, but rather give different perspectives on and approaches to each topic. We also analyzed aggregated student usage and performance data from MyMathLab for the previous edition of this text. The results of this analysis helped improve the quality and quantity of the exercises.

- Added short URLs to historical links that allow students to navigate directly to online information.

- Added new marginal notes throughout to guide the reader through the process of problem solution and to emphasize that each step in a mathematical argument is rigorously justified.

New to MyMathLab

Many improvements have been made to the overall functionality of MyMathLab (MML) since the previous edition. Beyond that, we have also increased and improved the content specific to this text.

- Instructors now have more exercises than ever to choose from in assigning homework. There are approximately 8080 assignable exercises in MML.

- The MML exercise-scoring engine has been updated to allow for more robust coverage of certain topics, including differential equations.

- A full suite of Interactive Figures have been added to support teaching and learning. The figures are designed to be used in lecture, as well as by students independently. The figures are editable using the freely available GeoGebra software. The figures were created by Marc Renault (Shippensburg University), Kevin Hopkins (Southwest Baptist University), Steve Phelps (University of Cincinnati), and Tim Brzezinski (Berlin High School, CT).

- Enhanced Sample Assignments include just-in-time prerequisite review, help keep skills fresh with distributed practice of key concepts (based on research by Jeff Hieb of University of Louisville), and provide opportunities to work exercises without learning aids (to help students develop confidence in their ability to solve problems independently).

- Additional Conceptual Questions augment text exercises to focus on deeper, theoretical understanding of the key concepts in calculus. These questions were written by faculty at Cornell University under an NSF grant. They are also assignable through Learning Catalytics.

- An Integrated Review version of the MML course contains pre-made quizzes to assess the prerequisite skills needed for each chapter, plus personalized remediation for any gaps in skills that are identified.

- Setup & Solve exercises now appear in many sections. These exercises require students to show how they set up a problem as well as the solution, better mirroring what is required of students on tests.

- Over 200 new instructional videos by Greg Wisloski and Dan Radelet (both of Indiana University of PA) augment the already robust collection within the course. These videos support the overall approach of the text—specifically, they go beyond routine procedures to show students how to generalize and connect key concepts.

Content Enhancements

Chapter 1

- Shortened 1.4 to focus on issues arising in use of mathematical software and potential pitfalls. Removed peripheral material on regression, along with associated exercises.

- Clarified explanation of definition of exponential function in 1.5.

- Replaced $\sin^{-1}$ notation for the inverse sine function with arcsin as default notation in 1.6, and similarly for other trig functions.

- Added new Exercises: **1.1:** 59–62, **1.2:** 21–22; **1.3:** 64–65, **1.6:** 61–64, 79cd; **PE:** 29–32.

Chapter 2

- Added definition of average speed in 2.1.

- Clarified definition of limits to allow for arbitrary domains. The definition of limits is now consistent with the definition in multivariable domains later in the text and with more general mathematical usage.

- Reworded limit and continuity definitions to remove implication symbols and improve comprehension.

- Added new Example 7 in 2.4 to illustrate limits of ratios of trig functions.

- Rewrote 2.5 Example 11 to solve the equation by finding a zero, consistent with previous discussion.

- Added new Exercises: **2.1:** 15–18; **2.2:** 3h–k, 4f–I; **2.4:** 19–20, 45–46; **2.5:** 31–32; **2.6:** 69–74; **PE:** 57–58; **AAE:** 35–38.

Chapter 3

- Clarified relation of slope and rate of change.

- Added new Figure 3.9 using the square root function to illustrate vertical tangent lines.

- Added figure of $x \sin (1/x)$ in 3.2 to illustrate how oscillation can lead to nonexistence of a derivative of a continuous function.

- Revised product rule to make order of factors consistent throughout text, including later dot product and cross product formulas.

- Added new Exercises: **3.2:** 36, 43–44; **3.3:** 65–66; **3.5:** 43–44, 61bc; **3.6:** 79–80, 111–113; **3.7:** 27–28; **3.8:** 97–100; **3.9:** 43–46; **3.10:** 47; **AAE:** 14–15, 26–27.

Chapter 4

- Added summary to 4.1.

- Added new Example 3 with new Figure 4.27 and Example 12 with new Figure 4.35 to give basic and advanced examples of concavity.

- Added new Exercises: **4.1:** 53–56, 67–70; **4.3:** 45–46, 67–68; **4.4:** 107–112; **4.6:** 37–42; **4.7:** 7–10; **4.8:** 115–118; **PE:** 1–16, 101–102; **AAE:** 19–20, 38–39. Moved Exercises 4.1: 53–68 to PE.

Chapter 5

- Improved discussion in 5.4 and added new Figure 5.18 to illustrate the Mean Value Theorem.

- Added new Exercises: **5.2:** 33–36; **5.4:** 71–72; **5.6:** 47–48; **PE:** 43–44, 75–76.

Chapter 6

- Clarified cylindrical shell method.

- Converted 6.5 Example 4 to metric units.

- Added introductory discussion of mass distribution along a line, with figure, in 6.6.

- Added new Exercises: **6.1:** 15–16; **6.2:** 49–50; **6.3:** 13–14; **6.5:** 1–2; **6.6:** 1–6, 21–22; **PE:** 17–18, 23–24, 37–38

Chapter 7

- Clarified discussion of separable differential equations in 7.2.

- Added new Exercises: **7.1:** 61–62, 73; **PE:** 41–42.

Chapter 8

- Updated 8.2 Integration by Parts discussion to emphasize $u(x)v'(x)\,dx$ form rather than $u\,dv$. Rewrote Examples 1–3 accordingly.

- Removed discussion of tabular integration and associated exercises.

- Updated discussion in 8.5 on how to find constants in the method of partial fractions.

- Updated notation in 8.8 to align with standard usage in statistics.

- Added new Exercises: **8.1:** 41–44; **8.2:** 53–56, 72–73; **8.3:** 75–76; **8.4:** 49–52; **8.5:** 51–66, 73–74; **8.8:** 35–38, 77–78; **PE:** 69–88.

Chapter 9

- Added new Example 3 with Figure 9.3 to illustrate how to construct a slope field.

- Added new Exercises: **9.1:** 11–14; **PE:** 17–22, 43–44.

Chapter 10

- Clarified the differences between a sequence and a series.

- Added new Figure 10.9 to illustrate sum of a series as area of a histogram.

- Added to 10.3 a discussion on the importance of bounding errors in approximations.

- Added new Figure 10.13 illustrating how to use integrals to bound remainder terms of partial sums.

- Rewrote Theorem 10 in 10.4 to bring out similarity to the integral comparison test.

- Added new Figure 10.16 to illustrate the differing behaviors of the harmonic and alternating harmonic series.

- Renamed the nth-Term Test the "nth-Term Test for Divergence" to emphasize that it says nothing about convergence.

- Added new Figure 10.19 to illustrate polynomials converging to $\ln (1 + x)$, which illustrates convergence on the half-open interval $(-1, 1]$.

- Used red dots and intervals to indicate intervals and points where divergence occurs, and blue to indicate convergence, throughout Chapter 10.

- Added new Figure 10.21 to show the six different possibilities for an interval of convergence.

- Added new Exercises: **10.1:** 27–30, 72–77; **10.2:** 19–22, 73–76, 105; **10.3:** 11–12, 39–42; **10.4:** 55–56; **10.5:** 45–46, 65–66; **10.6:** 57–82; **10.7:** 61–65; **10.8:** 23–24, 39–40; **10.9:** 11–12, 37–38; **PE:** 41–44, 97–102.

Chapter 11

- Added new Example 1 and Figure 11.2 in 11.1 to give a straightforward first example of a parametrized curve.

- Updated area formulas for polar coordinates to include conditions for positive r and nonoverlapping θ.

- Added new Example 3 and Figure 11.37 in 11.4 to illustrate intersections of polar curves.

- Added new Exercises: **11.1:** 19–28; **11.2:** 49–50; **11.4:** 21–24.

Chapter 12

- Added new Figure 12.13(b) to show the effect of scaling a vector.

- Added new Example 7 and Figure 12.26 in 12.3 to illustrate projection of a vector.

- Added discussion on general quadric surfaces in 12.6, with new Example 4 and new Figure 12.48 illustrating the description of an ellipsoid not centered at the origin via completing the square.

- Added new Exercises: **12.1:** 31–34, 59–60, 73–76; **12.2:** 43–44; **12.3:** 17–18; **12.4:** 51–57; **12.5:** 49–52.

Chapter 13

- Added sidebars on how to pronounce Greek letters such as kappa, tau, etc.

- Added new Exercises: **13.1:** 1–4, 27–36; **13.2:** 15–16, 19–20; **13.4:** 27–28; **13.6:** 1–2.

Chapter 14

- Elaborated on discussion of open and closed regions in 14.1.

- Standardized notation for evaluating partial derivatives, gradients, and directional derivatives at a point, throughout the chapter.

- Renamed "branch diagrams" as "dependency diagrams," which clarifies that they capture dependence of variables.

- Added new Exercises: **14.2:** 51–54; **14.3:** 51–54, 59–60, 71–74, 103–104; **14.4:** 20–30, 43–46, 57–58; **14.5:** 41–44; **14.6:** 9–10, 61; **14.7:** 61–62.

Chapter 15

- Added new Figure 15.21b to illustrate setting up limits of a double integral.

- Added new 15.5 Example 1, modified Examples 2 and 3, and added new Figures 15.31, 15.32, and 15.33 to give basic examples of setting up limits of integration for a triple integral.

- Added new material on joint probability distributions as an application of multivariable integration.

- Added new Examples 5, 6 and 7 to Section 15.6.

- Added new Exercises: **15.1:** 15–16, 27–28; **15.6:** 39–44; **15.7:** 1–22.

Chapter 16

- Added new Figure 16.4 to illustrate a line integral of a function.

- Added new Figure 16.17 to illustrate a gradient field.

- Added new Figure 16.19 to illustrate a line integral of a vector field.

- Clarified notation for line integrals in 16.2.

- Added discussion of the sign of potential energy in 16.3.

- Rewrote solution of Example 3 in 16.4 to clarify connection to Green's Theorem.

- Updated discussion of surface orientation in 16.6 along with Figure 16.52.

- Added new Exercises: **16.2:** 37–38, 41–46; **16.4:** 1–6; **16.6:** 49–50; **16.7:** 1–6; **16.8:** 1–4.

Appendices: Rewrote Appendix A7 on complex numbers.

Continuing Features

Rigor The level of rigor is consistent with that of earlier editions. We continue to distinguish between formal and informal discussions and to point out their differences. Starting with a more intuitive, less formal approach helps students understand a new or difficult concept so they can then appreciate its full mathematical precision and outcomes. We pay attention to defining ideas carefully and to proving theorems appropriate for calculus students, while mentioning deeper or subtler issues they would study in a more advanced course. Our organization and distinctions between informal and formal discussions give the instructor a degree of flexibility in the amount and depth of coverage of the various topics. For example, while we do not prove the Intermediate Value Theorem or the Extreme Value Theorem for continuous functions on a closed finite interval, we do state these theorems precisely, illustrate their meanings in numerous examples, and use them to prove other important results. Furthermore, for those instructors who desire greater depth of coverage, in Appendix 6 we discuss the reliance of these theorems on the completeness of the real numbers.

Writing Exercises Writing exercises placed throughout the text ask students to explore and explain a variety of calculus concepts and applications. In addition, the end of each chapter contains a list of questions for students to review and summarize what they have learned. Many of these exercises make good writing assignments.

End-of-Chapter Reviews and Projects In addition to problems appearing after each section, each chapter culminates with review questions, practice exercises covering the entire chapter, and a series of Additional and Advanced Exercises with more challenging or synthesizing problems. Most chapters also include descriptions of several **Technology Application Projects** that can be worked by individual students or groups of students over a longer period of time. These projects require the use of *Mathematica* or *Maple*, along with pre-made files that are available for download within MyMathLab.

Writing and Applications This text continues to be easy to read, conversational, and mathematically rich. Each new topic is motivated by clear, easy-to-understand examples and is then reinforced by its application to real-world problems of immediate interest to students. A hallmark of this book has been the application of calculus to science and engineering. These applied problems have been updated, improved, and extended continually over the last several editions.

Technology In a course using the text, technology can be incorporated according to the taste of the instructor. Each section contains exercises requiring the use of technology; these are marked with a $\boxed{T}$ if suitable for calculator or computer use, or they are labeled **Computer Explorations** if a computer algebra system (CAS, such as *Maple* or *Mathematica*) is required.

Additional Resources

MyMathLab® Online Course (access code required)

Built around Pearson's best-selling content, MyMathLab is an online homework, tutorial, and assessment program designed to work with this text to engage students and improve results. MyMathLab can be successfully implemented in any classroom environment—lab-based, hybrid, fully online, or traditional.

Used by more than 37 million students worldwide, MyMathLab delivers consistent, measurable gains in student learning outcomes, retention, and subsequent course success. Visit **www.mymathlab.com/results** to learn more.

Preparedness   One of the biggest challenges in calculus courses is making sure students are adequately prepared with the prerequisite skills needed to successfully complete their course work. MyMathLab supports students with just-in-time remediation and key-concept review.

- **Integrated Review Course** can be used for just-in-time prerequisite review. These courses contain pre-made quizzes to assess the prerequisite skills needed for each chapter, plus personalized remediation for any gaps in skills that are identified.

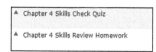

▲ Chapter 4 Skills Check Quiz

▲ Chapter 4 Skills Review Homework

Motivation   Students are motivated to succeed when they're engaged in the learning experience and understand the relevance and power of mathematics. MyMathLab's online homework offers students immediate feedback and tutorial assistance that motivates them to do more, which means they retain more knowledge and improve their test scores.

- **Exercises with immediate feedback**—the over 8080 assignable exercises for this text regenerate algorithmically to give students unlimited opportunity for practice and mastery. MyMathLab provides helpful feedback when students enter incorrect answers and includes optional learning aids such as Help Me Solve This, View an Example, videos, and an eText.

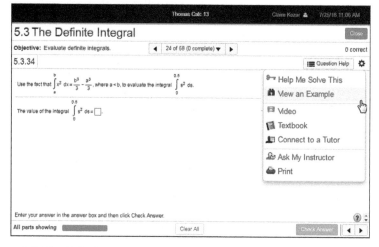

- **Setup and Solve Exercises** ask students to first describe how they will set up and approach the problem. This reinforces students' conceptual understanding of the process they are applying and promotes long-term retention of the skill.

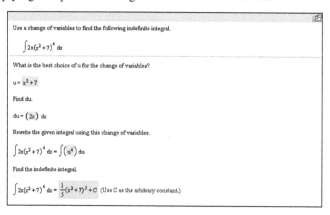

- **Additional Conceptual Questions** focus on deeper, theoretical understanding of the key concepts in calculus. These questions were written by faculty at Cornell University under an NSF grant and are also assignable through Learning Catalytics.

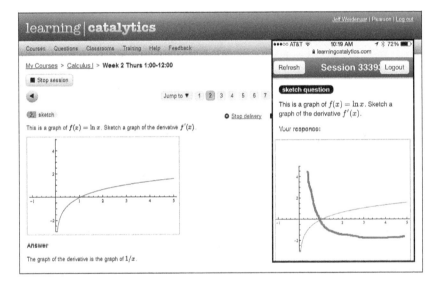

- **Learning Catalytics™** is a student response tool that uses students' smartphones, tablets, or laptops to engage them in more interactive tasks and thinking during lecture. Learning Catalytics fosters student engagement and peer-to-peer learning with real-time analytics. Learning Catalytics is available to all MyMathLab users.

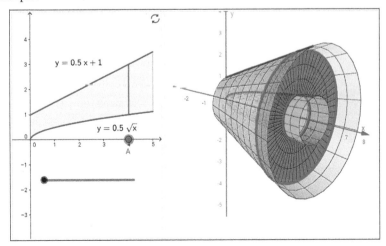

Learning and Teaching Tools

- **Interactive Figures** illustrate key concepts and allow manipulation for use as teaching and learning tools. We also include videos that use the Interactive Figures to explain key concepts.

- **Instructional videos**—hundreds of videos are available as learning aids within exercises and for self-study. The Guide to Video-Based Assignments makes it easy to assign videos for homework by showing which MyMathLab exercises correspond to each video.

- **The complete eText** is available to students through their MyMathLab courses for the lifetime of the edition, giving students unlimited access to the eText within any course using that edition of the text.

- **Enhanced Sample Assignments** These assignments include just-in-time prerequisite review, help keep skills fresh with distributed practice of key concepts, and provide opportunities to work exercises without learning aids so students can check their understanding.

- **PowerPoint Presentations** that cover each section of the book are available for download.

- **Mathematica manual and projects, Maple manual and projects, TI Graphing Calculator manual**—These manuals cover *Maple* 17, *Mathematica 8,* and the TI-84 Plus and TI-89, respectively. Each provides detailed guidance for integrating the software package or graphing calculator throughout the course, including syntax and commands.

- **Accessibility** and achievement go hand in hand. MyMathLab is compatible with the JAWS screen reader, and it enables students to read and interact with multiple-choice and free-response problem types via keyboard controls and math notation input. MyMathLab also works with screen enlargers, including ZoomText, MAGic, and SuperNova. And, all MyMathLab videos have closed-captioning. More information is available at **http://mymathlab.com/accessibility**.

- **A comprehensive gradebook** with enhanced reporting functionality allows you to efficiently manage your course.

  - **The Reporting Dashboard** offers insight as you view, analyze, and report learning outcomes. Student performance data is presented at the class, section, and program levels in an accessible, visual manner so you'll have the information you need to keep your students on track.

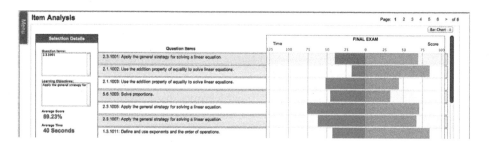

  - **Item Analysis** tracks class-wide understanding of particular exercises so you can refine your class lectures or adjust the course/department syllabus. Just-in-time teaching has never been easier!

MyMathLab comes from an experienced partner with educational expertise and an eye on the future. Whether you are just getting started with MyMathLab, or have a question along the way, we're here to help you learn about our technologies and how to incorporate them into your course. To learn more about how MyMathLab helps students succeed, visit **www.mymathlab.com** or contact your Pearson rep.

Instructor's Solutions Manual (downloadable)
ISBN: 0-13-443932-5 | 978-0-13-443932-7

The Instructor's Solutions Manual contains complete worked-out solutions to all the exercises in *Thomas' Calculus: Early Transcendentals*. It can be downloaded from within MyMathLab or the Pearson Instructor Resource Center, **www.pearsonhighered.com/irc**.

Student's Solutions Manual
Single Variable Calculus (Chapters 1–11), ISBN: 0-13-443933-3 | 978-0-13-443933-4
Multivariable Calculus (Chapters 10–16), ISBN: 0-13-443916-3 | 978-0-13-443916-7

The Student's Solutions Manual contains worked-out solutions to all the odd-numbered exercises in *Thomas' Calculus: Early Transcendentals*. These manuals are available in print and can be downloaded from within MyMathLab.

Just-In-Time Algebra and Trigonometry for Early Transcendentals Calculus, Fourth Edition
ISBN 0-321-67103-1 | 978-0-321-67103-5

Sharp algebra and trigonometry skills are critical to mastering calculus, and *Just-in-Time Algebra and Trigonometry for Early Transcendentals Calculus* by Guntram Mueller and Ronald I. Brent is designed to bolster these skills while students study calculus. As students make their way through calculus, this brief supplementary text is with them every step of the way, showing them the necessary algebra or trigonometry topics and pointing out potential problem spots. The easy-to-use table of contents has topics arranged in the order in which students will need them as they study calculus. This supplement is available in printed form only (note that MyMathLab contains a separate diagnostic and remediation system for gaps in algebra and trigonometry skills).

Technology Manuals and Projects (downloadable)
Maple Manual and Projects by Marie Vanisko, Carroll College
Mathematica Manual and Projects by Marie Vanisko, Carroll College
TI Graphing Calculator Manual by Elaine McDonald-Newman, Sonoma State University

These manuals and projects cover Maple 17, Mathematica 9, and the TI-84 Plus and TI-89. Each manual provides detailed guidance for integrating a specific software package or graphing calculator throughout the course, including syntax and commands. The projects include instructions and ready-made application files for Maple and Mathematica. These materials are available to download within MyMathLab.

TestGen®
ISBN: 0-13-443936-8 | 978-0-13-443936-5

TestGen® (**www.pearsoned.com/testgen**) enables instructors to build, edit, print, and administer tests using a computerized bank of questions developed to cover all the objectives of the text. TestGen is algorithmically based, allowing instructors to create multiple but equivalent versions of the same question or test with the click of a button. Instructors can also modify test bank questions or add new questions. The software and test bank are available for download from Pearson Education's online catalog, **www.pearsonhighered.com**.

PowerPoint® Lecture Slides
ISBN: 0-13-443943-0 | 978-0-13-443943-3

These classroom presentation slides were created for the *Thomas' Calculus* series. Key graphics from the book are included to help bring the concepts alive in the classroom. These files are available to qualified instructors through the Pearson Instructor Resource Center, **www.pearsonhighered.com/irc**, and within MyMathLab.

Acknowledgments

We are grateful to Duane Kouba, who created many of the new exercises. We would also like to express our thanks to the people who made many valuable contributions to this edition as it developed through its various stages:

Accuracy Checkers

Thomas Wegleitner
Jennifer Blue
Lisa Collette

Reviewers for the Fourteenth Edition

Alessandro Arsie, *University of Toledo*
Doug Baldwin, *SUNY Geneseo*
Steven Heilman, *UCLA*
David Horntrop, *New Jersey Institute of Technology*
Eric B. Kahn, *Bloomsburg University*
Colleen Kirk, *California Polytechnic State University*
Mark McConnell, *Princeton University*

Niels Martin Møller, *Princeton University*
James G. O'Brien, *Wentworth Institute of Technology*
Alan Saleski, *Loyola University Chicago*
Alan Von Hermann, *Santa Clara University*
Don Gayan Wilathgamuwa, *Montana State University*
James Wilson, *Iowa State University*

The following faculty members provided direction on the development of the MyMathLab course for this edition.

Charles Obare, *Texas State Technical College, Harlingen*
Elmira Yakutova-Lorentz, *Eastern Florida State College*
C. Sohn, *SUNY Geneseo*
Ksenia Owens, *Napa Valley College*
Ruth Mortha, *Malcolm X College*
George Reuter, *SUNY Geneseo*
Daniel E. Osborne, *Florida A&M University*
Luis Rodriguez, *Miami Dade College*
Abbas Meigooni, *Lincoln Land Community College*
Nader Yassin, *Del Mar College*
Arthur J. Rosenthal, *Salem State University*
Valerie Bouagnon, *DePaul University*
Brooke P. Quinlan, *Hillsborough Community College*
Shuvra Gupta, *Iowa State University*
Alexander Casti, *Farleigh Dickinson University*
Sharda K. Gudehithlu, *Wilbur Wright College*
Deanna Robinson, *McLennan Community College*

Kai Chuang, *Central Arizona College*
Vandana Srivastava, *Pitt Community College*
Brian Albright, *Concordia University*
Brian Hayes, *Triton College*
Gabriel Cuarenta, *Merced College*
John Beyers, *University of Maryland University College*
Daniel Pellegrini, *Triton College*
Debra Johnsen, *Orangeburg Calhoun Technical College*
Olga Tsukernik, *Rochester Institute of Technology*
Jorge Sarmiento, *County College of Morris*
Val Mohanakumar, *Hillsborough Community College*
MK Panahi, *El Centro College*
Sabrina Ripp, *Tulsa Community College*
Mona Panchal, *East Los Angeles College*
Gail Illich, *McLennan Community College*
Mark Farag, *Farleigh Dickinson University*
Selena Mohan, *Cumberland County College*

Dedication

We regret that prior to the writing of this edition our coauthor Maurice Weir passed away. Maury was dedicated to achieving the highest possible standards in the presentation of mathematics. He insisted on clarity, rigor, and readability. Maury was a role model to his students, his colleagues, and his coauthors. He was very proud of his daughters, Maia Coyle and Renee Waina, and of his grandsons, Matthew Ryan and Andrew Dean Waina. He will be greatly missed.

10

Infinite Sequences and Series

OVERVIEW In this chapter we introduce the topic of *infinite series*. Such series give us precise ways to express many numbers and functions, both familiar and new, as arithmetic sums with infinitely many terms. For example, we will learn that

$$\frac{\pi}{4} = 1 - \frac{1}{3} + \frac{1}{5} - \frac{1}{7} + \frac{1}{9} - \cdots$$

and

$$\cos x = 1 - \frac{x^2}{2} + \frac{x^4}{24} - \frac{x^6}{720} + \frac{x^8}{40,320} - \cdots.$$

We need to develop a method to make sense of such expressions. Everyone knows how to add two numbers together, or even several. But how do you add together infinitely many numbers? Or, when adding together functions, how do you add infinitely many powers of x? In this chapter we answer these questions, which are part of the theory of infinite sequences and series. As with the differential and integral calculus, limits play a major role in the development of infinite series.

One common and important application of series occurs when making computations with complicated functions. A hard-to-compute function is replaced by an expression that looks like an "infinite degree polynomial," an infinite series in powers of x, as we see with the cosine function given above. Using the first few terms of this infinite series can allow for highly accurate approximations of functions by polynomials, enabling us to work with more general functions than those we encountered before. These new functions are commonly obtained as solutions to differential equations arising in important applications of mathematics to science and engineering.

The terms "sequence" and "series" are sometimes used interchangeably in spoken language. In mathematics, however, each has a distinct meaning. A sequence is a type of infinite list, whereas a series is an infinite sum. To understand the infinite sums described by series, we are led to first study infinite sequences.

10.1 Sequences

HISTORICAL ESSAY
Sequences and Series
bit.ly/2NSNyUt

Sequences are fundamental to the study of infinite series and to many aspects of mathematics. We saw one example of a sequence when we studied Newton's Method in Section 4.7. Newton's Method produces a sequence of approximations x_n that become closer and closer to the root of a differentiable function. Now we will explore general sequences of numbers and the conditions under which they converge to a finite number.

Representing Sequences

A sequence is a list of numbers

$$a_1, a_2, a_3, \ldots, a_n, \ldots$$

in a given order. Each of a_1, a_2, a_3 and so on represents a number. These are the **terms** of the sequence. For example, the sequence

$$2, 4, 6, 8, 10, 12, \ldots, 2n, \ldots$$

has first term $a_1 = 2$, second term $a_2 = 4$, and nth term $a_n = 2n$. The integer n is called the **index** of a_n, and indicates where a_n occurs in the list. Order is important. The sequence $2, 4, 6, 8 \ldots$ is not the same as the sequence $4, 2, 6, 8 \ldots$.
We can think of the sequence

$$a_1, a_2, a_3, \ldots, a_n, \ldots$$

as a function that sends 1 to a_1, 2 to a_2, 3 to a_3, and in general sends the positive integer n to the nth term a_n. More precisely, an **infinite sequence** of numbers is a function whose domain is the set of positive integers. For example, the function associated with the sequence

$$2, 4, 6, 8, 10, 12, \ldots, 2n, \ldots$$

sends 1 to $a_1 = 2$, 2 to $a_2 = 4$, and so on. The general behavior of this sequence is described by the formula $a_n = 2n$.
We can change the index to start at any given number n. For example, the sequence

$$12, 14, 16, 18, 20, 22 \ldots$$

is described by the formula $a_n = 10 + 2n$, if we start with $n = 1$. It can also be described by the simpler formula $b_n = 2n$, where the index n starts at 6 and increases. To allow such simpler formulas, we let the first index of the sequence be any appropriate integer. In the sequence above, $\{a_n\}$ starts with a_1 while $\{b_n\}$ starts with b_6.
Sequences can be described by writing rules that specify their terms, such as

$$a_n = \sqrt{n}, \qquad b_n = (-1)^{n+1}\frac{1}{n}, \qquad c_n = \frac{n-1}{n}, \qquad d_n = (-1)^{n+1},$$

or by listing terms:

$$\{a_n\} = \left\{ \sqrt{1}, \sqrt{2}, \sqrt{3}, \ldots, \sqrt{n}, \ldots \right\}$$

$$\{b_n\} = \left\{ 1, -\frac{1}{2}, \frac{1}{3}, -\frac{1}{4}, \ldots, (-1)^{n+1}\frac{1}{n}, \ldots \right\}$$

$$\{c_n\} = \left\{ 0, \frac{1}{2}, \frac{2}{3}, \frac{3}{4}, \frac{4}{5}, \ldots, \frac{n-1}{n}, \ldots \right\}$$

$$\{d_n\} = \left\{ 1, -1, 1, -1, 1, -1, \ldots, (-1)^{n+1}, \ldots \right\}.$$

We also sometimes write a sequence using its rule, as with

$$\{a_n\} = \left\{ \sqrt{n} \right\}_{n=1}^{\infty}$$

and

$$\{b_n\} = \left\{ (-1)^{n+1}\frac{1}{n} \right\}_{n=1}^{\infty}.$$

Figure 10.1 shows two ways to represent sequences graphically. The first marks the first few points from $a_1, a_2, a_3, \ldots, a_n, \ldots$ on the real axis. The second method shows the graph of the function defining the sequence. The function is defined only on integer inputs, and the graph consists of some points in the xy-plane located at $(1, a_1)$, $(2, a_2), \ldots, (n, a_n), \ldots$.

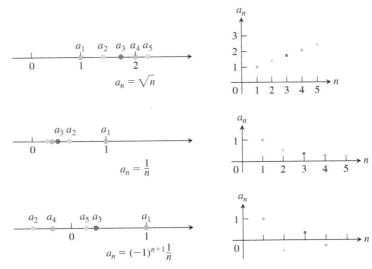

FIGURE 10.1 Sequences can be represented as points on the real line or as points in the plane where the horizontal axis n is the index number of the term and the vertical axis a_n is its value.

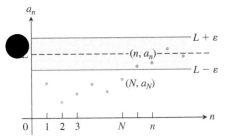

FIGURE 10.2 In the representation of a sequence as points in the plane, $a_n \to L$ if $y = L$ is a horizontal asymptote of the sequence of points $\{(n, a_n)\}$. In this figure, all the a_n's after a_N lie within ε of L.

Convergence and Divergence

Sometimes the numbers in a sequence approach a single value as the index n increases. This happens in the sequence

$$\left\{ 1, \frac{1}{2}, \frac{1}{3}, \frac{1}{4}, \ldots, \frac{1}{n}, \ldots \right\}$$

whose terms approach 0 as n gets large, and in the sequence

$$\left\{ 0, \frac{1}{2}, \frac{2}{3}, \frac{3}{4}, \frac{4}{5}, \ldots, 1 - \frac{1}{n}, \ldots \right\}$$

whose terms approach 1. On the other hand, sequences like

$$\left\{ \sqrt{1}, \sqrt{2}, \sqrt{3}, \ldots, \sqrt{n}, \ldots \right\}$$

have terms that get larger than any number as n increases, and sequences like

$$\left\{ 1, -1, 1, -1, 1, -1, \ldots, (-1)^{n+1}, \ldots \right\}$$

bounce back and forth between 1 and -1, never converging to a single value. The following definition captures the meaning of having a sequence converge to a limiting value. It says that if we go far enough out in the sequence, by taking the index n to be larger than some value N, the difference between a_n and the limit of the sequence becomes less than any preselected number $\varepsilon > 0$.

> **DEFINITIONS** The sequence $\{a_n\}$ **converges** to the number L if for every positive number ε there corresponds an integer N such that
> $$|a_n - L| < \varepsilon \qquad \text{whenever} \qquad n > N.$$
> If no such number L exists, we say that $\{a_n\}$ **diverges**.
>   If $\{a_n\}$ converges to L, we write $\lim_{n\to\infty} a_n = L$, or simply $a_n \to L$, and call L the **limit** of the sequence (Figure 10.2).

The definition is very similar to the definition of the limit of a function $f(x)$ as x tends to ∞ ($\lim_{x\to\infty} f(x)$ in Section 2.6). We will exploit this connection to calculate limits of sequences.

EXAMPLE 1   Show that

(a) $\lim\limits_{n\to\infty}\dfrac{1}{n}=0$ **(b)** $\lim\limits_{n\to\infty}k=k$ (any constant k)

Solution

(a) Let $\varepsilon>0$ be given. We must show that there exists an integer N such that

$$\left|\frac{1}{n}-0\right|<\varepsilon\qquad\text{whenever}\qquad n>N.$$

The inequality $|1/n-0|<\varepsilon$ will hold if $1/n<\varepsilon$ or $n>1/\varepsilon$. If N is any integer greater than $1/\varepsilon$, the inequality will hold for all $n>N$. This proves that $\lim_{n\to\infty}1/n=0$.

(b) Let $\varepsilon>0$ be given. We must show that there exists an integer N such that

$$|k-k|<\varepsilon\qquad\text{whenever}\qquad n>N.$$

Since $k-k=0$, we can use any positive integer for N and the inequality $|k-k|<\varepsilon$ will hold. This proves that $\lim_{n\to\infty}k=k$ for any constant k. ∎

EXAMPLE 2   Show that the sequence $\{1,-1,1,-1,1,-1,\ldots,(-1)^{n+1},\ldots\}$ diverges.

Solution   Suppose the sequence converges to some number L. Then the numbers in the sequence eventually get arbitrarily close to the limit L. This can't happen if they keep oscillating between 1 and -1. We can see this by choosing $\varepsilon=1/2$ in the definition of the limit. Then all terms a_n of the sequence with index n larger than some N must lie within $\varepsilon=1/2$ of L. Since the number 1 appears repeatedly as every other term of the sequence, we must have that the number 1 lies within the distance $\varepsilon=1/2$ of L. It follows that $|L-1|<1/2$, or equivalently, $1/2<L<3/2$. Likewise, the number -1 appears repeatedly in the sequence with arbitrarily high index. So we must also have that $|L-(-1)|<1/2$, or equivalently, $-3/2<L<-1/2$. But the number L cannot lie in both of the intervals $(1/2,3/2)$ and $(-3/2,-1/2)$ because they have no overlap. Therefore, no such limit L exists and so the sequence diverges.

Note that the same argument works for any positive number ε smaller than 1, not just $1/2$. ∎

The sequence $\{\sqrt{n}\}$ also diverges, but for a different reason. As n increases, its terms become larger than any fixed number. We describe the behavior of this sequence by writing

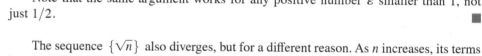

In writing infinity as the limit of a sequence, we are not saying that the differences between the terms a_n and ∞ become small as n increases. Nor are we asserting that there is some number infinity that the sequence approaches. We are merely using a notation that captures the idea that a_n eventually gets and stays larger than any fixed number as n gets large (see Figure 10.3a). The terms of a sequence might also decrease to negative infinity, as in Figure 10.3b.

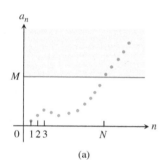

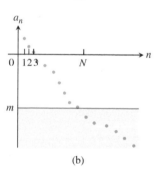

FIGURE 10.3 (a) The sequence diverges to ∞ because no matter what number M is chosen, the terms of the sequence after some index N all lie in the yellow band above M. (b) The sequence diverges to $-\infty$ because all terms after some index N lie below any chosen number m.

> DEFINITION   The sequence $\{a_n\}$ **diverges to infinity** if for every number M there is an integer N such that for all n larger than N, $a_n>M$. If this condition holds we write
>
> $$\lim_{n\to\infty}a_n=\infty\qquad\text{or}\qquad a_n\to\infty.$$
>
> Similarly, if for every number m there is an integer N such that for all $n>N$ we have $a_n<m$, then we say $\{a_n\}$ **diverges to negative infinity** and write
>
> $$\lim_{n\to\infty}a_n=-\infty\qquad\text{or}\qquad a_n\to-\infty.$$

A sequence may diverge without diverging to infinity or negative infinity, as we saw in Example 2. The sequences $\{1, -2, 3, -4, 5, -6, 7, -8, \dots\}$ and $\{1, 0, 2, 0, 3, 0, \dots\}$ are also examples of such divergence.

The convergence or divergence of a sequence is not affected by the values of any number of its initial terms (whether we omit or change the first 10, 1000, or even the first million terms does not matter). From Figure 10.2, we can see that only the part of the sequence that remains after discarding some initial number of terms determines whether the sequence has a limit and the value of that limit when it does exist.

Calculating Limits of Sequences

Since sequences are functions with domain restricted to the positive integers, it is not surprising that the theorems on limits of functions given in Chapter 2 have versions for sequences.

> **THEOREM 1**   Let $\{a_n\}$ and $\{b_n\}$ be sequences of real numbers, and let A and B be real numbers. The following rules hold if $\lim_{n\to\infty} a_n = A$ and $\lim_{n\to\infty} b_n = B$.
>
> 1. *Sum Rule:* $\qquad\qquad\qquad$ $\lim_{n\to\infty}(a_n + b_n) = A + B$
> 2. *Difference Rule:* $\qquad\quad$ $\lim_{n\to\infty}(a_n - b_n) = A - B$
> 3. *Constant Multiple Rule:* $\quad$ $\lim_{n\to\infty}(k \cdot b_n) = k \cdot B$ (any number k)
> 4. *Product Rule:* $\qquad\qquad$ $\lim_{n\to\infty}(a_n \cdot b_n) = A \cdot B$
> 5. *Quotient Rule:* $\qquad\qquad$ $\lim_{n\to\infty}\dfrac{a_n}{b_n} = \dfrac{A}{B}$ if $B \neq 0$

The proof is similar to that of Theorem 1 of Section 2.2 and is omitted.

EXAMPLE 3   By combining Theorem 1 with the limits of Example 1, we have:

(a) $\lim\limits_{n\to\infty}\left(-\dfrac{1}{n}\right) = -1 \cdot \lim\limits_{n\to\infty}\dfrac{1}{n} = -1 \cdot 0 = 0$ $\qquad$ Constant Multiple Rule and Example 1a

(b) $\lim\limits_{n\to\infty}\left(\dfrac{n-1}{n}\right) = \lim\limits_{n\to\infty}\left(1 - \dfrac{1}{n}\right) = \lim\limits_{n\to\infty}1 - \lim\limits_{n\to\infty}\dfrac{1}{n} = 1 - 0 = 1$ $\quad$ Difference Rule and Example 1a

(c) $\lim\limits_{n\to\infty}\dfrac{5}{n^2} = 5 \cdot \lim\limits_{n\to\infty}\dfrac{1}{n} \cdot \lim\limits_{n\to\infty}\dfrac{1}{n} = 5 \cdot 0 \cdot 0 = 0$ $\qquad$ Product Rule

(d) $\lim\limits_{n\to\infty}\dfrac{4 - 7n^6}{n^6 + 3} = \lim\limits_{n\to\infty}\dfrac{(4/n^6) - 7}{1 + (3/n^6)} = \dfrac{0 - 7}{1 + 0} = -7.$ $\qquad$ Divide numerator and denominator by n^6 and use the Sum and Quotient Rules.

Be cautious in applying Theorem 1. It does not say, for example, that each of the sequences $\{a_n\}$ and $\{b_n\}$ have limits if their sum $\{a_n + b_n\}$ has a limit. For instance, $\{a_n\} = \{1, 2, 3, \dots\}$ and $\{b_n\} = \{-1, -2, -3, \dots\}$ both diverge, but their sum $\{a_n + b_n\} = \{0, 0, 0, \dots\}$ clearly converges to 0.

One consequence of Theorem 1 is that every nonzero multiple of a divergent sequence $\{a_n\}$ diverges. Suppose, to the contrary, that $\{ca_n\}$ converges for some number $c \neq 0$. Then, by taking $k = 1/c$ in the Constant Multiple Rule in Theorem 1, we see that the sequence

$$\left\{\frac{1}{c} \cdot ca_n\right\} = \{a_n\}$$

converges. Thus, $\{ca_n\}$ cannot converge unless $\{a_n\}$ also converges. If $\{a_n\}$ does not converge, then $\{ca_n\}$ does not converge.

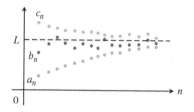

FIGURE 10.4 The terms of sequence $\{b_n\}$ are sandwiched between those of $\{a_n\}$ and $\{c_n\}$, forcing them to the same common limit L.

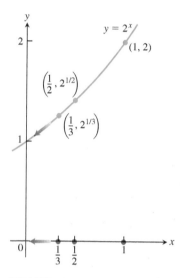

FIGURE 10.5 As $n \to \infty$, $1/n \to 0$ and $2^{1/n} \to 2^0$ (Example 6). The terms of $\{1/n\}$ are shown on the x-axis; the terms of $\{2^{1/n}\}$ are shown as the y-values on the graph of $f(x) = 2^x$.

The next theorem is the sequence version of the Sandwich Theorem in Section 2.2. You are asked to prove the theorem in Exercise 119. (See Figure 10.4.)

THEOREM 2—The Sandwich Theorem for Sequences
Let $\{a_n\}$, $\{b_n\}$, and $\{c_n\}$ be sequences of real numbers. If $a_n \le b_n \le c_n$ holds for all n beyond some index N, and if $\lim_{n\to\infty} a_n = \lim_{n\to\infty} c_n = L$, then $\lim_{n\to\infty} b_n = L$ also.

An immediate consequence of Theorem 2 is that, if $|b_n| \le c_n$ and $c_n \to 0$, then $b_n \to 0$ because $-c_n \le b_n \le c_n$. We use this fact in the next example.

EXAMPLE 4   Since $1/n \to 0$, we know that

(a) $\dfrac{\cos n}{n} \to 0$ because $-\dfrac{1}{n} \le \dfrac{\cos n}{n} \le \dfrac{1}{n};$

(b) $\dfrac{1}{2^n} \to 0$ because $0 \le \dfrac{1}{2^n} \le \dfrac{1}{n};$

(c) $(-1)^n \dfrac{1}{n} \to 0$ because $-\dfrac{1}{n} \le (-1)^n \dfrac{1}{n} \le \dfrac{1}{n}.$

(d) If $|a_n| \to 0$, then $a_n \to 0$ because $-|a_n| \le a_n \le |a_n|.$ ∎

The application of Theorems 1 and 2 is broadened by a theorem stating that applying a continuous function to a convergent sequence produces a convergent sequence. We state the theorem, leaving the proof as an exercise (Exercise 120).

THEOREM 3—The Continuous Function Theorem for Sequences
Let $\{a_n\}$ be a sequence of real numbers. If $a_n \to L$ and if f is a function that is continuous at L and defined at all a_n, then $f(a_n) \to f(L)$.

EXAMPLE 5   Show that $\sqrt{(n+1)/n} \to 1$.

Solution   We know that $(n+1)/n \to 1$. Taking $f(x) = \sqrt{x}$ and $L = 1$ in Theorem 3 gives $\sqrt{(n+1)/n} \to \sqrt{1} = 1$.   ∎

EXAMPLE 6   The sequence $\{1/n\}$ converges to 0. By taking $a_n = 1/n$, $f(x) = 2^x$, and $L = 0$ in Theorem 3, we see that $2^{1/n} = f(1/n) \to f(L) = 2^0 = 1$. The sequence $\{2^{1/n}\}$ converges to 1 (Figure 10.5).   ∎

Using L'Hôpital's Rule

The next theorem formalizes the connection between $\lim_{n\to\infty} a_n$ and $\lim_{x\to\infty} f(x)$. It enables us to use l'Hôpital's Rule to find the limits of some sequences.

THEOREM 4   Suppose that $f(x)$ is a function defined for all $x \ge n_0$ and that $\{a_n\}$ is a sequence of real numbers such that $a_n = f(n)$ for $n \ge n_0$. Then

$$\lim_{n\to\infty} a_n = L \qquad \text{whenever} \qquad \lim_{x\to\infty} f(x) = L.$$

Proof   Suppose that $\lim_{x\to\infty} f(x) = L$. Then for each positive number ε there is a number M such that

$$|f(x) - L| < \varepsilon \qquad \text{whenever} \qquad x > M.$$

Let N be an integer greater than M and greater than or equal to n_0. Since $a_n = f(n)$, it follows that for all $n > N$ we have

$$|a_n - L| = |f(n) - L| < \varepsilon. \qquad \blacksquare$$

EXAMPLE 7   Show that

$$\lim_{n\to\infty} \frac{\ln n}{n} = 0.$$

Solution   The function $(\ln x)/x$ is defined for all $x \geq 1$ and agrees with the given sequence at positive integers. Therefore, by Theorem 4, $\lim_{n\to\infty}(\ln n)/n$ will equal $\lim_{x\to\infty}(\ln x)/x$ if the latter exists. A single application of l'Hôpital's Rule shows that

$$\lim_{x\to\infty} \frac{\ln x}{x} = \lim_{x\to\infty} \frac{1/x}{1} = \frac{0}{1} = 0.$$

We conclude that $\lim_{n\to\infty}(\ln n)/n = 0$. $\qquad \blacksquare$

When we use l'Hôpital's Rule to find the limit of a sequence, we often treat n as a continuous real variable and differentiate directly with respect to n. This saves us from having to rewrite the formula for a_n as we did in Example 7.

EXAMPLE 8   Does the sequence whose nth term is

$$a_n = \left(\frac{n + 1}{n - 1}\right)^n$$

converge? If so, find $\lim_{n\to\infty} a_n$.

Solution   The limit leads to the indeterminate form 1^∞. We can apply l'Hôpital's Rule if we first change the form to $\infty \cdot 0$ by taking the natural logarithm of a_n:

$$\ln a_n = \ln\left(\frac{n + 1}{n - 1}\right)^n = n \ln\left(\frac{n + 1}{n - 1}\right).$$

Then,

$$\lim_{n\to\infty} \ln a_n = \lim_{n\to\infty} n \ln\left(\frac{n + 1}{n - 1}\right) \qquad \infty \cdot 0 \text{ form}$$

$$= \lim_{n\to\infty} \frac{\ln\left(\dfrac{n + 1}{n - 1}\right)}{1/n} \qquad \frac{0}{0} \text{ form}$$

$$= \lim_{n\to\infty} \frac{-2/(n^2 - 1)}{-1/n^2} \qquad \text{L'Hôpital's Rule: differentiate numerator and denominator.}$$

$$= \lim_{n\to\infty} \frac{2n^2}{n^2 - 1} = 2. \qquad \text{Simplify and evaluate.}$$

Since $\ln a_n \to 2$ and $f(x) = e^x$ is continuous, Theorem 3 tells us that

$$a_n = e^{\ln a_n} \to e^2.$$

The sequence $\{a_n\}$ converges to e^2. $\qquad \blacksquare$

Commonly Occurring Limits

The next theorem gives some limits that arise frequently.

Factorial Notation

The notation $n!$ ("n factorial")
means the product $1 \cdot 2 \cdot 3 \cdots n$
of the integers from 1 to n.
Notice that $(n + 1)! = (n + 1) \cdot n!$.
Thus, $4! = 1 \cdot 2 \cdot 3 \cdot 4 = 24$ and
$5! = 1 \cdot 2 \cdot 3 \cdot 4 \cdot 5 = 5 \cdot 4! = 120$. We
define $0!$ to be 1. Factorials grow even
faster than exponentials, as the table
suggests. The values in the table are
rounded.

n	e^n	$n!$
1	3	1
5	148	120
10	22,026	3,628,800
20	4.9×10^8	2.4×10^{18}

> **THEOREM 5** The following six sequences converge to the limits listed below:
>
> **1.** $\lim\limits_{n \to \infty} \dfrac{\ln n}{n} = 0$
> **2.** $\lim\limits_{n \to \infty} \sqrt[n]{n} = 1$
>
> **3.** $\lim\limits_{n \to \infty} x^{1/n} = 1 \quad (x > 0)$
> **4.** $\lim\limits_{n \to \infty} x^n = 0 \quad (|x| < 1)$
>
> **5.** $\lim\limits_{n \to \infty} \left(1 + \dfrac{x}{n}\right)^n = e^x \quad$ (any x)
> **6.** $\lim\limits_{n \to \infty} \dfrac{x^n}{n!} = 0 \quad$ (any x)
>
> In Formulas (3) through (6), x remains fixed as $n \to \infty$.

Proof The first limit was computed in Example 7. The next two can be proved by taking logarithms and applying Theorem 4 (Exercises 117 and 118). The remaining proofs are given in Appendix 5. ∎

EXAMPLE 9 These are examples of the limits in Theorem 5.

(a) $\dfrac{\ln(n^2)}{n} = \dfrac{2 \ln n}{n} \to 2 \cdot 0 = 0$ Formula 1

(b) $\sqrt[n]{n^2} = n^{2/n} = (n^{1/n})^2 \to (1)^2 = 1$ Formula 2

(c) $\sqrt[n]{3n} = 3^{1/n}(n^{1/n}) \to 1 \cdot 1 = 1$ Formula 3 with $x = 3$ and Formula 2

(d) $\left(-\dfrac{1}{2}\right)^n \to 0$ Formula 4 with $x = -\frac{1}{2}$

(e) $\left(\dfrac{n-2}{n}\right)^n = \left(1 + \dfrac{-2}{n}\right)^n \to e^{-2}$ Formula 5 with $x = -2$

(f) $\dfrac{100^n}{n!} \to 0$ Formula 6 with $x = 100$ ∎

Recursive Definitions

So far, we have calculated each a_n directly from the value of n. But sequences are often defined **recursively** by giving

1. The value(s) of the initial term or terms, and
2. A rule, called a **recursion formula**, for calculating any later term from terms that precede it.

EXAMPLE 10

(a) The statements $a_1 = 1$ and $a_n = a_{n-1} + 1$ for $n > 1$ define the sequence $1, 2, 3, \ldots, n, \ldots$ of positive integers. With $a_1 = 1$, we have $a_2 = a_1 + 1 = 2$, $a_3 = a_2 + 1 = 3$, and so on.

(b) The statements $a_1 = 1$ and $a_n = n \cdot a_{n-1}$ for $n > 1$ define the sequence $1, 2, 6, 24, \ldots, n!, \ldots$ of factorials. With $a_1 = 1$, we have $a_2 = 2 \cdot a_1 = 2$, $a_3 = 3 \cdot a_2 = 6$, $a_4 = 4 \cdot a_3 = 24$, and so on.

(c) The statements $a_1 = 1$, $a_2 = 1$, and $a_{n+1} = a_n + a_{n-1}$ for $n > 2$ define the sequence $1, 1, 2, 3, 5, \ldots$ of **Fibonacci numbers**. With $a_1 = 1$ and $a_2 = 1$, we have $a_3 = 1 + 1 = 2$, $a_4 = 2 + 1 = 3$, $a_5 = 3 + 2 = 5$, and so on.

(d) As we can see by applying Newton's method (see Exercise 145), the statements $x_0 = 1$ and $x_{n+1} = x_n - \left[(\sin x_n - x_n{}^2)/(\cos x_n - 2x_n) \right]$ for $n > 0$ define a sequence that, when it converges, gives a solution to the equation $\sin x - x^2 = 0$. ∎

Bounded Monotonic Sequences

Two concepts that play a key role in determining the convergence of a sequence are those of a *bounded* sequence and a *monotonic* sequence.

> **DEFINITION** A sequence $\{a_n\}$ is **bounded from above** if there exists a number M such that $a_n \leq M$ for all n. The number M is an **upper bound** for $\{a_n\}$. If M is an upper bound for $\{a_n\}$ but no number less than M is an upper bound for $\{a_n\}$, then M is the **least upper bound** for $\{a_n\}$.
>
> A sequence $\{a_n\}$ is **bounded from below** if there exists a number m such that $a_n \geq m$ for all n. The number m is a **lower bound** for $\{a_n\}$. If m is a lower bound for $\{a_n\}$ but no number greater than m is a lower bound for $\{a_n\}$, then m is the **greatest lower bound** for $\{a_n\}$.
>
> If $\{a_n\}$ is bounded from above and below, then $\{a_n\}$ is **bounded**. If $\{a_n\}$ is not bounded, then we say that $\{a_n\}$ is an **unbounded** sequence.

EXAMPLE 11

(a) The sequence $1, 2, 3, \ldots, n, \ldots$ has no upper bound because it eventually surpasses every number M. However, it is bounded below by every real number less than or equal to 1. The number $m = 1$ is the greatest lower bound of the sequence.

(b) The sequence $\dfrac{1}{2}, \dfrac{2}{3}, \dfrac{3}{4}, \ldots, \dfrac{n}{n+1}, \ldots$ is bounded above by every real number greater than or equal to 1. The upper bound $M = 1$ is the least upper bound (Exercise 137). The sequence is also bounded below by every number less than or equal to $\dfrac{1}{2}$, which is its greatest lower bound. ∎

Convergent sequences are bounded

If a sequence $\{a_n\}$ converges to the number L, then by definition there is a number N such that $|a_n - L| < 1$ if $n > N$. That is,

$$L - 1 < a_n < L + 1 \quad \text{for } n > N.$$

If M is a number larger than $L + 1$ and all of the finitely many numbers $a_1, a_2, \ldots, a_N$, then for every index n we have $a_n \leq M$ so that $\{a_n\}$ is bounded from above. Similarly, if m is a number smaller than $L - 1$ and all of the numbers $a_1, a_2, \ldots, a_N$, then m is a lower bound of the sequence. Therefore, all convergent sequences are bounded.

Although it is true that every convergent sequence is bounded, there are bounded sequences that fail to converge. One example is the bounded sequence $\{(-1)^{n+1}\}$ discussed in Example 2. The problem here is that some bounded sequences bounce around in the band determined by any lower bound m and any upper bound M (Figure 10.6). An important type of sequence that does not behave that way is one for which each term is at least as large, or at least as small, as its predecessor.

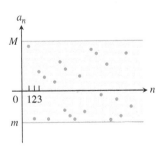

FIGURE 10.6 Some bounded sequences bounce around between their bounds and fail to converge to any limiting value.

> **DEFINITIONS** A sequence $\{a_n\}$ is **nondecreasing** if $a_n \leq a_{n+1}$ for all n. That is, $a_1 \leq a_2 \leq a_3 \leq \ldots$. The sequence is **nonincreasing** if $a_n \geq a_{n+1}$ for all n. The sequence $\{a_n\}$ is **monotonic** if it is either nondecreasing or nonincreasing.

EXAMPLE 12

(a) The sequence $1, 2, 3, \ldots, n, \ldots$ is nondecreasing.

(b) The sequence $\dfrac{1}{2}, \dfrac{2}{3}, \dfrac{3}{4}, \ldots, \dfrac{n}{n+1}, \ldots$ is nondecreasing.

(c) The sequence $1, \dfrac{1}{2}, \dfrac{1}{4}, \dfrac{1}{8}, \ldots, \dfrac{1}{2^n}, \ldots$ is nonincreasing.

(d) The constant sequence $3, 3, 3, \ldots, 3, \ldots$ is both nondecreasing and nonincreasing.

(e) The sequence $1, -1, 1, -1, 1, -1, \ldots$ is not monotonic.   ■

A nondecreasing sequence that is bounded from above always has a least upper bound. Likewise, a nonincreasing sequence bounded from below always has a greatest lower bound. These results are based on the *completeness property* of the real numbers, discussed in Appendix 6. We now prove that if L is the least upper bound of a nondecreasing sequence then the sequence converges to L, and that if L is the greatest lower bound of a nonincreasing sequence then the sequence converges to L.

> **THEOREM 6—The Monotonic Sequence Theorem**
> If a sequence $\{a_n\}$ is both bounded and monotonic, then the sequence converges.

Proof   Suppose $\{a_n\}$ is nondecreasing, L is its least upper bound, and we plot the points $(1, a_1), (2, a_2), \ldots, (n, a_n), \ldots$ in the xy-plane. If M is an upper bound of the sequence, all these points will lie on or below the line $y = M$ (Figure 10.7). The line $y = L$ is the lowest such line. None of the points (n, a_n) lies above $y = L$, but some do lie above any lower line $y = L - \varepsilon$, if ε is a positive number (because $L - \varepsilon$ is not an upper bound). The sequence converges to L because

a. $a_n \leq L$ for *all* values of n, and

b. given any $\varepsilon > 0$, there exists at least one integer N for which $a_N > L - \varepsilon$.

The fact that $\{a_n\}$ is nondecreasing tells us further that

$$a_n \geq a_N > L - \varepsilon \qquad \text{for all } n \geq N.$$

Thus, *all* the numbers a_n beyond the Nth number lie within ε of L. This is precisely the condition for L to be the limit of the sequence $\{a_n\}$.

The proof for nonincreasing sequences bounded from below is similar.   ■

It is important to realize that Theorem 6 does not say that convergent sequences are monotonic. The sequence $\{(-1)^{n+1}/n\}$ converges and is bounded, but it is not monotonic since it alternates between positive and negative values as it tends toward zero. What the theorem does say is that a nondecreasing sequence converges when it is bounded from above, but it diverges to infinity otherwise.

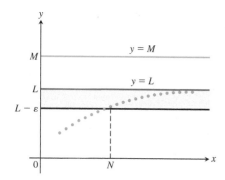

FIGURE 10.7   If the terms of a nondecreasing sequence have an upper bound M, they have a limit $L \leq M$.

EXERCISES 10.1

Finding Terms of a Sequence

Each of Exercises 1–6 gives a formula for the nth term a_n of a sequence $\{a_n\}$. Find the values of $a_1, a_2, a_3,$ and a_4.

1. $a_n = \dfrac{1-n}{n^2}$

2. $a_n = \dfrac{1}{n!}$

3. $a_n = \dfrac{(-1)^{n+1}}{2n-1}$

4. $a_n = 2 + (-1)^n$

5. $a_n = \dfrac{2^n}{2^{n+1}}$

6. $a_n = \dfrac{2^n - 1}{2^n}$

Each of Exercises 7–12 gives the first term or two of a sequence along with a recursion formula for the remaining terms. Write out the first ten terms of the sequence.

7. $a_1 = 1$, $a_{n+1} = a_n + (1/2^n)$

8. $a_1 = 1$, $a_{n+1} = a_n/(n+1)$

9. $a_1 = 2$, $a_{n+1} = (-1)^{n+1} a_n / 2$

10. $a_1 = -2$, $a_{n+1} = n a_n / (n + 1)$

11. $a_1 = a_2 = 1$, $a_{n+2} = a_{n+1} + a_n$

12. $a_1 = 2$, $a_2 = -1$, $a_{n+2} = a_{n+1} / a_n$

Finding a Sequence's Formula

In Exercises 13–30, find a formula for the nth term of the sequence.

13. $1, -1, 1, -1, 1, \ldots$ 1's with alternating signs

14. $-1, 1, -1, 1, -1, \ldots$ 1's with alternating signs

15. $1, -4, 9, -16, 25, \ldots$ Squares of the positive integers, with alternating signs

16. $1, -\dfrac{1}{4}, \dfrac{1}{9}, -\dfrac{1}{16}, \dfrac{1}{25}, \ldots$ Reciprocals of squares of the positive integers, with alternating signs

17. $\dfrac{1}{9}, \dfrac{2}{12}, \dfrac{2^2}{15}, \dfrac{2^3}{18}, \dfrac{2^4}{21}, \ldots$ Powers of 2 divided by multiples of 3

18. $-\dfrac{3}{2}, -\dfrac{1}{6}, \dfrac{1}{12}, \dfrac{3}{20}, \dfrac{5}{30}, \ldots$ Integers differing by 2 divided by products of consecutive integers

19. $0, 3, 8, 15, 24, \ldots$ Squares of the positive integers diminished by 1

20. $-3, -2, -1, 0, 1, \ldots$ Integers, beginning with -3

21. $1, 5, 9, 13, 17, \ldots$ Every other odd positive integer

22. $2, 6, 10, 14, 18, \ldots$ Every other even positive integer

23. $\dfrac{5}{1}, \dfrac{8}{2}, \dfrac{11}{6}, \dfrac{14}{24}, \dfrac{17}{120}, \ldots$ Integers differing by 3 divided by factorials

24. $\dfrac{1}{25}, \dfrac{8}{125}, \dfrac{27}{625}, \dfrac{64}{3125}, \dfrac{125}{15,625}, \ldots$ Cubes of positive integers divided by powers of 5

25. $1, 0, 1, 0, 1, \ldots$ Alternating 1's and 0's

26. $0, 1, 1, 2, 2, 3, 3, 4, \ldots$ Each positive integer repeated

27. $\dfrac{1}{2} - \dfrac{1}{3}, \dfrac{1}{3} - \dfrac{1}{4}, \dfrac{1}{4} - \dfrac{1}{5}, \dfrac{1}{5} - \dfrac{1}{6}, \ldots$

28. $\sqrt{5} - \sqrt{4}, \sqrt{6} - \sqrt{5}, \sqrt{7} - \sqrt{6}, \sqrt{8} - \sqrt{7}, \ldots$

29. $\sin\left(\dfrac{\sqrt{2}}{1+4}\right), \sin\left(\dfrac{\sqrt{3}}{1+9}\right), \sin\left(\dfrac{\sqrt{4}}{1+16}\right), \sin\left(\dfrac{\sqrt{5}}{1+25}\right), \ldots$

30. $\sqrt{\dfrac{5}{8}}, \sqrt{\dfrac{7}{11}}, \sqrt{\dfrac{9}{14}}, \sqrt{\dfrac{11}{17}}, \ldots$

Convergence and Divergence

Which of the sequences $\{a_n\}$ in Exercises 31–100 converge, and which diverge? Find the limit of each convergent sequence.

31. $a_n = 2 + (0.1)^n$

32. $a_n = \dfrac{n + (-1)^n}{n}$

33. $a_n = \dfrac{1 - 2n}{1 + 2n}$

34. $a_n = \dfrac{2n + 1}{1 - 3\sqrt{n}}$

35. $a_n = \dfrac{1 - 5n^4}{n^4 + 8n^3}$

36. $a_n = \dfrac{n + 3}{n^2 + 5n + 6}$

37. $a_n = \dfrac{n^2 - 2n + 1}{n - 1}$

38. $a_n = \dfrac{1 - n^3}{70 - 4n^2}$

39. $a_n = 1 + (-1)^n$

40. $a_n = (-1)^n \left(1 - \dfrac{1}{n}\right)$

41. $a_n = \left(\dfrac{n + 1}{2n}\right)\left(1 - \dfrac{1}{n}\right)$

42. $a_n = \left(2 - \dfrac{1}{2^n}\right)\left(3 + \dfrac{1}{2^n}\right)$

43. $a_n = \dfrac{(-1)^{n+1}}{2n - 1}$

44. $a_n = \left(-\dfrac{1}{2}\right)^n$

45. $a_n = \sqrt{\dfrac{2n}{n + 1}}$

46. $a_n = \dfrac{1}{(0.9)^n}$

47. $a_n = \sin\left(\dfrac{\pi}{2} + \dfrac{1}{n}\right)$

48. $a_n = n\pi \cos(n\pi)$

49. $a_n = \dfrac{\sin n}{n}$

50. $a_n = \dfrac{\sin^2 n}{2^n}$

51. $a_n = \dfrac{n}{2^n}$

52. $a_n = \dfrac{3^n}{n^3}$

53. $a_n = \dfrac{\ln(n + 1)}{\sqrt{n}}$

54. $a_n = \dfrac{\ln n}{\ln 2n}$

55. $a_n = 8^{1/n}$

56. $a_n = (0.03)^{1/n}$

57. $a_n = \left(1 + \dfrac{7}{n}\right)^n$

58. $a_n = \left(1 - \dfrac{1}{n}\right)^n$

59. $a_n = \sqrt[n]{10n}$

60. $a_n = \sqrt[n]{n^2}$

61. $a_n = \left(\dfrac{3}{n}\right)^{1/n}$

62. $a_n = (n + 4)^{1/(n+4)}$

63. $a_n = \dfrac{\ln n}{n^{1/n}}$

64. $a_n = \ln n - \ln(n + 1)$

65. $a_n = \sqrt[n]{4^n n}$

66. $u_n = \sqrt[n]{3^{2n+1}}$

67. $a_n = \dfrac{n!}{n^n}$ (*Hint:* Compare with $1/n$.)

68. $a_n = \dfrac{(-4)^n}{n!}$

69. $a_n = \dfrac{n!}{10^{6n}}$

70. $a_n = \dfrac{n!}{2^n \cdot 3^n}$

71. $a_n = \left(\dfrac{1}{n}\right)^{1/(\ln n)}$

72. $a_n = \dfrac{(n + 1)!}{(n + 3)!}$

73. $a_n = \dfrac{(2n + 2)!}{(2n - 1)!}$

74. $a_n = \dfrac{3e^n + e^{-n}}{e^n + 3e^{-n}}$

75. $a_n = \dfrac{e^{-2n} - 2e^{-3n}}{e^{-2n} - e^{-n}}$

76. $a_n = \left(1 - \dfrac{1}{2}\right) + \left(\dfrac{1}{2} - \dfrac{1}{3}\right) + \left(\dfrac{1}{3} - \dfrac{1}{4}\right) + \cdots$
$$+ \left(\dfrac{1}{n - 2} - \dfrac{1}{n - 1}\right) + \left(\dfrac{1}{n - 1} - \dfrac{1}{n}\right)$$

77. $a_n = (\ln 3 - \ln 2) + (\ln 4 - \ln 3) + (\ln 5 - \ln 4) + \cdots$
$$+ (\ln(n - 1) - \ln(n - 2)) + (\ln n - \ln(n - 1))$$

78. $a_n = \ln\left(1 + \dfrac{1}{n}\right)^n$

79. $a_n = \left(\dfrac{3n + 1}{3n - 1}\right)^n$

80. $a_n = \left(\dfrac{n}{n + 1}\right)^n$

81. $a_n = \left(\dfrac{x^n}{2n + 1}\right)^{1/n}$, $x > 0$

82. $a_n = \left(1 - \dfrac{1}{n^2}\right)^n$

83. $a_n = \dfrac{3^n \cdot 6^n}{2^{-n} \cdot n!}$

84. $a_n = \dfrac{(10/11)^n}{(9/10)^n + (11/12)^n}$

85. $a_n = \tanh n$

86. $a_n = \sinh(\ln n)$

87. $a_n = \dfrac{n^2}{2n - 1} \sin \dfrac{1}{n}$

88. $a_n = n\left(1 - \cos\dfrac{1}{n}\right)$

89. $a_n = \sqrt{n}\sin\dfrac{1}{\sqrt{n}}$

90. $a_n = (3^n + 5^n)^{1/n}$

91. $a_n = \tan^{-1} n$

92. $a_n = \dfrac{1}{\sqrt{n}}\tan^{-1} n$

93. $a_n = \left(\dfrac{1}{3}\right)^n + \dfrac{1}{\sqrt{2^n}}$

94. $a_n = \sqrt[n]{n^2 + n}$

95. $a_n = \dfrac{(\ln n)^{200}}{n}$

96. $a_n = \dfrac{(\ln n)^5}{\sqrt{n}}$

97. $a_n = n - \sqrt{n^2 - n}$

98. $a_n = \dfrac{1}{\sqrt{n^2 - 1} - \sqrt{n^2 + n}}$

99. $a_n = \dfrac{1}{n}\displaystyle\int_1^n \dfrac{1}{x}\,dx$

100. $a_n = \displaystyle\int_1^n \dfrac{1}{x^p}\,dx,\quad p > 1$

Recursively Defined Sequences

In Exercises 101–108, assume that each sequence converges and find its limit.

101. $a_1 = 2,\quad a_{n+1} = \dfrac{72}{1 + a_n}$

102. $a_1 = -1,\quad a_{n+1} = \dfrac{a_n + 6}{a_n + 2}$

103. $a_1 = -4,\quad a_{n+1} = \sqrt{8 + 2a_n}$

104. $a_1 = 0,\quad a_{n+1} = \sqrt{8 + 2a_n}$

105. $a_1 = 5,\quad a_{n+1} = \sqrt{5a_n}$

106. $a_1 = 3,\quad a_{n+1} = 12 - \sqrt{a_n}$

107. $2, 2 + \dfrac{1}{2}, 2 + \cfrac{1}{2 + \cfrac{1}{2}}, 2 + \cfrac{1}{2 + \cfrac{1}{2 + \cfrac{1}{2}}}, \ldots$

108. $\sqrt{1}, \sqrt{1 + \sqrt{1}}, \sqrt{1 + \sqrt{1 + \sqrt{1}}},$
$\sqrt{1 + \sqrt{1 + \sqrt{1 + \sqrt{1}}}}, \ldots$

Theory and Examples

109. The first term of a sequence is $x_1 = 1$. Each succeeding term is the sum of all those that come before it:

$$x_{n+1} = x_1 + x_2 + \cdots + x_n.$$

Write out enough early terms of the sequence to deduce a general formula for x_n that holds for $n \ge 2$.

110. A sequence of rational numbers is described as follows:

$$\dfrac{1}{1}, \dfrac{3}{2}, \dfrac{7}{5}, \dfrac{17}{12}, \ldots, \dfrac{a}{b}, \dfrac{a + 2b}{a + b}, \ldots.$$

Here the numerators form one sequence, the denominators form a second sequence, and their ratios form a third sequence. Let x_n and y_n be, respectively, the numerator and the denominator of the nth fraction $r_n = x_n/y_n$.

  a. Verify that $x_1{}^2 - 2y_1{}^2 = -1$, $x_2{}^2 - 2y_2{}^2 = +1$ and, more generally, that if $a^2 - 2b^2 = -1$ or $+1$, then

$$(a + 2b)^2 - 2(a + b)^2 = +1 \quad \text{or} \quad -1,$$

   respectively.

  b. The fractions $r_n = x_n/y_n$ approach a limit as n increases. What is that limit? (*Hint:* Use part (a) to show that $r_n{}^2 - 2 = \pm(1/y_n)^2$ and that y_n is not less than n.)

111. Newton's method The following sequences come from the recursion formula for Newton's method,

$$x_{n+1} = x_n - \dfrac{f(x_n)}{f'(x_n)}.$$

Do the sequences converge? If so, to what value? In each case, begin by identifying the function f that generates the sequence.

  a. $x_0 = 1,\quad x_{n+1} = x_n - \dfrac{x_n^2 - 2}{2x_n} = \dfrac{x_n}{2} + \dfrac{1}{x_n}$

  b. $x_0 = 1,\quad x_{n+1} = x_n - \dfrac{\tan x_n - 1}{\sec^2 x_n}$

  c. $x_0 = 1,\quad x_{n+1} = x_n - 1$

112. a. Suppose that $f(x)$ is differentiable for all x in $[0, 1]$ and that $f(0) = 0$. Define sequence $\{a_n\}$ by the rule $a_n = nf(1/n)$. Show that $\lim_{n \to \infty} a_n = f'(0)$. Use the result in part (a) to find the limits of the following sequences $\{a_n\}$.

  b. $a_n = n\tan^{-1}\dfrac{1}{n}$

  c. $a_n = n(e^{1/n} - 1)$

  d. $a_n = n\ln\left(1 + \dfrac{2}{n}\right)$

113. Pythagorean triples A triple of positive integers a, b, and c is called a **Pythagorean triple** if $a^2 + b^2 = c^2$. Let a be an odd positive integer and let

$$b = \left\lfloor \dfrac{a^2}{2} \right\rfloor \quad \text{and} \quad c = \left\lceil \dfrac{a^2}{2} \right\rceil$$

be, respectively, the integer floor and ceiling for $a^2/2$.

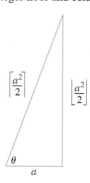

  a. Show that $a^2 + b^2 = c^2$. (*Hint:* Let $a = 2n + 1$ and express b and c in terms of n.)

  b. By direct calculation, or by appealing to the accompanying figure, find

$$\lim_{a \to \infty} \dfrac{\left\lfloor \dfrac{a^2}{2} \right\rfloor}{\left\lceil \dfrac{a^2}{2} \right\rceil}.$$

114. The nth root of $n!$

a. Show that $\lim_{n\to\infty}(2n\pi)^{1/(2n)} = 1$ and hence, using Stirling's approximation (Chapter 8, Additional Exercise 52a), that

$$\sqrt[n]{n!} \approx \frac{n}{e} \quad \text{for large values of } n.$$

T b. Test the approximation in part (a) for $n = 40, 50, 60, \ldots$, as far as your calculator will allow.

115. a. Assuming that $\lim_{n\to\infty}(1/n^c) = 0$ if c is any positive constant, show that

$$\lim_{n\to\infty}\frac{\ln n}{n^c} = 0$$

if c is any positive constant.

b. Prove that $\lim_{n\to\infty}(1/n^c) = 0$ if c is any positive constant. (*Hint:* If $\varepsilon = 0.001$ and $c = 0.04$, how large should N be to ensure that $|1/n^c - 0| < \varepsilon$ if $n > N$?)

116. The zipper theorem Prove the "zipper theorem" for sequences: If $\{a_n\}$ and $\{b_n\}$ both converge to L, then the sequence

$$a_1, b_1, a_2, b_2, \ldots, a_n, b_n, \ldots$$

converges to L.

117. Prove that $\lim_{n\to\infty}\sqrt[n]{n} = 1$.

118. Prove that $\lim_{n\to\infty}x^{1/n} = 1, (x > 0)$.

119. Prove Theorem 2.   **120.** Prove Theorem 3.

In Exercises 121–124, determine if the sequence is monotonic and if it is bounded.

121. $a_n = \dfrac{3n+1}{n+1}$ **122.** $a_n = \dfrac{(2n+3)!}{(n+1)!}$

123. $a_n = \dfrac{2^n 3^n}{n!}$ **124.** $a_n = 2 - \dfrac{2}{n} - \dfrac{1}{2^n}$

Which of the sequences in Exercises 125–134 converge, and which diverge? Give reasons for your answers.

125. $a_n = 1 - \dfrac{1}{n}$ **126.** $a_n = n - \dfrac{1}{n}$

127. $a_n = \dfrac{2^n - 1}{2^n}$ **128.** $a_n = \dfrac{2^n - 1}{3^n}$

129. $a_n = ((-1)^n + 1)\left(\dfrac{n+1}{n}\right)$

130. The first term of a sequence is $x_1 = \cos(1)$. The next terms are $x_2 = x_1$ or $\cos(2)$, whichever is larger; and $x_3 = x_2$ or $\cos(3)$, whichever is larger (farther to the right). In general,

$$x_{n+1} = \max\{x_n, \cos(n+1)\}.$$

131. $a_n = \dfrac{1 + \sqrt{2n}}{\sqrt{n}}$ **132.** $a_n = \dfrac{n+1}{n}$

133. $a_n = \dfrac{4^{n+1} + 3^n}{4^n}$ **134.** $a_1 = 1, \quad a_{n+1} = 2a_n - 3$

In Exercises 135–136, use the definition of convergence to prove the given limit.

135. $\lim_{n\to\infty}\dfrac{\sin n}{n} = 0$ **136.** $\lim_{n\to\infty}\left(1 - \dfrac{1}{n^2}\right) = 1$

137. The sequence $\{n/(n+1)\}$ has a least upper bound of 1 Show that if M is a number less than 1, then the terms of $\{n/(n+1)\}$ eventually exceed M. That is, if $M < 1$ there is an integer N such that $n/(n+1) > M$ whenever $n > N$. Since $n/(n+1) < 1$ for every n, this proves that 1 is a least upper bound for $\{n/(n+1)\}$.

138. Uniqueness of least upper bounds Show that if M_1 and M_2 are least upper bounds for the sequence $\{a_n\}$, then $M_1 = M_2$. That is, a sequence cannot have two different least upper bounds.

139. Is it true that a sequence $\{a_n\}$ of positive numbers must converge if it is bounded from above? Give reasons for your answer.

140. Prove that if $\{a_n\}$ is a convergent sequence, then to every positive number ε there corresponds an integer N such that

$$|a_m - a_n| < \varepsilon \quad \text{whenever} \quad m > N \quad \text{and} \quad n > N.$$

141. Uniqueness of limits Prove that limits of sequences are unique. That is, show that if L_1 and L_2 are numbers such that $a_n \to L_1$ and $a_n \to L_2$, then $L_1 = L_2$.

142. Limits and subsequences If the terms of one sequence appear in another sequence in their given order, we call the first sequence a **subsequence** of the second. Prove that if two sub-sequences of a sequence $\{a_n\}$ have different limits $L_1 \neq L_2$, then $\{a_n\}$ diverges.

143. For a sequence $\{a_n\}$ the terms of even index are denoted by a_{2k} and the terms of odd index by a_{2k+1}. Prove that if $a_{2k} \to L$ and $a_{2k+1} \to L$, then $a_n \to L$.

144. Prove that a sequence $\{a_n\}$ converges to 0 if and only if the sequence of absolute values $\{|a_n|\}$ converges to 0.

145. Sequences generated by Newton's method Newton's method, applied to a differentiable function $f(x)$, begins with a starting value x_0 and constructs from it a sequence of numbers $\{x_n\}$ that under favorable circumstances converges to a zero of f. The recursion formula for the sequence is

$$x_{n+1} = x_n - \frac{f(x_n)}{f'(x_n)}.$$

a. Show that the recursion formula for $f(x) = x^2 - a, a > 0$, can be written as $x_{n+1} = (x_n + a/x_n)/2$.

T b. Starting with $x_0 = 1$ and $a = 3$, calculate successive terms of the sequence until the display begins to repeat. What number is being approximated? Explain.

T **146. A recursive definition of $\pi/2$** If you start with $x_1 = 1$ and define the subsequent terms of $\{x_n\}$ by the rule $x_n = x_{n-1} + \cos x_{n-1}$, you generate a sequence that converges rapidly to $\pi/2$. (**a**) Try it. (**b**) Use the accompanying figure to explain why the convergence is so rapid.

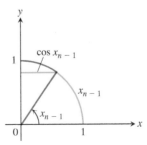

COMPUTER EXPLORATIONS

Use a CAS to perform the following steps for the sequences in Exercises 147–158.

    a. Calculate and then plot the first 25 terms of the sequence. Does the sequence appear to be bounded from above or below? Does it appear to converge or diverge? If it does converge, what is the limit L?

    b. If the sequence converges, find an integer N such that $|a_n - L| \leq 0.01$ for $n \geq N$. How far in the sequence do you have to get for the terms to lie within 0.0001 of L?

147. $a_n = \sqrt[n]{n}$

148. $a_n = \left(1 + \dfrac{0.5}{n}\right)^n$

149. $a_1 = 1, \quad a_{n+1} = a_n + \dfrac{1}{5^n}$

150. $a_1 = 1, \quad a_{n+1} = a_n + (-2)^n$

151. $a_n = \sin n$

152. $a_n = n \sin \dfrac{1}{n}$

153. $a_n = \dfrac{\sin n}{n}$

154. $a_n = \dfrac{\ln n}{n}$

155. $a_n = (0.9999)^n$

156. $a_n = (123456)^{1/n}$

157. $a_n = \dfrac{8^n}{n!}$

158. $a_n = \dfrac{n^{41}}{19^n}$

10.2 Infinite Series

An *infinite series* is the sum of an infinite sequence of numbers

$$a_1 + a_2 + a_3 + \cdots + a_n + \cdots$$

The goal of this section is to understand the meaning of such an infinite sum and to develop methods to calculate it. Since there are infinitely many terms to add in an infinite series, we cannot just keep adding to see what comes out. Instead we look at the result of summing just the first n terms of the sequence. The sum of the first n terms

$$s_n = a_1 + a_2 + a_3 + \cdots + a_n$$

is an ordinary finite sum and can be calculated by normal addition. It is called the *nth partial sum*. As n gets larger, we expect the partial sums to get closer and closer to a limiting value in the same sense that the terms of a sequence approach a limit, as discussed in Section 10.1.

    For example, to assign meaning to an expression like

$$1 + \frac{1}{2} + \frac{1}{4} + \frac{1}{8} + \frac{1}{16} + \cdots$$

we add the terms one at a time from the beginning and look for a pattern in how these partial sums grow.

Partial sum		Value	Suggestive expression for partial sum
First:	$s_1 = 1$	1	$2 - 1$
Second:	$s_2 = 1 + \dfrac{1}{2}$	$\dfrac{3}{2}$	$2 - \dfrac{1}{2}$
Third:	$s_3 = 1 + \dfrac{1}{2} + \dfrac{1}{4}$	$\dfrac{7}{4}$	$2 - \dfrac{1}{4}$
$\vdots$	$\vdots$	$\vdots$	$\vdots$
nth:	$s_n = 1 + \dfrac{1}{2} + \dfrac{1}{4} + \cdots + \dfrac{1}{2^{n-1}}$	$\dfrac{2^n - 1}{2^{n-1}}$	$2 - \dfrac{1}{2^{n-1}}$

Indeed there is a pattern. The partial sums form a sequence whose nth term is

$$s_n = 2 - \frac{1}{2^{n-1}}.$$

This sequence of partial sums converges to 2 because $\lim_{n\to\infty}(1/2^{n-1}) = 0$. We say

$$\text{“the sum of the infinite series } 1 + \frac{1}{2} + \frac{1}{4} + \cdots + \frac{1}{2^{n-1}} + \cdots \text{ is 2.”}$$

Is the sum of any finite number of terms in this series equal to 2? No. Can we actually add an infinite number of terms one by one? No. But we can still define their sum by defining it to be the limit of the sequence of partial sums as $n \to \infty$, in this case 2 (Figure 10.8). Our knowledge of sequences and limits enables us to break away from the confines of finite sums.

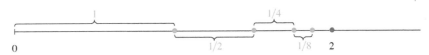

FIGURE 10.8 As the lengths $1, 1/2, 1/4, 1/8, \ldots$ are added one by one, the sum approaches 2.

HISTORICAL BIOGRAPHY
Blaise Pascal
(1623–1662)
bit.ly/2ItkU6x

DEFINITIONS Given a sequence of numbers $\{a_n\}$, an expression of the form

$$a_1 + a_2 + a_3 + \cdots + a_n + \cdots$$

is an **infinite series**. The number a_n is the ***n*th term** of the series. The sequence $\{s_n\}$ defined by

$$\begin{aligned} s_1 &= a_1 \\ s_2 &= a_1 + a_2 \\ &\vdots \\ s_n &= a_1 + a_2 + \cdots + a_n = \sum_{k=1}^{n} a_k \\ &\vdots \end{aligned}$$

is the **sequence of partial sums** of the series, the number s_n being the ***n*th partial sum**. If the sequence of partial sums converges to a limit L, we say that the series **converges** and that its **sum** is L. In this case, we also write

$$a_1 + a_2 + \cdots + a_n + \cdots = \sum_{n=1}^{\infty} a_n = L.$$

If the sequence of partial sums of the series does not converge, we say that the series **diverges**.

We can represent each term in an infinite series by the area of a rectangle. If all the terms a_n in the series are positive, then the series converges if the total area is finite, and diverges otherwise. Figure 10.9a shows an example where the series converges and Figure 10.9b shows an example where it diverges. The convergence of the total area is related to the convergence or divergence of improper integrals, as we found in Section 8.8. We make this connection explicit in the next section, where we develop an important test for convergence of series, the Integral Test.

When we begin to study a given series $a_1 + a_2 + \cdots + a_n + \cdots$, we might not know whether it converges or diverges. In either case, it is convenient to use sigma notation to write the series as

$$\sum_{n=1}^{\infty} a_n, \qquad \sum_{k=1}^{\infty} a_k, \qquad \text{or} \qquad \sum a_n$$

A useful shorthand when summation from 1 to ∞ is understood

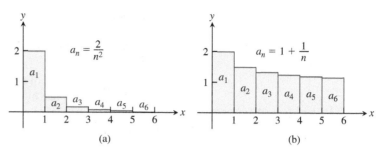

FIGURE 10.9 The sum of a series with positive terms can be interpreted as a total area of an infinite collection of rectangles. The series converges when the total area of the rectangles is finite (a) and diverges when the total area is unbounded (b). Note that the total area can be infinite even if the area of the rectangles is decreasing.

Geometric Series

Geometric series are series of the form

$$a + ar + ar^2 + \cdots + ar^{n-1} + \cdots = \sum_{n=1}^{\infty} ar^{n-1}$$

in which a and r are fixed real numbers and $a \neq 0$. The series can also be written as $\sum_{n=0}^{\infty} ar^n$. The **ratio** r can be positive, as in

$$1 + \frac{1}{2} + \frac{1}{4} + \cdots + \left(\frac{1}{2}\right)^{n-1} + \cdots, \qquad r = 1/2, a = 1$$

or negative, as in

$$1 - \frac{1}{3} + \frac{1}{9} - \cdots + \left(-\frac{1}{3}\right)^{n-1} + \cdots. \qquad r = -1/3, a = 1$$

If $r = 1$, the nth partial sum of the geometric series is

$$s_n = a + a(1) + a(1)^2 + \cdots + a(1)^{n-1} = na,$$

and the series diverges because $\lim_{n\to\infty} s_n = \pm\infty$, depending on the sign of a. If $r = -1$, the series diverges because the nth partial sums alternate between a and 0 and never approach a single limit. If $|r| \neq 1$, we can determine the convergence or divergence of the series in the following way:

$$s_n = a + ar + ar^2 + \cdots + ar^{n-1} \qquad \text{Write the } n\text{th partial sum.}$$
$$rs_n = ar + ar^2 + \cdots + ar^{n-1} + ar^n \qquad \text{Multiply } s_n \text{ by } r.$$
$$s_n - rs_n = a - ar^n \qquad \text{Subtract } rs_n \text{ from } s_n. \text{ Most of the terms on the right cancel.}$$
$$s_n(1 - r) = a(1 - r^n) \qquad \text{Factor.}$$
$$s_n = \frac{a(1 - r^n)}{1 - r}, \qquad (r \neq 1). \qquad \text{We can solve for } s_n \text{ if } r \neq 1.$$

If $|r| < 1$, then $r^n \to 0$ as $n \to \infty$ (as in Section 10.1), so $s_n \to a/(1 - r)$ in this case. On the other hand, if $|r| > 1$, then $|r^n| \to \infty$ and the series diverges.

If $|r| < 1$, the geometric series $a + ar + ar^2 + \cdots + ar^{n-1} + \cdots$ converges to $a/(1 - r)$:

$$\sum_{n=1}^{\infty} ar^{n-1} = \frac{a}{1 - r}, \qquad |r| < 1.$$

If $|r| \geq 1$, the series diverges.

The formula $a/(1 - r)$ for the sum of a geometric series applies *only* when the summation index begins with $n = 1$ in the expression $\sum_{n=1}^{\infty} ar^{n-1}$ (or with the index $n = 0$ if we write the series as $\sum_{n=0}^{\infty} ar^n$).

EXAMPLE 1   The geometric series with $a = 1/9$ and $r = 1/3$ is

$$\frac{1}{9} + \frac{1}{27} + \frac{1}{81} + \cdots = \sum_{n=1}^{\infty} \frac{1}{9}\left(\frac{1}{3}\right)^{n-1} = \frac{1/9}{1 - (1/3)} = \frac{1}{6}. \qquad \blacksquare$$

EXAMPLE 2   The series

$$\sum_{n=0}^{\infty} \frac{(-1)^n 5}{4^n} = 5 - \frac{5}{4} + \frac{5}{16} - \frac{5}{64} + \cdots$$

is a geometric series with $a = 5$ and $r = -1/4$. It converges to

$$\frac{a}{1 - r} = \frac{5}{1 + (1/4)} = 4. \qquad \blacksquare$$

EXAMPLE 3   You drop a ball from a meters above a flat surface. Each time the ball hits the surface after falling a distance h, it rebounds a distance rh, where r is positive but less than 1. Find the total distance the ball travels up and down (Figure 10.10).

Solution   The total distance is

$$s = a + \underbrace{2ar + 2ar^2 + 2ar^3 + \cdots}_{\text{This sum is } 2ar/(1-r).} = a + \frac{2ar}{1 - r} = a\frac{1 + r}{1 - r}.$$

If $a = 6$ m and $r = 2/3$, for instance, the distance is

$$s = 6 \cdot \frac{1 + (2/3)}{1 - (2/3)} = 6\left(\frac{5/3}{1/3}\right) = 30 \text{ m.} \qquad \blacksquare$$

EXAMPLE 4   Express the repeating decimal $5.232323\ldots$ as the ratio of two integers.

Solution   From the definition of a decimal number, we get a geometric series

$$5.232323\ldots = 5 + \frac{23}{100} + \frac{23}{(100)^2} + \frac{23}{(100)^3} + \cdots$$

$$= 5 + \frac{23}{100}\underbrace{\left(1 + \frac{1}{100} + \left(\frac{1}{100}\right)^2 + \cdots\right)}_{1/(1 - 0.01)} \quad \begin{array}{l} a = 1, \\ r = 1/100 \end{array}$$

$$= 5 + \frac{23}{100}\left(\frac{1}{0.99}\right) = 5 + \frac{23}{99} = \frac{518}{99} \qquad \blacksquare$$

Unfortunately, formulas like the one for the sum of a convergent geometric series are rare and we usually have to settle for an estimate of a series' sum (more about this later). The next example, however, is another case in which we can find the sum exactly.

EXAMPLE 5   Find the sum of the "telescoping" series $\sum_{n=1}^{\infty} \frac{1}{n(n + 1)}$.

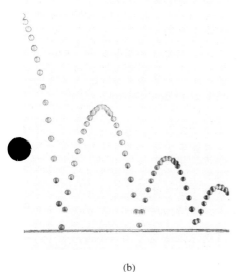

(a)

(b)

FIGURE 10.10   (a) Example 3 shows how to use a geometric series to calculate the total vertical distance traveled by a bouncing ball if the height of each rebound is reduced by the factor r. (b) A stroboscopic photo of a bouncing ball. (*Source: PSSC Physics*, 2nd ed., Reprinted by permission of Educational Development Center, Inc.)

Solution   We look for a pattern in the sequence of partial sums that might lead to a formula for s_k. The key observation is the partial fraction decomposition

$$\frac{1}{n(n + 1)} = \frac{1}{n} - \frac{1}{n + 1},$$

so

$$\sum_{n=1}^{k} \frac{1}{n(n + 1)} = \sum_{n=1}^{k} \left(\frac{1}{n} - \frac{1}{n + 1}\right)$$

and

$$s_k = \left(\frac{1}{1} - \frac{1}{2}\right) + \left(\frac{1}{2} - \frac{1}{3}\right) + \left(\frac{1}{3} - \frac{1}{4}\right) + \cdots + \left(\frac{1}{k} - \frac{1}{k + 1}\right).$$

Removing parentheses and canceling adjacent terms of opposite sign collapses the sum to

$$s_k = 1 - \frac{1}{k + 1}.$$

We now see that $s_k \to 1$ as $k \to \infty$. The series converges, and its sum is 1:

$$\sum_{n=1}^{\infty} \frac{1}{n(n + 1)} = 1.$$

The nth-Term Test for a Divergent Series

One reason that a series may fail to converge is that its terms don't become small.

EXAMPLE 6   The series

$$\sum_{n=1}^{\infty} \frac{n + 1}{n} = \frac{2}{1} + \frac{3}{2} + \frac{4}{3} + \cdots + \frac{n + 1}{n} + \cdots$$

diverges because the partial sums eventually outgrow every preassigned number. Each term is greater than 1, so the sum of n terms is greater than n.

We now show that $\lim_{n \to \infty} a_n$ must equal zero if the series $\sum_{n=1}^{\infty} a_n$ converges. To see why, let S represent the series' sum and $s_n = a_1 + a_2 + \cdots + a_n$ the nth partial sum. When n is large, both s_n and s_{n-1} are close to S, so their difference, a_n, is close to zero. More formally,

$$a_n = s_n - s_{n-1} \quad \rightarrow \quad S - S = 0. \qquad \text{Difference Rule for sequences}$$

This establishes the following theorem.

Caution

Theorem 7 *does not say* that $\sum_{n=1}^{\infty} a_n$ converges if $a_n \to 0$. It is possible for a series to diverge when $a_n \to 0$. (See Example 8.)

THEOREM 7   If $\displaystyle\sum_{n=1}^{\infty} a_n$ converges, then $a_n \to 0$.

Theorem 7 leads to a test for detecting the kind of divergence that occurred in Example 6.

The nth-Term Test for Divergence

$\displaystyle\sum_{n=1}^{\infty} a_n$ diverges if $\displaystyle\lim_{n \to \infty} a_n$ fails to exist or is different from zero.

EXAMPLE 7   The following are all examples of divergent series.

(a) $\displaystyle\sum_{n=1}^{\infty} n^2$ diverges because $n^2 \to \infty$.

(b) $\displaystyle\sum_{n=1}^{\infty} \frac{n+1}{n}$ diverges because $\dfrac{n+1}{n} \to 1$. $\quad \lim_{n\to\infty} a_n \neq 0$

(c) $\displaystyle\sum_{n=1}^{\infty} (-1)^{n+1}$ diverges because $\lim_{n\to\infty} (-1)^{n+1}$ does not exist.

(d) $\displaystyle\sum_{n=1}^{\infty} \frac{-n}{2n+5}$ diverges because $\lim_{n\to\infty} \dfrac{-n}{2n+5} = -\dfrac{1}{2} \neq 0$. ■

EXAMPLE 8   The series

$$1 + \frac{1}{2} + \frac{1}{2} + \frac{1}{4} + \frac{1}{4} + \frac{1}{4} + \frac{1}{4} + \cdots + \underbrace{\frac{1}{2^n} + \frac{1}{2^n} + \cdots + \frac{1}{2^n}}_{} + \cdots$$

$\underbrace{}_{\text{2 terms}}\quad \underbrace{}_{\text{4 terms}}\qquad\quad \underbrace{}_{2^n \text{ terms}}$

diverges because the terms can be grouped into infinitely many clusters each of which adds to 1, so the partial sums increase without bound. However, the terms of the series form a sequence that converges to 0. Example 1 of Section 10.3 shows that the harmonic series $\sum 1/n$ also behaves in this manner. ■

Combining Series

Whenever we have two convergent series, we can add them term by term, subtract them term by term, or multiply them by constants to make new convergent series.

THEOREM 8   If $\sum a_n = A$ and $\sum b_n = B$ are convergent series, then

1. *Sum Rule:* $\qquad\qquad\qquad \sum(a_n + b_n) = \sum a_n + \sum b_n = A + B$
2. *Difference Rule:* $\qquad\qquad \sum(a_n - b_n) = \sum a_n - \sum b_n = A - B$
3. *Constant Multiple Rule:* $\qquad \sum k a_n = k \sum a_n = kA \qquad$ (any number k).

Proof   The three rules for series follow from the analogous rules for sequences in Theorem 1, Section 10.1. To prove the Sum Rule for series, let

$$A_n = a_1 + a_2 + \cdots + a_n, \quad B_n = b_1 + b_2 + \cdots + b_n.$$

Then the partial sums of $\sum(a_n + b_n)$ are

$$\begin{aligned} s_n &= (a_1 + b_1) + (a_2 + b_2) + \cdots + (a_n + b_n) \\ &= (a_1 + \cdots + a_n) + (b_1 + \cdots + b_n) \\ &= A_n + B_n. \end{aligned}$$

Since $A_n \to A$ and $B_n \to B$, we have $s_n \to A + B$ by the Sum Rule for sequences. The proof of the Difference Rule is similar.

To prove the Constant Multiple Rule for series, observe that the partial sums of $\sum k a_n$ form the sequence

$$s_n = k a_1 + k a_2 + \cdots + k a_n = k(a_1 + a_2 + \cdots + a_n) = k A_n,$$

which converges to kA by the Constant Multiple Rule for sequences. ■

As corollaries of Theorem 8, we have the following results. We omit the proofs.

---

1. Every nonzero constant multiple of a divergent series diverges.
2. If $\sum a_n$ converges and $\sum b_n$ diverges, then $\sum(a_n + b_n)$ and $\sum(a_n - b_n)$ both diverge.

Caution Remember that $\Sigma(a_n + b_n)$ can converge *even if* both Σa_n and Σb_n diverge. For example, $\Sigma a_n = 1 + 1 + 1 + \cdots$ and $\Sigma b_n = (-1) + (-1) + (-1) + \cdots$ diverge, whereas $\Sigma(a_n + b_n) = 0 + 0 + 0 + \cdots$ converges to 0.

EXAMPLE 9 Find the sums of the following series.

(a) $\displaystyle\sum_{n=1}^{\infty} \frac{3^{n-1} - 1}{6^{n-1}} = \sum_{n=1}^{\infty} \left(\frac{1}{2^{n-1}} - \frac{1}{6^{n-1}} \right)$

$\displaystyle = \sum_{n=1}^{\infty} \frac{1}{2^{n-1}} - \sum_{n=1}^{\infty} \frac{1}{6^{n-1}}$ Difference Rule

$\displaystyle = \frac{1}{1 - (1/2)} - \frac{1}{1 - (1/6)}$ Geometric series with $a = 1$ and $r = 1/2, 1/6$

$\displaystyle = 2 - \frac{6}{5} = \frac{4}{5}$

(b) $\displaystyle\sum_{n=0}^{\infty} \frac{4}{2^n} = 4 \sum_{n=0}^{\infty} \frac{1}{2^n}$ Constant Multiple Rule

$\displaystyle = 4 \left(\frac{1}{1 - (1/2)} \right)$ Geometric series with $a = 1, r = 1/2$

$= 8$ ∎

Adding or Deleting Terms

We can add a finite number of terms to a series or delete a finite number of terms without altering the series' convergence or divergence, although in the case of convergence this will usually change the sum. If $\sum_{n=1}^{\infty} a_n$ converges, then $\sum_{n=k}^{\infty} a_n$ converges for any $k > 1$ and

$$\sum_{n=1}^{\infty} a_n = a_1 + a_2 + \cdots + a_{k-1} + \sum_{n=k}^{\infty} a_n.$$

Conversely, if $\sum_{n=k}^{\infty} a_n$ converges for any $k > 1$, then $\sum_{n=1}^{\infty} a_n$ converges. Thus,

$$\sum_{n=1}^{\infty} \frac{1}{5^n} = \frac{1}{5} + \frac{1}{25} + \frac{1}{125} + \sum_{n=4}^{\infty} \frac{1}{5^n}$$

and

$$\sum_{n=4}^{\infty} \frac{1}{5^n} = \left(\sum_{n=1}^{\infty} \frac{1}{5^n} \right) - \frac{1}{5} - \frac{1}{25} - \frac{1}{125}.$$

The convergence or divergence of a series is not affected by its first few terms. Only the "tail" of the series, the part that remains when we sum beyond some finite number of initial terms, influences whether it converges or diverges.

Reindexing

As long as we preserve the order of its terms, we can reindex any series without altering its convergence. To raise the starting value of the index h units, replace the n in the formula for a_n by $n - h$:

$$\sum_{n=1}^{\infty} a_n = \sum_{n=1+h}^{\infty} a_{n-h} = a_1 + a_2 + a_3 + \cdots.$$

To lower the starting value of the index h units, replace the n in the formula for a_n by $n + h$:

$$\sum_{n=1}^{\infty} a_n = \sum_{n=1-h}^{\infty} a_{n+h} = a_1 + a_2 + a_3 + \cdots.$$

We saw this reindexing in starting a geometric series with the index $n = 0$ instead of the index $n = 1$, but we can use any other starting index value as well. We usually give preference to indexings that lead to simple expressions.

EXAMPLE 10   We can write the geometric series

$$\sum_{n=1}^{\infty} \frac{1}{2^{n-1}} = 1 + \frac{1}{2} + \frac{1}{4} + \cdots$$

as

$$\sum_{n=0}^{\infty} \frac{1}{2^n}, \qquad \sum_{n=5}^{\infty} \frac{1}{2^{n-5}}, \qquad \text{or even} \qquad \sum_{n=-4}^{\infty} \frac{1}{2^{n+4}}.$$

The partial sums remain the same no matter what indexing we choose to use.   ∎

EXERCISES 10.2

Finding *n*th Partial Sums

In Exercises 1–6, find a formula for the *n*th partial sum of each series and use it to find the series' sum if the series converges.

1. $2 + \frac{2}{3} + \frac{2}{9} + \frac{2}{27} + \cdots + \frac{2}{3^{n-1}} + \cdots$

2. $\frac{9}{100} + \frac{9}{100^2} + \frac{9}{100^3} + \cdots + \frac{9}{100^n} + \cdots$

3. $1 - \frac{1}{2} + \frac{1}{4} - \frac{1}{8} + \cdots + (-1)^{n-1}\frac{1}{2^{n-1}} + \cdots$

4. $1 - 2 + 4 - 8 + \cdots + (-1)^{n-1} 2^{n-1} + \cdots$

5. $\frac{1}{2 \cdot 3} + \frac{1}{3 \cdot 4} + \frac{1}{4 \cdot 5} + \cdots + \frac{1}{(n + 1)(n + 2)} + \cdots$

6. $\frac{5}{1 \cdot 2} + \frac{5}{2 \cdot 3} + \frac{5}{3 \cdot 4} + \cdots + \frac{5}{n(n + 1)} + \cdots$

Series with Geometric Terms

In Exercises 7–14, write out the first eight terms of each series to show how the series starts. Then find the sum of the series or show that it diverges.

7. $\sum_{n=0}^{\infty} \frac{(-1)^n}{4^n}$

8. $\sum_{n=2}^{\infty} \frac{1}{4^n}$

9. $\sum_{n=1}^{\infty} \left(1 - \frac{7}{4^n}\right)$

10. $\sum_{n=0}^{\infty} (-1)^n \frac{5}{4^n}$

11. $\sum_{n=0}^{\infty} \left(\frac{5}{2^n} + \frac{1}{3^n}\right)$

12. $\sum_{n=0}^{\infty} \left(\frac{5}{2^n} - \frac{1}{3^n}\right)$

13. $\sum_{n=0}^{\infty} \left(\frac{1}{2^n} + \frac{(-1)^n}{5^n}\right)$

14. $\sum_{n=0}^{\infty} \left(\frac{2^{n+1}}{5^n}\right)$

In Exercises 15–22, determine if the geometric series converges or diverges. If a series converges, find its sum.

15. $1 + \left(\frac{2}{5}\right) + \left(\frac{2}{5}\right)^2 + \left(\frac{2}{5}\right)^3 + \left(\frac{2}{5}\right)^4 + \cdots$

16. $1 + (-3) + (-3)^2 + (-3)^3 + (-3)^4 + \cdots$

17. $\left(\frac{1}{8}\right) + \left(\frac{1}{8}\right)^2 + \left(\frac{1}{8}\right)^3 + \left(\frac{1}{8}\right)^4 + \left(\frac{1}{8}\right)^5 + \cdots$

18. $\left(\frac{-2}{3}\right)^2 + \left(\frac{-2}{3}\right)^3 + \left(\frac{-2}{3}\right)^4 + \left(\frac{-2}{3}\right)^5 + \left(\frac{-2}{3}\right)^6 + \cdots$

19. $1 - \left(\frac{2}{e}\right) + \left(\frac{2}{e}\right)^2 - \left(\frac{2}{e}\right)^3 + \left(\frac{2}{e}\right)^4 - \cdots$

20. $\left(\frac{1}{3}\right)^{-2} - \left(\frac{1}{3}\right)^{-1} + 1 - \left(\frac{1}{3}\right) + \left(\frac{1}{3}\right)^2 - \cdots$

21. $1 + \left(\frac{10}{9}\right)^2 + \left(\frac{10}{9}\right)^4 + \left(\frac{10}{9}\right)^6 + \left(\frac{10}{9}\right)^8 + \cdots$

22. $\frac{9}{4} - \frac{27}{8} + \frac{81}{16} - \frac{243}{32} + \frac{729}{64} - \cdots$

Repeating Decimals

Express each of the numbers in Exercises 23–30 as the ratio of two integers.

23. $0.\overline{23} = 0.23\,23\,23\ldots$

24. $0.\overline{234} = 0.234\,234\,234\ldots$

25. $0.\overline{7} = 0.7777\ldots$

26. $0.\overline{d} = 0.dddd\ldots,$ where d is a digit

27. $0.0\overline{6} = 0.06666\ldots$

28. $1.\overline{414} = 1.414\,414\,414\ldots$

29. $1.24\overline{123} = 1.24\,123\,123\,123\ldots$

30. $3.\overline{142857} = 3.142857\,142857\ldots$

Using the *n*th-Term Test

In Exercises 31–38, use the *n*th-Term Test for divergence to show that the series is divergent, or state that the test is inconclusive.

31. $\sum_{n=1}^{\infty} \frac{n}{n + 10}$

32. $\sum_{n=1}^{\infty} \frac{n(n + 1)}{(n + 2)(n + 3)}$

33. $\sum_{n=0}^{\infty} \frac{1}{n + 4}$

34. $\sum_{n=1}^{\infty} \frac{n}{n^2 + 3}$

35. $\displaystyle\sum_{n=1}^{\infty} \cos\frac{1}{n}$

36. $\displaystyle\sum_{n=0}^{\infty} \frac{e^n}{e^n + n}$

37. $\displaystyle\sum_{n=1}^{\infty} \ln\frac{1}{n}$

38. $\displaystyle\sum_{n=0}^{\infty} \cos n\pi$

Telescoping Series

In Exercises 39–44, find a formula for the nth partial sum of the series and use it to determine if the series converges or diverges. If a series converges, find its sum.

39. $\displaystyle\sum_{n=1}^{\infty} \left(\frac{1}{n} - \frac{1}{n+1}\right)$

40. $\displaystyle\sum_{n=1}^{\infty} \left(\frac{3}{n^2} - \frac{3}{(n+1)^2}\right)$

41. $\displaystyle\sum_{n=1}^{\infty} \left(\ln\sqrt{n+1} - \ln\sqrt{n}\right)$

42. $\displaystyle\sum_{n=1}^{\infty} (\tan(n) - \tan(n-1))$

43. $\displaystyle\sum_{n=1}^{\infty} \left(\cos^{-1}\left(\frac{1}{n+1}\right) - \cos^{-1}\left(\frac{1}{n+2}\right)\right)$

44. $\displaystyle\sum_{n=1}^{\infty} \left(\sqrt{n+4} - \sqrt{n+3}\right)$

Find the sum of each series in Exercises 45–52.

45. $\displaystyle\sum_{n=1}^{\infty} \frac{4}{(4n-3)(4n+1)}$

46. $\displaystyle\sum_{n=1}^{\infty} \frac{6}{(2n-1)(2n+1)}$

47. $\displaystyle\sum_{n=1}^{\infty} \frac{40n}{(2n-1)^2(2n+1)^2}$

48. $\displaystyle\sum_{n=1}^{\infty} \frac{2n+1}{n^2(n+1)^2}$

49. $\displaystyle\sum_{n=1}^{\infty} \left(\frac{1}{\sqrt{n}} - \frac{1}{\sqrt{n+1}}\right)$

50. $\displaystyle\sum_{n=1}^{\infty} \left(\frac{1}{2^{1/n}} - \frac{1}{2^{1/(n+1)}}\right)$

51. $\displaystyle\sum_{n=1}^{\infty} \left(\frac{1}{\ln(n+2)} - \frac{1}{\ln(n+1)}\right)$

52. $\displaystyle\sum_{n=1}^{\infty} (\tan^{-1}(n) - \tan^{-1}(n+1))$

Convergence or Divergence

Which series in Exercises 53–76 converge, and which diverge? Give reasons for your answers. If a series converges, find its sum.

53. $\displaystyle\sum_{n=0}^{\infty} \left(\frac{1}{\sqrt{2}}\right)^n$

54. $\displaystyle\sum_{n=0}^{\infty} \left(\sqrt{2}\right)^n$

55. $\displaystyle\sum_{n=1}^{\infty} (-1)^{n+1}\frac{3}{2^n}$

56. $\displaystyle\sum_{n=1}^{\infty} (-1)^{n+1} n$

57. $\displaystyle\sum_{n=0}^{\infty} \cos\left(\frac{n\pi}{2}\right)$

58. $\displaystyle\sum_{n=0}^{\infty} \frac{\cos n\pi}{5^n}$

59. $\displaystyle\sum_{n=0}^{\infty} e^{-2n}$

60. $\displaystyle\sum_{n=1}^{\infty} \ln\frac{1}{3^n}$

61. $\displaystyle\sum_{n=1}^{\infty} \frac{2}{10^n}$

62. $\displaystyle\sum_{n=0}^{\infty} \frac{1}{x^n}, \quad |x| > 1$

63. $\displaystyle\sum_{n=0}^{\infty} \frac{2^n - 1}{3^n}$

64. $\displaystyle\sum_{n=1}^{\infty} \left(1 - \frac{1}{n}\right)^n$

65. $\displaystyle\sum_{n=0}^{\infty} \frac{n!}{1000^n}$

66. $\displaystyle\sum_{n=1}^{\infty} \frac{n^n}{n!}$

67. $\displaystyle\sum_{n=1}^{\infty} \frac{2^n + 3^n}{4^n}$

68. $\displaystyle\sum_{n=1}^{\infty} \frac{2^n + 4^n}{3^n + 4^n}$

69. $\displaystyle\sum_{n=1}^{\infty} \ln\left(\frac{n}{n+1}\right)$

70. $\displaystyle\sum_{n=1}^{\infty} \ln\left(\frac{n}{2n+1}\right)$

71. $\displaystyle\sum_{n=0}^{\infty} \left(\frac{e}{\pi}\right)^n$

72. $\displaystyle\sum_{n=0}^{\infty} \frac{e^{n\pi}}{\pi^{ne}}$

73. $\displaystyle\sum_{n=1}^{\infty} \left(\frac{n}{n+1} - \frac{n+2}{n+3}\right)$

74. $\displaystyle\sum_{n=2}^{\infty} \left(\sin\left(\frac{\pi}{n}\right) - \sin\left(\frac{\pi}{n-1}\right)\right)$

75. $\displaystyle\sum_{n=1}^{\infty} \left(\cos\left(\frac{\pi}{n}\right) + \sin\left(\frac{\pi}{n}\right)\right)$

76. $\displaystyle\sum_{n=0}^{\infty} \left(\ln(4e^n - 1) - \ln(2e^n + 1)\right)$

Geometric Series with a Variable x

In each of the geometric series in Exercises 77–80, write out the first few terms of the series to find a and r, and find the sum of the series. Then express the inequality $|r| < 1$ in terms of x and find the values of x for which the inequality holds and the series converges.

77. $\displaystyle\sum_{n=0}^{\infty} (-1)^n x^n$

78. $\displaystyle\sum_{n=0}^{\infty} (-1)^n x^{2n}$

79. $\displaystyle\sum_{n=0}^{\infty} 3\left(\frac{x-1}{2}\right)^n$

80. $\displaystyle\sum_{n=0}^{\infty} \frac{(-1)^n}{2}\left(\frac{1}{3+\sin x}\right)^n$

In Exercises 81–86, find the values of x for which the given geometric series converges. Also, find the sum of the series (as a function of x) for those values of x.

81. $\displaystyle\sum_{n=0}^{\infty} 2^n x^n$

82. $\displaystyle\sum_{n=0}^{\infty} (-1)^n x^{-2n}$

83. $\displaystyle\sum_{n=0}^{\infty} (-1)^n (x+1)^n$

84. $\displaystyle\sum_{n=0}^{\infty} \left(-\frac{1}{2}\right)^n (x-3)^n$

85. $\displaystyle\sum_{n=0}^{\infty} \sin^n x$

86. $\displaystyle\sum_{n=0}^{\infty} (\ln x)^n$

Theory and Examples

87. The series in Exercise 5 can also be written as

$$\sum_{n=1}^{\infty} \frac{1}{(n+1)(n+2)} \quad \text{and} \quad \sum_{n=-1}^{\infty} \frac{1}{(n+3)(n+4)}.$$

Write it as a sum beginning with **(a)** $n = -2$, **(b)** $n = 0$, **(c)** $n = 5$.

88. The series in Exercise 6 can also be written as

$$\sum_{n=1}^{\infty} \frac{5}{n(n+1)} \quad \text{and} \quad \sum_{n=0}^{\infty} \frac{5}{(n+1)(n+2)}.$$

Write it as a sum beginning with **(a)** $n = -1$, **(b)** $n = 3$, **(c)** $n = 20$.

89. Make up an infinite series of nonzero terms whose sum is

  a. 1   **b.** -3 **c.** 0.

90. (*Continuation of Exercise 89.*) Can you make an infinite series of nonzero terms that converges to any number you want? Explain.

91. Show by example that $\sum(a_n/b_n)$ may diverge even though $\sum a_n$ and $\sum b_n$ converge and no b_n equals 0.

92. Find convergent geometric series $A = \Sigma a_n$ and $B = \Sigma b_n$ that illustrate the fact that $\Sigma a_n b_n$ may converge without being equal to AB.

93. Show by example that $\Sigma(a_n/b_n)$ may converge to something other than A/B even when $A = \Sigma a_n$, $B = \Sigma b_n \neq 0$, and no b_n equals 0.

94. If Σa_n converges and $a_n > 0$ for all n, can anything be said about $\Sigma(1/a_n)$? Give reasons for your answer.

95. What happens if you add a finite number of terms to a divergent series or delete a finite number of terms from a divergent series? Give reasons for your answer.

96. If Σa_n converges and Σb_n diverges, can anything be said about their term-by-term sum $\Sigma(a_n + b_n)$? Give reasons for your answer.

97. Make up a geometric series Σar^{n-1} that converges to the number 5 if

   a. $a = 2$ **b.** $a = 13/2$.

98. Find the value of b for which

$$1 + e^b + e^{2b} + e^{3b} + \cdots = 9.$$

99. For what values of r does the infinite series

$$1 + 2r + r^2 + 2r^3 + r^4 + 2r^5 + r^6 + \cdots$$

converge? Find the sum of the series when it converges.

100. The accompanying figure shows the first five of a sequence of squares. The outermost square has an area of 4 m^2. Each of the other squares is obtained by joining the midpoints of the sides of the squares before it. Find the sum of the areas of all the squares.

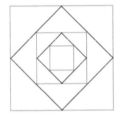

101. Drug dosage A patient takes a 300 mg tablet for the control of high blood pressure every morning at the same time. The concentration of the drug in the patient's system decays exponentially at a constant hourly rate of $k = 0.12$.

   a. How many milligrams of the drug are in the patient's system just before the second tablet is taken? Just before the third tablet is taken?

   b. In the long run, after taking the medication for at least six months, what quantity of drug is in the patient's body just before taking the next regularly scheduled morning tablet?

102. Show that the error $(L - s_n)$ obtained by replacing a convergent geometric series with one of its partial sums s_n is $ar^n/(1 - r)$.

103. The Cantor set To construct this set, we begin with the closed interval $[0, 1]$. From that interval, remove the middle open interval $(1/3, 2/3)$, leaving the two closed intervals $[0, 1/3]$ and $[2/3, 1]$. At the second step we remove the open middle third interval from each of those remaining. From $[0, 1/3]$ we remove the open interval $(1/9, 2/9)$, and from $[2/3, 1]$ we remove $(7/9, 8/9)$, leaving behind the four closed intervals $[0, 1/9]$, $[2/9, 1/3]$, $[2/3, 7/9]$, and $[8/9, 1]$. At the next step, we remove the middle open third interval from each closed interval left behind, so $(1/27, 2/27)$ is removed from $[0, 1/9]$, leaving the closed intervals $[0, 1/27]$ and $[2/27, 1/9]$; $(7/27, 8/27)$ is removed from $[2/9, 1/3]$, leaving behind $[2/9, 7/27]$ and $[8/27, 1/3]$, and so forth. We continue this process repeatedly without stopping, at each step removing the open third interval from every closed interval remaining behind from the preceding step. The numbers remaining in the interval $[0, 1]$, after all open middle third intervals have been removed, are the points in the Cantor set (named after Georg Cantor, 1845–1918). The set has some interesting properties.

   a. The Cantor set contains infinitely many numbers in $[0, 1]$. List 12 numbers that belong to the Cantor set.

   b. Show, by summing an appropriate geometric series, that the total length of all the open middle third intervals that have been removed from $[0, 1]$ is equal to 1.

104. Helga von Koch's snowflake curve Helga von Koch's snowflake is a curve of infinite length that encloses a region of finite area. To see why this is so, suppose the curve is generated by starting with an equilateral triangle whose sides have length 1.

   a. Find the length L_n of the nth curve C_n and show that $\lim_{n \to \infty} L_n = \infty$.

   b. Find the area A_n of the region enclosed by C_n and show that $\lim_{n \to \infty} A_n = (8/5)A_1$.

C_1 C_2 C_3 C_4

105. The largest circle in the accompanying figure has radius 1. Consider the sequence of circles of maximum area inscribed in semicircles of diminishing size. What is the sum of the areas of all of the circles?

10.3 The Integral Test

The most basic question we can ask about a series is whether it converges. In this section we begin to study this question, starting with series that have nonnegative terms. Such a series converges if its sequence of partial sums is bounded. If we establish that a given series does converge, we generally do not have a formula available for its sum. So to get an estimate for the sum of a convergent series, we investigate the error involved when using a partial sum to approximate the total sum.

Nondecreasing Partial Sums

Suppose that $\sum_{n=1}^{\infty} a_n$ is an infinite series with $a_n \geq 0$ for all n. Then each partial sum is greater than or equal to its predecessor because $s_{n+1} = s_n + a_n$, so

$$s_1 \leq s_2 \leq s_3 \leq \cdots \leq s_n \leq s_{n+1} \leq \cdots .$$

Since the partial sums form a nondecreasing sequence, the Monotonic Sequence Theorem (Theorem 6, Section 10.1) gives the following result.

> **Corollary of Theorem 6**
> A series $\sum_{n=1}^{\infty} a_n$ of nonnegative terms converges if and only if its partial sums are bounded from above.

EXAMPLE 1   As an application of the above corollary, consider the **harmonic series**

$$\sum_{n=1}^{\infty} \frac{1}{n} = 1 + \frac{1}{2} + \frac{1}{3} + \cdots + \frac{1}{n} + \cdots .$$

Although the nth term $1/n$ does go to zero, the series diverges because there is no upper bound for its partial sums. To see why, group the terms of the series in the following way:

$$1 + \frac{1}{2} + \underbrace{\left(\frac{1}{3} + \frac{1}{4}\right)}_{>\frac{2}{4} = \frac{1}{2}} + \underbrace{\left(\frac{1}{5} + \frac{1}{6} + \frac{1}{7} + \frac{1}{8}\right)}_{>\frac{4}{8} = \frac{1}{2}} + \underbrace{\left(\frac{1}{9} + \frac{1}{10} + \cdots + \frac{1}{16}\right)}_{>\frac{8}{16} = \frac{1}{2}} + \cdots .$$

The sum of the first two terms is 1.5. The sum of the next two terms is $1/3 + 1/4$, which is greater than $1/4 + 1/4 = 1/2$. The sum of the next four terms is $1/5 + 1/6 + 1/7 + 1/8$, which is greater than $1/8 + 1/8 + 1/8 + 1/8 = 1/2$. The sum of the next eight terms is $1/9 + 1/10 + 1/11 + 1/12 + 1/13 + 1/14 + 1/15 + 1/16$, which is greater than $8/16 = 1/2$. The sum of the next 16 terms is greater than $16/32 = 1/2$, and so on. In general, the sum of 2^n terms ending with $1/2^{n+1}$ is greater than $2^n/2^{n+1} = 1/2$. If $n = 2^k$, the partial sum s_n is greater than $k/2$, so the sequence of partial sums is not bounded from above. The harmonic series diverges. ∎

The Integral Test

We now introduce the Integral Test with a series that is related to the harmonic series, but whose nth term is $1/n^2$ instead of $1/n$.

EXAMPLE 2   Does the following series converge?

$$\sum_{n=1}^{\infty} \frac{1}{n^2} = 1 + \frac{1}{4} + \frac{1}{9} + \frac{1}{16} + \cdots + \frac{1}{n^2} + \cdots$$

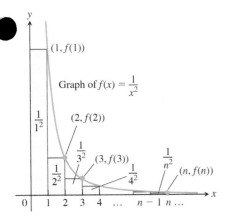

FIGURE 10.11 The sum of the areas of the rectangles under the graph of $f(x) = 1/x^2$ is less than the area under the graph (Example 2).

Solution We determine the convergence of $\sum_{n=1}^{\infty}(1/n^2)$ by comparing it with $\int_1^{\infty}(1/x^2)\,dx$. To carry out the comparison, we think of the terms of the series as values of the function $f(x) = 1/x^2$ and interpret these values as the areas of rectangles under the curve $y = 1/x^2$.

As Figure 10.11 shows,

$$s_n = \frac{1}{1^2} + \frac{1}{2^2} + \frac{1}{3^2} + \cdots + \frac{1}{n^2}$$

$$= f(1) + f(2) + f(3) + \cdots + f(n)$$

$$< f(1) + \int_1^n \frac{1}{x^2}\,dx \qquad \text{Rectangle areas sum to less than area under graph.}$$

$$< 1 + \int_1^{\infty} \frac{1}{x^2}\,dx \qquad \int_1^n (1/x^2)\,dx < \int_1^{\infty}(1/x^2)\,dx$$

$$< 1 + 1 = 2. \qquad \text{As in Section 8.8, Example 3,}\ \int_1^{\infty}(1/x^2)\,dx = 1.$$

Thus the partial sums of $\sum_{n=1}^{\infty}\left(1/n^2\right)$ are bounded from above (by 2) and the series converges. ∎

Caution
The series and integral need not have the same value in the convergent case. You will see in Example 6 that

$$\sum_{n=1}^{\infty}\left(1/n^2\right) \neq \int_1^{\infty}(1/x^2)\,dx = 1.$$

THEOREM 9—The Integral Test
Let $\{a_n\}$ be a sequence of positive terms. Suppose that $a_n = f(n)$, where f is a continuous, positive, decreasing function of x for all $x \geq N$ (N a positive integer). Then the series $\sum_{n=N}^{\infty} a_n$ and the integral $\int_N^{\infty} f(x)\,dx$ both converge or both diverge.

Proof We establish the test for the case $N = 1$. The proof for general N is similar.

We start with the assumption that f is a decreasing function with $f(n) = a_n$ for every n. This leads us to observe that the rectangles in Figure 10.12a, which have areas $a_1, a_2, \ldots, a_n$, collectively enclose more area than that under the curve $y = f(x)$ from $x = 1$ to $x = n + 1$. That is,

$$\int_1^{n+1} f(x)\,dx \leq a_1 + a_2 + \cdots + a_n.$$

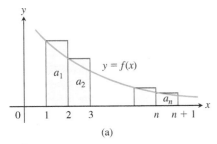

(a)

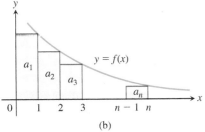

(b)

FIGURE 10.12 Subject to the conditions of the Integral Test, the series $\sum_{n=1}^{\infty} a_n$ and the integral $\int_1^{\infty} f(x)\,dx$ both converge or both diverge.

In Figure 10.12b the rectangles have been faced to the left instead of to the right. If we momentarily disregard the first rectangle of area a_1, we see that

$$a_2 + a_3 + \cdots + a_n \leq \int_1^n f(x)\,dx.$$

If we include a_1, we have

$$a_1 + a_2 + \cdots + a_n \leq a_1 + \int_1^n f(x)\,dx.$$

Combining these results gives

$$\int_1^{n+1} f(x)\,dx \leq a_1 + a_2 + \cdots + a_n \leq a_1 + \int_1^n f(x)\,dx.$$

These inequalities hold for each n, and continue to hold as $n \to \infty$.

If $\int_1^{\infty} f(x)\,dx$ is finite, the right-hand inequality shows that $\sum a_n$ is finite. If $\int_1^{\infty} f(x)\,dx$ is infinite, the left-hand inequality shows that $\sum a_n$ is infinite. Hence the series and the integral are either both finite or both infinite. ∎

The *p*-series $\displaystyle\sum_{n=1}^{\infty} \frac{1}{n^p}$ converges if $p > 1$, diverges if $p \leq 1$.

EXAMPLE 3   Show that the ***p*-series**

$$\sum_{n=1}^{\infty} \frac{1}{n^p} = \frac{1}{1^p} + \frac{1}{2^p} + \frac{1}{3^p} + \cdots + \frac{1}{n^p} + \cdots$$

(p a real constant) converges if $p > 1$, and diverges if $p \leq 1$.

Solution   If $p > 1$, then $f(x) = 1/x^p$ is a positive decreasing function of x. Since

$$\int_1^{\infty} \frac{1}{x^p}\, dx = \int_1^{\infty} x^{-p}\, dx = \lim_{b \to \infty} \left[\frac{x^{-p+1}}{-p+1} \right]_1^b \qquad \text{Evaluate the improper integral}$$

$$= \frac{1}{1-p} \lim_{b \to \infty} \left(\frac{1}{b^{p-1}} - 1 \right)$$

$$= \frac{1}{1-p}(0 - 1) = \frac{1}{p-1}, \qquad \begin{array}{l} b^{p-1} \to \infty \text{ as } b \to \infty \\ \text{because } p - 1 > 0. \end{array}$$

the series converges by the Integral Test. We emphasize that the sum of the *p*-series is *not* $1/(p-1)$. The series converges, but we don't know the value it converges to.

If $p \leq 0$, the series diverges by the nth-term test. If $0 < p < 1$, then $1 - p > 0$ and

$$\int_1^{\infty} \frac{1}{x^p}\, dx = \frac{1}{1-p} \lim_{b \to \infty} (b^{1-p} - 1) = \infty.$$

Therefore, the series diverges by the Integral Test.

If $p = 1$, we have the (divergent) harmonic series

$$1 + \frac{1}{2} + \frac{1}{3} + \cdots + \frac{1}{n} + \cdots.$$

In summary, we have convergence for $p > 1$ but divergence for all other values of p. ∎

The *p*-series with $p = 1$ is the **harmonic series** (Example 1). The *p*-Series Test shows that the harmonic series is just *barely* divergent; if we increase p to 1.000000001, for instance, the series converges!

The slowness with which the partial sums of the harmonic series approach infinity is impressive. For instance, it takes more than 178 million terms of the harmonic series to move the partial sums beyond 20. (See also Exercise 49b.)

EXAMPLE 4   The series $\sum_{n=1}^{\infty}(1/(n^2 + 1))$ is not a *p*-series, but it converges by the Integral Test. The function $f(x) = 1/(x^2 + 1)$ is positive, continuous, and decreasing for $x \geq 1$, and

$$\int_1^{\infty} \frac{1}{x^2 + 1}\, dx = \lim_{b \to \infty} \left[\arctan x \right]_1^b$$

$$= \lim_{b \to \infty} \left[\arctan b - \arctan 1 \right]$$

$$= \frac{\pi}{2} - \frac{\pi}{4} = \frac{\pi}{4}.$$

The Integral Test tells us that the series converges, but it does *not* say that $\pi/4$ or any other number is the sum of the series. ∎

EXAMPLE 5   Determine the convergence or divergence of the series.

(a) $\displaystyle\sum_{n=1}^{\infty} ne^{-n^2}$ **(b)** $\displaystyle\sum_{n=1}^{\infty} \frac{1}{2^{\ln n}}$

Solutions

(a) We apply the Integral Test and find that

$$\int_1^\infty \frac{x}{e^{x^2}}\, dx = \frac{1}{2}\int_1^\infty \frac{du}{e^u} \qquad u = x^2, du = 2x\, dx$$

$$= \lim_{b\to\infty}\left[-\frac{1}{2}e^{-u}\right]_1^b$$

$$= \lim_{b\to\infty}\left(-\frac{1}{2e^b} + \frac{1}{2e}\right) = \frac{1}{2e}.$$

Since the integral converges, the series also converges.

(b) Again applying the Integral Test,

$$\int_1^\infty \frac{dx}{2^{\ln x}} = \int_0^\infty \frac{e^u\, du}{2^u} \qquad u = \ln x, x = e^u, dx = e^u\, du$$

$$= \int_0^\infty \left(\frac{e}{2}\right)^u du$$

$$= \lim_{b\to\infty}\frac{1}{\ln\left(\frac{e}{2}\right)}\left(\left(\frac{e}{2}\right)^b - 1\right) = \infty. \qquad (e/2) > 1$$

The improper integral diverges, so the series diverges also.   ▪

Error Estimation

For some convergent series, such as the geometric series or the telescoping series in Example 5 of Section 10.2, we can actually find the total sum of the series. That is, we can find the limiting value S of the sequence of partial sums. For most convergent series, however, we cannot easily find the total sum. Nevertheless, we can *estimate* the sum by adding the first n terms to get s_n, but we need to know how far off s_n is from the total sum S. An approximation to a function or to a number is more useful when it is accompanied by a bound on the size of the worst possible error that could occur. With such an error bound we can try to make an estimate or approximation that is close enough for the problem at hand. Without a bound on the error size, we are just guessing and hoping that we are close to the actual answer. We now show a way to bound the error size using integrals.

Suppose that a series Σa_n with positive terms is shown to be convergent by the Integral Test, and we want to estimate the size of the **remainder** R_n measuring the difference between the total sum S of the series and its nth partial sum s_n. That is, we wish to estimate

$$R_n = S - s_n = a_{n+1} + a_{n+2} + a_{n+3} + \cdots.$$

To get a lower bound for the remainder, we compare the sum of the areas of the rectangles with the area under the curve $y = f(x)$ for $x \geq n$ (see Figure 10.13a). We see that

$$R_n = a_{n+1} + a_{n+2} + a_{n+3} + \cdots \geq \int_{n+1}^\infty f(x)\, dx.$$

Similarly, from Figure 10.13b, we find an upper bound with

$$R_n = a_{n+1} + a_{n+2} + a_{n+3} + \cdots \leq \int_n^\infty f(x)\, dx.$$

These comparisons prove the following result, giving bounds on the size of the remainder.

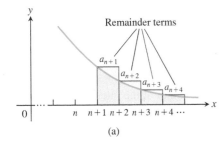

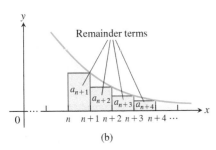

FIGURE 10.13 The remainder when using n terms is (a) larger than the integral of f over $[n + 1, \infty)$. (b) smaller than the integral of f over $[n, \infty)$.

> **Bounds for the Remainder in the Integral Test**
> Suppose $\{a_k\}$ is a sequence of positive terms with $a_k = f(k)$, where f is a continuous positive decreasing function of x for all $x \geq n$, and that Σa_n converges to S. Then the remainder $R_n = S - s_n$ satisfies the inequalities
>
> $$\int_{n+1}^{\infty} f(x)\, dx \leq R_n \leq \int_{n}^{\infty} f(x)\, dx. \tag{1}$$

If we add the partial sum s_n to each side of the inequalities in (1), we get

$$s_n + \int_{n+1}^{\infty} f(x)\, dx \leq S \leq s_n + \int_{n}^{\infty} f(x)\, dx \tag{2}$$

since $s_n + R_n = S$. The inequalities in (2) are useful for estimating the error in approximating the sum of a series known to converge by the Integral Test. The error can be no larger than the length of the interval containing S, with endpoints given by (2).

EXAMPLE 6 Estimate the sum of the series $\Sigma(1/n^2)$ using the inequalities in (2) and $n = 10$.

Solution We have that

$$\int_{n}^{\infty} \frac{1}{x^2}\, dx = \lim_{b \to \infty}\left[-\frac{1}{x}\right]_{n}^{b} = \lim_{b \to \infty}\left(-\frac{1}{b} + \frac{1}{n}\right) = \frac{1}{n}.$$

Using this result with the inequalities in (2), we get

$$s_{10} + \frac{1}{11} \leq S \leq s_{10} + \frac{1}{10}.$$

Taking $s_{10} = 1 + (1/4) + (1/9) + (1/16) + \cdots + (1/100) \approx 1.54977$, these last inequalities give

$$1.64068 \leq S \leq 1.64977.$$

If we approximate the sum S by the midpoint of this interval, we find that

$$\sum_{n=1}^{\infty} \frac{1}{n^2} \approx 1.6452.$$

The p-series for $p = 2$
$$\sum_{n=1}^{\infty} \frac{1}{n^2} = \frac{\pi^2}{6} \approx 1.64493$$

The error in this approximation is then less than half the length of the interval, so the error is less than 0.005. Using a trigonometric *Fourier series* (studied in advanced calculus), it can be shown that S is equal to $\pi^2/6 \approx 1.64493$. ∎

EXERCISES 10.3

Applying the Integral Test
Use the Integral Test to determine if the series in Exercises 1–12 converge or diverge. Be sure to check that the conditions of the Integral Test are satisfied.

1. $\displaystyle\sum_{n=1}^{\infty} \frac{1}{n^2}$

2. $\displaystyle\sum_{n=1}^{\infty} \frac{1}{n^{0.2}}$

3. $\displaystyle\sum_{n=1}^{\infty} \frac{1}{n^2 + 4}$

4. $\displaystyle\sum_{n=1}^{\infty} \frac{1}{n + 4}$

5. $\displaystyle\sum_{n=1}^{\infty} e^{-2n}$

6. $\displaystyle\sum_{n=2}^{\infty} \frac{1}{n(\ln n)^2}$

7. $\displaystyle\sum_{n=1}^{\infty} \frac{n}{n^2 + 4}$

8. $\displaystyle\sum_{n=2}^{\infty} \frac{\ln(n^2)}{n}$

9. $\displaystyle\sum_{n=1}^{\infty} \frac{n^2}{e^{n/3}}$

10. $\displaystyle\sum_{n=2}^{\infty} \frac{n - 4}{n^2 - 2n + 1}$

11. $\displaystyle\sum_{n=1}^{\infty} \frac{7}{\sqrt{n} + 4}$

12. $\displaystyle\sum_{n=2}^{\infty} \frac{1}{5n + 10\sqrt{n}}$

Determining Convergence or Divergence

Which of the series in Exercises 13–46 converge, and which diverge? Give reasons for your answers. (When you check an answer, remember that there may be more than one way to determine the series' convergence or divergence.)

13. $\sum_{n=1}^{\infty} \dfrac{1}{10^n}$ **14.** $\sum_{n=1}^{\infty} e^{-n}$ **15.** $\sum_{n=1}^{\infty} \dfrac{n}{n+1}$

16. $\sum_{n=1}^{\infty} \dfrac{5}{n+1}$ **17.** $\sum_{n=1}^{\infty} \dfrac{3}{\sqrt{n}}$ **18.** $\sum_{n=1}^{\infty} \dfrac{-2}{n\sqrt{n}}$

19. $\sum_{n=1}^{\infty} -\dfrac{1}{8^n}$ **20.** $\sum_{n=1}^{\infty} \dfrac{-8}{n}$ **21.** $\sum_{n=2}^{\infty} \dfrac{\ln n}{n}$

22. $\sum_{n=2}^{\infty} \dfrac{\ln n}{\sqrt{n}}$ **23.** $\sum_{n=1}^{\infty} \dfrac{2^n}{3^n}$ **24.** $\sum_{n=1}^{\infty} \dfrac{5^n}{4^n+3}$

25. $\sum_{n=0}^{\infty} \dfrac{-2}{n+1}$ **26.** $\sum_{n=1}^{\infty} \dfrac{1}{2n-1}$ **27.** $\sum_{n=1}^{\infty} \dfrac{2^n}{n+1}$

28. $\sum_{n=1}^{\infty} \left(1 + \dfrac{1}{n}\right)^n$ **29.** $\sum_{n=2}^{\infty} \dfrac{\sqrt{n}}{\ln n}$ **30.** $\sum_{n=1}^{\infty} \dfrac{1}{\sqrt{n}(\sqrt{n}+1)}$

31. $\sum_{n=1}^{\infty} \dfrac{1}{(\ln 2)^n}$ **32.** $\sum_{n=1}^{\infty} \dfrac{1}{(\ln 3)^n}$

33. $\sum_{n=3}^{\infty} \dfrac{(1/n)}{(\ln n)\sqrt{\ln^2 n - 1}}$ **34.** $\sum_{n=1}^{\infty} \dfrac{1}{n(1 + \ln^2 n)}$

35. $\sum_{n=1}^{\infty} n \sin \dfrac{1}{n}$ **36.** $\sum_{n=1}^{\infty} n \tan \dfrac{1}{n}$

37. $\sum_{n=1}^{\infty} \dfrac{e^n}{1 + e^{2n}}$ **38.** $\sum_{n=1}^{\infty} \dfrac{2}{1 + e^n}$

39. $\sum_{n=1}^{\infty} \dfrac{e^n}{10 + e^n}$ **40.** $\sum_{n=1}^{\infty} \dfrac{e^n}{(10 + e^n)^2}$

41. $\sum_{n=2}^{\infty} \dfrac{\sqrt{n+2} - \sqrt{n+1}}{\sqrt{n+1}\sqrt{n+2}}$ **42.** $\sum_{n=3}^{\infty} \dfrac{7}{\sqrt{n+1}\ln\sqrt{n+1}}$

43. $\sum_{n=1}^{\infty} \dfrac{8 \tan^{-1} n}{1 + n^2}$ **44.** $\sum_{n=1}^{\infty} \dfrac{n}{n^2 + 1}$

45. $\sum_{n=1}^{\infty} \operatorname{sech} n$ **46.** $\sum_{n=1}^{\infty} \operatorname{sech}^2 n$

Theory and Examples

For what values of a, if any, do the series in Exercises 47 and 48 converge?

47. $\sum_{n=1}^{\infty} \left(\dfrac{a}{n+2} - \dfrac{1}{n+4}\right)$ **48.** $\sum_{n=3}^{\infty} \left(\dfrac{1}{n-1} - \dfrac{2a}{n+1}\right)$

49. a. Draw illustrations like those in Figures 10.12a and 10.12b to show that the partial sums of the harmonic series satisfy the inequalities

$$\ln(n+1) = \int_1^{n+1} \dfrac{1}{x}\,dx \le 1 + \dfrac{1}{2} + \cdots + \dfrac{1}{n}$$

$$\le 1 + \int_1^n \dfrac{1}{x}\,dx = 1 + \ln n.$$

T **b.** There is absolutely no empirical evidence for the divergence of the harmonic series even though we know it diverges. The

partial sums just grow too slowly. To see what we mean, suppose you had started with $s_1 = 1$ the day the universe was formed, 13 billion years ago, and added a new term every *second*. About how large would the partial sum s_n be today, assuming a 365-day year?

50. Are there any values of x for which $\sum_{n=1}^{\infty}(1/nx)$ converges? Give reasons for your answer.

51. Is it true that if $\sum_{n=1}^{\infty} a_n$ is a divergent series of positive numbers, then there is also a divergent series $\sum_{n=1}^{\infty} b_n$ of positive numbers with $b_n < a_n$ for every n? Is there a "smallest" divergent series of positive numbers? Give reasons for your answers.

52. (*Continuation of Exercise 51.*) Is there a "largest" convergent series of positive numbers? Explain.

53. $\sum_{n=1}^{\infty} \left(1/\sqrt{n+1}\right)$ **diverges**

a. Use the accompanying graph to show that the partial sum $s_{50} = \sum_{n=1}^{50}\left(1/\sqrt{n+1}\right)$ satisfies

$$\int_1^{51} \dfrac{1}{\sqrt{x+1}}\,dx < s_{50} < \int_0^{50} \dfrac{1}{\sqrt{x+1}}\,dx.$$

Conclude that $11.5 < s_{50} < 12.3$.

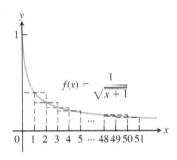

b. What should n be in order that the partial sum

$$s_n = \sum_{i=1}^{n}\left(1/\sqrt{i+1}\right) \text{ satisfy } s_n > 1000?$$

54. $\sum_{n=1}^{\infty} (1/n^4)$ **converges**

a. Use the accompanying graph to find an upper bound for the error if $s_{30} = \sum_{n=1}^{30}(1/n^4)$ is used to estimate the value of $\sum_{n=1}^{\infty}(1/n^4)$.

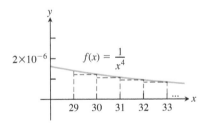

b. Find n so that the partial sum $s_n = \sum_{i=1}^{n}(1/i^4)$ estimates the value of $\sum_{n=1}^{\infty}(1/n^4)$ with an error of at most 0.000001.

55. Estimate the value of $\sum_{n=1}^{\infty}(1/n^3)$ to within 0.01 of its exact value.

56. Estimate the value of $\sum_{n=2}^{\infty}(1/(n^2+4))$ to within 0.1 of its exact value.

57. How many terms of the convergent series $\sum_{n=1}^{\infty}(1/n^{1.1})$ should be used to estimate its value with error at most 0.00001?

58. How many terms of the convergent series $\sum_{n=4}^{\infty} 1/(n(\ln n)^3)$ should be used to estimate its value with error at most 0.01?

59. The Cauchy condensation test The Cauchy condensation test says: Let $\{a_n\}$ be a nonincreasing sequence ($a_n \geq a_{n+1}$ for all n) of positive terms that converges to 0. Then $\sum a_n$ converges if and only if $\sum 2^n a_{2^n}$ converges. For example, $\sum(1/n)$ diverges because $\sum 2^n \cdot (1/2^n) = \sum 1$ diverges. Show why the test works.

60. Use the Cauchy condensation test from Exercise 59 to show that

a. $\displaystyle\sum_{n=2}^{\infty} \frac{1}{n \ln n}$ diverges;

b. $\displaystyle\sum_{n=1}^{\infty} \frac{1}{n^p}$ converges if $p > 1$ and diverges if $p \leq 1$.

61. Logarithmic p-series

a. Show that the improper integral

$$\int_2^{\infty} \frac{dx}{x(\ln x)^p} \quad (p \text{ a positive constant})$$

converges if and only if $p > 1$.

b. What implications does the fact in part (a) have for the convergence of the series

$$\sum_{n=2}^{\infty} \frac{1}{n(\ln n)^p}?$$

Give reasons for your answer.

62. (*Continuation of Exercise 61.*) Use the result in Exercise 61 to determine which of the following series converge and which diverge. Support your answer in each case.

a. $\displaystyle\sum_{n=2}^{\infty} \frac{1}{n(\ln n)}$
b. $\displaystyle\sum_{n=2}^{\infty} \frac{1}{n(\ln n)^{1.01}}$

c. $\displaystyle\sum_{n=2}^{\infty} \frac{1}{n \ln(n^3)}$
d. $\displaystyle\sum_{n=2}^{\infty} \frac{1}{n(\ln n)^3}$

63. Euler's constant Graphs like those in Figure 10.12 suggest that as n increases there is little change in the difference between the sum

$$1 + \frac{1}{2} + \cdots + \frac{1}{n}$$

and the integral

$$\ln n = \int_1^n \frac{1}{x} dx.$$

To explore this idea, carry out the following steps.

a. By taking $f(x) = 1/x$ in the proof of Theorem 9, show that

$$\ln(n+1) \leq 1 + \frac{1}{2} + \cdots + \frac{1}{n} \leq 1 + \ln n$$

or

$$0 < \ln(n+1) - \ln n \leq 1 + \frac{1}{2} + \cdots + \frac{1}{n} - \ln n \leq 1.$$

Thus, the sequence

$$a_n = 1 + \frac{1}{2} + \cdots + \frac{1}{n} - \ln n$$

is bounded from below and from above.

b. Show that

$$\frac{1}{n+1} < \int_n^{n+1} \frac{1}{x} dx = \ln(n+1) - \ln n,$$

and use this result to show that the sequence $\{a_n\}$ in part (a) is decreasing.

Since a decreasing sequence that is bounded from below converges, the numbers a_n defined in part (a) converge:

$$1 + \frac{1}{2} + \cdots + \frac{1}{n} - \ln n \to \gamma.$$

The number γ, whose value is $0.5772\ldots$, is called *Euler's constant*.

64. Use the Integral Test to show that the series

$$\sum_{n=0}^{\infty} e^{-n^2}$$

converges.

65. a. For the series $\sum(1/n^3)$, use the inequalities in Equation (2) with $n = 10$ to find an interval containing the sum S.

b. As in Example 5, use the midpoint of the interval found in part (a) to approximate the sum of the series. What is the maximum error for your approximation?

66. Repeat Exercise 65 using the series $\sum(1/n^4)$.

67. Area Consider the sequence $\{1/n\}_{n=1}^{\infty}$. On each subinterval $(1/(n+1), 1/n)$ within the interval $[0, 1]$, erect the rectangle with area a_n having height $1/n$ and width equal to the length of the subinterval. Find the total area $\sum a_n$ of all the rectangles. (*Hint:* Use the result of Example 5 in Section 10.2.)

68. Area Repeat Exercise 67, using trapezoids instead of rectangles. That is, on the subinterval $(1/(n+1), 1/n)$, let a_n denote the area of the trapezoid having heights $y = 1/(n+1)$ at $x = 1/(n+1)$ and $y = 1/n$ at $x = 1/n$.

10.4 Comparison Tests

We have seen how to determine the convergence of geometric series, p-series, and a few others. We can test the convergence of many more series by comparing their terms to those of a series whose convergence is already known.

> **THEOREM 10—Direct Comparison Test**
>
> Let Σa_n and Σb_n be two series with $0 \le a_n \le b_n$ for all n. Then
>
> **1.** If Σb_n converges, then Σa_n also converges.
>
> **2.** If Σa_n diverges, then Σb_n also diverges.

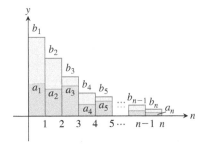

FIGURE 10.14 If the total area Σb_n of the taller b_n rectangles is finite, then so is the total area Σa_n of the shorter a_n rectangles.

Proof   The series Σa_n and Σb_n have nonnegative terms. The Corollary of Theorem 6 stated in Section 10.3 tells us that the series Σa_n and Σb_n converge if and only if their partial sums are bounded from above.

In Part (1) we assume that Σb_n converges to some number M. The partial sums $\sum_{n=1}^{N} a_n$ are all bounded from above by $M = \Sigma b_n$, since

$$s_N = a_1 + a_2 + \cdots + a_N \le b_1 + b_2 + \cdots + b_N \le \sum_{n=1}^{\infty} b_n = M.$$

Since the partial sums of Σa_n are bounded from above, the Corollary of Theorem 6 implies that Σa_n converges. We conclude that when Σb_n converges, then so does Σa_n. Figure 10.12 illustrates this result, with each term of each series interpreted as the area of a rectangle.

In Part (2), where we assume that Σa_n diverges, the partial sums of $\sum_{n=1}^{\infty} b_n$ are not bounded from above. If they were, the partial sums for Σa_n would also be bounded from above, since

$$a_1 + a_2 + \cdots + a_N \le b_1 + b_2 + \cdots + b_N,$$

and this would mean that Σa_n converges. We conclude that if Σa_n diverges, then so does Σb_n. ∎

EXAMPLE 1   We apply Theorem 10 to several series.

(a) The series

$$\sum_{n=1}^{\infty} \frac{5}{5n - 1}$$

diverges because its nth term

$$\frac{5}{5n - 1} = \frac{1}{n - \frac{1}{5}} > \frac{1}{n}$$

is greater than the nth term of the divergent harmonic series.

(b) The series

$$\sum_{n=0}^{\infty} \frac{1}{n!} = 1 + \frac{1}{1!} + \frac{1}{2!} + \frac{1}{3!} + \cdots$$

converges because its terms are all positive and less than or equal to the corresponding terms of

$$1 + \sum_{n=0}^{\infty} \frac{1}{2^n} = 1 + 1 + \frac{1}{2} + \frac{1}{2^2} + \cdots.$$

The geometric series on the left converges and we have

$$1 + \sum_{n=0}^{\infty} \frac{1}{2^n} = 1 + \frac{1}{1 - (1/2)} = 3.$$

The fact that 3 is an upper bound for the partial sums of $\sum_{n=0}^{\infty} (1/n!)$ does not mean that the series converges to 3. As we will see in Section 10.9, the series converges to e.

HISTORICAL BIOGRAPHY

Albert of Saxony
(ca. 1316–1390)
bit.ly/2DMOp6u

(c) The series

$$5 + \frac{2}{3} + \frac{1}{7} + 1 + \frac{1}{2 + \sqrt{1}} + \frac{1}{4 + \sqrt{2}} + \frac{1}{8 + \sqrt{3}} + \cdots + \frac{1}{2^n + \sqrt{n}} + \cdots$$

converges. To see this, we ignore the first three terms and compare the remaining terms with those of the convergent geometric series $\sum_{n=0}^{\infty}(1/2^n)$. The term $1/(2^n + \sqrt{n})$ of the truncated sequence is less than the corresponding term $1/2^n$ of the geometric series. We see that term by term we have the comparison

$$1 + \frac{1}{2 + \sqrt{1}} + \frac{1}{4 + \sqrt{2}} + \frac{1}{8 + \sqrt{3}} + \cdots \leq 1 + \frac{1}{2} + \frac{1}{4} + \frac{1}{8} + \cdots.$$

So the truncated series and the original series converge by an application of the Direct Comparison Test. ∎

The Limit Comparison Test

We now introduce a comparison test that is particularly useful for series in which a_n is a rational function of n.

THEOREM 11—Limit Comparison Test
Suppose that $a_n > 0$ and $b_n > 0$ for all $n \geq N$ (N an integer).

1. If $\lim_{n\to\infty} \frac{a_n}{b_n} = c$ and $c > 0$, then $\sum a_n$ and $\sum b_n$ both converge or both diverge.

2. If $\lim_{n\to\infty} \frac{a_n}{b_n} = 0$ and $\sum b_n$ converges, then $\sum a_n$ converges.

3. If $\lim_{n\to\infty} \frac{a_n}{b_n} = \infty$ and $\sum b_n$ diverges, then $\sum a_n$ diverges.

Proof We will prove Part 1. Parts 2 and 3 are left as Exercises 57a and b.
Since $c/2 > 0$, there exists an integer N such that

$$\left| \frac{a_n}{b_n} - c \right| < \frac{c}{2} \qquad \text{whenever} \qquad n > N. \qquad \text{Limit definition with } \varepsilon = c/2, L = c, \text{ and } a_n \text{ replaced by } a_n/b_n$$

Thus, for $n > N$,

$$-\frac{c}{2} < \frac{a_n}{b_n} - c < \frac{c}{2},$$

$$\frac{c}{2} < \frac{a_n}{b_n} < \frac{3c}{2},$$

$$\left(\frac{c}{2}\right) b_n < a_n < \left(\frac{3c}{2}\right) b_n.$$

If $\sum b_n$ converges, then $\sum (3c/2)b_n$ converges and $\sum a_n$ converges by the Direct Comparison Test. If $\sum b_n$ diverges, then $\sum (c/2)b_n$ diverges and $\sum a_n$ diverges by the Direct Comparison Test. ∎

EXAMPLE 2 Which of the following series converge, and which diverge?

(a) $\frac{3}{4} + \frac{5}{9} + \frac{7}{16} + \frac{9}{25} + \cdots = \sum_{n=1}^{\infty} \frac{2n+1}{(n+1)^2} = \sum_{n=1}^{\infty} \frac{2n+1}{n^2 + 2n + 1}$

(b) $\dfrac{1}{1} + \dfrac{1}{3} + \dfrac{1}{7} + \dfrac{1}{15} + \cdots = \displaystyle\sum_{n=1}^{\infty} \dfrac{1}{2^n - 1}$

(c) $\dfrac{1 + 2\ln 2}{9} + \dfrac{1 + 3\ln 3}{14} + \dfrac{1 + 4\ln 4}{21} + \cdots = \displaystyle\sum_{n=2}^{\infty} \dfrac{1 + n\ln n}{n^2 + 5}$

Solution We apply the Limit Comparison Test to each series.

(a) Let $a_n = (2n + 1)/(n^2 + 2n + 1)$. For large n, we expect a_n to behave like $2n/n^2 = 2/n$ since the leading terms dominate for large n, so we let $b_n = 1/n$. Since

$$\sum_{n=1}^{\infty} b_n = \sum_{n=1}^{\infty} \dfrac{1}{n} \text{ diverges}$$

and

$$\lim_{n\to\infty} \dfrac{a_n}{b_n} = \lim_{n\to\infty} \dfrac{2n^2 + n}{n^2 + 2n + 1} = 2,$$

$\sum a_n$ diverges by Part 1 of the Limit Comparison Test. We could just as well have taken $b_n = 2/n$, but $1/n$ is simpler.

(b) Let $a_n = 1/(2^n - 1)$. For large n, we expect a_n to behave like $1/2^n$, so we let $b_n = 1/2^n$. Since

$$\sum_{n=1}^{\infty} b_n = \sum_{n=1}^{\infty} \dfrac{1}{2^n} \text{ converges}$$

and

$$\lim_{n\to\infty} \dfrac{a_n}{b_n} = \lim_{n\to\infty} \dfrac{2^n}{2^n - 1} = \lim_{n\to\infty} \dfrac{1}{1 - (1/2^n)} = 1,$$

$\sum a_n$ converges by Part 1 of the Limit Comparison Test.

(c) Let $a_n = (1 + n\ln n)/(n^2 + 5)$. For large n, we expect a_n to behave like $(n\ln n)/n^2 = (\ln n)/n$, which is greater than $1/n$ for $n \geq 3$, so we let $b_n = 1/n$. Since

$$\sum_{n=2}^{\infty} b_n = \sum_{n=2}^{\infty} \dfrac{1}{n} \text{ diverges}$$

and

$$\lim_{n\to\infty} \dfrac{a_n}{b_n} = \lim_{n\to\infty} \dfrac{n + n^2\ln n}{n^2 + 5} = \infty,$$

$\sum a_n$ diverges by Part 3 of the Limit Comparison Test. ∎

EXAMPLE 3   Does $\displaystyle\sum_{n=1}^{\infty} \dfrac{\ln n}{n^{3/2}}$ converge?

Solution Because $\ln n$ grows more slowly than n^c for any positive constant c (Section 10.1, Exercise 115), we can compare the series to a convergent p-series. To get the p-series, we see that

$$\dfrac{\ln n}{n^{3/2}} < \dfrac{n^{1/4}}{n^{3/2}} = \dfrac{1}{n^{5/4}}$$

for n sufficiently large. Then taking $a_n = (\ln n)/n^{3/2}$ and $b_n = 1/n^{5/4}$, we have

$$\lim_{n\to\infty} \frac{a_n}{b_n} = \lim_{n\to\infty} \frac{\ln n}{n^{1/4}}$$

$$= \lim_{n\to\infty} \frac{1/n}{(1/4)n^{-3/4}} \qquad \text{l'Hôpital's Rule}$$

$$= \lim_{n\to\infty} \frac{4}{n^{1/4}} = 0.$$

Since $\Sigma b_n = \Sigma(1/n^{5/4})$ is a p-series with $p > 1$, it converges. Therefore Σa_n converges by Part 2 of the Limit Comparison Test. ∎

EXERCISES 10.4

Direct Comparison Test

In Exercises 1–8, use the Direct Comparison Test to determine if each series converges or diverges.

1. $\displaystyle\sum_{n=1}^{\infty} \frac{1}{n^2 + 30}$ **2.** $\displaystyle\sum_{n=1}^{\infty} \frac{n-1}{n^4 + 2}$ **3.** $\displaystyle\sum_{n=2}^{\infty} \frac{1}{\sqrt{n}-1}$

4. $\displaystyle\sum_{n=2}^{\infty} \frac{n+2}{n^2 - n}$ **5.** $\displaystyle\sum_{n=1}^{\infty} \frac{\cos^2 n}{n^{3/2}}$ **6.** $\displaystyle\sum_{n=1}^{\infty} \frac{1}{n3^n}$

7. $\displaystyle\sum_{n=1}^{\infty} \sqrt{\frac{n+4}{n^4 + 4}}$ **8.** $\displaystyle\sum_{n=1}^{\infty} \frac{\sqrt{n}+1}{\sqrt{n^2 + 3}}$

Limit Comparison Test

In Exercises 9–16, use the Limit Comparison Test to determine if each series converges or diverges.

9. $\displaystyle\sum_{n=1}^{\infty} \frac{n-2}{n^3 - n^2 + 3}$

(*Hint:* Limit Comparison with $\Sigma_{n=1}^{\infty} (1/n^2)$)

10. $\displaystyle\sum_{n=1}^{\infty} \sqrt{\frac{n+1}{n^2 + 2}}$

(*Hint:* Limit Comparison with $\Sigma_{n=1}^{\infty} (1/\sqrt{n})$)

11. $\displaystyle\sum_{n=2}^{\infty} \frac{n(n+1)}{(n^2+1)(n-1)}$ **12.** $\displaystyle\sum_{n=1}^{\infty} \frac{2^n}{3 + 4^n}$

13. $\displaystyle\sum_{n=1}^{\infty} \frac{5^n}{\sqrt{n}\,4^n}$ **14.** $\displaystyle\sum_{n=1}^{\infty} \left(\frac{2n+3}{5n+4}\right)^n$

15. $\displaystyle\sum_{n=2}^{\infty} \frac{1}{\ln n}$

(*Hint:* Limit Comparison with $\Sigma_{n=2}^{\infty} (1/n)$)

16. $\displaystyle\sum_{n=1}^{\infty} \ln\left(1 + \frac{1}{n^2}\right)$

(*Hint:* Limit Comparison with $\Sigma_{n=1}^{\infty} (1/n^2)$)

Determining Convergence or Divergence

Which of the series in Exercises 17–56 converge, and which diverge? Use any method, and give reasons for your answers.

17. $\displaystyle\sum_{n=1}^{\infty} \frac{1}{2\sqrt{n} + \sqrt[3]{n}}$ **18.** $\displaystyle\sum_{n=1}^{\infty} \frac{3}{n + \sqrt{n}}$ **19.** $\displaystyle\sum_{n=1}^{\infty} \frac{\sin^2 n}{2^n}$

20. $\displaystyle\sum_{n=1}^{\infty} \frac{1 + \cos n}{n^2}$ **21.** $\displaystyle\sum_{n=1}^{\infty} \frac{2n}{3n - 1}$ **22.** $\displaystyle\sum_{n=1}^{\infty} \frac{n+1}{n^2\sqrt{n}}$

23. $\displaystyle\sum_{n=1}^{\infty} \frac{10n + 1}{n(n+1)(n+2)}$ **24.** $\displaystyle\sum_{n=3}^{\infty} \frac{5n^3 - 3n}{n^2(n-2)(n^2+5)}$

25. $\displaystyle\sum_{n=1}^{\infty} \left(\frac{n}{3n+1}\right)^n$ **26.** $\displaystyle\sum_{n=1}^{\infty} \frac{1}{\sqrt{n^3 + 2}}$ **27.** $\displaystyle\sum_{n=3}^{\infty} \frac{1}{\ln(\ln n)}$

28. $\displaystyle\sum_{n=1}^{\infty} \frac{(\ln n)^2}{n^3}$ **29.** $\displaystyle\sum_{n=2}^{\infty} \frac{1}{\sqrt{n}\,\ln n}$ **30.** $\displaystyle\sum_{n=1}^{\infty} \frac{(\ln n)^2}{n^{3/2}}$

31. $\displaystyle\sum_{n=1}^{\infty} \frac{1}{1 + \ln n}$ **32.** $\displaystyle\sum_{n=2}^{\infty} \frac{\ln(n+1)}{n+1}$ **33.** $\displaystyle\sum_{n=2}^{\infty} \frac{1}{n\sqrt{n^2 - 1}}$

34. $\displaystyle\sum_{n=1}^{\infty} \frac{\sqrt{n}}{n^2 + 1}$ **35.** $\displaystyle\sum_{n=1}^{\infty} \frac{1-n}{n2^n}$ **36.** $\displaystyle\sum_{n=1}^{\infty} \frac{n + 2^n}{n^2 2^n}$

37. $\displaystyle\sum_{n=1}^{\infty} \frac{1}{3^{n-1} + 1}$ **38.** $\displaystyle\sum_{n=1}^{\infty} \frac{3^{n-1} + 1}{3^n}$ **39.** $\displaystyle\sum_{n=1}^{\infty} \frac{n+1}{n^2 + 3n} \cdot \frac{1}{5n}$

40. $\displaystyle\sum_{n=1}^{\infty} \frac{2^n + 3^n}{3^n + 4^n}$ **41.** $\displaystyle\sum_{n=1}^{\infty} \frac{2^n - n}{n2^n}$ **42.** $\displaystyle\sum_{n=1}^{\infty} \frac{\ln n}{\sqrt{n}\,e^n}$

43. $\displaystyle\sum_{n=2}^{\infty} \frac{1}{n!}$

(*Hint:* First show that $(1/n!) \le (1/n(n-1))$ for $n \ge 2$.)

44. $\displaystyle\sum_{n=1}^{\infty} \frac{(n-1)!}{(n+2)!}$ **45.** $\displaystyle\sum_{n=1}^{\infty} \sin\frac{1}{n}$ **46.** $\displaystyle\sum_{n=1}^{\infty} \tan\frac{1}{n}$

47. $\displaystyle\sum_{n=1}^{\infty} \frac{\tan^{-1} n}{n^{1.1}}$ **48.** $\displaystyle\sum_{n=1}^{\infty} \frac{\sec^{-1} n}{n^{1.3}}$ **49.** $\displaystyle\sum_{n=1}^{\infty} \frac{\coth n}{n^2}$

50. $\displaystyle\sum_{n=1}^{\infty} \frac{\tanh n}{n^2}$ **51.** $\displaystyle\sum_{n=1}^{\infty} \frac{1}{n\sqrt[n]{n}}$ **52.** $\displaystyle\sum_{n=1}^{\infty} \frac{\sqrt[n]{n}}{n^2}$

53. $\displaystyle\sum_{n=1}^{\infty} \frac{1}{1 + 2 + 3 + \cdots + n}$ **54.** $\displaystyle\sum_{n=1}^{\infty} \frac{1}{1 + 2^2 + 3^2 + \cdots + n^2}$

55. $\displaystyle\sum_{n=2}^{\infty} \frac{n}{(\ln n)^2}$ **56.** $\displaystyle\sum_{n=2}^{\infty} \frac{(\ln n)^2}{n}$

Theory and Examples

57. Prove **(a)** Part 2 and **(b)** Part 3 of the Limit Comparison Test.

58. If $\sum_{n=1}^{\infty} a_n$ is a convergent series of nonnegative numbers, can anything be said about $\sum_{n=1}^{\infty} (a_n/n)$? Explain.

59. Suppose that $a_n > 0$ and $b_n > 0$ for $n \geq N$ (N an integer). If $\lim_{n\to\infty}(a_n/b_n) = \infty$ and $\sum a_n$ converges, can anything be said about $\sum b_n$? Give reasons for your answer.

60. Prove that if $\sum a_n$ is a convergent series of nonnegative terms, then $\sum a_n^2$ converges.

61. Suppose that $a_n > 0$ and $\lim_{n\to\infty} a_n = \infty$. Prove that $\sum a_n$ diverges.

62. Suppose that $a_n > 0$ and $\lim_{n\to\infty} n^2 a_n = 0$. Prove that $\sum a_n$ converges.

63. Show that $\sum_{n=2}^{\infty} \left((\ln n)^q/n^p \right)$ converges for $-\infty < q < \infty$ and $p > 1$.

(*Hint:* Limit Comparison with $\sum_{n=2}^{\infty} 1/n^r$ for $1 < r < p$.)

64. (*Continuation of Exercise 63.*) Show that $\sum_{n=2}^{\infty} \left((\ln n)^q/n^p \right)$ diverges for $-\infty < q < \infty$ and $0 < p < 1$.

(*Hint:* Limit Comparison with an appropriate p-series.)

65. Decimal numbers Any real number in the interval $[0, 1]$ can be represented by a decimal (not necessarily unique) as

$$0.d_1 d_2 d_3 d_4 \ldots = \frac{d_1}{10} + \frac{d_2}{10^2} + \frac{d_3}{10^3} + \frac{d_4}{10^4} + \cdots,$$

where d_i is one of the integers $0, 1, 2, 3, \ldots, 9$. Prove that the series on the right-hand side always converges.

66. If $\sum a_n$ is a convergent series of positive terms, prove that $\sum \sin(a_n)$ converges.

In Exercises 67–72, use the results of Exercises 63 and 64 to determine if each series converges or diverges.

67. $\displaystyle\sum_{n=2}^{\infty} \frac{(\ln n)^3}{n^4}$

68. $\displaystyle\sum_{n=2}^{\infty} \sqrt{\frac{\ln n}{n}}$

69. $\displaystyle\sum_{n=2}^{\infty} \frac{(\ln n)^{1000}}{n^{1.001}}$

70. $\displaystyle\sum_{n=2}^{\infty} \frac{(\ln n)^{1/5}}{n^{0.99}}$

71. $\displaystyle\sum_{n=2}^{\infty} \frac{1}{n^{1.1}(\ln n)^3}$

72. $\displaystyle\sum_{n=2}^{\infty} \frac{1}{\sqrt{n} \cdot \ln n}$

COMPUTER EXPLORATIONS

73. It is not yet known whether the series

$$\sum_{n=1}^{\infty} \frac{1}{n^3 \sin^2 n}$$

converges or diverges. Use a CAS to explore the behavior of the series by performing the following steps.

a. Define the sequence of partial sums

$$s_k = \sum_{n=1}^{k} \frac{1}{n^3 \sin^2 n}.$$

What happens when you try to find the limit of s_k as $k \to \infty$? Does your CAS find a closed form answer for this limit?

b. Plot the first 100 points (k, s_k) for the sequence of partial sums. Do they appear to converge? What would you estimate the limit to be?

c. Next plot the first 200 points (k, s_k). Discuss the behavior in your own words.

d. Plot the first 400 points (k, s_k). What happens when $k = 355$? Calculate the number $355/113$. Explain from you calculation what happened at $k = 355$. For what values of k would you guess this behavior might occur again?

74. a. Use Theorem 8 to show that

$$S = \sum_{n=1}^{\infty} \frac{1}{n(n+1)} + \sum_{n=1}^{\infty} \left(\frac{1}{n^2} - \frac{1}{n(n+1)} \right)$$

where $S = \sum_{n=1}^{\infty} (1/n^2)$, the sum of a convergent p-series.

b. From Example 5, Section 10.2, show that

$$S = 1 + \sum_{n=1}^{\infty} \frac{1}{n^2(n+1)}.$$

c. Explain why taking the first M terms in the series in part (b) gives a better approximation to S than taking the first M terms in the original series $\sum_{n=1}^{\infty}(1/n^2)$.

d. We know the exact value of S is $\pi^2/6$. Which of the sums

$$\sum_{n=1}^{1000000} \frac{1}{n^2} \quad \text{or} \quad 1 + \sum_{n=1}^{1000} \frac{1}{n^2(n+1)}$$

gives a better approximation to S?

10.5 Absolute Convergence; The Ratio and Root Tests

When some of the terms of a series are positive and others are negative, the series may or may not converge. For example, the geometric series

$$5 - \frac{5}{4} + \frac{5}{16} - \frac{5}{64} + \cdots = \sum_{n=0}^{\infty} 5\left(\frac{-1}{4}\right)^n \tag{1}$$

converges (since $|r| = \frac{1}{4} < 1$), whereas the different geometric series

$$1 - \frac{5}{4} + \frac{25}{16} - \frac{125}{64} + \cdots = \sum_{n=0}^{\infty} \left(\frac{-5}{4}\right)^n \tag{2}$$

diverges (since $|r| = 5/4 > 1$). In series (1), there is some cancelation in the partial sums, which may be assisting the convergence property of the series. However, if we make all of the terms positive in series (1) to form the new series

$$5 + \frac{5}{4} + \frac{5}{16} + \frac{5}{64} + \cdots = \sum_{n=0}^{\infty} \left| 5\left(\frac{-1}{4}\right)^n \right| = \sum_{n=0}^{\infty} 5\left(\frac{1}{4}\right)^n,$$

we see that it still converges. For a general series with both positive and negative terms, we can apply the tests for convergence studied before to the series of absolute values of its terms. In doing so, we are led naturally to the following concept.

DEFINITION A series $\sum a_n$ **converges absolutely** (is **absolutely convergent**) if the corresponding series of absolute values, $\sum |a_n|$, converges.

So the geometric series (1) is absolutely convergent. We observed, too, that it is also convergent. This situation is always true: An absolutely convergent series is convergent as well, which we now prove.

Caution
Be careful when using Theorem 12. A convergent series need *not* converge absolutely, as you will see in the next section.

THEOREM 12—The Absolute Convergence Test

If $\sum_{n=1}^{\infty} |a_n|$ converges, then $\sum_{n=1}^{\infty} a_n$ converges.

Proof   For each n,

$$-|a_n| \le a_n \le |a_n|, \qquad \text{so} \qquad 0 \le a_n + |a_n| \le 2|a_n|.$$

If $\sum_{n=1}^{\infty} |a_n|$ converges, then $\sum_{n=1}^{\infty} 2|a_n|$ converges and, by the Direct Comparison Test, the nonnegative series $\sum_{n=1}^{\infty} (a_n + |a_n|)$ converges. The equality $a_n = (a_n + |a_n|) - |a_n|$ now lets us express $\sum_{n=1}^{\infty} a_n$ as the difference of two convergent series:

$$\sum_{n=1}^{\infty} a_n = \sum_{n=1}^{\infty} (a_n + |a_n| - |a_n|) = \sum_{n=1}^{\infty} (a_n + |a_n|) - \sum_{n=1}^{\infty} |a_n|.$$

Therefore, $\sum_{n=1}^{\infty} a_n$ converges.   ∎

EXAMPLE 1   This example gives two series that converge absolutely.

(a) For $\sum_{n=1}^{\infty} (-1)^{n+1} \frac{1}{n^2} = 1 - \frac{1}{4} + \frac{1}{9} - \frac{1}{16} + \cdots$, the corresponding series of absolute values is the convergent series

$$\sum_{n=1}^{\infty} \frac{1}{n^2} = 1 + \frac{1}{4} + \frac{1}{9} + \frac{1}{16} + \cdots.$$

The original series converges because it converges absolutely.

(b) For $\sum_{n=1}^{\infty} \frac{\sin n}{n^2} = \frac{\sin 1}{1} + \frac{\sin 2}{4} + \frac{\sin 3}{9} + \cdots$, which contains both positive and negative terms, the corresponding series of absolute values is

$$\sum_{n=1}^{\infty} \left| \frac{\sin n}{n^2} \right| = \frac{|\sin 1|}{1} + \frac{|\sin 2|}{4} + \cdots,$$

which converges by comparison with $\sum_{n=1}^{\infty} (1/n^2)$ because $|\sin n| \le 1$ for every n. The original series converges absolutely; therefore it converges.   ∎

The Ratio Test

The Ratio Test measures the rate of growth (or decline) of a series by examining the ratio a_{n+1}/a_n. For a geometric series $\sum ar^n$, this rate is a constant $((ar^{n+1})/(ar^n) = r)$, and the series converges if and only if its ratio is less than 1 in absolute value. The Ratio Test is a powerful rule extending that result.

ρ is the Greek lowercase letter rho, which is pronounced "row."

THEOREM 13—The Ratio Test

Let $\sum a_n$ be any series and suppose that

$$\lim_{n\to\infty} \left| \frac{a_{n+1}}{a_n} \right| = \rho.$$

Then **(a)** the series *converges absolutely* if $\rho < 1$, **(b)** the series *diverges* if $\rho > 1$ or ρ is infinite, **(c)** the test is *inconclusive* if $\rho = 1$.

Proof

(a) $\rho < 1$. Let r be a number between ρ and 1. Then the number $\varepsilon = r - \rho$ is positive. Since

$$\left| \frac{a_{n+1}}{a_n} \right| \to \rho,$$

$|a_{n+1}/a_n|$ must lie within ε of ρ when n is large enough, say, for all $n \geq N$. In particular,

$$\left| \frac{a_{n+1}}{a_n} \right| < \rho + \varepsilon = r, \qquad \text{when } n \geq N.$$

Hence

$$|a_{N+1}| < r|a_N|,$$
$$|a_{N+2}| < r|a_{N+1}| < r^2|a_N|,$$
$$|a_{N+3}| < r|a_{N+2}| < r^3|a_N|,$$
$$\vdots$$
$$|a_{N+m}| < r|a_{N+m-1}| < r^m|a_N|.$$

Therefore,

$$\sum_{m=N}^{\infty} |a_m| = \sum_{m=0}^{\infty} |a_{N+m}| \leq \sum_{m=0}^{\infty} |a_N|\, r^m = |a_N| \sum_{m=0}^{\infty} r^m.$$

The geometric series on the right-hand side converges because $0 < r < 1$, so the series of absolute values $\sum_{m=N}^{\infty} |a_m|$ converges by the Direct Comparison Test. Because adding or deleting finitely many terms in a series does not affect its convergence or divergence property, the series $\sum_{n=1}^{\infty} |a_n|$ also converges. That is, the series $\sum a_n$ is absolutely convergent.

(b) $1 < \rho \leq \infty$. From some index M on,

$$\left| \frac{a_{n+1}}{a_n} \right| > 1 \qquad \text{and} \qquad |a_M| < |a_{M+1}| < |a_{M+2}| < \cdots.$$

The terms of the series do not approach zero as n becomes infinite, and the series diverges by the nth-Term Test.

Chapter 10 Infinite Sequences and Series

(c) $\rho = 1$. The two series

$$\sum_{n=1}^{\infty} \frac{1}{n} \quad \text{and} \quad \sum_{n=1}^{\infty} \frac{1}{n^2}$$

show that some other test for convergence must be used when $\rho = 1$.

$$\text{For } \sum_{n=1}^{\infty} \frac{1}{n}: \quad \left| \frac{a_{n+1}}{a_n} \right| = \frac{1/(n+1)}{1/n} = \frac{n}{n+1} \to 1.$$

$$\text{For } \sum_{n=1}^{\infty} \frac{1}{n^2}: \quad \left| \frac{a_{n+1}}{a_n} \right| = \frac{1/(n+1)^2}{1/n^2} = \left(\frac{n}{n+1} \right)^2 \to 1^2 = 1.$$

In both cases, $\rho = 1$, yet the first series diverges, whereas the second converges. ∎

The Ratio Test is often effective when the terms of a series contain factorials of expressions involving n or expressions raised to a power involving n.

EXAMPLE 2   Investigate the convergence of the following series.

(a) $\displaystyle\sum_{n=0}^{\infty} \frac{2^n + 5}{3^n}$ (b) $\displaystyle\sum_{n=1}^{\infty} \frac{(2n)!}{n!n!}$ (c) $\displaystyle\sum_{n=1}^{\infty} \frac{4^n n! n!}{(2n)!}$

Solution   We apply the Ratio Test to each series.

(a) For the series $\sum_{n=0}^{\infty} (2^n + 5)/3^n$,

$$\left| \frac{a_{n+1}}{a_n} \right| = \frac{(2^{n+1} + 5)/3^{n+1}}{(2^n + 5)/3^n} = \frac{1}{3} \cdot \frac{2^{n+1} + 5}{2^n + 5} = \frac{1}{3} \cdot \left(\frac{2 + 5 \cdot 2^{-n}}{1 + 5 \cdot 2^{-n}} \right) \to \frac{1}{3} \cdot \frac{2}{1} = \frac{2}{3}.$$

The series converges absolutely (and thus converges) because $\rho = 2/3$ is less than 1. This does *not* mean that $2/3$ is the sum of the series. In fact,

$$\sum_{n=0}^{\infty} \frac{2^n + 5}{3^n} = \sum_{n=0}^{\infty} \left(\frac{2}{3} \right)^n + \sum_{n=0}^{\infty} \frac{5}{3^n} = \frac{1}{1 - (2/3)} + \frac{5}{1 - (1/3)} = \frac{21}{2}.$$

(b) If $a_n = \dfrac{(2n)!}{n!n!}$, then $a_{n+1} = \dfrac{(2n+2)!}{(n+1)!(n+1)!}$ and

$$\left| \frac{a_{n+1}}{a_n} \right| = \frac{n!n!(2n+2)(2n+1)(2n)!}{(n+1)!(n+1)!(2n)!}$$

$$= \frac{(2n+2)(2n+1)}{(n+1)(n+1)} = \frac{4n+2}{n+1} \to 4.$$

The series diverges because $\rho = 4$ is greater than 1.

(c) If $a_n = 4^n n! n! / (2n)!$, then

$$\left| \frac{a_{n+1}}{a_n} \right| = \frac{4^{n+1}(n+1)!(n+1)!}{(2n+2)(2n+1)(2n)!} \cdot \frac{(2n)!}{4^n n! n!}$$

$$= \frac{4(n+1)(n+1)}{(2n+2)(2n+1)} = \frac{2(n+1)}{2n+1} \to 1.$$

Because the limit is $\rho = 1$, we cannot decide from the Ratio Test whether the series converges. However, when we notice that $a_{n+1}/a_n = (2n+2)/(2n+1)$, we conclude that a_{n+1} is always greater than a_n because $(2n+2)/(2n+1)$ is always greater than 1. Therefore, all terms are greater than or equal to $a_1 = 2$, and the nth term does not approach zero as $n \to \infty$. The series diverges. ∎

The Root Test

The convergence tests we have so far for Σa_n work best when the formula for a_n is relatively simple. However, consider the series with the terms

$$a_n = \begin{cases} n/2^n, & n \text{ odd} \\ 1/2^n, & n \text{ even.} \end{cases}$$

To investigate convergence we write out several terms of the series:

$$\sum_{n=1}^{\infty} a_n = \frac{1}{2^1} + \frac{1}{2^2} + \frac{3}{2^3} + \frac{1}{2^4} + \frac{5}{2^5} + \frac{1}{2^6} + \frac{7}{2^7} + \cdots$$

$$= \frac{1}{2} + \frac{1}{4} + \frac{3}{8} + \frac{1}{16} + \frac{5}{32} + \frac{1}{64} + \frac{7}{128} + \cdots.$$

Clearly, this is not a geometric series. The nth term approaches zero as $n \to \infty$, so the nth-Term Test does not tell us if the series diverges. The Integral Test does not look promising. The Ratio Test produces

$$\left| \frac{a_{n+1}}{a_n} \right| = \begin{cases} \dfrac{1}{2n}, & n \text{ odd} \\[2mm] \dfrac{n+1}{2}, & n \text{ even} \end{cases}$$

As $n \to \infty$, the ratio is alternately small and large and therefore has no limit. However, we will see that the following test establishes that the series converges.

THEOREM 14—The Root Test

Let $\sum a_n$ be any series and suppose that

$$\lim_{n \to \infty} \sqrt[n]{|a_n|} = \rho.$$

Then (a) the series *converges absolutely* if $\rho < 1$, (b) the series *diverges* if $\rho > 1$ or ρ is infinite, (c) the test is *inconclusive* if $\rho = 1$.

Proof

(a) **$\rho < 1$.** Choose an $\varepsilon > 0$ so small that $\rho + \varepsilon < 1$. Since $\sqrt[n]{|a_n|} \to \rho$, the terms $\sqrt[n]{|a_n|}$ eventually get to within ε of ρ. So there exists an index M such that

$$\sqrt[n]{|a_n|} < \rho + \varepsilon \qquad \text{when } n \geq M.$$

Then it is also true that

$$|a_n| < (\rho + \varepsilon)^n \qquad \text{for } n \geq M.$$

Now, $\sum_{n=M}^{\infty} (\rho + \varepsilon)^n$ is a geometric series with ratio $(\rho + \varepsilon) < 1$ and therefore converges. By the Direct Comparison Test, $\sum_{n=M}^{\infty} |a_n|$ converges. Adding finitely many terms to a series does not affect its convergence or divergence, so the series

$$\sum_{n=1}^{\infty} |a_n| = |a_1| + \cdots + |a_{M-1}| + \sum_{n=M}^{\infty} |a_n|$$

also converges. Therefore, Σa_n converges absolutely.

(b) **$1 < \rho \leq \infty$.** For all indices beyond some integer M, we have $\sqrt[n]{|a_n|} > 1$, so that $|a_n| > 1$ for $n > M$. The terms of the series do not converge to zero. The series diverges by the nth-Term Test.

(c) **$\rho = 1$.** The series $\sum_{n=1}^{\infty} (1/n)$ and $\sum_{n=1}^{\infty} (1/n^2)$ show that the test is not conclusive when $\rho = 1$. The first series diverges and the second converges, but in both cases $\sqrt[n]{|a_n|} \to 1$. ∎

EXAMPLE 3 Consider again the series with terms $a_n = \begin{cases} n/2^n, & n \text{ odd} \\ 1/2^n, & n \text{ even}. \end{cases}$

Does Σa_n converge?

Solution We apply the Root Test, finding that

$$\sqrt[n]{|a_n|} = \begin{cases} \sqrt[n]{n}/2, & n \text{ odd} \\ 1/2, & n \text{ even}. \end{cases}$$

Therefore,

$$\frac{1}{2} \le \sqrt[n]{|a_n|} \le \frac{\sqrt[n]{n}}{2}.$$

Since $\sqrt[n]{n} \to 1$ (Section 10.1, Theorem 5), we have $\lim_{n\to\infty} \sqrt[n]{|a_n|} = 1/2$ by the Sandwich Theorem. The limit is less than 1, so the series converges absolutely by the Root Test. ∎

EXAMPLE 4 Which of the following series converge, and which diverge?

(a) $\displaystyle\sum_{n=1}^{\infty} \frac{n^2}{2^n}$ (b) $\displaystyle\sum_{n=1}^{\infty} \frac{2^n}{n^3}$ (c) $\displaystyle\sum_{n=1}^{\infty} \left(\frac{1}{1+n}\right)^n$

Solution We apply the Root Test to each series, noting that each series has positive terms.

(a) $\displaystyle\sum_{n=1}^{\infty} \frac{n^2}{2^n}$ converges because $\sqrt[n]{\dfrac{n^2}{2^n}} = \dfrac{\sqrt[n]{n^2}}{\sqrt[n]{2^n}} = \dfrac{(\sqrt[n]{n})^2}{2} \to \dfrac{1^2}{2} < 1.$

(b) $\displaystyle\sum_{n=1}^{\infty} \frac{2^n}{n^3}$ diverges because $\sqrt[n]{\dfrac{2^n}{n^3}} = \dfrac{2}{(\sqrt[n]{n})^3} \to \dfrac{2}{1^3} > 1.$

(c) $\displaystyle\sum_{n=1}^{\infty} \left(\frac{1}{1+n}\right)^n$ converges because $\sqrt[n]{\left(\dfrac{1}{1+n}\right)^n} = \dfrac{1}{1+n} \to 0 < 1.$

EXERCISES 10.5

Using the Ratio Test

In Exercises 1–8, use the Ratio Test to determine if each series converges absolutely or diverges.

1. $\displaystyle\sum_{n=1}^{\infty} \frac{2^n}{n!}$

2. $\displaystyle\sum_{n=1}^{\infty} (-1)^n \frac{n+2}{3^n}$

3. $\displaystyle\sum_{n=1}^{\infty} \frac{(n-1)!}{(n+1)^2}$

4. $\displaystyle\sum_{n=1}^{\infty} \frac{2^{n+1}}{n3^{n-1}}$

5. $\displaystyle\sum_{n=1}^{\infty} \frac{n^4}{(-4)^n}$

6. $\displaystyle\sum_{n=2}^{\infty} \frac{3^{n+2}}{\ln n}$

7. $\displaystyle\sum_{n=1}^{\infty} (-1)^n \frac{n^2(n+2)!}{n!\,3^{2n}}$

8. $\displaystyle\sum_{n=1}^{\infty} \frac{n5^n}{(2n+3)\ln(n+1)}$

Using the Root Test

In Exercises 9–16, use the Root Test to determine if each series converges absolutely or diverges.

9. $\displaystyle\sum_{n=1}^{\infty} \frac{7}{(2n+5)^n}$

10. $\displaystyle\sum_{n=1}^{\infty} \frac{4^n}{(3n)^n}$

11. $\displaystyle\sum_{n=1}^{\infty} \left(\frac{4n+3}{3n-5}\right)^n$

12. $\displaystyle\sum_{n=1}^{\infty} \left(-\ln\left(e^2+\frac{1}{n}\right)\right)^{n+1}$

13. $\displaystyle\sum_{n=1}^{\infty} \frac{-8}{(3+(1/n))^{2n}}$

14. $\displaystyle\sum_{n=1}^{\infty} \sin^n\left(\frac{1}{\sqrt{n}}\right)$

15. $\displaystyle\sum_{n=1}^{\infty} (-1)^n \left(1-\frac{1}{n}\right)^{n^2}$

(*Hint:* $\lim_{n\to\infty} (1+x/n)^n = e^x$)

16. $\displaystyle\sum_{n=2}^{\infty} \frac{(-1)^n}{n^{1+n}}$

Determining Convergence or Divergence

In Exercises 17–46, use any method to determine if the series converges or diverges. Give reasons for your answer.

17. $\displaystyle\sum_{n=1}^{\infty} \frac{n^{\sqrt{2}}}{2^n}$

18. $\displaystyle\sum_{n=1}^{\infty} (-1)^n\, n^2 e^{-n}$

19. $\displaystyle\sum_{n=1}^{\infty} n!(-e)^{-n}$

20. $\displaystyle\sum_{n=1}^{\infty} \frac{n!}{10^n}$

21. $\displaystyle\sum_{n=1}^{\infty} \frac{n^{10}}{10^n}$

22. $\displaystyle\sum_{n=1}^{\infty} \left(\frac{n-2}{n}\right)^n$

23. $\displaystyle\sum_{n=1}^{\infty} \frac{2 + (-1)^n}{1.25^n}$

24. $\displaystyle\sum_{n=1}^{\infty} \frac{(-2)^n}{3^n}$

25. $\displaystyle\sum_{n=1}^{\infty} (-1)^n \left(1 - \frac{3}{n}\right)^n$

26. $\displaystyle\sum_{n=1}^{\infty} \left(1 - \frac{1}{3n}\right)^n$

27. $\displaystyle\sum_{n=1}^{\infty} \frac{\ln n}{n^3}$

28. $\displaystyle\sum_{n=1}^{\infty} \frac{(-\ln n)^n}{n^n}$

29. $\displaystyle\sum_{n=1}^{\infty} \left(\frac{1}{n} - \frac{1}{n^2}\right)$

30. $\displaystyle\sum_{n=1}^{\infty} \left(\frac{1}{n} - \frac{1}{n^2}\right)^n$

31. $\displaystyle\sum_{n=1}^{\infty} \frac{e^n}{n^e}$

32. $\displaystyle\sum_{n=1}^{\infty} \frac{n \ln n}{(-2)^n}$

33. $\displaystyle\sum_{n=1}^{\infty} \frac{(n+1)(n+2)}{n!}$

34. $\displaystyle\sum_{n=1}^{\infty} e^{-n}(n^3)$

35. $\displaystyle\sum_{n=1}^{\infty} \frac{(n+3)!}{3!n!3^n}$

36. $\displaystyle\sum_{n=1}^{\infty} \frac{n2^n(n+1)!}{3^n n!}$

37. $\displaystyle\sum_{n=1}^{\infty} \frac{n!}{(2n+1)!}$

38. $\displaystyle\sum_{n=1}^{\infty} \frac{n!}{(-n)^n}$

39. $\displaystyle\sum_{n=2}^{\infty} \frac{-n}{(\ln n)^n}$

40. $\displaystyle\sum_{n=2}^{\infty} \frac{n}{(\ln n)^{(n/2)}}$

41. $\displaystyle\sum_{n=1}^{\infty} \frac{n! \ln n}{n(n+2)!}$

42. $\displaystyle\sum_{n=1}^{\infty} \frac{(-3)^n}{n^3 2^n}$

43. $\displaystyle\sum_{n=1}^{\infty} \frac{(n!)^2}{(2n)!}$

44. $\displaystyle\sum_{n=1}^{\infty} \frac{(2n+3)(2^n+3)}{3^n+2}$

45. $\displaystyle\sum_{n=3}^{\infty} \frac{2^n}{n^2}$

46. $\displaystyle\sum_{n=3}^{\infty} \frac{2^{n^2}}{n^{2^n}}$

Recursively Defined Terms Which of the series $\sum_{n=1}^{\infty} a_n$ defined by the formulas in Exercises 47–56 converge, and which diverge? Give reasons for your answers.

47. $a_1 = 2, \quad a_{n+1} = \dfrac{1 + \sin n}{n} a_n$

48. $a_1 = 1, \quad a_{n+1} = \dfrac{1 + \tan^{-1} n}{n} a_n$

49. $a_1 = \dfrac{1}{3}, \quad a_{n+1} = \dfrac{3n - 1}{2n + 5} a_n$

50. $a_1 = 3, \quad a_{n+1} = \dfrac{n}{n + 1} a_n$

51. $a_1 = 2, \quad a_{n+1} = \dfrac{2}{n} a_n$

52. $a_1 = 5, \quad a_{n+1} = \dfrac{\sqrt[n]{n}}{2} a_n$

53. $a_1 = 1, \quad a_{n+1} = \dfrac{1 + \ln n}{n} a_n$

54. $a_1 = \dfrac{1}{2}, \quad a_{n+1} = \dfrac{n + \ln n}{n + 10} a_n$

55. $a_1 = \dfrac{1}{3}, \quad a_{n+1} = \sqrt[n]{a_n}$

56. $a_1 = \dfrac{1}{2}, \quad a_{n+1} = (a_n)^{n+1}$

Convergence or Divergence
Which of the series in Exercises 57–64 converge, and which diverge? Give reasons for your answers.

57. $\displaystyle\sum_{n=1}^{\infty} \frac{2^n n! n!}{(2n)!}$

58. $\displaystyle\sum_{n=1}^{\infty} \frac{(-1)^n (3n)!}{n!(n+1)!(n+2)!}$

59. $\displaystyle\sum_{n=1}^{\infty} \frac{(n!)^n}{(n^n)^2}$

60. $\displaystyle\sum_{n=1}^{\infty} (-1)^n \frac{(n!)^n}{n^{(n^2)}}$

61. $\displaystyle\sum_{n=1}^{\infty} \frac{n^n}{2^{(n^2)}}$

62. $\displaystyle\sum_{n=1}^{\infty} \frac{n^n}{(2^n)^2}$

63. $\displaystyle\sum_{n=1}^{\infty} \frac{1 \cdot 3 \cdot \cdots \cdot (2n-1)}{4^n 2^n n!}$

64. $\displaystyle\sum_{n=1}^{\infty} \frac{1 \cdot 3 \cdot \cdots \cdot (2n-1)}{[2 \cdot 4 \cdot \cdots \cdot (2n)](3^n + 1)}$

65. Assume that b_n is a sequence of positive numbers converging to $4/5$. Determine if the following series converge or diverge.

a. $\displaystyle\sum_{n=1}^{\infty} (b_n)^{1/n}$

b. $\displaystyle\sum_{n=1}^{\infty} \left(\frac{5}{4}\right)^n (b_n)$

c. $\displaystyle\sum_{n=1}^{\infty} (b_n)^n$

d. $\displaystyle\sum_{n=1}^{\infty} \frac{1000^n}{n! + b_n}$

66. Assume that b_n is a sequence of positive numbers converging to $1/3$. Determine if the following series converge or diverge.

a. $\displaystyle\sum_{n=1}^{\infty} \frac{b_{n+1} b_n}{n \, 4^n}$

b. $\displaystyle\sum_{n=1}^{\infty} \frac{n^n}{n! \, b_1^2 b_2^2 \cdots b_n^2}$

Theory and Examples
67. Neither the Ratio Test nor the Root Test helps with p-series. Try them on

$$\sum_{n=1}^{\infty} \frac{1}{n^p}$$

and show that both tests fail to provide information about convergence.

68. Show that neither the Ratio Test nor the Root Test provides information about the convergence of

$$\sum_{n=2}^{\infty} \frac{1}{(\ln n)^p} \qquad (p \text{ constant}).$$

69. Let $a_n = \begin{cases} n/2^n, & \text{if } n \text{ is a prime number} \\ 1/2^n, & \text{otherwise.} \end{cases}$

Does $\sum a_n$ converge? Give reasons for your answer.

70. Show that $\sum_{n=1}^{\infty} 2^{(n^2)}/n!$ diverges. Recall from the Laws of Exponents that $2^{(n^2)} = (2^n)^n$.

10.6 Alternating Series and Conditional Convergence

A series in which the terms are alternately positive and negative is an **alternating series**. Here are three examples:

$$1 - \frac{1}{2} + \frac{1}{3} - \frac{1}{4} + \frac{1}{5} - \cdots + \frac{(-1)^{n+1}}{n} + \cdots \qquad (1)$$

$$-2 + 1 - \frac{1}{2} + \frac{1}{4} - \frac{1}{8} + \cdots + \frac{(-1)^n 4}{2^n} + \cdots \qquad (2)$$

$$1 - 2 + 3 - 4 + 5 - 6 + \cdots + (-1)^{n+1}n + \cdots \qquad (3)$$

We see from these examples that the nth term of an alternating series is of the form

$$a_n = (-1)^{n+1}u_n \qquad \text{or} \qquad a_n = (-1)^n u_n$$

where $u_n = |a_n|$ is a positive number.

Series (1), called the **alternating harmonic series**, converges, as we will see in a moment. Series (2), a geometric series with ratio $r = -1/2$, converges to $-2/[1 + (1/2)] = -4/3$. Series (3) diverges because the nth term does not approach zero.

We prove the convergence of the alternating harmonic series by applying the Alternating Series Test. This test is for *convergence* of an alternating series and cannot be used to conclude that such a series diverges. If we multiply $(u_1 - u_2 + u_3 - u_4 + \cdots)$ by -1, we see that the test is also valid for the alternating series $-u_1 + u_2 - u_3 + u_4 - \cdots$, as with the one in Series (2) given above.

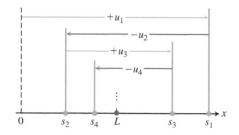

FIGURE 10.15 The partial sums of an alternating series that satisfies the hypotheses of Theorem 15 for $N = 1$ straddle the limit from the beginning.

THEOREM 15—The Alternating Series Test

The series

$$\sum_{n=1}^{\infty} (-1)^{n+1}u_n = u_1 - u_2 + u_3 - u_4 + \cdots$$

converges if the following conditions are satisfied:

1. The u_n's are all positive.
2. The u_n's are eventually nonincreasing: $u_n \geq u_{n+1}$ for all $n \geq N$, for some integer N.
3. $u_n \to 0$.

Proof We look at the case where $u_1, u_2, u_3, u_4, \ldots$ is nonincreasing, so that $N = 1$. If n is an even integer, say $n = 2m$, then the sum of the first n terms is

$$s_{2m} = (u_1 - u_2) + (u_3 - u_4) + \cdots + (u_{2m-1} - u_{2m})$$
$$= u_1 - (u_2 - u_3) - (u_4 - u_5) - \cdots - (u_{2m-2} - u_{2m-1}) - u_{2m}.$$

The first equality shows that s_{2m} is the sum of m nonnegative terms, since each term in parentheses is positive or zero. Hence $s_{2m+2} \geq s_{2m}$, and the sequence $\{s_{2m}\}$ is nondecreasing. The second equality shows that $s_{2m} \leq u_1$. Since $\{s_{2m}\}$ is nondecreasing and bounded from above, it has a limit, say

$$\lim_{m \to \infty} s_{2m} = L. \qquad \text{Theorem 6} \qquad (4)$$

If n is an odd integer, say $n = 2m + 1$, then the sum of the first n terms is $s_{2m+1} = s_{2m} + u_{2m+1}$. Since $u_n \to 0$,

$$\lim_{m \to \infty} u_{2m+1} = 0$$

and, as $m \to \infty$,

$$s_{2m+1} = s_{2m} + u_{2m+1} \to L + 0 = L. \tag{5}$$

Combining the results of Equations (4) and (5) gives $\lim_{n\to\infty} s_n = L$ (Section 10.1, Exercise 143). ∎

EXAMPLE 1   The alternating harmonic series

$$\sum_{n=1}^{\infty} (-1)^{n+1} \frac{1}{n} = 1 - \frac{1}{2} + \frac{1}{3} - \frac{1}{4} + \cdots$$

clearly satisfies the three requirements of Theorem 15 with $N = 1$; it therefore converges by the Alternating Series Test. Notice that the test gives no information about what the sum of the series might be. Figure 10.16 shows histograms of the partial sums of the divergent harmonic series and those of the convergent alternating harmonic series. It turns out that the alternating harmonic series converges to $\ln 2$. ∎

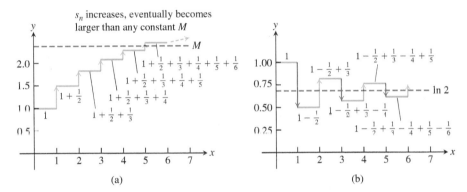

FIGURE 10.16   (a) The harmonic series diverges, with partial sums that eventually exceed any constant. (b) The alternating harmonic series converges to $\ln 2 \approx .693$.

Rather than directly verifying the definition $u_n \ge u_{n+1}$, a second way to show that the sequence $\{u_n\}$ is nonincreasing is to define a differentiable function $f(x)$ satisfying $f(n) = u_n$. That is, the values of f match the values of the sequence at every positive integer n. If $f'(x) \le 0$ for all x greater than or equal to some positive integer N, then $f(x)$ is nonincreasing for $x \ge N$. It follows that $f(n) \ge f(n + 1)$, or $u_n \ge u_{n+1}$, for $n \ge N$.

EXAMPLE 2   We show that the sequence $u_n = 10n/(n^2 + 16)$ is eventually nonincreasing. Define $f(x) = 10x/(x^2 + 16)$. Then from the Derivative Quotient Rule,

$$f'(x) = \frac{10(16 - x^2)}{(x^2 + 16)^2} \le 0 \qquad \text{whenever } x \ge 4.$$

It follows that $u_n \ge u_{n+1}$ for $n \ge 4$. That is, the sequence $\{u_n\}$ is nonincreasing for $n \ge 4$. ∎

A graphical interpretation of the partial sums (Figure 10.15) shows how an alternating series converges to its limit L when the three conditions of Theorem 15 are satisfied with $N = 1$. Starting from the origin of the x-axis, we lay off the positive distance $s_1 = u_1$. To find the point corresponding to $s_2 = u_1 - u_2$, we back up a distance equal to u_2. Since $u_2 \le u_1$, we do not back up any farther than the origin. We continue in this seesaw fashion, backing up or going forward as the signs in the series demand. But for $n \ge N$, each forward or backward step is shorter than (or at most the same size as) the preceding step because $u_{n+1} \le u_n$. And since the nth term approaches zero as n increases, the size of step we take forward or backward gets smaller and smaller. We oscillate back and forth across the limit L, and the amplitude of oscillation approaches zero. The limit L lies between any two successive sums s_n and s_{n+1} and hence differs from s_n by an amount less than u_{n+1}. Because

$$|L - s_n| < u_{n+1} \qquad \text{for } n \geq N,$$

we can make useful estimates of the sums of convergent alternating series.

THEOREM 16—The Alternating Series Estimation Theorem

If the alternating series $\sum_{n=1}^{\infty}(-1)^{n+1}u_n$ satisfies the three conditions of Theorem 15, then for $n \geq N$,

$$s_n = u_1 - u_2 + \cdots + (-1)^{n+1}u_n$$

approximates the sum L of the series with an error whose absolute value is less than u_{n+1}, the absolute value of the first unused term. Furthermore, the sum L lies between any two successive partial sums s_n and s_{n+1}, and the remainder, $L - s_n$, has the same sign as the first unused term.

We leave the verification of the sign of the remainder for Exercise 87.

EXAMPLE 3 We try Theorem 16 on a series whose sum we know:

$$\sum_{n=0}^{\infty}(-1)^n\frac{1}{2^n} = 1 - \frac{1}{2} + \frac{1}{4} - \frac{1}{8} + \frac{1}{16} - \frac{1}{32} + \frac{1}{64} - \frac{1}{128} + \frac{1}{256} - \cdots.$$

The theorem says that if we truncate the series after the eighth term, we throw away a total that is positive and less than $1/256$. The sum of the first eight terms is $s_8 = 0.6640625$ and the sum of the first nine terms is $s_9 = 0.66796875$. The sum of the geometric series is

$$\frac{1}{1 - (-1/2)} = \frac{1}{3/2} = \frac{2}{3},$$

and we note that $0.6640625 < (2/3) < 0.66796875$. The difference, $(2/3) - 0.6640625 = 0.0026041666\ldots$, is positive and is less than $(1/256) = 0.00390625$. ∎

Conditional Convergence

If we replace all the negative terms in the alternating series in Example 3, changing them to positive terms instead, we obtain the geometric series $\sum 1/2^n$. The original series and the new series of absolute values both converge (although to different sums). For an absolutely convergent series, changing infinitely many of the negative terms in the series to positive values does not change its property of still being a convergent series. Other convergent series may behave differently. The convergent alternating harmonic series has infinitely many negative terms, but if we change its negative terms to positive values, the resulting series is the divergent harmonic series. So the presence of infinitely many negative terms is essential to the convergence of the alternating harmonic series. The following terminology distinguishes these two types of convergent series.

DEFINITION A series that is convergent but not absolutely convergent is called **conditionally convergent**.

The alternating harmonic series is conditionally convergent, or **converges conditionally**. The next example extends that result to the alternating p-series.

EXAMPLE 4   If p is a positive constant, the sequence $\{1/n^p\}$ is a decreasing sequence with limit zero. Therefore, the alternating p-series

$$\sum_{n=1}^{\infty} \frac{(-1)^{n-1}}{n^p} = 1 - \frac{1}{2^p} + \frac{1}{3^p} - \frac{1}{4^p} + \cdots, \quad p > 0$$

converges.

If $p > 1$, the series converges absolutely as an ordinary p-series. If $0 < p \le 1$, the series converges conditionally by the alternating series test. For instance,

$$\text{Absolute convergence } (p = 3/2): \quad 1 - \frac{1}{2^{3/2}} + \frac{1}{3^{3/2}} - \frac{1}{4^{3/2}} + \cdots$$

$$\text{Conditional convergence } (p = 1/2): \quad 1 - \frac{1}{\sqrt{2}} + \frac{1}{\sqrt{3}} - \frac{1}{\sqrt{4}} + \cdots \qquad \blacksquare$$

We need to be careful when using a conditionally convergent series. We have seen with the alternating harmonic series that altering the signs of infinitely many terms of a conditionally convergent series can change its convergence status. Even more, simply changing the order of occurrence of infinitely many of its terms can also have a significant effect, as we now discuss.

Rearranging Series

We can always rearrange the terms of a *finite* collection of numbers without changing their sum. The same result is true for an infinite series that is absolutely convergent (see Exercise 96 for an outline of the proof).

THEOREM 17—The Rearrangement Theorem for Absolutely Convergent Series

If $\sum_{n=1}^{\infty} a_n$ converges absolutely, and $b_1, b_2, \ldots, b_n, \ldots$ is any arrangement of the sequence $\{a_n\}$, then $\sum b_n$ converges absolutely and

$$\sum_{n=1}^{\infty} b_n = \sum_{n=1}^{\infty} a_n.$$

On the other hand, if we rearrange the terms of a conditionally convergent series, we can get different results. In fact, for any real number r, a given conditionally convergent series can be rearranged so its sum is equal to r. (We omit the proof of this fact.) Here's an example of summing the terms of a conditionally convergent series with different orderings, with each ordering giving a different value for the sum.

EXAMPLE 5   We know that the alternating harmonic series $\sum_{n=1}^{\infty} (-1)^{n+1}/n$ converges to some number L. Moreover, by Theorem 16, L lies between the successive partial sums $s_2 = 1/2$ and $s_3 = 5/6$, so $L \ne 0$. If we multiply the series by 2 we obtain

$$2L = 2 \sum_{n=1}^{\infty} \frac{(-1)^{n+1}}{n} = 2\left(1 - \frac{1}{2} + \frac{1}{3} - \frac{1}{4} + \frac{1}{5} - \frac{1}{6} + \frac{1}{7} - \frac{1}{8} + \frac{1}{9} - \frac{1}{10} + \frac{1}{11} - \cdots\right)$$

$$= 2 - 1 + \frac{2}{3} - \frac{1}{2} + \frac{2}{5} - \frac{1}{3} + \frac{2}{7} - \frac{1}{4} + \frac{2}{9} - \frac{1}{5} + \frac{2}{11} - \cdots.$$

Now we change the order of this last sum by grouping each pair of terms with the same odd denominator, but leaving the negative terms with the even denominators as they are placed (so the denominators are the positive integers in their natural order). This rearrangement gives

$$(2 - 1) - \frac{1}{2} + \left(\frac{2}{3} - \frac{1}{3}\right) - \frac{1}{4} + \left(\frac{2}{5} - \frac{1}{5}\right) - \frac{1}{6} + \left(\frac{2}{7} - \frac{1}{7}\right) - \frac{1}{8} + \cdots$$

$$= \left(1 - \frac{1}{2} + \frac{1}{3} - \frac{1}{4} + \frac{1}{5} - \frac{1}{6} + \frac{1}{7} - \frac{1}{8} + \frac{1}{9} - \frac{1}{10} + \frac{1}{11} - \cdots\right)$$

$$= \sum_{n=1}^{\infty} \frac{(-1)^{n+1}}{n} = L.$$

So by rearranging the terms of the conditionally convergent series $\sum_{n=1}^{\infty} 2(-1)^{n+1}/n$, the series becomes $\sum_{n=1}^{\infty}(-1)^{n+1}/n$, which is the alternating harmonic series itself. If the two series are the same, it would imply that $2L = L$, which is clearly false since $L \neq 0$. ∎

Example 5 shows that we cannot rearrange the terms of a conditionally convergent series and expect the new series to be the same as the original one. When we use a conditionally convergent series, the terms must be added together in the order they are given to obtain a correct result. In contrast, Theorem 17 guarantees that the terms of an absolutely convergent series can be summed in any order without affecting the result.

Summary of Tests to Determine Convergence or Divergence

We have developed a variety of tests to determine convergence or divergence for an infinite series of constants. There are other tests we have not presented which are sometimes given in more advanced courses. Here is a summary of the tests we have considered.

1. **The nth-Term Test for Divergence:** Unless $a_n \to 0$, the series diverges.
2. **Geometric series:** $\sum ar^n$ converges if $|r| < 1$; otherwise it diverges.
3. **p-series:** $\sum 1/n^p$ converges if $p > 1$; otherwise it diverges.
4. **Series with nonnegative terms:** Try the Integral Test or try comparing to a known series with the Direct Comparison Test or the Limit Comparison Test. Try the Ratio or Root Test.
5. **Series with some negative terms:** Does $\sum |a_n|$ converge by the Ratio or Root Test, or by another of the tests listed above? Remember, absolute convergence implies convergence.
6. **Alternating series:** $\sum a_n$ converges if the series satisfies the conditions of the Alternating Series Test.

EXERCISES 10.6

Determining Convergence or Divergence
In Exercises 1–14, determine if the alternating series converges or diverges. Some of the series do not satisfy the conditions of the Alternating Series Test.

1. $\sum_{n=1}^{\infty} (-1)^{n+1} \frac{1}{\sqrt{n}}$

2. $\sum_{n=1}^{\infty} (-1)^{n+1} \frac{1}{n^{3/2}}$

3. $\sum_{n=1}^{\infty} (-1)^{n+1} \frac{1}{n3^n}$

4. $\sum_{n=2}^{\infty} (-1)^n \frac{4}{(\ln n)^2}$

5. $\sum_{n=1}^{\infty} (-1)^n \frac{n}{n^2 + 1}$

6. $\sum_{n=1}^{\infty} (-1)^{n+1} \frac{n^2 + 5}{n^2 + 4}$

7. $\sum_{n=1}^{\infty} (-1)^{n+1} \frac{2^n}{n^2}$

8. $\sum_{n=1}^{\infty} (-1)^n \frac{10^n}{(n + 1)!}$

9. $\sum_{n=1}^{\infty} (-1)^{n+1} \left(\frac{n}{10}\right)^n$

10. $\sum_{n=2}^{\infty} (-1)^{n+1} \frac{1}{\ln n}$

11. $\displaystyle\sum_{n=1}^{\infty}(-1)^{n+1}\frac{\ln n}{n}$

12. $\displaystyle\sum_{n=1}^{\infty}(-1)^{n}\ln\left(1+\frac{1}{n}\right)$

13. $\displaystyle\sum_{n=1}^{\infty}(-1)^{n+1}\frac{\sqrt{n}+1}{n+1}$

14. $\displaystyle\sum_{n=1}^{\infty}(-1)^{n+1}\frac{3\sqrt{n}+1}{\sqrt{n}+1}$

Absolute and Conditional Convergence

Which of the series in Exercises 15–48 converge absolutely, which converge, and which diverge? Give reasons for your answers.

15. $\displaystyle\sum_{n=1}^{\infty}(-1)^{n+1}(0.1)^{n}$

16. $\displaystyle\sum_{n=1}^{\infty}(-1)^{n+1}\frac{(0.1)^{n}}{n}$

17. $\displaystyle\sum_{n=1}^{\infty}(-1)^{n}\frac{1}{\sqrt{n}}$

18. $\displaystyle\sum_{n=1}^{\infty}\frac{(-1)^{n}}{1+\sqrt{n}}$

19. $\displaystyle\sum_{n=1}^{\infty}(-1)^{n+1}\frac{n}{n^{3}+1}$

20. $\displaystyle\sum_{n=1}^{\infty}(-1)^{n+1}\frac{n!}{2^{n}}$

21. $\displaystyle\sum_{n=1}^{\infty}(-1)^{n}\frac{1}{n+3}$

22. $\displaystyle\sum_{n=1}^{\infty}(-1)^{n}\frac{\sin n}{n^{2}}$

23. $\displaystyle\sum_{n=1}^{\infty}(-1)^{n+1}\frac{3+n}{5+n}$

24. $\displaystyle\sum_{n=1}^{\infty}\frac{(-2)^{n+1}}{n+5^{n}}$

25. $\displaystyle\sum_{n=1}^{\infty}(-1)^{n+1}\frac{1+n}{n^{2}}$

26. $\displaystyle\sum_{n=1}^{\infty}(-1)^{n+1}\left(\sqrt[n]{10}\right)$

27. $\displaystyle\sum_{n=1}^{\infty}(-1)^{n}n^{2}(2/3)^{n}$

28. $\displaystyle\sum_{n=2}^{\infty}(-1)^{n+1}\frac{1}{n\ln n}$

29. $\displaystyle\sum_{n=1}^{\infty}(-1)^{n}\frac{\tan^{-1}n}{n^{2}+1}$

30. $\displaystyle\sum_{n=1}^{\infty}(-1)^{n}\frac{\ln n}{n-\ln n}$

31. $\displaystyle\sum_{n=1}^{\infty}(-1)^{n}\frac{n}{n+1}$

32. $\displaystyle\sum_{n=1}^{\infty}(-5)^{-n}$

33. $\displaystyle\sum_{n=1}^{\infty}\frac{(-100)^{n}}{n!}$

34. $\displaystyle\sum_{n=1}^{\infty}\frac{(-1)^{n-1}}{n^{2}+2n+1}$

35. $\displaystyle\sum_{n=1}^{\infty}\frac{\cos n\pi}{n\sqrt{n}}$

36. $\displaystyle\sum_{n=1}^{\infty}\frac{\cos n\pi}{n}$

37. $\displaystyle\sum_{n=1}^{\infty}\frac{(-1)^{n}(n+1)^{n}}{(2n)^{n}}$

38. $\displaystyle\sum_{n=1}^{\infty}\frac{(-1)^{n+1}(n!)^{2}}{(2n)!}$

39. $\displaystyle\sum_{n=1}^{\infty}(-1)^{n}\frac{(2n)!}{2^{n}n!n!}$

40. $\displaystyle\sum_{n=1}^{\infty}(-1)^{n}\frac{(n!)^{2}3^{n}}{(2n+1)!}$

41. $\displaystyle\sum_{n=1}^{\infty}(-1)^{n}\left(\sqrt{n+1}-\sqrt{n}\right)$

42. $\displaystyle\sum_{n=1}^{\infty}(-1)^{n}\left(\sqrt{n^{2}+n}-n\right)$

43. $\displaystyle\sum_{n=1}^{\infty}(-1)^{n}\left(\sqrt{n+\sqrt{n}}-\sqrt{n}\right)$

44. $\displaystyle\sum_{n=1}^{\infty}\frac{(-1)^{n}}{\sqrt{n}+\sqrt{n+1}}$

45. $\displaystyle\sum_{n=1}^{\infty}(-1)^{n}\operatorname{sech}n$

46. $\displaystyle\sum_{n=1}^{\infty}(-1)^{n}\operatorname{csch}n$

47. $\dfrac{1}{4}-\dfrac{1}{6}+\dfrac{1}{8}-\dfrac{1}{10}+\dfrac{1}{12}-\dfrac{1}{14}+\cdots$

48. $1+\dfrac{1}{4}-\dfrac{1}{9}-\dfrac{1}{16}+\dfrac{1}{25}+\dfrac{1}{36}-\dfrac{1}{49}-\dfrac{1}{64}+\cdots$

Error Estimation

In Exercises 49–52, estimate the magnitude of the error involved in using the sum of the first four terms to approximate the sum of the entire series.

49. $\displaystyle\sum_{n=1}^{\infty}(-1)^{n+1}\frac{1}{n}$

50. $\displaystyle\sum_{n=1}^{\infty}(-1)^{n+1}\frac{1}{10^{n}}$

51. $\displaystyle\sum_{n=1}^{\infty}(-1)^{n+1}\frac{(0.01)^{n}}{n}$ As you will see in Section 10.7, the sum is $\ln(1.01)$.

52. $\dfrac{1}{1+t}=\displaystyle\sum_{n=0}^{\infty}(-1)^{n}t^{n},\quad 0<t<1$

In Exercises 53–56, determine how many terms should be used to estimate the sum of the entire series with an error of less than 0.001.

53. $\displaystyle\sum_{n=1}^{\infty}(-1)^{n}\frac{1}{n^{2}+3}$

54. $\displaystyle\sum_{n=1}^{\infty}(-1)^{n+1}\frac{n}{n^{2}+1}$

55. $\displaystyle\sum_{n=1}^{\infty}(-1)^{n+1}\frac{1}{(n+3\sqrt{n})^{3}}$

56. $\displaystyle\sum_{n=1}^{\infty}(-1)^{n}\frac{1}{\ln(\ln(n+2))}$

In Exercises 57–82, use any method to determine whether the series converges or diverges. Give reasons for your answer.

57. $\displaystyle\sum_{n=1}^{\infty}\frac{3^{n}}{n^{n}}$

58. $\displaystyle\sum_{n=1}^{\infty}\frac{3^{n}}{n^{3}}$

59. $\displaystyle\sum_{n=1}^{\infty}\left(\frac{1}{n+2}-\frac{1}{n+3}\right)$

60. $\displaystyle\sum_{n=1}^{\infty}\left(\frac{1}{2n+1}-\frac{1}{2n+2}\right)$

61. $\displaystyle\sum_{n=0}^{\infty}(-1)^{n}\frac{(n+2)!}{(2n)!}$

62. $\displaystyle\sum_{n=2}^{\infty}\frac{(3n)!}{(n!)^{3}}$

63. $\displaystyle\sum_{n=1}^{\infty}n^{-2/\sqrt{5}}$

64. $\displaystyle\sum_{n=2}^{\infty}\frac{3}{10+n^{4/3}}$

65. $\displaystyle\sum_{n=1}^{\infty}\left(1-\frac{2}{n}\right)^{n^{2}}$

66. $\displaystyle\sum_{n=0}^{\infty}\left(\frac{n+1}{n+2}\right)^{n}$

67. $\displaystyle\sum_{n=1}^{\infty}\frac{n-2}{n^{2}+3n}\left(-\frac{2}{3}\right)^{n}$

68. $\displaystyle\sum_{n=0}^{\infty}\frac{n+1}{(n+2)!}\left(\frac{3}{2}\right)^{n}$

69. $\dfrac{1}{2}-\dfrac{1}{2}+\dfrac{1}{2}-\dfrac{1}{2}+\dfrac{1}{2}-\dfrac{1}{2}+\cdots$

70. $1-\dfrac{1}{8}+\dfrac{1}{64}-\dfrac{1}{512}+\dfrac{1}{4096}-\cdots$

71. $\displaystyle\sum_{n=3}^{\infty}\sin\left(\frac{1}{\sqrt{n}}\right)$

72. $\displaystyle\sum_{n=1}^{\infty}\tan(n^{1/n})$

73. $\displaystyle\sum_{n=2}^{\infty}\frac{n}{\ln n}$

74. $\displaystyle\sum_{n=2}^{\infty}\frac{1}{n\sqrt{\ln n}}$

75. $\displaystyle\sum_{n=2}^{\infty}\ln\left(\frac{n+2}{n+1}\right)$

76. $\displaystyle\sum_{n=2}^{\infty}\left(\frac{\ln n}{n}\right)^{3}$

77. $\displaystyle\sum_{n=2}^{\infty}\frac{1}{1+2+2^{2}+\cdots+2^{n}}$

78. $\displaystyle\sum_{n=2}^{\infty}\frac{1+3+3^{2}+\cdots+3^{n-1}}{1+2+3+\cdots+n}$

79. $\sum_{n=0}^{\infty} (-1)^n \dfrac{e^n}{e^n + e^{n^2}}$

80. $\sum_{n=0}^{\infty} \dfrac{(2n+3)(2^n+3)}{3^n+2}$

81. $\sum_{n=1}^{\infty} \dfrac{n^2 3^n}{3 \cdot 5 \cdot 7 \cdots (2n+1)}$

82. $\sum_{n=1}^{\infty} \dfrac{4 \cdot 6 \cdot 8 \cdots (2n)}{5^{n+1}(n+2)!}$

T Approximate the sums in Exercises 83 and 84 with an error of magnitude less than 5×10^{-6}.

83. $\sum_{n=0}^{\infty} (-1)^n \dfrac{1}{(2n)!}$ As you will see in Section 10.9, the sum is cos 1, the cosine of 1 radian.

84. $\sum_{n=0}^{\infty} (-1)^n \dfrac{1}{n!}$ As you will see in Section 10.9 the sum is e^{-1}.

Theory and Examples

85. a. The series

$$\frac{1}{3} - \frac{1}{2} + \frac{1}{9} - \frac{1}{4} + \frac{1}{27} - \frac{1}{8} + \cdots + \frac{1}{3^n} - \frac{1}{2^n} + \cdots$$

does not meet one of the conditions of Theorem 15. Which one?

b. Use Theorem 17 to find the sum of the series in part (a).

T **86.** The limit L of an alternating series that satisfies the conditions of Theorem 15 lies between the values of any two consecutive partial sums. This suggests using the average

$$\frac{s_n + s_{n+1}}{2} = s_n + \frac{1}{2}(-1)^{n+2} a_{n+1}$$

to estimate L. Compute

$$s_{20} + \frac{1}{2} \cdot \frac{1}{21}$$

as an approximation to the sum of the alternating harmonic series. The exact sum is $\ln 2 = 0.69314718 \ldots$.

87. The sign of the remainder of an alternating series that satisfies the conditions of Theorem 15   Prove the assertion in Theorem 16 that whenever an alternating series satisfying the conditions of Theorem 15 is approximated with one of its partial sums, then the remainder (sum of the unused terms) has the same sign as the first unused term. (*Hint:* Group the remainder's terms in consecutive pairs.)

88. Show that the sum of the first $2n$ terms of the series

$$1 - \frac{1}{2} + \frac{1}{2} - \frac{1}{3} + \frac{1}{3} - \frac{1}{4} + \frac{1}{4} - \frac{1}{5} + \frac{1}{5} - \frac{1}{6} + \cdots$$

is the same as the sum of the first n terms of the series

$$\frac{1}{1 \cdot 2} + \frac{1}{2 \cdot 3} + \frac{1}{3 \cdot 4} + \frac{1}{4 \cdot 5} + \frac{1}{5 \cdot 6} + \cdots.$$

Do these series converge? What is the sum of the first $2n + 1$ terms of the first series? If the series converge, what is their sum?

89. Show that if $\sum_{n=1}^{\infty} a_n$ diverges, then $\sum_{n=1}^{\infty} |a_n|$ diverges.

90. Show that if $\sum_{n=1}^{\infty} a_n$ converges absolutely, then

$$\left| \sum_{n=1}^{\infty} a_n \right| \le \sum_{n=1}^{\infty} |a_n|.$$

91. Show that if $\sum_{n=1}^{\infty} a_n$ and $\sum_{n=1}^{\infty} b_n$ both converge absolutely, then so do the following.

a. $\sum_{n=1}^{\infty} (a_n + b_n)$ **b.** $\sum_{n=1}^{\infty} (a_n - b_n)$

c. $\sum_{n=1}^{\infty} k a_n$ (k any number)

92. Show by example that $\sum_{n=1}^{\infty} a_n b_n$ may diverge even if $\sum_{n=1}^{\infty} a_n$ and $\sum_{n=1}^{\infty} b_n$ both converge.

93. If $\sum a_n$ converges absolutely, prove that $\sum a_n^2$ converges.

94. Does the series

$$\sum_{n=1}^{\infty} \left(\frac{1}{n} - \frac{1}{n^2} \right)$$

converge or diverge? Justify your answer.

T **95.** In the alternating harmonic series, suppose the goal is to arrange the terms to get a new series that converges to $-1/2$. Start the new arrangement with the first negative term, which is $-1/2$. Whenever you have a sum that is less than or equal to $-1/2$, start introducing positive terms, taken in order, until the new total is greater than $-1/2$. Then add negative terms until the total is less than or equal to $-1/2$ again. Continue this process until your partial sums have been above the target at least three times and finish at or below it. If s_n is the sum of the first n terms of your new series, plot the points (n, s_n) to illustrate how the sums are behaving.

96. Outline of the proof of the Rearrangement Theorem (Theorem 17)

a. Let ε be a positive real number, let $L = \sum_{n=1}^{\infty} a_n$, and let $s_k = \sum_{n=1}^{k} a_n$. Show that for some index N_1 and for some index $N_2 \ge N_1$,

$$\sum_{n=N_1}^{\infty} |a_n| < \frac{\varepsilon}{2} \quad \text{and} \quad |s_{N_2} - L| < \frac{\varepsilon}{2}.$$

Since all the terms $a_1, a_2, \ldots, a_{N_2}$ appear somewhere in the sequence $\{b_n\}$, there is an index $N_3 \ge N_2$ such that if $n \ge N_3$, then $\left(\sum_{k=1}^{n} b_k \right) - s_{N_2}$ is at most a sum of terms a_m with $m \ge N_1$. Therefore, if $n \ge N_3$,

$$\left| \sum_{k=1}^{n} b_k - L \right| \le \left| \sum_{k=1}^{n} b_k - s_{N_2} \right| + |s_{N_2} - L|$$

$$\le \sum_{k=N_1}^{\infty} |a_k| + |s_{N_2} - L| < \varepsilon.$$

b. The argument in part (a) shows that if $\sum_{n=1}^{\infty} a_n$ converges absolutely then $\sum_{n=1}^{\infty} b_n$ converges and $\sum_{n=1}^{\infty} b_n = \sum_{n=1}^{\infty} a_n$. Now show that because $\sum_{n=1}^{\infty} a_n$ converges, $\sum_{n=1}^{\infty} b_n$ converges to $\sum_{n=1}^{\infty} a_n$.

10.7 Power Series

Now that we can test many infinite series of numbers for convergence, we can study sums that look like "infinite polynomials." We call these sums *power series* because they are defined as infinite series of powers of some variable, in our case x. Like polynomials, power series can be added, subtracted, multiplied, differentiated, and integrated to give new power series. With power series we can extend the methods of calculus to a vast array of functions, making the techniques of calculus applicable in an even wider setting.

Power Series and Convergence

We begin with the formal definition, which specifies the notation and terminology used for power series.

DEFINITIONS   **A power series about $x = 0$** is a series of the form

$$\sum_{n=0}^{\infty} c_n x^n = c_0 + c_1 x + c_2 x^2 + \cdots + c_n x^n + \cdots. \tag{1}$$

A power series about $x = a$ is a series of the form

$$\sum_{n=0}^{\infty} c_n(x - a)^n = c_0 + c_1(x - a) + c_2(x - a)^2 + \cdots + c_n(x - a)^n + \cdots \tag{2}$$

in which the **center** a and the **coefficients** $c_0, c_1, c_2, \ldots, c_n, \ldots$ are constants.

Equation (1) is the special case obtained by taking $a = 0$ in Equation (2). We will see that a power series defines a function $f(x)$ on a certain interval where it converges. Moreover, this function will be shown to be continuous and differentiable over the interior of that interval.

EXAMPLE 1   Taking all the coefficients to be 1 in Equation (1) gives the geometric power series

$$\sum_{n=0}^{\infty} x^n = 1 + x + x^2 + \cdots + x^n + \cdots.$$

This is the geometric series with first term 1 and ratio x. It converges to $1/(1 - x)$ for $|x| < 1$. We express this fact by writing

Power Series for $\dfrac{1}{1 - x}$

$$\frac{1}{1 - x} = \sum_{n=0}^{\infty} x^n, \quad |x| < 1$$

$$\frac{1}{1 - x} = 1 + x + x^2 + \cdots + x^n + \cdots, \quad -1 < x < 1. \tag{3}$$

∎

Up to now, we have used Equation (3) as a formula for the sum of the series on the right. We now change the focus: We think of the partial sums of the series on the right as polynomials $P_n(x)$ that approximate the function on the left. For values of x near zero, we need take only a few terms of the series to get a good approximation. As we move toward $x = 1$, or -1, we must take more terms. Figure 10.17 shows the graphs of $f(x) = 1/(1 - x)$ and the approximating polynomials $y_n = P_n(x)$ for $n = 0, 1, 2,$ and 8. The function $f(x) = 1/(1 - x)$ is not continuous on intervals containing $x = 1$, where it has a vertical asymptote. The approximations do not apply when $x \geq 1$.

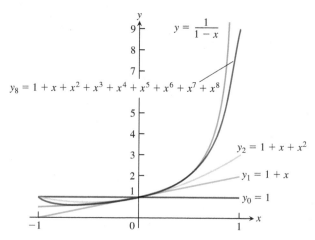

FIGURE 10.17 The graphs of $f(x) = 1/(1 - x)$ in Example 1 and four of its polynomial approximations.

EXAMPLE 2 The power series

$$1 - \frac{1}{2}(x - 2) + \frac{1}{4}(x - 2)^2 + \cdots + \left(-\frac{1}{2}\right)^n (x - 2)^n + \cdots \qquad (4)$$

matches Equation (2) with $a = 2$, $c_0 = 1$, $c_1 = -1/2$, $c_2 = 1/4, \ldots, c_n = (-1/2)^n$. This is a geometric series with first term 1 and ratio $r = -\dfrac{x - 2}{2}$. The series converges for $\left|\dfrac{x - 2}{2}\right| < 1$, which simplifies to $0 < x < 4$. The sum is

$$\frac{1}{1 - r} = \frac{1}{1 + \dfrac{x - 2}{2}} = \frac{2}{x},$$

so

$$\frac{2}{x} = 1 - \frac{(x - 2)}{2} + \frac{(x - 2)^2}{4} - \cdots + \left(-\frac{1}{2}\right)^n (x - 2)^n + \cdots, \qquad 0 < x < 4.$$

Series (4) generates useful polynomial approximations of $f(x) = 2/x$ for values of x near 2:

$$P_0(x) = 1$$

$$P_1(x) = 1 - \frac{1}{2}(x - 2) = 2 - \frac{x}{2}$$

$$P_2(x) = 1 - \frac{1}{2}(x - 2) + \frac{1}{4}(x - 2)^2 = 3 - \frac{3x}{2} + \frac{x^2}{4},$$

and so on (Figure 10.18). ∎

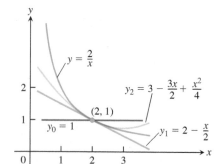

FIGURE 10.18 The graphs of $f(x) = 2/x$ and its first three polynomial approximations (Example 2).

The following example illustrates how we test a power series for convergence by using the Ratio Test to see where it converges and diverges.

EXAMPLE 3 For what values of x do the following power series converge?

(a) $\displaystyle\sum_{n=1}^{\infty} (-1)^{n-1} \frac{x^n}{n} = x - \frac{x^2}{2} + \frac{x^3}{3} - \cdots$

(b) $\displaystyle\sum_{n=1}^{\infty} (-1)^{n-1} \frac{x^{2n-1}}{2n - 1} = x - \frac{x^3}{3} + \frac{x^5}{5} - \cdots$

(c) $\displaystyle\sum_{n=0}^{\infty} \frac{x^n}{n!} = 1 + x + \frac{x^2}{2!} + \frac{x^3}{3!} + \cdots$

(d) $\displaystyle\sum_{n=0}^{\infty} n! x^n = 1 + x + 2!x^2 + 3!x^3 + \cdots$

Solution Apply the Ratio Test to the series $\sum |u_n|$, where u_n is the nth term of the power series in question.

(a) $\left| \dfrac{u_{n+1}}{u_n} \right| = \left| \dfrac{x^{n+1}}{n+1} \cdot \dfrac{n}{x} \right| = \dfrac{n}{n+1}|x| \to |x|.$

By the Ratio Test, the series converges absolutely for $|x| < 1$ and diverges for $|x| > 1$. At $x = 1$, we get the alternating harmonic series $1 - 1/2 + 1/3 - 1/4 + \cdots$, which converges. At $x = -1$, we get $-1 - 1/2 - 1/3 - 1/4 - \cdots$, the negative of the harmonic series, which diverges. Series (a) converges for $-1 < x \leq 1$ and diverges elsewhere.

We will see in Example 6 that this series converges to the function $\ln(1 + x)$ on the interval $(-1, 1]$ (see Figure 10.19).

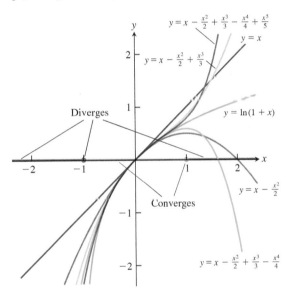

FIGURE 10.19 The power series $x - \dfrac{x^2}{2} + \dfrac{x^3}{3} - \dfrac{x^4}{4} + \cdots$ converges on the interval $(-1, 1]$.

(b) $\left| \dfrac{u_{n+1}}{u_n} \right| = \left| \dfrac{x^{2n+1}}{2n+1} \cdot \dfrac{2n-1}{x^{2n-1}} \right| = \dfrac{2n-1}{2n+1} x^2 \to x^2.$ $\quad$ 2(n + 1) - 1 = 2n + 1

By the Ratio Test, the series converges absolutely for $x^2 < 1$ and diverges for $x^2 > 1$. At $x = 1$ the series becomes $1 - 1/3 + 1/5 - 1/7 + \cdots$, which converges by the Alternating Series Theorem. It also converges at $x = -1$ because it is again an alternating series that satisfies the conditions for convergence. The value at $x = -1$ is the negative of the value at $x = 1$. Series (b) converges for $-1 \leq x \leq 1$ and diverges elsewhere.

(c) $\left| \dfrac{u_{n+1}}{u_n} \right| = \left| \dfrac{x^{n+1}}{(n+1)!} \cdot \dfrac{n!}{x^n} \right| = \dfrac{|x|}{n+1} \to 0$ for every x. $\quad \dfrac{n!}{(n+1)!} = \dfrac{1 \cdot 2 \cdot 3 \cdots n}{1 \cdot 2 \cdot 3 \cdots n \cdot (n+1)}$

The series converges absolutely for all x.

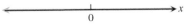

(d) $\left|\dfrac{u_{n+1}}{u_n}\right| = \left|\dfrac{(n+1)!x^{n+1}}{n!x^n}\right| = (n+1)|x| \to \infty$ unless $x = 0$.

The series diverges for all values of x except $x = 0$.

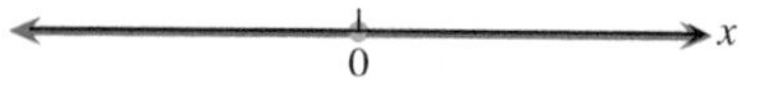

The previous example illustrated how a power series might converge. The next result shows that if a power series converges at more than one value, then it converges over an entire interval of values. The interval might be finite or infinite and contain one, both, or none of its endpoints. We will see that each endpoint of a finite interval must be tested independently for convergence or divergence.

THEOREM 18—The Convergence Theorem for Power Series

If the power series

$$\sum_{n=0}^{\infty} a_n x^n = a_0 + a_1 x + a_2 x^2 + \cdots \text{ converges at } x = c \neq 0, \text{ then it converges}$$

absolutely for all x with $|x| < |c|$. If the series diverges at $x = d$, then it diverges for all x with $|x| > |d|$.

Proof   The proof uses the Direct Comparison Test, with the given series compared to a converging geometric series.

Suppose the series $\sum_{n=0}^{\infty} a_n c^n$ converges. Then $\lim_{n\to\infty} a_n c^n = 0$ by the nth-Term Test. Hence, there is an integer N such that $|a_n c^n| < 1$ for all $n > N$, so that

$$|a_n| < \frac{1}{|c|^n} \qquad \text{for } n > N. \tag{5}$$

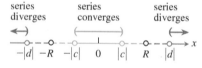

series diverges · series converges · series diverges

FIGURE 10.20 Convergence of $\sum a_n x^n$ at $x = c$ implies absolute convergence on the interval $-|c| < x < |c|$; divergence at $x = d$ implies divergence for $|x| > |d|$. The corollary to Theorem 18 asserts the existence of a radius of convergence $R \geq 0$. For $|x| < R$ the series converges absolutely and for $|x| > R$ it diverges.

Now take any x such that $|x| < |c|$, so that $|x|/|c| < 1$. Multiplying both sides of Equation (5) by $|x|^n$ gives

$$|a_n||x|^n < \frac{|x|^n}{|c|^n} \qquad \text{for } n > N.$$

Since $|x/c| < 1$, it follows that the geometric series $\sum_{n=0}^{\infty} |x/c|^n$ converges. By the Direct Comparison Test (Theorem 10), the series $\sum_{n=0}^{\infty} |a_n||x^n|$ converges, so the original power series $\sum_{n=0}^{\infty} a_n x^n$ converges absolutely for $-|c| < x < |c|$ as claimed by the theorem. (See Figure 10.20.)

Now suppose that the series $\sum_{n=0}^{\infty} a_n x^n$ diverges at $x = d$. If x is a number with $|x| > |d|$ and the series converges at x, then the first half of the theorem shows that the series also converges at d, contrary to our assumption. So the series diverges for all x with $|x| > |d|$. ∎

To simplify the notation, Theorem 18 deals with the convergence of series of the form $\sum a_n x^n$. For series of the form $\sum a_n (x - a)^n$ we can replace $x - a$ by x' and apply the results to the series $\sum a_n (x')^n$.

The Radius of Convergence of a Power Series

The theorem we have just proved and the examples we have studied lead to the conclusion that a power series $\sum c_n (x - a)^n$ behaves in one of three possible ways. It might converge only at $x = a$, or converge everywhere, or converge on some interval of radius R centered

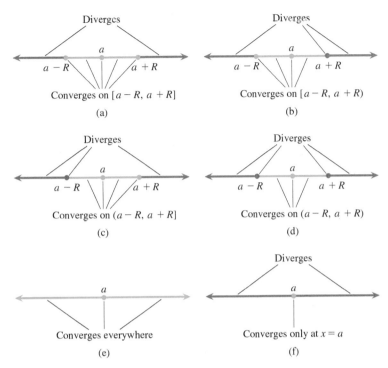

FIGURE 10.21 The six possibilities for an interval of convergence.

at $x = a$. We prove this as a Corollary to Theorem 18. When we also consider the convergence at the endpoints of an interval, there are six different possibilities. These are shown in Figure 10.21.

Corollary to Theorem 18

The convergence of the series $\sum c_n(x - a)^n$ is described by one of the following three cases:

1. There is a positive number R such that the series diverges for x with $|x - a| > R$ but converges absolutely for x with $|x - a| < R$. The series may or may not converge at either of the endpoints $x = a - R$ and $x = a + R$.

2. The series converges absolutely for every x $(R = \infty)$.

3. The series converges at $x = a$ and diverges elsewhere $(R = 0)$.

Proof We first consider the case where $a = 0$, so that we have a power series $\sum_{n=0}^{\infty} c_n x^n$ centered at 0. If the series converges everywhere we are in Case 2. If it converges only at $x = 0$ then we are in Case 3. Otherwise there is a nonzero number d such that $\sum_{n=0}^{\infty} c_n d^n$ diverges. Let S be the set of values of x for which $\sum_{n=0}^{\infty} c_n x^n$ converges. The set S does not include any x with $|x| > |d|$, since Theorem 18 implies the series diverges at all such values. So the set S is bounded. By the Completeness Property of the Real Numbers (Appendix 6) S has a least upper bound R. (This is the smallest number with the property that all elements of S are less than or equal to R.) Since we are not in Case 3, the series converges at some number $b \neq 0$ and, by Theorem 18, also on the open interval $(-|b|, |b|)$. Therefore, $R > 0$.

If $|x| < R$ then there is a number c in S with $|x| < c < R$, since otherwise R would not be the least upper bound for S. The series converges at c since $c \in S$, so by Theorem 18 the series converges absolutely at x.

Now suppose $|x| > R$. If the series converges at x, then Theorem 18 implies it converges absolutely on the open interval $(-|x|, |x|)$, so that S contains this interval. Since R is an upper bound for S, it follows that $|x| \le R$, which is a contradiction. So if $|x| > R$ then the series diverges. This proves the theorem for power series centered at $a = 0$.

For a power series centered at an arbitrary point $x = a$, set $x' = x - a$ and repeat the argument above, replacing x with x'. Since $x' = 0$ when $x = a$, convergence of the series $\sum_{n=0}^{\infty} |c_n(x')^n|$ on a radius R open interval centered at $x' = 0$ corresponds to convergence of the series $\sum_{n=0}^{\infty} |c_n(x - a)^n|$ on a radius R open interval centered at $x = a$. ∎

R is called the **radius of convergence** of the power series, and the interval of radius R centered at $x = a$ is called the **interval of convergence**. The interval of convergence may be open, closed, or half-open, depending on the particular series. At points x with $|x - a| < R$, the series converges absolutely. If the series converges for all values of x, we say its radius of convergence is infinite. If it converges only at $x = a$, we say its radius of convergence is zero.

How to Test a Power Series for Convergence

1. Use the Ratio Test (or Root Test) to find the largest open interval where the series converges absolutely,

$$|x - a| < R \qquad \text{or} \qquad a - R < x < a + R.$$

2. If R is finite, test for convergence or divergence at each endpoint, as in Examples 3a and b. Use a Comparison Test, the Integral Test, or the Alternating Series Test.

3. If R is finite, the series diverges for $|x - a| > R$ (it does not even converge conditionally) because the nth term does not approach zero for those values of x.

Operations on Power Series

On the intersection of their intervals of convergence, two power series can be added and subtracted term by term just like series of constants (Theorem 8). They can be multiplied just as we multiply polynomials, but we often limit the computation of the product to the first few terms, which are the most important. The following result gives a formula for the coefficients in the product, but we omit the proof. (Power series can also be divided in a way similar to division of polynomials, but we do not give a formula for the general coefficient here.)

THEOREM 19—Series Multiplication for Power Series

If $A(x) = \sum_{n=0}^{\infty} a_n x^n$ and $B(x) = \sum_{n=0}^{\infty} b_n x^n$ converge absolutely for $|x| < R$, and

$$c_n = a_0 b_n + a_1 b_{n-1} + a_2 b_{n-2} + \cdots + a_{n-1} b_1 + a_n b_0 = \sum_{k=0}^{n} a_k b_{n-k},$$

then $\sum_{n=0}^{\infty} c_n x^n$ converges absolutely to $A(x)B(x)$ for $|x| < R$:

$$\left(\sum_{n=0}^{\infty} a_n x^n\right)\left(\sum_{n=0}^{\infty} b_n x^n\right) = \sum_{n=0}^{\infty} c_n x^n.$$

Finding the general coefficient c_n in the product of two power series can be very tedious and the term may be unwieldy. The following computation provides an illustration of a product where we find the first few terms by multiplying the terms of the second series by each term of the first series:

$$\left(\sum_{n=0}^{\infty} x^n\right) \cdot \left(\sum_{n=0}^{\infty} (-1)^n \frac{x^{n+1}}{n+1}\right)$$

$$= (1 + x + x^2 + \cdots)\left(x - \frac{x^2}{2} + \frac{x^3}{3} - \cdots\right) \qquad \text{Multiply second series \ldots}$$

$$= \underbrace{\left(x - \frac{x^2}{2} + \frac{x^3}{3} - \cdots\right)}_{\text{by }1} + \underbrace{\left(x^2 - \frac{x^3}{2} + \frac{x^4}{3} - \cdots\right)}_{\text{by }x} + \underbrace{\left(x^3 - \frac{x^4}{2} + \frac{x^5}{3} - \cdots\right)}_{\text{by }x^2} + \cdots$$

$$= x + \frac{x^2}{2} + \frac{5x^3}{6} + \frac{7x^4}{12} \cdots. \qquad \text{and gather the first three powers.}$$

We can also substitute a function $f(x)$ for x in a convergent power series.

THEOREM 20   If $\sum_{n=0}^{\infty} a_n x^n$ converges absolutely for $|x| < R$ and f is a continuous function, then $\sum_{n=0}^{\infty} a_n(f(x))^n$ converges absolutely on the set of points x where $|f(x)| < R$.

Since $1/(1-x) = \sum_{n=0}^{\infty} x^n$ converges absolutely for $|x| < 1$, it follows from Theorem 20 that $1/(1 - 4x^2) = \sum_{n=0}^{\infty} (4x^2)^n$ converges absolutely when x satisfies $|4x^2| < 1$ or equivalently when $|x| < 1/2$.

Theorem 21 says that a power series can be differentiated term by term at each interior point of its interval of convergence. A proof is outlined in Exercise 64.

THEOREM 21—Term-by-Term Differentiation

If $\sum c_n(x-a)^n$ has radius of convergence $R > 0$, it defines a function

$$f(x) = \sum_{n=0}^{\infty} c_n(x-a)^n \qquad \text{on the interval} \qquad a - R < x < a + R.$$

This function f has derivatives of all orders inside the interval, and we obtain the derivatives by differentiating the original series term by term:

$$f'(x) = \sum_{n=1}^{\infty} n c_n(x-a)^{n-1},$$

$$f''(x) = \sum_{n=2}^{\infty} n(n-1) c_n(x-a)^{n-2},$$

and so on. Each of these derived series converges at every point of the interval $a - R < x < a + R.$

EXAMPLE 4   Find series for $f'(x)$ and $f''(x)$ if

$$f(x) = \frac{1}{1-x} = 1 + x + x^2 + x^3 + x^4 + \cdots + x^n + \cdots$$

$$= \sum_{n=0}^{\infty} x^n, \qquad -1 < x < 1.$$

Solution We differentiate the power series on the right term by term:

$$f'(x) = \frac{1}{(1-x)^2} = 1 + 2x + 3x^2 + 4x^3 + \cdots + nx^{n-1} + \cdots$$

$$= \sum_{n=1}^{\infty} nx^{n-1}, \quad -1 < x < 1;$$

$$f''(x) = \frac{2}{(1-x)^3} = 2 + 6x + 12x^2 + \cdots + n(n-1)x^{n-2} + \cdots$$

$$= \sum_{n=2}^{\infty} n(n-1)x^{n-2}, \quad -1 < x < 1. \qquad \blacksquare$$

Caution Term-by-term differentiation might not work for other kinds of series. For example, the trigonometric series

$$\sum_{n=1}^{\infty} \frac{\sin(n!x)}{n^2}$$

converges for all x. But if we differentiate term by term we get the series

$$\sum_{n=1}^{\infty} \frac{n! \cos(n!x)}{n^2},$$

which diverges for all x. This is not a power series since it is not a sum of positive integer powers of x. ●

It is also true that a power series can be integrated term by term throughout its interval of convergence. The proof is outlined in Exercise 65.

THEOREM 22—Term-by-Term Integration
Suppose that

$$f(x) = \sum_{n=0}^{\infty} c_n(x - a)^n$$

converges for $a - R < x < a + R (R > 0)$. Then

$$\sum_{n=0}^{\infty} c_n \frac{(x - a)^{n+1}}{n + 1}$$

converges for $a - R < x < a + R$ and

$$\int f(x)\, dx = \sum_{n=0}^{\infty} c_n \frac{(x - a)^{n+1}}{n + 1} + C$$

for $a - R < x < a + R$.

EXAMPLE 5 Identify the function

$$f(x) = \sum_{n=0}^{\infty} \frac{(-1)^n x^{2n+1}}{2n + 1} = x - \frac{x^3}{3} + \frac{x^5}{5} - \cdots, \quad -1 \le x \le 1.$$

Solution We differentiate the original series term by term and get

$$f'(x) = 1 - x^2 + x^4 - x^6 + \cdots, \quad -1 < x < 1. \qquad \text{Theorem 21}$$

This is a geometric series with first term 1 and ratio $-x^2$, so

$$f'(x) = \frac{1}{1 - (-x^2)} = \frac{1}{1 + x^2}.$$

We can now integrate $f'(x) = 1/(1 + x^2)$ to get

$$\int f'(x)\,dx = \int \frac{dx}{1 + x^2} = \tan^{-1}x + C.$$

The series for $f(x)$ is zero when $x = 0$, so $C = 0$. Hence

$$f(x) = x - \frac{x^3}{3} + \frac{x^5}{5} - \frac{x^7}{7} + \cdots = \tan^{-1}x, \qquad -1 < x < 1. \tag{6}$$

It can be shown that the series also converges to $\tan^{-1}x$ at the endpoints $x = \pm 1$, but we omit the proof. ∎

The Number π as a Series

$$\frac{\pi}{4} = \tan^{-1}1 = \sum_{n=0}^{\infty} \frac{(-1)^n}{2n + 1}$$

Notice that the original series in Example 5 converges at both endpoints of the original interval of convergence, but Theorem 22 can only guarantee the convergence of the differentiated series inside the interval.

EXAMPLE 6   The series

$$\frac{1}{1 + t} = 1 - t + t^2 - t^3 + \cdots$$

converges on the open interval $-1 < t < 1$. Therefore,

$$\ln(1 + x) = \int_0^x \frac{1}{1 + t}\,dt = t - \frac{t^2}{2} + \frac{t^3}{3} - \frac{t^4}{4} + \cdots \Big]_0^x \qquad \text{Theorem 22}$$

$$= x - \frac{x^2}{2} + \frac{x^3}{3} - \frac{x^4}{4} + \cdots$$

or

$$\ln(1 + x) = \sum_{n=1}^{\infty} \frac{(-1)^{n-1} x^n}{n}, \qquad -1 < x < 1.$$

Alternating Harmonic Series Sum

$$\ln 2 = \sum_{n=1}^{\infty} \frac{(-1)^{n-1}}{n}$$

It can also be shown that the series converges at $x = 1$ to the number $\ln 2$, but that was not guaranteed by the theorem. A proof of this is outlined in Exercise 61. ∎

EXERCISES 10.7

Intervals of Convergence

In Exercises 1–36, **(a)** find the series' radius and interval of convergence. For what values of x does the series converge **(b)** absolutely, **(c)** conditionally?

1. $\sum_{n=0}^{\infty} x^n$

2. $\sum_{n=0}^{\infty} (x + 5)^n$

3. $\sum_{n=0}^{\infty} (-1)^n (4x + 1)^n$

4. $\sum_{n=1}^{\infty} \frac{(3x - 2)^n}{n}$

5. $\sum_{n=0}^{\infty} \frac{(x - 2)^n}{10^n}$

6. $\sum_{n=0}^{\infty} (2x)^n$

7. $\sum_{n=0}^{\infty} \frac{nx^n}{n + 2}$

8. $\sum_{n=1}^{\infty} \frac{(-1)^n (x + 2)^n}{n}$

9. $\sum_{n=1}^{\infty} \frac{x^n}{n\sqrt{n}\,3^n}$

10. $\sum_{n=1}^{\infty} \frac{(x - 1)^n}{\sqrt{n}}$

11. $\sum_{n=0}^{\infty} \frac{(-1)^n x^n}{n!}$

12. $\sum_{n=0}^{\infty} \frac{3^n x^n}{n!}$

13. $\sum_{n=1}^{\infty} \frac{4^n x^{2n}}{n}$

14. $\sum_{n=1}^{\infty} \frac{(x - 1)^n}{n^3 3^n}$

15. $\sum_{n=0}^{\infty} \frac{x^n}{\sqrt{n^2 + 3}}$

16. $\sum_{n=0}^{\infty} \frac{(-1)^n x^{n+1}}{\sqrt{n + 3}}$

17. $\sum_{n=0}^{\infty} \frac{n(x+3)^n}{5^n}$

18. $\sum_{n=0}^{\infty} \frac{nx^n}{4^n(n^2+1)}$

19. $\sum_{n=0}^{\infty} \frac{\sqrt{n}x^n}{3^n}$

20. $\sum_{n=1}^{\infty} \sqrt[n]{n}(2x+5)^n$

21. $\sum_{n=1}^{\infty} (2+(-1)^n)\cdot(x+1)^{n-1}$

22. $\sum_{n=1}^{\infty} \frac{(-1)^n 3^{2n}(x-2)^n}{3n}$

23. $\sum_{n=1}^{\infty} \left(1+\frac{1}{n}\right)^n x^n$

24. $\sum_{n=1}^{\infty} (\ln n)x^n$

25. $\sum_{n=1}^{\infty} n^n x^n$

26. $\sum_{n=0}^{\infty} n!(x-4)^n$

27. $\sum_{n=1}^{\infty} \frac{(-1)^{n+1}(x+2)^n}{n2^n}$

28. $\sum_{n=0}^{\infty} (-2)^n(n+1)(x-1)^n$

29. $\sum_{n=2}^{\infty} \frac{x^n}{n(\ln n)^2}$ Get the information you need about $\sum 1/(n(\ln n)^2)$ from Section 10.3, Exercise 61.

30. $\sum_{n=2}^{\infty} \frac{x^n}{n\ln n}$ Get the information you need about $\sum 1/(n\ln n)$ from Section 10.3, Exercise 60.

31. $\sum_{n=1}^{\infty} \frac{(4x-5)^{2n+1}}{n^{3/2}}$

32. $\sum_{n=1}^{\infty} \frac{(3x+1)^{n+1}}{2n+2}$

33. $\sum_{n=1}^{\infty} \frac{1}{2\cdot4\cdot6\cdots(2n)} x^n$

34. $\sum_{n=1}^{\infty} \frac{3\cdot5\cdot7\cdots(2n+1)}{n^2\cdot2^n} x^{n+1}$

35. $\sum_{n=1}^{\infty} \frac{1+2+3+\cdots+n}{1^2+2^2+3^2+\cdots+n^2} x^n$

36. $\sum_{n=1}^{\infty} (\sqrt{n+1}-\sqrt{n})(x-3)^n$

In Exercises 37–40, find the series' radius of convergence.

37. $\sum_{n=1}^{\infty} \frac{n!}{3\cdot6\cdot9\cdots3n} x^n$

38. $\sum_{n=1}^{\infty} \left(\frac{2\cdot4\cdot6\cdots(2n)}{2\cdot5\cdot8\cdots(3n-1)}\right)^2 x^n$

39. $\sum_{n=1}^{\infty} \frac{(n!)^2}{2^n(2n)!} x^n$

40. $\sum_{n=1}^{\infty} \left(\frac{n}{n+1}\right)^{n^2} x^n$

(*Hint:* Apply the Root Test.)

In Exercises 41–48, use Theorem 20 to find the series' interval of convergence and, within this interval, the sum of the series as a function of x.

41. $\sum_{n=0}^{\infty} 3^n x^n$

42. $\sum_{n=0}^{\infty} (e^x-4)^n$

43. $\sum_{n=0}^{\infty} \frac{(x-1)^{2n}}{4^n}$

44. $\sum_{n=0}^{\infty} \frac{(x+1)^{2n}}{9^n}$

45. $\sum_{n=0}^{\infty} \left(\frac{\sqrt{x}}{2}-1\right)^n$

46. $\sum_{n=0}^{\infty} (\ln x)^n$

47. $\sum_{n=0}^{\infty} \left(\frac{x^2+1}{3}\right)^n$

48. $\sum_{n=0}^{\infty} \left(\frac{x^2-1}{2}\right)^n$

Using the Geometric Series

49. In Example 2 we represented the function $f(x)=2/x$ as a power series about $x=2$. Use a geometric series to represent $f(x)$ as a power series about $x=1$, and find its interval of convergence.

50. Use a geometric series to represent each of the given functions as a power series about $x=0$, and find their intervals of convergence.

   a. $f(x)=\frac{5}{3-x}$ b. $g(x)=\frac{3}{x-2}$

51. Represent the function $g(x)$ in Exercise 50 as a power series about $x=5$, and find the interval of convergence.

52. a. Find the interval of convergence of the power series

$$\sum_{n=0}^{\infty} \frac{8}{4^{n+2}} x^n.$$

   b. Represent the power series in part (a) as a power series about $x=3$ and identify the interval of convergence of the new series. (Later in the chapter you will understand why the new interval of convergence does not necessarily include all of the numbers in the original interval of convergence.)

Theory and Examples

53. For what values of x does the series

$$1-\frac{1}{2}(x-3)+\frac{1}{4}(x-3)^2+\cdots+\left(-\frac{1}{2}\right)^n(x-3)^n+\cdots$$

converge? What is its sum? What series do you get if you differentiate the given series term by term? For what values of x does the new series converge? What is its sum?

54. If you integrate the series in Exercise 53 term by term, what new series do you get? For what values of x does the new series converge, and what is another name for its sum?

55. The series

$$\sin x = x-\frac{x^3}{3!}+\frac{x^5}{5!}-\frac{x^7}{7!}+\frac{x^9}{9!}-\frac{x^{11}}{11!}+\cdots$$

converges to $\sin x$ for all x.

   a. Find the first six terms of a series for $\cos x$. For what values of x should the series converge?

   b. By replacing x by $2x$ in the series for $\sin x$, find a series that converges to $\sin 2x$ for all x.

   c. Using the result in part (a) and series multiplication, calculate the first six terms of a series for $2\sin x\cos x$. Compare your answer with the answer in part (b).

56. The series

$$e^x = 1+x+\frac{x^2}{2!}+\frac{x^3}{3!}+\frac{x^4}{4!}+\frac{x^5}{5!}+\cdots$$

converges to e^x for all x.

   a. Find a series for $(d/dx)e^x$. Do you get the series for e^x? Explain your answer.

b. Find a series for $\int e^x \, dx$. Do you get the series for e^x? Explain your answer.

c. Replace x by $-x$ in the series for e^x to find a series that converges to e^{-x} for all x. Then multiply the series for e^x and e^{-x} to find the first six terms of a series for $e^{-x} \cdot e^x$.

57. The series

$$\tan x = x + \frac{x^3}{3} + \frac{2x^5}{15} + \frac{17x^7}{315} + \frac{62x^9}{2835} + \cdots$$

converges to $\tan x$ for $-\pi/2 < x < \pi/2$.

a. Find the first five terms of the series for $\ln|\sec x|$. For what values of x should the series converge?

b. Find the first five terms of the series for $\sec^2 x$. For what values of x should this series converge?

c. Check your result in part (b) by squaring the series given for $\sec x$ in Exercise 58.

58. The series

$$\sec x = 1 + \frac{x^2}{2} + \frac{5}{24}x^4 + \frac{61}{720}x^6 + \frac{277}{8064}x^8 + \cdots$$

converges to $\sec x$ for $-\pi/2 < x < \pi/2$.

a. Find the first five terms of a power series for the function $\ln|\sec x + \tan x|$. For what values of x should the series converge?

b. Find the first four terms of a series for $\sec x \tan x$. For what values of x should the series converge?

c. Check your result in part (b) by multiplying the series for $\sec x$ by the series given for $\tan x$ in Exercise 57.

59. Uniqueness of convergent power series

a. Show that if two power series $\sum_{n=0}^{\infty} a_n x^n$ and $\sum_{n=0}^{\infty} b_n x^n$ are convergent and equal for all values of x in an open interval $(-c, c)$, then $a_n = b_n$ for every n. (*Hint:* Let $f(x) = \sum_{n=0}^{\infty} a_n x^n = \sum_{n=0}^{\infty} b_n x^n$. Differentiate term by term to show that a_n and b_n both equal $f^{(n)}(0)/(n!)$.)

b. Show that if $\sum_{n=0}^{\infty} a_n x^n = 0$ for all x in an open interval $(-c, c)$, then $a_n = 0$ for every n.

60. The sum of the series $\sum_{n=0}^{\infty} (n^2/2^n)$ To find the sum of this series, express $1/(1 - x)$ as a geometric series, differentiate both sides of the resulting equation with respect to x, multiply both sides of the result by x, differentiate again, multiply by x again, and set x equal to $1/2$. What do you get?

61. The sum of the alternating harmonic series This exercise will show that

$$\sum_{n=1}^{\infty} \frac{(-1)^{n+1}}{n} = \ln 2.$$

Let h_n be the nth partial sum of the harmonic series, and let s_n be the nth partial sum of the alternating harmonic series.

a. Use mathematical induction or algebra to show that

$$s_{2n} = h_{2n} - h_n.$$

b. Use the results in Exercise 63 in Section 10.3 to conclude that

$$\lim_{n \to \infty} (h_n - \ln n) = \gamma$$

and

$$\lim_{n \to \infty} (h_{2n} - \ln 2n) = \gamma,$$

where γ is Euler's constant.

c. Use these facts to show that

$$\sum_{n=1}^{\infty} \frac{(-1)^{n+1}}{n} = \lim_{n \to \infty} s_{2n} = \ln 2.$$

62. Assume that the series $\sum a_n x^n$ converges for $x = 4$ and diverges for $x = 7$. Answer true (T), false (F), or not enough information given (N) for the following statements about the series.

a. Converges absolutely for $x = -4$

b. Diverges for $x = 5$

c. Converges absolutely for $x = -8.5$

d. Converges for $x = -2$

e. Diverges for $x = 8$

f. Diverges for $x = -6$

g. Converges absolutely for $x = 0$

h. Converges absolutely for $x = -7.1$

63. Assume that the series $\sum a_n(x - 2)^n$ converges for $x = -1$ and diverges for $x = 6$. Answer true (T), false (F), or not enough information given (N) for the following statements about the series.

a. Converges absolutely for $x = 1$

b. Diverges for $x = -6$

c. Diverges for $x = 2$

d. Converges for $x = 0$

e. Converges absolutely for $x = 5$

f. Diverges for $x = 4.9$

g. Diverges for $x = 5.1$

h. Converges absolutely for $x = 4$

64. Proof of Theorem 21 Assume that $a = 0$ in Theorem 21 and that $f(x) = \sum_{n=0}^{\infty} c_n x^n$ converges for $-R < x < R$. Let $g(x) = \sum_{n=1}^{\infty} n c_n x^{n-1}$. This exercise will prove that $f'(x) = g(x)$, that is, $\lim_{h \to 0} \dfrac{f(x + h) - f(x)}{h} = g(x)$.

a. Use the Ratio Test to show that $g(x)$ converges for $-R < x < R$.

b. Use the Mean Value Theorem to show that

$$\frac{(x + h)^n - x^n}{h} = n c_n^{n-1}$$

for some c_n between x and $x + h$ for $n = 1, 2, 3, \dots$.

c. Show that

$$\left| g(x) - \frac{f(x + h) - f(x)}{h} \right| = \left| \sum_{n=2}^{\infty} n a_n \left(x^{n-1} - c_n^{n-1} \right) \right|$$

d. Use the Mean Value Theorem to show that

$$\frac{x^{n-1} - c_n^{n-1}}{x - c_n} = (n - 1) d_{n-1}^{n-2}$$

for some d_{n-1} between x and c_n for $n = 2, 3, 4, \dots$.

e. Explain why $|x - c_n| < h$ and why
$|d_{n-1}| \le \alpha = \max\{|x|, |x + h|\}$.

f. Show that
$$\left| g(x) - \frac{f(x + h) - f(x)}{h} \right| \le |h| \sum_{n=2}^{\infty} |n(n - 1)a_n \alpha^{n-2}|$$

g. Show that $\sum_{n=2}^{\infty} n(n - 1)\alpha^{n-2}$ converges for $-R < x < R$.

h. Let $h \to 0$ in part (f) to conclude that
$$\lim_{h \to 0} \frac{f(x + h) - f(x)}{h} = g(x).$$

65. Proof of Theorem 22 Assume that $a = 0$ in Theorem 22 and that $f(x) = \sum_{n=0}^{\infty} c_n x^n$ converges for $-R < x < R$. Let $g(x) = \sum_{n=0}^{\infty} \frac{c_n}{n + 1} x^{n+1}$. This exercise will prove that $g'(x) = f(x)$.

a. Use the Ratio Test to show that $g(x)$ converges for $-R < x < R$.

b. Use Theorem 21 to show that $g'(x) = f(x)$, that is,
$$\int f(x)\, dx = g(x) + C.$$

10.8 Taylor and Maclaurin Series

We have seen how geometric series can be used to generate a power series for functions such as $f(x) = 1/(1 - x)$ or $g(x) = 3/(x - 2)$. Now we expand our capability to represent a function with a power series. This section shows how functions that are infinitely differentiable generate power series called *Taylor series*. In many cases, these series provide useful polynomial approximations of the original functions. Because approximation by polynomials is extremely useful to both mathematicians and scientists, Taylor series are an important application of the theory of infinite series.

Series Representations

We know from Theorem 21 that within its interval of convergence I the sum of a power series is a continuous function with derivatives of all orders. But what about the other way around? If a function $f(x)$ has derivatives of all orders on an interval, can it be expressed as a power series on at least part of that interval? And if it can, what are its coefficients?

We can answer the last question readily if we assume that $f(x)$ is the sum of a power series about $x = a$,
$$f(x) = \sum_{n=0}^{\infty} a_n(x - a)^n$$
$$= a_0 + a_1(x - a) + a_2(x - a)^2 + \cdots + a_n(x - a)^n + \cdots$$

with a positive radius of convergence. By repeated term-by-term differentiation within the interval of convergence I, we obtain
$$f'(x) = a_1 + 2a_2(x - a) + 3a_3(x - a)^2 + \cdots + na_n(x - a)^{n-1} + \cdots,$$
$$f''(x) = 1 \cdot 2a_2 + 2 \cdot 3a_3(x - a) + 3 \cdot 4a_4(x - a)^2 + \cdots,$$
$$f'''(x) = 1 \cdot 2 \cdot 3a_3 + 2 \cdot 3 \cdot 4a_4(x - a) + 3 \cdot 4 \cdot 5a_5(x - a)^2 + \cdots,$$

with the nth derivative being
$$f^{(n)}(x) = n!a_n + \text{a sum of terms with } (x - a) \text{ as a factor.}$$

Since these equations all hold at $x = a$, we have
$$f'(a) = a_1, \quad f''(a) = 1 \cdot 2a_2, \quad f'''(a) = 1 \cdot 2 \cdot 3a_3,$$

and, in general,
$$f^{(n)}(a) = n!a_n.$$

These formulas reveal a pattern in the coefficients of any power series $\sum_{n=0}^{\infty} a_n(x-a)^n$ that converges to the values of f on I ("represents f on I"). If there *is* such a series (still an open question), then there is only one such series, and its nth coefficient is

$$a_n = \frac{f^{(n)}(a)}{n!}.$$

If f has a series representation, then the series must be

$$f(x) = f(a) + f'(a)(x-a) + \frac{f''(a)}{2!}(x-a)^2$$
$$+ \cdots + \frac{f^{(n)}(a)}{n!}(x-a)^n + \cdots. \tag{1}$$

But if we start with an arbitrary function f that is infinitely differentiable on an interval containing $x = a$ and use it to generate the series in Equation (1), does the series converge to $f(x)$ at each x in the interval of convergence? The answer is maybe—for some functions it will but for other functions it will not (as we will see in Example 4).

Taylor and Maclaurin Series

The series on the right-hand side of Equation (1) is the most important and useful series we will study in this chapter.

> **DEFINITIONS** Let f be a function with derivatives of all orders throughout some interval containing a as an interior point. Then the **Taylor series generated by f at $x = a$** is
>
> $$\sum_{k=0}^{\infty} \frac{f^{(k)}(a)}{k!}(x-a)^k = f(a) + f'(a)(x-a) + \frac{f''(a)}{2!}(x-a)^2$$
> $$+ \cdots + \frac{f^{(n)}(a)}{n!}(x-a)^n + \cdots.$$
>
> The **Maclaurin series of f** is the Taylor series generated by f at $x = 0$, or
>
> $$\sum_{k=0}^{\infty} \frac{f^{(k)}(0)}{k!}x^k = f(0) + f'(0)x + \frac{f''(0)}{2!}x^2 + \cdots + \frac{f^{(n)}(0)}{n!}x^n + \cdots.$$

The Maclaurin series generated by f is often just called the Taylor series of f.

EXAMPLE 1 Find the Taylor series generated by $f(x) = 1/x$ at $a = 2$. Where, if anywhere, does the series converge to $1/x$?

Solution We need to find $f(2), f'(2), f''(2), \ldots$. Taking derivatives we get

$$f(x) = x^{-1}, \quad f'(x) = -x^{-2}, \quad f''(x) = 2!x^{-3}, \ldots, f^{(n)}(x) = (-1)^n n! x^{-(n+1)},$$

so that

$$f(2) = 2^{-1} = \frac{1}{2}, \quad f'(2) = -\frac{1}{2^2}, \quad \frac{f''(2)}{2!} = 2^{-3} = \frac{1}{2^3}, \ldots, \frac{f^{(n)}(2)}{n!} = \frac{(-1)^n}{2^{n+1}}.$$

The Taylor series is

$$f(2) + f'(2)(x-2) - \frac{f''(2)}{2!}(x-2)^2 + \cdots + \frac{f^{(n)}(2)}{n!}(x-2)^n + \cdots$$
$$= \frac{1}{2} - \frac{(x-2)}{2^2} + \frac{(x-2)^2}{2^3} - \cdots + (-1)^n\frac{(x-2)^n}{2^{n+1}} + \cdots.$$

This is a geometric series with first term $1/2$ and ratio $r = -(x - 2)/2$. It converges absolutely for $|x - 2| < 2$ and its sum is

$$\frac{1/2}{1 + (x - 2)/2} = \frac{1}{2 + (x - 2)} = \frac{1}{x}.$$

In this example the Taylor series generated by $f(x) = 1/x$ at $a = 2$ converges to $1/x$ for $|x - 2| < 2$ or $0 < x < 4$. ∎

Taylor Polynomials

The linearization of a differentiable function f at a point a is the polynomial of degree one given by

$$P_1(x) = f(a) + f'(a)(x - a).$$

In Section 3.11 we used this linearization to approximate $f(x)$ at values of x near a. If f has derivatives of higher order at a, then it has higher-order polynomial approximations as well, one for each available derivative. These polynomials are called the Taylor polynomials of f.

DEFINITION Let f be a function with derivatives of order k for $k = 1, 2, \ldots, N$ in some interval containing a as an interior point. Then for any integer n from 0 through N, the **Taylor polynomial of order n** generated by f at $x = a$ is the polynomial

$$P_n(x) = f(a) + f'(a)(x - a) + \frac{f''(a)}{2!}(x - a)^2 + \cdots$$

$$+ \frac{f^{(k)}(a)}{k!}(x - a)^k + \cdots + \frac{f^{(n)}(a)}{n!}(x - a)^n.$$

We speak of a Taylor polynomial of *order n* rather than *degree n* because $f^{(n)}(a)$ may be zero. The first two Taylor polynomials of $f(x) = \cos x$ at $x = 0$, for example, are $P_0(x) = 1$ and $P_1(x) = 1$. The first-order Taylor polynomial has degree zero, not one.

Just as the linearization of f at $x = a$ provides the best linear approximation of f in the neighborhood of a, the higher-order Taylor polynomials provide the "best" polynomial approximations of their respective degrees. (See Exercise 44.)

EXAMPLE 2 Find the Taylor series and the Taylor polynomials generated by $f(x) = e^x$ at $x = 0$.

Solution Since $f^{(n)}(x) = e^x$ and $f^{(n)}(0) = 1$ for every $n = 0, 1, 2, \ldots$, the Taylor series generated by f at $x = 0$ (see Figure 10.22) is

$$f(0) + f'(0)x + \frac{f''(0)}{2!}x^2 + \cdots + \frac{f^{(n)}(0)}{n!}x^n + \cdots$$

$$= 1 + x + \frac{x^2}{2} + \cdots + \frac{x^n}{n!} + \cdots$$

$$= \sum_{k=0}^{\infty} \frac{x^k}{k!}.$$

This is also the Maclaurin series for e^x. In the next section we will see that the series converges to e^x at every x.

The Taylor polynomial of order n at $x = 0$ is

$$P_n(x) = 1 + x + \frac{x^2}{2} + \cdots + \frac{x^n}{n!}.$$

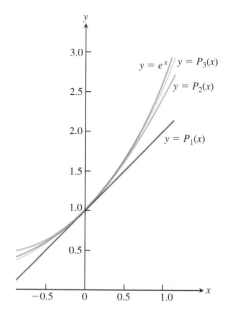

FIGURE 10.22 The graph of $f(x) = e^x$ and its Taylor polynomials

$P_1(x) = 1 + x$

$P_2(x) = 1 + x + (x^2/2!)$

$P_3(x) = 1 + x + (x^2/2!) + (x^3/3!)$.

Notice the very close agreement near the center $x = 0$ (Example 2).

EXAMPLE 3   Find the Taylor series and Taylor polynomials generated by $f(x) = \cos x$ at $x = 0$.

Solution   The cosine and its derivatives are

$$
\begin{aligned}
f(x) &= &\cos x, &\qquad f'(x) &= &-\sin x, \\
f''(x) &= &-\cos x, &\qquad f^{(3)}(x) &= &\sin x, \\
&\vdots & & &\vdots & \\
f^{(2n)}(x) &= (-1)^n \cos x, &\qquad f^{(2n+1)}(x) &= (-1)^{n+1} \sin x.
\end{aligned}
$$

At $x = 0$, the cosines are 1 and the sines are 0, so

$$f^{(2n)}(0) = (-1)^n, \qquad f^{(2n+1)}(0) = 0.$$

The Taylor series generated by f at 0 is

$$f(0) + f'(0)x + \frac{f''(0)}{2!}x^2 + \frac{f'''(0)}{3!}x^3 + \cdots + \frac{f^{(n)}(0)}{n!}x^n + \cdots$$

$$= 1 + 0 \cdot x - \frac{x^2}{2!} + 0 \cdot x^3 + \frac{x^4}{4!} + \cdots + (-1)^n \frac{x^{2n}}{(2n)!} + \cdots$$

$$= \sum_{k=0}^{\infty} \frac{(-1)^k x^{2k}}{(2k)!}.$$

This is also the Maclaurin series for $\cos x$. Notice that only even powers of x occur in the Taylor series generated by the cosine function, which is consistent with the fact that it is an even function. In Section 10.9, we will see that the series converges to $\cos x$ at every x.

Because $f^{(2n+1)}(0) = 0$, the Taylor polynomials of orders $2n$ and $2n + 1$ are identical:

$$P_{2n}(x) = P_{2n+1}(x) = 1 - \frac{x^2}{2!} + \frac{x^4}{4!} - \cdots + (-1)^n \frac{x^{2n}}{(2n)!}.$$

Figure 10.23 shows how well these polynomials approximate $f(x) = \cos x$ near $x = 0$. Only the right-hand portions of the graphs are given because the graphs are symmetric about the y-axis. ∎

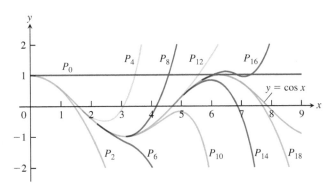

FIGURE 10.23   The polynomials

$$P_{2n}(x) = \sum_{k=0}^{n} \frac{(-1)^k x^{2k}}{(2k)!}$$

converge to $\cos x$ as $n \to \infty$. We can deduce the behavior of $\cos x$ arbitrarily far away solely from knowing the values of the cosine and its derivatives at $x = 0$ (Example 3).

EXAMPLE 4   It can be shown (though not easily) that

$$f(x) = \begin{cases} 0, & x = 0 \\ e^{-1/x^2}, & x \neq 0 \end{cases}$$

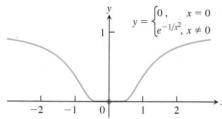

FIGURE 10.24 The graph of the continuous extension of $y = e^{-1/x^2}$ is so flat at the origin that all of its derivatives there are zero (Example 4). Therefore its Taylor series, which is zero everywhere, is not the function itself.

(Figure 10.24) has derivatives of all orders at $x = 0$ and that $f^{(n)}(0) = 0$ for all n. This means that the Taylor series generated by f at $x = 0$ is

$$f(0) + f'(0)x + \frac{f''(0)}{2!}x^2 + \cdots + \frac{f^{(n)}(0)}{n!}x^n + \cdots$$

$$= 0 + 0 \cdot x + 0 \cdot x^2 + \cdots + 0 \cdot x^n + \cdots$$

$$= 0 + 0 + \cdots + 0 + \cdots .$$

The series converges for every x (its sum is 0) but converges to $f(x)$ only at $x = 0$. That is, the Taylor series generated by $f(x)$ in this example is *not* equal to the function $f(x)$ over the entire interval of convergence. ∎

Two questions still remain.

1. For what values of x can we normally expect a Taylor series to converge to its generating function?

2. How accurately do a function's Taylor polynomials approximate the function on a given interval?

The answers are provided by a theorem of Taylor in the next section.

EXERCISES 10.8

Finding Taylor Polynomials

In Exercises 1–10, find the Taylor polynomials of orders 0, 1, 2, and 3 generated by f at a.

1. $f(x) = e^{2x}, \quad a = 0$
2. $f(x) = \sin x, \quad a = 0$
3. $f(x) = \ln x, \quad a = 1$
4. $f(x) = \ln(1 + x), \quad a = 0$
5. $f(x) = 1/x, \quad a = 2$
6. $f(x) = 1/(x + 2), \quad a = 0$
7. $f(x) = \sin x, \quad a = \pi/4$
8. $f(x) = \tan x, \quad a = \pi/4$
9. $f(x) = \sqrt{x}, \quad a = 4$
10. $f(x) = \sqrt{1 - x}, \quad a = 0$

Finding Taylor Series at $x = 0$ (Maclaurin Series)

Find the Maclaurin series for the functions in Exercises 11–24.

11. e^{-x}
12. xe^x
13. $\dfrac{1}{1 + x}$
14. $\dfrac{2 + x}{1 - x}$
15. $\sin 3x$
16. $\sin \dfrac{x}{2}$
17. $7 \cos(-x)$
18. $5 \cos \pi x$
19. $\cosh x = \dfrac{e^x + e^{-x}}{2}$
20. $\sinh x = \dfrac{e^x - e^{-x}}{2}$
21. $x^4 - 2x^3 - 5x + 4$
22. $\dfrac{x^2}{x + 1}$
23. $x \sin x$
24. $(x + 1) \ln(x + 1)$

Finding Taylor and Maclaurin Series

In Exercises 25–34, find the Taylor series generated by f at $x = a$.

25. $f(x) = x^3 - 2x + 4, \quad a = 2$
26. $f(x) = 2x^3 + x^2 + 3x - 8, \quad a = 1$
27. $f(x) = x^4 + x^2 + 1, \quad a = -2$
28. $f(x) = 3x^5 - x^4 + 2x^3 + x^2 - 2, \quad a = -1$
29. $f(x) = 1/x^2, \quad a = 1$
30. $f(x) = 1/(1 - x)^3, \quad a = 0$
31. $f(x) = e^x, \quad a = 2$
32. $f(x) = 2^x, \quad a = 1$
33. $f(x) = \cos(2x + (\pi/2)), \quad a = \pi/4$
34. $f(x) = \sqrt{x + 1}, \quad a = 0$

In Exercises 35–38, find the first three nonzero terms of the Maclaurin series for each function and the values of x for which the series converges absolutely.

35. $f(x) = \cos x - (2/(1 - x))$
36. $f(x) = (1 - x + x^2)e^x$
37. $f(x) = (\sin x) \ln(1 + x)$
38. $f(x) = x \sin^2 x$
39. $f(x) = x^4 e^{x^2}$
40. $f(x) = \dfrac{x^3}{1 + 2x}$

Theory and Examples

41. Use the Taylor series generated by e^x at $x = a$ to show that

$$e^x = e^a \left[1 + (x - a) + \frac{(x - a)^2}{2!} + \cdots \right].$$

42. (*Continuation of Exercise 41.*) Find the Taylor series generated by e^x at $x = 1$. Compare your answer with the formula in Exercise 41.

43. Let $f(x)$ have derivatives through order n at $x = a$. Show that the Taylor polynomial of order n and its first n derivatives have the same values that f and its first n derivatives have at $x = a$.

44. Approximation properties of Taylor polynomials Suppose that $f(x)$ is differentiable on an interval centered at $x = a$ and that $g(x) = b_0 + b_1(x - a) + \cdots + b_n(x - a)^n$ is a polynomial of degree n with constant coefficients $b_0, \ldots, b_n$. Let $E(x) = f(x) - g(x)$. Show that if we impose on g the conditions

i) $E(a) = 0$ The approximation error is zero at $x = a$.

ii) $\lim\limits_{x \to a} \dfrac{E(x)}{(x - a)^n} = 0$,   The error is negligible when compared to $(x - a)^n$.

then

$$g(x) = f(a) + f'(a)(x - a) + \frac{f''(a)}{2!}(x - a)^2 + \cdots$$

$$+ \frac{f^{(n)}(a)}{n!}(x - a)^n.$$

Thus, the Taylor polynomial $P_n(x)$ is the only polynomial of degree less than or equal to n whose error is both zero at $x = a$ and negligible when compared with $(x - a)^n$.

Quadratic Approximations The Taylor polynomial of order 2 generated by a twice-differentiable function $f(x)$ at $x = a$ is called the *quadratic approximation* of f at $x = a$. In Exercises 45–50, find the **(a)** linearization (Taylor polynomial of order 1) and **(b)** quadratic approximation of f at $x = 0$.

45. $f(x) = \ln(\cos x)$ **46.** $f(x) = e^{\sin x}$

47. $f(x) = 1/\sqrt{1 - x^2}$ **48.** $f(x) = \cosh x$

49. $f(x) = \sin x$ **50.** $f(x) = \tan x$

10.9 Convergence of Taylor Series

In the last section we asked when a Taylor series for a function can be expected to converge to the function that generates it. The finite-order Taylor polynomials that approximate the Taylor series provide estimates for the generating function. In order for these estimates to be useful, we need a way to control the possible errors we may encounter when approximating a function with its finite-order Taylor polynomials. How do we bound such possible errors? We answer the question in this section with the following theorem.

THEOREM 23—Taylor's Theorem

If f and its first n derivatives $f', f'', \ldots, f^{(n)}$ are continuous on the closed interval between a and b, and $f^{(n)}$ is differentiable on the open interval between a and b, then there exists a number c between a and b such that

$$f(b) = f(a) + f'(a)(b - a) + \frac{f''(a)}{2!}(b - a)^2 + \cdots$$

$$+ \frac{f^{(n)}(a)}{n!}(b - a)^n + \frac{f^{(n+1)}(c)}{(n + 1)!}(b - a)^{n+1}.$$

Taylor's Theorem is a generalization of the Mean Value Theorem (Exercise 49). There is a proof of Taylor's Theorem at the end of this section.

When we apply Taylor's Theorem, we usually want to hold a fixed and treat b as an independent variable. Taylor's formula is easier to use in circumstances like these if we change b to x. Here is a version of the theorem with this change.

Taylor's Formula

If f has derivatives of all orders in an open interval I containing a, then for each positive integer n and for each x in I,

$$f(x) = f(a) + f'(a)(x - a) + \frac{f''(a)}{2!}(x - a)^2 + \cdots$$

$$+ \frac{f^{(n)}(a)}{n!}(x - a)^n + R_n(x), \tag{1}$$

where

$$R_n(x) = \frac{f^{(n+1)}(c)}{(n + 1)!}(x - a)^{n+1} \qquad \text{for some } c \text{ between } a \text{ and } x. \tag{2}$$

When we state Taylor's theorem this way, it says that for each $x \in I$,

$$f(x) = P_n(x) + R_n(x).$$

The function $R_n(x)$ is determined by the value of the $(n + 1)$st derivative $f^{(n+1)}$ at a point c that depends on both a and x, and that lies somewhere between them. For any value of n we want, the equation gives both a polynomial approximation of f of that order and a formula for the error involved in using that approximation over the interval I.

Equation (1) is called **Taylor's formula**. The function $R_n(x)$ is called the **remainder of order n** or the **error term** for the approximation of f by $P_n(x)$ over I.

If $R_n(x) \to 0$ as $n \to \infty$ for all $x \in I$, we say that the Taylor series generated by f at $x = a$ **converges** to f on I, and we write

$$f(x) = \sum_{k=0}^{\infty} \frac{f^{(k)}(a)}{k!} (x - a)^k.$$

Often we can estimate R_n without knowing the value of c, as the following example illustrates.

EXAMPLE 1   Show that the Taylor series generated by $f(x) = e^x$ at $x = 0$ converges to $f(x)$ for every real value of x.

Solution   The function has derivatives of all orders throughout the interval $I = (-\infty, \infty)$. Equations (1) and (2) with $f(x) = e^x$ and $a = 0$ give

$$e^x = 1 + x + \frac{x^2}{2!} + \cdots + \frac{x^n}{n!} + R_n(x) \qquad \text{Polynomial from Section 10.8, Example 2}$$

and

$$R_n(x) = \frac{e^c}{(n + 1)!} x^{n+1} \qquad \text{for some } c \text{ between 0 and } x.$$

Since e^x is an increasing function of x, e^c lies between $e^0 = 1$ and e^x. When x is negative, so is c, and $e^c < 1$. When x is zero, $e^x = 1$ so that $R_n(x) = 0$. When x is positive, so is c, and $e^c < e^x$. Thus, for $R_n(x)$ given as above,

$$|R_n(x)| \le \frac{|x|^{n+1}}{(n + 1)!} \qquad \text{when } x \le 0, \qquad e^c < 1 \text{ since } c < 0$$

and

$$|R_n(x)| < e^x \frac{x^{n+1}}{(n + 1)!} \qquad \text{when } x > 0. \qquad e^c < e^x \text{ since } c < x$$

Finally, because

$$\lim_{n \to \infty} \frac{x^{n+1}}{(n + 1)!} = 0 \qquad \text{for every } x, \qquad \text{Section 10.1, Theorem 5}$$

$\lim_{n \to \infty} R_n(x) = 0$, and the series converges to e^x for every x. Thus,

$$e^x = \sum_{k=0}^{\infty} \frac{x^k}{k!} = 1 + x + \frac{x^2}{2!} + \cdots + \frac{x^k}{k!} + \cdots. \qquad (3)$$

The Number e as a Series

$$e = \sum_{n=0}^{\infty} \frac{1}{n!}$$

We can use the result of Example 1 with $x = 1$ to write

$$e = 1 + 1 + \frac{1}{2!} + \cdots + \frac{1}{n!} + R_n(1),$$

where for some c between 0 and 1,

$$R_n(1) = e^c \frac{1}{(n+1)!} < \frac{3}{(n+1)!}. \qquad e^c < e^1 < 3$$

Estimating the Remainder

It is often possible to estimate $R_n(x)$ as we did in Example 1. This method of estimation is so convenient that we state it as a theorem for future reference.

THEOREM 24—The Remainder Estimation Theorem

If there is a positive constant M such that $|f^{(n+1)}(t)| \leq M$ for all t between x and a, inclusive, then the remainder term $R_n(x)$ in Taylor's Theorem satisfies the inequality

$$|R_n(x)| \leq M \frac{|x-a|^{n+1}}{(n+1)!}.$$

If this inequality holds for every n and the other conditions of Taylor's Theorem are satisfied by f, then the series converges to $f(x)$.

The next two examples use Theorem 24 to show that the Taylor series generated by the sine and cosine functions do in fact converge to the functions themselves.

EXAMPLE 2   Show that the Taylor series for $\sin x$ at $x = 0$ converges for all x.

Solution   The function and its derivatives are

$$f(x) = \quad \sin x, \qquad f'(x) = \quad \cos x,$$

$$f''(x) = \quad -\sin x, \qquad f'''(x) = \quad -\cos x,$$

$$\vdots \qquad\qquad \vdots$$

$$f^{(2k)}(x) = (-1)^k \sin x, \qquad f^{(2k+1)}(x) = (-1)^k \cos x,$$

so

$$f^{(2k)}(0) = 0 \quad \text{and} \quad f^{(2k+1)}(0) = (-1)^k.$$

The series has only odd-powered terms and, for $n = 2k + 1$, Taylor's Theorem gives

$$\sin x = x - \frac{x^3}{3!} + \frac{x^5}{5!} - \cdots + \frac{(-1)^k x^{2k+1}}{(2k+1)!} + R_{2k+1}(x).$$

All the derivatives of $\sin x$ have absolute values less than or equal to 1, so we can apply the Remainder Estimation Theorem with $M = 1$ to obtain

$$|R_{2k+1}(x)| \leq 1 \cdot \frac{|x|^{2k+2}}{(2k+2)!}.$$

From Theorem 5, Rule 6, we have $(|x|^{2k+2}/(2k+2)!) \to 0$ as $k \to \infty$, whatever the value of x, so $R_{2k+1}(x) \to 0$ and the Maclaurin series for $\sin x$ converges to $\sin x$ for every x. Thus,

$$\sin x = \sum_{k=0}^{\infty} \frac{(-1)^k x^{2k+1}}{(2k+1)!} = x - \frac{x^3}{3!} + \frac{x^5}{5!} - \frac{x^7}{7!} + \cdots. \tag{4}$$

$$\sin x = x - \frac{x^3}{3!} + \frac{x^5}{5!} - \frac{x^7}{7!} + \cdots$$

EXAMPLE 3   Show that the Taylor series for $\cos x$ at $x = 0$ converges to $\cos x$ for every value of x.

Solution   We add the remainder term to the Taylor polynomial for $\cos x$ (Section 10.8, Example 3) to obtain Taylor's formula for $\cos x$ with $n = 2k$:

$$\cos x = 1 - \frac{x^2}{2!} + \frac{x^4}{4!} - \cdots + (-1)^k \frac{x^{2k}}{(2k)!} + R_{2k}(x).$$

Because the derivatives of the cosine have absolute value less than or equal to 1, the Remainder Estimation Theorem with $M = 1$ gives

$$|R_{2k}(x)| \le 1 \cdot \frac{|x|^{2k+1}}{(2k+1)!}.$$

For every value of x, $R_{2k}(x) \to 0$ as $k \to \infty$. Therefore, the series converges to $\cos x$ for every value of x. Thus,

$$\boxed{\cos x = 1 - \frac{x^2}{2!} + \frac{x^4}{4!} - \frac{x^6}{6!} + \cdots}$$

$$\cos x = \sum_{k=0}^{\infty} \frac{(-1)^k x^{2k}}{(2k)!} = 1 - \frac{x^2}{2!} + \frac{x^4}{4!} - \frac{x^6}{6!} + \cdots. \tag{5}$$

∎

Using Taylor Series

Since every Taylor series is a power series, the operations of adding, subtracting, and multiplying Taylor series are all valid on the intersection of their intervals of convergence.

EXAMPLE 4   Using known series, find the first few terms of the Taylor series for the given function by using power series operations.

(a) $\frac{1}{3}(2x + x \cos x)$ **(b)** $e^x \cos x$

Solution

(a) $\frac{1}{3}(2x + x \cos x) = \frac{2}{3}x + \frac{1}{3}x\left(1 - \frac{x^2}{2!} + \frac{x^4}{4!} - \cdots + (-1)^k \frac{x^{2k}}{(2k)!} + \cdots\right)$ Taylor series for $\cos x$

$$= \frac{2}{3}x + \frac{1}{3}x - \frac{x^3}{3!} + \frac{x^5}{3 \cdot 4!} - \cdots = x - \frac{x^3}{6} + \frac{x^5}{72} - \cdots$$

(b) $e^x \cos x = \left(1 + x + \frac{x^2}{2!} + \frac{x^3}{3!} + \frac{x^4}{4!} + \cdots\right)\left(1 - \frac{x^2}{2!} + \frac{x^4}{4!} - \cdots\right)$ Multiply the first series by each term of the second series.

$$= \left(1 + x + \frac{x^2}{2!} + \frac{x^3}{3!} + \frac{x^4}{4!} + \cdots\right) - \left(\frac{x^2}{2!} + \frac{x^3}{2!} + \frac{x^4}{2!2!} + \frac{x^5}{2!3!} + \cdots\right)$$

$$+ \left(\frac{x^4}{4!} + \frac{x^5}{4!} + \frac{x^6}{2!4!} + \cdots\right) + \cdots$$

$$= 1 + x - \frac{x^3}{3} - \frac{x^4}{6} + \cdots$$

∎

By Theorem 20, we can use the Taylor series of the function f to find the Taylor series of $f(u(x))$ where $u(x)$ is any continuous function. The Taylor series resulting from this substitution will converge for all x such that $u(x)$ lies within the interval of convergence of the Taylor series of f. For instance, we can find the Taylor series for $\cos 2x$ by substituting $2x$ for x in the Taylor series for $\cos x$:

$$\cos 2x = \sum_{k=0}^{\infty} \frac{(-1)^k (2x)^{2k}}{(2k)!} = 1 - \frac{(2x)^2}{2!} + \frac{(2x)^4}{4!} - \frac{(2x)^6}{6!} + \cdots \qquad \text{Eq. (5) with } 2x \text{ for } x$$

$$= 1 - \frac{2^2 x^2}{2!} + \frac{2^4 x^4}{4!} - \frac{2^6 x^6}{6!} + \cdots$$

$$= \sum_{k=0}^{\infty} (-1)^k \frac{2^{2k} x^{2k}}{(2k)!}.$$

EXAMPLE 5 For what values of x can we replace $\sin x$ by $x - (x^3/3!)$ and obtain an error whose magnitude is no greater than 3×10^{-4}?

Solution Here we can take advantage of the fact that the Taylor series for $\sin x$ is an alternating series for every nonzero value of x. According to the Alternating Series Estimation Theorem (Section 10.6), the error in truncating

$$\sin x = x - \frac{x^3}{3!} \bigg| + \frac{x^5}{5!} - \frac{x^7}{7!} + \cdots$$

after $(x^3/3!)$ is no greater than

$$\left| \frac{x^5}{5!} \right| = \frac{|x|^5}{120}.$$

Therefore the error will be less than or equal to 3×10^{-4} if

$$\frac{|x|^5}{120} < 3 \times 10^{-4} \qquad \text{or} \qquad |x| < \sqrt[5]{360 \times 10^{-4}} \approx 0.514. \qquad \begin{array}{l} \text{Rounded down,} \\ \text{to be safe} \end{array}$$

The Alternating Series Estimation Theorem tells us something that the Remainder Estimation Theorem does not: namely, that the estimate $x - (x^3/3!)$ for $\sin x$ is an underestimate when x is positive, because then $x^5/120$ is positive.

Figure 10.25 shows the graph of $\sin x$, along with the graphs of a number of its approximating Taylor polynomials. The graph of $P_3(x) = x - (x^3/3!)$ is almost indistinguishable from the sine curve when $0 \le x \le 1$. ∎

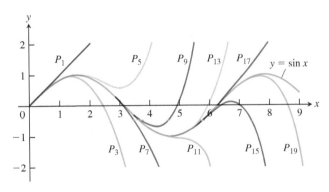

FIGURE 10.25 The polynomials

$$P_{2n+1}(x) = \sum_{k=0}^{n} \frac{(-1)^k x^{2k+1}}{(2k+1)!}$$

converge to $\sin x$ as $n \to \infty$. Notice how closely $P_3(x)$ approximates the sine curve for $x \le 1$ (Example 5).

A Proof of Taylor's Theorem

We prove Taylor's theorem assuming $a < b$. The proof for $a > b$ is nearly the same.

The Taylor polynomial

$$P_n(x) = f(a) + f'(a)(x - a) + \frac{f''(a)}{2!}(x - a)^2 + \cdots + \frac{f^{(n)}(a)}{n!}(x - a)^n$$

and its first n derivatives match the function f and its first n derivatives at $x = a$. We do not disturb that matching if we add another term of the form $K(x - a)^{n+1}$, where K is any constant, because such a term and its first n derivatives are all equal to zero at $x = a$. The new function

$$\phi_n(x) = P_n(x) + K(x - a)^{n+1}$$

and its first n derivatives still agree with f and its first n derivatives at $x = a$.

We now choose the particular value of K that makes the curve $y = \phi_n(x)$ agree with the original curve $y = f(x)$ at $x = b$. In symbols,

$$f(b) = P_n(b) + K(b - a)^{n+1}, \quad \text{or} \quad K = \frac{f(b) - P_n(b)}{(b - a)^{n+1}}. \tag{6}$$

With K defined by Equation (6), the function

$$F(x) = f(x) - \phi_n(x)$$

measures the difference between the original function f and the approximating function ϕ_n for each x in $[a, b]$.

We now use Rolle's Theorem (Section 4.2). First, because $F(a) = F(b) = 0$ and both F and F' are continuous on $[a, b]$, we know that

$$F'(c_1) = 0 \quad \text{for some } c_1 \text{ in } (a, b).$$

Next, because $F'(a) = F'(c_1) = 0$ and both F' and F'' are continuous on $[a, c_1]$, we know that

$$F''(c_2) = 0 \quad \text{for some } c_2 \text{ in } (a, c_1).$$

Rolle's Theorem, applied successively to $F'', F''', \ldots, F^{(n-1)}$, implies the existence of

$$c_3 \quad \text{in } (a, c_2) \qquad \text{such that } F'''(c_3) = 0,$$
$$c_4 \quad \text{in } (a, c_3) \qquad \text{such that } F^{(4)}(c_4) = 0,$$
$$\vdots$$
$$c_n \quad \text{in } (a, c_{n-1}) \quad \text{such that } F^{(n)}(c_n) = 0.$$

Finally, because $F^{(n)}$ is continuous on $[a, c_n]$ and differentiable on (a, c_n), and $F^{(n)}(a) = F^{(n)}(c_n) = 0$, Rolle's Theorem implies that there is a number c_{n+1} in (a, c_n) such that

$$F^{(n+1)}(c_{n+1}) = 0. \tag{7}$$

If we differentiate $F(x) = f(x) - P_n(x) - K(x - a)^{n+1}$ a total of $n + 1$ times, we get

$$F^{(n+1)}(x) = f^{(n+1)}(x) - 0 - (n + 1)!K. \tag{8}$$

Equations (7) and (8) together give

$$K = \frac{f^{(n+1)}(c)}{(n + 1)!} \qquad \text{for some number } c = c_{n+1} \text{ in } (a, b). \tag{9}$$

Equations (6) and (9) give

$$f(b) = P_n(b) + \frac{f^{(n+1)}(c)}{(n + 1)!}(b - a)^{n+1}.$$

This concludes the proof.

EXERCISES 10.9

Finding Taylor Series

Use substitution (as in Example 4) to find the Taylor series at $x = 0$ of the functions in Exercises 1–12.

1. e^{-5x} 2. $e^{-x/2}$ 3. $5\sin(-x)$

4. $\sin\left(\dfrac{\pi x}{2}\right)$ 5. $\cos 5x^2$ 6. $\cos\left(x^{2/3}/\sqrt{2}\right)$

7. $\ln(1 + x^2)$ 8. $\tan^{-1}(3x^4)$ 9. $\dfrac{1}{1 + \frac{3}{4}x^3}$

10. $\dfrac{1}{2 - x}$ 11. $\ln(3 + 6x)$ 12. $e^{-x^2 + \ln 5}$

Use power series operations to find the Taylor series at $x = 0$ for the functions in Exercises 13–30.

13. xe^x 14. $x^2\sin x$ 15. $\dfrac{x^2}{2} - 1 + \cos x$

16. $\sin x - x + \dfrac{x^3}{3!}$ 17. $x\cos \pi x$ 18. $x^2\cos(x^2)$

19. $\cos^2 x$ (Hint: $\cos^2 x = (1 + \cos 2x)/2$.)

20. $\sin^2 x$ 21. $\dfrac{x^2}{1 - 2x}$ 22. $x\ln(1 + 2x)$

23. $\dfrac{1}{(1 - x)^2}$ 24. $\dfrac{2}{(1 - x)^3}$ 25. $x\tan^{-1}x^2$

26. $\sin x \cdot \cos x$ 27. $e^x + \dfrac{1}{1 + x}$ 28. $\cos x - \sin x$

29. $\dfrac{x}{3}\ln(1 + x^2)$ 30. $\ln(1 + x) - \ln(1 - x)$

Find the first four nonzero terms in the Maclaurin series for the functions in Exercises 31–38.

31. $e^x\sin x$ 32. $\dfrac{\ln(1 + x)}{1 - x}$ 33. $(\tan^{-1}x)^2$

34. $\cos^2 x \cdot \sin x$ 35. $e^{\sin x}$ 36. $\sin(\tan^{-1}x)$

37. $\cos(e^x - 1)$ 38. $\cos\sqrt{x} + \ln(\cos x)$

Error Estimates

39. Estimate the error if $P_3(x) = x - (x^3/6)$ is used to estimate the value of $\sin x$ at $x = 0.1$.

40. Estimate the error if $P_4(x) = 1 + x + (x^2/2) + (x^3/6) + (x^4/24)$ is used to estimate the value of e^x at $x = 1/2$.

41. For approximately what values of x can you replace $\sin x$ by $x - (x^3/6)$ with an error of magnitude no greater than 5×10^{-4}? Give reasons for your answer.

42. If $\cos x$ is replaced by $1 - (x^2/2)$ and $|x| < 0.5$, what estimate can be made of the error? Does $1 - (x^2/2)$ tend to be too large, or too small? Give reasons for your answer.

43. How close is the approximation $\sin x = x$ when $|x| < 10^{-3}$? For which of these values of x is $x < \sin x$?

44. The estimate $\sqrt{1 + x} = 1 + (x/2)$ is used when x is small. Estimate the error when $|x| < 0.01$.

45. The approximation $e^x = 1 + x + (x^2/2)$ is used when x is small. Use the Remainder Estimation Theorem to estimate the error when $|x| < 0.1$.

46. (Continuation of Exercise 45.) When $x < 0$, the series for e^x is an alternating series. Use the Alternating Series Estimation Theorem to estimate the error that results from replacing e^x by $1 + x + (x^2/2)$ when $-0.1 < x < 0$. Compare your estimate with the one you obtained in Exercise 45.

Theory and Examples

47. Use the identity $\sin^2 x = (1 - \cos 2x)/2$ to obtain the Maclaurin series for $\sin^2 x$. Then differentiate this series to obtain the Maclaurin series for $2\sin x\cos x$. Check that this is the series for $\sin 2x$.

48. (Continuation of Exercise 47.) Use the identity $\cos^2 x = \cos 2x + \sin^2 x$ to obtain a power series for $\cos^2 x$.

49. **Taylor's Theorem and the Mean Value Theorem** Explain how the Mean Value Theorem (Section 4.2, Theorem 4) is a special case of Taylor's Theorem.

50. **Linearizations at inflection points** Show that if the graph of a twice-differentiable function $f(x)$ has an inflection point at $x = a$, then the linearization of f at $x = a$ is also the quadratic approximation of f at $x = a$. This explains why tangent lines fit so well at inflection points.

51. **The (second) second derivative test** Use the equation

$$f(x) = f(a) + f'(a)(x - a) + \dfrac{f''(c_2)}{2}(x - a)^2$$

to establish the following test.

Let f have continuous first and second derivatives and suppose that $f'(a) = 0$. Then

a. f has a local maximum at a if $f'' \leq 0$ throughout an interval whose interior contains a;

b. f has a local minimum at a if $f'' \geq 0$ throughout an interval whose interior contains a.

52. **A cubic approximation** Use Taylor's formula with $a = 0$ and $n = 3$ to find the standard cubic approximation of $f(x) = 1/(1 - x)$ at $x = 0$. Give an upper bound for the magnitude of the error in the approximation when $|x| \leq 0.1$.

53. a. Use Taylor's formula with $n = 2$ to find the quadratic approximation of $f(x) = (1 + x)^k$ at $x = 0$ (k a constant).

b. If $k = 3$, for approximately what values of x in the interval $[0, 1]$ will the error in the quadratic approximation be less than $1/100$?

54. **Improving approximations of π**

a. Let P be an approximation of π accurate to n decimals. Show that $P + \sin P$ gives an approximation correct to $3n$ decimals. (Hint: Let $P = \pi + x$.)

b. Try it with a calculator.

55. **The Taylor series generated by $f(x) = \sum_{n=0}^{\infty}a_n x^n$ is $\sum_{n=0}^{\infty}a_n x^n$** A function defined by a power series $\sum_{n=0}^{\infty}a_n x^n$ with a radius of convergence $R > 0$ has a Taylor series that converges to the function at every point of $(-R, R)$. Show this by showing that the Taylor series generated by $f(x) = \sum_{n=0}^{\infty}a_n x^n$ is the series $\sum_{n=0}^{\infty}a_n x^n$ itself.

An immediate consequence of this is that series like

$$x\sin x = x^2 - \dfrac{x^4}{3!} + \dfrac{x^6}{5!} - \dfrac{x^8}{7!} + \cdots$$

and

$$x^2 e^x = x^2 + x^3 + \frac{x^4}{2!} + \frac{x^5}{3!} + \cdots,$$

obtained by multiplying Taylor series by powers of x, as well as series obtained by integration and differentiation of convergent power series, are themselves the Taylor series generated by the functions they represent.

56. Taylor series for even functions and odd functions (*Continuation of Section 10.7, Exercise 59.*) Suppose that $f(x) = \sum_{n=0}^{\infty} a_n x^n$ converges for all x in an open interval $(-R, R)$. Show that

a. If f is even, then $a_1 = a_3 = a_5 = \cdots = 0$, i.e., the Taylor series for f at $x = 0$ contains only even powers of x.

b. If f is odd, then $a_0 = a_2 = a_4 = \cdots = 0$, i.e., the Taylor series for f at $x = 0$ contains only odd powers of x.

COMPUTER EXPLORATIONS

Taylor's formula with $n = 1$ and $a = 0$ gives the linearization of a function at $x = 0$. With $n = 2$ and $n = 3$ we obtain the standard quadratic and cubic approximations. In these exercises we explore the errors associated with these approximations. We seek answers to two questions:

a. For what values of x can the function be replaced by each approximation with an error less than 10^{-2}?

b. What is the maximum error we could expect if we replace the function by each approximation over the specified interval?

Using a CAS, perform the following steps to aid in answering questions (a) and (b) for the functions and intervals in Exercises 57–62.

Step 1: Plot the function over the specified interval.

Step 2: Find the Taylor polynomials $P_1(x)$, $P_2(x)$, and $P_3(x)$ at $x = 0$.

Step 3: Calculate the $(n + 1)$st derivative $f^{(n+1)}(c)$ associated with the remainder term for each Taylor polynomial. Plot the derivative as a function of c over the specified interval and estimate its maximum absolute value, M.

Step 4: Calculate the remainder $R_n(x)$ for each polynomial. Using the estimate M from Step 3 in place of $f^{(n+1)}(c)$, plot $R_n(x)$ over the specified interval. Then estimate the values of x that answer question (a).

Step 5: Compare your estimated error with the actual error $E_n(x) = |f(x) - P_n(x)|$ by plotting $E_n(x)$ over the specified interval. This will help answer question (b).

Step 6: Graph the function and its three Taylor approximations together. Discuss the graphs in relation to the information discovered in Steps 4 and 5.

57. $f(x) = \dfrac{1}{\sqrt{1 + x}}, \quad |x| \le \dfrac{3}{4}$

58. $f(x) = (1 + x)^{3/2}, \quad -\dfrac{1}{2} \le x \le 2$

59. $f(x) = \dfrac{x}{x^2 + 1}, \quad |x| \le 2$

60. $f(x) = (\cos x)(\sin 2x), \quad |x| \le 2$

61. $f(x) = e^{-x} \cos 2x, \quad |x| \le 1$

62. $f(x) = e^{x/3} \sin 2x, \quad |x| \le 2$

10.10 Applications of Taylor Series

We can use Taylor series to solve problems that would otherwise be intractable. For example, many functions have antiderivatives that cannot be expressed using familiar functions. In this section we show how to evaluate integrals of such functions by giving them as Taylor series. We also show how to use Taylor series to evaluate limits that lead to indeterminate forms and how Taylor series can be used to extend the exponential function from real to complex numbers. We begin with a discussion of the binomial series, which comes from the Taylor series of the function $f(x) = (1 + x)^m$, and conclude the section with Table 10.1, which lists some commonly used Taylor series.

The Binomial Series for Powers and Roots

The Taylor series generated by $f(x) = (1 + x)^m$, when m is constant, is

$$1 + mx + \frac{m(m - 1)}{2!}x^2 + \frac{m(m - 1)(m - 2)}{3!}x^3 + \cdots$$

$$+ \frac{m(m - 1)(m - 2) \cdots (m - k + 1)}{k!}x^k + \cdots. \quad (1)$$

This series, called the **binomial series**, converges absolutely for $|x| < 1$. To derive the series, we first list the function and its derivatives:

$$f(x) = (1 + x)^m$$
$$f'(x) = m(1 + x)^{m-1}$$
$$f''(x) = m(m - 1)(1 + x)^{m-2}$$
$$f'''(x) = m(m - 1)(m - 2)(1 + x)^{m-3}$$
$$\vdots$$
$$f^{(k)}(x) = m(m - 1)(m - 2) \cdots (m - k + 1)(1 + x)^{m-k}.$$

We then evaluate these at $x = 0$ and substitute into the Taylor series formula to obtain Series (1).

If m is an integer greater than or equal to zero, the series stops after $(m + 1)$ terms because the coefficients from $k = m + 1$ on are zero.

If m is not a positive integer or zero, the series is infinite and converges for $|x| < 1$. To see why, let u_k be the term involving x^k. Then apply the Ratio Test for absolute convergence to see that

$$\left| \frac{u_{k+1}}{u_k} \right| = \left| \frac{m - k}{k + 1} x \right| \to |x| \qquad \text{as } k \to \infty.$$

Our derivation of the binomial series shows only that it is generated by $(1 + x)^m$ and converges for $|x| < 1$. The derivation does not show that the series converges to $(1 + x)^m$. It does, but we leave the proof to Exercise 58. The following formulation gives a succinct way to express the series.

The Binomial Series

For $-1 < x < 1$,

$$(1 + x)^m = 1 + \sum_{k=1}^{\infty} \binom{m}{k} x^k,$$

where we define

$$\binom{m}{1} = m, \qquad \binom{m}{2} = \frac{m(m - 1)}{2!},$$

and

$$\binom{m}{k} = \frac{m(m - 1)(m - 2) \cdots (m - k + 1)}{k!} \qquad \text{for } k \geq 3.$$

EXAMPLE 1   If $m = -1$,

$$\binom{-1}{1} = -1, \qquad \binom{-1}{2} = \frac{-1(-2)}{2!} = 1,$$

and

$$\binom{-1}{k} = \frac{-1(-2)(-3) \cdots (-1 - k + 1)}{k!} = (-1)^k \left(\frac{k!}{k!} \right) = (-1)^k.$$

With these coefficient values and with x replaced by $-x$, the binomial series formula gives the familiar geometric series

$$(1 + x)^{-1} = 1 + \sum_{k=1}^{\infty} (-1)^k x^k = 1 - x + x^2 - x^3 + \cdots + (-1)^k x^k + \cdots. \qquad \blacksquare$$

EXAMPLE 2 We know from Section 3.11, Example 1, that $\sqrt{1 + x} \approx 1 + (x/2)$ for $|x|$ small. With $m = 1/2$, the binomial series gives quadratic and higher-order approximations as well, along with error estimates that come from the Alternating Series Estimation Theorem:

$$(1 + x)^{1/2} = 1 + \frac{x}{2} + \frac{\left(\frac{1}{2}\right)\left(-\frac{1}{2}\right)}{2!}x^2 + \frac{\left(\frac{1}{2}\right)\left(-\frac{1}{2}\right)\left(-\frac{3}{2}\right)}{3!}x^3$$

$$+ \frac{\left(\frac{1}{2}\right)\left(-\frac{1}{2}\right)\left(-\frac{3}{2}\right)\left(-\frac{5}{2}\right)}{4!}x^4 + \cdots$$

$$= 1 + \frac{x}{2} - \frac{x^2}{8} + \frac{x^3}{16} - \frac{5x^4}{128} + \cdots.$$

Substitution for x gives still other approximations. For example,

$$\sqrt{1 - x^2} \approx 1 - \frac{x^2}{2} - \frac{x^4}{8} \qquad \text{for } |x^2| \text{ small}$$

$$\sqrt{1 - \frac{1}{x}} \approx 1 - \frac{1}{2x} - \frac{1}{8x^2} \qquad \text{for } \left|\frac{1}{x}\right| \text{ small, that is, } |x| \text{ large.} \qquad \blacksquare$$

Evaluating Nonelementary Integrals

Sometimes we can use a familiar Taylor series to find the sum of a given power series in terms of a known function. For example,

$$x^2 - \frac{x^6}{3!} + \frac{x^{10}}{5!} - \frac{x^{14}}{7!} + \cdots = (x^2) - \frac{(x^2)^3}{3!} + \frac{(x^2)^5}{5!} - \frac{(x^2)^7}{7!} + \cdots = \sin x^2.$$

Additional examples are provided in Exercises 59–62.

Taylor series can be used to express nonelementary integrals in terms of series. Integrals like $\int \sin x^2 \, dx$ arise in the study of the diffraction of light.

EXAMPLE 3 Express $\int \sin x^2 \, dx$ as a power series.

Solution From the series for $\sin x$ we substitute x^2 for x to obtain

$$\sin x^2 = x^2 - \frac{x^6}{3!} + \frac{x^{10}}{5!} - \frac{x^{14}}{7!} + \frac{x^{18}}{9!} - \cdots.$$

Therefore,

$$\int \sin x^2 \, dx = C + \frac{x^3}{3} - \frac{x^7}{7 \cdot 3!} + \frac{x^{11}}{11 \cdot 5!} - \frac{x^{15}}{15 \cdot 7!} + \frac{x^{19}}{19 \cdot 9!} - \cdots. \qquad \blacksquare$$

EXAMPLE 4 Estimate $\int_0^1 \sin x^2 \, dx$ with an error of less than 0.001.

Solution From the indefinite integral in Example 3, we easily find that

$$\int_0^1 \sin x^2 \, dx = \frac{1}{3} - \frac{1}{7 \cdot 3!} + \frac{1}{11 \cdot 5!} - \frac{1}{15 \cdot 7!} + \frac{1}{19 \cdot 9!} - \cdots.$$

The series on the right-hand side alternates, and we find by numerical evaluations that

$$\frac{1}{11 \cdot 5!} \approx 0.00076$$

is the first term to be numerically less than 0.001. The sum of the preceding two terms gives

$$\int_0^1 \sin x^2 \, dx \approx \frac{1}{3} - \frac{1}{42} \approx 0.310.$$

With two more terms we could estimate

$$\int_0^1 \sin x^2 \, dx \approx 0.310268$$

with an error of less than 10^{-6}. With only one term beyond that we have

$$\int_0^1 \sin x^2 \, dx \approx \frac{1}{3} - \frac{1}{42} + \frac{1}{1320} - \frac{1}{75600} + \frac{1}{6894720} \approx 0.310268303,$$

with an error of about 1.08×10^{-9}. To guarantee this accuracy with the error formula for the Trapezoidal Rule would require using about 8000 subintervals.   ■

Arctangents

In Section 10.7, Example 5, we found a series for $\tan^{-1} x$ by differentiating to get

$$\frac{d}{dx} \tan^{-1} x = \frac{1}{1 + x^2} = 1 - x^2 + x^4 - x^6 + \cdots$$

and then integrating to get

$$\tan^{-1} x = x - \frac{x^3}{3} + \frac{x^5}{5} - \frac{x^7}{7} + \cdots .$$

However, we did not prove the term-by-term integration theorem on which this conclusion depended. We now derive the series again by integrating both sides of the finite formula

$$\frac{1}{1 + t^2} = 1 - t^2 + t^4 - t^6 + \cdots + (-1)^n t^{2n} + \frac{(-1)^{n+1} t^{2n+2}}{1 + t^2}, \tag{2}$$

in which the last term comes from adding the remaining terms as a geometric series with first term $a = (-1)^{n+1} t^{2n+2}$ and ratio $r = -t^2$. Integrating both sides of Equation (2) from $t = 0$ to $t = x$ gives

$$\tan^{-1} x = x - \frac{x^3}{3} + \frac{x^5}{5} - \frac{x^7}{7} + \cdots + (-1)^n \frac{x^{2n+1}}{2n + 1} + R_n(x),$$

where

$$R_n(x) = \int_0^x \frac{(-1)^{n+1} t^{2n+2}}{1 + t^2} \, dt.$$

The denominator of the integrand is greater than or equal to 1; hence

$$|R_n(x)| \le \int_0^{|x|} t^{2n+2} \, dt = \frac{|x|^{2n+3}}{2n + 3}.$$

If $|x| \le 1$, the right side of this inequality approaches zero as $n \to \infty$. Therefore $\lim_{n \to \infty} R_n(x) = 0$ if $|x| \le 1$ and

$$\tan^{-1} x = \sum_{n=0}^{\infty} \frac{(-1)^n x^{2n+1}}{2n + 1}, \qquad |x| \le 1.$$

$$\tag{3}$$

$$\tan^{-1} x = x - \frac{x^3}{3} + \frac{x^5}{5} - \frac{x^7}{7} + \cdots, \qquad |x| \le 1.$$

We take this route instead of finding the Taylor series directly because the formulas for the higher-order derivatives of $\tan^{-1} x$ are unmanageable. When we put $x = 1$ in Equation (3), we get **Leibniz's formula**:

$$\frac{\pi}{4} = 1 - \frac{1}{3} + \frac{1}{5} - \frac{1}{7} + \frac{1}{9} - \cdots + \frac{(-1)^n}{2n + 1} + \cdots.$$

Because this series converges very slowly, it is not used in approximating π to many decimal places. The series for $\tan^{-1} x$ converges most rapidly when x is near zero. For that reason, people who use the series for $\tan^{-1} x$ to compute π use various trigonometric identities.

For example, if

$$\alpha = \tan^{-1} \frac{1}{2} \quad \text{and} \quad \beta = \tan^{-1} \frac{1}{3},$$

then

$$\tan(\alpha + \beta) = \frac{\tan \alpha + \tan \beta}{1 - \tan \alpha \tan \beta} = \frac{\frac{1}{2} + \frac{1}{3}}{1 - \frac{1}{6}} = 1 = \tan \frac{\pi}{4}$$

and therefore

$$\frac{\pi}{4} = \alpha + \beta = \tan^{-1} \frac{1}{2} + \tan^{-1} \frac{1}{3}.$$

Now Equation (3) may be used with $x = 1/2$ to evaluate $\tan^{-1}(1/2)$ and with $x = 1/3$ to give $\tan^{-1}(1/3)$. The sum of these results, multiplied by 4, gives π.

Evaluating Indeterminate Forms

We can sometimes evaluate indeterminate forms by expressing the functions involved as Taylor series.

EXAMPLE 5   Evaluate

$$\lim_{x \to 1} \frac{\ln x}{x - 1}.$$

Solution   We represent $\ln x$ as a Taylor series in powers of $x - 1$. This can be accomplished by calculating the Taylor series generated by $\ln x$ at $x = 1$ directly or by replacing x by $x - 1$ in the series for $\ln(1 + x)$ in Section 10.7, Example 6. Either way, we obtain

$$\ln x = (x - 1) - \frac{1}{2}(x - 1)^2 + \cdots,$$

from which we find that

$$\lim_{x \to 1} \frac{\ln x}{x - 1} = \lim_{x \to 1} \left(1 - \frac{1}{2}(x - 1) + \cdots \right) = 1.$$

Of course, this particular limit can be evaluated using l'Hôpital's Rule just as well.   ∎

EXAMPLE 6   Evaluate

$$\lim_{x \to 0} \frac{\sin x - \tan x}{x^3}.$$

Solution The Taylor series for $\sin x$ and $\tan x$, to terms in x^5, are

$$\sin x = x - \frac{x^3}{3!} + \frac{x^5}{5!} - \cdots, \qquad \tan x = x + \frac{x^3}{3} + \frac{2x^5}{15} + \cdots.$$

Subtracting the series term by term, it follows that

$$\sin x - \tan x = -\frac{x^3}{2} - \frac{x^5}{8} - \cdots = x^3\left(-\frac{1}{2} - \frac{x^2}{8} - \cdots\right).$$

Division of both sides by x^3 and taking limits then gives

$$\lim_{x\to 0}\frac{\sin x - \tan x}{x^3} = \lim_{x\to 0}\left(-\frac{1}{2} - \frac{x^2}{8} - \cdots\right) = -\frac{1}{2}.$$

If we apply series to calculate $\lim_{x\to 0}((1/\sin x) - (1/x))$, we not only find the limit successfully but also discover an approximation formula for $\csc x$.

EXAMPLE 7 Find $\displaystyle\lim_{x\to 0}\left(\frac{1}{\sin x} - \frac{1}{x}\right)$.

Solution Using algebra and the Taylor series for $\sin x$, we have

$$\frac{1}{\sin x} - \frac{1}{x} = \frac{x - \sin x}{x\sin x} = \frac{x - \left(x - \frac{x^3}{3!} + \frac{x^5}{5!} - \cdots\right)}{x\cdot\left(x - \frac{x^3}{3!} + \frac{x^5}{5!} - \cdots\right)}$$

$$= \frac{x^3\left(\frac{1}{3!} - \frac{x^2}{5!} + \cdots\right)}{x^2\left(1 - \frac{x^2}{3!} + \cdots\right)} = x\cdot\frac{\frac{1}{3!} - \frac{x^2}{5!} + \cdots}{1 - \frac{x^2}{3!} + \cdots}.$$

Therefore,

$$\lim_{x\to 0}\left(\frac{1}{\sin x} - \frac{1}{x}\right) = \lim_{x\to 0}\left(x\cdot\frac{\frac{1}{3!} - \frac{x^2}{5!} + \cdots}{1 - \frac{x^2}{3!} + \cdots}\right) = 0.$$

From the quotient on the right, we can see that if $|x|$ is small, then

$$\frac{1}{\sin x} - \frac{1}{x} \approx x\cdot\frac{1}{3!} = \frac{x}{6} \qquad \text{or} \qquad \csc x \approx \frac{1}{x} + \frac{x}{6}.$$

Euler's Identity

A complex number is a number of the form $a + bi$, where a and b are real numbers and $i = \sqrt{-1}$ (see Appendix 7). If we substitute $x = i\theta$ (θ real) in the Taylor series for e^x and use the relations

$$i^2 = -1, \qquad i^3 = i^2 i = -i, \qquad i^4 = i^2 i^2 = 1, \qquad i^5 = i^4 i = i,$$

and so on, to simplify the result, we obtain

$$e^{i\theta} = 1 + \frac{i\theta}{1!} + \frac{i^2\theta^2}{2!} + \frac{i^3\theta^3}{3!} + \frac{i^4\theta^4}{4!} + \frac{i^5\theta^5}{5!} + \frac{i^6\theta^6}{6!} + \cdots$$

$$= \left(1 - \frac{\theta^2}{2!} + \frac{\theta^4}{4!} - \frac{\theta^6}{6!} + \cdots\right) + i\left(\theta - \frac{\theta^3}{3!} + \frac{\theta^5}{5!} - \cdots\right) = \cos\theta + i\sin\theta.$$

This does not *prove* that $e^{i\theta} = \cos\theta + i\sin\theta$ because we have not yet defined what it means to raise e to an imaginary power. Rather, it tells us how to define $e^{i\theta}$ so that its properties are consistent with the properties of the exponential function for real numbers.

DEFINITION

For any real number θ, $e^{i\theta} = \cos\theta + i\sin\theta$.   (4)

Equation (4), called **Euler's identity**, enables us to define e^{a+bi} to be $e^a \cdot e^{bi}$ for any complex number $a + bi$. So

$$e^{a+ib} = e^a(\cos b + i\sin b).$$

One consequence of this identity is the equation

$$e^{i\pi} = -1.$$

When written in the form $e^{i\pi} + 1 = 0$, this equation combines five of the most important constants in mathematics.

TABLE 10.1 Frequently Used Taylor Series

$$\frac{1}{1-x} = 1 + x + x^2 + \cdots + x^n + \cdots = \sum_{n=0}^{\infty} x^n, \qquad |x| < 1$$

$$\frac{1}{1+x} = 1 - x + x^2 - \cdots + (-x)^n + \cdots = \sum_{n=0}^{\infty} (-1)^n x^n, \qquad |x| < 1$$

$$e^x = 1 + x + \frac{x^2}{2!} + \cdots + \frac{x^n}{n!} + \cdots = \sum_{n=0}^{\infty} \frac{x^n}{n!}, \qquad |x| < \infty$$

$$\sin x = x - \frac{x^3}{3!} + \frac{x^5}{5!} - \cdots + (-1)^n \frac{x^{2n+1}}{(2n+1)!} + \cdots = \sum_{n=0}^{\infty} \frac{(-1)^n x^{2n+1}}{(2n+1)!}, \qquad |x| < \infty$$

$$\cos x = 1 - \frac{x^2}{2!} + \frac{x^4}{4!} - \cdots + (-1)^n \frac{x^{2n}}{(2n)!} + \cdots = \sum_{n=0}^{\infty} \frac{(-1)^n x^{2n}}{(2n)!}, \qquad |x| < \infty$$

$$\ln(1+x) = x - \frac{x^2}{2} + \frac{x^3}{3} - \cdots + (-1)^{n-1} \frac{x^n}{n} + \cdots = \sum_{n=1}^{\infty} \frac{(-1)^{n-1} x^n}{n}, \qquad -1 < x \leq 1$$

$$\tan^{-1} x = x - \frac{x^3}{3} + \frac{x^5}{5} - \cdots + (-1)^n \frac{x^{2n+1}}{2n+1} + \cdots = \sum_{n=0}^{\infty} \frac{(-1)^n x^{2n+1}}{2n+1}, \qquad |x| \leq 1$$

EXERCISES 10.10

Binomial Series

Find the first four terms of the binomial series for the functions in Exercises 1–10.

1. $(1 + x)^{1/2}$
2. $(1 + x)^{1/3}$
3. $(1 - x)^{-3}$
4. $(1 - 2x)^{1/2}$
5. $\left(1 + \dfrac{x}{2}\right)^{-2}$
6. $\left(1 - \dfrac{x}{3}\right)^{4}$
7. $(1 + x^3)^{-1/2}$
8. $(1 + x^2)^{-1/3}$
9. $\left(1 + \dfrac{1}{x}\right)^{1/2}$
10. $\dfrac{x}{\sqrt[3]{1 + x}}$

Find the binomial series for the functions in Exercises 11–14.

11. $(1 + x)^4$
12. $(1 + x^2)^3$
13. $(1 - 2x)^3$
14. $\left(1 - \dfrac{x}{2}\right)^{4}$

Approximations and Nonelementary Integrals

T In Exercises 15–18, use series to estimate the integrals' values with an error of magnitude less than 10^{-5}. (The answer section gives the integrals' values rounded to seven decimal places.)

15. $\displaystyle\int_0^{0.6} \sin x^2 \, dx$
16. $\displaystyle\int_0^{0.4} \dfrac{e^{-x} - 1}{x} \, dx$
17. $\displaystyle\int_0^{0.5} \dfrac{1}{\sqrt{1 + x^4}} dx$
18. $\displaystyle\int_0^{0.35} \sqrt[3]{1 + x^2} \, dx$

T Use series to approximate the values of the integrals in Exercises 19–22 with an error of magnitude less than 10^{-8}.

19. $\displaystyle\int_0^{0.1} \dfrac{\sin x}{x} dx$
20. $\displaystyle\int_0^{0.1} e^{-x^2} \, dx$
21. $\displaystyle\int_0^{0.1} \sqrt{1 + x^4} \, dx$
22. $\displaystyle\int_0^{1} \dfrac{1 - \cos x}{x^2} dx$

23. Estimate the error if $\cos t^2$ is approximated by $1 - \dfrac{t^4}{2} + \dfrac{t^8}{4!}$ in the integral $\int_0^1 \cos t^2 \, dt$.

24. Estimate the error if $\cos \sqrt{t}$ is approximated by $1 - \dfrac{t}{2} + \dfrac{t^2}{4!} - \dfrac{t^3}{6!}$ in the integral $\int_0^1 \cos \sqrt{t} \, dt$.

In Exercises 25–28, find a polynomial that will approximate $F(x)$ throughout the given interval with an error of magnitude less than 10^{-3}.

25. $F(x) = \displaystyle\int_0^x \sin t^2 \, dt, \quad [0, 1]$
26. $F(x) = \displaystyle\int_0^x t^2 e^{-t^2} \, dt, \quad [0, 1]$
27. $F(x) = \displaystyle\int_0^x \tan^{-1} t \, dt, \qquad$ (a) $[0, 0.5]$ (b) $[0, 1]$
28. $F(x) = \displaystyle\int_0^x \dfrac{\ln(1 + t)}{t} dt, \qquad$ (a) $[0, 0.5]$ (b) $[0, 1]$

Indeterminate Forms

Use series to evaluate the limits in Exercises 29–40.

29. $\displaystyle\lim_{x\to 0} \dfrac{e^x - (1 + x)}{x^2}$
30. $\displaystyle\lim_{x\to 0} \dfrac{e^x - e^{-x}}{x}$
31. $\displaystyle\lim_{t\to 0} \dfrac{1 - \cos t - (t^2/2)}{t^4}$
32. $\displaystyle\lim_{\theta\to 0} \dfrac{\sin \theta - \theta + (\theta^3/6)}{\theta^5}$
33. $\displaystyle\lim_{y\to 0} \dfrac{y - \tan^{-1} y}{y^3}$
34. $\displaystyle\lim_{y\to 0} \dfrac{\tan^{-1} y - \sin y}{y^3 \cos y}$
35. $\displaystyle\lim_{x\to\infty} x^2 \left(e^{-1/x^2} - 1\right)$
36. $\displaystyle\lim_{x\to\infty} (x + 1) \sin \dfrac{1}{x + 1}$
37. $\displaystyle\lim_{x\to 0} \dfrac{\ln(1 + x^2)}{1 - \cos x}$
38. $\displaystyle\lim_{x\to 2} \dfrac{x^2 - 4}{\ln(x - 1)}$
39. $\displaystyle\lim_{x\to 0} \dfrac{\sin 3x^2}{1 - \cos 2x}$
40. $\displaystyle\lim_{x\to 0} \dfrac{\ln(1 + x^3)}{x \cdot \sin x^2}$

Using Table 10.1

In Exercises 41–52, use Table 10.1 to find the sum of each series.

41. $1 + 1 + \dfrac{1}{2!} + \dfrac{1}{3!} + \dfrac{1}{4!} + \cdots$
42. $\left(\dfrac{1}{4}\right)^3 + \left(\dfrac{1}{4}\right)^4 + \left(\dfrac{1}{4}\right)^5 + \left(\dfrac{1}{4}\right)^6 + \cdots$
43. $1 - \dfrac{3^2}{4^2 \cdot 2!} + \dfrac{3^4}{4^4 \cdot 4!} - \dfrac{3^6}{4^6 \cdot 6!} + \cdots$
44. $\dfrac{1}{2} - \dfrac{1}{2 \cdot 2^2} + \dfrac{1}{3 \cdot 2^3} - \dfrac{1}{4 \cdot 2^4} + \cdots$
45. $\dfrac{\pi}{3} - \dfrac{\pi^3}{3^3 \cdot 3!} + \dfrac{\pi^5}{3^5 \cdot 5!} - \dfrac{\pi^7}{3^7 \cdot 7!} + \cdots$
46. $\dfrac{2}{3} - \dfrac{2^3}{3^3 \cdot 3} + \dfrac{2^5}{3^5 \cdot 5} - \dfrac{2^7}{3^7 \cdot 7} + \cdots$
47. $x^3 + x^4 + x^5 + x^6 + \cdots$
48. $1 - \dfrac{3^2 x^2}{2!} + \dfrac{3^4 x^4}{4!} - \dfrac{3^6 x^6}{6!} + \cdots$
49. $x^3 - x^5 + x^7 - x^9 + x^{11} - \cdots$
50. $x^2 - 2x^3 + \dfrac{2^2 x^4}{2!} - \dfrac{2^3 x^5}{3!} + \dfrac{2^4 x^6}{4!} - \cdots$
51. $-1 + 2x - 3x^2 + 4x^3 - 5x^4 + \cdots$
52. $1 + \dfrac{x}{2} + \dfrac{x^2}{3} + \dfrac{x^3}{4} + \dfrac{x^4}{5} + \cdots$

Theory and Examples

53. Replace x by $-x$ in the Taylor series for $\ln(1 + x)$ to obtain a series for $\ln(1 - x)$. Then subtract this from the Taylor series for $\ln(1 + x)$ to show that for $|x| < 1$,
$$\ln \dfrac{1 + x}{1 - x} = 2\left(x + \dfrac{x^3}{3} + \dfrac{x^5}{5} + \cdots\right).$$

54. How many terms of the Taylor series for $\ln(1 + x)$ should you add to be sure of calculating $\ln(1.1)$ with an error of magnitude less than 10^{-8}? Give reasons for your answer.

55. According to the Alternating Series Estimation Theorem, how many terms of the Taylor series for $\tan^{-1} 1$ would you have to add to be sure of finding $\pi/4$ with an error of magnitude less than 10^{-3}? Give reasons for your answer.

56. Show that the Taylor series for $f(x) = \tan^{-1} x$ diverges for $|x| > 1$.

T **57. Estimating Pi**   About how many terms of the Taylor series for $\tan^{-1} x$ would you have to use to evaluate each term on the right-hand side of the equation

$$\pi = 48 \tan^{-1}\frac{1}{18} + 32 \tan^{-1}\frac{1}{57} - 20 \tan^{-1}\frac{1}{239}$$

with an error of magnitude less than 10^{-6}? In contrast, the convergence of $\sum_{n=1}^{\infty}(1/n^2)$ to $\pi^2/6$ is so slow that even 50 terms will not yield two-place accuracy.

58. Use the following steps to prove that the binomial series in Equation (1) converges to $(1 + x)^m$.

   a. Differentiate the series

$$f(x) = 1 + \sum_{k=1}^{\infty}\binom{m}{k}x^k$$

   to show that

$$f'(x) = \frac{mf(x)}{1 + x}, \quad -1 < x < 1.$$

   b. Define $g(x) = (1 + x)^{-m} f(x)$ and show that $g'(x) = 0$.

   c. From part (b), show that

$$f(x) = (1 + x)^m.$$

59. a. Use the binomial series and the fact that

$$\frac{d}{dx}\sin^{-1}x = (1 - x^2)^{-1/2}$$

   to generate the first four nonzero terms of the Taylor series for $\sin^{-1} x$. What is the radius of convergence?

   b. Series for $\cos^{-1}x$ Use your result in part (a) to find the first five nonzero terms of the Taylor series for $\cos^{-1}x$.

60. a. Series for $\sinh^{-1}x$ Find the first four nonzero terms of the Taylor series for

$$\sinh^{-1}x = \int_0^x \frac{dt}{\sqrt{1 + t^2}}.$$

T **b.** Use the first *three* terms of the series in part (a) to estimate $\sinh^{-1} 0.25$. Give an upper bound for the magnitude of the estimation error.

61. Obtain the Taylor series for $1/(1 + x)^2$ from the series for $-1/(1 + x)$.

62. Use the Taylor series for $1/(1 - x^2)$ to obtain a series for $2x/(1 - x^2)^2$.

T **63. Estimating Pi**   The English mathematician Wallis discovered the formula

$$\frac{\pi}{4} = \frac{2 \cdot 4 \cdot 4 \cdot 6 \cdot 6 \cdot 8 \cdot \cdots}{3 \cdot 3 \cdot 5 \cdot 5 \cdot 7 \cdot 7 \cdot \cdots}.$$

Find π to two decimal places with this formula.

64. The complete elliptic integral of the first kind is the integral

$$K = \int_0^{\pi/2} \frac{d\theta}{\sqrt{1 - k^2 \sin^2\theta}},$$

where $0 < k < 1$ is constant.

   a. Show that the first four terms of the binomial series for $1/\sqrt{1 - x}$ are

$$(1 - x)^{-1/2} = 1 + \frac{1}{2}x + \frac{1 \cdot 3}{2 \cdot 4}x^2 + \frac{1 \cdot 3 \cdot 5}{2 \cdot 4 \cdot 6}x^3 + \cdots.$$

   b. From part (a) and the reduction integral Formula 67 at the back of the book, show that

$$K = \frac{\pi}{2}\left[1 + \left(\frac{1}{2}\right)^2 k^2 + \left(\frac{1 \cdot 3}{2 \cdot 4}\right)^2 k^4 + \left(\frac{1 \cdot 3 \cdot 5}{2 \cdot 4 \cdot 6}\right)^2 k^6 + \cdots\right].$$

65. Series for $\sin^{-1}x$ Integrate the binomial series for $(1 - x^2)^{-1/2}$ to show that for $|x| < 1$,

$$\sin^{-1}x = x + \sum_{n=1}^{\infty}\frac{1 \cdot 3 \cdot 5 \cdot \cdots \cdot (2n - 1)}{2 \cdot 4 \cdot 6 \cdot \cdots \cdot (2n)}\frac{x^{2n+1}}{2n + 1}.$$

66. Series for $\tan^{-1}x$ for $|x| > 1$ Derive the series

$$\tan^{-1}x = \frac{\pi}{2} - \frac{1}{x} + \frac{1}{3x^3} - \frac{1}{5x^5} + \cdots, \quad x > 1$$

$$\tan^{-1}x = -\frac{\pi}{2} - \frac{1}{x} + \frac{1}{3x^3} - \frac{1}{5x^5} + \cdots, \quad x < -1,$$

by integrating the series

$$\frac{1}{1 + t^2} = \frac{1}{t^2} \cdot \frac{1}{1 + (1/t^2)} = \frac{1}{t^2} - \frac{1}{t^4} + \frac{1}{t^6} - \frac{1}{t^8} + \cdots$$

in the first case from x to ∞ and in the second case from $-\infty$ to x.

Euler's Identity

67. Use Equation (4) to write the following powers of e in the form $a + bi$.

   a. $e^{-i\pi}$ **b.** $e^{i\pi/4}$ **c.** $e^{-i\pi/2}$

68. Use Equation (4) to show that

$$\cos\theta = \frac{e^{i\theta} + e^{-i\theta}}{2} \quad \text{and} \quad \sin\theta = \frac{e^{i\theta} - e^{-i\theta}}{2i}.$$

69. Establish the equations in Exercise 68 by combining the formal Taylor series for $e^{i\theta}$ and $e^{-i\theta}$.

70. Show that

   a. $\cosh i\theta = \cos\theta$,   **b.** $\sinh i\theta = i\sin\theta$.

71. By multiplying the Taylor series for e^x and $\sin x$, find the terms through x^5 of the Taylor series for $e^x \sin x$. This series is the imaginary part of the series for

$$e^x \cdot e^{ix} = e^{(1+i)x}.$$

Use this fact to check your answer. For what values of x should the series for $e^x \sin x$ converge?

72. When a and b are real, we define $e^{(a+ib)x}$ with the equation

$$e^{(a+ib)x} = e^{ax} \cdot e^{ibx} = e^{ax}(\cos bx + i\sin bx).$$

Differentiate the right-hand side of this equation to show that

$$\frac{d}{dx}e^{(a+ib)x} = (a + ib)e^{(a+ib)x}.$$

Thus the familiar rule $(d/dx)e^{kx} = ke^{kx}$ holds for k complex as well as real.

73. Use the definition of $e^{i\theta}$ to show that for any real numbers θ, θ_1, and θ_2,
a. $e^{i\theta_1}e^{i\theta_2} = e^{i(\theta_1+\theta_2)}$,
b. $e^{-i\theta} = 1/e^{i\theta}$.

74. Two complex numbers $a + ib$ and $c + id$ are equal if and only if $a = c$ and $b = d$. Use this fact to evaluate

$$\int e^{ax}\cos bx\, dx \quad \text{and} \quad \int e^{ax}\sin bx\, dx$$

from

$$\int e^{(a+ib)x}\, dx = \frac{a - ib}{a^2 + b^2}\, e^{(a+ib)x} + C,$$

where $C = C_1 + iC_2$ is a complex constant of integration.

CHAPTER 10 Questions to Guide Your Review

1. What is an infinite sequence? What does it mean for such a sequence to converge? To diverge? Give examples.

2. What is a monotonic sequence? Under what circumstances does such a sequence have a limit? Give examples.

3. What theorems are available for calculating limits of sequences? Give examples.

4. What theorem sometimes enables us to use l'Hôpital's Rule to calculate the limit of a sequence? Give an example.

5. What are the six commonly occurring limits in Theorem 5 that arise frequently when you work with sequences and series?

6. What is an infinite series? What does it mean for such a series to converge? To diverge? Give examples.

7. What is a geometric series? When does such a series converge? Diverge? When it does converge, what is its sum? Give examples.

8. Besides geometric series, what other convergent and divergent series do you know?

9. What is the nth-Term Test for Divergence? What is the idea behind the test?

10. What can be said about term-by-term sums and differences of convergent series? About constant multiples of convergent and divergent series?

11. What happens if you add a finite number of terms to a convergent series? A divergent series? What happens if you delete a finite number of terms from a convergent series? A divergent series?

12. How do you reindex a series? Why might you want to do this?

13. Under what circumstances will an infinite series of nonnegative terms converge? Diverge? Why study series of nonnegative terms?

14. What is the Integral Test? What is the reasoning behind it? Give an example of its use.

15. When do p-series converge? Diverge? How do you know? Give examples of convergent and divergent p-series.

16. What are the Direct Comparison Test and the Limit Comparison Test? What is the reasoning behind these tests? Give examples of their use.

7. What are the Ratio and Root Tests? Do they always give you the information you need to determine convergence or divergence? Give examples.

18. What is absolute convergence? Conditional convergence? How are the two related?

19. What is an alternating series? What theorem is available for determining the convergence of such a series?

20. How can you estimate the error involved in approximating the sum of an alternating series with one of the series' partial sums? What is the reasoning behind the estimate?

21. What do you know about rearranging the terms of an absolutely convergent series? Of a conditionally convergent series?

22. What is a power series? How do you test a power series for convergence? What are the possible outcomes?

23. What are the basic facts about
a. sums, differences, and products of power series?
b. substitution of a function for x in a power series?
c. term-by-term differentiation of power series?
d. term-by-term integration of power series?
e. Give examples.

24. What is the Taylor series generated by a function $f(x)$ at a point $x = a$? What information do you need about f to construct the series? Give an example.

25. What is a Maclaurin series?

26. Does a Taylor series always converge to its generating function? Explain.

27. What are Taylor polynomials? Of what use are they?

28. What is Taylor's formula? What does it say about the errors involved in using Taylor polynomials to approximate functions? In particular, what does Taylor's formula say about the error in a linearization? A quadratic approximation?

29. What is the binomial series? On what interval does it converge? How is it used?

30. How can you sometimes use power series to estimate the values of nonelementary definite integrals? To find limits?

31. What are the Taylor series for $1/(1 - x)$, $1/(1 + x)$, e^x, $\sin x$, $\cos x$, $\ln(1 + x)$, and $\tan^{-1}x$? How do you estimate the errors involved in replacing these series with their partial sums?

CHAPTER 10 Practice Exercises

Determining Convergence of Sequences

Which of the sequences whose nth terms appear in Exercises 1–18 converge, and which diverge? Find the limit of each convergent sequence.

1. $a_n = 1 + \dfrac{(-1)^n}{n}$

2. $a_n = \dfrac{1 - (-1)^n}{\sqrt{n}}$

3. $a_n = \dfrac{1 - 2^n}{2^n}$

4. $a_n = 1 + (0.9)^n$

5. $a_n = \sin \dfrac{n\pi}{2}$

6. $a_n = \sin n\pi$

7. $a_n = \dfrac{\ln(n^2)}{n}$

8. $a_n = \dfrac{\ln(2n + 1)}{n}$

9. $a_n = \dfrac{n + \ln n}{n}$

10. $a_n = \dfrac{\ln(2n^3 + 1)}{n}$

11. $a_n = \left(\dfrac{n - 5}{n}\right)^n$

12. $a_n = \left(1 + \dfrac{1}{n}\right)^{-n}$

13. $a_n = \sqrt[n]{\dfrac{3^n}{n}}$

14. $a_n = \left(\dfrac{3}{n}\right)^{1/n}$

15. $a_n = n(2^{1/n} - 1)$

16. $a_n = \sqrt[n]{2n + 1}$

17. $a_n = \dfrac{(n + 1)!}{n!}$

18. $a_n = \dfrac{(-4)^n}{n!}$

Convergent Series

Find the sums of the series in Exercises 19–24.

19. $\displaystyle\sum_{n=3}^{\infty} \dfrac{1}{(2n - 3)(2n - 1)}$

20. $\displaystyle\sum_{n=2}^{\infty} \dfrac{-2}{n(n + 1)}$

21. $\displaystyle\sum_{n=1}^{\infty} \dfrac{9}{(3n - 1)(3n + 2)}$

22. $\displaystyle\sum_{n=3}^{\infty} \dfrac{-8}{(4n - 3)(4n + 1)}$

23. $\displaystyle\sum_{n=0}^{\infty} e^{-n}$

24. $\displaystyle\sum_{n=1}^{\infty} (-1)^n \dfrac{3}{4^n}$

Determining Convergence of Series

Which of the series in Exercises 25–44 converge absolutely, which converge conditionally, and which diverge? Give reasons for your answers.

25. $\displaystyle\sum_{n=1}^{\infty} \dfrac{1}{\sqrt{n}}$

26. $\displaystyle\sum_{n=1}^{\infty} \dfrac{-5}{n}$

27. $\displaystyle\sum_{n=1}^{\infty} \dfrac{(-1)^n}{\sqrt{n}}$

28. $\displaystyle\sum_{n=1}^{\infty} \dfrac{1}{2n^3}$

29. $\displaystyle\sum_{n=1}^{\infty} \dfrac{(-1)^n}{\ln(n + 1)}$

30. $\displaystyle\sum_{n=2}^{\infty} \dfrac{1}{n(\ln n)^2}$

31. $\displaystyle\sum_{n=1}^{\infty} \dfrac{\ln n}{n^3}$

32. $\displaystyle\sum_{n=3}^{\infty} \dfrac{\ln n}{\ln(\ln n)}$

33. $\displaystyle\sum_{n=1}^{\infty} \dfrac{(-1)^n}{n\sqrt{n^2 + 1}}$

34. $\displaystyle\sum_{n=1}^{\infty} \dfrac{(-1)^n 3n^2}{n^3 + 1}$

35. $\displaystyle\sum_{n=1}^{\infty} \dfrac{n + 1}{n!}$

36. $\displaystyle\sum_{n=1}^{\infty} \dfrac{(-1)^n(n^2 + 1)}{2n^2 + n - 1}$

37. $\displaystyle\sum_{n=1}^{\infty} \dfrac{(-3)^n}{n!}$

38. $\displaystyle\sum_{n=1}^{\infty} \dfrac{2^n 3^n}{n^n}$

39. $\displaystyle\sum_{n=1}^{\infty} \dfrac{1}{\sqrt{n(n + 1)(n + 2)}}$

40. $\displaystyle\sum_{n=2}^{\infty} \dfrac{1}{n\sqrt{n^2 - 1}}$

41. $1 - \left(\dfrac{1}{\sqrt{3}}\right)^2 + \left(\dfrac{1}{\sqrt{3}}\right)^4 - \left(\dfrac{1}{\sqrt{3}}\right)^6 + \left(\dfrac{1}{\sqrt{3}}\right)^8 - \cdots$

42. $\displaystyle\sum_{n=0}^{\infty} \dfrac{(-1)^n}{e^{-n} + 1}$

43. $\displaystyle\sum_{n=0}^{\infty} \dfrac{1}{1 + r + r^2 + \cdots + r^n}, \quad \text{for } -1 < r < 1$

44. $\displaystyle\sum_{n=1}^{\infty} \dfrac{(-1)^n}{\sqrt{n + 100} - \sqrt{n}}$

Power Series

In Exercises 45–54, (a) find the series' radius and interval of convergence. Then identify the values of x for which the series converges (b) absolutely and (c) conditionally.

45. $\displaystyle\sum_{n=1}^{\infty} \dfrac{(x + 4)^n}{n3^n}$

46. $\displaystyle\sum_{n=1}^{\infty} \dfrac{(x - 1)^{2n-2}}{(2n - 1)!}$

47. $\displaystyle\sum_{n=1}^{\infty} \dfrac{(-1)^{n-1}(3x - 1)^n}{n^2}$

48. $\displaystyle\sum_{n=0}^{\infty} \dfrac{(n + 1)(2x + 1)^n}{(2n + 1)2^n}$

49. $\displaystyle\sum_{n=1}^{\infty} \dfrac{x^n}{n^n}$

50. $\displaystyle\sum_{n=1}^{\infty} \dfrac{x^n}{\sqrt{n}}$

51. $\displaystyle\sum_{n=0}^{\infty} \dfrac{(n + 1)x^{2n-1}}{3^n}$

52. $\displaystyle\sum_{n=0}^{\infty} \dfrac{(-1)^n(x - 1)^{2n+1}}{2n + 1}$

53. $\displaystyle\sum_{n=1}^{\infty} (\operatorname{csch} n)x^n$

54. $\displaystyle\sum_{n=1}^{\infty} (\coth n)x^n$

Maclaurin Series

Each of the series in Exercises 55–60 is the value of the Taylor series at $x = 0$ of a function $f(x)$ at a particular point. What function and what point? What is the sum of the series?

55. $1 - \dfrac{1}{4} + \dfrac{1}{16} - \cdots + (-1)^n \dfrac{1}{4^n} + \cdots$

56. $\dfrac{2}{3} - \dfrac{4}{18} + \dfrac{8}{81} - \cdots + (-1)^{n-1} \dfrac{2^n}{n3^n} + \cdots$

57. $\pi - \dfrac{\pi^3}{3!} + \dfrac{\pi^5}{5!} - \cdots + (-1)^n \dfrac{\pi^{2n+1}}{(2n + 1)!} + \cdots$

58. $1 - \dfrac{\pi^2}{9 \cdot 2!} + \dfrac{\pi^4}{81 \cdot 4!} - \cdots + (-1)^n \dfrac{\pi^{2n}}{3^{2n}(2n)!} + \cdots$

59. $1 + \ln 2 + \dfrac{(\ln 2)^2}{2!} + \cdots + \dfrac{(\ln 2)^n}{n!} + \cdots$

60. $\dfrac{1}{\sqrt{3}} - \dfrac{1}{9\sqrt{3}} + \dfrac{1}{45\sqrt{3}} - \cdots$

$+ (-1)^{n-1} \dfrac{1}{(2n - 1)(\sqrt{3})^{2n-1}} + \cdots$

Find Taylor series at $x = 0$ for the functions in Exercises 61–68.

61. $\dfrac{1}{1 - 2x}$

62. $\dfrac{1}{1 + x^3}$

63. $\sin \pi x$

64. $\sin \dfrac{2x}{3}$

65. $\cos \left(x^{5/3} \right)$

66. $\cos \dfrac{x^3}{\sqrt{5}}$

67. $e^{(\pi x/2)}$

68. e^{-x^2}

Taylor Series

In Exercises 69–72, find the first four nonzero terms of the Taylor series generated by f at $x = a$.

69. $f(x) = \sqrt{3 + x^2}$ at $x = -1$

70. $f(x) = 1/(1 - x)$ at $x = 2$

71. $f(x) = 1/(x + 1)$ at $x = 3$

72. $f(x) = 1/x$ at $x = a > 0$

Nonelementary Integrals

Use series to approximate the values of the integrals in Exercises 73–76 with an error of magnitude less than 10^{-8}. (The answer section gives the integrals' values rounded to 10 decimal places.)

73. $\displaystyle\int_0^{1/2} e^{-x^3}\,dx$

74. $\displaystyle\int_0^1 x \sin\left(x^3\right)\,dx$

75. $\displaystyle\int_0^{1/2} \dfrac{\tan^{-1} x}{x}\,dx$

76. $\displaystyle\int_0^{1/64} \dfrac{\tan^{-1} x}{\sqrt{x}}\,dx$

Using Series to Find Limits

In Exercises 77–82:

a. Use power series to evaluate the limit.

T b. Then use a grapher to support your calculation.

77. $\displaystyle\lim_{x \to 0} \dfrac{7 \sin x}{e^{2x} - 1}$

78. $\displaystyle\lim_{\theta \to 0} \dfrac{e^\theta - e^{-\theta} - 2\theta}{\theta - \sin \theta}$

79. $\displaystyle\lim_{t \to 0} \left(\dfrac{1}{2 - 2\cos t} - \dfrac{1}{t^2} \right)$

80. $\displaystyle\lim_{h \to 0} \dfrac{(\sin h)/h - \cos h}{h^2}$

81. $\displaystyle\lim_{z \to 0} \dfrac{1 - \cos^2 z}{\ln(1 - z) + \sin z}$

82. $\displaystyle\lim_{y \to 0} \dfrac{y^2}{\cos y - \cosh y}$

Theory and Examples

83. Use a series representation of $\sin 3x$ to find values of r and s for which

$$\lim_{x \to 0} \left(\dfrac{\sin 3x}{x^3} + \dfrac{r}{x^2} + s \right) = 0.$$

T **84.** Compare the accuracies of the approximations $\sin x \approx x$ and $\sin x \approx 6x/(6 + x^2)$ by comparing the graphs of $f(x) = \sin x - x$ and $g(x) = \sin x - (6x/(6 + x^2))$. Describe what you find.

85. Find the radius of convergence of the series

$$\sum_{n=1}^{\infty} \dfrac{2 \cdot 5 \cdot 8 \cdot \cdots \cdot (3n - 1)}{2 \cdot 4 \cdot 6 \cdot \cdots \cdot (2n)} x^n.$$

86. Find the radius of convergence of the series

$$\sum_{n=1}^{\infty} \dfrac{3 \cdot 5 \cdot 7 \cdot \cdots \cdot (2n + 1)}{4 \cdot 9 \cdot 14 \cdot \cdots \cdot (5n - 1)} (x - 1)^n.$$

87. Find a closed-form formula for the nth partial sum of the series $\sum_{n=2}^{\infty} \ln\left(1 - (1/n^2)\right)$ and use it to determine the convergence or divergence of the series.

88. Evaluate $\sum_{k=2}^{\infty} \left(1/(k^2 - 1)\right)$ by finding the limits as $n \to \infty$ of the series' nth partial sum.

89. a. Find the interval of convergence of the series

$$y = 1 + \dfrac{1}{6}x^3 + \dfrac{1}{180}x^6 + \cdots$$

$$+ \dfrac{1 \cdot 4 \cdot 7 \cdot \cdots \cdot (3n - 2)}{(3n)!} x^{3n} + \cdots.$$

b. Show that the function defined by the series satisfies a differential equation of the form

$$\dfrac{d^2 y}{dx^2} = x^a y + b$$

and find the values of the constants a and b.

90. a. Find the Maclaurin series for the function $x^2/(1 + x)$.

b. Does the series converge at $x = 1$? Explain.

91. If $\sum_{n=1}^{\infty} a_n$ and $\sum_{n=1}^{\infty} b_n$ are convergent series of nonnegative numbers, can anything be said about $\sum_{n=1}^{\infty} a_n b_n$? Give reasons for your answer.

92. If $\sum_{n=1}^{\infty} a_n$ and $\sum_{n=1}^{\infty} b_n$ are divergent series of nonnegative numbers, can anything be said about $\sum_{n=1}^{\infty} a_n b_n$? Give reasons for your answer.

93. Prove that the sequence $\{x_n\}$ and the series $\sum_{k=1}^{\infty} (x_{k+1} - x_k)$ both converge or both diverge.

94. Prove that $\sum_{n=1}^{\infty} (a_n/(1 + a_n))$ converges if $a_n > 0$ for all n and $\sum_{n=1}^{\infty} a_n$ converges.

95. Suppose that $a_1, a_2, a_3, \ldots, a_n$ are positive numbers satisfying the following conditions:

i) $a_1 \geq a_2 \geq a_3 \geq \cdots$;

ii) the series $a_2 + a_4 + a_8 + a_{16} + \cdots$ diverges.

Show that the series

$$\dfrac{a_1}{1} + \dfrac{a_2}{2} + \dfrac{a_3}{3} + \cdots$$

diverges.

96. Use the result in Exercise 95 to show that

$$1 + \sum_{n=2}^{\infty} \dfrac{1}{n \ln n}$$

diverges.

97. Show that if $a_n > 0$ and $\sum_{n=1}^{\infty} a_n$ converges, then $\sum_{n=1}^{\infty} \dfrac{\sqrt{a_n}}{n}$ converges.

98. Determine whether $\sum_{n=1}^{\infty} b_n$ converges or diverges.

    a. $b_1 = 1$, $\ b_{n+1} = (-1)^n \dfrac{n+1}{3n+2} b_n$

    b. $b_1 = 3$, $\ b_{n+1} = \dfrac{n}{\ln n} b_n$

99. Assume that $b_n > 0$ and $\sum_{n=1}^{\infty} b_n$ converges. What, if anything, can be said about the following series?

    a. $\sum_{n=1}^{\infty} \tan(b_n)$

    b. $\sum_{n=1}^{\infty} \ln(1 + b_n)$

    c. $\sum_{n=1}^{\infty} \ln(2 + b_n)$

100. Consider the convergent series $\sum_{n=1}^{\infty} \dfrac{(-1)^n}{e^n + e^{cn}}$, where c is a constant. What should c be so that the first 10 terms of the series estimate the sum of the entire series with an error of less than 0.00001?

101. Assume that the following sequence has a limit L. Find the value of L.

$$4^{1/3}, \ (4(4^{1/3}))^{1/3}, \ (4(4(4^{1/3}))^{1/3})^{1/3}, \ (4(4(4(4^{1/3}))^{1/3})^{1/3})^{1/3}, \ldots$$

102. Consider the infinite sequence of shaded right triangles in the accompanying diagram. Compute the total area of the triangles.

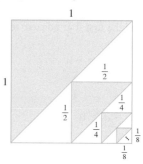

Determining Convergence of Series

Which of the series $\sum_{n=1}^{\infty} a_n$ defined by the formulas in Exercises 1–4 converge, and which diverge? Give reasons for your answers.

1. $\sum_{n=1}^{\infty} \dfrac{1}{(3n-2)^{n+(1/2)}}$

2. $\sum_{n=1}^{\infty} \dfrac{(\tan^{-1} n)^2}{n^2 + 1}$

3. $\sum_{n=1}^{\infty} (-1)^n \tanh n$

4. $\sum_{n=2}^{\infty} \dfrac{\log_n(n!)}{n^3}$

Which of the series $\sum_{n=1}^{\infty} a_n$ defined by the formulas in Exercises 5–8 converge, and which diverge? Give reasons for your answers.

5. $a_1 = 1$, $\ a_{n+1} = \dfrac{n(n+1)}{(n+2)(n+3)} a_n$

    (*Hint:* Write out several terms, see which factors cancel, and then generalize.)

6. $a_1 = a_2 = 7$, $\ a_{n+1} = \dfrac{n}{(n-1)(n+1)} a_n$ if $n \geq 2$

7. $a_1 = a_2 = 1$, $\ a_{n+1} = \dfrac{1}{1 + a_n}$ if $n \geq 2$

8. $a_n = 1/3^n$ if n is odd, $a_n = n/3^n$ if n is even

Choosing Centers for Taylor Series

Taylor's formula

$$f(x) = f(a) + f'(a)(x-a) + \dfrac{f''(a)}{2!}(x-a)^2 + \cdots$$

$$+ \dfrac{f^{(n)}(a)}{n!}(x-a)^n + \dfrac{f^{(n+1)}(c)}{(n+1)!}(x-a)^{n+1}$$

expresses the value of f at x in terms of the values of f and its derivatives at $x = a$. In numerical computations, we therefore need a to be a

point where we know the values of f and its derivatives. We also need a to be close enough to the values of f we are interested in to make $(x-a)^{n+1}$ so small we can neglect the remainder.

In Exercises 9–14, what Taylor series would you choose to represent the function near the given value of x? (There may be more than one good answer.) Write out the first four nonzero terms of the series you choose.

    9. $\cos x$ near $x = 1$

    10. $\sin x$ near $x = 6.3$

    11. e^x near $x = 0.4$

    12. $\ln x$ near $x = 1.3$

    13. $\cos x$ near $x = 69$

    14. $\tan^{-1} x$ near $x = 2$

Theory and Examples

15. Let a and b be constants with $0 < a < b$. Does the sequence $\{(a^n + b^n)^{1/n}\}$ converge? If it does converge, what is the limit?

16. Find the sum of the infinite series

$$1 + \dfrac{2}{10} + \dfrac{3}{10^2} + \dfrac{7}{10^3} + \dfrac{2}{10^4} + \dfrac{3}{10^5} + \dfrac{7}{10^6} + \dfrac{2}{10^7}$$

$$+ \dfrac{3}{10^8} + \dfrac{7}{10^9} + \cdots.$$

17. Evaluate

$$\sum_{n=0}^{\infty} \int_n^{n+1} \dfrac{1}{1+x^2} \, dx.$$

18. Find all values of x for which

$$\sum_{n=1}^{\infty} \dfrac{nx^n}{(n+1)(2x+1)^n}$$

converges absolutely.

19. a. Does the value of

$$\lim_{n\to\infty}\left(1 - \frac{\cos(a/n)}{n}\right)^n, \quad a \text{ constant},$$

appear to depend on the value of a? If so, how?

b. Does the value of

$$\lim_{n\to\infty}\left(1 - \frac{\cos(a/n)}{bn}\right)^n, \quad a \text{ and } b \text{ constant}, b \ne 0,$$

appear to depend on the value of b? If so, how?

c. Use calculus to confirm your findings in parts (a) and (b).

20. Show that if $\sum_{n=1}^{\infty} a_n$ converges, then

$$\sum_{n=1}^{\infty}\left(\frac{1 + \sin(a_n)}{2}\right)^n$$

converges.

21. Find a value for the constant b that will make the radius of convergence of the power series

$$\sum_{n=2}^{\infty}\frac{b^n x^n}{\ln n}$$

equal to 5.

22. How do you know that the functions $\sin x$, $\ln x$, and e^x are not polynomials? Give reasons for your answer.

23. Find the value of a for which the limit

$$\lim_{x\to 0}\frac{\sin(ax) - \sin x - x}{x^3}$$

is finite and evaluate the limit.

24. Find values of a and b for which

$$\lim_{x\to 0}\frac{\cos(ax) - b}{2x^2} = -1.$$

25. Raabe's (or Gauss's) Test   The following test, which we state without proof, is an extension of the Ratio Test.

Raabe's Test: If $\sum_{n=1}^{\infty} u_n$ is a series of positive constants and there exist constants C, K, and N such that

$$\frac{u_n}{u_{n+1}} = 1 + \frac{C}{n} + \frac{f(n)}{n^2},$$

where $|f(n)| < K$ for $n \ge N$, then $\sum_{n=1}^{\infty} u_n$ converges if $C > 1$ and diverges if $C \le 1$.

Show that the results of Raabe's Test agree with what you know about the series $\sum_{n=1}^{\infty}(1/n^2)$ and $\sum_{n=1}^{\infty}(1/n)$.

26. (*Continuation of Exercise 25.*) Suppose that the terms of $\sum_{n=1}^{\infty} u_n$ are defined recursively by the formulas

$$u_1 = 1, \quad u_{n+1} = \frac{(2n - 1)^2}{(2n)(2n + 1)}u_n.$$

Apply Raabe's Test to determine whether the series converges.

27. If $\sum_{n=1}^{\infty} a_n$ converges, and if $a_n \ne 1$ and $a_n > 0$ for all n,

a. Show that $\sum_{n=1}^{\infty} a_n^2$ converges.

b. Does $\sum_{n=1}^{\infty} a_n/(1 - a_n)$ converge? Explain.

28. (*Continuation of Exercise 27.*) If $\sum_{n=1}^{\infty} a_n$ converges, and if $1 > a_n > 0$ for all n, show that $\sum_{n=1}^{\infty}\ln(1 - a_n)$ converges.

(*Hint:* First show that $|\ln(1 - a_n)| \le a_n/(1 - a_n)$.)

29. Nicole Oresme's Theorem   Prove Nicole Oresme's Theorem that

$$1 + \frac{1}{2}\cdot 2 + \frac{1}{4}\cdot 3 + \cdots + \frac{n}{2^{n-1}} + \cdots = 4.$$

(*Hint:* Differentiate both sides of the equation $1/(1 - x) = 1 + \sum_{n=1}^{\infty} x^n$.)

30. a. Show that

$$\sum_{n=1}^{\infty}\frac{n(n + 1)}{x^n} = \frac{2x^2}{(x - 1)^3}$$

for $|x| > 1$ by differentiating the identity

$$\sum_{n=1}^{\infty} x^{n+1} = \frac{x^2}{1 - x}$$

twice, multiplying the result by x, and then replacing x by $1/x$.

b. Use part (a) to find the real solution greater than 1 of the equation

$$x = \sum_{n=1}^{\infty}\frac{n(n + 1)}{x^n}.$$

31. Quality control

a. Differentiate the series

$$\frac{1}{1 - x} = 1 + x + x^2 + \cdots + x^n + \cdots$$

to obtain a series for $1/(1 - x)^2$.

b. In one throw of two dice, the probability of getting a roll of 7 is $p = 1/6$. If you throw the dice repeatedly, the probability that a 7 will appear for the first time at the nth throw is $q^{n-1}p$, where $q = 1 - p = 5/6$. The expected number of throws until a 7 first appears is $\sum_{n=1}^{\infty} nq^{n-1}p$. Find the sum of this series.

c. As an engineer applying statistical control to an industrial operation, you inspect items taken at random from the assembly line. You classify each sampled item as either "good" or "bad." If the probability of an item's being good is p and of an item's being bad is $q = 1 - p$, the probability that the first bad item found is the nth one inspected is $p^{n-1}q$. The average number inspected up to and including the first bad item found is $\sum_{n=1}^{\infty} np^{n-1}q$. Evaluate this sum, assuming $0 < p < 1$.

32. Expected value   Suppose that a random variable X may assume the values 1, 2, 3, ..., with probabilities $p_1, p_2, p_3, \ldots$, where p_k is the probability that X equals k ($k = 1, 2, 3, \ldots$). Suppose also that $p_k \ge 0$ and that $\sum_{k=1}^{\infty} p_k = 1$. The **expected value** of X, denoted by $E(X)$, is the number $\sum_{k=1}^{\infty} kp_k$, provided the series converges. In each of the following cases, show that $\sum_{k=1}^{\infty} p_k = 1$ and find $E(X)$ if it exists. (*Hint:* See Exercise 31.)

a. $p_k = 2^{-k}$ **b.** $p_k = \frac{5^{k-1}}{6^k}$

c. $p_k = \frac{1}{k(k + 1)} = \frac{1}{k} - \frac{1}{k + 1}$

T **33. Safe and effective dosage** The concentration in the blood resulting from a single dose of a drug normally decreases with time as
the drug is eliminated from the body. Doses may therefore need to
be repeated periodically to keep the concentration from dropping
below some particular level. One model for the effect of repeated
doses gives the residual concentration just before the $(n + 1)$st
dose as

$$R_n = C_0 e^{-kt_0} + C_0 e^{-2kt_0} + \cdots + C_0 e^{-nkt_0},$$

where $C_0 =$ the change in concentration achievable by a single
dose (mg/mL), $k =$ the *elimination constant* (h^{-1}), and $t_0 =$ time
between doses (h). See the accompanying figure.

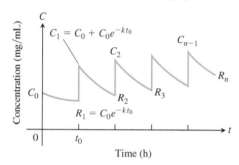

a. Write R_n in closed from as a single fraction, and find
$R = \lim_{n \to \infty} R_n$.

b. Calculate R_1 and R_{10} for $C_0 = 1$ mg/mL, $k = 0.1\ h^{-1}$, and
$t_0 = 10$ h. How good an estimate of R is R_{10}?

c. If $k = 0.01\ h^{-1}$ and $t_0 = 10$ h, find the smallest n such that
$R_n > (1/2)R$. Use $C_0 = 1$ mg/mL.
(*Source: Prescribing Safe and Effective Dosage,* B. Horelick and
S. Koont, COMAP, Inc., Lexington, MA.)

34. Time between drug doses (*Continuation of Exercise 33.*) If
a drug is known to be ineffective below a concentration C_L and
harmful above some higher concentration C_H, one need to find
values of C_0 and t_0 that will produce a concentration that is safe

(not above C_H) but effective (not below C_L). See the accompanying figure. We therefore want to find values for C_0 and t_0 for which

$$R = C_L \quad \text{and} \quad C_0 + R = C_H.$$

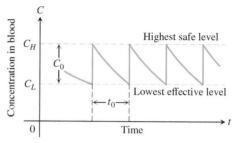

Thus $C_0 = C_H - C_L$. When these values are substituted in the
equation for R obtained in part (a) of Exercise 33, the resulting
equation simplifies to

$$t_0 = \frac{1}{k} \ln \frac{C_H}{C_L}.$$

To reach an effective level rapidly, one might administer a "loading" dose that would produce a concentration of C_H mg/mL. This
could be followed every t_0 hours by a dose that raises the concentration by $C_0 = C_H - C_L$ mg/mL.

a. Verify the preceding equation for t_0.

b. If $k = 0.05\ h^{-1}$ and the highest safe concentration is e times
the lowest effective concentration, find the length of time
between doses that will ensure safe and effective concentra-
tions.

c. Given $C_H = 2$ mg/mL, $C_L = 0.5$ mg/mL, and $k = 0.02\ h^{-1}$,
determine a scheme for administering the drug.

d. Suppose that $k = 0.2\ h^{-1}$ and that the smallest effective
concentration is 0.03 mg/mL. A single dose that produces
a concentration of 0.1 mg/mL is administered. About how
long will the drug remain effective?

CHAPTER 10 Technology Application Projects

Mathematica/Maple Projects

Projects can be found within MyMathLab.

• *Bouncing Ball*
The model predicts the height of a bouncing ball, and the time until it stops bouncing.

• *Taylor Polynomial Approximations of a Function*
A graphical animation shows the convergence of the Taylor polynomials to functions having derivatives of all orders over an interval in their
domains.

11

Parametric Equations and Polar Coordinates

OVERVIEW   In this chapter we study new ways to define curves in the plane. Instead of thinking of a curve as the graph of a function or equation, we think of it as the path of a moving particle whose position is changing over time. Then each of the x- and y-coordinates of the particle's position becomes a function of a third variable t. We can also change the way in which points in the plane themselves are described by using *polar coordinates* rather than the rectangular or Cartesian system. Both of these new tools are useful for describing motion, like that of planets and satellites, or projectiles moving in the plane or space.

11.1   Parametrizations of Plane Curves

Parametric Equations

Figure 11.1 shows the path of a moving particle in the xy-plane. Notice that the path fails the vertical line test, so it cannot be described as the graph of a function of the variable x. However, we can sometimes describe the path by a pair of equations, $x = f(t)$ and $y = g(t)$, where f and g are continuous functions. When studying motion, t usually denotes time. Equations like these can describe more general curves than those described by a single function, and they provide not only the graph of the path traced out but also the location of the particle $(x, y) = (f(t), g(t))$ at any time t.

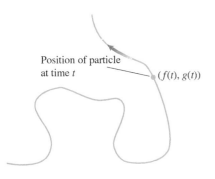

Position of particle
at time t $(f(t), g(t))$

FIGURE 11.1   The curve or path traced by a particle moving in the xy-plane is not always the graph of a function or single equation.

DEFINITION   If x and y are given as functions

$$x = f(t), \qquad y = g(t)$$

over an interval I of t-values, then the set of points $(x, y) = (f(t), g(t))$ defined by these equations is a **parametric curve**. The equations are **parametric equations** for the curve.

The variable t is a **parameter** for the curve, and its domain I is the **parameter interval**. If I is a closed interval, $a \leq t \leq b$, the point $(f(a), g(a))$ is the **initial point** of the curve and $(f(b), g(b))$ is the **terminal point**. When we give parametric equations and a parameter interval for a curve, we say that we have **parametrized** the curve. The equations and interval together constitute a **parametrization** of the curve. A given curve can be represented by different sets of parametric equations. (See Exercises 29 and 30.)

EXAMPLE 1   Sketch the curve defined by the parametric equations

$$x = \sin \pi t/2, \qquad y = t, \qquad 0 \leq t \leq 6.$$

Solution We make a table of values (Table 11.1), plot the points (x, y), and draw a smooth curve through them (Figure 11.2). If we think of the curve as the path of a moving particle, the particle starts at time $t = 0$ at the initial point $(0, 0)$ and then moves upward in a wavy path until at time $t = 6$ it reaches the terminal point $(0, 6)$. The direction of motion is shown by the arrows in Figure 11.2.

TABLE 11.1 **Values of** $x = \sin \pi t/2$ **and** $y = t$ **for selected values of** t.

t	x	y
0	0	0
1	1	1
2	0	2
3	−1	3
4	0	4
5	1	5
6	0	6

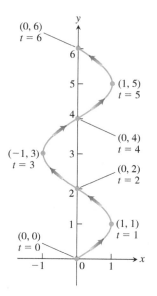

FIGURE 11.2 The curve given by the parametric equations $x = \sin \pi t/2$ and $y = t$ (Example 1).

EXAMPLE 2 Sketch the curve defined by the parametric equations

$$x = t^2, \qquad y = t + 1, \qquad -\infty < t < \infty.$$

Solution We make a table of values (Table 11.2), plot the points (x, y), and draw a smooth curve through them (Figure 11.3). We think of the curve as the path that a particle moves along the curve in the direction of the arrows. Although the time intervals in the table are equal, the consecutive points plotted along the curve are not at equal arc length distances. The reason for this is that the particle slows down as it gets nearer to the y-axis along the lower branch of the curve as t increases, and then speeds up after reaching the y-axis at $(0, 1)$ and moving along the upper branch. Since the interval of values for t is all real numbers, there is no initial point and no terminal point for the curve.

TABLE 11.2 **Values of** $x = t^2$ **and** $y = t + 1$ **for selected values of** t.

t	x	y
−3	9	−2
−2	4	−1
−1	1	0
0	0	1
1	1	2
2	4	3
3	9	4

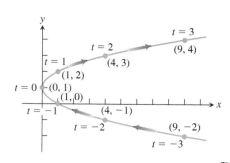

FIGURE 11.3 The curve given by the parametric equations $x = t^2$ and $y = t + 1$ (Example 2).

For this example we can use algebraic manipulation to eliminate the parameter t and obtain an algebraic equation for the curve in terms of x and y alone. We solve $y = t + 1$ for t and substitute the resulting equation $t = y - 1$ into the equation for x, which yields

$$x = t^2 = (y - 1)^2 = y^2 - 2y + 1.$$

The equation $x = y^2 - 2y + 1$ represents a parabola, as displayed in Figure 11.3. It is sometimes quite difficult, or even impossible, to eliminate the parameter from a pair of parametric equations, as we did here. ∎

EXAMPLE 3   Graph the parametric curves

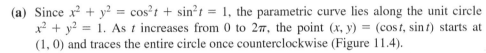

(a) $x = \cos t,$ $y = \sin t,$ $0 \le t \le 2\pi.$
(b) $x = a \cos t,$ $y = a \sin t,$ $0 \le t \le 2\pi.$

Solution

(a) Since $x^2 + y^2 = \cos^2 t + \sin^2 t = 1$, the parametric curve lies along the unit circle $x^2 + y^2 = 1$. As t increases from 0 to 2π, the point $(x, y) = (\cos t, \sin t)$ starts at $(1, 0)$ and traces the entire circle once counterclockwise (Figure 11.4).

(b) For $x = a \cos t, y = a \sin t, 0 \le t \le 2\pi$, we have $x^2 + y^2 = a^2 \cos^2 t + a^2 \sin^2 t = a^2$. The parametrization describes a motion that begins at the point $(a, 0)$ and traverses the circle $x^2 + y^2 = a^2$ once counterclockwise, returning to $(a, 0)$ at $t = 2\pi$. The graph is a circle centered at the origin with radius $r = |a|$ and coordinate points $(a \cos t, a \sin t)$. ∎

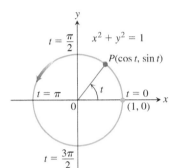

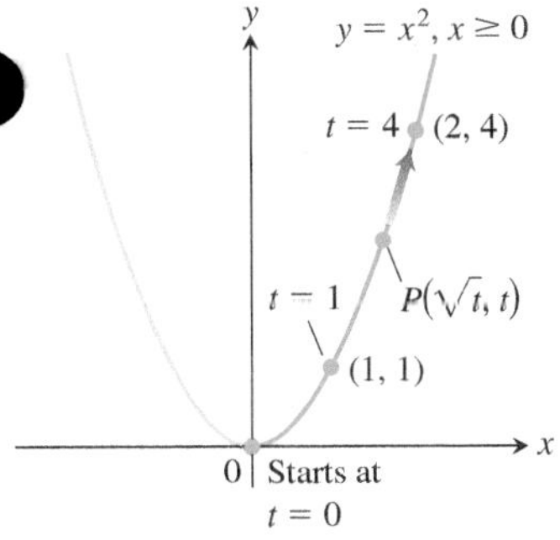

FIGURE 11.4   The equations $x = \cos t$ and $y = \sin t$ describe motion on the circle $x^2 + y^2 = 1$. The arrow shows the direction of increasing t (Example 3).

EXAMPLE 4   The position $P(x, y)$ of a particle moving in the xy-plane is given by the equations and parameter interval

$$x = \sqrt{t}, \qquad y = t, \qquad t \ge 0.$$

Identify the path traced by the particle and describe the motion.

Solution   We try to identify the path by eliminating t between the equations $x = \sqrt{t}$ and $y = t$, which might produce a re-cognizable algebraic relation between x and y. We find that

$$y = t = \left(\sqrt{t}\right)^2 = x^2.$$

Thus, the particle's position coordinates satisfy the equation $y = x^2$, so the particle moves along the parabola $y = x^2$.

It would be a mistake, however, to conclude that the particle's path is the entire parabola $y = x^2$; it is only half the parabola. The particle's x-coordinate is never negative. The particle starts at $(0, 0)$ when $t = 0$ and rises into the first quadrant as t increases (Figure 11.5). The parameter interval is $[0, \infty)$ and there is no terminal point. ∎

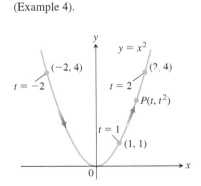

FIGURE 11.5   The equations $x = \sqrt{t}$ and $y = t$ and the interval $t \ge 0$ describe the path of a particle that traces the right-hand half of the parabola $y = x^2$ (Example 4).

The graph of any function $y = f(x)$ can always be given a **natural parametrization** $x = t$ and $y = f(t)$. The domain of the parameter in this case is the same as the domain of the function f.

EXAMPLE 5   A parametrization of the graph of the function $f(x) = x^2$ is given by

$$x = t, \qquad y = f(t) = t^2, \qquad -\infty < t < \infty.$$

When $t \ge 0$, this parametrization gives the same path in the xy-plane as we had in Example 4. However, since the parameter t here can now also be negative, we obtain the left-hand part of the parabola as well; that is, we have the entire parabolic curve. For this parametrization, there is no starting point and no terminal point (Figure 11.6). ∎

FIGURE 11.6   The path defined by $x = t, y = t^2, -\infty < t < \infty$ is the entire parabola $y = x^2$ (Example 5).

Notice that a parametrization also specifies *when* a particle moving along the curve is *located* at a specific point along the curve. In Example 4, the point (2, 4) is reached when $t = 4$; in Example 5, it is reached "earlier" when $t = 2$. You can see the implications of this aspect of parametrizations when considering the possibility of two objects coming into collision: they have to be at the exact same location point $P(x, y)$ for some (possibly different) values of their respective parameters. We will say more about this aspect of parametrizations when we study motion in Chapter 13.

EXAMPLE 6   Find a parametrization for the line through the point (a, b) having slope m.

Solution   A Cartesian equation of the line is $y - b = m(x - a)$. If we define the parameter t by $t = x - a$, we find that $x = a + t$ and $y - b = mt$. That is,

$$x = a + t, \qquad y = b + mt, \qquad -\infty < t < \infty$$

parametrizes the line. This parametrization differs from the one we would obtain by the natural parametrization in Example 5 when $t = x$. However, both parametrizations describe the same line. ∎

TABLE 11.3 Values of $x = t + (1/t)$ and $y = t - (1/t)$ for selected values of t.

t	$1/t$	x	y
0.1	10.0	10.1	−9.9
0.2	5.0	5.2	−4.8
0.4	2.5	2.9	−2.1
1.0	1.0	2.0	0.0
2.0	0.5	2.5	1.5
5.0	0.2	5.2	4.8
10.0	0.1	10.1	9.9

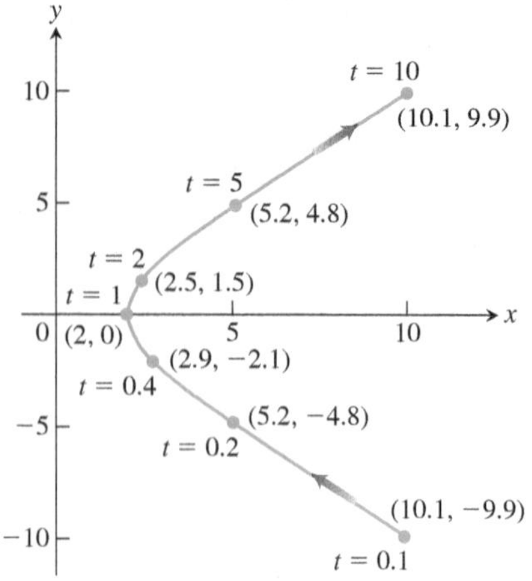

FIGURE 11.7 The curve for $x = t + (1/t)$, $y = t - (1/t)$, $t > 0$ in Example 7. (The part shown is for $0.1 \le t \le 10$.)

EXAMPLE 7   Sketch and identify the path traced by the point $P(x, y)$ if

$$x = t + \frac{1}{t}, \qquad y = t - \frac{1}{t}, \qquad t > 0.$$

Solution   We make a brief table of values in Table 11.3, plot the points, and draw a smooth curve through them, as we did in Example 1. Next we eliminate the parameter t from the equations. The procedure is more complicated than in Example 2. Taking the difference between x and y as given by the parametric equations, we find that

$$x - y = \left(t + \frac{1}{t}\right) - \left(t - \frac{1}{t}\right) = \frac{2}{t}.$$

If we add the two parametric equations, we get

$$x + y = \left(t + \frac{1}{t}\right) + \left(t - \frac{1}{t}\right) = 2t.$$

We can then eliminate the parameter t by multiplying these last equations together:

$$(x - y)(x + y) = \left(\frac{2}{t}\right)(2t) = 4.$$

Expanding the expression on the left-hand side, we obtain a standard equation for a hyperbola (reviewed in Section 11.6):

$$x^2 - y^2 = 4. \tag{1}$$

Thus the coordinates of all the points $P(x, y)$ described by the parametric equations satisfy Equation (1). However, Equation (1) does not require that the x-coordinate be positive. So there are points (x, y) on the hyperbola that do not satisfy the parametric equation $x = t + (1/t), t > 0$. In fact, the parametric equations do not yield any points on the left branch of the hyperbola given by Equation (1), points where the x-coordinate would be negative. For small positive values of t, the path lies in the fourth quadrant and rises into the first quadrant as t increases, crossing the x-axis when $t = 1$ (see Figure 11.7). The parameter domain is $(0, \infty)$ and there is no starting point and no terminal point for the path. ∎

Examples 4, 5, and 6 illustrate that a given curve, or portion of it, can be represented by different parametrizations. In the case of Example 7, we can also represent the right-hand branch of the hyperbola by the parametrization

$$x = \sqrt{4 + t^2}, \qquad y = t, \qquad -\infty < t < \infty,$$

which is obtained by solving Equation (1) for $x \geq 0$ and letting y be the parameter. Still another parametrization for the right-hand branch of the hyperbola given by Equation (1) is

$$x = 2 \sec t, \qquad y = 2 \tan t, \qquad -\frac{\pi}{2} < t < \frac{\pi}{2}.$$

This parametrization follows from the trigonometric identity $\sec^2 t - \tan^2 t = 1$, because

$$x^2 - y^2 = 4 \sec^2 t - 4 \tan^2 t = 4(\sec^2 t - \tan^2 t) = 4.$$

As t runs between $-\pi/2$ and $\pi/2$, $x = \sec t$ remains positive and $y = \tan t$ runs between $-\infty$ and ∞, so P traverses the hyperbola's right-hand branch. It comes in along the branch's lower half as $t \to 0^-$, reaches $(2, 0)$ at $t = 0$, and moves out into the first quadrant as t increases steadily toward $\pi/2$. This is the same branch of the hyperbola shown in Figure 11.7.

HISTORICAL BIOGRAPHY

Christian Huygens
(1629–1695)
bit.ly/2P6tPxq

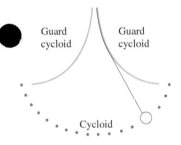

FIGURE 11.8 In Huygens' pendulum clock, the bob swings in a cycloid, so the frequency is independent of the amplitude.

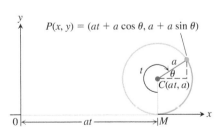

FIGURE 11.9 The position of $P(x, y)$ on the rolling wheel at angle t (Example 8).

Cycloids

The problem with a pendulum clock whose bob swings in a circular arc is that the frequency of the swing depends on the amplitude of the swing. The wider the swing, the longer it takes the bob to return to center (its lowest position).

This does not happen if the bob can be made to swing in a *cycloid*. In 1673, Christian Huygens designed a pendulum clock whose bob would swing in a cycloid, a curve we define in Example 8. He hung the bob from a fine wire constrained by guards that caused it to draw up as it swung away from center (Figure 11.8). We describe the path parametrically in the next example.

EXAMPLE 8 A wheel of radius a rolls along a horizontal straight line. Find parametric equations for the path traced by a point P on the wheel's circumference. The path is called a **cycloid**.

Solution We take the line to be the x-axis, mark a point P on the wheel, start the wheel with P at the origin, and roll the wheel to the right. As parameter, we use the angle t through which the wheel turns, measured in radians. Figure 11.9 shows the wheel a short while later when its base lies at units from the origin. The wheel's center C lies at (at, a) and the coordinates of P are

$$x = at + a \cos \theta, \qquad y = a + a \sin \theta.$$

To express θ in terms of t, we observe that $t + \theta = 3\pi/2$ in the figure, so that

$$\theta = \frac{3\pi}{2} - t.$$

This makes

$$\cos \theta = \cos\left(\frac{3\pi}{2} - t\right) = -\sin t, \qquad \sin \theta = \sin\left(\frac{3\pi}{2} - t\right) = -\cos t.$$

The equations we seek are

$$x = at - a \sin t, \qquad y = a - a \cos t.$$

These are usually written with the a factored out:

$$x = a(t - \sin t), \qquad y = a(1 - \cos t). \tag{2}$$

Figure 11.10 shows the first arch of the cycloid and part of the next.

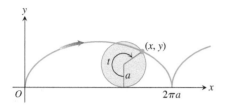

FIGURE 11.10 The cycloid curve $x = a(t - \sin t)$, $y = a(1 - \cos t)$, for $t \geq 0$.

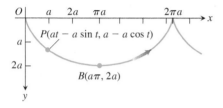

FIGURE 11.11 Turning Figure 11.10 upside down, the y-axis points downward, indicating the direction of the gravitational force. Equations (2) still describe the curve parametrically.

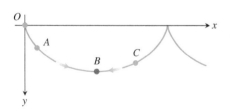

FIGURE 11.12 The cycloid is the unique curve which minimizes the time it takes for a frictionless bead to slide from point O to B.

FIGURE 11.13 Beads released simultaneously on the upside-down cycloid at O, A, and C will reach B at the same time.

Brachistochrones and Tautochrones

If we turn Figure 11.10 upside down, Equations (2) still apply and the resulting curve (Figure 11.11) has two interesting physical properties. The first relates to the origin O and the point B at the bottom of the first arch. Among all smooth curves joining these points, the cycloid is the curve along which a frictionless bead, subject only to the force of gravity, will slide from O to B the fastest. This makes the cycloid a **brachistochrone** ("brah-*kiss*-toe-krone"), or shortest-time curve for these points. The second property is that even if you start the bead partway down the curve toward B, it will still take the bead the same amount of time to reach B. This makes the cycloid a **tautochrone** ("*taw*-toe-krone"), or same-time curve for O and B.

Are there any other brachistochrones joining O and B, or is the cycloid the only one? We can formulate this as a mathematical question in the following way. At the start, the kinetic energy of the bead is zero, since its velocity (speed) is zero. The work done by gravity in moving the bead from $(0, 0)$ to any other point (x, y) in the plane is mgy, and this must equal the change in kinetic energy. (See Exercise 25 in Section 6.5.) That is,

$$mgy = \frac{1}{2}mv^2 - \frac{1}{2}m(0)^2.$$

Thus, the speed of the bead when it reaches (x, y) has to be $v = \sqrt{2gy}$. That is,

$$\frac{ds}{dT} = \sqrt{2gy} \qquad \text{\small ds is the arc length differential along the bead's path and T represents time.}$$

or

$$dT = \frac{ds}{\sqrt{2gy}} = \frac{\sqrt{1 + (dy/dx)^2}\,dx}{\sqrt{2gy}}. \tag{3}$$

The time T_f it takes the bead to slide along a particular path $y = f(x)$ from O to $B(a\pi, 2a)$ is

$$T_f = \int_{x=0}^{x=a\pi} \sqrt{\frac{1 + (dy/dx)^2}{2gy}}\,dx. \tag{4}$$

What curves $y = f(x)$, if any, minimize the value of this integral?

At first sight, we might guess that the straight line joining O and B would give the shortest time, but perhaps not. There might be some advantage in having the bead fall vertically at first to build up its speed faster. With a higher speed, the bead could travel a longer path and still reach B first. Indeed, this is the right idea. The solution, from a branch of mathematics known as the *calculus of variations*, is that the original cycloid from O to B is the one and only brachistochrone for O and B (Figure 11.12).

In the next section we show how to find the arc length differential ds for a parametrized curve. Once we know how to find ds, we can calculate the time given by the right-hand side of Equation (4) for the cycloid. This calculation gives the amount of time it takes a frictionless bead to slide down the cycloid to B after it is released from rest at O. The time turns out to be equal to $\pi\sqrt{a/g}$, where a is the radius of the wheel defining the particular cycloid. Moreover, if we start the bead at some lower point on the cycloid, corresponding to a parameter value $t_0 > 0$, we can integrate the parametric form of $ds/\sqrt{2gy}$ in Equation (3) over the interval $[t_0, \pi]$ to find the time it takes the bead to reach the point B. That calculation results in the same time $T = \pi\sqrt{a/g}$. It takes the bead the same amount of time to reach B no matter where it starts, which makes the cycloid a tautochrone. Beads starting simultaneously from O, A, and C in Figure 11.13, for instance, will all reach B at exactly the same time. This is the reason why Huygens' pendulum clock in Figure 11.8 is independent of the amplitude of the swing.

EXERCISES 11.1

Finding Cartesian from Parametric Equations

Exercises 1–18 give parametric equations and parameter intervals for the motion of a particle in the xy-plane. Identify the particle's path by finding a Cartesian equation for it. Graph the Cartesian equation. (The graphs will vary with the equation used.) Indicate the portion of the graph traced by the particle and the direction of motion.

1. $x = 3t, \quad y = 9t^2, \quad -\infty < t < \infty$

2. $x = -\sqrt{t}, \quad y = t, \quad t \geq 0$

3. $x = 2t - 5, \quad y = 4t - 7, \quad -\infty < t < \infty$

4. $x = 3 - 3t, \quad y = 2t, \quad 0 \leq t \leq 1$

5. $x = \cos 2t, \quad y = \sin 2t, \quad 0 \leq t \leq \pi$

6. $x = \cos(\pi - t), \quad y = \sin(\pi - t), \quad 0 \leq t \leq \pi$

7. $x = 4\cos t, \quad y = 2\sin t, \quad 0 \leq t \leq 2\pi$

8. $x = 4\sin t, \quad y = 5\cos t, \quad 0 \leq t \leq 2\pi$

9. $x = \sin t, \quad y = \cos 2t, \quad -\dfrac{\pi}{2} \leq t \leq \dfrac{\pi}{2}$

10. $x = 1 + \sin t, \quad y = \cos t - 2, \quad 0 \leq t \leq \pi$

11. $x = t^2, \quad y = t^6 - 2t^4, \quad -\infty < t < \infty$

12. $x = \dfrac{t}{t-1}, \quad y = \dfrac{t-2}{t+1}, \quad -1 < t < 1$

13. $x = t, \quad y = \sqrt{1 - t^2}, \quad -1 \leq t \leq 0$

14. $x = \sqrt{t + 1}, \quad y = \sqrt{t}, \quad t \geq 0$

15. $x = \sec^2 t - 1, \quad y = \tan t, \quad -\pi/2 < t < \pi/2$

16. $x = -\sec t, \quad y = \tan t, \quad -\pi/2 < t < \pi/2$

17. $x = -\cosh t, \quad y = \sinh t, \quad -\infty < t < \infty$

18. $x = 2\sinh t, \quad y = 2\cosh t, \quad -\infty < t < \infty$

In Exercises 19–24, match the parametric equations with the parametric curves labeled A through F.

19. $x = 1 - \sin t, \quad y = 1 + \cos t$

20. $x = \cos t, \quad y = 2\sin t$

21. $x = \dfrac{1}{4}t\cos t, \quad y = \dfrac{1}{4}t\sin t$

22. $x = \sqrt{t}, \quad y = \sqrt{t}\cos t$

23. $x = \ln t, \quad y = 3e^{-t/2}$

24. $x = \cos t, \quad y = \sin 3t$

A.

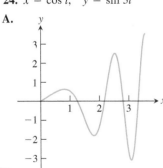

B.

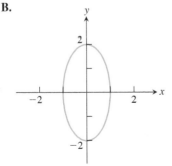

C.   **D.**

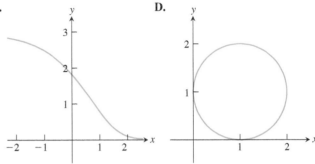

E.   **F.**

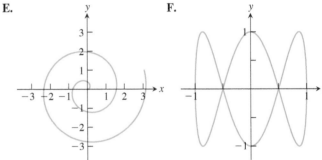

In Exercises 25–28, use the given graphs of $x = f(t)$ and $y = g(t)$ to sketch the corresponding parametric curve in the xy-plane.

25.

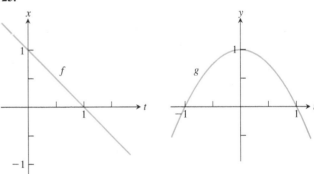

26.

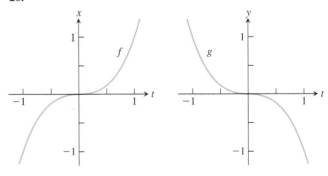

27.

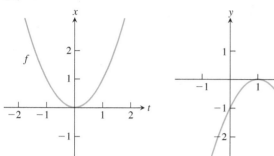

28.

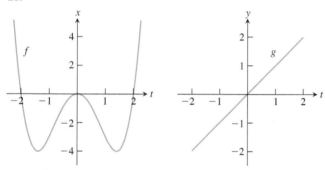

Finding Parametric Equations

29. Find parametric equations and a parameter interval for the motion of a particle that starts at $(a, 0)$ and traces the circle $x^2 + y^2 = a^2$

   a. once clockwise.

   b. once counterclockwise.

   c. twice clockwise.

   d. twice counterclockwise.

   (There are many ways to do these, so your answers may not be the same as the ones in the back of the book.)

30. Find parametric equations and a parameter interval for the motion of a particle that starts at $(a, 0)$ and traces the ellipse $(x^2/a^2) + (y^2/b^2) = 1$

   a. once clockwise.

   b. once counterclockwise.

   c. twice clockwise.

   d. twice counterclockwise.

   (As in Exercise 29, there are many correct answers.)

In Exercises 31–36, find a parametrization for the curve.

31. the line segment with endpoints $(-1, -3)$ and $(4, 1)$

32. the line segment with endpoints $(-1, 3)$ and $(3, -2)$

33. the lower half of the parabola $x - 1 = y^2$

34. the left half of the parabola $y = x^2 + 2x$

35. the ray (half line) with initial point $(2, 3)$ that passes through the point $(-1, -1)$

36. the ray (half line) with initial point $(-1, 2)$ that passes through the point $(0, 0)$

37. Find parametric equations and a parameter interval for the motion of a particle starting at the point $(2, 0)$ and tracing the top half of the circle $x^2 + y^2 = 4$ four times.

38. Find parametric equations and a parameter interval for the motion of a particle that moves along the graph of $y = x^2$ in the following way: Beginning at $(0, 0)$ it moves to $(3, 9)$, and then travels back and forth from $(3, 9)$ to $(-3, 9)$ infinitely many times.

39. Find parametric equations for the semicircle

$$x^2 + y^2 = a^2, \quad y > 0,$$

using as parameter the slope $t = dy/dx$ of the tangent to the curve at (x, y).

40. Find parametric equations for the circle

$$x^2 + y^2 = a^2,$$

using as parameter the arc length s measured counterclockwise from the point $(a, 0)$ to the point (x, y).

41. Find a parametrization for the line segment joining points $(0, 2)$ and $(4, 0)$ using the angle θ in the accompanying figure as the parameter.

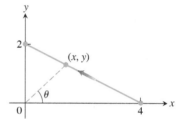

42. Find a parametrization for the curve $y = \sqrt{x}$ with terminal point $(0, 0)$ using the angle θ in the accompanying figure as the parameter.

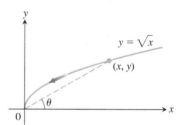

43. Find a parametrization for the circle $(x - 2)^2 + y^2 = 1$ starting at $(1, 0)$ and moving clockwise once around the circle, using the central angle θ in the accompanying figure as the parameter.

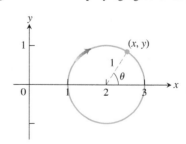

44. Find a parametrization for the circle $x^2 + y^2 = 1$ starting at $(1, 0)$ and moving counterclockwise to the terminal point $(0, 1)$, using the angle θ in the accompanying figure as the parameter.

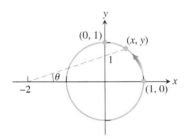

45. The witch of Maria Agnesi The bell-shaped witch of Maria Agnesi can be constructed in the following way. Start with a circle of radius 1, centered at the point $(0, 1)$, as shown in the accompanying figure. Choose a point A on the line $y = 2$ and connect it to the origin with a line segment. Call the point where the segment crosses the circle B. Let P be the point where the vertical line through A crosses the horizontal line through B. The witch is the curve traced by P as A moves along the line $y = 2$. Find parametric equations and a parameter interval for the witch by expressing the coordinates of P in terms of t, the radian measure of the angle that segment OA makes with the positive x-axis. The following equalities (which you may assume) will help.

a. $x = AQ$ **b.** $y = 2 - AB \sin t$

c. $AB \cdot OA = (AQ)^2$

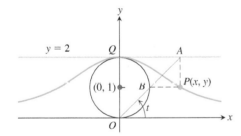

46. Hypocycloid When a circle rolls on the inside of a fixed circle, any point P on the circumference of the rolling circle describes a *hypocycloid*. Let the fixed circle be $x^2 + y^2 = a^2$, let the radius of the rolling circle be b, and let the initial position of the tracing point P be $A(a, 0)$. Find parametric equations for the hypocycloid, using as the parameter the angle θ from the positive x-axis to the line joining the circles' centers. In particular, if $b = a/4$, as in the accompanying figure, show that the hypocycloid is the astroid

$$x = a \cos^3 \theta, \quad y = a \sin^3 \theta.$$

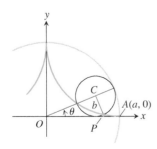

47. As the point N moves along the line $y = a$ in the accompanying figure, P moves in such a way that $OP = MN$. Find parametric equations for the coordinates of P as functions of the angle t that the line ON makes with the positive y-axis.

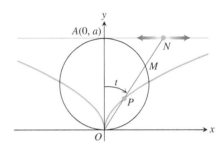

48. Trochoids A wheel of radius a rolls along a horizontal straight line without slipping. Find parametric equations for the curve traced out by a point P on a spoke of the wheel b units from its center. As parameter, use the angle θ through which the wheel turns. The curve is called a *trochoid*, which is a cycloid when $b = a$.

Distance Using Parametric Equations

49. Find the point on the parabola $x = t, y = t^2, -\infty < t < \infty$, closest to the point $(2, 1/2)$. (*Hint:* Minimize the square of the distance as a function of t.)

50. Find the point on the ellipse $x = 2 \cos t, y = \sin t, 0 \le t \le 2\pi$ closest to the point $(3/4, 0)$. (*Hint:* Minimize the square of the distance as a function of t.)

T GRAPHER EXPLORATIONS

If you have a parametric equation grapher, graph the equations over the given intervals in Exercises 51–58.

51. Ellipse $x = 4 \cos t, \quad y = 2 \sin t, \quad$ over

     a. $0 \le t \le 2\pi$

     b. $0 \le t \le \pi$

     c. $-\pi/2 \le t \le \pi/2$.

52. Hyperbola branch $x = \sec t$ (enter as $1/\cos(t)$), $y = \tan t$ (enter as $\sin(t)/\cos(t)$), over

     a. $-1.5 \le t \le 1.5$

     b. $-0.5 \le t \le 0.5$

     c. $-0.1 \le t \le 0.1$.

53. Parabola $x = 2t + 3, \quad y = t^2 - 1, \quad -2 \le t \le 2$

54. Cycloid $x = t - \sin t, \quad y = 1 - \cos t, \quad$ over

     a. $0 \le t \le 2\pi$

     b. $0 \le t \le 4\pi$

     c. $\pi \le t \le 3\pi$.

55. Deltoid

$$x = 2 \cos t + \cos 2t, \quad y = 2 \sin t - \sin 2t; \quad 0 \le t \le 2\pi$$

What happens if you replace 2 with -2 in the equations for x and y? Graph the new equations and find out.

56. A nice curve

$$x = 3 \cos t + \cos 3t, \quad y = 3 \sin t - \sin 3t; \quad 0 \le t \le 2\pi$$

What happens if you replace 3 with −3 in the equations for x and y? Graph the new equations and find out.

57. a. Epicycloid

$$x = 9 \cos t - \cos 9t, \quad y = 9 \sin t - \sin 9t; \quad 0 \le t \le 2\pi$$

b. Hypocycloid

$$x = 8 \cos t + 2 \cos 4t, \quad y = 8 \sin t - 2 \sin 4t; \quad 0 \le t \le 2\pi$$

c. Hypotrochoid

$$x = \cos t + 5 \cos 3t, \quad y = 6 \cos t - 5 \sin 3t; \quad 0 \le t \le 2\pi$$

58. a. $x = 6 \cos t + 5 \cos 3t, \quad y = 6 \sin t - 5 \sin 3t;$
$0 \le t \le 2\pi$

b. $x = 6 \cos 2t + 5 \cos 6t, \quad y = 6 \sin 2t - 5 \sin 6t;$
$0 \le t \le \pi$

c. $x = 6 \cos t + 5 \cos 3t, \quad y = 6 \sin 2t - 5 \sin 3t;$
$0 \le t \le 2\pi$

d. $x = 6 \cos 2t + 5 \cos 6t, \quad y = 6 \sin 4t - 5 \sin 6t;$
$0 \le t \le \pi$

11.2 Calculus with Parametric Curves

In this section we apply calculus to parametric curves. Specifically, we find slopes, lengths, and areas associated with parametrized curves.

Tangents and Areas

A parametrized curve $x = f(t)$ and $y = g(t)$ is **differentiable** at t if f and g are differentiable at t. At a point on a differentiable parametrized curve where y is also a differentiable function of x, the derivatives dy/dt, dx/dt, and dy/dx are related by the Chain Rule:

$$\frac{dy}{dt} = \frac{dy}{dx} \cdot \frac{dx}{dt}.$$

If $dx/dt \ne 0$, we may divide both sides of this equation by dx/dt to solve for dy/dx.

Parametric Formula for dy/dx

If all three derivatives exist and $dx/dt \ne 0$, then

$$\frac{dy}{dx} = \frac{dy/dt}{dx/dt}. \tag{1}$$

If parametric equations define y as a twice-differentiable function of x, we can apply Equation (1) to the function $dy/dx = y'$ to calculate d^2y/dx^2 as a function of t:

$$\frac{d^2y}{dx^2} = \frac{d}{dx}(y') = \frac{dy'/dt}{dx/dt}. \qquad \text{Eq. (1) with } y' \text{ in place of } y$$

Parametric Formula for d^2y/dx^2

If the equations $x = f(t)$, $y = g(t)$ define y as a twice-differentiable function of x, then at any point where $dx/dt \ne 0$ and $y' = dy/dx$,

$$\frac{d^2y}{dx^2} = \frac{dy'/dt}{dx/dt}. \tag{2}$$

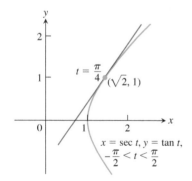

FIGURE 11.14 The curve in Example 1 is the right-hand branch of the hyperbola $x^2 - y^2 = 1$.

EXAMPLE 1   Find the tangent to the curve

$$x = \sec t, \qquad y = \tan t, \qquad -\frac{\pi}{2} < t < \frac{\pi}{2},$$

at the point $\left(\sqrt{2}, 1\right)$, where $t = \pi/4$ (Figure 11.14).

Solution The slope of the curve at t is

$$\frac{dy}{dx} = \frac{dy/dt}{dx/dt} = \frac{\sec^2 t}{\sec t \tan t} = \frac{\sec t}{\tan t}.$$ Eq. (1)

Setting t equal to $\pi/4$ gives

$$\frac{dy}{dx}\bigg|_{t=\pi/4} = \frac{\sec(\pi/4)}{\tan(\pi/4)} = \frac{\sqrt{2}}{1} = \sqrt{2}.$$

The tangent line is

$$y - 1 = \sqrt{2}\,(x - \sqrt{2})$$
$$y = \sqrt{2}\,x - 2 + 1$$
$$y = \sqrt{2}\,x - 1.$$

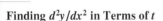

Finding d^2y/dx^2 in Terms of t

1. Express $y' = dy/dx$ in terms of t.
2. Find dy'/dt.
3. Divide dy'/dt by dx/dt.

EXAMPLE 2 Find d^2y/dx^2 as a function of t if $x = t - t^2$ and $y = t - t^3$.

Solution

1. Express $y' = dy/dx$ in terms of t.

$$y' = \frac{dy}{dx} = \frac{dy/dt}{dx/dt} = \frac{1 - 3t^2}{1 - 2t}$$

2. Differentiate y' with respect to t.

$$\frac{dy'}{dt} = \frac{d}{dt}\left(\frac{1 - 3t^2}{1 - 2t}\right) = \frac{2 - 6t + 6t^2}{(1 - 2t)^2}$$ Derivative Quotient Rule

3. Divide dy'/dt by dx/dt.

$$\frac{d^2y}{dx^2} = \frac{dy'/dt}{dx/dt} = \frac{(2 - 6t + 6t^2)/(1 - 2t)^2}{1 - 2t} = \frac{2 - 6t + 6t^2}{(1 - 2t)^3}$$ Eq. (2)

EXAMPLE 3 Find the area enclosed by the astroid (Figure 11.15)

$$x = \cos^3 t, \qquad y = \sin^3 t, \qquad 0 \le t \le 2\pi.$$

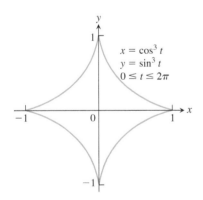

FIGURE 11.15 The astroid in Example 3.

Solution By symmetry, the enclosed area is 4 times the area beneath the curve in the first quadrant where $0 \le t \le \pi/2$. We can apply the definite integral formula for area studied in Chapter 5, using substitution to express the curve and differential dx in terms of the parameter t. Thus,

$$A = 4\int_0^1 y\, dx$$ 4 times area under y from $x = 0$ to $x = 1$

$$= 4\int_0^{\pi/2} (\sin^3 t)(3\cos^2 t \sin t)\, dt$$ Substitution for y and dx

$$= 12\int_0^{\pi/2} \left(\frac{1 - \cos 2t}{2}\right)^2 \left(\frac{1 + \cos 2t}{2}\right) dt$$ $\sin^4 t = \left(\dfrac{1 - \cos 2t}{2}\right)^2$

$$= \frac{3}{2}\int_0^{\pi/2} (1 - 2\cos 2t + \cos^2 2t)(1 + \cos 2t)\, dt$$ Expand squared term.

$$= \frac{3}{2}\int_0^{\pi/2} (1 - \cos 2t - \cos^2 2t + \cos^3 2t)\, dt$$ Multiply terms.

$$= \frac{3}{2}\left[\int_0^{\pi/2}(1-\cos 2t)\,dt - \int_0^{\pi/2}\cos^2 2t\,dt + \int_0^{\pi/2}\cos^3 2t\,dt\right]$$

$$= \frac{3}{2}\left[\left(t - \frac{1}{2}\sin 2t\right) - \frac{1}{2}\left(t + \frac{1}{4}\sin 2t\right) + \frac{1}{2}\left(\sin 2t - \frac{1}{3}\sin^3 2t\right)\right]_0^{\pi/2} \qquad \text{Section 8.2, Example 3}$$

$$= \frac{3}{2}\left[\left(\frac{\pi}{2} - 0 - 0 - 0\right) - \frac{1}{2}\left(\frac{\pi}{2} + 0 - 0 - 0\right) + \frac{1}{2}(0 - 0 - 0 + 0)\right] \qquad \text{Evaluate.}$$

$$= \frac{3\pi}{8}. \qquad \blacksquare$$

Length of a Parametrically Defined Curve

Let C be a curve given parametrically by the equations

$$x = f(t) \qquad \text{and} \qquad y = g(t), \qquad a \le t \le b.$$

We assume the functions f and g are **continuously differentiable** (meaning they have continuous first derivatives) on the interval $[a, b]$. We also assume that the derivatives $f'(t)$ and $g'(t)$ are not simultaneously zero, which prevents the curve C from having any corners or cusps. Such a curve is called a **smooth curve**. We subdivide the path (or arc) AB into n pieces at points $A = P_0, P_1, P_2, \ldots, P_n = B$ (Figure 11.16). These points correspond to a partition of the interval $[a, b]$ by $a = t_0 < t_1 < t_2 < \cdots < t_n = b$, where $P_k = (f(t_k), g(t_k))$. Join successive points of this subdivision by straight-line segments (Figure 11.16). A representative line segment has length

$$\begin{aligned}L_k &= \sqrt{(\Delta x_k)^2 + (\Delta y_k)^2}\\ &= \sqrt{[f(t_k) - f(t_{k-1})]^2 + [g(t_k) - g(t_{k-1})]^2}\end{aligned}$$

(see Figure 11.17). If Δt_k is small, the length L_k is approximately the length of arc $P_{k-1}P_k$. By the Mean Value Theorem there are numbers t_k^* and t_k^{**} in $[t_{k-1}, t_k]$ such that

$$\Delta x_k = f(t_k) - f(t_{k-1}) = f'(t_k^*)\,\Delta t_k,$$
$$\Delta y_k = g(t_k) - g(t_{k-1}) = g'(t_k^{**})\,\Delta t_k.$$

Assuming the path from A to B is traversed exactly once as t increases from $t = a$ to $t = b$, with no doubling back or retracing, an approximation to the (yet to be defined) "length" of the curve AB is the sum of all the lengths L_k:

$$\sum_{k=1}^{n} L_k = \sum_{k=1}^{n}\sqrt{(\Delta x_k)^2 + (\Delta y_k)^2}$$

$$= \sum_{k=1}^{n}\sqrt{[f'(t_k^*)]^2 + [g'(t_k^{**})]^2}\,\Delta t_k.$$

Although this last sum on the right is not exactly a Riemann sum (because f' and g' are evaluated at different points), it can be shown that its limit, as the norm of the partition tends to zero and the number of segments $n \to \infty$, is the definite integral

$$\lim_{\|P\| \to 0}\sum_{k=1}^{n}\sqrt{[f'(t_k^*)]^2 + [g'(t_k^{**})]^2}\,\Delta t_k = \int_a^b \sqrt{[f'(t)]^2 + [g'(t)]^2}\,dt.$$

Therefore, it is reasonable to define the length of the curve from A to B to be this integral.

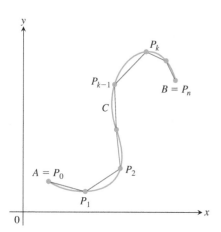

FIGURE 11.16 The length of the smooth curve C from A to B is approximated by the sum of the lengths of the polygonal path (straight-line segments) starting at $A = P_0$, then to P_1, and so on, ending at $B = P_n$.

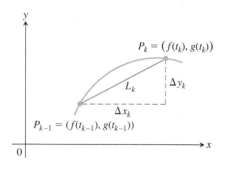

FIGURE 11.17 The arc $P_{k-1}P_k$ is approximated by the straight-line segment shown here, which has length $L_k = \sqrt{(\Delta x_k)^2 + (\Delta y_k)^2}$.

> **DEFINITION**   If a curve C is defined parametrically by $x = f(t)$ and $y = g(t)$, $a \leq t \leq b$, where f' and g' are continuous and not simultaneously zero on $[a, b]$, and C is traversed exactly once as t increases from $t = a$ to $t = b$, then **the length of C** is the definite integral
>
> $$L = \int_a^b \sqrt{[f'(t)]^2 + [g'(t)]^2}\, dt.$$

If $x = f(t)$ and $y = g(t)$, then using the Leibniz notation we can write the formula for arc length this way:

$$L = \int_a^b \sqrt{\left(\frac{dx}{dt}\right)^2 + \left(\frac{dy}{dt}\right)^2}\, dt. \tag{3}$$

A smooth curve C does not double back or reverse the direction of motion over the time interval $[a, b]$ since $(f')^2 + (g')^2 > 0$ throughout the interval. At a point where a curve does start to double back on itself, either the curve fails to be differentiable or both derivatives must simultaneously equal zero. We will examine this phenomenon in Chapter 13, where we study tangent vectors to curves.

If there are two different parametrizations for a curve C whose length we want to find, it does not matter which one we use. However, the parametrization we choose must meet the conditions stated in the definition of the length of C (see Exercise 41 for an example).

EXAMPLE 4   Using the definition, find the length of the circle of radius r defined parametrically by

$$x = r \cos t \quad \text{and} \quad y = r \sin t, \quad 0 \leq t \leq 2\pi.$$

Solution   As t varies from 0 to 2π, the circle is traversed exactly once, so the circumference is

$$L = \int_0^{2\pi} \sqrt{\left(\frac{dx}{dt}\right)^2 + \left(\frac{dy}{dt}\right)^2}\, dt.$$

We find

$$\frac{dx}{dt} = -r \sin t, \quad \frac{dy}{dt} = r \cos t$$

and

$$\left(\frac{dx}{dt}\right)^2 + \left(\frac{dy}{dt}\right)^2 = r^2(\sin^2 t + \cos^2 t) = r^2.$$

Therefore, the total arc length is

$$L = \int_0^{2\pi} \sqrt{r^2}\, dt = r\Big[t\Big]_0^{2\pi} = 2\pi r.$$

EXAMPLE 5   Find the length of the astroid (Figure 11.15)

$$x = \cos^3 t, \quad y = \sin^3 t, \quad 0 \leq t \leq 2\pi.$$

Solution   Because of the curve's symmetry with respect to the coordinate axes, its length is four times the length of the first-quadrant portion. We have

$$x = \cos^3 t, \qquad\qquad y = \sin^3 t$$

$$\left(\frac{dx}{dt}\right)^2 = [\,3\cos^2 t(-\sin t)\,]^2 = 9\cos^4 t \sin^2 t$$

$$\left(\frac{dy}{dt}\right)^2 = [\,3\sin^2 t(\cos t)\,]^2 = 9\sin^4 t \cos^2 t$$

$$\sqrt{\left(\frac{dx}{dt}\right)^2 + \left(\frac{dy}{dt}\right)^2} = \sqrt{9\cos^2 t \sin^2 t(\underbrace{\cos^2 t + \sin^2 t}_{1})}$$

$$= \sqrt{9\cos^2 t \sin^2 t}$$

$$= 3\,|\cos t \sin t| \qquad \cos t \sin t \geq 0 \text{ for } 0 \leq t \leq \pi/2$$

$$= 3\cos t \sin t.$$

Therefore,

$$\text{Length of first-quadrant portion} = \int_0^{\pi/2} 3\cos t \sin t\, dt$$

$$= \frac{3}{2}\int_0^{\pi/2} \sin 2t\, dt \qquad \cos t \sin t = (1/2)\sin 2t$$

$$= -\frac{3}{4}\cos 2t\,\Big]_0^{\pi/2} = \frac{3}{2}.$$

The length of the astroid is four times this: $4(3/2) = 6$.   ∎

EXAMPLE 6   Find the perimeter of the ellipse $\dfrac{x^2}{a^2} + \dfrac{y^2}{b^2} = 1$.

Solution   Parametrically, we represent the ellipse by the equations $x = a \sin t$ and $y = b \cos t$, $a > b$ and $0 \leq t \leq 2\pi$. Then,

$$\left(\frac{dx}{dt}\right)^2 + \left(\frac{dy}{dt}\right)^2 = a^2 \cos^2 t + b^2 \sin^2 t$$

$$= a^2 - (a^2 - b^2)\sin^2 t$$

$$= a^2[\,1 - e^2 \sin^2 t\,] \qquad e = \sqrt{1 - \frac{b^2}{a^2}} \text{ (eccentricity,}$$
$$\text{not the number } 2.71828\ldots)$$

From Equation (3), the perimeter is given by

$$P = 4a\int_0^{\pi/2} \sqrt{1 - e^2 \sin^2 t}\, dt.$$

(We investigate the meaning of the eccentricity e in Section 11.7.) The integral for P is nonelementary and is known as the *complete elliptic integral of the second kind*. We can compute its value to within any degree of accuracy using infinite series in the following way. From the binomial expansion for $\sqrt{1 - x^2}$ in Section 10.10, we have

$$\sqrt{1 - e^2 \sin^2 t} = 1 - \frac{1}{2}e^2 \sin^2 t - \frac{1}{2\cdot 4}e^4 \sin^4 t - \cdots, \qquad |e \sin t| \leq e < 1$$

Then to each term in this last expression we apply the integral Formula 157 (at the back of the book) for $\int_0^{\pi/2} \sin^n t \, dt$ when n is even, giving the perimeter

$$P = 4a \int_0^{\pi/2} \sqrt{1 - e^2 \sin^2 t} \, dt$$

$$= 4a \left[\frac{\pi}{2} - \left(\frac{1}{2} e^2 \right) \left(\frac{1}{2} \cdot \frac{\pi}{2} \right) - \left(\frac{1}{2 \cdot 4} e^4 \right) \left(\frac{1 \cdot 3}{2 \cdot 4} \cdot \frac{\pi}{2} \right) - \left(\frac{1 \cdot 3}{2 \cdot 4 \cdot 6} e^6 \right) \left(\frac{1 \cdot 3 \cdot 5}{2 \cdot 4 \cdot 6} \cdot \frac{\pi}{2} \right) - \cdots \right]$$

$$= 2\pi a \left[1 - \left(\frac{1}{2} \right)^2 e^2 - \left(\frac{1 \cdot 3}{2 \cdot 4} \right)^2 \frac{e^4}{3} - \left(\frac{1 \cdot 3 \cdot 5}{2 \cdot 4 \cdot 6} \right)^2 \frac{e^6}{5} - \cdots \right].$$

Since $e < 1$, the series on the right-hand side converges by comparison with the geometric series $\sum_{n-1}^{\infty} (e^2)^n$. We do not have an explicit value for P, but we can estimate it as closely as we like by summing finitely many terms from the infinite series. ∎

Length of a Curve $y = f(x)$

We will show that the length formula in Section 6.3 is a special case of Equation (3). Given a continuously differentiable function $y = f(x)$, $a \le x \le b$, we can assign $x = t$ as a parameter. The graph of the function f is then the curve C defined parametrically by

$$x = t \qquad \text{and} \qquad y = f(t), \qquad a \le t \le b,$$

which is a special case of what we have considered in this chapter. We have

$$\frac{dx}{dt} = 1 \qquad \text{and} \qquad \frac{dy}{dt} = f'(t).$$

From Equation (1),

$$\frac{dy}{dx} = \frac{dy/dt}{dx/dt} = f'(t),$$

giving

$$\left(\frac{dx}{dt} \right)^2 + \left(\frac{dy}{dt} \right)^2 = 1 + [f'(t)]^2$$

$$= 1 + [f'(x)]^2. \qquad t = x$$

Substitution into Equation (3) gives exactly the arc length formula for the graph of $y = f(x)$ that we found in Section 6.3.

The Arc Length Differential

As in Section 6.3, we define the arc length function for a parametrically defined curve $x = f(t)$ and $y = g(t)$, $a \le t \le b$, by

$$s(t) = \int_a^t \sqrt{[f'(z)]^2 + [g'(z)]^2} \, dz.$$

Then, by the Fundamental Theorem of Calculus,

$$\frac{ds}{dt} = \sqrt{[f'(t)]^2 + [g'(t)]^2} = \sqrt{\left(\frac{dx}{dt} \right)^2 + \left(\frac{dy}{dt} \right)^2}.$$

The differential of arc length is

$$ds = \sqrt{\left(\frac{dx}{dt} \right)^2 + \left(\frac{dy}{dt} \right)^2} \, dt. \tag{4}$$

Equation (4) is often abbreviated as

$$ds = \sqrt{dx^2 + dy^2}.$$

Just as in Section 6.3, we can integrate the differential ds between appropriate limits to find the total length of a curve.

Here's an example where we use the arc length differential to find the centroid of an arc.

EXAMPLE 7   Find the centroid of the first-quadrant arc of the astroid in Example 5.

Solution   We take the curve's density to be $\delta = 1$ and calculate the curve's mass and moments about the coordinate axes as we did in Section 6.6.

The distribution of mass is symmetric about the line $y = x$, so $\bar{x} = \bar{y}$. A typical segment of the curve (Figure 11.18) has mass

$$dm = 1 \cdot ds = \sqrt{\left(\frac{dx}{dt}\right)^2 + \left(\frac{dy}{dt}\right)^2}\, dt = 3 \cos t \sin t \, dt. \qquad \text{From Example 5}$$

The curve's mass is

$$M = \int_0^{\pi/2} dm = \int_0^{\pi/2} 3 \cos t \sin t \, dt = \frac{3}{2}. \qquad \text{Again from Example 5}$$

The curve's moment about the x-axis is

$$M_x = \int \tilde{y}\, dm = \int_0^{\pi/2} \sin^3 t \cdot 3 \cos t \sin t \, dt$$

$$= 3 \int_0^{\pi/2} \sin^4 t \cos t \, dt = 3 \cdot \frac{\sin^5 t}{5}\bigg]_0^{\pi/2} = \frac{3}{5}.$$

It follows that

$$\bar{y} = \frac{M_x}{M} = \frac{3/5}{3/2} = \frac{2}{5}.$$

The centroid is the point $(2/5, 2/5)$.

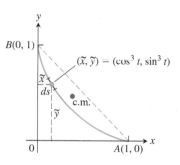

FIGURE 11.18 The centroid (c.m.) of the astroid arc in Example 7.

EXAMPLE 8   Find the time T_c it takes for a frictionless bead to slide along the cycloid $x = a(t - \sin t)$, $y = a(1 - \cos t)$ from $t = 0$ to $t = \pi$ (see Figure 11.13).

Solution   From Equation (3) in Section 11.1, we want to find the time

$$T_c = \int_{t=0}^{t=\pi} \frac{ds}{\sqrt{2gy}}.$$

We need to express ds parametrically in terms of the parameter t. For the cycloid, $dx/dt = a(1 - \cos t)$ and $dy/dt = a \sin t$, so

$$ds = \sqrt{\left(\frac{dx}{dt}\right)^2 + \left(\frac{dy}{dt}\right)^2}\, dt$$

$$= \sqrt{a^2 (1 - 2\cos t + \cos^2 t + \sin^2 t)}\, dt$$

$$= \sqrt{a^2 (2 - 2\cos t)}\, dt.$$

Substituting for ds and y in the integrand, it follows that

$$T_c = \int_0^{\pi} \sqrt{\frac{a^2(2 - 2\cos t)}{2ga(1 - \cos t)}}\, dt \qquad y = a(1 - \cos t)$$

$$= \int_0^{\pi} \sqrt{\frac{a}{g}}\, dt = \pi\sqrt{\frac{a}{g}}.$$

This is the amount of time it takes the frictionless bead to slide down the cycloid to B after it is released from rest at O (see Figure 11.13).

Areas of Surfaces of Revolution

In Section 6.4 we found integral formulas for the area of a surface when a curve is revolved about a coordinate axis. Specifically, we found that the surface area is $S = \int 2\pi y \, ds$ for revolution about the x-axis, and $S = \int 2\pi x \, ds$ for revolution about the y-axis. If the curve is parametrized by the equations $x = f(t)$ and $y = g(t), a \leq t \leq b$, where f and g are continuously differentiable and $(f')^2 + (g')^2 > 0$ on $[a, b]$, then the arc length differential ds is given by Equation (4). This observation leads to the following formulas for area of surfaces of revolution for smooth parametrized curves.

Area of Surface of Revolution for Parametrized Curves

If a smooth curve $x = f(t), y = g(t), a \leq t \leq b$, is traversed exactly once as t increases from a to b, then the areas of the surfaces generated by revolving the curve about the coordinate axes are as follows.

1. Revolution about the x-axis ($y \geq 0$):

$$S = \int_a^b 2\pi y \sqrt{\left(\frac{dx}{dt}\right)^2 + \left(\frac{dy}{dt}\right)^2} \, dt \qquad (5)$$

2. Revolution about the y-axis ($x \geq 0$):

$$S = \int_a^b 2\pi x \sqrt{\left(\frac{dx}{dt}\right)^2 + \left(\frac{dy}{dt}\right)^2} \, dt \qquad (6)$$

As with length, we can calculate surface area from any convenient parametrization that meets the stated criteria.

EXAMPLE 9   The standard parametrization of the circle of radius 1 centered at the point (0, 1) in the xy-plane is

$$x = \cos t, \qquad y = 1 + \sin t, \qquad 0 \leq t \leq 2\pi.$$

Use this parametrization to find the area of the surface swept out by revolving the circle about the x-axis (Figure 11.19).

Solution   We evaluate the formula

$$S = \int_a^b 2\pi y \sqrt{\left(\frac{dx}{dt}\right)^2 + \left(\frac{dy}{dt}\right)^2} \, dt \qquad \text{Eq. (5) for revolution about the } x\text{-axis: } y = 1 + \sin t \geq 0$$

$$= \int_0^{2\pi} 2\pi(1 + \sin t) \sqrt{(-\sin t)^2 + (\cos t)^2} \, dt$$

$$= 2\pi \int_0^{2\pi} (1 + \sin t) \, dt$$

$$= 2\pi \Big[t - \cos t \Big]_0^{2\pi} = 4\pi^2.$$

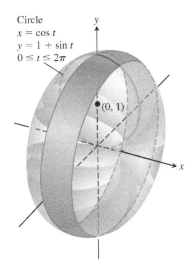

Circle
$x = \cos t$
$y = 1 + \sin t$
$0 \leq t \leq 2\pi$

(0, 1)

FIGURE 11.19   In Example 9 we calculate the area of the surface of revolution swept out by this parametrized curve.

EXERCISES **11.2**

Tangents to Parametrized Curves

In Exercises 1–14, find an equation for the line tangent to the curve at the point defined by the given value of t. Also, find the value of d^2y/dx^2 at this point.

1. $x = 2 \cos t$, $y = 2 \sin t$, $t = \pi/4$

2. $x = \sin 2\pi t$, $y = \cos 2\pi t$, $t = -1/6$

3. $x = 4 \sin t$, $y = 2 \cos t$, $t = \pi/4$

4. $x = \cos t$, $y = \sqrt{3} \cos t$, $t = 2\pi/3$

5. $x = t$, $y = \sqrt{t}$, $t = 1/4$

6. $x = \sec^2 t - 1$, $y = \tan t$, $t = -\pi/4$

7. $x = \sec t$, $y = \tan t$, $t = \pi/6$

8. $x = -\sqrt{t+1}$, $y = \sqrt{3t}$, $t = 3$

9. $x = 2t^2 + 3$, $y = t^4$, $t = -1$

10. $x = 1/t$, $y = -2 + \ln t$, $t = 1$

11. $x = t - \sin t$, $y = 1 - \cos t$, $t = \pi/3$

12. $x = \cos t$, $y = 1 + \sin t$, $t = \pi/2$

13. $x = \dfrac{1}{t+1}$, $y = \dfrac{t}{t-1}$, $t = 2$

14. $x = t + e^t$, $y = 1 - e^t$, $t = 0$

Implicitly Defined Parametrizations

Assuming that the equations in Exercises 15–20 define x and y implicitly as differentiable functions $x = f(t)$, $y = g(t)$, find the slope of the curve $x = f(t)$, $y = g(t)$ at the given value of t.

15. $x^3 + 2t^2 = 9$, $2y^3 - 3t^2 = 4$, $t = 2$

16. $x = \sqrt{5 - \sqrt{t}}$, $y(t - 1) = \sqrt{t}$, $t = 4$

17. $x + 2x^{3/2} = t^2 + t$, $y\sqrt{t+1} + 2t\sqrt{y} = 4$, $t = 0$

18. $x \sin t + 2x = t$, $t \sin t - 2t = y$, $t = \pi$

19. $x = t^3 + t$, $y + 2t^3 = 2x + t^2$, $t = 1$

20. $t = \ln(x - t)$, $y = te^t$, $t = 0$

Area

21. Find the area under one arch of the cycloid

$$x = a(t - \sin t), \quad y = a(1 - \cos t).$$

22. Find the area enclosed by the y-axis and the curve

$$x = t - t^2, \quad y = 1 + e^{-t}.$$

23. Find the area enclosed by the ellipse

$$x = a \cos t, \quad y = b \sin t, \quad 0 \le t \le 2\pi.$$

24. Find the area under $y = x^3$ over $[0, 1]$ using the following parametrizations.

  a. $x = t^2$, $y = t^6$ b. $x = t^3$, $y = t^9$

Lengths of Curves

Find the lengths of the curves in Exercises 25–30.

25. $x = \cos t$, $y = t + \sin t$, $0 \le t \le \pi$

26. $x = t^3$, $y = 3t^2/2$, $0 \le t \le \sqrt{3}$

27. $x = t^2/2$, $y = (2t + 1)^{3/2}/3$, $0 \le t \le 4$

28. $x = (2t + 3)^{3/2}/3$, $y = t + t^2/2$, $0 \le t \le 3$

29. $x = 8 \cos t + 8t \sin t$
    $y = 8 \sin t - 8t \cos t$,
    $0 \le t \le \pi/2$

30. $x = \ln(\sec t + \tan t) - \sin t$
    $y = \cos t$, $0 \le t \le \pi/3$

Surface Area

Find the areas of the surfaces generated by revolving the curves in Exercises 31–34 about the indicated axes.

31. $x = \cos t$, $y = 2 + \sin t$, $0 \le t \le 2\pi$; x-axis

32. $x = (2/3)t^{3/2}$, $y = 2\sqrt{t}$, $0 \le t \le \sqrt{3}$; y-axis

33. $x = t + \sqrt{2}$, $y = (t^2/2) + \sqrt{2}t$, $-\sqrt{2} \le t \le \sqrt{2}$; y-axis

34. $x = \ln(\sec t + \tan t) - \sin t$, $y = \cos t$, $0 \le t \le \pi/3$; x-axis

35. **A cone frustum**   The line segment joining the points $(0, 1)$ and $(2, 2)$ is revolved about the x-axis to generate a frustum of a cone. Find the surface area of the frustum using the parametrization $x = 2t, y = t + 1, 0 \le t \le 1$. Check your result with the geometry formula: Area $= \pi(r_1 + r_2)$(slant height).

36. **A cone**   The line segment joining the origin to the point (h, r) is revolved about the x-axis to generate a cone of height h and base radius r. Find the cone's surface area with the parametric equations $x = ht, y = rt, 0 \le t \le 1$. Check your result with the geometry formula: Area $= \pi r$(slant height).

Centroids

37. Find the coordinates of the centroid of the curve

$$x = \cos t + t \sin t, \quad y = \sin t - t \cos t, \quad 0 \le t \le \pi/2.$$

38. Find the coordinates of the centroid of the curve

$$x = e^t \cos t, \quad y = e^t \sin t, \quad 0 \le t \le \pi.$$

39. Find the coordinates of the centroid of the curve

$$x = \cos t, \quad y = t + \sin t, \quad 0 \le t \le \pi.$$

T 40. Most centroid calculations for curves are done with a calculator or computer that has an integral evaluation program. As a case in point, find, to the nearest hundredth, the coordinates of the centroid of the curve

$$x = t^3, \quad y = 3t^2/2, \quad 0 \le t \le \sqrt{3}.$$

Theory and Examples

41. **Length is independent of parametrization**   To illustrate the fact that the numbers we get for length do not depend on the way we parametrize our curves (except for the mild restrictions preventing doubling back mentioned earlier), calculate the length of the semicircle $y = \sqrt{1 - x^2}$ with these two different parametrizations:

  a. $x = \cos 2t$, $y = \sin 2t$, $0 \le t \le \pi/2$.

  b. $x = \sin \pi t$, $y = \cos \pi t$, $-1/2 \le t \le 1/2$.

42. a. Show that the Cartesian formula

$$L = \int_c^d \sqrt{1 + \left(\frac{dx}{dy}\right)^2} \, dy$$

for the length of the curve $x = g(y)$, $c \le y \le d$ (Section 6.3, Equation 4), is a special case of the parametric length formula

$$L = \int_a^b \sqrt{\left(\frac{dx}{dt}\right)^2 + \left(\frac{dy}{dt}\right)^2}\, dt.$$

Use this result to find the length of each curve.

b. $x = y^{3/2}$, $0 \le y \le 4/3$

c. $x = \frac{3}{2}y^{2/3}$, $0 \le y \le 1$

43. The curve with parametric equations

$$x = (1 + 2\sin\theta)\cos\theta, \quad y = (1 + 2\sin\theta)\sin\theta$$

is called a *limaçon* and is shown in the accompanying figure. Find the points (x, y) and the slopes of the tangent lines at these points for

a. $\theta = 0$.   **b.** $\theta = \pi/2$.   **c.** $\theta = 4\pi/3$.

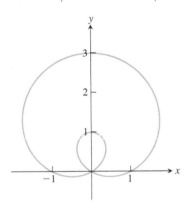

44. The curve with parametric equations

$$x = t, \quad y = 1 - \cos t, \quad 0 \le t \le 2\pi$$

is called a *sinusoid* and is shown in the accompanying figure. Find the point (x, y) where the slope of the tangent line is

a. largest.   **b.** smallest.

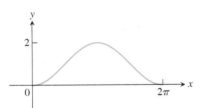

T The curves in Exercises 45 and 46 are called *Bowditch curves* or *Lissajous figures.* In each case, find the point in the interior of the first quadrant where the tangent to the curve is horizontal, and find the equations of the two tangents at the origin.

45.
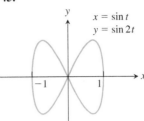

46.
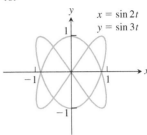

47. Cycloid

a. Find the length of one arch of the cycloid

$$x = a(t - \sin t), \quad y = a(1 - \cos t).$$

b. Find the area of the surface generated by revolving one arch of the cycloid in part (a) about the x-axis for $a = 1$.

48. Volume   Find the volume swept out by revolving the region bounded by the x-axis and one arch of the cycloid

$$x = t - \sin t, \quad y = 1 - \cos t$$

about the x-axis.

49. Find the volume swept out by revolving the region bounded by the x-axis and the graph of

$$x = 2t, \quad y = t(2 - t)$$

about the x-axis.

50. Find the volume swept out by revolving the region bounded by the y-axis and the graph of

$$x = t(1 - t), \quad y = 1 + t^2$$

about the y-axis.

COMPUTER EXPLORATIONS

In Exercises 51–54, use a CAS to perform the following steps for the given curve over the closed interval.

a. Plot the curve together with the polygonal path approximations for $n = 2, 4, 8$ partition points over the interval. (See Figure 11.16.)

b. Find the corresponding approximation to the length of the curve by summing the lengths of the line segments.

c. Evaluate the length of the curve using an integral. Compare your approximations for $n = 2, 4, 8$ with the actual length given by the integral. How does the actual length compare with the approximations as n increases? Explain your answer.

51. $x = \frac{1}{3}t^3$, $y = \frac{1}{2}t^2$, $0 \le t \le 1$

52. $x = 2t^3 - 16t^2 + 25t + 5$, $y = t^2 + t - 3$, $0 \le t \le 6$

53. $x = t - \cos t$, $y = 1 + \sin t$, $-\pi \le t \le \pi$

54. $x = e^t \cos t$, $y = e^t \sin t$, $0 \le t \le \pi$

11.3 Polar Coordinates

In this section we study polar coordinates and their relation to Cartesian coordinates. You will see that polar coordinates are very useful for calculating many multiple integrals studied in Chapter 15. They are also useful in describing the paths of planets and satellites.

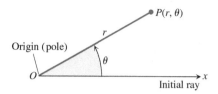

FIGURE 11.20 To define polar coordinates for the plane, we start with an origin, called the pole, and an initial ray.

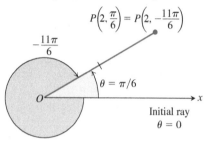

FIGURE 11.21 Polar coordinates are not unique.

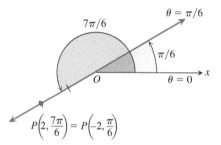

FIGURE 11.22 Polar coordinates can have negative r-values.

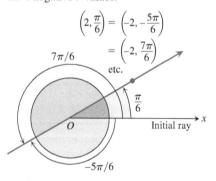

FIGURE 11.23 The point $P(2, \pi/6)$ has infinitely many polar coordinate pairs (Example 1).

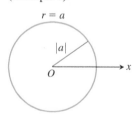

FIGURE 11.24 The polar equation for a circle is $r = a$.

Definition of Polar Coordinates

To define polar coordinates, we first fix an **origin** O (called the **pole**) and an **initial ray** from O (Figure 11.20). Usually the positive x-axis is chosen as the initial ray. Then each point P can be located by assigning to it a **polar coordinate pair** (r, θ) in which r gives the directed distance from O to P and θ gives the directed angle from the initial ray to ray OP. So we label the point P as

$$P(r, \theta)$$

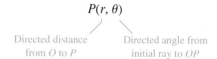

As in trigonometry, θ is positive when measured counterclockwise and negative when measured clockwise. The angle associated with a given point is not unique. While a point in the plane has just one pair of Cartesian coordinates, it has infinitely many pairs of polar coordinates. For instance, the point 2 units from the origin along the ray $\theta = \pi/6$ has polar coordinates $r = 2$, $\theta = \pi/6$. It also has coordinates $r = 2, \theta = -11\pi/6$ (Figure 11.21). In some situations we allow r to be negative. That is why we use directed distance in defining $P(r, \theta)$. The point $P(2, 7\pi/6)$ can be reached by turning $7\pi/6$ radians counterclockwise from the initial ray and going forward 2 units (Figure 11.22). It can also be reached by turning $\pi/6$ radians counterclockwise from the initial ray and going *backward* 2 units. So the point also has polar coordinates $r = -2, \theta = \pi/6$.

EXAMPLE 1 Find all the polar coordinates of the point $P(2, \pi/6)$.

Solution We sketch the initial ray of the coordinate system, draw the ray from the origin that makes an angle of $\pi/6$ radians with the initial ray, and mark the point $(2, \pi/6)$ (Figure 11.23). We then find the angles for the other coordinate pairs of P in which $r = 2$ and $r = -2$.

For $r = 2$, the complete list of angles is

$$\frac{\pi}{6}, \quad \frac{\pi}{6} \pm 2\pi, \quad \frac{\pi}{6} \pm 4\pi, \quad \frac{\pi}{6} \pm 6\pi, \dots.$$

For $r = -2$, the angles are

$$-\frac{5\pi}{6}, \quad -\frac{5\pi}{6} \pm 2\pi, \quad -\frac{5\pi}{6} \pm 4\pi, \quad -\frac{5\pi}{6} \pm 6\pi, \dots.$$

The corresponding coordinate pairs of P are

$$\left(2, \frac{\pi}{6} + 2n\pi\right), \qquad n = 0, \pm 1, \pm 2, \dots$$

and

$$\left(-2, -\frac{5\pi}{6} + 2n\pi\right), \qquad n = 0, \pm 1, \pm 2, \dots.$$

When $n = 0$, the formulas give $(2, \pi/6)$ and $(-2, -5\pi/6)$. When $n = 1$, they give $(2, 13\pi/6)$ and $(-2, 7\pi/6)$, and so on. ∎

Polar Equations and Graphs

If we hold r fixed at a constant value $r = a \neq 0$, the point $P(r, \theta)$ will lie $|a|$ units from the origin O. As θ varies over any interval of length 2π, P then traces a circle of radius $|a|$ centered at O (Figure 11.24).

If we hold θ fixed at a constant value $\theta = \theta_0$ and let r vary between $-\infty$ and ∞, the point $P(r, \theta)$ traces the line through O that makes an angle of measure θ_0 with the initial ray. (See Figure 11.22 for an example.)

(a)

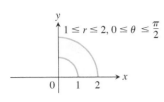

(b)

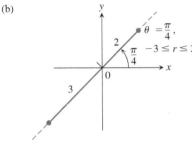

(c)

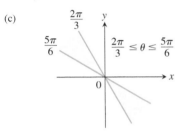

FIGURE 11.25 The graphs of typical inequalities in r and θ (Example 3).

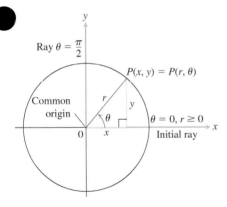

FIGURE 11.26 The usual way to relate polar and Cartesian coordinates.

EXAMPLE 2   A circle or line can have more than one polar equation.

(a) $r = 1$ and $r = -1$ are equations for the circle of radius 1 centered at O.

(b) $\theta = \pi/6$, $\theta = 7\pi/6$, and $\theta = -5\pi/6$ are equations for the line in Figure 11.23.   ■

Equations of the form $r = a$ and $\theta = \theta_0$ can be combined to define regions, segments, and rays.

EXAMPLE 3   Graph the sets of points whose polar coordinates satisfy the following conditions.

(a) $1 \le r \le 2$ and $0 \le \theta \le \dfrac{\pi}{2}$

(b) $-3 \le r \le 2$ and $\theta = \dfrac{\pi}{4}$

(c) $\dfrac{2\pi}{3} \le \theta \le \dfrac{5\pi}{6}$ (no restriction on r)

Solution   The graphs are shown in Figure 11.25.   ■

Relating Polar and Cartesian Coordinates

When we use both polar and Cartesian coordinates in a plane, we place the two origins together and let the initial polar ray be the positive x-axis. The ray $\theta = \pi/2$, $r > 0$, becomes the positive y-axis (Figure 11.26). The two coordinate systems are then related by the following equations.

Equations Relating Polar and Cartesian Coordinates

$$x = r \cos \theta, \qquad y = r \sin \theta, \qquad r^2 = x^2 + y^2, \qquad \tan \theta = \frac{y}{x}$$

The first two of these equations uniquely determine the Cartesian coordinates x and y given the polar coordinates r and θ. On the other hand, if x and y are given, the third equation gives two possible choices for r (a positive and a negative value). For each $(x, y) \neq (0, 0)$, there is a unique $\theta \in [0, 2\pi)$ satisfying the first two equations, each then giving a polar coordinate representation of the Cartesian point (x, y). The other polar coordinate representations for the point can be determined from these two, as in Example 1.

EXAMPLE 4   Here are some plane curves expressed in terms of both polar coordinate and Cartesian coordinate equations.

Polar equation	Cartesian equivalent
$r \cos \theta = 2$	$x = 2$
$r^2 \cos \theta \sin \theta = 4$	$xy = 4$
$r^2 \cos^2\theta - r^2 \sin^2\theta = 1$	$x^2 - y^2 = 1$
$r = 1 + 2r \cos \theta$	$y^2 - 3x^2 - 4x - 1 = 0$
$r = 1 - \cos \theta$	$x^4 + y^4 + 2x^2y^2 + 2x^3 + 2xy^2 - y^2 = 0$

Some curves are more simply expressed with polar coordinates; others are not.   ■

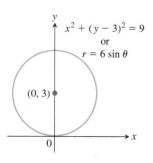

FIGURE 11.27 The circle in Example 5.

EXAMPLE 5   Find a polar equation for the circle $x^2 + (y - 3)^2 = 9$ (Figure 11.27).

Solution   We apply the equations relating polar and Cartesian coordinates:

$$x^2 + (y - 3)^2 = 9$$
$$x^2 + y^2 - 6y + 9 = 9 \qquad \text{Expand } (y-3)^2.$$
$$x^2 + y^2 - 6y = 0 \qquad \text{Cancelation}$$
$$r^2 - 6r \sin \theta = 0 \qquad x^2 + y^2 = r^2, \; y = r \sin \theta$$
$$r = 0 \quad \text{or} \quad r - 6 \sin \theta = 0$$
$$r = 6 \sin \theta \qquad \text{Includes both possibilities} \qquad \blacksquare$$

EXAMPLE 6   Replace the following polar equations by equivalent Cartesian equations and identify their graphs.

(a) $r \cos \theta = -4$

(b) $r^2 = 4r \cos \theta$

(c) $r = \dfrac{4}{2 \cos \theta - \sin \theta}$

Solution   We use the substitutions $r \cos \theta = x$, $r \sin \theta = y$, and $r^2 = x^2 + y^2$.

(a) $r \cos \theta = -4$

The Cartesian equation: $r \cos \theta = -4$
$$x = -4 \qquad \text{Substitute.}$$

The graph: Vertical line through $x = -4$ on the x-axis

(b) $r^2 = 4r \cos \theta$

The Cartesian equation: $r^2 = 4r \cos \theta$
$$x^2 + y^2 = 4x \qquad \text{Substitute.}$$
$$x^2 - 4x + y^2 = 0$$
$$x^2 - 4x + 4 + y^2 = 4 \qquad \text{Complete the square.}$$
$$(x - 2)^2 + y^2 = 4 \qquad \text{Factor.}$$

The graph: Circle, radius 2, center $(h, k) = (2, 0)$

(c) $r = \dfrac{4}{2 \cos \theta - \sin \theta}$

The Cartesian equation: $r(2 \cos \theta - \sin \theta) = 4$
$$2r \cos \theta - r \sin \theta = 4 \qquad \text{Multiply by } r.$$
$$2x - y = 4 \qquad \text{Substitute.}$$
$$y = 2x - 4 \qquad \text{Solve for } y.$$

The graph: Line, slope $m = 2$, y-intercept $b = -4$ $\blacksquare$

EXERCISES 11.3

Polar Coordinates

1. Which polar coordinate pairs label the same point?

a. $(3, 0)$ **b.** $(-3, 0)$ **c.** $(2, 2\pi/3)$

d. $(2, 7\pi/3)$ **e.** $(-3, \pi)$ **f.** $(2, \pi/3)$

g. $(-3, 2\pi)$ **h.** $(-2, -\pi/3)$

2. Which polar coordinate pairs label the same point?

a. $(-2, \pi/3)$ **b.** $(2, -\pi/3)$ **c.** (r, θ)

d. $(r, \theta + \pi)$ **e.** $(-r, \theta)$ **f.** $(2, -2\pi/3)$

g. $(-r, \theta + \pi)$ **h.** $(-2, 2\pi/3)$

3. Plot the following points (given in polar coordinates). Then find all the polar coordinates of each point.

   a. $(2, \pi/2)$ **b.** $(2, 0)$

   c. $(-2, \pi/2)$ **d.** $(-2, 0)$

4. Plot the following points (given in polar coordinates). Then find all the polar coordinates of each point.

   a. $(3, \pi/4)$ **b.** $(-3, \pi/4)$

   c. $(3, -\pi/4)$ **d.** $(-3, -\pi/4)$

Polar to Cartesian Coordinates

5. Find the Cartesian coordinates of the points in Exercise 1.

6. Find the Cartesian coordinates of the following points (given in polar coordinates).

   a. $\left(\sqrt{2}, \pi/4\right)$ **b.** $(1, 0)$

   c. $(0, \pi/2)$ **d.** $\left(-\sqrt{2}, \pi/4\right)$

   e. $(-3, 5\pi/6)$ **f.** $(5, \tan^{-1}(4/3))$

   g. $(-1, 7\pi)$ **h.** $\left(2\sqrt{3}, 2\pi/3\right)$

Cartesian to Polar Coordinates

7. Find the polar coordinates, $0 \le \theta < 2\pi$ and $r \ge 0$, of the following points given in Cartesian coordinates.

   a. $(1, 1)$ **b.** $(-3, 0)$

   c. $\left(\sqrt{3}, -1\right)$ **d.** $(-3, 4)$

8. Find the polar coordinates, $-\pi \le \theta < \pi$ and $r \ge 0$, of the following points given in Cartesian coordinates.

   a. $(-2, -2)$ **b.** $(0, 3)$

   c. $\left(-\sqrt{3}, 1\right)$ **d.** $(5, -12)$

9. Find the polar coordinates, $0 \le \theta < 2\pi$ and $r \le 0$, of the following points given in Cartesian coordinates.

   a. $(3, 3)$ **b.** $(-1, 0)$

   c. $\left(-1, \sqrt{3}\right)$ **d.** $(4, -3)$

10. Find the polar coordinates, $-\pi \le \theta < \pi$ and $r \le 0$, of the following points given in Cartesian coordinates.

   a. $(-2, 0)$ **b.** $(1, 0)$

   c. $(0, -3)$ **d.** $\left(\dfrac{\sqrt{3}}{2}, \dfrac{1}{2}\right)$

Graphing Sets of Polar Coordinate Points

Graph the sets of points whose polar coordinates satisfy the equations and inequalities in Exercises 11–26.

11. $r = 2$ **12.** $0 \le r \le 2$

13. $r \ge 1$ **14.** $1 \le r \le 2$

15. $0 \le \theta \le \pi/6, \;\; r \ge 0$ **16.** $\theta = 2\pi/3, \;\; r \le -2$

17. $\theta = \pi/3, \;\; -1 \le r \le 3$ **18.** $\theta = 11\pi/4, \;\; r \ge -1$

19. $\theta = \pi/2, \;\; r \ge 0$ **20.** $\theta = \pi/2, \;\; r \le 0$

21. $0 \le \theta \le \pi, \;\; r = 1$ **22.** $0 \le \theta \le \pi, \;\; r = -1$

23. $\pi/4 \le \theta \le 3\pi/4, \;\; 0 \le r \le 1$

24. $-\pi/4 \le \theta \le \pi/4, \;\; -1 \le r \le 1$

25. $-\pi/2 \le \theta \le \pi/2, \;\; 1 \le r \le 2$

26. $0 \le \theta \le \pi/2, \;\; 1 \le |r| \le 2$

Polar to Cartesian Equations

Replace the polar equations in Exercises 27–52 with equivalent Cartesian equations. Then describe or identify the graph.

27. $r \cos \theta = 2$ **28.** $r \sin \theta = -1$

29. $r \sin \theta = 0$ **30.** $r \cos \theta = 0$

31. $r = 4 \csc \theta$ **32.** $r = -3 \sec \theta$

33. $r \cos \theta + r \sin \theta = 1$ **34.** $r \sin \theta = r \cos \theta$

35. $r^2 = 1$ **36.** $r^2 = 4r \sin \theta$

37. $r = \dfrac{5}{\sin \theta - 2 \cos \theta}$ **38.** $r^2 \sin 2\theta = 2$

39. $r = \cot \theta \csc \theta$ **40.** $r = 4 \tan \theta \sec \theta$

41. $r = \csc \theta \, e^{r \cos \theta}$ **42.** $r \sin \theta = \ln r + \ln \cos \theta$

43. $r^2 + 2r^2 \cos \theta \sin \theta = 1$ **44.** $\cos^2 \theta = \sin^2 \theta$

45. $r^2 = -4r \cos \theta$ **46.** $r^2 = -6r \sin \theta$

47. $r = 8 \sin \theta$ **48.** $r = 3 \cos \theta$

49. $r = 2 \cos \theta + 2 \sin \theta$ **50.** $r = 2 \cos \theta - \sin \theta$

51. $r \sin \left(\theta + \dfrac{\pi}{6}\right) = 2$

52. $r \sin \left(\dfrac{2\pi}{3} - \theta\right) = 5$

Cartesian to Polar Equations

Replace the Cartesian equations in Exercises 53–66 with equivalent polar equations.

53. $x = 7$ **54.** $y = 1$ **55.** $x = y$

56. $x - y = 3$ **57.** $x^2 + y^2 = 4$ **58.** $x^2 - y^2 = 1$

59. $\dfrac{x^2}{9} + \dfrac{y^2}{4} = 1$ **60.** $xy = 2$

61. $y^2 = 4x$ **62.** $x^2 + xy + y^2 = 1$

63. $x^2 + (y - 2)^2 = 4$ **64.** $(x - 5)^2 + y^2 = 25$

65. $(x - 3)^2 + (y + 1)^2 = 4$ **66.** $(x + 2)^2 + (y - 5)^2 = 16$

67. Find all polar coordinates of the origin.

68. Vertical and horizontal lines

   a. Show that every vertical line in the xy-plane has a polar equation of the form $r = a \sec \theta$.

   b. Find the analogous polar equation for horizontal lines in the xy-plane.

11.4 Graphing Polar Coordinate Equations

It is often helpful to graph an equation expressed in polar coordinates in the Cartesian xy-plane. This section describes some techniques for graphing these equations using symmetries and tangents to the graph.

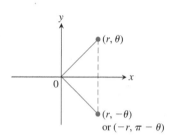

(a) About the x-axis

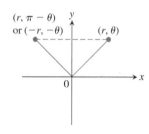

(b) About the y-axis

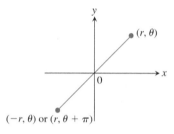

(c) About the origin

FIGURE 11.28 Three tests for symmetry in polar coordinates.

Symmetry

The following list shows how to test for three standard types of symmetries when using polar coordinates. These symmetries are illustrated in Figure 11.28.

Symmetry Tests for Polar Graphs in the Cartesian *xy*-Plane

1. *Symmetry about the x-axis:* If the point (r, θ) lies on the graph, then the point $(r, -\theta)$ or $(-r, \pi - \theta)$ lies on the graph (Figure 11.28a).

2. *Symmetry about the y-axis:* If the point (r, θ) lies on the graph, then the point $(r, \pi - \theta)$ or $(-r, -\theta)$ lies on the graph (Figure 11.28b).

3. *Symmetry about the origin:* If the point (r, θ) lies on the graph, then the point $(-r, \theta)$ or $(r, \theta + \pi)$ lies on the graph (Figure 11.28c).

Slope

The slope of a polar curve $r = f(\theta)$ in the *xy*-plane is dy/dx, but this is **not** given by the formula $r' = df/d\theta$. To see why, think of the graph of f as the graph of the parametric equations

$$x = r \cos \theta = f(\theta) \cos \theta, \qquad y = r \sin \theta = f(\theta) \sin \theta.$$

If f is a differentiable function of θ, then so are x and y and, when $dx/d\theta \neq 0$, we can calculate dy/dx from the parametric formula

$$\frac{dy}{dx} = \frac{dy/d\theta}{dx/d\theta} \qquad \text{Section 11.2, Eq. (1) with } t = \theta$$

$$= \frac{\dfrac{d}{d\theta}(f(\theta) \sin \theta)}{\dfrac{d}{d\theta}(f(\theta) \cos \theta)} \qquad \text{Substitute}$$

$$= \frac{\dfrac{df}{d\theta} \sin \theta + f(\theta) \cos \theta}{\dfrac{df}{d\theta} \cos \theta - f(\theta) \sin \theta} \qquad \text{Product Rule for derivatives}$$

Therefore we see that dy/dx is not the same as $df/d\theta$.

Slope of the Curve $r = f(\theta)$ in the Cartesian *xy*-Plane

$$\left.\frac{dy}{dx}\right|_{(r, \theta)} = \frac{f'(\theta) \sin \theta + f(\theta) \cos \theta}{f'(\theta) \cos \theta - f(\theta) \sin \theta} \qquad (1)$$

provided $dx/d\theta \neq 0$ at (r, θ).

If the curve $r = f(\theta)$ passes through the origin at $\theta = \theta_0$, then $f(\theta_0) = 0$, and the slope equation gives

$$\left.\frac{dy}{dx}\right|_{(0, \theta_0)} = \frac{f'(\theta_0) \sin \theta_0}{f'(\theta_0) \cos \theta_0} = \tan \theta_0.$$

That is, the slope at $(0, \theta_0)$ is $\tan \theta_0$. The reason we say "slope at $(0, \theta_0)$" and not just "slope at the origin" is that a polar curve may pass through the origin (or any point) more than once, with different slopes at different θ-values. This is not the case in our first example, however.

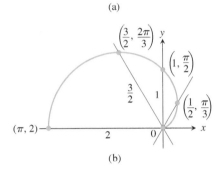

θ	$r = 1 - \cos\theta$
0	0
$\dfrac{\pi}{3}$	$\dfrac{1}{2}$
$\dfrac{\pi}{2}$	1
$\dfrac{2\pi}{3}$	$\dfrac{3}{2}$
π	2

(a)

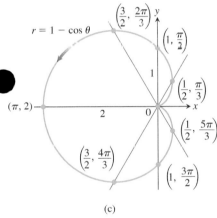

(b)

(c)

FIGURE 11.29 The steps in graphing the cardioid $r = 1 - \cos\theta$ (Example 1). The arrow shows the direction of increasing θ.

EXAMPLE 1   Graph the curve $r = 1 - \cos\theta$ in the Cartesian xy-plane.

Solution   The curve is symmetric about the x-axis because

$$(r, \theta) \text{ on the graph} \Rightarrow r = 1 - \cos\theta$$
$$\Rightarrow r = 1 - \cos(-\theta) \qquad \cos\theta = \cos(-\theta)$$
$$\Rightarrow (r, -\theta) \text{ on the graph.}$$

As θ increases from 0 to π, $\cos\theta$ decreases from 1 to -1, and $r = 1 - \cos\theta$ increases from a minimum value of 0 to a maximum value of 2. As θ continues on from π to 2π, $\cos\theta$ increases from -1 back to 1 and r decreases from 2 back to 0. The curve starts to repeat when $\theta = 2\pi$ because the cosine has period 2π.

The curve leaves the origin with slope $\tan(0) = 0$ and returns to the origin with slope $\tan(2\pi) = 0$.

We make a table of values from $\theta = 0$ to $\theta = \pi$, plot the points, draw a smooth curve through them with a horizontal tangent at the origin, and reflect the curve across the x-axis to complete the graph (Figure 11.29). The curve is called a *cardioid* because of its heart shape. ∎

EXAMPLE 2   Graph the curve $r^2 = 4\cos\theta$ in the Cartesian xy-plane.

Solution   The equation $r^2 = 4\cos\theta$ requires $\cos\theta \geq 0$, so we get the entire graph by running θ from $-\pi/2$ to $\pi/2$. The curve is symmetric about the x-axis because

$$(r, \theta) \text{ on the graph} \Rightarrow r^2 = 4\cos\theta$$
$$\Rightarrow r^2 = 4\cos(-\theta) \qquad \cos\theta = \cos(-\theta)$$
$$\Rightarrow (r, -\theta) \text{ on the graph.}$$

The curve is also symmetric about the origin because

$$(r, \theta) \text{ on the graph} \Rightarrow r^2 = 4\cos\theta$$
$$\Rightarrow (-r)^2 = 4\cos\theta$$
$$\Rightarrow (-r, \theta) \text{ on the graph.}$$

Together, these two symmetries imply symmetry about the y-axis.

The curve passes through the origin when $\theta = -\pi/2$ and $\theta = \pi/2$. It has a vertical tangent both times because $\tan\theta$ is infinite.

For each value of θ in the interval between $-\pi/2$ and $\pi/2$, the formula $r^2 = 4\cos\theta$ gives two values of r:

$$r = \pm 2\sqrt{\cos\theta}.$$

We make a short table of values, plot the corresponding points, and use information about symmetry and tangents to guide us in connecting the points with a smooth curve (Figure 11.30).

θ	$\cos\theta$	$r = \pm 2\sqrt{\cos\theta}$
0	1	± 2
$\pm\dfrac{\pi}{6}$	$\dfrac{\sqrt{3}}{2}$	$\approx \pm 1.9$
$\pm\dfrac{\pi}{4}$	$\dfrac{1}{\sqrt{2}}$	$\approx \pm 1.7$
$\pm\dfrac{\pi}{3}$	$\dfrac{1}{2}$	$\approx \pm 1.4$
$\pm\dfrac{\pi}{2}$	0	0

(a)

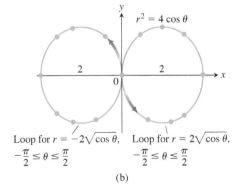

Loop for $r = -2\sqrt{\cos\theta}$,   Loop for $r = 2\sqrt{\cos\theta}$,
$-\dfrac{\pi}{2} \leq \theta \leq \dfrac{\pi}{2}$ $-\dfrac{\pi}{2} \leq \theta \leq \dfrac{\pi}{2}$

(b)

FIGURE 11.30 The graph of $r^2 = 4\cos\theta$. The arrows show the direction of increasing θ. The values of r in the table are rounded (Example 2). ∎

(a)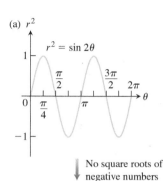

(b)

(c)

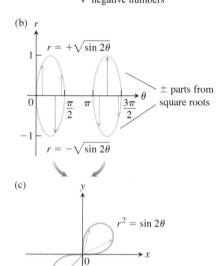

FIGURE 11.31 To plot $r = f(\theta)$ in the Cartesian $r\theta$-plane in (b), we first plot $r^2 = \sin 2\theta$ in the $r^2\theta$-plane in (a) and then ignore the values of θ for which $\sin 2\theta$ is negative. The radii from the sketch in (b) cover the polar graph of the lemniscate in (c) twice (Example 3).

Converting a Graph from the $r\theta$- to xy-Plane

One way to graph a polar equation $r = f(\theta)$ in the xy-plane is to make a table of (r, θ)-values, plot the corresponding points there, and connect them in order of increasing θ. This can work well if enough points have been plotted to reveal all the loops and dimples in the graph. Another method of graphing is to

1. first graph the function $r = f(\theta)$ in the *Cartesian* $r\theta$-plane,
2. then use that Cartesian graph as a "table" and guide to sketch the *polar* coordinate graph in the xy-plane.

This method is sometimes better than simple point plotting because the first Cartesian graph shows at a glance where r is positive, negative, and nonexistent, as well as where r is increasing and decreasing. Here is an example.

EXAMPLE 3   Graph the *lemniscate* curve $r^2 = \sin 2\theta$ in the Cartesian xy-plane.

Solution   For this example it will be easier to first plot r^2, instead of r, as a function of θ in the Cartesian $r^2\theta$-plane (see Figure 11.31a). We pass from there to the graph of $r = \pm\sqrt{\sin 2\theta}$ in the $r\theta$-plane (Figure 11.31b), and then draw the polar graph (Figure 11.31c). The graph in Figure 11.31b "covers" the final polar graph in Figure 11.31c twice. We could have managed with either loop alone, with the two upper halves, or with the two lower halves. The double covering does no harm, however, and we actually learn a little more about the behavior of the function this way.   ■

USING TECHNOLOGY   Graphing Polar Curves Parametrically

For complicated polar curves we may need to use a graphing calculator or computer to graph the curve. If the device does not plot polar graphs directly, we can convert $r = f(\theta)$ into parametric form using the equations

$$x = r \cos \theta = f(\theta) \cos \theta, \qquad y = r \sin \theta = f(\theta) \sin \theta.$$

Then we use the device to draw a parametrized curve in the Cartesian xy-plane.

EXERCISES 11.4

Symmetries and Polar Graphs
Identify the symmetries of the curves in Exercises 1–12. Then sketch the curves in the xy-plane.

1. $r = 1 + \cos \theta$ 2. $r = 2 - 2 \cos \theta$

3. $r = 1 - \sin \theta$ 4. $r = 1 + \sin \theta$

5. $r = 2 + \sin \theta$ 6. $r = 1 + 2 \sin \theta$

7. $r = \sin (\theta/2)$ 8. $r = \cos (\theta/2)$

9. $r^2 = \cos \theta$ 10. $r^2 = \sin \theta$

11. $r^2 = -\sin \theta$ 12. $r^2 = -\cos \theta$

Graph the lemniscates in Exercises 13–16. What symmetries do these curves have?

13. $r^2 = 4 \cos 2\theta$ 14. $r^2 = 4 \sin 2\theta$

15. $r^2 = -\sin 2\theta$ 16. $r^2 = -\cos 2\theta$

Slopes of Polar Curves in the xy-Plane
Find the slopes of the curves in Exercises 17–20 at the given points. Sketch the curves along with their tangents at these points.

17. **Cardioid** $r = -1 + \cos \theta; \quad \theta = \pm \pi/2$

18. **Cardioid** $r = -1 + \sin \theta; \quad \theta = 0, \pi$

19. **Four-leaved rose** $r = \sin 2\theta; \quad \theta = \pm \pi/4, \pm 3\pi/4$

20. **Four-leaved rose** $r = \cos 2\theta; \quad \theta = 0, \pm \pi/2, \pi$

Concavity of Polar Curves in the *xy*-Plane

Equation (1) gives the formula for the derivative y' of a polar curve $r = f(\theta)$. The second derivative is $\dfrac{d^2y}{dx^2} = \dfrac{dy'/d\theta}{dx/d\theta}$ (see Equation (2) in Section 11.2). Find the slope and concavity of the curves in Exercises 21–24 at the given points.

21. $r = \sin\theta, \quad \theta = \pi/6, \pi/3$ 22. $r = e^\theta, \quad \theta = 0, \pi$

23. $r = \theta, \quad \theta = 0, \pi/2$ 24. $r = 1/\theta, \quad \theta = -\pi, 1$

Graphing Limaçons

Graph the limaçons in Exercises 25–28. Limaçon ("*lee*-ma-sahn") is Old French for "snail." You will understand the name when you graph the limaçons in Exercise 25. Equations for limaçons have the form $r = a \pm b \cos\theta$ or $r = a \pm b \sin\theta$. There are four basic shapes.

25. **Limaçons with an inner loop**

   a. $r = \dfrac{1}{2} + \cos\theta$ b. $r = \dfrac{1}{2} + \sin\theta$

26. **Cardioids**

   a. $r = 1 - \cos\theta$ b. $r = -1 + \sin\theta$

27. **Dimpled limaçons**

   a. $r = \dfrac{3}{2} + \cos\theta$ b. $r = \dfrac{3}{2} - \sin\theta$

28. **Oval limaçons**

   a. $r = 2 + \cos\theta$ b. $r = -2 + \sin\theta$

Graphing Polar Regions and Curves in the *xy*-Plane

29. Sketch the region defined by the inequalities $-1 \le r \le 2$ and $-\pi/2 \le \theta \le \pi/2$.

30. Sketch the region defined by the inequalities $0 \le r \le 2\sec\theta$ and $-\pi/4 \le \theta \le \pi/4$.

In Exercises 31 and 32, sketch the region defined by the inequality.

31. $0 \le r \le 2 - 2\cos\theta$ 32. $0 \le r^2 \le \cos\theta$

T 33. Which of the following has the same graph as $r = 1 - \cos\theta$?

   a. $r = -1 - \cos\theta$ b. $r = 1 + \cos\theta$

   Confirm your answer with algebra.

T 34. Which of the following has the same graph as $r = \cos 2\theta$?

   a. $r = -\sin(2\theta + \pi/2)$ b. $r = -\cos(\theta/2)$

   Confirm your answer with algebra.

T 35. **A rose within a rose**   Graph the equation $r = 1 - 2\sin 3\theta$.

T 36. **The nephroid of Freeth**   Graph the nephroid of Freeth:

$$r = 1 + 2\sin\frac{\theta}{2}.$$

T 37. **Roses**   Graph the roses $r = \cos m\theta$ for $m = 1/3, 2, 3,$ and 7.

T 38. **Spirals**   Polar coordinates are just the thing for defining spirals. Graph the following spirals.

   a. $r = \theta$

   b. $r = -\theta$

   c. *A logarithmic spiral:* $r = e^{\theta/10}$

   d. *A hyperbolic spiral:* $r = 8/\theta$

   e. *An equilateral hyperbola:* $r = \pm 10/\sqrt{\theta}$

   (Use different colors for the two branches.)

T 39. Graph the equation $r = \sin\left(\frac{8}{7}\theta\right)$ for $0 \le \theta \le 14\pi$.

T 40. Graph the equation

$$r = \sin^2(2.3\theta) + \cos^4(2.3\theta)$$

for $0 \le \theta \le 10\pi$.

11.5 Areas and Lengths in Polar Coordinates

This section shows how to calculate areas of plane regions and lengths of curves in polar coordinates.

Area in the Plane

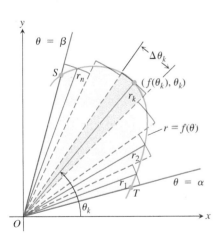

FIGURE 11.32 To derive a formula for the area of region *OTS*, we approximate the region with fan-shaped circular sectors.

The region *OTS* in Figure 11.32 is bounded by the rays $\theta = \alpha$ and $\theta = \beta$ and the curve $r = f(\theta)$. We approximate the region with n nonoverlapping fan-shaped circular sectors based on a partition P of angle *TOS*. The typical sector has radius $r_k = f(\theta_k)$ and central angle of radian measure $\Delta\theta_k$. Its area is $\Delta\theta_k/2\pi$ times the area of a circle of radius r_k, or

$$A_k = \frac{1}{2}r_k^2\,\Delta\theta_k = \frac{1}{2}\left(f(\theta_k)\right)^2\Delta\theta_k.$$

The area of region *OTS* is approximately

$$\sum_{k=1}^{n} A_k = \sum_{k=1}^{n}\frac{1}{2}\left(f(\theta_k)\right)^2\Delta\theta_k.$$

If f is continuous, we expect the approximations to improve as the norm of the partition P goes to zero, where the norm of P is the largest value of $\Delta\theta_k$. We are therefore led to the following formula for the region's area:

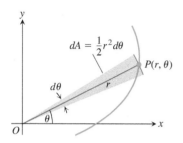

FIGURE 11.33 The area differential dA for the curve $r = f(\theta)$.

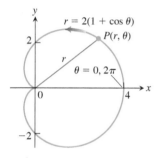

FIGURE 11.34 The cardioid in Example 1.

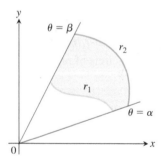

FIGURE 11.35 The area of the shaded region is calculated by subtracting the area of the region between r_1 and the origin from the area of the region between r_2 and the origin.

$$A = \lim_{\|P\| \to 0} \sum_{k=1}^{n} \frac{1}{2}\big(f(\theta_k)\big)^2 \, \Delta\theta_k = \int_{\alpha}^{\beta} \frac{1}{2}\big(f(\theta)\big)^2 \, d\theta.$$

Area of the Fan-Shaped Region Between the Origin and the Curve
$r = f(\theta)$ when $\alpha \le \theta \le \beta, r \ge 0$, and $\beta - \alpha \le 2\pi$.

$$A = \int_{\alpha}^{\beta} \frac{1}{2} r^2 \, d\theta$$

This is the integral of the **area differential** (Figure 11.33)

$$dA = \frac{1}{2} r^2 \, d\theta = \frac{1}{2}\big(f(\theta)\big)^2 \, d\theta.$$

In the area formula above, we assumed that $r \ge 0$ and that the region does not sweep out an angle of more than 2π. This avoids issues with negatively signed areas or with regions that overlap themselves. More general regions can usually be handled by subdividing them into regions of this type if necessary.

EXAMPLE 1   Find the area of the region in the xy-plane enclosed by the cardioid $r = 2(1 + \cos\theta)$.

Solution   We graph the cardioid (Figure 11.34) and determine that the radius OP sweeps out the region exactly once as θ runs from 0 to 2π. The area is therefore

$$\int_{\theta=0}^{\theta=2\pi} \frac{1}{2} r^2 \, d\theta = \int_{0}^{2\pi} \frac{1}{2} \cdot 4(1 + \cos\theta)^2 \, d\theta$$

$$= \int_{0}^{2\pi} 2(1 + 2\cos\theta + \cos^2\theta) \, d\theta$$

$$= \int_{0}^{2\pi} \left(2 + 4\cos\theta + 2 \cdot \frac{1 + \cos 2\theta}{2}\right) d\theta$$

$$= \int_{0}^{2\pi} (3 + 4\cos\theta + \cos 2\theta) \, d\theta$$

$$= \left[3\theta + 4\sin\theta + \frac{\sin 2\theta}{2}\right]_{0}^{2\pi} = 6\pi - 0 = 6\pi. \qquad \blacksquare$$

To find the area of a region like the one in Figure 11.35, which lies between two polar curves $r_1 = r_1(\theta)$ and $r_2 = r_2(\theta)$ from $\theta = \alpha$ to $\theta = \beta$, we subtract the integral of $(1/2)r_1^2 \, d\theta$ from the integral of $(1/2)r_2^2 \, d\theta$. This leads to the following formula.

Area of the Region $0 \le r_1(\theta) \le r \le r_2(\theta), \alpha \le \theta \le \beta$, and $\beta - \alpha \le 2\pi$.

$$A = \int_{\alpha}^{\beta} \frac{1}{2} r_2^2 \, d\theta - \int_{\alpha}^{\beta} \frac{1}{2} r_1^2 \, d\theta = \int_{\alpha}^{\beta} \frac{1}{2}\big(r_2^2 - r_1^2\big) \, d\theta \qquad (1)$$

EXAMPLE 2   Find the area of the region that lies inside the circle $r = 1$ and outside the cardioid $r = 1 - \cos\theta$.

Solution   We sketch the region to determine its boundaries and find the limits of integration (Figure 11.36). The outer curve is $r_2 = 1$, the inner curve is $r_1 = 1 - \cos\theta$, and θ runs from $-\pi/2$ to $\pi/2$. The area, from Equation (1), is

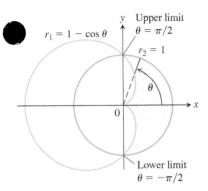

FIGURE 11.36 The region and limits of integration in Example 2.

$$A = \int_{-\pi/2}^{\pi/2} \frac{1}{2} \left(r_2{}^2 - r_1{}^2 \right) d\theta \qquad \text{Eq. (1)}$$

$$= 2 \int_0^{\pi/2} \frac{1}{2} \left(r_2{}^2 - r_1{}^2 \right) d\theta \qquad \text{Symmetry}$$

$$= \int_0^{\pi/2} (1 - (1 - 2\cos\theta + \cos^2\theta)) \, d\theta \qquad r_2 = 1 \text{ and } r_1 = 1 - \cos\theta$$

$$= \int_0^{\pi/2} (2\cos\theta - \cos^2\theta) \, d\theta = \int_0^{\pi/2} \left(2\cos\theta - \frac{1 + \cos 2\theta}{2} \right) d\theta$$

$$= \left[2\sin\theta - \frac{\theta}{2} - \frac{\sin 2\theta}{4} \right]_0^{\pi/2} = 2 - \frac{\pi}{4}. \qquad \blacksquare$$

The fact that we can represent a point in different ways in polar coordinates requires extra care in deciding when a point lies on the graph of a polar equation and in determining the points in which polar graphs intersect. (We needed intersection points in Example 2.) In Cartesian coordinates, we can always find the points where two curves cross by solving their equations simultaneously. In polar coordinates, the story is different. Simultaneous solution may reveal some intersection points without revealing others, so it is sometimes difficult to find all points of intersection of two polar curves. One way to identify all the points of intersection is to graph the equations.

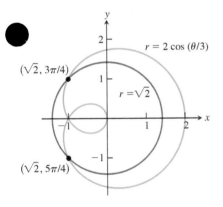

FIGURE 11.37 The curves $r = 2\cos(\theta/3)$ and $r = \sqrt{2}$ intersect at two points (Example 3).

EXAMPLE 3   Find all of the points where the curve $r = 2\cos(\theta/3)$ intersects the circle of radius $\sqrt{2}$ centered at the origin.

Solution   Note that the function $r = 2\cos(\theta/3)$ takes both positive and negative values. Therefore, when we look for the points where this curve intersects the circle, it is important to take into account that the circle is described both by the equation $r = \sqrt{2}$ *and* the equation $r = -\sqrt{2}$.

Solving $2\cos(\theta/3) = \sqrt{2}$ for θ yields:

$$2\cos(\theta/3) = \sqrt{2}, \quad \cos(\theta/3) = \sqrt{2}/2, \quad \theta/3 = \pi/4, \quad \theta = 3\pi/4.$$

This gives us one point, $\left(\sqrt{2}, 3\pi/4 \right)$, where the two curves intersect. However, as we can see by looking at the graphs in Figure 11.37, there is a second intersection point. To find the second point, we solve $2\cos(\theta/3) = -\sqrt{2}$ for θ:

$$2\cos(\theta/3) = -\sqrt{2}, \quad \cos(\theta/3) = -\sqrt{2}/2, \quad \theta/3 = 3\pi/4, \quad \theta = 9\pi/4.$$

The second intersection point is located at $\left(-\sqrt{2}, 9\pi/4 \right)$. We can specify this point in polar coordinates using a positive value of r and an angle between 0 and 2π. In polar coordinates, adding multiples of 2π to θ gives a second description of the same point in the plane. Similarly, changing the sign of r, while at the same time adding or subtracting π to θ, also gives a description of the same point. So in polar coordinates $\left(-\sqrt{2}, 9\pi/4 \right)$ describes the same point in the plane as $\left(-\sqrt{2}, \pi/4 \right)$ and also as $\left(\sqrt{2}, 5\pi/4 \right)$. The second intersection point is located at $\left(\sqrt{2}, 5\pi/4 \right)$. $\blacksquare$

Length of a Polar Curve

We can obtain a polar coordinate formula for the length of a curve $r = f(\theta)$, $\alpha \le \theta \le \beta$, by parametrizing the curve as

$$x = r\cos\theta = f(\theta)\cos\theta, \qquad y = r\sin\theta = f(\theta)\sin\theta, \qquad \alpha \le \theta \le \beta. \qquad (2)$$

The parametric length formula, Equation (3) from Section 11.2, then gives the length as

$$L = \int_\alpha^\beta \sqrt{\left(\frac{dx}{d\theta}\right)^2 + \left(\frac{dy}{d\theta}\right)^2}\, d\theta.$$

This equation becomes

$$L = \int_\alpha^\beta \sqrt{r^2 + \left(\frac{dr}{d\theta}\right)^2}\, d\theta$$

when Equations (2) are substituted for x and y (Exercise 29).

Length of a Polar Curve

If $r = f(\theta)$ has a continuous first derivative for $\alpha \le \theta \le \beta$ and if the point $P(r, \theta)$ traces the curve $r = f(\theta)$ exactly once as θ runs from α to β, then the length of the curve is

$$L = \int_\alpha^\beta \sqrt{r^2 + \left(\frac{dr}{d\theta}\right)^2}\, d\theta. \qquad (3)$$

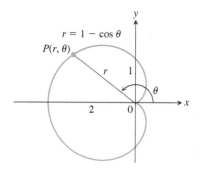

FIGURE 11.38 Calculating the length of a cardioid (Example 4).

EXAMPLE 4 Find the length of the cardioid $r = 1 - \cos\theta$.

Solution We sketch the cardioid to determine the limits of integration (Figure 11.38). The point $P(r, \theta)$ traces the curve once, counterclockwise as θ runs from 0 to 2π, so these are the values we take for α and β.

With

$$r = 1 - \cos\theta, \qquad \frac{dr}{d\theta} = \sin\theta,$$

we have

$$r^2 + \left(\frac{dr}{d\theta}\right)^2 = (1 - \cos\theta)^2 + (\sin\theta)^2$$

$$= 1 - 2\cos\theta + \underbrace{\cos^2\theta + \sin^2\theta}_{1} = 2 - 2\cos\theta$$

and

$$L = \int_\alpha^\beta \sqrt{r^2 + \left(\frac{dr}{d\theta}\right)^2}\, d\theta = \int_0^{2\pi} \sqrt{2 - 2\cos\theta}\, d\theta$$

$$= \int_0^{2\pi} \sqrt{4\sin^2\frac{\theta}{2}}\, d\theta \qquad 1 - \cos\theta = 2\sin^2(\theta/2)$$

$$= \int_0^{2\pi} 2\left|\sin\frac{\theta}{2}\right|\, d\theta$$

$$= \int_0^{2\pi} 2\sin\frac{\theta}{2}\, d\theta \qquad \sin(\theta/2) \ge 0 \quad \text{for} \quad 0 \le \theta \le 2\pi$$

$$= \left[-4\cos\frac{\theta}{2}\right]_0^{2\pi} = 4 + 4 = 8.$$

EXERCISES 11.5

Finding Polar Areas

Find the areas of the regions in Exercises 1–8.

1. Bounded by the spiral $r = \theta$ for $0 \le \theta \le \pi$

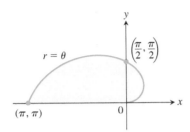

2. Bounded by the circle $r = 2 \sin \theta$ for $\pi/4 \le \theta \le \pi/2$

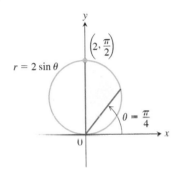

3. Inside the oval limaçon $r = 4 + 2 \cos \theta$

4. Inside the cardioid $r = a(1 + \cos \theta)$, $a > 0$

5. Inside one leaf of the four-leaved rose $r = \cos 2\theta$

6. Inside one leaf of the three-leaved rose $r = \cos 3\theta$

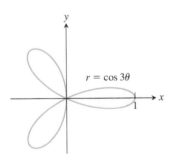

7. Inside one loop of the lemniscate $r^2 = 4 \sin 2\theta$

8. Inside the six-leaved rose $r^2 = 2 \sin 3\theta$

Find the areas of the regions in Exercises 9–18.

9. Shared by the circles $r = 2 \cos \theta$ and $r = 2 \sin \theta$

10. Shared by the circles $r = 1$ and $r = 2 \sin \theta$

11. Shared by the circle $r = 2$ and the cardioid $r = 2(1 - \cos \theta)$

12. Shared by the cardioids $r = 2(1 + \cos \theta)$ and $r = 2(1 - \cos \theta)$

13. Inside the lemniscate $r^2 = 6 \cos 2\theta$ and outside the circle $r = \sqrt{3}$

14. Inside the circle $r = 3a \cos \theta$ and outside the cardioid $r = a(1 + \cos \theta)$, $a > 0$

15. Inside the circle $r = -2 \cos \theta$ and outside the circle $r = 1$

16. Inside the circle $r = 6$ above the line $r = 3 \csc \theta$

17. Inside the circle $r = 4 \cos \theta$ and to the right of the vertical line $r = \sec \theta$

18. Inside the circle $r = 4 \sin \theta$ and below the horizontal line $r = 3 \csc \theta$

19. a. Find the area of the shaded region in the accompanying figure.

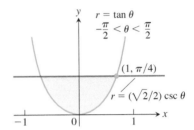

b. It looks as if the graph of $r = \tan \theta$, $-\pi/2 < \theta < \pi/2$, could be asymptotic to the lines $x = 1$ and $x = -1$. Is it? Give reasons for your answer.

20. The area of the region that lies inside the cardioid curve $r = \cos \theta + 1$ and outside the circle $r = \cos \theta$ is not

$$\frac{1}{2} \int_0^{2\pi} \left[(\cos \theta + 1)^2 - \cos^2 \theta \right] d\theta = \pi.$$

Why not? What *is* the area? Give reasons for your answers.

Finding Lengths of Polar Curves

Find the lengths of the curves in Exercises 21–28.

21. The spiral $r = \theta^2$, $0 \le \theta \le \sqrt{5}$

22. The spiral $r = e^\theta/\sqrt{2}$, $0 \le \theta \le \pi$

23. The cardioid $r = 1 + \cos \theta$

24. The curve $r = a \sin^2 (\theta/2)$, $0 \le \theta \le \pi$, $a > 0$

25. The parabolic segment $r = 6/(1 + \cos \theta)$, $0 \le \theta \le \pi/2$

26. The parabolic segment $r = 2/(1 - \cos \theta)$, $\pi/2 \le \theta \le \pi$

27. The curve $r = \cos^3 (\theta/3)$, $0 \le \theta \le \pi/4$

28. The curve $r = \sqrt{1 + \sin 2\theta}$, $0 \le \theta \le \pi\sqrt{2}$

29. The length of the curve $r = f(\theta)$, $\alpha \le \theta \le \beta$ Assuming that the necessary derivatives are continuous, show how the substitutions

$$x = f(\theta) \cos \theta, \quad y = f(\theta) \sin \theta$$

(Equations 2 in the text) transform

$$L = \int_\alpha^\beta \sqrt{\left(\frac{dx}{d\theta}\right)^2 + \left(\frac{dy}{d\theta}\right)^2}\, d\theta$$

into

$$L = \int_\alpha^\beta \sqrt{r^2 + \left(\frac{dr}{d\theta}\right)^2}\, d\theta.$$

30. Circumferences of circles As usual, when faced with a new formula, it is a good idea to try it on familiar objects to be sure it gives results consistent with past experience. Use the length formula in Equation (3) to calculate the circumferences of the following circles ($a > 0$).

a. $r = a$ **b.** $r = a \cos \theta$ **c.** $r = a \sin \theta$

Theory and Examples

31. Average value If f is continuous, the average value of the polar coordinate r over the curve $r = f(\theta)$, $\alpha \le \theta \le \beta$, with respect to θ is given by the formula

$$r_{av} = \frac{1}{\beta - \alpha} \int_{\alpha}^{\beta} f(\theta) \, d\theta.$$

Use this formula to find the average value of r with respect to θ over the following curves ($a > 0$).

a. The cardioid $r = a(1 - \cos \theta)$

b. The circle $r = a$

c. The circle $r = a \cos \theta$, $-\pi/2 \le \theta \le \pi/2$

32. $r = f(\theta)$ *vs.* $r = 2f(\theta)$ Can anything be said about the relative lengths of the curves $r = f(\theta)$, $\alpha \le \theta \le \beta$, and $r = 2f(\theta)$, $\alpha \le \theta \le \beta$? Give reasons for your answer.

11.6 Conic Sections

HISTORICAL BIOGRAPHY
Gregory St. Vincent
(1584–1667)
bit.ly/2xLhZ55

In this section we define and review parabolas, ellipses, and hyperbolas geometrically and derive their standard Cartesian equations. These curves are called *conic sections* or *conics* because they are formed by cutting a double cone with a plane (Figure 11.39). This

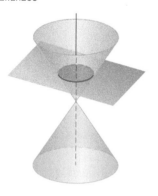

Circle: plane perpendicular to cone axis

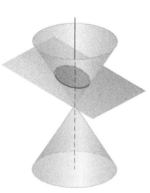

Ellipse: plane oblique to cone axis

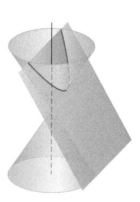

Parabola: plane parallel to side of cone

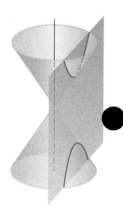

Hyperbola: plane parallel to cone axis

(a)

Point: plane through cone vertex only

Single line: plane tangent to cone

Pair of intersecting lines

(b)

FIGURE 11.39 The standard conic sections (a) are the curves in which a plane cuts a *double* cone. Hyperbolas come in two parts, called *branches*. The point and lines obtained by passing the plane through the cone's vertex (b) are *degenerate* conic sections.

geometric method was the only way that conic sections could be described by Greek mathematicians, since they did not have our tools of Cartesian or polar coordinates. In the next section we express the conics in polar coordinates.

Parabolas

> **DEFINITIONS** A set that consists of all the points in a plane equidistant from a given fixed point and a given fixed line in the plane is a **parabola**. The fixed point is the **focus** of the parabola. The fixed line is the **directrix**.

If the focus F lies on the directrix L, the parabola is the line through F perpendicular to L. We consider this to be a degenerate case and assume henceforth that F does not lie on L.

A parabola has its simplest equation when its focus and directrix straddle one of the coordinate axes. For example, suppose that the focus lies at the point $F(0, p)$ on the positive y-axis and that the directrix is the line $y = -p$ (Figure 11.40). In the notation of the figure, a point $P(x, y)$ lies on the parabola if and only if $PF = PQ$. From the distance formula,

$$PF = \sqrt{(x - 0)^2 + (y - p)^2} = \sqrt{x^2 + (y - p)^2}$$
$$PQ = \sqrt{(x - x)^2 + (y - (-p))^2} = \sqrt{(y + p)^2}.$$

When we equate these expressions, square, and simplify, we get

$$y = \frac{x^2}{4p} \quad \text{or} \quad x^2 = 4py. \qquad \text{Standard form} \tag{1}$$

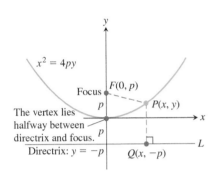

$x^2 = 4py$

Focus $F(0, p)$

The vertex lies halfway between directrix and focus.

Directrix: $y = -p$ $Q(x, -p)$

$P(x, y)$

FIGURE 11.40 The standard form of the parabola $x^2 = 4py$, $p > 0$.

These equations reveal the parabola's symmetry about the y-axis. We call the y-axis the **axis** of the parabola (short for "axis of symmetry").

The point where a parabola crosses its axis is the **vertex**. The vertex of the parabola $x^2 = 4py$ lies at the origin (Figure 11.40). The positive number p is the parabola's **focal length**.

If the parabola opens downward, with its focus at $(0, -p)$ and its directrix the line $y = p$, then Equations (1) become

$$y = -\frac{x^2}{4p} \quad \text{and} \quad x^2 = -4py.$$

By interchanging the variables x and y, we obtain similar equations for parabolas opening to the right or to the left (Figure 11.41).

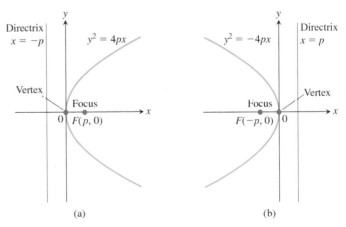

(a)   (b)

FIGURE 11.41 (a) The parabola $y^2 = 4px$. (b) The parabola $y^2 = -4px$.

EXAMPLE 1   Find the focus and directrix of the parabola $y^2 = 10x$.

Solution   We find the value of p in the standard equation $y^2 = 4px$:

$$4p = 10, \quad \text{so} \quad p = \frac{10}{4} = \frac{5}{2}.$$

Then we find the focus and directrix for this value of p:

$$\text{Focus:} \quad (p, 0) = \left(\frac{5}{2}, 0\right)$$

$$\text{Directrix:} \quad x = -p \quad \text{or} \quad x = -\frac{5}{2}. \quad \blacksquare$$

Ellipses

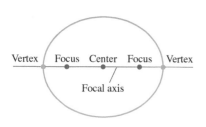

FIGURE 11.42   Points on the focal axis of an ellipse.

> **DEFINITIONS**   An **ellipse** is the set of points in a plane whose distances from two fixed points in the plane have a constant sum. The two fixed points are the foci of the ellipse.
>
> The line through the foci of an ellipse is the ellipse's **focal axis**. The point on the axis halfway be-tween the foci is the **center**. The points where the focal axis and ellipse cross are the ellipse's **vertices** (Figure 11.42).

If the foci are $F_1(-c, 0)$ and $F_2(c, 0)$ (Figure 11.43), and $PF_1 + PF_2$ is denoted by $2a$, then the coordinates of a point P on the ellipse satisfy the equation

$$\sqrt{(x + c)^2 + y^2} + \sqrt{(x - c)^2 + y^2} = 2a.$$

To simplify this equation, we move the second radical to the right-hand side, square, isolate the remaining radical, and square again, obtaining

$$\frac{x^2}{a^2} + \frac{y^2}{a^2 - c^2} = 1. \tag{2}$$

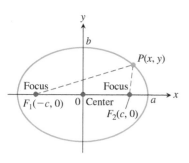

FIGURE 11.43   The ellipse defined by the equation $PF_1 + PF_2 = 2a$ is the graph of the equation $(x^2/a^2) + (y^2/b^2) = 1$, where $b^2 = a^2 - c^2$.

Since $PF_1 + PF_2$ is greater than the length F_1F_2 (by the triangle inequality for triangle PF_1F_2), the number $2a$ is greater than $2c$. Accordingly, $a > c$ and the number $a^2 - c^2$ in Equation (2) is positive.

The algebraic steps leading to Equation (2) can be reversed to show that every point P whose coordinates satisfy an equation of this form with $0 < c < a$ also satisfies the equation $PF_1 + PF_2 = 2a$. A point therefore lies on the ellipse if and only if its coordinates satisfy Equation (2).

If we let b denote the positive square root of $a^2 - c^2$,

$$b = \sqrt{a^2 - c^2}, \tag{3}$$

then $a^2 - c^2 = b^2$ and Equation (2) takes the form

$$\frac{x^2}{a^2} + \frac{y^2}{b^2} = 1. \tag{4}$$

Equation (4) reveals that this ellipse is symmetric with respect to the origin and both coordinate axes. It lies inside the rectangle bounded by the lines $x = \pm a$ and $y = \pm b$. It crosses the axes at the points $(\pm a, 0)$ and $(0, \pm b)$. The tangents at these points are perpendicular to the axes because

$$\frac{dy}{dx} = -\frac{b^2 x}{a^2 y}, \qquad \text{Obtained from Eq. (4)} \\ \text{by implicit differentiation}$$

which is zero if $x = 0$ and infinite if $y = 0$.

The **major axis** of the ellipse in Equation (4) is the line segment of length $2a$ joining the points $(\pm a, 0)$. The **minor axis** is the line segment of length $2b$ joining the points $(0, \pm b)$. The number a itself is the **semimajor axis**, the number b the **semiminor axis**. The number c, found from Equation (3) as

$$c = \sqrt{a^2 - b^2},$$

is the **center-to-focus distance** of the ellipse. If $a = b$ then the ellipse is a circle.

EXAMPLE 2   The ellipse

$$\frac{x^2}{16} + \frac{y^2}{9} = 1 \tag{5}$$

shown in Figure 11.44 has

Semimajor axis: $\quad a = \sqrt{16} = 4, \qquad$ Semiminor axis: $\quad b = \sqrt{9} = 3,$

Center-to-focus distance: $\quad c = \sqrt{16 - 9} = \sqrt{7},$

Foci: $\quad (\pm c, 0) = \left(\pm \sqrt{7}, 0 \right),$

Vertices: $\quad (\pm a, 0) = (\pm 4, 0),$

Center: $\quad (0, 0).$ ∎

If we interchange x and y in Equation (5), we have the equation

$$\frac{x^2}{9} + \frac{y^2}{16} = 1. \tag{6}$$

The major axis of this ellipse is now vertical instead of horizontal, with the foci and vertices on the y-axis. We can determine which way the major axis runs simply by finding the intercepts of the ellipse with the coordinate axes. The longer of the two axes of the ellipse is the major axis.

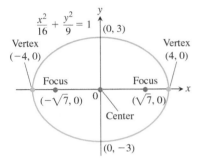

$\dfrac{x^2}{16} + \dfrac{y^2}{9} = 1$

Vertex $(-4, 0)$

Focus $(-\sqrt{7}, 0)$

$(0, 3)$

Vertex $(4, 0)$

Focus $(\sqrt{7}, 0)$

Center

$(0, -3)$

FIGURE 11.44   An ellipse with its major axis horizontal (Example 2).

Standard-Form Equations for Ellipses Centered at the Origin

Foci on the x-axis: $\quad \dfrac{x^2}{a^2} + \dfrac{y^2}{b^2} = 1 \quad (a > b)$

Center-to-focus distance: $\quad c = \sqrt{a^2 - b^2}$

Foci: $\quad (\pm c, 0)$

Vertices: $\quad (\pm a, 0)$

Foci on the y-axis: $\quad \dfrac{x^2}{b^2} + \dfrac{y^2}{a^2} = 1 \quad (a > b)$

Center-to-focus distance: $\quad c = \sqrt{a^2 - b^2}$

Foci: $\quad (0, \pm c)$

Vertices: $\quad (0, \pm a)$

In each case, a is the semimajor axis and b is the semiminor axis.

Hyperbolas

DEFINITIONS   A **hyperbola** is the set of points in a plane whose distances from two fixed points in the plane have a constant difference. The two fixed points are the foci of the hyperbola.

The line through the foci of a hyperbola is the **focal axis**. The point on the axis halfway between the foci is the hyperbola's **center**. The points where the focal axis and hyperbola cross are the **vertices** (Figure 11.45).

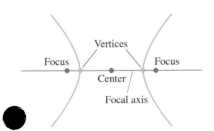

Vertices

Focus

Focus

Center

Focal axis

FIGURE 11.45   Points on the focal axis of a hyperbola.

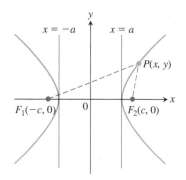

FIGURE 11.46 Hyperbolas have two branches. For points on the right-hand branch of the hyperbola shown here, $PF_1 - PF_2 = 2a$. For points on the left-hand branch, $PF_2 - PF_1 = 2a$. We then let $b = \sqrt{c^2 - a^2}$.

If the foci are $F_1(-c, 0)$ and $F_2(c, 0)$ (Figure 11.46) and the constant difference is $2a$, then a point (x, y) lies on the hyperbola if and only if

$$\sqrt{(x + c)^2 + y^2} - \sqrt{(x - c)^2 + y^2} = \pm 2a. \qquad (7)$$

To simplify this equation, we move the second radical to the right-hand side, square, isolate the remaining radical, and square again, obtaining

$$\frac{x^2}{a^2} + \frac{y^2}{a^2 - c^2} = 1. \qquad (8)$$

So far, this looks just like the equation for an ellipse. But now $a^2 - c^2$ is negative because $2a$, being the difference of two sides of triangle PF_1F_2, is less than $2c$, the third side.

The algebraic steps leading to Equation (8) can be reversed to show that every point P whose coordinates satisfy an equation of this form with $0 < a < c$ also satisfies Equation (7). A point therefore lies on the hyperbola if and only if its coordinates satisfy Equation (8).

If we let b denote the positive square root of $c^2 - a^2$,

$$b = \sqrt{c^2 - a^2}, \qquad (9)$$

then $a^2 - c^2 = -b^2$ and Equation (8) takes the compact form

$$\frac{x^2}{a^2} - \frac{y^2}{b^2} = 1. \qquad (10)$$

The differences between Equation (10) and the equation for an ellipse (Equation 4) are the minus sign and the new relation

$$c^2 = a^2 + b^2. \qquad \text{From Eq. (9)}$$

Like the ellipse, the hyperbola is symmetric with respect to the origin and coordinate axes. It crosses the x-axis at the points $(\pm a, 0)$. The tangents at these points are vertical because

$$\frac{dy}{dx} = \frac{b^2 x}{a^2 y} \qquad \begin{array}{l} \text{Obtained from Eq. (10) by} \\ \text{implicit differentiation} \end{array}$$

and this is infinite when $y = 0$. The hyperbola has no y-intercepts; in fact, no part of the curve lies between the lines $x = -a$ and $x = a$.

The lines

$$y = \pm \frac{b}{a} x$$

are the two **asymptotes** of the hyperbola defined by Equation (10). The fastest way to find the equations of the asymptotes is to replace the 1 in Equation (10) by 0 and solve the new equation for y:

$$\underbrace{\frac{x^2}{a^2} - \frac{y^2}{b^2} = 1}_{\text{hyperbola}} \rightarrow \underbrace{\frac{x^2}{a^2} - \frac{y^2}{b^2} = 0}_{\text{0 for 1}} \rightarrow \underbrace{y = \pm \frac{b}{a} x.}_{\text{asymptotes}}$$

EXAMPLE 3   The equation

$$\frac{x^2}{4} - \frac{y^2}{5} = 1 \qquad (11)$$

is Equation (10) with $a^2 = 4$ and $b^2 = 5$ (Figure 11.47). We have

Center-to-focus distance: $c = \sqrt{a^2 + b^2} = \sqrt{4 + 5} = 3,$

Foci: $(\pm c, 0) = (\pm 3, 0),$ Vertices: $(\pm a, 0) = (\pm 2, 0),$

Center: $(0, 0),$

Asymptotes: $\dfrac{x^2}{4} - \dfrac{y^2}{5} = 0$ or $y = \pm \dfrac{\sqrt{5}}{2} x.$

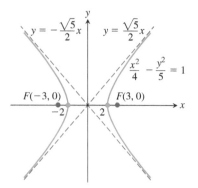

FIGURE 11.47 The hyperbola and its asymptotes in Example 3.

If we interchange x and y in Equation (11), the foci and vertices of the resulting hyperbola will lie along the y-axis. We still find the asymptotes in the same way as before, but now their equations will be $y = \pm 2x/\sqrt{5}$.

Standard-Form Equations for Hyperbolas Centered at the Origin

Foci on the x-axis: $\dfrac{x^2}{a^2} - \dfrac{y^2}{b^2} = 1$

    Center-to-focus distance: $c = \sqrt{a^2 + b^2}$

    Foci: $(\pm c, 0)$

    Vertices: $(\pm a, 0)$

    Asymptotes: $\dfrac{x^2}{a^2} - \dfrac{y^2}{b^2} = 0$ or $y = \pm \dfrac{b}{a}x$

Foci on the y-axis: $\dfrac{y^2}{a^2} - \dfrac{x^2}{b^2} = 1$

    Center-to-focus distance: $c = \sqrt{a^2 + b^2}$

    Foci: $(0, \pm c)$

    Vertices: $(0, \pm a)$

    Asymptotes: $\dfrac{y^2}{a^2} - \dfrac{x^2}{b^2} = 0$ or $y = \pm \dfrac{a}{b}x$

Notice the difference in the asymptote equations (b/a in the first, a/b in the second).

We shift conics using the principles reviewed in Section 1.2, replacing x by $x + h$ and y by $y + k$.

EXAMPLE 4 Show that the equation $x^2 - 4y^2 + 2x + 8y - 7 = 0$ represents a hyperbola. Find its center, asymptotes, and foci.

Solution We reduce the equation to standard form by completing the square in x and y as follows:

$$(x^2 + 2x) - 4(y^2 - 2y) = 7$$
$$(x^2 + 2x + 1) - 4(y^2 - 2y + 1) = 7 + 1 - 4$$
$$\frac{(x + 1)^2}{4} - (y - 1)^2 = 1.$$

This is the standard form Equation (10) of a hyperbola with x replaced by $x + 1$ and y replaced by $y - 1$. The hyperbola is shifted one unit to the left and one unit upward, and it has center $x + 1 = 0$ and $y - 1 = 0$, or $x = -1$ and $y = 1$. Moreover,

$$a^2 = 4, \qquad b^2 = 1, \qquad c^2 = a^2 + b^2 = 5,$$

so the asymptotes are the two lines

$$\frac{x + 1}{2} - (y - 1) = 0 \qquad \text{and} \qquad \frac{x + 1}{2} + (y - 1) = 0,$$

or

$$y - 1 = \pm \frac{1}{2}(x + 1).$$

The shifted foci have coordinates $\left(-1 \pm \sqrt{5}, 1\right)$.

EXERCISES 11.6

Identifying Graphs

Match the parabolas in Exercises 1–4 with the following equations:

$$x^2 = 2y, \quad x^2 = -6y, \quad y^2 = 8x, \quad y^2 = -4x.$$

Then find each parabola's focus and directrix.

1.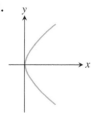

2.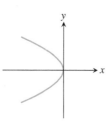

3.

4.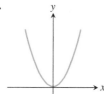

Match each conic section in Exercises 5–8 with one of these equations:

$$\frac{x^2}{4} + \frac{y^2}{9} = 1, \qquad \frac{x^2}{2} + y^2 = 1,$$

$$\frac{y^2}{4} - x^2 = 1, \qquad \frac{x^2}{4} - \frac{y^2}{9} = 1.$$

Then find the conic section's foci and vertices. If the conic section is a hyperbola, find its asymptotes as well.

5.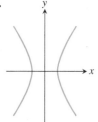

6.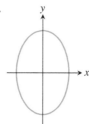

7.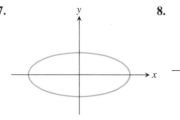

8.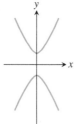

Parabolas

Exercises 9–16 give equations of parabolas. Find each parabola's focus and directrix. Then sketch the parabola. Include the focus and directrix in your sketch.

9. $y^2 = 12x$ **10.** $x^2 = 6y$ **11.** $x^2 = -8y$

12. $y^2 = -2x$ **13.** $y = 4x^2$ **14.** $y = -8x^2$

15. $x = -3y^2$ **16.** $x = 2y^2$

Ellipses

Exercises 17–24 give equations for ellipses. Put each equation in standard form. Then sketch the ellipse. Include the foci in your sketch.

17. $16x^2 + 25y^2 = 400$ **18.** $7x^2 + 16y^2 = 112$

19. $2x^2 + y^2 = 2$ **20.** $2x^2 + y^2 = 4$

21. $3x^2 + 2y^2 = 6$ **22.** $9x^2 + 10y^2 = 90$

23. $6x^2 + 9y^2 = 54$ **24.** $169x^2 + 25y^2 = 4225$

Exercises 25 and 26 give information about the foci and vertices of ellipses centered at the origin of the xy-plane. In each case, find the ellipse's standard-form equation from the given information.

25. Foci: $\left(\pm\sqrt{2}, 0\right)$ Vertices: $(\pm 2, 0)$

26. Foci: $(0, \pm 4)$ Vertices: $(0, \pm 5)$

Hyperbolas

Exercises 27–34 give equations for hyperbolas. Put each equation in standard form and find the hyperbola's asymptotes. Then sketch the hyperbola. Include the asymptotes and foci in your sketch.

27. $x^2 - y^2 = 1$ **28.** $9x^2 - 16y^2 = 144$

29. $y^2 - x^2 = 8$ **30.** $y^2 - x^2 = 4$

31. $8x^2 - 2y^2 = 16$ **32.** $y^2 - 3x^2 = 3$

33. $8y^2 - 2x^2 = 16$ **34.** $64x^2 - 36y^2 = 2304$

Exercises 35–38 give information about the foci, vertices, and asymptotes of hyperbolas centered at the origin of the xy-plane. In each case, find the hyperbola's standard-form equation from the information given.

35. Foci: $\left(0, \pm\sqrt{2}\right)$
    Asymptotes: $y = \pm x$

36. Foci: $(\pm 2, 0)$
    Asymptotes: $y = \pm\dfrac{1}{\sqrt{3}}x$

37. Vertices: $(\pm 3, 0)$
    Asymptotes: $y = \pm\dfrac{4}{3}x$

38. Vertices: $(0, \pm 2)$
    Asymptotes: $y = \pm\dfrac{1}{2}x$

Shifting Conic Sections

You may wish to review Section 1.2 before solving Exercises 39–56.

39. The parabola $y^2 = 8x$ is shifted down 2 units and right 1 unit to generate the parabola $(y + 2)^2 = 8(x - 1)$.

  a. Find the new parabola's vertex, focus, and directrix.

  b. Plot the new vertex, focus, and directrix, and sketch in the parabola.

40. The parabola $x^2 = -4y$ is shifted left 1 unit and up 3 units to generate the parabola $(x + 1)^2 = -4(y - 3)$.

  a. Find the new parabola's vertex, focus, and directrix.

  b. Plot the new vertex, focus, and directrix, and sketch in the parabola.

41. The ellipse $(x^2/16) + (y^2/9) = 1$ is shifted 4 units to the right and 3 units up to generate the ellipse

$$\frac{(x-4)^2}{16} + \frac{(y-3)^2}{9} = 1.$$

   a. Find the foci, vertices, and center of the new ellipse.

   b. Plot the new foci, vertices, and center, and sketch in the new ellipse.

42. The ellipse $(x^2/9) + (y^2/25) = 1$ is shifted 3 units to the left and 2 units down to generate the ellipse

$$\frac{(x+3)^2}{9} + \frac{(y+2)^2}{25} = 1.$$

   a. Find the foci, vertices, and center of the new ellipse.

   b. Plot the new foci, vertices, and center, and sketch in the new ellipse.

43. The hyperbola $(x^2/16) - (y^2/9) = 1$ is shifted 2 units to the right to generate the hyperbola

$$\frac{(x-2)^2}{16} - \frac{y^2}{9} = 1.$$

   a. Find the center, foci, vertices, and asymptotes of the new hyperbola.

   b. Plot the new center, foci, vertices, and asymptotes, and sketch in the hyperbola.

44. The hyperbola $(y^2/4) - (x^2/5) = 1$ is shifted 2 units down to generate the hyperbola

$$\frac{(y+2)^2}{4} - \frac{x^2}{5} = 1.$$

   a. Find the center, foci, vertices, and asymptotes of the new hyperbola.

   b. Plot the new center, foci, vertices, and asymptotes, and sketch in the hyperbola.

Exercises 45–48 give equations for parabolas and tell how many units up or down and to the right or left each parabola is to be shifted. Find an equation for the new parabola, and find the new vertex, focus, and directrix.

45. $y^2 = 4x$, left 2, down 3 **46.** $y^2 = -12x$, right 4, up 3

47. $x^2 = 8y$, right 1, down 7 **48.** $x^2 = 6y$, left 3, down 2

Exercises 49–52 give equations for ellipses and tell how many units up or down and to the right or left each ellipse is to be shifted. Find an equation for the new ellipse, and find the new foci, vertices, and center.

49. $\dfrac{x^2}{6} + \dfrac{y^2}{9} = 1$, left 2, down 1

50. $\dfrac{x^2}{2} + y^2 = 1$, right 3, up 4

51. $\dfrac{x^2}{3} + \dfrac{y^2}{2} = 1$, right 2, up 3

52. $\dfrac{x^2}{16} + \dfrac{y^2}{25} = 1$, left 4, down 5

Exercises 53–56 give equations for hyperbolas and tell how many units up or down and to the right or left each hyperbola is to be shifted. Find an equation for the new hyperbola, and find the new center, foci, vertices, and asymptotes.

53. $\dfrac{x^2}{4} - \dfrac{y^2}{5} = 1$, right 2, up 2

54. $\dfrac{x^2}{16} - \dfrac{y^2}{9} = 1$, left 2, down 1

55. $y^2 - x^2 = 1$, left 1, down 1

56. $\dfrac{y^2}{3} - x^2 = 1$, right 1, up 3

Find the center, foci, vertices, asymptotes, and radius, as appropriate, of the conic sections in Exercises 57–68.

57. $x^2 + 4x + y^2 = 12$

58. $2x^2 + 2y^2 - 28x + 12y + 114 = 0$

59. $x^2 + 2x + 4y - 3 = 0$ **60.** $y^2 - 4y - 8x - 12 = 0$

61. $x^2 + 5y^2 + 4x = 1$ **62.** $9x^2 + 6y^2 + 36y = 0$

63. $x^2 + 2y^2 - 2x - 4y = -1$

64. $4x^2 + y^2 + 8x - 2y = -1$

65. $x^2 - y^2 - 2x + 4y = 4$ **66.** $x^2 - y^2 + 4x - 6y = 6$

67. $2x^2 - y^2 + 6y = 3$ **68.** $y^2 - 4x^2 + 16x = 24$

Theory and Examples

69. If lines are drawn parallel to the coordinate axes through a point P on the parabola $y^2 = kx, k > 0$, the parabola partitions the rectangular region bounded by these lines and the coordinate axes into two smaller regions, A and B.

   a. If the two smaller regions are revolved about the y-axis, show that they generate solids whose volumes have the ratio 4:1.

   b. What is the ratio of the volumes generated by revolving the regions about the x-axis?

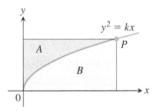

70. Suspension bridge cables hang in parabolas The suspension bridge cable shown in the accompanying figure supports a uniform load of w pounds per horizontal foot. It can be shown that if H is the horizontal tension of the cable at the origin, then the curve of the cable satisfies the equation

$$\frac{dy}{dx} = \frac{w}{H}x.$$

Show that the cable hangs in a parabola by solving this differential equation subject to the initial condition that $y = 0$ when $x = 0$.

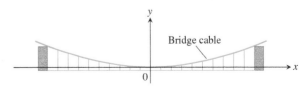

71. The width of a parabola at the focus Show that the number $4p$ is the *width* of the parabola $x^2 = 4py$ ($p > 0$) at the focus by showing that the line $y = p$ cuts the parabola at points that are $4p$ units apart.

72. The asymptotes of $(x^2/a^2) - (y^2/b^2) = 1$ Show that the vertical distance between the line $y = (b/a)x$ and the upper half of the right-hand branch $y = (b/a)\sqrt{x^2 - a^2}$ of the hyperbola $(x^2/a^2) - (y^2/b^2) = 1$ approaches 0 by showing that

$$\lim_{x\to\infty}\left(\frac{b}{a}x - \frac{b}{a}\sqrt{x^2 - a^2}\right) = \frac{b}{a}\lim_{x\to\infty}\left(x - \sqrt{x^2 - a^2}\right) = 0.$$

Similar results hold for the remaining portions of the hyperbola and the lines $y = \pm(b/a)x$.

73. Area Find the dimensions of the rectangle of largest area that can be inscribed in the ellipse $x^2 + 4y^2 = 4$ with its sides parallel to the coordinate axes. What is the area of the rectangle?

74. Volume Find the volume of the solid generated by revolving the region enclosed by the ellipse $9x^2 + 4y^2 = 36$ about the (a) x-axis, (b) y-axis.

75. Volume The "triangular" region in the first quadrant bounded by the x-axis, the line $x = 4$, and the hyperbola $9x^2 - 4y^2 = 36$ is revolved about the x-axis to generate a solid. Find the volume of the solid.

76. Tangents Show that the tangents to the curve $y^2 = 4px$ from any point on the line $x = -p$ are perpendicular.

77. Tangents Find equations for the tangents to the circle $(x - 2)^2 + (y - 1)^2 = 5$ at the points where the circle crosses the coordinate axes.

78. Volume The region bounded on the left by the y-axis, on the right by the hyperbola $x^2 - y^2 = 1$, and above and below by the lines $y = \pm 3$ is revolved about the y-axis to generate a solid. Find the volume of the solid.

79. Centroid Find the centroid of the region that is bounded below by the x-axis and above by the ellipse $(x^2/9) + (y^2/16) = 1$.

80. Surface area The curve $y = \sqrt{x^2 + 1}$, $0 \le x \le \sqrt{2}$, which is part of the upper branch of the hyperbola $y^2 - x^2 = 1$, is revolved about the x-axis to generate a surface. Find the area of the surface.

81. The reflective property of parabolas The accompanying figure shows a typical point $P(x_0, y_0)$ on the parabola $y^2 = 4px$. The line L is tangent to the parabola at P. The parabola's focus lies at $F(p, 0)$. The ray L' extending from P to the right is parallel to the x-axis. We show that light from F to P will be reflected out along L' by showing that β equals α. Establish this equality by taking the following steps.

a. Show that $\tan \beta = 2p/y_0$.

b. Show that $\tan \phi = y_0/(x_0 - p)$.

c. Use the identity

$$\tan \alpha = \frac{\tan \phi - \tan \beta}{1 + \tan \phi \tan \beta}$$

to show that $\tan \alpha = 2p/y_0$.

Since α and β are both acute, $\tan \beta = \tan \alpha$ implies $\beta = \alpha$.

This reflective property of parabolas is used in applications like car headlights, radio telescopes, and satellite TV dishes.

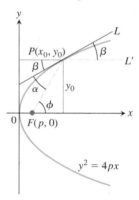

11.7 Conics in Polar Coordinates

Polar coordinates are especially important in astronomy and astronautical engineering because satellites, moons, planets, and comets all move approximately along ellipses, parabolas, and hyperbolas that can be described with a single relatively simple polar coordinate equation. We develop that equation here after first introducing the idea of a conic section's *eccentricity*. The eccentricity reveals the conic section's type (circle, ellipse, parabola, or hyperbola) and the degree to which it is "squashed" or flattened.

Eccentricity

Although the center-to-focus distance c does not appear in the standard Cartesian equation

$$\frac{x^2}{a^2} + \frac{y^2}{b^2} = 1, \quad (a > b)$$

for an ellipse, we can still determine c from the equation $c = \sqrt{a^2 - b^2}$. If we fix a and vary c over the interval $0 \le c \le a$, the resulting ellipses will vary in shape. They are circles if $c = 0$ (so that $a = b$) and flatten, becoming more oblong, as c increases. If $c = a$, the foci and vertices overlap and the ellipse degenerates into a line segment. Thus we are led to consider the ratio $e = c/a$. We use this ratio for hyperbolas as well, except in this case c equals $\sqrt{a^2 + b^2}$ instead of $\sqrt{a^2 - b^2}$. We refer to this ratio as the *eccentricity* of the ellipse or hyperbola.

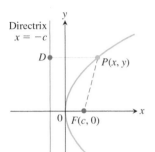

FIGURE 11.48 The distance from the focus F to any point P on a parabola equals the distance from P to the nearest point D on the directrix, so $PF = PD$.

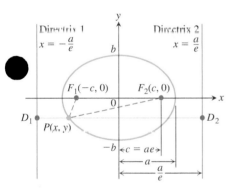

FIGURE 11.49 The foci and directrices of the ellipse $(x^2/a^2) + (y^2/b^2) = 1$. Directrix 1 corresponds to focus F_1 and directrix 2 to focus F_2.

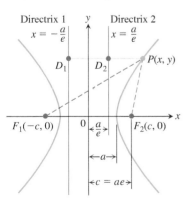

IGURE 11.50 The foci and directrices of the hyperbola $(x^2/a^2) - (y^2/b^2) = 1$. No matter where P lies on the hyperbola, $PF_1 = e \cdot PD_1$ and $PF_2 = e \cdot PD_2$.

> **DEFINITION**
> The **eccentricity** of the ellipse $(x^2/a^2) + (y^2/b^2) = 1 \ (a > b)$ is
> $$e = \frac{c}{a} = \frac{\sqrt{a^2 - b^2}}{a}.$$
> The **eccentricity** of the hyperbola $(x^2/a^2) - (y^2/b^2) = 1$ is
> $$e = \frac{c}{a} = \frac{\sqrt{a^2 + b^2}}{a}.$$
> The **eccentricity** of a parabola is $e = 1$.

Whereas a parabola has one focus and one directrix, each **ellipse** has two foci and two **directrices**. These are the lines perpendicular to the major axis at distances $\pm a/e$ from the center. From Figure 11.48 we see that a parabola has the property

$$PF = 1 \cdot PD \tag{1}$$

for any point P on it, where F is the focus and D is the point nearest P on the directrix. For an ellipse, it can be shown that the equations that replace Equation (1) are

$$PF_1 = e \cdot PD_1, \qquad PF_2 = e \cdot PD_2. \tag{2}$$

Here, e is the eccentricity, P is any point on the ellipse, F_1 and F_2 are the foci, and D_1 and D_2 are the points on the directrices nearest P (Figure 11.49).

In both Equations (2) the directrix and focus must correspond; that is, if we use the distance from P to F_1, we must also use the distance from P to the directrix at the same end of the ellipse. The directrix $x = -a/e$ corresponds to $F_1(-c, 0)$, and the directrix $x = a/e$ corresponds to $F_2(c, 0)$.

As with the ellipse, it can be shown that the lines $x = \pm a/e$ act as **directrices** for the **hyperbola** and that

$$PF_1 = e \cdot PD_1 \qquad \text{and} \qquad PF_2 = e \cdot PD_2. \tag{3}$$

Here P is any point on the hyperbola, F_1 and F_2 are the foci, and D_1 and D_2 are the points nearest P on the directrices (Figure 11.50).

In both the ellipse and the hyperbola, the eccentricity is the ratio of the distance between the foci to the distance between the vertices (because $c/a = 2c/2a$).

$$\text{Eccentricity} = \frac{\text{distance between foci}}{\text{distance between vertices}}$$

In an ellipse, the foci are closer together than the vertices and the ratio is less than 1. In a hyperbola, the foci are further apart than the vertices and the ratio is greater than 1.

The "focus–directrix" equation $PF = e \cdot PD$ unites the parabola, ellipse, and hyperbola in the following way. Suppose that the distance PF of a point P from a fixed point F (the focus) is a constant multiple of its distance from a fixed line (the directrix). That is, suppose

$$PF = e \cdot PD, \tag{4}$$

where e is the constant of proportionality. Then the path traced by P is

(a) a *parabola* if $e = 1$,

(b) an *ellipse* of eccentricity e if $e < 1$, and

(c) a *hyperbola* of eccentricity e if $e > 1$.

As e increases ($e \rightarrow 1^-$), ellipses become more oblong, and ($e \rightarrow \infty$) hyperbolas flatten toward two lines parallel to the directrix. There are no coordinates in Equation (4), and when we try to translate it into Cartesian coordinate form, it translates in different ways depending on the size of e. However, as we are about to see, in polar coordinates the equation $PF = e \cdot PD$ translates into a single equation regardless of the value of e.

Given the focus and corresponding directrix of a hyperbola centered at the origin and with foci on the x-axis, we can use the dimensions shown in Figure 11.50 to find e. Knowing e, we can derive a Cartesian equation for the hyperbola from the equation $PF = e \cdot PD$, as in the next example. We can find equations for ellipses centered at the origin and with foci on the x-axis in a similar way, using the dimensions shown in Figure 11.49.

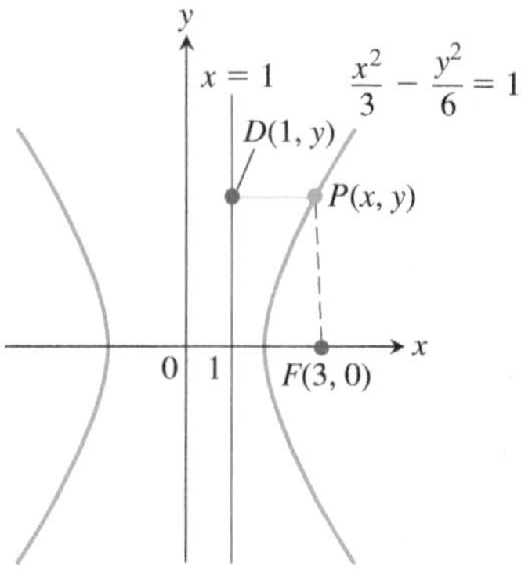

EXAMPLE 1   Find a Cartesian equation for the hyperbola centered at the origin that has a focus at $(3, 0)$ and the line $x = 1$ as the corresponding directrix.

Solution   We first use the dimensions shown in Figure 11.50 to find the hyperbola's eccentricity. The focus is (see Figure 11.51)

$$(c, 0) = (3, 0), \quad \text{so} \quad c = 3.$$

Again from Figure 11.50, the directrix is the line

$$x = \frac{a}{e} = 1, \quad \text{so} \quad a = e.$$

When combined with the equation $e = c/a$ that defines eccentricity, these results give

$$e = \frac{c}{a} = \frac{3}{e}, \quad \text{so} \quad e^2 = 3 \quad \text{and} \quad e = \sqrt{3}.$$

Knowing e, we can now derive the equation we want from the equation $PF = e \cdot PD$. In the coordinates of Figure 11.51, we have

$$PF = e \cdot PD \qquad \text{Eq. (4)}$$
$$\sqrt{(x-3)^2 + (y-0)^2} = \sqrt{3}\,|x - 1| \qquad e = \sqrt{3}$$
$$x^2 - 6x + 9 + y^2 = 3(x^2 - 2x + 1) \qquad \text{Square both sides.}$$
$$2x^2 - y^2 = 6 \qquad \text{Simplify.}$$
$$\frac{x^2}{3} - \frac{y^2}{6} = 1.$$

$\blacksquare$

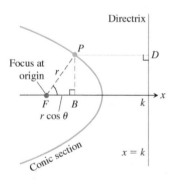

FIGURE 11.51   The hyperbola and directrix in Example 1.

Polar Equations

To find a polar equation for an ellipse, parabola, or hyperbola, we place one focus at the origin and the corresponding directrix to the right of the origin along the vertical line $x = k$ (Figure 11.52). In polar coordinates, this makes

$$PF = r$$

and

$$PD = k - FB = k - r\cos\theta.$$

The conic's focus–directrix equation $PF = e \cdot PD$ then becomes

$$r = e(k - r\cos\theta),$$

which can be solved for r to obtain the following expression.

FIGURE 11.52   If a conic section is put in the position with its focus placed at the origin and a directrix perpendicular to the initial ray and right of the origin, we can find its polar equation from the conic's focus–directrix equation.

Polar Equation for a Conic with Eccentricity e

$$r = \frac{ke}{1 + e\cos\theta}, \tag{5}$$

where $x = k > 0$ is the vertical directrix.

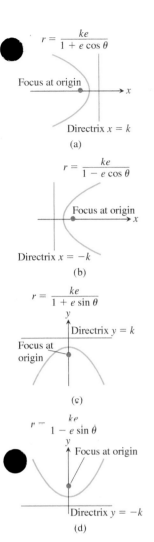

$$r = \frac{ke}{1 + e \cos \theta}$$

Focus at origin

Directrix $x = k$

(a)

$$r = \frac{ke}{1 - e \cos \theta}$$

Focus at origin

Directrix $x = -k$

(b)

$$r = \frac{ke}{1 + e \sin \theta}$$

Directrix $y = k$

Focus at origin

(c)

$$r = \frac{ke}{1 - e \sin \theta}$$

Focus at origin

Directrix $y = -k$

(d)

FIGURE 11.53 Equations for conic sections with eccentricity $e > 0$ but different locations of the directrix. The graphs here show a parabola, so $e = 1$.

EXAMPLE 2 Here are polar equations for three conics. The eccentricity values identifying the conic are the same for both polar and Cartesian coordinates.

$$e = \frac{1}{2}: \quad \text{ellipse} \quad r = \frac{k}{2 + \cos \theta}$$

$$e = 1: \quad \text{parabola} \quad r = \frac{k}{1 + \cos \theta}$$

$$e = 2: \quad \text{hyperbola} \quad r = \frac{2k}{1 + 2 \cos \theta}$$

You may see variations of Equation (5), depending on the location of the directrix. If the directrix is the line $x = -k$ to the left of the origin (the origin is still a focus), we replace Equation (5) with

$$r = \frac{ke}{1 - e \cos \theta}.$$

The denominator now has a $(-)$ instead of a $(+)$. If the directrix is either of the lines $y = k$ or $y = -k$, the equations have sines in them instead of cosines, as shown in Figure 11.53.

EXAMPLE 3 Find an equation for the hyperbola with eccentricity $3/2$ and directrix $x = 2$.

Solution We use Equation (5) with $k = 2$ and $e = 3/2$:

$$r = \frac{2(3/2)}{1 + (3/2) \cos \theta} \quad \text{or} \quad r = \frac{6}{2 + 3 \cos \theta}.$$

EXAMPLE 4 Find the directrix of the parabola $r = \frac{25}{10 + 10 \cos \theta}$.

Solution We divide the numerator and denominator by 10 to put the equation in standard polar form:

$$r = \frac{5/2}{1 + \cos \theta}.$$

This is the equation

$$r = \frac{ke}{1 + e \cos \theta}$$

with $k = 5/2$ and $e = 1$. The equation of the directrix is $x = 5/2$.

From the ellipse diagram in Figure 11.54, we see that k is related to the eccentricity e and the semimajor axis a by the equation

$$k = \frac{a}{e} - ea.$$

From this, we find that $ke = a(1 - e^2)$. Replacing ke in Equation (5) by $a(1 - e^2)$ gives the standard polar equation for an ellipse.

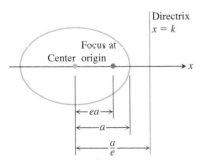

Directrix $x = k$

Focus at Center origin

FIGURE 11.54 In an ellipse with semimajor axis a, the focus–directrix distance is $k = (a/e) - ea$, so $ke = a(1 - e^2)$.

> **Polar Equation for the Ellipse with Eccentricity e and Semimajor Axis a**
>
> $$r = \frac{a(1 - e^2)}{1 + e \cos \theta} \qquad (6)$$

Notice that when $e = 0$, Equation (6) becomes $r = a$, which represents a circle.

Lines

Suppose the perpendicular from the origin to line L meets L at the point $P_0(r_0, \theta_0)$, with $r_0 \geq 0$ (Figure 11.55). Then, if $P(r, \theta)$ is any other point on L, the points P, P_0, and O are the vertices of a right triangle, from which we can read the relation

$$r_0 = r \cos (\theta - \theta_0).$$

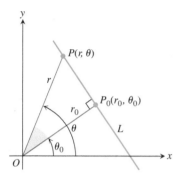

FIGURE 11.55 We can obtain a polar equation for line L by reading the relation $r_0 = r \cos (\theta - \theta_0)$ from the right triangle OP_0P.

> **The Standard Polar Equation for Lines**
> If the point $P_0(r_0, \theta_0)$ is the foot of the perpendicular from the origin to the line L, and $r_0 \geq 0$, then an equation for L is
>
> $$r \cos (\theta - \theta_0) = r_0. \qquad (7)$$

For example, if $\theta_0 = \pi/3$ and $r_0 = 2$, we find that

$$r \cos \left(\theta - \frac{\pi}{3} \right) = 2$$

$$r \left(\cos \theta \cos \frac{\pi}{3} + \sin \theta \sin \frac{\pi}{3} \right) = 2$$

$$\frac{1}{2} r \cos \theta + \frac{\sqrt{3}}{2} r \sin \theta = 2, \qquad \text{or} \qquad x + \sqrt{3}\, y = 4.$$

Circles

To find a polar equation for the circle of radius a centered at $P_0(r_0, \theta_0)$, we let $P(r, \theta)$ be a point on the circle and apply the Law of Cosines to triangle OP_0P (Figure 11.56). This gives

$$a^2 = r_0^2 + r^2 - 2r_0 r \cos (\theta - \theta_0).$$

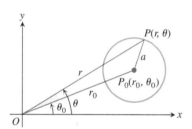

FIGURE 11.56 We can get a polar equation for this circle by applying the Law of Cosines to triangle OP_0P.

If the circle passes through the origin, then $r_0 = a$ and this equation simplifies to

$$a^2 = a^2 + r^2 - 2ar \cos (\theta - \theta_0)$$
$$r^2 = 2ar \cos (\theta - \theta_0)$$
$$r = 2a \cos (\theta - \theta_0).$$

If the circle's center lies on the positive x-axis, $\theta_0 = 0$ and we get the further simplification

$$r = 2a \cos \theta. \qquad (8)$$

If the center lies on the positive y-axis, $\theta = \pi/2$, $\cos (\theta - \pi/2) = \sin \theta$, and the equation $r = 2a \cos (\theta - \theta_0)$ becomes

$$r = 2a \sin \theta. \qquad (9)$$

Equations for circles through the origin centered on the negative x- and y-axes can be obtained by replacing r with $-r$ in the above equations.

EXAMPLE 5   Here are several polar equations given by Equations (8) and (9) for circles through the origin and having centers that lie on the x- or y-axis.

Radius	Center (polar coordinates)	Polar equation
3	$(3, 0)$	$r = 6 \cos \theta$
2	$(2, \pi/2)$	$r = 4 \sin \theta$
1/2	$(-1/2, 0)$	$r = -\cos \theta$
1	$(-1, \pi/2)$	$r = -2 \sin \theta$

EXERCISES 11.7

Ellipses and Eccentricity

In Exercises 1–8, find the eccentricity of the ellipse. Then find and graph the ellipse's foci and directrices.

1. $16x^2 + 25y^2 = 400$ **2.** $7x^2 + 16y^2 = 112$

3. $2x^2 + y^2 = 2$ **4.** $2x^2 + y^2 = 4$

5. $3x^2 + 2y^2 = 6$ **6.** $9x^2 + 10y^2 = 90$

7. $6x^2 + 9y^2 = 54$ **8.** $169x^2 + 25y^2 = 4225$

Exercises 9–12 give the foci or vertices and the eccentricities of ellipses centered at the origin of the xy-plane. In each case, find the ellipse's standard-form equation in Cartesian coordinates.

9. Foci: $(0, \pm 3)$ **10.** Foci: $(\pm 8, 0)$

    Eccentricity: 0.5   Eccentricity: 0.2

11. Vertices: $(0, \pm 70)$ **12.** Vertices: $(\pm 10, 0)$

    Eccentricity: 0.1   Eccentricity: 0.24

Exercises 13–16 give foci and corresponding directrices of ellipses centered at the origin of the xy-plane. In each case, use the dimensions in Figure 11.49 to find the eccentricity of the ellipse. Then find the ellipse's standard-form equation in Cartesian coordinates.

13. Focus: $\left(\sqrt{5}, 0 \right)$ **14.** Focus: $(4, 0)$

    Directrix: $x = \dfrac{9}{\sqrt{5}}$ Directrix: $x = \dfrac{16}{3}$

15. Focus: $(-4, 0)$ **16.** Focus: $\left(-\sqrt{2}, 0 \right)$

    Directrix: $x = -16$ Directrix: $x = -2\sqrt{2}$

Hyperbolas and Eccentricity

In Exercises 17–24, find the eccentricity of the hyperbola. Then find and graph the hyperbola's foci and directrices.

17. $x^2 - y^2 = 1$ **18.** $9x^2 - 16y^2 = 144$

19. $y^2 - x^2 = 8$ **20.** $y^2 - x^2 = 4$

21. $8x^2 - 2y^2 = 16$ **22.** $y^2 - 3x^2 = 3$

23. $8y^2 - 2x^2 = 16$ **24.** $64x^2 - 36y^2 = 2304$

Exercises 25–28 give the eccentricities and the vertices or foci of hyperbolas centered at the origin of the xy-plane. In each case, find the hyperbola's standard-form equation in Cartesian coordinates.

25. Eccentricity: 3   **26.** Eccentricity: 2

    Vertices: $(0, \pm 1)$ Vertices: $(\pm 2, 0)$

27. Eccentricity: 3   **28.** Eccentricity: 1.25

    Foci: $(\pm 3, 0)$ Foci: $(0, \pm 5)$

Eccentricities and Directrices

Exercises 29–36 give the eccentricities of conic sections with one focus at the origin along with the directrix corresponding to that focus. Find a polar equation for each conic section.

29. $e = 1, \quad x = 2$ **30.** $e = 1, \quad y = 2$

31. $e = 5, \quad y = -6$ **32.** $e = 2, \quad x = 4$

33. $e = 1/2, \quad x = 1$ **34.** $e = 1/4, \quad x = -2$

35. $e = 1/5, \quad y = -10$ **36.** $e = 1/3, \quad y = 6$

Parabolas and Ellipses

Sketch the parabolas and ellipses in Exercises 37–44. Include the directrix that corresponds to the focus at the origin. Label the vertices with appropriate polar coordinates. Label the centers of the ellipses as well.

37. $r = \dfrac{1}{1 + \cos \theta}$ **38.** $r = \dfrac{6}{2 + \cos \theta}$

39. $r = \dfrac{25}{10 - 5 \cos \theta}$ **40.** $r = \dfrac{4}{2 - 2 \cos \theta}$

41. $r = \dfrac{400}{16 + 8 \sin \theta}$ **42.** $r = \dfrac{12}{3 + 3 \sin \theta}$

43. $r = \dfrac{8}{2 - 2 \sin \theta}$ **44.** $r = \dfrac{4}{2 - \sin \theta}$

Lines

Sketch the lines in Exercises 45–48 and find Cartesian equations for them.

45. $r \cos \left(\theta - \dfrac{\pi}{4} \right) = \sqrt{2}$ **46.** $r \cos \left(\theta + \dfrac{3\pi}{4} \right) = 1$

47. $r \cos \left(\theta - \dfrac{2\pi}{3} \right) = 3$ **48.** $r \cos \left(\theta + \dfrac{\pi}{3} \right) = 2$

Find a polar equation in the form $r \cos (\theta - \theta_0) = r_0$ for each of the lines in Exercises 49–52.

49. $\sqrt{2}x + \sqrt{2}y = 6$ **50.** $\sqrt{3}x - y = 1$

51. $y = -5$ **52.** $x = -4$

Circles

Sketch the circles in Exercises 53–56. Give polar coordinates for their centers and identify their radii.

53. $r = 4 \cos \theta$ **54.** $r = 6 \sin \theta$

55. $r = -2 \cos \theta$ **56.** $r = -8 \sin \theta$

Find polar equations for the circles in Exercises 57–64. Sketch each circle in the coordinate plane and label it with both its Cartesian and polar equations.

57. $(x - 6)^2 + y^2 = 36$ **58.** $(x + 2)^2 + y^2 = 4$

59. $x^2 + (y - 5)^2 = 25$ **60.** $x^2 + (y + 7)^2 = 49$

61. $x^2 + 2x + y^2 = 0$ **62.** $x^2 - 16x + y^2 = 0$

63. $x^2 + y^2 + y = 0$ **64.** $x^2 + y^2 - \dfrac{4}{3}y = 0$

Examples of Polar Equations

T Graph the lines and conic sections in Exercises 65–74.

65. $r = 3 \sec (\theta - \pi/3)$ **66.** $r = 4 \sec (\theta + \pi/6)$

67. $r = 4 \sin \theta$ **68.** $r = -2 \cos \theta$

69. $r = 8/(4 + \cos \theta)$ **70.** $r = 8/(4 + \sin \theta)$

71. $r = 1/(1 - \sin \theta)$ **72.** $r = 1/(1 + \cos \theta)$

73. $r = 1/(1 + 2 \sin \theta)$ **74.** $r = 1/(1 + 2 \cos \theta)$

75. Perihelion and aphelion   A planet travels about its sun in an ellipse whose semimajor axis has length a. (See accompanying figure.)

a. Show that $r = a(1 - e)$ when the planet is closest to the sun and that $r = a(1 + e)$ when the planet is farthest from the sun.

b. Use the data in the table in Exercise 76 to find how close each planet in our solar system comes to the sun and how far away each planet gets from the sun.

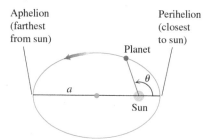

Aphelion (farthest from sun)   Perihelion (closest to sun)

Planet a Sun θ

76. Planetary orbits   Use the data in the table below and Equation (6) to find polar equations for the orbits of the planets.

Planet	Semimajor axis (astronomical units)	Eccentricity
Mercury	0.3871	0.2056
Venus	0.7233	0.0068
Earth	1.000	0.0167
Mars	1.524	0.0934
Jupiter	5.203	0.0484
Saturn	9.539	0.0543
Uranus	19.18	0.0460
Neptune	30.06	0.0082

CHAPTER 11   Questions to Guide Your Review

1. What is a parametrization of a curve in the xy-plane? Does a function $y = f(x)$ always have a parametrization? Are parametrizations of a curve unique? Give examples.

2. Give some typical parametrizations for lines, circles, parabolas, ellipses, and hyperbolas. How might the parametrized curve differ from the graph of its Cartesian equation?

3. What is a cycloid? What are typical parametric equations for cycloids? What physical properties account for the importance of cycloids?

4. What is the formula for the slope dy/dx of a parametrized curve $x = f(t)$, $y = g(t)$? When does the formula apply? When can you expect to be able to find d^2y/dx^2 as well? Give examples.

5. How can you sometimes find the area bounded by a parametrized curve and one of the coordinate axes?

6. How do you find the length of a smooth parametrized curve $x = f(t)$, $y = g(t)$, $a \le t \le b$? What does smoothness have to do with length? What else do you need to know about the parametrization in order to find the curve's length? Give examples.

7. What is the arc length function for a smooth parametrized curve? What is its arc length differential?

8. Under what conditions can you find the area of the surface generated by revolving a curve $x = f(t)$, $y = g(t)$, $a \le t \le b$, about the x-axis? the y-axis? Give examples.

9. What are polar coordinates? What equations relate polar coordinates to Cartesian coordinates? Why might you want to change from one coordinate system to the other?

10. What consequence does the lack of uniqueness of polar coordinates have for graphing? Give an example.

11. How do you graph equations in polar coordinates? Include in your discussion symmetry, slope, behavior at the origin, and the use of Cartesian graphs. Give examples.

12. How do you find the area of a region $0 \le r_1(\theta) \le r \le r_2(\theta)$, $\alpha \le \theta \le \beta$, in the polar coordinate plane? Give examples.

13. Under what conditions can you find the length of a curve $r = f(\theta)$, $\alpha \le \theta \le \beta$, in the polar coordinate plane? Give an example of a typical calculation.

14. What is a parabola? What are the Cartesian equations for parabolas whose vertices lie at the origin and whose foci lie on the coordinate axes? How can you find the focus and directrix of such a parabola from its equation?

15. What is an ellipse? What are the Cartesian equations for ellipses centered at the origin with foci on one of the coordinate axes?

How can you find the foci, vertices, and directrices of such an ellipse from its equation?

16. What is a hyperbola? What are the Cartesian equations for hyperbolas centered at the origin with foci on one of the coordinate axes? How can you find the foci, vertices, and directrices of such an ellipse from its equation?

17. What is the eccentricity of a conic section? How can you classify conic sections by eccentricity? How does eccentricity change the shape of ellipses and hyperbolas?

18. Explain the equation $PF = e \cdot PD$.

19. What are the standard equations for lines and conic sections in polar coordinates? Give examples.

CHAPTER 11 Practice Exercises

Identifying Parametric Equations in the Plane

Exercises 1–6 give parametric equations and parameter intervals for the motion of a particle in the xy-plane. Identify the particle's path by finding a Cartesian equation for it. Graph the Cartesian equation and indicate the direction of motion and the portion traced by the particle.

1. $x = t/2$, $y = t + 1$, $-\infty < t < \infty$
2. $x = \sqrt{t}$, $y = 1 - \sqrt{t}$; $t \ge 0$
3. $x = (1/2) \tan t$, $y = (1/2) \sec t$; $-\pi/2 < t < \pi/2$
4. $x = -2 \cos t$, $y = 2 \sin t$; $0 \le t \le \pi$
5. $x = -\cos t$, $y = \cos^2 t$; $0 \le t \le \pi$
6. $x = 4 \cos t$, $y = 9 \sin t$; $0 \le t \le 2\pi$

Finding Parametric Equations and Tangent Lines

7. Find parametric equations and a parameter interval for the motion of a particle in the xy-plane that traces the ellipse $16x^2 + 9y^2 = 144$ once counterclockwise. (There are many ways to do this.)

8. Find parametric equations and a parameter interval for the motion of a particle that starts at the point $(-2, 0)$ in the xy-plane and traces the circle $x^2 + y^2 = 4$ three times clockwise. (There are many ways to do this.)

In Exercises 9 and 10, find an equation for the line in the xy-plane that is tangent to the curve at the point corresponding to the given value of t. Also, find the value of d^2y/dx^2 at this point.

9. $x = (1/2) \tan t$, $y = (1/2) \sec t$; $t = \pi/3$
10. $x = 1 + 1/t^2$, $y = 1 - 3/t$; $t = 2$
11. Eliminate the parameter to express the curve in the form $y = f(x)$.
    a. $x = 4t^2$, $y = t^3 - 1$
    b. $x = \cos t$, $y = \tan t$
12. Find parametric equations for the given curve.
    a. Line through $(1, -2)$ with slope 3
    b. $(x - 1)^2 + (y + 2)^2 = 9$
    c. $y = 4x^2 - x$
    d. $9x^2 + 4y^2 = 36$

Lengths of Curves

Find the lengths of the curves in Exercises 13–19.

13. $y = x^{1/2} - (1/3)x^{3/2}$, $1 \le x \le 4$
14. $x = y^{2/3}$, $1 \le y \le 8$
15. $y = (5/12)x^{6/5} - (5/8)x^{4/5}$, $1 \le x \le 32$
16. $x = (y^3/12) + (1/y)$, $1 \le y \le 2$
17. $x = 5 \cos t - \cos 5t$, $y = 5 \sin t - \sin 5t$, $0 \le t \le \pi/2$
18. $x = t^3 - 6t^2$, $y = t^3 + 6t^2$, $0 \le t \le 1$
19. $x = 3 \cos \theta$, $y = 3 \sin \theta$, $0 \le \theta \le \dfrac{3\pi}{2}$

20. Find the length of the enclosed loop $x = t^2$, $y = (t^3/3) - t$ shown here. The loop starts at $t = -\sqrt{3}$ and ends at $t = \sqrt{3}$.

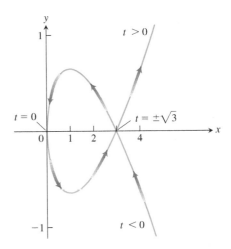

Surface Areas

Find the areas of the surfaces generated by revolving the curves in Exercises 21 and 22 about the indicated axes.

21. $x = t^2/2$, $y = 2t$, $0 \le t \le \sqrt{5}$; x-axis
22. $x = t^2 + 1/(2t)$, $y = 4\sqrt{t}$, $1/\sqrt{2} \le t \le 1$; y-axis

Polar to Cartesian Equations

Sketch the lines in Exercises 23–28. Also, find a Cartesian equation for each line.

23. $r \cos\left(\theta + \dfrac{\pi}{3}\right) = 2\sqrt{3}$ **24.** $r \cos\left(\theta - \dfrac{3\pi}{4}\right) = \dfrac{\sqrt{2}}{2}$

25. $r = 2 \sec \theta$ **26.** $r = -\sqrt{2} \sec \theta$

27. $r = -(3/2) \csc \theta$ **28.** $r = \left(3\sqrt{3}\right) \csc \theta$

Find Cartesian equations for the circles in Exercises 29–32. Sketch each circle in the coordinate plane and label it with both its Cartesian and polar equations.

29. $r = -4 \sin \theta$ **30.** $r = 3\sqrt{3} \sin \theta$

31. $r = 2\sqrt{2} \cos \theta$ **32.** $r = -6 \cos \theta$

Cartesian to Polar Equations

Find polar equations for the circles in Exercises 33–36. Sketch each circle in the coordinate plane and label it with both its Cartesian and polar equations.

33. $x^2 + y^2 + 5y = 0$ **34.** $x^2 + y^2 - 2y = 0$

35. $x^2 + y^2 - 3x = 0$ **36.** $x^2 + y^2 + 4x = 0$

Graphs in Polar Coordinates

Sketch the regions defined by the polar coordinate inequalities in Exercises 37 and 38.

37. $0 \le r \le 6 \cos \theta$ **38.** $-4 \sin \theta \le r \le 0$

Match each graph in Exercises 39–46 with the appropriate equation (a)–(l). There are more equations than graphs, so some equations will not be matched.

a. $r = \cos 2\theta$ **b.** $r \cos \theta = 1$ **c.** $r = \dfrac{6}{1 - 2\cos\theta}$

d. $r = \sin 2\theta$ **e.** $r = \theta$ **f.** $r^2 = \cos 2\theta$

g. $r = 1 + \cos\theta$ **h.** $r = 1 - \sin\theta$ **i.** $r = \dfrac{2}{1 - \cos\theta}$

j. $r^2 = \sin 2\theta$ **k.** $r = -\sin\theta$ **l.** $r = 2\cos\theta + 1$

39. Four-leaved rose   **40.** Spiral

     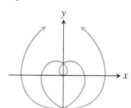

41. Limaçon   **42.** Lemniscate

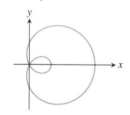   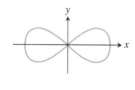

43. Circle   **44.** Cardioid

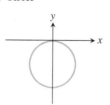   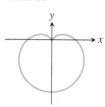

45. Parabola   **46.** Lemniscate

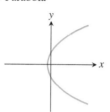   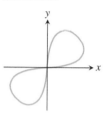

Area in Polar Coordinates

Find the areas of the regions in the polar coordinate plane described in Exercises 47–50.

47. Enclosed by the limaçon $r = 2 - \cos\theta$

48. Enclosed by one leaf of the three-leaved rose $r = \sin 3\theta$

49. Inside the "figure eight" $r = 1 + \cos 2\theta$ and outside the circle $r = 1$

50. Inside the cardioid $r = 2(1 + \sin\theta)$ and outside the circle $r = 2\sin\theta$

Length in Polar Coordinates

Find the lengths of the curves given by the polar coordinate equations in Exercises 51–54.

51. $r = -1 + \cos\theta$

52. $r = 2\sin\theta + 2\cos\theta, \quad 0 \le \theta \le \pi/2$

53. $r = 8\sin^3(\theta/3), \quad 0 \le \theta \le \pi/4$

54. $r = \sqrt{1 + \cos 2\theta}, \quad -\pi/2 \le \theta \le \pi/2$

Graphing Conic Sections

Sketch the parabolas in Exercises 55–58. Include the focus and directrix in each sketch.

55. $x^2 = -4y$ **56.** $x^2 = 2y$

57. $y^2 = 3x$ **58.** $y^2 = -(8/3)x$

Find the eccentricities of the ellipses and hyperbolas in Exercises 59–62. Sketch each conic section. Include the foci, vertices, and asymptotes (as appropriate) in your sketch.

59. $16x^2 + 7y^2 = 112$ **60.** $x^2 + 2y^2 = 4$

61. $3x^2 - y^2 = 3$ **62.** $5y^2 - 4x^2 = 20$

Exercises 63–68 give equations for conic sections and tell how many units up or down and to the right or left each curve is to be shifted. Find an equation for the new conic section, and find the new foci, vertices, centers, and asymptotes, as appropriate. If the curve is a parabola, find the new directrix as well.

63. $x^2 = -12y$,   right 2, up 3   **64.** $y^2 = 10x$,   left 1/2, down 1

65. $\dfrac{x^2}{9} + \dfrac{y^2}{25} = 1$,   left 3, down 5

66. $\dfrac{x^2}{169} + \dfrac{y^2}{144} = 1$,   right 5, up 12

67. $\dfrac{y^2}{8} - \dfrac{x^2}{2} = 1$,   right 2, up $2\sqrt{2}$

68. $\dfrac{x^2}{36} - \dfrac{y^2}{64} = 1$,   left 10, down 3

Identifying Conic Sections
Complete the squares to identify the conic sections in Exercises 69–76. Find their foci, vertices, centers, and asymptotes (as appropriate). If the curve is a parabola, find its directrix as well.

69. $x^2 - 4x - 4y^2 = 0$ **70.** $4x^2 - y^2 + 4y = 8$

71. $y^2 - 2y + 16x = -49$ **72.** $x^2 - 2x + 8y = -17$

73. $9x^2 + 16y^2 + 54x - 64y = -1$

74. $25x^2 + 9y^2 - 100x + 54y = 44$

75. $x^2 + y^2 - 2x - 2y = 0$ **76.** $x^2 + y^2 + 4x + 2y = 1$

Conics in Polar Coordinates
Sketch the conic sections whose polar coordinate equations are given in Exercises 77–80. Give polar coordinates for the vertices and, in the case of ellipses, for the centers as well.

77. $r = \dfrac{2}{1 + \cos\theta}$ **78.** $r = \dfrac{8}{2 + \cos\theta}$

79. $r = \dfrac{6}{1 - 2\cos\theta}$ **80.** $r = \dfrac{12}{3 + \sin\theta}$

Exercises 81–84 give the eccentricities of conic sections with one focus at the origin of the polar coordinate plane, along with the directrix for that focus. Find a polar equation for each conic section.

81. $e = 2$, $r\cos\theta = 2$ **82.** $e = 1$, $r\cos\theta = -4$

83. $e = 1/2$, $r\sin\theta = 2$ **84.** $e = 1/3$, $r\sin\theta = -6$

Theory and Examples
85. Find the volume of the solid generated by revolving the region enclosed by the ellipse $9x^2 + 4y^2 = 36$ about **(a)** the x-axis, **(b)** the y-axis.

86. The "triangular" region in the first quadrant bounded by the x-axis, the line $x = 4$, and the hyperbola $9x^2 - 4y^2 = 36$ is revolved about the x-axis to generate a solid. Find the volume of the solid.

87. Show that the equations $x = r\cos\theta$, $y = r\sin\theta$ transform the polar equation

$$r = \dfrac{k}{1 + e\cos\theta}$$

into the Cartesian equation

$$(1 - e^2)x^2 + y^2 + 2kex - k^2 = 0.$$

88. Archimedes spirals The graph of an equation of the form $r = a\theta$, where a is a nonzero constant, is called an *Archimedes spiral*. Is there anything special about the widths between the successive turns of such a spiral?

CHAPTER 11   Additional and Advanced Exercises

Finding Conic Sections
1. Find an equation for the parabola with focus $(4, 0)$ and directrix $x = 3$. Sketch the parabola together with its vertex, focus, and directrix.

2. Find the vertex, focus, and directrix of the parabola
$$x^2 - 6x - 12y + 9 = 0.$$

3. Find an equation for the curve traced by the point $P(x, y)$ if the distance from P to the vertex of the parabola $x^2 = 4y$ is twice the distance from P to the focus. Identify the curve.

4. A line segment of length $a + b$ runs from the x-axis to the y-axis. The point P on the segment lies a units from one end and b units from the other end. Show that P traces an ellipse as the ends of the segment slide along the axes.

5. The vertices of an ellipse of eccentricity 0.5 lie at the points $(0, \pm 2)$. Where do the foci lie?

6. Find an equation for the ellipse of eccentricity $2/3$ that has the line $x = 2$ as a directrix and the point $(4, 0)$ as the corresponding focus.

7. One focus of a hyperbola lies at the point $(0, -7)$ and the corresponding directrix is the line $y = -1$. Find an equation for the hyperbola if its eccentricity is **(a)** 2, **(b)** 5.

8. Find an equation for the hyperbola with foci $(0, -2)$ and $(0, 2)$ that passes through the point $(12, 7)$.

9. Show that the line
$$b^2xx_1 + a^2yy_1 - a^2b^2 = 0$$
is tangent to the ellipse $b^2x^2 + a^2y^2 - a^2b^2 = 0$ at the point (x_1, y_1) on the ellipse.

10. Show that the line
$$b^2xx_1 - a^2yy_1 - a^2b^2 = 0$$
is tangent to the hyperbola $b^2x^2 - a^2y^2 - a^2b^2 = 0$ at the point (x_1, y_1) on the hyperbola.

Equations and Inequalities
What points in the xy-plane satisfy the equations and inequalities in Exercises 11–16? Draw a figure for each exercise.

11. $(x^2 - y^2 - 1)(x^2 + y^2 - 25)(x^2 + 4y^2 - 4) = 0$

12. $(x + y)(x^2 + y^2 - 1) = 0$

13. $(x^2/9) + (y^2/16) \le 1$

14. $(x^2/9) - (y^2/16) \le 1$

15. $(9x^2 + 4y^2 - 36)(4x^2 + 9y^2 - 16) \le 0$

16. $(9x^2 + 4y^2 - 36)(4x^2 + 9y^2 - 16) > 0$

Polar Coordinates

17. a. Find an equation in polar coordinates for the curve

$$x = e^{2t} \cos t, \quad y = e^{2t} \sin t; \quad -\infty < t < \infty.$$

b. Find the length of the curve from $t = 0$ to $t = 2\pi$.

18. Find the length of the curve $r = 2 \sin^3(\theta/3), 0 \le \theta \le 3\pi$, in the polar coordinate plane.

Exercises 19–22 give the eccentricities of conic sections with one focus at the origin of the polar coordinate plane, along with the directrix for that focus. Find a polar equation for each conic section.

19. $e = 2, \quad r \cos \theta = 2$ **20.** $e = 1, \quad r \cos \theta = -4$

21. $e = 1/2, \quad r \sin \theta = 2$ **22.** $e = 1/3, \quad r \sin \theta = -6$

Theory and Examples

23. Epicycloids When a circle rolls externally along the circumference of a second, fixed circle, any point P on the circumference of the rolling circle describes an *epicycloid*, as shown here. Let the fixed circle have its center at the origin O and have radius a.

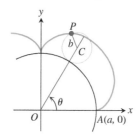

Let the radius of the rolling circle be b and let the initial position of the tracing point P be $A(a, 0)$. Find parametric equations for the epicycloid, using as the parameter the angle θ from the positive x-axis to the line through the circles' centers.

24. Find the centroid of the region enclosed by the x-axis and the cycloid arch

$$x = a(t - \sin t), \quad y = a(1 - \cos t); \quad 0 \le t \le 2\pi.$$

The Angle Between the Radius Vector and the Tangent Line to a Polar Coordinate Curve In Cartesian coordinates, when we want to discuss the direction of a curve at a point, we use the angle ϕ measured counterclockwise from the positive x-axis to the tangent line. In polar coordinates, it is more convenient to calculate the angle ψ from the *radius vector* to the tangent line (see the accompanying figure). The angle ϕ can then be calculated from the relation

$$\phi = \theta + \psi, \tag{1}$$

which comes from applying the Exterior Angle Theorem to the triangle in the accompanying figure.

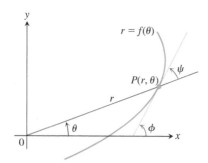

Suppose the equation of the curve is given in the form $r = f(\theta)$, where $f(\theta)$ is a differentiable function of θ. Then

$$x = r \cos \theta \quad \text{and} \quad y = r \sin \theta \tag{2}$$

are differentiable functions of θ with

$$\frac{dx}{d\theta} = -r \sin \theta + \cos \theta \frac{dr}{d\theta},$$

$$\frac{dy}{d\theta} = r \cos \theta + \sin \theta \frac{dr}{d\theta}. \tag{3}$$

Since $\psi = \phi - \theta$ from (1),

$$\tan \psi = \tan(\phi - \theta) = \frac{\tan \phi - \tan \theta}{1 + \tan \phi \tan \theta}.$$

Furthermore,

$$\tan \phi = \frac{dy}{dx} = \frac{dy/d\theta}{dx/d\theta}$$

because $\tan \phi$ is the slope of the curve at P. Also,

$$\tan \theta = \frac{y}{x}.$$

Hence

$$\tan \psi = \frac{\dfrac{dy/d\theta}{dx/d\theta} - \dfrac{y}{x}}{1 + \dfrac{y}{x}\dfrac{dy/d\theta}{dx/d\theta}} = \frac{x\dfrac{dy}{d\theta} - y\dfrac{dx}{d\theta}}{x\dfrac{dx}{d\theta} + y\dfrac{dy}{d\theta}}. \tag{4}$$

The numerator in the last expression in Equation (4) is found from Equations (2) and (3) to be

$$x\frac{dy}{d\theta} - y\frac{dx}{d\theta} = r^2.$$

Similarly, the denominator is

$$x\frac{dx}{d\theta} + y\frac{dy}{d\theta} = r\frac{dr}{d\theta}.$$

When we substitute these into Equation (4), we obtain

$$\tan \psi = \frac{r}{dr/d\theta}. \tag{5}$$

This is the equation we use for finding ψ as a function of θ.

25. Show, by reference to a figure, that the angle β between the tangents to two curves at a point of intersection may be found from the formula

$$\tan \beta = \frac{\tan \psi_2 - \tan \psi_1}{1 + \tan \psi_2 \tan \psi_1}. \tag{6}$$

When will the two curves intersect at right angles?

26. Find the value of $\tan \psi$ for the curve $r = \sin^4(\theta/4)$.

27. Find the angle between the radius vector to the curve $r = 2a \sin 3\theta$ and its tangent when $\theta = \pi/6$.

28. a. Graph the hyperbolic spiral $r\theta = 1$. What appears to happen to ψ as the spiral winds in around the origin?

   b. Confirm your finding in part (a) analytically.

29. The circles $r = \sqrt{3} \cos \theta$ and $r - \sin \theta$ intersect at the point $\left(\sqrt{3}/2, \pi/3\right)$. Show that their tangents are perpendicular there.

30. Find the angle at which the cardioid $r = a(1 - \cos \theta)$ crosses the ray $\theta = \pi/2$.

CHAPTER 11   Technology Application Projects

Mathematica/Maple Projects

Projects can be found within MyMathLab.

- *Radar Tracking of a Moving Object*
  Part I: Convert from polar to Cartesian coordinates.

- *Parametric and Polar Equations with a Figure Skater*
  Part I: Visualize position, velocity, and acceleration to analyze motion defined by parametric equations.
  Part II: Find and analyze the equations of motion for a figure skater tracing a polar plot.

12

Vectors and the Geometry of Space

OVERVIEW In this chapter we begin the study of multivariable calculus. To apply calculus in many real-world situations, we introduce three-dimensional coordinate systems and vectors. We establish coordinates in space by adding a third axis that measures distance above and below the xy-plane. Then we define vectors, which provide simple ways to define equations for lines, planes, curves, and surfaces in space.

12.1 Three-Dimensional Coordinate Systems

To locate a point in space, we use three mutually perpendicular coordinate axes, arranged as in Figure 12.1. The axes shown there make a *right-handed* coordinate frame. When you hold your right hand so that the fingers curl from the positive x-axis toward the positive y-axis, your thumb points along the positive z-axis. So when you look down on the xy-plane from the positive direction of the z-axis, positive angles in the plane are measured counterclockwise from the positive x-axis and around the positive z-axis. (In a *left-handed* coordinate frame, the z-axis would point downward in Figure 12.1 and angles in the plane would be positive when measured clockwise from the positive x-axis. Right-handed and left-handed coordinate frames are not equivalent.)

The Cartesian coordinates (x, y, z) of a point P in space are the values at which the planes through P perpendicular to the axes cut the axes. Cartesian coordinates for space are also called **rectangular coordinates** because the axes that define them meet at right angles. Points on the x-axis have y- and z-coordinates equal to zero. That is, they have coordinates of the form $(x, 0, 0)$. Similarly, points on the y-axis have coordinates of the form $(0, y, 0)$, and points on the z-axis have coordinates of the form $(0, 0, z)$.

The planes determined by the coordinates axes are the **xy-plane**, whose standard equation is $z = 0$; the **yz-plane**, whose standard equation is $x = 0$; and the **xz-plane**, whose standard equation is $y = 0$. They meet at the **origin** $(0, 0, 0)$ (Figure 12.2). The origin is also identified by simply 0 or sometimes the letter O.

The three **coordinate planes** $x = 0$, $y = 0$, and $z = 0$ divide space into eight cells called **octants**. The octant in which the point coordinates are all positive is called the **first octant**; there is no convention for numbering the other seven octants.

The points in a plane perpendicular to the x-axis all have the same x-coordinate, this being the number at which that plane cuts the x-axis. The y- and z-coordinates can be any numbers. Similarly, the points in a plane perpendicular to the y-axis have a common y-coordinate and the points in a plane perpendicular to the z-axis have a common z-coordinate. To write equations for these planes, we name the common coordinate's value. The plane $x = 2$ is the plane perpendicular to the x-axis at $x = 2$. The plane $y = 3$ is the plane perpendicular to the y-axis at $y = 3$. The plane $z = 5$ is the plane perpendicular to the z-axis at $z = 5$. Figure 12.3 shows the planes $x = 2$, $y = 3$, and $z = 5$, together with their intersection point $(2, 3, 5)$.

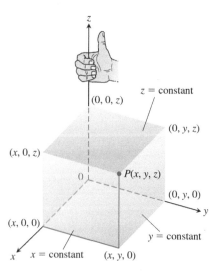

FIGURE 12.1 The Cartesian coordinate system is right-handed.

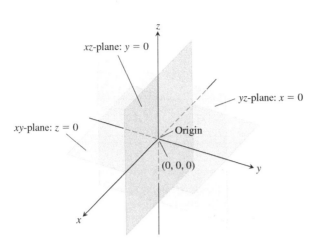

FIGURE 12.2 The planes $x = 0$, $y = 0$, and $z = 0$ divide space into eight octants.

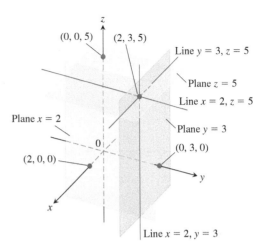

FIGURE 12.3 The planes $x = 2$, $y = 3$, and $z = 5$ determine three lines through the point $(2, 3, 5)$.

The planes $x = 2$ and $y = 3$ in Figure 12.3 intersect in a line parallel to the z-axis. This line is described by the *pair* of equations $x = 2$, $y = 3$. A point (x, y, z) lies on the line if and only if $x = 2$ and $y = 3$. Similarly, the line of intersection of the planes $y = 3$ and $z = 5$ is described by the equation pair $y = 3$, $z = 5$. This line runs parallel to the x-axis. The line of intersection of the planes $x = 2$ and $z = 5$, parallel to the y-axis, is described by the equation pair $x = 2$, $z = 5$.

In the following examples, we match coordinate equations and inequalities with the sets of points they define in space.

EXAMPLE 1 We interpret these equations and inequalities geometrically.

(a) $z \geq 0$ — The half-space consisting of the points on and above the xy-plane.

(b) $x = -3$ — The plane perpendicular to the x-axis at $x = -3$. This plane lies parallel to the yz-plane and 3 units behind it.

(c) $z = 0, x \leq 0, y \geq 0$ — The second quadrant of the xy-plane.

(d) $x \geq 0, y \geq 0, z \geq 0$ — The first octant.

(e) $-1 \leq y \leq 1$ — The slab between the planes $y = -1$ and $y = 1$ (planes included).

(f) $y = -2, z = 2$ — The line in which the planes $y = -2$ and $z = 2$ intersect. Alternatively, the line through the point $(0, -2, 2)$ parallel to the x-axis. ∎

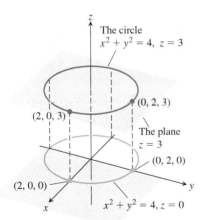

FIGURE 12.4 The circle $x^2 + y^2 = 4$ in the plane $z = 3$ (Example 2).

EXAMPLE 2 What points (x, y, z) satisfy the equations

$$x^2 + y^2 = 4 \qquad \text{and} \qquad z = 3?$$

Solution The points lie in the horizontal plane $z = 3$ and, in this plane, make up the circle $x^2 + y^2 = 4$. We call this set of points "the circle $x^2 + y^2 = 4$ in the plane $z = 3$" or, more simply, "the circle $x^2 + y^2 = 4$, $z = 3$" (Figure 12.4). ∎

Distance and Spheres in Space

The formula for the distance between two points in the xy-plane extends to points in space.

The Distance Between $P_1(x_1, y_1, z_1)$ and $P_2(x_2, y_2, z_2)$

$$|P_1 P_2| = \sqrt{(x_2 - x_1)^2 + (y_2 - y_1)^2 + (z_2 - z_1)^2}$$

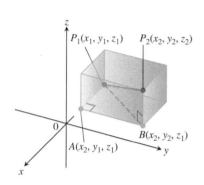

FIGURE 12.5 We find the distance between P_1 and P_2 by applying the Pythagorean theorem to the right triangles P_1AB and P_1BP_2.

Proof   We construct a rectangular box with faces parallel to the coordinate planes and the points P_1 and P_2 at opposite corners of the box (Figure 12.5). If $A(x_2, y_1, z_1)$ and $B(x_2, y_2, z_1)$ are the vertices of the box indicated in the figure, then the three box edges P_1A, AB, and BP_2 have lengths

$$|P_1 A| = |x_2 - x_1|, \qquad |AB| = |y_2 - y_1|, \qquad |BP_2| = |z_2 - z_1|.$$

Because triangles $P_1 B P_2$ and $P_1 AB$ are both right-angled, two applications of the Pythagorean theorem give

$$|P_1 P_2|^2 = |P_1 B|^2 + |BP_2|^2 \qquad \text{and} \qquad |P_1 B|^2 = |P_1 A|^2 + |AB|^2$$

(see Figure 12.5). So

$$
\begin{aligned}
|P_1 P_2|^2 &= |P_1 B|^2 + |BP_2|^2 \\
&= |P_1 A|^2 + |AB|^2 + |BP_2|^2 \qquad \text{\small Substitute } |P_1 B|^2 = |P_1 A|^2 + |AB|^2. \\
&= |x_2 - x_1|^2 + |y_2 - y_1|^2 + |z_2 - z_1|^2 \\
&= (x_2 - x_1)^2 + (y_2 - y_1)^2 + (z_2 - z_1)^2.
\end{aligned}
$$

Therefore

$$|P_1 P_2| = \sqrt{(x_2 - x_1)^2 + (y_2 - y_1)^2 + (z_2 - z_1)^2}. \qquad \blacksquare$$

EXAMPLE 3   The distance between $P_1(2, 1, 5)$ and $P_2(-2, 3, 0)$ is

$$
\begin{aligned}
|P_1 P_2| &= \sqrt{(-2 - 2)^2 + (3 - 1)^2 + (0 - 5)^2} \\
&= \sqrt{16 + 4 + 25} \\
&= \sqrt{45} \approx 6.708. \qquad \blacksquare
\end{aligned}
$$

We can use the distance formula to write equations for spheres in space (Figure 12.6). A point $P(x, y, z)$ lies on the sphere of radius a centered at $P_0(x_0, y_0, z_0)$ precisely when $|P_0 P| = a$ or

$$(x - x_0)^2 + (y - y_0)^2 + (z - z_0)^2 = a^2.$$

The Standard Equation for the Sphere of Radius a and Center (x_0, y_0, z_0)

$$(x - x_0)^2 + (y - y_0)^2 + (z - z_0)^2 = a^2$$

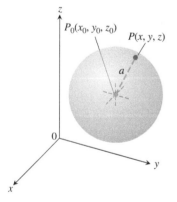

FIGURE 12.6   The sphere of radius a centered at the point (x_0, y_0, z_0).

EXAMPLE 4   Find the center and radius of the sphere

$$x^2 + y^2 + z^2 + 3x - 4z + 1 = 0.$$

Solution   We find the center and radius of a sphere the way we find the center and radius of a circle: Complete the squares on the x-, y-, and z-terms as necessary and write each

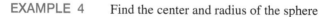

quadratic as a squared linear expression. Then, from the equation in standard form, read off the center and radius. For the sphere here, we have

$$x^2 + y^2 + z^2 + 3x - 4z + 1 = 0$$
$$(x^2 + 3x) + y^2 + (z^2 - 4z) = -1$$
$$\left(x^2 + 3x + \left(\frac{3}{2}\right)^2\right) + y^2 + \left(z^2 - 4z + \left(\frac{-4}{2}\right)^2\right) = -1 + \left(\frac{3}{2}\right)^2 + \left(\frac{-4}{2}\right)^2$$
$$\left(x + \frac{3}{2}\right)^2 + y^2 + (z - 2)^2 = -1 + \frac{9}{4} + 4 = \frac{21}{4}.$$

From this standard form, we read that $x_0 = -3/2$, $y_0 = 0$, $z_0 = 2$, and $a = \sqrt{21}/2$. The center is $(-3/2, 0, 2)$. The radius is $\sqrt{21}/2$. ∎

EXAMPLE 5 Here are some geometric interpretations of inequalities and equations involving spheres.

(a) $x^2 + y^2 + z^2 < 4$ The interior of the sphere $x^2 + y^2 + z^2 = 4$.

(b) $x^2 + y^2 + z^2 \le 4$ The solid ball bounded by the sphere $x^2 + y^2 + z^2 = 4$. Alternatively, the sphere $x^2 + y^2 + z^2 = 4$ together with its interior.

(c) $x^2 + y^2 + z^2 > 4$ The exterior of the sphere $x^2 + y^2 + z^2 = 4$.

(d) $x^2 + y^2 + z^2 = 4, z \le 0$ The lower hemisphere cut from the sphere $x^2 + y^2 + z^2 = 4$ by the xy-plane (the plane $z = 0$). ∎

Just as polar coordinates give another way to locate points in the xy-plane (Section 11.3), alternative coordinate systems, different from the Cartesian coordinate system developed here, exist for three-dimensional space. We examine two of these coordinate systems in Section 15.7.

EXERCISES 12.1

Geometric Interpretations of Equations
In Exercises 1–16, give a geometric description of the set of points in space whose coordinates satisfy the given pairs of equations.

1. $x = 2, \quad y = 3$
2. $x = -1, \quad z = 0$
3. $y = 0, \quad z = 0$
4. $x = 1, \quad y = 0$
5. $x^2 + y^2 = 4, \quad z = 0$
6. $x^2 + y^2 = 4, \quad z = -2$
7. $x^2 + z^2 = 4, \quad y = 0$
8. $y^2 + z^2 = 1, \quad x = 0$
9. $x^2 + y^2 + z^2 = 1, \quad x = 0$
10. $x^2 + y^2 + z^2 = 25, \quad y = -4$
11. $x^2 + y^2 + (z + 3)^2 = 25, \quad z = 0$
12. $x^2 + (y - 1)^2 + z^2 = 4, \quad y = 0$
13. $x^2 + y^2 = 4, \quad z = y$
14. $x^2 + y^2 + z^2 = 4, \quad y = x$
15. $y = x^2, \quad z = 0$
16. $z = y^2, \quad x = 1$

Geometric Interpretations of Inequalities and Equations
In Exercises 17–24, describe the sets of points in space whose coordinates satisfy the given inequalities or combinations of equations and inequalities.

17. a. $x \ge 0, \quad y \ge 0, \quad z = 0$ b. $x \ge 0, \quad y \le 0, \quad z = 0$
18. a. $0 \le x \le 1$ b. $0 \le x \le 1, \quad 0 \le y \le 1$
    c. $0 \le x \le 1, \quad 0 \le y \le 1, \quad 0 \le z \le 1$
19. a. $x^2 + y^2 + z^2 \le 1$ b. $x^2 + y^2 + z^2 > 1$
20. a. $x^2 + y^2 \le 1, \quad z = 0$ b. $x^2 + y^2 \le 1, \quad z = 3$
    c. $x^2 + y^2 \le 1, \quad$ no restriction on z
21. a. $1 \le x^2 + y^2 + z^2 \le 4$ b. $x^2 + y^2 + z^2 \le 1, \quad z \ge 0$
22. a. $x = y, \quad z = 0$ b. $x = y, \quad$ no restriction on z
23. a. $y \ge x^2, \quad z \ge 0$ b. $x \le y^2, \quad 0 \le z \le 2$
24. a. $z = 1 - y, \quad$ no restriction on x
    b. $z = y^3, \quad x = 2$

Distance

In Exercises 25–30, find the distance between points P_1 and P_2.

25. $P_1(1, 1, 1)$, $\quad P_2(3, 3, 0)$

26. $P_1(-1, 1, 5)$, $\quad P_2(2, 5, 0)$

27. $P_1(1, 4, 5)$, $\quad P_2(4, -2, 7)$

28. $P_1(3, 4, 5)$, $\quad P_2(2, 3, 4)$

29. $P_1(0, 0, 0)$, $\quad P_2(2, -2, -2)$

30. $P_1(5, 3, -2)$, $\quad P_2(0, 0, 0)$

31. Find the distance from the point $(3, -4, 2)$ to the
    a. xy-plane   **b.** yz-plane   **c.** xz-plane

32. Find the distance from the point $(-2, 1, 4)$ to the
    a. plane $x = 3$ **b.** plane $y = -5$ **c.** plane $z = -1$

33. Find the distance from the point $(4, 3, 0)$ to the
    a. x-axis   **b.** y-axis   **c.** z-axis

34. Find the distance from the
    a. x-axis to the plane $z = 3$.
    b. origin to the plane $2 = z - x$.
    c. point $(0, 4, 0)$ to the plane $y = x$.

In Exercises 35–44, describe the given set with a single equation or with a pair of equations.

35. The plane perpendicular to the
    a. x-axis at $(3, 0, 0)$ **b.** y-axis at $(0, -1, 0)$
    c. z-axis at $(0, 0, -2)$

36. The plane through the point $(3, -1, 2)$ perpendicular to the
    a. x-axis   **b.** y-axis   **c.** z-axis

37. The plane through the point $(3, -1, 1)$ parallel to the
    a. xy-plane   **b.** yz-plane   **c.** xz-plane

38. The circle of radius 2 centered at $(0, 0, 0)$ and lying in the
    a. xy-plane   **b.** yz-plane   **c.** xz-plane

39. The circle of radius 2 centered at $(0, 2, 0)$ and lying in the
    a. xy-plane   **b.** yz-plane   **c.** plane $y = 2$

40. The circle of radius 1 centered at $(-3, 4, 1)$ and lying in a plane parallel to the
    a. xy-plane   **b.** yz-plane   **c.** xz-plane

41. The line through the point $(1, 3, -1)$ parallel to the
    a. x-axis   **b.** y-axis   **c.** z-axis

42. The set of points in space equidistant from the origin and the point $(0, 2, 0)$

43. The circle in which the plane through the point $(1, 1, 3)$ perpendicular to the z-axis meets the sphere of radius 5 centered at the origin

44. The set of points in space that lie 2 units from the point $(0, 0, 1)$ and, at the same time, 2 units from the point $(0, 0, -1)$

Inequalities to Describe Sets of Points

Write inequalities to describe the sets in Exercises 45–50.

45. The slab bounded by the planes $z = 0$ and $z = 1$ (planes included)

46. The solid cube in the first octant bounded by the coordinate planes and the planes $x = 2$, $y = 2$, and $z = 2$

47. The half-space consisting of the points on and below the xy-plane

48. The upper hemisphere of the sphere of radius 1 centered at the origin

49. The **(a)** interior and **(b)** exterior of the sphere of radius 1 centered at the point $(1, 1, 1)$

50. The closed region bounded by the spheres of radius 1 and radius 2 centered at the origin. (*Closed* means the spheres are to be included. Had we wanted the spheres left out, we would have asked for the *open* region bounded by the spheres. This is analogous to the way we use *closed* and *open* to describe intervals: *closed* means endpoints included, *open* means endpoints left out. Closed sets include boundaries; open sets leave them out.)

Spheres

Find the center C and the radius a for the spheres in Exercises 51–60.

51. $(x + 2)^2 + y^2 + (z - 2)^2 = 8$

52. $(x - 1)^2 + \left(y + \dfrac{1}{2}\right)^2 + (z + 3)^2 = 25$

53. $(x - \sqrt{2})^2 + (y - \sqrt{2})^2 + (z + \sqrt{2})^2 = 2$

54. $x^2 + \left(y + \dfrac{1}{3}\right)^2 + \left(z - \dfrac{1}{3}\right)^2 = \dfrac{16}{9}$

55. $x^2 + y^2 + z^2 + 4x - 4z = 0$

56. $x^2 + y^2 + z^2 - 6y + 8z = 0$

57. $2x^2 + 2y^2 + 2z^2 + x + y + z = 9$

58. $3x^2 + 3y^2 + 3z^2 + 2y - 2z = 9$

59. $x^2 + y^2 + z^2 - 4x + 6y - 10z = 11$

60. $(x - 1)^2 + (y - 2)^2 + (z + 1)^2 = 103 + 2x + 4y - 2z$

Find equations for the spheres whose centers and radii are given in Exercises 61–64.

	Center	Radius
61.	$(1, 2, 3)$	$\sqrt{14}$
62.	$(0, -1, 5)$	2
63.	$\left(-1, \dfrac{1}{2}, -\dfrac{2}{3}\right)$	$\dfrac{4}{9}$
64.	$(0, -7, 0)$	7

Theory and Examples

65. Find a formula for the distance from the point $P(x, y, z)$ to the
    a. x-axis.   **b.** y-axis.   **c.** z-axis.

66. Find a formula for the distance from the point $P(x, y, z)$ to the
    a. xy-plane.   **b.** yz-plane.   **c.** xz-plane.

67. Find the perimeter of the triangle with vertices $A(-1, 2, 1)$, $B(1, -1, 3)$, and $C(3, 4, 5)$.

68. Show that the point $P(3, 1, 2)$ is equidistant from the points $A(2, -1, 3)$ and $B(4, 3, 1)$.

69. Find an equation for the set of all points equidistant from the planes $y = 3$ and $y = -1$.

70. Find an equation for the set of all points equidistant from the point $(0, 0, 2)$ and the xy-plane.

71. Find the point on the sphere $x^2 + (y - 3)^2 + (z + 5)^2 = 4$ nearest
    a. the xy-plane.   b. the point $(0, 7, -5)$.

72. Find the point equidistant from the points $(0, 0, 0)$, $(0, 4, 0)$, $(3, 0, 0)$, and $(2, 2, -3)$.

73. Find an equation for the set of points equidistant from the point $(0, 0, 2)$ and the x-axis.

74. Find an equation for the set of points equidistant from the y-axis and the plane $z = 6$.

75. Find an equation for the set of points equidistant from the
    a. xy-plane and yz-plane.
    b. x-axis and y-axis.

76. Find all points that simultaneously lie 3 units from each of the points $(2, 0, 0)$, $(0, 2, 0)$, and $(0, 0, 2)$.

12.2 Vectors

Some of the things we measure are determined simply by their magnitudes. To record mass, length, or time, for example, we need only write down a number and name an appropriate unit of measure. We need more information to describe a force, displacement, or velocity. To describe a force, we need to record the direction in which it acts as well as how large it is. To describe a body's displacement, we have to say in what direction it moved as well as how far. To describe a body's velocity, we have to know its direction of motion, as well as how fast it is going. In this section we show how to represent things that have both magnitude and direction in the plane or in space.

Component Form

A quantity such as force, displacement, or velocity is called a **vector** and is represented by a **directed line segment** (Figure 12.7). The arrow points in the direction of the action and its length gives the magnitude of the action in terms of a suitably chosen unit. For example, a force vector points in the direction in which the force acts and its length is a measure of the force's strength; a velocity vector points in the direction of motion and its length is the speed of the moving object. Figure 12.8 displays the velocity vector **v** at a specific location for a particle moving along a path in the plane or in space. (This application of vectors is studied in Chapter 13.)

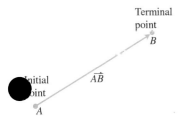

FIGURE 12.7 The directed line segment $\overrightarrow{AB}$ is called a vector.

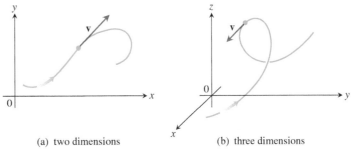

(a) two dimensions   (b) three dimensions

FIGURE 12.8 The velocity vector of a particle moving along a path (a) in the plane (b) in space. The arrowhead on the path indicates the direction of motion of the particle.

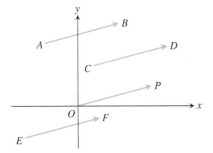

FIGURE 12.9 The four arrows in the plane (directed line segments) shown here have the same length and direction. They therefore represent the same vector, and we write $\overrightarrow{AB} = \overrightarrow{CD} = \overrightarrow{OP} = \overrightarrow{EF}$.

> **DEFINITIONS** The vector represented by the directed line segment $\overrightarrow{AB}$ has **initial point** A and **terminal point** B and its **length** is denoted by $|\overrightarrow{AB}|$. Two vectors are **equal** if they have the same length and direction.

The arrows we use when we draw vectors are understood to represent the same vector if they have the same length, are parallel, and point in the same direction (Figure 12.9) regardless of the initial point.

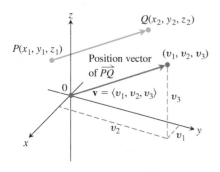

FIGURE 12.10 A vector $\overrightarrow{PQ}$ in standard position has its initial point at the origin. The directed line segments $\overrightarrow{PQ}$ and $\mathbf{v}$ are parallel and have the same length.

HISTORICAL BIOGRAPHY

Carl Friedrich Gauss
(1777–1855)
bit.ly/2xZSPyS

In textbooks, vectors are usually written in lowercase, boldface letters, for example $\mathbf{u}$, $\mathbf{v}$, and $\mathbf{w}$. Sometimes we use uppercase boldface letters, such as $\mathbf{F}$, to denote a force vector. In handwritten form, it is customary to draw small arrows above the letters, for example $\vec{u}$, $\vec{v}$, $\vec{w}$, and $\vec{F}$.

We need a way to represent vectors algebraically so that we can be more precise about the direction of a vector. Let $\mathbf{v} = \overrightarrow{PQ}$. There is one directed line segment equal to $\overrightarrow{PQ}$ whose initial point is the origin (Figure 12.10). It is the representative of $\mathbf{v}$ in **standard position** and is the vector we normally use to represent $\mathbf{v}$. We can specify $\mathbf{v}$ by writing the coordinates of its terminal point (v_1, v_2, v_3) when $\mathbf{v}$ is in standard position. If $\mathbf{v}$ is a vector in the plane its terminal point (v_1, v_2) has two coordinates.

> **DEFINITION** If $\mathbf{v}$ is a **two-dimensional** vector in the plane equal to the vector with initial point at the origin and terminal point (v_1, v_2), then the **component form** of $\mathbf{v}$ is
>
> $$\mathbf{v} = \langle v_1, v_2 \rangle.$$
>
> If $\mathbf{v}$ is a **three-dimensional** vector equal to the vector with initial point at the origin and terminal point (v_1, v_2, v_3), then the **component form** of $\mathbf{v}$ is
>
> $$\mathbf{v} = \langle v_1, v_2, v_3 \rangle.$$

So a two-dimensional vector is an ordered pair $\mathbf{v} = \langle v_1, v_2 \rangle$ of real numbers, and a three-dimensional vector is an ordered triple $\mathbf{v} = \langle v_1, v_2, v_3 \rangle$ of real numbers. The numbers v_1, v_2, and v_3 are the **components** of $\mathbf{v}$.

If $\mathbf{v} = \langle v_1, v_2, v_3 \rangle$ is represented by the directed line segment $\overrightarrow{PQ}$, where the initial point is $P(x_1, y_1, z_1)$ and the terminal point is $Q(x_2, y_2, z_2)$, then $x_1 + v_1 = x_2$, $y_1 + v_2 = y_2$, and $z_1 + v_3 = z_2$ (see Figure 12.10). Thus, $v_1 = x_2 - x_1$, $v_2 = y_2 - y_1$, and $v_3 = z_2 - z_1$ are the components of $\overrightarrow{PQ}$.

In summary, given the points $P(x_1, y_1, z_1)$ and $Q(x_2, y_2, z_2)$, the standard position vector $\mathbf{v} = \langle v_1, v_2, v_3 \rangle$ equal to $\overrightarrow{PQ}$ is

$$\mathbf{v} = \langle x_2 - x_1, y_2 - y_1, z_2 - z_1 \rangle.$$

If $\mathbf{v}$ is two-dimensional with $P(x_1, y_1)$ and $Q(x_2, y_2)$ as points in the plane, then $\mathbf{v} = \langle x_2 - x_1, y_2 - y_1 \rangle$. There is no third component for planar vectors. With this understanding, we will develop the algebra of three-dimensional vectors and simply drop the third component when the vector is two-dimensional (a planar vector).

Two vectors are equal if and only if their standard position vectors are identical. Thus $\langle u_1, u_2, u_3 \rangle$ and $\langle v_1, v_2, v_3 \rangle$ are **equal** if and only if $u_1 = v_1$, $u_2 = v_2$, and $u_3 = v_3$.

The **magnitude** or **length** of the vector $\overrightarrow{PQ}$ is the length of any of its equivalent directed line segment representations. In particular, if $\mathbf{v} = \langle x_2 - x_1, y_2 - y_1, z_2 - z_1 \rangle$ is the standard position vector for $\overrightarrow{PQ}$, then the distance formula gives the magnitude or length of $\mathbf{v}$, denoted by the symbol $|\mathbf{v}|$ or $\|\mathbf{v}\|$.

> The **magnitude** or **length** of the vector $\mathbf{v} = \overrightarrow{PQ}$ is the nonnegative number
>
> $$|\mathbf{v}| = \sqrt{v_1^2 + v_2^2 + v_3^2} = \sqrt{(x_2 - x_1)^2 + (y_2 - y_1)^2 + (z_2 - z_1)^2}$$
>
> (see Figure 12.10).

The only vector with length 0 is the **zero vector** $\mathbf{0} = \langle 0, 0 \rangle$ or $\mathbf{0} = \langle 0, 0, 0 \rangle$. This vector is also the only vector with no specific direction.

EXAMPLE 1   Find the **(a)** component form and **(b)** length of the vector with initial point $P(-3, 4, 1)$ and terminal point $Q(-5, 2, 2)$.

Solution

(a) The standard position vector $\mathbf{v}$ representing $\overrightarrow{PQ}$ has components

$$v_1 = x_2 - x_1 = -5 - (-3) = -2, \qquad v_2 = y_2 - y_1 = 2 - 4 = -2,$$

and

$$v_3 = z_2 - z_1 = 2 - 1 = 1.$$

The component form of $\overrightarrow{PQ}$ is

$$\mathbf{v} = \langle -2, -2, 1 \rangle.$$

(b) The length or magnitude of $\mathbf{v} = \overrightarrow{PQ}$ is

$$|\mathbf{v}| = \sqrt{(-2)^2 + (-2)^2 + (1)^2} = \sqrt{9} = 3.$$

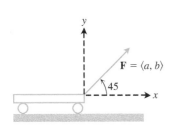

FIGURE 12.11   The force pulling the cart forward is represented by the vector **F** whose horizontal component is the effective force (Example 2).

EXAMPLE 2   A small cart is being pulled along a smooth horizontal floor with a 20-lb force **F** making a 45° angle to the floor (Figure 12.11). What is the *effective* force moving the cart forward?

Solution   The effective force is the horizontal component of $\mathbf{F} = \langle a, b \rangle$, given by

$$a = |\mathbf{F}| \cos 45° = (20)\left(\frac{\sqrt{2}}{2}\right) \approx 14.14 \text{ lb}.$$

Notice that **F** is a two-dimensional vector.

Vector Algebra Operations

Two principal operations involving vectors are *vector addition* and *scalar multiplication*. A **scalar** is simply a real number, and is called such when we want to draw attention to the differences between numbers and vectors. Scalars can be positive, negative, or zero and are used to "scale" a vector by multiplication.

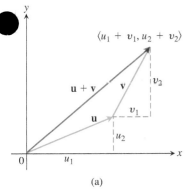

(a)

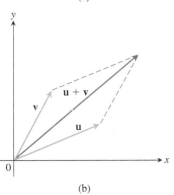

(b)

FIGURE 12.12   (a) Geometric interpretation of the vector sum. (b) The parallelogram law of vector addition in which both vectors are in standard position.

DEFINITIONS   Let $\mathbf{u} = \langle u_1, u_2, u_3 \rangle$ and $\mathbf{v} = \langle v_1, v_2, v_3 \rangle$ be vectors with k a scalar.

Addition: $\mathbf{u} + \mathbf{v} = \langle u_1 + v_1, u_2 + v_2, u_3 + v_3 \rangle$

Scalar multiplication: $k\mathbf{u} = \langle ku_1, ku_2, ku_3 \rangle$

We add vectors by adding the corresponding components of the vectors. We multiply a vector by a scalar by multiplying each component by the scalar. The definitions also apply to planar vectors, except in that case there are only two components, $\langle u_1, u_2 \rangle$ and $\langle v_1, v_2 \rangle$.

The definition of vector addition is illustrated geometrically for planar vectors in Figure 12.12a, where the initial point of one vector is placed at the terminal point of the other. Another interpretation is shown in Figure 12.12b (called the **parallelogram law** of addition), where the sum, called the **resultant vector**, is the diagonal of the parallelogram. In physics, forces add vectorially as do velocities, accelerations, and so on. So the force acting on a particle subject to two gravitational forces, for example, is obtained by adding the two force vectors.

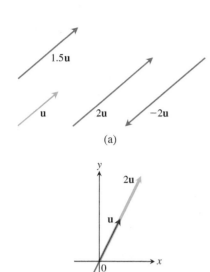

(a)

(b)

FIGURE 12.13 (a) Scalar multiples of **u**. (b) Scalar multiples of a vector **u** in standard position.

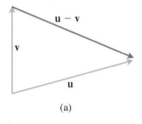

(a)

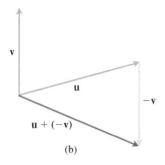

(b)

FIGURE 12.14 (a) The vector **u** − **v**, when added to **v**, gives **u**. (b) **u** − **v** = **u** + (−**v**).

Figure 12.13 displays a geometric interpretation of the product $k\mathbf{u}$ of the scalar k and vector **u**. If $k > 0$, then $k\mathbf{u}$ has the same direction as **u**; if $k < 0$, then the direction of $k\mathbf{u}$ is opposite to that of **u**. Comparing the lengths of **u** and $k\mathbf{u}$, we see that

$$|k\mathbf{u}| = \sqrt{(ku_1)^2 + (ku_2)^2 + (ku_3)^2} = \sqrt{k^2(u_1^2 + u_2^2 + u_3^2)}$$
$$= \sqrt{k^2}\sqrt{u_1^2 + u_2^2 + u_3^2} = |k||\mathbf{u}|.$$

The length of $k\mathbf{u}$ is the absolute value of the scalar k times the length of **u**. The vector $(-1)\mathbf{u} = -\mathbf{u}$ has the same length as **u** but points in the opposite direction.

The **difference u − v** of two vectors is defined by

$$\mathbf{u} - \mathbf{v} = \mathbf{u} + (-\mathbf{v}).$$

If $\mathbf{u} = \langle u_1, u_2, u_3 \rangle$ and $\mathbf{v} = \langle v_1, v_2, v_3 \rangle$, then

$$\mathbf{u} - \mathbf{v} = \langle u_1 - v_1, u_2 - v_2, u_3 - v_3 \rangle.$$

Note that $(\mathbf{u} - \mathbf{v}) + \mathbf{v} = \mathbf{u}$, so adding the vector $(\mathbf{u} - \mathbf{v})$ to **v** gives **u** (Figure 12.14a). Figure 12.14b shows the difference **u** − **v** as the sum **u** + (−**v**).

EXAMPLE 3 Let $\mathbf{u} = \langle -1, 3, 1 \rangle$ and $\mathbf{v} = \langle 4, 7, 0 \rangle$. Find the components of

(a) $2\mathbf{u} + 3\mathbf{v}$ **(b)** $\mathbf{u} - \mathbf{v}$ **(c)** $\left|\dfrac{1}{2}\mathbf{u}\right|$.

Solution

(a) $2\mathbf{u} + 3\mathbf{v} = 2\langle -1, 3, 1 \rangle + 3\langle 4, 7, 0 \rangle = \langle -2, 6, 2 \rangle + \langle 12, 21, 0 \rangle = \langle 10, 27, 2 \rangle$

(b) $\mathbf{u} - \mathbf{v} = \langle -1, 3, 1 \rangle - \langle 4, 7, 0 \rangle = \langle -1 - 4, 3 - 7, 1 - 0 \rangle = \langle -5, -4, 1 \rangle$

(c) $\left|\dfrac{1}{2}\mathbf{u}\right| = \left|\left\langle -\dfrac{1}{2}, \dfrac{3}{2}, \dfrac{1}{2} \right\rangle\right| = \sqrt{\left(-\dfrac{1}{2}\right)^2 + \left(\dfrac{3}{2}\right)^2 + \left(\dfrac{1}{2}\right)^2} = \dfrac{1}{2}\sqrt{11}.$

Vector operations have many of the properties of ordinary arithmetic.

Properties of Vector Operations

Let **u**, **v**, **w** be vectors and a, b be scalars.

1. $\mathbf{u} + \mathbf{v} = \mathbf{v} + \mathbf{u}$
2. $(\mathbf{u} + \mathbf{v}) + \mathbf{w} = \mathbf{u} + (\mathbf{v} + \mathbf{w})$
3. $\mathbf{u} + \mathbf{0} = \mathbf{u}$
4. $\mathbf{u} + (-\mathbf{u}) = \mathbf{0}$
5. $0\mathbf{u} = \mathbf{0}$
6. $1\mathbf{u} = \mathbf{u}$
7. $a(b\mathbf{u}) = (ab)\mathbf{u}$
8. $a(\mathbf{u} + \mathbf{v}) = a\mathbf{u} + a\mathbf{v}$
9. $(a + b)\mathbf{u} = a\mathbf{u} + b\mathbf{u}$

These properties are readily verified using the definitions of vector addition and multiplication by a scalar. For instance, to establish Property 1, we have

$$\mathbf{u} + \mathbf{v} = \langle u_1, u_2, u_3 \rangle + \langle v_1, v_2, v_3 \rangle$$
$$= \langle u_1 + v_1, u_2 + v_2, u_3 + v_3 \rangle$$
$$= \langle v_1 + u_1, v_2 + u_2, v_3 + u_3 \rangle$$
$$= \langle v_1, v_2, v_3 \rangle + \langle u_1, u_2, u_3 \rangle$$
$$= \mathbf{v} + \mathbf{u}.$$

When three or more space vectors lie in the same plane, we say they are **coplanar** vectors. For example, the vectors **u**, **v**, and **u** + **v** are always coplanar.

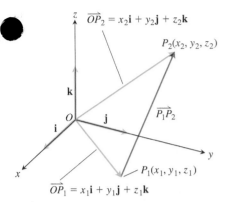

FIGURE 12.15 The vector from P_1 to P_2 is $\overrightarrow{P_1P_2} = (x_2 - x_1)\mathbf{i} + (y_2 - y_1)\mathbf{j} + (z_2 - z_1)\mathbf{k}$.

Unit Vectors

A vector $\mathbf{v}$ of length 1 is called a **unit vector**. The **standard unit vectors** are

$$\mathbf{i} = \langle 1, 0, 0 \rangle, \quad \mathbf{j} = \langle 0, 1, 0 \rangle, \quad \text{and} \quad \mathbf{k} = \langle 0, 0, 1 \rangle.$$

Any vector $\mathbf{v} = \langle v_1, v_2, v_3 \rangle$ can be written as a *linear combination* of the standard unit vectors as follows:

$$\begin{aligned}
\mathbf{v} = \langle v_1, v_2, v_3 \rangle &= \langle v_1, 0, 0 \rangle + \langle 0, v_2, 0 \rangle + \langle 0, 0, v_3 \rangle \\
&= v_1 \langle 1, 0, 0 \rangle + v_2 \langle 0, 1, 0 \rangle + v_3 \langle 0, 0, 1 \rangle \\
&= v_1 \mathbf{i} + v_2 \mathbf{j} + v_3 \mathbf{k}.
\end{aligned}$$

We call the scalar (or number) v_1 the **i-component** of the vector $\mathbf{v}$, v_2 the **j-component**, and v_3 the **k-component**. As shown in Figure 12.15, the component form for the vector from $P_1(x_1, y_1, z_1)$ to $P_2(x_2, y_2, z_2)$ is

$$\overrightarrow{P_1P_2} = (x_2 - x_1)\mathbf{i} + (y_2 - y_1)\mathbf{j} + (z_2 - z_1)\mathbf{k}.$$

If $\mathbf{v} \neq \mathbf{0}$, then its length $|\mathbf{v}|$ is not zero and

$$\left| \frac{1}{|\mathbf{v}|} \mathbf{v} \right| = \frac{1}{|\mathbf{v}|} |\mathbf{v}| = 1.$$

That is, $\mathbf{v}/|\mathbf{v}|$ is a unit vector in the direction of $\mathbf{v}$, called **the direction** of the nonzero vector $\mathbf{v}$.

EXAMPLE 4 Find a unit vector $\mathbf{u}$ in the direction of the vector from $P_1(1, 0, 1)$ to $P_2(3, 2, 0)$.

Solution We divide $\overrightarrow{P_1P_2}$ by its length:

$$\overrightarrow{P_1P_2} = (3 - 1)\mathbf{i} + (2 - 0)\mathbf{j} + (0 - 1)\mathbf{k} = 2\mathbf{i} + 2\mathbf{j} - \mathbf{k}$$

$$\left| \overrightarrow{P_1P_2} \right| = \sqrt{(2)^2 + (2)^2 + (-1)^2} = \sqrt{4 + 4 + 1} = \sqrt{9} = 3$$

$$\mathbf{u} = \frac{\overrightarrow{P_1P_2}}{\left| \overrightarrow{P_1P_2} \right|} = \frac{2\mathbf{i} + 2\mathbf{j} - \mathbf{k}}{3} = \frac{2}{3}\mathbf{i} + \frac{2}{3}\mathbf{j} - \frac{1}{3}\mathbf{k}.$$

This unit vector $\mathbf{u}$ is the direction of $\overrightarrow{P_1P_2}$. ∎

EXAMPLE 5 If $\mathbf{v} = 3\mathbf{i} - 4\mathbf{j}$ is a velocity vector, express $\mathbf{v}$ as a product of its speed times its direction of motion.

Solution Speed is the magnitude (length) of $\mathbf{v}$:

$$|\mathbf{v}| = \sqrt{(3)^2 + (-4)^2} = \sqrt{9 + 16} = 5.$$

The unit vector $\mathbf{v}/|\mathbf{v}|$ is the direction of $\mathbf{v}$:

$$\frac{\mathbf{v}}{|\mathbf{v}|} = \frac{3\mathbf{i} - 4\mathbf{j}}{5} = \frac{3}{5}\mathbf{i} - \frac{4}{5}\mathbf{j}.$$

So

$$\mathbf{v} = 3\mathbf{i} - 4\mathbf{j} = 5\left(\frac{3}{5}\mathbf{i} - \frac{4}{5}\mathbf{j} \right).$$

Length   Direction of motion
(speed)

In summary, we can express any nonzero vector $\mathbf{v}$ in terms of its two important features, length and direction, by writing $\mathbf{v} = |\mathbf{v}| \dfrac{\mathbf{v}}{|\mathbf{v}|}$.

If $\mathbf{v} \neq \mathbf{0}$, then

1. $\dfrac{\mathbf{v}}{|\mathbf{v}|}$ is a unit vector called the direction of $\mathbf{v}$;

2. the equation $\mathbf{v} = |\mathbf{v}| \dfrac{\mathbf{v}}{|\mathbf{v}|}$ expresses $\mathbf{v}$ as its length times its direction.

EXAMPLE 6   A force of 6 newtons is applied in the direction of the vector $\mathbf{v} = 2\mathbf{i} + 2\mathbf{j} - \mathbf{k}$. Express the force $\mathbf{F}$ as a product of its magnitude and direction.

Solution   The force vector has magnitude 6 and direction $\dfrac{\mathbf{v}}{|\mathbf{v}|}$, so

$$\mathbf{F} = 6\,\frac{\mathbf{v}}{|\mathbf{v}|} = 6\,\frac{2\mathbf{i} + 2\mathbf{j} - \mathbf{k}}{\sqrt{2^2 + 2^2 + (-1)^2}} = 6\,\frac{2\mathbf{i} + 2\mathbf{j} - \mathbf{k}}{3}$$

$$= 6\left(\frac{2}{3}\mathbf{i} + \frac{2}{3}\mathbf{j} - \frac{1}{3}\mathbf{k}\right).$$

Midpoint of a Line Segment

Vectors are often useful in geometry. For example, the coordinates of the midpoint of a line segment are found by averaging.

The **midpoint** M of the line segment joining points $P_1(x_1, y_1, z_1)$ and $P_2(x_2, y_2, z_2)$ is the point

$$\left(\frac{x_1 + x_2}{2}, \frac{y_1 + y_2}{2}, \frac{z_1 + z_2}{2}\right).$$

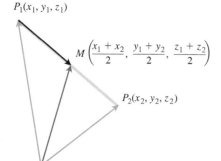

FIGURE 12.16 The coordinates of the midpoint are the averages of the coordinates of P_1 and P_2.

To see why, observe (Figure 12.16) that

$$\overrightarrow{OM} = \overrightarrow{OP_1} + \frac{1}{2}\,(\overrightarrow{P_1 P_2}) = \overrightarrow{OP_1} + \frac{1}{2}\,(\overrightarrow{OP_2} - \overrightarrow{OP_1})$$

$$= \frac{1}{2}\,(\overrightarrow{OP_1} + \overrightarrow{OP_2})$$

$$= \frac{x_1 + x_2}{2}\mathbf{i} + \frac{y_1 + y_2}{2}\mathbf{j} + \frac{z_1 + z_2}{2}\mathbf{k}.$$

EXAMPLE 7   The midpoint of the segment joining $P_1(3, -2, 0)$ and $P_2(7, 4, 4)$ is

$$\left(\frac{3 + 7}{2}, \frac{-2 + 4}{2}, \frac{0 + 4}{2}\right) = (5, 1, 2).$$

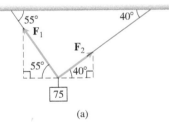

FIGURE 12.17 Vectors representing the velocities of the airplane **u** and tailwind **v** in Example 8.

Applications

An important application of vectors occurs in navigation.

EXAMPLE 8   A jet airliner, flying due east at 500 mph in still air, encounters a 70-mph tailwind blowing in the direction 60° north of east. The airplane holds its compass heading due east but, because of the wind, acquires a new ground speed and direction. What are they?

Solution   If **u** is the velocity of the airplane alone and **v** is the velocity of the tailwind, then $|\mathbf{u}| = 500$ and $|\mathbf{v}| = 70$ (Figure 12.17). The velocity of the airplane with respect to the ground is given by the magnitude and direction of the resultant vector **u** + **v**. If we let the positive x-axis represent east and the positive y-axis represent north, then the component forms of **u** and **v** are

$$\mathbf{u} = \langle 500, 0 \rangle \qquad \text{and} \qquad \mathbf{v} = \langle 70 \cos 60°, 70 \sin 60° \rangle = \langle 35, 35\sqrt{3} \rangle.$$

Therefore,

$$\mathbf{u} + \mathbf{v} = \langle 535, 35\sqrt{3} \rangle = 535\mathbf{i} + 35\sqrt{3}\,\mathbf{j}$$
$$|\mathbf{u} + \mathbf{v}| = \sqrt{535^2 + (35\sqrt{3})^2} \approx 538.4$$

and

$$\theta = \tan^{-1}\frac{35\sqrt{3}}{535} \approx 6.5°. \qquad \text{Figure 12.17}$$

The new ground speed of the airplane is about 538.4 mph, and its new direction is about 6.5° north of east.   ◼

Another important application occurs in physics and engineering when several forces are acting on a single object.

EXAMPLE 9   A 75-N weight is suspended by two wires, as shown in Figure 12.18a. Find the forces $\mathbf{F}_1$ and $\mathbf{F}_2$ acting in both wires.

Solution   The force vectors $\mathbf{F}_1$ and $\mathbf{F}_2$ have magnitudes $|\mathbf{F}_1|$ and $|\mathbf{F}_2|$ and components that are measured in newtons. The resultant force is the sum $\mathbf{F}_1 + \mathbf{F}_2$ and must be equal in magnitude and acting in the opposite (or upward) direction to the weight vector **w** (see Figure 12.18b). It follows from the figure that

$$\mathbf{F}_1 = \langle -|\mathbf{F}_1| \cos 55°, |\mathbf{F}_1| \sin 55° \rangle \qquad \text{and} \qquad \mathbf{F}_2 = \langle |\mathbf{F}_2| \cos 40°, |\mathbf{F}_2| \sin 40° \rangle.$$

Since $\mathbf{F}_1 + \mathbf{F}_2 = \langle 0, 75 \rangle$, the resultant vector leads to the system of equations

$$-|\mathbf{F}_1| \cos 55° + |\mathbf{F}_2| \cos 40° = 0$$
$$|\mathbf{F}_1| \sin 55° + |\mathbf{F}_2| \sin 40° = 75.$$

Solving for $|\mathbf{F}_2|$ in the first equation and substituting the result into the second equation, we get

$$|\mathbf{F}_2| = \frac{|\mathbf{F}_1| \cos 55°}{\cos 40°} \qquad \text{and} \qquad |\mathbf{F}_1| \sin 55° + \frac{|\mathbf{F}_1| \cos 55°}{\cos 40°} \sin 40° = 75.$$

It follows that

$$|\mathbf{F}_1| = \frac{75}{\sin 55° + \cos 55° \tan 40°} \approx 57.67 \text{ N},$$

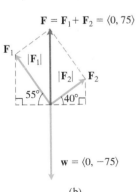

FIGURE 12.18 The suspended weight in Example 9.

and

$$|\mathbf{F}_2| = \frac{75 \cos 55°}{\sin 55° \cos 40° + \cos 55° \sin 40°}$$

$$= \frac{75 \cos 55°}{\sin (55° + 40°)} \approx 43.18 \text{ N.}$$

The force vectors are then

$$\mathbf{F}_1 = \langle -|\mathbf{F}_1| \cos 55°, |\mathbf{F}_1| \sin 55° \rangle \approx \langle -33.08, 47.24 \rangle$$

and

$$\mathbf{F}_2 = \langle |\mathbf{F}_2| \cos 40°, |\mathbf{F}_2| \sin 40° \rangle \approx \langle 33.08, 27.76 \rangle.$$

EXERCISES 12.2

Vectors in the Plane

In Exercises 1–8, let $\mathbf{u} = \langle 3, -2 \rangle$ and $\mathbf{v} = \langle -2, 5 \rangle$. Find the (a) component form and (b) magnitude (length) of the vector.

1. $3\mathbf{u}$

2. $-2\mathbf{v}$

3. $\mathbf{u} + \mathbf{v}$

4. $\mathbf{u} - \mathbf{v}$

5. $2\mathbf{u} - 3\mathbf{v}$

6. $-2\mathbf{u} + 5\mathbf{v}$

7. $\frac{3}{5}\mathbf{u} + \frac{4}{5}\mathbf{v}$

8. $-\frac{5}{13}\mathbf{u} + \frac{12}{13}\mathbf{v}$

In Exercises 9–16, find the component form of the vector.

9. The vector $\overrightarrow{PQ}$, where $P = (1, 3)$ and $Q = (2, -1)$

10. The vector $\overrightarrow{OP}$ where O is the origin and P is the midpoint of segment RS, where $R = (2, -1)$ and $S = (-4, 3)$

11. The vector from the point $A = (2, 3)$ to the origin

12. The sum of $\overrightarrow{AB}$ and $\overrightarrow{CD}$, where $A = (1, -1), B = (2, 0)$, $C = (-1, 3)$, and $D = (-2, 2)$

13. The unit vector that makes an angle $\theta = 2\pi/3$ with the positive x-axis

14. The unit vector that makes an angle $\theta = -3\pi/4$ with the positive x-axis

15. The unit vector obtained by rotating the vector $\langle 0, 1 \rangle$ 120° counterclockwise about the origin

16. The unit vector obtained by rotating the vector $\langle 1, 0 \rangle$ 135° counterclockwise about the origin

Vectors in Space

In Exercises 17–22, express each vector in the form $\mathbf{v} = v_1\mathbf{i} + v_2\mathbf{j} + v_3\mathbf{k}$.

17. $\overrightarrow{P_1P_2}$ if P_1 is the point $(5, 7, -1)$ and P_2 is the point $(2, 9, -2)$

18. $\overrightarrow{P_1P_2}$ if P_1 is the point $(1, 2, 0)$ and P_2 is the point $(-3, 0, 5)$

19. $\overrightarrow{AB}$ if A is the point $(-7, -8, 1)$ and B is the point $(-10, 8, 1)$

20. $\overrightarrow{AB}$ if A is the point $(1, 0, 3)$ and B is the point $(-1, 4, 5)$

21. $5\mathbf{u} - \mathbf{v}$ if $\mathbf{u} = \langle 1, 1, -1 \rangle$ and $\mathbf{v} = \langle 2, 0, 3 \rangle$

22. $-2\mathbf{u} + 3\mathbf{v}$ if $\mathbf{u} = \langle -1, 0, 2 \rangle$ and $\mathbf{v} = \langle 1, 1, 1 \rangle$

Geometric Representations

In Exercises 23 and 24, copy vectors $\mathbf{u}$, $\mathbf{v}$, and $\mathbf{w}$ head to tail as needed to sketch the indicated vector.

23.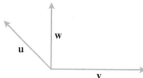

a. $\mathbf{u} + \mathbf{v}$

b. $\mathbf{u} + \mathbf{v} + \mathbf{w}$

c. $\mathbf{u} - \mathbf{v}$

d. $\mathbf{u} - \mathbf{w}$

24.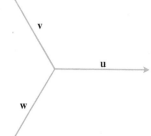

a. $\mathbf{u} - \mathbf{v}$

b. $\mathbf{u} - \mathbf{v} + \mathbf{w}$

c. $2\mathbf{u} - \mathbf{v}$

d. $\mathbf{u} + \mathbf{v} + \mathbf{w}$

Length and Direction

In Exercises 25–30, express each vector as a product of its length and direction.

25. $2\mathbf{i} + \mathbf{j} - 2\mathbf{k}$

26. $9\mathbf{i} - 2\mathbf{j} + 6\mathbf{k}$

27. $5\mathbf{k}$

28. $\frac{3}{5}\mathbf{i} + \frac{4}{5}\mathbf{k}$

29. $\frac{1}{\sqrt{6}}\mathbf{i} - \frac{1}{\sqrt{6}}\mathbf{j} - \frac{1}{\sqrt{6}}\mathbf{k}$

30. $\frac{\mathbf{i}}{\sqrt{3}} + \frac{\mathbf{j}}{\sqrt{3}} + \frac{\mathbf{k}}{\sqrt{3}}$

31. Find the vectors whose lengths and directions are given. Try to do the calculations without writing.

Length	Direction
a. 2	$\mathbf{i}$
b. $\sqrt{3}$	$-\mathbf{k}$
c. $\dfrac{1}{2}$	$\dfrac{3}{5}\mathbf{j} + \dfrac{4}{5}\mathbf{k}$
d. 7	$\dfrac{6}{7}\mathbf{i} - \dfrac{2}{7}\mathbf{j} + \dfrac{3}{7}\mathbf{k}$

32. Find the vectors whose lengths and directions are given. Try to do the calculations without writing.

Length	Direction
a. 7	$-\mathbf{j}$
b. $\sqrt{2}$	$-\dfrac{3}{5}\mathbf{i} - \dfrac{4}{5}\mathbf{k}$
c. $\dfrac{13}{12}$	$\dfrac{3}{13}\mathbf{i} - \dfrac{4}{13}\mathbf{j} - \dfrac{12}{13}\mathbf{k}$
d. $a > 0$	$\dfrac{1}{\sqrt{2}}\mathbf{i} + \dfrac{1}{\sqrt{3}}\mathbf{j} - \dfrac{1}{\sqrt{6}}\mathbf{k}$

33. Find a vector of magnitude 7 in the direction of $\mathbf{v} = 12\mathbf{i} - 5\mathbf{k}$.

34. Find a vector of magnitude 3 in the direction opposite to the direction of $\mathbf{v} = (1/2)\mathbf{i} - (1/2)\mathbf{j} - (1/2)\mathbf{k}$.

Direction and Midpoints

In Exercises 35–38, find **a.** the direction of $\overrightarrow{P_1 P_2}$ and **b.** the midpoint of line segment $P_1 P_2$.

35. $P_1(-1, 1, 5)$ $P_2(2, 5, 0)$

36. $P_1(1, 4, 5)$ $P_2(4, -2, 7)$

37. $P_1(3, 4, 5)$ $P_2(2, 3, 4)$

38. $P_1(0, 0, 0)$ $P_2(2, -2, -2)$

39. If $\overrightarrow{AB} = \mathbf{i} + 4\mathbf{j} - 2\mathbf{k}$ and B is the point $(5, 1, 3)$, find A.

40. If $\overrightarrow{AB} = -7\mathbf{i} + 3\mathbf{j} + 8\mathbf{k}$ and A is the point $(-2, -3, 6)$, find B.

Theory and Applications

41. Linear combination Let $\mathbf{u} = 2\mathbf{i} + \mathbf{j}, \mathbf{v} = \mathbf{i} + \mathbf{j}$, and $\mathbf{w} = \mathbf{i} - \mathbf{j}$. Find scalars a and b such that $\mathbf{u} = a\mathbf{v} + b\mathbf{w}$.

42. Linear combination Let $\mathbf{u} = \mathbf{i} - 2\mathbf{j}, \mathbf{v} = 2\mathbf{i} + 3\mathbf{j}$, and $\mathbf{w} = \mathbf{i} + \mathbf{j}$. Write $\mathbf{u} = \mathbf{u}_1 + \mathbf{u}_2$, where $\mathbf{u}_1$ is parallel to $\mathbf{v}$ and $\mathbf{u}_2$ is parallel to $\mathbf{w}$. (See Exercise 41.)

43. Linear combination Let $\mathbf{u} = \langle 1, 2, 1 \rangle$, $\mathbf{v} = \langle 1, -1, -1 \rangle$, $\mathbf{w} = \langle 1, 1, -1 \rangle$, and $\mathbf{z} = \langle 2, -3, -4 \rangle$. Find scalars a, b, and c such that $\mathbf{z} = a\mathbf{u} + b\mathbf{v} + c\mathbf{w}$.

44. Linear combination Let $\mathbf{u} = \langle 1, 2, 2 \rangle$, $\mathbf{v} = \langle 1, -1, -1 \rangle$, $\mathbf{w} = \langle 1, 3, -1 \rangle$, and $\mathbf{z} = \langle 2, 11, 8 \rangle$. Write $\mathbf{z} = \mathbf{u}_1 + \mathbf{u}_2 + \mathbf{u}_3$, where $\mathbf{u}_1$ is parallel to $\mathbf{u}$, $\mathbf{u}_2$ is parallel to $\mathbf{v}$, and $\mathbf{u}_3$ is parallel to $\mathbf{w}$. What are $\mathbf{u}_1, \mathbf{u}_2, \mathbf{u}_3$?

45. Velocity An airplane is flying in the direction 25° west of north at 800 km/h. Find the component form of the velocity of the airplane, assuming that the positive x-axis represents due east and the positive y-axis represents due north.

46. (*Continuation of Example 8.*) What speed and direction should the jetliner in Example 8 have in order for the resultant vector to be 500 mph due east?

47. Consider a 100-N weight suspended by two wires as shown in the accompanying figure. Find the magnitudes and components of the force vectors $\mathbf{F}_1$ and $\mathbf{F}_2$.

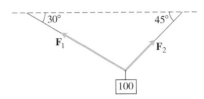

48. Consider a 50-N weight suspended by two wires as shown in the accompanying figure. If the magnitude of vector $\mathbf{F}_1$ is 35 N, find angle α and the magnitude of vector $\mathbf{F}_2$.

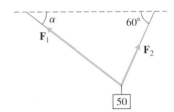

49. Consider a w-N weight suspended by two wires as shown in the accompanying figure. If the magnitude of vector $\mathbf{F}_2$ is 100 N, find w and the magnitude of vector $\mathbf{F}_1$.

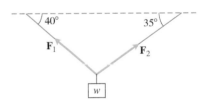

50. Consider a 25-N weight suspended by two wires as shown in the accompanying figure. If the magnitudes of vectors $\mathbf{F}_1$ and $\mathbf{F}_2$ are both 75 N, then angles α and β are equal. Find α.

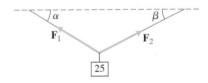

51. Location A bird flies from its nest 5 km in the direction 60° north of east, where it stops to rest on a tree. It then flies 10 km in the direction due southeast and lands atop a telephone pole. Place an xy-coordinate system so that the origin is the bird's nest, the x-axis points east, and the y-axis points north.

a. At what point is the tree located?

b. At what point is the telephone pole?

52. Use similar triangles to find the coordinates of the point Q that divides the segment from $P_1(x_1, y_1, z_1)$ to $P_2(x_2, y_2, z_2)$ into two lengths whose ratio is $p/q = r$.

53. Medians of a triangle Suppose that A, B, and C are the corner points of the thin triangular plate of constant density shown here.

a. Find the vector from C to the midpoint M of side AB.

b. Find the vector from C to the point that lies two-thirds of the way from C to M on the median CM.

c. Find the coordinates of the point in which the medians of $\triangle ABC$ intersect. According to Exercise 27, Section 6.6, this point is the plate's center of mass. (See the figure.)

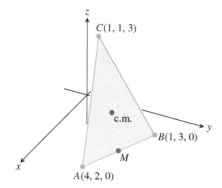

54. Find the vector from the origin to the point of intersection of the medians of the triangle whose vertices are

$$A(1, -1, 2), \quad B(2, 1, 3), \quad \text{and} \quad C(-1, 2, -1).$$

55. Let $ABCD$ be a general, not necessarily planar, quadrilateral in space. Show that the two segments joining the midpoints of opposite sides of $ABCD$ bisect each other. (*Hint:* Show that the segments have the same midpoint.)

56. Vectors are drawn from the center of a regular n-sided polygon in the plane to the vertices of the polygon. Show that the sum of the vectors is zero. (*Hint:* What happens to the sum if you rotate the polygon about its center?)

57. Suppose that A, B, and C are vertices of a triangle and that a, b, and c are, respectively, the midpoints of the opposite sides. Show that $\overrightarrow{Aa} + \overrightarrow{Bb} + \overrightarrow{Cc} = 0$.

58. Unit vectors in the plane Show that a unit vector in the plane can be expressed as $\mathbf{u} = (\cos\theta)\mathbf{i} + (\sin\theta)\mathbf{j}$, obtained by rotating $\mathbf{i}$ through an angle θ in the counterclockwise direction. Explain why this form gives *every* unit vector in the plane.

59. Consider a triangle whose vertices are $A(2, -3, 4)$, $B(1, 0, -1)$, and $C(3, 1, 2)$.

a. Find $\overrightarrow{AB} + \overrightarrow{BC} + \overrightarrow{CA}$.   b. Find $\overrightarrow{BA} + \overrightarrow{AC} + \overrightarrow{CB}$.

12.3 The Dot Product

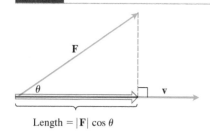

FIGURE 12.19 The magnitude of the force $\mathbf{F}$ in the direction of vector $\mathbf{v}$ is the length $|\mathbf{F}|\cos\theta$ of the projection of $\mathbf{F}$ onto $\mathbf{v}$.

If a force $\mathbf{F}$ is applied to a particle moving along a path, we often need to know the magnitude of the force in the direction of motion. If $\mathbf{v}$ is parallel to the tangent line to the path at the point where $\mathbf{F}$ is applied, then we want the magnitude of $\mathbf{F}$ in the direction of $\mathbf{v}$. Figure 12.19 shows that the scalar quantity we seek is the length $|\mathbf{F}|\cos\theta$, where θ is the angle between the two vectors $\mathbf{F}$ and $\mathbf{v}$.

In this section we show how to calculate easily the angle between two vectors directly from their components. A key part of the calculation is an expression called the *dot product*. Dot products are also called *inner* or *scalar* products because the product results in a scalar, not a vector. After investigating the dot product, we apply it to finding the projection of one vector onto another (as displayed in Figure 12.19) and to finding the work done by a constant force acting through a displacement.

Angle Between Vectors

When two nonzero vectors $\mathbf{u}$ and $\mathbf{v}$ are placed so their initial points coincide, they form an angle θ of measure $0 \le \theta \le \pi$ (Figure 12.20). If the vectors do not lie along the same line, the angle θ is measured in the plane containing both of them. If they do lie along the same line, the angle between them is 0 if they point in the same direction and π if they point in opposite directions. The angle θ is the **angle between $\mathbf{u}$ and $\mathbf{v}$**. Theorem 1 gives a formula to determine this angle.

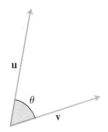

FIGURE 12.20 The angle between $\mathbf{u}$ and $\mathbf{v}$ given by Theorem 1 lies in the interval $[0, \pi]$.

> **THEOREM 1—Angle Between Two Vectors**
> The angle θ between two nonzero vectors $\mathbf{u} = \langle u_1, u_2, u_3 \rangle$ and $\mathbf{v} = \langle v_1, v_2, v_3 \rangle$ is given by
>
> $$\theta = \cos^{-1}\left(\frac{u_1 v_1 + u_2 v_2 + u_3 v_3}{|\mathbf{u}||\mathbf{v}|}\right).$$

We use the law of cosines to prove Theorem 1, but before doing so, we focus attention on the expression $u_1v_1 + u_2v_2 + u_3v_3$ in the calculation for θ. This expression is the sum of the products of the corresponding components of the vectors $\mathbf{u}$ and $\mathbf{v}$.

DEFINITION   The **dot product u · v** ("**u dot v**") of vectors $\mathbf{u} = \langle u_1, u_2, u_3 \rangle$ and $\mathbf{v} = \langle v_1, v_2, v_3 \rangle$ is the scalar

$$\mathbf{u} \cdot \mathbf{v} = u_1v_1 + u_2v_2 + u_3v_3.$$

EXAMPLE 1   We illustrate the definition.

(a) $\langle 1, -2, -1 \rangle \cdot \langle -6, 2, -3 \rangle = (1)(-6) + (-2)(2) + (-1)(-3)$

$$= -6 - 4 + 3 = -7$$

(b) $\left(\dfrac{1}{2}\mathbf{i} + 3\mathbf{j} + \mathbf{k}\right) \cdot (4\mathbf{i} - \mathbf{j} + 2\mathbf{k}) = \left(\dfrac{1}{2}\right)(4) + (3)(-1) + (1)(2) = 1$ ∎

The dot product of a pair of two-dimensional vectors is defined in a similar fashion:

$$\langle u_1, u_2 \rangle \cdot \langle v_1, v_2 \rangle = u_1v_1 + u_2v_2.$$

We will see throughout the remainder of the book that the dot product is a key tool for many important geometric and physical calculations in space (and the plane).

Proof of Theorem 1   Applying the law of cosines (Equation (8), Section 1.3) to the triangle in Figure 12.21, we find that

$$|\mathbf{w}|^2 = |\mathbf{u}|^2 + |\mathbf{v}|^2 - 2|\mathbf{u}||\mathbf{v}|\cos\theta \qquad \text{Law of cosines}$$

$$2|\mathbf{u}||\mathbf{v}|\cos\theta = |\mathbf{u}|^2 + |\mathbf{v}|^2 - |\mathbf{w}|^2.$$

Because $\mathbf{w} = \mathbf{u} - \mathbf{v}$, the component form of $\mathbf{w}$ is $\langle u_1 - v_1, u_2 - v_2, u_3 - v_3 \rangle$. So

$$|\mathbf{u}|^2 = \left(\sqrt{u_1{}^2 + u_2{}^2 + u_3{}^2}\right)^2 = u_1{}^2 + u_2{}^2 + u_3{}^2$$

$$|\mathbf{v}|^2 = \left(\sqrt{v_1{}^2 + v_2{}^2 + v_3{}^2}\right)^2 = v_1{}^2 + v_2{}^2 + v_3{}^2$$

$$|\mathbf{w}|^2 = \left(\sqrt{(u_1 - v_1)^2 + (u_2 - v_2)^2 + (u_3 - v_3)^2}\right)^2$$

$$= (u_1 - v_1)^2 + (u_2 - v_2)^2 + (u_3 - v_3)^2$$

$$= u_1{}^2 - 2u_1v_1 + v_1{}^2 + u_2{}^2 - 2u_2v_2 + v_2{}^2 + u_3{}^2 - 2u_3v_3 + v_3{}^2$$

and

$$|\mathbf{u}|^2 + |\mathbf{v}|^2 - |\mathbf{w}|^2 = 2(u_1v_1 + u_2v_2 + u_3v_3).$$

Therefore,

$$2|\mathbf{u}||\mathbf{v}|\cos\theta = |\mathbf{u}|^2 + |\mathbf{v}|^2 - |\mathbf{w}|^2 = 2(u_1v_1 + u_2v_2 + u_3v_3)$$

$$|\mathbf{u}||\mathbf{v}|\cos\theta = u_1v_1 + u_2v_2 + u_3v_3$$

$$\cos\theta = \frac{u_1v_1 + u_2v_2 + u_3v_3}{|\mathbf{u}||\mathbf{v}|}.$$

Since $0 \leq \theta < \pi$, we have $\theta = \cos^{-1}\left(\dfrac{u_1v_1 + u_2v_2 + u_3v_3}{|\mathbf{u}||\mathbf{v}|}\right)$.   ∎

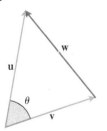

FIGURE 12.21   The parallelogram law of addition of vectors gives $\mathbf{w} = \mathbf{u} - \mathbf{v}$.

Dot Product and Angles

The angle between two nonzero vectors **u** and **v** is $\theta = \cos^{-1}\left(\dfrac{\mathbf{u} \cdot \mathbf{v}}{|\mathbf{u}||\mathbf{v}|}\right)$.

The dot product of two vectors **u** and **v** is given by $\mathbf{u} \cdot \mathbf{v} = |\mathbf{u}||\mathbf{v}|\cos\theta$.

EXAMPLE 2   Find the angle between $\mathbf{u} = \mathbf{i} - 2\mathbf{j} - 2\mathbf{k}$ and $\mathbf{v} = 6\mathbf{i} + 3\mathbf{j} + 2\mathbf{k}$.

Solution   We use the formula above:

$$\mathbf{u} \cdot \mathbf{v} = (1)(6) + (-2)(3) + (-2)(2) = 6 - 6 - 4 = -4$$
$$|\mathbf{u}| = \sqrt{(1)^2 + (-2)^2 + (-2)^2} = \sqrt{9} = 3$$
$$|\mathbf{v}| = \sqrt{(6)^2 + (3)^2 + (2)^2} = \sqrt{49} = 7$$
$$\theta = \cos^{-1}\left(\frac{\mathbf{u} \cdot \mathbf{v}}{|\mathbf{u}||\mathbf{v}|}\right) = \cos^{-1}\left(\frac{-4}{(3)(7)}\right) \approx 1.76 \text{ radians or } 100.98°. \quad \blacksquare$$

The angle formula applies to two-dimensional vectors as well. Note that the angle θ is acute if $\mathbf{u} \cdot \mathbf{v} > 0$ and obtuse if $\mathbf{u} \cdot \mathbf{v} < 0$.

EXAMPLE 3   Find the angle θ in the triangle ABC determined by the vertices $A = (0, 0)$, $B = (3, 5)$, and $C = (5, 2)$ (Figure 12.22).

Solution   The angle θ is the angle between the vectors $\vec{CA}$ and $\vec{CB}$. The component forms of these two vectors are

$$\vec{CA} = \langle -5, -2 \rangle \quad \text{and} \quad \vec{CB} = \langle -2, 3 \rangle.$$

First we calculate the dot product and magnitudes of these two vectors.

$$\vec{CA} \cdot \vec{CB} = (-5)(-2) + (-2)(3) = 4$$
$$|\vec{CA}| = \sqrt{(-5)^2 + (-2)^2} = \sqrt{29}$$
$$|\vec{CB}| = \sqrt{(-2)^2 + (3)^2} = \sqrt{13}$$

Then applying the angle formula, we have

$$\theta = \cos^{-1}\left(\frac{\vec{CA} \cdot \vec{CB}}{|\vec{CA}||\vec{CB}|}\right) = \cos^{-1}\left(\frac{4}{(\sqrt{29})(\sqrt{13})}\right)$$
$$\approx 78.1° \text{ or } 1.36 \text{ radians.} \quad \blacksquare$$

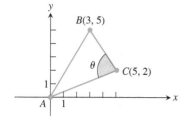

FIGURE 12.22   The triangle in Example 3.

Orthogonal Vectors

Two nonzero vectors **u** and **v** are perpendicular if the angle between them is $\pi/2$. For such vectors, we have $\mathbf{u} \cdot \mathbf{v} = 0$ because $\cos(\pi/2) = 0$. The converse is also true. If **u** and **v** are nonzero vectors with $\mathbf{u} \cdot \mathbf{v} = |\mathbf{u}||\mathbf{v}|\cos\theta = 0$, then $\cos\theta = 0$ and $\theta = \cos^{-1} 0 = \pi/2$. The following definition also allows for one or both of the vectors to be the zero vector.

DEFINITION   Vectors **u** and **v** are orthogonal if $\mathbf{u} \cdot \mathbf{v} = 0$.

EXAMPLE 4   To determine if two vectors are orthogonal, calculate their dot product.

(a) $\mathbf{u} = \langle 3, -2 \rangle$ and $\mathbf{v} = \langle 4, 6 \rangle$ are orthogonal because $\mathbf{u} \cdot \mathbf{v} = (3)(4) + (-2)(6) = 0$.

(b) $\mathbf{u} = 3\mathbf{i} - 2\mathbf{j} + \mathbf{k}$ and $\mathbf{v} = 2\mathbf{j} + 4\mathbf{k}$ are orthogonal because

$$\mathbf{u} \cdot \mathbf{v} = (3)(0) + (-2)(2) + (1)(4) = 0.$$

(c) $\mathbf{0}$ is orthogonal to every vector $\mathbf{u}$ since

$$\mathbf{0} \cdot \mathbf{u} = \langle 0, 0, 0 \rangle \cdot \langle u_1, u_2, u_3 \rangle$$
$$= (0)(u_1) + (0)(u_2) + (0)(u_3) = 0.$$ ∎

Dot Product Properties and Vector Projections

The dot product obeys many of the laws that hold for ordinary products of real numbers (scalars).

Properties of the Dot Product
If $\mathbf{u}$, $\mathbf{v}$, and $\mathbf{w}$ are any vectors and c is a scalar, then

1. $\mathbf{u} \cdot \mathbf{v} = \mathbf{v} \cdot \mathbf{u}$
2. $(c\mathbf{u}) \cdot \mathbf{v} = \mathbf{u} \cdot (c\mathbf{v}) = c(\mathbf{u} \cdot \mathbf{v})$
3. $\mathbf{u} \cdot (\mathbf{v} + \mathbf{w}) = \mathbf{u} \cdot \mathbf{v} + \mathbf{u} \cdot \mathbf{w}$
4. $\mathbf{u} \cdot \mathbf{u} = |\mathbf{u}|^2$
5. $\mathbf{0} \cdot \mathbf{u} = 0.$

Proofs of Properties 1 and 3   The properties are easy to prove using the definition. For instance, here are the proofs of Properties 1 and 3.

1. $\mathbf{u} \cdot \mathbf{v} = u_1v_1 + u_2v_2 + u_3v_3 = v_1u_1 + v_2u_2 + v_3u_3 = \mathbf{v} \cdot \mathbf{u}$
3. $\mathbf{u} \cdot (\mathbf{v} + \mathbf{w}) = \langle u_1, u_2, u_3 \rangle \cdot \langle v_1 + w_1, v_2 + w_2, v_3 + w_3 \rangle$
$$= u_1(v_1 + w_1) + u_2(v_2 + w_2) + u_3(v_3 + w_3)$$
$$= u_1v_1 + u_1w_1 + u_2v_2 + u_2w_2 + u_3v_3 + u_3w_3$$
$$= (u_1v_1 + u_2v_2 + u_3v_3) + (u_1w_1 + u_2w_2 + u_3w_3)$$
$$= \mathbf{u} \cdot \mathbf{v} + \mathbf{u} \cdot \mathbf{w}$$ ∎

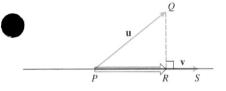
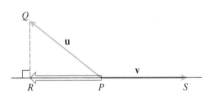

FIGURE 12.23   The vector projection of $\mathbf{u}$ onto $\mathbf{v}$.

We now return to the problem of projecting one vector onto another, posed in the opening to this section. The **vector projection** of $\mathbf{u} = \overrightarrow{PQ}$ onto a nonzero vector $\mathbf{v} = \overrightarrow{PS}$ (Figure 12.23) is the vector $\overrightarrow{PR}$ determined by dropping a perpendicular from Q to the line PS. The notation for this vector is

$$\text{proj}_{\mathbf{v}}\ \mathbf{u}   \qquad \text{("the vector projection of } \mathbf{u} \text{ onto } \mathbf{v}\text{")}.$$

If $\mathbf{u}$ represents a force, then $\text{proj}_{\mathbf{v}}\ \mathbf{u}$ represents the effective force in the direction of $\mathbf{v}$ (Figure 12.24).

If the angle θ between $\mathbf{u}$ and $\mathbf{v}$ is acute, $\text{proj}_{\mathbf{v}}\ \mathbf{u}$ has length $|\mathbf{u}|\cos\theta$ and direction $\mathbf{v}/|\mathbf{v}|$ (Figure 12.25). If θ is obtuse, $\cos\theta < 0$ and $\text{proj}_{\mathbf{v}}\ \mathbf{u}$ has length $-|\mathbf{u}|\cos\theta$ and direction $-\mathbf{v}/|\mathbf{v}|$. In both cases,

$$\text{proj}_{\mathbf{v}}\ \mathbf{u} = \left(|\mathbf{u}|\cos\theta\right)\frac{\mathbf{v}}{|\mathbf{v}|}$$

$$= \left(\frac{\mathbf{u} \cdot \mathbf{v}}{|\mathbf{v}|}\right)\frac{\mathbf{v}}{|\mathbf{v}|}   \qquad |\mathbf{u}|\cos\theta = \frac{|\mathbf{u}||\mathbf{v}|\cos\theta}{|\mathbf{v}|} = \frac{\mathbf{u} \cdot \mathbf{v}}{|\mathbf{v}|}$$

$$= \left(\frac{\mathbf{u} \cdot \mathbf{v}}{|\mathbf{v}|^2}\right)\mathbf{v}.$$

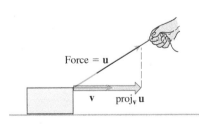

FIGURE 12.24   If we pull on the box with force $\mathbf{u}$, the effective force moving the box forward in the direction $\mathbf{v}$ is the projection of $\mathbf{u}$ onto $\mathbf{v}$.

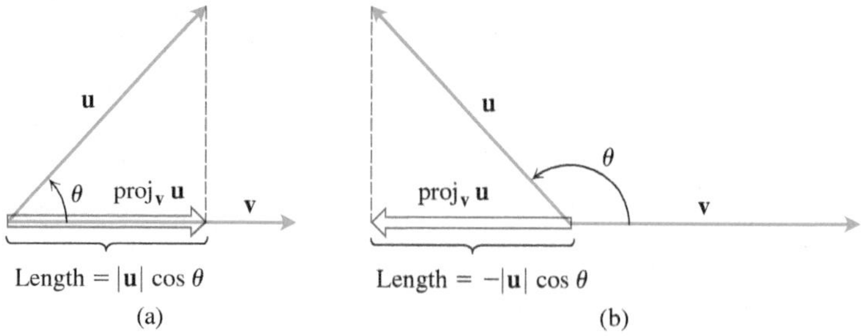

FIGURE 12.25 The length of $\text{proj}_v\, \mathbf{u}$ is (a) $|\mathbf{u}|\cos\theta$ if $\cos\theta \geq 0$ and (b) $-|\mathbf{u}|\cos\theta$ if $\cos\theta < 0$.

The number $|\mathbf{u}|\cos\theta$ is called the **scalar component of u in the direction of v** (or of **u** onto **v**). To summarize,

---

The vector projection of **u** onto **v** is the vector

$$\text{proj}_v\, \mathbf{u} = \left(\frac{\mathbf{u}\cdot\mathbf{v}}{|\mathbf{v}|^2}\right)\mathbf{v} = \left(\frac{\mathbf{u}\cdot\mathbf{v}}{|\mathbf{v}|}\right)\frac{\mathbf{v}}{|\mathbf{v}|}. \tag{1}$$

The scalar component of **u** in the direction of **v** is the scalar

$$|\mathbf{u}|\cos\theta = \frac{\mathbf{u}\cdot\mathbf{v}}{|\mathbf{v}|} = \mathbf{u}\cdot\frac{\mathbf{v}}{|\mathbf{v}|}. \tag{2}$$

Note that both the vector projection of **u** onto **v** and the scalar component of **u** in the direction of **v** depend only on the direction of the vector **v** and not its length. This is because in both cases we take the dot product of **u** with the direction vector $\mathbf{v}/|\mathbf{v}|$, which is the direction of **v**, and for the projection we go on to multiply the result by the direction vector.

EXAMPLE 5 Find the vector projection of $\mathbf{u} = 6\mathbf{i} + 3\mathbf{j} + 2\mathbf{k}$ onto $\mathbf{v} = \mathbf{i} - 2\mathbf{j} - 2\mathbf{k}$ and the scalar component of **u** in the direction of **v**.

Solution We find $\text{proj}_v\, \mathbf{u}$ from Equation (1):

$$\text{proj}_v\, \mathbf{u} = \frac{\mathbf{u}\cdot\mathbf{v}}{|\mathbf{v}|^2}\mathbf{v} = \frac{\mathbf{u}\cdot\mathbf{v}}{\mathbf{v}\cdot\mathbf{v}}\mathbf{v} = \frac{6 - 6 - 4}{1 + 4 + 4}(\mathbf{i} - 2\mathbf{j} - 2\mathbf{k})$$

$$= -\frac{4}{9}(\mathbf{i} - 2\mathbf{j} - 2\mathbf{k}) = -\frac{4}{9}\mathbf{i} + \frac{8}{9}\mathbf{j} + \frac{8}{9}\mathbf{k}.$$

We find the scalar component of **u** in the direction of **v** from Equation (2):

$$|\mathbf{u}|\cos\theta = \mathbf{u}\cdot\frac{\mathbf{v}}{|\mathbf{v}|} = (6\mathbf{i} + 3\mathbf{j} + 2\mathbf{k})\cdot\left(\frac{1}{3}\mathbf{i} - \frac{2}{3}\mathbf{j} - \frac{2}{3}\mathbf{k}\right)$$

$$= 2 - 2 - \frac{4}{3} = -\frac{4}{3}. \qquad\blacksquare$$

Equations (1) and (2) also apply to two-dimensional vectors. We demonstrate this in the next example.

EXAMPLE 6 Find the vector projection of a force $\mathbf{F} = 5\mathbf{i} + 2\mathbf{j}$ onto $\mathbf{v} = \mathbf{i} - 3\mathbf{j}$ and the scalar component of **F** in the direction of **v**.

Solution   The vector projection is

$$\text{proj}_{\mathbf{v}}\,\mathbf{F} = \left(\frac{\mathbf{F}\cdot\mathbf{v}}{|\mathbf{v}|^2}\right)\mathbf{v} = \left(\frac{\mathbf{F}\cdot\mathbf{v}}{\mathbf{v}\cdot\mathbf{v}}\right)\mathbf{v}$$

$$= \frac{5-6}{1+9}(\mathbf{i}-3\mathbf{j}) = -\frac{1}{10}(\mathbf{i}-3\mathbf{j})$$

$$= -\frac{1}{10}\mathbf{i} + \frac{3}{10}\mathbf{j}.$$

The scalar component of $\mathbf{F}$ in the direction of $\mathbf{v}$ is

$$|\mathbf{F}|\cos\theta = \frac{\mathbf{F}\cdot\mathbf{v}}{|\mathbf{v}|} = \frac{5-6}{\sqrt{1+9}} = -\frac{1}{\sqrt{10}}.$$ ∎

EXAMPLE 7   Verify that the vector $\mathbf{u} - \text{proj}_{\mathbf{v}}\,\mathbf{u}$ is orthogonal to the projection vector $\text{proj}_{\mathbf{v}}\,\mathbf{u}$.

Solution   The vector $\text{proj}_{\mathbf{v}}\,\mathbf{u} = \left(\dfrac{\mathbf{u}\cdot\mathbf{v}}{|\mathbf{v}|^2}\right)\mathbf{v}$ is parallel to $\mathbf{v}$. So it suffices to show that the vector $\mathbf{u} - \text{proj}_{\mathbf{v}}\,\mathbf{u}$ is orthogonal to $\mathbf{v}$. We verify orthogonality by showing that the dot product of $\mathbf{u} - \text{proj}_{\mathbf{v}}\,\mathbf{u}$ with $\mathbf{v}$ is zero:

$$(\mathbf{u} - \text{proj}_{\mathbf{v}}\,\mathbf{u})\cdot\mathbf{v} = \mathbf{u}\cdot\mathbf{v} - \left(\frac{\mathbf{u}\cdot\mathbf{v}}{|\mathbf{v}|^2}\mathbf{v}\right)\cdot\mathbf{v} \qquad \text{Definition of } \text{proj}_{\mathbf{v}}\,\mathbf{u}$$

$$= \mathbf{u}\cdot\mathbf{v} - \frac{\mathbf{u}\cdot\mathbf{v}}{|\mathbf{v}|^2}\mathbf{v}\cdot\mathbf{v} \qquad \text{Dot product property (2)}$$

$$= \mathbf{u}\cdot\mathbf{v} - \frac{\mathbf{u}\cdot\mathbf{v}}{|\mathbf{v}|^2}|\mathbf{v}|^2 \qquad \mathbf{v}\cdot\mathbf{v} = |\mathbf{v}|^2$$

$$= \mathbf{u}\cdot\mathbf{v} - \mathbf{u}\cdot\mathbf{v} = 0.$$ ∎

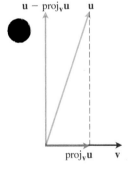

FIGURE 12.26   The vector $\mathbf{u}$ is the sum of two perpendicular vectors: a vector $\text{proj}_{\mathbf{v}}\,\mathbf{u}$, parallel to $\mathbf{v}$, and a vector $\mathbf{u} - \text{proj}_{\mathbf{v}}\,\mathbf{u}$, perpendicular to $\mathbf{v}$.

Example 7 verifies that the vector $\mathbf{u} - \text{proj}_{\mathbf{v}}\,\mathbf{u}$ is orthogonal to the projection vector $\text{proj}_{\mathbf{v}}\,\mathbf{u}$ (which has the same direction as $\mathbf{v}$). So the equation

$$\mathbf{u} = \text{proj}_{\mathbf{v}}\,\mathbf{u} + (\mathbf{u} - \text{proj}_{\mathbf{v}}\,\mathbf{u}) = \underbrace{\left(\frac{\mathbf{u}\cdot\mathbf{v}}{|\mathbf{v}|^2}\right)\mathbf{v}}_{\text{Parallel to } \mathbf{v}} + \underbrace{\left(\mathbf{u} - \left(\frac{\mathbf{u}\cdot\mathbf{v}}{|\mathbf{v}|^2}\right)\mathbf{v}\right)}_{\text{Orthogonal to } \mathbf{v}}$$

expresses $\mathbf{u}$ as a sum of orthogonal vectors (see Figure 12.26).

Work

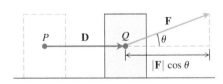

FIGURE 12.27   The work done by a constant force $\mathbf{F}$ during a displacement $\mathbf{D}$ is $(|\mathbf{F}|\cos\theta)|\mathbf{D}|$, which is the dot product $\mathbf{F}\cdot\mathbf{D}$.

In Chapter 6, we calculated the work done by a constant force of magnitude F in moving an object through a distance d as $W = Fd$. That formula holds only if the force is directed along the line of motion. If a force $\mathbf{F}$ moving an object through a displacement $\mathbf{D} = \overrightarrow{PQ}$ has some other direction, the work is performed by the component of $\mathbf{F}$ in the direction of $\mathbf{D}$. If θ is the angle between $\mathbf{F}$ and $\mathbf{D}$ (Figure 12.27), then

$$\text{Work} = \left(\begin{array}{c}\text{scalar component of } \mathbf{F}\\ \text{in the direction of } \mathbf{D}\end{array}\right)(\text{length of } \mathbf{D})$$

$$= (|\mathbf{F}|\cos\theta)|\mathbf{D}|$$

$$= \mathbf{F}\cdot\mathbf{D}.$$

> **DEFINITION** The **work** done by a constant force $\mathbf{F}$ acting through a displacement $\mathbf{D} = \overrightarrow{PQ}$ is
>
> $$W = \mathbf{F} \cdot \mathbf{D}.$$

EXAMPLE 8   If $|\mathbf{F}| = 40$ N (newtons), $|\mathbf{D}| = 3$ m, and $\theta = 60°$, the work done by $\mathbf{F}$ in acting from P to Q is

$$
\begin{aligned}
\text{Work} &= \mathbf{F} \cdot \mathbf{D} && \text{Definition}\\
&= |\mathbf{F}||\mathbf{D}| \cos \theta \\
&= (40)(3) \cos 60° && \text{Given values}\\
&= (120)(1/2) = 60 \text{ J (joules)}.
\end{aligned}
$$

We encounter more challenging work problems in Chapter 16 when we learn to find the work done by a variable force along a more general *path* in space.

EXERCISES 12.3

Dot Product and Projections
In Exercises 1–8, find

  a. $\mathbf{v} \cdot \mathbf{u}, |\mathbf{v}|, |\mathbf{u}|$

  b. the cosine of the angle between $\mathbf{v}$ and $\mathbf{u}$

  c. the scalar component of $\mathbf{u}$ in the direction of $\mathbf{v}$

  d. the vector $\text{proj}_{\mathbf{v}} \mathbf{u}$.

1. $\mathbf{v} = 2\mathbf{i} - 4\mathbf{j} + \sqrt{5}\mathbf{k}, \quad \mathbf{u} = -2\mathbf{i} + 4\mathbf{j} - \sqrt{5}\mathbf{k}$

2. $\mathbf{v} = (3/5)\mathbf{i} + (4/5)\mathbf{k}, \quad \mathbf{u} = 5\mathbf{i} + 12\mathbf{j}$

3. $\mathbf{v} = 10\mathbf{i} + 11\mathbf{j} - 2\mathbf{k}, \quad \mathbf{u} = 3\mathbf{j} + 4\mathbf{k}$

4. $\mathbf{v} = 2\mathbf{i} + 10\mathbf{j} - 11\mathbf{k}, \quad \mathbf{u} = 2\mathbf{i} + 2\mathbf{j} + \mathbf{k}$

5. $\mathbf{v} = 5\mathbf{j} - 3\mathbf{k}, \quad \mathbf{u} = \mathbf{i} + \mathbf{j} + \mathbf{k}$

6. $\mathbf{v} = -\mathbf{i} + \mathbf{j}, \quad \mathbf{u} = \sqrt{2}\mathbf{i} + \sqrt{3}\mathbf{j} + 2\mathbf{k}$

7. $\mathbf{v} = 5\mathbf{i} + \mathbf{j}, \quad \mathbf{u} = 2\mathbf{i} + \sqrt{17}\mathbf{j}$

8. $\mathbf{v} = \left\langle \dfrac{1}{\sqrt{2}}, \dfrac{1}{\sqrt{3}} \right\rangle, \quad \mathbf{u} = \left\langle \dfrac{1}{\sqrt{2}}, -\dfrac{1}{\sqrt{3}} \right\rangle$

Angle Between Vectors
T Find the angles between the vectors in Exercises 9–12 to the nearest hundredth of a radian.

9. $\mathbf{u} = 2\mathbf{i} + \mathbf{j}, \quad \mathbf{v} = \mathbf{i} + 2\mathbf{j} - \mathbf{k}$

10. $\mathbf{u} = 2\mathbf{i} - 2\mathbf{j} + \mathbf{k}, \quad \mathbf{v} = 3\mathbf{i} + 4\mathbf{k}$

11. $\mathbf{u} = \sqrt{3}\mathbf{i} - 7\mathbf{j}, \quad \mathbf{v} = \sqrt{3}\mathbf{i} + \mathbf{j} - 2\mathbf{k}$

12. $\mathbf{u} = \mathbf{i} + \sqrt{2}\mathbf{j} - \sqrt{2}\mathbf{k}, \quad \mathbf{v} = -\mathbf{i} + \mathbf{j} + \mathbf{k}$

13. Triangle Find the measures of the angles of the triangle whose vertices are $A = (-1, 0)$, $B = (2, 1)$, and $C = (1, -2)$.

14. Rectangle Find the measures of the angles between the diagonals of the rectangle whose vertices are $A = (1, 0)$, $B = (0, 3)$, $C = (3, 4)$, and $D = (4, 1)$.

15. Direction angles and direction cosines The *direction angles* α, β, and γ of a vector $\mathbf{v} = a\mathbf{i} + b\mathbf{j} + c\mathbf{k}$ are defined as follows:

α is the angle between $\mathbf{v}$ and the positive x-axis ($0 \le \alpha \le \pi$)

β is the angle between $\mathbf{v}$ and the positive y-axis ($0 \le \beta \le \pi$)

γ is the angle between $\mathbf{v}$ and the positive z-axis ($0 \le \gamma \le \pi$).

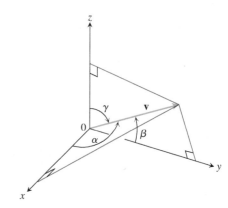

  a. Show that

$$\cos \alpha = \frac{a}{|\mathbf{v}|}, \qquad \cos \beta = \frac{b}{|\mathbf{v}|}, \qquad \cos \gamma = \frac{c}{|\mathbf{v}|},$$

and $\cos^2 \alpha + \cos^2 \beta + \cos^2 \gamma = 1$. These cosines are called the *direction cosines* of $\mathbf{v}$.

  b. Unit vectors are built from direction cosines Show that if $\mathbf{v} = a\mathbf{i} + b\mathbf{j} + c\mathbf{k}$ is a unit vector, then a, b, and c are the direction cosines of $\mathbf{v}$.

16. Water main construction A water main is to be constructed with a 20% grade in the north direction and a 10% grade in the east direction. Determine the angle θ required in the water main for the turn from north to east.

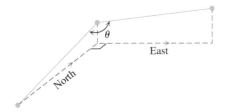

For Exercises 17 and 18, find the acute angle between the given lines by using vectors parallel to the lines.

17. $y = x, \quad y = 2x + 3$

18. $2 - x + 2y = 0, \quad 3x - 4y = -12$

Theory and Examples

19. Sums and differences In the accompanying figure, it looks as if $v_1 + v_2$ and $v_1 - v_2$ are orthogonal. Is this mere coincidence, or are there circumstances under which we may expect the sum of two vectors to be orthogonal to their difference? Give reasons for your answer.

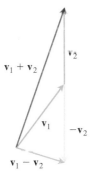

20. Orthogonality on a circle Suppose that AB is the diameter of a circle with center O and that C is a point on one of the two arcs joining A and B. Show that $\overrightarrow{CA}$ and $\overrightarrow{CB}$ are orthogonal.

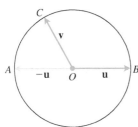

21. Diagonals of a rhombus Show that the diagonals of a rhombus (parallelogram with sides of equal length) are perpendicular.

22. Perpendicular diagonals Show that squares are the only rectangles with perpendicular diagonals.

23. When parallelograms are rectangles Prove that a parallelogram is a rectangle if and only if its diagonals are equal in length. (This fact is often exploited by carpenters.)

24. Diagonal of parallelogram Show that the indicated diagonal of the parallelogram determined by vectors u and v bisects the angle between u and v if $|u| = |v|$.

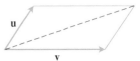

25. Projectile motion A gun with muzzle velocity of 1200 ft/sec is fired at an angle of 8° above the horizontal. Find the horizontal and vertical components of the velocity.

26. Inclined plane Suppose that a box is being towed up an inclined plane as shown in the figure. Find the force w needed to make the component of the force parallel to the inclined plane equal to 2.5 lb.

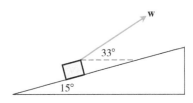

27. a. Cauchy-Schwartz inequality Since $u \cdot v = |u||v| \cos \theta$, show that the inequality $|u \cdot v| \le |u||v|$ holds for any vectors u and v.

   b. Under what circumstances, if any, does $|u \cdot v|$ equal $|u||v|$? Give reasons for your answer.

28. Dot multiplication is positive definite Show that dot multiplication of vectors is *positive definite*; that is, show that $u \cdot u \ge 0$ for every vector u and that $u \cdot u = 0$ if and only if $u = 0$.

29. Orthogonal unit vectors If u_1 and u_2 are orthogonal unit vectors and $v = au_1 + bu_2$, find $v \cdot u_1$.

30. Cancelation in dot products In real-number multiplication, if $uv_1 = uv_2$ and $u \ne 0$, we can cancel the u and conclude that $v_1 = v_2$. Does the same rule hold for the dot product? That is, if $u \cdot v_1 = u \cdot v_2$ and $u \ne 0$, can you conclude that $v_1 = v_2$? Give reasons for your answer.

31. If u and v are orthogonal, show that $\text{proj}_v\, u = 0$.

32. A force $F = 2i + j - 3k$ is applied to a spacecraft with velocity vector $v = 3i - j$. Express F as a sum of a vector parallel to v and a vector orthogonal to v.

Equations for Lines in the Plane

33. Line perpendicular to a vector Show that $v = ai + bj$ is perpendicular to the line $ax + by = c$ by establishing that the slope of the vector v is the negative reciprocal of the slope of the given line.

34. Line parallel to a vector Show that the vector $v = ai + bj$ is parallel to the line $bx - ay = c$ by establishing that the slope of the line segment representing v is the same as the slope of the given line.

In Exercises 35–38, use the result of Exercise 33 to find an equation for the line through P perpendicular to $\mathbf{v}$. Then sketch the line. Include $\mathbf{v}$ in your sketch *as a vector starting at the origin*.

35. $P(2, 1)$, $\quad \mathbf{v} = \mathbf{i} + 2\mathbf{j}$ $\qquad$ **36.** $P(-1, 2)$, $\quad \mathbf{v} = -2\mathbf{i} - \mathbf{j}$

37. $P(-2, -7)$, $\quad \mathbf{v} = -2\mathbf{i} + \mathbf{j}$ $\quad$ **38.** $P(11, 10)$, $\quad \mathbf{v} = 2\mathbf{i} - 3\mathbf{j}$

In Exercises 39–42, use the result of Exercise 34 to find an equation for the line through P parallel to $\mathbf{v}$. Then sketch the line. Include $\mathbf{v}$ in your sketch *as a vector starting at the origin*.

39. $P(-2, 1)$, $\quad \mathbf{v} = \mathbf{i} - \mathbf{j}$ $\qquad$ **40.** $P(0, -2)$, $\quad \mathbf{v} = 2\mathbf{i} + 3\mathbf{j}$

41. $P(1, 2)$, $\quad \mathbf{v} = -\mathbf{i} - 2\mathbf{j}$ $\qquad$ **42.** $P(1, 3)$, $\quad \mathbf{v} = 3\mathbf{i} - 2\mathbf{j}$

Work

43. Work along a line Find the work done by a force $\mathbf{F} = 5\mathbf{i}$ (magnitude 5 N) in moving an object along the line from the origin to the point $(1, 1)$ (distance in meters).

44. Locomotive The Union Pacific's *Big Boy* locomotive could pull 6000-ton trains with a tractive effort (pull) of 602,148 N (135,375 lb). At this level of effort, about how much work did *Big Boy* do on the (approximately straight) 605-km journey from San Francisco to Los Angeles?

45. Inclined plane How much work does it take to slide a crate 20 m along a loading dock by pulling on it with a 200-N force at an angle of 30° from the horizontal?

46. Sailboat The wind passing over a boat's sail exerted a 1000-lb magnitude force $\mathbf{F}$ as shown here. How much work did the wind perform in moving the boat forward 1 mi? Answer in foot-pounds.

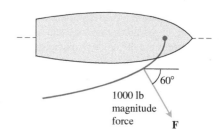

Angles Between Lines in the Plane

The **acute angle between intersecting lines** that do not cross at right angles is the same as the angle determined by vectors normal to the lines or by the vectors parallel to the lines.

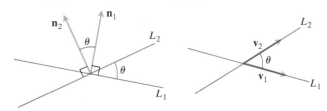

Use this fact and the results of Exercise 33 or 34 to find the acute angles between the lines in Exercises 47–52.

47. $3x + y = 5$, $\quad 2x - y = 4$

48. $y = \sqrt{3}x - 1$, $\quad y = -\sqrt{3}x + 2$

49. $\sqrt{3}x - y = -2$, $\quad x - \sqrt{3}y = 1$

50. $x + \sqrt{3}y = 1$, $\quad \left(1 - \sqrt{3}\right)x + \left(1 + \sqrt{3}\right)y = 8$

51. $3x - 4y = 3$, $\quad x - y = 7$

52. $12x + 5y = 1$, $\quad 2x - 2y = 3$

12.4 The Cross Product

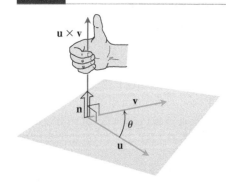

FIGURE 12.28 The construction of $\mathbf{u} \times \mathbf{v}$.

In studying lines in the plane, when we needed to describe how a line was tilting, we used the notions of slope and angle of inclination. In space, we want a way to describe how a *plane* is tilting. We accomplish this by multiplying two vectors in the plane together to get a third vector perpendicular to the plane. The direction of this third vector tells us the "inclination" of the plane. The product we use to multiply the vectors together is the *vector* or *cross product*, the second of the two vector multiplication methods. The cross product gives us a simple way to find a variety of geometric quantities, including volumes, areas, and perpendicular vectors. We study the cross product in this section.

The Cross Product of Two Vectors in Space

We start with two nonzero vectors $\mathbf{u}$ and $\mathbf{v}$ in space. Two vectors are *parallel* if one is a nonzero multiple of the other. If $\mathbf{u}$ and $\mathbf{v}$ are not parallel, they determine a plane. The vectors in this plane are linear combinations of $\mathbf{u}$ and $\mathbf{v}$, so they can be written as a sum $a\mathbf{u} + b\mathbf{v}$. We select the unit vector $\mathbf{n}$ perpendicular to the plane by the **right-hand rule**. This means that we choose $\mathbf{n}$ to be the unit (normal) vector that points the way your right thumb points when your fingers curl through the angle θ from $\mathbf{u}$ to $\mathbf{v}$ (Figure 12.28). Then we define a new vector as follows.

> **DEFINITION** The **cross product** $\mathbf{u} \times \mathbf{v}$ ("u cross v") is the vector
>
> $$\mathbf{u} \times \mathbf{v} = (|\mathbf{u}||\mathbf{v}| \sin \theta) \, \mathbf{n}.$$

Unlike the dot product, the cross product is a vector. For this reason it is also called the **vector product** of **u** and **v**, and can be applied *only* to vectors in space. The vector **u** × **v** is orthogonal to both **u** and **v** because it is a scalar multiple of **n**.

There is a straightforward way to calculate the cross product of two vectors from their components. The method does not require that we know the angle between them (as suggested by the definition), but we postpone that calculation momentarily so we can focus first on the properties of the cross product.

Since the sines of 0 and π are both zero, it makes sense to define the cross product of two parallel nonzero vectors to be **0**. If one or both of **u** and **v** are zero, we also define **u** × **v** to be zero. This way, the cross product of two vectors **u** and **v** is zero if and only if **u** and **v** are parallel or one or both of them are zero.

Parallel Vectors

Nonzero vectors **u** and **v** are parallel if and only if **u** × **v** = **0**.

The cross product obeys the following laws.

Properties of the Cross Product

If **u**, **v**, and **w** are any vectors and r, s are scalars, then

1. $(r\mathbf{u}) \times (s\mathbf{v}) = (rs)(\mathbf{u} \times \mathbf{v})$ 2. $\mathbf{u} \times (\mathbf{v} + \mathbf{w}) = \mathbf{u} \times \mathbf{v} + \mathbf{u} \times \mathbf{w}$

3. $\mathbf{v} \times \mathbf{u} = -(\mathbf{u} \times \mathbf{v})$ 4. $(\mathbf{v} + \mathbf{w}) \times \mathbf{u} = \mathbf{v} \times \mathbf{u} + \mathbf{w} \times \mathbf{u}$

5. $\mathbf{0} \times \mathbf{u} = \mathbf{0}$ 6. $\mathbf{u} \times (\mathbf{v} \times \mathbf{w}) = (\mathbf{u} \cdot \mathbf{w})\mathbf{v} - (\mathbf{u} \cdot \mathbf{v})\mathbf{w}$

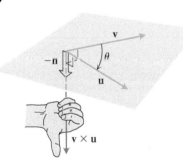

FIGURE 12.29 The construction of **v** × **u**.

To visualize Property 3, for example, notice that when the fingers of your right hand curl through the angle θ from **v** to **u**, your thumb points the opposite way; the unit vector we choose in forming **v** × **u** is the negative of the one we choose in forming **u** × **v** (Figure 12.29).

Property 1 can be verified by applying the definition of cross product to both sides of the equation and comparing the results. Property 2 is proved in Appendix 8. Property 4 follows by multiplying both sides of the equation in Property 2 by -1 and reversing the order of the products using Property 3. Property 5 is a definition. As a rule, cross product multiplication is *not associative* so $(\mathbf{u} \times \mathbf{v}) \times \mathbf{w}$ does not generally equal $\mathbf{u} \times (\mathbf{v} \times \mathbf{w})$. (See Additional Exercise 17.)

When we apply the definition and Property 3 to calculate the pairwise cross products of **i**, **j**, and **k**, we find (Figure 12.30)

$$\mathbf{i} \times \mathbf{j} = -(\mathbf{j} \times \mathbf{i}) = \mathbf{k}$$
$$\mathbf{j} \times \mathbf{k} = -(\mathbf{k} \times \mathbf{j}) = \mathbf{i}$$
$$\mathbf{k} \times \mathbf{i} = -(\mathbf{i} \times \mathbf{k}) = \mathbf{j}$$

and

$$\mathbf{i} \times \mathbf{i} = \mathbf{j} \times \mathbf{j} = \mathbf{k} \times \mathbf{k} = \mathbf{0}.$$

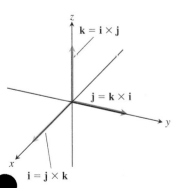

FIGURE 12.30 The pairwise cross products of **i**, **j**, and **k**.

|**u** × **v**| Is the Area of a Parallelogram

Because **n** is a unit vector, the magnitude of **u** × **v** is

---

$$|\mathbf{u} \times \mathbf{v}| = |\mathbf{u}||\mathbf{v}|\,|\sin\theta|\,|\mathbf{n}| = |\mathbf{u}||\mathbf{v}|\sin\theta.$$

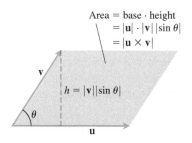

Area = base · height
= $|\mathbf{u}| \cdot |\mathbf{v}| \, |\sin\theta|$
= $|\mathbf{u} \times \mathbf{v}|$

$h = |\mathbf{v}||\sin\theta|$

FIGURE 12.31 The parallelogram determined by **u** and **v**.

This is the area of the parallelogram determined by **u** and **v** (Figure 12.31), $|\mathbf{u}|$ being the base of the parallelogram and $|\mathbf{v}||\sin\theta|$ the height.

Determinant Formula for u × v

Our next objective is to calculate $\mathbf{u} \times \mathbf{v}$ from the components of **u** and **v** relative to a Cartesian coordinate system.

Suppose that

$$\mathbf{u} = u_1\mathbf{i} + u_2\mathbf{j} + u_3\mathbf{k} \quad \text{and} \quad \mathbf{v} = v_1\mathbf{i} + v_2\mathbf{j} + v_3\mathbf{k}.$$

Then the distributive laws and the rules for multiplying **i**, **j**, and **k** tell us that

$$\begin{aligned}
\mathbf{u} \times \mathbf{v} &= (u_1\mathbf{i} + u_2\mathbf{j} + u_3\mathbf{k}) \times (v_1\mathbf{i} + v_2\mathbf{j} + v_3\mathbf{k}) \\
&= u_1v_1\mathbf{i} \times \mathbf{i} + u_1v_2\mathbf{i} \times \mathbf{j} + u_1v_3\mathbf{i} \times \mathbf{k} \\
&\quad + u_2v_1\mathbf{j} \times \mathbf{i} + u_2v_2\mathbf{j} \times \mathbf{j} + u_2v_3\mathbf{j} \times \mathbf{k} \\
&\quad + u_3v_1\mathbf{k} \times \mathbf{i} + u_3v_2\mathbf{k} \times \mathbf{j} + u_3v_3\mathbf{k} \times \mathbf{k} \\
&= (u_2v_3 - u_3v_2)\mathbf{i} - (u_1v_3 - u_3v_1)\mathbf{j} + (u_1v_2 - u_2v_1)\mathbf{k}.
\end{aligned}$$

The component terms in the last line are hard to remember, but they are the same as the terms in the expansion of the symbolic determinant

$$\begin{vmatrix} \mathbf{i} & \mathbf{j} & \mathbf{k} \\ u_1 & u_2 & u_3 \\ v_1 & v_2 & v_3 \end{vmatrix}.$$

So we restate the calculation in this easy-to-remember form.

Determinants

2×2 and 3×3 determinants are evaluated as follows:

$$\begin{vmatrix} a & b \\ c & d \end{vmatrix} = ad - bc$$

$$\begin{vmatrix} a_1 & a_2 & a_3 \\ b_1 & b_2 & b_3 \\ c_1 & c_2 & c_3 \end{vmatrix} = a_1 \begin{vmatrix} b_2 & b_3 \\ c_2 & c_3 \end{vmatrix} - a_2 \begin{vmatrix} b_1 & b_3 \\ c_1 & c_3 \end{vmatrix} + a_3 \begin{vmatrix} b_1 & b_2 \\ c_1 & c_2 \end{vmatrix}$$

Calculating the Cross Product as a Determinant

If $\mathbf{u} = u_1\mathbf{i} + u_2\mathbf{j} + u_3\mathbf{k}$ and $\mathbf{v} = v_1\mathbf{i} + v_2\mathbf{j} + v_3\mathbf{k}$, then

$$\mathbf{u} \times \mathbf{v} = \begin{vmatrix} \mathbf{i} & \mathbf{j} & \mathbf{k} \\ u_1 & u_2 & u_3 \\ v_1 & v_2 & v_3 \end{vmatrix}.$$

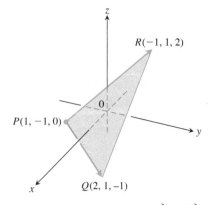

FIGURE 12.32 The vector $\overrightarrow{PQ} \times \overrightarrow{PR}$ is perpendicular to the plane of triangle PQR (Example 2). The area of triangle PQR is half of $|\overrightarrow{PQ} \times \overrightarrow{PR}|$ (Example 3).

EXAMPLE 1 Find $\mathbf{u} \times \mathbf{v}$ and $\mathbf{v} \times \mathbf{u}$ if $\mathbf{u} = 2\mathbf{i} + \mathbf{j} + \mathbf{k}$ and $\mathbf{v} = -4\mathbf{i} + 3\mathbf{j} + \mathbf{k}$.

Solution We expand the symbolic determinant:

$$\mathbf{u} \times \mathbf{v} = \begin{vmatrix} \mathbf{i} & \mathbf{j} & \mathbf{k} \\ 2 & 1 & 1 \\ -4 & 3 & 1 \end{vmatrix} = \begin{vmatrix} 1 & 1 \\ 3 & 1 \end{vmatrix} \mathbf{i} - \begin{vmatrix} 2 & 1 \\ -4 & 1 \end{vmatrix} \mathbf{j} + \begin{vmatrix} 2 & 1 \\ -4 & 3 \end{vmatrix} \mathbf{k}$$

$$= -2\mathbf{i} - 6\mathbf{j} + 10\mathbf{k}$$

$$\mathbf{v} \times \mathbf{u} = -(\mathbf{u} \times \mathbf{v}) = 2\mathbf{i} + 6\mathbf{j} - 10\mathbf{k} \qquad \text{Property 3}$$

EXAMPLE 2 Find a vector perpendicular to the plane of $P(1, -1, 0)$, $Q(2, 1, -$ and $R(-1, 1, 2)$ (Figure 12.32).

Solution   The vector $\overrightarrow{PQ} \times \overrightarrow{PR}$ is perpendicular to the plane because it is perpendicular to both vectors. In terms of components,

$$\overrightarrow{PQ} = (2 - 1)\mathbf{i} + (1 + 1)\mathbf{j} + (-1 - 0)\mathbf{k} = \mathbf{i} + 2\mathbf{j} - \mathbf{k}$$
$$\overrightarrow{PR} = (-1 - 1)\mathbf{i} + (1 + 1)\mathbf{j} + (2 - 0)\mathbf{k} = -2\mathbf{i} + 2\mathbf{j} + 2\mathbf{k}$$

$$\overrightarrow{PQ} \times \overrightarrow{PR} = \begin{vmatrix} \mathbf{i} & \mathbf{j} & \mathbf{k} \\ 1 & 2 & -1 \\ -2 & 2 & 2 \end{vmatrix} = \begin{vmatrix} 2 & -1 \\ 2 & 2 \end{vmatrix}\mathbf{i} - \begin{vmatrix} 1 & -1 \\ -2 & 2 \end{vmatrix}\mathbf{j} + \begin{vmatrix} 1 & 2 \\ -2 & 2 \end{vmatrix}\mathbf{k}$$

$$= 6\mathbf{i} + 6\mathbf{k}. \qquad \blacksquare$$

EXAMPLE 3   Find the area of the triangle with vertices $P(1, -1, 0)$, $Q(2, 1, -1)$, and $R(-1, 1, 2)$ (Figure 12.32).

Solution   The area of the parallelogram determined by P, Q, and R is

$$|\overrightarrow{PQ} \times \overrightarrow{PR}| = |6\mathbf{i} + 6\mathbf{k}| \qquad \text{Values from Example 2}$$
$$= \sqrt{(6)^2 + (6)^2} = \sqrt{2 \cdot 36} = 6\sqrt{2}.$$

The triangle's area is half of this, or $3\sqrt{2}$. $\qquad \blacksquare$

EXAMPLE 4   Find a unit vector perpendicular to the plane of $P(1, -1, 0)$, $Q(2, 1, -1)$, and $R(-1, 1, 2)$.

Solution   Since $\overrightarrow{PQ} \times \overrightarrow{PR}$ is perpendicular to the plane, its direction $\mathbf{n}$ is a unit vector perpendicular to the plane. Taking values from Examples 2 and 3, we have

$$\mathbf{n} = \frac{\overrightarrow{PQ} \times \overrightarrow{PR}}{|\overrightarrow{PQ} \times \overrightarrow{PR}|} = \frac{6\mathbf{i} + 6\mathbf{k}}{6\sqrt{2}} = \frac{1}{\sqrt{2}}\mathbf{i} + \frac{1}{\sqrt{2}}\mathbf{k}. \qquad \blacksquare$$

For ease in calculating the cross product using determinants, we usually write vectors in the form $\mathbf{v} = v_1\mathbf{i} + v_2\mathbf{j} + v_3\mathbf{k}$ rather than as ordered triples $\mathbf{v} = \langle v_1, v_2, v_3 \rangle$.

Torque

When we turn a bolt by applying a force $\mathbf{F}$ to a wrench (Figure 12.33), we produce a torque that causes the bolt to rotate. The **torque vector** points in the direction of the axis of the bolt according to the right-hand rule (so the rotation is counterclockwise when viewed from the *tip* of the vector). The magnitude of the torque depends on how far out on the wrench the force is applied and on how much of the force is perpendicular to the wrench at the point of application. The number we use to measure the torque's magnitude is the product of the length of the lever arm $\mathbf{r}$ and the scalar component of $\mathbf{F}$ perpendicular to $\mathbf{r}$. In the notation of Figure 12.33,

$$\text{Magnitude of torque vector} = |\mathbf{r}||\mathbf{F}| \sin\theta,$$

or $|\mathbf{r} \times \mathbf{F}|$. If we let $\mathbf{n}$ be a unit vector along the axis of the bolt in the direction of the torque, then a complete description of the torque vector is $\mathbf{r} \times \mathbf{F}$, or

$$\text{Torque vector} = \mathbf{r} \times \mathbf{F} = (|\mathbf{r}||\mathbf{F}| \sin\theta)\,\mathbf{n}.$$

Recall that we defined $\mathbf{u} \times \mathbf{v}$ to be $\mathbf{0}$ when $\mathbf{u}$ and $\mathbf{v}$ are parallel. This is consistent with the torque interpretation as well. If the force $\mathbf{F}$ in Figure 12.33 is parallel to the wrench, meaning that we are trying to turn the bolt by pushing or pulling along the line of the wrench's handle, the torque produced is zero.

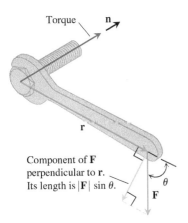

FIGURE 12.33   The torque vector describes the tendency of the force $\mathbf{F}$ to drive the bolt forward.

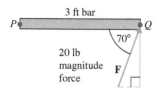

FIGURE 12.34 The magnitude of the torque exerted by **F** at P is about 56.4 ft-lb (Example 5). The bar rotates counter-clockwise around P.

EXAMPLE 5   The magnitude of the torque generated by force **F** at the pivot point P in Figure 12.34 is

$$\left| \overrightarrow{PQ} \times \mathbf{F} \right| = \left| \overrightarrow{PQ} \right| |\mathbf{F}| \sin 70° \approx (3)(20)(0.94) \approx 56.4 \text{ ft-lb}.$$

In this example the torque vector is pointing out of the page toward you.   ∎

Triple Scalar or Box Product

The product $(\mathbf{u} \times \mathbf{v}) \cdot \mathbf{w}$ is called the **triple scalar product** of **u**, **v**, and **w** (in that order). As you can see from the formula

$$|(\mathbf{u} \times \mathbf{v}) \cdot \mathbf{w}| = |\mathbf{u} \times \mathbf{v}||\mathbf{w}||\cos \theta|,$$

the absolute value of this product is the volume of the parallelepiped (parallelogram-sided box) determined by **u**, **v**, and **w** (Figure 12.35). The number $|\mathbf{u} \times \mathbf{v}|$ is the area of the base parallelogram. The number $|\mathbf{w}||\cos \theta|$ is the parallelepiped's height. Because of this geometry, $(\mathbf{u} \times \mathbf{v}) \cdot \mathbf{w}$ is also called the **box product** of **u**, **v**, and **w**.

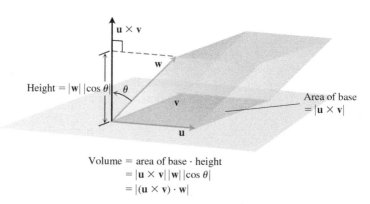

$$\text{Volume} = \text{area of base} \cdot \text{height}$$
$$= |\mathbf{u} \times \mathbf{v}||\mathbf{w}||\cos \theta|$$
$$= |(\mathbf{u} \times \mathbf{v}) \cdot \mathbf{w}|$$

FIGURE 12.35 The number $|(\mathbf{u} \times \mathbf{v}) \cdot \mathbf{w}|$ is the volume of a parallelepiped.

By treating the planes of **v** and **w** and of **w** and **u** as the base planes of the parallelepiped determined by **u**, **v**, and **w**, we see that

$$(\mathbf{u} \times \mathbf{v}) \cdot \mathbf{w} = (\mathbf{v} \times \mathbf{w}) \cdot \mathbf{u} = (\mathbf{w} \times \mathbf{u}) \cdot \mathbf{v}.$$

Since the dot product is commutative, we also have

$$(\mathbf{u} \times \mathbf{v}) \cdot \mathbf{w} = \mathbf{u} \cdot (\mathbf{v} \times \mathbf{w}).$$

The dot and cross may be interchanged in a triple scalar product without altering its value.

The triple scalar product can be evaluated as a determinant:

$$(\mathbf{u} \times \mathbf{v}) \cdot \mathbf{w} = \left(\begin{vmatrix} u_2 & u_3 \\ v_2 & v_3 \end{vmatrix} \mathbf{i} - \begin{vmatrix} u_1 & u_3 \\ v_1 & v_3 \end{vmatrix} \mathbf{j} + \begin{vmatrix} u_1 & u_2 \\ v_1 & v_2 \end{vmatrix} \mathbf{k} \right) \cdot \mathbf{w}$$

$$= w_1 \begin{vmatrix} u_2 & u_3 \\ v_2 & v_3 \end{vmatrix} - w_2 \begin{vmatrix} u_1 & u_3 \\ v_1 & v_3 \end{vmatrix} + w_3 \begin{vmatrix} u_1 & u_2 \\ v_1 & v_2 \end{vmatrix}$$

$$= \begin{vmatrix} u_1 & u_2 & u_3 \\ v_1 & v_2 & v_3 \\ w_1 & w_2 & w_3 \end{vmatrix}.$$

Calculating the Triple Scalar Product as a Determinant

$$(\mathbf{u} \times \mathbf{v}) \cdot \mathbf{w} = \begin{vmatrix} u_1 & u_2 & u_3 \\ v_1 & v_2 & v_3 \\ w_1 & w_2 & w_3 \end{vmatrix}$$

$$\begin{vmatrix} u_1 & u_2 & u_3 \\ v_1 & v_2 & v_3 \\ w_1 & w_2 & w_3 \end{vmatrix} = \pm \begin{vmatrix} w_1 & w_2 & w_3 \\ v_1 & v_2 & v_3 \\ u_1 & u_2 & u_3 \end{vmatrix}$$

Any two rows of a matrix can be interchanged without changing the absolute value of the determinant. So we can take the vectors **u**, **v**, **w** in any order when calculating the absolute value of the triple product.

EXAMPLE 6 Find the volume of the box (parallelepiped) determined by $\mathbf{u} = \mathbf{i} + 2\mathbf{j} - \mathbf{k}$, $\mathbf{v} = -2\mathbf{i} + 3\mathbf{k}$, and $\mathbf{w} = 7\mathbf{j} - 4\mathbf{k}$.

Solution Using the rule for calculating a 3×3 determinant, we find

$$(\mathbf{u} \times \mathbf{v}) \cdot \mathbf{w} = \begin{vmatrix} 1 & 2 & -1 \\ -2 & 0 & 3 \\ 0 & 7 & -4 \end{vmatrix} = (1)\begin{vmatrix} 0 & 3 \\ 7 & -4 \end{vmatrix} - (2)\begin{vmatrix} -2 & 3 \\ 0 & -4 \end{vmatrix} + (-1)\begin{vmatrix} -2 & 0 \\ 0 & 7 \end{vmatrix} = -23.$$

The volume is $|(\mathbf{u} \times \mathbf{v}) \cdot \mathbf{w}| = 23$ units cubed. ∎

EXERCISES 12.4

Cross Product Calculations
In Exercises 1–8, find the length and direction (when defined) of $\mathbf{u} \times \mathbf{v}$ and $\mathbf{v} \times \mathbf{u}$.

1. $\mathbf{u} = 2\mathbf{i} - 2\mathbf{j} - \mathbf{k}$, $\mathbf{v} = \mathbf{i} - \mathbf{k}$
2. $\mathbf{u} = 2\mathbf{i} + 3\mathbf{j}$, $\mathbf{v} = -\mathbf{i} + \mathbf{j}$
3. $\mathbf{u} = 2\mathbf{i} - 2\mathbf{j} + 4\mathbf{k}$, $\mathbf{v} = -\mathbf{i} + \mathbf{j} - 2\mathbf{k}$
4. $\mathbf{u} = \mathbf{i} + \mathbf{j} - \mathbf{k}$, $\mathbf{v} = 0$
5. $\mathbf{u} = 2\mathbf{i}$, $\mathbf{v} = -3\mathbf{j}$
6. $\mathbf{u} = \mathbf{i} \times \mathbf{j}$, $\mathbf{v} = \mathbf{j} \times \mathbf{k}$
7. $\mathbf{u} = -8\mathbf{i} - 2\mathbf{j} - 4\mathbf{k}$, $\mathbf{v} = 2\mathbf{i} + 2\mathbf{j} + \mathbf{k}$
8. $\mathbf{u} = \frac{3}{2}\mathbf{i} - \frac{1}{2}\mathbf{j} + \mathbf{k}$, $\mathbf{v} = \mathbf{i} + \mathbf{j} + 2\mathbf{k}$

In Exercises 9–14, sketch the coordinate axes and then include the vectors **u**, **v**, and $\mathbf{u} \times \mathbf{v}$ as vectors starting at the origin.

9. $\mathbf{u} = \mathbf{i}$, $\mathbf{v} = \mathbf{j}$
10. $\mathbf{u} = \mathbf{i} - \mathbf{k}$, $\mathbf{v} = \mathbf{j}$
11. $\mathbf{u} = \mathbf{i} - \mathbf{k}$, $\mathbf{v} = \mathbf{j} + \mathbf{k}$
12. $\mathbf{u} = 2\mathbf{i} - \mathbf{j}$, $\mathbf{v} = \mathbf{i} + 2\mathbf{j}$
13. $\mathbf{u} = \mathbf{i} + \mathbf{j}$, $\mathbf{v} = \mathbf{i} - \mathbf{j}$
14. $\mathbf{u} = \mathbf{j} + 2\mathbf{k}$, $\mathbf{v} = \mathbf{i}$

Triangles in Space
In Exercises 15–18,

  a. Find the area of the triangle determined by the points P, Q, and R.

  b. Find a unit vector perpendicular to plane PQR.

15. $P(1, -1, 2)$, $Q(2, 0, -1)$, $R(0, 2, 1)$
16. $P(1, 1, 1)$, $Q(2, 1, 3)$, $R(3, -1, 1)$
17. $P(2, -2, 1)$, $Q(3, -1, 2)$, $R(3, -1, 1)$
18. $P(-2, 2, 0)$, $Q(0, 1, -1)$, $R(-1, 2, -2)$

Triple Scalar Products
In Exercises 19–22, verify that $(\mathbf{u} \times \mathbf{v}) \cdot \mathbf{w} = (\mathbf{v} \times \mathbf{w}) \cdot \mathbf{u} = (\mathbf{w} \times \mathbf{u}) \cdot \mathbf{v}$ and find the volume of the parallelepiped (box) determined by **u**, **v**, and **w**.

	u	**v**	**w**
19.	$2\mathbf{i}$	$2\mathbf{j}$	$2\mathbf{k}$
20.	$\mathbf{i} - \mathbf{j} + \mathbf{k}$	$2\mathbf{i} + \mathbf{j} - 2\mathbf{k}$	$-\mathbf{i} + 2\mathbf{j} - \mathbf{k}$
21.	$2\mathbf{i} + \mathbf{j}$	$2\mathbf{i} - \mathbf{j} + \mathbf{k}$	$\mathbf{i} + 2\mathbf{k}$
22.	$\mathbf{i} + \mathbf{j} - 2\mathbf{k}$	$-\mathbf{i} - \mathbf{k}$	$2\mathbf{i} + 4\mathbf{j} - 2\mathbf{k}$

Theory and Examples
23. **Parallel and perpendicular vectors** Let $\mathbf{u} = 5\mathbf{i} - \mathbf{j} + \mathbf{k}$, $\mathbf{v} = \mathbf{j} - 5\mathbf{k}$, $\mathbf{w} = -15\mathbf{i} + 3\mathbf{j} - 3\mathbf{k}$. Which vectors, if any, are (a) perpendicular? (b) Parallel? Give reasons for your answers.

24. **Parallel and perpendicular vectors** Let $\mathbf{u} = \mathbf{i} + 2\mathbf{j} - \mathbf{k}$, $\mathbf{v} = -\mathbf{i} + \mathbf{j} + \mathbf{k}$, $\mathbf{w} = \mathbf{i} + \mathbf{k}$, $\mathbf{r} = -(\pi/2)\mathbf{i} - \pi\mathbf{j} + (\pi/2)\mathbf{k}$. Which vectors, if any, are (a) perpendicular? (b) Parallel? Give reasons for your answers.

In Exercises 25 and 26, find the magnitude of the torque exerted by $\mathbf{F}$ on the bolt at P if $|\overrightarrow{PQ}| = 8$ in. and $|\mathbf{F}| = 30$ lb. Answer in foot-pounds.

25. 26.

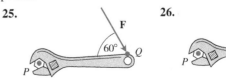

27. Which of the following are *always true*, and which are *not always true*? Give reasons for your answers.

a. $|\mathbf{u}| = \sqrt{\mathbf{u} \cdot \mathbf{u}}$

b. $\mathbf{u} \cdot \mathbf{u} = |\mathbf{u}|$

c. $\mathbf{u} \times \mathbf{0} = \mathbf{0} \times \mathbf{u} = \mathbf{0}$

d. $\mathbf{u} \times (-\mathbf{u}) = \mathbf{0}$

e. $\mathbf{u} \times \mathbf{v} = \mathbf{v} \times \mathbf{u}$

f. $\mathbf{u} \times (\mathbf{v} + \mathbf{w}) = \mathbf{u} \times \mathbf{v} + \mathbf{u} \times \mathbf{w}$

g. $(\mathbf{u} \times \mathbf{v}) \cdot \mathbf{v} = 0$

h. $(\mathbf{u} \times \mathbf{v}) \cdot \mathbf{w} = \mathbf{u} \cdot (\mathbf{v} \times \mathbf{w})$

28. Which of the following are *always true*, and which are *not always true*? Give reasons for your answers.

a. $\mathbf{u} \cdot \mathbf{v} = \mathbf{v} \cdot \mathbf{u}$

b. $\mathbf{u} \times \mathbf{v} = -(\mathbf{v} \times \mathbf{u})$

c. $(-\mathbf{u}) \times \mathbf{v} = -(\mathbf{u} \times \mathbf{v})$

d. $(c\mathbf{u}) \cdot \mathbf{v} = \mathbf{u} \cdot (c\mathbf{v}) = c(\mathbf{u} \cdot \mathbf{v})$ (any number c)

e. $c(\mathbf{u} \times \mathbf{v}) = (c\mathbf{u}) \times \mathbf{v} = \mathbf{u} \times (c\mathbf{v})$ (any number c)

f. $\mathbf{u} \cdot \mathbf{u} = |\mathbf{u}|^2$

g. $(\mathbf{u} \times \mathbf{u}) \cdot \mathbf{u} = 0$

h. $(\mathbf{u} \times \mathbf{v}) \cdot \mathbf{u} = \mathbf{v} \cdot (\mathbf{u} \times \mathbf{v})$

29. Given nonzero vectors $\mathbf{u}$, $\mathbf{v}$, and $\mathbf{w}$, use dot product and cross product notation, as appropriate, to describe the following.

a. The vector projection of $\mathbf{u}$ onto $\mathbf{v}$

b. A vector orthogonal to $\mathbf{u}$ and $\mathbf{v}$

c. A vector orthogonal to $\mathbf{u} \times \mathbf{v}$ and $\mathbf{w}$

d. The volume of the parallelepiped determined by $\mathbf{u}$, $\mathbf{v}$, and $\mathbf{w}$

e. A vector orthogonal to $\mathbf{u} \times \mathbf{v}$ and $\mathbf{u} \times \mathbf{w}$

f. A vector of length $|\mathbf{u}|$ in the direction of $\mathbf{v}$

30. Compute $(\mathbf{i} \times \mathbf{j}) \times \mathbf{j}$ and $\mathbf{i} \times (\mathbf{j} \times \mathbf{j})$. What can you conclude about the associativity of the cross product?

31. Let $\mathbf{u}$, $\mathbf{v}$, and $\mathbf{w}$ be vectors. Which of the following make sense, and which do not? Give reasons for your answers.

a. $(\mathbf{u} \times \mathbf{v}) \cdot \mathbf{w}$ b. $\mathbf{u} \times (\mathbf{v} \cdot \mathbf{w})$

c. $\mathbf{u} \times (\mathbf{v} \times \mathbf{w})$ d. $\mathbf{u} \cdot (\mathbf{v} \cdot \mathbf{w})$

32. Cross products of three vectors Show that except in degenerate cases, $(\mathbf{u} \times \mathbf{v}) \times \mathbf{w}$ lies in the plane of $\mathbf{u}$ and $\mathbf{v}$, whereas $\mathbf{u} \times (\mathbf{v} \times \mathbf{w})$ lies in the plane of $\mathbf{v}$ and $\mathbf{w}$. What *are* the degenerate cases?

33. Cancelation in cross products If $\mathbf{u} \times \mathbf{v} = \mathbf{u} \times \mathbf{w}$ and $\mathbf{u} \neq \mathbf{0}$, then does $\mathbf{v} = \mathbf{w}$? Give reasons for your answer.

34. Double cancelation If $\mathbf{u} \neq \mathbf{0}$ and if $\mathbf{u} \times \mathbf{v} = \mathbf{u} \times \mathbf{w}$ and $\mathbf{u} \cdot \mathbf{v} = \mathbf{u} \cdot \mathbf{w}$, then does $\mathbf{v} = \mathbf{w}$? Give reasons for your answer.

Area of a Parallelogram

Find the areas of the parallelograms whose vertices are given in Exercises 35–40.

35. $A(1, 0)$, $B(0, 1)$, $C(-1, 0)$, $D(0, -1)$

36. $A(0, 0)$, $B(7, 3)$, $C(9, 8)$, $D(2, 5)$

37. $A(-1, 2)$, $B(2, 0)$, $C(7, 1)$, $D(4, 3)$

38. $A(-6, 0)$, $B(1, -4)$, $C(3, 1)$, $D(-4, 5)$

39. $A(0, 0, 0)$, $B(3, 2, 4)$, $C(5, 1, 4)$, $D(2, -1, 0)$

40. $A(1, 0, -1)$, $B(1, 7, 2)$, $C(2, 4, -1)$, $D(0, 3, 2)$

Area of a Triangle

Find the areas of the triangles whose vertices are given in Exercises 41–47.

41. $A(0, 0)$, $B(-2, 3)$, $C(3, 1)$

42. $A(-1, -1)$, $B(3, 3)$, $C(2, 1)$

43. $A(-5, 3)$, $B(1, -2)$, $C(6, -2)$

44. $A(-6, 0)$, $B(10, -5)$, $C(-2, 4)$

45. $A(1, 0, 0)$, $B(0, 2, 0)$, $C(0, 0, -1)$

46. $A(0, 0, 0)$, $B(-1, 1, -1)$, $C(3, 0, 3)$

47. $A(1, -1, 1)$, $B(0, 1, 1)$, $C(1, 0, -1)$

48. Find the volume of a parallelepiped if four of its eight vertices are $A(0, 0, 0), B(1, 2, 0), C(0, -3, 2)$, and $D(3, -4, 5)$.

49. Triangle area Find a 2×2 determinant formula for the area of the triangle in the xy-plane with vertices at $(0, 0)$, (a_1, a_2), and (b_1, b_2). Explain your work.

50. Triangle area Find a concise 3×3 determinant formula that gives the area of a triangle in the xy-plane having vertices (a_1, a_2), (b_1, b_2), and (c_1, c_2).

Volume of a Tetrahedron

Using the methods of Section 6.1, where volume is computed by integrating cross-sectional area, it can be shown that the volume of a tetrahedron formed by three vectors is equal to $\frac{1}{6}$ the volume of the parallelepiped formed by the three vectors. Find the volumes of the tetrahedra whose vertices are given in Exercises 51–54.

51. $A(0, 0, 0)$, $B(2, 0, 0)$, $C(0, 3, 0)$, $D(0, 0, 4)$

52. $A(0, 0, 0)$, $B(1, 0, 2)$, $C(0, 2, 1)$, $D(3, 4, 0)$

53. $A(1, -1, 0)$, $B(0, 2, -2)$, $C(-3, 0, 3)$, $D(0, 4, 4)$

54. $A(-1, 2, 3)$, $B(2, 0, 1)$, $C(1, -3, 2)$, $D(-2, 1, -1)$

In Exercises 55–57, determine whether the given points are coplanar.

55. $A(1, 1, 1)$, $B(-1, 0, 4)$, $C(0, 2, 1)$, $D(2, -2, 3)$

56. $A(0, 0, 4)$, $B(6, 2, 0)$, $C(2, -1, 1)$, $D(-3, -4, 3)$

57. $A(0, 1, 2)$, $B(-1, 1, 0)$, $C(2, 0, -1)$, $D(1, -1, 1)$

12.5 Lines and Planes in Space

This section shows how to use scalar and vector products to write equations for lines, line segments, and planes in space. We will use these representations throughout the rest of the book in studying the calculus of curves and surfaces in space.

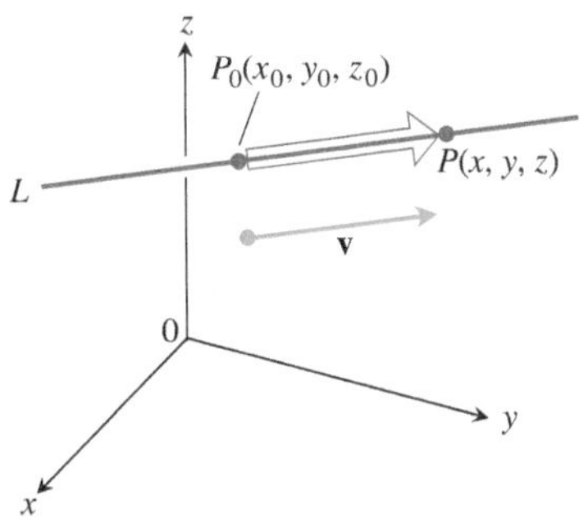

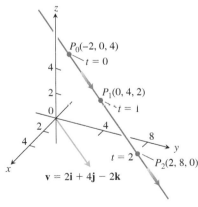

FIGURE 12.36 A point P lies on L through P_0 parallel to $\mathbf{v}$ if and only if $\overrightarrow{P_0P}$ is a scalar multiple of $\mathbf{v}$.

Lines and Line Segments in Space

In the plane, a line is determined by a point and a number giving the slope of the line. In space a line is determined by a point and a *vector* giving the direction of the line.

Suppose that L is a line in space passing through a point $P_0(x_0, y_0, z_0)$ parallel to a vector $\mathbf{v} = v_1\mathbf{i} + v_2\mathbf{j} + v_3\mathbf{k}$. Then L is the set of all points $P(x, y, z)$ for which $\overrightarrow{P_0P}$ is parallel to $\mathbf{v}$ (Figure 12.36). Thus, $\overrightarrow{P_0P} = t\mathbf{v}$ for some scalar parameter t. The value of t depends on the location of the point P along the line, and the domain of t is $(-\infty, \infty)$. The expanded form of the equation $\overrightarrow{P_0P} = t\mathbf{v}$ is

$$(x - x_0)\mathbf{i} + (y - y_0)\mathbf{j} + (z - z_0)\mathbf{k} = t(v_1\mathbf{i} + v_2\mathbf{j} + v_3\mathbf{k}),$$

which can be rewritten as

$$x\mathbf{i} + y\mathbf{j} + z\mathbf{k} = x_0\mathbf{i} + y_0\mathbf{j} + z_0\mathbf{k} + t(v_1\mathbf{i} + v_2\mathbf{j} + v_3\mathbf{k}). \tag{1}$$

If $\mathbf{r}(t)$ is the position vector of a point $P(x, y, z)$ on the line and $\mathbf{r}_0$ is the position vector of the point $P_0(x_0, y_0, z_0)$, then Equation (1) gives the following vector form for the equation of a line in space.

> **Vector Equation for a Line**
> **A vector equation for the line L through $P_0(x_0, y_0, z_0)$ parallel to $\mathbf{v}$ is**
> $$\mathbf{r}(t) = \mathbf{r}_0 + t\mathbf{v}, \qquad -\infty < t < \infty, \tag{2}$$
> where $\mathbf{r}$ is the position vector of a point $P(x, y, z)$ on L and $\mathbf{r}_0$ is the position vector of $P_0(x_0, y_0, z_0)$.

Equating the corresponding components of the two sides of Equation (1) gives three scalar equations involving the parameter t:

$$x = x_0 + tv_1, \qquad y = y_0 + tv_2, \qquad z = z_0 + tv_3.$$

These equations give us the standard parametrization of the line for the parameter interval $-\infty < t < \infty$.

> **Parametric Equations for a Line**
> **The standard parametrization of the line through $P_0(x_0, y_0, z_0)$ parallel to $\mathbf{v} = v_1\mathbf{i} + v_2\mathbf{j} + v_3\mathbf{k}$ is**
> $$x = x_0 + tv_1, \quad y = y_0 + tv_2, \quad z = z_0 + tv_3, \quad -\infty < t < \infty \tag{3}$$

EXAMPLE 1   Find parametric equations for the line through $(-2, 0, 4)$ parallel to $\mathbf{v} = 2\mathbf{i} + 4\mathbf{j} - 2\mathbf{k}$ (Figure 12.37).

Solution   With $P_0(x_0, y_0, z_0)$ equal to $(-2, 0, 4)$ and $v_1\mathbf{i} + v_2\mathbf{j} + v_3\mathbf{k}$ equal to $2\mathbf{i} + 4\mathbf{j} - 2\mathbf{k}$, Equations (3) become

$$x = -2 + 2t, \qquad y = 4t, \qquad z = 4 - 2t. \qquad \blacksquare$$

EXAMPLE 2   Find parametric equations for the line through $P(-3, 2, -3)$ and $Q(1, -1, 4)$.

Solution   The vector

$$\overrightarrow{PQ} = (1 - (-3))\mathbf{i} + (-1 - 2)\mathbf{j} + (4 - (-3))\mathbf{k} = 4\mathbf{i} - 3\mathbf{j} + 7\mathbf{k}$$

FIGURE 12.37 Selected points and parameter values on the line in Example 1. The arrows show the direction of increasing t.

is parallel to the line, and Equations (3) with $(x_0, y_0, z_0) = (-3, 2, -3)$ give

$$x = -3 + 4t, \qquad y = 2 - 3t, \qquad z = -3 + 7t.$$

We could have chosen $Q(1, -1, 4)$ as the "base point" and written

$$x = 1 + 4t, \qquad y = -1 - 3t, \qquad z = 4 + 7t.$$

These equations serve as well as the first; they simply place you at a different point on the line for a given value of t. ∎

Notice that parametrizations are not unique. Not only can the "base point" change, but so can the parameter. The equations $x = -3 + 4t^3$, $y = 2 - 3t^3$, and $z = -3 + 7t^3$ also parametrize the line in Example 2.

To parametrize a line segment joining two points, we first parametrize the line through the points. We then find the t-values for the endpoints and restrict t to lie in the closed interval bounded by these values. The line equations together with this added restriction parametrize the segment.

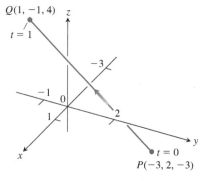

$Q(1, -1, 4)$

$t = 1$

$t = 0$

$P(-3, 2, -3)$

FIGURE 12.38 Example 3 derives a parametrization of line segment PQ. The arrow shows the direction of increasing t.

EXAMPLE 3 Parametrize the line segment joining the points $P(-3, 2, -3)$ and $Q(1, -1, 4)$ (Figure 12.38).

Solution We begin with equations for the line through P and Q, taking them, in this case, from Example 2:

$$x = -3 + 4t, \qquad y = 2 - 3t, \qquad z = -3 + 7t.$$

We observe that the point

$$(x, y, z) = (-3 + 4t, 2 - 3t, -3 + 7t)$$

on the line passes through $P(-3, 2, -3)$ at $t = 0$ and $Q(1, -1, 4)$ at $t = 1$. We add the restriction $0 \le t \le 1$ to parametrize the segment:

$$x = -3 + 4t, \qquad y = 2 - 3t, \qquad z = -3 + 7t, \qquad 0 \le t \le 1. \qquad ∎$$

The vector form (Equation (2)) for a line in space is more revealing if we think of a line as the path of a particle starting at position $P_0(x_0, y_0, z_0)$ and moving in the direction of vector $\mathbf{v}$. Rewriting Equation (2), we have

$$\mathbf{r}(t) = \mathbf{r}_0 + t\mathbf{v}$$

$$= \mathbf{r}_0 + t|\mathbf{v}| \frac{\mathbf{v}}{|\mathbf{v}|}. \qquad (4)$$

Initial position   Time   Speed   Direction

In other words, the position of the particle at time t is its initial position plus its distance moved (speed × time) in the direction $\mathbf{v}/|\mathbf{v}|$ of its straight-line motion.

EXAMPLE 4 A helicopter is to fly directly from a helipad at the origin in the direction of the point $(1, 1, 1)$ at a speed of 60 ft/sec. What is the position of the helicopter after 10 sec?

Solution We place the origin at the starting position (helipad) of the helicopter. Then the unit vector

$$\mathbf{u} = \frac{1}{\sqrt{3}}\mathbf{i} + \frac{1}{\sqrt{3}}\mathbf{j} + \frac{1}{\sqrt{3}}\mathbf{k}$$

gives the flight direction of the helicopter. From Equation (4), the position of the helicopter at any time t is

$$\mathbf{r}(t) = \mathbf{r}_0 + t(\text{speed})\mathbf{u}$$

$$= \mathbf{0} + t(60)\left(\frac{1}{\sqrt{3}}\mathbf{i} + \frac{1}{\sqrt{3}}\mathbf{j} + \frac{1}{\sqrt{3}}\mathbf{k}\right)$$

$$= 20\sqrt{3}\,t(\mathbf{i} + \mathbf{j} + \mathbf{k}).$$

When $t = 10$ sec,

$$\mathbf{r}(10) = 200\sqrt{3}\,(\mathbf{i} + \mathbf{j} + \mathbf{k})$$

$$= \langle 200\sqrt{3}, 200\sqrt{3}, 200\sqrt{3}\rangle.$$

After 10 sec of flight from the origin toward $(1, 1, 1)$, the helicopter is located at the point $(200\sqrt{3}, 200\sqrt{3}, 200\sqrt{3})$ in space. It has traveled a distance of $(60 \text{ ft/sec})(10 \text{ sec}) = 600$ ft, which is the length of the vector $\mathbf{r}(10)$. ∎

The Distance from a Point to a Line in Space

To find the distance from a point S to a line that passes through a point P parallel to a vector $\mathbf{v}$, we find the absolute value of the scalar component of $\overrightarrow{PS}$ in the direction of a vector normal to the line (Figure 12.39). In the notation of the figure, the absolute value of the scalar component is $|\overrightarrow{PS}| \sin \theta$, which is $\dfrac{|\overrightarrow{PS}||\mathbf{v}| \sin \theta}{|\mathbf{v}|} = \dfrac{|\overrightarrow{PS} \times \mathbf{v}|}{|\mathbf{v}|}$.

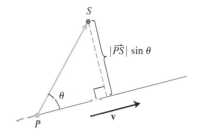

FIGURE 12.39 The distance from S to the line through P parallel to $\mathbf{v}$ is $|\overrightarrow{PS}| \sin \theta$, where θ is the angle between $\overrightarrow{PS}$ and $\mathbf{v}$.

Distance from a Point S to a Line Through P Parallel to $\mathbf{v}$

$$d = \frac{|\overrightarrow{PS} \times \mathbf{v}|}{|\mathbf{v}|} \qquad (5)$$

EXAMPLE 5 Find the distance from the point $S(1, 1, 5)$ to the line

$$L: \quad x = 1 + t, \quad y = 3 - t, \quad z = 2t.$$

Solution We see from the equations for L that L passes through $P(1, 3, 0)$ parallel to $\mathbf{v} = \mathbf{i} - \mathbf{j} + 2\mathbf{k}$. With

$$\overrightarrow{PS} = (1 - 1)\mathbf{i} + (1 - 3)\mathbf{j} + (5 - 0)\mathbf{k} = -2\mathbf{j} + 5\mathbf{k}$$

and

$$\overrightarrow{PS} \times \mathbf{v} = \begin{vmatrix} \mathbf{i} & \mathbf{j} & \mathbf{k} \\ 0 & -2 & 5 \\ 1 & -1 & 2 \end{vmatrix} = \mathbf{i} + 5\mathbf{j} + 2\mathbf{k},$$

Equation (5) gives

$$d = \frac{|\overrightarrow{PS} \times \mathbf{v}|}{|\mathbf{v}|} = \frac{\sqrt{1 + 25 + 4}}{\sqrt{1 + 1 + 4}} = \frac{\sqrt{30}}{\sqrt{6}} = \sqrt{5}. \qquad ∎$$

An Equation for a Plane in Space

A plane in space is determined by knowing a point on the plane and its "tilt" or orientation. This "tilt" is defined by specifying a vector that is perpendicular or normal to the plane.

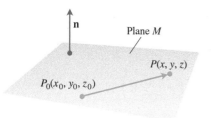

FIGURE 12.40 The standard equation for a plane in space is defined in terms of a vector normal to the plane: A point P lies in the plane through P_0 normal to $\mathbf{n}$ if and only if $\mathbf{n} \cdot \overrightarrow{P_0P} = 0$.

Suppose that plane M passes through a point $P_0(x_0, y_0, z_0)$ and is normal to the nonzero vector $\mathbf{n} = A\mathbf{i} + B\mathbf{j} + C\mathbf{k}$. A vector from P_0 to any point P on the plane is orthogonal to $\mathbf{n}$. Then M is the set of all points $P(x, y, z)$ for which $\overrightarrow{P_0P}$ is orthogonal to $\mathbf{n}$ (Figure 12.40). Thus, the dot product $\mathbf{n} \cdot \overrightarrow{P_0P} = 0$. This equation is equivalent to

$$(A\mathbf{i} + B\mathbf{j} + C\mathbf{k}) \cdot [(x - x_0)\mathbf{i} + (y - y_0)\mathbf{j} + (z - z_0)\mathbf{k}] = 0,$$

so the plane M consists of the points (x, y, z) satisfying

$$A(x - x_0) + B(y - y_0) + C(z - z_0) = 0.$$

Equation for a Plane

The plane through $P_0(x_0, y_0, z_0)$ normal to $\mathbf{n} = A\mathbf{i} + B\mathbf{j} + C\mathbf{k}$ has

Vector equation: $\mathbf{n} \cdot \overrightarrow{P_0P} = 0$

Component equation: $A(x - x_0) + B(y - y_0) + C(z - z_0) = 0$

Component equation simplified: $Ax + By + Cz = D$, where
$$D = Ax_0 + By_0 + Cz_0$$

EXAMPLE 6 Find an equation for the plane through $P_0(-3, 0, 7)$ perpendicular to $\mathbf{n} = 5\mathbf{i} + 2\mathbf{j} - \mathbf{k}$.

Solution The component equation is

$$5(x - (-3)) + 2(y - 0) + (-1)(z - 7) = 0.$$

Simplifying, we obtain

$$5x + 15 + 2y - z + 7 = 0$$
$$5x + 2y - z = -22.$$

Notice in Example 6 how the components of $\mathbf{n} = 5\mathbf{i} + 2\mathbf{j} - \mathbf{k}$ became the coefficients of x, y, and z in the equation $5x + 2y - z = -22$. The vector $\mathbf{n} = A\mathbf{i} + B\mathbf{j} + C\mathbf{k}$ is normal to the plane $Ax + By + Cz = D$.

EXAMPLE 7 Find an equation for the plane through $A(0, 0, 1)$, $B(2, 0, 0)$, and $C(0, 3, 0)$.

Solution We find a vector normal to the plane and use it with one of the points (it does not matter which) to write an equation for the plane.

The cross product

$$\overrightarrow{AB} \times \overrightarrow{AC} = \begin{vmatrix} \mathbf{i} & \mathbf{j} & \mathbf{k} \\ 2 & 0 & -1 \\ 0 & 3 & -1 \end{vmatrix} = 3\mathbf{i} + 2\mathbf{j} + 6\mathbf{k}$$

is normal to the plane. We substitute the components of this vector and the coordinates of $A(0, 0, 1)$ into the component form of the equation to obtain

$$3(x - 0) + 2(y - 0) + 6(z - 1) = 0$$
$$3x + 2y + 6z = 6.$$

Lines of Intersection

Just as lines are parallel if and only if they have the same direction, two planes are **parallel** if and only if their normals are parallel, or $\mathbf{n}_1 = k\mathbf{n}_2$ for some scalar k. Two planes that are not parallel intersect in a line.

EXAMPLE 8   Find a vector parallel to the line of intersection of the planes $3x - 6y - 2z = 15$ and $2x + y - 2z = 5$.

Solution   The line of intersection of two planes is perpendicular to both planes' normal vectors $\mathbf{n}_1$ and $\mathbf{n}_2$ (Figure 12.41) and therefore parallel to $\mathbf{n}_1 \times \mathbf{n}_2$. Turning this around, $\mathbf{n}_1 \times \mathbf{n}_2$ is a vector parallel to the planes' line of intersection. In our case,

$$\mathbf{n}_1 \times \mathbf{n}_2 = \begin{vmatrix} \mathbf{i} & \mathbf{j} & \mathbf{k} \\ 3 & -6 & -2 \\ 2 & 1 & -2 \end{vmatrix} = 14\mathbf{i} + 2\mathbf{j} + 15\mathbf{k}.$$

Any nonzero scalar multiple of $\mathbf{n}_1 \times \mathbf{n}_2$ will do as well.   ∎

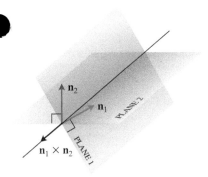

FIGURE 12.41   How the line of intersection of two planes is related to the planes' normal vectors (Example 8).

EXAMPLE 9   Find parametric equations for the line in which the planes $3x - 6y - 2z = 15$ and $2x + y - 2z = 5$ intersect.

Solution   We find a vector parallel to the line and a point on the line and use Equations (3).

Example 8 identifies $\mathbf{v} = 14\mathbf{i} + 2\mathbf{j} + 15\mathbf{k}$ as a vector parallel to the line. To find a point on the line, we can take any point common to the two planes. Substituting $z = 0$ in the plane equations and solving for x and y simultaneously identifies one of these points as $(3, -1, 0)$. The line is

$$x = 3 + 14t, \qquad y = -1 + 2t, \qquad z = 15t.$$

The choice $z = 0$ is arbitrary and we could have chosen $z = 1$ or $z = -1$ just as well. Or we could have let $x = 0$ and solved for y and z. The different choices would simply give different parametrizations of the same line.   ∎

Sometimes we want to know where a line and a plane intersect. For example, if we are looking at a flat plate and a line segment passes through it, we may be interested in knowing what portion of the line segment is hidden from our view by the plate. This application is used in computer graphics (Exercise 78).

EXAMPLE 10   Find the point where the line

$$x = \frac{8}{3} + 2t, \qquad y = -2t, \qquad z = 1 + t$$

intersects the plane $3x + 2y + 6z = 6$.

Solution   The point

$$\left(\frac{8}{3} + 2t, -2t, 1 + t \right)$$

lies in the plane if its coordinates satisfy the equation of the plane, that is, if

$$3\left(\frac{8}{3} + 2t \right) + 2(-2t) + 6(1 + t) = 6$$

$$8 + 6t - 4t + 6 + 6t = 6$$

$$8t = -8$$

$$t = -1.$$

The point of intersection is

$$(x, y, z)\big|_{t=-1} = \left(\frac{8}{3} - 2, 2, 1 - 1 \right) = \left(\frac{2}{3}, 2, 0 \right).$$ ∎

The Distance from a Point to a Plane

If P is a point on a plane with normal $\mathbf{n}$, then the distance from any point S to the plane is the length of the vector projection of $\overrightarrow{PS}$ onto $\mathbf{n}$, as given in the following formula.

Distance from a Point S to a Plane with Normal $\mathbf{n}$ at Point P

$$d = \left| \overrightarrow{PS} \cdot \frac{\mathbf{n}}{|\mathbf{n}|} \right| \qquad (6)$$

EXAMPLE 11   Find the distance from $S(1, 1, 3)$ to the plane $3x + 2y + 6z = 6$.

Solution   We find a point P in the plane and calculate the length of the vector projection of $\overrightarrow{PS}$ onto a vector $\mathbf{n}$ normal to the plane (Figure 12.42). The coefficients in the equation $3x + 2y + 6z = 6$ give

$$\mathbf{n} = 3\mathbf{i} + 2\mathbf{j} + 6\mathbf{k}.$$

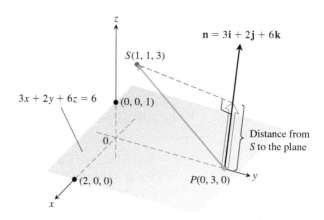

FIGURE 12.42   The distance from S to the plane is the length of the vector projection of $\overrightarrow{PS}$ onto $\mathbf{n}$ (Example 11).

The points on the plane easiest to find from the plane's equation are the intercepts. If we take P to be the y-intercept $(0, 3, 0)$, then

$$\overrightarrow{PS} = (1 - 0)\mathbf{i} + (1 - 3)\mathbf{j} + (3 - 0)\mathbf{k} = \mathbf{i} - 2\mathbf{j} + 3\mathbf{k},$$
$$|\mathbf{n}| = \sqrt{(3)^2 + (2)^2 + (6)^2} = \sqrt{49} = 7.$$

Therefore, the distance from S to the plane is

$$d = \left| \overrightarrow{PS} \cdot \frac{\mathbf{n}}{|\mathbf{n}|} \right| \qquad \text{Length of } \mathrm{proj}_{\mathbf{n}} \overrightarrow{PS}$$

$$= \left| (\mathbf{i} - 2\mathbf{j} + 3\mathbf{k}) \cdot \left(\frac{3}{7}\mathbf{i} + \frac{2}{7}\mathbf{j} + \frac{6}{7}\mathbf{k} \right) \right|$$

$$= \left| \frac{3}{7} - \frac{4}{7} + \frac{18}{7} \right| = \frac{17}{7}. \qquad \blacksquare$$

Angles Between Planes

The angle between two intersecting planes is defined to be the acute angle between their normal vectors (Figure 12.43).

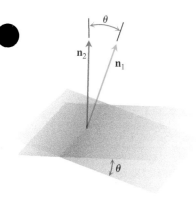

FIGURE 12.43 The angle between two planes is obtained from the angle between their normals.

EXAMPLE 12 Find the angle between the planes $3x - 6y - 2z = 15$ and $2x + y - 2z = 5$.

Solution The vectors

$$\mathbf{n}_1 = 3\mathbf{i} - 6\mathbf{j} - 2\mathbf{k}, \qquad \mathbf{n}_2 = 2\mathbf{i} + \mathbf{j} - 2\mathbf{k}$$

are normals to the planes. The angle between them is

$$\theta = \cos^{-1}\left(\frac{\mathbf{n}_1 \cdot \mathbf{n}_2}{|\mathbf{n}_1||\mathbf{n}_2|}\right)$$

$$= \cos^{-1}\left(\frac{4}{21}\right) \approx 1.38 \text{ radians.} \qquad \text{About 79 degrees} \qquad \blacksquare$$

EXERCISES 12.5

Lines and Line Segments

Find parametric equations for the lines in Exercises 1–12.

1. The line through the point $P(3, -4, -1)$ parallel to the vector $\mathbf{i} + \mathbf{j} + \mathbf{k}$

2. The line through $P(1, 2, -1)$ and $Q(-1, 0, 1)$

3. The line through $P(-2, 0, 3)$ and $Q(3, 5, -2)$

4. The line through $P(1, 2, 0)$ and $Q(1, 1, -1)$

5. The line through the origin parallel to the vector $2\mathbf{j} + \mathbf{k}$

6. The line through the point $(3, -2, 1)$ parallel to the line $x = 1 + 2t, y = 2 - t, z = 3t$

7. The line through $(1, 1, 1)$ parallel to the z-axis

8. The line through $(2, 4, 5)$ perpendicular to the plane $3x + 7y - 5z = 21$

9. The line through $(0, -7, 0)$ perpendicular to the plane $x + 2y + 2z = 13$

10. The line through $(2, 3, 0)$ perpendicular to the vectors $\mathbf{u} = \mathbf{i} + 2\mathbf{j} + 3\mathbf{k}$ and $\mathbf{v} = 3\mathbf{i} + 4\mathbf{j} + 5\mathbf{k}$

11. The x-axis   **12.** The z-axis

Find parametrizations for the line segments joining the points in Exercises 13–20. Draw coordinate axes and sketch each segment, indicating the direction of increasing t for your parametrization.

13. $(0, 0, 0)$, $(1, 1, 3/2)$ **14.** $(0, 0, 0)$, $(1, 0, 0)$

15. $(1, 0, 0)$, $(1, 1, 0)$ **16.** $(1, 1, 0)$, $(1, 1, 1)$

17. $(0, 1, 1)$, $(0, -1, 1)$ **18.** $(0, 2, 0)$, $(3, 0, 0)$

19. $(2, 0, 2)$, $(0, 2, 0)$ **20.** $(1, 0, -1)$, $(0, 3, 0)$

Planes

Find equations for the planes in Exercises 21–26.

21. The plane through $P_0(0, 2, -1)$ normal to $\mathbf{n} = 3\mathbf{i} - 2\mathbf{j} - \mathbf{k}$

22. The plane through $(1, -1, 3)$ parallel to the plane

$$3x + y + z = 7$$

23. The plane through $(1, 1, -1)$, $(2, 0, 2)$, and $(0, -2, 1)$

24. The plane through $(2, 4, 5)$, $(1, 5, 7)$, and $(-1, 6, 8)$

25. The plane through $P_0(2, 4, 5)$ perpendicular to the line

$$x = 5 + t, \quad y = 1 + 3t, \quad z = 4t$$

26. The plane through $A(1, -2, 1)$ perpendicular to the vector from the origin to A

27. Find the point of intersection of the lines $x = 2t + 1$, $y = 3t + 2$, $z = 4t + 3$, and $x = s + 2, y = 2s + 4, z = -4s - 1$, and then find the plane determined by these lines.

28. Find the point of intersection of the lines $x = t, y = -t + 2$, $z = t + 1$, and $x = 2s + 2, y = s + 3, z = 5s + 6$, and then find the plane determined by these lines.

In Exercises 29 and 30, find the plane containing the intersecting lines.

29. $L1: x = -1 + t, \quad y = 2 + t, \quad z = 1 - t; \quad -\infty < t < \infty$
$L2: x = 1 - 4s, \quad y = 1 + 2s, \quad z = 2 - 2s; \quad -\infty < s < \infty$

30. $L1: x = t, \quad y = 3 - 3t, \quad z = -2 - t; \quad -\infty < t < \infty$
$L2: x = 1 + s, \quad y = 4 + s, \quad z = -1 + s; \quad -\infty < s < \infty$

31. Find a plane through $P_0(2, 1, -1)$ and perpendicular to the line of intersection of the planes $2x + y - z = 3, x + 2y + z = 2$.

32. Find a plane through the points $P_1(1, 2, 3)$, $P_2(3, 2, 1)$ and perpendicular to the plane $4x - y + 2z = 7$.

Distances

In Exercises 33–38, find the distance from the point to the line.

33. $(0, 0, 12)$; $x = 4t, \quad y = -2t, \quad z = 2t$

34. $(0, 0, 0)$; $x = 5 + 3t, \quad y = 5 + 4t, \quad z = -3 - 5t$

35. $(2, 1, 3)$; $x = 2 + 2t, \quad y = 1 + 6t, \quad z = 3$

36. $(2, 1, -1)$; $x = 2t, \quad y = 1 + 2t, \quad z = 2t$

37. $(3, -1, 4)$; $x = 4 - t, \quad y = 3 + 2t, \quad z = -5 + 3t$

38. $(-1, 4, 3)$; $x = 10 + 4t, \quad y = -3, \quad z = 4t$

In Exercises 39–44, find the distance from the point to the plane.

39. $(2, -3, 4)$, $x + 2y + 2z = 13$

40. $(0, 0, 0)$, $3x + 2y + 6z = 6$

41. $(0, 1, 1)$, $4y + 3z = -12$

42. $(2, 2, 3)$, $2x + y + 2z = 4$

43. $(0, -1, 0)$, $2x + y + 2z = 4$

44. $(1, 0, -1)$, $-4x + y + z = 4$

45. Find the distance from the plane $x + 2y + 6z = 1$ to the plane $x + 2y + 6z = 10$.

46. Find the distance from the line $x = 2 + t, y = 1 + t$, $z = -(1/2) - (1/2)t$ to the plane $x + 2y + 6z = 10$.

Angles

Find the angles between the planes in Exercises 47 and 48.

47. $x + y = 1$, $2x + y - 2z = 2$

48. $5x + y - z = 10$, $x - 2y + 3z = -1$

Find the acute angles between the intersecting lines in Exercises 49 and 50.

49. $x = t, y = 2t, z = -t$ and $x = 1 - t, y = 5 + t, z = 2t$

50. $x = 2 + t, y = 4t + 2, z = 1 + t$ and
$x = 3t - 2, y = -2, z = 2 - 2t$

Find the acute angles between the lines and planes in Exercises 51 and 52.

51. $x = 1 - t, y = 3t, z = 1 + t$; $2x - y + 3z = 6$

52. $x = 2, y = 3 + 2t, z = 1 - 2t$; $x - y + z = 0$

[T] Use a calculator to find the acute angles between the planes in Exercises 53–56 to the nearest hundredth of a radian.

53. $2x + 2y + 2z = 3$, $2x - 2y - z = 5$

54. $x + y + z = 1$, $z = 0$ (the xy-plane)

55. $2x + 2y - z = 3$, $x + 2y + z = 2$

56. $4y + 3z = -12$, $3x + 2y + 6z = 6$

Intersecting Lines and Planes

In Exercises 57–60, find the point in which the line meets the plane.

57. $x = 1 - t$, $y = 3t$, $z = 1 + t$; $2x - y + 3z = 6$

58. $x = 2$, $y = 3 + 2t$, $z = -2 - 2t$; $6x + 3y - 4z = -12$

59. $x = 1 + 2t$, $y = 1 + 5t$, $z = 3t$; $x + y + z = 2$

60. $x = -1 + 3t$, $y = -2$, $z = 5t$; $2x - 3z = 7$

Find parametrizations for the lines in which the planes in Exercises 61–64 intersect.

61. $x + y + z = 1$, $x + y = 2$

62. $3x - 6y - 2z = 3$, $2x + y - 2z = 2$

63. $x - 2y + 4z = 2$, $x + y - 2z = 5$

64. $5x - 2y = 11$, $4y - 5z = -17$

Given two lines in space, either they are parallel, they intersect, or they are skew (lie in parallel planes). In Exercises 65 and 66, determine whether the lines, taken two at a time, are parallel, intersect, or are skew. If they intersect, find the point of intersection. Otherwise, find the distance between the two lines.

65. $L1: x = 3 + 2t$, $y = -1 + 4t$, $z = 2 - t$; $-\infty < t < \infty$
    $L2: x = 1 + 4s, y = 1 + 2s, z = -3 + 4s$; $-\infty < s < \infty$
    $L3: x = 3 + 2r, y = 2 + r, z = -2 + 2r$; $-\infty < r < \infty$

66. $L1: x = 1 + 2t$, $y = -1 - t$, $z = 3t$; $-\infty < t < \infty$
    $L2: x = 2 - s$, $y = 3s$, $z = 1 + s$; $-\infty < s < \infty$
    $L3: x = 5 + 2r$, $y = 1 - r$, $z = 8 + 3r$; $-\infty < r < \infty$

Theory and Examples

67. Use Equations (3) to generate a parametrization of the line through $P(2, -4, 7)$ parallel to $\mathbf{v}_1 = 2\mathbf{i} - \mathbf{j} + 3\mathbf{k}$. Then generate another parametrization of the line using the point $P_2(-2, -2, 1)$ and the vector $\mathbf{v}_2 = -\mathbf{i} + (1/2)\mathbf{j} - (3/2)\mathbf{k}$.

68. Use the component form to generate an equation for the plane through $P_1(4, 1, 5)$ normal to $\mathbf{n}_1 = \mathbf{i} - 2\mathbf{j} + \mathbf{k}$. Then generate another equation for the same plane using the point $P_2(3, -2, 0)$ and the normal vector $\mathbf{n}_2 = -\sqrt{2}\mathbf{i} + 2\sqrt{2}\mathbf{j} - \sqrt{2}\mathbf{k}$.

69. Find the points in which the line $x = 1 + 2t, y = -1 - t$, $z = 3t$ meets the coordinate planes. Describe the reasoning behind your answer.

70. Find equations for the line in the plane $z = 3$ that makes an angle of $\pi/6$ rad with $\mathbf{i}$ and an angle of $\pi/3$ rad with $\mathbf{j}$. Describe the reasoning behind your answer.

71. Is the line $x = 1 - 2t, y = 2 + 5t, z = -3t$ parallel to the plane $2x + y - z = 8$? Give reasons for your answer.

72. How can you tell when two planes $A_1x + B_1y + C_1z = D_1$ and $A_2x + B_2y + C_2z = D_2$ are parallel? Perpendicular? Give reasons for your answer.

73. Find two different planes whose intersection is the line $x = 1 + t, y = 2 - t, z = 3 + 2t$. Write equations for each plane in the form $Ax + By + Cz = D$.

74. Find a plane through the origin that is perpendicular to the plane $M: 2x + 3y + z = 12$ in a right angle. How do you know that your plane is perpendicular to M?

75. The graph of $(x/a) + (y/b) + (z/c) = 1$ is a plane for any nonzero numbers $a, b,$ and c. Which planes have an equation of this form?

76. Suppose L_1 and L_2 are disjoint (nonintersecting) nonparallel lines. Is it possible for a nonzero vector to be perpendicular to both L_1 and L_2? Give reasons for your answer.

77. Perspective in computer graphics   In computer graphics and perspective drawing, we need to represent objects seen by the eye in space as images on a two-dimensional plane. Suppose that the eye is at $E(x_0, 0, 0)$ as shown here and that we want to represent a point $P_1(x_1, y_1, z_1)$ as a point on the yz-plane. We do this by projecting P_1 onto the plane with a ray from E. The point P_1 will be portrayed as the point $P(0, y, z)$. The problem for us as graphics designers is to find y and z given E and P_1.

a. Write a vector equation that holds between $\overrightarrow{EP}$ and $\overrightarrow{EP_1}$. Use the equation to express y and z in terms of $x_0, x_1, y_1,$ and z_1.

b. Test the formulas obtained for y and z in part (a) by investigating their behavior at $x_1 = 0$ and $x_1 = x_0$ and by seeing what happens as $x_0 \to \infty$. What do you find?

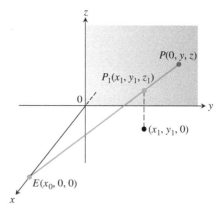

78. Hidden lines in computer graphics Here is another typical problem in computer graphics. Your eye is at (4, 0, 0). You are looking at a triangular plate whose vertices are at (1, 0, 1), (1, 1, 0), and (−2, 2, 2). The line segment from (1, 0, 0) to (0, 2, 2) passes through the plate. What portion of the line segment is hidden from your view by the plate? (This is an exercise in finding intersections of lines and planes.)

12.6 Cylinders and Quadric Surfaces

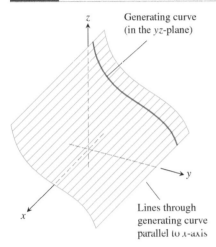

FIGURE 12.44 A cylinder and generating curve.

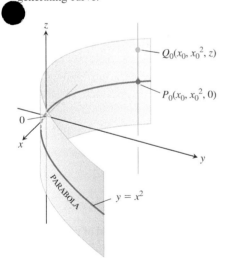

FIGURE 12.45 Every point of the cylinder in Example 1 has coordinates of the form (x_0, x_0^2, z). We call it "the cylinder $y = x^2$."

Up to now, we have studied two special types of surfaces: spheres and planes. In this section, we extend our inventory to include a variety of cylinders and quadric surfaces. Quadric surfaces are surfaces defined by second-degree equations in x, y, and z. Spheres are quadric surfaces, but there are others of equal interest which will be needed in Chapters 14–16.

Cylinders

A **cylinder** is a surface that is generated by moving a straight line along a given planar curve while holding the line parallel to a given fixed line. The curve is called a **generating curve** for the cylinder (Figure 12.44). In solid geometry, where *cylinder* means *circular cylinder*, the generating curves are circles, but now we allow generating curves of any kind. The cylinder in our first example is generated by a parabola.

EXAMPLE 1   Find an equation for the cylinder made by the lines parallel to the z-axis that pass through the parabola $y = x^2$, $z = 0$ (Figure 12.45).

Solution   The point $P_0(x_0, x_0^2, 0)$ lies on the parabola $y = x^2$ in the xy-plane. Then, for any value of z, the point $Q(x_0, x_0^2, z)$ lies on the cylinder because it lies on the line $x = x_0$, $y = x_0^2$ through P_0 parallel to the z-axis. Conversely, any point $Q(x_0, x_0^2, z)$ whose y-coordinate is the square of its x-coordinate lies on the cylinder because it lies on the line $x = x_0$, $y = x_0^2$ through P_0 parallel to the z-axis (Figure 12.45).

Regardless of the value of z, therefore, the points on the surface are the points whose coordinates satisfy the equation $y = x^2$. This makes $y = x^2$ an equation for the cylinder. Because of this, we call the cylinder "the cylinder $y = x^2$." ∎

As Example 1 suggests, any curve $f(x, y) = c$ in the xy-plane defines a cylinder parallel to the z-axis whose equation is also $f(x, y) = c$. For instance, the equation $x^2 + y^2 = 1$ defines the circular cylinder made by the lines parallel to the z-axis that pass through the circle $x^2 + y^2 = 1$ in the xy-plane.

In a similar way, any curve $g(x, z) = c$ in the xz-plane defines a cylinder parallel to the y-axis whose space equation is also $g(x, z) = c$. Any curve $h(y, z) = c$ defines a cylinder parallel to the x-axis whose space equation is also $h(y, z) = c$. The axis of a cylinder need not be parallel to a coordinate axis, however.

Quadric Surfaces

A **quadric surface** is the graph in space of a second-degree equation in x, y, and z. We first focus on quadric surfaces given by the equation

$$Ax^2 + By^2 + Cz^2 + Dz = E,$$

where A, B, C, D, and E are constants. The basic quadric surfaces are **ellipsoids**, **paraboloids**, **elliptical cones**, and **hyperboloids**. Spheres are special cases of ellipsoids. We present a few examples illustrating how to sketch a quadric surface, and then give a summary table of graphs of the basic types.

EXAMPLE 2   The **ellipsoid**

$$\frac{x^2}{a^2} + \frac{y^2}{b^2} + \frac{z^2}{c^2} = 1$$

(Figure 12.46) cuts the coordinate axes at $(\pm a, 0, 0)$, $(0, \pm b, 0)$, and $(0, 0, \pm c)$. It lies within the rectangular box defined by the inequalities $|x| \le a$, $|y| \le b$, and $|z| \le c$. The surface is symmetric with respect to each of the coordinate planes because each variable in the defining equation is squared.

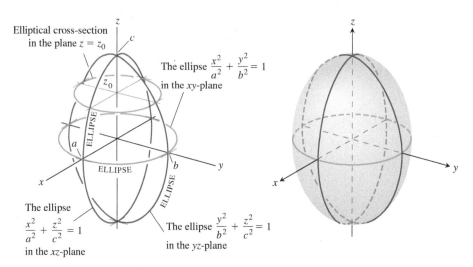

FIGURE 12.46 The ellipsoid

$$\frac{x^2}{a^2} + \frac{y^2}{b^2} + \frac{z^2}{c^2} = 1$$

in Example 2 has elliptical cross-sections in each of the three coordinate planes.

The curves in which the three coordinate planes cut the surface are ellipses. For example,

$$\frac{x^2}{a^2} + \frac{y^2}{b^2} = 1 \qquad \text{when} \qquad z = 0.$$

The curve cut from the surface by the plane $z = z_0$, $|z_0| < c$, is the ellipse

$$\frac{x^2}{a^2(1 - (z_0/c)^2)} + \frac{y^2}{b^2(1 - (z_0/c)^2)} = 1.$$

If any two of the semiaxes a, b, and c are equal, the surface is an **ellipsoid of revolution**. If all three are equal, the surface is a sphere. ∎

EXAMPLE 3   The **hyperbolic paraboloid**

$$\frac{y^2}{b^2} - \frac{x^2}{a^2} = \frac{z}{c}, \qquad c > 0$$

has symmetry with respect to the planes $x = 0$ and $y = 0$ (Figure 12.47). The cross-sections in these planes are

$$x = 0: \quad \text{the parabola } z = \frac{c}{b^2}y^2. \tag{1}$$

$$y = 0: \quad \text{the parabola } z = -\frac{c}{a^2}x^2. \tag{2}$$

In the plane $x = 0$, the parabola opens upward from the origin. The parabola in the plane $y = 0$ opens downward.

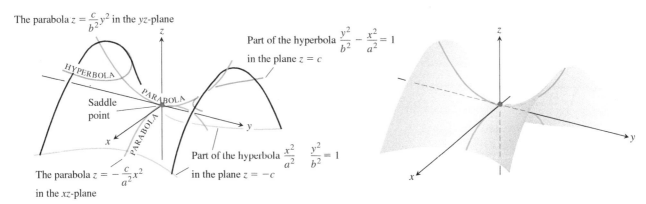

The parabola $z = \dfrac{c}{b^2}y^2$ in the yz-plane

Part of the hyperbola $\dfrac{y^2}{b^2} - \dfrac{x^2}{a^2} = 1$ in the plane $z = c$

HYPERBOLA

PARABOLA

Saddle point

PARABOLA

Part of the hyperbola $\dfrac{x^2}{a^2} - \dfrac{y^2}{b^2} = 1$ in the plane $z = -c$

The parabola $z = -\dfrac{c}{a^2}x^2$ in the xz-plane

FIGURE 12.47 The hyperbolic paraboloid $(y^2/b^2) - (x^2/a^2) = z/c, c > 0$. The cross-sections in planes perpendicular to the z-axis above and below the xy-plane are hyperbolas. The cross-sections in planes perpendicular to the other axes are parabolas.

If we cut the surface by a plane $z = z_0 > 0$, the cross-section is a hyperbola,

$$\frac{y^2}{b^2} - \frac{x^2}{a^2} = \frac{z_0}{c},$$

with its focal axis parallel to the y-axis and its vertices on the parabola in Equation (1). If z_0 is negative, the focal axis is parallel to the x-axis and the vertices lie on the parabola in Equation (2).

Near the origin, the surface is shaped like a saddle or mountain pass. To a person traveling along the surface in the yz-plane the origin looks like a minimum. To a person traveling the xz-plane the origin looks like a maximum. Such a point is called a **saddle point** of a surface. We will say more about saddle points in Section 14.7. ∎

Table 12.1 shows graphs of the six basic types of quadric surfaces. Each surface shown is symmetric with respect to the z-axis, but other coordinate axes can serve as well (with appropriate changes to the equation).

General Quadric Surfaces

The quadric surfaces we have considered have symmetries relative to the x-, y-, or z-axes. The general equation of second degree in three variables x, y, z is

$$Ax^2 + By^2 + Cz^2 + Dxy + Exz + Fyz + Gx + Hy + Iz + J = 0,$$

where $A, B, C, D, E, F, G, H, I,$ and J are constants. This equation leads to surfaces similar to those in Table 12.1, but in general these surfaces might be translated and rotated relative to the x-, y-, and z-axes. Terms of the type $Gx, Hy,$ or Iz in the above formula lead to translations, which can be seen by a process of completing the square.

EXAMPLE 4   Identify the surface given by the equation

$$x^2 + y^2 + 4z^2 - 2x + 4y + 1 = 0.$$

Solution   We complete the squares to simplify the expression:

$$x^2 + y^2 + 4z^2 - 2x + 4y + 1 = (x - 1)^2 - 1 + (y + 2)^2 - 4 + 4z^2 + 1$$
$$= (x - 1)^2 + (y + 2)^2 + 4z^2 - 4.$$

TABLE 12.1 **Graphs of Quadric Surfaces**

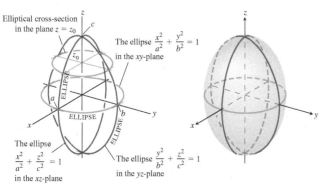

ELLIPSOID $\dfrac{x^2}{a^2} + \dfrac{y^2}{b^2} + \dfrac{z^2}{c^2} = 1$

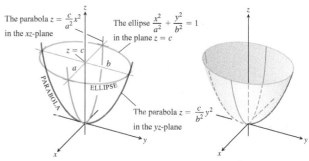

ELLIPTICAL PARABOLOID $\dfrac{x^2}{a^2} + \dfrac{y^2}{b^2} = \dfrac{z}{c}$

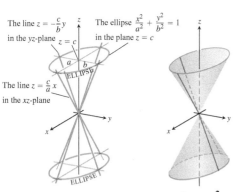

ELLIPTICAL CONE $\dfrac{x^2}{a^2} + \dfrac{y^2}{b^2} = \dfrac{z^2}{c^2}$

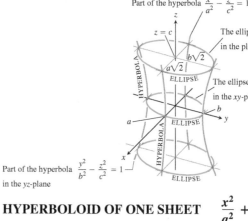

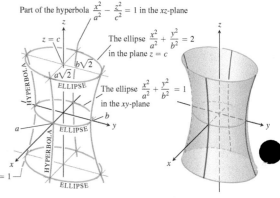

HYPERBOLOID OF ONE SHEET $\dfrac{x^2}{a^2} + \dfrac{y^2}{b^2} - \dfrac{z^2}{c^2} = 1$

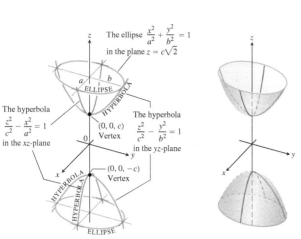

HYPERBOLOID OF TWO SHEETS $\dfrac{z^2}{c^2} - \dfrac{x^2}{a^2} - \dfrac{y^2}{b^2} = 1$

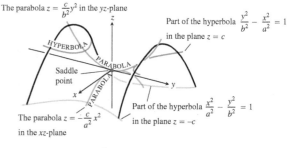

HYPERBOLIC PARABOLOID $\dfrac{y^2}{b^2} - \dfrac{x^2}{a^2} = \dfrac{z}{c}, c > 0$

We can rewrite the original equation as

$$\frac{(x-1)^2}{4} + \frac{(y+2)^2}{4} + \frac{z^2}{1} = 1.$$

This is the equation of an ellipsoid whose three semiaxes have lengths 2, 2, and 1 and which is centered at the point $(1, -2, 0)$, as shown in Figure 12.48. ∎

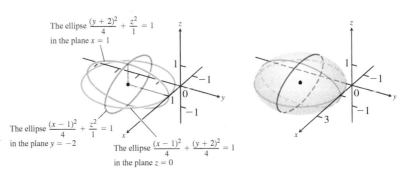

FIGURE 12.48 An ellipsoid centered at the point $(1, -2, 0)$.

Matching Equations with Surfaces

Exercises 1–12, match the equation with the surface it defines. Also, identify each surface by type (paraboloid, ellipsoid, etc.). The surfaces are labeled (a)–(l).

1. $x^2 + y^2 + 4z^2 = 10$

2. $z^2 + 4y^2 - 4x^2 = 4$

3. $9y^2 + z^2 - 16$

4. $y^2 + z^2 = x^2$

5. $x = y^2 - z^2$

6. $x = -y^2 - z^2$

7. $x^2 + 2z^2 = 8$

8. $z^2 + x^2 - y^2 = 1$

9. $x = z^2 - y^2$

10. $z = -4x^2 - y^2$

11. $x^2 + 4z^2 = y^2$

12. $9x^2 + 4y^2 + 2z^2 = 36$

a.

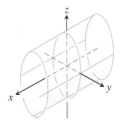

b.

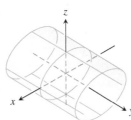

c.

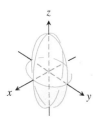

d.

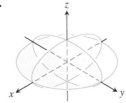

e.

f.

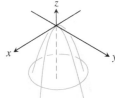

g.

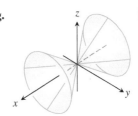

h.

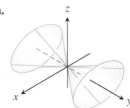

i.

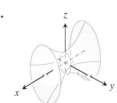

j.

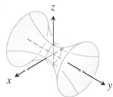

k.

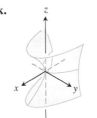

l.

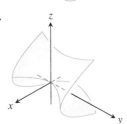

Drawing

Sketch the surfaces in Exercises 13–44.

CYLINDERS

13. $x^2 + y^2 = 4$

14. $z = y^2 - 1$

15. $x^2 + 4z^2 = 16$

16. $4x^2 + y^2 = 36$

ELLIPSOIDS

17. $9x^2 + y^2 + z^2 = 9$

18. $4x^2 + 4y^2 + z^2 = 16$

19. $4x^2 + 9y^2 + 4z^2 = 36$

20. $9x^2 + 4y^2 + 36z^2 = 36$

PARABOLOIDS AND CONES

21. $z = x^2 + 4y^2$

22. $z = 8 - x^2 - y^2$

23. $x = 4 - 4y^2 - z^2$

24. $y = 1 - x^2 - z^2$

25. $x^2 + y^2 = z^2$

26. $4x^2 + 9z^2 = 9y^2$

HYPERBOLOIDS

27. $x^2 + y^2 - z^2 = 1$

28. $y^2 + z^2 - x^2 = 1$

29. $z^2 - x^2 - y^2 = 1$

30. $(y^2/4) - (x^2/4) - z^2 = 1$

HYPERBOLIC PARABOLOIDS

31. $y^2 - x^2 = z$

32. $x^2 - y^2 = z$

ASSORTED

33. $z = 1 + y^2 - x^2$

34. $4x^2 + 4y^2 = z^2$

35. $y = -(x^2 + z^2)$

36. $16x^2 + 4y^2 = 1$

37. $x^2 + y^2 - z^2 = 4$

38. $x^2 + z^2 = y$

39. $x^2 + z^2 = 1$

40. $16y^2 + 9z^2 = 4x^2$

41. $z = -(x^2 + y^2)$

42. $y^2 - x^2 - z^2 = 1$

43. $4y^2 + z^2 - 4x^2 = 4$

44. $x^2 + y^2 = z$

Theory and Examples

45. a. Express the area A of the cross-section cut from the ellipsoid

$$x^2 + \frac{y^2}{4} + \frac{z^2}{9} = 1$$

by the plane $z = c$ as a function of c. (The area of an ellipse with semiaxes a and b is πab.)

b. Use slices perpendicular to the z-axis to find the volume of the ellipsoid in part (a).

c. Now find the volume of the ellipsoid

$$\frac{x^2}{a^2} + \frac{y^2}{b^2} + \frac{z^2}{c^2} = 1.$$

Does your formula give the volume of a sphere of radius a if $a = b = c$?

46. The barrel shown here is shaped like an ellipsoid with equal pieces cut from the ends by planes perpendicular to the z-axis. The cross-sections perpendicular to the z-axis are circular. The barrel is $2h$ units high, its midsection radius is R, and its end radii are both r. Find a formula for the barrel's volume. Then check two things. First, suppose the sides of the barrel are straightened to turn the barrel into a cylinder of radius R and height $2h$. Does your formula give the cylinder's volume? Second, suppose $r = 0$ and $h = R$ so the barrel is a sphere. Does your formula give the sphere's volume?

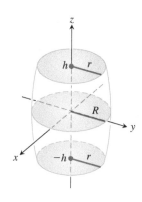

47. Show that the volume of the segment cut from the paraboloid

$$\frac{x^2}{a^2} + \frac{y^2}{b^2} = \frac{z}{c}$$

by the plane $z = h$ equals half the segment's base times its altitude.

48. a. Find the volume of the solid bounded by the hyperboloid

$$\frac{x^2}{a^2} + \frac{y^2}{b^2} - \frac{z^2}{c^2} = 1$$

and the planes $z = 0$ and $z = h$, $h > 0$.

b. Express your answer in part (a) in terms of h and the areas A_0 and A_h of the regions cut by the hyperboloid from the planes $z = 0$ and $z = h$.

c. Show that the volume in part (a) is also given by the formula

$$V = \frac{h}{6}(A_0 + 4A_m + A_h),$$

where A_m is the area of the region cut by the hyperboloid from the plane $z = h/2$.

Viewing Surfaces

T Plot the surfaces in Exercises 49–52 over the indicated domains. If you can, rotate the surface into different viewing positions.

49. $z = y^2$, $\quad -2 \le x \le 2$, $\quad -0.5 \le y \le 2$

50. $z = 1 - y^2$, $\quad -2 \le x \le 2$, $\quad -2 \le y \le 2$

51. $z = x^2 + y^2$, $\quad -3 \le x \le 3$, $\quad -3 \le y \le 3$

52. $z = x^2 + 2y^2$ over

   a. $-3 \le x \le 3$, $\quad -3 \le y \le 3$

   b. $-1 \le x \le 1$, $\quad -2 \le y \le 3$

   c. $-2 \le x \le 2$, $\quad -2 \le y \le 2$

   d. $-2 \le x \le 2$, $\quad -1 \le y \le 1$

COMPUTER EXPLORATIONS

Use a CAS to plot the surfaces in Exercises 53–58. Identify the type of quadric surface from your graph.

53. $\dfrac{x^2}{9} + \dfrac{y^2}{36} = 1 - \dfrac{z^2}{25}$

54. $\dfrac{x^2}{9} - \dfrac{z^2}{9} = 1 - \dfrac{y^2}{16}$

55. $5x^2 = z^2 - 3y^2$

56. $\dfrac{y^2}{16} = 1 - \dfrac{x^2}{9} + z$

57. $\dfrac{x^2}{9} - 1 = \dfrac{y^2}{16} + \dfrac{z^2}{2}$

58. $y - \sqrt{4 - z^2} = 0$

CHAPTER 12   Questions to Guide Your Review

1. When do directed line segments in the plane represent the same vector?

2. How are vectors added and subtracted geometrically? Algebraically?

3. How do you find a vector's magnitude and direction?

4. If a vector is multiplied by a positive scalar, how is the result related to the original vector? What if the scalar is zero? Negative?

5. Define the *dot product (scalar product)* of two vectors. Which algebraic laws are satisfied by dot products? Give examples. When is the dot product of two vectors equal to zero?

6. What geometric interpretation does the dot product have? Give examples.

7. What is the vector projection of a vector **u** onto a vector **v**? Give an example of a useful application of a vector projection.

8. Define the *cross product (vector product)* of two vectors. Which algebraic laws are satisfied by cross products, and which are not? Give examples. When is the cross product of two vectors equal to zero?

9. What geometric or physical interpretations do cross products have? Give examples.

10. What is the determinant formula for calculating the cross product of two vectors relative to the Cartesian **i, j, k**-coordinate system? Use it in an example.

11. How do you find equations for lines, line segments, and planes in space? Give examples. Can you express a line in space by a single equation? A plane?

12. How do you find the distance from a point to a line in space? From a point to a plane? Give examples.

13. What are box products? What significance do they have? How are they evaluated? Give an example.

14. How do you find equations for spheres in space? Give examples.

15. How do you find the intersection of two lines in space? A line and a plane? Two planes? Give examples.

16. What is a cylinder? Give examples of equations that define cylinders in Cartesian coordinates.

17. What are quadric surfaces? Give examples of different kinds of ellipsoids, paraboloids, cones, and hyperboloids (equations and sketches).

CHAPTER 12   Practice Exercises

Vector Calculations in Two Dimensions

In Exercises 1–4, let $\mathbf{u} = \langle -3, 4 \rangle$ and $\mathbf{v} = \langle 2, -5 \rangle$. Find (a) the component form of the vector and (b) its magnitude.

1. $3\mathbf{u} - 4\mathbf{v}$ 2. $\mathbf{u} + \mathbf{v}$

3. $-2\mathbf{u}$ 4. $5\mathbf{v}$

In Exercises 5–8, find the component form of the vector.

5. The vector obtained by rotating $\langle 0, 1 \rangle$ through an angle of $2\pi/3$ radians

6. The unit vector that makes an angle of $\pi/6$ radian with the positive x-axis

7. The vector 2 units long in the direction $4\mathbf{i} - \mathbf{j}$

8. The vector 5 units long in the direction opposite to the direction of $(3/5)\mathbf{i} + (4/5)\mathbf{j}$

Express the vectors in Exercises 9–12 in terms of their lengths and directions.

9. $\sqrt{2}\mathbf{i} + \sqrt{2}\mathbf{j}$ 10. $-\mathbf{i} - \mathbf{j}$

11. Velocity vector $\mathbf{v} = (-2\sin t)\mathbf{i} + (2\cos t)\mathbf{j}$ when $t = \pi/2$.

12. Velocity vector $\mathbf{v} = (e^t \cos t - e^t \sin t)\mathbf{i} + (e^t \sin t + e^t \cos t)\mathbf{j}$ when $t = \ln 2$.

Vector Calculations in Three Dimensions

Express the vectors in Exercises 13 and 14 in terms of their lengths and directions.

13. $2\mathbf{i} - 3\mathbf{j} + 6\mathbf{k}$ 14. $\mathbf{i} + 2\mathbf{j} - \mathbf{k}$

15. Find a vector 2 units long in the direction of $\mathbf{v} = 4\mathbf{i} - \mathbf{j} + 4\mathbf{k}$.

16. Find a vector 5 units long in the direction opposite to the direction of $\mathbf{v} = (3/5)\mathbf{i} + (4/5)\mathbf{k}$.

In Exercises 17 and 18, find $|\mathbf{v}|, |\mathbf{u}|, \mathbf{v} \cdot \mathbf{u}, \mathbf{u} \cdot \mathbf{v}, \mathbf{v} \times \mathbf{u}, \mathbf{u} \times \mathbf{v}, |\mathbf{v} \times \mathbf{u}|$, the angle between **v** and **u**, the scalar component of **u** in the direction of **v**, and the vector projection of **u** onto **v**.

17. $\mathbf{v} = \mathbf{i} + \mathbf{j}$
    $\mathbf{u} = 2\mathbf{i} + \mathbf{j} - 2\mathbf{k}$

18. $\mathbf{v} = \mathbf{i} + \mathbf{j} + 2\mathbf{k}$
    $\mathbf{u} = -\mathbf{i} - \mathbf{k}$

In Exercises 19 and 20, find $\text{proj}_\mathbf{v} \, \mathbf{u}$.

19. $\mathbf{v} = 2\mathbf{i} + \mathbf{j} - \mathbf{k}$
    $\mathbf{u} = \mathbf{i} + \mathbf{j} - 5\mathbf{k}$

20. $\mathbf{u} = \mathbf{i} - 2\mathbf{j}$
    $\mathbf{v} = \mathbf{i} + \mathbf{j} + \mathbf{k}$

In Exercises 21 and 22, draw coordinate axes and then sketch **u**, **v**, and $\mathbf{u} \times \mathbf{v}$ as vectors at the origin.

21. $\mathbf{u} = \mathbf{i}, \quad \mathbf{v} = \mathbf{i} + \mathbf{j}$ 22. $\mathbf{u} = \mathbf{i} - \mathbf{j}, \quad \mathbf{v} = \mathbf{i} + \mathbf{j}$

23. If $|\mathbf{v}| = 2, |\mathbf{w}| = 3$, and the angle between **v** and **w** is $\pi/3$, find $|\mathbf{v} - 2\mathbf{w}|$.

24. For what value or values of a will the vectors $\mathbf{u} = 2\mathbf{i} + 4\mathbf{j} - 5\mathbf{k}$ and $\mathbf{v} = -4\mathbf{i} - 8\mathbf{j} + a\mathbf{k}$ be parallel?

In Exercises 25 and 26, find **(a)** the area of the parallelogram determined by vectors **u** and **v** and **(b)** the volume of the parallelepiped determined by the vectors **u**, **v**, and **w**.

25. $\mathbf{u} = \mathbf{i} + \mathbf{j} - \mathbf{k}$, $\mathbf{v} = 2\mathbf{i} + \mathbf{j} + \mathbf{k}$, $\mathbf{w} = -\mathbf{i} - 2\mathbf{j} + 3\mathbf{k}$

26. $\mathbf{u} = \mathbf{i} + \mathbf{j}$, $\mathbf{v} = \mathbf{j}$, $\mathbf{w} = \mathbf{i} + \mathbf{j} + \mathbf{k}$

Lines, Planes, and Distances

27. Suppose that **n** is normal to a plane and that **v** is parallel to the plane. Describe how you would find a vector **n** that is both perpendicular to **v** and parallel to the plane.

28. Find a vector in the plane parallel to the line $ax + by = c$.

In Exercises 29 and 30, find the distance from the point to the line.

29. $(2, 2, 0)$; $x = -t$, $y = t$, $z = -1 + t$

30. $(0, 4, 1)$; $x = 2 + t$, $y = 2 + t$, $z = t$

31. Parametrize the line that passes through the point $(1, 2, 3)$ parallel to the vector $\mathbf{v} = -3\mathbf{i} + 7\mathbf{k}$.

32. Parametrize the line segment joining the points $P(1, 2, 0)$ and $Q(1, 3, -1)$.

In Exercises 33 and 34, find the distance from the point to the plane.

33. $(6, 0, -6)$, $x - y = 4$

34. $(3, 0, 10)$, $2x + 3y + z = 2$

35. Find an equation for the plane that passes through the point $(3, -2, 1)$ normal to the vector $\mathbf{n} = 2\mathbf{i} + \mathbf{j} + \mathbf{k}$.

36. Find an equation for the plane that passes through the point $(-1, 6, 0)$ perpendicular to the line $x = -1 + t, y = 6 - 2t$, $z = 3t$.

In Exercises 37 and 38, find an equation for the plane through points P, Q, and R.

37. $P(1, -1, 2)$, $Q(2, 1, 3)$, $R(-1, 2, -1)$

38. $P(1, 0, 0)$, $Q(0, 1, 0)$, $R(0, 0, 1)$

39. Find the points in which the line $x = 1 + 2t, y = -1 - t, z = 3t$ meets the three coordinate planes.

40. Find the point in which the line through the origin perpendicular to the plane $2x - y - z = 4$ meets the plane $3x - 5y + 2z = 6$.

41. Find the acute angle between the planes $x = 7$ and $x + y + \sqrt{2}z = -3$.

42. Find the acute angle between the planes $x + y = 1$ and $y + z = 1$.

43. Find parametric equations for the line in which the planes $x + 2y + z = 1$ and $x - y + 2z = -8$ intersect.

44. Show that the line in which the planes

$$x + 2y - 2z = 5 \quad \text{and} \quad 5x - 2y - z = 0$$

intersect is parallel to the line

$$x = -3 + 2t, \quad y = 3t, \quad z = 1 + 4t.$$

45. The planes $3x + 6z = 1$ and $2x + 2y - z = 3$ intersect in a line.

  a. Show that the planes are orthogonal.

  b. Find equations for the line of intersection.

46. Find an equation for the plane that passes through the point $(1, 2, 3)$ parallel to $\mathbf{u} = 2\mathbf{i} + 3\mathbf{j} + \mathbf{k}$ and $\mathbf{v} = \mathbf{i} - \mathbf{j} + 2\mathbf{k}$.

47. Is $\mathbf{v} = 2\mathbf{i} - 4\mathbf{j} + \mathbf{k}$ related in any special way to the plane $2x + y = 5$? Give reasons for your answer.

48. The equation $\mathbf{n} \cdot \overrightarrow{P_0 P} = 0$ represents the plane through P_0 normal to **n**. What set does the inequality $\mathbf{n} \cdot \overrightarrow{P_0 P} > 0$ represent?

49. Find the distance from the point $P(1, 4, 0)$ to the plane through $A(0, 0, 0)$, $B(2, 0, -1)$, and $C(2, -1, 0)$.

50. Find the distance from the point $(2, 2, 3)$ to the plane $2x + 3y + 5z = 0$.

51. Find a vector parallel to the plane $2x - y - z = 4$ and orthogonal to $\mathbf{i} + \mathbf{j} + \mathbf{k}$.

52. Find a unit vector orthogonal to **A** in the plane of **B** and **C** if $\mathbf{A} = 2\mathbf{i} - \mathbf{j} + \mathbf{k}, \mathbf{B} = \mathbf{i} + 2\mathbf{j} + \mathbf{k}$, and $\mathbf{C} = \mathbf{i} + \mathbf{j} - 2\mathbf{k}$.

53. Find a vector of magnitude 2 parallel to the line of intersection of the planes $x + 2y + z - 1 = 0$ and $x - y + 2z + 7 = 0$.

54. Find the point in which the line through the origin perpendicular to the plane $2x - y - z = 4$ meets the plane $3x - 5y + 2z = 6$.

55. Find the point in which the line through $P(3, 2, 1)$ normal to the plane $2x - y + 2z = -2$ meets the plane.

56. What angle does the line of intersection of the planes $2x + y - z = 0$ and $x + y + 2z = 0$ make with the positive x-axis?

57. The line

$$L: \quad x = 3 + 2t, \quad y = 2t, \quad z = t$$

intersects the plane $x + 3y - z = -4$ in a point P. Find the coordinates of P and find equations for the line in the plane through P perpendicular to L.

58. Show that for every real number k the plane

$$x - 2y + z + 3 + k(2x - y - z + 1) = 0$$

contains the line of intersection of the planes

$$x - 2y + z + 3 = 0 \quad \text{and} \quad 2x - y - z + 1 = 0.$$

59. Find an equation for the plane through $A(-2, 0, -3)$ and $B(1, -2, 1)$ that lies parallel to the line through $C(-2, -13/5, 26/5)$ and $D(16/5, -13/5, 0)$.

60. Is the line $x = 1 + 2t, y = -2 + 3t, z = -5t$ related in any way to the plane $-4x - 6y + 10z = 9$? Give reasons for your answer.

61. Which of the following are equations for the plane through the points $P(1, 1, -1)$, $Q(3, 0, 2)$, and $R(-2, 1, 0)$?

  a. $(2\mathbf{i} - 3\mathbf{j} + 3\mathbf{k}) \cdot ((x + 2)\mathbf{i} + (y - 1)\mathbf{j} + z\mathbf{k}) = 0$

  b. $x = 3 - t$, $y = -11t$, $z = 2 - 3t$

  c. $(x + 2) + 11(y - 1) = 3z$

  d. $(2\mathbf{i} - 3\mathbf{j} + 3\mathbf{k}) \times ((x + 2)\mathbf{i} + (y - 1)\mathbf{j} + z\mathbf{k}) = \mathbf{0}$

  e. $(2\mathbf{i} - \mathbf{j} + 3\mathbf{k}) \times (-3\mathbf{i} + \mathbf{k}) \cdot ((x + 2)\mathbf{i} + (y - 1)\mathbf{j} + z\mathbf{k})$
    $= 0$

62. The parallelogram shown here has vertices at $A(2, -1, 4)$, $B(1, 0, -1)$, $C(1, 2, 3)$, and D. Find

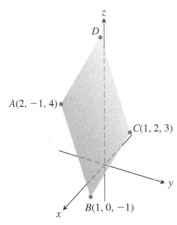

a. the coordinates of D.

b. the cosine of the interior angle at B.

c. the vector projection of $\overrightarrow{BA}$ onto $\overrightarrow{BC}$.

d. the area of the parallelogram.

e. an equation for the plane of the parallelogram.

f. the areas of the orthogonal projections of the parallelogram on the three coordinate planes.

63. Distance between skew lines Find the distance between the line L_1 through the points $A(1, 0, -1)$ and $B(-1, 1, 0)$ and the line L_2 through the points $C(3, 1, -1)$ and $D(4, 5, -2)$. The distance is to be measured along the line perpendicular to the two lines. First find a vector **n** perpendicular to both lines. Then project $\overrightarrow{AC}$ onto **n**.

64. (*Continuation of Exercise 63.*) Find the distance between the line through $A(4, 0, 2)$ and $B(2, 4, 1)$ and the line through $C(1, 3, 2)$ and $D(2, 2, 4)$.

Quadric Surfaces

Identify and sketch the surfaces in Exercises 65–76.

65. $x^2 + y^2 + z^2 = 4$

66. $x^2 + (y - 1)^2 + z^2 = 1$

67. $4x^2 + 4y^2 + z^2 = 4$

68. $36x^2 + 9y^2 + 4z^2 = 36$

69. $z = -(x^2 + y^2)$

70. $y = -(x^2 + z^2)$

71. $x^2 + y^2 = z^2$

72. $x^2 + z^2 = y^2$

73. $x^2 + y^2 - z^2 = 4$

74. $4y^2 + z^2 - 4x^2 = 4$

75. $y^2 - x^2 - z^2 = 1$

76. $z^2 - x^2 - y^2 = 1$

CHAPTER 12 Additional and Advanced Exercises

1. Submarine hunting Two surface ships on maneuvers are trying to determine a submarine's course and speed to prepare for an aircraft intercept. As shown here, ship A is located at $(4, 0, 0)$, whereas ship B is located at $(0, 5, 0)$. All coordinates are given in thousands of feet. Ship A locates the submarine in the direction of the vector $2\mathbf{i} + 3\mathbf{j} - (1/3)\mathbf{k}$, and ship B locates it in the direction of the vector $18\mathbf{i} - 6\mathbf{j} - \mathbf{k}$. Four minutes ago, the submarine was located at $(2, -1, -1/3)$. The aircraft is due in 20 min. Assuming that the submarine moves in a straight line at a constant speed, to what position should the surface ships direct the aircraft?

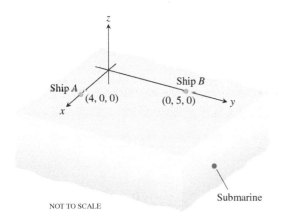

NOT TO SCALE

2. A helicopter rescue Two helicopters, H_1 and H_2, are traveling together. At time $t = 0$, they separate and follow different straight-line paths given by

$$H_1: \quad x = 6 + 40t, \quad y = -3 + 10t, \quad z = -3 + 2t$$
$$H_2: \quad x = 6 + 110t, \quad y = -3 + 4t, \quad z = -3 + t.$$

Time t is measured in hours, and all coordinates are measured in miles. Due to system malfunctions, H_2 stops its flight at $(446, 13, 1)$ and, in a negligible amount of time, lands at $(446, 13, 0)$. Two hours later, H_1 is advised of this fact and heads toward H_2 at 150 mph. How long will it take H_1 to reach H_2?

3. Torque The operator's manual for the Toro® 21-in. lawnmower says "tighten the spark plug to 15 ft-lb (20.4 N · m)." If you are installing the plug with a 10.5-in. socket wrench that places the center of your hand 9 in. from the axis of the spark plug, about how hard should you pull? Answer in pounds.

9 in.

4. Rotating body   The line through the origin and the point $A(1, 1, 1)$ is the axis of rotation of a rigid body rotating with a constant angular speed of $3/2$ rad/sec. The rotation appears to be clockwise when we look toward the origin from A. Find the velocity $\mathbf{v}$ of the point of the body that is at the position $B(1, 3, 2)$.

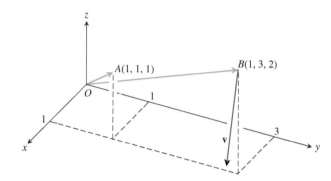

5. Consider the weight suspended by two wires in each diagram. Find the magnitudes and components of vectors $\mathbf{F}_1$ and $\mathbf{F}_2$, and angles α and β.

a.

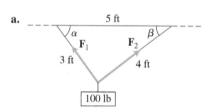

b.

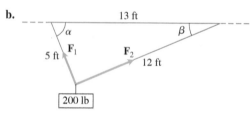

(*Hint:* This triangle is a right triangle.)

6. Consider a weight of w N suspended by two wires in the diagram, where $\mathbf{T}_1$ and $\mathbf{T}_2$ are force vectors directed along the wires.

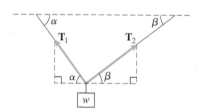

a. Find the vectors $\mathbf{T}_1$ and $\mathbf{T}_2$ and show that their magnitudes are

$$|\mathbf{T}_1| = \frac{w \cos \beta}{\sin (\alpha + \beta)}$$

and

$$|\mathbf{T}_2| = \frac{w \cos \alpha}{\sin (\alpha + \beta)}.$$

b. For a fixed β determine the value of α which minimizes the magnitude $|\mathbf{T}_1|$.

c. For a fixed α determine the value of β which minimizes the magnitude $|\mathbf{T}_2|$.

7. Determinants and planes

a. Show that

$$\begin{vmatrix} x_1 - x & y_1 - y & z_1 - z \\ x_2 - x & y_2 - y & z_2 - z \\ x_3 - x & y_3 - y & z_3 - z \end{vmatrix} = 0$$

is an equation for the plane through the three noncollinear points $P_1(x_1, y_1, z_1)$, $P_2(x_2, y_2, z_2)$, and $P_3(x_3, y_3, z_3)$.

b. What set of points in space is described by the equation

$$\begin{vmatrix} x & y & z & 1 \\ x_1 & y_1 & z_1 & 1 \\ x_2 & y_2 & z_2 & 1 \\ x_3 & y_3 & z_3 & 1 \end{vmatrix} = 0 ?$$

8. Determinants and lines   Show that the lines

$$x = a_1 s + b_1, \quad y = a_2 s + b_2, \quad z = a_3 s + b_3, \quad -\infty < s < \infty$$

and

$$x = c_1 t + d_1, \quad y = c_2 t + d_2, \quad z = c_3 t + d_3, \quad -\infty < t < \infty,$$

intersect or are parallel if and only if

$$\begin{vmatrix} a_1 & c_1 & b_1 - d_1 \\ a_2 & c_2 & b_2 - d_2 \\ a_3 & c_3 & b_3 - d_3 \end{vmatrix} = 0.$$

9. Consider a regular tetrahedron of side length 2.

a. Use vectors to find the angle θ formed by the base of the tetrahedron and any one of its other edges.

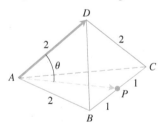

b. Use vectors to find the angle θ formed by any two adjacent faces of the tetrahedron. This angle is commonly referred to as a dihedral angle.

10. In the figure here, D is the midpoint of side AB of triangle ABC, and E is one-third of the way between C and B. Use vectors to prove that F is the midpoint of line segment CD.

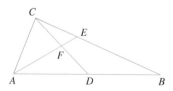

11. Use vectors to show that the distance from $P_1(x_1, y_1)$ to the line $ax + by = c$ is

$$d = \frac{|ax_1 + by_1 - c|}{\sqrt{a^2 + b^2}}.$$

12. a. Use vectors to show that the distance from $P_1(x_1, y_1, z_1)$ to the plane $Ax + By + Cz = D$ is

$$d = \frac{|Ax_1 + By_1 + Cz_1 - D|}{\sqrt{A^2 + B^2 + C^2}}.$$

   b. Find an equation for the sphere that is tangent to the planes $x + y + z = 3$ and $x + y + z = 9$ if the planes $2x - y = 0$ and $3x - z = 0$ pass through the center of the sphere.

13. a. **Distance between parallel planes** Show that the distance between the parallel planes $Ax + By + Cz = D_1$ and $Ax + By + Cz = D_2$ is

$$d = \frac{|D_1 - D_2|}{|A\mathbf{i} + B\mathbf{j} + C\mathbf{k}|}.$$

   b. Find the distance between the planes $2x + 3y - z = 6$ and $2x + 3y - z = 12$.

   c. Find an equation for the plane parallel to the plane $2x - y + 2z = -4$ if the point $(3, 2, -1)$ is equidistant from the two planes.

   d. Write equations for the planes that lie parallel to and 5 units away from the plane $x - 2y + z = 3$.

14. Prove that four points A, B, C, and D are coplanar (lie in a common plane) if and only if $\overrightarrow{AD} \cdot (\overrightarrow{AB} \times \overrightarrow{BC}) = 0$.

15. **The projection of a vector on a plane** Let P be a plane in space and let $\mathbf{v}$ be a vector. The vector projection of $\mathbf{v}$ onto the plane P, $\text{proj}_P \mathbf{v}$, can be defined informally as follows. Suppose the sun is shining so that its rays are normal to the plane P. Then $\text{proj}_P \mathbf{v}$ is the "shadow" of $\mathbf{v}$ onto P. If P is the plane $x + 2y + 6z = 6$ and $\mathbf{v} = \mathbf{i} + \mathbf{j} + \mathbf{k}$, find $\text{proj}_P \mathbf{v}$.

16. The accompanying figure shows nonzero vectors $\mathbf{v}$, $\mathbf{w}$, and $\mathbf{z}$, with $\mathbf{z}$ orthogonal to the line L, and $\mathbf{v}$ and $\mathbf{w}$ making equal angles β with L. Assuming $|\mathbf{v}| = |\mathbf{w}|$, find $\mathbf{w}$ in terms of $\mathbf{v}$ and $\mathbf{z}$.

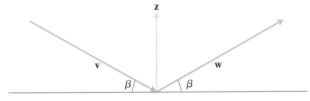

17. **Triple vector products** The *triple vector products* $(\mathbf{u} \times \mathbf{v}) \times \mathbf{w}$ and $\mathbf{u} \times (\mathbf{v} \times \mathbf{w})$ are usually not equal, although the formulas for evaluating them from components are similar:

$$(\mathbf{u} \times \mathbf{v}) \times \mathbf{w} = (\mathbf{u} \cdot \mathbf{w})\mathbf{v} - (\mathbf{v} \cdot \mathbf{w})\mathbf{u}.$$

$$\mathbf{u} \times (\mathbf{v} \times \mathbf{w}) = (\mathbf{u} \cdot \mathbf{w})\mathbf{v} - (\mathbf{u} \cdot \mathbf{v})\mathbf{w}.$$

Verify each formula for the following vectors by evaluating its two sides and comparing the results.

u	v	w
a. $2\mathbf{i}$	$2\mathbf{j}$	$2\mathbf{k}$
b. $\mathbf{i} - \mathbf{j} + \mathbf{k}$	$2\mathbf{i} + \mathbf{j} - 2\mathbf{k}$	$-\mathbf{i} + 2\mathbf{j} - \mathbf{k}$
c. $2\mathbf{i} + \mathbf{j}$	$2\mathbf{i} - \mathbf{j} + \mathbf{k}$	$\mathbf{i} + 2\mathbf{k}$
d. $\mathbf{i} + \mathbf{j} - 2\mathbf{k}$	$-\mathbf{i} - \mathbf{k}$	$2\mathbf{i} + 4\mathbf{j} - 2\mathbf{k}$

18. **Cross and dot products** Show that if $\mathbf{u}$, $\mathbf{v}$, $\mathbf{w}$, and $\mathbf{r}$ are any vectors, then

   a. $\mathbf{u} \times (\mathbf{v} \times \mathbf{w}) + \mathbf{v} \times (\mathbf{w} \times \mathbf{u}) + \mathbf{w} \times (\mathbf{u} \times \mathbf{v}) = \mathbf{0}$

   b. $\mathbf{u} \times \mathbf{v} = (\mathbf{u} \cdot \mathbf{v} \times \mathbf{i})\mathbf{i} + (\mathbf{u} \cdot \mathbf{v} \times \mathbf{j})\mathbf{j} + (\mathbf{u} \cdot \mathbf{v} \times \mathbf{k})\mathbf{k}$

   c. $(\mathbf{u} \times \mathbf{v}) \cdot (\mathbf{w} \times \mathbf{r}) = \begin{vmatrix} \mathbf{u} \cdot \mathbf{w} & \mathbf{v} \cdot \mathbf{w} \\ \mathbf{u} \cdot \mathbf{r} & \mathbf{v} \cdot \mathbf{r} \end{vmatrix}.$

19. **Cross and dot products** Prove or disprove the formula

$$\mathbf{u} \times (\mathbf{u} \times (\mathbf{u} \times \mathbf{v})) \cdot \mathbf{w} = -|\mathbf{u}|^2 \mathbf{u} \cdot \mathbf{v} \times \mathbf{w}.$$

20. By forming the cross product of two appropriate vectors, derive the trigonometric identity

$$\sin(A - B) = \sin A \cos B - \cos A \sin B.$$

21. Use vectors to prove that

$$(a^2 + b^2)(c^2 + d^2) \geq (ac + bd)^2$$

for any four numbers a, b, c, and d. (*Hint:* Let $\mathbf{u} = a\mathbf{i} + b\mathbf{j}$ and $\mathbf{v} = c\mathbf{i} + d\mathbf{j}$.)

22. **Dot multiplication is positive definite** Show that dot multiplication of vectors is *positive definite;* that is, show $\mathbf{u} \cdot \mathbf{u} \geq 0$ for every vector $\mathbf{u}$ and that $\mathbf{u} \cdot \mathbf{u} = 0$ if and only if $\mathbf{u} = \mathbf{0}$.

23. Show that $|\mathbf{u} + \mathbf{v}| \leq |\mathbf{u}| + |\mathbf{v}|$ for any vectors $\mathbf{u}$ and $\mathbf{v}$.

24. Show that $\mathbf{w} = |\mathbf{v}|\mathbf{u} + |\mathbf{u}|\mathbf{v}$ bisects the angle between $\mathbf{u}$ and $\mathbf{v}$.

25. Show that $|\mathbf{v}|\mathbf{u} + |\mathbf{u}|\mathbf{v}$ and $|\mathbf{v}|\mathbf{u} - |\mathbf{u}|\mathbf{v}$ are orthogonal.

CHAPTER 12   Technology Application Projects

Mathematica/Maple Projects

Projects can be found within MyMathLab.

- *Using Vectors to Represent Lines and Find Distances*
  Parts I and II: Learn the advantages of interpreting lines as vectors.
  Part III: Use vectors to find the distance from a point to a line.

- *Putting a Scene in Three Dimensions onto a Two-Dimensional Canvas*
  Use the concept of planes in space to obtain a two-dimensional image.

- *Getting Started in Plotting in 3D*
  Part I: Use the vector definition of lines and planes to generate graphs and equations, and to compare different forms for the equations of a single line.
  Part II: Plot functions that are defined implicitly.

13

Vector-Valued Functions and Motion in Space

OVERVIEW In this chapter we introduce the calculus of vector-valued functions. The domains of these functions are sets of real numbers, as before, but their ranges consist of vectors instead of scalars. When a vector-valued function changes, the change can occur in both magnitude and direction, so the derivative is itself a vector. The integral of a vector-valued function is also a vector. We use the calculus of these functions to describe the paths and motions of objects moving in a plane or in space, so their velocities and accelerations are given by vectors. We also introduce new concepts that quantify the way that the path of an object moving in space can twist and turn.

13.1 Curves in Space and Their Tangents

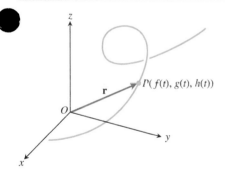

FIGURE 13.1 The position vector $\mathbf{r} = \overrightarrow{OP}$ of a particle moving through space is a function of time.

When a particle moves through space during a time interval I, we think of the particle's coordinates as functions defined on I:

$$x = f(t), \qquad y = g(t), \qquad z = h(t), \qquad t \in I. \tag{1}$$

The points $(x, y, z) = (f(t), g(t), h(t))$, $t \in I$, make up the **curve** in space that we call the particle's **path**. The equations and interval in Equation (1) parametrize the curve.

A curve in space can also be represented in vector form. The vector

$$\mathbf{r}(t) = \overrightarrow{OP} = f(t)\mathbf{i} + g(t)\mathbf{j} + h(t)\mathbf{k} \tag{2}$$

from the origin to the particle's position $P(f(t), g(t), h(t))$ at time t is the particle's position vector (Figure 13.1). The functions f, g, and h are the **component functions** (or components) of the position vector. We think of the particle's path as the curve traced by $\mathbf{r}$ during the time interval I. Figure 13.2 displays several space curves generated by a computer graphing program.

Equation (2) defines $\mathbf{r}$ as a vector function of the real variable t on the interval I. More generally, a **vector-valued function** or **vector function** on a domain set D is a rule that assigns a vector in space to each element in D. For now, the domains will be intervals of real numbers, and the graph of the function represents a curve in space. Later, in Chapter 16, the domains will be regions in the plane, and in that setting the graph will represent a surface in space. Vector functions on a domain in the plane or in space also give rise to "vector fields," which are important to the study of fluid flows, gravitational fields, and electromagnetic phenomena. We investigate vector fields and their applications in Chapter 16.

Real-valued functions are called **scalar functions** to distinguish them from vector functions. The components of $\mathbf{r}$ in Equation (2) are scalar functions of t. The domain of a vector-valued function is the common domain of its components.

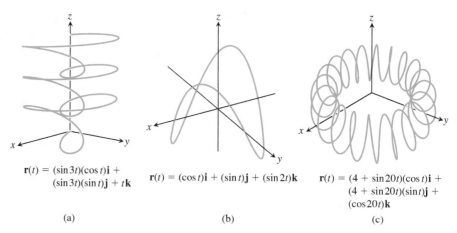

$\mathbf{r}(t) = (\sin 3t)(\cos t)\mathbf{i} + (\sin 3t)(\sin t)\mathbf{j} + t\mathbf{k}$

$\mathbf{r}(t) = (\cos t)\mathbf{i} + (\sin t)\mathbf{j} + (\sin 2t)\mathbf{k}$

$\mathbf{r}(t) = (4 + \sin 20t)(\cos t)\mathbf{i} + (4 + \sin 20t)(\sin t)\mathbf{j} + (\cos 20t)\mathbf{k}$

(a)

(b)

(c)

FIGURE 13.2 Space curves are defined by the position vectors $\mathbf{r}(t)$.

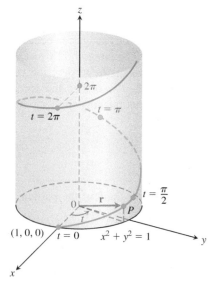

FIGURE 13.3 The upper half of the helix $\mathbf{r}(t) = (\cos t)\mathbf{i} + (\sin t)\mathbf{j} + t\mathbf{k}$ (Example 1).

EXAMPLE 1 Graph the vector function

$$\mathbf{r}(t) = (\cos t)\mathbf{i} + (\sin t)\mathbf{j} + t\mathbf{k}.$$

Solution This vector function $\mathbf{r}(t)$ is defined for all real values of t. The curve traced by $\mathbf{r}$ winds around the circular cylinder $x^2 + y^2 = 1$ (Figure 13.3). The curve lies on the cylinder because the $\mathbf{i}$- and $\mathbf{j}$-components of $\mathbf{r}$, being the x- and y-coordinates of the tip of $\mathbf{r}$, satisfy the cylinder's equation:

$$x^2 + y^2 = (\cos t)^2 + (\sin t)^2 = 1.$$

The curve rises as the $\mathbf{k}$-component $z = t$ increases. Each time t increases by 2π, the curve completes one turn around the cylinder. The curve is called a **helix** (from an old Greek word for "spiral"). The equations

$$x = \cos t, \qquad y = \sin t, \qquad z = t$$

parametrize the helix. The domain is the largest set of points t for which all three equations are defined, or $-\infty < t < \infty$ for this example. Figure 13.4 shows more helices. ∎

$\mathbf{r}(t) = (\cos t)\mathbf{i} + (\sin t)\mathbf{j} + t\mathbf{k}$

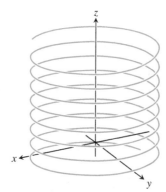

$\mathbf{r}(t) = (\cos t)\mathbf{i} + (\sin t)\mathbf{j} + 0.3t\mathbf{k}$

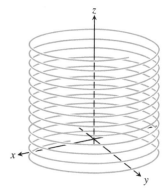

$\mathbf{r}(t) = (\cos 5t)\mathbf{i} + (\sin 5t)\mathbf{j} + t\mathbf{k}$

FIGURE 13.4 Helices spiral upward around a cylinder, like coiled springs.

Limits and Continuity

The way we define limits of vector-valued functions is similar to the way we define limits of real-valued functions.

> **DEFINITION**   Let $\mathbf{r}(t) = f(t)\mathbf{i} + g(t)\mathbf{j} + h(t)\mathbf{k}$ be a vector function with domain D, and let $\mathbf{L}$ be a vector. We say that $\mathbf{r}$ has **limit L** as t approaches t_0 and write
>
> $$\lim_{t \to t_0} \mathbf{r}(t) = \mathbf{L}$$
>
> if, for every number $\varepsilon > 0$, there exists a corresponding number $\delta > 0$ such that for all $t \in D$
>
> $$\left| \mathbf{r}(t) - \mathbf{L} \right| < \varepsilon \quad \text{whenever} \quad 0 < \left| t - t_0 \right| < \delta.$$

If $\mathbf{L} = L_1\mathbf{i} + L_2\mathbf{j} + L_3\mathbf{k}$, then it can be shown that $\lim_{t \to t_0} \mathbf{r}(t) = \mathbf{L}$ precisely when

$$\lim_{t \to t_0} f(t) = L_1, \qquad \lim_{t \to t_0} g(t) = L_2, \qquad \text{and} \qquad \lim_{t \to t_0} h(t) = L_3.$$

We omit the proof. The equation

> To calculate the limit of a vector function, we find the limit of each component scalar function.

$$\lim_{t \to t_0} \mathbf{r}(t) = \left(\lim_{t \to t_0} f(t) \right)\mathbf{i} + \left(\lim_{t \to t_0} \mathbf{g(t)} \right)\mathbf{j} + \left(\lim_{t \to t_0} \mathbf{h(t)} \right)\mathbf{k} \qquad (3)$$

provides a practical way to calculate limits of vector functions.

EXAMPLE 2   If $\mathbf{r}(t) = (\cos t)\mathbf{i} + (\sin t)\mathbf{j} + t\mathbf{k}$, then

$$\lim_{t \to \pi/4} \mathbf{r}(t) = \left(\lim_{t \to \pi/4} \cos t \right)\mathbf{i} + \left(\lim_{t \to \pi/4} \sin t \right)\mathbf{j} + \left(\lim_{t \to \pi/4} t \right)\mathbf{k}$$

$$= \frac{\sqrt{2}}{2}\mathbf{i} + \frac{\sqrt{2}}{2}\mathbf{j} + \frac{\pi}{4}\mathbf{k}. \qquad \blacksquare$$

We define continuity for vector functions the same way we define continuity for scalar functions defined over an interval.

> **DEFINITION**   A vector function $\mathbf{r}(t)$ is **continuous at a point** $t = t_0$ in its domain if $\lim_{t \to t_0} \mathbf{r}(t) = \mathbf{r}(t_0)$. The function is **continuous** if it is continuous at every point in its domain.

From Equation (3), we see that $\mathbf{r}(t)$ is continuous at $t = t_0$ if and only if each component function is continuous there (Exercise 45).

EXAMPLE 3

(a) All the space curves shown in Figures 13.2 and 13.4 are continuous because their component functions are continuous at every value of t in $(-\infty, \infty)$.

(b) The function

$$\mathbf{g}(t) = (\cos t)\mathbf{i} + (\sin t)\mathbf{j} + \lfloor t \rfloor \mathbf{k}$$

is discontinuous at every integer, because the greatest integer function $\lfloor t \rfloor$ is discontinuous at every integer. $\blacksquare$

Derivatives and Motion

Suppose that $\mathbf{r}(t) = f(t)\mathbf{i} + g(t)\mathbf{j} + h(t)\mathbf{k}$ is the position vector of a particle moving along a curve in space and that f, g, and h are differentiable functions of t. Then the difference between the particle's positions at time t and time $t + \Delta t$ is the vector

$$\Delta\mathbf{r} = \mathbf{r}(t + \Delta t) - \mathbf{r}(t)$$

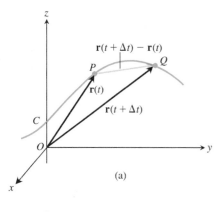

(a)

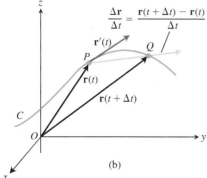

(b)

FIGURE 13.5 As $\Delta t \to 0$, the point Q approaches the point P along the curve C. In the limit, the vector $\overrightarrow{PQ}\,/\Delta t$ becomes the tangent vector $\mathbf{r}'(t)$.

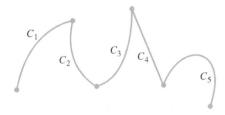

FIGURE 13.6 A piecewise smooth curve made up of five smooth curves connected end to end in a continuous fashion. The curve here is not smooth at the points joining the five smooth curves.

(Figure 13.5a). In terms of components,

$$
\begin{aligned}
\Delta\mathbf{r} &= \mathbf{r}(t + \Delta t) - \mathbf{r}(t) \\
&= [\, f(t + \Delta t)\mathbf{i} + g(t + \Delta t)\mathbf{j} + h(t + \Delta t)\mathbf{k}\,] - [\, f(t)\mathbf{i} + g(t)\mathbf{j} + h(t)\mathbf{k}\,] \\
&= [\, f(t + \Delta t) - f(t)\,]\mathbf{i} + [\, g(t + \Delta t) - g(t)\,]\mathbf{j} + [\, h(t + \Delta t) - h(t)\,]\mathbf{k}.
\end{aligned}
$$

As Δt approaches zero, three things seem to happen simultaneously. First, Q approaches P along the curve. Second, the secant line PQ seems to approach a limiting position tangent to the curve at P. Third, the quotient $\Delta\mathbf{r}/\Delta t$ (Figure 13.5b) approaches the limit

$$
\lim_{\Delta t \to 0} \frac{\Delta\mathbf{r}}{\Delta t} = \left[\lim_{\Delta t \to 0} \frac{f(t + \Delta t) - f(t)}{\Delta t} \right]\mathbf{i} + \left[\lim_{\Delta t \to 0} \frac{g(t + \Delta t) - g(t)}{\Delta t} \right]\mathbf{j}
$$

$$
+ \left[\lim_{\Delta t \to 0} \frac{h(t + \Delta t) - h(t)}{\Delta t} \right]\mathbf{k}
$$

$$
= \left[\frac{df}{dt} \right]\mathbf{i} + \left[\frac{dg}{dt} \right]\mathbf{j} + \left[\frac{dh}{dt} \right]\mathbf{k}.
$$

These observations lead us to the following definition.

DEFINITION   The vector function $\mathbf{r}(t) = f(t)\mathbf{i} + g(t)\mathbf{j} + h(t)\mathbf{k}$ has a **derivative (is differentiable) at** t if f, g, and h have derivatives at t. The derivative is the vector function

$$
\mathbf{r}'(t) = \frac{d\mathbf{r}}{dt} = \lim_{\Delta t \to 0} \frac{\mathbf{r}(t + \Delta t) - \mathbf{r}(t)}{\Delta t} = \frac{df}{dt}\mathbf{i} + \frac{dg}{dt}\mathbf{j} + \frac{dh}{dt}\mathbf{k}.
$$

A vector function $\mathbf{r}$ is **differentiable** if it is differentiable at every point of its domain. The curve traced by $\mathbf{r}$ is **smooth** if $d\mathbf{r}/dt$ is continuous and never $\mathbf{0}$, that is, if f, g, and h have continuous first derivatives that are not simultaneously 0.

The geometric significance of the definition of derivative is shown in Figure 13.5. The points P and Q have position vectors $\mathbf{r}(t)$ and $\mathbf{r}(t + \Delta t)$, and the vector $\overrightarrow{PQ}$ is represented by $\mathbf{r}(t + \Delta t) - \mathbf{r}(t)$. For $\Delta t > 0$, the scalar multiple $(1/\Delta t)(\mathbf{r}(t + \Delta t) - \mathbf{r}(t))$ points in the same direction as the vector $\overrightarrow{PQ}$. As $\Delta t \to 0$, this vector approaches a vector that is tangent to the curve at P (Figure 13.5b). The vector $\mathbf{r}'(t)$, when different from the zero vector $\mathbf{0}$, is defined to be the vector **tangent** to the curve at P. The **tangent line** to the curve at a point $(f(t_0), g(t_0), h(t_0))$ is defined to be the line through the point parallel to $\mathbf{r}'(t_0)$. We require $d\mathbf{r}/dt \neq \mathbf{0}$ for a smooth curve to make sure the curve has a continuously turning tangent at each point. On a smooth curve, there are no sharp corners or cusps.

A curve that is made up of a finite number of smooth curves pieced together in a continuous fashion is called **piecewise smooth** (Figure 13.6).

Look once again at Figure 13.5. We drew the figure for Δt positive, so $\Delta\mathbf{r}$ points forward, in the direction of the motion. The vector $\Delta\mathbf{r}/\Delta t$, having the same direction as $\Delta\mathbf{r}$, points forward too. Had Δt been negative, $\Delta\mathbf{r}$ would have pointed backward, against the direction of motion. The quotient $\Delta\mathbf{r}/\Delta t$, however, being a negative scalar multiple of $\Delta\mathbf{r}$, would once again have pointed forward. No matter how $\Delta\mathbf{r}$ points, $\Delta\mathbf{r}/\Delta t$ points forward and we expect the vector $d\mathbf{r}/dt = \lim_{\Delta t \to 0} \Delta\mathbf{r}/\Delta t$, when different from $\mathbf{0}$, to do the same. This means that the derivative $d\mathbf{r}/dt$, which is the rate of change of position with respect to time, always points in the direction of motion. For a smooth curve, $d\mathbf{r}/dt$ is never zero; the particle does not stop or reverse direction.

> DEFINITIONS   If **r** is the position vector of a particle moving along a smooth curve in space, then
>
> $$\mathbf{v}(t) = \frac{d\mathbf{r}}{dt}$$
>
> is the particle's **velocity vector**, tangent to the curve. At any time t, the direction of **v** is the **direction of motion**, the magnitude of **v** is the particle's **speed**, and the derivative $\mathbf{a} = d\mathbf{v}/dt$, when it exists, is the particle's **acceleration vector**. In summary,
>
> **1.** Velocity is the derivative of position: $\mathbf{v} = \dfrac{d\mathbf{r}}{dt}$.
>
> **2.** Speed is the magnitude of velocity: Speed $= |\mathbf{v}|$.
>
> **3.** Acceleration is the derivative of velocity: $\mathbf{a} = \dfrac{d\mathbf{v}}{dt} = \dfrac{d^2\mathbf{r}}{dt^2}$.
>
> **4.** The unit vector $\mathbf{v}/|\mathbf{v}|$ is the direction of motion at time t.

EXAMPLE 4   Find the velocity, speed, and acceleration of a particle whose motion in space is given by the position vector $\mathbf{r}(t) = 2 \cos t\,\mathbf{i} + 2 \sin t\,\mathbf{j} + 5 \cos^2 t\,\mathbf{k}$. Sketch the velocity vector $\mathbf{v}(7\pi/4)$.

Solution   The velocity and acceleration vectors at time t are

$$\mathbf{v}(t) = \mathbf{r}'(t) = -2 \sin t\,\mathbf{i} + 2 \cos t\,\mathbf{j} - 10 \cos t \sin t\,\mathbf{k}$$
$$= -2 \sin t\,\mathbf{i} + 2 \cos t\,\mathbf{j} - 5 \sin 2t\,\mathbf{k},$$
$$\mathbf{a}(t) = \mathbf{r}''(t) = -2 \cos t\,\mathbf{i} - 2 \sin t\,\mathbf{j} - 10 \cos 2t\,\mathbf{k},$$

and the speed is

$$|\mathbf{v}(t)| = \sqrt{(-2 \sin t)^2 + (2 \cos t)^2 + (-5 \sin 2t)^2} = \sqrt{4 + 25 \sin^2 2t}.$$

When $t = 7\pi/4$, we have

$$\mathbf{v}\left(\frac{7\pi}{4}\right) = \sqrt{2}\,\mathbf{i} + \sqrt{2}\,\mathbf{j} + 5\mathbf{k}, \qquad \mathbf{a}\left(\frac{7\pi}{4}\right) = -\sqrt{2}\,\mathbf{i} + \sqrt{2}\,\mathbf{j}, \qquad \left|\mathbf{v}\left(\frac{7\pi}{4}\right)\right| = \sqrt{29}.$$

A sketch of the curve of motion, and the velocity vector when $t = 7\pi/4$, can be seen in Figure 13.7.   ∎

We can express the velocity of a moving particle as the product of its speed and direction:

$$\text{Velocity} = |\mathbf{v}|\left(\frac{\mathbf{v}}{|\mathbf{v}|}\right) = (\text{speed})(\text{direction}).$$

Differentiation Rules

Because the derivatives of vector functions may be computed component by component, the rules for differentiating vector functions have the same form as the rules for differentiating scalar functions.

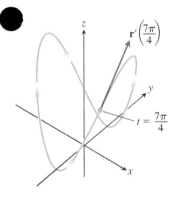

FIGURE 13.7   The curve and the velocity vector when $t = 7\pi/4$ for the motion given in Example 4.

> **Differentiation Rules for Vector Functions**
>
> Let $\mathbf{u}$ and $\mathbf{v}$ be differentiable vector functions of t, $\mathbf{C}$ a constant vector, c any scalar, and f any differentiable scalar function.
>
> **1.** *Constant Function Rule:* $\quad \dfrac{d}{dt}\mathbf{C} = \mathbf{0}$
>
> **2.** *Scalar Multiple Rules:* $\quad \dfrac{d}{dt}\big[c\mathbf{u}(t)\big] = c\mathbf{u}'(t)$
>
> $\dfrac{d}{dt}\big[f(t)\mathbf{u}(t)\big] = f'(t)\mathbf{u}(t) + f(t)\mathbf{u}'(t)$
>
> **3.** *Sum Rule:* $\quad \dfrac{d}{dt}\big[\mathbf{u}(t) + \mathbf{v}(t)\big] = \mathbf{u}'(t) + \mathbf{v}'(t)$
>
> **4.** *Difference Rule:* $\quad \dfrac{d}{dt}\big[\mathbf{u}(t) - \mathbf{v}(t)\big] = \mathbf{u}'(t) - \mathbf{v}'(t)$
>
> **5.** *Dot Product Rule:* $\quad \dfrac{d}{dt}\big[\mathbf{u}(t) \cdot \mathbf{v}(t)\big] = \mathbf{u}'(t) \cdot \mathbf{v}(t) + \mathbf{u}(t) \cdot \mathbf{v}'(t)$
>
> **6.** *Cross Product Rule:* $\quad \dfrac{d}{dt}\big[\mathbf{u}(t) \times \mathbf{v}(t)\big] = \mathbf{u}'(t) \times \mathbf{v}(t) + \mathbf{u}(t) \times \mathbf{v}'(t)$
>
> **7.** *Chain Rule:* $\quad \dfrac{d}{dt}\big[\mathbf{u}(f(t))\big] = f'(t)\mathbf{u}'(f(t))$

When you use the Cross Product Rule, remember to preserve the order of the factors. If $\mathbf{u}$ comes first on the left side of the equation, it must also come first on the right or the signs will be wrong.

We will prove the product rules and Chain Rule but leave the rules for constants, scalar multiples, sums, and differences as exercises.

Proof of the Dot Product Rule   Suppose that

$$\mathbf{u} = u_1(t)\mathbf{i} + u_2(t)\mathbf{j} + u_3(t)\mathbf{k}$$

and

$$\mathbf{v} = v_1(t)\mathbf{i} + v_2(t)\mathbf{j} + v_3(t)\mathbf{k}.$$

Then

$$\frac{d}{dt}(\mathbf{u} \cdot \mathbf{v}) = \frac{d}{dt}(u_1v_1 + u_2v_2 + u_3v_3)$$

$$= \underbrace{u_1'v_1 + u_2'v_2 + u_3'v_3}_{\mathbf{u}' \cdot \mathbf{v}} + \underbrace{u_1v_1' + u_2v_2' + u_3v_3'}_{\mathbf{u} \cdot \mathbf{v}'}. \quad \blacksquare$$

Proof of the Cross Product Rule   We model the proof after the proof of the Product Rule for scalar functions. According to the definition of derivative,

$$\frac{d}{dt}(\mathbf{u} \times \mathbf{v}) = \lim_{h \to 0} \frac{\mathbf{u}(t+h) \times \mathbf{v}(t+h) - \mathbf{u}(t) \times \mathbf{v}(t)}{h}.$$

To change this fraction into an equivalent one that contains the difference quotients for the derivatives of $\mathbf{u}$ and $\mathbf{v}$, we subtract and add $\mathbf{u}(t) \times \mathbf{v}(t+h)$ in the numerator. Then

$$\frac{d}{dt}(\mathbf{u} \times \mathbf{v})$$

$$= \lim_{h \to 0} \frac{\mathbf{u}(t+h) \times \mathbf{v}(t+h) - \mathbf{u}(t) \times \mathbf{v}(t+h) + \mathbf{u}(t) \times \mathbf{v}(t+h) - \mathbf{u}(t) \times \mathbf{v}(t)}{h}$$

$$= \lim_{h \to 0} \left[\frac{\mathbf{u}(t+h) - \mathbf{u}(t)}{h} \times \mathbf{v}(t+h) + \mathbf{u}(t) \times \frac{\mathbf{v}(t+h) - \mathbf{v}(t)}{h} \right]$$

$$= \lim_{h \to 0} \frac{\mathbf{u}(t+h) - \mathbf{u}(t)}{h} \times \lim_{h \to 0} \mathbf{v}(t+h) + \lim_{h \to 0} \mathbf{u}(t) \times \lim_{h \to 0} \frac{\mathbf{v}(t+h) - \mathbf{v}(t)}{h}.$$

The last of these equalities holds because the limit of the cross product of two vector functions is the cross product of their limits if the latter exist (Exercise 46). As h approaches zero, $\mathbf{v}(t + h)$ approaches $\mathbf{v}(t)$ because $\mathbf{v}$, being differentiable at t, is continuous at t (Exercise 47). The two fractions approach the values of $d\mathbf{u}/dt$ and $d\mathbf{v}/dt$ at t. In short,

$$\frac{d}{dt}(\mathbf{u} \times \mathbf{v}) = \frac{d\mathbf{u}}{dt} \times \mathbf{v} + \mathbf{u} \times \frac{d\mathbf{v}}{dt}. \qquad \blacksquare$$

Proof of the Chain Rule   Suppose that $\mathbf{u}(s) = a(s)\mathbf{i} + b(s)\mathbf{j} + c(s)\mathbf{k}$ is a differentiable vector function of s and that $s = f(t)$ is a differentiable scalar function of t. Then a, b, and c are differentiable functions of t, and the Chain Rule for differentiable real-valued functions gives

$$\frac{d}{dt}[\mathbf{u}(s)] = \frac{da}{dt}\mathbf{i} + \frac{db}{dt}\mathbf{j} + \frac{dc}{dt}\mathbf{k}$$

$$= \frac{da}{ds}\frac{ds}{dt}\mathbf{i} + \frac{db}{ds}\frac{ds}{dt}\mathbf{j} + \frac{dc}{ds}\frac{ds}{dt}\mathbf{k}$$

$$= \frac{ds}{dt}\left(\frac{da}{ds}\mathbf{i} + \frac{db}{ds}\mathbf{j} + \frac{dc}{ds}\mathbf{k}\right)$$

$$= \frac{ds}{dt}\frac{d\mathbf{u}}{ds}$$

$$= f'(t)\mathbf{u}'(f(t)). \qquad s = f(t) \qquad \blacksquare$$

As an algebraic convenience, we sometimes write the product of a scalar c and a vector $\mathbf{v}$ as $\mathbf{v}c$ instead of $c\mathbf{v}$. This permits us, for instance, to write the Chain Rule in a familiar form:

$$\frac{d\mathbf{u}}{dt} = \frac{d\mathbf{u}}{ds}\frac{ds}{dt},$$

where $s = f(t)$.

Vector Functions of Constant Length

When we track a particle moving on a sphere centered at the origin (Figure 13.8), the position vector has a constant length equal to the radius of the sphere. The velocity vector $d\mathbf{r}/dt$, tangent to the path of motion, is tangent to the sphere and hence perpendicular to $\mathbf{r}$. This is always the case for a differentiable vector function of constant length. The vector and its first derivative are orthogonal. By direct calculation,

$$\mathbf{r}(t) \cdot \mathbf{r}(t) = |\mathbf{r}(t)|^2 = c^2 \qquad |\mathbf{r}(t)| = c \text{ is constant.}$$

$$\frac{d}{dt}[\mathbf{r}(t) \cdot \mathbf{r}(t)] = 0 \qquad \text{Differentiate both sides.}$$

$$\mathbf{r}'(t) \cdot \mathbf{r}(t) + \mathbf{r}(t) \cdot \mathbf{r}'(t) = 0 \qquad \text{Rule 5 with } \mathbf{r}(t) = \mathbf{u}(t) = \mathbf{v}(t)$$

$$2\mathbf{r}'(t) \cdot \mathbf{r}(t) = 0.$$

Thus the vectors $\mathbf{r}'(t)$ and $\mathbf{r}(t)$ are orthogonal because their dot product is 0. In summary, the following holds.

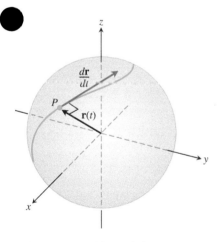

FIGURE 13.8 If a particle moves on a sphere in such a way that its position $\mathbf{r}$ is a differentiable function of time, then $\mathbf{r} \cdot (d\mathbf{r}/dt) = 0$.

If $\mathbf{r}$ is a differentiable vector function of t and the length of $\mathbf{r}(t)$ is constant, then

$$\mathbf{r} \cdot \frac{d\mathbf{r}}{dt} = 0. \qquad (4)$$

We will use this observation repeatedly in Section 13.4. The converse is also true (see Exercise 41).

EXERCISES 13.1

In Exercises 1–4, find the given limits.

1. $\lim\limits_{t \to \pi} \left[\left(\sin \dfrac{t}{2} \right) \mathbf{i} + \left(\cos \dfrac{2}{3}t \right) \mathbf{j} + \left(\tan \dfrac{5}{4}t \right) \mathbf{k} \right]$

2. $\lim\limits_{t \to -1} \left[t^3 \mathbf{i} + \left(\sin \dfrac{\pi}{2}t \right) \mathbf{j} + (\ln(t + 2)) \mathbf{k} \right]$

3. $\lim\limits_{t \to 1} \left[\left(\dfrac{t^2 - 1}{\ln t} \right) \mathbf{i} - \left(\dfrac{\sqrt{t} - 1}{1 - t} \right) \mathbf{j} + (\tan^{-1} t) \mathbf{k} \right]$

4. $\lim\limits_{t \to 0} \left[\left(\dfrac{\sin t}{t} \right) \mathbf{i} + \left(\dfrac{\tan^2 t}{\sin 2t} \right) \mathbf{j} - \left(\dfrac{t^3 - 8}{t + 2} \right) \mathbf{k} \right]$

Motion in the Plane

In Exercises 5–8, $\mathbf{r}(t)$ is the position of a particle in the xy-plane at time t. Find an equation in x and y whose graph is the path of the particle. Then find the particle's velocity and acceleration vectors at the given value of t.

5. $\mathbf{r}(t) = (t + 1)\mathbf{i} + (t^2 - 1)\mathbf{j}, \quad t = 1$

6. $\mathbf{r}(t) = \dfrac{t}{t + 1}\mathbf{i} + \dfrac{1}{t}\mathbf{j}, \quad t = -\dfrac{1}{2}$

7. $\mathbf{r}(t) = e^t\mathbf{i} + \dfrac{2}{9}e^{2t}\mathbf{j}, \quad t = \ln 3$

8. $\mathbf{r}(t) = (\cos 2t)\mathbf{i} + (3 \sin 2t)\mathbf{j}, \quad t = 0$

Exercises 9–12 give the position vectors of particles moving along various curves in the xy-plane. In each case, find the particle's velocity and acceleration vectors at the stated times and sketch them as vectors on the curve.

9. Motion on the circle $x^2 + y^2 = 1$

$\quad \mathbf{r}(t) = (\sin t)\mathbf{i} + (\cos t)\mathbf{j}; \quad t = \pi/4 \text{ and } \pi/2$

10. Motion on the circle $x^2 + y^2 = 16$

$\quad \mathbf{r}(t) = \left(4 \cos \dfrac{t}{2} \right) \mathbf{i} + \left(4 \sin \dfrac{t}{2} \right) \mathbf{j}; \quad t = \pi \text{ and } 3\pi/2$

11. Motion on the cycloid $x = t - \sin t, \quad y = 1 - \cos t$

$\quad \mathbf{r}(t) = (t - \sin t)\mathbf{i} + (1 - \cos t)\mathbf{j}; \quad t = \pi \text{ and } 3\pi/2$

12. Motion on the parabola $y = x^2 + 1$

$\quad \mathbf{r}(t) = t\mathbf{i} + (t^2 + 1)\mathbf{j}; \quad t = -1, 0, \text{ and } 1$

Motion in Space

In Exercises 13–18, $\mathbf{r}(t)$ is the position of a particle in space at time t. Find the particle's velocity and acceleration vectors. Then find the particle's speed and direction of motion at the given value of t. Write the particle's velocity at that time as the product of its speed and direction.

13. $\mathbf{r}(t) = (t + 1)\mathbf{i} + (t^2 - 1)\mathbf{j} + 2t\mathbf{k}, \quad t = 1$

14. $\mathbf{r}(t) = (1 + t)\mathbf{i} + \dfrac{t^2}{\sqrt{2}}\mathbf{j} + \dfrac{t^3}{3}\mathbf{k}, \quad t = 1$

15. $\mathbf{r}(t) = (2 \cos t)\mathbf{i} + (3 \sin t)\mathbf{j} + 4t\mathbf{k}, \quad t = \pi/2$

16. $\mathbf{r}(t) = (\sec t)\mathbf{i} + (\tan t)\mathbf{j} + \dfrac{4}{3}t\mathbf{k}, \quad t = \pi/6$

17. $\mathbf{r}(t) = (2 \ln(t + 1))\mathbf{i} + t^2\mathbf{j} + \dfrac{t^2}{2}\mathbf{k}, \quad t = 1$

18. $\mathbf{r}(t) = e^{-t}\mathbf{i} + (2 \cos 3t)\mathbf{j} + (2 \sin 3t)\mathbf{k}, \quad t = 0$

In Exercises 19–22, $\mathbf{r}(t)$ is the position of a particle in space at time t. Find the angle between the velocity and acceleration vectors at time $t = 0$.

19. $\mathbf{r}(t) = (3t + 1)\mathbf{i} + \sqrt{3}t\mathbf{j} + t^2\mathbf{k}$

20. $\mathbf{r}(t) = \left(\dfrac{\sqrt{2}}{2}t \right)\mathbf{i} + \left(\dfrac{\sqrt{2}}{2}t - 16t^2 \right)\mathbf{j}$

21. $\mathbf{r}(t) = (\ln(t^2 + 1))\mathbf{i} + (\tan^{-1} t)\mathbf{j} + \sqrt{t^2 + 1}\,\mathbf{k}$

22. $\mathbf{r}(t) = \dfrac{4}{9}(1 + t)^{3/2}\mathbf{i} + \dfrac{4}{9}(1 - t)^{3/2}\mathbf{j} + \dfrac{1}{3}t\mathbf{k}$

Tangents to Curves

As mentioned in the text, the **tangent line** to a smooth curve $\mathbf{r}(t) = f(t)\mathbf{i} + g(t)\mathbf{j} + h(t)\mathbf{k}$ at $t = t_0$ is the line that passes through the point $(f(t_0), g(t_0), h(t_0))$ parallel to $\mathbf{v}(t_0)$, the curve's velocity vector at t_0. In Exercises 23–26, find parametric equations for the line that is tangent to the given curve at the given parameter value $t = t_0$.

23. $\mathbf{r}(t) = (\sin t)\mathbf{i} + (t^2 - \cos t)\mathbf{j} + e^t\mathbf{k}, \quad t_0 = 0$

24. $\mathbf{r}(t) = t^2\mathbf{i} + (2t - 1)\mathbf{j} + t^3\mathbf{k}, \quad t_0 = 2$

25. $\mathbf{r}(t) = \ln t\,\mathbf{i} + \dfrac{t - 1}{t + 2}\mathbf{j} + t \ln t\,\mathbf{k}, \quad t_0 = 1$

26. $\mathbf{r}(t) = (\cos t)\mathbf{i} + (\sin t)\mathbf{j} + (\sin 2t)\mathbf{k}, \quad t_0 = \dfrac{\pi}{2}$

In Exercises 27–30, find the value(s) of t so that the tangent line to the given curve contains the given point.

27. $\mathbf{r}(t) = t^2\mathbf{i} + (1 + t)\mathbf{j} + (2t - 3)\mathbf{k}; \quad (-8, 2, -1)$

28. $\mathbf{r}(t) = t\mathbf{i} + 3\mathbf{j} + \left(\dfrac{2}{3}t^{3/2} \right)\mathbf{k}; \quad (0, 3, -8/3)$

29. $\mathbf{r}(t) = 2t\mathbf{i} + t^2\mathbf{j} - t^2\mathbf{k}; \quad (0, -4, 4)$

30. $\mathbf{r}(t) = -t\mathbf{i} + t^2\mathbf{j} + (\ln t)\mathbf{k}; \quad (2, -5, -3)$

In Exercises 31–36, $\mathbf{r}(t)$ is the position of a particle in space at time t. Match each position function with one of the graphs A–F.

31. $\mathbf{r}(t) = (t \cos t)\mathbf{i} + (t \sin t)\mathbf{j} + t\mathbf{k}$

32. $\mathbf{r}(t) = (\cos t)\mathbf{i} + (\sin t)\mathbf{j} + (\sin 2t)\mathbf{k}$

33. $\mathbf{r}(t) = t^2\mathbf{i} + (t^2 + 1)\mathbf{j} + t^4\mathbf{k}$

34. $\mathbf{r}(t) = t\mathbf{i} + (\ln t)\mathbf{j} + (\sin t)\mathbf{k}$

35. $\mathbf{r}(t) = t\mathbf{i} + (\cos t)\mathbf{j} + (\sin t)\mathbf{k}$

36. $\mathbf{r}(t) = (t \sin t)\mathbf{i} + (t \cos t)\mathbf{j} + \left(\dfrac{t}{t^2 + 1}\right)\mathbf{k}$

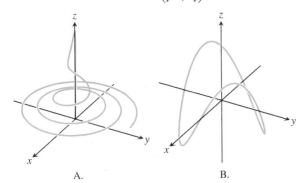

A.   B.

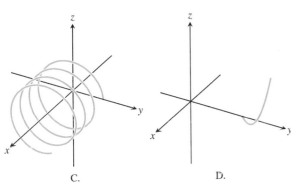

C.   D.

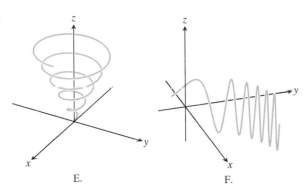

E.   F.

Theory and Examples

37. Motion along a circle Each of the following equations in parts (a)–(e) describes the motion of a particle having the same path, namely the unit circle $x^2 + y^2 = 1$. Although the path of each particle in parts (a)–(e) is the same, the behavior, or "dynamics," of each particle is different. For each particle, answer the following questions.

   i) Does the particle have constant speed? If so, what is its constant speed?

   ii) Is the particle's acceleration vector always orthogonal to its velocity vector?

   iii) Does the particle move clockwise or counterclockwise around the circle?

   iv) Does the particle begin at the point $(1, 0)$?

   a. $\mathbf{r}(t) = (\cos t)\mathbf{i} + (\sin t)\mathbf{j}, \quad t \geq 0$

   b. $\mathbf{r}(t) = \cos(2t)\mathbf{i} + \sin(2t)\mathbf{j}, \quad t \geq 0$

   c. $\mathbf{r}(t) = \cos(t - \pi/2)\mathbf{i} + \sin(t - \pi/2)\mathbf{j}, \quad t \geq 0$

   d. $\mathbf{r}(t) = (\cos t)\mathbf{i} - (\sin t)\mathbf{j}, \quad t \geq 0$

   e. $\mathbf{r}(t) = \cos(t^2)\mathbf{i} + \sin(t^2)\mathbf{j}, \quad t \geq 0$

38. Motion along a circle Show that the vector-valued function

$$\mathbf{r}(t) = (2\mathbf{i} + 2\mathbf{j} + \mathbf{k})$$
$$+ \cos t\left(\frac{1}{\sqrt{2}}\mathbf{i} - \frac{1}{\sqrt{2}}\mathbf{j}\right) + \sin t\left(\frac{1}{\sqrt{3}}\mathbf{i} + \frac{1}{\sqrt{3}}\mathbf{j} + \frac{1}{\sqrt{3}}\mathbf{k}\right)$$

describes the motion of a particle moving in the circle of radius 1 centered at the point $(2, 2, 1)$ and lying in the plane $x + y - 2z = 2$.

39. Motion along a parabola A particle moves along the top of the parabola $y^2 = 2x$ from left to right at a constant speed of 5 units per second. Find the velocity of the particle as it moves through the point $(2, 2)$.

40. Motion along a cycloid A particle moves in the xy-plane in such a way that its position at time t is

$$\mathbf{r}(t) = (t - \sin t)\mathbf{i} + (1 - \cos t)\mathbf{j}.$$

T **a.** Graph $\mathbf{r}(t)$. The resulting curve is a cycloid.

   b. Find the maximum and minimum values of $|\mathbf{v}|$ and $|\mathbf{a}|$. (*Hint:* Find the extreme values of $|\mathbf{v}|^2$ and $|\mathbf{a}|^2$ first and take square roots later.)

41. Let $\mathbf{r}$ be a differentiable vector function of t. Show that if $\mathbf{r} \cdot (d\mathbf{r}/dt) = 0$ for all t, then $|\mathbf{r}|$ is constant.

42. Derivatives of triple scalar products

   a. Show that if $\mathbf{u}$, $\mathbf{v}$, and $\mathbf{w}$ are differentiable vector functions of t, then

$$\frac{d}{dt}(\mathbf{u} \cdot \mathbf{v} \times \mathbf{w}) = \frac{d\mathbf{u}}{dt} \cdot \mathbf{v} \times \mathbf{w} + \mathbf{u} \cdot \frac{d\mathbf{v}}{dt} \times \mathbf{w} + \mathbf{u} \cdot \mathbf{v} \times \frac{d\mathbf{w}}{dt}.$$

   b. Show that

$$\frac{d}{dt}\left(\mathbf{r} \cdot \frac{d\mathbf{r}}{dt} \times \frac{d^2\mathbf{r}}{dt^2}\right) = \mathbf{r} \cdot \left(\frac{d\mathbf{r}}{dt} \times \frac{d^3\mathbf{r}}{dt^3}\right).$$

   (*Hint:* Differentiate on the left and look for vectors whose products are zero.)

43. Prove the two Scalar Multiple Rules for vector functions.

44. Prove the Sum and Difference Rules for vector functions.

45. Component test for continuity at a point Show that the vector function $\mathbf{r}$ defined by $\mathbf{r}(t) = f(t)\mathbf{i} + g(t)\mathbf{j} + h(t)\mathbf{k}$ is continuous at $t = t_0$ if and only if f, g, and h are continuous at t_0.

46. Limits of cross products of vector functions   Suppose that $\mathbf{r}_1(t) = f_1(t)\mathbf{i} + f_2(t)\mathbf{j} + f_3(t)\mathbf{k}$, $\mathbf{r}_2(t) = g_1(t)\mathbf{i} + g_2(t)\mathbf{j} + g_3(t)\mathbf{k}$, $\lim_{t \to t_0} \mathbf{r}_1(t) = \mathbf{A}$, and $\lim_{t \to t_0} \mathbf{r}_2(t) = \mathbf{B}$. Use the determinant formula for cross products and the Limit Product Rule for scalar functions to show that

$$\lim_{t \to t_0} (\mathbf{r}_1(t) \times \mathbf{r}_2(t)) = \mathbf{A} \times \mathbf{B}.$$

47. Differentiable vector functions are continuous   Show that if $\mathbf{r}(t) = f(t)\mathbf{i} + g(t)\mathbf{j} + h(t)\mathbf{k}$ is differentiable at $t = t_0$, then it is continuous at t_0 as well.

48. Constant Function Rule   Prove that if $\mathbf{u}$ is the vector function with the constant value $\mathbf{C}$, then $d\mathbf{u}/dt = \mathbf{0}$.

COMPUTER EXPLORATIONS

Use a CAS to perform the following steps in Exercises 49–52.

  a. Plot the space curve traced out by the position vector $\mathbf{r}$.

  b. Find the components of the velocity vector $d\mathbf{r}/dt$.

  c. Evaluate $d\mathbf{r}/dt$ at the given point t_0 and determine the equation of the tangent line to the curve at $\mathbf{r}(t_0)$.

  d. Plot the tangent line together with the curve over the given interval.

49. $\mathbf{r}(t) = (\sin t - t \cos t)\mathbf{i} + (\cos t + t \sin t)\mathbf{j} + t^2\mathbf{k}$, $0 \le t \le 6\pi$, $t_0 = 3\pi/2$

50. $\mathbf{r}(t) = \sqrt{2}t\,\mathbf{i} + e^t\mathbf{j} + e^{-t}\mathbf{k}$, $-2 \le t \le 3$, $t_0 = 1$

51. $\mathbf{r}(t) = (\sin 2t)\mathbf{i} + (\ln(1 + t))\mathbf{j} + t\mathbf{k}$, $0 \le t \le 4\pi$, $t_0 = \pi/4$

52. $\mathbf{r}(t) = (\ln(t^2 + 2))\mathbf{i} + (\tan^{-1} 3t)\mathbf{j} + \sqrt{t^2 + 1}\,\mathbf{k}$, $-3 \le t \le 5$, $t_0 = 3$

In Exercises 53 and 54, you will explore graphically the behavior of the helix

$$\mathbf{r}(t) = (\cos at)\mathbf{i} + (\sin at)\mathbf{j} + bt\mathbf{k}$$

as you change the values of the constants a and b. Use a CAS to perform the steps in each exercise.

53. Set $b = 1$. Plot the helix $\mathbf{r}(t)$ together with the tangent line to the curve at $t = 3\pi/2$ for $a = 1, 2, 4,$ and 6 over the interval $0 \le t \le 4\pi$. Describe in your own words what happens to the graph of the helix and the position of the tangent line as a increases through these positive values.

54. Set $a = 1$. Plot the helix $\mathbf{r}(t)$ together with the tangent line to the curve at $t = 3\pi/2$ for $b = 1/4, 1/2, 2,$ and 4 over the interval $0 \le t \le 4\pi$. Describe in your own words what happens to the graph of the helix and the position of the tangent line as b increases through these positive values.

13.2 Integrals of Vector Functions; Projectile Motion

In this section we investigate integrals of vector functions and their application to motion along a path in space or in the plane.

Integrals of Vector Functions

A differentiable vector function $\mathbf{R}(t)$ is an **antiderivative** of a vector function $\mathbf{r}(t)$ on an interval I if $d\mathbf{R}/dt = \mathbf{r}$ at each point of I. If $\mathbf{R}$ is an antiderivative of $\mathbf{r}$ on I, it can be shown, working one component at a time, that every antiderivative of $\mathbf{r}$ on I has the form $\mathbf{R} + \mathbf{C}$ for some constant vector $\mathbf{C}$ (Exercise 45). The set of all antiderivatives of $\mathbf{r}$ on I is the **indefinite integral** of $\mathbf{r}$ on I.

> **DEFINITION**   The **indefinite integral** of $\mathbf{r}$ with respect to t is the set of all antiderivatives of $\mathbf{r}$, denoted by $\int \mathbf{r}(t)\,dt$. If $\mathbf{R}$ is any antiderivative of $\mathbf{r}$, then
>
> $$\int \mathbf{r}(t)\,dt = \mathbf{R}(t) + \mathbf{C}.$$

The usual arithmetic rules for indefinite integrals apply.

EXAMPLE 1   To integrate a vector function, we integrate each of its components.

$$\int ((\cos t)\mathbf{i} + \mathbf{j} - 2t\mathbf{k})\,dt = \left(\int \cos t\,dt\right)\mathbf{i} + \left(\int dt\right)\mathbf{j} - \left(\int 2t\,dt\right)\mathbf{k} \qquad (1)$$

$$= (\sin t + C_1)\mathbf{i} + (t + C_2)\mathbf{j} - (t^2 + C_3)\mathbf{k} \qquad (2)$$

$$= (\sin t)\mathbf{i} + t\mathbf{j} - t^2\mathbf{k} + \mathbf{C} \qquad \text{\scriptsize $C = C_1\mathbf{i} + C_2\mathbf{j} - C_3\mathbf{k}$}$$

As in the integration of scalar functions, we recommend that you skip the steps in Equations (1) and (2) and go directly to the final form. Find an antiderivative for each component and add a *constant vector* at the end. ∎

Definite integrals of vector functions are best defined in terms of components. The definition is consistent with how we compute limits and derivatives of vector functions.

DEFINITION   If the components of $\mathbf{r}(t) = f(t)\mathbf{i} + g(t)\mathbf{j} + h(t)\mathbf{k}$ are integrable over $[a, b]$, then so is $\mathbf{r}$, and the **definite integral** of $\mathbf{r}$ from a to b is

$$\int_a^b \mathbf{r}(t)\, dt = \left(\int_a^b f(t)\, dt\right)\mathbf{i} + \left(\int_a^b g(t)\, dt\right)\mathbf{j} + \left(\int_a^b h(t)\, dt\right)\mathbf{k}.$$

EXAMPLE 2   As in Example 1, we integrate each component.

$$\int_0^\pi ((\cos t)\mathbf{i} + \mathbf{j} - 2t\mathbf{k})\, dt = \left(\int_0^\pi \cos t\, dt\right)\mathbf{i} + \left(\int_0^\pi dt\right)\mathbf{j} - \left(\int_0^\pi 2t\, dt\right)\mathbf{k}$$

$$= \Big[\sin t\Big]_0^\pi \mathbf{i} + \Big[t\Big]_0^\pi \mathbf{j} - \Big[t^2\Big]_0^\pi \mathbf{k}$$

$$= [0 - 0]\mathbf{i} + [\pi - 0]\mathbf{j} - [\pi^2 - 0^2]\mathbf{k}$$

$$= \pi\mathbf{j} - \pi^2\mathbf{k}$$

∎

The Fundamental Theorem of Calculus for continuous vector functions says that

$$\int_a^b \mathbf{r}(t)\, dt = \mathbf{R}(t)\Big]_a^b = \mathbf{R}(b) - \mathbf{R}(a)$$

where $\mathbf{R}$ is any antiderivative of $\mathbf{r}$, so that $\mathbf{R}'(t) = \mathbf{r}(t)$ (Exercise 46). Notice that an antiderivative of a vector function is also a vector function, whereas a definite integral of a vector function is a single constant vector.

EXAMPLE 3   Suppose we do not know the path of a hang glider, but only its acceleration vector $\mathbf{a}(t) = -(3\cos t)\mathbf{i} - (3\sin t)\mathbf{j} + 2\mathbf{k}$. We also know that initially (at time $t = 0$) the glider departed from the point $(4, 0, 0)$ with velocity $\mathbf{v}(0) = 3\mathbf{j}$. Find the glider's position as a function of t.

Solution   Our goal is to find $\mathbf{r}(t)$ knowing

$$\text{The differential equation:} \quad \mathbf{a} = \frac{d^2\mathbf{r}}{dt^2} = -(3\cos t)\mathbf{i} - (3\sin t)\mathbf{j} + 2\mathbf{k}$$

$$\text{The initial conditions:} \quad \mathbf{v}(0) = 3\mathbf{j} \quad \text{and} \quad \mathbf{r}(0) = 4\mathbf{i} + 0\mathbf{j} + 0\mathbf{k}.$$

Integrating both sides of the differential equation with respect to t gives

$$\mathbf{v}(t) = -(3\sin t)\mathbf{i} + (3\cos t)\mathbf{j} + 2t\mathbf{k} + \mathbf{C}_1.$$

We use $\mathbf{v}(0) = 3\mathbf{j}$ to find $\mathbf{C}_1$:

$$3\mathbf{j} = -(3\sin 0)\mathbf{i} + (3\cos 0)\mathbf{j} + (0)\mathbf{k} + \mathbf{C}_1$$

$$3\mathbf{j} = 3\mathbf{j} + \mathbf{C}_1$$

$$\mathbf{C}_1 = \mathbf{0}.$$

The glider's velocity as a function of time is

$$\frac{d\mathbf{r}}{dt} = \mathbf{v}(t) = -(3\sin t)\mathbf{i} + (3\cos t)\mathbf{j} + 2t\mathbf{k}.$$

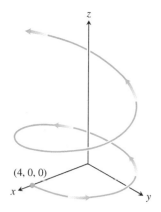

FIGURE 13.9 The path of the hang glider in Example 3. Although the path spirals around the z-axis, it is not a helix.

Integrating both sides of this last differential equation gives

$$\mathbf{r}(t) = (3\cos t)\mathbf{i} + (3\sin t)\mathbf{j} + t^2\mathbf{k} + \mathbf{C}_2.$$

We then use the initial condition $\mathbf{r}(0) = 4\mathbf{i}$ to find $\mathbf{C}_2$:

$$4\mathbf{i} = (3\cos 0)\mathbf{i} + (3\sin 0)\mathbf{j} + (0^2)\mathbf{k} + \mathbf{C}_2$$
$$4\mathbf{i} = 3\mathbf{i} + (0)\mathbf{j} + (0)\mathbf{k} + \mathbf{C}_2$$
$$\mathbf{C}_2 = \mathbf{i}.$$

The glider's position as a function of t is

$$\mathbf{r}(t) = (1 + 3\cos t)\mathbf{i} + (3\sin t)\mathbf{j} + t^2\mathbf{k}.$$

This is the path of the glider shown in Figure 13.9. Although the path resembles that of a helix due to its spiraling nature around the z-axis, it is not a helix because of the way it is rising. (We say more about this in Section 13.5.) ∎

The Vector and Parametric Equations for Ideal Projectile Motion

A classic example of integrating vector functions is the derivation of the equations for the motion of a projectile. In physics, projectile motion describes how an object fired at some angle from an initial position, and acted upon by only the force of gravity, moves in a vertical coordinate plane. In the classic example, we ignore the effects of any frictional drag on the object, which may vary with its speed and altitude, and also the fact that the force of gravity changes slightly with the projectile's changing height. In addition, we ignore the long-distance effects of Earth turning beneath the projectile, such as in a rocket launch or the firing of a projectile from a cannon. Ignoring these effects gives us a reasonable approximation of the motion in most cases.

To derive equations for projectile motion, we assume that the projectile behaves like a particle moving in a vertical coordinate plane and that the only force acting on the projectile during its flight is the constant force of gravity, which always points straight down. We assume that the projectile is launched from the origin at time $t = 0$ into the first quadrant with an initial velocity $\mathbf{v}_0$ (Figure 13.10). If $\mathbf{v}_0$ makes an angle α with the horizontal, then

$$\mathbf{v}_0 = (|\mathbf{v}_0|\cos \alpha)\mathbf{i} + (|\mathbf{v}_0|\sin \alpha)\mathbf{j}.$$

If we use the simpler notation v_0 for the initial speed $|\mathbf{v}_0|$, then

$$\mathbf{v}_0 = (v_0 \cos \alpha)\mathbf{i} + (v_0 \sin \alpha)\mathbf{j}. \tag{3}$$

The projectile's initial position is

$$\mathbf{r}_0 = 0\mathbf{i} + 0\mathbf{j} = \mathbf{0}. \tag{4}$$

Newton's second law of motion says that the force acting on the projectile is equal to the projectile's mass m times its acceleration, or $m(d^2\mathbf{r}/dt^2)$ if $\mathbf{r}$ is the projectile's position vector and t is time. If the force is solely the gravitational force $-mg\mathbf{j}$, then

$$m\frac{d^2\mathbf{r}}{dt^2} = -mg\mathbf{j} \quad \text{and} \quad \frac{d^2\mathbf{r}}{dt^2} = -g\mathbf{j},$$

where g is the acceleration due to gravity. We find $\mathbf{r}$ as a function of t by solving the following initial value problem.

Differential equation: $\quad \dfrac{d^2\mathbf{r}}{dt^2} = -g\mathbf{j}$

Initial conditions: $\quad \mathbf{r} = \mathbf{r}_0 \quad \text{and} \quad \dfrac{d\mathbf{r}}{dt} = \mathbf{v}_0 \quad \text{when } t = 0$

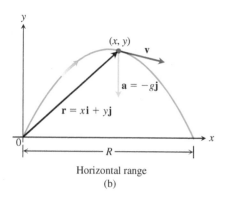

FIGURE 13.10 (a) Position, velocity, acceleration, and launch angle at $t = 0$. (b) Position, velocity, and acceleration at a later time t.

The first integration gives

$$\frac{d\mathbf{r}}{dt} = -(gt)\mathbf{j} + \mathbf{v}_0.$$

A second integration gives

$$\mathbf{r} = -\frac{1}{2}gt^2\mathbf{j} + \mathbf{v}_0 t + \mathbf{r}_0.$$

Substituting the values of $\mathbf{v}_0$ and $\mathbf{r}_0$ from Equations (3) and (4) gives

$$\mathbf{r} = -\frac{1}{2}gt^2\mathbf{j} + \underbrace{(v_0 \cos \alpha)t\mathbf{i} + (v_0 \sin \alpha)t\mathbf{j}}_{\mathbf{v}_0 t} + \mathbf{0}.$$

Collecting terms, we obtain the following.

Ideal Projectile Motion Equation

$$\mathbf{r} = (v_0 \cos \alpha)t\mathbf{i} + \left((v_0 \sin \alpha)t - \frac{1}{2}gt^2\right)\mathbf{j}. \qquad (5)$$

Equation (5) is the *vector equation* of the path for ideal projectile motion. The angle α is the projectile's launch angle (**firing angle, angle of elevation**), and v_0, as we said before, is the projectile's **initial speed**. The components of $\mathbf{r}$ give the parametric equations

$$x = (v_0 \cos \alpha)t \qquad \text{and} \qquad y = (v_0 \sin \alpha)t - \frac{1}{2}gt^2, \qquad (6)$$

where x is the distance downrange and y is the height of the projectile at time $t \geq 0$.

EXAMPLE 4 A projectile is fired from the origin over horizontal ground at an initial speed of 500 m/sec and a launch angle of 60°. Where will the projectile be 10 sec later?

Solution We use Equation (5) with $v_0 = 500$, $\alpha = 60°$, $g = 9.8$, and $t = 10$ to find the projectile's components 10 sec after firing.

$$\mathbf{r} = (v_0 \cos \alpha)t\mathbf{i} + \left((v_0 \sin \alpha)t - \frac{1}{2}gt^2\right)\mathbf{j}$$

$$= (500)\left(\frac{1}{2}\right)(10)\mathbf{i} + \left((500)\left(\frac{\sqrt{3}}{2}\right)10 - \left(\frac{1}{2}\right)(9.8)(100)\right)\mathbf{j}$$

$$\approx 2500\mathbf{i} + 3840\mathbf{j}$$

Ten seconds after firing, the projectile is about 3840 m above ground and 2500 m downrange from the origin. ∎

Ideal projectiles move along parabolas, as we now deduce from Equations (6). If we substitute $t = x/(v_0 \cos \alpha)$ from the first equation into the second, we obtain the Cartesian coordinate equation

$$y = -\left(\frac{g}{2v_0{}^2 \cos^2 \alpha}\right)x^2 + (\tan \alpha)x.$$

This equation has the form $y = ax^2 + bx$, so its graph is a parabola.

A projectile reaches its highest point when its vertical velocity component is zero. When fired over horizontal ground, the projectile lands when its vertical component equals zero in Equation (5), and the **range** R is the distance from the origin to the point of impact. We summarize the results here, which you are asked to verify in Exercise 31.

Height, Flight Time, and Range for Ideal Projectile Motion
For ideal projectile motion when an object is launched from the origin over a horizontal surface with initial speed v_0 and launch angle α:

$$\textit{Maximum height:} \qquad y_{\max} = \frac{(v_0 \sin \alpha)^2}{2g}$$

$$\textit{Flight time:} \qquad t = \frac{2v_0 \sin \alpha}{g}$$

$$\textit{Range:} \qquad R = \frac{v_0{}^2}{g} \sin 2\alpha.$$

If we fire our ideal projectile from the point (x_0, y_0) instead of the origin (Figure 13.11), the position vector for the path of motion is

$$\mathbf{r} = (x_0 + (v_0 \cos \alpha)t)\mathbf{i} + \left(y_0 + (v_0 \sin \alpha)t - \frac{1}{2}gt^2 \right)\mathbf{j}, \qquad (7)$$

as you are asked to show in Exercise 33.

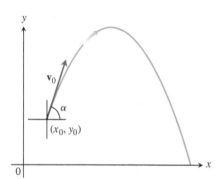

FIGURE 13.11 The path of a projectile fired from (x_0, y_0) with an initial velocity $\mathbf{v}_0$ at an angle of α degrees with the horizontal.

Projectile Motion with Wind Gusts

The next example shows how to account for another force acting on a projectile, due to a gust of wind. We assume that the path of the baseball in Example 5 lies in a vertical plane.

EXAMPLE 5   A baseball is hit when it is 3 ft above the ground. It leaves the bat with initial speed of 152 ft/sec, making an angle of 20° with the horizontal. At the instant the ball is hit, an instantaneous gust of wind blows in the horizontal direction directly opposite the direction the ball is taking toward the outfield, adding a component of $-8.8\mathbf{i}$ (ft/sec) to the ball's initial velocity (8.8 ft/sec = 6 mph).

(a) Find a vector equation (position vector) for the path of the baseball.

(b) How high does the baseball go, and when does it reach maximum height?

(c) Assuming that the ball is not caught, find its range and flight time.

Solution

(a) Using Equation (3) and accounting for the gust of wind, the initial velocity of the baseball is

$$\mathbf{v}_0 = (v_0 \cos \alpha)\mathbf{i} + (v_0 \sin \alpha)\mathbf{j} - 8.8\mathbf{i}$$
$$= (152 \cos 20°)\mathbf{i} + (152 \sin 20°)\mathbf{j} - (8.8)\mathbf{i}$$
$$= (152 \cos 20° - 8.8)\mathbf{i} + (152 \sin 20°)\mathbf{j}.$$

The initial position is $\mathbf{r}_0 = 0\mathbf{i} + 3\mathbf{j}$. Integration of $d^2\mathbf{r}/dt^2 = -g\mathbf{j}$ gives

$$\frac{d\mathbf{r}}{dt} = -(gt)\mathbf{j} + \mathbf{v}_0.$$

A second integration gives

$$\mathbf{r} = -\frac{1}{2}gt^2\mathbf{j} + \mathbf{v}_0 t + \mathbf{r}_0.$$

Substituting the values of $\mathbf{v}_0$ and $\mathbf{r}_0$ into the last equation gives the position vector of the baseball.

$$\begin{aligned}
\mathbf{r} &= -\frac{1}{2}gt^2\mathbf{j} + \mathbf{v}_0 t + \mathbf{r}_0 \\
&= -16t^2\mathbf{j} + (152 \cos 20° - 8.8)t\mathbf{i} + (152 \sin 20°)t\mathbf{j} + 3\mathbf{j} \\
&= (152 \cos 20° - 8.8)t\mathbf{i} + \left(3 + (152 \sin 20°)t - 16t^2\right)\mathbf{j}.
\end{aligned}$$

(b) The baseball reaches its highest point when the vertical component of velocity is zero, or

$$\frac{dy}{dt} = 152 \sin 20° - 32t = 0.$$

Solving for t we find

$$t = \frac{152 \sin 20°}{32} \approx 1.62 \text{ sec.}$$

Substituting this time into the vertical component for $\mathbf{r}$ gives the maximum height

$$y_{max} = 3 + (152 \sin 20°)(1.62) - 16(1.62)^2 \approx 45.2 \text{ ft.}$$

That is, the maximum height of the baseball is about 45.2 ft, reached about 1.6 sec after leaving the bat.

(c) To find when the baseball lands, we set the vertical component for $\mathbf{r}$ equal to 0 and solve for t:

$$\begin{aligned}
3 + (152 \sin 20°)t - 16t^2 &= 0 \\
3 + (51.99)t - 16t^2 &= 0.
\end{aligned}$$

The solution values are about $t = 3.3$ sec and $t = -0.06$ sec. Substituting the positive time into the horizontal component for $\mathbf{r}$, we find the range

$$R = (152 \cos 20° - 8.8)(3.3) \approx 442 \text{ ft.}$$

Thus, the horizontal range is about 442 ft, and the flight time is about 3.3 sec. ∎

In Exercises 41 and 42, we consider projectile motion when there is air resistance slowing down the flight.

EXERCISES 13.2

Integrating Vector-Valued Functions
Evaluate the integrals in Exercises 1–10.

1. $\displaystyle\int_0^1 \left[t^3\mathbf{i} + 7\mathbf{j} + (t+1)\mathbf{k} \right] dt$

2. $\displaystyle\int_1^2 \left[(6-6t)\mathbf{i} + 3\sqrt{t}\mathbf{j} + \left(\frac{4}{t^2}\right)\mathbf{k} \right] dt$

3. $\displaystyle\int_{-\pi/4}^{\pi/4} \left[(\sin t)\mathbf{i} + (1 + \cos t)\mathbf{j} + (\sec^2 t)\mathbf{k} \right] dt$

4. $\displaystyle\int_0^{\pi/3} \left[(\sec t \tan t)\mathbf{i} + (\tan t)\mathbf{j} + (2 \sin t \cos t)\mathbf{k} \right] dt$

5. $\displaystyle\int_1^4 \left[\frac{1}{t}\mathbf{i} + \frac{1}{5-t}\mathbf{j} + \frac{1}{2t}\mathbf{k} \right] dt$

6. $\displaystyle\int_0^1 \left[\frac{2}{\sqrt{1-t^2}}\mathbf{i} + \frac{\sqrt{3}}{1+t^2}\mathbf{k} \right] dt$

7. $\displaystyle\int_0^1 \left[te^{t^2}\mathbf{i} + e^{-t}\mathbf{j} + \mathbf{k} \right] dt$

8. $\displaystyle\int_1^{\ln 3} \left[te^t\mathbf{i} + e^t\mathbf{j} + \ln t\, \mathbf{k} \right] dt$

9. $\int_0^{\pi/2} \left[\cos t \, \mathbf{i} - \sin 2t \, \mathbf{j} + \sin^2 t \, \mathbf{k} \right] dt$

10. $\int_0^{\pi/4} \left[\sec t \, \mathbf{i} + \tan^2 t \, \mathbf{j} - t \sin t \, \mathbf{k} \right] dt$

Initial Value Problems

Solve the initial value problems in Exercises 11–20 for **r** as a vector function of t.

11. Differential equation: $\dfrac{d\mathbf{r}}{dt} = -t\mathbf{i} - t\mathbf{j} - t\mathbf{k}$

     Initial condition: $\mathbf{r}(0) = \mathbf{i} + 2\mathbf{j} + 3\mathbf{k}$

12. Differential equation: $\dfrac{d\mathbf{r}}{dt} = (180t)\mathbf{i} + (180t - 16t^2)\mathbf{j}$

     Initial condition: $\mathbf{r}(0) = 100\mathbf{j}$

13. Differential equation: $\dfrac{d\mathbf{r}}{dt} = \dfrac{3}{2}(t+1)^{1/2}\mathbf{i} + e^{-t}\mathbf{j} + \dfrac{1}{t+1}\mathbf{k}$

     Initial condition: $\mathbf{r}(0) = \mathbf{k}$

14. Differential equation: $\dfrac{d\mathbf{r}}{dt} = (t^3 + 4t)\mathbf{i} + t\mathbf{j} + 2t^2\mathbf{k}$

     Initial condition: $\mathbf{r}(0) = \mathbf{i} + \mathbf{j}$

15. Differential equation:

$$\dfrac{d\mathbf{r}}{dt} = (\tan t)\mathbf{i} + \left(\cos\left(\dfrac{1}{2}t\right)\right)\mathbf{j} - (\sec 2t)\mathbf{k}$$

     Initial condition: $\mathbf{r}(0) = 3\mathbf{i} - 2\mathbf{j} + \mathbf{k}$

16. Differential equation:

$$\dfrac{d\mathbf{r}}{dt} = \left(\dfrac{t}{t^2+2}\right)\mathbf{i} - \left(\dfrac{t^2+1}{t-2}\right)\mathbf{j} + \left(\dfrac{t^2+4}{t^2+3}\right)\mathbf{k}$$

     Initial condition: $\mathbf{r}(0) = \mathbf{i} - \mathbf{j} + \mathbf{k}$

17. Differential equation: $\dfrac{d^2\mathbf{r}}{dt^2} = -32\mathbf{k}$

     Initial conditions: $\mathbf{r}(0) = 100\mathbf{k}$ and

          $\left.\dfrac{d\mathbf{r}}{dt}\right|_{t=0} = 8\mathbf{i} + 8\mathbf{j}$

18. Differential equation: $\dfrac{d^2\mathbf{r}}{dt^2} = -(\mathbf{i} + \mathbf{j} + \mathbf{k})$

     Initial conditions: $\mathbf{r}(0) = 10\mathbf{i} + 10\mathbf{j} + 10\mathbf{k}$ and

          $\left.\dfrac{d\mathbf{r}}{dt}\right|_{t=0} = \mathbf{0}$

19. Differential equation: $\dfrac{d^2\mathbf{r}}{dt^2} = e^t\mathbf{i} - e^{-t}\mathbf{j} + 4e^{2t}\mathbf{k}$

     Initial conditions: $\mathbf{r}(0) = 3\mathbf{i} + \mathbf{j} + 2\mathbf{k}$ and

          $\left.\dfrac{d\mathbf{r}}{dt}\right|_{t=0} = -\mathbf{i} + 4\mathbf{j}$

20. Differential equation:

$$\dfrac{d^2\mathbf{r}}{dt^2} = (\sin t)\mathbf{i} - (\cos t)\mathbf{j} + (4\sin t \cos t)\mathbf{k}$$

     Initial conditions: $\mathbf{r}(0) = \mathbf{i} - \mathbf{k}$ and

          $\left.\dfrac{d\mathbf{r}}{dt}\right|_{t=0} = \mathbf{i}$

Motion Along a Straight Line

21. At time $t = 0$, a particle is located at the point (1, 2, 3). It travels in a straight line to the point (4, 1, 4), has speed 2 at (1, 2, 3) and constant acceleration $3\mathbf{i} - \mathbf{j} + \mathbf{k}$. Find an equation for the position vector $\mathbf{r}(t)$ of the particle at time t.

22. A particle traveling in a straight line is located at the point $(1, -1, 2)$ and has speed 2 at time $t = 0$. The particle moves toward the point (3, 0, 3) with constant acceleration $2\mathbf{i} + \mathbf{j} + \mathbf{k}$. Find its position vector $\mathbf{r}(t)$ at time t.

Projectile Motion

Projectile flights in the following exercises are to be treated as ideal unless stated otherwise. All launch angles are assumed to be measured from the horizontal. All projectiles are assumed to be launched from the origin over a horizontal surface unless stated otherwise.

23. Travel time   A projectile is fired at a speed of 840 m/sec at an angle of 60°. How long will it take to get 21 km downrange?

24. Range and height versus speed

   a. Show that doubling a projectile's initial speed at a given launch angle multiplies its range by 4.

   b. By about what percentage should you increase the initial speed to double the height and range?

25. Flight time and height   A projectile is fired with an initial speed of 500 m/sec at an angle of elevation of 45°.

   a. When and how far away will the projectile strike?

   b. How high overhead will the projectile be when it is 5 km downrange?

   c. What is the greatest height reached by the projectile?

26. Throwing a baseball   A baseball is thrown from the stands 32 ft above the field at an angle of 30° up from the horizontal. When and how far away will the ball strike the ground if its initial speed is 32 ft/sec?

27. Firing golf balls   A spring gun at ground level fires a golf ball at an angle of 45°. The ball lands 10 m away.

   a. What was the ball's initial speed?

   b. For the same initial speed, find the two firing angles that make the range 6 m.

28. Beaming electrons   An electron in a TV tube is beamed horizontally at a speed of 5×10^6 m/sec toward the face of the tube 40 cm away. About how far will the electron drop before it hits?

29. Equal-range firing angles   What two angles of elevation will enable a projectile to reach a target 16 km downrange on the same level as the gun if the projectile's initial speed is 400 m/sec?

30. Finding muzzle speed   Find the muzzle speed of a gun whose maximum range is 24.5 km.

31. Verify the results given in the text (following Example 4) for the maximum height, flight time, and range for ideal projectile motion.

32. Colliding marbles The accompanying figure shows an experiment with two marbles. Marble A was launched toward marble B with launch angle α and initial speed v_0. At the same instant, marble B was released to fall from rest at $R \tan \alpha$ units directly above a spot R units downrange from A. The marbles were found to collide regardless of the value of v_0. Was this mere coincidence, or must this happen? Give reasons for your answer.

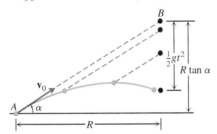

33. Firing from (x_0, y_0) Derive the equations

$$x = x_0 + (v_0 \cos \alpha)t,$$

$$y = y_0 + (v_0 \sin \alpha)t - \frac{1}{2}gt^2$$

(see Equation (7) in the text) by solving the following initial value problem for a vector $\mathbf{r}$ in the plane.

Differential equation: $\dfrac{d^2\mathbf{r}}{dt^2} = -g\mathbf{j}$

Initial conditions: $\mathbf{r}(0) = x_0\mathbf{i} + y_0\mathbf{j}$

$\dfrac{d\mathbf{r}}{dt}(0) = (v_0 \cos \alpha)\mathbf{i} + (v_0 \sin \alpha)\mathbf{j}$

34. Where trajectories crest For a projectile fired from the ground at launch angle α with initial speed v_0, consider α as a variable and v_0 as a fixed constant. For each α, $0 < \alpha < \pi/2$, we obtain a parabolic trajectory as shown in the accompanying figure. Show that the points in the plane that give the maximum heights of these parabolic trajectories all lie on the ellipse

$$x^2 + 4\left(y - \frac{v_0^2}{4g}\right)^2 = \frac{v_0^4}{4g^2},$$

where $x \geq 0$.

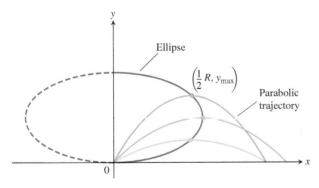

35. Launching downhill An ideal projectile is launched straight down an inclined plane as shown in the accompanying figure.

a. Show that the greatest downhill range is achieved when the initial velocity vector bisects angle AOR.

b. If the projectile were fired uphill instead of down, what launch angle would maximize its range? Give reasons for your answer.

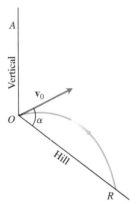

36. Elevated green A golf ball is hit with an initial speed of 116 ft/sec at an angle of elevation of 45° from the tee to a green that is elevated 45 ft above the tee as shown in the diagram. Assuming that the pin, 369 ft downrange, does not get in the way, where will the ball land in relation to the pin?

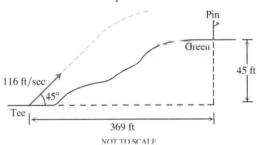

37. Volleyball A volleyball is hit when it is 4 ft above the ground and 12 ft from a 6-ft-high net. It leaves the point of impact with an initial velocity of 35 ft/sec at an angle of 27° and slips by the opposing team untouched.

a. Find a vector equation for the path of the volleyball.

b. How high does the volleyball go, and when does it reach maximum height?

c. Find its range and flight time.

d. When is the volleyball 7 ft above the ground? How far (ground distance) is the volleyball from where it will land?

e. Suppose that the net is raised to 8 ft. Does this change things? Explain.

38. Shot put In Moscow in 1987, Natalya Lisouskaya set a women's world record by putting an 8 lb 13 oz shot 73 ft 10 in. Assuming that she launched the shot at a 40° angle to the horizontal from 6.5 ft above the ground, what was the shot's initial speed?

39. Model train The accompanying multiflash photograph shows a model train engine moving at a constant speed on a straight horizontal track. As the engine moved along, a marble was fired into the air by a spring in the engine's smokestack. The marble, which continued to move with the same forward speed as the engine, rejoined the engine 1 sec after it was fired. Measure the angle the marble's path made with the horizontal and use the information to find how high the marble went and how fast the engine was moving.

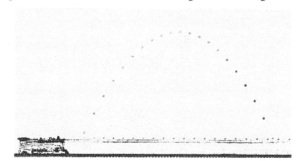

Source: PSSC Physics, 2nd ed., Reprinted by permission of Educational Development Center, Inc.

40. Hitting a baseball under a wind gust A baseball is hit when it is 2.5 ft above the ground. It leaves the bat with an initial velocity of 145 ft/sec at a launch angle of 23°. At the instant the ball is hit, an instantaneous gust of wind blows against the ball, adding a component of $-14\mathbf{i}$ (ft/sec) to the ball's initial velocity. A 15-ft-high fence lies 300 ft from home plate in the direction of the flight.

a. Find a vector equation for the path of the baseball.

b. How high does the baseball go, and when does it reach maximum height?

c. Find the range and flight time of the baseball, assuming that the ball is not caught.

d. When is the baseball 20 ft high? How far (ground distance) is the baseball from home plate at that height?

e. Has the batter hit a home run? Explain.

Projectile Motion with Linear Drag

The main force affecting the motion of a projectile, other than gravity, is air resistance. This slowing down force is **drag force**, and it acts in a direction *opposite* to the velocity of the projectile (see accompanying figure). For projectiles moving through the air at relatively low speeds, however, the drag force is (very nearly) proportional to the speed (to the first power) and so is called **linear**.

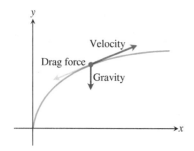

41. Linear drag Derive the equations

$$x = \frac{v_0}{k}(1 - e^{-kt})\cos\alpha$$

$$y = \frac{v_0}{k}(1 - e^{-kt})(\sin\alpha) + \frac{g}{k^2}(1 - kt - e^{-kt})$$

by solving the following initial value problem for a vector **r** in the plane.

Differential equation: $\dfrac{d^2\mathbf{r}}{dt^2} = -g\mathbf{j} - k\mathbf{v} = -g\mathbf{j} - k\dfrac{d\mathbf{r}}{dt}$

Initial conditions: $\mathbf{r}(0) = \mathbf{0}$

$$\frac{d\mathbf{r}}{dt}\bigg|_{t=0} = \mathbf{v}_0 = (v_0\cos\alpha)\mathbf{i} + (v_0\sin\alpha)\mathbf{j}$$

The **drag coefficient** k is a positive constant representing resistance due to air density, v_0 and α are the projectile's initial speed and launch angle, and g is the acceleration of gravity.

42. Hitting a baseball with linear drag Consider the baseball problem in Example 5 when there is linear drag (see Exercise 41). Assume a drag coefficient $k = 0.12$, but no gust of wind.

a. From Exercise 41, find a vector form for the path of the baseball.

b. How high does the baseball go, and when does it reach maximum height?

c. Find the range and flight time of the baseball.

d. When is the baseball 30 ft high? How far (ground distance) is the baseball from home plate at that height?

e. A 10-ft-high outfield fence is 340 ft from home plate in the direction of the flight of the baseball. The outfielder can jump and catch any ball up to 11 ft off the ground to stop it from going over the fence. Has the batter hit a home run?

Theory and Examples

43. Establish the following properties of integrable vector functions.

a. The *Constant Scalar Multiple Rule:*

$$\int_a^b k\mathbf{r}(t)\,dt = k\int_a^b \mathbf{r}(t)\,dt \quad \text{(any scalar } k\text{)}$$

The *Rule for Negatives,*

$$\int_a^b (-\mathbf{r}(t))\,dt = -\int_a^b \mathbf{r}(t)\,dt,$$

is obtained by taking $k = -1$.

b. The *Sum and Difference Rules:*

$$\int_a^b (\mathbf{r}_1(t) \pm \mathbf{r}_2(t))\,dt = \int_a^b \mathbf{r}_1(t)\,dt \pm \int_a^b \mathbf{r}_2(t)\,dt$$

c. The *Constant Vector Multiple Rules:*

$$\int_a^b \mathbf{C}\cdot\mathbf{r}(t)\,dt = \mathbf{C}\cdot\int_a^b \mathbf{r}(t)\,dt \quad \text{(any constant vector } \mathbf{C}\text{)}$$

and

$$\int_a^b \mathbf{C}\times\mathbf{r}(t)\,dt = \mathbf{C}\times\int_a^b \mathbf{r}(t)\,dt \quad \text{(any constant vector } \mathbf{C}\text{)}$$

44. Products of scalar and vector functions Suppose that the scalar function $u(t)$ and the vector function $\mathbf{r}(t)$ are both defined for $a \le t \le b$.

   a. Show that $u\mathbf{r}$ is continuous on $[a, b]$ if u and $\mathbf{r}$ are continuous on $[a, b]$.

   b. If u and $\mathbf{r}$ are both differentiable on $[a, b]$, show that $u\mathbf{r}$ is differentiable on $[a, b]$ and that

$$\frac{d}{dt}(u\mathbf{r}) = u\frac{d\mathbf{r}}{dt} + \mathbf{r}\frac{du}{dt}.$$

45. Antiderivatives of vector functions

   a. Use Corollary 2 of the Mean Value Theorem for scalar functions to show that if two vector functions $\mathbf{R}_1(t)$ and $\mathbf{R}_2(t)$ have identical derivatives on an interval I, then the functions differ by a constant vector value throughout I.

   b. Use the result in part (a) to show that if $\mathbf{R}(t)$ is any antiderivative of $\mathbf{r}(t)$ on I, then any other antiderivative of $\mathbf{r}$ on I equals $\mathbf{R}(t) + \mathbf{C}$ for some constant vector $\mathbf{C}$.

46. The Fundamental Theorem of Calculus The Fundamental Theorem of Calculus for scalar functions of a real variable holds for vector functions of a real variable as well. Prove this by using the theorem for scalar functions to show first that if a vector function $\mathbf{r}(t)$ is continuous for $a \le t \le b$, then

$$\frac{d}{dt}\int_a^t \mathbf{r}(\tau)\, d\tau = \mathbf{r}(t)$$

at every point t of (a, b). Then use the conclusion in part (b) of Exercise 45 to show that if $\mathbf{R}$ is any antiderivative of $\mathbf{r}$ on $[a, b]$ then

$$\int_a^b \mathbf{r}(t)\, dt = \mathbf{R}(b) - \mathbf{R}(a).$$

47. Hitting a baseball with linear drag under a wind gust Consider again the baseball problem in Example 5. This time assume a drag coefficient of 0.08 *and* an instantaneous gust of wind that adds a component of $-17.6\mathbf{i}$ (ft/sec) to the initial velocity at the instant the baseball is hit.

   a. Find a vector equation for the path of the baseball.

   b. How high does the baseball go, and when does it reach maximum height?

   c. Find the range and flight time of the baseball.

   d. When is the baseball 35 ft high? How far (ground distance) is the baseball from home plate at that height?

   e. A 20-ft-high outfield fence is 380 ft from home plate in the direction of the flight of the baseball. Has the batter hit a home run? If "yes," what change in the horizontal component of the ball's initial velocity would have kept the ball in the park? If "no," what change would have allowed it to be a home run?

48. Height versus time Show that a projectile attains three-quarters of its maximum height in half the time it takes to reach the maximum height.

13.3 Arc Length in Space

In this and the next two sections, we study the mathematical features of a curve's shape that describe the sharpness of its turning and its twisting.

Arc Length Along a Space Curve

Base point

FIGURE 13.12 Smooth curves can be scaled like number lines, the coordinate of each point being its directed distance along the curve from a preselected base point.

One of the features of smooth space and plane curves is that they have a measurable length. This enables us to locate points along these curves by giving their directed distance s along the curve from some base point, the way we locate points on coordinate axes by giving their directed distance from the origin (Figure 13.12). This is what we did for plane curves in Section 13.2.

To measure distance along a smooth curve in space, we add a z-term to the formula we use for curves in the plane.

> **DEFINITION** The **length** of a smooth curve $\mathbf{r}(t) = x(t)\mathbf{i} + y(t)\mathbf{j} + z(t)\mathbf{k}$, $a \le t \le b$, that is traced exactly once as t increases from $t = a$ to $t = b$, is
>
> $$L = \int_a^b \sqrt{\left(\frac{dx}{dt}\right)^2 + \left(\frac{dy}{dt}\right)^2 + \left(\frac{dz}{dt}\right)^2}\, dt. \qquad (1)$$

Just as for plane curves, we can calculate the length of a curve in space from any convenient parametrization that meets the stated conditions. We omit the proof.

The square root in Equation (1) is $|\mathbf{v}|$, the length of a velocity vector $d\mathbf{r}/dt$. This enables us to write the formula for length a shorter way.

Arc Length Formula

$$L = \int_a^b |\mathbf{v}| \, dt \qquad (2)$$

EXAMPLE 1 A glider is soaring upward along the helix $\mathbf{r}(t) = (\cos t)\mathbf{i} + (\sin t)\mathbf{j} + t\mathbf{k}$. How long is the glider's path from $t = 0$ to $t = 2\pi$?

Solution The path segment during this time corresponds to one full turn of the helix (Figure 13.13). The length of this portion of the curve is

$$L = \int_a^b |\mathbf{v}| \, dt = \int_0^{2\pi} \sqrt{(-\sin t)^2 + (\cos t)^2 + (1)^2} \, dt$$

$$= \int_0^{2\pi} \sqrt{2} \, dt = 2\pi\sqrt{2} \text{ units of length.}$$

This is $\sqrt{2}$ times the circumference of the circle in the xy-plane over which the helix stands. ∎

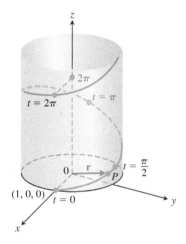

FIGURE 13.13 The helix in Example 1, $\mathbf{r}(t) = (\cos t)\mathbf{i} + (\sin t)\mathbf{j} + t\mathbf{k}$.

If we choose a base point $P(t_0)$ on a smooth curve C parametrized by t, each value of t determines a point $P(t) = (x(t), y(t), z(t))$ on C and a "directed distance"

$$s(t) = \int_{t_0}^t |\mathbf{v}(\tau)| \, d\tau,$$

measured along C from the base point (Figure 13.14). This is the arc length function we defined in Section 11.2 for plane curves that have no z-component. If $t > t_0$, $s(t)$ is the distance along the curve from $P(t_0)$ to $P(t)$. If $t < t_0$, $s(t)$ is the negative of the distance. Each value of s determines a point on C, and this parametrizes C with respect to s. We call s an **arc length parameter** for the curve. The parameter's value increases in the direction of increasing t. We will see that the arc length parameter is particularly effective for investigating the turning and twisting nature of a space curve.

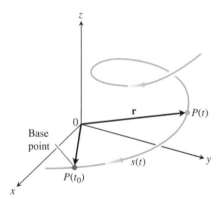

FIGURE 13.14 The directed distance along the curve from $P(t_0)$ to any point $P(t)$ is $s(t) = \int_{t_0}^t |\mathbf{v}(\tau)| \, d\tau$.

Arc Length Parameter with Base Point $P(t_0)$

$$s(t) = \int_{t_0}^t \sqrt{[x'(\tau)]^2 + [y'(\tau)]^2 + [z'(\tau)]^2} \, d\tau = \int_{t_0}^t |\mathbf{v}(\tau)| \, d\tau \qquad (3)$$

τ is the Greek letter tau (rhymes with "now")

We use the Greek letter τ ("tau") as the variable of integration in Equation (3) because the letter t is already in use as the upper limit.

If a curve $\mathbf{r}(t)$ is already given in terms of some parameter t and $s(t)$ is the arc length function given by Equation (3), then we may be able to solve for t as a function of s: $t = t(s)$. Then the curve can be reparametrized in terms of s by substituting for t: $\mathbf{r} = \mathbf{r}(t(s))$. The new parametrization identifies a point on the curve with its directed distance along the curve from the base point.

EXAMPLE 2 This is an example for which we can actually find the arc length parametrization of a curve. If $t_0 = 0$, the arc length parameter along the helix

$$\mathbf{r}(t) = (\cos t)\mathbf{i} + (\sin t)\mathbf{j} + t\mathbf{k}$$

from t_0 to t is

$$s(t) = \int_{t_0}^{t} |\mathbf{v}(\tau)| \, d\tau \qquad \text{Eq. (3)}$$

$$= \int_{0}^{t} \sqrt{2} \, d\tau \qquad \text{Value from Example 1}$$

$$= \sqrt{2}\, t.$$

Solving this equation for t gives $t = s/\sqrt{2}$. Substituting into the position vector $\mathbf{r}$ gives the following arc length parametrization for the helix:

$$\mathbf{r}(t(s)) = \left(\cos \frac{s}{\sqrt{2}} \right) \mathbf{i} + \left(\sin \frac{s}{\sqrt{2}} \right) \mathbf{j} + \frac{s}{\sqrt{2}} \mathbf{k}. \qquad ■$$

Unlike Example 2, the arc length parametrization is generally difficult to find analytically for a curve already given in terms of some other parameter t. Fortunately, however, we rarely need an exact formula for $s(t)$ or its inverse $t(s)$.

Speed on a Smooth Curve

Since the derivatives beneath the radical in Equation (3) are continuous (the curve is smooth), the Fundamental Theorem of Calculus tells us that s is a differentiable function of t with derivative

$$\frac{ds}{dt} = |\mathbf{v}(t)|. \qquad (4)$$

Equation (4) says that the speed with which a particle moves along its path is the magnitude of $\mathbf{v}$, consistent with what we know.

Although the base point $P(t_0)$ plays a role in defining s in Equation (3), it plays no role in Equation (4). The rate at which a moving particle covers distance along its path is independent of how far away it is from the base point.

Notice that $ds/dt > 0$ since, by definition, $|\mathbf{v}|$ is never zero for a smooth curve. We see once again that s is an increasing function of t.

Unit Tangent Vector

We already know the velocity vector $\mathbf{v} = d\mathbf{r}/dt$ is tangent to the curve $\mathbf{r}(t)$ and that the vector

$$\mathbf{T} = \frac{\mathbf{v}}{|\mathbf{v}|}$$

is therefore a unit vector tangent to the (smooth) curve, called the **unit tangent vector** (Figure 13.15). The unit tangent vector $\mathbf{T}$ is a differentiable function of t whenever $\mathbf{v}$ is a differentiable function of t. As we will see in Section 13.5, $\mathbf{T}$ is one of three unit vectors in a traveling reference frame that is used to describe the motion of objects traveling in three dimensions.

EXAMPLE 3 Find the unit tangent vector of the curve

$$\mathbf{r}(t) = (1 + 3\cos t)\mathbf{i} + (3\sin t)\mathbf{j} + t^2\mathbf{k}$$

representing the path of the glider in Example 3, Section 13.2.

Solution In that example, we found

$$\mathbf{v} = \frac{d\mathbf{r}}{dt} = -(3\sin t)\mathbf{i} + (3\cos t)\mathbf{j} + 2t\mathbf{k}$$

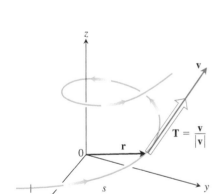

FIGURE 13.15 We find the unit tangent vector $\mathbf{T}$ by dividing $\mathbf{v}$ by its length $|\mathbf{v}|$.

and

$$|\mathbf{v}| = \sqrt{9 + 4t^2}.$$

Thus,

$$\mathbf{T} = \frac{\mathbf{v}}{|\mathbf{v}|} = -\frac{3\sin t}{\sqrt{9 + 4t^2}}\mathbf{i} + \frac{3\cos t}{\sqrt{9 + 4t^2}}\mathbf{j} + \frac{2t}{\sqrt{9 + 4t^2}}\mathbf{k}. \qquad \blacksquare$$

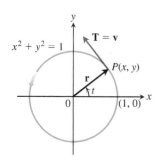

FIGURE 13.16 Counterclockwise motion around the unit circle.

For the counterclockwise motion

$$\mathbf{r}(t) = (\cos t)\mathbf{i} + (\sin t)\mathbf{j}$$

around the unit circle, we see that

$$\mathbf{v} = (-\sin t)\mathbf{i} + (\cos t)\mathbf{j}$$

is already a unit vector, so $\mathbf{T} = \mathbf{v}$ and $\mathbf{T}$ is orthogonal to $\mathbf{r}$ (Figure 13.16).

The velocity vector is the change in the position vector $\mathbf{r}$ with respect to time t, but how does the position vector change with respect to arc length? More precisely, what is the derivative $d\mathbf{r}/ds$? Since $ds/dt > 0$ for the curves we are considering, s is one-to-one and has an inverse that gives t as a differentiable function of s (Section 3.8). The derivative of the inverse is

$$\frac{dt}{ds} = \frac{1}{ds/dt} = \frac{1}{|\mathbf{v}|}.$$

This makes $\mathbf{r}$ a differentiable function of s whose derivative can be calculated with the Chain Rule to be

$$\frac{d\mathbf{r}}{ds} = \frac{d\mathbf{r}}{dt}\frac{dt}{ds} = \mathbf{v}\frac{1}{|\mathbf{v}|} = \frac{\mathbf{v}}{|\mathbf{v}|} = \mathbf{T}. \qquad (5)$$

This equation says that $d\mathbf{r}/ds$ is the unit tangent vector in the direction of the velocity vector $\mathbf{v}$ (Figure 13.15).

EXERCISES 13.3

Finding Tangent Vectors and Lengths

In Exercises 1–8, find the curve's unit tangent vector. Also, find the length of the indicated portion of the curve.

1. $\mathbf{r}(t) = (2\cos t)\mathbf{i} + (2\sin t)\mathbf{j} + \sqrt{5}t\mathbf{k}, \quad 0 \le t \le \pi$

2. $\mathbf{r}(t) = (6\sin 2t)\mathbf{i} + (6\cos 2t)\mathbf{j} + 5t\mathbf{k}, \quad 0 \le t \le \pi$

3. $\mathbf{r}(t) = t\mathbf{i} + (2/3)t^{3/2}\mathbf{k}, \quad 0 \le t \le 8$

4. $\mathbf{r}(t) = (2 + t)\mathbf{i} - (t + 1)\mathbf{j} + t\mathbf{k}, \quad 0 \le t \le 3$

5. $\mathbf{r}(t) = (\cos^3 t)\mathbf{j} + (\sin^3 t)\mathbf{k}, \quad 0 \le t \le \pi/2$

6. $\mathbf{r}(t) = 6t^3\mathbf{i} - 2t^3\mathbf{j} - 3t^3\mathbf{k}, \quad 1 \le t \le 2$

7. $\mathbf{r}(t) = (t\cos t)\mathbf{i} + (t\sin t)\mathbf{j} + \left(2\sqrt{2}/3\right)t^{3/2}\mathbf{k}, \quad 0 \le t \le \pi$

8. $\mathbf{r}(t) = (t\sin t + \cos t)\mathbf{i} + (t\cos t - \sin t)\mathbf{j}, \quad \sqrt{2} \le t \le 2$

9. Find the point on the curve

$$\mathbf{r}(t) = (5\sin t)\mathbf{i} + (5\cos t)\mathbf{j} + 12t\mathbf{k}$$

at a distance 26π units along the curve from the point $(0, 5, 0)$ in the direction of increasing arc length.

10. Find the point on the curve

$$\mathbf{r}(t) = (12\sin t)\mathbf{i} - (12\cos t)\mathbf{j} + 5t\mathbf{k}$$

at a distance 13π units along the curve from the point $(0, -12, 0)$ in the direction opposite to the direction of increasing arc length.

Arc Length Parameter

In Exercises 11–14, find the arc length parameter along the curve from the point where $t = 0$ by evaluating the integral

$$s = \int_0^t |\mathbf{v}(\tau)| \, d\tau$$

from Equation (3). Then find the length of the indicated portion of the curve.

11. $\mathbf{r}(t) = (4\cos t)\mathbf{i} + (4\sin t)\mathbf{j} + 3t\mathbf{k}, \quad 0 \le t \le \pi/2$

12. $\mathbf{r}(t) = (\cos t + t\sin t)\mathbf{i} + (\sin t - t\cos t)\mathbf{j}, \quad \pi/2 \le t \le \pi$

13. $\mathbf{r}(t) = (e^t\cos t)\mathbf{i} + (e^t\sin t)\mathbf{j} + e^t\mathbf{k}, \quad -\ln 4 \le t \le 0$

14. $\mathbf{r}(t) = (1 + 2t)\mathbf{i} + (1 + 3t)\mathbf{j} + (6 - 6t)\mathbf{k}, \quad -1 \le t \le 0$

Theory and Examples

15. Arc length Find the length of the curve

$$\mathbf{r}(t) = \left(\sqrt{2}t\right)\mathbf{i} + \left(\sqrt{2}t\right)\mathbf{j} + (1 - t^2)\mathbf{k}$$

from $(0, 0, 1)$ to $\left(\sqrt{2}, \sqrt{2}, 0\right)$.

16. Length of helix The length $2\pi\sqrt{2}$ of the turn of the helix in Example 1 is also the length of the diagonal of a square 2π units on a side. Show how to obtain this square by cutting away and flattening a portion of the cylinder around which the helix winds.

17. Ellipse

  a. Show that the curve $\mathbf{r}(t) = (\cos t)\mathbf{i} + (\sin t)\mathbf{j} + (1 - \cos t)\mathbf{k}$, $0 \leq t \leq 2\pi$, is an ellipse by showing that it is the intersection of a right circular cylinder and a plane. Find equations for the cylinder and plane.

  b. Sketch the ellipse on the cylinder. Add to your sketch the unit tangent vectors at $t = 0$, $\pi/2$, π, and $3\pi/2$.

  c. Show that the acceleration vector always lies parallel to the plane (orthogonal to a vector normal to the plane). Thus, if you draw the acceleration as a vector attached to the ellipse, it will lie in the plane of the ellipse. Add the acceleration vectors for $t = 0$, $\pi/2$, π, and $3\pi/2$ to your sketch.

  d. Write an integral for the length of the ellipse. Do not try to evaluate the integral; it is nonelementary.

  T **e. Numerical integrator** Estimate the length of the ellipse to two decimal places.

18. Length is independent of parametrization To illustrate that the length of a smooth space curve does not depend on the parametrization you use to compute it, calculate the length of one turn of the helix in Example 1 with the following parametrizations.

  a. $\mathbf{r}(t) = (\cos 4t)\mathbf{i} + (\sin 4t)\mathbf{j} + 4t\mathbf{k}$, $0 \leq t \leq \pi/2$

  b. $\mathbf{r}(t) = [\cos (t/2)]\mathbf{i} + [\sin (t/2)]\mathbf{j} + (t/2)\mathbf{k}$, $0 \leq t \leq 4\pi$

  c. $\mathbf{r}(t) = (\cos t)\mathbf{i} - (\sin t)\mathbf{j} - t\mathbf{k}$, $-2\pi \leq t \leq 0$

19. The involute of a circle If a string wound around a fixed circle is unwound while held taut in the plane of the circle, its end P traces an *involute* of the circle. In the accompanying figure, the circle in question is the circle $x^2 + y^2 = 1$ and the tracing point starts at $(1, 0)$. The unwound portion of the string is tangent to the circle at Q, and t is the radian measure of the angle from the positive x-axis to segment OQ. Derive the parametric equations

$$x = \cos t + t \sin t, \quad y = \sin t - t \cos t, \quad t > 0$$

of the point $P(x, y)$ for the involute.

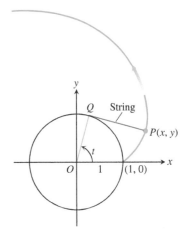

20. (*Continuation of Exercise 19.*) Find the unit tangent vector to the involute of the circle at the point $P(x, y)$.

21. Distance along a line Show that if $\mathbf{u}$ is a unit vector, then the arc length parameter along the line $\mathbf{r}(t) = P_0 + t\mathbf{u}$ from the point $P_0(x_0, y_0, z_0)$ where $t = 0$, is t itself.

22. Use Simpson's Rule with $n = 10$ to approximate the length of arc of $\mathbf{r}(t) = t\mathbf{i} + t^2\mathbf{j} + t^3\mathbf{k}$ from the origin to the point $(2, 4, 8)$.

13.4 Curvature and Normal Vectors of a Curve

In this section we study how a curve turns or bends. To gain perspective, we look first at curves in the coordinate plane. Then we consider curves in space.

Curvature of a Plane Curve

As a particle moves along a smooth curve in the plane, $\mathbf{T} = d\mathbf{r}/ds$ turns as the curve bends. Since $\mathbf{T}$ is a unit vector, its length remains constant and only its direction changes as the particle moves along the curve. The rate at which $\mathbf{T}$ turns per unit of length along the curve is called the *curvature* (Figure 13.17). The traditional symbol for the curvature function is the Greek letter κ ("kappa").

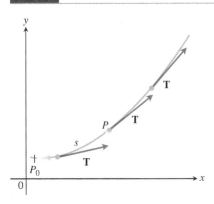

FIGURE 13.17 As P moves along the curve in the direction of increasing arc length, the unit tangent vector turns. The value of $|d\mathbf{T}/ds|$ at P is called the *curvature* of the curve at P.

> **DEFINITION** If $\mathbf{T}$ is the unit vector of a smooth curve, the **curvature** function of the curve is
> $$\kappa = \left|\frac{d\mathbf{T}}{ds}\right|.$$

If $|d\mathbf{T}/ds|$ is large, $\mathbf{T}$ turns sharply as the particle passes through P, and the curvature at P is large. If $|d\mathbf{T}/ds|$ is close to zero, $\mathbf{T}$ turns more slowly and the curvature at P is smaller.

If a smooth curve $\mathbf{r}(t)$ is already given in terms of some parameter t other than the arc length parameter s, we can calculate the curvature as

κ is the Greek letter kappa

$$\kappa = \left|\frac{d\mathbf{T}}{ds}\right| = \left|\frac{d\mathbf{T}}{dt}\frac{dt}{ds}\right| \qquad \text{Chain Rule}$$

$$= \frac{1}{|ds/dt|}\left|\frac{d\mathbf{T}}{dt}\right|$$

$$= \frac{1}{|\mathbf{v}|}\left|\frac{d\mathbf{T}}{dt}\right|. \qquad \frac{ds}{dt} = |\mathbf{v}|$$

Formula for Calculating Curvature

If $\mathbf{r}(t)$ is a smooth curve, then the curvature is the scalar function

$$\kappa = \frac{1}{|\mathbf{v}|}\left|\frac{d\mathbf{T}}{dt}\right|, \tag{1}$$

where $\mathbf{T} = \mathbf{v}/|\mathbf{v}|$ is the unit tangent vector.

Testing the definition, we see in Examples 1 and 2 below that the curvature is constant for straight lines and circles.

EXAMPLE 1   A straight line is parametrized by $\mathbf{r}(t) = \mathbf{C} + t\mathbf{v}$ for constant vectors $\mathbf{C}$ and $\mathbf{v}$. Thus, $\mathbf{r}'(t) = \mathbf{v}$, and the unit tangent vector $\mathbf{T} = \mathbf{v}/|\mathbf{v}|$ is a constant vector that always points in the same direction and has derivative $\mathbf{0}$ (Figure 13.18). It follows that, for any value of the parameter t, the curvature of the straight line is zero:

FIGURE 13.18   Along a straight line, $\mathbf{T}$ always points in the same direction. The curvature, $|d\mathbf{T}/ds|$, is zero (Example 1).

$$\kappa = \frac{1}{|\mathbf{v}|}\left|\frac{d\mathbf{T}}{dt}\right| = \frac{1}{|\mathbf{v}|}|\mathbf{0}| = 0. \qquad \blacksquare$$

EXAMPLE 2   Here we find the curvature of a circle. We begin with the parametrization

$$\mathbf{r}(t) = (a \cos t)\mathbf{i} + (a \sin t)\mathbf{j}$$

of a circle of radius a. Then,

$$\mathbf{v} = \frac{d\mathbf{r}}{dt} = -(a \sin t)\mathbf{i} + (a \cos t)\mathbf{j}$$

$$|\mathbf{v}| = \sqrt{(-a \sin t)^2 + (a \cos t)^2} = \sqrt{a^2} = |a| = a. \qquad \text{Since } |a| > 0, |a| = a.$$

From this we find

$$\mathbf{T} = \frac{\mathbf{v}}{|\mathbf{v}|} = -(\sin t)\mathbf{i} + (\cos t)\mathbf{j}$$

$$\frac{d\mathbf{T}}{dt} = -(\cos t)\mathbf{i} - (\sin t)\mathbf{j}$$

$$\left|\frac{d\mathbf{T}}{dt}\right| = \sqrt{\cos^2 t + \sin^2 t} = 1.$$

Hence, for any value of the parameter t, the curvature of the circle is

$$\kappa = \frac{1}{|\mathbf{v}|}\left|\frac{d\mathbf{T}}{dt}\right| = \frac{1}{a}(1) = \frac{1}{a} = \frac{1}{\text{radius}}.$$

Although the formula for calculating κ in Equation (1) is also valid for space curves, in the next section we find a computational formula that is usually more convenient to apply.

Among the vectors orthogonal to the unit tangent vector $\mathbf{T}$, there is one of particular significance because it points in the direction in which the curve is turning. Since $\mathbf{T}$ has constant length (because its length is always 1), the derivative $d\mathbf{T}/ds$ is orthogonal to $\mathbf{T}$ (Equation 4, Section 13.1). Therefore, if we divide $d\mathbf{T}/ds$ by its length κ, we obtain a *unit* vector $\mathbf{N}$ orthogonal to $\mathbf{T}$ (Figure 13.19).

> **DEFINITION** At a point where $\kappa \neq 0$, the **principal unit normal** vector for a smooth curve in the plane is
> $$\mathbf{N} = \frac{1}{\kappa}\frac{d\mathbf{T}}{ds}.$$

The vector $d\mathbf{T}/ds$ points in the direction in which $\mathbf{T}$ turns as the curve bends. Therefore, if we face in the direction of increasing arc length, the vector $d\mathbf{T}/ds$ points toward the right if $\mathbf{T}$ turns clockwise and toward the left if $\mathbf{T}$ turns counterclockwise. In other words, the principal normal vector $\mathbf{N}$ will point toward the concave side of the curve (Figure 13.19).

If a smooth curve $\mathbf{r}(t)$ is already given in terms of some parameter t other than the arc length parameter s, we can use the Chain Rule to calculate $\mathbf{N}$ directly.

$$\mathbf{N} = \frac{d\mathbf{T}/ds}{|d\mathbf{T}/ds|}$$
$$= \frac{(d\mathbf{T}/dt)(dt/ds)}{|d\mathbf{T}/dt||dt/ds|}$$
$$= \frac{d\mathbf{T}/dt}{|d\mathbf{T}/dt|}. \qquad \frac{dt}{ds} = \frac{1}{ds/dt} > 0 \text{ cancels.}$$

This formula enables us to find $\mathbf{N}$ without having to find κ and s first.

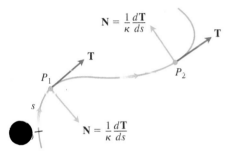

FIGURE 13.19 The vector $d\mathbf{T}/ds$, normal to the curve, always points in the direction in which $\mathbf{T}$ is turning. The unit normal vector $\mathbf{N}$ is the direction of $d\mathbf{T}/ds$.

> **Formula for Calculating N**
> If $\mathbf{r}(t)$ is a smooth curve, then the principal unit normal is
> $$\mathbf{N} = \frac{d\mathbf{T}/dt}{|d\mathbf{T}/dt|}, \qquad (2)$$
> where $\mathbf{T} = \mathbf{v}/|\mathbf{v}|$ is the unit tangent vector.

EXAMPLE 3 Find $\mathbf{T}$ and $\mathbf{N}$ for the circular motion
$$\mathbf{r}(t) = (\cos 2t)\mathbf{i} + (\sin 2t)\mathbf{j}.$$

Solution We first find $\mathbf{T}$:
$$\mathbf{v} = -(2\sin 2t)\mathbf{i} + (2\cos 2t)\mathbf{j}$$
$$|\mathbf{v}| = \sqrt{4\sin^2 2t + 4\cos^2 2t} = 2$$
$$\mathbf{T} = \frac{\mathbf{v}}{|\mathbf{v}|} = -(\sin 2t)\mathbf{i} + (\cos 2t)\mathbf{j}.$$

From this we find

$$\frac{d\mathbf{T}}{dt} = -(2 \cos 2t)\mathbf{i} - (2 \sin 2t)\mathbf{j}$$

$$\left|\frac{d\mathbf{T}}{dt}\right| = \sqrt{4 \cos^2 2t + 4 \sin^2 2t} = 2$$

and

$$\mathbf{N} = \frac{d\mathbf{T}/dt}{|d\mathbf{T}/dt|} = -(\cos 2t)\mathbf{i} - (\sin 2t)\mathbf{j}. \qquad \text{Eq. (2)}$$

Notice that $\mathbf{T} \cdot \mathbf{N} = 0$, verifying that $\mathbf{N}$ is orthogonal to $\mathbf{T}$. Notice too, that for the circular motion here, $\mathbf{N}$ points from $\mathbf{r}(t)$ toward the circle's center at the origin. ∎

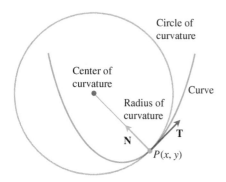

FIGURE 13.20 The center of the osculating circle at $P(x, y)$ lies toward the inner side of the curve.

Circle of Curvature for Plane Curves

The **circle of curvature** or **osculating circle** at a point P on a plane curve where $\kappa \neq 0$ is the circle in the plane of the curve that

1. is tangent to the curve at P (has the same tangent line the curve has)
2. has the same curvature the curve has at P
3. has center that lies toward the concave or inner side of the curve (as in Figure 13.20).

The **radius of curvature** of the curve at P is the radius of the circle of curvature, which, according to Example 2, is

$$\text{Radius of curvature} = \rho = \frac{1}{\kappa}.$$

To find ρ, we find κ and take the reciprocal. The **center of curvature** of the curve at P is the center of the circle of curvature.

EXAMPLE 4   Find and graph the osculating circle of the parabola $y = x^2$ at the origin.

Solution   We parametrize the parabola using the parameter $t = x$ (Section 11.1, Example 5):

$$\mathbf{r}(t) = t\mathbf{i} + t^2\mathbf{j}.$$

First we find the curvature of the parabola at the origin, using Equation (1):

$$\mathbf{v} = \frac{d\mathbf{r}}{dt} = \mathbf{i} + 2t\mathbf{j}$$

$$|\mathbf{v}| = \sqrt{1 + 4t^2}$$

so that

$$\mathbf{T} = \frac{\mathbf{v}}{|\mathbf{v}|} = (1 + 4t^2)^{-1/2}\mathbf{i} + 2t(1 + 4t^2)^{-1/2}\mathbf{j}.$$

From this we find

$$\frac{d\mathbf{T}}{dt} = -4t(1 + 4t^2)^{-3/2}\mathbf{i} + \left[2(1 + 4t^2)^{-1/2} - 8t^2(1 + 4t^2)^{-3/2}\right]\mathbf{j}.$$

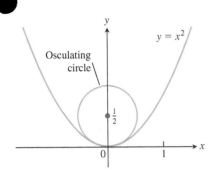

FIGURE 13.21 The osculating circle for the parabola $y = x^2$ at the origin (Example 4).

At the origin, $t = 0$, so the curvature is

$$\kappa(0) = \frac{1}{|\mathbf{v}(0)|}\left|\frac{d\mathbf{T}}{dt}(0)\right| \qquad \text{Eq. (1)}$$

$$= \frac{1}{\sqrt{1}}|0\mathbf{i} + 2\mathbf{j}|$$

$$= (1)\sqrt{0^2 + 2^2} = 2.$$

Therefore, the radius of curvature is $1/\kappa = 1/2$. At the origin we have $t = 0$ and $\mathbf{T} = \mathbf{i}$, so $\mathbf{N} = \mathbf{j}$. Thus the center of the circle is $(0, 1/2)$. The equation of the osculating circle is

$$(x - 0)^2 + \left(y - \frac{1}{2}\right)^2 = \left(\frac{1}{2}\right)^2.$$

You can see from Figure 13.21 that the osculating circle is a better approximation to the parabola at the origin than is the tangent line approximation $y = 0$. ∎

Curvature and Normal Vectors for Space Curves

If a smooth curve in space is specified by the position vector $\mathbf{r}(t)$ as a function of some parameter t, and if s is the arc length parameter of the curve, then the unit tangent vector $\mathbf{T}$ is $d\mathbf{r}/ds = \mathbf{v}/|\mathbf{v}|$. The **curvature** in space is then defined to be

$$\kappa = \left|\frac{d\mathbf{T}}{ds}\right| = \frac{1}{|\mathbf{v}|}\left|\frac{d\mathbf{T}}{dt}\right| \qquad (3)$$

just as for plane curves. The vector $d\mathbf{T}/ds$ is orthogonal to $\mathbf{T}$, and we define the **principal unit normal** to be

$$\mathbf{N} = \frac{1}{\kappa}\frac{d\mathbf{T}}{ds} = \frac{d\mathbf{T}/dt}{|d\mathbf{T}/dt|}. \qquad (4)$$

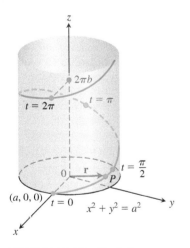

FIGURE 13.22 The helix

$$\mathbf{r}(t) = (a \cos t)\mathbf{i} + (a \sin t)\mathbf{j} + bt\mathbf{k},$$

drawn with a and b positive and $t \geq 0$ (Example 5).

EXAMPLE 5 Find the curvature for the helix (Figure 13.22)

$$\mathbf{r}(t) = (a \cos t)\mathbf{i} + (a \sin t)\mathbf{j} + bt\mathbf{k}, \qquad a, b \geq 0, \qquad a^2 + b^2 \neq 0.$$

Solution We calculate $\mathbf{T}$ from the velocity vector $\mathbf{v}$:

$$\mathbf{v} = -(a \sin t)\mathbf{i} + (a \cos t)\mathbf{j} + b\mathbf{k}$$

$$|\mathbf{v}| = \sqrt{a^2 \sin^2 t + a^2 \cos^2 t + b^2} = \sqrt{a^2 + b^2}$$

$$\mathbf{T} = \frac{\mathbf{v}}{|\mathbf{v}|} = \frac{1}{\sqrt{a^2 + b^2}}\left[-(a \sin t)\mathbf{i} + (a \cos t)\mathbf{j} + b\mathbf{k}\right].$$

Then using Equation (3),

$$\kappa = \frac{1}{|\mathbf{v}|}\left|\frac{d\mathbf{T}}{dt}\right|$$

$$= \frac{1}{\sqrt{a^2 + b^2}}\left|\frac{1}{\sqrt{a^2 + b^2}}\left[-(a \cos t)\mathbf{i} - (a \sin t)\mathbf{j}\right]\right|$$

$$= \frac{a}{a^2 + b^2}|-(\cos t)\mathbf{i} - (\sin t)\mathbf{j}|$$

$$= \frac{a}{a^2 + b^2}\sqrt{(\cos t)^2 + (\sin t)^2} = \frac{a}{a^2 + b^2}.$$

From this equation, we see that increasing b for a fixed a decreases the curvature. Decreasing a for a fixed b eventually decreases the curvature as well.

If $b = 0$, the helix reduces to a circle of radius a and its curvature reduces to $1/a$, as it should. If $a = 0$, the helix becomes the z-axis, and its curvature reduces to 0, again as it should. ∎

EXAMPLE 6   Find **N** for the helix in Example 5 and describe how the vector is pointing.

Solution We have

$$\frac{d\mathbf{T}}{dt} = -\frac{1}{\sqrt{a^2 + b^2}} \left[(a \cos t)\mathbf{i} + (a \sin t)\mathbf{j} \right] \qquad \text{Example 5}$$

$$\left| \frac{d\mathbf{T}}{dt} \right| = \frac{1}{\sqrt{a^2 + b^2}} \sqrt{a^2 \cos^2 t + a^2 \sin^2 t} = \frac{a}{\sqrt{a^2 + b^2}}$$

$$\mathbf{N} = \frac{d\mathbf{T}/dt}{|d\mathbf{T}/dt|} \qquad \text{Eq. (4)}$$

$$= -\frac{\sqrt{a^2 + b^2}}{a} \cdot \frac{1}{\sqrt{a^2 + b^2}} \left[(a \cos t)\mathbf{i} + (a \sin t)\mathbf{j} \right]$$

$$= -(\cos t)\mathbf{i} - (\sin t)\mathbf{j}.$$

Thus, **N** is parallel to the xy-plane and always points toward the z-axis. ∎

EXERCISES 13.4

Plane Curves
Find **T**, **N**, and κ for the plane curves in Exercises 1–4.

1. $\mathbf{r}(t) = t\mathbf{i} + (\ln \cos t)\mathbf{j}, \quad -\pi/2 < t < \pi/2$

2. $\mathbf{r}(t) = (\ln \sec t)\mathbf{i} + t\mathbf{j}, \quad -\pi/2 < t < \pi/2$

3. $\mathbf{r}(t) = (2t + 3)\mathbf{i} + (5 - t^2)\mathbf{j}$

4. $\mathbf{r}(t) = (\cos t + t \sin t)\mathbf{i} + (\sin t - t \cos t)\mathbf{j}, \quad t > 0$

5. **A formula for the curvature of the graph of a function in the xy-plane**

   a. The graph $y = f(x)$ in the xy-plane automatically has the parametrization $x = x, y = f(x)$, and the vector formula $\mathbf{r}(x) = x\mathbf{i} + f(x)\mathbf{j}$. Use this formula to show that if f is a twice-differentiable function of x, then

   $$\kappa(x) = \frac{|f''(x)|}{\left[1 + (f'(x))^2 \right]^{3/2}}.$$

   b. Use the formula for κ in part (a) to find the curvature of $y = \ln (\cos x), -\pi/2 < x < \pi/2$. Compare your answer with the answer in Exercise 1.

   c. Show that the curvature is zero at a point of inflection.

6. **A formula for the curvature of a parametrized plane curve**

   a. Show that the curvature of a smooth curve $\mathbf{r}(t) = f(t)\mathbf{i} + g(t)\mathbf{j}$ defined by twice-differentiable functions $x = f(t)$ and $y = g(t)$ is given by the formula

   $$\kappa = \frac{|\dot{x}\ddot{y} - \dot{y}\ddot{x}|}{(\dot{x}^2 + \dot{y}^2)^{3/2}}.$$

   The dots in the formula denote differentiation with respect to t, one derivative for each dot. Apply the formula to find the curvatures of the following curves.

   b. $\mathbf{r}(t) = t\mathbf{i} + (\ln \sin t)\mathbf{j}, \quad 0 < t < \pi$

   c. $\mathbf{r}(t) = [\tan^{-1}(\sinh t)]\mathbf{i} + (\ln \cosh t)\mathbf{j}$

7. **Normals to plane curves**

   a. Show that $\mathbf{n}(t) = -g'(t)\mathbf{i} + f'(t)\mathbf{j}$ and $-\mathbf{n}(t) = g'(t)\mathbf{i} - f'(t)\mathbf{j}$ are both normal to the curve $\mathbf{r}(t) = f(t)\mathbf{i} + g(t)\mathbf{j}$ at the point $(f(t), g(t))$.

   To obtain **N** for a particular plane curve, we can choose the one of **n** or $-\mathbf{n}$ from part (a) that points toward the concave side of the curve, and make it into a unit vector. (See Figure 13.19.) Apply this method to find **N** for the following curves.

   b. $\mathbf{r}(t) = t\mathbf{i} + e^{2t}\mathbf{j}$

   c. $\mathbf{r}(t) = \sqrt{4 - t^2}\mathbf{i} + t\mathbf{j}, \quad -2 \leq t \leq 2$

8. (*Continuation of Exercise 7.*)

   a. Use the method of Exercise 7 to find **N** for the curve $\mathbf{r}(t) = t\mathbf{i} + (1/3)t^3\mathbf{j}$ when $t < 0$; when $t > 0$.

   b. Calculate **N** for $t \neq 0$ directly from **T** using Equation (4) for the curve in part (a). Does **N** exist at $t = 0$? Graph the curve and explain what is happening to **N** as t passes from negative to positive values.

Space Curves
Find **T**, **N**, and κ for the space curves in Exercises 9–16.

9. $\mathbf{r}(t) = (3 \sin t)\mathbf{i} + (3 \cos t)\mathbf{j} + 4t\mathbf{k}$

10. $\mathbf{r}(t) = (\cos t + t \sin t)\mathbf{i} + (\sin t - t \cos t)\mathbf{j} + 3\mathbf{k}$

11. $\mathbf{r}(t) = (e^t \cos t)\mathbf{i} + (e^t \sin t)\mathbf{j} + 2\mathbf{k}$

12. $\mathbf{r}(t) = (6 \sin 2t)\mathbf{i} + (6 \cos 2t)\mathbf{j} + 5t\mathbf{k}$

13. $\mathbf{r}(t) = (t^3/3)\mathbf{i} + (t^2/2)\mathbf{j}, \quad t > 0$

14. $\mathbf{r}(t) = (\cos^3 t)\mathbf{i} + (\sin^3 t)\mathbf{j}, \quad 0 < t < \pi/2$

15. $\mathbf{r}(t) = t\mathbf{i} + (a\cosh(t/a))\mathbf{j}, \quad a > 0$

16. $\mathbf{r}(t) = (\cosh t)\mathbf{i} - (\sinh t)\mathbf{j} + t\mathbf{k}$

More on Curvature

17. Show that the parabola $y = ax^2, a \neq 0$, has its largest curvature at its vertex and has no minimum curvature. (*Note:* Since the curvature of a curve remains the same if the curve is translated or rotated, this result is true for any parabola.)

18. Show that the ellipse $x = a\cos t, y = b\sin t, a > b > 0$, has its largest curvature on its major axis and its smallest curvature on its minor axis. (As in Exercise 17, the same is true for any ellipse.)

19. **Maximizing the curvature of a helix** In Example 5, we found the curvature of the helix $\mathbf{r}(t) = (a\cos t)\mathbf{i} + (a\sin t)\mathbf{j} + bt\mathbf{k}$ $(a, b \geq 0)$ to be $\kappa = a/(a^2 + b^2)$. What is the largest value κ can have for a given value of b? Give reasons for your answer.

20. **Total curvature** We find the **total curvature** of the portion of a smooth curve that runs from $s = s_0$ to $s = s_1 > s_0$ by integrating κ from s_0 to s_1. If the curve has some other parameter, say t, then the total curvature is

$$K = \int_{s_0}^{s_1} \kappa \, ds = \int_{t_0}^{t_1} \kappa \frac{ds}{dt} dt = \int_{t_0}^{t_1} \kappa |\mathbf{v}| \, dt,$$

where t_0 and t_1 correspond to s_0 and s_1. Find the total curvatures of

a. The portion of the helix $\mathbf{r}(t) = (3\cos t)\mathbf{i} + (3\sin t)\mathbf{j} + t\mathbf{k}$, $0 \leq t \leq 4\pi$.

b. The parabola $y = x^2, -\infty < x < \infty$.

21. Find an equation for the circle of curvature of the curve $\mathbf{r}(t) = t\mathbf{i} + (\sin t)\mathbf{j}$ at the point $(\pi/2, 1)$. (The curve parametrizes the graph of $y = \sin x$ in the xy-plane.)

22. Find an equation for the circle of curvature of the curve $\mathbf{r}(t) = (2\ln t)\mathbf{i} - [t + (1/t)]\mathbf{j}, e^{-2} \leq t \leq e^2$, at the point $(0, -2)$, where $t = 1$.

T The formula

$$\kappa(x) = \frac{|f''(x)|}{[1 + (f'(x))^2]^{3/2}},$$

derived in Exercise 5, expresses the curvature $\kappa(x)$ of a twice-differentiable plane curve $y = f(x)$ as a function of x. Find the curvature function of each of the curves in Exercises 23–26. Then graph $f(x)$ together with $\kappa(x)$ over the given interval. You will find some surprises.

23. $y = x^2, \quad -2 \leq x \leq 2$ 24. $y = x^4/4, \quad -2 \leq x \leq 2$

25. $y = \sin x, \quad 0 \leq x \leq 2\pi$ 26. $y = e^x, \quad -1 \leq x \leq 2$

In Exercises 27 and 28, determine the maximum curvature for the graph of each function.

27. $f(x) = \ln x$ 28. $f(x) = \dfrac{x}{x + 1}$ for $x > -1$

29. **Osculating circle** Show that the center of the osculating circle for the parabola $y = x^2$ at the point (a, a^2) is located at

$$\left(-4a^3, 3a^2 + \frac{1}{2} \right).$$

30. **Osculating circle** Find a parametrization of the osculating circle for the parabola $y = x^2$ when $x = 1$.

COMPUTER EXPLORATIONS

In Exercises 31–38 you will use a CAS to explore the osculating circle at a point P on a plane curve where $\kappa \neq 0$. Use a CAS to perform the following steps:

a. Plot the plane curve given in parametric or function form over the specified interval to see what it looks like.

b. Calculate the curvature κ of the curve at the given value t_0 using the appropriate formula from Exercise 5 or 6. Use the parametrization $x = t$ and $y = f(t)$ if the curve is given as a function $y = f(x)$.

c. Find the unit normal vector $\mathbf{N}$ at t_0. Notice that the signs of the components of $\mathbf{N}$ depend on whether the unit tangent vector $\mathbf{T}$ is turning clockwise or counterclockwise at $t = t_0$. (See Exercise 7.)

d. If $\mathbf{C} = a\mathbf{i} + b\mathbf{j}$ is the vector from the origin to the center (a, b) of the osculating circle, find the center $\mathbf{C}$ from the vector equation

$$\mathbf{C} = \mathbf{r}(t_0) + \frac{1}{\kappa(t_0)} \mathbf{N}(t_0).$$

The point $P(x_0, y_0)$ on the curve is given by the position vector $\mathbf{r}(t_0)$.

e. Plot implicitly the equation $(x - a)^2 + (y - b)^2 = 1/\kappa^2$ of the osculating circle. Then plot the curve and osculating circle together. You may need to experiment with the size of the viewing window, but be sure the axes are equally scaled.

31. $\mathbf{r}(t) = (3\cos t)\mathbf{i} + (5\sin t)\mathbf{j}, \quad 0 \leq t \leq 2\pi, \quad t_0 = \pi/4$

32. $\mathbf{r}(t) = (\cos^3 t)\mathbf{i} + (\sin^3 t)\mathbf{j}, \quad 0 \leq t \leq 2\pi, \quad t_0 = \pi/4$

33. $\mathbf{r}(t) = t^2\mathbf{i} + (t^3 - 3t)\mathbf{j}, \quad -4 \leq t \leq 4, \quad t_0 = 3/5$

34. $\mathbf{r}(t) = (t^3 - 2t^2 - t)\mathbf{i} + \dfrac{3t}{\sqrt{1 + t^2}}\mathbf{j}, \quad -2 \leq t \leq 5, \quad t_0 = 1$

35. $\mathbf{r}(t) = (2t - \sin t)\mathbf{i} + (2 - 2\cos t)\mathbf{j}, \quad 0 \leq t \leq 3\pi,$ $t_0 = 3\pi/2$

36. $\mathbf{r}(t) = (e^{-t}\cos t)\mathbf{i} + (e^{-t}\sin t)\mathbf{j}, \quad 0 \leq t \leq 6\pi, \quad t_0 = \pi/4$

37. $y = x^2 - x, \quad -2 \leq x \leq 5, \quad x_0 = 1$

38. $y = x(1 - x)^{2/5}, \quad -1 \leq x \leq 2, \quad x_0 = 1/2$

13.5 Tangential and Normal Components of Acceleration

If you are traveling along a curve in space, the Cartesian $\mathbf{i}$, $\mathbf{j}$, and $\mathbf{k}$ coordinate system for representing the vectors describing your motion may not be very relevant to you. Instead, the vectors that represent your forward direction (the unit tangent vector $\mathbf{T}$), the direction in which your path is turning (the unit normal vector $\mathbf{N}$), and the tendency of your motion

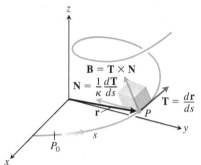

FIGURE 13.23 The **TNB** frame of mutually orthogonal unit vectors traveling along a curve in space.

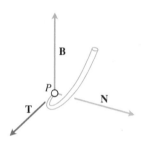

FIGURE 13.24 The vectors **T**, **N**, and **B** (in that order) make a right-handed frame of mutually orthogonal unit vectors in space.

to "twist" out of the plane created by these vectors in the direction perpendicular to this plane (defined by the *unit binormal vector* $\mathbf{B} = \mathbf{T} \times \mathbf{N}$) are likely to be more important. Expressing the acceleration vector along the curve as a linear combination of this **TNB** frame of mutually orthogonal unit vectors traveling with the motion (Figure 13.23) can reveal much about the nature of your path and your motion along it.

The TNB Frame

The **binormal vector** of a curve in space is $\mathbf{B} = \mathbf{T} \times \mathbf{N}$, which is a unit vector that is orthogonal to both **T** and **N** (Figure 13.24). Together **T**, **N**, and **B** define a moving right-handed vector frame that plays a significant role in calculating the paths of particles moving through space. It is called the **Frenet** ("fre-*nay*") **frame** (after Jean-Frédéric Frenet, 1816–1900), or the **TNB frame**.

Tangential and Normal Components of Acceleration

When an object is accelerated by gravity, brakes, or rocket motors, we often need to know how much of the acceleration acts in the direction of motion, which is the direction of the tangent vector **T**. We can calculate this using the Chain Rule to rewrite **v** as

$$\mathbf{v} = \frac{d\mathbf{r}}{dt} = \frac{d\mathbf{r}}{ds}\frac{ds}{dt} = \mathbf{T}\frac{ds}{dt}.$$

Then we differentiate both ends of this string of equalities to get

$$\mathbf{a} = \frac{d\mathbf{v}}{dt} = \frac{d}{dt}\left(\mathbf{T}\frac{ds}{dt}\right) = \frac{d^2s}{dt^2}\mathbf{T} + \frac{ds}{dt}\frac{d\mathbf{T}}{dt}$$

$$= \frac{d^2s}{dt^2}\mathbf{T} + \frac{ds}{dt}\left(\frac{d\mathbf{T}}{ds}\frac{ds}{dt}\right) = \frac{d^2s}{dt^2}\mathbf{T} + \frac{ds}{dt}\left(\kappa\mathbf{N}\frac{ds}{dt}\right) \qquad \frac{d\mathbf{T}}{ds} = \kappa\mathbf{N}$$

$$= \frac{d^2s}{dt^2}\mathbf{T} + \kappa\left(\frac{ds}{dt}\right)^2\mathbf{N}.$$

DEFINITION If the acceleration vector is written as

$$\mathbf{a} = a_T\mathbf{T} + a_N\mathbf{N}, \tag{1}$$

then

$$a_T = \frac{d^2s}{dt^2} = \frac{d}{dt}|\mathbf{v}| \qquad \text{and} \qquad a_N = \kappa\left(\frac{ds}{dt}\right)^2 = \kappa|\mathbf{v}|^2 \tag{2}$$

are the **tangential** and **normal** scalar components of acceleration.

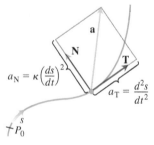

FIGURE 13.25 The tangential and normal components of acceleration. The acceleration **a** always lies in the plane of **T** and **N** and is orthogonal to **B**.

Notice that the binormal vector **B** does not appear in Equation (1). No matter how the path of the moving object we are watching may appear to twist and turn in space, the acceleration **a** *always lies in the plane of* **T** and **N** orthogonal to **B**. The equation also tells us exactly how much of the acceleration takes place tangent to the motion (d^2s/dt^2) and how much takes place normal to the motion $[\kappa(ds/dt)^2]$ (Figure 13.25).

What information can we discover from Equations (2)? By definition, acceleration **a** is the rate of change of velocity **v**, and in general, both the length and direction of **v** change as an object moves along its path. The tangential component of acceleration a_T measures the rate of change of the *length* of **v** (that is, the change in the speed). The normal component of acceleration a_N measures the rate of change of the *direction* of **v**.

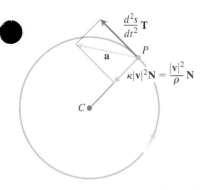

FIGURE 13.26 The tangential and normal components of the acceleration of an object that is speeding up as it moves counterclockwise around a circle of radius ρ.

Notice that the normal scalar component of the acceleration is the curvature times the *square* of the speed. This explains why you have to hold on when your car makes a sharp (large κ), high-speed (large $|\mathbf{v}|$) turn. If you double the speed of your car, you will experience four times the normal component of acceleration for the same curvature.

If an object moves in a circle at a constant speed, d^2s/dt^2 is zero and all the acceleration points along $\mathbf{N}$ toward the circle's center. If the object is speeding up or slowing down, $\mathbf{a}$ has a nonzero tangential component (Figure 13.26).

To calculate a_N, we usually use the formula $a_N = \sqrt{|\mathbf{a}|^2 - a_T{}^2}$, which comes from solving the equation $|\mathbf{a}|^2 = \mathbf{a} \cdot \mathbf{a} = a_T{}^2 + a_N{}^2$ for a_N. With this formula, we can find a_N without having to calculate κ first.

Formula for Calculating the Normal Component of Acceleration

$$a_N = \sqrt{|\mathbf{a}|^2 - a_T{}^2} \qquad (3)$$

EXAMPLE 1   Without finding $\mathbf{T}$ and $\mathbf{N}$, write the acceleration of the motion

$$\mathbf{r}(t) = (\cos t + t \sin t)\mathbf{i} + (\sin t - t \cos t)\mathbf{j}, \qquad t > 0$$

in the form $\mathbf{a} = a_T\mathbf{T} + a_N\mathbf{N}$. (The path of the motion is the involute of the circle in Figure 13.27. See also Section 13.3, Exercise 19.)

Solution   We use the first of Equations (2) to find a_T:

$$\mathbf{v} = \frac{d\mathbf{r}}{dt} = (-\sin t + \sin t + t \cos t)\mathbf{i} + (\cos t - \cos t + t \sin t)\mathbf{j}$$

$$= (t \cos t)\mathbf{i} + (t \sin t)\mathbf{j}$$

$$|\mathbf{v}| = \sqrt{t^2 \cos^2 t + t^2 \sin^2 t} = \sqrt{t^2} = |t| = t \qquad\qquad t > 0$$

$$a_T = \frac{d}{dt}|\mathbf{v}| = \frac{d}{dt}(t) = 1. \qquad\qquad\qquad \text{Eq. (2)}$$

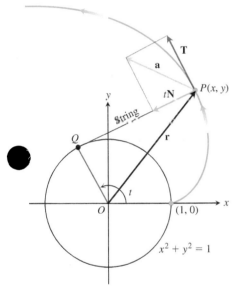

FIGURE 13.27 The tangential and normal components of the acceleration of the motion $\mathbf{r}(t) = (\cos t + t \sin t)\mathbf{i} + (\sin t - t \cos t)\mathbf{j}$, for $t > 0$. If a string wound around a fixed circle is unwound while held taut in the plane of the circle, its end P traces an involute of the circle (Example 1).

Knowing a_T, we use Equation (3) to find a_N:

$$\mathbf{a} = (\cos t - t \sin t)\mathbf{i} + (\sin t + t \cos t)\mathbf{j}$$

$$|\mathbf{a}|^2 = t^2 + 1 \qquad\qquad\qquad \text{After some algebra}$$

$$a_N = \sqrt{|\mathbf{a}|^2 - a_T{}^2}$$

$$ = \sqrt{(t^2 + 1) - (1)} = \sqrt{t^2} = t.$$

We then use Equation (1) to find $\mathbf{a}$:

$$\mathbf{a} = a_T\mathbf{T} + a_N\mathbf{N} = (1)\mathbf{T} + (t)\mathbf{N} = \mathbf{T} + t\mathbf{N}. \qquad\qquad \blacksquare$$

Torsion

How does $d\mathbf{B}/ds$ behave in relation to $\mathbf{T}$, $\mathbf{N}$, and $\mathbf{B}$? From the rule for differentiating a cross product in Section 13.1, we have

$$\frac{d\mathbf{B}}{ds} = \frac{d(\mathbf{T} \times \mathbf{N})}{ds} = \frac{d\mathbf{T}}{ds} \times \mathbf{N} + \mathbf{T} \times \frac{d\mathbf{N}}{ds}.$$

Since $\mathbf{N}$ is the direction of $d\mathbf{T}/ds$, $(d\mathbf{T}/ds) \times \mathbf{N} = \mathbf{0}$ and

$$\frac{d\mathbf{B}}{ds} = \mathbf{0} + \mathbf{T} \times \frac{d\mathbf{N}}{ds} = \mathbf{T} \times \frac{d\mathbf{N}}{ds}.$$

From this we see that $d\mathbf{B}/ds$ is orthogonal to $\mathbf{T}$, since a cross product is orthogonal to its factors.

Since $d\mathbf{B}/ds$ is also orthogonal to $\mathbf{B}$ (the latter has constant length), it follows that $d\mathbf{B}/ds$ is orthogonal to the plane of $\mathbf{B}$ and $\mathbf{T}$. In other words, $d\mathbf{B}/ds$ is parallel to $\mathbf{N}$, so $d\mathbf{B}/ds$ is a scalar multiple of $\mathbf{N}$. In symbols,

$$\frac{d\mathbf{B}}{ds} = -\tau\mathbf{N}.$$

The negative sign in this equation is traditional. The scalar τ is called the *torsion* along the curve. Notice that

$$\frac{d\mathbf{B}}{ds} \cdot \mathbf{N} = -\tau\mathbf{N} \cdot \mathbf{N} = -\tau(1) = -\tau.$$

We use this equation for our next definition.

DEFINITION   Let $\mathbf{B} = \mathbf{T} \times \mathbf{N}$. The **torsion** function of a smooth curve is

$$\tau = -\frac{d\mathbf{B}}{ds} \cdot \mathbf{N}. \qquad (4)$$

Unlike the curvature κ, which is never negative, the torsion τ may be positive, negative, or zero.

The three planes determined by $\mathbf{T}$, $\mathbf{N}$, and $\mathbf{B}$ are named and shown in Figure 13.28. The curvature $\kappa = |d\mathbf{T}/ds|$ can be thought of as the rate at which the normal plane turns as the point P moves along its path. Similarly, the torsion $\tau = -(d\mathbf{B}/ds) \cdot \mathbf{N}$ is the rate at which the osculating plane turns about $\mathbf{T}$ as P moves along the curve. Torsion measures how the curve twists.

Look at Figure 13.29. If P is a train climbing up a curved track, the rate at which the headlight turns from side to side per unit distance is the curvature of the track. The rate at which the engine tends to twist out of the plane formed by $\mathbf{T}$ and $\mathbf{N}$ is the torsion. It can be shown that a space curve is a helix if and only if it has constant nonzero curvature and constant nonzero torsion.

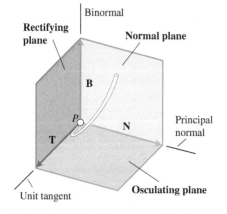

FIGURE 13.28 The names of the three planes determined by $\mathbf{T}$, $\mathbf{N}$, and $\mathbf{B}$.

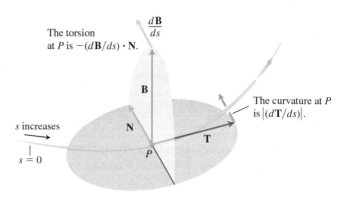

FIGURE 13.29 Every moving body travels with a **TNB** frame that characterizes the geometry of its path of motion.

Formulas for Computing Curvature and Torsion

We now give easy-to-use formulas for computing the curvature and torsion of a smooth curve. From Equations (1) and (2), we have

$$\mathbf{v} \times \mathbf{a} = \left(\frac{ds}{dt}\mathbf{T}\right) \times \left[\frac{d^2s}{dt^2}\mathbf{T} + \kappa\left(\frac{ds}{dt}\right)^2\mathbf{N}\right] \qquad \begin{array}{l}\mathbf{v} = d\mathbf{r}/dt = \\ (ds/dt)\mathbf{T}\end{array}$$

$$= \left(\frac{ds}{dt}\frac{d^2s}{dt^2}\right)(\mathbf{T} \times \mathbf{T}) + \kappa\left(\frac{ds}{dt}\right)^3(\mathbf{T} \times \mathbf{N})$$

$$= \kappa\left(\frac{ds}{dt}\right)^3\mathbf{B}. \qquad \begin{array}{l}\mathbf{T} \times \mathbf{T} = 0 \text{ and} \\ \mathbf{T} \times \mathbf{N} = \mathbf{B}\end{array}$$

It follows that

$$|\mathbf{v} \times \mathbf{a}| = \kappa\left|\frac{ds}{dt}\right|^3|\mathbf{B}| = \kappa|\mathbf{v}|^3. \qquad \frac{ds}{dt} = |\mathbf{v}| \quad \text{and} \quad |\mathbf{B}| = 1$$

Solving for κ gives the following formula.

Vector Formula for Curvature

$$\kappa = \frac{|\mathbf{v} \times \mathbf{a}|}{|\mathbf{v}|^3} \tag{5}$$

Equation (5) calculates the curvature, a geometric property of the curve, from the velocity and acceleration of any vector representation of the curve in which $|\mathbf{v}|$ is different from zero. From any formula for motion along a curve, no matter how variable the motion may be (as long as $\mathbf{v}$ is never zero), we can calculate a geometric property of the curve that seems to have nothing to do with the way the curve is parametrically defined.

The most widely used formula for torsion, derived in more advanced texts, is given in a determinant form.

Formula for Torsion

$$\tau = \frac{\begin{vmatrix} \dot{x} & \dot{y} & \dot{z} \\ \ddot{x} & \ddot{y} & \ddot{z} \\ \dddot{x} & \dddot{y} & \dddot{z} \end{vmatrix}}{|\mathbf{v} \times \mathbf{a}|^2} \quad (\text{if } \mathbf{v} \times \mathbf{a} \neq \mathbf{0}) \tag{6}$$

Newton's Dot Notation for Derivatives
The dots in Equation (6) denote differentiation with respect to t, one derivative for each dot. Thus, $\dot{x}$ ("x dot") means dx/dt, $\ddot{x}$ ("x double dot") means d^2x/dt^2, and $\dddot{x}$ ("x triple dot") means d^3x/dt^3. Similarly, $\dot{y} = dy/dt$, and so on.

This formula calculates the torsion directly from the derivatives of the component functions $x = f(t)$, $y = g(t)$, $z = h(t)$ that make up $\mathbf{r}$. The determinant's first row comes from $\mathbf{v}$, the second row comes from $\mathbf{a}$, and the third row comes from $\dot{\mathbf{a}} = d\mathbf{a}/dt$. This formula for torsion is traditionally written using Newton's dot notation for derivatives.

EXAMPLE 2   Use Equations (5) and (6) to find the curvature κ and torsion τ for the helix

$$\mathbf{r}(t) = (a\cos t)\mathbf{i} + (a\sin t)\mathbf{j} + bt\mathbf{k}, \qquad a, b \geq 0, \qquad a^2 + b^2 \neq 0.$$

Solution We calculate the curvature with Equation (5):

$$\mathbf{v} = -(a \sin t)\mathbf{i} + (a \cos t)\mathbf{j} + b\mathbf{k}$$

$$\mathbf{a} = -(a \cos t)\mathbf{i} - (a \sin t)\mathbf{j}$$

$$\mathbf{v} \times \mathbf{a} = \begin{vmatrix} \mathbf{i} & \mathbf{j} & \mathbf{k} \\ -a \sin t & a \cos t & b \\ -a \cos t & -a \sin t & 0 \end{vmatrix}$$

$$= (ab \sin t)\mathbf{i} - (ab \cos t)\mathbf{j} + a^2\mathbf{k}$$

$$\kappa = \frac{|\mathbf{v} \times \mathbf{a}|}{|\mathbf{v}|^3} = \frac{\sqrt{a^2b^2 + a^4}}{(a^2 + b^2)^{3/2}} = \frac{a\sqrt{a^2 + b^2}}{(a^2 + b^2)^{3/2}} = \frac{a}{a^2 + b^2}. \qquad (7)$$

Notice that Equation (7) agrees with the result in Example 5 in Section 13.4, where we calculated the curvature directly from its definition.

To evaluate Equation (6) for the torsion, we find the entries in the determinant by differentiating $\mathbf{r}$ with respect to t. We already have $\mathbf{v}$ and $\mathbf{a}$, and

$$\dot{\mathbf{a}} = \frac{d\mathbf{a}}{dt} = (a \sin t)\mathbf{i} - (a \cos t)\mathbf{j}.$$

Hence,

$$\tau = \frac{\begin{vmatrix} \dot{x} & \dot{y} & \dot{z} \\ \ddot{x} & \ddot{y} & \ddot{z} \\ \dddot{x} & \dddot{y} & \dddot{z} \end{vmatrix}}{|\mathbf{v} \times \mathbf{a}|^2} = \frac{\begin{vmatrix} -a \sin t & a \cos t & b \\ -a \cos t & -a \sin t & 0 \\ a \sin t & -a \cos t & 0 \end{vmatrix}}{(a\sqrt{a^2 + b^2})^2} \qquad \text{\small Value of } |\mathbf{v} \times \mathbf{a}| \text{ from Eq. (7)}$$

$$= \frac{b(a^2 \cos^2 t + a^2 \sin^2 t)}{a^2(a^2 + b^2)}$$

$$= \frac{b}{a^2 + b^2}.$$

From this last equation we see that the torsion of a helix about a circular cylinder is constant. In fact, constant curvature and constant torsion characterize the helix among all curves in space. ∎

Computation Formulas for Curves in Space

Unit tangent vector: $\qquad\qquad\qquad \mathbf{T} = \dfrac{\mathbf{v}}{|\mathbf{v}|}$

Principal unit normal vector: $\qquad \mathbf{N} = \dfrac{d\mathbf{T}/dt}{|d\mathbf{T}/dt|}$

Binormal vector: $\qquad\qquad\qquad\quad \mathbf{B} = \mathbf{T} \times \mathbf{N}$

Curvature: $\qquad\qquad\qquad\qquad\quad \kappa = \left|\dfrac{d\mathbf{T}}{ds}\right| = \dfrac{|\mathbf{v} \times \mathbf{a}|}{|\mathbf{v}|^3}$

Torsion: $\qquad\qquad\qquad\qquad\quad \tau = -\dfrac{d\mathbf{B}}{ds} \cdot \mathbf{N} = \dfrac{\begin{vmatrix} \dot{x} & \dot{y} & \dot{z} \\ \ddot{x} & \ddot{y} & \ddot{z} \\ \dddot{x} & \dddot{y} & \dddot{z} \end{vmatrix}}{|\mathbf{v} \times \mathbf{a}|^2}$

Tangential and normal scalar
components of acceleration: $\qquad \mathbf{a} = a_{\mathrm{T}}\mathbf{T} + a_{\mathrm{N}}\mathbf{N}$

$\qquad\qquad\qquad\qquad\qquad\qquad\qquad a_{\mathrm{T}} = \dfrac{d}{dt}|\mathbf{v}|$

$\qquad\qquad\qquad\qquad\qquad\qquad\qquad a_{\mathrm{N}} = \kappa|\mathbf{v}|^2 = \sqrt{|\mathbf{a}|^2 - a_{\mathrm{T}}^2}$

Finding Tangential and Normal Components

In Exercises 1 and 2, write $\mathbf{a}$ in the form $\mathbf{a} = a_\mathrm{T}\mathbf{T} + a_\mathrm{N}\mathbf{N}$ without finding $\mathbf{T}$ and $\mathbf{N}$.

1. $\mathbf{r}(t) = (a\cos t)\mathbf{i} + (a\sin t)\mathbf{j} + bt\,\mathbf{k}$

2. $\mathbf{r}(t) = (1 + 3t)\mathbf{i} + (t - 2)\mathbf{j} - 3t\,\mathbf{k}$

In Exercises 3–6, write $\mathbf{a}$ in the form $\mathbf{a} = a_\mathrm{T}\mathbf{T} + a_\mathrm{N}\mathbf{N}$ at the given value of t without finding $\mathbf{T}$ and $\mathbf{N}$.

3. $\mathbf{r}(t) = (t + 1)\mathbf{i} + 2t\mathbf{j} + t^2\mathbf{k}, \quad t = 1$

4. $\mathbf{r}(t) = (t\cos t)\mathbf{i} + (t\sin t)\mathbf{j} + t^2\mathbf{k}, \quad t = 0$

5. $\mathbf{r}(t) = t^2\mathbf{i} + (t + (1/3)t^3)\mathbf{j} + (t - (1/3)t^3)\mathbf{k}, \quad t = 0$

6. $\mathbf{r}(t) = (e^t\cos t)\mathbf{i} + (e^t\sin t)\mathbf{j} + \sqrt{2}e^t\mathbf{k}, \quad t = 0$

Finding the TNB Frame

In Exercises 7 and 8, find $\mathbf{r}$, $\mathbf{T}$, $\mathbf{N}$, and $\mathbf{B}$ at the given value of t. Then find equations for the osculating, normal, and rectifying planes at that value of t.

7. $\mathbf{r}(t) = (\cos t)\mathbf{i} + (\sin t)\mathbf{j} - \mathbf{k}, \quad t = \pi/4$

8. $\mathbf{r}(t) = (\cos t)\mathbf{i} + (\sin t)\mathbf{j} + t\,\mathbf{k}, \quad t = 0$

In Exercises 9–16 of Section 13.4, you found $\mathbf{T}$, $\mathbf{N}$, and κ. Now, in the following Exercises 9–16, find $\mathbf{B}$ and τ for these space curves.

9. $\mathbf{r}(t) = (3\sin t)\mathbf{i} + (3\cos t)\mathbf{j} + 4t\mathbf{k}$

10. $\mathbf{r}(t) = (\cos t + t\sin t)\mathbf{i} + (\sin t - t\cos t)\mathbf{j} + 3\mathbf{k}$

11. $\mathbf{r}(t) = (e^t\cos t)\mathbf{i} + (e^t\sin t)\mathbf{j} + 2\mathbf{k}$

12. $\mathbf{r}(t) = (6\sin 2t)\mathbf{i} + (6\cos 2t)\mathbf{j} + 5t\mathbf{k}$

13. $\mathbf{r}(t) = (t^3/3)\mathbf{i} + (t^2/2)\mathbf{j}, \quad t > 0$

14. $\mathbf{r}(t) = (\cos^3 t)\mathbf{i} + (\sin^3 t)\mathbf{j}, \quad 0 < t < \pi/2$

15. $\mathbf{r}(t) = t\mathbf{i} + (a\cosh(t/a))\mathbf{j}, \quad a > 0$

16. $\mathbf{r}(t) = (\cosh t)\mathbf{i} - (\sinh t)\mathbf{j} + t\mathbf{k}$

Physical Applications

17. The speedometer on your car reads a steady 35 mph. Could you be accelerating? Explain.

18. Can anything be said about the acceleration of a particle that is moving at a constant speed? Give reasons for your answer.

19. Can anything be said about the speed of a particle whose acceleration is always orthogonal to its velocity? Give reasons for your answer.

20. An object of mass m travels along the parabola $y = x^2$ with a constant speed of 10 units/sec. What is the force on the object due to its acceleration at $(0, 0)$? at $(2^{1/2}, 2)$? Write your answers in terms of $\mathbf{i}$ and $\mathbf{j}$. (Remember Newton's law, $\mathbf{F} = m\mathbf{a}$.)

Theory and Examples

21. Show that κ and τ are both zero for the line
$$\mathbf{r}(t) = (x_0 + At)\mathbf{i} + (y_0 + Bt)\mathbf{j} + (z_0 + Ct)\mathbf{k}.$$

22. Show that a moving particle will move in a straight line if the normal component of its acceleration is zero.

23. A sometime shortcut to curvature If you already know $|a_\mathrm{N}|$ and $|\mathbf{v}|$, then the formula $a_\mathrm{N} = \kappa|\mathbf{v}|^2$ gives a convenient way to find the curvature. Use it to find the curvature and radius of curvature of the curve
$$\mathbf{r}(t) = (\cos t + t\sin t)\mathbf{i} + (\sin t - t\cos t)\mathbf{j}, \quad t > 0.$$
(Take a_N and $|\mathbf{v}|$ from Example 1.)

24. What can be said about the torsion of a smooth plane curve $\mathbf{r}(t) = f(t)\mathbf{i} + g(t)\mathbf{j}$? Give reasons for your answer.

25. Differentiable curves with zero torsion lie in planes That a sufficiently differentiable curve with zero torsion lies in a plane is a special case of the fact that a particle whose velocity remains perpendicular to a fixed vector $\mathbf{C}$ moves in a plane perpendicular to $\mathbf{C}$. This, in turn, can be viewed as the following result.

Suppose $\mathbf{r}(t) = f(t)\mathbf{i} + g(t)\mathbf{j} + h(t)\mathbf{k}$ is twice differentiable for all t in an interval $[a, b]$, that $\mathbf{r} = 0$ when $t = a$, and that $\mathbf{v} \cdot \mathbf{k} = 0$ for all t in $[a, b]$. Show that $h(t) = 0$ for all t in $[a, b]$. (*Hint:* Start with $\mathbf{a} = d^2\mathbf{r}/dt^2$ and apply the initial conditions in reverse order.)

26. A formula that calculates τ from B and v If we start with the definition $\tau = -(d\mathbf{B}/ds) \cdot \mathbf{N}$ and apply the Chain Rule to rewrite $d\mathbf{B}/ds$ as
$$\frac{d\mathbf{B}}{ds} = \frac{d\mathbf{B}}{dt}\frac{dt}{ds} = \frac{d\mathbf{B}}{dt}\frac{1}{|\mathbf{v}|},$$
we arrive at the formula
$$\tau = -\frac{1}{|\mathbf{v}|}\left(\frac{d\mathbf{B}}{dt} \cdot \mathbf{N}\right).$$
Use the formula to find the torsion of the helix in Example 2.

COMPUTER EXPLORATIONS

Rounding the answers to four decimal places, use a CAS to find $\mathbf{v}$, $\mathbf{a}$, speed, $\mathbf{T}$, $\mathbf{N}$, $\mathbf{B}$, κ, τ, and the tangential and normal components of acceleration for the curves in Exercises 27–30 at the given values of t.

27. $\mathbf{r}(t) = (t\cos t)\mathbf{i} + (t\sin t)\mathbf{j} + t\mathbf{k}, \quad t = \sqrt{3}$

28. $\mathbf{r}(t) = (e^t\cos t)\mathbf{i} + (e^t\sin t)\mathbf{j} + e^t\mathbf{k}, \quad t = \ln 2$

29. $\mathbf{r}(t) = (t - \sin t)\mathbf{i} + (1 - \cos t)\mathbf{j} + \sqrt{-t}\mathbf{k}, \quad t = -3\pi$

30. $\mathbf{r}(t) = (3t - t^2)\mathbf{i} + (3t^2)\mathbf{j} + (3t + t^3)\mathbf{k}, \quad t = 1$

13.6   Velocity and Acceleration in Polar Coordinates

In this section we derive equations for velocity and acceleration in polar coordinates. These equations are useful for calculating the paths of planets and satellites in space, and we use them to examine Kepler's three laws of planetary motion.

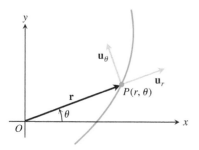

FIGURE 13.30 The length of **r** is the positive polar coordinate r of the point P. Thus, $\mathbf{u}_r$, which is $\mathbf{r}/|\mathbf{r}|$, is also $\mathbf{r}/r$. Equations (1) express $\mathbf{u}_r$ and $\mathbf{u}_\theta$ in terms of **i** and **j**.

Motion in Polar and Cylindrical Coordinates

When a particle at $P(r, \theta)$ moves along a curve in the polar coordinate plane, we express its position, velocity, and acceleration in terms of the moving unit vectors

$$\mathbf{u}_r = (\cos \theta)\mathbf{i} + (\sin \theta)\mathbf{j}, \qquad \mathbf{u}_\theta = -(\sin \theta)\mathbf{i} + (\cos \theta)\mathbf{j}, \qquad (1)$$

shown in Figure 13.30. The vector $\mathbf{u}_r$ points along the position vector $\overrightarrow{OP}$, so $\mathbf{r} = r\mathbf{u}_r$. The vector $\mathbf{u}_\theta$, orthogonal to $\mathbf{u}_r$, points in the direction of increasing θ.

We find from Equations (1) that

$$\frac{d\mathbf{u}_r}{d\theta} = -(\sin \theta)\mathbf{i} + (\cos \theta)\mathbf{j} = \mathbf{u}_\theta$$

$$\frac{d\mathbf{u}_\theta}{d\theta} = -(\cos \theta)\mathbf{i} - (\sin \theta)\mathbf{j} = -\mathbf{u}_r.$$

When we differentiate $\mathbf{u}_r$ and $\mathbf{u}_\theta$ with respect to t to find how they change with time, the Chain Rule gives

$$\dot{\mathbf{u}}_r = \frac{d\mathbf{u}_r}{d\theta}\dot{\theta} = \dot{\theta}\mathbf{u}_\theta, \qquad \dot{\mathbf{u}}_\theta = \frac{d\mathbf{u}_\theta}{d\theta}\dot{\theta} = -\dot{\theta}\mathbf{u}_r. \qquad (2)$$

Hence, we can express the velocity vector in terms of $\mathbf{u}_r$ and $\mathbf{u}_\theta$ as

$$\mathbf{v} = \dot{\mathbf{r}} = \frac{d}{dt}(r\mathbf{u}_r) = \dot{r}\mathbf{u}_r + r\dot{\mathbf{u}}_r = \dot{r}\mathbf{u}_r + r\dot{\theta}\mathbf{u}_\theta.$$

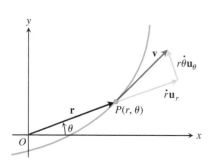

FIGURE 13.31 In polar coordinates, the velocity vector is

$$\mathbf{v} = \dot{r}\mathbf{u}_r + r\dot{\theta}\mathbf{u}_\theta.$$

See Figure 13.31. As in the previous section, we use Newton's dot notation for time derivatives to keep the formulas as simple as we can: $\dot{\mathbf{u}}_r$ means $d\mathbf{u}_r/dt$, $\dot{\theta}$ means $d\theta/dt$, and so on.

The acceleration is

$$\mathbf{a} = \dot{\mathbf{v}} = (\ddot{r}\mathbf{u}_r + \dot{r}\dot{\mathbf{u}}_r) + (\dot{r}\dot{\theta}\mathbf{u}_\theta + r\ddot{\theta}\mathbf{u}_\theta + r\dot{\theta}\dot{\mathbf{u}}_\theta).$$

When Equations (2) are used to evaluate $\dot{\mathbf{u}}_r$ and $\dot{\mathbf{u}}_\theta$ and the components are separated, the equation for acceleration in terms of $\mathbf{u}_r$ and $\mathbf{u}_\theta$ becomes

$$\mathbf{a} = (\ddot{r} - r\dot{\theta}^2)\mathbf{u}_r + (r\ddot{\theta} + 2\dot{r}\dot{\theta})\mathbf{u}_\theta.$$

To extend these equations of motion to space, we add $z\mathbf{k}$ to the right-hand side of the equation $\mathbf{r} = r\mathbf{u}_r$. Then, in these *cylindrical coordinates*, we have

Position:	$\mathbf{r} = r\mathbf{u}_r + z\mathbf{k}$
Velocity:	$\mathbf{v} = \dot{r}\mathbf{u}_r + r\dot{\theta}\mathbf{u}_\theta + \dot{z}\mathbf{k}$
Acceleration:	$\mathbf{a} = (\ddot{r} - r\dot{\theta}^2)\mathbf{u}_r + (r\ddot{\theta} + 2\dot{r}\dot{\theta})\mathbf{u}_\theta + \ddot{z}\mathbf{k}$

(3)

The vectors $\mathbf{u}_r$, $\mathbf{u}_\theta$, and **k** make a right-handed frame (Figure 13.32) in which

$$\mathbf{u}_r \times \mathbf{u}_\theta = \mathbf{k}, \qquad \mathbf{u}_\theta \times \mathbf{k} = \mathbf{u}_r, \qquad \mathbf{k} \times \mathbf{u}_r = \mathbf{u}_\theta.$$

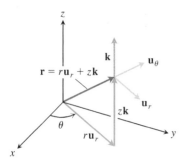

FIGURE 13.32 Position vector and basic unit vectors in cylindrical coordinates. Notice that $|\mathbf{r}| \neq r$ if $z \neq 0$ since $|\mathbf{r}| = \sqrt{r^2 + z^2}$.

Planets Move in Planes

Newton's law of gravitation says that if **r** is the radius vector from the center of a sun of mass M to the center of a planet of mass m, then the force **F** of the gravitational attraction between the planet and sun is

$$\mathbf{F} = -\frac{GmM}{|\mathbf{r}|^2}\frac{\mathbf{r}}{|\mathbf{r}|}$$

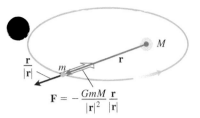

FIGURE 13.33 The force of gravity is directed along the line joining the centers of mass.

$$\mathbf{F} = -\frac{GmM}{|\mathbf{r}|^2}\frac{\mathbf{r}}{|\mathbf{r}|}$$

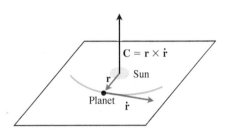

FIGURE 13.34 A planet that obeys Newton's laws of gravitation and motion travels in the plane through the sun's center of mass perpendicular to $\mathbf{C} = \mathbf{r} \times \dot{\mathbf{r}}$.

(Figure 13.33). The number G is the **universal gravitational constant**. If we measure mass in kilograms, force in newtons, and distance in meters, G is about 6.6738×10^{-11} Nm^2 kg^{-2}.

Combining the gravitation law with Newton's second law, $\mathbf{F} = m\ddot{\mathbf{r}}$, for the force acting on the planet gives

$$m\ddot{\mathbf{r}} = -\frac{GmM}{|\mathbf{r}|^2}\frac{\mathbf{r}}{|\mathbf{r}|},$$

$$\ddot{\mathbf{r}} = -\frac{GM}{|\mathbf{r}|^2}\frac{\mathbf{r}}{|\mathbf{r}|}.$$

The planet is therefore accelerated toward the sun's center of mass at all times.

Since $\ddot{\mathbf{r}}$ is a scalar multiple of $\mathbf{r}$, we have

$$\mathbf{r} \times \ddot{\mathbf{r}} = \mathbf{0}.$$

From this last equation,

$$\frac{d}{dt}(\mathbf{r} \times \dot{\mathbf{r}}) = \dot{\mathbf{r}} \times \dot{\mathbf{r}} + \mathbf{r} \times \ddot{\mathbf{r}} = \mathbf{r} \times \ddot{\mathbf{r}} = \mathbf{0}.$$

$$\underbrace{}_{()}$$

It follows that

$$\mathbf{r} \times \dot{\mathbf{r}} = \mathbf{C} \qquad (4)$$

for some constant vector $\mathbf{C}$.

Equation (4) tells us that $\mathbf{r}$ and $\dot{\mathbf{r}}$ always lie in a plane perpendicular to $\mathbf{C}$. Hence, the planet moves in a fixed plane through the center of its sun (Figure 13.34). We next see how Kepler's laws describe the motion in a precise way.

Kepler's First Law (Ellipse Law)

Kepler's first law says that a planet's path is an ellipse with the sun at one focus. The eccentricity of the ellipse is

$$e = \frac{r_0 v_0{}^2}{GM} - 1 \qquad (5)$$

and the polar equation (see Section 11.7, Equation (5)) is

$$r = \frac{(1+e)r_0}{1 + e\cos\theta}. \qquad (6)$$

Here v_0 is the speed when the planet is positioned at its minimum distance r_0 from the sun. We omit the lengthy proof. The sun's mass M is 1.99×10^{30} kg.

Kepler's Second Law (Equal Area Law)

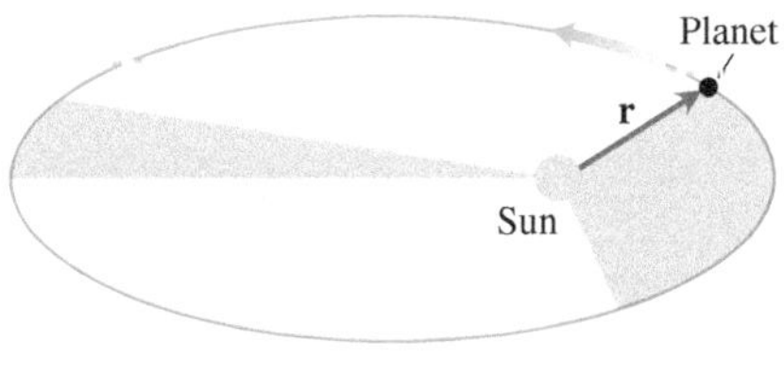

Kepler's second law says that the radius vector from the sun to a planet (the vector $\mathbf{r}$ in our model) sweeps out equal areas in equal times, as displayed in Figure 13.35. In that figure, we assume the plane of the planet is the xy-plane, so the unit vector in the direction of $\mathbf{C}$ is $\mathbf{k}$. We introduce polar coordinates in the plane, choosing as initial line $\theta = 0$, the direction $\mathbf{r}$ when $|\mathbf{r}| = r$ is a minimum value. Then at $t = 0$, we have $r(0) = r_0$ being a minimum so

$$\dot{r}|_{t=0} = \frac{dr}{dt}\bigg|_{t=0} = 0 \quad \text{and} \quad v_0 = |\mathbf{v}|_{t=0} = [r\dot{\theta}]_{t=0}. \qquad \text{Eq. (3), } \dot{z} = 0$$

FIGURE 13.35 The line joining a planet to its sun sweeps over equal areas in equal times.

To derive Kepler's second law, we use Equation (3) to evaluate the cross product $\mathbf{C} = \mathbf{r} \times \dot{\mathbf{r}}$ from Equation (4):

$$
\begin{aligned}
\mathbf{C} = \mathbf{r} \times \dot{\mathbf{r}} &= \mathbf{r} \times \mathbf{v} \\
&= r\mathbf{u}_r \times (\dot{r}\mathbf{u}_r + r\dot{\theta}\mathbf{u}_\theta) \qquad \text{Eq. (3), } \dot{z} = 0 \\
&= r\dot{r}\underbrace{(\mathbf{u}_r \times \mathbf{u}_r)}_{0} + r(r\dot{\theta})\underbrace{(\mathbf{u}_r \times \mathbf{u}_\theta)}_{\mathbf{k}} \\
&= r(r\dot{\theta})\mathbf{k}.
\end{aligned}
\tag{7}
$$

Setting t equal to zero shows that

$$\mathbf{C} = \left[r(r\dot{\theta}) \right]_{t=0} \mathbf{k} = r_0 v_0 \mathbf{k}.$$

Substituting this value for $\mathbf{C}$ in Equation (7) gives

$$r_0 v_0 \mathbf{k} = r^2 \dot{\theta}\mathbf{k}, \quad \text{or} \quad r^2\dot{\theta} = r_0 v_0.$$

This is where the area comes in. The area differential in polar coordinates is

$$dA = \frac{1}{2}r^2 \, d\theta$$

(Section 11.5). Accordingly, dA/dt has the constant value

$$\frac{dA}{dt} = \frac{1}{2}r^2\dot{\theta} = \frac{1}{2}r_0 v_0. \tag{8}$$

So dA/dt is constant, giving Kepler's second law.

HISTORICAL BIOGRAPHY
Johannes Kepler
(1571–1630)
bit.ly/2zIRwXb

Kepler's Third Law (Time–Distance Law)

The time T it takes a planet to go around its sun once is the planet's **orbital period**. *Kepler's third law* says that T and the orbit's semimajor axis a are related by the equation

$$\frac{T^2}{a^3} = \frac{4\pi^2}{GM}.$$

Since the right-hand side of this equation is constant within a given solar system, the ratio of T^2 to a^3 is *the same for every planet in the system.*

Here is a partial derivation of Kepler's third law. The area enclosed by the planet's elliptical orbit is calculated as follows:

$$
\begin{aligned}
\text{Area} &= \int_0^T dA \\
&= \int_0^T \frac{1}{2}r_0 v_0 \, dt \qquad \text{Eq. (8)} \\
&= \frac{1}{2}Tr_0 v_0.
\end{aligned}
$$

If b is the semiminor axis, the area of the ellipse is πab, so

$$T = \frac{2\pi ab}{r_0 v_0} = \frac{2\pi a^2}{r_0 v_0}\sqrt{1 - e^2}. \qquad \text{For any ellipse, } b = a\sqrt{1 - e^2}. \tag{9}$$

It remains only to express a and e in terms of r_0, v_0, G, and M. Equation (5) does this for e. For a, we observe that setting θ equal to π in Equation (6) gives

$$r_{\max} = r_0\frac{1 + e}{1 - e}.$$

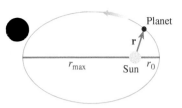

FIGURE 13.36 The length of the major axis of the ellipse is $2a = r_0 + r_{max}$.

Hence, from Figure 13.36,

$$2a = r_0 + r_{max} = \frac{2r_0}{1 - e} = \frac{2r_0 GM}{2GM - r_0 v_0^2}. \tag{10}$$

Squaring both sides of Equation (9) and substituting the results of Equations (5) and (10) produces Kepler's third law (Exercise 11).

EXERCISES 13.6

In Exercises 1–7, find the velocity and acceleration vectors in terms of $\mathbf{u}_r$ and $\mathbf{u}_\theta$.

1. $r = \theta$ and $\dfrac{d\theta}{dt} = 2$

2. $r = \dfrac{1}{\theta}$ and $\dfrac{d\theta}{dt} = t^2$

3. $r = a(1 - \cos\theta)$ and $\dfrac{d\theta}{dt} = 3$

4. $r = a\sin 2\theta$ and $\dfrac{d\theta}{dt} = 2t$

5. $r = e^{a\theta}$ and $\dfrac{d\theta}{dt} = 2$

6. $r = a(1 + \sin t)$ and $\theta = 1 - e^{-t}$

7. $r = 2\cos 4t$ and $\theta = 2t$

8. **Type of orbit**   For what values of v_0 in Equation (5) is the orbit in Equation (6) a circle? An ellipse? A parabola? A hyperbola?

9. **Circular orbits**   Show that a planet in a circular orbit moves with a constant speed. (*Hint:* This is a consequence of one of Kepler's laws.)

10. Suppose that $\mathbf{r}$ is the position vector of a particle moving along a plane curve and dA/dt is the rate at which the vector sweeps out area. Without introducing coordinates, and assuming the necessary derivatives exist, give a geometric argument based on increments and limits for the validity of the equation

$$\frac{dA}{dt} = \frac{1}{2}|\mathbf{r} \times \dot{\mathbf{r}}|.$$

11. **Kepler's third law**   Complete the derivation of Kepler's third law (the part following Equation (10)).

12. Do the data in the accompanying table support Kepler's third law? Give reasons for your answer.

Planet	Semimajor axis a $(10^{10}$ m)	Period T (years)
Mercury	5.79	0.241
Venus	10.81	0.615
Mars	22.78	1.881
Saturn	142.70	29.457

13. **Earth's major axis**   Estimate the length of the major axis of Earth's orbit if its orbital period is 365.256 days.

14. Estimate the length of the major axis of the orbit of Uranus if its orbital period is 84 years.

15. The eccentricity of Earth's orbit is $e = 0.0167$, so the orbit is nearly circular, with radius approximately 150×10^6 km. Find the rate dA/dt in units of km^2/sec satisfying Kepler's second law.

16. **Jupiter's orbital period**   Estimate the orbital period of Jupiter, assuming that $a = 77.8 \times 10^{10}$ m.

17. **Mass of Jupiter**   Io is one of the moons of Jupiter. It has a semimajor axis of 0.042×10^{10} m and an orbital period of 1.769 days. Use these data to estimate the mass of Jupiter.

18. **Distance from Earth to the moon**   The period of the moon's rotation around Earth is 2.36055×10^6 sec. Estimate the distance to the moon.

CHAPTER 13   Questions to Guide Your Review

1. State the rules for differentiating and integrating vector functions. Give examples.

2. How do you define and calculate the velocity, speed, direction of motion, and acceleration of a body moving along a sufficiently differentiable space curve? Give an example.

3. What is special about the derivatives of vector functions of constant length? Give an example.

4. What are the vector and parametric equations for ideal projectile motion? How do you find a projectile's maximum height, flight time, and range? Give examples.

5. How do you define and calculate the length of a segment of a smooth space curve? Give an example. What mathematical assumptions are involved in the definition?

6. How do you measure distance along a smooth curve in space from a preselected base point? Give an example.

7. What is a differentiable curve's unit tangent vector? Give an example.

8. Define curvature, circle of curvature (osculating circle), center of curvature, and radius of curvature for twice-differentiable curves in the plane. Give examples. What curves have zero curvature? Constant curvature?

9. What is a plane curve's principal normal vector? When is it defined? Which way does it point? Give an example.

10. How do you define $\mathbf{N}$ and κ for curves in space? How are these quantities related? Give examples.

11. What is a curve's binormal vector? Give an example. How is this vector related to the curve's torsion? Give an example.

12. What formulas are available for writing a moving object's acceleration as a sum of its tangential and normal components? Give an example. Why might one want to write the acceleration this way? What if the object moves at a constant speed? At a constant speed around a circle?

13. State Kepler's laws.

CHAPTER 13 Practice Exercises

Motion in the Plane

In Exercises 1 and 2, graph the curves and sketch their velocity and acceleration vectors at the given values of t. Then write $\mathbf{a}$ in the form $\mathbf{a} = a_T\mathbf{T} + a_N\mathbf{N}$ without finding $\mathbf{T}$ and $\mathbf{N}$, and find the value of κ at the given values of t.

1. $\mathbf{r}(t) = (4\cos t)\mathbf{i} + \left(\sqrt{2}\sin t\right)\mathbf{j}, \quad t = 0$ and $\pi/4$

2. $\mathbf{r}(t) = \left(\sqrt{3}\sec t\right)\mathbf{i} + \left(\sqrt{3}\tan t\right)\mathbf{j}, \quad t = 0$

3. The position of a particle in the plane at time t is

$$\mathbf{r} = \frac{1}{\sqrt{1 + t^2}}\mathbf{i} + \frac{t}{\sqrt{1 + t^2}}\mathbf{j}.$$

   Find the particle's highest speed.

4. Suppose $\mathbf{r}(t) = (e^t\cos t)\mathbf{i} + (e^t\sin t)\mathbf{j}$. Show that the angle between $\mathbf{r}$ and $\mathbf{a}$ never changes. What *is* the angle?

5. **Finding curvature** At point P, the velocity and acceleration of a particle moving in the plane are $\mathbf{v} = 3\mathbf{i} + 4\mathbf{j}$ and $\mathbf{a} = 5\mathbf{i} + 15\mathbf{j}$. Find the curvature of the particle's path at P.

6. Find the point on the curve $y = e^x$ where the curvature is greatest.

7. A particle moves around the unit circle in the xy-plane. Its position at time t is $\mathbf{r} = x\mathbf{i} + y\mathbf{j}$, where x and y are differentiable functions of t. Find dy/dt if $\mathbf{v}\cdot\mathbf{i} = y$. Is the motion clockwise or counterclockwise?

8. You send a message through a pneumatic tube that follows the curve $9y = x^3$ (distance in meters). At the point $(3, 3)$, $\mathbf{v}\cdot\mathbf{i} = 4$ and $\mathbf{a}\cdot\mathbf{i} = -2$. Find the values of $\mathbf{v}\cdot\mathbf{j}$ and $\mathbf{a}\cdot\mathbf{j}$ at $(3, 3)$.

9. **Characterizing circular motion** A particle moves in the plane so that its velocity and position vectors are always orthogonal. Show that the particle moves in a circle centered at the origin.

10. **Speed along a cycloid** A circular wheel with radius 1 ft and center C rolls to the right along the x-axis at a half-turn per second. (See the accompanying figure.) At time t seconds, the position vector of the point P on the wheel's circumference is

$$\mathbf{r} = (\pi t - \sin \pi t)\mathbf{i} + (1 - \cos \pi t)\mathbf{j}.$$

   a. Sketch the curve traced by P during the interval $0 \le t \le 3$.

   b. Find $\mathbf{v}$ and $\mathbf{a}$ at $t = 0, 1, 2$, and 3 and add these vectors to your sketch.

   c. At any given time, what is the forward speed of the topmost point of the wheel? Of C?

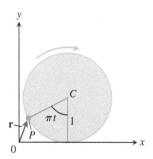

Projectile Motion

11. **Shot put** A shot leaves the thrower's hand 6.5 ft above the ground at a 45° angle at 44 ft/sec. Where is it 3 sec later?

12. **Javelin** A javelin leaves the thrower's hand 7 ft above the ground at a 45° angle at 80 ft/sec. How high does it go?

13. A golf ball is hit with an initial speed v_0 at an angle α to the horizontal from a point that lies at the foot of a straight-sided hill that is inclined at an angle ϕ to the horizontal, where

$$0 < \phi < \alpha < \frac{\pi}{2}.$$

   Show that the ball lands at a distance

$$\frac{2v_0^2 \cos \alpha}{g \cos^2 \phi} \sin (\alpha - \phi),$$

   measured up the face of the hill. Hence, show that the greatest range that can be achieved for a given v_0 occurs when $\alpha = (\phi/2) + (\pi/4)$, i.e., when the initial velocity vector bisects the angle between the vertical and the hill.

14. **Javelin** In Potsdam in 1988, Petra Felke of (then) East Germany set a women's world record by throwing a javelin 262 ft 5 in.

   a. Assuming that Felke launched the javelin at a 40° angle to the horizontal 6.5 ft above the ground, what was the javelin's initial speed?

   b. How high did the javelin go?

Motion in Space

nd the lengths of the curves in Exercises 15 and 16.

15. $\mathbf{r}(t) = (2 \cos t)\mathbf{i} + (2 \sin t)\mathbf{j} + t^2\mathbf{k}, \quad 0 \le t \le \pi/4$

16. $\mathbf{r}(t) = (3 \cos t)\mathbf{i} + (3 \sin t)\mathbf{j} + 2t^{3/2}\mathbf{k}, \quad 0 \le t \le 3$

In Exercises 17–20, find $\mathbf{T}$, $\mathbf{N}$, $\mathbf{B}$, κ, and τ at the given value of t.

17. $\mathbf{r}(t) = \frac{4}{9}(1 + t)^{3/2}\mathbf{i} + \frac{4}{9}(1 - t)^{3/2}\mathbf{j} + \frac{1}{3}t\mathbf{k}, \quad t = 0$

18. $\mathbf{r}(t) = (e^t \sin 2t)\mathbf{i} + (e^t \cos 2t)\mathbf{j} + 2e^t\mathbf{k}, \quad t = 0$

19. $\mathbf{r}(t) = t\mathbf{i} + \frac{1}{2}e^{2t}\mathbf{j}, \quad t = \ln 2$

20. $\mathbf{r}(t) = (3 \cosh 2t)\mathbf{i} + (3 \sinh 2t)\mathbf{j} + 6t\mathbf{k}, \quad t = \ln 2$

In Exercises 21 and 22, write $\mathbf{a}$ in the form $\mathbf{a} = a_\mathbf{T}\mathbf{T} + a_\mathbf{N}\mathbf{N}$ at $t = 0$ without finding $\mathbf{T}$ and $\mathbf{N}$.

21. $\mathbf{r}(t) = (2 + 3t + 3t^2)\mathbf{i} + (4t + 4t^2)\mathbf{j} - (6 \cos t)\mathbf{k}$

22. $\mathbf{r}(t) = (2 + t)\mathbf{i} + (t + 2t^2)\mathbf{j} + (1 + t^2)\mathbf{k}$

23. Find $\mathbf{T}$, $\mathbf{N}$, $\mathbf{B}$, κ, and τ as functions of t if

$$\mathbf{r}(t) = (\sin t)\mathbf{i} + \left(\sqrt{2} \cos t\right)\mathbf{j} + (\sin t)\mathbf{k}.$$

24. At what times in the interval $0 \le t \le \pi$ are the velocity and acceleration vectors of the motion $\mathbf{r}(t) = \mathbf{i} + (5 \cos t)\mathbf{j} + (3 \sin t)\mathbf{k}$ orthogonal?

25. The position of a particle moving in space at time $t \ge 0$ is

$$\mathbf{r}(t) = 2\mathbf{i} + \left(4 \sin \frac{t}{2}\right)\mathbf{j} + \left(3 - \frac{t}{\pi}\right)\mathbf{k}.$$

Find the first time $\mathbf{r}$ is orthogonal to the vector $\mathbf{i} - \mathbf{j}$.

26. Find equations for the osculating, normal, and rectifying planes of the curve $\mathbf{r}(t) = t\mathbf{i} + t^2\mathbf{j} + t^3\mathbf{k}$ at the point $(1, 1, 1)$.

27. Find parametric equations for the line that is tangent to the curve $\mathbf{r}(t) = e^t\mathbf{i} + (\sin t)\mathbf{j} + \ln(1 - t)\mathbf{k}$ at $t = 0$.

28. Find parametric equations for the line tangent to the helix $\mathbf{r}(t) = \left(\sqrt{2} \cos t\right)\mathbf{i} + \left(\sqrt{2} \sin t\right)\mathbf{j} + t\mathbf{k}$ at the point where $t = \pi/4$.

Theory and Examples

29. Synchronous curves By eliminating α from the ideal projectile equations

$$x = (v_0 \cos \alpha)t, \quad y = (v_0 \sin \alpha)t - \frac{1}{2}gt^2,$$

show that $x^2 + (y + gt^2/2)^2 = v_0^2 t^2$. This shows that projectiles launched simultaneously from the origin at the same initial speed will, at any given instant, all lie on the circle of radius $v_0 t$ centered at $(0, -gt^2/2)$, regardless of their launch angle. These circles are the *synchronous curves* of the launching.

30. Radius of curvature Show that the radius of curvature of a twice-differentiable plane curve $\mathbf{r}(t) = f(t)\mathbf{i} + g(t)\mathbf{j}$ is given by the formula

$$\rho = \frac{\dot{x}^2 + \dot{y}^2}{\sqrt{\ddot{x}^2 + \ddot{y}^2 - \ddot{s}^2}}, \quad \text{where} \quad \ddot{s} = \frac{d}{dt}\sqrt{\dot{x}^2 + \dot{y}^2}.$$

31. An alternative definition of curvature in the plane An alternative definition gives the curvature of a sufficiently differentiable plane curve to be $|d\phi/ds|$, where ϕ is the angle between $\mathbf{T}$ and $\mathbf{i}$ (Figure 13.37a). Figure 13.37b shows the distance s measured counterclockwise around the circle $x^2 + y^2 = a^2$ from the point $(a, 0)$ to a point P, along with the angle ϕ at P. Calculate the circle's curvature using the alternative definition. (*Hint:* $\phi = \theta + \pi/2$.)

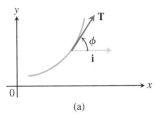

(a)

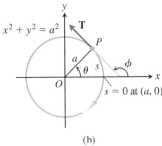

(h)

FIGURE 13.37 Figures for Exercise 31.

32. The view from *Skylab 4* What percentage of Earth's surface area could the astronauts see when *Skylab 4* was at its apogee height, 437 km above the surface? To find out, model the visible surface as the surface generated by revolving the circular arc GT, shown here, about the y-axis. Then carry out these steps:

1. Use similar triangles in the figure to show that $y_0/6380 = 6380/(6380 + 437)$. Solve for y_0.

2. To four significant digits, calculate the visible area as

$$VA = \int_{y_0}^{6380} 2\pi x \sqrt{1 + \left(\frac{dx}{dy}\right)^2}\, dy.$$

3. Express the result as a percentage of Earth's surface area.

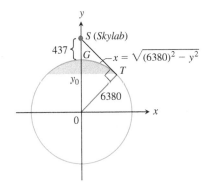

CHAPTER 13   Additional and Advanced Exercises

Applications

1. A frictionless particle P, starting from rest at time $t = 0$ at the point $(a, 0, 0)$, slides down the helix

$$\mathbf{r}(\theta) = (a \cos \theta)\mathbf{i} + (a \sin \theta)\mathbf{j} + b\theta\mathbf{k} \quad (a, b > 0)$$

under the influence of gravity, as in the accompanying figure. The θ in this equation is the cylindrical coordinate θ and the helix is the curve $r = a$, $z = b\theta$, $\theta \geq 0$, in cylindrical coordinates. We assume θ to be a differentiable function of t for the motion. The law of conservation of energy tells us that the particle's speed after it has fallen straight down a distance z is $\sqrt{2gz}$, where g is the constant acceleration of gravity.

   a. Find the angular velocity $d\theta/dt$ when $\theta = 2\pi$.

   b. Express the particle's θ- and z-coordinates as functions of t.

   c. Express the tangential and normal components of the velocity $d\mathbf{r}/dt$ and acceleration $d^2\mathbf{r}/dt^2$ as functions of t. Does the acceleration have any nonzero component in the direction of the binormal vector $\mathbf{B}$?

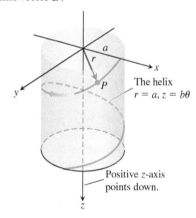

The helix
$r = a$, $z = b\theta$

Positive z-axis
points down.

2. Suppose the curve in Exercise 1 is replaced by the conical helix $r = a\theta$, $z = b\theta$ shown in the accompanying figure.

   a. Express the angular velocity $d\theta/dt$ as a function of θ.

   b. Express the distance the particle travels along the helix as a function of θ.

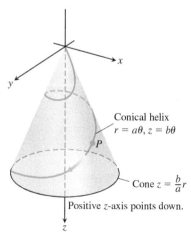

Conical helix
$r = a\theta$, $z = b\theta$

Cone $z = \dfrac{b}{a}r$

Positive z-axis points down.

Motion in Polar and Cylindrical Coordinates

3. Deduce from the orbit equation

$$r = \frac{(1 + e)r_0}{1 + e \cos \theta}$$

   that a planet is closest to its sun when $\theta = 0$ and show that $r = r_0$ at that time.

4. **A Kepler equation**   The problem of locating a planet in its orbit at a given time and date eventually leads to solving "Kepler" equations of the form

$$f(x) = x - 1 - \frac{1}{2}\sin x = 0.$$

   a. Show that this particular equation has a solution between $x = 0$ and $x = 2$.

   b. With your computer or calculator in radian mode, use Newton's method to find the solution to as many places as you can.

5. In Section 13.6, we found the velocity of a particle moving in the plane to be

$$\mathbf{v} = \dot{x}\mathbf{i} + \dot{y}\mathbf{j} = \dot{r}\mathbf{u}_r + r\dot{\theta}\mathbf{u}_\theta.$$

   a. Express $\dot{x}$ and $\dot{y}$ in terms of $\dot{r}$ and $r\dot{\theta}$ by evaluating the dot products $\mathbf{v} \cdot \mathbf{i}$ and $\mathbf{v} \cdot \mathbf{j}$.

   b. Express $\dot{r}$ and $r\dot{\theta}$ in terms of $\dot{x}$ and $\dot{y}$ by evaluating the dot products $\mathbf{v} \cdot \mathbf{u}_r$ and $\mathbf{v} \cdot \mathbf{u}_\theta$.

6. Express the curvature of a twice-differentiable curve $r = f(\theta)$ in the polar coordinate plane in terms of f and its derivatives.

7. A slender rod through the origin of the polar coordinate plane rotates (in the plane) about the origin at the rate of 3 rad/min. A beetle starting from the point $(2, 0)$ crawls along the rod toward the origin at the rate of 1 in./min.

   a. Find the beetle's acceleration and velocity in polar form when it is halfway to (1 in. from) the origin.

   b. To the nearest tenth of an inch, what will be the length of the path the beetle has traveled by the time it reaches the origin?

8. **Arc length in cylindrical coordinates**

   a. Show that when you express $ds^2 = dx^2 + dy^2 + dz^2$ in terms of cylindrical coordinates, you get $ds^2 = dr^2 + r^2\, d\theta^2 + dz^2$.

   b. Interpret this result geometrically in terms of the edges and a diagonal of a box. Sketch the box.

   c. Use the result in part (a) to find the length of the curve $r = e^\theta$, $z = e^\theta$, $0 \leq \theta \leq \ln 8$.

9. **Unit vectors for position and motion in cylindrical coordinates**   When the position of a particle moving in space is given in cylindrical coordinates, the unit vectors we use to describe its position and motion are

$$\mathbf{u}_r = (\cos \theta)\mathbf{i} + (\sin \theta)\mathbf{j}, \qquad \mathbf{u}_\theta = -(\sin \theta)\mathbf{i} + (\cos \theta)\mathbf{j},$$

and **k** (see accompanying figure). The particle's position vector is then $\mathbf{r} = r\mathbf{u}_r + z\mathbf{k}$, where r is the positive polar distance coordinate of the particle's position.

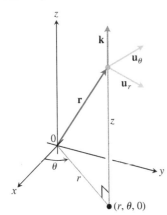

a. Show that $\mathbf{u}_r$, $\mathbf{u}_\theta$, and **k**, in this order, form a right-handed frame of unit vectors.

b. Show that

$$\frac{d\mathbf{u}_r}{d\theta} = \mathbf{u}_\theta \quad \text{and} \quad \frac{d\mathbf{u}_\theta}{d\theta} = -\mathbf{u}_r.$$

c. Assuming that the necessary derivatives with respect to t exist, express $\mathbf{v} = \dot{\mathbf{r}}$ and $\mathbf{a} = \ddot{\mathbf{r}}$ in terms of $\mathbf{u}_r$, $\mathbf{u}_\theta$, **k**, $\dot{r}$, and $\dot{\theta}$.

d. Conservation of angular momentum Let $\mathbf{r}(t)$ denote the position in space of a moving object at time t. Suppose the force acting on the object at time t is

$$\mathbf{F}(t) = -\frac{c}{|\mathbf{r}(t)|^3}\mathbf{r}(t),$$

where c is a constant. In physics the **angular momentum** of an object at time t is defined to be $\mathbf{L}(t) = \mathbf{r}(t) \times m\mathbf{v}(t)$, where m is the mass of the object and $\mathbf{v}(t)$ is the velocity. Prove that angular momentum is a conserved quantity; i.e., prove that $\mathbf{L}(t)$ is a constant vector, independent of time. Remember Newton's law $\mathbf{F} = m\mathbf{a}$. (This is a calculus problem, not a physics problem.)

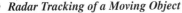

CHAPTER 13 Technology Application Projects

Mathematica/Maple Projects
Projects can be found within MyMathLab.

- *Radar Tracking of a Moving Object*
  Visualize position, velocity, and acceleration vectors to analyze motion.

- *Parametric and Polar Equations with a Figure Skater*
  Visualize position, velocity, and acceleration vectors to analyze motion.

- *Moving in Three Dimensions*
  Compute distance traveled, speed, curvature, and torsion for motion along a space curve. Visualize and compute the tangential, normal, and binormal vectors associated with motion along a space curve.

14

Partial Derivatives

OVERVIEW The volume of a right circular cylinder is a function $V = \pi r^2 h$ of its radius and its height, so it is a function $V(r, h)$ of two variables r and h. The speed of sound through seawater is primarily a function of salinity S and temperature T. The monthly payment on a home mortgage is a function of the principal borrowed P, the interest rate i, and the term t of the loan. These are examples of functions that depend on more than one independent variable.

In this chapter we extend the ideas of single-variable differential calculus to functions of several variables. Their derivatives are more varied and interesting because of the different ways the variables can interact. The applications of these derivatives are also more varied than for single-variable calculus, and in the next chapter we will see that the same is true for integrals involving several variables.

14.1 Functions of Several Variables

Real-valued functions of several independent real variables are defined analogously to functions of a single variable. Points in the domain are now ordered pairs (triples, quadruples, n-tuples) of real numbers, and values in the range are real numbers.

> **DEFINITIONS** Suppose D is a set of n-tuples of real numbers $(x_1, x_2, \ldots, x_n)$. A **real-valued function** f on D is a rule that assigns a unique (single) real number
>
> $$w = f(x_1, x_2, \ldots, x_n)$$
>
> to each element in D. The set D is the function's **domain**. The set of w-values taken on by f is the function's **range**. The symbol w is the **dependent variable** of f, and f is said to be a function of the n **independent variables** x_1 to x_n. We also call the x_j's the function's **input variables** and call w the function's **output variable**.

If f is a function of two independent variables, we usually call the independent variables x and y and the dependent variable z, and we picture the domain of f as a region in the xy-plane (Figure 14.1). If f is a function of three independent variables, we call the independent variables x, y, and z and the dependent variable w, and we picture the domain as a region in space.

In applications, we tend to use letters that remind us of what the variables stand for. To say that the volume of a right circular cylinder is a function of its radius and height, we might write $V = f(r, h)$. To be more specific, we might replace the notation $f(r, h)$ by the formula

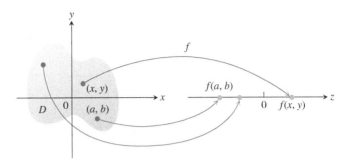

FIGURE 14.1 An arrow diagram for the function $z = f(x, y)$.

that calculates the value of V from the values of r and h, and write $V = \pi r^2 h$. In either case, r and h would be the independent variables and V the dependent variable of the function.

As usual, we evaluate functions defined by formulas by substituting the values of the independent variables in the formula and calculating the corresponding value of the dependent variable. For example, the value of $f(x, y, z) = \sqrt{x^2 + y^2 + z^2}$ at the point $(3, 0, 4)$ is

$$f(3, 0, 4) = \sqrt{(3)^2 + (0)^2 + (4)^2} = \sqrt{25} = 5.$$

Domains and Ranges

In defining a function of more than one variable, we follow the usual practice of excluding inputs that lead to complex numbers or division by zero. If $f(x, y) = \sqrt{y - x^2}$, then y cannot be less than x^2. If $f(x, y) = 1/(xy)$, then xy cannot be zero. The domain of a function is assumed to be the largest set for which the defining rule generates real numbers, unless the domain is otherwise specified explicitly. The range consists of the set of output values for the dependent variable.

EXAMPLE 1

(a) These are functions of two variables. Note the restrictions that may apply to their domains in order to obtain a real value for the dependent variable z.

Function	Domain	Range
$z = \sqrt{y - x^2}$	$y \geq x^2$	$[0, \infty)$
$z = \dfrac{1}{xy}$	$xy \neq 0$	$(-\infty, 0) \cup (0, \infty)$
$z = \sin xy$	Entire plane	$[-1, 1]$

(b) These are functions of three variables with restrictions on some of their domains.

Function	Domain	Range
$w = \sqrt{x^2 + y^2 + z^2}$	Entire space	$[0, \infty)$
$w = \dfrac{1}{x^2 + y^2 + z^2}$	$(x, y, z) \neq (0, 0, 0)$	$(0, \infty)$
$w = xy \ln z$	Half-space $z > 0$	$(-\infty, \infty)$

Functions of Two Variables

Regions in the plane can have interior points and boundary points just like intervals on the real line. Closed intervals $[a, b]$ include their boundary points, open intervals (a, b) don't include their boundary points, and intervals such as $[a, b)$ are neither open nor closed.

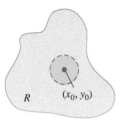

(a) Interior point

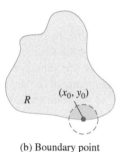

(b) Boundary point

FIGURE 14.2 Interior points and boundary points of a plane region R. An interior point is necessarily a point of R. A boundary point of R need not belong to R.

DEFINITIONS A point (x_0, y_0) in a region (set) R in the xy-plane is an **interior point** of R if it is the center of a disk of positive radius that lies entirely in R (Figure 14.2). A point (x_0, y_0) is a **boundary point** of R if every disk centered at (x_0, y_0) contains points that lie outside of R as well as points that lie in R. (The boundary point itself need not belong to R.)

The interior points of a region, as a set, make up the **interior** of the region. The region's boundary points make up its **boundary**. A region is **open** if it consists entirely of interior points. A region is **closed** if it contains all its boundary points (Figure 14.3).

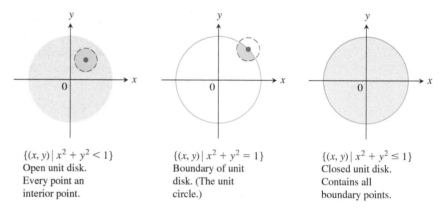

$\{(x, y) \mid x^2 + y^2 < 1\}$
Open unit disk.
Every point an
interior point.

$\{(x, y) \mid x^2 + y^2 = 1\}$
Boundary of unit
disk. (The unit
circle.)

$\{(x, y) \mid x^2 + y^2 \leq 1\}$
Closed unit disk.
Contains all
boundary points.

FIGURE 14.3 Interior points and boundary points of the unit disk in the plane.

As with a half-open interval of real numbers $[a, b)$, some regions in the plane are neither open nor closed. If you start with the open disk in Figure 14.3 and add to it some, but not all, of its boundary points, the resulting set is neither open nor closed. The boundary points that *are* there keep the set from being open. The absence of the remaining boundary points keeps the set from being closed. Two interesting examples are the empty set and the entire plane. The empty set has no interior points and no boundary points. This implies that the empty set is open (because it does not contain points that are not interior points), and at the same time it is closed (because there are no boundary points that it fails to contain). The entire xy-plane is also both open and closed: open because every point in the plane is an interior point, and closed because it has no boundary points. The empty set and the entire plane are the only subsets of the plane that are both open and closed. Other sets may be open, or closed, or neither.

DEFINITIONS A region in the plane is **bounded** if it lies inside a disk of finite radius. A region is **unbounded** if it is not bounded.

Examples of *bounded* sets in the plane include line segments, triangles, interiors of triangles, rectangles, circles, and disks. Examples of *unbounded* sets in the plane include lines, coordinate axes, the graphs of functions defined on infinite intervals, quadrants, half-planes, and the plane itself.

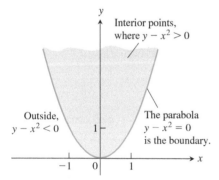

FIGURE 14.4 The domain of $f(x, y)$ in Example 2 consists of the shaded region and its bounding parabola.

EXAMPLE 2 Describe the domain of the function $f(x, y) = \sqrt{y - x^2}$.

Solution Since f is defined only where $y - x^2 \geq 0$, the domain is the closed unbounded region shown in Figure 14.4. The parabola $y = x^2$ is the boundary of the domain. The points above the parabola make up the domain's interior. ∎

Graphs, Level Curves, and Contours of Functions of Two Variables

There are two standard ways to picture the values of a function $f(x, y)$. One is to draw and label curves in the domain on which f has a constant value. The other is to sketch the surface $z = f(x, y)$ in space.

> **DEFINITIONS** The set of points in the plane where a function $f(x, y)$ has a constant value $f(x, y) = c$ is called a **level curve** of f. The set of all points $(x, y, f(x, y))$ in space, for (x, y) in the domain of f, is called the **graph** of f. The graph of f is also called the **surface** $z = f(x, y)$.

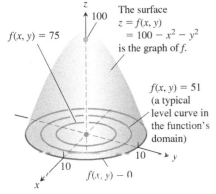

FIGURE 14.5 The graph and selected level curves of the function $f(x, y)$ in Example 3. The level curves lie in the xy-plane, which is the domain of the function $f(x, y)$.

EXAMPLE 3 Graph $f(x, y) = 100 - x^2 - y^2$ and plot the level curves $f(x, y) = 0$, $f(x, y) = 51$, and $f(x, y) = 75$ in the domain of f in the plane.

Solution The domain of f is the entire xy-plane, and the range of f is the set of real numbers less than or equal to 100. The graph is the paraboloid $z = 100 - x^2 - y^2$, the positive portion of which is shown in Figure 14.5.

The level curve $f(x, y) = 0$ is the set of points in the xy-plane at which

$$f(x, y) = 100 - x^2 - y^2 = 0, \quad \text{or} \quad x^2 + y^2 = 100,$$

which is the circle of radius 10 centered at the origin. Similarly, the level curves $f(x, y) = 51$ and $f(x, y) = 75$ (Figure 14.5) are the circles

$$f(x, y) = 100 - x^2 - y^2 = 51, \quad \text{or} \quad x^2 + y^2 = 49$$
$$f(x, y) = 100 - x^2 - y^2 = 75, \quad \text{or} \quad x^2 + y^2 = 25.$$

The level curve $f(x, y) = 100$ consists of the origin alone. (It is still a level curve.)

If $x^2 + y^2 > 100$, then the values of $f(x, y)$ are negative. For example, the circle $x^2 + y^2 = 144$, which is the circle centered at the origin with radius 12, gives the constant value $f(x, y) = -44$ and is a level curve of f. ∎

The curve in space in which the plane $z = c$ cuts a surface $z = f(x, y)$ is made up of the points that represent the function value $f(x, y) = c$. It is called the **contour curve** $f(x, y) = c$ to distinguish it from the level curve $f(x, y) = c$ in the domain of f. Figure 14.6 shows the contour curve $f(x, y) = 75$ on the surface $z = 100 - x^2 - y^2$ defined by the function $f(x, y) = 100 - x^2 - y^2$. The contour curve lies directly above the circle $x^2 + y^2 = 25$, which is the level curve $f(x, y) = 75$ in the function's domain.

The distinction between level curves and contour curves is often overlooked, and it is common to call both types of curves by the same name, relying on context to make it clear which type of curve is meant. On most maps, for example, the curves that represent constant elevation (height above sea level) are called contours, not level curves (Figure 14.7).

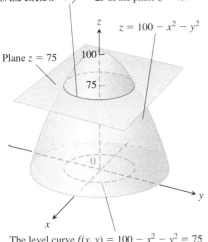

The contour curve $f(x, y) = 100 - x^2 - y^2 = 75$ is the circle $x^2 + y^2 = 25$ in the plane $z = 75$.

The level curve $f(x, y) = 100 - x^2 - y^2 = 75$ is the circle $x^2 + y^2 = 25$ in the xy-plane.

FIGURE 14.6 A plane $z = c$ parallel to the xy-plane intersecting a surface $z = f(x, y)$ produces a contour curve.

Functions of Three Variables

In the plane, the points where a function of two independent variables has a constant value $f(x, y) = c$ make a curve in the function's domain. In space, the points where a function of three independent variables has a constant value $f(x, y, z) = c$ make a surface in the function's domain.

> **DEFINITION** The set of points (x, y, z) in space where a function of three independent variables has a constant value $f(x, y, z) = c$ is called a **level surface** of f.

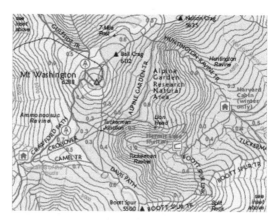

FIGURE 14.7 Contours on Mt. Washington in New Hampshire. (Reprinted by permission of the Appalachian Mountain Club.)

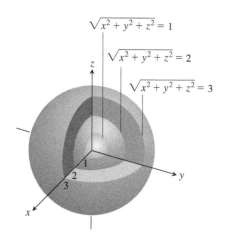

$$\sqrt{x^2 + y^2 + z^2} = 1$$
$$\sqrt{x^2 + y^2 + z^2} = 2$$
$$\sqrt{x^2 + y^2 + z^2} = 3$$

FIGURE 14.8 The level surfaces of $f(x, y, z) = \sqrt{x^2 + y^2 + z^2}$ are concentric spheres (Example 4).

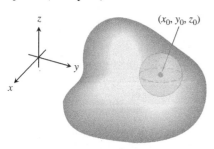

(a) Interior point

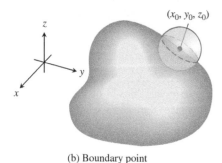

(b) Boundary point

FIGURE 14.9 Interior points and boundary points of a region in space. As with regions in the plane, a boundary point need not belong to the space region R.

Since the graphs of functions of three variables consist of points $(x, y, z, f(x, y, z))$ lying in a four-dimensional space, we cannot sketch them effectively in our three-dimensional frame of reference. We can see how the function behaves, however, by looking at its three-dimensional level surfaces.

EXAMPLE 4 Describe the level surfaces of the function

$$f(x, y, z) = \sqrt{x^2 + y^2 + z^2}.$$

Solution The value of f is the distance from the origin to the point (x, y, z). Each level surface $\sqrt{x^2 + y^2 + z^2} = c$, $c > 0$, is a sphere of radius c centered at the origin. Figure 14.8 shows a cutaway view of three of these spheres. The level surface $\sqrt{x^2 + y^2 + z^2} = 0$ consists of the origin alone.

We are not graphing the function here; we are looking at level surfaces in the function's domain. The level surfaces show how the function's values change as we move through its domain. If we remain on a sphere of radius c centered at the origin, the function maintains a constant value, namely c. If we move from a point on one sphere to a point on another, the function's value changes. It increases if we move away from the origin and decreases if we move toward the origin. The way the values change depends on the direction we take. The dependence of change on direction is important. We return to it in Section 14.5. ∎

The definitions of interior, boundary, open, closed, bounded, and unbounded for regions in space are similar to those for regions in the plane. To accommodate the extra dimension, we use solid balls of positive radius instead of disks.

DEFINITIONS A point (x_0, y_0, z_0) in a region R in space is an **interior point** of R if it is the center of a solid ball that lies entirely in R (Figure 14.9a). A point (x_0, y_0, z_0) is a **boundary point** of R if every solid ball centered at (x_0, y_0, z_0) contains points that lie outside of R as well as points that lie inside R (Figure 14.9b). The **interior** of R is the set of interior points of R. The **boundary** of R is the set of boundary points of R.

A region is **open** if it consists entirely of interior points. A region is **closed** if it contains its entire boundary.

Examples of *open* sets in space include the interior of a sphere, the open half-space $z > 0$, the first octant (where x, y, and z are all positive), and space itself. Examples of *closed* sets in space include lines, planes, and the closed half-space $z \geq 0$. A solid sphere with part of its boundary removed or a solid cube with a missing face, edge, or corner point is *neither open nor closed*.

Functions of more than three independent variables are also important. For example, the temperature on a surface in space may depend not only on the location of the point $P(x, y, z)$ on the surface but also on the time t when it is visited, so we would write $T = f(x, y, z, t)$.

Computer Graphing

Three-dimensional graphing software makes it possible to graph functions of two variables. We can often get information more quickly from a graph than from a formula, since the surfaces reveal increasing and decreasing behavior, and high points or low points.

EXAMPLE 5   The temperature w beneath the Earth's surface is a function of the depth x beneath the surface and the time t of the year. If we measure x in feet and t as the number of days elapsed from the expected date of the yearly highest surface temperature, we can model the variation in temperature with the function

$$w = \cos(1.7 \times 10^{-2}t - 0.2x)e^{-0.2x}.$$

(The temperature at 0 ft is scaled to vary from $+1$ to -1, so that the variation at x feet can be interpreted as a fraction of the variation at the surface.)

Figure 14.10 shows a graph of the function. At a depth of 15 ft, the variation (change in vertical amplitude in the figure) is about 5% of the surface variation. At 25 ft, there is almost no variation during the year.

The graph also shows that the temperature 15 ft below the surface is about half a year out of phase with the surface temperature. When the temperature is lowest on the surface (late January, say), it is at its highest 15 ft below. Fifteen feet below the ground, the seasons are reversed. ∎

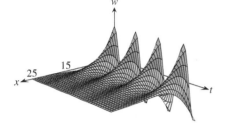

FIGURE 14.10   This graph shows the seasonal variation of the temperature below ground as a fraction of surface temperature (Example 5).

Figure 14.11 shows computer-generated graphs of a number of functions of two variables together with their level curves.

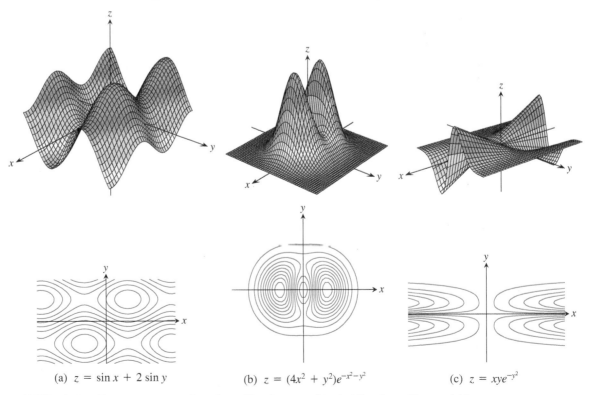

(a) $z = \sin x + 2 \sin y$

(b) $z = (4x^2 + y^2)e^{-x^2-y^2}$

(c) $z = xye^{-y^2}$

FIGURE 14.11   Computer-generated graphs and level curves of typical functions of two variables.

Domain, Range, and Level Curves

In Exercises 1–4, find the specific function values.

1. $f(x, y) = x^2 + xy^3$

   a. $f(0, 0)$ **b.** $f(-1, 1)$

   c. $f(2, 3)$ **d.** $f(-3, -2)$

2. $f(x, y) = \sin(xy)$

   a. $f\left(2, \dfrac{\pi}{6}\right)$ **b.** $f\left(-3, \dfrac{\pi}{12}\right)$

   c. $f\left(\pi, \dfrac{1}{4}\right)$ **d.** $f\left(-\dfrac{\pi}{2}, -7\right)$

3. $f(x, y, z) = \dfrac{x - y}{y^2 + z^2}$

   a. $f(3, -1, 2)$ **b.** $f\left(1, \dfrac{1}{2}, -\dfrac{1}{4}\right)$

   c. $f\left(0, -\dfrac{1}{3}, 0\right)$ **d.** $f(2, 2, 100)$

4. $f(x, y, z) = \sqrt{49 - x^2 - y^2 - z^2}$

   a. $f(0, 0, 0)$ **b.** $f(2, -3, 6)$

   c. $f(-1, 2, 3)$ **d.** $f\left(\dfrac{4}{\sqrt{2}}, \dfrac{5}{\sqrt{2}}, \dfrac{6}{\sqrt{2}}\right)$

In Exercises 5–12, find and sketch the domain for each function.

5. $f(x, y) = \sqrt{y - x - 2}$

6. $f(x, y) = \ln(x^2 + y^2 - 4)$

7. $f(x, y) = \dfrac{(x - 1)(y + 2)}{(y - x)(y - x^3)}$

8. $f(x, y) = \dfrac{\sin(xy)}{x^2 + y^2 - 25}$

9. $f(x, y) = \cos^{-1}(y - x^2)$

10. $f(x, y) = \ln(xy + x - y - 1)$

11. $f(x, y) = \sqrt{(x^2 - 4)(y^2 - 9)}$

12. $f(x, y) = \dfrac{1}{\ln(4 - x^2 - y^2)}$

In Exercises 13–16, find and sketch the level curves $f(x, y) = c$ on the same set of coordinate axes for the given values of c. We refer to these level curves as a contour map.

13. $f(x, y) = x + y - 1, \quad c = -3, -2, -1, 0, 1, 2, 3$

14. $f(x, y) = x^2 + y^2, \quad c = 0, 1, 4, 9, 16, 25$

15. $f(x, y) = xy, \quad c = -9, -4, -1, 0, 1, 4, 9$

16. $f(x, y) = \sqrt{25 - x^2 - y^2}, \quad c = 0, 1, 2, 3, 4$

In Exercises 17–30, **(a)** find the function's domain, **(b)** find the function's range, **(c)** describe the function's level curves, **(d)** find the boundary of the function's domain, **(e)** determine if the domain is an open region, a closed region, or neither, and **(f)** decide if the domain is bounded or unbounded.

17. $f(x, y) = y - x$ **18.** $f(x, y) = \sqrt{y - x}$

19. $f(x, y) = 4x^2 + 9y^2$ **20.** $f(x, y) = x^2 - y^2$

21. $f(x, y) = xy$ **22.** $f(x, y) = y/x^2$

23. $f(x, y) = \dfrac{1}{\sqrt{16 - x^2 - y^2}}$ **24.** $f(x, y) = \sqrt{9 - x^2 - y^2}$

25. $f(x, y) = \ln(x^2 + y^2)$ **26.** $f(x, y) = e^{-(x^2+y^2)}$

27. $f(x, y) = \sin^{-1}(y - x)$ **28.** $f(x, y) = \tan^{-1}\left(\dfrac{y}{x}\right)$

29. $f(x, y) = \ln(x^2 + y^2 - 1)$ **30.** $f(x, y) = \ln(9 - x^2 - y^2)$

Matching Surfaces with Level Curves

Exercises 31–36 show level curves for six functions. The graphs of these functions are given on the next page (items a–f), as are their equations (items g–l). Match each set of level curves with the appropriate graph and appropriate equation.

31.

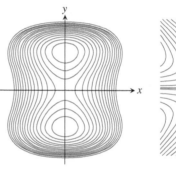

32.

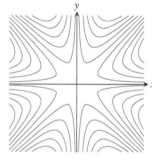

33.

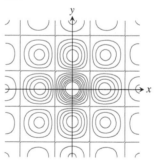

34.

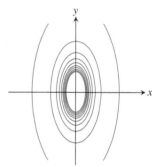

35.

36.

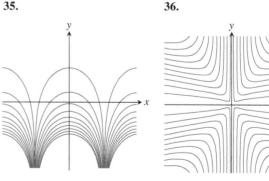

a.

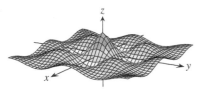

b.

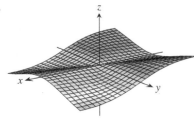

c.

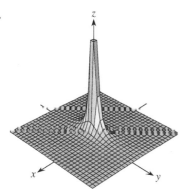

d.

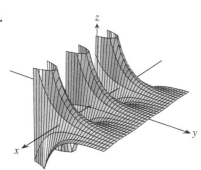

e.

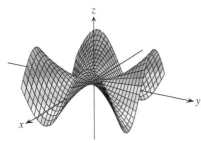

f.

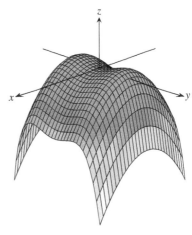

g. $z = -\dfrac{xy^2}{x^2 + y^2}$ **h.** $z = y^2 - y^4 - x^2$

i. $z = (\cos x)(\cos y)\, e^{-\sqrt{x^2+y^2}/4}$

j. $z = e^{-y}\cos x$ **k.** $z = \dfrac{1}{4x^2 + y^2}$

l. $z = \dfrac{xy(x^2 - y^2)}{x^2 + y^2}$

Functions of Two Variables

Display the values of the functions in Exercises 37–48 in two ways: (a) by sketching the surface $z = f(x, y)$ and (b) by drawing an assortment of level curves in the function's domain. Label each level curve with its function value.

37. $f(x, y) = y^2$ **38.** $f(x, y) = \sqrt{x}$

39. $f(x, y) = x^2 + y^2$ **40.** $f(x, y) = \sqrt{x^2 + y^2}$

41. $f(x, y) = x^2 - y$ **42.** $f(x, y) = 4 - x^2 - y^2$

43. $f(x, y) = 4x^2 + y^2$ **44.** $f(x, y) = 6 - 2x - 3y$

45. $f(x, y) = 1 - |y|$ **46.** $f(x, y) = 1 - |x| - |y|$

47. $f(x, y) = \sqrt{x^2 + y^2 + 4}$ **48.** $f(x, y) = \sqrt{x^2 + y^2 - 4}$

Finding Level Curves

In Exercises 49–52, find an equation for and sketch the graph of the level curve of the function $f(x, y)$ that passes through the given point.

49. $f(x, y) = 16 - x^2 - y^2, \quad \left(2\sqrt{2}, \sqrt{2}\right)$

50. $f(x, y) = \sqrt{x^2 - 1}, \quad (1, 0)$

51. $f(x, y) = \sqrt{x + y^2 - 3}, \quad (3, -1)$

52. $f(x, y) = \dfrac{2y - x}{x + y + 1}, \quad (-1, 1)$

Sketching Level Surfaces

In Exercises 53–60, sketch a typical level surface for the function.

53. $f(x, y, z) = x^2 + y^2 + z^2$ **54.** $f(x, y, z) = \ln(x^2 + y^2 + z^2)$

55. $f(x, y, z) = x + z$ **56.** $f(x, y, z) = z$

57. $f(x, y, z) = x^2 + y^2$ **58.** $f(x, y, z) = y^2 + z^2$

59. $f(x, y, z) = z - x^2 - y^2$

60. $f(x, y, z) = (x^2/25) + (y^2/16) + (z^2/9)$

Finding Level Surfaces

In Exercises 61–64, find an equation for the level surface of the function through the given point.

61. $f(x, y, z) = \sqrt{x - y} - \ln z$, $(3, -1, 1)$

62. $f(x, y, z) = \ln(x^2 + y + z^2)$, $(-1, 2, 1)$

63. $g(x, y, z) = \sqrt{x^2 + y^2 + z^2}$, $\left(1, -1, \sqrt{2}\right)$

64. $g(x, y, z) = \dfrac{x - y + z}{2x + y - z}$, $(1, 0, -2)$

In Exercises 65–68, find and sketch the domain of f. Then find an equation for the level curve or surface of the function passing through the given point.

65. $f(x, y) = \displaystyle\sum_{n=0}^{\infty} \left(\dfrac{x}{y}\right)^n$, $(1, 2)$

66. $g(x, y, z) = \displaystyle\sum_{n=0}^{\infty} \dfrac{(x + y)^n}{n! z^n}$, $(\ln 4, \ln 9, 2)$

67. $f(x, y) = \displaystyle\int_x^y \dfrac{d\theta}{\sqrt{1 - \theta^2}}$, $(0, 1)$

68. $g(x, y, z) = \displaystyle\int_x^y \dfrac{dt}{1 + t^2} + \int_0^z \dfrac{d\theta}{\sqrt{4 - \theta^2}}$, $\left(0, 1, \sqrt{3}\right)$

COMPUTER EXPLORATIONS

Use a CAS to perform the following steps for each of the functions in Exercises 69–72.

a. Plot the surface over the given rectangle.

b. Plot several level curves in the rectangle.

c. Plot the level curve of f through the given point.

69. $f(x, y) = x \sin \dfrac{y}{2} + y \sin 2x$, $0 \le x \le 5\pi$, $0 \le y \le 5\pi$, $P(3\pi, 3\pi)$

70. $f(x, y) = (\sin x)(\cos y) e^{\sqrt{x^2 + y^2}/8}$, $0 \le x \le 5\pi$, $0 \le y \le 5\pi$, $P(4\pi, 4\pi)$

71. $f(x, y) = \sin(x + 2 \cos y)$, $-2\pi \le x \le 2\pi$, $-2\pi \le y \le 2\pi$, $P(\pi, \pi)$

72. $f(x, y) = e^{(x^{0.1} - y)} \sin(x^2 + y^2)$, $0 \le x \le 2\pi$, $-2\pi \le y \le \pi$, $P(\pi, -\pi)$

Use a CAS to plot the implicitly defined level surfaces in Exercises 73–76.

73. $4 \ln(x^2 + y^2 + z^2) = 1$ **74.** $x^2 + z^2 = 1$

75. $x + y^2 - 3z^2 = 1$

76. $\sin\left(\dfrac{x}{2}\right) - (\cos y)\sqrt{x^2 + z^2} = 2$

Parametrized Surfaces   Just as you describe curves in the plane parametrically with a pair of equations $x = f(t), y = g(t)$ defined on some parameter interval I, you can sometimes describe surfaces in space with a triple of equations $x = f(u, v), y = g(u, v), z = h(u, v)$ defined on some parameter rectangle $a \le u \le b, c \le v \le d$. Many computer algebra systems permit you to plot such surfaces in *parametric mode*. (Parametrized surfaces are discussed in detail in Section 16.5.) Use a CAS to plot the surfaces in Exercises 77–80. Also plot several level curves in the *xy*-plane.

77. $x = u \cos v$, $y = u \sin v$, $z = u$, $0 \le u \le 2$, $0 \le v \le 2\pi$

78. $x = u \cos v$, $y = u \sin v$, $z = v$, $0 \le u \le 2$, $0 \le v \le 2\pi$

79. $x = (2 + \cos u) \cos v$, $y = (2 + \cos u) \sin v$, $z = \sin u$, $0 \le u \le 2\pi$, $0 \le v \le 2\pi$

80. $x = 2 \cos u \cos v$, $y = 2 \cos u \sin v$, $z = 2 \sin u$, $0 \le u \le 2\pi$, $0 \le v \le \pi$

14.2 Limits and Continuity in Higher Dimensions

In this section we develop limits and continuity for multivariable functions. The theory is similar to that developed for single-variable functions, but since we now have more than one independent variable, there is additional complexity that requires some new ideas.

Limits for Functions of Two Variables

If the values of $f(x, y)$ lie arbitrarily close to a fixed real number L for all points (x, y) sufficiently close to a point (x_0, y_0), we say that f approaches the limit L as (x, y) approaches (x_0, y_0). This is similar to the informal definition for the limit of a function of a single variable. Notice, however, that when (x_0, y_0) lies in the interior of f's domain, (x, y) can approach (x_0, y_0) from any direction, not just from the left or the right. For the limit to exist, the same limiting value must be obtained whatever direction of approach is taken. We illustrate this issue in several examples following the definition.

DEFINITION   We say that a function $f(x, y)$ approaches the **limit** L as (x, y) approaches (x_0, y_0), and write

$$\lim_{(x, y) \to (x_0, y_0)} f(x, y) = L$$

if, for every number $\varepsilon > 0$, there exists a corresponding number $\delta > 0$ such that for all (x, y) in the domain of f,

$$\left| f(x, y) - L \right| < \varepsilon \qquad \text{whenever} \qquad 0 < \sqrt{(x - x_0)^2 + (y - y_0)^2} < \delta.$$

The definition of limit says that the distance between $f(x, y)$ and L becomes arbitrarily small whenever the distance from (x, y) to (x_0, y_0) is made sufficiently small (but not 0). The definition applies to interior points (x_0, y_0) as well as boundary points of the domain of f, although a boundary point need not lie within the domain. The points (x, y) that approach (x_0, y_0) are always taken to be in the domain of f. See Figure 14.12.

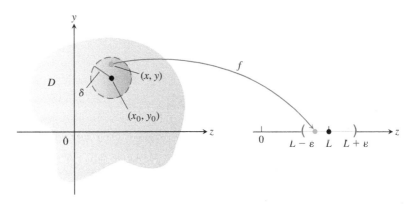

FIGURE 14.12   In the limit definition, δ is the radius of a disk centered at (x_0, y_0). For all points (x, y) within this disk, the function values $f(x, y)$ lie inside the corresponding interval $(L - \varepsilon, L + \varepsilon)$.

As for functions of a single variable, it can be shown that

$$\lim_{(x, y) \to (x_0, y_0)} x = x_0$$

$$\lim_{(x, y) \to (x_0, y_0)} y = y_0$$

$$\lim_{(x, y) \to (x_0, y_0)} k = k \qquad \text{(any number } k\text{)}.$$

For example, in the first limit statement above, $f(x, y) = x$ and $L = x_0$. Using the definition of limit, suppose that $\varepsilon > 0$ is chosen. If we let δ equal this ε, we see that if

$$0 < \sqrt{(x - x_0)^2 + (y - y_0)^2} < \delta = \varepsilon,$$

then

$$\sqrt{(x - x_0)^2} < \delta \qquad {\scriptstyle (x - x_0)^2 \,\le\, (x - x_0)^2 + (y - y_0)^2}$$

$$\left| x - x_0 \right| < \varepsilon \qquad {\scriptstyle \sqrt{a^2} = |a|}$$

$$\left| f(x, y) - x_0 \right| < \varepsilon. \qquad {\scriptstyle x = f(x, y)}$$

That is,

$$\left| f(x, y) - x_0 \right| < \varepsilon \qquad \text{whenever} \qquad 0 < \sqrt{(x - x_0)^2 + (y - y_0)^2} < \delta.$$

So a δ has been found satisfying the requirement of the definition, and therefore we have proved that

$$\lim_{(x, y) \to (x_0, y_0)} f(x, y) = \lim_{(x, y) \to (x_0, y_0)} x = x_0.$$

As with single-variable functions, the limit of the sum of two functions is the sum of their limits (when they both exist), with similar results for the limits of the differences, constant multiples, products, quotients, powers, and roots. These facts are summarized in Theorem 1.

THEOREM 1—Properties of Limits of Functions of Two Variables
The following rules hold if L, M, and k are real numbers and

$$\lim_{(x,y)\to(x_0,y_0)} f(x,y) = L \quad \text{and} \quad \lim_{(x,y)\to(x_0,y_0)} g(x,y) = M.$$

1. *Sum Rule:* $\lim_{(x,y)\to(x_0,y_0)} (f(x,y) + g(x,y)) = L + M$

2. *Difference Rule:* $\lim_{(x,y)\to(x_0,y_0)} (f(x,y) - g(x,y)) = L - M$

3. *Constant Multiple Rule:* $\lim_{(x,y)\to(x_0,y_0)} kf(x,y) = kL$ (any number k)

4. *Product Rule:* $\lim_{(x,y)\to(x_0,y_0)} (f(x,y) \cdot g(x,y)) = L \cdot M$

5. *Quotient Rule:* $\lim_{(x,y)\to(x_0,y_0)} \dfrac{f(x,y)}{g(x,y)} = \dfrac{L}{M}, \quad M \neq 0$

6. *Power Rule:* $\lim_{(x,y)\to(x_0,y_0)} [f(x,y)]^n = L^n$, n a positive integer

7. *Root Rule:* $\lim_{(x,y)\to(x_0,y_0)} \sqrt[n]{f(x,y)} = \sqrt[n]{L} = L^{1/n}$,

n a positive integer, and if n is even, we assume that $L > 0$.

Although we will not prove Theorem 1 here, we give an informal discussion of why it is true. If (x, y) is sufficiently close to (x_0, y_0), then $f(x, y)$ is close to L and $g(x, y)$ is close to M (from the informal interpretation of limits). It is then reasonable that $f(x, y) + g(x, y)$ is close to $L + M$; $f(x, y) - g(x, y)$ is close to $L - M$; $kf(x, y)$ is close to kL; $f(x, y)g(x, y)$ is close to LM; and $f(x, y)/g(x, y)$ is close to L/M if $M \neq 0$.

When we apply Theorem 1 to polynomials and rational functions, we obtain the useful result that the limits of these functions as $(x, y) \to (x_0, y_0)$ can be calculated by evaluating the functions at (x_0, y_0). The only requirement is that the rational functions be defined at (x_0, y_0).

EXAMPLE 1   In this example, we can combine the three simple results following the limit definition with the results in Theorem 1 to calculate the limits. We simply substitute the x- and y-values of the point being approached into the functional expression to find the limiting value.

(a) $\lim_{(x,y)\to(0,1)} \dfrac{x - xy + 3}{x^2 y + 5xy - y^3} = \dfrac{0 - (0)(1) + 3}{(0)^2(1) + 5(0)(1) - (1)^3} = -3$

(b) $\lim_{(x,y)\to(3,-4)} \sqrt{x^2 + y^2} = \sqrt{(3)^2 + (-4)^2} = \sqrt{25} = 5$ ∎

EXAMPLE 2   Find $\lim_{(x,y)\to(0,0)} \dfrac{x^2 - xy}{\sqrt{x} - \sqrt{y}}$.

Solution Since the denominator $\sqrt{x} - \sqrt{y}$ approaches 0 as $(x, y) \to (0, 0)$, we cannot use the Quotient Rule from Theorem 1. If we multiply numerator and denominator by $\sqrt{x} + \sqrt{y}$, however, we produce an equivalent fraction whose limit we *can* find:

$$\lim_{(x, y) \to (0,0)} \frac{x^2 - xy}{\sqrt{x} - \sqrt{y}} = \lim_{(x, y) \to (0,0)} \frac{(x^2 - xy)(\sqrt{x} + \sqrt{y})}{(\sqrt{x} - \sqrt{y})(\sqrt{x} + \sqrt{y})} \qquad \text{Multiply by a form equal to 1.}$$

$$= \lim_{(x, y) \to (0,0)} \frac{x(x - y)(\sqrt{x} + \sqrt{y})}{x - y} \qquad \text{Algebra}$$

$$= \lim_{(x, y) \to (0,0)} x(\sqrt{x} + \sqrt{y}) \qquad \text{Cancel the nonzero factor } (x - y).$$

$$= 0(\sqrt{0} + \sqrt{0}) = 0 \qquad \text{Known limit values}$$

We can cancel the factor $(x - y)$ because the path $y = x$ (where we would have $x - y = 0$) is *not* in the domain of the function

$$f(x, y) = \frac{x^2 - xy}{\sqrt{x} - \sqrt{y}}. \qquad \blacksquare$$

EXAMPLE 3 Find $\displaystyle\lim_{(x, y) \to (0,0)} \frac{4xy^2}{x^2 + y^2}$ if it exists.

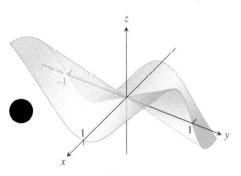

FIGURE 14.13 The surface graph shows the limit of the function in Example 3 must be 0, if it exists.

Solution We first observe that along the line $x = 0$, the function always has value 0 when $y \neq 0$. Likewise, along the line $y = 0$, the function has value 0 provided $x \neq 0$. So if the limit does exist as (x, y) approaches $(0, 0)$, the value of the limit must be 0 (see Figure 14.13). To see if this is true, we apply the definition of limit.

Let $\varepsilon > 0$ be given, but arbitrary. We want to find a $\delta > 0$ such that

$$\left| \frac{4xy^2}{x^2 + y^2} - 0 \right| < \varepsilon \qquad \text{whenever} \qquad 0 < \sqrt{x^2 + y^2} < \delta$$

or

$$\frac{4|x|y^2}{x^2 + y^2} < \varepsilon \qquad \text{whenever} \qquad 0 < \sqrt{x^2 + y^2} < \delta.$$

Since $y^2 \leq x^2 + y^2$ we have that

$$\frac{4|x|y^2}{x^2 + y^2} \leq 4|x| = 4\sqrt{x^2} \leq 4\sqrt{x^2 + y^2}. \qquad \frac{y^2}{x^2 + y^2} \leq 1$$

So if we choose $\delta = \varepsilon/4$ and let $0 < \sqrt{x^2 + y^2} < \delta$, we get

$$\left| \frac{4xy^2}{x^2 + y^2} - 0 \right| \leq 4\sqrt{x^2 + y^2} < 4\delta = 4\left(\frac{\varepsilon}{4}\right) = \varepsilon.$$

It follows from the definition that

$$\lim_{(x, y) \to (0,0)} \frac{4xy^2}{x^2 + y^2} = 0. \qquad \blacksquare$$

EXAMPLE 4 If $f(x, y) = \dfrac{y}{x}$, does $\displaystyle\lim_{(x,\,y)\to(0,\,0)} f(x, y)$ exist?

Solution The domain of f does not include the y-axis, so we do not consider any points (x, y) where $x = 0$ in the approach toward the origin $(0, 0)$. Along the x-axis, the value of the function is $f(x, 0) = 0$ for all $x \neq 0$. So if the limit does exist as $(x, y) \to (0, 0)$, the value of the limit must be $L = 0$. On the other hand, along the line $y = x$, the value of the function is $f(x, x) = x/x = 1$ for all $x \neq 0$. That is, the function f approaches the value 1 along the line $y = x$. This means that for every disk of radius δ centered at $(0, 0)$, the disk will contain points $(x, 0)$ on the x-axis where the value of the function is 0, and also points (x, x) along the line $y = x$ where the value of the function is 1. So no matter how small we choose δ as the radius of the disk in Figure 14.12, there will be points within the disk for which the function values differ by 1. Therefore, the limit cannot exist because we can take ε to be any number less than 1 in the limit definition and deny that $L = 0$ or 1, or any other real number. The limit does not exist because we have different limiting values along different paths approaching the point $(0, 0)$. ∎

Continuity

As with functions of a single variable, continuity is defined in terms of limits.

DEFINITION A function $f(x, y)$ is **continuous at the point** (x_0, y_0) if

1. f is defined at (x_0, y_0),

2. $\displaystyle\lim_{(x,\,y)\to(x_0,\,y_0)} f(x, y)$ exists,

3. $\displaystyle\lim_{(x,\,y)\to(x_0,\,y_0)} f(x, y) = f(x_0, y_0).$

A function is **continuous** if it is continuous at every point of its domain.

As with the definition of limit, the definition of continuity applies at boundary points as well as interior points of the domain of f. The only requirement is that each point (x, y) near (x_0, y_0) be in the domain of f.

A consequence of Theorem 1 is that algebraic combinations of continuous functions are continuous at every point at which all the functions involved are defined. This means that sums, differences, constant multiples, products, quotients, and powers of continuous functions are continuous where defined. In particular, polynomials and rational functions of two variables are continuous at every point at which they are defined.

EXAMPLE 5 Show that

$$f(x, y) = \begin{cases} \dfrac{2xy}{x^2 + y^2}, & (x, y) \neq (0, 0) \\ 0, & (x, y) = (0, 0) \end{cases}$$

is continuous at every point except the origin (Figure 14.14).

Solution The function f is continuous at every point (x, y) except $(0, 0)$ because its values at points other than $(0, 0)$ are given by a rational function of x and y, and therefore at those points the limiting value is simply obtained by substituting the values of x and y into that rational expression.

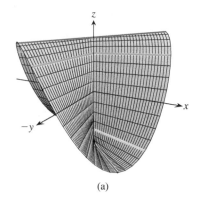

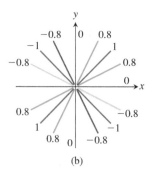

FIGURE 14.14 (a) The graph of

$$f(x, y) = \begin{cases} \dfrac{2xy}{x^2 + y^2}, & (x, y) \neq (0, 0) \\ 0, & (x, y) = (0, 0). \end{cases}$$

The function is continuous at every point except the origin. (b) The values of f are different constants along each line $y = mx$, $x \neq 0$ (Example 5).

At $(0, 0)$, the value of f is defined, but f has no limit as $(x, y) \to (0, 0)$. The reason is that different paths of approach to the origin can lead to different results, as we now see.

For every value of m, the function f has a constant value on the "punctured" line $y = mx$, $x \neq 0$, because

$$f(x, y)\bigg|_{y=mx} = \frac{2xy}{x^2 + y^2}\bigg|_{y=mx} = \frac{2x(mx)}{x^2 + (mx)^2} = \frac{2mx^2}{x^2 + m^2x^2} = \frac{2m}{1 + m^2}.$$

Therefore, f has this number as its limit as (x, y) approaches $(0, 0)$ along the line:

$$\lim_{\substack{(x, y) \to (0,0) \\ \text{along } y = mx}} f(x, y) = \lim_{(x, y) \to (0,0)} \left[f(x, y)\bigg|_{y=mx} \right] = \frac{2m}{1 + m^2}.$$

This limit changes with each value of the slope m. There is therefore no single number we may call the limit of f as (x, y) approaches the origin. The limit fails to exist, and the function is not continuous at the origin. ∎

Examples 4 and 5 illustrate an important point about limits of functions of two or more variables. For a limit to exist at a point, the limit must be the same along every approach path. This result is analogous to the single-variable case where both the left- and right-sided limits had to have the same value. For functions of two or more variables, if we ever find paths with different limits, we know the function has no limit at the point they approach.

Two-Path Test for Nonexistence of a Limit

If a function $f(x, y)$ has different limits along two different paths in the domain of f as (x, y) approaches (x_0, y_0), then $\lim_{(x, y) \to (x_0, y_0)} f(x, y)$ does not exist.

EXAMPLE 6   Show that the function

$$f(x, y) = \frac{2x^2y}{x^4 + y^2}$$

(Figure 14.15) has no limit as (x, y) approaches $(0, 0)$.

Solution   The limit cannot be found by direct substitution, which gives the indeterminate form $0/0$. We examine the values of f along parabolic curves that end at $(0, 0)$. Along the curve $y = kx^2$, $x \neq 0$, the function has the constant value

$$f(x, y)\bigg|_{y=kx^2} = \frac{2x^2y}{x^4 + y^2}\bigg|_{y=kx^2} = \frac{2x^2(kx^2)}{x^4 + (kx^2)^2} = \frac{2kx^4}{x^4 + k^2x^4} = \frac{2k}{1 + k^2}.$$

Therefore,

$$\lim_{\substack{(x, y) \to (0,0) \\ \text{along } y = kx^2}} f(x, y) = \lim_{(x, y) \to (0,0)} \left[f(x, y)\bigg|_{y=kx^2} \right] = \frac{2k}{1 + k^2}.$$

This limit varies with the path of approach. If (x, y) approaches $(0, 0)$ along the parabola $y = x^2$, for instance, $k = 1$ and the limit is 1. If (x, y) approaches $(0, 0)$ along the x-axis, $k = 0$ and the limit is 0. By the two-path test, f has no limit as (x, y) approaches $(0, 0)$. ∎

It can be shown that the function in Example 6 has limit 0 along every straight line path $y = mx$ (Exercise 57). This implies the following observation:

Having the same limit along all straight lines approaching (x_0, y_0) does not imply that a limit exists at (x_0, y_0).

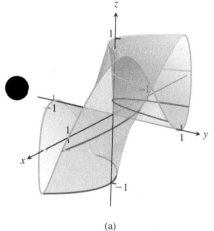

(a)

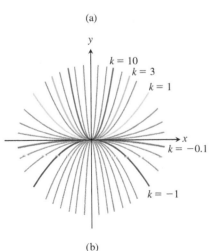

(b)

FIGURE 14.15   (a) The graph of $f(x, y) = 2x^2y/(x^4 + y^2)$. (b) Along each path $y = kx^2$ the value of f is constant, but varies with k (Example 6).

Whenever it is correctly defined, the composition of continuous functions is also continuous. The only requirement is that each function be continuous where it is applied. The proof, omitted here, is similar to that for functions of a single variable (Theorem 9 in Section 2.5).

Continuity of Compositions

If f is continuous at (x_0, y_0) and g is a single-variable function continuous at $f(x_0, y_0)$, then the composition $h = g \circ f$ defined by $h(x, y) = g(f(x, y))$ is continuous at (x_0, y_0).

For example, the composite functions

$$e^{x-y}, \qquad \cos \frac{xy}{x^2 + 1}, \qquad \ln (1 + x^2 y^2)$$

are continuous at every point (x, y).

Functions of More Than Two Variables

The definitions of limit and continuity for functions of two variables and the conclusions about limits and continuity for sums, products, quotients, powers, and compositions all extend to functions of three or more variables. Functions like

$$\ln (x + y + z) \qquad \text{and} \qquad \frac{y \sin z}{x - 1}$$

are continuous throughout their domains, and limits like

$$\lim_{P \to (1,0,-1)} \frac{e^{x+z}}{z^2 + \cos \sqrt{xy}} = \frac{e^{1-1}}{(-1)^2 + \cos 0} = \frac{1}{2},$$

where P denotes the point (x, y, z), may be found by direct substitution.

Extreme Values of Continuous Functions on Closed, Bounded Sets

The Extreme Value Theorem (Theorem 1, Section 4.1) states that a function of a single variable that is continuous at every point of a closed, bounded interval $[a, b]$ takes on an absolute maximum value and an absolute minimum value at least once in $[a, b]$. The same holds true of a function $z = f(x, y)$ that is continuous on a closed, bounded set R in the plane (like a line segment, a disk, or a filled-in triangle). The function takes on an absolute maximum value at some point in R and an absolute minimum value at some point in R. The function may take on a maximum or minimum value more than once over R.

Similar results hold for functions of three or more variables. A continuous function $w = f(x, y, z)$ must take on absolute maximum and minimum values on any closed, bounded set (such as a solid ball or cube, spherical shell, or rectangular solid) on which it is defined. We will learn how to find these extreme values in Section 14.7.

EXERCISES 14.2

Limits with Two Variables
Find the limits in Exercises 1–12.

1. $\displaystyle \lim_{(x, y) \to (0,0)} \frac{3x^2 - y^2 + 5}{x^2 + y^2 + 2}$

2. $\displaystyle \lim_{(x, y) \to (0,4)} \frac{x}{\sqrt{y}}$

3. $\displaystyle \lim_{(x, y) \to (3,4)} \sqrt{x^2 + y^2 - 1}$

4. $\displaystyle \lim_{(x, y) \to (2,-3)} \left(\frac{1}{x} + \frac{1}{y} \right)^2$

5. $\displaystyle \lim_{(x, y) \to (0,\pi/4)} \sec x \tan y$

6. $\displaystyle \lim_{(x, y) \to (0,0)} \cos \frac{x^2 + y^3}{x + y + 1}$

7. $\lim\limits_{(x,y)\to(0,\ln 2)} e^{x-y}$

8. $\lim\limits_{(x,y)\to(1,1)} \ln|1 + x^2 y^2|$

9. $\lim\limits_{(x,y)\to(0,0)} \dfrac{e^y \sin x}{x}$

10. $\lim\limits_{(x,y)\to(1/27,\,\pi^3)} \cos\sqrt[3]{xy}$

11. $\lim\limits_{(x,y)\to(1,\,\pi/6)} \dfrac{x \sin y}{x^2 + 1}$

12. $\lim\limits_{(x,y)\to(\pi/2,0)} \dfrac{\cos y + 1}{y - \sin x}$

Limits of Quotients

Find the limits in Exercises 13–24 by rewriting the fractions first.

13. $\lim\limits_{\substack{(x,y)\to(1,1) \\ x\ne y}} \dfrac{x^2 - 2xy + y^2}{x - y}$

14. $\lim\limits_{\substack{(x,y)\to(1,1) \\ x\ne y}} \dfrac{x^2 - y^2}{x - y}$

15. $\lim\limits_{\substack{(x,y)\to(1,1) \\ x\ne 1}} \dfrac{xy - y - 2x + 2}{x - 1}$

16. $\lim\limits_{\substack{(x,y)\to(2,-4) \\ x\ne -4,\, x\ne x^2}} \dfrac{y + 4}{x^2 y - xy + 4x^2 - 4x}$

17. $\lim\limits_{\substack{(x,y)\to(0,0) \\ x\ne y}} \dfrac{x - y + 2\sqrt{x} - 2\sqrt{y}}{\sqrt{x} - \sqrt{y}}$

18. $\lim\limits_{\substack{(x,y)\to(2,2) \\ x+y\ne 4}} \dfrac{x + y - 4}{\sqrt{x + y} - 2}$

19. $\lim\limits_{\substack{(x,y)\to(2,0) \\ 2x-y\ne 4}} \dfrac{\sqrt{2x - y} - 2}{2x - y - 4}$

20. $\lim\limits_{\substack{(x,y)\to(4,3) \\ x\ne y+1}} \dfrac{\sqrt{x} - \sqrt{y + 1}}{x - y - 1}$

21. $\lim\limits_{(x,y)\to(0,0)} \dfrac{\sin(x^2 + y^2)}{x^2 + y^2}$

22. $\lim\limits_{(x,y)\to(0,0)} \dfrac{1 - \cos(xy)}{xy}$

23. $\lim\limits_{(x,y)\to(1,-1)} \dfrac{x^3 + y^3}{x + y}$

24. $\lim\limits_{(x,y)\to(2,2)} \dfrac{x - y}{x^4 - y^4}$

Limits with Three Variables

Find the limits in Exercises 25–30.

25. $\lim\limits_{P\to(1,3,4)} \left(\dfrac{1}{x} + \dfrac{1}{y} + \dfrac{1}{z}\right)$

26. $\lim\limits_{P\to(1,-1,-1)} \dfrac{2xy + yz}{x^2 + z^2}$

27. $\lim\limits_{P\to(\pi,\pi,0)} (\sin^2 x + \cos^2 y + \sec^2 z)$

28. $\lim\limits_{P\to(-1/4,\pi/2,2)} \tan^{-1} xyz$

29. $\lim\limits_{P\to(\pi,0,3)} ze^{-2y} \cos 2x$

30. $\lim\limits_{P\to(2,-3,6)} \ln\sqrt{x^2 + y^2 + z^2}$

Continuity for Two Variables

At what points (x, y) in the plane are the functions in Exercises 31–34 continuous?

31. a. $f(x, y) = \sin(x + y)$ **b.** $f(x, y) = \ln(x^2 + y^2)$

32. a. $f(x, y) = \dfrac{x + y}{x - y}$ **b.** $f(x, y) = \dfrac{y}{x^2 + 1}$

33. a. $g(x, y) = \sin\dfrac{1}{xy}$ **b.** $g(x, y) = \dfrac{x + y}{2 + \cos x}$

34. a. $g(x, y) = \dfrac{x^2 + y^2}{x^2 - 3x + 2}$ **b.** $g(x, y) = \dfrac{1}{x^2 - y}$

Continuity for Three Variables

At what points (x, y, z) in space are the functions in Exercises 35–40 continuous?

35. a. $f(x, y, z) = x^2 + y^2 - 2z^2$
   b. $f(x, y, z) = \sqrt{x^2 + y^2 - 1}$

36. a. $f(x, y, z) = \ln xyz$ **b.** $f(x, y, z) = e^{x+y} \cos z$

37. a. $h(x, y, z) = xy \sin\dfrac{1}{z}$ **b.** $h(x, y, z) = \dfrac{1}{x^2 + z^2 - 1}$

38. a. $h(x, y, z) = \dfrac{1}{|y| + |z|}$ **b.** $h(x, y, z) = \dfrac{1}{|xy| + |z|}$

39. a. $h(x, y, z) = \ln(z - x^2 - y^2 - 1)$
   b. $h(x, y, z) = \dfrac{1}{z - \sqrt{x^2 + y^2}}$

40. a. $h(x, y, z) = \sqrt{4 - x^2 - y^2 - z^2}$
   b. $h(x, y, z) = \dfrac{1}{4 - \sqrt{x^2 + y^2 + z^2 - 9}}$

No Limit Exists at the Origin

By considering different paths of approach, show that the functions in Exercises 41–48 have no limit as $(x, y) \to (0, 0)$.

41. $f(x, y) = -\dfrac{x}{\sqrt{x^2 + y^2}}$ **42.** $f(x, y) = \dfrac{x^4}{x^4 + y^2}$

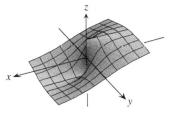
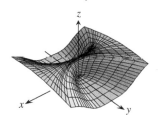

43. $f(x, y) = \dfrac{x^4 - y^2}{x^4 + y^2}$ **44.** $f(x, y) = \dfrac{xy}{|xy|}$

45. $g(x, y) = \dfrac{x - y}{x + y}$ **46.** $g(x, y) = \dfrac{x^2 - y}{x - y}$

47. $h(x, y) = \dfrac{x^2 + y}{y}$ **48.** $h(x, y) = \dfrac{x^2 y}{x^4 + y^2}$

Theory and Examples

In Exercises 49–54, show that the limits do not exist.

49. $\lim\limits_{(x,y)\to(1,1)} \dfrac{xy^2 - 1}{y - 1}$ **50.** $\lim\limits_{(x,y)\to(1,-1)} \dfrac{xy + 1}{x^2 - y^2}$

51. $\lim\limits_{(x,y)\to(0,1)} \dfrac{x \ln y}{x^2 + (\ln y)^2}$ **52.** $\lim\limits_{(x,y)\to(1,0)} \dfrac{xe^y - 1}{xe^y - 1 + y}$

53. $\lim\limits_{(x,y)\to(0,0)} \dfrac{y + \sin x}{x + \sin y}$ **54.** $\lim\limits_{(x,y)\to(1,1)} \dfrac{\tan y - y \tan x}{y - x}$

55. Let $f(x, y) = \begin{cases} 1, & y \geq x^4 \\ 1, & y \leq 0 \\ 0, & \text{otherwise.} \end{cases}$

Find each of the following limits, or explain that the limit does not exist.

a. $\displaystyle\lim_{(x, y)\to(0,1)} f(x, y)$

b. $\displaystyle\lim_{(x, y)\to(2,3)} f(x, y)$

c. $\displaystyle\lim_{(x, y)\to(0,0)} f(x, y)$

56. Let $f(x, y) = \begin{cases} x^2, & x \geq 0 \\ x^3, & x < 0 \end{cases}$.

Find the following limits.

a. $\displaystyle\lim_{(x, y)\to(3,-2)} f(x, y)$

b. $\displaystyle\lim_{(x, y)\to(-2, 1)} f(x, y)$

c. $\displaystyle\lim_{(x, y)\to(0,0)} f(x, y)$

57. Show that the function in Example 6 has limit 0 along every straight line approaching $(0, 0)$.

58. If $f(x_0, y_0) = 3$, what can you say about

$$\lim_{(x, y)\to(x_0, y_0)} f(x, y)$$

if f is continuous at (x_0, y_0)? If f is not continuous at (x_0, y_0)? Give reasons for your answers.

The Sandwich Theorem for functions of two variables states that if $g(x, y) \leq f(x, y) \leq h(x, y)$ for all $(x, y) \neq (x_0, y_0)$ in a disk centered at (x_0, y_0) and if g and h have the same finite limit L as $(x, y) \to (x_0, y_0)$, then

$$\lim_{(x, y)\to(x_0, y_0)} f(x, y) = L.$$

Use this result to support your answers to the questions in Exercises 59–62.

59. Does knowing that

$$1 - \frac{x^2 y^2}{3} < \frac{\tan^{-1} xy}{xy} < 1$$

tell you anything about

$$\lim_{(x, y)\to(0,0)} \frac{\tan^{-1} xy}{xy} ?$$

Give reasons for your answer.

60. Does knowing that

$$2|xy| - \frac{x^2 y^2}{6} < 4 - 4\cos\sqrt{|xy|} < 2|xy|$$

tell you anything about

$$\lim_{(x, y)\to(0,0)} \frac{4 - 4\cos\sqrt{|xy|}}{|xy|} ?$$

Give reasons for your answer.

61. Does knowing that $\left|\sin(1/x)\right| \leq 1$ tell you anything about

$$\lim_{(x, y)\to(0,0)} y \sin\frac{1}{x} ?$$

Give reasons for your answer.

62. Does knowing that $\left|\cos(1/y)\right| \leq 1$ tell you anything about

$$\lim_{(x, y)\to(0,0)} x \cos\frac{1}{y} ?$$

Give reasons for your answer.

63. (*Continuation of Example 5.*)

a. Reread Example 5. Then substitute $m = \tan\theta$ into the formula

$$f(x, y)\Big|_{y=mx} = \frac{2m}{1 + m^2}$$

and simplify the result to show how the value of f varies with the line's angle of inclination.

b. Use the formula you obtained in part (a) to show that the limit of f as $(x, y) \to (0, 0)$ along the line $y = mx$ varies from -1 to 1 depending on the angle of approach.

64. **Continuous extension**   Define $f(0, 0)$ in a way that extends

$$f(x, y) = xy\frac{x^2 - y^2}{x^2 + y^2}$$

to be continuous at the origin.

Changing Variables to Polar Coordinates

If you cannot make any headway with $\lim_{(x, y)\to(0,0)} f(x, y)$ in rectangular coordinates, try changing to polar coordinates. Substitute $x = r\cos\theta$, $y = r\sin\theta$, and investigate the limit of the resulting expression as $r \to 0$. In other words, try to decide whether there exists a number L satisfying the following criterion:

Given $\varepsilon > 0$, there exists a $\delta > 0$ such that for all r and θ,

$$|r| < \delta \quad \Rightarrow \quad |f(r, \theta) - L| < \varepsilon. \tag{1}$$

If such an L exists, then

$$\lim_{(x, y)\to(0,0)} f(x, y) = \lim_{r\to0} f(r\cos\theta, r\sin\theta) = L.$$

For instance,

$$\lim_{(x, y)\to(0,0)} \frac{x^3}{x^2 + y^2} = \lim_{r\to0} \frac{r^3\cos^3\theta}{r^2} = \lim_{r\to0} r\cos^3\theta = 0.$$

To verify the last of these equalities, we need to show that Equation (1) is satisfied with $f(r, \theta) = r\cos^3\theta$ and $L = 0$. That is, we need to show that given any $\varepsilon > 0$, there exists a $\delta > 0$ such that for all r and θ,

$$|r| < \delta \quad \Rightarrow \quad |r\cos^3\theta - 0| < \varepsilon.$$

Since

$$|r\cos^3\theta| = |r||\cos^3\theta| \leq |r|\cdot 1 = |r|,$$

the implication holds for all r and θ if we take $\delta = \varepsilon$.

In contrast,

$$\frac{x^2}{x^2 + y^2} = \frac{r^2\cos^2\theta}{r^2} = \cos^2\theta$$

takes on all values from 0 to 1 regardless of how small $|r|$ is, so that $\lim_{(x, y) \to (0,0)} x^2/(x^2 + y^2)$ does not exist.

In each of these instances, the existence or nonexistence of the limit as $r \to 0$ is fairly clear. Shifting to polar coordinates does not always help, however, and may even tempt us to false conclusions. For example, the limit may exist along every straight line (or ray) $\theta = $ constant and yet fail to exist in the broader sense. Example 5 illustrates this point. In polar coordinates, $f(x, y) = (2x^2y)/(x^4 + y^2)$ becomes

$$f(r \cos \theta, r \sin \theta) = \frac{r \cos \theta \sin 2\theta}{r^2 \cos^4 \theta + \sin^2 \theta}$$

for $r \neq 0$. If we hold θ constant and let $r \to 0$, the limit is 0. On the path $y = x^2$, however, we have $r \sin \theta = r^2 \cos^2 \theta$ and

$$f(r \cos \theta, r \sin \theta) = \frac{r \cos \theta \sin 2\theta}{r^2 \cos^4 \theta + (r \cos^2 \theta)^2}$$

$$= \frac{2r \cos^2 \theta \sin \theta}{2r^2 \cos^4 \theta} = \frac{r \sin \theta}{r^2 \cos^2 \theta} = 1.$$

In Exercises 65–70, find the limit of f as $(x, y) \to (0, 0)$ or show that the limit does not exist.

65. $f(x, y) = \dfrac{x^3 - xy^2}{x^2 + y^2}$

66. $f(x, y) = \cos\left(\dfrac{x^3 - y^3}{x^2 + y^2}\right)$

67. $f(x, y) = \dfrac{y^2}{x^2 + y^2}$

68. $f(x, y) = \dfrac{2x}{x^2 + x + y^2}$

69. $f(x, y) = \tan^{-1}\left(\dfrac{|x| + |y|}{x^2 + y^2}\right)$

70. $f(x, y) = \dfrac{x^2 - y^2}{x^2 + y^2}$

In Exercises 71 and 72, define $f(0, 0)$ in a way that extends f to be continuous at the origin.

71. $f(x, y) = \ln\left(\dfrac{3x^2 - x^2y^2 + 3y^2}{x^2 + y^2}\right)$

72. $f(x, y) = \dfrac{3x^2y}{x^2 + y^2}$

Using the Limit Definition

Each of Exercises 73–78 gives a function $f(x, y)$ and a positive number ε. In each exercise, show that there exists a $\delta > 0$ such that for all (x, y),

$$\sqrt{x^2 + y^2} < \delta \quad \Rightarrow \quad |f(x, y) - f(0, 0)| < \varepsilon.$$

73. $f(x, y) = x^2 + y^2, \quad \varepsilon = 0.01$

74. $f(x, y) = y/(x^2 + 1), \quad \varepsilon = 0.05$

75. $f(x, y) = (x + y)/(x^2 + 1), \quad \varepsilon = 0.01$

76. $f(x, y) = (x + y)/(2 + \cos x), \quad \varepsilon = 0.02$

77. $f(x, y) = \dfrac{xy^2}{x^2 + y^2}$ and $f(0, 0) = 0, \quad \varepsilon = 0.04$

78. $f(x, y) = \dfrac{x^3 + y^4}{x^2 + y^2}$ and $f(0, 0) = 0, \quad \varepsilon = 0.02$

Each of Exercises 79–82 gives a function $f(x, y, z)$ and a positive number ε. In each exercise, show that there exists a $\delta > 0$ such that for all (x, y, z),

$$\sqrt{x^2 + y^2 + z^2} < \delta \quad \Rightarrow \quad |f(x, y, z) - f(0, 0, 0)| < \varepsilon.$$

79. $f(x, y, z) = x^2 + y^2 + z^2, \quad \varepsilon = 0.015$

80. $f(x, y, z) = xyz, \quad \varepsilon = 0.008$

81. $f(x, y, z) = \dfrac{x + y + z}{x^2 + y^2 + z^2 + 1}, \quad \varepsilon = 0.015$

82. $f(x, y, z) = \tan^2 x + \tan^2 y + \tan^2 z, \quad \varepsilon = 0.03$

83. Show that $f(x, y, z) = x + y - z$ is continuous at every point (x_0, y_0, z_0).

84. Show that $f(x, y, z) = x^2 + y^2 + z^2$ is continuous at the origin.

14.3 Partial Derivatives

The calculus of several variables is similar to single-variable calculus applied to several variables one at a time. When we hold all but one of the independent variables of a function constant and differentiate with respect to that one variable, we get a "partial" derivative. This section shows how partial derivatives are defined and interpreted geometrically, and how to calculate them by applying the rules for differentiating functions of a single variable. The idea of *differentiability* for functions of several variables requires more than the existence of the partial derivatives because a point can be approached from many different directions. However, we will see that differentiable functions of several variables behave similarly to differentiable single-variable functions. In particular, they are continuous and can be well approximated by linear functions.

Partial Derivatives of a Function of Two Variables

If (x_0, y_0) is a point in the domain of a function $f(x, y)$, the vertical plane $y = y_0$ will cut the surface $z = f(x, y)$ in the curve $z = f(x, y_0)$ (Figure 14.16). This curve is the graph of

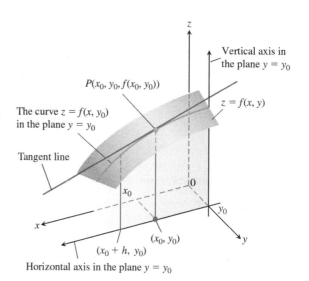

FIGURE 14.16 The intersection of the plane $y = y_0$ with the surface $z = f(x, y)$, viewed from above the first quadrant of the xy-plane.

the function $z = f(x, y_0)$ in the plane $y = y_0$. The horizontal coordinate in this plane is x; the vertical coordinate is z. The y-value is held constant at y_0, so y is not a variable.

We define the partial derivative of f with respect to x at the point (x_0, y_0) as the ordinary derivative of $f(x, y_0)$ with respect to x at the point $x = x_0$. To distinguish partial derivatives from ordinary derivatives we use the symbol ∂ rather than the d previously used. In the definition, h represents a real number, positive or negative.

DEFINITION The **partial derivative of $f(x, y)$ with respect to x at the point** (x_0, y_0) is

$$\left.\frac{\partial f}{\partial x}\right|_{(x_0, y_0)} = \lim_{h \to 0} \frac{f(x_0 + h, y_0) - f(x_0, y_0)}{h},$$

provided the limit exists.

The partial derivative of $f(x, y)$ with respect to x at the point (x_0, y_0) is the same as the ordinary derivative of $f(x, y_0)$ at the point x_0:

$$\left.\frac{\partial f}{\partial x}\right|_{(x_0, y_0)} = \left.\frac{d}{dx} f(x, y_0)\right|_{x = x_0}.$$

A variety of notations are used to denote the partial derivative at a point (x_0, y_0), including

$$\frac{\partial f}{\partial x}(x_0, y_0), \qquad f_x(x_0, y_0), \qquad \text{and} \qquad \left.\frac{\partial z}{\partial x}\right|_{(x_0, y_0)}.$$

When we do not specify a specific point (x_0, y_0) at which the partial derivative is being evaluated, then the partial derivative becomes a *function* whose domain is the points where the partial derivative exists. Notations for this function include

$$\frac{\partial f}{\partial x}, \qquad f_x, \qquad \text{and} \qquad \frac{\partial z}{\partial x}.$$

The slope of the curve $z = f(x, y_0)$ at the point $P(x_0, y_0, f(x_0, y_0))$ in the plane $y = y_0$ is the value of the partial derivative of f with respect to x at (x_0, y_0). (In Figure 14.16 this slope is negative.) The tangent line to the curve at P is the line in the plane $y = y_0$ that passes through P with this slope. The partial derivative $\partial f / \partial x$ at (x_0, y_0) gives the rate of change of f with respect to x when y is held fixed at the value y_0.

The definition of the partial derivative of $f(x, y)$ with respect to y at a point (x_0, y_0) is similar to the definition of the partial derivative of f with respect to x. We hold x fixed at the value x_0 and take the ordinary derivative of $f(x_0, y)$ with respect to y at y_0.

DEFINITION The **partial derivative of $f(x, y)$ with respect to y at the point (x_0, y_0)** is

$$\frac{\partial f}{\partial y}\bigg|_{(x_0, y_0)} = \frac{d}{dy} f(x_0, y)\bigg|_{y=y_0} = \lim_{h \to 0} \frac{f(x_0, y_0 + h) - f(x_0, y_0)}{h},$$

provided the limit exists.

The slope of the curve $z = f(x_0, y)$ at the point $P(x_0, y_0, f(x_0, y_0))$ in the vertical plane $x = x_0$ (Figure 14.17) is the partial derivative of f with respect to y at (x_0, y_0). The tangent line to the curve at P is the line in the plane $x = x_0$ that passes through P with this slope. The partial derivative gives the rate of change of f with respect to y at (x_0, y_0) when x is held fixed at the value x_0.

The partial derivative with respect to y is denoted the same way as the partial derivative with respect to x:

$$\frac{\partial f}{\partial y}(x_0, y_0), \qquad f_y(x_0, y_0), \qquad \frac{\partial f}{\partial y}, \qquad f_y.$$

Notice that we now have two tangent lines associated with the surface $z = f(x, y)$ at the point $P(x_0, y_0, f(x_0, y_0))$ (Figure 14.18). Is the plane they determine tangent to the surface at P? We will see that it is for the *differentiable* functions defined at the end of this section, and we will learn how to find the tangent plane in Section 14.6. First we have to better understand partial derivatives.

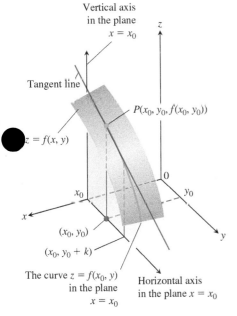

FIGURE 14.17 The intersection of the plane $x = x_0$ with the surface $z = f(x, y)$, viewed from above the first quadrant of the xy-plane.

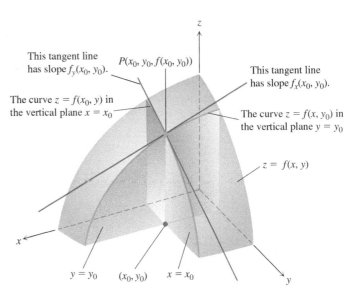

FIGURE 14.18 Figures 14.16 and 14.17 combined. The tangent lines at the point $(x_0, y_0, f(x_0, y_0))$ determine a plane that, in this picture at least, appears to be tangent to the surface.

Calculations

The definitions of $\partial f/\partial x$ and $\partial f/\partial y$ give us two different ways of differentiating f at a point: with respect to x in the usual way while treating y as a constant and with respect to y in the usual way while treating x as a constant. As the following examples show, the values of these partial derivatives are usually different at a given point (x_0, y_0).

EXAMPLE 1   Find the values of $\partial f/\partial x$ and $\partial f/\partial y$ at the point $(4, -5)$ if

$$f(x, y) = x^2 + 3xy + y - 1.$$

Solution   To find $\partial f/\partial x$, we treat y as a constant and differentiate with respect to x:

$$\frac{\partial f}{\partial x} = \frac{\partial}{\partial x}(x^2 + 3xy + y - 1) = 2x + 3 \cdot 1 \cdot y + 0 - 0 = 2x + 3y.$$

The value of $\partial f/\partial x$ at $(4, -5)$ is $2(4) + 3(-5) = -7$.
    To find $\partial f/\partial y$, we treat x as a constant and differentiate with respect to y:

$$\frac{\partial f}{\partial y} = \frac{\partial}{\partial y}(x^2 + 3xy + y - 1) = 0 + 3 \cdot x \cdot 1 + 1 - 0 = 3x + 1.$$

The value of $\partial f/\partial y$ at $(4, -5)$ is $3(4) + 1 = 13$.   ∎

EXAMPLE 2   Find $\partial f/\partial y$ as a function if $f(x, y) = y \sin xy$.

Solution   We treat x as a constant and f as a product of y and $\sin xy$:

$$\frac{\partial f}{\partial y} = \frac{\partial}{\partial y}(y \sin xy) = y \frac{\partial}{\partial y} \sin xy + (\sin xy) \frac{\partial}{\partial y}(y)$$

$$= (y \cos xy) \frac{\partial}{\partial y}(xy) + \sin xy = xy \cos xy + \sin xy.$$ ∎

EXAMPLE 3   Find f_x and f_y as functions if

$$f(x, y) = \frac{2y}{y + \cos x}.$$

Solution   We treat f as a quotient. With y held constant, we use the quotient rule to get

$$f_x = \frac{\partial}{\partial x}\left(\frac{2y}{y + \cos x}\right) = \frac{(y + \cos x)\frac{\partial}{\partial x}(2y) - 2y \frac{\partial}{\partial x}(y + \cos x)}{(y + \cos x)^2}$$

$$= \frac{(y + \cos x)(0) - 2y(-\sin x)}{(y + \cos x)^2} = \frac{2y \sin x}{(y + \cos x)^2}.$$

With x held constant and again applying the quotient rule, we get

$$f_y = \frac{\partial}{\partial y}\left(\frac{2y}{y + \cos x}\right) = \frac{(y + \cos x)\frac{\partial}{\partial y}(2y) - 2y \frac{\partial}{\partial y}(y + \cos x)}{(y + \cos x)^2}$$

$$= \frac{(y + \cos x)(2) - 2y(1)}{(y + \cos x)^2} = \frac{2 \cos x}{(y + \cos x)^2}.$$ ∎

    Implicit differentiation works for partial derivatives the way it works for ordinary derivatives, as the next example illustrates.

EXAMPLE 4   Find $\partial z/\partial x$ assuming that the equation

$$yz - \ln z = x + y$$

defines z as a function of the two independent variables x and y and the partial derivative exists.

Solution   We differentiate both sides of the equation with respect to x, holding y constant and treating z as a differentiable function of x:

$$\frac{\partial}{\partial x}(yz) - \frac{\partial}{\partial x}\ln z = \frac{\partial x}{\partial x} + \frac{\partial y}{\partial x}$$

$$y\frac{\partial z}{\partial x} - \frac{1}{z}\frac{\partial z}{\partial x} = 1 + 0 \qquad \text{With } y \text{ constant, } \frac{\partial}{\partial x}(yz) = y\frac{\partial z}{\partial x}.$$

$$\left(y - \frac{1}{z}\right)\frac{\partial z}{\partial x} = 1$$

$$\frac{\partial z}{\partial x} = \frac{z}{yz - 1}.$$ ∎

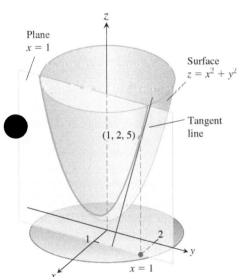

Plane
$x = 1$

Surface
$z = x^2 + y^2$

$(1, 2, 5)$

Tangent line

$x = 1$

FIGURE 14.19   The tangent to the curve of intersection of the plane $x = 1$ and surface $z = x^2 + y^2$ at the point $(1, 2, 5)$ (Example 5).

EXAMPLE 5   The plane $x = 1$ intersects the paraboloid $z = x^2 + y^2$ in a parabola. Find the slope of the tangent to the parabola at $(1, 2, 5)$ (Figure 14.19).

Solution   The parabola lies in a plane parallel to the yz-plane, and the slope is the value of the partial derivative $\partial z/\partial y$ at $(1, 2)$:

$$\left.\frac{\partial z}{\partial y}\right|_{(1,2)} = \left.\frac{\partial}{\partial y}(x^2 + y^2)\right|_{(1,2)} = \left.2y\right|_{(1,2)} = 2(2) = 4.$$

As a check, we can treat the parabola as the graph of the single-variable function $z = (1)^2 + y^2 = 1 + y^2$ in the plane $x = 1$ and ask for the slope at $y = 2$. The slope, calculated now as an ordinary derivative, is

$$\left.\frac{dz}{dy}\right|_{y=2} = \left.\frac{d}{dy}(1 + y^2)\right|_{y=2} = \left.2y\right|_{y=2} = 4.$$ ∎

Functions of More Than Two Variables

The definitions of the partial derivatives of functions of more than two independent variables are similar to the definitions for functions of two variables. They are ordinary derivatives with respect to one variable, taken while the other independent variables are held constant.

EXAMPLE 6   If x, y, and z are independent variables and

$$f(x, y, z) = x \sin(y + 3z),$$

then

$$\frac{\partial f}{\partial z} = \frac{\partial}{\partial z}[x \sin(y + 3z)] = x\frac{\partial}{\partial z}\sin(y + 3z) \qquad x \text{ held constant}$$

$$= x \cos(y + 3z)\frac{\partial}{\partial z}(y + 3z) \qquad \text{Chain rule}$$

$$= 3x \cos(y + 3z). \qquad y \text{ held constant}$$ ∎

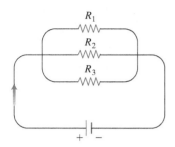

FIGURE 14.20 Resistors arranged this way are said to be connected in parallel (Example 7). Each resistor lets a portion of the current through. Their equivalent resistance R is calculated with the formula

$$\frac{1}{R} = \frac{1}{R_1} + \frac{1}{R_2} + \frac{1}{R_3}.$$

EXAMPLE 7   If resistors of R_1, R_2, and R_3 ohms are connected in parallel to make an R-ohm resistor, the value of R can be found from the equation

$$\frac{1}{R} = \frac{1}{R_1} + \frac{1}{R_2} + \frac{1}{R_3}$$

(Figure 14.20). Find the value of $\partial R/\partial R_2$ when $R_1 = 30$, $R_2 = 45$, and $R_3 = 90$ ohms.

Solution   To find $\partial R/\partial R_2$, we treat R_1 and R_3 as constants and, using implicit differentiation, differentiate both sides of the equation with respect to R_2:

$$\frac{\partial}{\partial R_2}\left(\frac{1}{R}\right) = \frac{\partial}{\partial R_2}\left(\frac{1}{R_1} + \frac{1}{R_2} + \frac{1}{R_3}\right)$$

$$-\frac{1}{R^2}\frac{\partial R}{\partial R_2} = 0 - \frac{1}{R_2{}^2} + 0$$

$$\frac{\partial R}{\partial R_2} = \frac{R^2}{R_2{}^2} = \left(\frac{R}{R_2}\right)^2.$$

When $R_1 = 30$, $R_2 = 45$, and $R_3 = 90$,

$$\frac{1}{R} = \frac{1}{30} + \frac{1}{45} + \frac{1}{90} = \frac{3 + 2 + 1}{90} = \frac{6}{90} = \frac{1}{15},$$

so $R = 15$ and

$$\frac{\partial R}{\partial R_2} = \left(\frac{15}{45}\right)^2 = \left(\frac{1}{3}\right)^2 = \frac{1}{9}.$$

Thus at the given values, a small change in the resistance R_2 leads to a change in R about 1/9th as large.

Partial Derivatives and Continuity

A function $f(x, y)$ can have partial derivatives with respect to both x and y at a point without the function being continuous there. This is different from functions of a single variable, where the existence of a derivative implies continuity. If the partial derivatives of $f(x, y)$ exist and are continuous throughout a disk centered at (x_0, y_0), however, then f is continuous at (x_0, y_0), as we see at the end of this section.

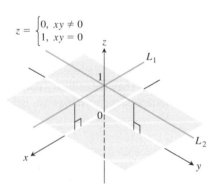

FIGURE 14.21   The graph of

$$f(x, y) = \begin{cases} 0, & xy \neq 0 \\ 1, & xy = 0 \end{cases}$$

consists of the lines L_1 and L_2 (lying 1 unit above the xy-plane) and the four open quadrants of the xy-plane. The function has partial derivatives at the origin but is not continuous there (Example 8).

EXAMPLE 8   Let

$$f(x, y) = \begin{cases} 0, & xy \neq 0 \\ 1, & xy = 0 \end{cases}$$

(Figure 14.21).

(a) Find the limit of f as (x, y) approaches $(0, 0)$ along the line $y = x$.

(b) Prove that f is not continuous at the origin.

(c) Show that both partial derivatives $\partial f/\partial x$ and $\partial f/\partial y$ exist at the origin.

Solution

(a) Since $f(x, y)$ is constantly zero along the line $y = x$ (except at the origin), we have

$$\lim_{(x, y)\to(0,0)} f(x, y)\Big|_{y=x} = \lim_{(x, y)\to(0,0)} 0 = 0.$$

(b) Since $f(0, 0) = 1$, the limit in part (a) is not equal to $f(0, 0)$, which proves that f is not continuous at $(0, 0)$.

(c) To find $\partial f/\partial x$ at $(0, 0)$, we hold y fixed at $y = 0$. Then $f(x, y) = 1$ for all x, and the graph of f is the line L_1 in Figure 14.21. The slope of this line at any x is $\partial f/\partial x = 0$. In particular, $\partial f/\partial x = 0$ at $(0, 0)$. Similarly, $\partial f/\partial y$ is the slope of line L_2 at any y, so $\partial f/\partial y = 0$ at $(0, 0)$. ∎

What Example 8 suggests is that we need a stronger requirement for differentiability in higher dimensions than the mere existence of the partial derivatives. We define differentiability for functions of two variables (which is slightly more complicated than for single-variable functions) at the end of this section and then revisit the connection to continuity.

Second-Order Partial Derivatives

When we differentiate a function $f(x, y)$ twice, we produce its second-order derivatives. These derivatives are usually denoted by

$$\frac{\partial^2 f}{\partial x^2} \text{ or } f_{xx}, \qquad \frac{\partial^2 f}{\partial y^2} \text{ or } f_{yy},$$

$$\frac{\partial^2 f}{\partial x\,\partial y} \text{ or } f_{yx}, \qquad \text{and} \qquad \frac{\partial^2 f}{\partial y\,\partial x} \text{ or } f_{xy}.$$

The defining equations are

$$\frac{\partial^2 f}{\partial x^2} = \frac{\partial}{\partial x}\left(\frac{\partial f}{\partial x}\right), \qquad \frac{\partial^2 f}{\partial x\,\partial y} = \frac{\partial}{\partial x}\left(\frac{\partial f}{\partial y}\right),$$

and so on. Notice the order in which the mixed partial derivatives are taken:

$$\frac{\partial^2 f}{\partial x\,\partial y} \qquad \text{Differentiate first with respect to } y, \text{ then with respect to } x.$$

$$f_{yx} = (f_y)_x \qquad \text{Means the same thing.}$$

HISTORICAL BIOGRAPHY
Pierre-Simon Laplace
(1749–1827)
bit.ly/2NdfIV5

EXAMPLE 9 If $f(x, y) = x \cos y + ye^x$, find the second-order derivatives

$$\frac{\partial^2 f}{\partial x^2}, \qquad \frac{\partial^2 f}{\partial y\,\partial x}, \qquad \frac{\partial^2 f}{\partial y^2}, \qquad \text{and} \qquad \frac{\partial^2 f}{\partial x\,\partial y}.$$

Solution The first step is to calculate both first partial derivatives.

$$\frac{\partial f}{\partial x} = \frac{\partial}{\partial x}(x \cos y + ye^x) \qquad\qquad \frac{\partial f}{\partial y} = \frac{\partial}{\partial y}(x \cos y + ye^x)$$

$$= \cos y + ye^x \qquad\qquad\qquad = -x \sin y + e^x$$

Now we find both partial derivatives of each first partial:

$$\frac{\partial^2 f}{\partial y\,\partial x} = \frac{\partial}{\partial y}\left(\frac{\partial f}{\partial x}\right) = -\sin y + e^x \qquad\qquad \frac{\partial^2 f}{\partial x\,\partial y} = \frac{\partial}{\partial x}\left(\frac{\partial f}{\partial y}\right) = -\sin y + e^x$$

$$\frac{\partial^2 f}{\partial x^2} = \frac{\partial}{\partial x}\left(\frac{\partial f}{\partial x}\right) = ye^x. \qquad\qquad \frac{\partial^2 f}{\partial y^2} = \frac{\partial}{\partial y}\left(\frac{\partial f}{\partial y}\right) = -x \cos y.$$

The Mixed Derivative Theorem

You may have noticed that the "mixed" second-order partial derivatives

$$\frac{\partial^2 f}{\partial y\, \partial x} \quad \text{and} \quad \frac{\partial^2 f}{\partial x\, \partial y}$$

in Example 9 are equal. This is not a coincidence. They must be equal whenever f, f_x, f_y, f_{xy}, and f_{yx} are continuous, as stated in the following theorem. However, the mixed derivatives can be different when the continuity conditions are not satisfied (see Exercise 82).

THEOREM 2—The Mixed Derivative Theorem

If $f(x, y)$ and its partial derivatives f_x, f_y, f_{xy}, and f_{yx} are defined throughout an open region containing a point (a, b) and are all continuous at (a, b), then

$$f_{xy}(a, b) = f_{yx}(a, b).$$

HISTORICAL BIOGRAPHY

Alexis Clairaut

(1713–1765)

`bit.ly/2Rfl7yc`

Theorem 2 is also known as Clairaut's Theorem, named after the French mathematician Alexis Clairaut, who discovered it. A proof is given in Appendix 9. Theorem 2 says that to calculate a mixed second-order derivative, we may differentiate in either order, provided the continuity conditions are satisfied. This ability to proceed in different order sometimes simplifies our calculations.

EXAMPLE 10   Find $\dfrac{\partial^2 w}{\partial x\, \partial y}$ if

$$w = xy + \frac{e^y}{y^2 + 1}.$$

Solution   The symbol $\partial^2 w / \partial x\, \partial y$ tells us to differentiate first with respect to y and then with respect to x. However, if we interchange the order of differentiation and differentiate first with respect to x we get the answer more quickly. In two steps,

$$\frac{\partial w}{\partial x} = y \quad \text{and} \quad \frac{\partial^2 w}{\partial y\, \partial x} = 1.$$

If we differentiate first with respect to y, we obtain $\partial^2 w / \partial x\, \partial y = 1$ as well. We can differentiate in either order because the conditions of Theorem 2 hold for w at all points (x_0, y_0).

Partial Derivatives of Still Higher Order

Although we will deal mostly with first- and second-order partial derivatives, because these appear the most frequently in applications, there is no theoretical limit to how many times we can differentiate a function as long as the derivatives involved exist. Thus, we get third- and fourth-order derivatives denoted by symbols like

$$\frac{\partial^3 f}{\partial x\, \partial y^2} = f_{yyx},$$

$$\frac{\partial^4 f}{\partial x^2\, \partial y^2} = f_{yyxx},$$

and so on. As with second-order derivatives, the order of differentiation is immaterial as long as all the derivatives through the order in question are continuous.

EXAMPLE 11   Find f_{yxyz} if $f(x, y, z) = 1 - 2xy^2z + x^2y$.

Solution   We first differentiate with respect to the variable y, then x, then y again, and finally with respect to z:

$$f_y = -4xyz + x^2$$
$$f_{yx} = -4yz + 2x$$
$$f_{yxy} = -4z$$
$$f_{yxyz} = -4.$$

Differentiability

The concept of *differentiability* for functions of several variables is more complicated than for single-variable functions because a point in the domain can be approached along more than one path. In defining the partial derivatives for a function of two variables, we intersected the surface of the graph with vertical planes parallel to the xz- and yz-planes, creating a curve on each plane, called a *trace*. The partial derivatives were seen as the slopes of the two tangent lines to these trace curves at the point on the surface corresponding to the point (x_0, y_0) being approached in the domain. (See Figure 14.18.) For a differentiable function, it would seem reasonable to assume that if we were to rotate slightly one of these vertical planes, keeping it vertical but no longer parallel to its coordinate plane, then a smooth trace curve would appear on that plane that would have a tangent line at the point on the surface having a slope differing just slightly from what it was before (when the plane was parallel to its coordinate plane). However, the mere existence of the original partial derivative does not guarantee that result. Just as having a limit in the x- and y-coordinate directions alone does not imply the function itself has a limit at (x_0, y_0), as we saw in Figure 14.21, so is it the case that the existence of both partial derivatives is not enough by itself to ensure derivatives exist for trace curves in other vertical planes. For the existence of differentiability, a property is needed to ensure that no abrupt change occurs in the function resulting from small changes in the independent variables along any path approaching (x_0, y_0), paths along which *both* variables x and y are allowed to change, rather than just one of them at a time.

In our study of functions of a single variable, we found that if a function $y = f(x)$ is differentiable at $x = x_0$, then the change Δy resulting in the change of x from x_0 to $x_0 + \Delta x$ is close to the change ΔL along the tangent line (or linear approximation L of the function f at x_0). That is, from Equation (1) in Section 3.11,

$$\Delta y = f'(x_0)\Delta x + \varepsilon \Delta x$$

in which $\varepsilon \to 0$ as $\Delta x \to 0$. The extension of this result is what we use to *define* differentiability for functions of two variables.

DEFINITION   A function $z = f(x, y)$ is **differentiable at** (x_0, y_0) if $f_x(x_0, y_0)$ and $f_y(x_0, y_0)$ exist and $\Delta z = f(x_0 + \Delta x, y_0 + \Delta y) - f(x_0, y_0)$ satisfies an equation of the form

$$\Delta z = f_x(x_0, y_0)\Delta x + f_y(x_0, y_0)\Delta y + \varepsilon_1 \Delta x + \varepsilon_2 \Delta y$$

in which each of $\varepsilon_1, \varepsilon_2 \to 0$ as both $\Delta x, \Delta y \to 0$. We call f **differentiable** if it is differentiable at every point in its domain, and say that its graph is a **smooth surface**.

The following theorem (proved in Appendix 9) and its accompanying corollary tell us that functions with *continuous* first partial derivatives at (x_0, y_0) are differentiable there, and they are closely approximated locally by a linear function. We study this approximation in Section 14.6.

THEOREM 3—The Increment Theorem for Functions of Two Variables

Suppose that the first partial derivatives of $f(x, y)$ are defined throughout an open region R containing the point (x_0, y_0) and that f_x and f_y are continuous at (x_0, y_0). Then the change

$$\Delta z = f(x_0 + \Delta x, y_0 + \Delta y) - f(x_0, y_0)$$

in the value of f that results from moving from (x_0, y_0) to another point $(x_0 + \Delta x, y_0 + \Delta y)$ in R satisfies an equation of the form

$$\Delta z = f_x(x_0, y_0)\Delta x + f_y(x_0, y_0)\Delta y + \varepsilon_1 \Delta x + \varepsilon_2 \Delta y$$

in which each of $\varepsilon_1, \varepsilon_2 \to 0$ as both $\Delta x, \Delta y \to 0$.

Corollary of Theorem 3

If the partial derivatives f_x and f_y of a function $f(x, y)$ are continuous throughout an open region R, then f is differentiable at every point of R.

If $z = f(x, y)$ is differentiable, then the definition of differentiability ensures that $\Delta z = f(x_0 + \Delta x, y_0 + \Delta y) - f(x_0, y_0)$ approaches 0 as Δx and Δy approach 0. This tells us that a function of two variables is continuous at every point where it is differentiable.

THEOREM 4—Differentiability Implies Continuity

If a function $f(x, y)$ is differentiable at (x_0, y_0), then f is continuous at (x_0, y_0).

As we can see from Corollary 3 and Theorem 4, a function $f(x, y)$ must be continuous at a point (x_0, y_0) if f_x and f_y are continuous throughout an open region containing (x_0, y_0). Remember, however, that it is still possible for a function of two variables to be discontinuous at a point where its first partial derivatives exist, as we saw in Example 8. Existence alone of the partial derivatives at that point is not enough, but continuity of the partial derivatives guarantees differentiability.

EXERCISES 14.3

Calculating First-Order Partial Derivatives

In Exercises 1–22, find $\partial f/\partial x$ and $\partial f/\partial y$.

1. $f(x, y) = 2x^2 - 3y - 4$ 2. $f(x, y) = x^2 - xy + y^2$
3. $f(x, y) = (x^2 - 1)(y + 2)$
4. $f(x, y) = 5xy - 7x^2 - y^2 + 3x - 6y + 2$
5. $f(x, y) = (xy - 1)^2$ 6. $f(x, y) = (2x - 3y)^3$
7. $f(x, y) = \sqrt{x^2 + y^2}$ 8. $f(x, y) = (x^3 + (y/2))^{2/3}$
9. $f(x, y) = 1/(x + y)$ 10. $f(x, y) = x/(x^2 + y^2)$
11. $f(x, y) = (x + y)/(xy - 1)$ 12. $f(x, y) = \tan^{-1}(y/x)$
13. $f(x, y) = e^{(x+y+1)}$ 14. $f(x, y) = e^{-x}\sin(x + y)$
15. $f(x, y) = \ln(x + y)$ 16. $f(x, y) = e^{xy}\ln y$
17. $f(x, y) = \sin^2(x - 3y)$ 18. $f(x, y) = \cos^2(3x - y^2)$

19. $f(x, y) = x^y$ 20. $f(x, y) = \log_y x$
21. $f(x, y) = \int_x^y g(t)\,dt$ (g continuous for all t)
22. $f(x, y) = \sum_{n=0}^{\infty}(xy)^n$ ($|xy| < 1$)

In Exercises 23–34, find f_x, f_y, and f_z.
23. $f(x, y, z) = 1 + xy^2 - 2z^2$
24. $f(x, y, z) = xy + yz + xz$
25. $f(x, y, z) = x - \sqrt{y^2 + z^2}$
26. $f(x, y, z) = (x^2 + y^2 + z^2)^{-1/2}$
27. $f(x, y, z) = \sin^{-1}(xyz)$

28. $f(x, y, z) = \sec^{-1}(x + yz)$

29. $f(x, y, z) = \ln(x + 2y + 3z)$

30. $f(x, y, z) = yz \ln(xy)$

31. $f(x, y, z) = e^{-(x^2+y^2+z^2)}$

32. $f(x, y, z) = e^{-xyz}$

33. $f(x, y, z) = \tanh(x + 2y + 3z)$

34. $f(x, y, z) = \sinh(xy - z^2)$

In Exercises 35–40, find the partial derivative of the function with respect to each variable.

35. $f(t, \alpha) = \cos(2\pi t - \alpha)$

36. $g(u, v) = v^2 e^{(2u/v)}$

37. $h(\rho, \phi, \theta) = \rho \sin \phi \cos \theta$

38. $g(r, \theta, z) = r(1 - \cos \theta) - z$

39. Work done by the heart (Section 3.11, Exercise 59)

$$W(P, V, \delta, v, g) = PV + \frac{V\delta v^2}{2g}$$

40. Wilson lot size formula (Section 4.6, Exercise 59)

$$A(c, h, k, m, q) = \frac{km}{q} + cm + \frac{hq}{2}$$

Calculating Second-Order Partial Derivatives

Find all the second-order partial derivatives of the functions in Exercises 41–50.

41. $f(x, y) = x + y + xy$ **42.** $f(x, y) = \sin xy$

43. $g(x, y) = x^2y + \cos y + y \sin x$

44. $h(x, y) = xe^y + y + 1$ **45.** $r(x, y) = \ln(x + y)$

46. $s(x, y) = \tan^{-1}(y/x)$ **47.** $w = x^2 \tan(xy)$

48. $w = ye^{x^2 - y}$ **49.** $w = x \sin(x^2y)$

50. $w = \dfrac{x - y}{x^2 + y}$ **51.** $f(x, y) = x^2y^3 - x^4 + y^5$

52. $g(x, y) = \cos x^2 - \sin 3y$ **53.** $z = x \sin(2x - y^2)$

54. $z = xe^{x/y^2}$

Mixed Partial Derivatives

In Exercises 55–60, verify that $w_{xy} = w_{yx}$.

55. $w = \ln(2x + 3y)$ **56.** $w = e^x + x \ln y + y \ln x$

57. $w = xy^2 + x^2y^3 + x^3y^4$

58. $w = x \sin y + y \sin x + xy$

59. $\omega = \dfrac{x^2}{y^3}$ **60.** $\omega - \dfrac{3x - y}{x + y}$

61. Which order of differentiation will calculate f_{xy} faster: x first or y first? Try to answer without writing anything down.

   a. $f(x, y) = x \sin y + e^y$

   b. $f(x, y) = 1/x$

   c. $f(x, y) = y + (x/y)$

   d. $f(x, y) = y + x^2y + 4y^3 - \ln(y^2 + 1)$

   e. $f(x, y) = x^2 + 5xy + \sin x + 7e^x$

   f. $f(x, y) = x \ln xy$

62. The fifth-order partial derivative $\partial^5 f/\partial x^2 \partial y^3$ is zero for each of the following functions. To show this as quickly as possible, which variable would you differentiate with respect to first: x or y? Try to answer without writing anything down.

   a. $f(x, y) = y^2x^4e^x + 2$

   b. $f(x, y) = y^2 + y(\sin x - x^4)$

   c. $f(x, y) = x^2 + 5xy + \sin x + 7e^x$

   d. $f(x, y) = xe^{y^2/2}$

Using the Partial Derivative Definition

In Exercises 63–66, use the limit definition of partial derivative to compute the partial derivatives of the functions at the specified points.

63. $f(x, y) = 1 - x + y - 3x^2y$, $\dfrac{\partial f}{\partial x}$ and $\dfrac{\partial f}{\partial y}$ at $(1, 2)$

64. $f(x, y) = 4 + 2x - 3y - xy^2$, $\dfrac{\partial f}{\partial x}$ and $\dfrac{\partial f}{\partial y}$ at $(-2, 1)$

65. $f(x, y) = \sqrt{2x + 3y - 1}$, $\dfrac{\partial f}{\partial x}$ and $\dfrac{\partial f}{\partial y}$ at $(-2, 3)$

66. $f(x, y) = \begin{cases} \dfrac{\sin(x^3 + y^4)}{x^2 + y^2}, & (x, y) \neq (0, 0) \\ 0, & (x, y) = (0, 0), \end{cases}$

$\dfrac{\partial f}{\partial x}$ and $\dfrac{\partial f}{\partial y}$ at $(0, 0)$

67. Let $f(x, y) = 2x + 3y - 4$. Find the slope of the line tangent to this surface at the point $(2, -1)$ and lying in the **a.** plane $x = 2$. **b.** plane $y = -1$.

68. Let $f(x, y) = x^2 + y^3$. Find the slope of the line tangent to this surface at the point $(-1, 1)$ and lying in the **a.** plane $x = -1$ **b.** plane $y = 1$.

69. Three variables Let $w = f(x, y, z)$ be a function of three independent variables and write the formal definition of the partial derivative $\partial f/\partial z$ at (x_0, y_0, z_0). Use this definition to find $\partial f/\partial z$ at $(1, 2, 3)$ for $f(x, y, z) = x^2yz^2$.

70. Three variables Let $w = f(x, y, z)$ be a function of three independent variables and write the formal definition of the partial derivative $\partial f/\partial y$ at (x_0, y_0, z_0). Use this definition to find $\partial f/\partial y$ at $(-1, 0, 3)$ for $f(x, y, z) = -2xy^2 + yz^2$.

In Exercises 71–74, find a function $z = f(x, y)$ whose partial derivatives are as given, or explain why this is impossible.

71. $\dfrac{\partial f}{\partial x} = 3x^2y^2 - 2x$, $\dfrac{\partial f}{\partial y} = 2x^3y + 6y$

72. $\dfrac{\partial f}{\partial x} = 2xe^{xy^2} + x^2y^2e^{xy^2} + 3$, $\dfrac{\partial f}{\partial y} = 2x^3ye^{xy^2} - e^y$

73. $\dfrac{\partial f}{\partial x} = \dfrac{2y}{(x + y)^2}$, $\dfrac{\partial f}{\partial y} = \dfrac{2x}{(x + y)^2}$

74. $\dfrac{\partial f}{\partial x} = xy \cos(xy) + \sin(xy)$, $\dfrac{\partial f}{\partial y} = x \cos(xy)$

Differentiating Implicitly

75. Find the value of $\partial z/\partial x$ at the point $(1, 1, 1)$ if the equation

$$xy + z^3x - 2yz = 0$$

defines z as a function of the two independent variables x and y and the partial derivative exists.

76. Find the value of $\partial x / \partial z$ at the point $(1, -1, -3)$ if the equation

$$xz + y \ln x - x^2 + 4 = 0$$

defines x as a function of the two independent variables y and z and the partial derivative exists.

Exercises 77 and 78 are about the triangle shown here.

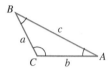

77. Express A implicitly as a function of a, b, and c and calculate $\partial A / \partial a$ and $\partial A / \partial b$.

78. Express a implicitly as a function of A, b, and B and calculate $\partial a / \partial A$ and $\partial a / \partial B$.

79. Two dependent variables Express v_x in terms of u and v if the equations $x = v \ln u$ and $y = u \ln v$ define u and v as functions of the independent variables x and y, and if v_x exists. (*Hint:* Differentiate both equations with respect to x and solve for v_x by eliminating u_x.)

80. Two dependent variables Find $\partial x / \partial u$ and $\partial y / \partial u$ if the equations $u = x^2 - y^2$ and $v = x^2 - y$ define x and y as functions of the independent variables u and v, and the partial derivatives exist. (See the hint in Exercise 79.) Then let $s = x^2 + y^2$ and find $\partial s / \partial u$.

Theory and Examples

81. Let $f(x, y) = \begin{cases} y^3, & y \geq 0 \\ -y^2, & y < 0. \end{cases}$

Find f_x, f_y, f_{xy}, and f_{yx}, and state the domain for each partial derivative.

82. Let $f(x, y) = \begin{cases} xy \dfrac{x^2 - y^2}{x^2 + y^2}, & \text{if } (x, y) \neq 0, \\ 0, & \text{if } (x, y) = 0. \end{cases}$

The graph of f is shown on page 813.

a. Show that $\dfrac{\partial f}{\partial y}(x, 0) = x$ for all x, and $\dfrac{\partial f}{\partial x}(0, y) = -y$ for all y.

b. Show that $\dfrac{\partial^2 f}{\partial y \, \partial x}(0, 0) \neq \dfrac{\partial^2 f}{\partial x \, \partial y}(0, 0)$.

The **three-dimensional Laplace equation**

$$\frac{\partial^2 f}{\partial x^2} + \frac{\partial^2 f}{\partial y^2} + \frac{\partial^2 f}{\partial z^2} = 0$$

is satisfied by steady-state temperature distributions $T = f(x, y, z)$ in space, by gravitational potentials, and by electrostatic potentials. The **two-dimensional Laplace equation**

$$\frac{\partial^2 f}{\partial x^2} + \frac{\partial^2 f}{\partial y^2} = 0,$$

obtained by dropping the $\partial^2 f / \partial z^2$ term from the previous equation, describes potentials and steady-state temperature distributions in a plane (see the accompanying figure). The plane (a) may be treated as a thin slice of the solid (b) perpendicular to the z-axis.

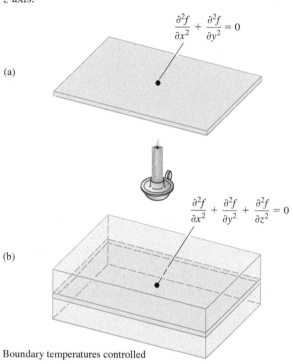

Boundary temperatures controlled

Show that each function in Exercises 83–90 satisfies a Laplace equation.

83. $f(x, y, z) = x^2 + y^2 - 2z^2$

84. $f(x, y, z) = 2z^3 - 3(x^2 + y^2)z$

85. $f(x, y) = e^{-2y} \cos 2x$

86. $f(x, y) = \ln \sqrt{x^2 + y^2}$

87. $f(x, y) = 3x + 2y - 4$

88. $f(x, y) = \tan^{-1} \dfrac{x}{y}$

89. $f(x, y, z) = (x^2 + y^2 + z^2)^{-1/2}$

90. $f(x, y, z) = e^{3x+4y} \cos 5z$

The wave equation If we stand on an ocean shore and take a snapshot of the waves, the picture shows a regular pattern of peaks and valleys in an instant of time. We see periodic vertical motion in space, with respect to distance. If we stand in the water, we can feel the rise and fall of the water as the waves go by. We see periodic vertical motion in time. In physics, this beautiful symmetry is expressed by the **one-dimensional wave equation**

$$\frac{\partial^2 w}{\partial t^2} = c^2 \frac{\partial^2 w}{\partial x^2},$$

where w is the wave height, x is the distance variable, t is the time variable, and c is the velocity with which the waves are propagated.

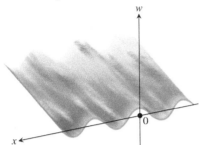

In our example, x is the distance across the ocean's surface, but in other applications, x might be the distance along a vibrating string, distance through air (sound waves), or distance through space (light waves). The number c varies with the medium and type of wave.

Show that the functions in Exercises 91–97 are all solutions of the wave equation.

91. $w = \sin(x + ct)$ **92.** $w = \cos(2x + 2ct)$

93. $w = \sin(x + ct) + \cos(2x + 2ct)$

94. $w = \ln(2x + 2ct)$ **95.** $w = \tan(2x - 2ct)$

96. $w = 5\cos(3x + 3ct) + e^{x+ct}$

97. $w = f(u)$, where f is a differentiable function of u, and $u = a(x + ct)$, where a is a constant

98. Does a function $f(x, y)$ with continuous first partial derivatives throughout an open region R have to be continuous on R? Give reasons for your answer.

99. If a function $f(x, y)$ has continuous second partial derivatives throughout an open region R, must the first-order partial derivatives of f be continuous on R? Give reasons for your answer.

100. The heat equation An important partial differential equation that describes the distribution of heat in a region at time t can be represented by the *one-dimensional heat equation*

$$\frac{\partial f}{\partial t} = \frac{\partial^2 f}{\partial x^2}.$$

Show that $u(x, t) = \sin(\alpha x) \cdot e^{-\beta t}$ satisfies the heat equation for constants α and β. What is the relationship between α and β for this function to be a solution?

101. Let $f(x, y) = \begin{cases} \dfrac{xy^2}{x^2 + y^4}, & (x, y) \neq (0, 0) \\ 0, & (x, y) = (0, 0). \end{cases}$

Show that $f_x(0, 0)$ and $f_y(0, 0)$ exist, but f is not differentiable at $(0, 0)$. (*Hint:* Use Theorem 4 and show that f is not continuous at $(0, 0)$.)

102. Let $f(x, y) = \begin{cases} 0, & x^2 < y < 2x^2 \\ 1, & \text{otherwise.} \end{cases}$

Show that $f_x(0, 0)$ and $f_y(0, 0)$ exist, but f is not differentiable at $(0, 0)$.

103. The Korteweg-deVries equation

This nonlinear differential equation, which describes wave motion on shallow water surfaces, is given by

$$4u_t + u_{xxx} + 12uu_x = 0.$$

Show that $u(x, t) = \operatorname{sech}^2(x - t)$ satisfies the Korteweg-deVries equation.

104. Show that $T = \dfrac{1}{\sqrt{x^2 + y^2}}$ satisfies the equation $T_{xx} + T_{yy} = T^3$.

14.4 The Chain Rule

The Chain Rule for functions of a single variable studied in Section 3.6 says that when $w = f(x)$ is a differentiable function of x and $x = g(t)$ is a differentiable function of t, w is a differentiable function of t and dw/dt can be calculated by the formula

$$\frac{dw}{dt} = \frac{dw}{dx}\frac{dx}{dt}.$$

To find dw/dt, we read down the route from w to t, multiplying derivatives along the way.

Chain Rule

$w = f(x)$ — Dependent variable

$\dfrac{dw}{dx}$

x — Intermediate variable

$\dfrac{dx}{dt}$

t — Independent variable

$\dfrac{dw}{dt} = \dfrac{dw}{dx}\dfrac{dx}{dt}$

For this composite function $w(t) = f(g(t))$, we can think of t as the independent variable and $x = g(t)$ as the "intermediate variable," because t determines the value of x which in turn gives the value of w from the function f. We display the Chain Rule in a "dependency diagram" in the margin. Such diagrams capture which variables depend on which.

For functions of several variables the Chain Rule has more than one form, which depends on how many independent and intermediate variables are involved. However, once the variables are taken into account, the Chain Rule works in the same way we just discussed.

Functions of Two Variables

The Chain Rule formula for a differentiable function $w = f(x, y)$ when $x = x(t)$ and $y = y(t)$ are both differentiable functions of t is given in the following theorem.

THEOREM 5—Chain Rule For Functions of One Independent Variable and Two Intermediate Variables

If $w = f(x, y)$ is differentiable and if $x = x(t)$, $y = y(t)$ are differentiable functions of t, then the composition $w = f(x(t), y(t))$ is a differentiable function of t and

$$\frac{dw}{dt} = f_x(x(t), y(t))x'(t) + f_y(x(t), y(t))y'(t),$$

or

$$\frac{dw}{dt} = \frac{\partial f}{\partial x}\frac{dx}{dt} + \frac{\partial f}{\partial y}\frac{dy}{dt}.$$

Each of $\dfrac{\partial f}{\partial x}$, $\dfrac{\partial w}{\partial x}$, f_x indicates the partial derivative of f with respect to x.

Proof   The proof consists of showing that if x and y are differentiable at $t = t_0$, then w is differentiable at t_0 and

$$\frac{dw}{dt}(t_0) = \frac{\partial w}{\partial x}(P_0)\frac{dx}{dt}(t_0) + \frac{\partial w}{\partial y}(P_0)\frac{dy}{dt}(t_0),$$

where $P_0 = (x(t_0), y(t_0))$.

Let Δx, Δy, and Δw be the increments that result from changing t from t_0 to $t_0 + \Delta t$. Since f is differentiable (see the definition in Section 14.3),

$$\Delta w = \frac{\partial w}{\partial x}(P_0)\Delta x + \frac{\partial w}{\partial y}(P_0)\Delta y + \varepsilon_1 \Delta x + \varepsilon_2 \Delta y,$$

where $\varepsilon_1, \varepsilon_2 \to 0$ as $\Delta x, \Delta y \to 0$. To find dw/dt, we divide this equation through by Δt and let Δt approach zero. The division gives

$$\frac{\Delta w}{\Delta t} = \frac{\partial w}{\partial x}(P_0)\frac{\Delta x}{\Delta t} + \frac{\partial w}{\partial y}(P_0)\frac{\Delta y}{\Delta t} + \varepsilon_1\frac{\Delta x}{\Delta t} + \varepsilon_2\frac{\Delta y}{\Delta t}.$$

Letting Δt approach zero gives

$$\frac{dw}{dt}(t_0) = \lim_{\Delta t \to 0}\frac{\Delta w}{\Delta t}$$

$$= \frac{\partial w}{\partial x}(P_0)\frac{dx}{dt}(t_0) + \frac{\partial w}{\partial y}(P_0)\frac{dy}{dt}(t_0) + 0 \cdot \frac{dx}{dt}(t_0) + 0 \cdot \frac{dy}{dt}(t_0). \qquad \blacksquare$$

Often we write $\partial w/\partial x$ for the partial derivative $\partial f/\partial x$, so we can rewrite the Chain Rule in Theorem 5 in the form

To remember the Chain Rule, picture the diagram below. To find dw/dt, start at w and read down each route to t, multiplying derivatives along the way. Then add the products.

Chain Rule

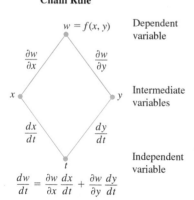

$$\frac{dw}{dt} = \frac{\partial w}{\partial x}\frac{dx}{dt} + \frac{\partial w}{\partial y}\frac{dy}{dt}.$$

However, the meaning of the dependent variable w is different on each side of the preceding equation. On the left-hand side, it refers to the composite function $w = f(x(t), y(t))$ as a function of the single variable t. On the right-hand side, it refers to the function $w = f(x, y)$ as a function of the two variables x and y. Moreover, the single derivatives dw/dt, dx/dt, and dy/dt are being evaluated at a point t_0, whereas the partial derivatives $\partial w/\partial x$ and $\partial w/\partial y$ are being evaluated at the point (x_0, y_0), with $x_0 = x(t_0)$ and $y_0 = y(t_0)$. With that understanding, we will use both of these forms interchangeably throughout the text whenever no confusion will arise.

The **dependency diagram** on the preceding page provides a convenient way to remember the Chain Rule. The "true" independent variable in the composite function is t, whereas x and y are *intermediate variables* (controlled by t) and w is the dependent variable.

A more precise notation for the Chain Rule shows where the various derivatives in Theorem 5 are evaluated:

$$\frac{dw}{dt}(t_0) = \frac{\partial f}{\partial x}(x_0, y_0)\frac{dx}{dt}(t_0) + \frac{\partial f}{\partial y}(x_0, y_0)\frac{dy}{dt}(t_0),$$

or, using another notation,

$$\left.\frac{dw}{dt}\right|_{t_0} = \left.\frac{\partial f}{\partial x}\right|_{(x_0, y_0)}\left.\frac{dx}{dt}\right|_{t_0} + \left.\frac{\partial f}{\partial y}\right|_{(x_0, y_0)}\left.\frac{dy}{dt}\right|_{t_0}.$$

EXAMPLE 1   Use the Chain Rule to find the derivative of

$$w = xy$$

with respect to t along the path $x = \cos t, y = \sin t$. What is the derivative's value at $t = \pi/2$?

Solution   We apply the Chain Rule to find dw/dt as follows:

$$\frac{dw}{dt} = \frac{\partial w}{\partial x}\frac{dx}{dt} + \frac{\partial w}{\partial y}\frac{dy}{dt}$$

$$= \frac{\partial(xy)}{\partial x}\frac{d}{dt}(\cos t) + \frac{\partial(xy)}{\partial y}\frac{d}{dt}(\sin t)$$

$$= (y)(-\sin t) + (x)(\cos t)$$

$$= (\sin t)(-\sin t) + (\cos t)(\cos t)$$

$$= -\sin^2 t + \cos^2 t$$

$$= \cos 2t.$$

In this example, we can check the result with a more direct calculation. As a function of t,

$$w = xy = \cos t \sin t = \frac{1}{2}\sin 2t,$$

so

$$\frac{dw}{dt} = \frac{d}{dt}\left(\frac{1}{2}\sin 2t\right) = \frac{1}{2}(2\cos 2t) = \cos 2t.$$

In either case, at the given value of t,

$$\left.\frac{dw}{dt}\right|_{t=\pi/2} = \cos\left(2\frac{\pi}{2}\right) = \cos\pi = -1. \qquad \blacksquare$$

Functions of Three Variables

You can probably predict the Chain Rule for functions of three intermediate variables, as it only involves adding the expected third term to the two-variable formula.

THEOREM 6—Chain Rule for Functions of One Independent Variable and Three Intermediate Variables

If $w = f(x, y, z)$ is differentiable and x, y, and z are differentiable functions of t, then w is a differentiable function of t and

$$\frac{dw}{dt} = \frac{\partial w}{\partial x}\frac{dx}{dt} + \frac{\partial w}{\partial y}\frac{dy}{dt} + \frac{\partial w}{\partial z}\frac{dz}{dt}.$$

Here we have three routes from w to t instead of two, but finding dw/dt is still the same. Read down each route, multiplying derivatives along the way; then add.

Chain Rule

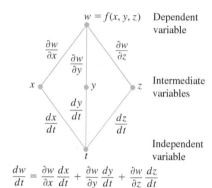

$$\frac{dw}{dt} = \frac{\partial w}{\partial x}\frac{dx}{dt} + \frac{\partial w}{\partial y}\frac{dy}{dt} + \frac{\partial w}{\partial z}\frac{dz}{dt}$$

The proof is identical with the proof of Theorem 5 except that there are now three intermediate variables instead of two. The dependency diagram we use for remembering the new equation is similar as well, with three routes from w to t.

EXAMPLE 2   Find dw/dt if

$$w = xy + z, \qquad x = \cos t, \qquad y = \sin t, \qquad z = t.$$

In this example the values of $w(t)$ are changing along the path of a helix (Section 13.1) as t changes. What is the derivative's value at $t = 0$?

Solution   Using the Chain Rule for three intermediate variables, we have

$$\frac{dw}{dt} = \frac{\partial w}{\partial x}\frac{dx}{dt} + \frac{\partial w}{\partial y}\frac{dy}{dt} + \frac{\partial w}{\partial z}\frac{dz}{dt}$$

$$= (y)(-\sin t) + (x)(\cos t) + (1)(1)$$

$$= (\sin t)(-\sin t) + (\cos t)(\cos t) + 1 \qquad \text{Substitute for intermediate variables.}$$

$$= -\sin^2 t + \cos^2 t + 1 = 1 + \cos 2t,$$

so

$$\left.\frac{dw}{dt}\right|_{t=0} = 1 + \cos(0) = 2.$$ ∎

For a physical interpretation of change along a curve, think of an object whose position is changing with time t. If $w = T(x, y, z)$ is the temperature at each point (x, y, z) along a curve C with parametric equations $x = x(t)$, $y = y(t)$, and $z = z(t)$, then the composite function $w = T(x(t), y(t), z(t))$ represents the temperature relative to t along the curve. The derivative dw/dt is then the instantaneous rate of change of temperature due to the motion along the curve, as calculated in Theorem 6.

Functions Defined on Surfaces

If we are interested in the temperature $w = f(x, y, z)$ at points (x, y, z) on the earth's surface, we might prefer to think of x, y, and z as functions of the variables r and s that give the points' longitudes and latitudes. If $x = g(r, s)$, $y = h(r, s)$, and $z = k(r, s)$, we could then express the temperature as a function of r and s with the composite function

$$w = f(g(r, s), h(r, s), k(r, s)).$$

Under the conditions stated below, w has partial derivatives with respect to both r and s that can be calculated in the following way.

THEOREM 7—Chain Rule for Two Independent Variables and Three Intermediate Variables

Suppose that $w = f(x, y, z)$, $x = g(r, s)$, $y = h(r, s)$, and $z = k(r, s)$. If all four functions are differentiable, then w has partial derivatives with respect to r and s, given by the formulas

$$\frac{\partial w}{\partial r} = \frac{\partial w}{\partial x}\frac{\partial x}{\partial r} + \frac{\partial w}{\partial y}\frac{\partial y}{\partial r} + \frac{\partial w}{\partial z}\frac{\partial z}{\partial r}$$

$$\frac{\partial w}{\partial s} = \frac{\partial w}{\partial x}\frac{\partial x}{\partial s} + \frac{\partial w}{\partial y}\frac{\partial y}{\partial s} + \frac{\partial w}{\partial z}\frac{\partial z}{\partial s}.$$

The first of these equations can be derived from the Chain Rule in Theorem 6 by holding s fixed and treating r as t. The second can be derived in the same way, holding r fixed and treating s as t. The dependency diagrams for both equations are shown in Figure 14.22.

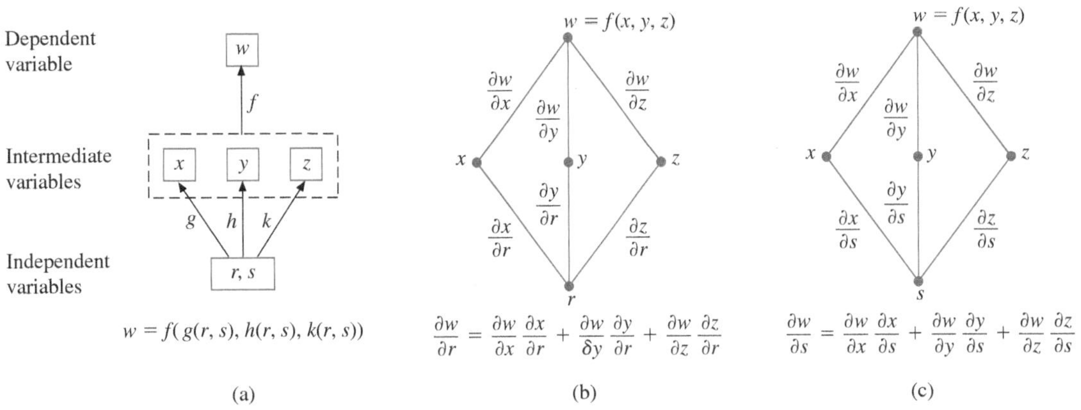

Dependent variable

Intermediate variables

Independent variables

$w = f(g(r, s), h(r, s), k(r, s))$

(a)

$w = f(x, y, z)$

$$\frac{\partial w}{\partial r} = \frac{\partial w}{\partial x}\frac{\partial x}{\partial r} + \frac{\partial w}{\partial y}\frac{\partial y}{\partial r} + \frac{\partial w}{\partial z}\frac{\partial z}{\partial r}$$

(b)

$w = f(x, y, z)$

$$\frac{\partial w}{\partial s} = \frac{\partial w}{\partial x}\frac{\partial x}{\partial s} + \frac{\partial w}{\partial y}\frac{\partial y}{\partial s} + \frac{\partial w}{\partial z}\frac{\partial z}{\partial s}$$

(c)

FIGURE 14.22 Composite function and dependency diagrams for Theorem 7.

EXAMPLE 3   Express $\partial w/\partial r$ and $\partial w/\partial s$ in terms of r and s if

$$w = x + 2y + z^2, \qquad x = \frac{r}{s}, \qquad y = r^2 + \ln s, \qquad z = 2r.$$

Solution   Using the formulas in Theorem 7, we find

$$\frac{\partial w}{\partial r} = \frac{\partial w}{\partial x}\frac{\partial x}{\partial r} + \frac{\partial w}{\partial y}\frac{\partial y}{\partial r} + \frac{\partial w}{\partial z}\frac{\partial z}{\partial r}$$

$$= (1)\left(\frac{1}{s}\right) + (2)(2r) + (2z)(2)$$

$$= \frac{1}{s} + 4r + (4r)(2) = \frac{1}{s} + 12r \qquad \text{Substitute for intermediate variable } z.$$

$$\frac{\partial w}{\partial s} = \frac{\partial w}{\partial x}\frac{\partial x}{\partial s} + \frac{\partial w}{\partial y}\frac{\partial y}{\partial s} + \frac{\partial w}{\partial z}\frac{\partial z}{\partial s}$$

$$= (1)\left(-\frac{r}{s^2}\right) + (2)\left(\frac{1}{s}\right) + (2z)(0) = \frac{2}{s} - \frac{r}{s^2}.$$

If f is a function of two intermediate variables instead of three, each equation in Theorem 7 becomes correspondingly one term shorter.

Chain Rule

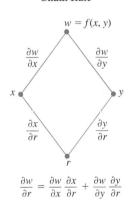

$w = f(x, y)$

$$\frac{\partial w}{\partial r} = \frac{\partial w}{\partial x}\frac{\partial x}{\partial r} + \frac{\partial w}{\partial y}\frac{\partial y}{\partial r}$$

FIGURE 14.23 Dependency diagram for the equation

$$\frac{\partial w}{\partial r} = \frac{\partial w}{\partial x}\frac{\partial x}{\partial r} + \frac{\partial w}{\partial y}\frac{\partial y}{\partial r}.$$

If $w = f(x, y)$, $x = g(r, s)$, and $y = h(r, s)$, then

$$\frac{\partial w}{\partial r} = \frac{\partial w}{\partial x}\frac{\partial x}{\partial r} + \frac{\partial w}{\partial y}\frac{\partial y}{\partial r} \qquad \text{and} \qquad \frac{\partial w}{\partial s} = \frac{\partial w}{\partial x}\frac{\partial x}{\partial s} + \frac{\partial w}{\partial y}\frac{\partial y}{\partial s}.$$

Figure 14.23 shows the dependency diagram for the first of these equations. The diagram for the second equation is similar; just replace r with s.

EXAMPLE 4   Express $\partial w/\partial r$ and $\partial w/\partial s$ in terms of r and s if

$$w = x^2 + y^2, \qquad x = r - s, \qquad y = r + s.$$

Solution The preceding discussion gives the following.

$$\frac{\partial w}{\partial r} = \frac{\partial w}{\partial x}\frac{\partial x}{\partial r} + \frac{\partial w}{\partial y}\frac{\partial y}{\partial r} \qquad\qquad \frac{\partial w}{\partial s} = \frac{\partial w}{\partial x}\frac{\partial x}{\partial s} + \frac{\partial w}{\partial y}\frac{\partial y}{\partial s}$$

$$= (2x)(1) + (2y)(1) \qquad\qquad = (2x)(-1) + (2y)(1)$$

$$= 2(r-s) + 2(r+s) \qquad\qquad = -2(r-s) + 2(r+s)$$

$$= 4r \qquad\qquad\qquad\qquad\qquad = 4s$$

Substitute for the intermediate variables.

If f is a function of a single intermediate variable x, our equations are even simpler.

If $w = f(x)$ and $x = g(r, s)$, then

$$\frac{\partial w}{\partial r} = \frac{dw}{dx}\frac{\partial x}{\partial r} \qquad \text{and} \qquad \frac{\partial w}{\partial s} = \frac{dw}{dx}\frac{\partial x}{\partial s}.$$

In this case, we use the ordinary (single-variable) derivative, dw/dx. The dependency diagram is shown in Figure 14.24.

Implicit Differentiation Revisited

The two-variable Chain Rule in Theorem 5 leads to a formula that takes some of the algebra out of implicit differentiation. Suppose that

1. The function $F(x, y)$ is differentiable and

2. The equation $F(x, y) = 0$ defines y implicitly as a differentiable function of x, say $y = h(x)$.

Since $w = F(x, y) = 0$, the derivative dw/dx must be zero. Computing the derivative from the Chain Rule (dependency diagram in Figure 14.25), we find

$$0 = \frac{dw}{dx} = F_x\frac{dx}{dx} + F_y\frac{dy}{dx} \qquad \text{\small Theorem 5 with}$$
$$\qquad\qquad\qquad\qquad\qquad\qquad \text{\small } t = x \text{ and } f = F$$

$$= F_x \cdot 1 + F_y \cdot \frac{dy}{dx}.$$

If $F_y = \partial w/\partial y \neq 0$, we can solve this equation for dy/dx to get

$$\frac{dy}{dx} = -\frac{F_x}{F_y}.$$

We state this result formally.

THEOREM 8—A Formula for Implicit Differentiation

Suppose that $F(x, y)$ is differentiable and that the equation $F(x, y) = 0$ defines y as a differentiable function of x. Then at any point where $F_y \neq 0$,

$$\frac{dy}{dx} = -\frac{F_x}{F_y}. \qquad\qquad (1)$$

Chain Rule

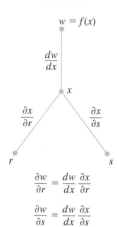

FIGURE 14.24 Dependency diagram for differentiating f as a composite function of r and s with one intermediate variable.

$w = F(x, y)$

$\frac{\partial w}{\partial x} = F_x$ $\qquad$ $F_y = \frac{\partial w}{\partial y}$

x $\qquad\qquad$ $y = h(x)$

$\frac{dx}{dx} = 1$ $\qquad$ $\frac{dy}{dx} = h'(x)$

x

$\frac{dw}{dx} = F_x \cdot 1 + F_y \cdot \frac{dy}{dx}$

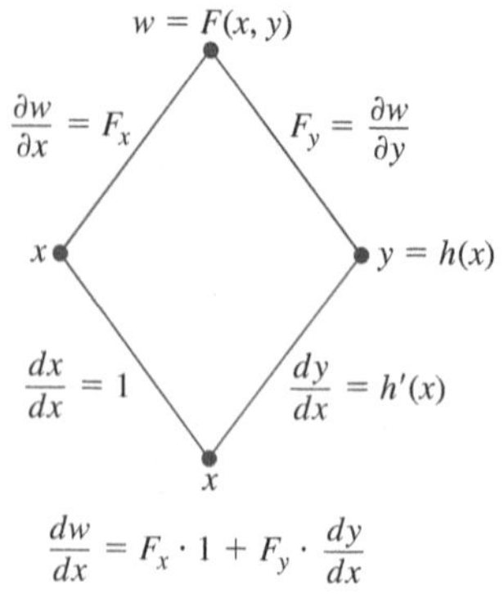

FIGURE 14.25 Dependency diagram for differentiating $w = F(x, y)$ with respect to x. Setting $dw/dx = 0$ leads to a simple computational formula for implicit differentiation (Theorem 8).

EXAMPLE 5   Use Theorem 8 to find dy/dx if $y^2 - x^2 - \sin xy = 0$.

Solution   Take $F(x, y) = y^2 - x^2 - \sin xy$. Then

$$\frac{dy}{dx} = -\frac{F_x}{F_y} = -\frac{-2x - y \cos xy}{2y - x \cos xy} = \frac{2x + y \cos xy}{2y - x \cos xy}.$$

This calculation is significantly shorter than a single-variable calculation using implicit differentiation.   ∎

The result in Theorem 8 is easily extended to three variables. Suppose that the equation $F(x, y, z) = 0$ defines the variable z implicitly as a function $z = f(x, y)$. Then for all (x, y) in the domain of f, we have $F(x, y, f(x, y)) = 0$. Assuming that F and f are differentiable functions, we can use the Chain Rule to differentiate the equation $F(x, y, z) = 0$ with respect to the independent variable x:

$$0 = \frac{\partial F}{\partial x}\frac{\partial x}{\partial x} + \frac{\partial F}{\partial y}\frac{\partial y}{\partial x} + \frac{\partial F}{\partial z}\frac{\partial z}{\partial x}$$

$$= F_x \cdot 1 + F_y \cdot 0 + F_z \cdot \frac{\partial z}{\partial x}, \qquad \begin{array}{l} y \text{ is constant when} \\ \text{differentiating with} \\ \text{respect to } x. \end{array}$$

so

$$F_x + F_z \frac{\partial z}{\partial x} = 0.$$

A similar calculation for differentiating with respect to the independent variable y gives

$$F_y + F_z \frac{\partial z}{\partial y} = 0.$$

Whenever $F_z \neq 0$, we can solve these last two equations for the partial derivatives of $z = f(x, y)$ to obtain

$$\frac{\partial z}{\partial x} = -\frac{F_x}{F_z} \quad \text{and} \quad \frac{\partial z}{\partial y} = -\frac{F_y}{F_z}. \tag{2}$$

An important result from advanced calculus, called the **Implicit Function Theorem**, states the conditions for which our results in Equations (2) are valid. If the partial derivatives F_x, F_y, and F_z are continuous throughout an open region R in space containing the point (x_0, y_0, z_0), and if for some constant c, $F(x_0, y_0, z_0) = c$ and $F_z(x_0, y_0, z_0) \neq 0$, then the equation $F(x, y, z) = c$ defines z implicitly as a differentiable function of x and y near (x_0, y_0, z_0), and the partial derivatives of z are given by Equations (2).

EXAMPLE 6   Find $\dfrac{\partial z}{\partial x}$ and $\dfrac{\partial z}{\partial y}$ at $(0, 0, 0)$ if $x^3 + z^2 + ye^{xz} + z \cos y = 0$.

Solution   Let $F(x, y, z) = x^3 + z^2 + ye^{xz} + z \cos y$. Then

$$F_x = 3x^2 + zye^{xz}, \qquad F_y = e^{xz} - z \sin y, \qquad \text{and} \qquad F_z = 2z + xye^{xz} + \cos y.$$

Since $F(0, 0, 0) = 0$, $F_z(0, 0, 0) = 1 \neq 0$, and all first partial derivatives are continuous, the Implicit Function Theorem says that $F(x, y, z) = 0$ defines z as a differentiable function of x and y near the point $(0, 0, 0)$. From Equations (2),

$$\frac{\partial z}{\partial x} = -\frac{F_x}{F_z} = -\frac{3x^2 + zye^{xz}}{2z + xye^{xz} + \cos y} \quad \text{and} \quad \frac{\partial z}{\partial y} = -\frac{F_y}{F_z} = -\frac{e^{xz} - z \sin y}{2z + xye^{xz} + \cos y}.$$

At $(0, 0, 0)$ we find

$$\frac{\partial z}{\partial x} = -\frac{0}{1} = 0 \quad \text{and} \quad \frac{\partial z}{\partial y} = -\frac{1}{1} = -1. \qquad \blacksquare$$

Functions of Many Variables

We have seen several different forms of the Chain Rule in this section, but each one is just a special case of one general formula. When solving particular problems, it may help to draw the appropriate dependency diagram by placing the dependent variable on top, the intermediate variables in the middle, and the selected independent variable at the bottom. To find the derivative of the dependent variable with respect to the selected independent variable, start at the dependent variable and read down each route of the dependency diagram to the independent variable, calculating and multiplying the derivatives along each route. Then add the products found for the different routes.

In general, suppose that $w = f(x, y, \ldots, v)$ is a differentiable function of the intermediate variables $x, y, \ldots, v$ (a finite set) and the $x, y, \ldots, v$ are differentiable functions of the independent variables $p, q, \ldots, t$ (another finite set). Then w is a differentiable function of the variables p through t, and the partial derivatives of w with respect to these variables are given by equations of the form

$$\frac{\partial w}{\partial p} = \frac{\partial w}{\partial x}\frac{\partial x}{\partial p} + \frac{\partial w}{\partial y}\frac{\partial y}{\partial p} + \cdots + \frac{\partial w}{\partial v}\frac{\partial v}{\partial p}.$$

The other equations are obtained by replacing p by $q, \ldots, t$, one at a time.

One way to remember this equation is to think of the right-hand side as the dot product of two vectors with components

$$\underbrace{\left(\frac{\partial w}{\partial x}, \frac{\partial w}{\partial y}, \ldots, \frac{\partial w}{\partial v}\right)}_{\substack{\text{Derivatives of } w \text{ with} \\ \text{respect to the} \\ \text{intermediate variables}}} \quad \text{and} \quad \underbrace{\left(\frac{\partial x}{\partial p}, \frac{\partial y}{\partial p}, \ldots, \frac{\partial v}{\partial p}\right)}_{\substack{\text{Derivatives of the intermediate} \\ \text{variables with respect to the} \\ \text{selected independent variable}}}.$$

EXERCISES 14.4

Chain Rule: One Independent Variable

In Exercises 1–6, (a) express dw/dt as a function of t, both by using the Chain Rule and by expressing w in terms of t and differentiating directly with respect to t. Then (b) evaluate dw/dt at the given value of t.

1. $w = x^2 + y^2$, $x = \cos t$, $y = \sin t$; $t = \pi$

2. $w = x^2 + y^2$, $x = \cos t + \sin t$, $y = \cos t - \sin t$; $t = 0$

3. $w = \frac{x}{z} + \frac{y}{z}$, $x = \cos^2 t$, $y = \sin^2 t$, $z = 1/t$; $t = 3$

4. $w = \ln(x^2 + y^2 + z^2)$, $x = \cos t$, $y = \sin t$, $z = 4\sqrt{t}$; $t = 3$

5. $w = 2ye^x - \ln z$, $x = \ln(t^2 + 1)$, $y = \tan^{-1} t$, $z = e^t$; $t = 1$

6. $w = z - \sin xy$, $x = t$, $y = \ln t$, $z = e^{t-1}$; $t = 1$

Chain Rule: Two and Three Independent Variables

In Exercises 7 and 8, (a) express $\partial z/\partial u$ and $\partial z/\partial v$ as functions of u and v both by using the Chain Rule and by expressing z directly in terms of u and v before differentiating. Then (b) evaluate $\partial z/\partial u$ and $\partial z/\partial v$ at the given point (u, v).

7. $z = 4e^x \ln y$, $x = \ln(u \cos v)$, $y = u \sin v$;
   $(u, v) = (2, \pi/4)$

8. $z = \tan^{-1}(x/y)$, $x = u\cos v$, $y = u\sin v$;
  $(u, v) = (1.3, \pi/6)$

In Exercises 9 and 10, **(a)** express $\partial w/\partial u$ and $\partial w/\partial v$ as functions of u and v both by using the Chain Rule and by expressing w directly in terms of u and v before differentiating. Then **(b)** evaluate $\partial w/\partial u$ and $\partial w/\partial v$ at the given point (u, v).

9. $w = xy + yz + xz$, $x = u + v$, $y = u - v$, $z = uv$;
  $(u, v) = (1/2, 1)$

10. $w = \ln(x^2 + y^2 + z^2)$, $x = ue^v \sin u$, $y = ue^v \cos u$,
  $z = ue^v$; $(u, v) = (-2, 0)$

In Exercises 11 and 12, **(a)** express $\partial u/\partial x$, $\partial u/\partial y$, and $\partial u/\partial z$ as functions of x, y, and z both by using the Chain Rule and by expressing u directly in terms of x, y, and z before differentiating. Then **(b)** evaluate $\partial u/\partial x$, $\partial u/\partial y$, and $\partial u/\partial z$ at the given point (x, y, z).

11. $u = \dfrac{p - q}{q - r}$, $p = x + y + z$, $q = x - y + z$,
  $r = x + y - z$; $(x, y, z) = \left(\sqrt{3}, 2, 1\right)$

12. $u = e^{qr}\sin^{-1} p$, $p = \sin x$, $q = z^2 \ln y$, $r = 1/z$;
  $(x, y, z) = (\pi/4, 1/2, -1/2)$

Using a Dependency Diagram

In Exercises 13–24, draw a dependency diagram and write a Chain Rule formula for each derivative.

13. $\dfrac{dz}{dt}$ for $z = f(x, y)$, $x = g(t)$, $y = h(t)$

14. $\dfrac{dz}{dt}$ for $z = f(u, v, w)$, $u = g(t)$, $v = h(t)$, $w = k(t)$

15. $\dfrac{\partial w}{\partial u}$ and $\dfrac{\partial w}{\partial v}$ for $w = h(x, y, z)$, $x = f(u, v)$, $y = g(u, v)$,
  $z = k(u, v)$

16. $\dfrac{\partial w}{\partial x}$ and $\dfrac{\partial w}{\partial y}$ for $w = f(r, s, t)$, $r = g(x, y)$, $s = h(x, y)$,
  $t = k(x, y)$

17. $\dfrac{\partial w}{\partial u}$ and $\dfrac{\partial w}{\partial v}$ for $w = g(x, y)$, $x = h(u, v)$, $y = k(u, v)$

18. $\dfrac{\partial w}{\partial x}$ and $\dfrac{\partial w}{\partial y}$ for $w = g(u, v)$, $u = h(x, y)$, $v = k(x, y)$

19. $\dfrac{\partial z}{\partial t}$ and $\dfrac{\partial z}{\partial s}$ for $z = f(x, y)$, $x = g(t, s)$, $y = h(t, s)$

20. $\dfrac{\partial y}{\partial r}$ for $y = f(u)$, $u = g(r, s)$

21. $\dfrac{\partial w}{\partial s}$ and $\dfrac{\partial w}{\partial t}$ for $w = g(u)$, $u = h(s, t)$

22. $\dfrac{\partial w}{\partial p}$ for $w = f(x, y, z, v)$, $x = g(p, q)$, $y = h(p, q)$,
  $z = j(p, q)$, $v = k(p, q)$

23. $\dfrac{\partial w}{\partial r}$ and $\dfrac{\partial w}{\partial s}$ for $w = f(x, y)$, $x = g(r)$, $y = h(s)$

24. $\dfrac{\partial w}{\partial s}$ for $w = g(x, y)$, $x = h(r, s, t)$, $y = k(r, s, t)$

Implicit Differentiation

Assuming that the equations in Exercises 25–30 define y as a differentiable function of x, use Theorem 8 to find the value of dy/dx at the given point.

25. $x^3 - 2y^2 + xy = 0$, $(1, 1)$

26. $xy + y^2 - 3x - 3 = 0$, $(-1, 1)$

27. $x^2 + xy + y^2 - 7 = 0$, $(1, 2)$

28. $xe^y + \sin xy + y - \ln 2 = 0$, $(0, \ln 2)$

29. $(x^3 - y^4)^6 + \ln(x^2 + y) = 1$, $(-1, 0)$

30. $xe^{x^2 y} - ye^x = x - y + 2$, $(1, 1)$

Find the values of $\partial z/\partial x$ and $\partial z/\partial y$ at the points in Exercises 31–34.

31. $z^3 - xy + yz + y^3 - 2 = 0$, $(1, 1, 1)$

32. $\dfrac{1}{x} + \dfrac{1}{y} + \dfrac{1}{z} - 1 = 0$, $(2, 3, 6)$

33. $\sin(x + y) + \sin(y + z) + \sin(x + z) = 0$, (π, π, π)

34. $xe^y + ye^z + 2\ln x - 2 - 3\ln 2 = 0$, $(1, \ln 2, \ln 3)$

Finding Partial Derivatives at Specified Points

35. Find $\partial w/\partial r$ when $r = 1, s = -1$ if $w = (x + y + z)^2$, $x = r - s$, $y = \cos(r + s), z = \sin(r + s)$.

36. Find $\partial w/\partial v$ when $u = -1, v = 2$ if $w = xy + \ln z$, $x = v^2/u, y = u + v, z = \cos u$.

37. Find $\partial w/\partial v$ when $u = 0, v = 0$ if $w = x^2 + (y/x)$, $x = u - 2v + 1, y = 2u + v - 2$.

38. Find $\partial z/\partial u$ when $u = 0, v = 1$ if $z = \sin xy + x\sin y$, $x = u^2 + v^2, y = uv$.

39. Find $\partial z/\partial u$ and $\partial z/\partial v$ when $u = \ln 2, v = 1$ if $z = 5\tan^{-1} x$ and $x = e^u + \ln v$.

40. Find $\partial z/\partial u$ and $\partial z/\partial v$ when $u = 1, v = -2$ if $z = \ln q$ and $q = \sqrt{v} + 3\tan^{-1} u$.

Theory and Examples

41. Assume that $w = f(s^3 + t^2)$ and $f'(x) = e^x$. Find $\dfrac{\partial w}{\partial t}$ and $\dfrac{\partial w}{\partial s}$.

42. Assume that $w = f\left(ts^2, \dfrac{s}{t}\right)$, $\dfrac{\partial f}{\partial x}(x, y) = xy$, and $\dfrac{\partial f}{\partial y}(x, y) = \dfrac{x^2}{2}$.
  Find $\dfrac{\partial w}{\partial t}$ and $\dfrac{\partial w}{\partial s}$.

43. Assume that $z = f(x, y)$, $x = g(t)$, $y = h(t)$, $f_x(2, -1) = 3$, and $f_y(2, -1) = -2$. If $g(0) = 2$, $h(0) = -1$, $g'(0) = 5$, and $h'(0) = -4$, find $\dfrac{dz}{dt}\Big|_{t=0}$.

44. Assume that $z = f(x, y)^2$, $x = g(t)$, $y = h(t)$, $f_x(1, 0) = -1$, $f_y(1, 0) = 1$, and $f(1, 0) = 2$. If $g(3) = 1$, $h(3) = 0$, $g'(3) = -3$, and $h'(3) = 4$, find $\dfrac{dz}{dt}\Big|_{t=3}$.

45. Assume that $z = f(w)$, $w = g(x, y)$, $x = 2r^3 - s^2$, and $y = re^s$. If $g_x(2, 1) = -3$, $g_y(2, 1) = 2$, $f'(7) = -1$, and $g(2, 1) = 7$, find $\dfrac{\partial z}{\partial r}\Big|_{r=1, s=0}$ and $\dfrac{\partial z}{\partial s}\Big|_{r=1, s=0}$.

46. Assume that $z = \ln(f(w))$, $w = g(x, y)$, $x = \sqrt{r - s}$, and $y = r^2 s$. If $g_x(2, -9) = -1$, $g_y(2, -9) = 3$, $f'(-2) = 2$, $f(-2) = 5$, and $g(2, -9) = -2$, find $\dfrac{\partial z}{\partial r}\Big|_{r=3, s=-1}$ and $\dfrac{\partial z}{\partial s}\Big|_{r=3, s=-1}$.

47. Changing voltage in a circuit The voltage V in a circuit that satisfies the law $V = IR$ is slowly dropping as the battery wears out. At the same time, the resistance R is increasing as the resistor heats up. Use the equation

$$\frac{dV}{dt} = \frac{\partial V}{\partial I}\frac{dI}{dt} + \frac{\partial V}{\partial R}\frac{dR}{dt}$$

to find how the current is changing at the instant when $R = 600$ ohms, $I = 0.04$ amp, $dR/dt = 0.5$ ohm/sec, and $dV/dt = -0.01$ volt/sec.

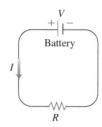

48. Changing dimensions in a box The lengths a, b, and c of the edges of a rectangular box are changing with time. At the instant in question, $a = 1$ m, $b = 2$ m, $c = 3$ m, $da/dt = db/dt = 1$ m/sec, and $dc/dt = -3$ m/sec. At what rates are the box's volume V and surface area S changing at that instant? Are the box's interior diagonals increasing in length or decreasing?

49. If $f(u, v, w)$ is differentiable and $u = x - y, v = y - z$, and $w = z - x$, show that

$$\frac{\partial f}{\partial x} + \frac{\partial f}{\partial y} + \frac{\partial f}{\partial z} = 0.$$

50. Polar coordinates Suppose that we substitute polar coordinates $x = r\cos\theta$ and $y = r\sin\theta$ in a differentiable function $w = f(x, y)$.

a. Show that

$$\frac{\partial w}{\partial r} = f_x\cos\theta + f_y\sin\theta$$

and

$$\frac{1}{r}\frac{\partial w}{\partial \theta} = -f_x\sin\theta + f_y\cos\theta.$$

b. Solve the equations in part (a) to express f_x and f_y in terms of $\partial w/\partial r$ and $\partial w/\partial\theta$.

c. Show that

$$(f_x)^2 + (f_y)^2 = \left(\frac{\partial w}{\partial r}\right)^2 + \frac{1}{r^2}\left(\frac{\partial w}{\partial\theta}\right)^2.$$

51. Laplace equations Show that if $w = f(u, v)$ satisfies the Laplace equation $f_{uu} + f_{vv} = 0$ and if $u = (x^2 - y^2)/2$ and $v = xy$, then w satisfies the Laplace equation $w_{xx} + w_{yy} = 0$.

52. Laplace equations Let $w = f(u) + g(v)$, where $u = x + iy$, $v = x - iy$, and $i = \sqrt{-1}$. Show that w satisfies the Laplace equation $w_{xx} + w_{yy} = 0$ if all the necessary functions are differentiable.

53. Extreme values on a helix Suppose that the partial derivatives of a function $f(x, y, z)$ at points on the helix $x = \cos t, y = \sin t$, $z = t$ are

$$f_x = \cos t, \qquad f_y = \sin t, \qquad f_z = t^2 + t - 2.$$

At what points on the curve, if any, can f take on extreme values?

54. A space curve Let $w = x^2 e^{2y} \cos 3z$. Find the value of dw/dt at the point $(1, \ln 2, 0)$ on the curve $x = \cos t, y = \ln(t + 2)$, $z = t$.

55. Temperature on a circle Let $T = f(x, y)$ be the temperature at the point (x, y) on the circle $x = \cos t, y = \sin t, 0 \le t \le 2\pi$ and suppose that

$$\frac{\partial T}{\partial x} = 8x - 4y, \qquad \frac{\partial T}{\partial y} = 8y - 4x.$$

a. Find where the maximum and minimum temperatures on the circle occur by examining the derivatives dT/dt and d^2T/dt^2.

b. Suppose that $T = 4x^2 - 4xy + 4y^2$. Find the maximum and minimum values of T on the circle.

56. Temperature on an ellipse Let $T = g(x, y)$ be the temperature at the point (x, y) on the ellipse

$$x = 2\sqrt{2}\cos t, \qquad y = \sqrt{2}\sin t, \qquad 0 \le t \le 2\pi,$$

and suppose that

$$\frac{\partial T}{\partial x} = y, \qquad \frac{\partial T}{\partial y} = x.$$

a. Locate the maximum and minimum temperatures on the ellipse by examining dT/dt and d^2T/dt^2.

b. Suppose that $T = xy - 2$. Find the maximum and minimum values of T on the ellipse.

57. The temperature $T = T(x, y)$ in °C at point (x, y) satisfies $T_x(1, 2) = 3$ and $T_y(1, 2) = -1$. If $x = e^{2t-2}$ cm and $y = 2 + \ln t$ cm, find the rate at which the temperature T changes when $t = 1$ sec.

58. A bug crawls on the surface $z = x^2 - y^2$ directly above a path in the xy-plane given by $x = f(t)$ and $y = g(t)$. If $f(2) = 4, f'(2) = -1, g(2) = -2$, and $g'(2) = -3$, then at what rate is the bug's elevation z changing when $t = 2$?

Differentiating Integrals Under mild continuity restrictions, it is true that if

$$F(x) = \int_a^b g(t, x)\,dt,$$

then $F'(x) = \displaystyle\int_a^b g_x(t, x)\,dt$. Using this fact and the Chain Rule, we can find the derivative of

$$F(x) = \int_a^{f(x)} g(t, x)\,dt$$

by letting

$$G(u, x) = \int_a^u g(t, x)\,dt,$$

where $u = f(x)$. Find the derivatives of the functions in Exercises 59 and 60.

59. $F(x) = \displaystyle\int_0^{x^2} \sqrt{t^4 + x^3}\,dt$

60. $F(x) = \displaystyle\int_{x^2}^1 \sqrt{t^3 + x^2}\,dt$

14.5 Directional Derivatives and Gradient Vectors

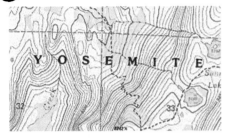

If you look at the map (Figure 14.26) showing contours within Yosemite National Park in California, you will notice that the streams flow perpendicular to the contours. The streams are following paths of steepest descent so the waters reach lower elevations as quickly as possible. Therefore, the fastest instantaneous rate of change in a stream's elevation above sea level has a particular direction. In this section, you will see why this direction, called the "downhill" direction, is perpendicular to the contours.

FIGURE 14.26 Contours within Yosemite National Park in California show streams, which follow paths of steepest descent, running perpendicular to the contours. (*Source:* Yosemite National Park Map from U.S. Geological Survey, http://www.usgs.gov)

Directional Derivatives in the Plane

We know from Section 14.4 that if $f(x, y)$ is differentiable, then the rate at which f changes with respect to t along a differentiable curve $x = g(t)$, $y = h(t)$ is

$$\frac{df}{dt} = \frac{\partial f}{\partial x}\frac{dx}{dt} + \frac{\partial f}{\partial y}\frac{dy}{dt}.$$

At any point $P_0(x_0, y_0) = P_0(g(t_0), h(t_0))$, this equation gives the rate of change of f with respect to increasing t and therefore depends, among other things, on the direction of motion along the curve. If the curve is a straight line and t is the arc length parameter along the line measured from P_0 in the direction of a given unit vector $\mathbf{u}$, then df/dt is the rate of change of f with respect to distance in its domain in the direction of $\mathbf{u}$. By varying $\mathbf{u}$, we find the rates at which f changes with respect to distance as we move through P_0 in different directions. We now define this idea more precisely.

Suppose that the function $f(x, y)$ is defined throughout a region R in the xy-plane, that $P_0(x_0, y_0)$ is a point in R, and that $\mathbf{u} = u_1\mathbf{i} + u_2\mathbf{j}$ is a unit vector. Then the equations

$$x = x_0 + su_1, \qquad y = y_0 + su_2$$

parametrize the line through P_0 parallel to $\mathbf{u}$. If the parameter s measures arc length from P_0 in the direction of $\mathbf{u}$, we find the rate of change of f at P_0 in the direction of $\mathbf{u}$ by calculating df/ds at P_0 (Figure 14.27).

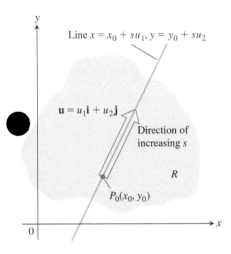

Line $x = x_0 + su_1, y = y_0 + su_2$

$\mathbf{u} = u_1\mathbf{i} + u_2\mathbf{j}$

Direction of increasing s

R

$P_0(x_0, y_0)$

FIGURE 14.27 The rate of change of f in the direction of $\mathbf{u}$ at a point P_0 is the rate at which f changes along this line at P_0.

DEFINITION The **derivative of f at $P_0(x_0, y_0)$ in the direction of the unit vector $\mathbf{u} = u_1\mathbf{i} + u_2\mathbf{j}$** is the number

$$\left(\frac{df}{ds}\right)_{\mathbf{u}, P_0} = \lim_{s \to 0} \frac{f(x_0 + su_1, y_0 + su_2) - f(x_0, y_0)}{s}, \tag{1}$$

provided the limit exists.

The **directional derivative** defined by Equation (1) is also denoted by

$$D_{\mathbf{u}}f(P_0) \qquad \text{or} \qquad D_{\mathbf{u}}f\big|_{P_0} \qquad \begin{array}{l}\text{"The derivative of } f \\ \text{in the direction of } \mathbf{u}, \\ \text{evaluated at } P_0\text{"}\end{array}$$

The partial derivatives $f_x(x_0, y_0)$ and $f_y(x_0, y_0)$ are the directional derivatives of f at P_0 in the $\mathbf{i}$ and $\mathbf{j}$ directions. This observation can be seen by comparing Equation (1) to the definitions of the two partial derivatives given in Section 14.3.

EXAMPLE 1   Using the definition, find the derivative of

$$f(x, y) = x^2 + xy$$

at $P_0(1, 2)$ in the direction of the unit vector $\mathbf{u} = \left(1/\sqrt{2}\right)\mathbf{i} + \left(1/\sqrt{2}\right)\mathbf{j}$.

Solution   Applying the definition in Equation (1), we obtain

$$\left(\frac{df}{ds}\right)_{\mathbf{u},\,P_0} = \lim_{s\to 0}\frac{f(x_0 + su_1, y_0 + su_2) - f(x_0, y_0)}{s} \qquad \text{Eq. (1)}$$

$$= \lim_{s\to 0}\frac{f\left(1 + s\cdot\dfrac{1}{\sqrt{2}},\, 2 + s\cdot\dfrac{1}{\sqrt{2}}\right) - f(1, 2)}{s} \qquad \text{Substitute.}$$

$$= \lim_{s\to 0}\frac{\left(1 + \dfrac{s}{\sqrt{2}}\right)^2 + \left(1 + \dfrac{s}{\sqrt{2}}\right)\left(2 + \dfrac{s}{\sqrt{2}}\right) - (1^2 + 1\cdot 2)}{s}$$

$$= \lim_{s\to 0}\frac{\left(1 + \dfrac{2s}{\sqrt{2}} + \dfrac{s^2}{2}\right) + \left(2 + \dfrac{3s}{\sqrt{2}} + \dfrac{s^2}{2}\right) - 3}{s}$$

$$= \lim_{s\to 0}\frac{\dfrac{5s}{\sqrt{2}} + s^2}{s} = \lim_{s\to 0}\left(\frac{5}{\sqrt{2}} + s\right) = \frac{5}{\sqrt{2}}.$$

The rate of change of $f(x, y) = x^2 + xy$ at $P_0(1, 2)$ in the direction $\mathbf{u}$ is $5/\sqrt{2}$.   ■

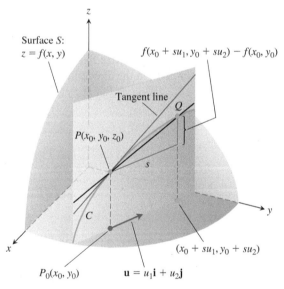

FIGURE 14.28   The slope of the trace curve C at P_0 is $\lim\limits_{Q\to P} \text{slope}\,(PQ)$; this is the directional derivative

$$\left(\frac{df}{ds}\right)_{\mathbf{u},\,P_0} = D_{\mathbf{u}}f\big|_{P_0}.$$

Interpretation of the Directional Derivative

The equation $z = f(x, y)$ represents a surface S in space. If $z_0 = f(x_0, y_0)$, then the point $P(x_0, y_0, z_0)$ lies on S. The vertical plane that passes through P and $P_0(x_0, y_0)$ parallel to $\mathbf{u}$ intersects S in a curve C (Figure 14.28). The rate of change of f in the direction of $\mathbf{u}$ is the slope of the tangent to C at P in the right-handed system formed by the vectors $\mathbf{u}$ and $\mathbf{k}$.

When $\mathbf{u} = \mathbf{i}$, the directional derivative at P_0 is $\partial f/\partial x$ evaluated at (x_0, y_0). When $\mathbf{u} = \mathbf{j}$, the directional derivative at P_0 is $\partial f/\partial y$ evaluated at (x_0, y_0). The directional derivative generalizes the two partial derivatives. We can now ask for the rate of change of f in any direction $\mathbf{u}$, not just the directions $\mathbf{i}$ and $\mathbf{j}$.

For a physical interpretation of the directional derivative, suppose that $T = f(x, y)$ is the temperature at each point (x, y) over a region in the plane. Then $f(x_0, y_0)$ is the temperature at the point $P_0(x_0, y_0)$ and $D_{\mathbf{u}}f\big|_{P_0}$ is the instantaneous rate of change of the temperature at P_0 stepping off in the direction $\mathbf{u}$.

Calculation and Gradients

We now develop an efficient formula to calculate the directional derivative for a differentiable function f. We begin with the line

$$x = x_0 + su_1, \qquad y = y_0 + su_2, \tag{2}$$

through $P_0(x_0, y_0)$, parametrized with the arc length parameter s increasing in the direction of the unit vector $\mathbf{u} = u_1\mathbf{i} + u_2\mathbf{j}$. Then by the Chain Rule we find

$$\left(\frac{df}{ds}\right)_{\mathbf{u},P_0} = \frac{\partial f}{\partial x}\bigg|_{P_0}\frac{dx}{ds} + \frac{\partial f}{\partial y}\bigg|_{P_0}\frac{dy}{ds} \quad \text{Chain Rule for differentiable } f$$

$$= \frac{\partial f}{\partial x}\bigg|_{P_0}u_1 + \frac{\partial f}{\partial y}\bigg|_{P_0}u_2 \quad \text{From Eqs. (2), } dx/ds = u_1 \text{ and } dy/ds = u_2$$

$$= \underbrace{\left[\frac{\partial f}{\partial x}\bigg|_{P_0}\mathbf{i} + \frac{\partial f}{\partial y}\bigg|_{P_0}\mathbf{j}\right]}_{\text{Gradient of } f \text{ at } P_0} \cdot \underbrace{\left[u_1\mathbf{i} + u_2\mathbf{j}\right]}_{\text{Direction } \mathbf{u}}. \tag{3}$$

Equation (3) says that the derivative of a differentiable function f in the direction of $\mathbf{u}$ at P_0 is the dot product of $\mathbf{u}$ with a special vector, which we now define.

DEFINITION The **gradient vector** (or **gradient**) of $f(x, y)$ is the vector

$$\nabla f = \frac{\partial f}{\partial x}\mathbf{i} + \frac{\partial f}{\partial y}\mathbf{j}.$$

The value of the gradient vector obtained by evaluating the partial derivatives at a point $P_0(x_0, y_0)$ is written

$$\nabla f\big|_{P_0} \qquad \text{or} \qquad \nabla f(x_0, y_0).$$

The notation ∇f is read "grad f" as well as "gradient of f" and "del f." The symbol ∇ by itself is read "del." Another notation for the gradient is grad f. Using the gradient notation, we restate Equation (3) as a theorem.

THEOREM 9—The Directional Derivative Is a Dot Product
If $f(x, y)$ is differentiable in an open region containing $P_0(x_0, y_0)$, then

$$\left(\frac{df}{ds}\right)_{\mathbf{u},P_0} = \nabla f\big|_{P_0} \cdot \mathbf{u}, \tag{4}$$

the dot product of the gradient ∇f at P_0 with the vector $\mathbf{u}$. In brief, $D_{\mathbf{u}}f = \nabla f \cdot \mathbf{u}$.

EXAMPLE 2   Find the derivative of $f(x, y) = xe^y + \cos(xy)$ at the point $(2, 0)$ in the direction of $\mathbf{v} = 3\mathbf{i} - 4\mathbf{j}$.

Solution   Recall that the direction of a vector $\mathbf{v}$ is the unit vector obtained by dividing $\mathbf{v}$ by its length:

$$\mathbf{u} = \frac{\mathbf{v}}{|\mathbf{v}|} = \frac{\mathbf{v}}{5} = \frac{3}{5}\mathbf{i} - \frac{4}{5}\mathbf{j}.$$

The partial derivatives of f are everywhere continuous and at $(2, 0)$ are given by

$$f_x(2, 0) = (e^y - y \sin(xy))\Big|_{(2, 0)} = e^0 - 0 = 1$$

$$f_y(2, 0) = (xe^y - x \sin(xy))\Big|_{(2, 0)} = 2e^0 - 2 \cdot 0 = 2.$$

The gradient of f at $(2, 0)$ is

$$\nabla f\big|_{(2,0)} = f_x(2, 0)\mathbf{i} + f_y(2, 0)\mathbf{j} = \mathbf{i} + 2\mathbf{j}$$

(Figure 14.29). The derivative of f at $(2, 0)$ in the direction of $\mathbf{v}$ is therefore

$$D_{\mathbf{u}}f\big|_{(2, 0)} = \nabla f\big|_{(2, 0)} \cdot \mathbf{u} \qquad \text{Eq. (4) with the } D_{\mathbf{u}}f\big|_{P_0} \text{ notation}$$

$$= (\mathbf{i} + 2\mathbf{j}) \cdot \left(\frac{3}{5}\mathbf{i} - \frac{4}{5}\mathbf{j}\right) = \frac{3}{5} - \frac{8}{5} = -1. \qquad \blacksquare$$

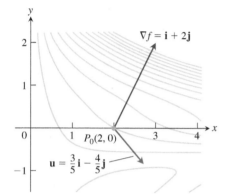

FIGURE 14.29   Picture ∇f as a vector in the domain of f. The figure shows a number of level curves of f. The rate at which f changes at $(2, 0)$ in the direction $\mathbf{u}$ is $\nabla f \cdot \mathbf{u} = -1$, which is the component of ∇f in the direction of unit vector $\mathbf{u}$ (Example 2).

Evaluating the dot product in the brief version of Equation (4) gives

$$D_{\mathbf{u}}f = \nabla f \cdot \mathbf{u} = |\nabla f||\mathbf{u}| \cos\theta = |\nabla f| \cos\theta,$$

where θ is the angle between the vectors $\mathbf{u}$ and ∇f, and reveals the following properties.

Properties of the Directional Derivative $D_{\mathbf{u}}f = \nabla f \cdot \mathbf{u} = |\nabla f| \cos\theta$

1. The function f increases most rapidly when $\cos\theta = 1$, which means that $\theta = 0$ and $\mathbf{u}$ is the direction of ∇f. That is, at each point P in its domain, f increases most rapidly in the direction of the gradient vector ∇f at P. The derivative in this direction is

$$D_{\mathbf{u}}f = |\nabla f| \cos(0) = |\nabla f|.$$

2. Similarly, f decreases most rapidly in the direction of $-\nabla f$. The derivative in this direction is $D_{\mathbf{u}}f = |\nabla f| \cos(\pi) = -|\nabla f|$.

3. Any direction $\mathbf{u}$ orthogonal to a gradient $\nabla f \neq 0$ is a direction of zero change in f because θ then equals $\pi/2$ and

$$D_{\mathbf{u}}f = |\nabla f| \cos(\pi/2) = |\nabla f| \cdot 0 = 0.$$

As we discuss later, these properties hold in three dimensions as well as two.

EXAMPLE 3   Find the directions in which $f(x, y) = (x^2/2) + (y^2/2)$

(a) increases most rapidly at the point $(1, 1)$, and

(b) decreases most rapidly at $(1, 1)$.

(c) What are the directions of zero change in f at $(1, 1)$?

Solution

(a) The function increases most rapidly in the direction of ∇f at $(1, 1)$. The gradient there is

$$\nabla f\big|_{(1, 1)} = (x\mathbf{i} + y\mathbf{j})\Big|_{(1, 1)} = \mathbf{i} + \mathbf{j}.$$

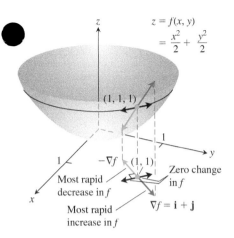

FIGURE 14.30 The direction in which $f(x, y)$ increases most rapidly at $(1, 1)$ is the direction of $\nabla f|_{(1,1)} = \mathbf{i} + \mathbf{j}$. It corresponds to the direction of steepest ascent on the surface at $(1, 1, 1)$ (Example 3).

Its direction is

$$\mathbf{u} = \frac{\mathbf{i} + \mathbf{j}}{|\mathbf{i} + \mathbf{j}|} = \frac{\mathbf{i} + \mathbf{j}}{\sqrt{(1)^2 + (1)^2}} = \frac{1}{\sqrt{2}}\mathbf{i} + \frac{1}{\sqrt{2}}\mathbf{j}.$$

(b) The function decreases most rapidly in the direction of $-\nabla f$ at $(1, 1)$, which is

$$-\mathbf{u} = -\frac{1}{\sqrt{2}}\mathbf{i} - \frac{1}{\sqrt{2}}\mathbf{j}.$$

(c) The directions of zero change at $(1, 1)$ are the directions orthogonal to ∇f:

$$\mathbf{n} = -\frac{1}{\sqrt{2}}\mathbf{i} + \frac{1}{\sqrt{2}}\mathbf{j} \quad \text{and} \quad -\mathbf{n} = \frac{1}{\sqrt{2}}\mathbf{i} - \frac{1}{\sqrt{2}}\mathbf{j}.$$

See Figure 14.30.   ■

Gradients and Tangents to Level Curves

If a differentiable function $f(x, y)$ has a constant value c along a smooth curve $\mathbf{r} = g(t)\mathbf{i} + h(t)\mathbf{j}$ (making the curve part of a level curve of f), then $f(g(t), h(t)) = c$. Differentiating both sides of this equation with respect to t leads to the equations

$$\frac{d}{dt} f(g(t), h(t)) = \frac{d}{dt}(c)$$

$$\frac{\partial f}{\partial x}\frac{dg}{dt} + \frac{\partial f}{\partial y}\frac{dh}{dt} = 0 \qquad \text{Chain Rule} \qquad (5)$$

$$\underbrace{\left(\frac{\partial f}{\partial x}\mathbf{i} + \frac{\partial f}{\partial y}\mathbf{j}\right)}_{\nabla f} \cdot \underbrace{\left(\frac{dg}{dt}\mathbf{i} + \frac{dh}{dt}\mathbf{j}\right)}_{\frac{d\mathbf{r}}{dt}} = 0.$$

Equation (5) says that ∇f is normal to the tangent vector $d\mathbf{r}/dt$, so it is normal to the curve. This is seen in Figure 14.31, where ∇f is a nonzero vector (it is possible for ∇f to be the zero vector).

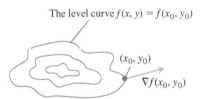

FIGURE 14.31 The gradient of a differentiable function of two variables at a point is always normal to the function's level curve through that point.

> At every point (x_0, y_0) in the domain of a differentiable function $f(x, y)$, the gradient of f is normal to the level curve through (x_0, y_0) (Figure 14.31).

Equation (5) validates our observation that streams flow perpendicular to the contours in topographical maps (see Figure 14.26). Since the downflowing stream will reach its destination in the fastest way, it must flow in the direction of the negative gradient vectors from Property 2 for the directional derivative. Equation (5) tells us these directions are perpendicular to the level curves.

This observation also enables us to find equations for tangent lines to level curves. They are the lines normal to the gradients. The line through a point $P_0(x_0, y_0)$ normal to a nonzero vector $\mathbf{N} = A\mathbf{i} + B\mathbf{j}$ has the equation

$$A(x - x_0) + B(y - y_0) = 0$$

(Exercise 39). If $\mathbf{N}$ is the gradient $\nabla f|_{(x_0, y_0)} = f_x(x_0, y_0)\mathbf{i} + f_y(x_0, y_0)\mathbf{j}$, and this gradient is not the zero vector, then this equation gives the following formula.

$$f_x(x_0, y_0)(x - x_0) + f_y(x_0, y_0)(y - y_0) = 0 \qquad (6)$$

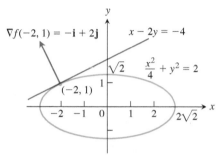

FIGURE 14.32 We can find the tangent to the ellipse $(x^2/4) + y^2 = 2$ by treating the ellipse as a level curve of the function $f(x, y) = (x^2/4) + y^2$ (Example 4).

EXAMPLE 4   Find an equation for the tangent to the ellipse

$$\frac{x^2}{4} + y^2 = 2$$

(Figure 14.32) at the point $(-2, 1)$.

Solution   The ellipse is a level curve of the function

$$f(x, y) = \frac{x^2}{4} + y^2.$$

The gradient of f at $(-2, 1)$ is

$$\nabla f\big|_{(-2, 1)} = \left(\frac{x}{2}\mathbf{i} + 2y\mathbf{j}\right)\bigg|_{(-2, 1)} = -\mathbf{i} + 2\mathbf{j}.$$

Because this gradient vector is nonzero, the tangent to the ellipse at $(-2, 1)$ is the line

$$(-1)(x + 2) + (2)(y - 1) = 0 \qquad \text{Eq. (6)}$$

$$x - 2y = -4. \qquad \text{Simplify.}$$

If we know the gradients of two functions f and g, we automatically know the gradients of their sum, difference, constant multiples, product, and quotient. You are asked to establish the following rules in Exercise 40. Notice that these rules have the same form as the corresponding rules for derivatives of single-variable functions.

Algebra Rules for Gradients

1. *Sum Rule:* $\nabla(f + g) = \nabla f + \nabla g$

2. *Difference Rule:* $\nabla(f - g) = \nabla f - \nabla g$

3. *Constant Multiple Rule:* $\nabla(kf) = k\nabla f$ (any number k)

4. *Product Rule:* $\nabla(fg) = f\nabla g + g\nabla f$

5. *Quotient Rule:* $\nabla\left(\dfrac{f}{g}\right) = \dfrac{g\nabla f - f\nabla g}{g^2}$ } Scalar multipliers on left of gradients

---

EXAMPLE 5   We illustrate two of the rules with

$$f(x, y) = x - y \qquad g(x, y) = 3y$$
$$\nabla f = \mathbf{i} - \mathbf{j} \qquad \nabla g = 3\mathbf{j}.$$

We have

1. $\nabla(f - g) = \nabla(x - 4y) = \mathbf{i} - 4\mathbf{j} = \nabla f - \nabla g$ Rule 2

2. $\nabla(fg) = \nabla(3xy - 3y^2) = 3y\mathbf{i} + (3x - 6y)\mathbf{j}$

and

$$f\nabla g + g\nabla f = (x - y)3\mathbf{j} + 3y(\mathbf{i} - \mathbf{j}) \qquad \text{Substitute.}$$
$$= 3y\mathbf{i} + (3x - 6y)\mathbf{j}. \qquad \text{Simplify.}$$

We have therefore verified for this example that $\nabla(fg) = f\nabla g + g\nabla f$.

Functions of Three Variables

For a differentiable function $f(x, y, z)$ and a unit vector $\mathbf{u} = u_1\mathbf{i} + u_2\mathbf{j} + u_3\mathbf{k}$ in space, we have

$$\nabla f = \frac{\partial f}{\partial x}\mathbf{i} + \frac{\partial f}{\partial y}\mathbf{j} + \frac{\partial f}{\partial z}\mathbf{k}$$

and

$$D_{\mathbf{u}}f = \nabla f \cdot \mathbf{u} = \frac{\partial f}{\partial x}u_1 + \frac{\partial f}{\partial y}u_2 + \frac{\partial f}{\partial z}u_3.$$

The directional derivative can once again be written in the form

$$D_{\mathbf{u}}f = \nabla f \cdot \mathbf{u} = |\nabla f||\mathbf{u}| \cos \theta = |\nabla f| \cos \theta,$$

so the properties listed earlier for functions of two variables extend to three variables. At any given point, f increases most rapidly in the direction of ∇f and decreases most rapidly in the direction of $-\nabla f$. In any direction orthogonal to ∇f, the derivative is zero.

EXAMPLE 6

(a) Find the derivative of $f(x, y, z) = x^3 - xy^2 - z$ at $P_0(1, 1, 0)$ in the direction of $\mathbf{v} = 2\mathbf{i} - 3\mathbf{j} + 6\mathbf{k}$.

(b) In what directions does f change most rapidly at P_0, and what are the rates of change in these directions?

Solution

(a) The direction of $\mathbf{v}$ is obtained by dividing $\mathbf{v}$ by its length:

$$|\mathbf{v}| = \sqrt{(2)^2 + (-3)^2 + (6)^2} = \sqrt{49} = 7$$

$$\mathbf{u} = \frac{\mathbf{v}}{|\mathbf{v}|} = \frac{2}{7}\mathbf{i} - \frac{3}{7}\mathbf{j} + \frac{6}{7}\mathbf{k}.$$

The partial derivatives of f at P_0 are

$$f_x = (3x^2 - y^2)\Big|_{(1, 1, 0)} = 2, \qquad f_y = -2xy\Big|_{(1, 1, 0)} = -2, \qquad f_z = -1\Big|_{(1, 1, 0)} = -1.$$

The gradient of f at P_0 is

$$\nabla f|_{(1,1,0)} = 2\mathbf{i} - 2\mathbf{j} - \mathbf{k}.$$

The derivative of f at P_0 in the direction of $\mathbf{v}$ is therefore

$$D_{\mathbf{u}}f|_{(1, 1, 0)} = \nabla f|_{(1, 1, 0)} \cdot \mathbf{u} = (2\mathbf{i} - 2\mathbf{j} - \mathbf{k}) \cdot \left(\frac{2}{7}\mathbf{i} - \frac{3}{7}\mathbf{j} + \frac{6}{7}\mathbf{k}\right)$$

$$= \frac{4}{7} + \frac{6}{7} - \frac{6}{7} = \frac{4}{7}.$$

(b) The function increases most rapidly in the direction of $\nabla f = 2\mathbf{i} - 2\mathbf{j} - \mathbf{k}$ and decreases most rapidly in the direction of $-\nabla f$. The rates of change in the directions are, respectively,

$$|\nabla f| = \sqrt{(2)^2 + (-2)^2 + (-1)^2} = \sqrt{9} = 3 \qquad \text{and} \qquad -|\nabla f| = -3. \quad \blacksquare$$

The Chain Rule for Paths

If $\mathbf{r}(t) = x(t)\mathbf{i} + y(t)\mathbf{j} + z(t)\mathbf{k}$ is a smooth path C, and $w = f(\mathbf{r}(t))$ is a scalar function evaluated along C, then according to the Chain Rule, Theorem 6 in Section 14.4,

$$\frac{dw}{dt} = \frac{\partial w}{\partial x}\frac{dx}{dt} + \frac{\partial w}{\partial y}\frac{dy}{dt} + \frac{\partial w}{\partial z}\frac{dz}{dt}.$$

The partial derivatives on the right-hand side of the above equation are evaluated along the curve $\mathbf{r}(t)$, and the derivatives of the intermediate variables are evaluated at t. If we express this equation using vector notation, we have

The Derivative Along a Path

$$\frac{d}{dt}f(\mathbf{r}(t)) = \nabla f(\mathbf{r}(t)) \cdot \mathbf{r}'(t). \tag{7}$$

What Equation (7) says is that the derivative of the composite function $f(\mathbf{r}(t))$ is the "derivative" (gradient) of the outside function f "times" (dot product) the derivative of the inside function $\mathbf{r}$. This is analogous to the "Outside-Inside" Rule for derivatives of composite functions studied in Section 3.6. That is, the multivariable Chain Rule for paths has exactly *the same form* as the rule for single-variable differential calculus when appropriate interpretations are given to the meanings of the terms and operations involved.

EXERCISES 14.5

Calculating Gradients
In Exercises 1–6, find the gradient of the function at the given point. Then sketch the gradient together with the level curve that passes through the point.

1. $f(x, y) = y - x$, $(2, 1)$
2. $f(x, y) = \ln(x^2 + y^2)$, $(1, 1)$
3. $g(x, y) = xy^2$, $(2, -1)$
4. $g(x, y) = \frac{x^2}{2} - \frac{y^2}{2}$, $(\sqrt{2}, 1)$
5. $f(x, y) = \sqrt{2x + 3y}$, $(-1, 2)$
6. $f(x, y) = \tan^{-1}\frac{\sqrt{x}}{y}$, $(4, -2)$

In Exercises 7–10, find ∇f at the given point.
7. $f(x, y, z) = x^2 + y^2 - 2z^2 + z \ln x$, $(1, 1, 1)$
8. $f(x, y, z) = 2z^3 - 3(x^2 + y^2)z + \tan^{-1} xz$, $(1, 1, 1)$
9. $f(x, y, z) = (x^2 + y^2 + z^2)^{-1/2} + \ln(xyz)$, $(-1, 2, -2)$
10. $f(x, y, z) = e^{x+y}\cos z + (y + 1)\sin^{-1} x$, $(0, 0, \pi/6)$

Finding Directional Derivatives
In Exercises 11–18, find the derivative of the function at P_0 in the direction of $\mathbf{u}$.
11. $f(x, y) = 2xy - 3y^2$, $P_0(5, 5)$, $\mathbf{u} = 4\mathbf{i} + 3\mathbf{j}$
12. $f(x, y) = 2x^2 + y^2$, $P_0(-1, 1)$, $\mathbf{u} = 3\mathbf{i} - 4\mathbf{j}$
13. $g(x, y) = \frac{x - y}{xy + 2}$, $P_0(1, -1)$, $\mathbf{u} = 12\mathbf{i} + 5\mathbf{j}$
14. $h(x, y) = \tan^{-1}(y/x) + \sqrt{3}\sin^{-1}(xy/2)$, $P_0(1, 1)$, $\mathbf{u} = 3\mathbf{i} - 2\mathbf{j}$

15. $f(x, y, z) = xy + yz + zx$, $P_0(1, -1, 2)$, $\mathbf{u} = 3\mathbf{i} + 6\mathbf{j} - 2\mathbf{k}$
16. $f(x, y, z) = x^2 + 2y^2 - 3z^2$, $P_0(1, 1, 1)$, $\mathbf{u} = \mathbf{i} + \mathbf{j} + \mathbf{k}$
17. $g(x, y, z) = 3e^x \cos yz$, $P_0(0, 0, 0)$, $\mathbf{u} = 2\mathbf{i} + \mathbf{j} - 2\mathbf{k}$
18. $h(x, y, z) = \cos xy + e^{yz} + \ln zx$, $P_0(1, 0, 1/2)$, $\mathbf{u} = \mathbf{i} + 2\mathbf{j} + 2\mathbf{k}$

In Exercises 19–24, find the directions in which the functions increase and decrease most rapidly at P_0. Then find the derivatives of the functions in these directions.
19. $f(x, y) = x^2 + xy + y^2$, $P_0(-1, 1)$
20. $f(x, y) = x^2y + e^{xy}\sin y$, $P_0(1, 0)$
21. $f(x, y, z) = (x/y) - yz$, $P_0(4, 1, 1)$
22. $g(x, y, z) = xe^y + z^2$, $P_0(1, \ln 2, 1/2)$
23. $f(x, y, z) = \ln xy + \ln yz + \ln xz$, $P_0(1, 1, 1)$
24. $h(x, y, z) = \ln(x^2 + y^2 - 1) + y + 6z$, $P_0(1, 1, 0)$

Tangent Lines to Level Curves
In Exercises 25–28, sketch the curve $f(x, y) = c$ together with ∇f and the tangent line at the given point. Then write an equation for the tangent line.
25. $x^2 + y^2 = 4$, $(\sqrt{2}, \sqrt{2})$
26. $x^2 - y = 1$, $(\sqrt{2}, 1)$
27. $xy = -4$, $(2, -2)$
28. $x^2 - xy + y^2 = 7$, $(-1, 2)$

Theory and Examples

29. Let $f(x, y) = x^2 - xy + y^2 - y$. Find the directions $\mathbf{u}$ and the values of $D_{\mathbf{u}} f(1, -1)$ for which

    a. $D_{\mathbf{u}} f(1, -1)$ is largest   **b.** $D_{\mathbf{u}} f(1, -1)$ is smallest

    c. $D_{\mathbf{u}} f(1, -1) = 0$ **d.** $D_{\mathbf{u}} f(1, -1) = 4$

    e. $D_{\mathbf{u}} f(1, -1) = -3$

30. Let $f(x, y) = \dfrac{(x - y)}{(x + y)}$. Find the directions $\mathbf{u}$ and the values of $D_{\mathbf{u}} f\left(-\dfrac{1}{2}, \dfrac{3}{2}\right)$ for which

    a. $D_{\mathbf{u}} f\left(-\dfrac{1}{2}, \dfrac{3}{2}\right)$ is largest   **b.** $D_{\mathbf{u}} f\left(-\dfrac{1}{2}, \dfrac{3}{2}\right)$ is smallest

    c. $D_{\mathbf{u}} f\left(-\dfrac{1}{2}, \dfrac{3}{2}\right) = 0$ **d.** $D_{\mathbf{u}} f\left(-\dfrac{1}{2}, \dfrac{3}{2}\right) = -2$

    e. $D_{\mathbf{u}} f\left(-\dfrac{1}{2}, \dfrac{3}{2}\right) = 1$

31. Zero directional derivative In what direction is the derivative of $f(x, y) = xy + y^2$ at $P(3, 2)$ equal to zero?

32. Zero directional derivative In what directions is the derivative of $f(x, y) = (x^2 - y^2)/(x^2 + y^2)$ at $P(1, 1)$ equal to zero?

33. Is there a direction $\mathbf{u}$ in which the rate of change of $f(x, y) = x^2 - 3xy + 4y^2$ at $P(1, 2)$ equals 14? Give reasons for your answer.

34. Changing temperature along a circle Is there a direction $\mathbf{u}$ in which the rate of change of the temperature function $T(x, y, z) = 2xy - yz$ (temperature in degrees Celsius, distance in feet) at $P(1, -1, 1)$ is $-3°C/ft$? Give reasons for your answer.

35. The derivative of $f(x, y)$ at $P_0(1, 2)$ in the direction of $\mathbf{i} + \mathbf{j}$ is $2\sqrt{2}$ and in the direction of $-2\mathbf{j}$ is -3. What is the derivative of f in the direction of $-\mathbf{i} - 2\mathbf{j}$? Give reasons for your answer.

36. The derivative of $f(x, y, z)$ at a point P is greatest in the direction of $\mathbf{v} = \mathbf{i} + \mathbf{j} - \mathbf{k}$. In this direction, the value of the derivative is $2\sqrt{3}$.

    a. What is ∇f at P? Give reasons for your answer.

    b. What is the derivative of f at P in the direction of $\mathbf{i} + \mathbf{j}$?

37. Directional derivatives and scalar components How is the derivative of a differentiable function $f(x, y, z)$ at a point P_0 in the direction of a unit vector $\mathbf{u}$ related to the scalar component of $(\nabla f)_{P_0}$ in the direction of $\mathbf{u}$? Give reasons for your answer.

38. Directional derivatives and partial derivatives Assuming that the necessary derivatives of $f(x, y, z)$ are defined, how are $D_{\mathbf{i}} f$, $D_{\mathbf{j}} f$, and $D_{\mathbf{k}} f$ related to f_x, f_y, and f_z? Give reasons for your answer.

39. Lines in the xy-plane Show that $A(x - x_0) + B(y - y_0) = 0$ is an equation for the line in the xy-plane through the point (x_0, y_0) normal to the vector $\mathbf{N} = A\mathbf{i} + B\mathbf{j}$.

40. The algebra rules for gradients Given a constant k and the gradients

$$\nabla f = \frac{\partial f}{\partial x}\mathbf{i} + \frac{\partial f}{\partial y}\mathbf{j} + \frac{\partial f}{\partial z}\mathbf{k},$$

$$\nabla g = \frac{\partial g}{\partial x}\mathbf{i} + \frac{\partial g}{\partial y}\mathbf{j} + \frac{\partial g}{\partial z}\mathbf{k},$$

establish the algebra rules for gradients.

In Exercises 41–44, find a parametric equation for the line that is perpendicular to the graph of the given equation at the given point.

41. $x^2 + y^2 = 25$, $(-3, 4)$

42. $x^2 + xy + y^2 = 3$, $(2, -1)$

43. $x^2 + y^2 + z^2 = 14$, $(3, -2, 1)$

44. $z = x^3 - xy^2$, $(-1, 1, 0)$

14.6

14.6 Tangent Planes and Differentials

In single-variable differential calculus we saw how the derivative defined the tangent line to the graph of a differentiable function at a point on the graph. The tangent line then provided for a linearization of the function at the point. In this section, we will see analogously how the gradient defines the *tangent plane* to the level surface of a function $w = f(x, y, z)$ at a point on the surface. The tangent plane then provides for a linearization of f at the point and defines the total differential of the function.

Tangent Planes and Normal Lines

If $\mathbf{r}(t) = x(t)\mathbf{i} + y(t)\mathbf{j} + z(t)\mathbf{k}$ is a smooth curve on the level surface $f(x, y, z) = c$ of a differentiable function f, we found in Equation (7) of the last section that

$$\frac{d}{dt} f(\mathbf{r}(t)) = \nabla f(\mathbf{r}(t)) \cdot \mathbf{r}'(t).$$

Since f is constant along the curve $\mathbf{r}$, the derivative on the left-hand side of the equation is 0, so the gradient ∇f is orthogonal to the curve's velocity vector $\mathbf{r}'$.

Now let us restrict our attention to the curves that pass through a point P_0 (Figure 14.33). All the velocity vectors at P_0 are orthogonal to ∇f at P_0, so the curves' tangent lines all lie in the plane through P_0 normal to ∇f. We now define this plane.

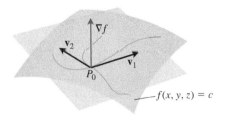

FIGURE 14.33 The gradient ∇f is orthogonal to the velocity vector of every smooth curve in the surface through P_0. The velocity vectors at P_0 therefore lie in a common plane, which we call the tangent plane at P_0.

DEFINITIONS   The **tangent plane** to the level surface $f(x, y, z) = c$ of a differentiable function f at a point P_0 where the gradient is not zero is the plane through P_0 normal to $\nabla f|_{P_0}$.

The **normal line** of the surface at P_0 is the line through P_0 parallel to $\nabla f|_{P_0}$.

The results of Section 12.5 imply that the tangent plane and normal line satisfy the following equations, as long as the gradient at the point P_0 is not the zero vector.

Tangent Plane to $f(x, y, z) = c$ at $P_0(x_0, y_0, z_0)$

$$f_x(P_0)(x - x_0) + f_y(P_0)(y - y_0) + f_z(P_0)(z - z_0) = 0 \tag{1}$$

Normal Line to $f(x, y, z) = c$ at $P_0(x_0, y_0, z_0)$

$$x = x_0 + f_x(P_0)t, \qquad y = y_0 + f_y(P_0)t, \qquad z = z_0 + f_z(P_0)t \tag{2}$$

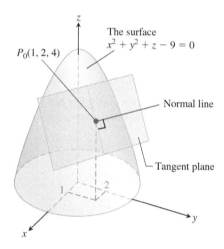

$P_0(1, 2, 4)$

The surface
$x^2 + y^2 + z - 9 = 0$

Normal line

Tangent plane

FIGURE 14.34 The tangent plane and normal line to this level surface at P_0 (Example 1).

EXAMPLE 1   Find the tangent plane and normal line of the level surface

$$f(x, y, z) = x^2 + y^2 + z - 9 = 0 \qquad \text{A circular paraboloid}$$

at the point $P_0(1, 2, 4)$.

Solution   The surface is shown in Figure 14.34.
    The tangent plane is the plane through P_0 perpendicular to the gradient of f at P_0. The gradient is

$$\nabla f|_{P_0} = (2x\mathbf{i} + 2y\mathbf{j} + \mathbf{k})\Big|_{(1, 2, 4)} = 2\mathbf{i} + 4\mathbf{j} + \mathbf{k}.$$

The tangent plane is therefore the plane

$$2(x - 1) + 4(y - 2) + (z - 4) = 0, \qquad \text{or} \qquad 2x + 4y + z = 14.$$

The line normal to the surface at P_0 is

$$x = 1 + 2t, \qquad y = 2 + 4t, \qquad z = 4 + t. \qquad \blacksquare$$

To find an equation for the plane tangent to a smooth surface $z = f(x, y)$ at a point $P_0(x_0, y_0, z_0)$ where $z_0 = f(x_0, y_0)$, we first observe that the equation $z = f(x, y)$ is equivalent to $f(x, y) - z = 0$. The surface $z = f(x, y)$ is therefore the zero level surface of the function $F(x, y, z) = f(x, y) - z$. The partial derivatives of F are

$$F_x = \frac{\partial}{\partial x}(f(x, y) - z) = f_x - 0 = f_x$$

$$F_y = \frac{\partial}{\partial y}(f(x, y) - z) = f_y - 0 = f_y$$

$$F_z = \frac{\partial}{\partial z}(f(x, y) - z) = 0 - 1 = -1.$$

The formula

$$F_x(P_0)(x - x_0) + F_y(P_0)(y - y_0) + F_z(P_0)(z - z_0) = 0$$

for the plane tangent to the level surface at P_0 therefore reduces to

$$f_x(x_0, y_0)(x - x_0) + f_y(x_0, y_0)(y - y_0) - (z - z_0) = 0,$$

as long as the gradient is not the zero vector at the point P_0.

Plane Tangent to a Surface $z = f(x, y)$ at $(x_0, y_0, f(x_0, y_0))$

The plane tangent to the surface $z = f(x, y)$ of a differentiable function f at the point $P_0(x_0, y_0, z_0) = (x_0, y_0, f(x_0, y_0))$ is

$$f_x(x_0, y_0)(x - x_0) + f_y(x_0, y_0)(y - y_0) - (z - z_0) = 0. \qquad (3)$$

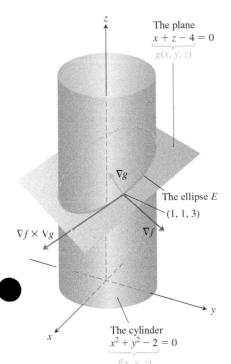

The plane
$x + z - 4 = 0$
$g(x, y, z)$

∇g

The ellipse E
$(1, 1, 3)$

$\nabla f \times \nabla g$ ∇f

The cylinder
$x^2 + y^2 - 2 = 0$
$f(x, y, z)$

FIGURE 14.35 This cylinder and plane intersect in an ellipse E (Example 3).

EXAMPLE 2   Find the plane tangent to the surface $z = x \cos y - ye^x$ at $(0, 0, 0)$.

Solution   We calculate the partial derivatives of $f(x, y) = x \cos y - ye^x$ and use Equation (3):

$$f_x(0, 0) = (\cos y - ye^x)\Big|_{(0, 0)} = 1 - 0 \cdot 1 = 1$$

$$f_y(0, 0) = (-x \sin y - e^x)\Big|_{(0, 0)} = 0 - 1 = -1.$$

The tangent plane is therefore

$$1 \cdot (x - 0) - 1 \cdot (y - 0) - (z - 0) = 0, \qquad \text{Eq. (3)}$$

or

$$x - y - z = 0.$$

EXAMPLE 3   The surfaces

$$f(x, y, z) = x^2 + y^2 - 2 = 0 \qquad \text{A cylinder}$$

and

$$g(x, y, z) = x + z - 4 = 0 \qquad \text{A plane}$$

meet in an ellipse E (Figure 14.35). Find parametric equations for the line tangent to E at the point $P_0(1, 1, 3)$.

Solution   The tangent line is orthogonal to both ∇f and ∇g at P_0, and therefore parallel to $\mathbf{v} = \nabla f \times \nabla g$. The components of $\mathbf{v}$ and the coordinates of P_0 give us equations for the line. We have

$$\nabla f\big|_{(1, 1, 3)} = (2x\mathbf{i} + 2y\mathbf{j})\Big|_{(1, 1, 3)} = 2\mathbf{i} + 2\mathbf{j}$$

$$\nabla g\big|_{(1, 1, 3)} = (\mathbf{i} + \mathbf{k})\Big|_{(1, 1, 3)} = \mathbf{i} + \mathbf{k}$$

$$\mathbf{v} = (2\mathbf{i} + 2\mathbf{j}) \times (\mathbf{i} + \mathbf{k}) = \begin{vmatrix} \mathbf{i} & \mathbf{j} & \mathbf{k} \\ 2 & 2 & 0 \\ 1 & 0 & 1 \end{vmatrix} = 2\mathbf{i} - 2\mathbf{j} - 2\mathbf{k}.$$

The tangent line to the ellipse of intersection is

$$x = 1 + 2t, \qquad y = 1 - 2t, \qquad z = 3 - 2t.$$

Estimating Change in a Specific Direction

The directional derivative plays a role similar to that of an ordinary derivative when we want to estimate how much the value of a function f changes if we move a small distance ds from a point P_0 to another point nearby. If f were a function of a single variable, we would have

$$df = f'(P_0)\,ds. \qquad \text{Ordinary derivative} \times \text{increment}$$

For a function of two or more variables, we use the formula

$$df = (\nabla f|_{P_0} \cdot \mathbf{u})\,ds, \qquad \text{Directional derivative} \times \text{increment}$$

where $\mathbf{u}$ is the direction of the motion away from P_0.

Estimating the Change in f in a Direction u
To estimate the change in the value of a differentiable function f when we move a small distance ds from a point P_0 in a particular direction $\mathbf{u}$, use the formula

$$df = \underbrace{(\nabla f|_{P_0} \cdot \mathbf{u})}_{\substack{\text{Directional} \\ \text{derivative}}} \ \underbrace{ds}_{\substack{\text{Distance} \\ \text{increment}}}$$

EXAMPLE 4   Estimate how much the value of

$$f(x, y, z) = y \sin x + 2yz$$

will change if the point $P(x, y, z)$ moves 0.1 unit from $P_0(0, 1, 0)$ straight toward $P_1(2, 2, -2)$.

Solution   We first find the derivative of f at P_0 in the direction of the vector $\overrightarrow{P_0P_1} = 2\mathbf{i} + \mathbf{j} - 2\mathbf{k}$. The direction of this vector is

$$\mathbf{u} = \frac{\overrightarrow{P_0P_1}}{|\overrightarrow{P_0P_1}|} = \frac{\overrightarrow{P_0P_1}}{3} = \frac{2}{3}\mathbf{i} + \frac{1}{3}\mathbf{j} - \frac{2}{3}\mathbf{k}.$$

The gradient of f at P_0 is

$$\nabla f|_{(0, 1, 0)} = ((y \cos x)\mathbf{i} + (\sin x + 2z)\mathbf{j} + 2y\mathbf{k})\Big|_{(0, 1, 0)} = \mathbf{i} + 2\mathbf{k}.$$

Therefore,

$$\nabla f|_{P_0} \cdot \mathbf{u} = (\mathbf{i} + 2\mathbf{k}) \cdot \left(\frac{2}{3}\mathbf{i} + \frac{1}{3}\mathbf{j} - \frac{2}{3}\mathbf{k}\right) = \frac{2}{3} - \frac{4}{3} = -\frac{2}{3}.$$

The change df in f that results from moving $ds = 0.1$ unit away from P_0 in the direction of $\mathbf{u}$ is approximately

$$df = (\nabla f|_{P_0} \cdot \mathbf{u})(ds) = \left(-\frac{2}{3}\right)(0.1) \approx -0.067 \text{ unit.}$$

See Figure 14.36.   ■

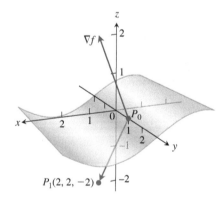

FIGURE 14.36   As $P(x, y, z)$ moves off the level surface at P_0 by 0.1 unit directly toward P_1, the function f changes value by approximately -0.067 unit (Example 4).

How to Linearize a Function of Two Variables

Functions of two variables can be quite complicated, and we sometimes need to approximate them with simpler ones that give the accuracy required for specific applications without being so difficult to work with. We do this in a way that is similar to the way we find linear replacements for functions of a single variable (Section 3.11).

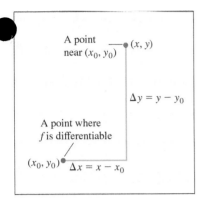

FIGURE 14.37 If f is differentiable at (x_0, y_0), then the value of f at any point (x, y) nearby is approximately $f(x_0, y_0) + f_x(x_0, y_0)\Delta x + f_y(x_0, y_0)\Delta y$.

Suppose the function we wish to approximate is $z = f(x, y)$ near a point (x_0, y_0) at which we know the values of f, f_x, and f_y and at which f is differentiable. If we move from (x_0, y_0) to any nearby point (x, y) by increments $\Delta x = x - x_0$ and $\Delta y = y - y_0$ (see Figure 14.37), then the definition of differentiability from Section 14.3 gives the change

$$f(x, y) - f(x_0, y_0) = f_x(x_0, y_0)\Delta x + f_y(x_0, y_0)\Delta y + \varepsilon_1\Delta x + \varepsilon_2\Delta y,$$

where $\varepsilon_1, \varepsilon_2 \to 0$ as $\Delta x, \Delta y \to 0$. If the increments Δx and Δy are small, the products $\varepsilon_1\Delta x$ and $\varepsilon_2\Delta y$ will eventually be smaller still and we have the approximation

$$f(x, y) \approx \underbrace{f(x_0, y_0) + f_x(x_0, y_0)(x - x_0) + f_y(x_0, y_0)(y - y_0)}_{L(x,\, y)}.$$

In other words, as long as Δx and Δy are small, f will have approximately the same value as the linear function L.

DEFINITIONS The **linearization** of a function $f(x, y)$ at a point (x_0, y_0) where f is differentiable is the function

$$L(x, y) = f(x_0, y_0) + f_x(x_0, y_0)(x - x_0) + f_y(x_0, y_0)(y - y_0).$$

The approximation

$$f(x, y) \approx L(x, y)$$

is the **standard linear approximation** of f at (x_0, y_0).

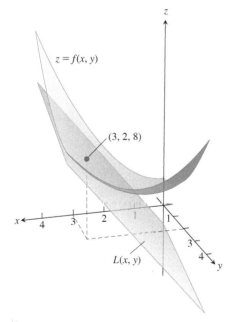

FIGURE 14.38 The tangent plane $L(x, y)$ represents the linearization of $f(x, y)$ in Example 5.

From Equation (3), we find that the plane $z = L(x, y)$ is tangent to the surface $z = f(x, y)$ at the point (x_0, y_0). Thus, the linearization of a function of two variables is a tangent-*plane* approximation in the same way that the linearization of a function of a single variable is a tangent-*line* approximation. (See Exercise 57.)

EXAMPLE 5 Find the linearization of

$$f(x, y) = x^2 - xy + \frac{1}{2}y^2 + 3$$

at the point $(3, 2)$.

Solution We first evaluate f, f_x, and f_y at the point $(x_0, y_0) = (3, 2)$:

$$f(3, 2) = \left.\left(x^2 - xy + \frac{1}{2}y^2 + 3\right)\right|_{(3, 2)} = 8$$

$$f_x(3, 2) = \left.\frac{\partial}{\partial x}\left(x^2 - xy + \frac{1}{2}y^2 + 3\right)\right|_{(3, 2)} = (2x - y)\Big|_{(3, 2)} = 4$$

$$f_y(3, 2) = \left.\frac{\partial}{\partial y}\left(x^2 - xy + \frac{1}{2}y^2 + 3\right)\right|_{(3, 2)} = (-x + y)\Big|_{(3, 2)} = -1,$$

giving

$$L(x, y) = f(x_0, y_0) + f_x(x_0, y_0)(x - x_0) + f_y(x_0, y_0)(y - y_0)$$
$$= 8 + (4)(x - 3) + (-1)(y - 2) = 4x - y - 2.$$

The linearization of f at $(3, 2)$ is $L(x, y) = 4x - y - 2$ (see Figure 14.38). ∎

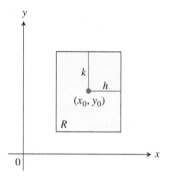

FIGURE 14.39 The rectangular region R: $|x - x_0| \leq h, |y - y_0| \leq k$ in the xy-plane.

When approximating a differentiable function $f(x, y)$ by its linearization $L(x, y)$ at (x_0, y_0), an important question is how accurate the approximation might be.

If we can find a common upper bound M for $|f_{xx}|$, $|f_{yy}|$, and $|f_{xy}|$ on a rectangle R centered at (x_0, y_0) (Figure 14.39), then we can bound the error E throughout R by using a simple formula (derived in Section 14.9). The **error** is defined by $E(x, y) = f(x, y) - L(x, y)$.

The Error in the Standard Linear Approximation

If f has continuous first and second partial derivatives throughout an open set containing a rectangle R centered at (x_0, y_0) and if M is any upper bound for the values of $|f_{xx}|$, $|f_{yy}|$, and $|f_{xy}|$ on R, then the error $E(x, y)$ incurred in replacing $f(x, y)$ on R by its linearization

$$L(x, y) = f(x_0, y_0) + f_x(x_0, y_0)(x - x_0) + f_y(x_0, y_0)(y - y_0)$$

satisfies the inequality

$$|E(x, y)| \leq \frac{1}{2} M \big(|x - x_0| + |y - y_0| \big)^2.$$

To make $|E(x, y)|$ small for a given M, we just make $|x - x_0|$ and $|y - y_0|$ small.

Differentials

Recall from Section 3.11 that for a function of a single variable, $y = f(x)$, we defined the change in f as x changes from a to $a + \Delta x$ by

$$\Delta f = f(a + \Delta x) - f(a)$$

and the differential of f as

$$df = f'(a)\Delta x.$$

We now consider the differential of a function of two variables.

Suppose a differentiable function $f(x, y)$ and its partial derivatives exist at a point (x_0, y_0). If we move to a nearby point $(x_0 + \Delta x, y_0 + \Delta y)$, the change in f is

$$\Delta f = f(x_0 + \Delta x, y_0 + \Delta y) - f(x_0, y_0).$$

A straightforward calculation based on the definition of $L(x, y)$, using the notation $x - x_0 = \Delta x$ and $y - y_0 = \Delta y$, shows that the corresponding change in L is

$$\Delta L = L(x_0 + \Delta x, y_0 + \Delta y) - L(x_0, y_0) = f_x(x_0, y_0)\Delta x + f_y(x_0, y_0)\Delta y.$$

The **differentials** dx and dy are independent variables, so they can be assigned any values. Often we take $dx = \Delta x = x - x_0$, and $dy = \Delta y = y - y_0$. We then have the following definition of the differential or *total* differential of f.

DEFINITION If we move from (x_0, y_0) to a point $(x_0 + dx, y_0 + dy)$ nearby, the resulting change

$$df = f_x(x_0, y_0)\, dx + f_y(x_0, y_0)\, dy$$

in the linearization of f is called the **total differential of f**.

EXAMPLE 6   Suppose that a cylindrical can is designed to have a radius of 1 in. and a height of 5 in., but that the radius and height are off by the amounts $dr = +0.03$ and $dh = -0.1$. Estimate the resulting absolute change in the volume of the can.

Solution   To estimate the absolute change in $V = \pi r^2 h$, we use

$$\Delta V \approx dV = V_r(r_0, h_0)\, dr + V_h(r_0, h_0)\, dh.$$

With $V_r = 2\pi rh$ and $V_h = \pi r^2$, we get

$$dV = 2\pi r_0 h_0\, dr + \pi r_0{}^2\, dh = 2\pi(1)(5)(0.03) + \pi(1)^2(-0.1)$$
$$= 0.3\pi - 0.1\pi = 0.2\pi \approx 0.63 \text{ in}^3$$

EXAMPLE 7   Your company manufactures stainless steel right circular cylindrical molasses storage tanks that are 25 ft high with a radius of 5 ft. How sensitive are the tanks' volumes to small variations in height and radius?

Solution   With $V = \pi r^2 h$, the total differential gives the approximation for the change in volume as

$$dV = V_r(5, 25)\, dr + V_h(5, 25)\, dh$$
$$= (2\pi rh)\Big|_{(5,\,25)} dr + (\pi r^2)\Big|_{(5,\,25)} dh$$
$$= 250\pi\, dr + 25\pi\, dh.$$

Thus, a 1-unit change in r will change V by about 250π units. A 1-unit change in h will change V by about 25π units. The tank's volume is 10 times more sensitive to a small change in r than it is to a small change of equal size in h. As a quality control engineer concerned with being sure the tanks have the correct volume, you would want to pay special attention to their radii.

In contrast, if the values of r and h are reversed to make $r = 25$ and $h = 5$, then the total differential in V becomes

$$dV = (2\pi rh)\Big|_{(25,\,5)} dr + (\pi r^2)\Big|_{(25,\,5)} dh = 250\pi\, dr + 625\pi\, dh.$$

Now the volume is more sensitive to changes in h than to changes in r (Figure 14.40).

The general rule is that functions are most sensitive to small changes in the variables that generate the largest partial derivatives.

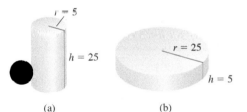

FIGURE 14.40   The volume of cylinder (a) is more sensitive to a small change in r than it is to an equally small change in h. The volume of cylinder (b) is more sensitive to small changes in h than it is to small changes in r (Example 7).

Functions of More Than Two Variables

Analogous results hold for differentiable functions of more than two variables.

1. The **linearization** of $f(x, y, z)$ at a point $P_0(x_0, y_0, z_0)$ is

$$L(x, y, z) = f(P_0) + f_x(P_0)(x - x_0) + f_y(P_0)(y - y_0) + f_z(P_0)(z - z_0).$$

2. Suppose that R is a closed rectangular solid centered at P_0 and lying in an open region on which the second partial derivatives of f are continuous. Suppose also that $|f_{xx}|, |f_{yy}|, |f_{zz}|, |f_{xy}|, |f_{xz}|,$ and $|f_{yz}|$ are all less than or equal to M throughout R. Then the **error** $E(x, y, z) = f(x, y, z) - L(x, y, z)$ in the approximation of f by L is bounded throughout R by the inequality

$$|E| \le \frac{1}{2}M\big(|x - x_0| + |y - y_0| + |z - z_0|\big)^2.$$

3. If the second partial derivatives of f are continuous and if x, y, and z change from x_0, y_0, and z_0 by small amounts dx, dy, and dz, the **total differential**

$$df = f_x(P_0)\, dx + f_y(P_0)\, dy + f_z(P_0)\, dz$$

gives a good approximation of the resulting change in f.

EXAMPLE 8   Find the linearization $L(x, y, z)$ of

$$f(x, y, z) = x^2 - xy + 3 \sin z$$

at the point $(x_0, y_0, z_0) = (2, 1, 0)$. Find an upper bound for the error incurred in replacing f by L on the rectangular region

$$R: \quad |x - 2| \le 0.01, \qquad |y - 1| \le 0.02, \qquad |z| \le 0.01.$$

Solution   Routine calculations give

$$f(2, 1, 0) = 2, \qquad f_x(2, 1, 0) = 3, \qquad f_y(2, 1, 0) = -2, \qquad f_z(2, 1, 0) = 3.$$

Thus,

$$L(x, y, z) = 2 + 3(x - 2) + (-2)(y - 1) + 3(z - 0) = 3x - 2y + 3z - 2.$$

Since

$$f_{xx} = 2, \qquad f_{yy} = 0, \qquad f_{zz} = -3 \sin z, \qquad f_{xy} = -1, \qquad f_{xz} = 0, \qquad f_{yz} = 0,$$

and $|-3 \sin z| \le 3 \sin 0.01 \approx 0.03$, we may take $M = 2$ as a bound on the second partials. Hence, the error incurred by replacing f by L on R satisfies

$$|E| \le \frac{1}{2}(2)(0.01 + 0.02 + 0.01)^2 = 0.0016.$$

EXERCISES 14.6

Tangent Planes and Normal Lines to Surfaces
In Exercises 1–10, find equations for the

(a) tangent plane and

(b) normal line at the point P_0 on the given surface.

1. $x^2 + y^2 + z^2 = 3$, $P_0(1, 1, 1)$

2. $x^2 + y^2 - z^2 = 18$, $P_0(3, 5, -4)$

3. $2z - x^2 = 0$, $P_0(2, 0, 2)$

4. $x^2 + 2xy - y^2 + z^2 = 7$, $P_0(1, -1, 3)$

5. $\cos \pi x - x^2 y + e^{xz} + yz = 4$, $P_0(0, 1, 2)$

6. $x^2 - xy - y^2 - z = 0$, $P_0(1, 1, -1)$

7. $x + y + z = 1$, $P_0(0, 1, 0)$

8. $x^2 + y^2 - 2xy - x + 3y - z = -4$, $P_0(2, -3, 18)$

9. $x \ln y + y \ln z = x$, $P_0(2, 1, e)$

10. $ye^x - ze^{y^2} = z$, $P_0(0, 0, 1)$

In Exercises 11–14, find an equation for the plane that is tangent to the given surface at the given point.

11. $z = \ln(x^2 + y^2)$, $(1, 0, 0)$

12. $z = e^{-(x^2+y^2)}$, $(0, 0, 1)$

13. $z = \sqrt{y - x}$, $(1, 2, 1)$

14. $z = 4x^2 + y^2$, $(1, 1, 5)$

Tangent Lines to Intersecting Surfaces
In Exercises 15–20, find parametric equations for the line tangent to the curve of intersection of the surfaces at the given point.

15. Surfaces: $x + y^2 + 2z = 4$, $x = 1$
 Point: $(1, 1, 1)$

16. Surfaces: $xyz = 1$, $x^2 + 2y^2 + 3z^2 = 6$
 Point: $(1, 1, 1)$

17. Surfaces: $x^2 + 2y + 2z = 4$, $y = 1$
 Point: $(1, 1, 1/2)$

18. Surfaces: $x + y^2 + z = 2$, $y = 1$
 Point: $(1/2, 1, 1/2)$

19. Surfaces: $x^3 + 3x^2y^2 + y^3 + 4xy - z^2 = 0$,
  $x^2 + y^2 + z^2 = 11$
 Point: $(1, 1, 3)$

20. Surfaces: $x^2 + y^2 = 4$, $x^2 + y^2 - z = 0$
 Point: $(\sqrt{2}, \sqrt{2}, 4)$

14.6 Tangent Planes and Differentials   **861**

Estimating Change

21. By about how much will

$$f(x, y, z) = \ln\sqrt{x^2 + y^2 + z^2}$$

change if the point $P(x, y, z)$ moves from $P_0(3, 4, 12)$ a distance of $ds = 0.1$ unit in the direction of $3\mathbf{i} + 6\mathbf{j} - 2\mathbf{k}$?

22. By about how much will

$$f(x, y, z) = e^x \cos yz$$

change as the point $P(x, y, z)$ moves from the origin a distance of $ds = 0.1$ unit in the direction of $2\mathbf{i} + 2\mathbf{j} - 2\mathbf{k}$?

23. By about how much will

$$g(x, y, z) = x + x \cos z - y \sin z + y$$

change if the point $P(x, y, z)$ moves from $P_0(2, -1, 0)$ a distance of $ds = 0.2$ unit toward the point $P_1(0, 1, 2)$?

24. By about how much will

$$h(x, y, z) = \cos(\pi xy) + xz^2$$

change if the point $P(x, y, z)$ moves from $P_0(-1, -1, -1)$ a distance of $ds = 0.1$ unit toward the origin?

25. Temperature change along a circle Suppose that the Celsius temperature at the point (x, y) in the xy-plane is $T(x, y) = x \sin 2y$ and that distance in the xy-plane is measured in meters. A particle is moving *clockwise* around the circle of radius 1 m centered at the origin at the constant rate of 2 m/sec.

a. How fast is the temperature experienced by the particle changing in degrees Celsius per meter at the point $P(1/2, \sqrt{3}/2)$?

b. How fast is the temperature experienced by the particle changing in degrees Celsius per second at P?

26. Changing temperature along a space curve The Celsius temperature in a region in space is given by $T(x, y, z) = 2x^2 - xyz$. A particle is moving in this region and its position at time t is given by $x = 2t^2$, $y = 3t$, $z = -t^2$, where time is measured in seconds and distance in meters.

a. How fast is the temperature experienced by the particle changing in degrees Celsius per meter when the particle is at the point $P(8, 6, -4)$?

b. How fast is the temperature experienced by the particle changing in degrees Celsius per second at P?

Finding Linearizations

In Exercises 27–32, find the linearization $L(x, y)$ of the function at each point.

27. $f(x, y) = x^2 + y^2 + 1$ at   **a.** $(0, 0)$,   **b.** $(1, 1)$

28. $f(x, y) = (x + y + 2)^2$ at   **a.** $(0, 0)$,   **b.** $(1, 2)$

29. $f(x, y) = 3x - 4y + 5$ at   **a.** $(0, 0)$,   **b.** $(1, 1)$

30. $f(x, y) = x^3 y^4$ at   **a.** $(1, 1)$,   **b.** $(0, 0)$

31. $f(x, y) = e^x \cos y$ at   **a.** $(0, 0)$,   **b.** $(0, \pi/2)$

32. $f(x, y) = e^{2y-x}$ at   **a.** $(0, 0)$,   **b.** $(1, 2)$

33. Wind chill factor Wind chill, a measure of the apparent temperature felt on exposed skin, is a function of air temperature and

wind speed. The precise formula, updated by the National Weather Service in 2001 and based on modern heat transfer theory, a human face model, and skin tissue resistance, is

$$W = W(v, T) = 35.74 + 0.6215\,T - 35.75\,v^{0.16}$$
$$+ 0.4275\,T \cdot v^{0.16},$$

where T is air temperature in °F and v is wind speed in mph. A partial wind chill chart is given.

		\multicolumn{9}{c}{T(°F)}
		30	25	20	15	10	5	0	−5	−10
	5	25	19	13	7	1	−5	−11	−16	−22
	10	21	15	9	3	−4	−10	−16	−22	−28
v	15	19	13	6	0	−7	−13	−19	−26	−32
(mph)	20	17	11	4	−2	−9	−15	−22	−29	−35
	25	16	9	3	−4	−11	−17	−24	−31	−37
	30	15	8	1	−5	−12	−19	−26	−33	−39
	35	14	7	0	−7	−14	−21	−27	−34	−41

a. Use the table to find $W(20, 25)$, $W(30, -10)$, and $W(15, 15)$.

b. Use the formula to find $W(10, -40)$, $W(50, -40)$, and $W(60, 30)$.

c. Find the linearization $L(v, T)$ of the function $W(v, T)$ at the point $(25, 5)$.

d. Use $L(v, T)$ in part (c) to estimate the following wind chill values.

   i) $W(24, 6)$ **ii)** $W(27, 2)$

   iii) $W(5, -10)$ (Explain why this value is much different from the value found in the table.)

34. Find the linearization $L(v, T)$ of the function $W(v, T)$ in Exercise 33 at the point $(50, -20)$. Use it to estimate the following wind chill values.

a. $W(49, -22)$

b. $W(53, -19)$

c. $W(60, -30)$

Bounding the Error in Linear Approximations

In Exercises 35–40, find the linearization $L(x, y)$ of the function $f(x, y)$ at P_0. Then find an upper bound for the magnitude $|E|$ of the error in the approximation $f(x, y) \approx L(x, y)$ over the rectangle R.

35. $f(x, y) = x^2 - 3xy + 5$ at $P_0(2, 1)$,
   $R: \quad |x - 2| \le 0.1, \quad |y - 1| \le 0.1$

36. $f(x, y) = (1/2)x^2 + xy + (1/4)y^2 + 3x - 3y + 4$ at $P_0(2, 2)$,
   $R: \quad |x - 2| \le 0.1, \quad |y - 2| \le 0.1$

37. $f(x, y) = 1 + y + x \cos y$ at $P_0(0, 0)$,
   $R: \quad |x| \le 0.2, \quad |y| \le 0.2$
   (Use $|\cos y| \le 1$ and $|\sin y| \le 1$ in estimating E.)

38. $f(x, y) = xy^2 + y \cos(x - 1)$ at $P_0(1, 2)$,
   $R: \quad |x - 1| \le 0.1, \quad |y - 2| \le 0.1$

39. $f(x, y) = e^x \cos y$ at $P_0(0, 0)$,

    R: $|x| \le 0.1$, $|y| \le 0.1$

    (Use $e^x \le 1.11$ and $|\cos y| \le 1$ in estimating E.)

40. $f(x, y) = \ln x + \ln y$ at $P_0(1, 1)$,

    R: $|x - 1| \le 0.2$, $|y - 1| \le 0.2$

Linearizations for Three Variables

Find the linearizations $L(x, y, z)$ of the functions in Exercises 41–46 at the given points.

41. $f(x, y, z) = xy + yz + xz$ at

    a. $(1, 1, 1)$ **b.** $(1, 0, 0)$ **c.** $(0, 0, 0)$

42. $f(x, y, z) = x^2 + y^2 + z^2$ at

    a. $(1, 1, 1)$ **b.** $(0, 1, 0)$ **c.** $(1, 0, 0)$

43. $f(x, y, z) = \sqrt{x^2 + y^2 + z^2}$ at

    a. $(1, 0, 0)$ **b.** $(1, 1, 0)$ **c.** $(1, 2, 2)$

44. $f(x, y, z) = (\sin xy)/z$ at

    a. $(\pi/2, 1, 1)$ **b.** $(2, 0, 1)$

45. $f(x, y, z) = e^x + \cos (y + z)$ at

    a. $(0, 0, 0)$ **b.** $\left(0, \dfrac{\pi}{2}, 0\right)$ **c.** $\left(0, \dfrac{\pi}{4}, \dfrac{\pi}{4}\right)$

46. $f(x, y, z) = \tan^{-1} (xyz)$ at

    a. $(1, 0, 0)$ **b.** $(1, 1, 0)$ **c.** $(1, 1, 1)$

In Exercises 47–50, find the linearization $L(x, y, z)$ of the function $f(x, y, z)$ at P_0. Then find an upper bound for the magnitude of the error E in the approximation $f(x, y, z) \approx L(x, y, z)$ over the region R.

47. $f(x, y, z) = xz - 3yz + 2$ at $P_0(1, 1, 2)$,

    R: $|x - 1| \le 0.01$, $|y - 1| \le 0.01$, $|z - 2| \le 0.02$

48. $f(x, y, z) = x^2 + xy + yz + (1/4)z^2$ at $P_0(1, 1, 2)$,

    R: $|x - 1| \le 0.01$, $|y - 1| \le 0.01$, $|z - 2| \le 0.08$

49. $f(x, y, z) = xy + 2yz - 3xz$ at $P_0(1, 1, 0)$,

    R: $|x - 1| \le 0.01$, $|y - 1| \le 0.01$, $|z| \le 0.01$

50. $f(x, y, z) = \sqrt{2} \cos x \sin (y + z)$ at $P_0(0, 0, \pi/4)$,

    R: $|x| \le 0.01$, $|y| \le 0.01$, $|z - \pi/4| \le 0.01$

Estimating Error; Sensitivity to Change

51. Estimating maximum error   Suppose that T is to be found from the formula $T = x (e^y + e^{-y})$, where x and y are found to be 2 and $\ln 2$ with maximum possible errors of $|dx| = 0.1$ and $|dy| = 0.02$. Estimate the maximum possible error in the computed value of T.

52. Variation in electrical resistance   The resistance R produced by wiring resistors of R_1 and R_2 ohms in parallel (see accompanying figure) can be calculated from the formula

$$\frac{1}{R} = \frac{1}{R_1} + \frac{1}{R_2}.$$

a. Show that

$$dR = \left(\frac{R}{R_1}\right)^2 dR_1 + \left(\frac{R}{R_2}\right)^2 dR_2.$$

b. You have designed a two-resistor circuit, like the one shown, to have resistances of $R_1 = 100$ ohms and $R_2 = 400$ ohms, but there is always some variation in manufacturing and the resistors received by your firm will probably not have these exact values. Will the value of R be more sensitive to variation in R_1 or to variation in R_2? Give reasons for your answer.

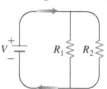

c. In another circuit like the one shown, you plan to change R_1 from 20 to 20.1 ohms and R_2 from 25 to 24.9 ohms. By about what percentage will this change R?

53. You plan to calculate the area of a long, thin rectangle from measurements of its length and width. Which dimension should you measure more carefully? Give reasons for your answer.

54. a. Around the point $(1, 0)$, is $f(x, y) = x^2(y + 1)$ more sensitive to changes in x or to changes in y? Give reasons for your answer.

    b. What ratio of dx to dy will make df equal zero at $(1, 0)$?

55. Value of a 2×2 determinant   If $|a|$ is much greater than $|b|, |c|,$ and $|d|$, to which of $a, b, c,$ and d is the value of the determinant

$$f(a, b, c, d) = \begin{vmatrix} a & b \\ c & d \end{vmatrix}$$

most sensitive? Give reasons for your answer.

56. The Wilson lot size formula   The Wilson lot size formula in economics says that the most economical quantity Q of goods (radios, shoes, brooms, whatever) for a store to order is given by the formula $Q = \sqrt{2KM/h}$, where K is the cost of placing the order, M is the number of items sold per week, and h is the weekly holding cost for each item (cost of space, utilities, security, and so on). To which of the variables $K, M,$ and h is Q most sensitive near the point $(K_0, M_0, h_0) = (2, 20, 0.05)$? Give reasons for your answer.

Theory and Examples

57. The linearization of $f(x, y)$ is a tangent-plane approximation   Show that the tangent plane at the point $P_0(x_0, y_0, f(x_0, y_0))$ on the surface $z = f(x, y)$ defined by a differentiable function f is the plane

$$f_x(x_0, y_0)(x - x_0) + f_y(x_0, y_0)(y - y_0) - (z - f(x_0, y_0)) = 0$$

or

$$z = f(x_0, y_0) + f_x(x_0, y_0)(x - x_0) + f_y(x_0, y_0)(y - y_0).$$

Thus, the tangent plane at P_0 is the graph of the linearization of f at P_0 (see accompanying figure).

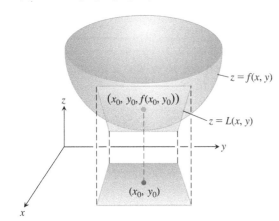

58. Change along the involute of a circle Find the derivative of $f(x, y) = x^2 + y^2$ in the direction of the unit tangent vector of the curve

$$\mathbf{r}(t) = (\cos t + t \sin t)\mathbf{i} + (\sin t - t \cos t)\mathbf{j}, \qquad t > 0.$$

59. Tangent curves A smooth curve is *tangent* to the surface at a point of intersection if its velocity vector is orthogonal to ∇f there.

Show that the curve

$$\mathbf{r}(t) = \sqrt{t}\mathbf{i} + \sqrt{t}\mathbf{j} + (2t - 1)\mathbf{k}$$

is tangent to the surface $x^2 + y^2 - z = 1$ when $t = 1$.

60. Normal curves A smooth curve is *normal* to a surface $f(x, y, z) = c$ at a point of intersection if the curve's velocity vector is a nonzero scalar multiple of ∇f at the point.

Show that the curve

$$\mathbf{r}(t) = \sqrt{t}\mathbf{i} + \sqrt{t}\mathbf{j} - \frac{1}{4}(t + 3)\mathbf{k}$$

is normal to the surface $x^2 + y^2 - z = 3$ when $t = 1$.

61. Consider a closed rectangular box with a square base, as shown in the figure. Assume x is measured with an error of at most 0.5% and y is measured with an error of at most 0.75%, so we have $|dx|/x < 0.005$ and $|dy|/y < 0.0075$.

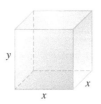

a. Use a differential to estimate the relative error $|dV|/V$ in computing the box's volume V.

b. Use a differential to estimate the relative error $|dS|/S$ in computing the box's surface area S.

Hint for **b:** $\dfrac{4x^2 + 4xy}{2x^2 + 4xy} \leq \dfrac{4x^2 + 8xy}{2x^2 + 4xy} = 2$ and

$$\dfrac{4xy}{2x^2 + 4xy} \leq \dfrac{2x^2 + 4xy}{2x^2 + 4xy} = 1.$$

14.7 Extreme Values and Saddle Points

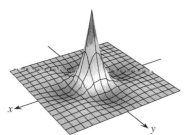

FIGURE 14.41 The function

$$z = (\cos x)(\cos y)e^{-\sqrt{x^2+y^2}}$$

has a maximum value of 1 and a minimum value of about -0.067 on the square region $|x| \leq 3\pi/2,\ |y| \leq 3\pi/2$.

Continuous functions of two variables assume extreme values on closed, bounded domains (see Figures 14.41 and 14.42). We see in this section that we can narrow the search for these extreme values by examining the functions' first partial derivatives. A function of two variables can assume extreme values only at boundary points of the domain or at interior domain points where both first partial derivatives are zero or where one or both of the first partial derivatives fail to exist. However, the vanishing of derivatives at an interior point (a, b) does not always signal the presence of an extreme value. The surface that is the graph of the function might be shaped like a saddle right above (a, b) and cross its tangent plane there.

Derivative Tests for Local Extreme Values

To find the local extreme values of a function of a single variable, we look for points where the graph has a horizontal tangent line. At such points, we then look for local maxima, local minima, and points of inflection. For a function $f(x, y)$ of two variables, we look for points where the surface $z = f(x, y)$ has a horizontal tangent *plane*. At such points, we then look for local maxima, local minima, and saddle points. We begin by defining maxima and minima.

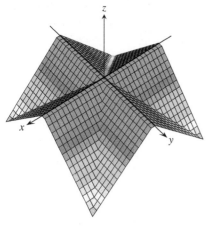

FIGURE 14.42 The "roof surface"

$$z = \frac{1}{2}\left(\left||x| - |y|\right| - |x| - |y|\right)$$

has a maximum value of 0 and a minimum value of $-a$ on the square region $|x| \le a$, $|y| \le a$.

DEFINITIONS Let $f(x, y)$ be defined on a region R containing the point (a, b). Then

1. $f(a, b)$ is a **local maximum** value of f if $f(a, b) \ge f(x, y)$ for all domain points (x, y) in an open disk centered at (a, b).
2. $f(a, b)$ is a **local minimum** value of f if $f(a, b) \le f(x, y)$ for all domain points (x, y) in an open disk centered at (a, b).

Local maxima correspond to mountain peaks on the surface $z = f(x, y)$ and local minima correspond to valley bottoms (Figure 14.43). At such points the tangent planes, when they exist, are horizontal. Local extrema are also called **relative extrema**.

As with functions of a single variable, the key to identifying the local extrema is the First Derivative Test, which we next state and prove.

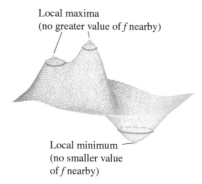

Local maxima
(no greater value of f nearby)

Local minimum
(no smaller value of f nearby)

FIGURE 14.43 A local maximum occurs at a mountain peak and a local minimum occurs at a valley low point.

THEOREM 10—First Derivative Test for Local Extreme Values
If $f(x, y)$ has a local maximum or minimum value at an interior point (a, b) of its domain and if the first partial derivatives exist there, then $f_x(a, b) = 0$ and $f_y(a, b) = 0$.

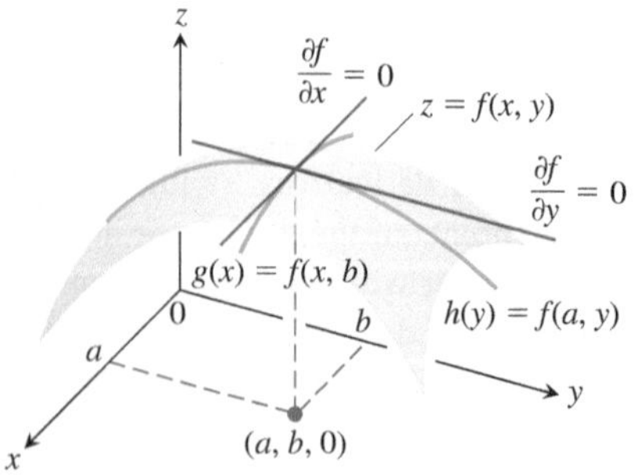

Proof If f has a local extremum at (a, b), then the function $g(x) = f(x, b)$ has a local extremum at $x = a$ (Figure 14.44). Therefore, $g'(a) = 0$ (Chapter 4, Theorem 2). Now $g'(a) = f_x(a, b)$, so $f_x(a, b) = 0$. A similar argument with the function $h(y) = f(a, y)$ shows that $f_y(a, b) = 0$. ∎

If we substitute the values $f_x(a, b) = 0$ and $f_y(a, b) = 0$ into the equation

$$f_x(a, b)(x - a) + f_y(a, b)(y - b) - (z - f(a, b)) = 0$$

for the tangent plane to the surface $z = f(x, y)$ at (a, b), the equation reduces to

$$0 \cdot (x - a) + 0 \cdot (y - b) - z + f(a, b) = 0$$

or

$$z = f(a, b).$$

Thus, Theorem 10 says that the surface does indeed have a horizontal tangent plane at a local extremum, provided there is a tangent plane there.

FIGURE 14.44 If a local maximum of f occurs at $x = a$, $y = b$, then the first partial derivatives $f_x(a, b)$ and $f_y(a, b)$ are both zero.

DEFINITION An interior point of the domain of a function $f(x, y)$ where both f_x and f_y are zero or where one or both of f_x and f_y do not exist is a **critical point** of f.

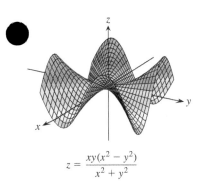

$$z = \frac{xy(x^2 - y^2)}{x^2 + y^2}$$

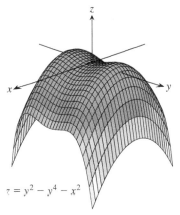

$z = y^2 - y^4 - x^2$

FIGURE 14.45 Saddle points at the origin.

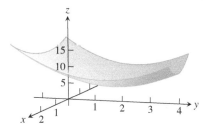

FIGURE 14.46 The graph of the function $f(x, y) = x^2 + y^2 - 4y + 9$ is a paraboloid which has a local minimum value of 5 at the point $(0, 2)$ (Example 1).

Theorem 10 says that the only points where a function $f(x, y)$ can assume extreme values are critical points and boundary points. As with differentiable functions of a single variable, not every critical point gives rise to a local extremum. A differentiable function of a single variable might have a point of inflection. A differentiable function of two variables might have a *saddle point*, with the graph of f crossing the tangent plane defined there.

> **DEFINITION** A differentiable function $f(x, y)$ has a **saddle point** at a critical point (a, b) if in every open disk centered at (a, b) there are domain points (x, y) where $f(x, y) > f(a, b)$ and domain points (x, y) where $f(x, y) < f(a, b)$. The corresponding point $(a, b, f(a, b))$ on the surface $z = f(x, y)$ is called a saddle point of the surface (Figure 14.45).

EXAMPLE 1 Find the local extreme values of $f(x, y) = x^2 + y^2 - 4y + 9$.

Solution The domain of f is the entire plane (so there are no boundary points) and the partial derivatives $f_x = 2x$ and $f_y = 2y - 4$ exist everywhere. Therefore, local extreme values can occur only where

$$f_x = 2x = 0 \qquad \text{and} \qquad f_y = 2y - 4 = 0.$$

The only possibility is the point $(0, 2)$, where the value of f is 5. Since $f(x, y) = x^2 + (y - 2)^2 + 5$ is never less than 5, we see that the critical point $(0, 2)$ gives a local minimum (Figure 14.46). ∎

EXAMPLE 2 Find the local extreme values (if any) of $f(x, y) = y^2 - x^2$.

Solution The domain of f is the entire plane (so there are no boundary points) and the partial derivatives $f_x = -2x$ and $f_y = 2y$ exist everywhere. Therefore, local extrema can occur only at the origin $(0, 0)$ where $f_x = 0$ and $f_y = 0$. Along the positive x-axis, however, f has the value $f(x, 0) = -x^2 < 0$; along the positive y-axis, f has the value $f(0, y) = y^2 > 0$. Therefore, every open disk in the xy-plane centered at $(0, 0)$ contains points where the function is positive and points where it is negative. The function has a saddle point at the origin and no local extreme values (Figure 14.47a). Figure 14.47b displays the level curves (they are hyperbolas) of f, and shows the function decreasing and increasing in an alternating fashion among the four groupings of hyperbolas. ∎

That $f_x = f_y = 0$ at an interior point (a, b) of R does not guarantee f has a local extreme value there. If f and its first and second partial derivatives are continuous on R, however, we may be able to learn more from the following theorem, proved in Section 14.9.

> **THEOREM 11—Second Derivative Test for Local Extreme Values**
> Suppose that $f(x, y)$ and its first and second partial derivatives are continuous throughout a disk centered at (a, b) and that $f_x(a, b) = f_y(a, b) = 0$. Then
>
> **i)** f has a **local maximum** at (a, b) if $f_{xx} < 0$ and $f_{xx}f_{yy} - f_{xy}^2 > 0$ at (a, b).
> **ii)** f has a **local minimum** at (a, b) if $f_{xx} > 0$ and $f_{xx}f_{yy} - f_{xy}^2 > 0$ at (a, b).
> **iii)** f has a **saddle point** at (a, b) if $f_{xx}f_{yy} - f_{xy}^2 < 0$ at (a, b).
> **iv)** the test is inconclusive at (a, b) if $f_{xx}f_{yy} - f_{xy}^2 = 0$ at (a, b). In this case, we must find some other way to determine the behavior of f at (a, b).

The expression $f_{xx}f_{yy} - f_{xy}^2$ is called the **discriminant** or **Hessian** of f. It is sometimes easier to remember it in determinant form,

$$f_{xx}f_{yy} - f_{xy}^2 = \begin{vmatrix} f_{xx} & f_{xy} \\ f_{xy} & f_{yy} \end{vmatrix}.$$

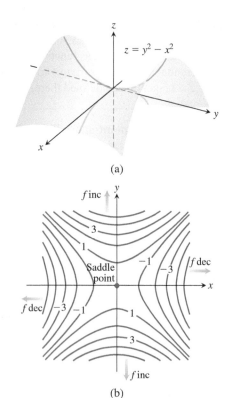

(a)

(b)

FIGURE 14.47 (a) The origin is a saddle point of the function $f(x, y) = y^2 - x^2$. There are no local extreme values (Example 2). (b) Level curves for the function f in Example 2.

Theorem 11 says that if the discriminant is positive at the point (a, b), then the surface curves the same way in all directions: downward if $f_{xx} < 0$, giving rise to a local maximum, and upward if $f_{xx} > 0$, giving a local minimum. On the other hand, if the discriminant is negative at (a, b), then the surface curves up in some directions and down in others, so we have a saddle point.

EXAMPLE 3 Find the local extreme values of the function

$$f(x, y) = xy - x^2 - y^2 - 2x - 2y + 4.$$

Solution The function is defined and differentiable for all x and y, and its domain has no boundary points. The function therefore has extreme values only at the points where f_x and f_y are simultaneously zero. This leads to

$$f_x = y - 2x - 2 = 0, \qquad f_y = x - 2y - 2 = 0,$$

or

$$x = y = -2.$$

Therefore, the point $(-2, -2)$ is the only point where f may take on an extreme value. To see if it does so, we calculate

$$f_{xx} = -2, \qquad f_{yy} = -2, \qquad f_{xy} = 1.$$

The discriminant of f at $(a, b) = (-2, -2)$ is

$$f_{xx}f_{yy} - f_{xy}{}^2 = (-2)(-2) - (1)^2 = 4 - 1 = 3.$$

The combination

$$f_{xx} < 0 \qquad \text{and} \qquad f_{xx}f_{yy} - f_{xy}{}^2 > 0$$

tells us that f has a local maximum at $(-2, -2)$. The value of f at this point is $f(-2, -2) = 8$. ∎

EXAMPLE 4 Find the local extreme values of $f(x, y) = 3y^2 - 2y^3 - 3x^2 + 6xy$.

Solution Since f is differentiable everywhere, it can assume extreme values only where

$$f_x = 6y - 6x = 0 \qquad \text{and} \qquad f_y = 6y - 6y^2 + 6x = 0.$$

From the first of these equations we find $x = y$, and substitution for y into the second equation then gives

$$6x - 6x^2 + 6x = 0 \qquad \text{or} \qquad 6x(2 - x) = 0.$$

The two critical points are therefore $(0, 0)$ and $(2, 2)$.

To classify the critical points, we calculate the second derivatives:

$$f_{xx} = -6, \qquad f_{yy} = 6 - 12y, \qquad f_{xy} = 6.$$

The discriminant is given by

$$f_{xx}f_{yy} - f_{xy}{}^2 = (-36 + 72y) - 36 = 72(y - 1).$$

At the critical point $(0, 0)$ we see that the value of the discriminant is the negative number -72, so the function has a saddle point at the origin. At the critical point $(2, 2)$ we see that the discriminant has the positive value 72. Combining this result with the negative value of the second partial $f_{xx} = -6$, Theorem 11 says that the critical point $(2, 2)$ gives a local maximum value of $f(2, 2) = 12 - 16 - 12 + 24 = 8$. A graph of the surface is shown in Figure 14.48. ∎

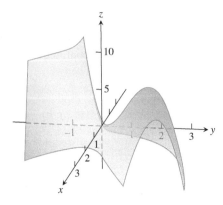

FIGURE 14.48 The surface $z = 3y^2 - 2y^3 - 3x^2 + 6xy$ has a saddle point at the origin and a local maximum at the point $(2, 2)$ (Example 4).

EXAMPLE 5   Find the critical points of the function $f(x, y) = 10xye^{-(x^2+y^2)}$ and use the Second Derivative Test to classify each point as one where a saddle, local minimum, or local maximum occurs.

Solution   First we find the partial derivatives f_x and f_y and set them simultaneously to zero in seeking the critical points:

$$f_x = 10ye^{-(x^2+y^2)} - 20x^2ye^{-(x^2+y^2)} = 10y(1 - 2x^2)e^{-(x^2+y^2)} = 0 \Rightarrow y = 0 \text{ or } 1 - 2x^2 = 0,$$
$$f_y = 10xe^{-(x^2+y^2)} - 20xy^2e^{-(x^2+y^2)} = 10x(1 - 2y^2)e^{-(x^2+y^2)} = 0 \Rightarrow x = 0 \text{ or } 1 - 2y^2 = 0.$$

Since both partial derivatives are continuous everywhere, the only critical points are

$$(0, 0), \left(\frac{1}{\sqrt{2}}, \frac{1}{\sqrt{2}}\right), \left(-\frac{1}{\sqrt{2}}, \frac{1}{\sqrt{2}}\right), \left(\frac{1}{\sqrt{2}}, -\frac{1}{\sqrt{2}}\right), \text{ and } \left(-\frac{1}{\sqrt{2}}, -\frac{1}{\sqrt{2}}\right).$$

Next we calculate the second partial derivatives in order to evaluate the discriminant at each critical point:

$$f_{xx} = -20xy(1 - 2x^2)e^{-(x^2+y^2)} - 40xye^{-(x^2+y^2)} = -20xy(3 - 2x^2)e^{-(x^2+y^2)},$$
$$f_{xy} = f_{yx} = 10(1 - 2x^2)e^{-(x^2+y^2)} - 20y^2(1 - 2x^2)e^{-(x^2+y^2)} = 10(1 - 2x^2)(1 - 2y^2)e^{-(x^2+y^2)},$$
$$f_{yy} = -20xy(1 - 2y^2)e^{-(x^2+y^2)} - 40xye^{-(x^2+y^2)} = -20xy(3 - 2y^2)e^{-(x^2+y^2)}.$$

The following table summarizes the values needed by the Second Derivative Test.

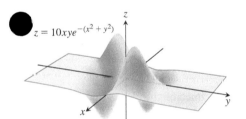

$z = 10xye^{-(x^2+y^2)}$

Critical Point	f_{xx}	f_{xy}	f_{yy}	Discriminant D
$(0, 0)$	0	10	0	-100
$\left(\dfrac{1}{\sqrt{2}}, \dfrac{1}{\sqrt{2}}\right)$	$-\dfrac{20}{e}$	0	$-\dfrac{20}{e}$	$\dfrac{400}{e^2}$
$\left(-\dfrac{1}{\sqrt{2}}, \dfrac{1}{\sqrt{2}}\right)$	$\dfrac{20}{e}$	0	$\dfrac{20}{e}$	$\dfrac{400}{e^2}$
$\left(\dfrac{1}{\sqrt{2}}, -\dfrac{1}{\sqrt{2}}\right)$	$\dfrac{20}{e}$	0	$\dfrac{20}{e}$	$\dfrac{400}{e^2}$
$\left(-\dfrac{1}{\sqrt{2}}, -\dfrac{1}{\sqrt{2}}\right)$	$-\dfrac{20}{e}$	0	$-\dfrac{20}{e}$	$\dfrac{400}{e^2}$

FIGURE 14.49   A graph of the function in Example 5.

From the table we find that $D < 0$ at the critical point $(0, 0)$, giving a saddle; $D > 0$ and $f_{xx} < 0$ at the critical points $\left(1/\sqrt{2}, 1/\sqrt{2}\right)$ and $\left(-1/\sqrt{2}, -1/\sqrt{2}\right)$, giving local maximum values there; and $D > 0$ and $f_{xx} > 0$ at the critical points $\left(-1/\sqrt{2}, 1/\sqrt{2}\right)$ and $\left(1/\sqrt{2}, -1/\sqrt{2}\right)$, each giving local minimum values. A graph of the surface is shown in Figure 14.49.   ∎

Absolute Maxima and Minima on Closed Bounded Regions

We organize the search for the absolute extrema of a continuous function $f(x, y)$ on a closed and bounded region R into three steps.

1. *List the interior points of R* where f may have local maxima and minima and evaluate f at these points. These are the critical points of f.

2. *List the boundary points of R* where f has local maxima and minima and evaluate f at these points. We show how to do this in the next example.

3. *Look through the lists* for the maximum and minimum values of f. These will be the absolute maximum and minimum values of f on R.

EXAMPLE 6   Find the absolute maximum and minimum values of

$$f(x, y) = 2 + 2x + 4y - x^2 - y^2$$

on the triangular region in the first quadrant bounded by the lines $x = 0$, $y = 0$, and $y = 9 - x$.

Solution   Since f is differentiable, the only places where f can assume these values are points inside the triangle where $f_x = f_y = 0$ and points on the boundary (Figure 14.50a).

(a) Interior points. For these we have

$$f_x = 2 - 2x = 0, \qquad f_y = 4 - 2y = 0,$$

yielding the single point $(x, y) = (1, 2)$. The value of f there is

$$f(1, 2) = 7.$$

(b) Boundary points. We take the triangle one side at a time:

i) On the segment OA, $y = 0$. The function

$$f(x, y) = f(x, 0) = 2 + 2x - x^2$$

may now be regarded as a function of x defined on the closed interval $0 \le x \le 9$. Its extreme values (as we know from Chapter 4) may occur at the endpoints

$$\begin{aligned} x &= 0 \qquad \text{where} \qquad f(0, 0) = 2 \\ x &= 9 \qquad \text{where} \qquad f(9, 0) = 2 + 18 - 81 = -61 \end{aligned}$$

or at the interior points where $f'(x, 0) = 2 - 2x = 0$. The only interior point where $f'(x, 0) = 0$ is $x = 1$, where

$$f(x, 0) = f(1, 0) = 3.$$

ii) On the segment OB, $x = 0$ and

$$f(x, y) = f(0, y) = 2 + 4y - y^2.$$

As in part i), we consider $f(0, y)$ as a function of y defined on the closed interval $[0, 9]$. Its extreme values can occur at the endpoints or at interior points where $f'(0, y) = 0$. Since $f'(0, y) = 4 - 2y$, the only interior point where $f'(0, y) = 0$ occurs at $(0, 2)$, with $f(0, 2) = 6$. So the candidates for this segment are

$$f(0, 0) = 2, \qquad f(0, 9) = -43, \qquad f(0, 2) = 6.$$

iii) We have already accounted for the values of f at the endpoints of AB, so we need only look at the interior points of the line segment AB. With $y = 9 - x$, we have

$$f(x, y) = 2 + 2x + 4(9 - x) - x^2 - (9 - x)^2 = -43 + 16x - 2x^2.$$

Setting $f'(x, 9 - x) = 16 - 4x = 0$ gives

$$x = 4.$$

At this value of x,

$$y = 9 - 4 = 5 \qquad \text{and} \qquad f(x, y) = f(4, 5) = -11.$$

Summary   We list all the function value candidates: 7, 2, −61, 3, −43, 6, −11. The maximum is 7, which f assumes at $(1, 2)$. The minimum is −61, which f assumes at $(9, 0)$. See Figure 14.50b. ■

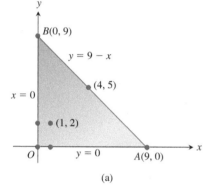

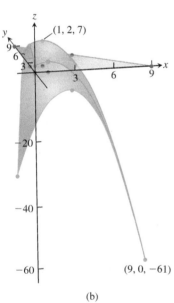

FIGURE 14.50 (a) This triangular region is the domain of the function in Example 6. (b) The graph of the function in Example 6. The blue points are the candidates for maxima or minima.

Solving extreme value problems with algebraic constraints on the variables usually requires the method of Lagrange multipliers introduced in the next section. But sometimes we can solve such problems directly, as in the next example.

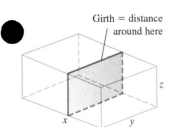

Girth = distance around here

FIGURE 14.51   The box in Example 7.

EXAMPLE 7   A delivery company accepts only rectangular boxes the sum of whose length and girth (perimeter of a cross-section) does not exceed 108 in. Find the dimensions of an acceptable box of largest volume.

Solution   Let x, y, and z represent the length, width, and height of the rectangular box, respectively. Then the girth is $2y + 2z$. We want to maximize the volume $V = xyz$ of the box (Figure 14.51) satisfying $x + 2y + 2z = 108$ (the largest box accepted by the delivery company). Thus, we can write the volume of the box as a function of two variables:

$$V(y, z) = (108 - 2y - 2z)yz \qquad \begin{matrix} V = xyz \text{ and} \\ x = 108 - 2y - 2z \end{matrix}$$
$$= 108yz - 2y^2z - 2yz^2.$$

Setting the first partial derivatives equal to zero,

$$V_y(y, z) = 108z - 4yz - 2z^2 = (108 - 4y - 2z)z = 0$$
$$V_z(y, z) = 108y - 2y^2 - 4yz = (108 - 2y - 4z)y = 0,$$

gives the critical points $(0, 0)$, $(0, 54)$, $(54, 0)$, and $(18, 18)$. The volume is zero at $(0, 0)$, $(0, 54)$, and $(54, 0)$, which are not maximum values. At the point $(18, 18)$, we apply the Second Derivative Test (Theorem 11):

$$V_{yy} = -4z, \qquad V_{zz} = -4y, \qquad V_{yz} = 108 - 4y - 4z.$$

Then

$$V_{yy}V_{zz} - V_{yz}{}^2 = 16yz - 16(27 - y - z)^2.$$

Thus,

$$V_{yy}(18, 18) = -4(18) < 0$$

and

$$\left(V_{yy}V_{zz} - V_{yz}{}^2\right)\Big|_{(18, 18)} = 16(18)(18) - 16(-9)^2 > 0,$$

so $(18, 18)$ gives a maximum volume. The dimensions of the package are $x = 108 - 2(18) - 2(18) = 36$ in., $y = 18$ in., and $z = 18$ in. The maximum volume is $V = (36)(18)(18) = 11,664$ in^3, or 6.75 ft^3.   ■

Despite the power of Theorem 11, we urge you to remember its limitations. It does not apply to boundary points of a function's domain, where it is possible for a function to have extreme values along with nonzero derivatives. Also, it does not apply to points where either f_x or f_y fails to exist.

Summary of Max-Min Tests

The extreme values of $f(x, y)$ can occur only at

i) boundary points of the domain of f

ii) critical points (interior points where $f_x = f_y = 0$ or points where f_x or f_y fails to exist)

If the first- and second-order partial derivatives of f are continuous throughout a disk centered at a point (a, b) and $f_x(a, b) = f_y(a, b) = 0$, the nature of $f(a, b)$ can be tested with the **Second Derivative Test**:

i) $f_{xx} < 0$ and $f_{xx}f_{yy} - f_{xy}{}^2 > 0$ at (a, b) ⟹ **local maximum**

ii) $f_{xx} > 0$ and $f_{xx}f_{yy} - f_{xy}{}^2 > 0$ at (a, b) ⟹ **local minimum**

iii) $f_{xx}f_{yy} - f_{xy}{}^2 < 0$ at (a, b) ⟹ **saddle point**

iv) $f_{xx}f_{yy} - f_{xy}{}^2 = 0$ at (a, b) ⟹ **test is inconclusive**

EXERCISES 14.7

Finding Local Extrema

Find all the local maxima, local minima, and saddle points of the functions in Exercises 1–30.

1. $f(x, y) = x^2 + xy + y^2 + 3x - 3y + 4$
2. $f(x, y) = 2xy - 5x^2 - 2y^2 + 4x + 4y - 4$
3. $f(x, y) = x^2 + xy + 3x + 2y + 5$
4. $f(x, y) = 5xy - 7x^2 + 3x - 6y + 2$
5. $f(x, y) = 2xy - x^2 - 2y^2 + 3x + 4$
6. $f(x, y) = x^2 - 4xy + y^2 + 6y + 2$
7. $f(x, y) = 2x^2 + 3xy + 4y^2 - 5x + 2y$
8. $f(x, y) = x^2 - 2xy + 2y^2 - 2x + 2y + 1$
9. $f(x, y) = x^2 - y^2 - 2x + 4y + 6$
10. $f(x, y) = x^2 + 2xy$
11. $f(x, y) = \sqrt{56x^2 - 8y^2 - 16x - 31} + 1 - 8x$
12. $f(x, y) = 1 - \sqrt[3]{x^2 + y^2}$
13. $f(x, y) = x^3 - y^3 - 2xy + 6$
14. $f(x, y) = x^3 + 3xy + y^3$
15. $f(x, y) = 6x^2 - 2x^3 + 3y^2 + 6xy$
16. $f(x, y) = x^3 + y^3 + 3x^2 - 3y^2 - 8$
17. $f(x, y) = x^3 + 3xy^2 - 15x + y^3 - 15y$
18. $f(x, y) = 2x^3 + 2y^3 - 9x^2 + 3y^2 - 12y$
19. $f(x, y) = 4xy - x^4 - y^4$
20. $f(x, y) = x^4 + y^4 + 4xy$
21. $f(x, y) = \dfrac{1}{x^2 + y^2 - 1}$
22. $f(x, y) = \dfrac{1}{x} + xy + \dfrac{1}{y}$
23. $f(x, y) = y \sin x$
24. $f(x, y) = e^{2x} \cos y$
25. $f(x, y) = e^{x^2 + y^2 - 4x}$
26. $f(x, y) = e^y - ye^x$
27. $f(x, y) = e^{-y}(x^2 + y^2)$
28. $f(x, y) = e^x(x^2 - y^2)$
29. $f(x, y) = 2 \ln x + \ln y - 4x - y$
30. $f(x, y) = \ln(x + y) + x^2 - y$

Finding Absolute Extrema

In Exercises 31–38, find the absolute maxima and minima of the functions on the given domains.

31. $f(x, y) = 2x^2 - 4x + y^2 - 4y + 1$ on the closed triangular plate bounded by the lines $x = 0$, $y = 2$, $y = 2x$ in the first quadrant
32. $D(x, y) = x^2 - xy + y^2 + 1$ on the closed triangular plate in the first quadrant bounded by the lines $x = 0$, $y = 4$, $y = x$
33. $f(x, y) = x^2 + y^2$ on the closed triangular plate bounded by the lines $x = 0$, $y = 0$, $y + 2x = 2$ in the first quadrant
34. $T(x, y) = x^2 + xy + y^2 - 6x$ on the rectangular plate $0 \le x \le 5$, $-3 \le y \le 3$

35. $T(x, y) = x^2 + xy + y^2 - 6x + 2$ on the rectangular plate $0 \le x \le 5$, $-3 \le y \le 0$
36. $f(x, y) = 48xy - 32x^3 - 24y^2$ on the rectangular plate $0 \le x \le 1$, $0 \le y \le 1$
37. $f(x, y) = (4x - x^2) \cos y$ on the rectangular plate $1 \le x \le 3$, $-\pi/4 \le y \le \pi/4$ (see accompanying figure)

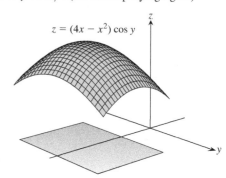

$z = (4x - x^2) \cos y$

38. $f(x, y) = 4x - 8xy + 2y + 1$ on the triangular plate bounded by the lines $x = 0$, $y = 0$, $x + y = 1$ in the first quadrant
39. Find two numbers a and b with $a \le b$ such that
$$\int_a^b (6 - x - x^2)\, dx$$
has its largest value.
40. Find two numbers a and b with $a \le b$ such that
$$\int_a^b (24 - 2x - x^2)^{1/3}\, dx$$
has its largest value.
41. **Temperatures** A flat circular plate has the shape of the region $x^2 + y^2 \le 1$. The plate, including the boundary where $x^2 + y^2 = 1$, is heated so that the temperature at the point (x, y) is
$$T(x, y) = x^2 + 2y^2 - x.$$
Find the temperatures at the hottest and coldest points on the plate.
42. Find the critical point of
$$f(x, y) = xy + 2x - \ln x^2 y$$
in the open first quadrant ($x > 0$, $y > 0$) and show that f takes on a minimum there.

Theory and Examples

43. Find the maxima, minima, and saddle points of $f(x, y)$, if any, given that
   a. $f_x = 2x - 4y$ and $f_y = 2y - 4x$
   b. $f_x = 2x - 2$ and $f_y = 2y - 4$
   c. $f_x = 9x^2 - 9$ and $f_y = 2y + 4$
   Describe your reasoning in each case.

44. The discriminant $f_{xx}f_{yy} - f_{xy}{}^2$ is zero at the origin for each of the following functions, so the Second Derivative Test fails there. Determine whether the function has a maximum, a minimum, or neither at the origin by imagining what the surface $z = f(x, y)$ looks like. Describe your reasoning in each case.

   a. $f(x, y) = x^2y^2$ **b.** $f(x, y) = 1 - x^2y^2$

   c. $f(x, y) = xy^2$ **d.** $f(x, y) = x^3y^2$

   e. $f(x, y) = x^3y^3$ **f.** $f(x, y) = x^4y^4$

45. Show that $(0, 0)$ is a critical point of $f(x, y) = x^2 + kxy + y^2$ no matter what value the constant k has. (*Hint:* Consider two cases: $k = 0$ and $k \neq 0$.)

46. For what values of the constant k does the Second Derivative Test guarantee that $f(x, y) = x^2 + kxy + y^2$ will have a saddle point at $(0, 0)$? A local minimum at $(0, 0)$? For what values of k is the Second Derivative Test inconclusive? Give reasons for your answers.

47. If $f_x(a, b) = f_y(a, b) = 0$, must f have a local maximum or minimum value at (a, b)? Give reasons for your answer.

48. Can you conclude anything about $f(a, b)$ if f and its first and second partial derivatives are continuous throughout a disk centered at the critical point (a, b) and $f_{xx}(a, b)$ and $f_{yy}(a, b)$ differ in sign? Give reasons for your answer.

49. Among all the points on the graph of $z = 10 - x^2 - y^2$ that lie above the plane $x + 2y + 3z = 0$, find the point farthest from the plane.

50. Find the point on the graph of $z = x^2 + y^2 + 10$ nearest the plane $x + 2y - z = 0$.

51. Find the point on the plane $3x + 2y + z = 6$ that is nearest the origin.

52. Find the minimum distance from the point $(2, -1, 1)$ to the plane $x + y - z = 2$.

53. Find three numbers whose sum is 9 and whose sum of squares is a minimum.

54. Find three positive numbers whose sum is 3 and whose product is a maximum.

55. Find the maximum value of $s = xy + yz + xz$ where $x + y + z = 6$.

56. Find the minimum distance from the cone $z = \sqrt{x^2 + y^2}$ to the point $(-6, 4, 0)$.

57. Find the dimensions of the rectangular box of maximum volume that can be inscribed inside the sphere $x^2 + y^2 + z^2 = 4$.

58. Among all closed rectangular boxes of volume 27 cm^3, what is the smallest surface area?

59. You are to construct an open rectangular box from 12 ft^2 of material. What dimensions will result in a box of maximum volume?

60. Consider the function $f(x, y) = x^2 + y^2 + 2xy - x - y + 1$ over the square $0 \le x \le 1$ and $0 \le y \le 1$.

   a. Show that f has an absolute minimum along the line segment $2x + 2y = 1$ in this square. What *is* the absolute minimum value?

   b. Find the absolute maximum value of f over the square.

61. Find the point on the graph of $y^2 - xz^2 = 4$ nearest the origin.

62. A rectangular box is inscribed in the region in the first octant bounded above by the plane with x-intercept 6, y-intercept 6, and z-intercept 6.

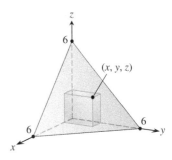

   a. Find an equation for the plane.

   b. Find the dimensions of the box of maximum volume.

Extreme Values on Parametrized Curves   To find the extreme values of a function $f(x, y)$ on a curve $x = x(t)$, $y = y(t)$, we treat f as a function of the single variable t and use the Chain Rule to find where df/dt is zero. As in any other single-variable case, the extreme values of f are then found among the values at the

   a. critical points (points where df/dt is zero or fails to exist), and

   b. endpoints of the parameter domain.

Find the absolute maximum and minimum values of the following functions on the given curves.

63. Functions:

   a. $f(x, y) = x + y$ **b.** $g(x, y) = xy$ **c.** $h(x, y) = 2x^2 + y^2$

   Curves:

    i) The semicircle $x^2 + y^2 = 4$, $y \ge 0$

    ii) The quarter circle $x^2 + y^2 = 4$, $x \ge 0$, $y \ge 0$

   Use the parametric equations $x = 2 \cos t$, $y = 2 \sin t$.

64. Functions:

   a. $f(x, y) = 2x + 3y$

   b. $g(x, y) = xy$

   c. $h(x, y) = x^2 + 3y^2$

   Curves:

    i) The semiellipse $(x^2/9) + (y^2/4) = 1$, $y \ge 0$

    ii) The quarter ellipse $(x^2/9) + (y^2/4) = 1$, $x \ge 0$, $y \ge 0$

   Use the parametric equations $x = 3 \cos t$, $y = 2 \sin t$.

65. Function: $f(x, y) = xy$

   Curves:

    i) The line $x = 2t$, $y = t + 1$

    ii) The line segment $x = 2t$, $y = t + 1$, $-1 \le t \le 0$

    iii) The line segment $x = 2t$, $y = t + 1$, $0 \le t \le 1$

66. Functions:

   a. $f(x, y) = x^2 + y^2$ **b.** $g(x, y) = 1/(x^2 + y^2)$

   Curves:

    i) The line $x = t$, $y = 2 - 2t$

    ii) The line segment $x = t$, $y = 2 - 2t$, $0 \le t \le 1$

67. Least squares and regression lines When we try to fit a line $y = mx + b$ to a set of numerical data points (x_1, y_1), $(x_2, y_2), \ldots, (x_n, y_n)$, we usually choose the line that minimizes the sum of the squares of the vertical distances from the points to the line. In theory, this means finding the values of m and b that minimize the value of the function

$$w = (mx_1 + b - y_1)^2 + \cdots + (mx_n + b - y_n)^2. \quad (1)$$

(See the accompanying figure.) Show that the values of m and b that do this are

$$m = \frac{\left(\sum x_k\right)\left(\sum y_k\right) - n\sum x_k y_k}{\left(\sum x_k\right)^2 - n\sum x_k^2}, \quad (2)$$

$$b = \frac{1}{n}\left(\sum y_k - m\sum x_k\right), \quad (3)$$

with all sums running from $k = 1$ to $k = n$. Many scientific calculators have these formulas built in, enabling you to find m and b with only a few keystrokes after you have entered the data.

The line $y = mx + b$ determined by these values of m and b is called the **least squares line**, **regression line**, or **trend line** for the data under study. Finding a least squares line lets you

1. summarize data with a simple expression,
2. predict values of y for other, experimentally untried values of x,
3. handle data analytically.

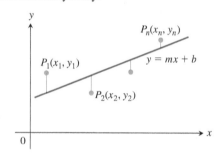

In Exercises 68–70, use Equations (2) and (3) to find the least squares line for each set of data points. Then use the linear equation you obtain to predict the value of y that would correspond to $x = 4$.

68. $(-2, 0)$, $(0, 2)$, $(2, 3)$

69. $(-1, 2)$, $(0, 1)$, $(3, -4)$

70. $(0, 0)$, $(1, 2)$, $(2, 3)$

COMPUTER EXPLORATIONS

In Exercises 71–76, you will explore functions to identify their local extrema. Use a CAS to perform the following steps:

  a. Plot the function over the given rectangle.

  b. Plot some level curves in the rectangle.

  c. Calculate the function's first partial derivatives and use the CAS equation solver to find the critical points. How do the critical points relate to the level curves plotted in part (b)? Which critical points, if any, appear to give a saddle point? Give reasons for your answer.

  d. Calculate the function's second partial derivatives and find the discriminant $f_{xx}f_{yy} - f_{xy}^2$.

  e. Using the max-min tests, classify the critical points found in part (c). Are your findings consistent with your discussion in part (c)?

71. $f(x, y) = x^2 + y^3 - 3xy$, $-5 \leq x \leq 5$, $-5 \leq y \leq 5$

72. $f(x, y) = x^3 - 3xy^2 + y^2$, $-2 \leq x \leq 2$, $-2 \leq y \leq 2$

73. $f(x, y) = x^4 + y^2 - 8x^2 - 6y + 16$, $-3 \leq x \leq 3$, $-6 \leq y \leq 6$

74. $f(x, y) = 2x^4 + y^4 - 2x^2 - 2y^2 + 3$, $-3/2 \leq x \leq 3/2$, $-3/2 \leq y \leq 3/2$

75. $f(x, y) = 5x^6 + 18x^5 - 30x^4 + 30xy^2 - 120x^3$, $-4 \leq x \leq 3$, $-2 \leq y \leq 2$

76. $f(x, y) = \begin{cases} x^5 \ln(x^2 + y^2), & (x, y) \neq (0, 0) \\ 0, & (x, y) = (0, 0) \end{cases}$, $-2 \leq x \leq 2$, $-2 \leq y \leq 2$

14.8 Lagrange Multipliers

HISTORICAL BIOGRAPHY
Joseph Louis Lagrange
(1736–1813)
bit.ly/2NdfJs7

Sometimes we need to find the extreme values of a function whose domain is constrained to lie within some particular subset of the plane—for example, a disk, a closed triangular region, or along a curve. We saw an instance of this situation in Example 6 of the previous section. Here we explore a powerful method for finding extreme values of constrained functions: the method of *Lagrange multipliers*.

Constrained Maxima and Minima

To gain some insight, we first consider a problem where a constrained minimum can be found by eliminating a variable.

EXAMPLE 1   Find the point $p(x, y, z)$ on the plane $2x + y - z - 5 = 0$ that is closest to the origin.

Solution   The problem asks us to find the minimum value of the function

$$|\overrightarrow{OP}| = \sqrt{(x - 0)^2 + (y - 0)^2 + (z - 0)^2} = \sqrt{x^2 + y^2 + z^2}$$

subject to the constraint that

$$2x + y - z - 5 = 0.$$

Since $|\overrightarrow{OP}|$ has a minimum value wherever the function

$$f(x, y, z) = x^2 + y^2 + z^2$$

has a minimum value, we may solve the problem by finding the minimum value of $f(x, y, z)$ subject to the constraint $2x + y - z - 5 = 0$ (thus avoiding square roots). If we regard x and y as the independent variables in this equation and write z as

$$z = 2x + y - 5,$$

our problem reduces to one of finding the points (x, y) at which the function

$$h(x, y) = f(x, y, 2x + y - 5) = x^2 + y^2 + (2x + y - 5)^2$$

has its minimum value or values. Since the domain of h is the entire xy-plane, the First Derivative Test of Section 14.7 tells us that any minima that h might have must occur at points where

$$h_x = 2x + 2(2x + y - 5)(2) = 0, \qquad h_y = 2y + 2(2x + y - 5) = 0.$$

This leads to

$$10x + 4y - 20, \qquad 4x + 4y = 10,$$

which has the solution

$$x = \frac{5}{3}, \qquad y = \frac{5}{6}.$$

We may apply a geometric argument together with the Second Derivative Test to show that these values minimize h. The z-coordinate of the corresponding point on the plane $z = 2x + y - 5$ is

$$z = 2\left(\frac{5}{3}\right) + \frac{5}{6} - 5 = -\frac{5}{6}.$$

Therefore, the point we seek is

$$\text{Closest point:} \qquad P\left(\frac{5}{3}, \frac{5}{6}, -\frac{5}{6}\right).$$

The distance from P to the origin is $5/\sqrt{6} \approx 2.04$. ∎

Attempts to solve a constrained maximum or minimum problem by substitution, as we might call the method of Example 1, do not always go smoothly.

EXAMPLE 2   Find the points on the hyperbolic cylinder $x^2 - z^2 - 1 = 0$ that are closest to the origin.

Solution 1   The cylinder is shown in Figure 14.52. We seek the points on the cylinder closest to the origin. These are the points whose coordinates minimize the value of the function

$$f(x, y, z) = x^2 + y^2 + z^2 \qquad \text{Square of the distance}$$

subject to the constraint that $x^2 - z^2 - 1 = 0$. If we regard x and y as independent variables in the constraint equation, then

$$z^2 = x^2 - 1$$

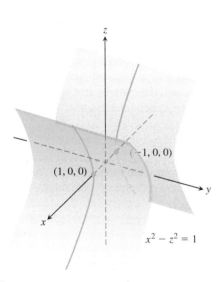

FIGURE 14.52   The hyperbolic cylinder $x^2 - z^2 - 1 = 0$ in Example 2.

The hyperbolic cylinder $x^2 - z^2 = 1$

On this part,
$x = \sqrt{z^2 + 1}$

On this part,
$x = -\sqrt{z^2 + 1}$

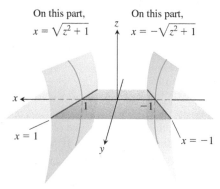

FIGURE 14.53 The region in the xy-plane from which the first two coordinates of the points (x, y, z) on the hyperbolic cylinder $x^2 - z^2 = 1$ are selected excludes the band $-1 < x < 1$ in the xy-plane (Example 2).

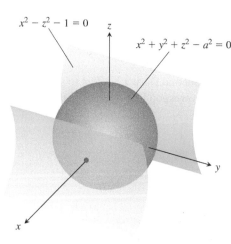

FIGURE 14.54 A sphere expanding like a soap bubble centered at the origin until it just touches the hyperbolic cylinder $x^2 - z^2 - 1 = 0$ (Example 2).

λ is the Greek letter lambda

and the values of $f(x, y, z) = x^2 + y^2 + z^2$ on the cylinder are given by the function

$$h(x, y) = x^2 + y^2 + (x^2 - 1) = 2x^2 + y^2 - 1.$$

To find the points on the cylinder whose coordinates minimize f, we look for the points in the xy-plane whose coordinates minimize h. The only extreme value of h occurs where

$$h_x = 4x = 0 \qquad \text{and} \qquad h_y = 2y = 0,$$

that is, at the point $(0, 0)$. But there are no points on the cylinder where both x and y are zero. What went wrong?

What happened was that the First Derivative Test found (as it should have) the point *in the domain of h* where h has a minimum value. We, on the other hand, want the points *on the cylinder* where h has a minimum value. Although the domain of h is the entire xy-plane, the domain from which we can select the first two coordinates of the points (x, y, z) on the cylinder is restricted to the "shadow" of the cylinder on the xy-plane; it does not include the band between the lines $x = -1$ and $x = 1$ (Figure 14.53).

We can avoid this problem if we treat y and z as independent variables (instead of x and y) and express x in terms of y and z as

$$x^2 = z^2 + 1.$$

With this substitution, $f(x, y, z) = x^2 + y^2 + z^2$ becomes

$$k(y, z) = (z^2 + 1) + y^2 + z^2 = 1 + y^2 + 2z^2$$

and we look for the points where k takes on its smallest value. The domain of k in the yz-plane now matches the domain from which we select the y- and z-coordinates of the points (x, y, z) on the cylinder. Hence, the points that minimize k in the plane will have corresponding points on the cylinder. The smallest values of k occur where

$$k_y = 2y = 0 \qquad \text{and} \qquad k_z = 4z = 0,$$

or where $y = z = 0$. This leads to

$$x^2 = z^2 + 1 = 1, \qquad x = \pm 1.$$

The corresponding points on the cylinder are $(\pm 1, 0, 0)$. We can see from the inequality

$$k(y, z) = 1 + y^2 + 2z^2 \geq 1$$

that the points $(\pm 1, 0, 0)$ give a minimum value for k. We can also see that the minimum distance from the origin to a point on the cylinder is 1 unit.

Solution 2   Another way to find the points on the cylinder closest to the origin is to imagine a small sphere centered at the origin expanding like a soap bubble until it just touches the cylinder (Figure 14.54). At each point of contact, the cylinder and sphere have the same tangent plane and normal line. Therefore, if the sphere and cylinder are represented as the level surfaces obtained by setting

$$f(x, y, z) = x^2 + y^2 + z^2 - a^2 \qquad \text{and} \qquad g(x, y, z) = x^2 - z^2 - 1$$

equal to 0, then the gradients ∇f and ∇g will be parallel where the surfaces touch. At any point of contact, we should therefore be able to find a scalar λ ("lambda") such that

$$\nabla f = \lambda \nabla g,$$

or

$$2x\mathbf{i} + 2y\mathbf{j} + 2z\mathbf{k} = \lambda(2x\mathbf{i} - 2z\mathbf{k}).$$

Thus, the coordinates x, y, and z of any point of tangency will have to satisfy the three scalar equations

$$2x = 2\lambda x, \qquad 2y = 0, \qquad 2z = -2\lambda z.$$

For what values of λ will a point (x, y, z) whose coordinates satisfy these scalar equations also lie on the surface $x^2 - z^2 - 1 = 0$? To answer this question, we use our knowledge that no point on the surface has a zero x-coordinate to conclude that $x \neq 0$. Hence, $2x = 2\lambda x$ only if

$$2 = 2\lambda, \quad \text{or} \quad \lambda = 1.$$

For $\lambda = 1$, the equation $2z = -2\lambda z$ becomes $2z = -2z$. If this equation is to be satisfied as well, z must be zero. Since $y = 0$ also (from the equation $2y = 0$), we conclude that the points we seek all have coordinates of the form

$$(x, 0, 0).$$

What points on the surface $x^2 - z^2 = 1$ have coordinates of this form? The answer is the points $(x, 0, 0)$ for which

$$x^2 - (0)^2 = 1, \quad x^2 = 1, \quad \text{or} \quad x = \pm 1.$$

The points on the cylinder closest to the origin are the points $(\pm 1, 0, 0)$. ∎

The Method of Lagrange Multipliers

In Solution 2 of Example 2, we used the **method of Lagrange multipliers**. The method says that the local extreme values of a function $f(x, y, z)$ whose variables are subject to a constraint $g(x, y, z) = 0$ are to be found on the surface $g = 0$ among the points where

$$\nabla f = \lambda \nabla g$$

for some scalar λ (called a **Lagrange multiplier**).

To explore the method further and see why it works, we first make the following observation, which we state as a theorem.

THEOREM 12—The Orthogonal Gradient Theorem

Suppose that $f(x, y, z)$ is differentiable in a region whose interior contains a smooth curve

$$C: \quad \mathbf{r}(t) = x(t)\mathbf{i} + y(t)\mathbf{j} + z(t)\mathbf{k}.$$

If P_0 is a point on C where f has a local maximum or minimum relative to its values on C, then ∇f is orthogonal to C at P_0.

Proof We show that ∇f is orthogonal to the curve's tangent vector $\mathbf{r}'$ at P_0. The values of f on C are given by the composition $f(x(t), y(t), z(t))$, whose derivative with respect to t is

$$\frac{df}{dt} = \frac{\partial f}{\partial x}\frac{dx}{dt} + \frac{\partial f}{\partial y}\frac{dy}{dt} + \frac{\partial f}{\partial z}\frac{dz}{dt} = \nabla f \cdot \mathbf{r}'.$$

At any point P_0 where f has a local maximum or minimum relative to its values on the curve, $df/dt = 0$, so

$$\nabla f \cdot \mathbf{r}' = 0.$$ ∎

By dropping the z-terms in Theorem 12, we obtain a similar result for functions of two variables.

COROLLARY At the points on a smooth curve $\mathbf{r}(t) = x(t)\mathbf{i} + y(t)\mathbf{j}$ where a differentiable function $f(x, y)$ takes on its local maxima and minima relative to its values on the curve, $\nabla f \cdot \mathbf{r}' = 0$.

Theorem 12 is the key to the method of Lagrange multipliers. Suppose that $f(x, y, z)$ and $g(x, y, z)$ are differentiable and that P_0 is a point on the surface $g(x, y, z) = 0$ where f has a local maximum or minimum value relative to its other values on the surface. We assume also that $\nabla g \neq \mathbf{0}$ at points on the surface $g(x, y, z) = 0$. Then f takes on a local maximum or minimum at P_0 relative to its values on every differentiable curve through P_0 on the surface $g(x, y, z) = 0$. Therefore, ∇f is orthogonal to the tangent vector of every such differentiable curve through P_0. So is ∇g, moreover (because ∇g is orthogonal to the level surface $g = 0$, as we saw in Section 14.5). Therefore, at P_0, ∇f is some scalar multiple λ of ∇g.

The Method of Lagrange Multipliers

Suppose that $f(x, y, z)$ and $g(x, y, z)$ are differentiable and $\nabla g \neq \mathbf{0}$ when $g(x, y, z) = 0$. To find the local maximum and minimum values of f subject to the constraint $g(x, y, z) = 0$ (if these exist), find the values of x, y, z, and λ that simultaneously satisfy the equations

$$\nabla f = \lambda \nabla g \qquad \text{and} \qquad g(x, y, z) = 0. \qquad (1)$$

For functions of two independent variables, the condition is similar, but without the variable z.

Some care must be used in applying this method. An extreme value may not actually exist (Exercise 45).

EXAMPLE 3   Find the greatest and smallest values that the function

$$f(x, y) = xy$$

takes on the ellipse (Figure 14.55)

$$\frac{x^2}{8} + \frac{y^2}{2} = 1.$$

Solution   We want to find the extreme values of $f(x, y) = xy$ subject to the constraint

$$g(x, y) = \frac{x^2}{8} + \frac{y^2}{2} - 1 = 0.$$

To do so, we first find the values of x, y, and λ for which

$$\nabla f = \lambda \nabla g \qquad \text{and} \qquad g(x, y) = 0.$$

The gradient equation in Equations (1) gives

$$y\mathbf{i} + x\mathbf{j} = \frac{\lambda}{4} x\mathbf{i} + \lambda y\mathbf{j},$$

from which we find

$$y = \frac{\lambda}{4} x, \qquad x = \lambda y, \qquad \text{and} \qquad y = \frac{\lambda}{4}(\lambda y) = \frac{\lambda^2}{4} y,$$

so that $y = 0$ or $\lambda = \pm 2$. We now consider these two cases.

Case 1:   If $y = 0$, then $x = y = 0$. But $(0, 0)$ is not on the ellipse. Hence, $y \neq 0$.

Case 2:   If $y \neq 0$, then $\lambda = \pm 2$ and $x = \pm 2y$. Substituting this in the equation $g(x, y) = 0$ gives

$$\frac{(\pm 2y)^2}{8} + \frac{y^2}{2} = 1, \qquad 4y^2 + 4y^2 = 8 \qquad \text{and} \qquad y = \pm 1.$$

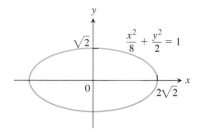

$$\frac{x^2}{8} + \frac{y^2}{2} = 1$$

FIGURE 14.55   Example 3 shows how to find the largest and smallest values of the product xy on this ellipse.

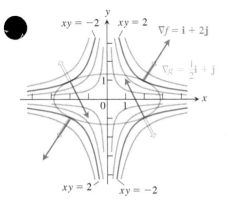

FIGURE 14.56 When subjected to the constraint $g(x, y) = x^2/8 + y^2/2 - 1 = 0$, the function $f(x, y) = xy$ takes on extreme values at the four points $(\pm 2, \pm 1)$. These are the points on the ellipse where ∇f (red) is a scalar multiple of ∇g (blue) (Example 3).

The function $f(x, y) = xy$ therefore takes on its extreme values on the ellipse at the four points $(\pm 2, 1), (\pm 2, -1)$. The extreme values are $xy = 2$ and $xy = -2$.

The Geometry of the Solution The level curves of the function $f(x, y) = xy$ are the hyperbolas $xy = c$ (Figure 14.56). The farther the hyperbolas lie from the origin, the larger the absolute value of f. We want to find the extreme values of $f(x, y)$, given that the point (x, y) also lies on the ellipse $x^2 + 4y^2 = 8$. Which hyperbolas intersecting the ellipse lie farthest from the origin? The hyperbolas that just graze the ellipse, the ones that are tangent to it, are farthest. At these points, any vector normal to the hyperbola is normal to the ellipse, so $\nabla f = y\mathbf{i} + x\mathbf{j}$ is a multiple ($\lambda = \pm 2$) of $\nabla g = (x/4)\mathbf{i} + y\mathbf{j}$. At the point $(2, 1)$, for example,

$$\nabla f = \mathbf{i} + 2\mathbf{j}, \qquad \nabla g = \frac{1}{2}\mathbf{i} + \mathbf{j}, \qquad \text{and} \qquad \nabla f = 2\nabla g.$$

At the point $(-2, 1)$,

$$\nabla f = \mathbf{i} - 2\mathbf{j}, \qquad \nabla g = -\frac{1}{2}\mathbf{i} + \mathbf{j}, \qquad \text{and} \qquad \nabla f = -2\nabla g. \qquad \blacksquare$$

EXAMPLE 4   Find the maximum and minimum values of the function $f(x, y) = 3x + 4y$ on the circle $x^2 + y^2 = 1$.

Solution   We model this as a Lagrange multiplier problem with

$$f(x, y) = 3x + 4y, \qquad g(x, y) = x^2 + y^2 - 1$$

and look for the values of x, y, and λ that satisfy the equations

$$\nabla f = \lambda \nabla g: \quad 3\mathbf{i} + 4\mathbf{j} = 2x\lambda\mathbf{i} + 2y\lambda\mathbf{j}$$
$$g(x, y) = 0: \quad x^2 + y^2 - 1 = 0.$$

The gradient equation in Equations (1) implies that $\lambda \neq 0$ and gives

$$x = \frac{3}{2\lambda}, \qquad y = \frac{2}{\lambda}.$$

These equations tell us, among other things, that x and y have the same sign. With these values for x and y, the equation $g(x, y) = 0$ gives

$$\left(\frac{3}{2\lambda}\right)^2 + \left(\frac{2}{\lambda}\right)^2 - 1 = 0,$$

so

$$\frac{9}{4\lambda^2} + \frac{4}{\lambda^2} = 1, \qquad 9 + 16 = 4\lambda^2, \qquad 4\lambda^2 = 25, \qquad \text{and} \qquad \lambda = \pm\frac{5}{2}.$$

Thus,

$$x = \frac{3}{2\lambda} = \pm\frac{3}{5}, \qquad y = \frac{2}{\lambda} = \pm\frac{4}{5},$$

and $f(x, y) = 3x + 4y$ has extreme values at $(x, y) = \pm(3/5, 4/5)$.

By calculating the value of $3x + 4y$ at the points $\pm(3/5, 4/5)$, we see that its maximum and minimum values on the circle $x^2 + y^2 = 1$ are

$$3\left(\frac{3}{5}\right) + 4\left(\frac{4}{5}\right) = \frac{25}{5} = 5 \qquad \text{and} \qquad 3\left(-\frac{3}{5}\right) + 4\left(-\frac{4}{5}\right) = -\frac{25}{5} = -5.$$

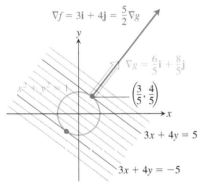

FIGURE 14.57 The function $f(x, y) = 3x + 4y$ takes on its largest value on the unit circle $g(x, y) = x^2 + y^2 - 1 = 0$ at the point $(3/5, 4/5)$ and its smallest value at the point $(-3/5, -4/5)$ (Example 4). At each of these points, ∇f is a scalar multiple of ∇g. The figure shows the gradients at the first point but not the second.

The Geometry of the Solution The level curves of $f(x, y) = 3x + 4y$ are the lines $3x + 4y = c$ (Figure 14.57). The farther the lines lie from the origin, the larger the absolute value of f. We want to find the extreme values of $f(x, y)$ given that the point (x, y)

also lies on the circle $x^2 + y^2 = 1$. Which lines intersecting the circle lie farthest from the origin? The lines tangent to the circle are farthest. At the points of tangency, any vector normal to the line is normal to the circle, so the gradient $\nabla f = 3\mathbf{i} + 4\mathbf{j}$ is a multiple ($\lambda = \pm 5/2$) of the gradient $\nabla g = 2x\mathbf{i} + 2y\mathbf{j}$. At the point $(3/5, 4/5)$, for example,

$$\nabla f = 3\mathbf{i} + 4\mathbf{j}, \qquad \nabla g = \frac{6}{5}\mathbf{i} + \frac{8}{5}\mathbf{j}, \qquad \text{and} \qquad \nabla f = \frac{5}{2}\nabla g. \qquad \blacksquare$$

Lagrange Multipliers with Two Constraints

Many problems require us to find the extreme values of a differentiable function $f(x, y, z)$ whose variables are subject to two constraints. If the constraints are

$$g_1(x, y, z) = 0 \qquad \text{and} \qquad g_2(x, y, z) = 0$$

and g_1 and g_2 are differentiable, with ∇g_1 not parallel to ∇g_2, we find the constrained local maxima and minima of f by introducing two Lagrange multipliers λ and μ (mu, pronounced "mew"). That is, we locate the points $P(x, y, z)$ where f takes on its constrained extreme values by finding the values of x, y, z, λ, and μ that simultaneously satisfy the three equations

$$\boxed{\nabla f = \lambda \nabla g_1 + \mu \nabla g_2, \qquad g_1(x, y, z) = 0, \qquad g_2(x, y, z) = 0 \qquad (2)}$$

Equations (2) have a nice geometric interpretation. The surfaces $g_1 = 0$ and $g_2 = 0$ (usually) intersect in a smooth curve, say C (Figure 14.58). Along this curve we seek the points where f has local maximum and minimum values relative to its other values on the curve. These are the points where ∇f is normal to C, as we saw in Theorem 12. But ∇g_1 and ∇g_2 are also normal to C at these points because C lies in the surfaces $g_1 = 0$ and $g_2 = 0$. Therefore, ∇f lies in the plane determined by ∇g_1 and ∇g_2, which means that $\nabla f = \lambda \nabla g_1 + \mu \nabla g_2$ for some λ and μ. Since the points we seek also lie in both surfaces, their coordinates must satisfy the equations $g_1(x, y, z) = 0$ and $g_2(x, y, z) = 0$, which are the remaining requirements in Equations (2).

EXAMPLE 5   The plane $x + y + z = 1$ cuts the cylinder $x^2 + y^2 = 1$ in an ellipse (Figure 14.59). Find the points on the ellipse that lie closest to and farthest from the origin.

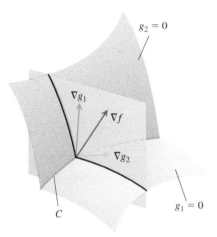

FIGURE 14.58 The vectors ∇g_1 and ∇g_2 lie in a plane perpendicular to the curve C because ∇g_1 is normal to the surface $g_1 = 0$ and ∇g_2 is normal to the surface $g_2 = 0$.

μ is the Greek letter mu, pronounced "mew"

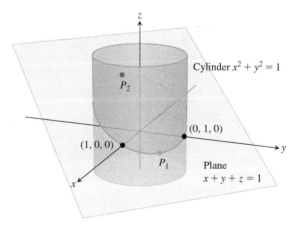

FIGURE 14.59 On the ellipse where the plane and cylinder meet, we find the points closest to and farthest from the origin (Example 5).

Solution   We find the extreme values of

$$f(x, y, z) = x^2 + y^2 + z^2$$

(the square of the distance from (x, y, z) to the origin) subject to the constraints

$$g_1(x, y, z) = x^2 + y^2 - 1 = 0 \qquad (3)$$

$$g_2(x, y, z) = x + y + z - 1 = 0. \qquad (4)$$

The gradient equation in Equations (2) then gives

$$\nabla f = \lambda \nabla g_1 + \mu \nabla g_2$$

$$2x\mathbf{i} + 2y\mathbf{j} + 2z\mathbf{k} = \lambda(2x\mathbf{i} + 2y\mathbf{j}) + \mu(\mathbf{i} + \mathbf{j} + \mathbf{k})$$

$$2x\mathbf{i} + 2y\mathbf{j} + 2z\mathbf{k} = (2\lambda x + \mu)\mathbf{i} + (2\lambda y + \mu)\mathbf{j} + \mu\mathbf{k}$$

or

$$2x = 2\lambda x + \mu, \qquad 2y = 2\lambda y + \mu, \qquad 2z = \mu. \qquad (5)$$

The scalar equations in Equations (5) yield

$$2x = 2\lambda x + 2z \Longrightarrow (1 - \lambda)x = z,$$

$$2y = 2\lambda y + 2z \Longrightarrow (1 - \lambda)y = z. \qquad (6)$$

Equations (6) are satisfied simultaneously if either $\lambda = 1$ and $z = 0$ or $\lambda \neq 1$ and $x = y = z/(1 - \lambda)$.

If $z = 0$, then solving Equations (3) and (4) simultaneously to find the corresponding points on the ellipse gives the two points $(1, 0, 0)$ and $(0, 1, 0)$. This makes sense when you look at Figure 14.59.

If $x = y$, then Equations (3) and (4) give

$$x^2 + x^2 - 1 = 0 \qquad\qquad x + x + z - 1 = 0$$

$$2x^2 = 1 \qquad\qquad z = 1 - 2x$$

$$x = \pm\frac{\sqrt{2}}{2} \qquad\qquad z = 1 \mp \sqrt{2}.$$

The corresponding points on the ellipse are

$$P_1 = \left(\frac{\sqrt{2}}{2}, \frac{\sqrt{2}}{2}, 1 - \sqrt{2}\right) \qquad \text{and} \qquad P_2 = \left(-\frac{\sqrt{2}}{2}, -\frac{\sqrt{2}}{2}, 1 + \sqrt{2}\right).$$

Here we need to be careful, however. Although P_1 and P_2 both give local maxima of f on the ellipse, P_2 is farther from the origin than P_1.

The points on the ellipse closest to the origin are $(1, 0, 0)$ and $(0, 1, 0)$. The point on the ellipse farthest from the origin is P_2. (See Figure 14.59.) ■

EXERCISES   14.8

Two Independent Variables with One Constraint

1. **Extrema on an ellipse**   Find the points on the ellipse $x^2 + 2y^2 = 1$ where $f(x, y) = xy$ has its extreme values.

2. **Extrema on a circle**   Find the extreme values of $f(x, y) = xy$ subject to the constraint $g(x, y) = x^2 + y^2 - 10 = 0$.

3. **Maximum on a line**   Find the maximum value of $f(x, y) = 49 - x^2 - y^2$ on the line $x + 3y = 10$.

4. **Extrema on a line**   Find the local extreme values of $f(x, y) = x^2y$ on the line $x + y = 3$.

5. **Constrained minimum**   Find the points on the curve $xy^2 = 54$ nearest the origin.

6. **Constrained minimum**   Find the points on the curve $x^2y = 2$ nearest the origin.

7. Use the method of Lagrange multipliers to find

   a. **Minimum on a hyperbola** The minimum value of $x + y$, subject to the constraints $xy = 16, x > 0, y > 0$

   b. **Maximum on a line** The maximum value of xy, subject to the constraint $x + y = 16$.

   Comment on the geometry of each solution.

8. **Extrema on a curve** Find the points on the curve $x^2 + xy + y^2 = 1$ in the xy-plane that are nearest to and farthest from the origin.

9. **Minimum surface area with fixed volume** Find the dimensions of the closed right circular cylindrical can of smallest surface area whose volume is 16π cm^3.

10. **Cylinder in a sphere** Find the radius and height of the open right circular cylinder of largest surface area that can be inscribed in a sphere of radius a. What *is* the largest surface area?

11. **Rectangle of greatest area in an ellipse** Use the method of Lagrange multipliers to find the dimensions of the rectangle of greatest area that can be inscribed in the ellipse $x^2/16 + y^2/9 = 1$ with sides parallel to the coordinate axes.

12. **Rectangle of longest perimeter in an ellipse** Find the dimensions of the rectangle of largest perimeter that can be inscribed in the ellipse $x^2/a^2 + y^2/b^2 = 1$ with sides parallel to the coordinate axes. What *is* the largest perimeter?

13. **Extrema on a circle** Find the maximum and minimum values of $x^2 + y^2$ subject to the constraint $x^2 - 2x + y^2 - 4y = 0$.

14. **Extrema on a circle** Find the maximum and minimum values of $3x - y + 6$ subject to the constraint $x^2 + y^2 = 4$.

15. **Ant on a metal plate** The temperature at a point (x, y) on a metal plate is $T(x, y) = 4x^2 - 4xy + y^2$. An ant on the plate walks around the circle of radius 5 centered at the origin. What are the highest and lowest temperatures encountered by the ant?

16. **Cheapest storage tank** Your firm has been asked to design a storage tank for liquid petroleum gas. The customer's specifications call for a cylindrical tank with hemispherical ends, and the tank is to hold 8000 m^3 of gas. The customer also wants to use the smallest amount of material possible in building the tank. What radius and height do you recommend for the cylindrical portion of the tank?

Three Independent Variables with One Constraint

17. **Minimum distance to a point** Find the point on the plane $x + 2y + 3z = 13$ closest to the point $(1, 1, 1)$.

18. **Maximum distance to a point** Find the point on the sphere $x^2 + y^2 + z^2 = 4$ farthest from the point $(1, -1, 1)$.

19. **Minimum distance to the origin** Find the minimum distance from the surface $x^2 - y^2 - z^2 = 1$ to the origin.

20. **Minimum distance to the origin** Find the point on the surface $z = xy + 1$ nearest the origin.

21. **Minimum distance to the origin** Find the points on the surface $z^2 = xy + 4$ closest to the origin.

22. **Minimum distance to the origin** Find the point(s) on the surface $xyz = 1$ closest to the origin.

23. **Extrema on a sphere** Find the maximum and minimum values of
$$f(x, y, z) = x - 2y + 5z$$
on the sphere $x^2 + y^2 + z^2 = 30$.

24. **Extrema on a sphere** Find the points on the sphere $x^2 + y^2 + z^2 = 25$ where $f(x, y, z) = x + 2y + 3z$ has its maximum and minimum values.

25. **Minimizing a sum of squares** Find three real numbers whose sum is 9 and the sum of whose squares is as small as possible.

26. **Maximizing a product** Find the largest product the positive numbers x, y, and z can have if $x + y + z^2 = 16$.

27. **Rectangular box of largest volume in a sphere** Find the dimensions of the closed rectangular box with maximum volume that can be inscribed in the unit sphere.

28. **Box with vertex on a plane** Find the volume of the largest closed rectangular box in the first octant having three faces in the coordinate planes and a vertex on the plane $x/a + y/b + z/c = 1$, where $a > 0, b > 0$, and $c > 0$.

29. **Hottest point on a space probe** A space probe in the shape of the ellipsoid
$$4x^2 + y^2 + 4z^2 = 16$$
enters Earth's atmosphere and its surface begins to heat. After 1 hour, the temperature at the point (x, y, z) on the probe's surface is
$$T(x, y, z) = 8x^2 + 4yz - 16z + 600.$$
Find the hottest point on the probe's surface.

30. **Extreme temperatures on a sphere** Suppose that the Celsius temperature at the point (x, y, z) on the sphere $x^2 + y^2 + z^2 = 1$ is $T = 400xyz^2$. Locate the highest and lowest temperatures on the sphere.

31. **Cobb-Douglas production function** During the 1920s, Charles Cobb and Paul Douglas modeled total production output P (of a firm, industry, or entire economy) as a function of labor hours involved x and capital invested y (which includes the monetary worth of all buildings and equipment). The Cobb-Douglas production function is given by
$$P(x, y) = kx^\alpha y^{1-\alpha},$$
where k and α are constants representative of a particular firm or economy.

   a. Show that a doubling of both labor and capital results in a doubling of production P.

   b. Suppose a particular firm has the production function for $k = 120$ and $\alpha = 3/4$. Assume that each unit of labor costs $250 and each unit of capital costs $400, and that the total expenses for all costs cannot exceed $100,000. Find the maximum production level for the firm.

32. (*Continuation of Exercise 31.*) If the cost of a unit of labor is c_1 and the cost of a unit of capital is c_2, and if the firm can spend only B dollars as its total budget, then production P is constrained by $c_1 x + c_2 y = B$. Show that the maximum production level subject to the constraint occurs at the point
$$x = \frac{\alpha B}{c_1} \quad \text{and} \quad y = \frac{(1 - \alpha)B}{c_2}.$$

33. **Maximizing a utility function: an example from economics** In economics, the usefulness or *utility* of amounts x and y of two capital goods G_1 and G_2 is sometimes measured by a function $U(x, y)$. For example, G_1 and G_2 might be two chemicals a pharmaceutical company needs to have on hand and $U(x, y)$ the gain

from manufacturing a product whose synthesis requires different amounts of the chemicals depending on the process used. If G_1 costs a dollars per kilogram, G_2 costs b dollars per kilogram, and the total amount allocated for the purchase of G_1 and G_2 together is c dollars, then the company's managers want to maximize $U(x, y)$ given that $ax + by = c$. Thus, they need to solve a typical Lagrange multiplier problem.

Suppose that

$$U(x, y) = xy + 2x$$

and that the equation $ax + by = c$ simplifies to

$$2x + y = 30.$$

Find the maximum value of U and the corresponding values of x and y subject to this latter constraint.

34. **Blood types** Human blood types are classified by three gene forms A, B, and O. Blood types AA, BB, and OO are *homozygous*, and blood types AB, AO, and BO are *heterozygous*. If p, q, and r represent the proportions of the three gene forms to the population, respectively, then the *Hardy-Weinberg Law* asserts that the proportion Q of heterozygous persons in any specific population is modeled by

$$Q(p, q, r) = 2(pq + pr + qr),$$

subject to $p + q + r = 1$. Find the maximum value of Q.

35. **Length of a beam** In Section 4.6, Exercise 45, we posed a problem of finding the length L of the shortest beam that can reach over a wall of height h to a tall building located k units from the wall. Use Lagrange multipliers to show that

$$L = (h^{2/3} + k^{2/3})^{3/2}.$$

36. **Locating a radio telescope** You are in charge of erecting a radio telescope on a newly discovered planet. To minimize interference, you want to place it where the magnetic field of the planet is weakest. The planet is spherical, with a radius of 6 units. Based on a coordinate system whose origin is at the center of the planet, the strength of the magnetic field is given by $M(x, y, z) = 6x - y^2 + xz + 60$. Where should you locate the radio telescope?

Extreme Values Subject to Two Constraints

37. Maximize the function $f(x, y, z) = x^2 + 2y - z^2$ subject to the constraints $2x - y = 0$ and $y + z = 0$.

38. Minimize the function $f(x, y, z) = x^2 + y^2 + z^2$ subject to the constraints $x + 2y + 3z = 6$ and $x + 3y + 9z = 9$.

39. **Minimum distance to the origin** Find the point closest to the origin on the line of intersection of the planes $y + 2z = 12$ and $x + y = 6$.

40. **Maximum value on line of intersection** Find the maximum value that $f(x, y, z) = x^2 + 2y - z^2$ can have on the line of intersection of the planes $2x - y = 0$ and $y + z = 0$.

41. **Extrema on a curve of intersection** Find the extreme values of $f(x, y, z) = x^2yz + 1$ on the intersection of the plane $z = 1$ with the sphere $x^2 + y^2 + z^2 = 10$.

42. a. **Maximum on line of intersection** Find the maximum value of $w = xyz$ on the line of intersection of the two planes $x + y + z = 40$ and $x + y - z = 0$.

 b. Give a geometric argument to support your claim that you have found a maximum, and not a minimum, value of w.

43. **Extrema on a circle of intersection** Find the extreme values of the function $f(x, y, z) = xy + z^2$ on the circle in which the plane $y - x = 0$ intersects the sphere $x^2 + y^2 + z^2 = 4$.

44. **Minimum distance to the origin** Find the point closest to the origin on the curve of intersection of the plane $2y + 4z = 5$ and the cone $z^2 = 4x^2 + 4y^2$.

Theory and Examples

45. **The condition $\nabla f = \lambda \nabla g$ is not sufficient** Although $\nabla f = \lambda \nabla g$ is a necessary condition for the occurrence of an extreme value of $f(x, y)$ subject to the conditions $g(x, y) = 0$ and $\nabla g \neq \mathbf{0}$, it does not in itself guarantee that one exists. As a case in point, try using the method of Lagrange multipliers to find a maximum value of $f(x, y) = x + y$ subject to the constraint that $xy = 16$. The method will identify the two points $(4, 4)$ and $(-4, -4)$ as candidates for the location of extreme values. Yet the sum $x + y$ has no maximum value on the hyperbola $xy = 16$. The farther you go from the origin on this hyperbola in the first quadrant, the larger the sum $f(x, y) = x + y$ becomes.

46. **A least squares plane** The plane $z = Ax + By + C$ is to be "fitted" to the following points (x_k, y_k, z_k):

$$(0, 0, 0), \quad (0, 1, 1), \quad (1, 1, 1), \quad (1, 0, -1).$$

Find the values of A, B, and C that minimize

$$\sum_{k=1}^{4} (Ax_k + By_k + C - z_k)^2,$$

the sum of the squares of the deviations.

47. a. **Maximum on a sphere** Show that the maximum value of $a^2b^2c^2$ on a sphere of radius r centered at the origin of a Cartesian abc-coordinate system is $(r^2/3)^3$.

 b. **Geometric and arithmetic means** Using part (a), show that for nonnegative numbers a, b, and c,

$$(abc)^{1/3} \leq \frac{a + b + c}{3};$$

 that is, the *geometric mean* of three nonnegative numbers is less than or equal to their *arithmetic mean*.

48. **Sum of products** Let $a_1, a_2, \ldots, a_n$ be n positive numbers. Find the maximum of $\sum_{i=1}^{n} a_i x_i$ subject to the constraint $\sum_{i=1}^{n} x_i^2 = 1$.

COMPUTER EXPLORATIONS

In Exercises 49–54, use a CAS to perform the following steps implementing the method of Lagrange multipliers for finding constrained extrema:

 a. Form the function $h = f - \lambda_1 g_1 - \lambda_2 g_2$, where f is the function to optimize subject to the constraints $g_1 = 0$ and $g_2 = 0$.

 b. Determine all the first partial derivatives of h, including the partials with respect to λ_1 and λ_2, and set them equal to 0.

 c. Solve the system of equations found in part (b) for all the unknowns, including λ_1 and λ_2.

 d. Evaluate f at each of the solution points found in part (c) and select the extreme value subject to the constraints asked for in the exercise.

49. Minimize $f(x, y, z) = xy + yz$ subject to the constraints $x^2 + y^2 - 2 = 0$ and $x^2 + z^2 - 2 = 0$.

50. Minimize $f(x, y, z) = xyz$ subject to the constraints $x^2 + y^2 - 1 = 0$ and $x - z = 0$.

51. Maximize $f(x, y, z) = x^2 + y^2 + z^2$ subject to the constraints $2y + 4z - 5 = 0$ and $4x^2 + 4y^2 - z^2 = 0$.

52. Minimize $f(x, y, z) = x^2 + y^2 + z^2$ subject to the constraints $x^2 - xy + y^2 - z^2 - 1 = 0$ and $x^2 + y^2 - 1 = 0$.

53. Minimize $f(x, y, z, w) = x^2 + y^2 + z^2 + w^2$ subject to the constraints $2x - y + z - w - 1 = 0$ and $x + y - z + w - 1 = 0$.

54. Determine the distance from the line $y = x + 1$ to the parabola $y^2 = x$. (*Hint:* Let (x, y) be a point on the line and (w, z) a point on the parabola. You want to minimize $(x - w)^2 + (y - z)^2$.)

14.9 Taylor's Formula for Two Variables

In this section we use Taylor's formula to derive the Second Derivative Test for local extreme values (Section 14.7) and the error formula for linearizations of functions of two independent variables (Section 14.6). The use of Taylor's formula in these derivations leads to an extension of the formula that provides polynomial approximations of all orders for functions of two independent variables.

Derivation of the Second Derivative Test

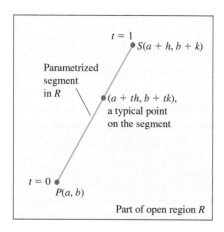

FIGURE 14.60 We begin the derivation of the Second Derivative Test at $P(a, b)$ by parametrizing a typical line segment from P to a point S nearby.

Let $f(x, y)$ have continuous first and second partial derivatives in an open region R containing a point $P(a, b)$ where $f_x = f_y = 0$ (Figure 14.60). Let h and k be increments small enough to put the point $S(a + h, b + k)$ and the line segment joining it to P inside R. We parametrize the segment PS as

$$x = a + th, \qquad y = b + tk, \qquad 0 \le t \le 1.$$

If $F(t) = f(a + th, b + tk)$, the Chain Rule gives

$$F'(t) = f_x \frac{dx}{dt} + f_y \frac{dy}{dt} = hf_x + kf_y.$$

Since f_x and f_y are differentiable (because they have continuous partial derivatives), F' is a differentiable function of t and

$$F'' = \frac{\partial F'}{\partial x}\frac{dx}{dt} + \frac{\partial F'}{\partial y}\frac{dy}{dt} = \frac{\partial}{\partial x}(hf_x + kf_y) \cdot h + \frac{\partial}{\partial y}(hf_x + kf_y) \cdot k$$

$$= h^2 f_{xx} + 2hk f_{xy} + k^2 f_{yy}. \qquad f_{xy} = f_{yx}$$

Since F and F' are continuous on $[0, 1]$ and F' is differentiable on $(0, 1)$, we can apply Taylor's formula with $n = 2$ and $a = 0$ to obtain

$$F(1) = F(0) + F'(0)(1 - 0) + F''(c)\frac{(1 - 0)^2}{2} \tag{1}$$

$$= F(0) + F'(0) + \frac{1}{2}F''(c)$$

for some c between 0 and 1. Writing Equation (1) in terms of f gives

$$f(a + h, b + k) = f(a, b) + hf_x(a, b) + kf_y(a, b)$$
$$+ \frac{1}{2}\left(h^2 f_{xx} + 2hk f_{xy} + k^2 f_{yy}\right)\Big|_{(a+ch, b+ck)}. \tag{2}$$

Since $f_x(a, b) = f_y(a, b) = 0$, this reduces to

$$f(a + h, b + k) - f(a, b) = \frac{1}{2}\left(h^2 f_{xx} + 2hk f_{xy} + k^2 f_{yy}\right)\Big|_{(a+ch, b+ck)}. \tag{3}$$

To determine whether f has an extremum at (a, b), we examine the sign of the difference $f(a + h, b + k) - f(a, b)$. By Equation (3), this is the same as the sign of

$$Q(c) = (h^2 f_{xx} + 2hk f_{xy} + k^2 f_{yy})\Big|_{(a+ch, \, b+ck)}.$$

Now, if $Q(0) \neq 0$, the sign of $Q(c)$ will be the same as the sign of $Q(0)$ for sufficiently small values of h and k. We can predict the sign of

$$Q(0) = h^2 f_{xx}(a, b) + 2hk f_{xy}(a, b) + k^2 f_{yy}(a, b) \tag{4}$$

from the signs of f_{xx} and $f_{xx}f_{yy} - f_{xy}^2$ at (a, b). Multiply both sides of Equation (4) by f_{xx} and rearrange the right-hand side to get

$$f_{xx}Q(0) = \left(hf_{xx} + kf_{xy}\right)^2 + \left(f_{xx}f_{yy} - f_{xy}^2\right)k^2. \tag{5}$$

From Equation (5) we see that

1. If $f_{xx} < 0$ and $f_{xx}f_{yy} - f_{xy}^2 > 0$ at (a, b), then $Q(0) < 0$ for all sufficiently small nonzero values of h and k, and f has a *local maximum* value at (a, b).

2. If $f_{xx} > 0$ and $f_{xx}f_{yy} - f_{xy}^2 > 0$ at (a, b), then $Q(0) > 0$ for all sufficiently small nonzero values of h and k, and f has a *local minimum* value at (a, b).

3. If $f_{xx}f_{yy} - f_{xy}^2 < 0$ at (a, b), there are combinations of arbitrarily small nonzero values of h and k for which $Q(0) > 0$, and other values for which $Q(0) < 0$. Arbitrarily close to the point $P_0(a, b, f(a, b))$ on the surface $z = f(x, y)$ there are points above P_0 and points below P_0, so f has a *saddle point* at (a, b).

4. If $f_{xx}f_{yy} - f_{xy}^2 = 0$ at (a, b), another test is needed. The possibility that $Q(0)$ equals zero prevents us from drawing conclusions about the sign of $Q(c)$.

The Error Formula for Linear Approximations

We want to show that the difference $E(x, y)$ between the values of a function $f(x, y)$ and its linearization $L(x, y)$ at (x_0, y_0) satisfies the inequality

$$|E(x, y)| \leq \frac{1}{2} M \left(|x - x_0| + |y - y_0|\right)^2.$$

The function f is assumed to have continuous second partial derivatives throughout an open set containing a closed rectangular region R centered at (x_0, y_0). The number M is an upper bound for $|f_{xx}|, |f_{yy}|$, and $|f_{xy}|$ on R.

The inequality we want comes from Equation (2). We substitute x_0 and y_0 for a and b, and $x - x_0$ and $y - y_0$ for h and k, respectively, and rearrange the result as

$$f(x, y) = \underbrace{f(x_0, y_0) + f_x(x_0, y_0)(x - x_0) + f_y(x_0, y_0)(y - y_0)}_{\text{linearization } L(x, y)}$$

$$\underbrace{+ \frac{1}{2} \left((x - x_0)^2 f_{xx} + 2(x - x_0)(y - y_0) f_{xy} + (y - y_0)^2 f_{yy}\right)\Big|_{(x_0 + c(x-x_0), \, y_0 + c(y-y_0))}}_{\text{error } E(x, y)}.$$

This equation reveals that

$$|E| \leq \frac{1}{2}\left(|x - x_0|^2 |f_{xx}| + 2|x - x_0||y - y_0||f_{xy}| + |y - y_0|^2 |f_{yy}|\right).$$

Hence, if M is an upper bound for the values of $|f_{xx}|, |f_{xy}|$, and $|f_{yy}|$ on R, then

$$|E| \leq \frac{1}{2}\left(|x - x_0|^2 M + 2|x - x_0||y - y_0|M + |y - y_0|^2 M\right)$$

$$= \frac{1}{2} M \left(|x - x_0| + |y - y_0|\right)^2.$$

Taylor's Formula for Functions of Two Variables

The formulas derived earlier for F' and F'' can be obtained by applying to $f(x, y)$ the differentiation operators

$$\left(h\frac{\partial}{\partial x} + k\frac{\partial}{\partial y}\right) \quad \text{and} \quad \left(h\frac{\partial}{\partial x} + k\frac{\partial}{\partial y}\right)^2 = h^2\frac{\partial^2}{\partial x^2} + 2hk\frac{\partial^2}{\partial x\,\partial y} + k^2\frac{\partial^2}{\partial y^2}.$$

These are the first two instances of a more general formula,

$$F^{(n)}(t) = \frac{d^n}{dt^n}F(t) = \left(h\frac{\partial}{\partial x} + k\frac{\partial}{\partial y}\right)^n f(x, y), \tag{6}$$

which says that applying d^n/dt^n to $F(t)$ gives the same result as applying the operator

$$\left(h\frac{\partial}{\partial x} + k\frac{\partial}{\partial y}\right)^n$$

to $f(x, y)$ after expanding it by the Binomial Theorem.

If the partial derivatives of f through order $n + 1$ are continuous throughout a rectangular region centered at (a, b), we may extend the Taylor formula for $F(t)$ to

$$F(t) = F(0) + F'(0)t + \frac{F''(0)}{2!}t^2 + \cdots + \frac{F^{(n)}(0)}{n!}t^{(n)} + \text{remainder},$$

and take $t = 1$ to obtain

$$F(1) = F(0) + F'(0) + \frac{F''(0)}{2!} + \cdots + \frac{F^{(n)}(0)}{n!} + \text{remainder}.$$

When we replace the first n derivatives on the right of this last series by their equivalent expressions from Equation (6) evaluated at $t = 0$ and add the appropriate remainder term, we arrive at the following formula.

Taylor's Formula for $f(x, y)$ at the Point (a, b)

Suppose $f(x, y)$ and its partial derivatives through order $n + 1$ are continuous throughout an open rectangular region R centered at a point (a, b). Then, throughout R,

$$f(a + h, b + k) = f(a, b) + (hf_x + kf_y)\Big|_{(a,b)} + \frac{1}{2!}(h^2f_{xx} + 2hkf_{xy} + k^2f_{yy})\Big|_{(a,b)}$$

$$+ \frac{1}{3!}(h^3f_{xxx} + 3h^2kf_{xxy} + 3hk^2f_{xyy} + k^3f_{yyy})\Big|_{(a,b)} + \cdots + \frac{1}{n!}\left(h\frac{\partial}{\partial x} + k\frac{\partial}{\partial y}\right)^n f\Big|_{(a,b)}$$

$$+ \frac{1}{(n+1)!}\left(h\frac{\partial}{\partial x} + k\frac{\partial}{\partial y}\right)^{n+1}f\Big|_{(a+ch,\,b+ck)}. \tag{7}$$

The first n derivative terms are evaluated at (a, b). The last term is evaluated at some point $(a + ch, b + ck)$ on the line segment joining (a, b) and $(a + h, b + k)$.

If $(a, b) = (0, 0)$ and we treat h and k as independent variables (denoting them now by x and y), then Equation (7) assumes the following form.

Taylor's Formula for $f(x, y)$ at the Origin

$$f(x, y) = f(0, 0) + xf_x + yf_y + \frac{1}{2!}(x^2 f_{xx} + 2xy f_{xy} + y^2 f_{yy})$$

$$+ \frac{1}{3!}(x^3 f_{xxx} + 3x^2 y f_{xxy} + 3xy^2 f_{xyy} + y^3 f_{yyy}) + \cdots + \frac{1}{n!}\left(x^n \frac{\partial^n f}{\partial x^n} + nx^{n-1}y \frac{\partial^n f}{\partial x^{n-1}\partial y} + \cdots + y^n \frac{\partial^n f}{\partial y^n}\right)$$

$$+ \frac{1}{(n+1)!}\left(x^{n+1}\frac{\partial^{n+1} f}{\partial x^{n+1}} + (n+1)x^n y \frac{\partial^{n+1} f}{\partial x^n \partial y} + \cdots + y^{n+1}\frac{\partial^{n+1} f}{\partial y^{n+1}}\right)\Bigg|_{(cx,\,cy)} \qquad (8)$$

The first n derivative terms are evaluated at $(0, 0)$. The last term is evaluated at a point on the line segment joining the origin and (x, y).

Taylor's formula provides polynomial approximations of two-variable functions. The first n derivative terms give the polynomial; the last term gives the approximation error. The first three terms of Taylor's formula give the function's linearization. To improve on the linearization, we add higher-power terms.

EXAMPLE 1   Find a quadratic approximation to $f(x, y) = \sin x \sin y$ near the origin. How accurate is the approximation if $|x| \le 0.1$ and $|y| \le 0.1$?

Solution   We take $n = 2$ in Equation (8):

$$f(x, y) = f(0, 0) + (xf_x + yf_y) + \frac{1}{2}(x^2 f_{xx} + 2xy f_{xy} + y^2 f_{yy})$$

$$+ \frac{1}{6}(x^3 f_{xxx} + 3x^2 y f_{xxy} + 3xy^2 f_{xyy} + y^3 f_{yyy})\Bigg|_{(cx,\,cy)}.$$

Calculating the values of the partial derivatives,

$$f(0, 0) = \sin x \sin y \Big|_{(0, 0)} = 0, \qquad f_{xx}(0, 0) = -\sin x \sin y \Big|_{(0, 0)} = 0,$$

$$f_x(0, 0) = \cos x \sin y \Big|_{(0, 0)} = 0, \qquad f_{xy}(0, 0) = \cos x \cos y \Big|_{(0, 0)} = 1,$$

$$f_y(0, 0) = \sin x \cos y \Big|_{(0, 0)} = 0, \qquad f_{yy}(0, 0) = -\sin x \sin y \Big|_{(0, 0)} = 0,$$

we have the result

$$\sin x \sin y \approx 0 + 0 + 0 + \frac{1}{2}(x^2(0) + 2xy(1) + y^2(0)), \quad \text{or} \quad \sin x \sin y \approx xy.$$

The error in the approximation is

$$E(x, y) = \frac{1}{6}(x^3 f_{xxx} + 3x^2 y f_{xxy} + 3xy^2 f_{xyy} + y^3 f_{yyy})\Bigg|_{(cx,\,cy)}.$$

The third derivatives never exceed 1 in absolute value because they are products of sines and cosines. Also, $|x| \le 0.1$ and $|y| \le 0.1$. Hence

$$|E(x, y)| \le \frac{1}{6}((0.1)^3 + 3(0.1)^3 + 3(0.1)^3 + (0.1)^3) = \frac{8}{6}(0.1)^3 \le 0.00134$$

(rounded up). The error will not exceed 0.00134 if $|x| \le 0.1$ and $|y| \le 0.1$.

Finding Quadratic and Cubic Approximations

In Exercises 1–10, use Taylor's formula for $f(x, y)$ at the origin to find quadratic and cubic approximations of f near the origin.

1. $f(x, y) = xe^y$

2. $f(x, y) = e^x \cos y$

3. $f(x, y) = y \sin x$

4. $f(x, y) = \sin x \cos y$

5. $f(x, y) = e^x \ln (1 + y)$

6. $f(x, y) = \ln (2x + y + 1)$

7. $f(x, y) = \sin (x^2 + y^2)$

8. $f(x, y) = \cos (x^2 + y^2)$

9. $f(x, y) = \dfrac{1}{1 - x - y}$

10. $f(x, y) = \dfrac{1}{1 - x - y + xy}$

11. Use Taylor's formula to find a quadratic approximation of $f(x, y) = \cos x \cos y$ at the origin. Estimate the error in the approximation if $|x| \le 0.1$ and $|y| \le 0.1$.

12. Use Taylor's formula to find a quadratic approximation of $e^x \sin y$ at the origin. Estimate the error in the approximation if $|x| \le 0.1$ and $|y| \le 0.1$.

14.10 Partial Derivatives with Constrained Variables

In finding partial derivatives of functions like $w = f(x, y)$, we have assumed x and y to be independent. In many applications, however, this is not the case. For example, the internal energy U of a gas may be expressed as a function $U = f(P, V, T)$ of pressure P, volume V, and temperature T. If the individual molecules of the gas do not interact, however, P, V, and T obey (and are constrained by) the ideal gas law

$$PV = nRT \qquad (n \text{ and } R \text{ constant}),$$

and fail to be independent. In this section we learn how to find partial derivatives in situations like this, which occur in economics, engineering, and physics.

Decide Which Variables Are Dependent and Which Are Independent

If the variables in a function $w = f(x, y, z)$ are constrained by a relation like the one imposed on x, y, and z by the equation $z = x^2 + y^2$, the geometric meanings and the numerical values of the partial derivatives of f will depend on which variables are chosen to be dependent and which are chosen to be independent. To see how this choice can affect the outcome, we consider the calculation of $\partial w / \partial x$ when $w = x^2 + y^2 + z^2$ and $z = x^2 + y^2$.

EXAMPLE 1   Find $\partial w / \partial x$ if $w = x^2 + y^2 + z^2$ and $z = x^2 + y^2$.

Solution   We are given two equations in the four unknowns x, y, z, and w. Like many such systems, this one can be solved for two of the unknowns (the dependent variables) in terms of the others (the independent variables). In being asked for $\partial w / \partial x$, we are told that w is to be a dependent variable and x an independent variable. The possible choices for the other variables come down to

	Dependent	Independent
Choice 1:	w, z	x, y
Choice 2:	w, y	x, z

In either case, we can express w explicitly in terms of the selected independent variables. We do this by using the second equation $z = x^2 + y^2$ to eliminate the remaining dependent variable in the first equation.

In the first case, the remaining dependent variable is z. We eliminate it from the first equation by replacing it by $x^2 + y^2$. The resulting expression for w is

$$w = x^2 + y^2 + z^2 = x^2 + y^2 + (x^2 + y^2)^2$$
$$= x^2 + y^2 + x^4 + 2x^2y^2 + y^4$$

and therefore

$$\frac{\partial w}{\partial x} = 2x + 4x^3 + 4xy^2. \tag{1}$$

This is the formula for $\partial w / \partial x$ when x and y are the independent variables.

In the second case, where the independent variables are x and z and the remaining dependent variable is y, we eliminate the dependent variable y in the expression for w by replacing y^2 in the second equation by $z - x^2$. This gives

$$w = x^2 + y^2 + z^2 = x^2 + (z - x^2) + z^2 = z + z^2$$

and therefore

$$\frac{\partial w}{\partial x} = 0. \tag{2}$$

This is the formula for $\partial w / \partial x$ when x and z are the independent variables.

The formulas for $\partial w / \partial x$ in Equations (1) and (2) are genuinely different. We cannot change either formula into the other by using the relation $z = x^2 + y^2$. There is not just one $\partial w / \partial x$, there are two, and we see that the original instruction to find $\partial w / \partial x$ was incomplete. *Which* $\partial w / \partial x$? we ask.

The geometric interpretations of Equations (1) and (2) help to explain why the equations differ. The function $w = x^2 + y^2 + z^2$ measures the square of the distance from the point (x, y, z) to the origin. The condition $z = x^2 + y^2$ says that the point (x, y, z) lies on the paraboloid of revolution shown in Figure 14.61. What does it mean to calculate $\partial w / \partial x$ at a point $P(x, y, z)$ that can move only on this surface? What is the value of $\partial w / \partial x$ when the coordinates of P are, say, $(1, 0, 1)$?

If we take x and y to be independent, then we find $\partial w / \partial x$ by holding y fixed (at $y = 0$ in this case) and letting x vary. Hence, P moves along the parabola $z = x^2$ in the xz-plane. As P moves on this parabola, w, which is the square of the distance from P to the origin, changes. We calculate $\partial w / \partial x$ in this case (our first solution above) to be

$$\frac{\partial w}{\partial x} = 2x + 4x^3 + 4xy^2.$$

At the point $P(1, 0, 1)$, the value of this derivative is

$$\frac{\partial w}{\partial x} = 2 + 4 + 0 = 6.$$

If we take x and z to be independent, then we find $\partial w / \partial x$ by holding z fixed while x varies. Since the z-coordinate of P is 1, varying x moves P along a circle in the plane $z = 1$. As P moves along this circle, its distance from the origin remains constant, and w, being the square of this distance, does not change. That is,

$$\frac{\partial w}{\partial x} = 0,$$

as we found in our second solution. ∎

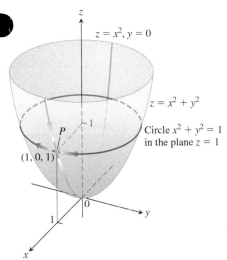

FIGURE 14.61 If P is constrained to lie on the paraboloid $z = x^2 + y^2$, the value of the partial derivative of $w = x^2 + y^2 + z^2$ with respect to x at P depends on the direction of motion (Example 1). (1) As x changes, with $y = 0$, P moves up or down the surface on the parabola $z = x^2$ in the xz-plane with $\partial w / \partial x = 2x + 4x^3$. (2) As x changes, with $z = 1$, P moves on the circle $x^2 + y^2 = 1, z = 1$, and $\partial w / \partial x = 0$.

How to Find $\partial w / \partial x$ When the Variables in $w = f(x, y, z)$ Are Constrained by Another Equation

As we saw in Example 1, a typical routine for finding $\partial w / \partial x$ when the variables in the function $w = f(x, y, z)$ are related by another equation has three steps. These steps apply to finding $\partial w / \partial y$ and $\partial w / \partial z$ as well.

1. *Decide* which variables are to be dependent and which are to be independent. (In practice, the decision is based on the physical or theoretical context of our work. In the exercises at the end of this section, we say which variables are which.)

2. *Eliminate* the other dependent variable(s) in the expression for w.

3. *Differentiate* as usual.

If we cannot carry out Step 2 after deciding which variables are dependent, we differentiate the equations as they are and try to solve for $\partial w / \partial x$ afterward. The next example shows how this is done.

EXAMPLE 2   Find $\partial w / \partial x$ at the point $(x, y, z) = (2, -1, 1)$ if

$$w = x^2 + y^2 + z^2, \qquad z^3 - xy + yz + y^3 = 1,$$

and x and y are the independent variables.

Solution   It is not convenient to eliminate z in the expression for w. We therefore differentiate both equations implicitly with respect to x, treating x and y as independent variables and w and z as dependent variables. This gives

$$\frac{\partial w}{\partial x} = 2x + 2z \frac{\partial z}{\partial x} \tag{3}$$

and

$$3z^2 \frac{\partial z}{\partial x} - y + y \frac{\partial z}{\partial x} + 0 = 0. \tag{4}$$

These equations may now be combined to express $\partial w / \partial x$ in terms of x, y, and z. We solve Equation (4) for $\partial z / \partial x$ to get

$$\frac{\partial z}{\partial x} = \frac{y}{y + 3z^2}$$

and substitute into Equation (3) to get

$$\frac{\partial w}{\partial x} = 2x + \frac{2yz}{y + 3z^2}.$$

The value of this derivative at $(x, y, z) = (2, -1, 1)$ is

$$\frac{\partial w}{\partial x} \bigg|_{(2, -1, 1)} = 2(2) + \frac{2(-1)(1)}{-1 + 3(1)^2} = 4 + \frac{-2}{2} = 3. \qquad \blacksquare$$

Notation

To show what variables are assumed to be independent in calculating a derivative, we can use the following notation:

$$\left(\frac{\partial w}{\partial x} \right)_y \qquad \partial w / \partial x \text{ with } x \text{ and } y \text{ independent}$$

$$\left(\frac{\partial f}{\partial y} \right)_{x, t} \qquad \partial f / \partial y \text{ with } y, x, \text{ and } t \text{ independent}$$

EXAMPLE 3   Find $(\partial w/\partial x)_{y,z}$ if $w = x^2 + y - z + \sin t$ and $x + y = t$.

Solution   With x, y, z independent, we have

$$t = x + y, \qquad w = x^2 + y - z + \sin (x + y)$$

$$\left(\frac{\partial w}{\partial x}\right)_{y,z} = 2x + 0 - 0 + \cos (x + y) \frac{\partial}{\partial x}(x + y)$$

$$= 2x + \cos (x + y).$$

Arrow Diagrams

In solving problems like the one in Example 3, it often helps to start with an arrow diagram that shows how the variables and functions are related. If

$$w = x^2 + y - z + \sin t \qquad \text{and} \qquad x + y = t$$

and we are asked to find $\partial w/\partial x$ when $x, y,$ and z are independent, the appropriate diagram is one like this:

$$\begin{pmatrix} x \\ y \\ z \end{pmatrix} \longrightarrow \begin{pmatrix} x \\ y \\ z \\ t \end{pmatrix} \longrightarrow \ w \qquad\qquad (5)$$

Independent   Intermediate   Dependent
variables   variables   variable

To avoid confusion between the independent and intermediate variables with the same symbolic names in the diagram, it is helpful to rename the intermediate variables (so they are seen as *functions* of the independent variables). Thus, let $u = x, v = y,$ and $s = z$ denote the renamed intermediate variables. With this notation, the arrow diagram becomes

$$\begin{pmatrix} x \\ y \\ z \end{pmatrix} \longrightarrow \begin{pmatrix} u \\ v \\ s \\ t \end{pmatrix} \longrightarrow \ w \qquad\qquad (6)$$

Independent   Intermediate   Dependent
variables   variables and   variable
       relations
$u = x$
$v = y$
$s = z$
$t = x + y$

The diagram shows the independent variables on the left, the intermediate variables and their relation to the independent variables in the middle, and the dependent variable on the right. The function w now becomes

$$w = u^2 + v - s + \sin t,$$

where

$$u = x, \qquad v = y, \qquad s = z, \qquad \text{and} \qquad t = x + y.$$

To find $\partial w/\partial x$, we apply the four-variable form of the Chain Rule to w, guided by the arrow diagram in Equation (6):

$$\frac{\partial w}{\partial x} = \frac{\partial w}{\partial u}\frac{\partial u}{\partial x} + \frac{\partial w}{\partial v}\frac{\partial v}{\partial x} + \frac{\partial w}{\partial s}\frac{\partial s}{\partial x} + \frac{\partial w}{\partial t}\frac{\partial t}{\partial x}$$

$$= (2u)(1) + (1)(0) + (-1)(0) + (\cos t)(1)$$

$$= 2u + \cos t$$

$$= 2x + \cos (x + y). \qquad\qquad \text{\small Substitute the original independent variables } u = x \text{ and } t = x + y$$

Finding Partial Derivatives with Constrained Variables

In Exercises 1–3, begin by drawing a diagram that shows the relations among the variables.

1. If $w = x^2 + y^2 + z^2$ and $z = x^2 + y^2$, find

 a. $\left(\dfrac{\partial w}{\partial y}\right)_z$ **b.** $\left(\dfrac{\partial w}{\partial z}\right)_x$ **c.** $\left(\dfrac{\partial w}{\partial z}\right)_y$.

2. If $w = x^2 + y - z + \sin t$ and $x + y = t$, find

 a. $\left(\dfrac{\partial w}{\partial y}\right)_{x, z}$ **b.** $\left(\dfrac{\partial w}{\partial y}\right)_{z, t}$ **c.** $\left(\dfrac{\partial w}{\partial z}\right)_{x, y}$

 d. $\left(\dfrac{\partial w}{\partial z}\right)_{y, t}$ **e.** $\left(\dfrac{\partial w}{\partial t}\right)_{x, z}$ **f.** $\left(\dfrac{\partial w}{\partial t}\right)_{y, z}$.

3. Let $U = f(P, V, T)$ be the internal energy of a gas that obeys the ideal gas law $PV = nRT$ (n and R constant). Find

 a. $\left(\dfrac{\partial U}{\partial P}\right)_V$ **b.** $\left(\dfrac{\partial U}{\partial T}\right)_V$.

4. Find

 a. $\left(\dfrac{\partial w}{\partial x}\right)_y$ **b.** $\left(\dfrac{\partial w}{\partial z}\right)_y$

at the point $(x, y, z) = (0, 1, \pi)$ if

$$w = x^2 + y^2 + z^2 \qquad \text{and} \qquad y \sin z + z \sin x = 0.$$

5. Find

 a. $\left(\dfrac{\partial w}{\partial y}\right)_x$ **b.** $\left(\dfrac{\partial w}{\partial y}\right)_z$

at the point $(w, x, y, z) = (4, 2, 1, -1)$ if

$$w = x^2 y^2 + yz - z^3 \qquad \text{and} \qquad x^2 + y^2 + z^2 = 6.$$

6. Find $(\partial u / \partial y)_x$ at the point $(u, v) = \left(\sqrt{2}, 1\right)$ if $x = u^2 + v^2$ and $y = uv$.

7. Suppose that $x^2 + y^2 = r^2$ and $x = r \cos \theta$, as in polar coordinates. Find

$$\left(\frac{\partial x}{\partial r}\right)_{\theta} \qquad \text{and} \qquad \left(\frac{\partial r}{\partial x}\right)_y.$$

8. Suppose that

$$w = x^2 - y^2 + 4z + t \qquad \text{and} \qquad x + 2z + t = 25.$$

Show that the equations

$$\frac{\partial w}{\partial x} = 2x - 1 \qquad \text{and} \qquad \frac{\partial w}{\partial x} = 2x - 2$$

each give $\partial w / \partial x$, depending on which variables are chosen to be dependent and which variables are chosen to be independent. Identify the independent variables in each case.

Theory and Examples

9. Establish the fact, widely used in hydrodynamics, that if $f(x, y, z) = 0$, then

$$\left(\frac{\partial x}{\partial y}\right)_z \left(\frac{\partial y}{\partial z}\right)_x \left(\frac{\partial z}{\partial x}\right)_y = -1.$$

(*Hint:* Express all the derivatives in terms of the formal partial derivatives $\partial f / \partial x$, $\partial f / \partial y$, and $\partial f / \partial z$.)

10. If $z = x + f(u)$, where $u = xy$, show that

$$x \frac{\partial z}{\partial x} - y \frac{\partial z}{\partial y} = x.$$

11. Suppose that the equation $g(x, y, z) = 0$ determines z as a differentiable function of the independent variables x and y and that $g_z \neq 0$. Show that

$$\left(\frac{\partial z}{\partial y}\right)_x = -\frac{\partial g / \partial y}{\partial g / \partial z}.$$

12. Suppose that $f(x, y, z, w) = 0$ and $g(x, y, z, w) = 0$ determine z and w as differentiable functions of the independent variables x and y, and suppose that

$$\frac{\partial f}{\partial z} \frac{\partial g}{\partial w} - \frac{\partial f}{\partial w} \frac{\partial g}{\partial z} \neq 0.$$

Show that

$$\left(\frac{\partial z}{\partial x}\right)_y = -\frac{\dfrac{\partial f}{\partial x}\dfrac{\partial g}{\partial w} - \dfrac{\partial f}{\partial w}\dfrac{\partial g}{\partial x}}{\dfrac{\partial f}{\partial z}\dfrac{\partial g}{\partial w} - \dfrac{\partial f}{\partial w}\dfrac{\partial g}{\partial z}}$$

and

$$\left(\frac{\partial w}{\partial y}\right)_x = -\frac{\dfrac{\partial f}{\partial z}\dfrac{\partial g}{\partial y} - \dfrac{\partial f}{\partial y}\dfrac{\partial g}{\partial z}}{\dfrac{\partial f}{\partial z}\dfrac{\partial g}{\partial w} - \dfrac{\partial f}{\partial w}\dfrac{\partial g}{\partial z}}.$$

CHAPTER 14 Questions to Guide Your Review

1. What is a real-valued function of two independent variables? Three independent variables? Give examples.

2. What does it mean for sets in the plane or in space to be open? Closed? Give examples. Give examples of sets that are neither open nor closed.

3. How can you display the values of a function $f(x, y)$ of two in-dependent variables graphically? How do you do the same for a function $f(x, y, z)$ of three independent variables?

4. What does it mean for a function $f(x, y)$ to have limit L as $(x, y) \rightarrow (x_0, y_0)$? What are the basic properties of limits of functions of two independent variables?

5. When is a function of two (three) independent variables continuous at a point in its domain? Give examples of functions that are continuous at some points but not others.

6. What can be said about algebraic combinations and compositions of continuous functions?

7. Explain the two-path test for nonexistence of limits.

8. How are the partial derivatives $\partial f/\partial x$ and $\partial f/\partial y$ of a function $f(x, y)$ defined? How are they interpreted and calculated?

9. How does the relation between first partial derivatives and continuity of functions of two independent variables differ from the relation between first derivatives and continuity for real-valued functions of a single independent variable? Give an example.

10. What is the Mixed Derivative Theorem for mixed second-order partial derivatives? How can it help in calculating partial derivatives of second and higher orders? Give examples.

11. What does it mean for a function $f(x, y)$ to be differentiable? What does the Increment Theorem say about differentiability?

12. How can you sometimes decide from examining f_x and f_y that a function $f(x, y)$ is differentiable? What is the relation between the differentiability of f and the continuity of f at a point?

13. What is the general Chain Rule? What form does it take for functions of two independent variables? Three independent variables? Functions defined on surfaces? How do you diagram these different forms? Give examples. What pattern enables one to remember all the different forms?

14. What is the derivative of a function $f(x, y)$ at a point P_0 in the direction of a unit vector $\mathbf{u}$? What rate does it describe? What geometric interpretation does it have? Give examples.

15. What is the gradient vector of a differentiable function $f(x, y)$? How is it related to the function's directional derivatives? State the analogous results for functions of three independent variables.

16. How do you find the tangent line at a point on a level curve of a differentiable function $f(x, y)$? How do you find the tangent plane and normal line at a point on a level surface of a differentiable function $f(x, y, z)$? Give examples.

17. How can you use directional derivatives to estimate change?

18. How do you linearize a function $f(x, y)$ of two independent variables at a point (x_0, y_0)? Why might you want to do this? How do you linearize a function of three independent variables?

19. What can you say about the accuracy of linear approximations of functions of two (three) independent variables?

20. If (x, y) moves from (x_0, y_0) to a point $(x_0 + dx, y_0 + dy)$ nearby, how can you estimate the resulting change in the value of a differentiable function $f(x, y)$? Give an example.

21. How do you define local maxima, local minima, and saddle points for a differentiable function $f(x, y)$? Give examples.

22. What derivative tests are available for determining the local extreme values of a function $f(x, y)$? How do they enable you to narrow your search for these values? Give examples.

23. How do you find the extrema of a continuous function $f(x, y)$ on a closed bounded region of the xy-plane? Give an example.

24. Describe the method of Lagrange multipliers and give examples.

25. How does Taylor's formula for a function $f(x, y)$ generate polynomial approximations and error estimates?

26. If $w = f(x, y, z)$, where the variables x, y, and z are constrained by an equation $g(x, y, z) = 0$, what is the meaning of the notation $(\partial w/\partial x)_y$? How can an arrow diagram help you calculate this partial derivative with constrained variables? Give examples.

CHAPTER 14   Practice Exercises

Domain, Range, and Level Curves

In Exercises 1–4, find the domain and range of the given function and identify its level curves. Sketch a typical level curve.

1. $f(x, y) = 9x^2 + y^2$ 2. $f(x, y) = e^{x+y}$

3. $g(x, y) = 1/xy$ 4. $g(x, y) = \sqrt{x^2 - y}$

In Exercises 5–8, find the domain and range of the given function and identify its level surfaces. Sketch a typical level surface.

5. $f(x, y, z) = x^2 + y^2 - z$ 6. $g(x, y, z) = x^2 + 4y^2 + 9z^2$

7. $h(x, y, z) = \dfrac{1}{x^2 + y^3 + z^2}$ 8. $k(x, y, z) = \dfrac{1}{x^2 + y^2 + z^2 + 1}$

Evaluating Limits

Find the limits in Exercises 9–14.

9. $\displaystyle\lim_{(x, y)\to(\pi, \ln 2)} e^y \cos x$ 10. $\displaystyle\lim_{(x, y)\to(0, 0)} \dfrac{2 + y}{x + \cos y}$

11. $\displaystyle\lim_{(x, y)\to(1, 1)} \dfrac{x - y}{x^2 - y^2}$ 12. $\displaystyle\lim_{(x, y)\to(1, 1)} \dfrac{x^3 y^3 - 1}{xy - 1}$

13. $\displaystyle\lim_{P\to(1, -1, e)} \ln|x + y + z|$ 14. $\displaystyle\lim_{P\to(1, -1, -1)} \tan^{-1}(x + y + z)$

By considering different paths of approach, show that the limits in Exercises 15 and 16 do not exist.

15. $\displaystyle\lim_{\substack{(x,y)\to(0,0)\\ y\neq x^2}} \dfrac{y}{x^2 - y}$ 16. $\displaystyle\lim_{\substack{(x,y)\to(0,0)\\ xy\neq 0}} \dfrac{x^2 + y^2}{xy}$

17. **Continuous extension**   Let $f(x, y) = (x^2 - y^2)/(x^2 + y^2)$ for $(x, y) \neq (0, 0)$. Is it possible to define $f(0, 0)$ in a way that makes f continuous at the origin? Why?

18. **Continuous extension**   Let

$$f(x, y) = \begin{cases} \dfrac{\sin(x - y)}{|x| + |y|}, & |x| + |y| \neq 0 \\ 0, & (x, y) = (0, 0). \end{cases}$$

Is f continuous at the origin? Why?

Partial Derivatives

In Exercises 19–24, find the partial derivative of the function with respect to each variable.

19. $g(r, \theta) = r \cos \theta + r \sin \theta$

20. $f(x, y) = \dfrac{1}{2} \ln(x^2 + y^2) + \tan^{-1}\dfrac{y}{x}$

21. $f(R_1, R_2, R_3) = \dfrac{1}{R_1} + \dfrac{1}{R_2} + \dfrac{1}{R_3}$

22. $h(x, y, z) = \sin(2\pi x + y - 3z)$

23. $P(n, R, T, V) = \dfrac{nRT}{V}$ (the ideal gas law)

24. $f(r, l, T, w) = \dfrac{1}{2rl}\sqrt{\dfrac{T}{\pi w}}$

Second-Order Partials

Find the second-order partial derivatives of the functions in Exercises 25–28.

25. $g(x, y) = y + \dfrac{x}{y}$

26. $g(x, y) = e^x + y \sin x$

27. $f(x, y) = x + xy - 5x^3 + \ln(x^2 + 1)$

28. $f(x, y) = y^2 - 3xy + \cos y + 7e^y$

Chain Rule Calculations

29. Find dw/dt at $t = 0$ if $w = \sin(xy + \pi)$, $x = e^t$, and $y = \ln(t + 1)$.

30. Find dw/dt at $t = 1$ if $w = xe^y + y \sin z - \cos z$, $x = 2\sqrt{t}$, $y = t - 1 + \ln t$, and $z = \pi t$.

31. Find $\partial w/\partial r$ and $\partial w/\partial s$ when $r = \pi$ and $s = 0$ if $w = \sin(2x - y)$, $x = r + \sin s$, $y = rs$.

32. Find $\partial w/\partial u$ and $\partial w/\partial v$ when $u = v = 0$ if $w = \ln\sqrt{1 + x^2} - \tan^{-1} x$ and $x = 2e^u \cos v$.

33. Find the value of the derivative of $f(x, y, z) = xy + yz + xz$ with respect to t on the curve $x = \cos t$, $y = \sin t$, $z = \cos 2t$ at $t = 1$.

34. Show that if $w = f(s)$ is any differentiable function of s and if $s = y + 5x$, then

$$\frac{\partial w}{\partial x} - 5\frac{\partial w}{\partial y} = 0.$$

Implicit Differentiation

Assuming that the equations in Exercises 35 and 36 define y as a differentiable function of x, find the value of dy/dx at point P.

35. $1 - x - y^2 - \sin xy = 0$, $\quad P(0, 1)$

36. $2xy + e^{x+y} - 2 = 0$, $\quad P(0, \ln 2)$

Directional Derivatives

In Exercises 37–40, find the directions in which f increases and decreases most rapidly at P_0 and find the derivative of f in each direction. Also, find the derivative of f at P_0 in the direction of the vector $\mathbf{v}$.

37. $f(x, y) = \cos x \cos y$, $\quad P_0(\pi/4, \pi/4)$, $\quad \mathbf{v} = 3\mathbf{i} + 4\mathbf{j}$

38. $f(x, y) = x^2 e^{-2y}$, $\quad P_0(1, 0)$, $\quad \mathbf{v} = \mathbf{i} + \mathbf{j}$

39. $f(x, y, z) = \ln(2x + 3y + 6z)$, $\quad P_0(-1, -1, 1)$, $\mathbf{v} = 2\mathbf{i} + 3\mathbf{j} + 6\mathbf{k}$

40. $f(x, y, z) = x^2 + 3xy - z^2 + 2y + z + 4$, $\quad P_0(0, 0, 0)$, $\mathbf{v} = \mathbf{i} + \mathbf{j} + \mathbf{k}$

41. Derivative in velocity direction Find the derivative of $f(x, y, z) = xyz$ in the direction of the velocity vector of the helix

$$\mathbf{r}(t) = (\cos 3t)\mathbf{i} + (\sin 3t)\mathbf{j} + 3t\mathbf{k}$$

at $t = \pi/3$.

42. Maximum directional derivative What is the largest value that the directional derivative of $f(x, y, z) = xyz$ can have at the point $(1, 1, 1)$?

43. Directional derivatives with given values At the point $(1, 2)$, the function $f(x, y)$ has a derivative of 2 in the direction toward $(2, 2)$ and a derivative of -2 in the direction toward $(1, 1)$.

a. Find $f_x(1, 2)$ and $f_y(1, 2)$.

b. Find the derivative of f at $(1, 2)$ in the direction toward the point $(4, 6)$.

44. Which of the following statements are true if $f(x, y)$ is differentiable at (x_0, y_0)? Give reasons for your answers.

a. If $\mathbf{u}$ is a unit vector, the derivative of f at (x_0, y_0) in the direction of $\mathbf{u}$ is $(f_x(x_0, y_0)\mathbf{i} + f_y(x_0, y_0)\mathbf{j}) \cdot \mathbf{u}$.

b. The derivative of f at (x_0, y_0) in the direction of $\mathbf{u}$ is a vector.

c. The directional derivative of f at (x_0, y_0) has its greatest value in the direction of ∇f.

d. At (x_0, y_0), vector ∇f is normal to the curve $f(x, y) = f(x_0, y_0)$.

Gradients, Tangent Planes, and Normal Lines

In Exercises 45 and 46, sketch the surface $f(x, y, z) = c$ together with ∇f at the given points.

45. $x^2 + y + z^2 = 0$; $\quad (0, -1, \pm 1)$, $\quad (0, 0, 0)$

46. $y^2 + z^2 = 4$; $\quad (2, \pm 2, 0)$, $\quad (2, 0, \pm 2)$

In Exercises 47 and 48, find an equation for the plane tangent to the level surface $f(x, y, z) = c$ at the point P_0. Also, find parametric equations for the line that is normal to the surface at P_0.

47. $x^2 - y - 5z = 0$, $\quad P_0(2, -1, 1)$

48. $x^2 + y^2 + z = 4$, $\quad P_0(1, 1, 2)$

In Exercises 49 and 50, find an equation for the plane tangent to the surface $z = f(x, y)$ at the given point.

49. $z = \ln(x^2 + y^2)$, $\quad (0, 1, 0)$

50. $z = 1/(x^2 + y^2)$, $\quad (1, 1, 1/2)$

In Exercises 51 and 52, find equations for the lines that are tangent and normal to the level curve $f(x, y) = c$ at the point P_0. Then sketch the lines and level curve together with ∇f at P_0.

51. $y - \sin x = 1$, $\quad P_0(\pi, 1)$

52. $\dfrac{y^2}{2} - \dfrac{x^2}{2} = \dfrac{3}{2}$, $\quad P_0(1, 2)$

Tangent Lines to Curves

In Exercises 53 and 54, find parametric equations for the line that is tangent to the curve of intersection of the surfaces at the given point.

53. Surfaces: $x^2 + 2y + 2z = 4$, $\quad y = 1$

Point: $\quad (1, 1, 1/2)$

54. Surfaces: $x + y^2 + z = 2$, $\quad y = 1$

Point: $\quad (1/2, 1, 1/2)$

Linearizations

In Exercises 55 and 56, find the linearization $L(x, y)$ of the function $f(x, y)$ at the point P_0. Then find an upper bound for the magnitude of the error E in the approximation $f(x, y) \approx L(x, y)$ over the rectangle R.

55. $f(x, y) = \sin x \cos y$, $\quad P_0(\pi/4, \pi/4)$

$$R: \quad \left|x - \frac{\pi}{4}\right| \le 0.1, \quad \left|y - \frac{\pi}{4}\right| \le 0.1$$

56. $f(x, y) = xy - 3y^2 + 2$, $P_0(1, 1)$
  R: $|x - 1| \leq 0.1$, $|y - 1| \leq 0.2$

Find the linearizations of the functions in Exercises 57 and 58 at the given points.

57. $f(x, y, z) = xy + 2yz - 3xz$ at $(1, 0, 0)$ and $(1, 1, 0)$

58. $f(x, y, z) = \sqrt{2} \cos x \sin(y + z)$ at $(0, 0, \pi/4)$ and $(\pi/4, \pi/4, 0)$

Estimates and Sensitivity to Change

59. Measuring the volume of a pipeline You plan to calculate the volume inside a stretch of pipeline that is about 36 in. in diameter and 1 mile long. With which measurement should you be more careful, the length or the diameter? Why?

60. Sensitivity to change Is $f(x, y) = x^2 - xy + y^2 - 3$ more sensitive to changes in x or to changes in y when it is near the point $(1, 2)$? How do you know?

61. Change in an electrical circuit Suppose that the current I (amperes) in an electrical circuit is related to the voltage V (volts) and the resistance R (ohms) by the equation $I = V/R$. If the voltage drops from 24 to 23 volts and the resistance drops from 100 to 80 ohms, will I increase or decrease? By about how much? Is the change in I more sensitive to change in the voltage or to change in the resistance? How do you know?

62. Maximum error in estimating the area of an ellipse If $a = 10$ cm and $b = 16$ cm to the nearest millimeter, what should you expect the maximum percentage error to be in the calculated area $A = \pi ab$ of the ellipse $x^2/a^2 + y^2/b^2 = 1$?

63. Error in estimating a product Let $y = uv$ and $z = u + v$, where u and v are positive independent variables.

  a. If u is measured with an error of 2% and v with an error of 3%, about what is the percentage error in the calculated value of y?

  b. Show that the percentage error in the calculated value of z is less than the percentage error in the value of y.

64. Cardiac index To make different people comparable in studies of cardiac output, researchers divide the measured cardiac output by the body surface area to find the *cardiac index* C:

$$C = \frac{\text{cardiac output}}{\text{body surface area}}.$$

The body surface area B of a person with weight w and height h is approximated by the formula

$$B = 71.84w^{0.425}h^{0.725},$$

which gives B in square centimeters when w is measured in kilograms and h in centimeters. You are about to calculate the cardiac index of a person 180 cm tall, weighing 70 kg, with cardiac output of 7 L/min. Which will have a greater effect on the calculation, a 1-kg error in measuring the weight or a 1-cm error in measuring the height?

Local Extrema

Test the functions in Exercises 65–70 for local maxima and minima and saddle points. Find each function's value at these points.

65. $f(x, y) = x^2 - xy + y^2 + 2x + 2y - 4$

66. $f(x, y) = 5x^2 + 4xy - 2y^2 + 4x - 4y$

67. $f(x, y) = 2x^3 + 3xy + 2y^3$

68. $f(x, y) = x^3 + y^3 - 3xy + 15$

69. $f(x, y) = x^3 + y^3 + 3x^2 - 3y^2$

70. $f(x, y) = x^4 - 8x^2 + 3y^2 - 6y$

Absolute Extrema

In Exercises 71–78, find the absolute maximum and minimum values of f on the region R.

71. $f(x, y) = x^2 + xy + y^2 - 3x + 3y$
  R: The triangular region cut from the first quadrant by the line $x + y = 4$

72. $f(x, y) = x^2 - y^2 - 2x + 4y + 1$
  R: The rectangular region in the first quadrant bounded by the coordinate axes and the lines $x = 4$ and $y = 2$

73. $f(x, y) = y^2 - xy - 3y + 2x$
  R: The square region enclosed by the lines $x = \pm 2$ and $y = \pm 2$

74. $f(x, y) = 2x + 2y - x^2 - y^2$
  R: The square region bounded by the coordinate axes and the lines $x = 2$, $y = 2$ in the first quadrant

75. $f(x, y) = x^2 - y^2 - 2x + 4y$
  R: The triangular region bounded below by the x-axis, above by the line $y = x + 2$, and on the right by the line $x = 2$

76. $f(x, y) = 4xy - x^4 - y^4 + 16$
  R: The triangular region bounded below by the line $y = -2$, above by the line $y = x$, and on the right by the line $x = 2$

77. $f(x, y) = x^3 + y^3 + 3x^2 - 3y^2$
  R: The square region enclosed by the lines $x = \pm 1$ and $y = \pm 1$

78. $f(x, y) = x^3 + 3xy + y^3 + 1$
  R: The square region enclosed by the lines $x = \pm 1$ and $y = \pm 1$

Lagrange Multipliers

79. Extrema on a circle Find the extreme values of $f(x, y) = x^3 + y^2$ on the circle $x^2 + y^2 = 1$.

80. Extrema on a circle Find the extreme values of $f(x, y) = xy$ on the circle $x^2 + y^2 = 1$.

81. Extrema in a disk Find the extreme values of $f(x, y) = x^2 + 3y^2 + 2y$ on the unit disk $x^2 + y^2 \leq 1$.

82. Extrema in a disk Find the extreme values of $f(x, y) = x^2 + y^2 - 3x - xy$ on the disk $x^2 + y^2 \leq 9$.

83. Extrema on a sphere Find the extreme values of $f(x, y, z) = x - y + z$ on the unit sphere $x^2 + y^2 + z^2 = 1$.

84. Minimum distance to origin Find the points on the surface $x^2 - zy = 4$ closest to the origin.

85. Minimizing cost of a box A closed rectangular box is to have volume V cm^3. The cost of the material used in the box is a cents/cm^2 for top and bottom, b cents/cm^2 for front and back, and c cents/cm^2 for the remaining sides. What dimensions minimize the total cost of materials?

86. Least volume Find the plane $x/a + y/b + z/c = 1$ that passes through the point $(2, 1, 2)$ and cuts off the least volume from the first octant.

87. Extrema on curve of intersecting surfaces Find the extreme values of $f(x, y, z) = x(y + z)$ on the curve of intersection of the right circular cylinder $x^2 + y^2 = 1$ and the hyperbolic cylinder $xz = 1$.

88. Minimum distance to origin on curve of intersecting plane and cone Find the point closest to the origin on the curve of intersection of the plane $x + y + z = 1$ and the cone $z^2 = 2x^2 + 2y^2$.

Theory and Examples

89. Let $w = f(r, \theta)$, $r = \sqrt{x^2 + y^2}$, and $\theta = \tan^{-1}(y/x)$. Find $\partial w/\partial x$ and $\partial w/\partial y$ and express your answers in terms of r and θ.

90. Let $z = f(u, v)$, $u = ax + by$, and $v = ax - by$. Express z_x and z_y in terms of f_u, f_v, and the constants a and b.

91. If a and b are constants, $w = u^3 + \tanh u + \cos u$, and $u = ax + by$, show that

$$a \frac{\partial w}{\partial y} = b \frac{\partial w}{\partial x}.$$

92. Using the Chain Rule If $w = \ln(x^2 + y^2 + 2z)$, $x = r + s$, $y = r - s$, and $z = 2rs$, find w_r and w_s by the Chain Rule. Then check your answer another way.

93. Angle between vectors The equations $e^u \cos v - x = 0$ and $e^u \sin v - y = 0$ define u and v as differentiable functions of x and y. Show that the angle between the vectors

$$\frac{\partial u}{\partial x}\mathbf{i} + \frac{\partial u}{\partial y}\mathbf{j} \quad \text{and} \quad \frac{\partial v}{\partial x}\mathbf{i} + \frac{\partial v}{\partial y}\mathbf{j}$$

is constant.

94. Polar coordinates and second derivatives Introducing polar coordinates $x = r \cos \theta$ and $y = r \sin \theta$ changes $f(x, y)$ to $g(r, \theta)$. Find the value of $\partial^2 g/\partial\theta^2$ at the point $(r, \theta) = (2, \pi/2)$, given that

$$\frac{\partial f}{\partial x} = \frac{\partial f}{\partial y} = \frac{\partial^2 f}{\partial x^2} = \frac{\partial^2 f}{\partial y^2} = 1$$

at that point.

95. Normal line parallel to a plane Find the points on the surface

$$(y + z)^2 + (z - x)^2 = 16$$

where the normal line is parallel to the yz-plane.

96. Tangent plane parallel to xy-plane Find the points on the surface

$$xy + yz + zx - x - z^2 = 0$$

where the tangent plane is parallel to the xy-plane.

97. When gradient is parallel to position vector Suppose that $\nabla f(x, y, z)$ is always parallel to the position vector $x\mathbf{i} + y\mathbf{j} + z\mathbf{k}$. Show that $f(0, 0, a) = f(0, 0, -a)$ for any a.

98. One-sided directional derivative in all directions, but no gradient The *one-sided directional derivative of f at $P(x_0, y_0, z_0)$ in the direction* $\mathbf{u} = u_1\mathbf{i} + u_2\mathbf{j} + u_3\mathbf{k}$ is the number

$$\lim_{s \to 0^+} \frac{f(x_0 + su_1, y_0 + su_2, z_0 + su_3) - f(x_0, y_0, z_0)}{s}.$$

Show that the one-sided directional derivative of

$$f(x, y, z) = \sqrt{x^2 + y^2 + z^2}$$

at the origin equals 1 in any direction but that f has no gradient vector at the origin.

99. Normal line through origin Show that the line normal to the surface $xy + z = 2$ at the point $(1, 1, 1)$ passes through the origin.

100. Tangent plane and normal line
  a. Sketch the surface $x^2 - y^2 + z^2 = 4$.
  b. Find a vector normal to the surface at $(2, -3, 3)$. Add the vector to your sketch.
  c. Find equations for the tangent plane and normal line at $(2, -3, 3)$.

Partial Derivatives with Constrained Variables

In Exercises 101 and 102, begin by drawing a diagram that shows the relations among the variables.

101. If $w = x^2 e^{yz}$ and $z = x^2 - y^2$ find
  a. $\left(\dfrac{\partial w}{\partial y}\right)_z$ b. $\left(\dfrac{\partial w}{\partial z}\right)_x$ c. $\left(\dfrac{\partial w}{\partial z}\right)_y$.

102. Let $U = f(P, V, T)$ be the internal energy of a gas that obeys the ideal gas law $PV = nRT$ (n and R constant). Find
  a. $\left(\dfrac{\partial U}{\partial T}\right)_P$ b. $\left(\dfrac{\partial U}{\partial V}\right)_T$.

CHAPTER 14 Additional and Advanced Exercises

Partial Derivatives

1. Function with saddle at the origin If you did Exercise 60 in Section 14.2, you know that the function

$$f(x, y) = \begin{cases} xy\dfrac{x^2 - y^2}{x^2 + y^2}, & (x, y) \neq (0, 0) \\ 0, & (x, y) = (0, 0) \end{cases}$$

(see the accompanying figure) is continuous at $(0, 0)$. Find $f_{xy}(0, 0)$ and $f_{yx}(0, 0)$.

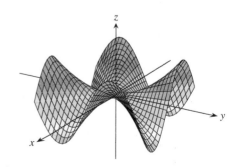

2. Finding a function from second partials Find a function $w = f(x, y)$ whose first partial derivatives are $\partial w/\partial x = 1 + e^x \cos y$ and $\partial w/\partial y = 2y - e^x \sin y$ and whose value at the point $(\ln 2, 0)$ is $\ln 2$.

3. A proof of Leibniz's Rule Leibniz's Rule says that if f is continuous on $[a, b]$ and if $u(x)$ and $v(x)$ are differentiable functions of x whose values lie in $[a, b]$, then

$$\frac{d}{dx} \int_{u(x)}^{v(x)} f(t)\, dt = f(v(x)) \frac{dv}{dx} - f(u(x)) \frac{du}{dx}.$$

Prove the rule by setting

$$g(u, v) = \int_u^v f(t)\, dt, \qquad u = u(x), \qquad v = v(x)$$

and calculating dg/dx with the Chain Rule.

4. Finding a function with constrained second partials Suppose that f is a twice-differentiable function of r, that $r = \sqrt{x^2 + y^2 + z^2}$, and that

$$f_{xx} + f_{yy} + f_{zz} = 0.$$

Show that for some constants a and b,

$$f(r) = \frac{a}{r} + b.$$

5. Homogeneous functions A function $f(x, y)$ is *homogeneous of degree n* (n a nonnegative integer) if $f(tx, ty) = t^n f(x, y)$ for all t, x, and y. For such a function (sufficiently differentiable), prove that

a. $x \dfrac{\partial f}{\partial x} + y \dfrac{\partial f}{\partial y} = nf(x, y)$

b. $x^2 \left(\dfrac{\partial^2 f}{\partial x^2}\right) + 2xy \left(\dfrac{\partial^2 f}{\partial x \partial y}\right) + y^2 \left(\dfrac{\partial^2 f}{\partial y^2}\right) = n(n-1)f.$

6. Surface in polar coordinates Let

$$f(r, \theta) = \begin{cases} \dfrac{\sin 6r}{6r}, & r \neq 0 \\ 1, & r = 0, \end{cases}$$

where r and θ are polar coordinates. Find

a. $\lim\limits_{r \to 0} f(r, \theta)$

b. $f_r(0, 0)$

c. $f_\theta(r, \theta), \quad r \neq 0.$

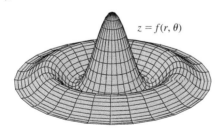
$z = f(r, \theta)$

Gradients and Tangents

7. Properties of position vectors Let $\mathbf{r} = x\mathbf{i} + y\mathbf{j} + z\mathbf{k}$ and let $r = |\mathbf{r}|$.

a. Show that $\nabla r = \mathbf{r}/r$.

b. Show that $\nabla(r^n) = nr^{n-2}\mathbf{r}$.

c. Find a function whose gradient equals $\mathbf{r}$.

d. Show that $\mathbf{r} \cdot d\mathbf{r} = r\, dr$.

e. Show that $\nabla(\mathbf{A} \cdot \mathbf{r}) = \mathbf{A}$ for any constant vector $\mathbf{A}$.

8. Gradient orthogonal to tangent Suppose that a differentiable function $f(x, y)$ has the constant value c along the differentiable curve $x = g(t), y = h(t)$; that is,

$$f(g(t), h(t)) = c$$

for all values of t. Differentiate both sides of this equation with respect to t to show that ∇f is orthogonal to the curve's tangent vector at every point on the curve.

9. Curve tangent to a surface Show that the curve

$$\mathbf{r}(t) = (\ln t)\mathbf{i} + (t \ln t)\mathbf{j} + t\mathbf{k}$$

is tangent to the surface

$$xz^2 - yz + \cos xy = 1$$

at $(0, 0, 1)$.

10. Curve tangent to a surface Show that the curve

$$\mathbf{r}(t) = \left(\frac{t^3}{4} - 2\right)\mathbf{i} + \left(\frac{4}{t} - 3\right)\mathbf{j} + \cos(t - 2)\mathbf{k}$$

is tangent to the surface

$$x^3 + y^3 + z^3 - xyz = 0$$

at $(0, -1, 1)$.

Extreme Values

11. Extrema on a surface Show that the only possible maxima and minima of z on the surface $z = x^3 + y^3 - 9xy + 27$ occur at $(0, 0)$ and $(3, 3)$. Show that neither a maximum nor a minimum occurs at $(0, 0)$. Determine whether z has a maximum or a minimum at $(3, 3)$.

12. Maximum in closed first quadrant Find the maximum value of $f(x, y) = 6xye^{-(2x+3y)}$ in the closed first quadrant (includes the nonnegative axes).

13. Minimum volume cut from first octant Find the minimum volume for a region bounded by the planes $x = 0, y = 0, z = 0$ and a plane tangent to the ellipsoid

$$\frac{x^2}{a^2} + \frac{y^2}{b^2} + \frac{z^2}{c^2} = 1$$

at a point in the first octant.

14. Minimum distance from a line to a parabola in xy-plane By minimizing the function $f(x, y, u, v) = (x - u)^2 + (y - v)^2$ subject to the constraints $y = x + 1$ and $u = v^2$, find the minimum distance in the xy-plane from the line $y = x + 1$ to the parabola $y^2 = x$.

Theory and Examples

15. Boundedness of first partials implies continuity Prove the following theorem: If $f(x, y)$ is defined in an open region R of the xy-plane and if f_x and f_y are bounded on R, then $f(x, y)$ is continuous on R. (The assumption of boundedness is essential.)

16. Suppose that $\mathbf{r}(t) = g(t)\mathbf{i} + h(t)\mathbf{j} + k(t)\mathbf{k}$ is a smooth curve in the domain of a differentiable function $f(x, y, z)$. Describe the relation between df/dt, ∇f, and $\mathbf{v} = d\mathbf{r}/dt$. What can be said about ∇f and $\mathbf{v}$ at interior points of the curve where f has extreme values relative to its other values on the curve? Give reasons for your answer.

17. Finding functions from partial derivatives Suppose that f and g are functions of x and y such that

$$\frac{\partial f}{\partial y} = \frac{\partial g}{\partial x} \quad \text{and} \quad \frac{\partial f}{\partial x} = \frac{\partial g}{\partial y},$$

and suppose that

$$\frac{\partial f}{\partial x} = 0, \quad f(1, 2) = g(1, 2) = 5, \quad \text{and} \quad f(0, 0) = 4.$$

Find $f(x, y)$ and $g(x, y)$.

18. Rate of change of the rate of change We know that if $f(x, y)$ is a function of two variables and if $\mathbf{u} = a\mathbf{i} + b\mathbf{j}$ is a unit vector, then $D_{\mathbf{u}}f(x, y) = f_x(x, y)a + f_y(x, y)b$ is the rate of change of $f(x, y)$ at (x, y) in the direction of $\mathbf{u}$. Give a similar formula for the rate of change *of the rate of change of* $f(x, y)$ at (x, y) in the direction $\mathbf{u}$.

19. Path of a heat-seeking particle A heat-seeking particle has the property that at any point (x, y) in the plane it moves in the direction of maximum temperature increase. If the temperature at (x, y) is $T(x, y) = -e^{-2y}\cos x$, find an equation $y = f(x)$ for the path of a heat-seeking particle at the point $(\pi/4, 0)$.

20. Velocity after a ricochet A particle traveling in a straight line with constant velocity $\mathbf{i} + \mathbf{j} - 5\mathbf{k}$ passes through the point $(0, 0, 30)$ and hits the surface $z = 2x^2 + 3y^2$. The particle ricochets off the surface, the angle of reflection being equal to the angle of incidence. Assuming no loss of speed, what is the velocity of the particle after the ricochet? Simplify your answer.

21. Directional derivatives tangent to a surface Let S be the surface that is the graph of $f(x, y) = 10 - x^2 - y^2$. Suppose that the temperature in space at each point (x, y, z) is $T(x, y, z) = x^2y + y^2z + 4x + 14y + z$.

a. Among all the possible directions tangential to the surface S at the point $(0, 0, 10)$, which direction will make the rate of change of temperature at $(0, 0, 10)$ a maximum?

b. Which direction tangential to S at the point $(1, 1, 8)$ will make the rate of change of temperature a maximum?

22. Drilling another borehole On a flat surface of land, geologists drilled a borehole straight down and hit a mineral deposit at 1000 ft. They drilled a second borehole 100 ft to the north of the first and hit the mineral deposit at 950 ft. A third borehole 100 ft east of the first borehole struck the mineral deposit at 1025 ft. The geologists have reasons to believe that the mineral deposit is in the shape of a dome, and for the sake of economy, they would like to find where the deposit is closest to the surface. Assuming the surface to be the xy-plane, in what direction from the first borehole would you suggest the geologists drill their fourth borehole?

The one-dimensional heat equation If $w(x, t)$ represents the temperature at position x at time t in a uniform wire with perfectly insulated sides, then the partial derivatives w_{xx} and w_t satisfy a differential equation of the form

$$w_{xx} = \frac{1}{c^2}\, w_t.$$

This equation is called the *one-dimensional heat equation*. The value of the positive constant c^2 is determined by the material from which the wire is made.

23. Find all solutions of the one-dimensional heat equation of the form $w = e^{rt}\sin \pi x$, where r is a constant.

24. Find all solutions of the one-dimensional heat equation that have the form $w = e^{rt}\sin kx$ and satisfy the conditions that $w(0, t) = 0$ and $w(L, t) = 0$. What happens to these solutions as $t \to \infty$?

CHAPTER 14   Technology Application Projects

Mathematica/Maple Projects

Projects can be found within MyMathLab.

- *Plotting Surfaces*
  Efficiently generate plots of surfaces, contours, and level curves.

- *Exploring the Mathematics Behind Skateboarding: Analysis of the Directional Derivative*
  The path of a skateboarder is introduced, first on a level plane, then on a ramp, and finally on a paraboloid. Compute, plot, and analyze the directional derivative in terms of the skateboarder.

- *Looking for Patterns and Applying the Method of Least Squares to Real Data*
  Fit a line to a set of numerical data points by choosing the line that minimizes the sum of the squares of the vertical distances from the points to the line.

- *Lagrange Goes Skateboarding: How High Does He Go?*
  Revisit and analyze the skateboarders' adventures for maximum and minimum heights from both a graphical and analytic perspective using Lagrange multipliers.

15

Multiple Integrals

OVERVIEW In this chapter we define the *double integral* of a function of two variables $f(x, y)$ over a region in the plane as the limit of approximating Riemann sums. Just as a single integral can represent signed area, so can a double integral represent signed volume. Double integrals can be evaluated using the Fundamental Theorem of Calculus studied in Section 5.4, but now the evaluations are done twice by integrating with respect to each of the variables x and y in turn. Double integrals can be used to find areas of more general regions in the plane than those encountered in Chapter 5. Moreover, just as the Substitution Rule could simplify finding single integrals, we can sometimes use polar coordinates to simplify computing a double integral. We study more general substitutions for evaluating double integrals as well.

We also define *triple integrals* for a function of three variables $f(x, y, z)$ over a region in space. Triple integrals can be used to find volumes of still more general regions in space, and their evaluation is like that of double integrals with yet a third evaluation. *Cylindrical* or *spherical coordinates* can sometimes be used to simplify the calculation of a triple integral, and we investigate those techniques. Double and triple integrals have a number of applications, such as calculating the average value of a multivariable function, and finding moments and centers of mass.

15.1 Double and Iterated Integrals over Rectangles

In Chapter 5 we defined the definite integral of a continuous function $f(x)$ over an interval $[a, b]$ as a limit of Riemann sums. In this section we extend this idea to define the *double integral* of a continuous function of two variables $f(x, y)$ over a bounded rectangle R in the plane. The Riemann sums for the integral of a single-variable function $f(x)$ are obtained by partitioning a finite interval into thin subintervals, multiplying the width of each subinterval by the value of f at a point c_k inside that subinterval, and then adding together all the products. A similar method of partitioning, multiplying, and summing is used to construct double integrals as limits of approximating Riemann sums.

Double Integrals

We begin our investigation of double integrals by considering the simplest type of planar region, a rectangle. We consider a function $f(x, y)$ defined on a rectangular region R,

$$R: \quad a \le x \le b, \quad c \le y \le d.$$

We subdivide R into small rectangles using a network of lines parallel to the x- and y-axes (Figure 15.1). The lines divide R into n rectangular pieces, where the number of such pieces n gets large as the width and height of each piece gets small. These rectangles form a **partition** of R. A small rectangular piece of width Δx and height Δy has area

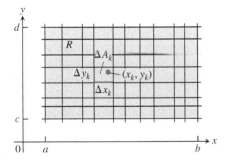

FIGURE 15.1 Rectangular grid partitioning the region R into small rectangles of area $\Delta A_k = \Delta x_k \, \Delta y_k$.

$\Delta A = \Delta x \Delta y$. If we number the small pieces partitioning R in some order, then their areas are given by numbers $\Delta A_1, \Delta A_2, \ldots, \Delta A_n$, where ΔA_k is the area of the kth small rectangle.

To form a Riemann sum over R, we choose a point (x_k, y_k) in the kth small rectangle, multiply the value of f at that point by the area ΔA_k, and add together the products:

$$S_n = \sum_{k=1}^{n} f(x_k, y_k)\, \Delta A_k.$$

Depending on how we pick (x_k, y_k) in the kth small rectangle, we may get different values for S_n.

We are interested in what happens to these Riemann sums as the widths and heights of all the small rectangles in the partition of R approach zero. The **norm** of a partition P, written $\|P\|$, is the largest width or height of any rectangle in the partition. If $\|P\| = 0.1$ then all the rectangles in the partition of R have width at most 0.1 and height at most 0.1. Sometimes the Riemann sums converge as the norm of P goes to zero, written $\|P\| \to 0$. The resulting limit is then written as

$$\lim_{\|P\| \to 0} \sum_{k=1}^{n} f(x_k, y_k)\, \Delta A_k.$$

As $\|P\| \to 0$ and the rectangles get narrow and short, their number n increases, so we can also write this limit as

$$\lim_{n \to \infty} \sum_{k=1}^{n} f(x_k, y_k)\, \Delta A_k,$$

with the understanding that $\|P\| \to 0$, and hence $\Delta A_k \to 0$, as $n \to \infty$.

Many choices are involved in a limit of this kind. The collection of small rectangles is determined by the grid of vertical and horizontal lines that determine a rectangular partition of R. In each of the resulting small rectangles there is a choice of an arbitrary point (x_k, y_k) at which f is evaluated. These choices together determine a single Riemann sum. To form a limit, we repeat the whole process again and again, choosing partitions whose rectangle widths and heights both go to zero and whose number goes to infinity.

When a limit of the sums S_n exists, giving the same limiting value no matter what choices are made, then the function f is said to be **integrable** and the limit is called the **double integral** of f over R, written as

$$\iint_R f(x, y)\, dA \qquad \text{or} \qquad \iint_R f(x, y)\, dx\, dy.$$

It can be shown that if $f(x, y)$ is a continuous function throughout R, then f is integrable, as in the single-variable case discussed in Chapter 5. Many discontinuous functions are also integrable, including functions that are discontinuous only on a finite number of points or smooth curves. We leave the proof of these facts to a more advanced text.

Double Integrals as Volumes

When $f(x, y)$ is a positive function over a rectangular region R in the xy-plane, we may interpret the double integral of f over R as the volume of the 3-dimensional solid region over the xy-plane bounded below by R and above by the surface $z = f(x, y)$ (Figure 15.2). Each term $f(x_k, y_k)\, \Delta A_k$ in the sum $S_n = \sum f(x_k, y_k)\, \Delta A_k$ is the volume of a vertical rectangular box that approximates the volume of the portion of the solid that stands directly above the base ΔA_k. The sum S_n thus approximates what we want to call the total volume of the solid. We *define* this volume to be

$$\text{Volume} = \lim_{n \to \infty} S_n = \iint_R f(x, y)\, dA,$$

where $\Delta A_k \to 0$ as $n \to \infty$.

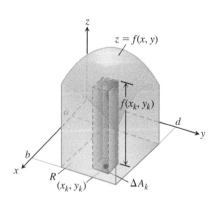

FIGURE 15.2 Approximating solids with rectangular boxes leads us to define the volumes of more general solids as double integrals. The volume of the solid shown here is the double integral of $f(x, y)$ over the base region R.

As you might expect, this more general method of calculating volume agrees with the methods in Chapter 6, but we do not prove this here. Figure 15.3 shows Riemann sum approximations to the volume becoming more accurate as the number n of boxes increases.

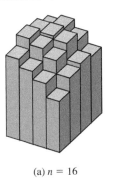  

(a) $n = 16$ (b) $n = 64$ (c) $n = 256$

FIGURE 15.3 As n increases, the Riemann sum approximations approach the total volume of the solid shown in Figure 15.2.

Fubini's Theorem for Calculating Double Integrals

Suppose that we wish to calculate the volume under the plane $z = 4 - x - y$ over the rectangular region $R: 0 \le x \le 2, 0 \le y \le 1$ in the xy-plane. If we apply the method of slicing from Section 6.1, with slices perpendicular to the x-axis (Figure 15.4), then the volume is

$$\int_{x=0}^{x=2} A(x)\, dx, \tag{1}$$

where $A(x)$ is the cross-sectional area at x. For each value of x, we may calculate $A(x)$ as the integral

$$A(x) = \int_{y=0}^{y=1} (4 - x - y)\, dy, \tag{2}$$

which is the area under the curve $z = 4 - x - y$ in the plane of the cross-section at x. In calculating $A(x)$, x is held fixed and the integration takes place with respect to y. Combining Equations (1) and (2), we see that the volume of the entire solid is

$$\text{Volume} = \int_{x=0}^{x=2} A(x)\, dx = \int_{x=0}^{x=2} \left(\int_{y=0}^{y=1} (4 - x - y)\, dy \right) dx$$

$$= \int_{x=0}^{x=2} \left[4y - xy - \frac{y^2}{2} \right]_{y=0}^{y=1} dx = \int_{x=0}^{x=2} \left(\frac{7}{2} - x \right) dx$$

$$= \left[\frac{7}{2}x - \frac{x^2}{2} \right]_0^2 = 5.$$

If we just wanted to write a formula for the volume, without carrying out any of the integrations, we could write

$$\text{Volume} = \int_0^2 \int_0^1 (4 - x - y)\, dy\, dx. \tag{3}$$

The expression on the right, called an **iterated** or **repeated integral**, says that the volume is obtained by integrating $4 - x - y$ with respect to y from $y = 0$ to $y = 1$ while holding x fixed, and then integrating the resulting expression in x from $x = 0$ to $x = 2$. The limits of integration 0 and 1 are associated with y, so they are placed on the integral closest to dy. The other limits of integration, 0 and 2, are associated with the variable x, so they are placed on the outside integral symbol that is paired with dx.

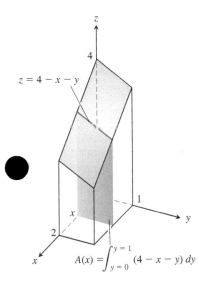

FIGURE 15.4 To obtain the cross-sectional area $A(x)$, we hold x fixed and integrate with respect to y.

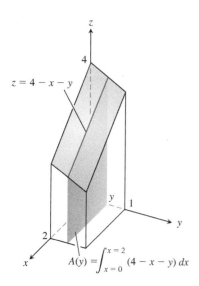

$z = 4 - x - y$

$A(y) = \int_{x=0}^{x=2} (4 - x - y)\, dx$

FIGURE 15.5 To obtain the cross-sectional area $A(y)$, we hold y fixed and integrate with respect to x.

What would have happened if we had calculated the volume by slicing with planes perpendicular to the y-axis (Figure 15.5)? As a function of y, the typical cross-sectional area is

$$A(y) = \int_{x=0}^{x=2} (4 - x - y)\, dx = \left[4x - \frac{x^2}{2} - xy \right]_{x=0}^{x=2} = 6 - 2y. \tag{4}$$

The volume of the entire solid is therefore

$$\text{Volume} = \int_{y=0}^{y=1} A(y)\, dy = \int_{y=0}^{y=1} (6 - 2y)\, dy = \left[6y - y^2 \right]_0^1 = 5,$$

in agreement with our earlier calculation.

Again, we may give a formula for the volume as an iterated integral by writing

$$\text{Volume} = \int_0^1 \int_0^2 (4 - x - y)\, dx\, dy.$$

The expression on the right says we can find the volume by integrating $4 - x - y$ with respect to x from $x = 0$ to $x = 2$ as in Equation (4) and integrating the result with respect to y from $y = 0$ to $y = 1$. In this iterated integral, the order of integration is first x and then y, the reverse of the order in Equation (3).

What do these two volume calculations with iterated integrals have to do with the double integral

$$\iint_R (4 - x - y)\, dA$$

over the rectangle $R: 0 \le x \le 2, 0 \le y \le 1$? The answer is that both iterated integrals give the value of the double integral. This is what we would reasonably expect, since the double integral measures the volume of the same region as the two iterated integrals. A theorem published in 1907 by Guido Fubini says that the double integral of any continuous function over a rectangle can be calculated as an iterated integral in either order of integration. (Fubini proved his theorem in greater generality, but this is what it says in our setting.)

THEOREM 1—Fubini's Theorem (First Form)
If $f(x, y)$ is continuous throughout the rectangular region $R: a \le x \le b$, $c \le y \le d$, then

$$\iint_R f(x, y)\, dA = \int_c^d \int_a^b f(x, y)\, dx\, dy = \int_a^b \int_c^d f(x, y)\, dy\, dx.$$

Fubini's Theorem says that double integrals over rectangles can be calculated as iterated integrals. Thus, we can evaluate a double integral by integrating with respect to one variable at a time using the Fundamental Theorem of Calculus.

Fubini's Theorem also says that we may calculate the double integral by integrating in *either* order, a genuine convenience. When we calculate a volume by slicing, we may use either planes perpendicular to the x-axis or planes perpendicular to the y-axis.

EXAMPLE 1   Calculate $\iint_R f(x, y)\, dA$ for

$$f(x, y) = 100 - 6x^2 y \quad \text{and} \quad R: \ 0 \le x \le 2, \ -1 \le y \le 1.$$

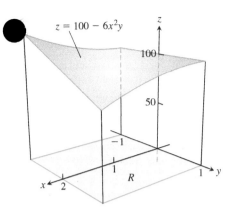

FIGURE 15.6 The double integral $\iint_R f(x, y)\, dA$ gives the volume under this surface over the rectangular region R (Example 1).

Solution   Figure 15.6 displays the volume beneath the surface. By Fubini's Theorem,

$$\iint_R f(x, y)\, dA = \int_{-1}^{1}\int_{0}^{2}(100 - 6x^2 y)\, dx\, dy = \int_{-1}^{1}\Big[100x - 2x^3 y\Big]_{x=0}^{x=2} dy$$

$$= \int_{-1}^{1}(200 - 16y)\, dy = \Big[200y - 8y^2\Big]_{-1}^{1} = 400.$$

Reversing the order of integration gives the same answer:

$$\int_{0}^{2}\int_{-1}^{1}(100 - 6x^2 y)\, dy\, dx = \int_{0}^{2}\Big[100y - 3x^2 y^2\Big]_{y=-1}^{y=1} dx$$

$$= \int_{0}^{2}\Big[(100 - 3x^2) - (-100 - 3x^2)\Big]\, dx$$

$$= \int_{0}^{2} 200\, dx = 400.$$

EXAMPLE 2   Find the volume of the region bounded above by the elliptical paraboloid $z = 10 + x^2 + 3y^2$ and below by the rectangle R: $0 \le x \le 1, 0 \le y \le 2$.

Solution   The surface and volume are shown in Figure 15.7. The volume is given by the double integral

$$V = \iint_R (10 + x^2 + 3y^2)\, dA = \int_{0}^{1}\int_{0}^{2}(10 + x^2 + 3y^2)\, dy\, dx$$

$$= \int_{0}^{1}\Big[10y + x^2 y + y^3\Big]_{y=0}^{y=2} dx$$

$$= \int_{0}^{1}(20 + 2x^2 + 8)\, dx = \Big[20x + \frac{2}{3}x^3 + 8x\Big]_{0}^{1} = \frac{86}{3}.$$

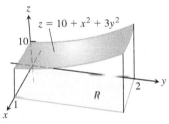

FIGURE 15.7 The double integral $\iint_R f(x, y)\, dA$ gives the volume under this surface over the rectangular region R (Example 2).

EXERCISES 15.1

Evaluating Iterated Integrals

In Exercises 1–14, evaluate the iterated integral.

1. $\displaystyle\int_{1}^{2}\int_{0}^{4} 2xy\, dy\, dx$

2. $\displaystyle\int_{0}^{2}\int_{-1}^{1}(x - y)\, dy\, dx$

3. $\displaystyle\int_{-1}^{0}\int_{-1}^{1}(x + y + 1)\, dx\, dy$

4. $\displaystyle\int_{0}^{1}\int_{0}^{1}\Big(1 - \frac{x^2 + y^2}{2}\Big)\, dx\, dy$

5. $\displaystyle\int_{0}^{3}\int_{0}^{2}(4 - y^2)\, dy\, dx$

6. $\displaystyle\int_{0}^{3}\int_{-2}^{0}(x^2 y - 2xy)\, dy\, dx$

7. $\displaystyle\int_{0}^{1}\int_{0}^{1}\frac{y}{1 + xy}\, dx\, dy$

8. $\displaystyle\int_{1}^{4}\int_{0}^{4}\Big(\frac{x}{2} + \sqrt{y}\Big)\, dx\, dy$

9. $\displaystyle\int_{0}^{\ln 2}\int_{1}^{\ln 5} e^{2x+y}\, dy\, dx$

10. $\displaystyle\int_{0}^{1}\int_{1}^{2} xye^x\, dy\, dx$

11. $\displaystyle\int_{-1}^{2}\int_{0}^{\pi/2} y \sin x\, dx\, dy$

12. $\displaystyle\int_{\pi}^{2\pi}\int_{0}^{\pi}(\sin x + \cos y)\, dx\, dy$

13. $\displaystyle\int_{1}^{4}\int_{1}^{e}\frac{\ln x}{xy}\, dx\, dy$

14. $\displaystyle\int_{-1}^{2}\int_{1}^{2} x \ln y\, dy\, dx$

15. Find all values of the constant c so that $\displaystyle\int_{0}^{1}\int_{0}^{c}(2x + y)\, dx\, dy = 3.$

16. Find all values of the constant c so that

$$\int_{-1}^{c}\int_{0}^{2}(xy + 1)\, dy\, dx = 4 + 4c.$$

Evaluating Double Integrals over Rectangles

In Exercises 17–24, evaluate the double integral over the given region R.

17. $\displaystyle\iint_R (6y^2 - 2x)\, dA,$ R: $0 \le x \le 1,\ \ 0 \le y \le 2$

18. $\displaystyle\iint_R \Big(\frac{\sqrt{x}}{y^2}\Big)\, dA,$ R: $0 \le x \le 4,\ \ 1 \le y \le 2$

19. $\displaystyle\iint_R xy \cos y \, dA,$ $R: -1 \le x \le 1, \quad 0 \le y \le \pi$

20. $\displaystyle\iint_R y \sin (x + y) \, dA,$ $R: -\pi \le x \le 0, \quad 0 \le y \le \pi$

21. $\displaystyle\iint_R e^{x-y} \, dA,$ $R: 0 \le x \le \ln 2, \quad 0 \le y \le \ln 2$

22. $\displaystyle\iint_R xye^{xy^2} \, dA,$ $R: 0 \le x \le 2, \quad 0 \le y \le 1$

23. $\displaystyle\iint_R \frac{xy^3}{x^2 + 1} \, dA,$ $R: 0 \le x \le 1, \quad 0 \le y \le 2$

24. $\displaystyle\iint_R \frac{y}{x^2y^2 + 1} \, dA,$ $R: 0 \le x \le 1, \quad 0 \le y \le 1$

In Exercises 25 and 26, integrate f over the given region.

25. **Square** $f(x, y) = 1/(xy)$ over the square $1 \le x \le 2$, $1 \le y \le 2$

26. **Rectangle** $f(x, y) = y \cos xy$ over the rectangle $0 \le x \le \pi$, $0 \le y \le 1$

In Exercises 27 and 28, sketch the solid whose volume is given by the specified integral.

27. $\displaystyle\int_0^1 \int_0^2 (9 - x^2 - y^2) \, dy \, dx$ 28. $\displaystyle\int_0^3 \int_1^4 (7 - x - y) \, dx \, dy$

29. Find the volume of the region bounded above by the paraboloid $z = x^2 + y^2$ and below by the square $R: -1 \le x \le 1$, $-1 \le y \le 1$.

30. Find the volume of the region bounded above by the elliptical paraboloid $z = 16 - x^2 - y^2$ and below by the square $R: 0 \le x \le 2, 0 \le y \le 2$.

31. Find the volume of the region bounded above by the plane $z = 2 - x - y$ and below by the square $R: 0 \le x \le 1$, $0 \le y \le 1$.

32. Find the volume of the region bounded above by the plane $z = y/2$ and below by the rectangle $R: 0 \le x \le 4, 0 \le y \le 2$.

33. Find the volume of the region bounded above by the surface $z = 2 \sin x \cos y$ and below by the rectangle $R: 0 \le x \le \pi/2$, $0 \le y \le \pi/4$.

34. Find the volume of the region bounded above by the surface $z = 4 - y^2$ and below by the rectangle $R: 0 \le x \le 1$, $0 \le y \le 2$.

35. Find a value of the constant k so that $\displaystyle\int_1^2 \int_0^3 kx^2y \, dx \, dy = 1$.

36. Evaluate $\displaystyle\int_{-1}^1 \int_0^{\pi/2} x \sin \sqrt{y} \, dy \, dx$.

37. Use Fubini's Theorem to evaluate

$$\int_0^2 \int_0^1 \frac{x}{1 + xy} \, dx \, dy.$$

38. Use Fubini's Theorem to evaluate

$$\int_0^1 \int_0^3 xe^{xy} \, dx \, dy.$$

T 39. Use a software application to compute the integrals

a. $\displaystyle\int_0^1 \int_0^2 \frac{y - x}{(x + y)^3} \, dx \, dy$ b. $\displaystyle\int_0^2 \int_0^1 \frac{y - x}{(x + y)^3} \, dy \, dx$

Explain why your results do not contradict Fubini's Theorem.

40. If $f(x, y)$ is continuous over $R: a \le x \le b, c \le y \le d$ and

$$F(x, y) = \int_a^x \int_c^y f(u, v) \, dv \, du$$

on the interior of R, find the second partial derivatives F_{xy} and F_{yx}.

15.2 Double Integrals over General Regions

In this section we define and evaluate double integrals over bounded regions in the plane which are more general than rectangles. These double integrals are also evaluated as iterated integrals, with the main practical problem being that of determining the limits of integration. Since the region of integration may have boundaries other than line segments parallel to the coordinate axes, the limits of integration often involve variables, not just constants.

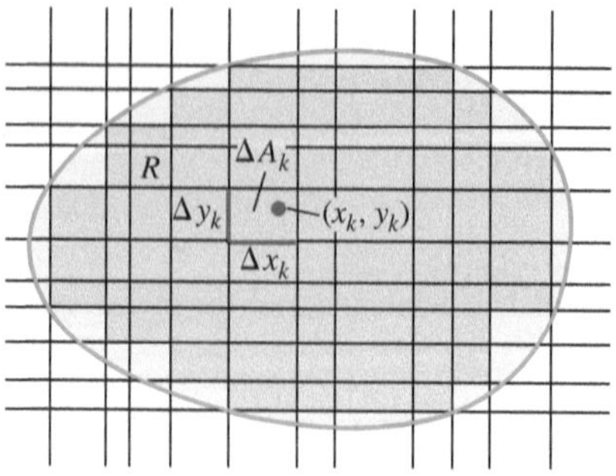

Double Integrals over Bounded, Nonrectangular Regions

FIGURE 15.8 A rectangular grid partitioning a bounded, nonrectangular region into rectangular cells.

To define the double integral of a function $f(x, y)$ over a bounded, nonrectangular region R, such as the one in Figure 15.8, we again begin by covering R with a grid of small rectangular cells whose union contains all points of R. This time, however, we cannot exactly fill R with a finite number of rectangles lying inside R, since its boundary is curved, and some of the small rectangles in the grid lie partly outside R. A partition of R is formed by taking the rectangles that lie completely inside it, not using any that are either partly or completely outside. For commonly arising regions, more and more of R is included as the norm of a partition (the largest width or height of any rectangle used) approaches zero.

Once we have a partition of R, we number the rectangles in some order from 1 to n and let ΔA_k be the area of the kth rectangle. We then choose a point (x_k, y_k) in the kth rectangle and form the Riemann sum

$$S_n = \sum_{k=1}^{n} f(x_k, y_k) \, \Delta A_k.$$

As the norm of the partition forming S_n goes to zero, $\|P\| \to 0$, the width and height of each enclosed rectangle go to zero, their area ΔA_k goes to zero, and their number goes to infinity. If $f(x, y)$ is a continuous function, then these Riemann sums converge to a limiting value, not dependent on any of the choices we made. This limit is called the **double integral** of $f(x, y)$ over R:

$$\lim_{\|P\| \to 0} \sum_{k=1}^{n} f(x_k, y_k) \, \Delta A_k = \iint_R f(x, y) \, dA.$$

The nature of the boundary of R introduces issues not found in integrals over an interval. When R has a curved boundary, the n rectangles of a partition lie inside R but do not cover all of R. In order for a partition to approximate R well, the parts of R covered by small rectangles lying partly outside R must become negligible as the norm of the partition approaches zero. This property of being nearly filled in by a partition of small norm is satisfied by all the regions that we will encounter. There is no problem with boundaries made from polygons, circles, ellipses, and from continuous graphs over an interval, joined end to end. A curve with a "fractal" type of shape would be problematic, but such curves arise rarely in most applications. A careful discussion of which type of regions R can be used for computing double integrals is left to a more advanced text.

Volumes

If $f(x, y)$ is positive and continuous over R, we define the volume of the solid region between R and the surface $z = f(x, y)$ to be $\iint_R f(x, y) \, dA$, as before (Figure 15.9).

If R is a region like the one shown in the xy-plane in Figure 15.10, bounded "above" and "below" by the curves $y = g_2(x)$ and $y = g_1(x)$ and on the sides by the lines

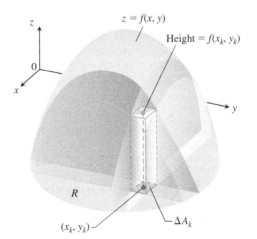

Volume $= \lim \sum f(x_k, y_k) \, \Delta A_k = \iint_R f(x, y) \, dA$

FIGURE 15.9 We define the volumes of solids with curved bases as a limit of approximating rectangular boxes.

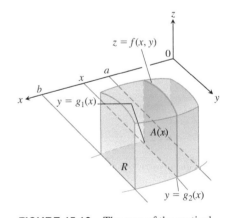

FIGURE 15.10 The area of the vertical slice shown here is $A(x)$. To calculate the volume of the solid, we integrate this area from $x = a$ to $x = b$:

$$\int_a^b A(x) \, dx = \int_a^b \int_{g_1(x)}^{g_2(x)} f(x, y) \, dy \, dx.$$

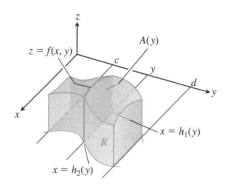

FIGURE 15.11 The volume of the solid shown here is

$$\int_c^d A(y)\, dy = \int_c^d \int_{h_1(y)}^{h_2(y)} f(x, y)\, dx\, dy.$$

For a given solid, Theorem 2 says we can calculate the volume as in Figure 15.10, or in the way shown here. Both calculations have the same result.

$x = a$, $x = b$, we may again calculate the volume by the method of slicing. We first calculate the cross-sectional area

$$A(x) = \int_{y=g_1(x)}^{y=g_2(x)} f(x, y)\, dy$$

and then integrate $A(x)$ from $x = a$ to $x = b$ to get the volume as an iterated integral:

$$V = \int_a^b A(x)\, dx = \int_a^b \int_{g_1(x)}^{g_2(x)} f(x, y)\, dy\, dx. \qquad (1)$$

Similarly, if R is a region like the one shown in Figure 15.11, bounded by the curves $x = h_2(y)$ and $x = h_1(y)$ and the lines $y = c$ and $y = d$, then the volume calculated by slicing is given by the iterated integral

$$\text{Volume} = \int_c^d \int_{h_1(y)}^{h_2(y)} f(x, y)\, dx\, dy. \qquad (2)$$

That the iterated integrals in Equations (1) and (2) both give the volume that we defined to be the double integral of f over R is a consequence of the following stronger form of Fubini's Theorem.

THEOREM 2—Fubini's Theorem (Stronger Form)
Let $f(x, y)$ be continuous on a region R.

1. If R is defined by $a \le x \le b$, $g_1(x) \le y \le g_2(x)$, with g_1 and g_2 continuous on $[a, b]$, then

$$\iint_R f(x, y)\, dA = \int_a^b \int_{g_1(x)}^{g_2(x)} f(x, y)\, dy\, dx.$$

2. If R is defined by $c \le y \le d$, $h_1(y) \le x \le h_2(y)$, with h_1 and h_2 continuous on $[c, d]$, then

$$\iint_R f(x, y)\, dA = \int_c^d \int_{h_1(y)}^{h_2(y)} f(x, y)\, dx\, dy.$$

EXAMPLE 1   Find the volume of the prism whose base is the triangle in the xy-plane bounded by the x-axis and the lines $y = x$ and $x = 1$ and whose top lies in the plane

$$z = f(x, y) = 3 - x - y.$$

Solution   See Figure 15.12. For any x between 0 and 1, y may vary from $y = 0$ to $y = x$ (Figure 15.12b). Hence,

$$V = \int_0^1 \int_0^x (3 - x - y)\, dy\, dx = \int_0^1 \left[3y - xy - \frac{y^2}{2} \right]_{y=0}^{y=x} dx$$

$$= \int_0^1 \left(3x - \frac{3x^2}{2} \right) dx = \left[\frac{3x^2}{2} - \frac{x^3}{2} \right]_{x=0}^{x=1} = 1.$$

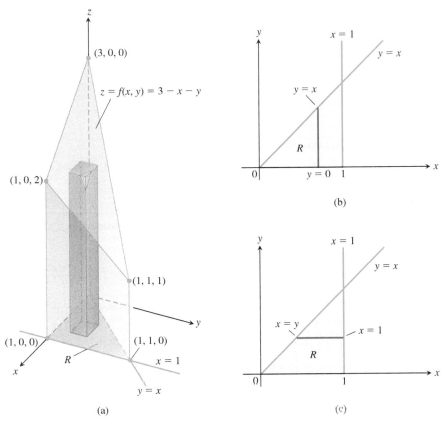

FIGURE 15.12 (a) Prism with a triangular base in the xy-plane. The volume of this prism is defined as a double integral over R. To evaluate it as an iterated integral, we may integrate first with respect to y and then with respect to x, or the other way around (Example 1). (b) Integration limits of

$$\int_{x=0}^{x=1}\int_{y=0}^{y=x} f(x, y) \, dy \, dx.$$

If we integrate first with respect to y, we integrate along a vertical line through R and then integrate from left to right to include all the vertical lines in R. (c) Integration limits of

$$\int_{y=0}^{y=1}\int_{x=y}^{x=1} f(x, y) \, dx \, dy.$$

If we integrate first with respect to x, we integrate along a horizontal line through R and then integrate from bottom to top to include all the horizontal lines in R.

When the order of integration is reversed (Figure 15.12c), the integral for the volume is

$$V = \int_0^1 \int_y^1 (3 - x - y) \, dx \, dy = \int_0^1 \left[3x - \frac{x^2}{2} - xy \right]_{x=y}^{x=1} dy$$

$$= \int_0^1 \left(3 - \frac{1}{2} - y - 3y + \frac{y^2}{2} + y^2 \right) dy$$

$$= \int_0^1 \left(\frac{5}{2} - 4y + \frac{3}{2}y^2 \right) dy = \left[\frac{5}{2}y - 2y^2 + \frac{y^3}{2} \right]_{y=0}^{y=1} = 1.$$

The two integrals are equal, as they should be.

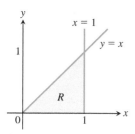

FIGURE 15.13 The region of integration in Example 2.

Although Fubini's Theorem assures us that a double integral may be calculated as an iterated integral in either order of integration, the value of one integral may be easier to find than the value of the other. The next example shows how this can happen.

EXAMPLE 2   Calculate

$$\iint_R \frac{\sin x}{x}\,dA,$$

where R is the triangle in the xy-plane bounded by the x-axis, the line $y = x$, and the line $x = 1$.

Solution   The region of integration is shown in Figure 15.13. If we integrate first with respect to y and next with respect to x, then because x is held fixed in the first integration, we find

$$\int_0^1 \left(\int_0^x \frac{\sin x}{x}\,dy \right) dx = \int_0^1 \left[y\,\frac{\sin x}{x} \right]_{y=0}^{y=x} dx = \int_0^1 \sin x\,dx = -\cos(1) + 1 \approx 0.46.$$

If we reverse the order of integration and attempt to calculate

$$\int_0^1 \int_y^1 \frac{\sin x}{x}\,dx\,dy,$$

we run into a problem because $\int ((\sin x)/x)\,dx$ cannot be expressed in terms of elementary functions (there is no simple antiderivative).

There is no general rule for predicting which order of integration will be the good one in circumstances like these. If the order you first choose doesn't work, try the other. Sometimes neither order will work, and then we may need to use numerical approximations. ■

Finding Limits of Integration

We now give a procedure for finding limits of integration that applies for many regions in the plane. Regions that are more complicated, and for which this procedure fails, can often be split up into pieces on which the procedure works.

Using Vertical Cross-Sections   When faced with evaluating $\iint_R f(x, y)\,dA$, integrating first with respect to y and then with respect to x, do the following three steps:

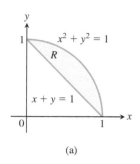

(a)

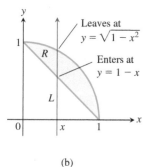

(b)

1. *Sketch.* Sketch the region of integration and label the bounding curves (Figure 15.14a).

2. *Find the y-limits of integration.* Imagine a vertical line L cutting through R in the direction of increasing y. Mark the y-values where L enters and leaves. These are the y-limits of integration and are usually functions of x (instead of constants) (Figure 15.14b).

3. *Find the x-limits of integration.* Choose x-limits that include all the vertical lines through R. The integral shown here (see Figure 15.14c) is

$$\iint_R f(x, y)\,dA = \int_{x=0}^{x=1} \int_{y=1-x}^{y=\sqrt{1-x^2}} f(x, y)\,dy\,dx.$$

Using Horizontal Cross-Sections   To evaluate the same double integral as an iterated integral with the order of integration reversed, use horizontal lines instead of vertical lines in Steps 2 and 3 (see Figure 15.15). The integral is

$$\iint_R f(x, y)\,dA = \int_0^1 \int_{1-y}^{\sqrt{1-y^2}} f(x, y)\,dx\,dy.$$

Smallest x Largest x
is $x = 0$ is $x = 1$

(c)

FIGURE 15.14 Finding the limits of integration when integrating first with respect to y and then with respect to x.

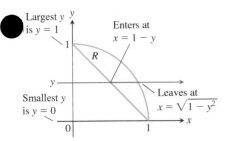

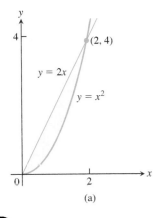

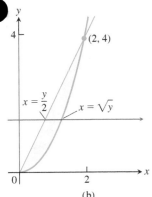

FIGURE 15.15 Finding the limits of integration when integrating first with respect to x and then with respect to y.

FIGURE 15.16 Region of integration for Example 3.

EXAMPLE 3 Sketch the region of integration for the integral

$$\int_0^2 \int_{x^2}^{2x} (4x + 2)\, dy\, dx$$

and write an equivalent integral with the order of integration reversed.

Solution The region of integration is given by the inequalities $x^2 \le y \le 2x$ and $0 \le x \le 2$. It is therefore the region bounded by the curves $y = x^2$ and $y = 2x$ between $x = 0$ and $x = 2$ (Figure 15.16a).

To find limits for integrating in the reverse order, we imagine a horizontal line passing from left to right through the region. It enters at $x = y/2$ and leaves at $x = \sqrt{y}$. To include all such lines, we let y run from $y = 0$ to $y = 4$ (Figure 15.16b). The integral is

$$\int_0^4 \int_{y/2}^{\sqrt{y}} (4x + 2)\, dx\, dy.$$

The common value of these integrals is 8. ∎

Properties of Double Integrals

Like single integrals, double integrals of continuous functions have algebraic properties that are useful in computations and applications.

If $f(x, y)$ and $g(x, y)$ are continuous on the bounded region R, then the following properties hold.

1. *Constant Multiple:* $\displaystyle\iint_R c\,f(x, y)\, dA = c \iint_R f(x, y)\, dA$ (any number c)

2. *Sum and Difference:*

$$\iint_R \big(f(x, y) \pm g(x, y)\big)\, dA = \iint_R f(x, y)\, dA \pm \iint_R g(x, y)\, dA$$

3. Domination:

   (a) $\displaystyle\iint_R f(x, y)\, dA \ge 0$ if $f(x, y) \ge 0$ on R

   (b) $\displaystyle\iint_R f(x, y)\, dA \ge \iint_R g(x, y)\, dA$ if $f(x, y) \ge g(x, y)$ on R

4. *Additivity:* If R is the union of two nonoverlapping regions R_1 and R_2, then

$$\iint_R f(x, y)\, dA = \iint_{R_1} f(x, y)\, dA + \iint_{R_2} f(x, y)\, dA$$

Property 4 assumes that the region of integration R is decomposed into nonoverlapping regions R_1 and R_2 with boundaries consisting of a finite number of line segments or smooth curves. Figure 15.17 illustrates an example of this property.

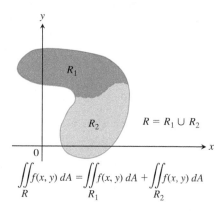

$$\iint_R f(x, y)\, dA = \iint_{R_1} f(x, y)\, dA + \iint_{R_2} f(x, y)\, dA$$

FIGURE 15.17 The Additivity Property for rectangular regions holds for regions bounded by smooth curves.

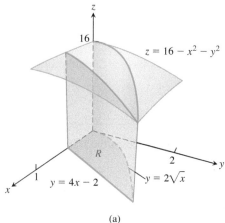

(a)

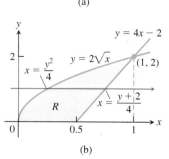

(b)

FIGURE 15.18 (a) The solid "wedge-like" region whose volume is found in Example 4. (b) The region of integration R showing the order $dx\, dy$.

The idea behind these properties is that integrals behave like sums. If the function $f(x, y)$ is replaced by its constant multiple $cf(x, y)$, then a Riemann sum for f,

$$S_n = \sum_{k=1}^{n} f(x_k, y_k)\, \Delta A_k,$$

is replaced by a Riemann sum for cf:

$$\sum_{k=1}^{n} cf(x_k, y_k)\, \Delta A_k = c \sum_{k=1}^{n} f(x_k, y_k)\, \Delta A_k = cS_n.$$

Taking limits as $n \to \infty$ shows that $c \lim_{n\to\infty} S_n = c \iint_R f\, dA$ and $\lim_{n\to\infty} cS_n = \iint_R cf\, dA$ are equal. It follows that the Constant Multiple Property carries over from sums to double integrals.

The other properties are also easy to verify for Riemann sums, and carry over to double integrals for the same reason. While this discussion gives the idea, an actual proof that these properties hold requires a more careful analysis of how Riemann sums converge.

EXAMPLE 4   Find the volume of the wedgelike solid that lies beneath the surface $z = 16 - x^2 - y^2$ and above the region R bounded by the curve $y = 2\sqrt{x}$, the line $y = 4x - 2$, and the x-axis.

Solution   Figure 15.18a shows the surface and the "wedgelike" solid whose volume we want to calculate. Figure 15.18b shows the region of integration in the xy-plane. If we integrate in the order $dy\, dx$ (first with respect to y and then with respect to x), two integrations will be required because y varies from $y = 0$ to $y = 2\sqrt{x}$ for $0 \le x \le 0.5$, and then varies from $y = 4x - 2$ to $y = 2\sqrt{x}$ for $0.5 \le x \le 1$. So we choose to integrate in the order $dx\, dy$, which requires only one double integral whose limits of integration are indicated in Figure 15.18b. The volume is then calculated as the iterated integral:

$$\iint_R (16 - x^2 - y^2)\, dA$$

$$= \int_0^2 \int_{y^2/4}^{(y+2)/4} (16 - x^2 - y^2)\, dx\, dy$$

$$= \int_0^2 \left[16x - \frac{x^3}{3} - xy^2 \right]_{x=y^2/4}^{x=(y+2)/4} dx$$

$$= \int_0^2 \left[4(y+2) - \frac{(y+2)^3}{3 \cdot 64} - \frac{(y+2)y^2}{4} - 4y^2 + \frac{y^6}{3 \cdot 64} + \frac{y^4}{4} \right] dy$$

$$= \left[\frac{191y}{24} + \frac{63y^2}{32} - \frac{145y^3}{96} - \frac{49y^4}{768} + \frac{y^5}{20} + \frac{y^7}{1344} \right]_0^2 = \frac{20803}{1680} \approx 12.4. \quad \blacksquare$$

Our development of the double integral has focused on its representation of the volume of the solid region between R and the surface $z = f(x, y)$ of a positive continuous function. Just as we saw with signed area in the case of single integrals, when $f(x_k, y_k)$ is negative, the product $f(x_k, y_k)\, \Delta A_k$ is the negative of the volume of the rectangular box shown in Figure 15.9 that was used to form the approximating Riemann sum. So for an arbitrary continuous function f defined over R, the limit of any Riemann sum represents the *signed* volume (not the total volume) of the solid region between R and the surface. The double integral has other interpretations as well, and in the next section we will see how it is used to calculate the area of a general region in the plane.

EXERCISES 15.2

Sketching Regions of Integration

In Exercises 1–8, sketch the described regions of integration.

1. $0 \le x \le 3$, $\quad 0 \le y \le 2x$
2. $-1 \le x \le 2$, $\quad x - 1 \le y \le x^2$
3. $-2 \le y \le 2$, $\quad y^2 \le x \le 4$
4. $0 \le y \le 1$, $\quad y \le x \le 2y$
5. $0 \le x \le 1$, $\quad e^x \le y \le e$
6. $1 \le x \le e^2$, $\quad 0 \le y \le \ln x$
7. $0 \le y \le 1$, $\quad 0 \le x \le \sin^{-1} y$
8. $0 \le y \le 8$, $\quad \frac{1}{4}y \le x \le y^{1/3}$

Finding Limits of Integration

In Exercises 9–18, write an iterated integral for $\iint_R dA$ over the described region R using (a) vertical cross-sections, (b) horizontal cross-sections.

9.

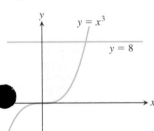

10.

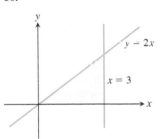

11.

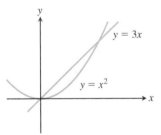

12.

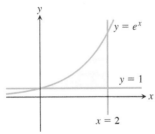

13. Bounded by $y = \sqrt{x}$, $y = 0$, and $x = 9$
14. Bounded by $y = \tan x$, $x = 0$, and $y = 1$
15. Bounded by $y = e^{-x}$, $y = 1$, and $x = \ln 3$
16. Bounded by $y = 0$, $x = 0$, $y = 1$, and $y = \ln x$
17. Bounded by $y = 3 - 2x$, $y = x$, and $x = 0$
18. Bounded by $y = x^2$ and $y = x + 2$

Finding Regions of Integration and Double Integrals

In Exercises 19–24, sketch the region of integration and evaluate the integral.

19. $\displaystyle\int_0^\pi \int_0^x x \sin y \, dy \, dx$ **20.** $\displaystyle\int_0^\pi \int_0^{\sin x} y \, dy \, dx$

21. $\displaystyle\int_1^{\ln 8} \int_0^{\ln y} e^{x+y} \, dx \, dy$ **22.** $\displaystyle\int_1^2 \int_y^{y^2} dx \, dy$

23. $\displaystyle\int_0^1 \int_0^{y^2} 3y^3 e^{xy} \, dx \, dy$ **24.** $\displaystyle\int_1^4 \int_0^{\sqrt{x}} \frac{3}{2} e^{y/\sqrt{x}} \, dy \, dx$

In Exercises 25–28, integrate f over the given region.

25. **Quadrilateral** $f(x, y) = x/y$ over the region in the first quadrant bounded by the lines $y = x$, $y = 2x$, $x = 1$, and $x = 2$
26. **Triangle** $f(x, y) = x^2 + y^2$ over the triangular region with vertices $(0, 0)$, $(1, 0)$, and $(0, 1)$
27. **Triangle** $f(u, v) = v - \sqrt{u}$ over the triangular region cut from the first quadrant of the uv-plane by the line $u + v = 1$
28. **Curved region** $f(s, t) = e^s \ln t$ over the region in the first quadrant of the st-plane that lies above the curve $s = \ln t$ from $t = 1$ to $t = 2$

Each of Exercises 29–32 gives an integral over a region in a Cartesian coordinate plane. Sketch the region and evaluate the integral.

29. $\displaystyle\int_{-2}^0 \int_v^{-v} 2 \, dp \, dv$ (the pv-plane)

30. $\displaystyle\int_0^1 \int_0^{\sqrt{1-s^2}} 8t \, dt \, ds$ (the st-plane)

31. $\displaystyle\int_{-\pi/3}^{\pi/3} \int_0^{\sec t} 3 \cos t \, du \, dt$ (the tu-plane)

32. $\displaystyle\int_0^{3/2} \int_1^{4-2u} \frac{4 - 2u}{v^2} \, dv \, du$ (the uv-plane)

Reversing the Order of Integration

In Exercises 33–46, sketch the region of integration and write an equivalent double integral with the order of integration reversed.

33. $\displaystyle\int_0^1 \int_2^{4-2x} dy \, dx$ **34.** $\displaystyle\int_0^2 \int_{y-2}^0 dx \, dy$

35. $\displaystyle\int_0^1 \int_y^{\sqrt{y}} dx \, dy$ **36.** $\displaystyle\int_0^1 \int_{1-x}^{1-x^2} dy \, dx$

37. $\displaystyle\int_0^1 \int_1^{e^x} dy \, dx$ **38.** $\displaystyle\int_0^{\ln 2} \int_{e^y}^2 dx \, dy$

39. $\displaystyle\int_0^{3/2} \int_0^{9-4x^2} 16x \, dy \, dx$ **40.** $\displaystyle\int_0^2 \int_0^{4-y^2} y \, dx \, dy$

41. $\displaystyle\int_0^1 \int_{-\sqrt{1-y^2}}^{\sqrt{1-y^2}} 3y \, dx \, dy$ **42.** $\displaystyle\int_0^2 \int_{-\sqrt{4-x^2}}^{\sqrt{4-x^2}} 6x \, dy \, dx$

43. $\displaystyle\int_1^e \int_0^{\ln x} xy \, dy \, dx$ **44.** $\displaystyle\int_0^{\pi/6} \int_{\sin x}^{1/2} xy^2 \, dy \, dx$

45. $\displaystyle\int_0^3 \int_1^{e^y} (x + y) \, dx \, dy$ **46.** $\displaystyle\int_0^{\sqrt{3}} \int_0^{\tan^{-1} y} \sqrt{xy} \, dx \, dy$

In Exercises 47–56, sketch the region of integration, reverse the order of integration, and evaluate the integral.

47. $\displaystyle\int_0^\pi\int_x^\pi \frac{\sin y}{y}\,dy\,dx$

48. $\displaystyle\int_0^2\int_x^2 2y^2\sin xy\,dy\,dx$

49. $\displaystyle\int_0^1\int_y^1 x^2 e^{xy}\,dx\,dy$

50. $\displaystyle\int_0^2\int_0^{4-x^2}\frac{xe^{2y}}{4-y}\,dy\,dx$

51. $\displaystyle\int_0^{2\sqrt{\ln 3}}\int_{y/2}^{\sqrt{\ln 3}} e^{x^2}\,dx\,dy$

52. $\displaystyle\int_0^3\int_{\sqrt{x/3}}^1 e^{y^3}\,dy\,dx$

53. $\displaystyle\int_0^{1/16}\int_{y^{1/4}}^{1/2}\cos(16\pi x^5)\,dx\,dy$

54. $\displaystyle\int_0^8\int_{\sqrt[3]{x}}^2 \frac{dy\,dx}{y^4+1}$

55. Square region $\iint_R (y-2x^2)\,dA$ where R is the region bounded by the square $|x|+|y|=1$

56. Triangular region $\iint_R xy\,dA$ where R is the region bounded by the lines $y=x$, $y=2x$, and $x+y=2$

Volume Beneath a Surface $z=f(x,y)$

57. Find the volume of the region bounded above by the paraboloid $z=x^2+y^2$ and below by the triangle enclosed by the lines $y=x$, $x=0$, and $x+y=2$ in the xy-plane.

58. Find the volume of the solid that is bounded above by the cylinder $z=x^2$ and below by the region enclosed by the parabola $y=2-x^2$ and the line $y=x$ in the xy-plane.

59. Find the volume of the solid whose base is the region in the xy-plane that is bounded by the parabola $y=4-x^2$ and the line $y=3x$, while the top of the solid is bounded by the plane $z=x+4$.

60. Find the volume of the solid in the first octant bounded by the coordinate planes, the cylinder $x^2+y^2=4$, and the plane $z+y=3$.

61. Find the volume of the solid in the first octant bounded by the coordinate planes, the plane $x=3$, and the parabolic cylinder $z=4-y^2$.

62. Find the volume of the solid cut from the first octant by the surface $z=4-x^2-y$.

63. Find the volume of the wedge cut from the first octant by the cylinder $z=12-3y^2$ and the plane $x+y=2$.

64. Find the volume of the solid cut from the square column $|x|+|y|\le1$ by the planes $z=0$ and $3x+z=3$.

65. Find the volume of the solid that is bounded on the front and back by the planes $x=2$ and $x=1$, on the sides by the cylinders $y=\pm1/x$, and above and below by the planes $z=x+1$ and $z=0$.

66. Find the volume of the solid bounded on the front and back by the planes $x=\pm\pi/3$, on the sides by the cylinders $y=\pm\sec x$, above by the cylinder $z=1+y^2$, and below by the xy-plane.

In Exercises 67 and 68, sketch the region of integration and the solid whose volume is given by the double integral.

67. $\displaystyle\int_0^3\int_0^{2-2x/3}\left(1-\frac13 x-\frac12 y\right)dy\,dx$

68. $\displaystyle\int_0^4\int_{-\sqrt{16-y^2}}^{\sqrt{16-y^2}}\sqrt{25-x^2-y^2}\,dx\,dy$

Integrals over Unbounded Regions

Improper double integrals can often be computed similarly to improper integrals of one variable. The first iteration of the following improper integrals is conducted just as if they were proper integrals. One then evaluates an improper integral of a single variable by taking appropriate limits, as in Section 8.8. Evaluate the improper integrals in Exercises 69–72 as iterated integrals.

69. $\displaystyle\int_1^\infty\int_{e^{-x}}^1 \frac{1}{x^3 y}\,dy\,dx$

70. $\displaystyle\int_{-1}^1\int_{-1/\sqrt{1-x^2}}^{1/\sqrt{1-x^2}}(2y+1)\,dy\,dx$

71. $\displaystyle\int_{-\infty}^\infty\int_{-\infty}^\infty \frac{1}{(x^2+1)(y^2+1)}\,dx\,dy$

72. $\displaystyle\int_0^\infty\int_0^\infty xe^{-(x+2y)}\,dx\,dy$

Approximating Integrals with Finite Sums

In Exercises 73 and 74, approximate the double integral of $f(x,y)$ over the region R partitioned by the given vertical lines $x=a$ and horizontal lines $y=c$. In each subrectangle, use (x_k,y_k) as indicated for your approximation.

$$\iint_R f(x,y)\,dA \approx \sum_{k=1}^n f(x_k,y_k)\,\Delta A_k$$

73. $f(x,y)=x+y$ over the region R bounded above by the semicircle $y=\sqrt{1-x^2}$ and below by the x-axis, using the partition $x=-1,-1/2,0,1/4,1/2,1$ and $y=0,1/2,1$ with (x_k,y_k) the lower left corner in the kth subrectangle (provided the subrectangle lies within R)

74. $f(x,y)=x+2y$ over the region R inside the circle $(x-2)^2+(y-3)^2=1$ using the partition $x=1,3/2,2,5/2,3$ and $y=2,5/2,3,7/2,4$ with (x_k,y_k) the center (centroid) in the kth subrectangle (provided the subrectangle lies within R)

Theory and Examples

75. Circular sector Integrate $f(x,y)=\sqrt{4-x^2}$ over the smaller sector cut from the disk $x^2+y^2\le4$ by the rays $\theta=\pi/6$ and $\theta=\pi/2$.

76. Unbounded region Integrate $f(x,y)=1/[(x^2-x)(y-1)^{2/3}]$ over the infinite rectangle $2\le x<\infty,0\le y\le2$.

77. Noncircular cylinder A solid right (noncircular) cylinder has its base R in the xy-plane and is bounded above by the paraboloid $z=x^2+y^2$. The cylinder's volume is

$$V=\int_0^1\int_0^y(x^2+y^2)\,dx\,dy+\int_1^2\int_0^{2-y}(x^2+y^2)\,dx\,dy.$$

Sketch the base region R and express the cylinder's volume as a single iterated integral with the order of integration reversed. Then evaluate the integral to find the volume.

78. Converting to a double integral   Evaluate the integral

$$\int_0^2 (\tan^{-1} \pi x - \tan^{-1} x) \, dx.$$

(*Hint:* Write the integrand as an integral.)

79. Maximizing a double integral   What region R in the xy-plane maximizes the value of

$$\iint_R (4 - x^2 - 2y^2) \, dA?$$

Give reasons for your answer.

80. Minimizing a double integral   What region R in the xy-plane minimizes the value of

$$\iint_R (x^2 + y^2 - 9) \, dA?$$

Give reasons for your answer.

81. Is it possible to evaluate the integral of a continuous function $f(x, y)$ over a rectangular region in the xy-plane and get different answers depending on the order of integration? Give reasons for your answer.

82. How would you evaluate the double integral of a continuous function $f(x, y)$ over the region R in the xy-plane enclosed by the triangle with vertices $(0, 1)$, $(2, 0)$, and $(1, 2)$? Give reasons for your answer.

83. Unbounded region   Prove that

$$\int_{-\infty}^{\infty} \int_{-\infty}^{\infty} e^{-x^2-y^2} \, dx \, dy = \lim_{b \to \infty} \int_{-b}^{b} \int_{-b}^{b} e^{-x^2-y^2} \, dx \, dy$$

$$= 4 \left(\int_0^{\infty} e^{-x^2} \, dx \right)^2.$$

84. Improper double integral   Evaluate the improper integral

$$\int_0^1 \int_0^3 \frac{x^2}{(y-1)^{2/3}} \, dy \, dx.$$

COMPUTER EXPLORATIONS

Use a CAS double-integral evaluator to estimate the values of the integrals in Exercises 85–88.

85. $\displaystyle\int_1^3 \int_1^x \frac{1}{xy} \, dy \, dx$ **86.** $\displaystyle\int_0^1 \int_0^1 e^{-(x^2+y^2)} \, dy \, dx$

87. $\displaystyle\int_0^1 \int_0^1 \tan^{-1} xy \, dy \, dx$

88. $\displaystyle\int_{-1}^1 \int_0^{\sqrt{1-x^2}} 3\sqrt{1 - x^2 - y^2} \, dy \, dx$

Use a CAS double-integral evaluator to find the integrals in Exercises 89–94. Then reverse the order of integration and evaluate, again with a CAS.

89. $\displaystyle\int_0^1 \int_{2y}^4 e^{x^2} \, dx \, dy$

90. $\displaystyle\int_0^3 \int_{x^2}^9 x \cos (y^2) \, dy \, dx$

91. $\displaystyle\int_0^2 \int_{y^3}^{4\sqrt{2y}} (x^2 y - xy^2) \, dx \, dy$

92. $\displaystyle\int_0^2 \int_0^{4-y^2} e^{xy} \, dx \, dy$

93. $\displaystyle\int_1^2 \int_0^{x^2} \frac{1}{x + y} \, dy \, dx$ **94.** $\displaystyle\int_1^2 \int_{y^3}^8 \frac{1}{\sqrt{x^2 + y^2}} \, dx \, dy$

15.3   Area by Double Integration

In this section we show how to use double integrals to calculate the areas of bounded regions in the plane, and to find the average value of a function of two variables.

Areas of Bounded Regions in the Plane

If we take $f(x, y) = 1$ in the definition of the double integral over a region R in the preceding section, the Riemann sums reduce to

$$S_n = \sum_{k=1}^n f(x_k, y_k) \, \Delta A_k = \sum_{k=1}^n \Delta A_k. \tag{1}$$

This is simply the sum of the areas of the small rectangles in the partition of R, and approximates what we would like to call the area of R. As the norm of a partition of R approaches zero, the height and width of all rectangles in the partition approach zero, and the coverage of R becomes increasingly complete (Figure 15.8). We define the area of R to be the limit

$$\lim_{\|P\| \to 0} \sum_{k=1}^n \Delta A_k = \iint_R dA. \tag{2}$$

> DEFINITION   The **area** of a closed, bounded plane region R is
> $$A = \iint_R dA.$$

As with the other definitions in this chapter, the definition here applies to a greater variety of regions than does the earlier single-variable definition of area, but it agrees with the earlier definition on regions to which they both apply. To evaluate the integral in the definition of area, we integrate the constant function $f(x, y) = 1$ over R.

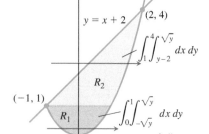

FIGURE 15.19   The region in Example 1.

EXAMPLE 1   Find the area of the region R bounded by $y = x$ and $y = x^2$ in the first quadrant.

Solution   We sketch the region (Figure 15.19), noting where the two curves intersect at the origin and $(1, 1)$, and calculate the area as

$$A = \int_0^1 \int_{x^2}^x dy\, dx = \int_0^1 \Big[y \Big]_{x^2}^x dx = \int_0^1 (x - x^2)\, dx = \left[\frac{x^2}{2} - \frac{x^3}{3} \right]_0^1 = \frac{1}{6}.$$

Notice that the single-variable integral $\int_0^1 (x - x^2)\, dx$, obtained from evaluating the inside iterated integral, is the integral for the area between these two curves using the method of Section 5.6.

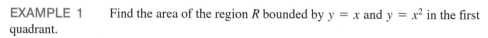

EXAMPLE 2   Find the area of the region R enclosed by the parabola $y = x^2$ and the line $y = x + 2$.

Solution   If we divide R into the regions R_1 and R_2 shown in Figure 15.20a, we may calculate the area as

$$A = \iint_{R_1} dA + \iint_{R_2} dA = \int_0^1 \int_{-\sqrt{y}}^{\sqrt{y}} dx\, dy + \int_1^4 \int_{y-2}^{\sqrt{y}} dx\, dy.$$

On the other hand, reversing the order of integration (Figure 15.20b) gives

$$A = \int_{-1}^2 \int_{x^2}^{x+2} dy\, dx.$$

This second result, which requires only one integral, is simpler to evaluate, giving

$$A = \int_{-1}^2 \Big[y \Big]_{x^2}^{x+2} dx = \int_{-1}^2 (x + 2 - x^2)\, dx = \left[\frac{x^2}{2} + 2x - \frac{x^3}{3} \right]_{-1}^2 = \frac{9}{2}. \quad \blacksquare$$

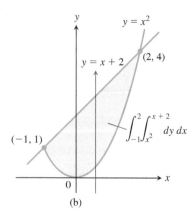

FIGURE 15.20   Calculating this area takes (a) two double integrals if the first integration is with respect to x, but (b) only one if the first integration is with respect to y (Example 2).

EXAMPLE 3   Find the area of the playing field described by $R: -2 \le x \le 2, -1 - \sqrt{4 - x^2} \le y \le 1 + \sqrt{4 - x^2}$, using

(a) Fubini's Theorem   **(b)** Simple geometry.

Solution   The region R is shown in Figure 15.21a.

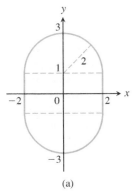

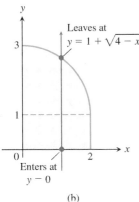

FIGURE 15.21 (a) The playing field described by the region R in Example 3. (b) First quadrant of the playing field.

(a) From the symmetries observed in the figure, we see that the area of R is 4 times its area in the first quadrant. As shown in Figure 15.21b, a vertical line at x enters this part of the region at $y = 0$ and exits at $y = 1 + \sqrt{4 - x^2}$. Therefore, using Fubini's Theorem, we have

$$A = \iint\limits_R dA = 4 \int_0^2 \int_0^{1+\sqrt{4-x^2}} dy \, dx$$

$$= 4 \int_0^2 \left(1 + \sqrt{4 - x^2} \right) dx$$

$$= 4 \left[x + \frac{x}{2}\sqrt{4 - x^2} + \frac{4}{2} \sin^{-1} \frac{x}{2} \right]_0^2 \qquad \text{Integral Table Formula 45}$$

$$= 4 \left(2 + 0 + 2 \cdot \frac{\pi}{2} - 0 \right) = 8 + 4\pi.$$

(b) The region R consists of a rectangle mounted on two sides by half disks of radius 2. The area can be computed by summing the area of the 4×2 rectangle and the area of a circle of radius 2, so

$$A = 8 + \pi 2^2 = 8 + 4\pi. \qquad \blacksquare$$

Average Value

The average value of an integrable function of one variable on a closed interval is the integral of the function over the interval divided by the length of the interval. For an integrable function of two variables defined on a bounded region in the plane, the average value is the integral over the region divided by the area of the region. This can be visualized by thinking of the region as being the base of a tank with vertical walls around the boundary of the region, and imagining that the tank is filled with water that is sloshing around. The value $f(x, y)$ is then the height of the water that is directly above the point (x, y). The average height of the water in the tank can be found by letting the water settle down to a constant height. This height is equal to the volume of water in the tank divided by the area of R. We therefore define the average value of an integrable function f over a region R as follows:

$$\textbf{Average value of } f \text{ over } R = \frac{1}{\text{area of } R} \iint\limits_R f \, dA. \qquad (3)$$

If f is the temperature of a thin plate covering R, then the double integral of f over R divided by the area of R is the plate's average temperature. If $f(x, y)$ is the distance from the point (x, y) to a fixed point P, then the average value of f over R is the average distance of points in R from P.

EXAMPLE 4 Find the average value of $f(x, y) = x \cos xy$ over the rectangle $R: 0 \le x \le \pi, 0 \le y \le 1$.

Solution The value of the integral of f over R is

$$\int_0^\pi \int_0^1 x \cos xy \, dy \, dx = \int_0^\pi \left[\sin xy \right]_{y=0}^{y=1} dx \qquad \int x \cos xy \, dy = \sin xy + C$$

$$= \int_0^\pi (\sin x - 0) \, dx = -\cos x \Big]_0^\pi = 1 + 1 = 2.$$

The area of R is π. The average value of f over R is $2/\pi$. $\blacksquare$

EXERCISES 15.3

Area by Double Integrals

In Exercises 1–12, sketch the region bounded by the given lines and curves. Then express the region's area as an iterated double integral and evaluate the integral.

1. The coordinate axes and the line $x + y = 2$
2. The lines $x = 0, y = 2x$, and $y = 4$
3. The parabola $x = -y^2$ and the line $y = x + 2$
4. The parabola $x = y - y^2$ and the line $y = -x$
5. The curve $y = e^x$ and the lines $y = 0, x = 0$, and $x = \ln 2$
6. The curves $y = \ln x$ and $y = 2 \ln x$ and the line $x = e$, in the first quadrant
7. The parabolas $x = y^2$ and $x = 2y - y^2$
8. The parabolas $x = y^2 - 1$ and $x = 2y^2 - 2$
9. The lines $y = x, y = x/3$, and $y = 2$
10. The lines $y = 1 - x$ and $y = 2$ and the curve $y = e^x$
11. The lines $y = 2x, y = x/2$, and $y = 3 - x$
12. The lines $y = x - 2$ and $y = -x$ and the curve $y = \sqrt{x}$

Identifying the Region of Integration

The integrals and sums of integrals in Exercises 13–18 give the areas of regions in the xy-plane. Sketch each region, label each bounding curve with its equation, and give the coordinates of the points where the curves intersect. Then find the area of the region.

13. $\int_0^6 \int_{y^2/3}^{2y} dx\, dy$

14. $\int_0^3 \int_{-x}^{x(2-x)} dy\, dx$

15. $\int_0^{\pi/4} \int_{\sin x}^{\cos x} dy\, dx$

16. $\int_{-1}^2 \int_{y^2}^{y+2} dx\, dy$

17. $\int_{-1}^0 \int_{-2x}^{1-x} dy\, dx + \int_0^2 \int_{-x/2}^{1-x} dy\, dx$

18. $\int_0^2 \int_{x^2-4}^0 dy\, dx + \int_0^4 \int_0^{\sqrt{x}} dy\, dx$

Finding Average Values

19. Find the average value of $f(x, y) = \sin(x + y)$ over
    a. the rectangle $0 \le x \le \pi, \ 0 \le y \le \pi$.
    b. the rectangle $0 \le x \le \pi, \ 0 \le y \le \pi/2$.

20. Which do you think will be larger, the average value of $f(x, y) = xy$ over the square $0 \le x \le 1, 0 \le y \le 1$, or the average value of f over the quarter circle $x^2 + y^2 \le 1$ in the first quadrant? Calculate them to find out.

21. Find the average height of the paraboloid $z = x^2 + y^2$ over the square $0 \le x \le 2, 0 \le y \le 2$.

22. Find the average value of $f(x, y) = 1/(xy)$ over the square $\ln 2 \le x \le 2 \ln 2, \ln 2 \le y \le 2 \ln 2$.

Theory and Examples

23. **Geometric area** Find the area of the region
    $$R: 0 \le x \le 2, \ 2 - x \le y \le \sqrt{4 - x^2},$$
    using (a) Fubini's Theorem, (b) simple geometry.

24. **Geometric area** Find the area of the circular washer with outer radius 2 and inner radius 1, using (a) Fubini's Theorem, (b) simple geometry.

25. **Bacterium population** If $f(x, y) = (10{,}000\,e^y)/(1 + |x|/2)$ represents the "population density" of a certain bacterium on the xy-plane, where x and y are measured in centimeters, find the total population of bacteria within the rectangle $-5 \le x \le 5$ and $-2 \le y \le 0$.

26. **Regional population** If $f(x, y) = 100\,(y + 1)$ represents the population density of a planar region on Earth, where x and y are measured in miles, find the number of people in the region bounded by the curves $x = y^2$ and $x = 2y - y^2$.

27. **Average temperature in Texas** According to the *Texas Almanac*, Texas has 254 counties and a National Weather Service station in each county. Assume that at time t_0, each of the 2 weather stations recorded the local temperature. Find a formula that would give a reasonable approximation of the average temperature in Texas at time t_0. Your answer should involve information that you would expect to be readily available in the *Texas Almanac*.

28. If $y = f(x)$ is a nonnegative continuous function over the closed interval $a \le x \le b$, show that the double integral definition of area for the closed plane region bounded by the graph of f, the vertical lines $x = a$ and $x = b$, and the x-axis agrees with the definition for area beneath the curve in Section 5.3.

29. Suppose $f(x, y)$ is continuous over a region R in the plane and that the area $A(R)$ of the region is defined. If there are constants m and M such that $m \le f(x, y) \le M$ for all $(x, y) \in R$, prove that
    $$mA(R) \le \iint_R f(x, y)\, dA \le MA(R).$$

30. Suppose $f(x, y)$ is continuous and nonnegative over a region R in the plane with a defined area $A(R)$. If $\iint_R f(x, y)\, dA = 0$, prove that $f(x, y) = 0$ at every point $(x, y) \in R$.

15.4 Double Integrals in Polar Form

Double integrals are sometimes easier to evaluate if we change to polar coordinates. This section shows how to accomplish the change and how to evaluate double integrals over regions whose boundaries are given by polar equations.

Integrals in Polar Coordinates

When we defined the double integral of a function over a region R in the xy-plane, we began by cutting R into rectangles whose sides were parallel to the coordinate axes. These were the natural shapes to use because their sides have either constant x-values or constant y-values. In polar coordinates, the natural shape is a "polar rectangle" whose sides have constant r- and θ-values. To avoid ambiguities when describing the region of integration with polar coordinates, we use polar coordinate points (r, θ) where $r \geq 0$.

Suppose that a function $f(r, \theta)$ is defined over a region R that is bounded by the rays $\theta = \alpha$ and $\theta = \beta$ and by the continuous curves $r = g_1(\theta)$ and $r = g_2(\theta)$. Suppose also that $0 \leq g_1(\theta) \leq g_2(\theta) \leq a$ for every value of θ between α and β. Then R lies in a fan-shaped region Q defined by the inequalities $0 \leq r \leq a$ and $\alpha \leq \theta \leq \beta$, where $0 \leq \beta - \alpha \leq 2\pi$. See Figure 15.22.

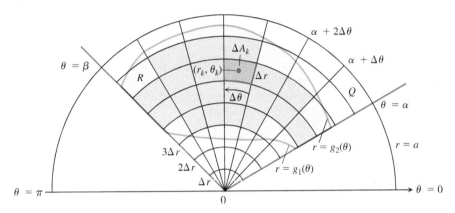

FIGURE 15.22 The region R: $g_1(\theta) \leq r \leq g_2(\theta)$, $\alpha \leq \theta \leq \beta$, is contained in the fan-shaped region Q: $0 \leq r \leq a$, $\alpha \leq \theta \leq \beta$, where $0 \leq \beta - \alpha \leq 2\pi$. The partition of Q by circular arcs and rays induces a partition of R.

We cover Q by a grid of circular arcs and rays. The arcs are cut from circles centered at the origin, with radii $\Delta r, 2\Delta r, \ldots, m\Delta r$, where $\Delta r = a/m$. The rays are given by

$$\theta = \alpha, \qquad \theta = \alpha + \Delta\theta, \qquad \theta = \alpha + 2\Delta\theta, \qquad \ldots, \qquad \theta = \alpha + m'\Delta\theta = \beta,$$

where $\Delta\theta = (\beta - \alpha)/m'$. The arcs and rays partition Q into small patches called "polar rectangles."

We number the polar rectangles that lie inside R (the order does not matter), calling their areas $\Delta A_1, \Delta A_2, \ldots, \Delta A_n$. We let (r_k, θ_k) be any point in the polar rectangle whose area is ΔA_k. We then form the sum

$$S_n = \sum_{k=1}^{n} f(r_k, \theta_k)\, \Delta A_k.$$

If f is continuous throughout R, this sum will approach a limit as we refine the grid to make Δr and $\Delta\theta$ go to zero. The limit is called the double integral of f over R. In symbols,

$$\lim_{n \to \infty} S_n = \iint\limits_{R} f(r, \theta)\, dA.$$

To evaluate this limit, we first have to write the sum S_n in a way that expresses ΔA_k in terms of Δr and $\Delta\theta$. For convenience we choose r_k to be the average of the radii of the inner and outer arcs bounding the kth polar rectangle ΔA_k. The radius of the inner arc bounding ΔA_k is then $r_k - (\Delta r/2)$ (Figure 15.23). The radius of the outer arc is $r_k + (\Delta r/2)$.

The area of a wedge-shaped sector of a circle having radius r and angle θ is

$$A = \frac{1}{2}\theta \cdot r^2,$$

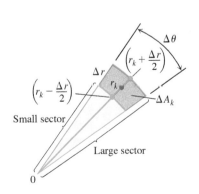

FIGURE 15.23 The observation that

$$\Delta A_k = \begin{pmatrix} \text{area of} \\ \text{large sector} \end{pmatrix} - \begin{pmatrix} \text{area of} \\ \text{small sector} \end{pmatrix}$$

leads to the formula $\Delta A_k = r_k\, \Delta r\, \Delta\theta$.

as can be seen by multiplying πr^2, the area of the circle, by $\theta/2\pi$, the fraction of the circle's area contained in the wedge. So the areas of the circular sectors subtended by these arcs at the origin are

$$\text{Inner radius:} \quad \frac{1}{2}\left(r_k - \frac{\Delta r}{2}\right)^2 \Delta\theta$$

$$\text{Outer radius:} \quad \frac{1}{2}\left(r_k + \frac{\Delta r}{2}\right)^2 \Delta\theta.$$

Therefore,

$$\Delta A_k = \text{area of large sector} - \text{area of small sector}$$

$$= \frac{\Delta\theta}{2}\left[\left(r_k + \frac{\Delta r}{2}\right)^2 - \left(r_k - \frac{\Delta r}{2}\right)^2\right] = \frac{\Delta\theta}{2}(2r_k\,\Delta r) = r_k\,\Delta r\,\Delta\theta.$$

Combining this result with the sum defining S_n gives

$$S_n = \sum_{k=1}^{n} f(r_k, \theta_k)\, r_k\, \Delta r\, \Delta\theta.$$

As $n \to \infty$ and the values of Δr and $\Delta\theta$ approach zero, these sums converge to the double integral

$$\lim_{n\to\infty} S_n = \iint_R f(r, \theta)\, r\, dr\, d\theta.$$

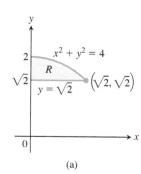

(a)

A version of Fubini's Theorem says that the limit approached by these sums can be evaluated by repeated single integrations with respect to r and θ as

$$\iint_R f(r, \theta)\, dA = \int_{\theta=\alpha}^{\theta=\beta} \int_{r=g_1(\theta)}^{r=g_2(\theta)} f(r, \theta)\, r\, dr\, d\theta.$$

Finding Limits of Integration

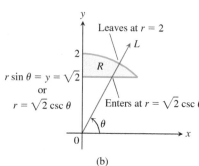

(b)

The procedure for finding limits of integration in rectangular coordinates also works for polar coordinates. We illustrate this using the region R shown in Figure 15.24. To evaluate $\iint_R f(r, \theta)\, dA$ in polar coordinates, integrating first with respect to r and then with respect to θ, take the following steps.

1. *Sketch.* Sketch the region and label the bounding curves (Figure 15.24a).
2. *Find the r-limits of integration.* Imagine a ray L from the origin cutting through R in the direction of increasing r. Mark the r-values where L enters and leaves R. These are the r-limits of integration. They usually depend on the angle θ that L makes with the positive x-axis (Figure 15.24b).
3. *Find the θ-limits of integration.* Find the smallest and largest θ-values that bound R. These are the θ-limits of integration (Figure 15.24c). The polar iterated integral is

$$\iint_R f(r, \theta)\, dA = \int_{\theta=\pi/4}^{\theta=\pi/2} \int_{r=\sqrt{2}\csc\theta}^{r=2} f(r, \theta)\, r\, dr\, d\theta.$$

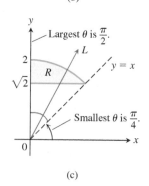

(c)

FIGURE 15.24 Finding the limits of integration in polar coordinates.

EXAMPLE 1   Find the limits of integration for integrating $f(r, \theta)$ over the region R that lies inside the cardioid $r = 1 + \cos\theta$ and outside the circle $r = 1$.

Solution

1. We first sketch the region and label the bounding curves (Figure 15.25).
2. Next we find the *r-limits of integration.* A typical ray from the origin enters R where $r = 1$ and leaves where $r = 1 + \cos\theta$.

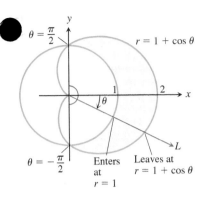

FIGURE 15.25 Finding the limits of integration in polar coordinates for the region in Example 1.

Area Differential in Polar Coordinates

$$dA = r \, dr \, d\theta$$

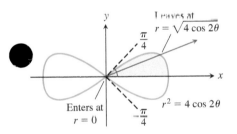

FIGURE 15.26 To integrate over the shaded region, we run r from 0 to $\sqrt{4 \cos 2\theta}$ and θ from 0 to $\pi/4$ (Example 2).

3. Finally we find the *θ-limits of integration*. The rays from the origin that intersect R run from $\theta = -\pi/2$ to $\theta = \pi/2$. The integral is

$$\int_{-\pi/2}^{\pi/2} \int_{1}^{1+\cos\theta} f(r, \theta) \, r \, dr \, d\theta.$$ ◼

If $f(r, \theta)$ is the constant function whose value is 1, then the integral of f over R is the area of R.

Area in Polar Coordinates

The area of a closed and bounded region R in the polar coordinate plane is

$$A = \iint_R r \, dr \, d\theta.$$

This formula for area is consistent with all earlier formulas.

EXAMPLE 2 Find the area enclosed by the lemniscate $r^2 = 4 \cos 2\theta$.

Solution We graph the lemniscate to determine the limits of integration (Figure 15.26) and see from the symmetry of the region that the total area is 4 times the first-quadrant portion.

$$A = 4 \int_0^{\pi/4} \int_0^{\sqrt{4\cos 2\theta}} r \, dr \, d\theta = 4 \int_0^{\pi/4} \left[\frac{r^2}{2} \right]_{r=0}^{r=\sqrt{4\cos 2\theta}} d\theta$$

$$= 4 \int_0^{\pi/4} 2 \cos 2\theta \, d\theta = 4 \sin 2\theta \Big]_0^{\pi/4} = 4.$$ ◼

Changing Cartesian Integrals into Polar Integrals

The procedure for changing a Cartesian integral $\iint_R f(x, y) \, dx \, dy$ into a polar integral has two steps. First substitute $x = r \cos\theta$ and $y = r \sin\theta$, and replace $dx \, dy$ by $r \, dr \, d\theta$ in the Cartesian integral. Then supply polar limits of integration for the boundary of R. The Cartesian integral then becomes

$$\iint_R f(x, y) \, dx \, dy = \iint_G f(r \cos\theta, r \sin\theta) \, r \, dr \, d\theta,$$

where G denotes the same region of integration, but now described in polar coordinates. This is like the substitution method in Chapter 5 except that there are now two variables to substitute for instead of one. Notice that the area differential $dx \, dy$ is not replaced by $dr \, d\theta$ but by $r \, dr \, d\theta$. A more general discussion of changes of variables (substitutions) in multiple integrals is given in Section 15.8.

EXAMPLE 3 Evaluate

$$\iint_R e^{x^2 + y^2} \, dy \, dx,$$

where R is the semicircular region bounded by the x-axis and the curve $y = \sqrt{1 - x^2}$ (Figure 15.27).

Solution In Cartesian coordinates, the integral in question is a nonelementary integral and there is no direct way to integrate $e^{x^2 + y^2}$ with respect to either x or y. Yet this integral and others like it are important in mathematics—in statistics, for example—and we need

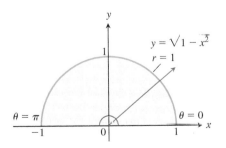

FIGURE 15.27 The semicircular region in Example 3 is the region

$$0 \leq r \leq 1, \qquad 0 \leq \theta \leq \pi.$$

to evaluate it. Polar coordinates make this possible. Substituting $x = r \cos \theta$ and $y = r \sin \theta$ and replacing $dy\, dx$ by $r\, dr\, d\theta$ give

$$\iint\limits_R e^{x^2+y^2}\, dy\, dx = \int_0^\pi \int_0^1 e^{r^2} r\, dr\, d\theta = \int_0^\pi \left[\frac{1}{2}e^{r^2}\right]_0^1 d\theta$$

$$= \int_0^\pi \frac{1}{2}(e-1)\, d\theta = \frac{\pi}{2}(e-1).$$

The r in the $r\, dr\, d\theta$ is what allowed us to integrate e^{r^2}. Without it, we would have been unable to find an antiderivative for the first (innermost) iterated integral. ∎

EXAMPLE 4 Evaluate the integral

$$\int_0^1 \int_0^{\sqrt{1-x^2}} (x^2 + y^2)\, dy\, dx.$$

Solution Integration with respect to y gives

$$\int_0^1 \left(x^2 \sqrt{1 - x^2} + \frac{(1 - x^2)^{3/2}}{3}\right) dx,$$

which is difficult to evaluate without tables. Things go better if we change the original integral to polar coordinates. The region of integration in Cartesian coordinates is given by the inequalities $0 \le y \le \sqrt{1 - x^2}$ and $0 \le x \le 1$, which correspond to the interior of the unit quarter circle $x^2 + y^2 = 1$ in the first quadrant. (See Figure 15.27, first quadrant.) Substituting the polar coordinates $x = r \cos \theta$, $y = r \sin \theta$, $0 \le \theta \le \pi/2$, and $0 \le r \le 1$, and replacing $dy\, dx$ by $r\, dr\, d\theta$ in the double integral, we get

$$\int_0^1 \int_0^{\sqrt{1-x^2}} (x^2 + y^2)\, dy\, dx = \int_0^{\pi/2} \int_0^1 (r^2)\, r\, dr\, d\theta$$

$$= \int_0^{\pi/2} \left[\frac{r^4}{4}\right]_{r=0}^{r=1} d\theta = \int_0^{\pi/2} \frac{1}{4} d\theta = \frac{\pi}{8}.$$

The polar coordinate transformation is effective here because $x^2 + y^2$ simplifies to r^2 and the limits of integration become constants. ∎

EXAMPLE 5 Find the volume of the solid region bounded above by the paraboloid $z = 9 - x^2 - y^2$ and below by the unit circle in the xy-plane.

Solution The region of integration R is bounded by the unit circle $x^2 + y^2 = 1$, which is described in polar coordinates by $r = 1, 0 \le \theta \le 2\pi$. The solid region is shown in Figure 15.28. The volume is given by the double integral

$$\iint\limits_R (9 - x^2 - y^2)\, dA = \int_0^{2\pi} \int_0^1 (9 - r^2)\, r\, dr\, d\theta \qquad r^2 = x^2 + y^2,\quad dA = r\, dr\, d\theta.$$

$$= \int_0^{2\pi} \int_0^1 (9r - r^3)\, dr\, d\theta$$

$$= \int_0^{2\pi} \left[\frac{9}{2}r^2 - \frac{1}{4}r^4\right]_{r=0}^{r=1} d\theta$$

$$= \frac{17}{4} \int_0^{2\pi} d\theta = \frac{17\pi}{2}.$$

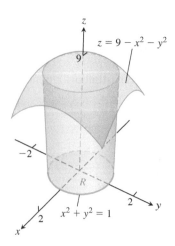

FIGURE 15.28 The solid region in Example 5.

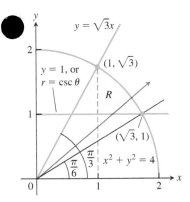

FIGURE 15.29 The region R in Example 6.

EXAMPLE 6 Using polar integration, find the area of the region R in the xy-plane enclosed by the circle $x^2 + y^2 = 4$, above the line $y = 1$, and below the line $y = \sqrt{3}x$.

Solution A sketch of the region R is shown in Figure 15.29. First we note that the line $y = \sqrt{3}x$ has slope $\sqrt{3} = \tan \theta$, so $\theta = \pi/3$. Next we observe that the line $y = 1$ intersects the circle $x^2 + y^2 = 4$ when $x^2 + 1 = 4$, or $x = \sqrt{3}$. Moreover, the radial line from the origin through the point $(\sqrt{3}, 1)$ has slope $1/\sqrt{3} = \tan \theta$, giving its angle of inclination as $\theta = \pi/6$. This information is shown in Figure 15.29.

Now, for the region R, as θ varies from $\pi/6$ to $\pi/3$, the polar coordinate r varies from the horizontal line $y = 1$ to the circle $x^2 + y^2 = 4$. Substituting $r \sin \theta$ for y in the equation for the horizontal line, we have $r \sin \theta = 1$, or $r = \csc \theta$, which is the polar equation of the line. The polar equation for the circle is $r = 2$. So in polar coordinates, for $\pi/6 \le \theta \le \pi/3$, r varies from $r = \csc \theta$ to $r = 2$. It follows that the iterated integral for the area is

$$\iint_R dA = \int_{\pi/6}^{\pi/3} \int_{\csc \theta}^{2} r \, dr \, d\theta$$

$$= \int_{\pi/6}^{\pi/3} \left[\frac{1}{2} r^2 \right]_{r=\csc \theta}^{r=2} d\theta$$

$$= \int_{\pi/6}^{\pi/3} \frac{1}{2} \left[4 - \csc^2 \theta \right] d\theta$$

$$= \frac{1}{2} \left[4\theta + \cot \theta \right]_{\pi/6}^{\pi/3}$$

$$= \frac{1}{2} \left(\frac{4\pi}{3} + \frac{1}{\sqrt{3}} \right) - \frac{1}{2} \left(\frac{4\pi}{6} + \sqrt{3} \right) = \frac{\pi}{3} - \frac{\sqrt{3}}{3}.$$

EXERCISES 15.4

Regions in Polar Coordinates

In Exercises 1–8, describe the given region in polar coordinates.

1.

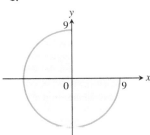

2.

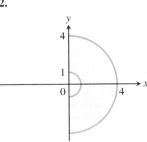

3.

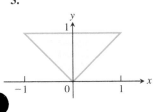

4.

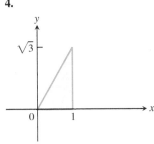

5.

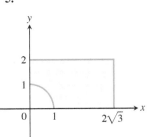

6.

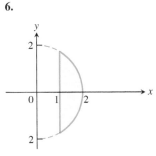

7. The region enclosed by the circle $x^2 + y^2 = 2x$

8. The region enclosed by the semicircle $x^2 + y^2 = 2y, y \ge 0$

Evaluating Polar Integrals

In Exercises 9–22, change the Cartesian integral into an equivalent polar integral. Then evaluate the polar integral.

9. $\displaystyle\int_{-1}^{1} \int_{0}^{\sqrt{1-x^2}} dy \, dx$

10. $\displaystyle\int_{0}^{1} \int_{0}^{\sqrt{1-y^2}} (x^2 + y^2) \, dx \, dy$

11. $\displaystyle\int_{0}^{2} \int_{0}^{\sqrt{4-y^2}} (x^2 + y^2) \, dx \, dy$

12. $\displaystyle\int_{-a}^{a}\int_{-\sqrt{a^2-x^2}}^{\sqrt{a^2-x^2}} dy\, dx$

13. $\displaystyle\int_{0}^{6}\int_{0}^{y} x\, dx\, dy$

14. $\displaystyle\int_{0}^{2}\int_{0}^{x} y\, dy\, dx$

15. $\displaystyle\int_{1}^{\sqrt{3}}\int_{1}^{x} dy\, dx$

16. $\displaystyle\int_{\sqrt{2}}^{2}\int_{\sqrt{4-y^2}}^{y} dx\, dy$

17. $\displaystyle\int_{-1}^{0}\int_{-\sqrt{1-x^2}}^{0} \frac{2}{1+\sqrt{x^2+y^2}}\, dy\, dx$

18. $\displaystyle\int_{-1}^{1}\int_{-\sqrt{1-x^2}}^{\sqrt{1-x^2}} \frac{2}{(1+x^2+y^2)^2}\, dy\, dx$

19. $\displaystyle\int_{0}^{\ln 2}\int_{0}^{\sqrt{(\ln 2)^2-y^2}} e^{\sqrt{x^2+y^2}}\, dx\, dy$

20. $\displaystyle\int_{-1}^{1}\int_{-\sqrt{1-y^2}}^{\sqrt{1-y^2}} \ln(x^2+y^2+1)\, dx\, dy$

21. $\displaystyle\int_{0}^{1}\int_{x}^{\sqrt{2-x^2}} (x+2y)\, dy\, dx$

22. $\displaystyle\int_{1}^{2}\int_{0}^{\sqrt{2x-x^2}} \frac{1}{(x^2+y^2)^2}\, dy\, dx$

In Exercises 23–26, sketch the region of integration and convert each polar integral or sum of integrals to a Cartesian integral or sum of integrals. Do not evaluate the integrals.

23. $\displaystyle\int_{0}^{\pi/2}\int_{0}^{1} r^3 \sin\theta\cos\theta\, dr\, d\theta$

24. $\displaystyle\int_{\pi/6}^{\pi/2}\int_{1}^{\csc\theta} r^2 \cos\theta\, dr\, d\theta$

25. $\displaystyle\int_{0}^{\pi/4}\int_{0}^{2\sec\theta} r^5 \sin^2\theta\, dr\, d\theta$

26. $\displaystyle\int_{0}^{\tan^{-1}\frac{4}{3}}\int_{0}^{3\sec\theta} r^7\, dr\, d\theta + \int_{\tan^{-1}\frac{4}{3}}^{\pi/2}\int_{0}^{4\csc\theta} r^7\, dr\, d\theta$

Area in Polar Coordinates

27. Find the area of the region cut from the first quadrant by the curve $r = 2(2-\sin 2\theta)^{1/2}$.

28. Cardioid overlapping a circle Find the area of the region that lies inside the cardioid $r = 1 + \cos\theta$ and outside the circle $r = 1$.

29. One leaf of a rose Find the area enclosed by one leaf of the rose $r = 12\cos 3\theta$.

30. Snail shell Find the area of the region enclosed by the positive x-axis and spiral $r = 4\theta/3$, $0 \le \theta \le 2\pi$. The region looks like a snail shell.

31. Cardioid in the first quadrant Find the area of the region cut from the first quadrant by the cardioid $r = 1 + \sin\theta$.

32. Overlapping cardioids Find the area of the region common to the interiors of the cardioids $r = 1 + \cos\theta$ and $r = 1 - \cos\theta$.

Average Values

In polar coordinates, the **average value** of a function over a region R (Section 15.3) is given by

$$\frac{1}{\text{Area}(R)} \iint_R f(r,\theta)\, r\, dr\, d\theta.$$

33. Average height of a hemisphere Find the average height of the hemispherical surface $z = \sqrt{a^2 - x^2 - y^2}$ above the disk $x^2 + y^2 \le a^2$ in the xy-plane.

34. Average height of a cone Find the average height of the (single) cone $z = \sqrt{x^2 + y^2}$ above the disk $x^2 + y^2 \le a^2$ in the xy-plane.

35. Average distance from interior of disk to center Find the average distance from a point $P(x, y)$ in the disk $x^2 + y^2 \le a^2$ to the origin.

36. Average distance squared from a point in a disk to a point in its boundary Find the average value of the *square* of the distance from the point $P(x, y)$ in the disk $x^2 + y^2 \le 1$ to the boundary point $A(1, 0)$.

Theory and Examples

37. Converting to a polar integral Integrate $f(x, y) = \left[\ln(x^2 + y^2)\right]/\sqrt{x^2 + y^2}$ over the region $1 \le x^2 + y^2 \le e$.

38. Converting to a polar integral Integrate $f(x, y) = \left[\ln(x^2 + y^2)\right]/(x^2 + y^2)$ over the region $1 \le x^2 + y^2 \le e^2$.

39. Volume of noncircular right cylinder The region that lies inside the cardioid $r = 1 + \cos\theta$ and outside the circle $r = 1$ is the base of a solid right cylinder. The top of the cylinder lies in the plane $z = x$. Find the cylinder's volume.

40. Volume of noncircular right cylinder The region enclosed by the lemniscate $r^2 = 2\cos 2\theta$ is the base of a solid right cylinder whose top is bounded by the sphere $z = \sqrt{2 - r^2}$. Find the cylinder's volume.

41. Converting to polar integrals

a. The usual way to evaluate the improper integral $I = \int_0^\infty e^{-x^2}\, dx$ is first to calculate its square:

$$I^2 = \left(\int_0^\infty e^{-x^2}\, dx\right)\left(\int_0^\infty e^{-y^2}\, dy\right) = \int_0^\infty\int_0^\infty e^{-(x^2+y^2)}\, dx\, dy.$$

Evaluate the last integral using polar coordinates and solve the resulting equation for I.

b. Evaluate

$$\lim_{x\to\infty}\text{erf}(x) = \lim_{x\to\infty}\int_0^x \frac{2e^{-t^2}}{\sqrt{\pi}}\, dt.$$

42. Converting to a polar integral Evaluate the integral

$$\int_0^\infty\int_0^\infty \frac{1}{(1+x^2+y^2)^2}\, dx\, dy.$$

43. Existence Integrate the function $f(x, y) = 1/(1 - x^2 - y^2)$ over the disk $x^2 + y^2 \leq 3/4$. Does the integral of $f(x, y)$ over the disk $x^2 + y^2 \leq 1$ exist? Give reasons for your answer.

44. Area formula in polar coordinates Use the double integral in polar coordinates to derive the formula

$$A = \int_{\alpha}^{\beta} \frac{1}{2} r^2 \, d\theta$$

for the area of the fan-shaped region between the origin and polar curve $r = f(\theta)$, $\alpha \leq \theta \leq \beta$.

45. Average distance to a given point inside a disk Let P_0 be a point inside a circle of radius a and let h denote the distance from P_0 to the center of the circle. Let d denote the distance from an arbitrary point P to P_0. Find the average value of d^2 over the region enclosed by the circle. (*Hint:* Simplify your work by placing the center of the circle at the origin and P_0 on the x-axis.)

46. Area Suppose that the area of a region in the polar coordinate plane is

$$A = \int_{\pi/4}^{3\pi/4} \int_{\csc\theta}^{2\sin\theta} r \, dr \, d\theta.$$

Sketch the region and find its area.

47. Evaluate the integral $\iint_R \sqrt{x^2 + y^2} \, dA$, where R is the region inside the upper semicircle of radius 2 centered at the origin, but outside the circle $x^2 + (y - 1)^2 = 1$.

48. Evaluate the integral $\iint_R (x^2 + y^2)^{-2} \, dA$, where R is the region inside the circle $x^2 + y^2 = 2$ for $x \leq -1$.

COMPUTER EXPLORATIONS

In Exercises 49–52, use a CAS to change the Cartesian integrals into an equivalent polar integral and evaluate the polar integral. Perform the following steps in each exercise.

a. Plot the Cartesian region of integration in the xy-plane.

b. Change each boundary curve of the Cartesian region in part (a) to its polar representation by solving its Cartesian equation for r and θ.

c. Using the results in part (b), plot the polar region of integration in the $r\theta$-plane.

d. Change the integrand from Cartesian to polar coordinates. Determine the limits of integration from your plot in part (c) and evaluate the polar integral using the CAS integration utility.

49. $\displaystyle\int_0^1 \int_x^1 \frac{y}{x^2 + y^2} \, dy \, dx$

50. $\displaystyle\int_0^1 \int_0^{x/2} \frac{x}{x^2 + y^2} \, dy \, dx$

51. $\displaystyle\int_0^1 \int_{-y/3}^{y/3} \frac{y}{\sqrt{x^2 + y^2}} \, dx \, dy$

52. $\displaystyle\int_0^1 \int_y^{2-y} \sqrt{x + y} \, dx \, dy$

15.5 Triple Integrals in Rectangular Coordinates

Just as double integrals allow us to deal with more general situations than could be handled by single integrals, triple integrals enable us to solve still more general problems. We use triple integrals to calculate the volumes of three-dimensional shapes and the average value of a function over a three-dimensional region. Triple integrals also arise in the study of vector fields and fluid flow in three dimensions, as we will see in Chapter 16.

Triple Integrals

If $F(x, y, z)$ is a function defined on a closed bounded region D in space, such as the region occupied by a solid ball or a lump of clay, then the integral of F over D may be defined in the following way. We partition a rectangular boxlike region containing D into rectangular cells by planes parallel to the coordinate axes (Figure 15.30). We number the cells that lie completely inside D from 1 to n in some order, the kth cell having dimensions Δx_k by Δy_k by Δz_k and volume $\Delta V_k = \Delta x_k \Delta y_k \Delta z_k$. We choose a point (x_k, y_k, z_k) in each cell and form the sum

$$S_n = \sum_{k=1}^{n} F(x_k, y_k, z_k) \, \Delta V_k. \tag{1}$$

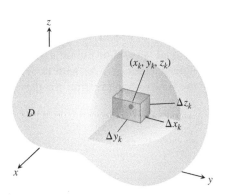

FIGURE 15.30 Partitioning a solid with rectangular cells of volume ΔV_k.

We are interested in what happens as D is partitioned by smaller and smaller cells, so that Δx_k, Δy_k, Δz_k, and the norm of the partition $\|P\|$, the largest value among Δx_k, Δy_k, Δz_k, all approach zero. When a single limiting value is attained, no matter how the partitions and points (x_k, y_k, z_k) are chosen, we say that F is **integrable** over D. As before, it can be shown that when F is continuous and the bounding surface of D is formed from finitely many smooth surfaces joined together along finitely many smooth curves, then F is integrable. As $\|P\| \to 0$ and the number of cells n goes to ∞, the sums S_n approach a limit. We call this limit the **triple integral of F over D** and write

$$\lim_{n \to \infty} S_n = \iiint_D F(x, y, z)\, dV \qquad \text{or} \qquad \lim_{\|P\| \to 0} S_n = \iiint_D F(x, y, z)\, dx\, dy\, dz.$$

The regions D over which continuous functions are integrable are those having "reasonably smooth" boundaries.

Volume of a Region in Space

If F is the constant function whose value is 1, then the sums in Equation (1) reduce to

$$S_n = \sum_{k=1}^{n} F(x_k, y_k, z_k)\, \Delta V_k = \sum_{k=1}^{n} 1 \cdot \Delta V_k = \sum_{k=1}^{n} \Delta V_k.$$

As Δx_k, Δy_k, and Δz_k approach zero, the cells ΔV_k become smaller and more numerous and fill up more and more of D. We therefore define the volume of D to be the triple integral

$$\lim_{n \to \infty} \sum_{k=1}^{n} \Delta V_k = \iiint_D dV.$$

DEFINITION   The **volume** of a closed, bounded region D in space is

$$V = \iiint_D dV.$$

This definition is in agreement with our previous definitions of volume, although we omit the verification of this fact. As we see in a moment, this integral enables us to calculate the volumes of solids enclosed by curved surfaces. These are more general solids than the ones encountered before (Chapter 6 and Section 15.2).

Finding Limits of Integration in the Order $dz\, dy\, dx$

We evaluate a triple integral by applying a three-dimensional version of Fubini's Theorem (Section 15.2) to evaluate it by three repeated single integrations. As with double integrals, there is a geometric procedure for finding the limits of integration for these iterated integrals.

To evaluate

$$\iiint_D F(x, y, z)\, dV$$

over a region D, integrate first with respect to z, then with respect to y, and finally with respect to x. (You might choose a different order of integration, but the procedure is similar, as we illustrate in Example 2.)

1. *Sketch.* Sketch the region D along with its "shadow" R (vertical projection) in the xy-plane. Label the upper and lower bounding surfaces of D and the upper and lower bounding curves of R.

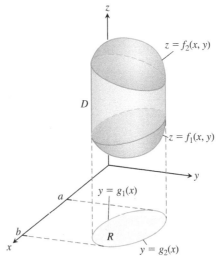

2. *Find the z-limits of integration.* Draw a line M passing through a typical point (x, y) in R parallel to the z-axis. As z increases, M enters D at $z = f_1(x, y)$ and leaves at $z = f_2(x, y)$. These are the z-limits of integration.

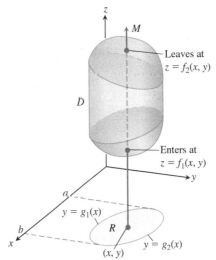

3. *Find the y-limits of integration.* Draw a line L through (x, y) parallel to the y-axis. As y increases, L enters R at $y = g_1(x)$ and leaves at $y = g_2(x)$. These are the y-limits of integration.

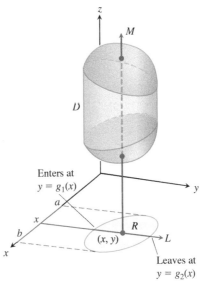

4. *Find the x-limits of integration.* Choose x-limits that include all lines through R parallel to the y-axis ($x = a$ and $x = b$ in the preceding figure). These are the x-limits of integration. The integral is

$$\int_{x=a}^{x=b} \int_{y=g_1(x)}^{y=g_2(x)} \int_{z=f_1(x,y)}^{z=f_2(x,y)} F(x, y, z) \, dz \, dy \, dx.$$

Follow similar procedures if you change the order of integration. The "shadow" of region D lies in the plane of the last two variables with respect to which the iterated integration takes place.

The preceding procedure applies whenever a solid region D is bounded above and below by a surface, and when the "shadow" region R is bounded by a lower and upper curve. It does not apply to regions with complicated holes through them, although sometimes such regions can be subdivided into simpler regions for which the procedure does apply.

We illustrate this method of finding the limits of integration in our first example.

EXAMPLE 1   Let S be the sphere of radius 5 centered at the origin, and let D be the region under the sphere that lies above the plane $z = 3$. Set up the limits of integration for evaluating the triple integral of a function $F(x, y, z)$ over the region D.

Solution   The region under the sphere that lies above the plane $z = 3$ is enclosed by the surfaces $x^2 + y^2 + z^2 = 25$ and $z = 3$.

To find the limits of integration, we first sketch the region, as shown in Figure 15.31. The "shadow region" R in the xy-plane is a circle of some radius centered at the origin. By considering a side view of the region D, we can determine that the radius of this circle is 4; see Figure 15.32a.

If we fix a point (x, y) in R and draw a vertical line M above (x, y), then we see that this line enters the region D at the height $z = 3$ and leaves the region at the height $z = \sqrt{25 - x^2 - y^2}$; see Figure 15.31. This gives us the z-limits of integration.

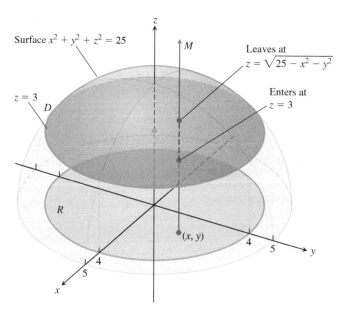

FIGURE 15.31   Finding the limits of integration for evaluating the triple integral of a function defined over the portion of the sphere of radius 5 that lies above the plane $z = 3$ (Example 1).

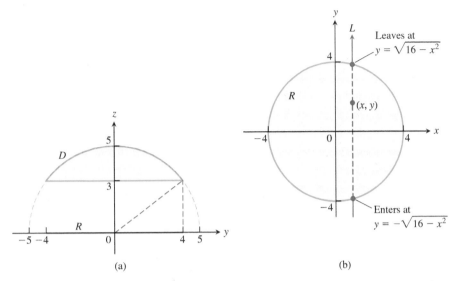

FIGURE 15.32 (a) Side view of the region from Example 1, looking down the x-axis. The dashed right triangle has a hypotenuse of length 5 and sides of lengths 3 and 4. In this side view, the shadow region R lies between -4 and 4 on the y-axis. (b) The "shadow region" R shown face-on in the xy-plane.

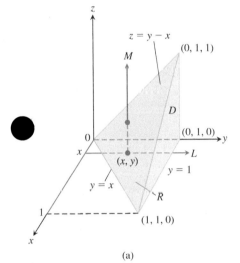

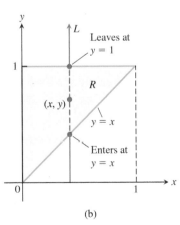

FIGURE 15.33 (a) The tetrahedron in Example 2 showing how the limits of integration are found for the order $dz\,dy\,dx$. (b) The "shadow region" R shown face-on in the xy-plane.

To find the y-limits of integration, we consider a line L that lies in the region R, passes through the point (x, y), and is parallel to the y-axis. For clarity we have separately pictured the region R and the line L in Figure 15.32b. The line L enters R when $y = -\sqrt{16 - x^2}$ and exits when $y = \sqrt{16 - x^2}$. This gives us the y-limits of integration.

Finally, as L sweeps across R from left to right, the value of x varies from $x = -4$ to $x = 4$. This gives us the x-limits of integration. Therefore, the triple integral of F over the region D is given by

$$\iiint_D F(x, y, z)\, dz\, dy\, dx = \int_{-4}^{4} \int_{-\sqrt{16-x^2}}^{\sqrt{16-x^2}} \int_{3}^{\sqrt{25-x^2-y^2}} F(x, y, z)\, dz\, dy\, dx. \quad \blacksquare$$

The region D in Example 1 has a great deal of symmetry, which makes visualization easier. Even without symmetry, the steps in finding the limits of integration are the same, as shown in the next example.

EXAMPLE 2 Set up the limits of integration for evaluating the triple integral of a function $F(x, y, z)$ over the tetrahedron D whose vertices are $(0, 0, 0)$, $(1, 1, 0)$, $(0, 1, 0)$, and $(0, 1, 1)$. Use the order of integration $dz\,dy\,dx$.

Solution The region D and its "shadow" R in the xy-plane are shown in Figure 15.33a. The "side" face of D is parallel to the xz-plane, the "back" face lies in the yz-plane, and the "top" face is contained in the plane $z = y - x$.

To find the z-limits of integration, fix a point (x, y) in the shadow region R, and consider the vertical line M that passes through (x, y) and is parallel to the z-axis. This line enters D at the height $z = 0$, and it exits at height $z = y - x$.

To find the y-limits of integration we again fix a point (x, y) in R, but now we consider a line L that lies in R, passes through (x, y), and is parallel to the y-axis. This line is shown in Figure 15.33a and also in the face-on view of R that is pictured in Figure 15.33b. The line L enters R when $y = x$ and exits when $y = 1$.

Finally, as L sweeps across R, the value of x varies from $x = 0$ to $x = 1$. Therefore, the triple integral of F over the region D is given by

$$\iiint\limits_{D} F(x, y, z)\, dz\, dy\, dx = \int_0^1 \int_x^1 \int_0^{y-x} F(x, y, z)\, dz\, dy\, dx.$$ ∎

In the next example we project the region D onto the xz-plane instead of the xy-plane, to show how to use a different order of integration.

EXAMPLE 3   Find the volume of the tetrahedron D from Example 2 by integrating $F(x, y, z) = 1$ over the region using the order $dz\, dy\, dx$. Then do the same calculation using the order $dy\, dz\, dx$.

Solution   Using the limits of integration that we found in Example 2, we calculate the volume of the tetrahedron as follows:

$$V = \int_0^1 \int_x^1 \int_0^{y-x} dz\, dy\, dx \qquad \text{Integrand is 1 when computing volume.}$$

$$= \int_0^1 \int_x^1 (y - x)\, dy\, dx \qquad \text{Integrate over } z \text{ and evaluate.}$$

$$= \int_0^1 \left[\frac{1}{2} y^2 - xy \right]_{y=x}^{y=1} dx \qquad \text{Integrate over } y.$$

$$= \int_0^1 \left(\frac{1}{2} - x + \frac{1}{2} x^2 \right) dx \qquad \text{Evaluate.}$$

$$= \left[\frac{1}{2} x - \frac{1}{2} x^2 + \frac{1}{6} x^3 \right]_0^1 \qquad \text{Integrate over } x.$$

$$= \frac{1}{6}. \qquad \text{Evaluate.}$$

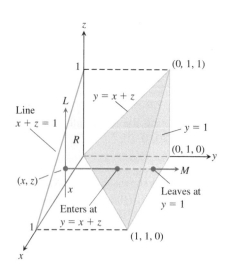

FIGURE 15.34   Finding the limits of integration for evaluating the triple integral of a function defined over the tetrahedron D (Example 3).

Now we will compute the volume using the order of integration $dy\, dz\, dx$. The procedure for finding the limits of integration is similar, except that we find the limits for y first, then for z, and then for x. The region D is the same tetrahedron as before, but now the "shadow region" R lies in the xz-plane, as shown in Figure 15.34.

To find the y-limits of integration, we fix a point (x, z) in the shadow R and consider the line M that passes through (x, z) and is parallel to the y-axis. As shown in Figure 15.34, this line enters D when $y = x + z$, and it leaves when $y = 1$.

Next we find the z-limits of integration. The line L that passes through a point (x, z) in R and is parallel to the z-axis enters R when $z = 0$ and exits when $z = 1 - x$ (see Figure 15.34).

Finally, as L sweeps across R, the value of x varies from $x = 0$ to $x = 1$. Therefore, the volume of the tetrahedron is

$$V = \int_0^1 \int_0^{1-x} \int_{x+z}^1 dy\, dz\, dx$$

$$= \int_0^1 \int_0^{1-x} (1 - x - z)\, dz\, dx$$

$$= \int_0^1 \left[(1 - x)z - \frac{1}{2} z^2 \right]_{z=0}^{z=1-x} dx$$

$$= \int_0^1 \left[(1 - x)^2 - \frac{1}{2}(1 - x)^2 \right] dx$$

$$= \frac{1}{2} \int_0^1 (1 - x)^2 \, dx$$

$$= -\frac{1}{6}(1 - x)^3 \Big]_0^1 = \frac{1}{6}.$$

Next we set up and evaluate a triple integral over a more complicated region.

EXAMPLE 4   Find the volume of the region D enclosed by the surfaces $z = x^2 + 3y^2$ and $z = 8 - x^2 - y^2$.

Solution   The volume is

$$V = \iiint_D dz \, dy \, dx,$$

the integral of $F(x, y, z) = 1$ over D. To find the limits of integration for evaluating the integral, we first sketch the region. The surfaces (Figure 15.35) intersect on the elliptical cylinder $x^2 + 3y^2 = 8 - x^2 - y^2$ or $x^2 + 2y^2 = 4, z > 0$. The boundary of the region R, the projection of D onto the xy-plane, is an ellipse with the same equation: $x^2 + 2y^2 = 4$. The "upper" boundary of R is the curve $y = \sqrt{(4 - x^2)/2}$. The lower boundary is the curve $y = -\sqrt{(4 - x^2)/2}$.

Now we find the z-limits of integration. The line M passing through a typical point (x, y) in R parallel to the z-axis enters D at $z = x^2 + 3y^2$ and leaves at $z = 8 - x^2 - y^2$.

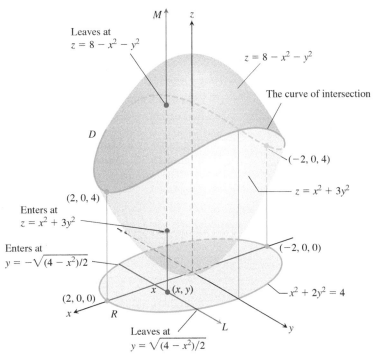

FIGURE 15.35   The volume of the region enclosed by two paraboloids, calculated in Example 4.

Next we find the y-limits of integration. The line L through (x, y) that lies parallel to the y-axis enters the region R when $y = -\sqrt{(4 - x^2)/2}$ and leaves when $y = \sqrt{(4 - x^2)/2}$

Finally we find the x-limits of integration. As L sweeps across R, the value of x varies from $x = -2$ at $(-2, 0, 0)$ to $x = 2$ at $(2, 0, 0)$. The volume of D is

$$V = \iiint_D dz\, dy\, dx \qquad \text{Integrand is 1 when computing volume.}$$

$$= \int_{-2}^{2} \int_{-\sqrt{(4-x^2)/2}}^{\sqrt{(4-x^2)/2}} \int_{x^2+3y^2}^{8-x^2-y^2} dz\, dy\, dx \qquad \text{Substitute limits of integration.}$$

$$= \int_{-2}^{2} \int_{-\sqrt{(4-x^2)/2}}^{\sqrt{(4-x^2)/2}} (8 - 2x^2 - 4y^2)\, dy\, dx \qquad \text{Integrate over } z \text{ and evaluate.}$$

$$= \int_{-2}^{2} \left[(8 - 2x^2)y - \frac{4}{3}y^3 \right]_{y=-\sqrt{(4-x^2)/2}}^{y=\sqrt{(4-x^2)/2}} dx \qquad \text{Integrate over } y.$$

$$= \int_{-2}^{2} \left(2(8 - 2x^2)\sqrt{\frac{4 - x^2}{2}} - \frac{8}{3}\left(\frac{4 - x^2}{2}\right)^{3/2} \right) dx \qquad \text{Evaluate.}$$

$$= \int_{-2}^{2} \left[8\left(\frac{4 - x^2}{2}\right)^{3/2} - \frac{8}{3}\left(\frac{4 - x^2}{2}\right)^{3/2} \right] dx = \frac{4\sqrt{2}}{3} \int_{-2}^{2} (4 - x^2)^{3/2}\, dx$$

$$= 8\pi\sqrt{2}. \qquad \text{After integration with the substitution } x = 2\sin u$$

Average Value of a Function in Space

The average value of a function F over a region D in space is defined by the formula

$$\textbf{Average value of } F \text{ over } D = \frac{1}{\text{volume of } D} \iiint_D F\, dV. \qquad (2)$$

For example, if $F(x, y, z) = \sqrt{x^2 + y^2 + z^2}$, then the average value of F over D is the average distance of points in D from the origin. If $F(x, y, z)$ is the temperature at (x, y, z) on a solid that occupies a region D in space, then the average value of F over D is the average temperature of the solid.

EXAMPLE 5   Find the average value of $F(x, y, z) = xyz$ throughout the cubical region D bounded by the coordinate planes and the planes $x = 2$, $y = 2$, and $z = 2$ in the first octant.

Solution   We sketch the cube with enough detail to show the limits of integration (Figure 15.36). We then use Equation (2) to calculate the average value of F over the cube.

The volume of the region D is $(2)(2)(2) = 8$. The value of the integral of F over the cube is

$$\int_0^2 \int_0^2 \int_0^2 xyz\, dx\, dy\, dz = \int_0^2 \int_0^2 \left[\frac{x^2}{2}yz \right]_{x=0}^{x=2} dy\, dz = \int_0^2 \int_0^2 2yz\, dy\, dz$$

$$= \int_0^2 \left[y^2 z \right]_{y=0}^{y=2} dz = \int_0^2 4z\, dz = \left[2z^2 \right]_0^2 = 8.$$

With these values, Equation (2) gives

$$\begin{array}{c} \text{Average value of} \\ xyz \text{ over the cube} \end{array} = \frac{1}{\text{volume}} \iiint_{\text{cube}} xyz\, dV = \left(\frac{1}{8}\right)(8) = 1.$$

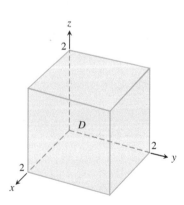

FIGURE 15.36   The region of integration in Example 5.

In evaluating the integral, we chose the order $dx\, dy\, dz$, but any of the other five possible orders would have done as well.

Properties of Triple Integrals

Triple integrals have the same algebraic properties as double and single integrals. Simply replace the double integrals in the four properties given in Section 15.2, page 907, with triple integrals.

EXERCISES 15.5

Triple Integrals in Different Iteration Orders

1. Evaluate the integral in Example 3 taking $F(x, y, z) = 1$ to find the volume of the tetrahedron in the order $dz\, dx\, dy$.

2. Volume of rectangular solid   Write six different iterated triple integrals for the volume of the rectangular solid in the first octant bounded by the coordinate planes and the planes $x = 1$, $y = 2$, and $z = 3$. Evaluate one of the integrals.

3. Volume of tetrahedron   Write six different iterated triple integrals for the volume of the tetrahedron cut from the first octant by the plane $6x + 3y + 2z = 6$. Evaluate one of the integrals.

4. Volume of solid   Write six different iterated triple integrals for the volume of the region in the first octant enclosed by the cylinder $x^2 + z^2 = 4$ and the plane $y = 3$. Evaluate one of the integrals.

5. Volume enclosed by paraboloids   Let D be the region bounded by the paraboloids $z = 8 - x^2 - y^2$ and $z = x^2 + y^2$. Write six different triple iterated integrals for the volume of D. Evaluate one of the integrals.

6. Volume inside paraboloid beneath a plane   Let D be the region bounded by the paraboloid $z = x^2 + y^2$ and the plane $z = 2y$. Write triple iterated integrals in the order $dz\, dx\, dy$ and $dz\, dy\, dx$ that give the volume of D. Do not evaluate either integral.

Evaluating Triple Iterated Integrals

Evaluate the integrals in Exercises 7–20.

7. $\displaystyle\int_0^1 \int_0^1 \int_0^1 (x^2 + y^2 + z^2)\, dz\, dy\, dx$

8. $\displaystyle\int_0^{\sqrt{2}} \int_0^{3y} \int_{x^2+3y^2}^{8-x^2-y^2} dz\, dx\, dy$ **9.** $\displaystyle\int_1^e \int_1^{e^2} \int_1^{e^3} \frac{1}{xyz}\, dx\, dy\, dz$

10. $\displaystyle\int_0^1 \int_0^{3-3x} \int_0^{3-3x-y} dz\, dy\, dx$ **11.** $\displaystyle\int_0^{\pi/6} \int_0^1 \int_{-2}^3 y \sin z\, dx\, dy\, dz$

12. $\displaystyle\int_1^1 \int_0^1 \int_0^2 (x + y + z)\, dy\, dx\, dz$

13. $\displaystyle\int_0^3 \int_0^{\sqrt{9-x^2}} \int_0^{\sqrt{9-x^2}} dz\, dy\, dx$ **14.** $\displaystyle\int_0^2 \int_{-\sqrt{4-y^2}}^{\sqrt{4-y^2}} \int_0^{2x+y} dz\, dx\, dy$

15. $\displaystyle\int_0^1 \int_0^{2-x} \int_0^{2-x-y} dz\, dy\, dx$ **16.** $\displaystyle\int_0^1 \int_0^{1-x^2} \int_3^{4-x^2-y} x\, dz\, dy\, dx$

17. $\displaystyle\int_0^\pi \int_0^\pi \int_0^\pi \cos(u + v + w)\, du\, dv\, dw$ (uvw-space)

18. $\displaystyle\int_0^1 \int_1^{\sqrt{e}} \int_1^e se^s \ln r \frac{(\ln t)^2}{t}\, dt\, dr\, ds$ (rst-space)

19. $\displaystyle\int_0^{\pi/4} \int_0^{\ln \sec v} \int_{-\infty}^{2t} e^x\, dx\, dt\, dv$ (tvx-space)

20. $\displaystyle\int_0^7 \int_0^2 \int_0^{\sqrt{4-q^2}} \frac{q}{r + 1}\, dp\, dq\, dr$ (pqr-space)

Finding Equivalent Iterated Integrals

21. Here is the region of integration of the integral

$$\int_{-1}^1 \int_{x^2}^1 \int_0^{1-y} dz\, dy\, dx.$$

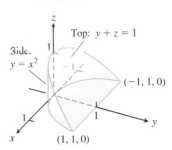

Rewrite the integral as an equivalent iterated integral in the order

a. $dy\, dz\, dx$ **b.** $dy\, dx\, dz$

c. $dx\, dy\, dz$ **d.** $dx\, dz\, dy$

e. $dz\, dx\, dy$.

22. Here is the region of integration of the integral

$$\int_0^1 \int_{-1}^0 \int_0^{y^2} dz\, dy\, dx.$$

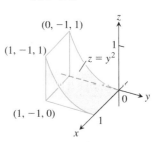

Rewrite the integral as an equivalent iterated integral in the order

a. $dy\, dz\, dx$ **b.** $dy\, dx\, dz$

c. $dx\, dy\, dz$ **d.** $dx\, dz\, dy$

e. $dz\, dx\, dy$.

Finding Volumes Using Triple Integrals

Find the volumes of the regions in Exercises 23–36.

23. The region between the cylinder $z = y^2$ and the xy-plane that is bounded by the planes $x = 0, x = 1, y = -1, y = 1$

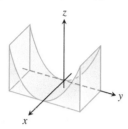

24. The region in the first octant bounded by the coordinate planes and the planes $x + z = 1, y + 2z = 2$

25. The region in the first octant bounded by the coordinate planes, the plane $y + z = 2$, and the cylinder $x = 4 - y^2$

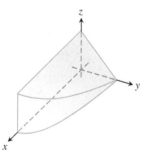

26. The wedge cut from the cylinder $x^2 + y^2 = 1$ by the planes $z = -y$ and $z = 0$

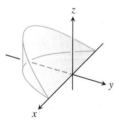

27. The tetrahedron in the first octant bounded by the coordinate planes and the plane passing through $(1, 0, 0), (0, 2, 0)$, and $(0, 0, 3)$

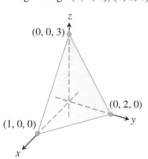

28. The region in the first octant bounded by the coordinate planes, the plane $y = 1 - x$, and the surface $z = \cos(\pi x/2), 0 \le x \le 1$

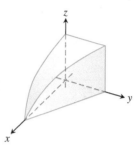

29. The region common to the interiors of the cylinders $x^2 + y^2 = 1$ and $x^2 + z^2 = 1$, one-eighth of which is shown in the accompanying figure

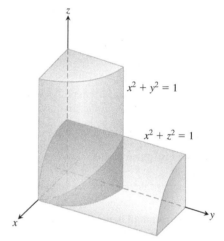

30. The region in the first octant bounded by the coordinate planes and the surface $z = 4 - x^2 - y$

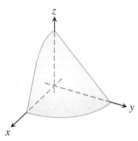

31. The region in the first octant bounded by the coordinate planes, the plane $x + y = 4$, and the cylinder $y^2 + 4z^2 = 16$

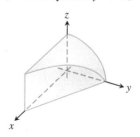

32. The region cut from the cylinder $x^2 + y^2 = 4$ by the plane $z = 0$ and the plane $x + z = 3$

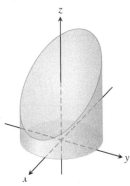

33. The region between the planes $x + y + 2z = 2$ and $2x + 2y + z = 4$ in the first octant

34. The finite region bounded by the planes $z = x$, $x + z = 8$, $z = y$, $y = 8$, and $z = 0$

35. The region cut from the solid elliptical cylinder $x^2 + 4y^2 \leq 4$ by the *xy*-plane and the plane $z = x + 2$

36. The region bounded in back by the plane $x = 0$, on the front and sides by the parabolic cylinder $x = 1 - y^2$, on the top by the paraboloid $z = x^2 + y^2$, and on the bottom by the *xy*-plane

Average Values

In Exercises 37–40, find the average value of $F(x, y, z)$ over the given region.

37. $F(x, y, z) = x^2 + 9$ over the cube in the first octant bounded by the coordinate planes and the planes $x = 2$, $y = 2$, and $z = 2$

38. $F(x, y, z) = x + y - z$ over the rectangular solid in the first octant bounded by the coordinate planes and the planes $x = 1$, $y = 1$, and $z = 2$

39. $F(x, y, z) = x^2 + y^2 + z^2$ over the cube in the first octant bounded by the coordinate planes and the planes $x = 1$, $y = 1$, and $z = 1$

40. $F(x, y, z) = xyz$ over the cube in the first octant bounded by the coordinate planes and the planes $x = 2$, $y = 2$, and $z = 2$

Changing the Order of Integration

Evaluate the integrals in Exercises 41–44 by changing the order of integration in an appropriate way.

41. $\int_0^4 \int_0^1 \int_{2y}^2 \frac{4 \cos (x^2)}{2\sqrt{z}} \, dx \, dy \, dz$

42. $\int_0^1 \int_0^1 \int_{x^2}^1 12xze^{zy^2} \, dy \, dx \, dz$

43. $\int_0^1 \int_{\sqrt[3]{z}}^1 \int_0^{\ln 3} \frac{\pi e^{2x} \sin \pi y^2}{y^2} \, dx \, dy \, dz$

44. $\int_0^2 \int_0^{4-x^2} \int_0^x \frac{\sin 2z}{4 - z} \, dy \, dz \, dx$

Theory and Examples

45. Finding an upper limit of an iterated integral Solve for a:

$$\int_0^1 \int_0^{4-a-x^2} \int_a^{4-x^2-y} dz \, dy \, dx = \frac{4}{15}.$$

46. Ellipsoid For what value of c is the volume of the ellipsoid $x^2 + (y/2)^2 + (z/c)^2 = 1$ equal to 8π?

47. Minimizing a triple integral What domain D in space minimizes the value of the integral

$$\iiint_D (4x^2 + 4y^2 + z^2 - 4) \, dV \, ?$$

Give reasons for your answer.

48. Maximizing a triple integral What domain D in space maximizes the value of the integral

$$\iiint_D (1 - x^2 - y^2 - z^2) \, dV \, ?$$

Give reasons for your answer.

COMPUTER EXPLORATIONS

In Exercises 49–52, use a CAS integration utility to evaluate the triple integral of the given function over the specified solid region.

49. $F(x, y, z) = x^2 y^2 z$ over the solid cylinder bounded by $x^2 + y^2 = 1$ and the planes $z = 0$ and $z = 1$

50. $F(x, y, z) = |xyz|$ over the solid bounded below by the paraboloid $z = x^2 + y^2$ and above by the plane $z = 1$

51. $F(x, y, z) = \dfrac{z}{(x^2 + y^2 + z^2)^{3/2}}$ over the solid bounded below by the cone $z = \sqrt{x^2 + y^2}$ and above by the plane $z = 1$

52. $F(x, y, z) = x^4 + y^2 + z^2$ over the solid sphere $x^2 + y^2 + z^2 \leq 1$

15.6 Applications

This section shows how to calculate the masses and moments of two- and three-dimensional objects in Cartesian coordinates. The definitions and ideas are similar to the single-variable case we studied in Section 6.6, but now we can consider more realistic situations. We also look at how multiple integrals are used to compute probabilities.

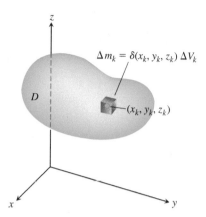

FIGURE 15.37 To define an object's mass, we first imagine it to be partitioned into a finite number of mass elements Δm_k.

Masses and First Moments

If $\delta(x, y, z)$ is the density (mass per unit volume) of an object occupying a region D in space, the integral of δ over D gives the **mass** of the object. To see why, imagine partitioning the object into n mass elements like the one in Figure 15.37. The object's mass is the limit

$$M = \lim_{n \to \infty} \sum_{k=1}^{n} \Delta m_k = \lim_{n \to \infty} \sum_{k=1}^{n} \delta(x_k, y_k, z_k) \, \Delta V_k = \iiint_D \delta(x, y, z) \, dV.$$

The *first moment* of a solid region D about a coordinate plane is defined as the triple integral over D of the distance from a point (x, y, z) in D to the plane multiplied by the density of the solid at that point. For instance, the first moment about the yz-plane is the integral

$$M_{yz} = \iiint_D x \, \delta(x, y, z) \, dV.$$

The *center of mass* is found from the first moments. For instance, the x-coordinate of the center of mass is $\bar{x} = M_{yz}/M$.

For a two-dimensional object, such as a thin, flat plate, we calculate first moments about the coordinate axes by simply dropping the z-coordinate. So the first moment about the y-axis is the double integral over the region R forming the plate of the distance from the axis multiplied by the density, or

$$M_y = \iint_R x \, \delta(x, y) \, dA.$$

Table 15.1 summarizes the formulas.

TABLE 15.1 Mass and first moment formulas

THREE-DIMENSIONAL SOLID

Mass: $M = \displaystyle\iiint_D \delta \, dV$ $\delta = \delta(x, y, z)$ is the density at (x, y, z).

First moments about the coordinate planes:

$$M_{yz} = \iiint_D x \, \delta \, dV, \qquad M_{xz} = \iiint_D y \, \delta \, dV, \qquad M_{xy} = \iiint_D z \, \delta \, dV$$

Center of mass: $\bar{x} = \dfrac{M_{yz}}{M}, \qquad \bar{y} = \dfrac{M_{xz}}{M}, \qquad \bar{z} = \dfrac{M_{xy}}{M}$

TWO-DIMENSIONAL PLATE

Mass: $M = \displaystyle\iint_R \delta \, dA$ $\delta = \delta(x, y)$ is the density at (x, y).

First moments: $M_y = \displaystyle\iint_R x \, \delta \, dA, \qquad M_x = \iint_R y \, \delta \, dA$

Center of mass: $\bar{x} = \dfrac{M_y}{M}, \qquad \bar{y} = \dfrac{M_x}{M}$

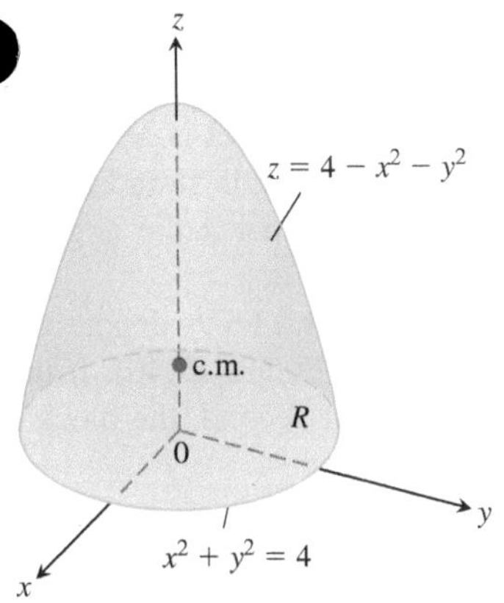

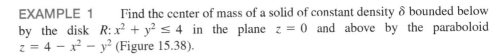

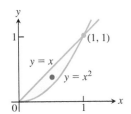

FIGURE 15.38 Finding the center of mass of a solid (Example 1).

EXAMPLE 1   Find the center of mass of a solid of constant density δ bounded below by the disk $R: x^2 + y^2 \le 4$ in the plane $z = 0$ and above by the paraboloid $z = 4 - x^2 - y^2$ (Figure 15.38).

Solution   By symmetry $\bar{x} = \bar{y} = 0$. To find $\bar{z}$, we first calculate

$$M_{xy} = \iint_R \int_{z=0}^{z=4-x^2-y^2} z\,\delta\,dz\,dy\,dx = \iint_R \left[\frac{z^2}{2}\right]_{z=0}^{z=4-x^2-y^2} \delta\,dy\,dx$$

$$= \frac{\delta}{2} \iint_R (4 - x^2 - y^2)^2 \,dy\,dx$$

$$= \frac{\delta}{2} \int_0^{2\pi} \int_0^2 (4 - r^2)^2 r\,dr\,d\theta \qquad \text{Polar coordinates simplify the integration.}$$

$$= \frac{\delta}{2} \int_0^{2\pi} \left[-\frac{1}{6}(4 - r^2)^3\right]_{r=0}^{r=2} d\theta = \frac{16\delta}{3} \int_0^{2\pi} d\theta = \frac{32\pi\delta}{3}.$$

A similar calculation gives the mass

$$M = \iint_R \int_0^{4-x^2-y^2} \delta\,dz\,dy\,dx = 8\pi\delta.$$

Therefore $\bar{z} = (M_{xy}/M) = 4/3$ and the center of mass is $(\bar{x}, \bar{y}, \bar{z}) = (0, 0, 4/3)$.   ■

When the density of a solid object or plate is constant (as in Example 1), the center of mass is called the **centroid** of the object. To find a centroid, we set δ equal to 1 and proceed to find $\bar{x}$, $\bar{y}$, and $\bar{z}$ as before, by dividing first moments by masses. These calculations are also valid for two-dimensional objects.

EXAMPLE 2   Find the centroid of the region in the first quadrant that is bounded above by the line $y = x$ and below by the parabola $y = x^2$.

Solution   We sketch the region and include enough detail to determine the limits of integration (Figure 15.39). We then set δ equal to 1 and evaluate the appropriate formulas from Table 15.1:

$$M = \int_0^1 \int_{x^2}^x 1\,dy\,dx = \int_0^1 \left[y\right]_{y=x^2}^{y=x} dx = \int_0^1 (x - x^2)\,dx = \left[\frac{x^2}{2} - \frac{x^3}{3}\right]_0^1 = \frac{1}{6}$$

$$M_x = \int_0^1 \int_{x^2}^x y\,dy\,dx = \int_0^1 \left[\frac{y^2}{2}\right]_{y=x^2}^{y=x} dx$$

$$= \int_0^1 \left(\frac{x^2}{2} - \frac{x^4}{2}\right) dx = \left[\frac{x^3}{6} - \frac{x^5}{10}\right]_0^1 = \frac{1}{15}$$

$$M_y = \int_0^1 \int_{x^2}^x x\,dy\,dx = \int_0^1 \left[xy\right]_{y=x^2}^{y=x} dx = \int_0^1 (x^2 - x^3)\,dx = \left[\frac{x^3}{3} - \frac{x^4}{4}\right]_0^1 = \frac{1}{12}.$$

From these values of M, M_x, and M_y, we find

$$\bar{x} = \frac{M_y}{M} = \frac{1/12}{1/6} = \frac{1}{2} \qquad \text{and} \qquad \bar{y} = \frac{M_x}{M} = \frac{1/15}{1/6} = \frac{2}{5}.$$

The centroid is the point $(1/2, 2/5)$.   ■

FIGURE 15.39   The centroid of this region is found in Example 2.

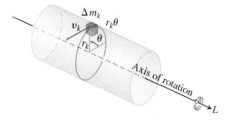

FIGURE 15.40 To find an integral for the amount of energy stored in a rotating shaft, we first imagine the shaft to be partitioned into small blocks. Each block has its own kinetic energy. We add the contributions of the individual blocks to find the kinetic energy of the shaft.

Moments of Inertia

An object's first moments (Table 15.1) tell us about balance and about the torque the object experiences about different axes in a gravitational field. If the object is a rotating shaft, however, we are more likely to be interested in how much energy is stored in the shaft or about how much energy is generated by a shaft rotating at a particular angular velocity. This is where the second moment or moment of inertia comes in.

Think of partitioning the shaft into small blocks of mass Δm_k and let r_k denote the distance from the kth block's center of mass to the axis of rotation (Figure 15.40). If the shaft rotates at a constant angular velocity of $\omega = d\theta/dt$ radians per second, the block's center of mass will trace its orbit at a linear speed of

$$v_k = \frac{d}{dt}(r_k\theta) = r_k \frac{d\theta}{dt} = r_k\omega.$$

The block's kinetic energy will be approximately

$$\frac{1}{2}\Delta m_k v_k^2 = \frac{1}{2}\Delta m_k (r_k\omega)^2 = \frac{1}{2}\omega^2 r_k^2 \, \Delta m_k.$$

The kinetic energy of the shaft will be approximately

$$\sum \frac{1}{2}\omega^2 r_k^2 \, \Delta m_k.$$

The integral approached by these sums as the shaft is partitioned into smaller and smaller blocks gives the shaft's kinetic energy:

$$KE_{shaft} = \int \frac{1}{2}\omega^2 r^2 \, dm = \frac{1}{2}\omega^2 \int r^2 \, dm. \tag{1}$$

The factor

$$I = \int r^2 \, dm$$

is the *moment of inertia* of the shaft about its axis of rotation, and we see from Equation (1) that the shaft's kinetic energy is

$$KE_{shaft} = \frac{1}{2}I\omega^2.$$

The moment of inertia of a shaft resembles in some ways the inertial mass of a locomotive. To start a locomotive with mass m moving at a linear velocity v, we need to provide a kinetic energy of $KE = (1/2)mv^2$. To stop the locomotive we have to remove this amount of energy. To start a shaft with moment of inertia I rotating at an angular velocity ω, we need to provide a kinetic energy of $KE = (1/2)I\omega^2$. To stop the shaft we have to take this amount of energy back out. The shaft's moment of inertia is analogous to the locomotive's mass. What makes the locomotive hard to start or stop is its mass. What makes the shaft hard to start or stop is its moment of inertia. The moment of inertia depends not only on the mass of the shaft but also on its distribution. Mass that is farther away from the axis of rotation contributes more to the moment of inertia.

We now derive a formula for the moment of inertia for a solid in space. If $r(x, y, z)$ is the distance from the point (x, y, z) in D to a line L, then the moment of inertia of the mass $\Delta m_k = \delta(x_k, y_k, z_k)\Delta V_k$ about the line L (as in Figure 15.40) is approximately $\Delta I_k = r^2(x_k, y_k, z_k) \Delta m_k$. **The moment of inertia about L of the entire object is**

$$I_L = \lim_{n\to\infty} \sum_{k=1}^{n} \Delta I_k = \lim_{n\to\infty} \sum_{k=1}^{n} r^2(x_k, y_k, z_k)\,\delta(x_k, y_k, z_k)\,\Delta V_k = \iiint_D r^2\delta \, dV.$$

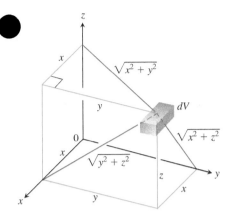

FIGURE 15.41 Distances from dV to the coordinate planes and axes.

If L is the x-axis, then $r^2 = y^2 + z^2$ (Figure 15.41) and

$$I_x = \iiint_D (y^2 + z^2)\, \delta(x, y, z)\, dV.$$

Similarly, if L is the y-axis or z-axis we have

$$I_y = \iiint_D (x^2 + z^2)\, \delta(x, y, z)\, dV \quad \text{and} \quad I_z = \iiint_D (x^2 + y^2)\, \delta(x, y, z)\, dV.$$

Table 15.2 summarizes the formulas for these moments of inertia (second moments because they invoke the *squares* of the distances). It shows the definition of the *polar moment* about the origin as well.

TABLE 15.2 Moments of inertia (second moments) formulas

THREE-DIMENSIONAL SOLID

About the x-axis: $I_x = \iiint (y^2 + z^2)\, \delta\, dV$ $\delta = \delta(x, y, z)$

About the y-axis: $I_y = \iiint (x^2 + z^2)\, \delta\, dV$

About the z-axis: $I_z = \iiint (x^2 + y^2)\, \delta\, dV$

About a line L: $I_L = \iiint r^2(x, y, z)\, \delta\, dV$ $r(x, y, z) = $ distance from the point (x, y, z) to line L

TWO-DIMENSIONAL PLATE

About the x-axis: $I_x = \iint y^2\, \delta\, dA$ $\delta = \delta(x, y)$

About the y-axis: $I_y = \iint x^2\, \delta\, dA$

About a line L: $I_L = \iint r^2(x, y)\, \delta\, dA$ $r(x, y) = $ distance from (x, y) to L

About the origin (polar moment): $I_0 = \iint (x^2 + y^2)\, \delta\, dA = I_x + I_y$

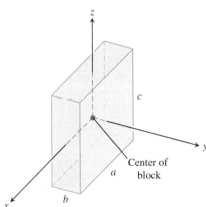

FIGURE 15.42 Finding I_x, I_y, and I_z for the block shown here. The origin lies at the center of the block (Example 3).

EXAMPLE 3   Find I_x, I_y, I_z for the rectangular solid of constant density δ shown in Figure 15.42.

Solution   The formula for I_x gives

$$I_x = \int_{-c/2}^{c/2} \int_{-b/2}^{b/2} \int_{-a/2}^{a/2} (y^2 + z^2)\, \delta\, dx\, dy\, dz.$$

We can avoid some of the work of integration by observing that $(y^2 + z^2)\delta$ is an even function of x, y, and z since δ is constant. The rectangular solid consists of eight symmetric pieces, one in each octant. We can evaluate the integral on one of these pieces and then multiply by 8 to get the total value.

$$I_x = 8 \int_0^{c/2} \int_0^{b/2} \int_0^{a/2} (y^2 + z^2)\, \delta\, dx\, dy\, dz = 4a\delta \int_0^{c/2} \int_0^{b/2} (y^2 + z^2)\, dy\, dz$$

$$= 4a\delta \int_0^{c/2} \left[\frac{y^3}{3} + z^2 y \right]_{y=0}^{y=b/2} dz$$

$$= 4a\delta \int_0^{c/2} \left(\frac{b^3}{24} + \frac{z^2 b}{2} \right) dz$$

$$= 4a\delta \left(\frac{b^3 c}{48} + \frac{c^3 b}{48} \right) = \frac{abc\delta}{12}(b^2 + c^2) = \frac{M}{12}(b^2 + c^2). \qquad M = abc\delta$$

Similarly,

$$I_y = \frac{M}{12}(a^2 + c^2) \qquad \text{and} \qquad I_z = \frac{M}{12}(a^2 + b^2). \qquad \blacksquare$$

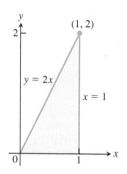

FIGURE 15.43 The triangular region covered by the plate in Example 4.

EXAMPLE 4   A thin plate covers the triangular region bounded by the x-axis and the lines $x = 1$ and $y = 2x$ in the first quadrant. The plate's density at the point (x, y) is $\delta(x, y) = 6x + 6y + 6$. Find the plate's moments of inertia about the coordinate axes and the origin.

Solution   We sketch the plate and put in enough detail to determine the limits of integration for the integrals we have to evaluate (Figure 15.43). The moment of inertia about the x-axis is

$$I_x = \int_0^1 \int_0^{2x} y^2 \delta(x, y)\, dy\, dx = \int_0^1 \int_0^{2x} (6xy^2 + 6y^3 + 6y^2)\, dy\, dx$$

$$= \int_0^1 \left[2xy^3 + \frac{3}{2} y^4 + 2y^3 \right]_{y=0}^{y=2x} dx = \int_0^1 (40x^4 + 16x^3)\, dx$$

$$= \left[8x^5 + 4x^4 \right]_0^1 = 12.$$

Similarly, the moment of inertia about the y-axis is

$$I_y = \int_0^1 \int_0^{2x} x^2 \delta(x, y)\, dy\, dx = \frac{39}{5}.$$

Notice that we integrate y^2 times density in calculating I_x and x^2 times density to find I_y.

Since we know I_x and I_y, we do not need to evaluate an integral to find I_0; we can use the equation $I_0 = I_x + I_y$ from Table 15.2 instead:

$$I_0 = 12 + \frac{39}{5} = \frac{60 + 39}{5} = \frac{99}{5}. \qquad \blacksquare$$

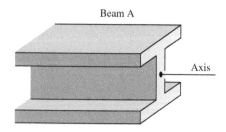

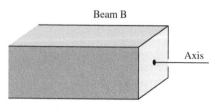

FIGURE 15.44 The greater the polar moment of inertia of the cross-section of a beam about the beam's longitudinal axis, the stiffer the beam. Beams A and B have the same cross-sectional area, but A is stiffer.

The moment of inertia also plays a role in determining how much a horizontal metal beam will bend under a load. The stiffness of the beam is a constant times I, the moment of inertia of a typical cross-section of the beam about the beam's longitudinal axis. The greater the value of I, the stiffer the beam and the less it will bend under a given load. That is why we use I-beams instead of beams whose cross-sections are square. The flanges at the top and bottom of the beam hold most of the beam's mass away from the longitudinal axis to increase the value of I (Figure 15.44).

Probability

In Chapter 8 we saw that the probability that a continuous random variable X takes values between a and b is found by integrating a probability density function f,

$$P(a \le X \le b) = \int_a^b f(x)\, dx.$$

A similar process applies to probabilities involving two continuous random variables. The probability that a pair of random variables (X, Y) takes values lying within a particular region is determined by a *joint probability density function* f. Integrating the joint probability density function over a region R in the plane gives the probability that the pair of random variables take values in that region:

$$P((X, Y) \in R) = \iint_R f(x, y)\, dx\, dy.$$

If the region is a rectangle, then this expression has the simple form

$$P(a \le X \le b \text{ and } c \le Y \le d) = \int_a^b \int_c^d f(x, y)\, dx\, dy.$$

A joint probability density function f is defined by three basic properties. The first property ensures that there are no negative probabilities, and the second implies that the total probability of all possible outcomes is one. The final property describes the connection of f to probabilities.

DEFINITION   A **joint probability density function** f is a function that satisfies three conditions:

1. $f(x, y) \ge 0$

2. $\displaystyle\int_{-\infty}^{\infty} \int_{-\infty}^{\infty} f(x, y)\, dx\, dy = 1$

3. $\displaystyle P((X, Y) \in R) = \iint_R f(x, y)\, dx\, dy.$

A pair of random variables has a *uniform* distribution on a region R with finite area A if $f(x, y) = 1/A$ for any $(x, y) \in R$, and $f(x, y) = 0$ otherwise.

EXAMPLE 5   A random number generator is used to generate two random real numbers X and Y in succession. The first number X is chosen between 0 and 10, and the second number Y is chosen between 0 and 5. The random number generation is done by a process that gives a uniform distribution. Find the joint probability density function f for the pair of numbers (X, Y) and use it to compute the probability that X is larger than Y.

Solution   The joint probability density function f is constant on the rectangle $0 \le x \le 10, 0 \le y \le 5$, because (X, Y) are uniformly distributed. The area of the rectangle is 50, so f takes the value $1/50$ inside this rectangle:

$$f(x, y) = \begin{cases} 1/50, & \text{if } 0 \le x \le 10 \text{ and } 0 \le y \le 5, \\ 0, & \text{otherwise.} \end{cases}$$

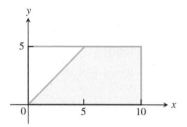

FIGURE 15.45 The pair of random variables X and Y take values anywhere in this rectangle with equal probability. In the shaded region we have $X > Y$.

To compute the probability that $X > Y$, we integrate the joint probability density function f over the region in the rectangle where $X > Y$. This region is bounded on the left by the line $x = y$ and on the right by the line $x = 10$. An integral over this region has limits of integration given by $y \leq x \leq 10, 0 \leq y \leq 5$ (see Figure 15.45). The probability is given by

$$P(X > Y) = \int_0^5 \int_y^{10} \frac{1}{50} \, dx \, dy = \frac{3}{4}.$$

There is a 75% probability that the first number is larger than the second. ∎

EXAMPLE 6 Using the joint probability density function

$$f(x, y) = \begin{cases} e^{-(x+y)}, & \text{if } 0 < x \text{ and } 0 < y \\ 0, & \text{otherwise} \end{cases}$$

find the probability that $1 < X < 2$ and $2 < Y < 3$.

Solution

$$P(1 < X < 2, 2 < Y < 3) = \int_2^3 \int_1^2 e^{-(x+y)} \, dx \, dy = e^{-5} + e^{-3} - 2e^{-4} \approx 0.01989.$$

There is slightly less than a 2% probability that X and Y fall within these bounds. ∎

Means and Expected Values

The mean, or expected value, of a random variable was seen in Section 8.9 to be

$$\mu = \int_{-\infty}^{\infty} x f(x) \, dx.$$

When X and Y have joint probability density function f, the expected value of X and the expected value of Y are

$$\mu_X = \int_{-\infty}^{\infty} \int_{-\infty}^{\infty} x f(x, y) \, dx \, dy \qquad \text{and} \qquad \mu_Y = \int_{-\infty}^{\infty} \int_{-\infty}^{\infty} y f(x, y) \, dx \, dy.$$

These indicate the average value expected for each of X and Y. The expected values μ_X and μ_Y are sometimes called the first moments of the distribution, because their defining formulas have the same form as those seen in Table 15.1 for the moments of a two-dimensional plate. The joint probability density function plays the role in computing μ_X that the mass density function plays in computing the x-coordinate $\bar{x}$ of the center of mass, and the same applies to μ_Y and $\bar{y}$. One can roughly think of the joint probability density function as measuring the probability concentration per unit area on the plane, just as density measures the mass per unit area for a plate.

EXAMPLE 7 Find the expected values μ_X and μ_Y for the joint probability density function in Example 5.

Solution For the joint probability density function in Example 5, we compute

$$\mu_X = \int_0^5 \int_0^{10} x(1/50) \, dx \, dy = 5$$

and

$$\mu_Y = \int_0^5 \int_0^{10} y(1/50) \, dx \, dy = 2.5.$$

The expected value of X is 5 and that of Y is 2.5. ∎

EXERCISES 15.6

Plates of Constant Density

1. **Finding a center of mass** Find the center of mass of a thin plate of density $\delta = 3$ bounded by the lines $x = 0$, $y = x$, and the parabola $y = 2 - x^2$ in the first quadrant.

2. **Finding moments of inertia** Find the moments of inertia about the coordinate axes of a thin rectangular plate of constant density δ gm/cm^2 bounded by the lines $x = 3$ and $y = 3$ in the first quadrant.

3. **Finding a centroid** Find the centroid of the region in the first quadrant bounded by the x-axis, the parabola $y^2 = 2x$, and the line $x + y = 4$.

4. **Finding a centroid** Find the centroid of the triangular region cut from the first quadrant by the line $x + y = 3$.

5. **Finding a centroid** Find the centroid of the region cut from the first quadrant by the circle $x^2 + y^2 = a^2$.

6. **Finding a centroid** Find the centroid of the region between the x-axis and the arch $y = \sin x$, $0 \le x \le \pi$.

7. **Finding moments of inertia** Find the moment of inertia about the x-axis of a thin plate of density $\delta = 1$ gm/cm^2 bounded by the circle $x^2 + y^2 = 4$. Then use your result to find I_y and I_0 for the plate.

8. **Finding a moment of inertia** Find the moment of inertia with respect to the y-axis of a thin sheet of constant density $\delta = 1$ gm/cm^2 bounded by the curve $y = (\sin^2 x)/x^2$ and the interval $\pi \le x \le 2\pi$ of the x-axis.

9. **The centroid of an infinite region** Find the centroid of the infinite region in the second quadrant enclosed by the coordinate axes and the curve $y = e^x$. (Use improper integrals in the mass-moment formulas.)

10. **The first moment of an infinite plate** Find the first moment about the y-axis of a thin plate of density $\delta(x, y) = 1$ covering the infinite region under the curve $y = e^{-x^2/2}$ in the first quadrant.

Plates with Varying Density

11. **Finding a moment of inertia** Find the moment of inertia about the x-axis of a thin plate bounded by the parabola $x = y - y^2$ and the line $x + y = 0$ if $\delta(x, y) = x + y$.

12. **Finding mass** Find the mass of a thin plate occupying the smaller region cut from the ellipse $x^2 + 4y^2 = 12$ by the parabola $x = 4y^2$ if $\delta(x, y) = 5x$ kg/m^2.

13. **Finding a center of mass** Find the center of mass of a thin triangular plate bounded by the y-axis and the lines $y = x$ and $y = 2 - x$ if $\delta(x, y) = 6x + 3y + 3$.

14. **Finding a center of mass and moment of inertia** Find the center of mass and moment of inertia about the x-axis of a thin plate bounded by the curves $x = y^2$ and $x = 2y - y^2$ if the density at the point (x, y) is $\delta(x, y) = y + 1$.

15. **Center of mass, moment of inertia** Find the center of mass and the moment of inertia about the y-axis of a thin rectangular plate cut from the first quadrant by the lines $x = 6$ and $y = 1$ if $\delta(x, y) = x + y + 1$.

16. **Center of mass, moment of inertia** Find the center of mass and the moment of inertia about the y-axis of a thin plate bounded by the line $y = 1$ and the parabola $y = x^2$ if the density is $\delta(x, y) = y + 1$.

17. **Center of mass, moment of inertia** Find the center of mass and the moment of inertia about the y-axis of a thin plate bounded by the x-axis, the lines $x = \pm 1$, and the parabola $y = x^2$ if $\delta(x, y) = 7y + 1$.

18. **Center of mass, moment of inertia** Find the center of mass and the moment of inertia about the x-axis of a thin rectangular plate bounded by the lines $x = 0, x = 20, y = -1$, and $y = 1$ if $\delta(x, y) = 1 + (x/20)$.

19. **Center of mass, moments of inertia** Find the center of mass, the moment of inertia about the coordinate axes, and the polar moment of inertia of a thin triangular plate bounded by the lines $y = x, y = -x$, and $y = 1$ if $\delta(x, y) = y + 1$ kg/m^2.

20. **Center of mass, moments of inertia** Repeat Exercise 19 for $\delta(x, y) = 3x^2 + 1$ kg/m^2.

Solids with Constant Density

21. **Moments of inertia** Find the moments of inertia of the rectangular solid shown here with respect to its edges by calculating I_x, I_y, and I_z.

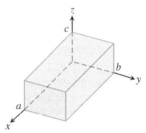

22. **Moments of inertia** The coordinate axes in the figure run through the centroid of a solid wedge parallel to the labeled edges. Find I_x, I_y, and I_z if $a = b = 6$ and $c = 4$.

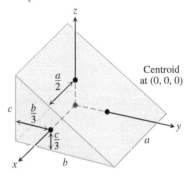

23. **Center of mass and moments of inertia** A solid "trough" of constant density is bounded below by the surface $z = 4y^2$, above by the plane $z = 4$, and on the ends by the planes $x = 1$ and $x = -1$. Find the center of mass and the moments of inertia with respect to the three axes.

24. Center of mass A solid of constant density is bounded below by the plane $z = 0$, on the sides by the elliptical cylinder $x^2 + 4y^2 = 4$, and above by the plane $z = 2 - x$ (see the accompanying figure).

a. Find $\bar{x}$ and $\bar{y}$.

b. Evaluate the integral

$$M_{xy} = \int_{-2}^{2} \int_{-(1/2)\sqrt{4-x^2}}^{(1/2)\sqrt{4-x^2}} \int_{0}^{2-x} z \, dz \, dy \, dx$$

using integral tables to carry out the final integration with respect to x. Then divide M_{xy} by M to verify that $\bar{z} = 5/4$.

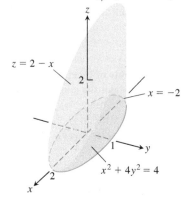

25. a. Center of mass Find the center of mass of a solid of constant density bounded below by the paraboloid $z = x^2 + y^2$ and above by the plane $z = 4$.

b. Find the plane $z = c$ that divides the solid into two parts of equal volume. This plane does not pass through the center of mass.

26. Moments A solid cube, 2 units on a side, is bounded by the planes $x = \pm 1, z = \pm 1, y = 3$, and $y = 5$. Find the center of mass and the moments of inertia about the coordinate axes.

27. Moment of inertia about a line A wedge like the one in Exercise 22 has $a = 4, b = 6$, and $c = 3$. Make a quick sketch to check for yourself that the square of the distance from a typical point (x, y, z) of the wedge to the line $L: z = 0, y = 6$ is $r^2 = (y - 6)^2 + z^2$. Then calculate the moment of inertia of the wedge about L.

28. Moment of inertia about a line A wedge like the one in Exercise 22 has $a = 4, b = 6$, and $c = 3$. Make a quick sketch to check for yourself that the square of the distance from a typical point (x, y, z) of the wedge to the line $L: x = 4, y = 0$ is $r^2 = (x - 4)^2 + y^2$. Then calculate the moment of inertia of the wedge about L.

Solids with Varying Density
In Exercises 29 and 30, find

a. the mass of the solid. **b.** the center of mass.

29. A solid region in the first octant is bounded by the coordinate planes and the plane $x + y + z = 2$. The density of the solid is $\delta(x, y, z) = 2x$ gm/cm^3.

30. A solid in the first octant is bounded by the planes $y = 0$ and $z = 0$ and by the surfaces $z = 4 - x^2$ and $x = y^2$ (see the accompanying figure). Its density function is $\delta(x, y, z) = kxy$, k a constant.

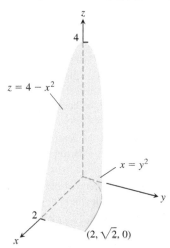

In Exercises 31 and 32, find

a. the mass of the solid.

b. the center of mass.

c. the moments of inertia about the coordinate axes.

31. A solid cube in the first octant is bounded by the coordinate planes and by the planes $x = 1, y = 1$, and $z = 1$. The density of the cube is $\delta(x, y, z) = x + y + z + 1$.

32. A wedge like the one in Exercise 22 has dimensions $a = 2$, $b = 6$, and $c = 3$. The density is $\delta(x, y, z) = x + 1$. Notice that if the density is constant, the center of mass will be $(0, 0, 0)$.

33. Mass Find the mass of the solid bounded by the planes $x + z = 1, x - z = -1, y = 0$, and the surface $y = \sqrt{z}$. The density of the solid is $\delta(x, y, z) = 2y + 5$ kg/m^3.

34. Mass Find the mass of the solid region bounded by the parabolic surfaces $z = 16 - 2x^2 - 2y^2$ and $z = 2x^2 + 2y^2$ if the density of the solid is $\delta(x, y, z) = \sqrt{x^2 + y^2}$.

Theory and Examples
The Parallel Axis Theorem Let $L_{c.m.}$ be a line through the center of mass of a body of mass m and let L be a parallel line h units away from $L_{c.m.}$. The *Parallel Axis Theorem* says that the moments of inertia $I_{c.m.}$ and I_L of the body about $L_{c.m.}$ and L satisfy the equation

$$I_L = I_{c.m.} + mh^2. \qquad (2)$$

As in the two-dimensional case, the theorem gives a quick way to calculate one moment when the other moment and the mass are known.

35. Proof of the Parallel Axis Theorem

a. Show that the first moment of a body in space about any plane through the body's center of mass is zero. (*Hint:* Place the body's center of mass

the body's center of mass at the origin and let the plane be the yz-plane. What does the formula $\bar{x} = M_{yz}/M$ then tell you?)

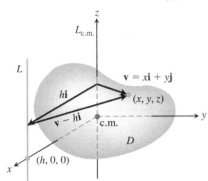

b. To prove the Parallel Axis Theorem, place the body with its center of mass at the origin, with the line $L_{c.m.}$ along the z-axis and the line L perpendicular to the xy-plane at the point $(h, 0, 0)$. Let D be the region of space occupied by the body. Then, in the notation of the figure,

$$I_L = \iiint_D |\mathbf{v} - h\mathbf{i}|^2 \, dm.$$

Expand the integrand in this integral and complete the proof.

36. The moment of inertia about a diameter of a solid sphere of constant density and radius a is $(2/5)ma^2$, where m is the mass of the sphere. Find the moment of inertia about a line tangent to the sphere.

37. The moment of inertia of the solid in Exercise 21 about the z-axis is $I_z = abc(a^2 + b^2)/3$.

a. Use Equation (2) to find the moment of inertia of the solid about the line parallel to the z-axis through the solid's center of mass.

b. Use Equation (2) and the result in part (a) to find the moment of inertia of the solid about the line $x = 0, y = 2b$.

38. If $a = b = 6$ and $c = 4$, the moment of inertia of the solid wedge in Exercise 22 about the x-axis is $I_x = 208$. Find the moment of inertia of the wedge about the line $y = 4, z = -4/3$ (the edge of the wedge's narrow end).

Joint Probability Density Functions

For Exercises 39–42, verify that f gives a joint probability density function. Then find the expected values μ_X and μ_Y.

39. $f(x, y) = \begin{cases} x + y, & \text{if } 0 \le x \le 1 \text{ and } 0 \le y \le 1, \\ 0, & \text{otherwise.} \end{cases}$

40. $f(x, y) = \begin{cases} 4xy, & \text{if } 0 \le x \le 1 \text{ and } 0 \le y \le 1, \\ 0, & \text{otherwise.} \end{cases}$

41. $f(x, y) = \begin{cases} 6x^2y, & \text{if } 0 \le x \le 1 \text{ and } 0 \le y \le 1, \\ 0, & \text{otherwise.} \end{cases}$

42. $f(x, y) = \begin{cases} \frac{3}{2}(x^2 + y^2), & \text{if } 0 \le x \le 1 \text{ and } 0 \le y \le 1, \\ 0, & \text{otherwise.} \end{cases}$

43. Suppose that f is a uniform joint probability density function on $0 \le x < 2, 0 \le y < 3$. What is the formula for f? What is the probability that $X < Y$?

44. The following formula defines a joint probability density function. What is the value of C? What are the expected values μ_X and μ_Y?

$$f(x, y) = \begin{cases} Cxy, & \text{if } 0 \le x \le 2 \text{ and } 0 \le y \le 3, \\ 0, & \text{otherwise.} \end{cases}$$

15.7 Triple Integrals in Cylindrical and Spherical Coordinates

When a calculation in physics, engineering, or geometry involves a cylinder, cone, or sphere, we can often simplify our work by using cylindrical or spherical coordinates, which are introduced in this section. The procedure for transforming to these coordinates and evaluating the resulting triple integrals is similar to the transformation to polar coordinates in the plane studied in Section 15.4.

Integration in Cylindrical Coordinates

We obtain cylindrical coordinates for space by combining polar coordinates in the xy-plane with the usual z-axis. This assigns to every point in space one or more coordinate triples of the form (r, θ, z), as shown in Figure 15.46. Here we require $r \ge 0$.

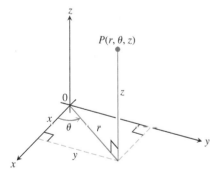

FIGURE 15.46 The cylindrical coordinates of a point in space are r, θ, and z.

> DEFINITION **Cylindrical coordinates** represent a point P in space by ordered triples (r, θ, z) in which $r \ge 0$,
>
> **1.** r and θ are polar coordinates for the vertical projection of P on the xy-plane
>
> **2.** z is the rectangular vertical coordinate.

The values of x, y, r, and θ in rectangular and cylindrical coordinates are related by the usual equations.

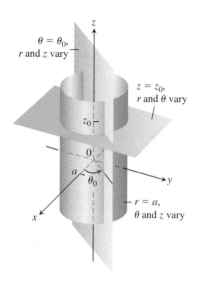

FIGURE 15.47 Constant-coordinate equations in cylindrical coordinates yield cylinders and planes.

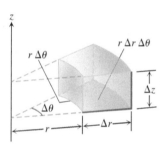

FIGURE 15.48 In cylindrical coordinates the volume of the wedge is approximated by the product $\Delta V = \Delta z \, r \, \Delta r \, \Delta \theta$.

Volume Differential in Cylindrical Coordinates

$$dV = dz \, r \, dr \, d\theta$$

Equations Relating Rectangular (x, y, z) and Cylindrical (r, θ, z) Coordinates

$$x = r \cos \theta, \qquad y = r \sin \theta, \qquad z = z,$$
$$r^2 = x^2 + y^2, \qquad \tan \theta = y/x$$

In cylindrical coordinates, the equation $r = a$ describes not just a circle in the xy-plane but an entire cylinder about the z-axis (Figure 15.47). The z-axis is given by $r = 0$. The equation $\theta = \theta_0$ describes the plane that contains the z-axis and makes an angle θ_0 with the positive x-axis. And, just as in rectangular coordinates, the equation $z = z_0$ describes a plane perpendicular to the z-axis.

Cylindrical coordinates are good for describing cylinders whose axes run along the z-axis and planes that either contain the z-axis or lie perpendicular to the z-axis. Surfaces like these have equations of constant coordinate value:

$$r = 4 \qquad \text{Cylinder, radius 4, axis the } z\text{-axis}$$
$$\theta = \frac{\pi}{3} \qquad \text{Plane containing the } z\text{-axis}$$
$$z = 2. \qquad \text{Plane perpendicular to the } z\text{-axis}$$

When computing triple integrals over a region D in cylindrical coordinates, we partition the region into n small cylindrical wedges, rather than into rectangular boxes. In the kth cylindrical wedge, r, θ and z change by Δr_k, $\Delta \theta_k$, and Δz_k, and the largest of these numbers among all the cylindrical wedges is called the **norm** of the partition. We define the triple integral as a limit of Riemann sums using these wedges. The volume of such a cylindrical wedge ΔV_k is obtained by taking the area ΔA_k of its base in the $r\theta$-plane and multiplying by the height Δz (Figure 15.48).

For a point (r_k, θ_k, z_k) in the center of the kth wedge, we calculated in polar coordinates that $\Delta A_k = r_k \, \Delta r_k \, \Delta \theta_k$. So $\Delta V_k = \Delta z_k \, r_k \, \Delta r_k \, \Delta \theta_k$ and a Riemann sum for f over D has the form

$$S_n = \sum_{k=1}^{n} f(r_k, \theta_k, z_k) \, \Delta z_k \, r_k \, \Delta r_k \, \Delta \theta_k.$$

The triple integral of a function f over D is obtained by taking a limit of such Riemann sums with partitions whose norms approach zero:

$$\lim_{n \to \infty} S_n = \iiint_D f \, dV = \iiint_D f \, dz \, r \, dr \, d\theta.$$

Triple integrals in cylindrical coordinates are then evaluated as iterated integrals, as in the following example. Although the definition of cylindrical coordinates makes sense without any restrictions on θ, in most situations when integrating, we will need to restrict θ to an interval of length 2π. So we impose the requirement that $\alpha \le \theta \le \beta$, where $0 \le \beta - \alpha \le 2\pi$.

EXAMPLE 1   Find the limits of integration in cylindrical coordinates for integrating a function $f(r, \theta, z)$ over the region D bounded below by the plane $z = 0$, laterally by the circular cylinder $x^2 + (y - 1)^2 = 1$, and above by the paraboloid $z = x^2 + y^2$.

Solution   The base of D is also the region's projection R on the xy-plane. The boundary of R is the circle $x^2 + (y - 1)^2 = 1$. Its polar coordinate equation is

$$x^2 + (y - 1)^2 = 1$$
$$x^2 + y^2 - 2y + 1 = 1$$
$$r^2 - 2r \sin \theta = 0$$
$$r = 2 \sin \theta.$$

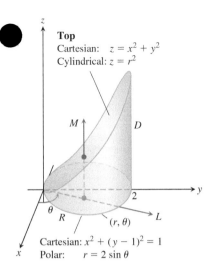

Top
Cartesian: $z = x^2 + y^2$
Cylindrical: $z = r^2$

Cartesian: $x^2 + (y-1)^2 = 1$
Polar: $r = 2 \sin \theta$

FIGURE 15.49 Finding the limits of integration for evaluating an integral in cylindrical coordinates (Example 1).

The region is sketched in Figure 15.49.

We find the limits of integration, starting with the z-limits. A line M through a typical point (r, θ) in R parallel to the z-axis enters D at $z = 0$ and leaves at $z = x^2 + y^2 = r^2$.

Next we find the r-limits of integration. A ray L through (r, θ) from the origin enters R at $r = 0$ and leaves at $r = 2 \sin \theta$.

Finally we find the θ-limits of integration. As L sweeps across R, the angle θ it makes with the positive x-axis runs from $\theta = 0$ to $\theta = \pi$. The integral is

$$\iiint_D f(r, \theta, z)\, dV = \int_0^\pi \int_0^{2 \sin \theta} \int_0^{r^2} f(r, \theta, z)\, dz\, r\, dr\, d\theta. \qquad \blacksquare$$

Example 1 illustrates a good procedure for finding limits of integration in cylindrical coordinates. The procedure is summarized as follows.

How to Integrate in Cylindrical Coordinates

To evaluate

$$\iiint_D f(r, \theta, z)\, dV$$

over a region D in space in cylindrical coordinates, integrating first with respect to z, then with respect to r, and finally with respect to θ, take the following steps.

1. *Sketch.* Sketch the region D along with its projection R on the xy-plane. Label the surfaces and curves that bound D and R.

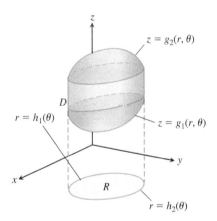

2. *Find the z-limits of integration.* Draw a line M through a typical point (r, θ) of R parallel to the z-axis. As z increases, M enters D at $z = g_1(r, \theta)$ and leaves at $z = g_2(r, \theta)$. These are the z-limits of integration.

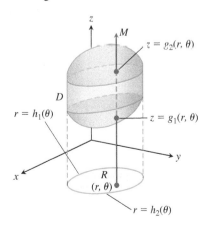

3. *Find the r-limits of integration.* Draw a ray L through (r, θ) from the origin. The ray enters R at $r = h_1(\theta)$ and leaves at $r = h_2(\theta)$. These are the r-limits of integration.

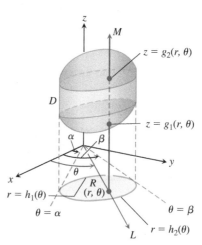

4. *Find the θ-limits of integration.* As L sweeps across R, the angle θ it makes with the positive x-axis runs from $\theta = \alpha$ to $\theta = \beta$. These are the θ-limits of integration. The integral is

$$\iiint_D f(r, \theta, z)\, dV = \int_{\theta=\alpha}^{\theta=\beta} \int_{r=h_1(\theta)}^{r=h_2(\theta)} \int_{z=g_1(r, \theta)}^{z=g_2(r, \theta)} f(r, \theta, z)\, dz\, r\, dr\, d\theta.$$

EXAMPLE 2 Find the centroid ($\delta = 1$) of the solid enclosed by the cylinder $x^2 + y^2 = 4$, bounded above by the paraboloid $z = x^2 + y^2$, and bounded below by the xy-plane.

Solution We sketch the solid, bounded above by the paraboloid $z = r^2$ and below by the plane $z = 0$ (Figure 15.50). Its base R is the disk $0 \le r \le 2$ in the xy-plane.

The solid's centroid $(\bar{x}, \bar{y}, \bar{z})$ lies on its axis of symmetry, here the z-axis. This makes $\bar{x} = \bar{y} = 0$. To find $\bar{z}$, we divide the first moment M_{xy} by the mass M.

To find the limits of integration for the mass and moment integrals, we continue with the four basic steps. We completed our initial sketch. The remaining steps give the limits of integration.

The z-limits. A line M through a typical point (r, θ) in the base parallel to the z-axis enters the solid at $z = 0$ and leaves at $z = r^2$.

The r-limits. A ray L through (r, θ) from the origin enters R at $r = 0$ and leaves at $r = 2$.

The θ-limits. As L sweeps over the base like a clock hand, the angle θ it makes with the positive x-axis runs from $\theta = 0$ to $\theta = 2\pi$. The value of M_{xy} is

$$M_{xy} = \int_0^{2\pi}\int_0^2\int_0^{r^2} z\, dz\, r\, dr\, d\theta = \int_0^{2\pi}\int_0^2 \left[\frac{z^2}{2}\right]_0^{r^2} r\, dr\, d\theta$$
$$= \int_0^{2\pi}\int_0^2 \frac{r^5}{2}\, dr\, d\theta = \int_0^{2\pi}\left[\frac{r^6}{12}\right]_0^2 d\theta = \int_0^{2\pi} \frac{16}{3}\, d\theta = \frac{32\pi}{3}.$$

The value of M is

$$M = \int_0^{2\pi}\int_0^2\int_0^{r^2} dz\, r\, dr\, d\theta = \int_0^{2\pi}\int_0^2 \left[z\right]_0^{r^2} r\, dr\, d\theta$$
$$= \int_0^{2\pi}\int_0^2 r^3\, dr\, d\theta = \int_0^{2\pi}\left[\frac{r^4}{4}\right]_0^2 d\theta = \int_0^{2\pi} 4\, d\theta = 8\pi.$$

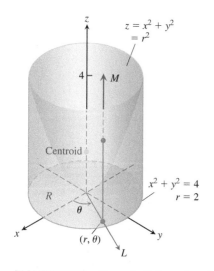

FIGURE 15.50 Example 2 shows how to find the centroid of this solid.

Therefore,

$$\bar{z} = \frac{M_{xy}}{M} = \frac{32\pi}{3} \frac{1}{8\pi} = \frac{4}{3},$$

and the centroid is $(0, 0, 4/3)$. Notice that the centroid lies on the z-axis, outside the solid. ∎

Spherical Coordinates and Integration

Spherical coordinates locate points in space with two angles and one distance, as shown in Figure 15.51. The first coordinate, $\rho = |\overrightarrow{OP}|$, is the point's distance from the origin and is never negative. The second coordinate, ϕ, is the angle $\overrightarrow{OP}$ makes with the positive z-axis. It is required to lie in the interval $[0, \pi]$. The third coordinate is the angle θ as measured in cylindrical coordinates.

ϕ is the Greek letter phi, pronounced "fee."

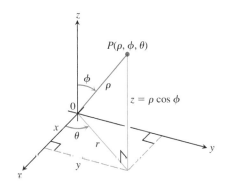

FIGURE 15.51 The spherical coordinates ρ, ϕ, and θ and their relation to x, y, z, and r.

> **DEFINITION** **Spherical coordinates** represent a point P in space by ordered triples (ρ, ϕ, θ) in which
>
> 1. ρ is the distance from P to the origin ($\rho \geq 0$).
> 2. ϕ is the angle $\overrightarrow{OP}$ makes with the positive z-axis ($0 \leq \phi \leq \pi$).
> 3. θ is the angle from cylindrical coordinates.

On maps of the earth, θ is related to the meridian of a point on the earth and ϕ to its latitude, while ρ is related to elevation above the earth's surface.

The equation $\rho = a$ describes the sphere of radius a centered at the origin (Figure 15.52). The equation $\phi = \phi_0$ describes a single cone whose vertex lies at the origin and whose axis lies along the z-axis. (We broaden our interpretation to include the xy-plane as the cone $\phi = \pi/2$.) If ϕ_0 is greater than $\pi/2$, the cone $\phi = \phi_0$ opens downward. The equation $\theta = \theta_0$ describes the half-plane that contains the z-axis and makes an angle θ_0 with the positive x-axis.

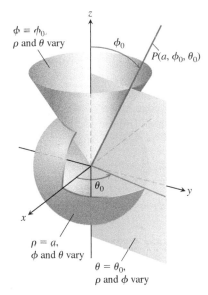

$\phi = \phi_0$, ρ and θ vary

$\rho = a$, ϕ and θ vary

$\theta = \theta_0$, ρ and ϕ vary

FIGURE 15.52 Constant-coordinate equations in spherical coordinates yield spheres, single cones, and half-planes.

> **Equations Relating Spherical Coordinates to Cartesian and Cylindrical Coordinates**
>
>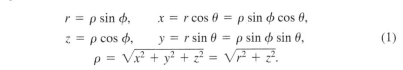
>
> $$r = \rho \sin \phi, \qquad x = r \cos \theta = \rho \sin \phi \cos \theta,$$
> $$z = \rho \cos \phi, \qquad y = r \sin \theta = \rho \sin \phi \sin \theta, \qquad (1)$$
> $$\rho = \sqrt{x^2 + y^2 + z^2} = \sqrt{r^2 + z^2}.$$

EXAMPLE 3 Find a spherical coordinate equation for the sphere $x^2 + y^2 + (z - 1)^2 = 1$.

Solution We use Equations (1) to substitute for x, y, and z:

$$x^2 + y^2 + (z - 1)^2 = 1$$
$$\rho^2 \sin^2 \phi \cos^2 \theta + \rho^2 \sin^2 \phi \sin^2 \theta + (\rho \cos \phi - 1)^2 = 1 \qquad \text{Eqs. (1)}$$
$$\rho^2 \sin^2 \phi \underbrace{(\cos^2 \theta + \sin^2 \theta)}_{1} + \rho^2 \cos^2 \phi - 2\rho \cos \phi + 1 = 1$$

$$\rho^2 \underbrace{(\sin^2 \phi + \cos^2 \phi)}_{1} = 2\rho \cos \phi$$

$$\rho^2 = 2\rho \cos \phi$$
$$\rho = 2 \cos \phi. \qquad \rho > 0$$

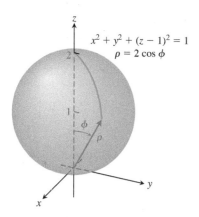

FIGURE 15.53   The sphere in Example 3.

$$x^2 + y^2 + (z - 1)^2 = 1$$
$$\rho = 2\cos\phi$$

The angle ϕ varies from 0 at the north pole of the sphere to $\pi/2$ at the south pole; the angle θ does not appear in the expression for ρ, reflecting the symmetry about the z-axis (see Figure 15.53).

EXAMPLE 4   Find a spherical coordinate equation for the cone $z = \sqrt{x^2 + y^2}$.

Solution 1   *Use geometry.* The cone is symmetric with respect to the z-axis and cuts the first quadrant of the yz-plane along the line $z = y$. The angle between the cone and the positive z-axis is therefore $\pi/4$ radians. The cone consists of the points whose spherical coordinates have ϕ equal to $\pi/4$, so its equation is $\phi = \pi/4$. (See Figure 15.54.)

Solution 2   *Use algebra.* If we use Equations (1) to substitute for x, y, and z we obtain the same result:

$$z = \sqrt{x^2 + y^2}$$
$$\rho\cos\phi = \sqrt{\rho^2\sin^2\phi} \qquad \text{Example 3}$$
$$\rho\cos\phi = \rho\sin\phi \qquad \rho > 0,\ \sin\phi \geq 0$$
$$\cos\phi = \sin\phi$$
$$\phi = \frac{\pi}{4}. \qquad 0 \leq \phi \leq \pi$$

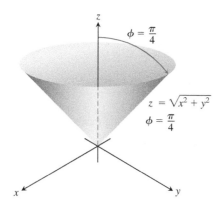

$$\phi = \frac{\pi}{4}$$

$$z = \sqrt{x^2 + y^2}$$
$$\phi = \frac{\pi}{4}$$

FIGURE 15.54   The cone in Example 4.

Spherical coordinates are useful for describing spheres centered at the origin, half-planes hinged along the z-axis, and cones whose vertices lie at the origin and whose axes lie along the z-axis. Surfaces like these have equations of constant coordinate value:

$$\rho = 4 \qquad \text{Sphere, radius 4, center at origin}$$

$$\phi = \frac{\pi}{3} \qquad \begin{array}{l}\text{Cone opening up from the origin, making} \\ \text{angle of } \pi/3 \text{ radians with the positive } z\text{-axis}\end{array}$$

$$\theta = \frac{\pi}{3}. \qquad \begin{array}{l}\text{Half-plane, hinged along the } z\text{-axis, making an} \\ \text{angle of } \pi/3 \text{ radians with the positive } x\text{-axis}\end{array}$$

When computing triple integrals over a region D in spherical coordinates, we partition the region into n spherical wedges. The size of the kth spherical wedge, which contains a point $(\rho_k, \phi_k, \theta_k)$, is given by the changes $\Delta\rho_k$, $\Delta\phi_k$, and $\Delta\theta_k$ in ρ, ϕ, and θ. Such a spherical wedge has one edge a circular arc of length $\rho_k\,\Delta\phi_k$, another edge a circular arc of length $\rho_k\sin\phi_k\,\Delta\theta_k$, and thickness $\Delta\rho_k$. The spherical wedge closely approximates a cube of these dimensions when $\Delta\rho_k$, $\Delta\phi_k$, and $\Delta\theta_k$ are all small (Figure 15.55). It can be shown that the volume of this spherical wedge ΔV_k is $\Delta V_k = \rho_k^2\sin\phi_k\,\Delta\rho_k\,\Delta\phi_k\,\Delta\theta_k$ for $(\rho_k, \phi_k, \theta_k)$, a point chosen inside the wedge.

The corresponding Riemann sum for a function $f(\rho, \phi, \theta)$ is

$$S_n = \sum_{k=1}^{n} f(\rho_k, \phi_k, \theta_k)\,\rho_k^2\sin\phi_k\,\Delta\rho_k\,\Delta\phi_k\,\Delta\theta_k.$$

As the norm of a partition approaches zero, and the spherical wedges get smaller, the Riemann sums have a limit when f is continuous:

$$\lim_{n\to\infty} S_n = \iiint_D f(\rho, \phi, \theta)\,dV = \iiint_D f(\rho, \phi, \theta)\,\rho^2\sin\phi\,d\rho\,d\phi\,d\theta.$$

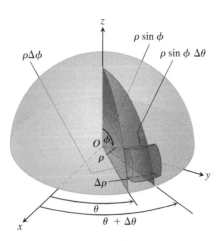

FIGURE 15.55   In spherical coordinates we use the volume of a spherical wedge, which closely approximates that of a cube.

Volume Differential in Spherical Coordinates

$$dV = \rho^2\sin\phi\,d\rho\,d\phi\,d\theta$$

To evaluate integrals in spherical coordinates, we usually integrate first with respect to ρ. The procedure for finding the limits of integration is as follows. We restrict our attention to integrating over domains that are solids of revolution about the z-axis (or portions thereof) and for which the limits for θ and ϕ are constant. As with cylindrical coordinates, we restrict θ in the form $\alpha \le \theta \le \beta$ and $0 \le \beta - \alpha \le 2\pi$.

How to Integrate in Spherical Coordinates

To evaluate

$$\iiint_D f(\rho, \phi, \theta) \, dV$$

over a region D in space in spherical coordinates, integrating first with respect to ρ, then with respect to ϕ, and finally with respect to θ, take the following steps.

1. *Sketch.* Sketch the region D along with its projection R on the xy-plane. Label the surfaces that bound D.

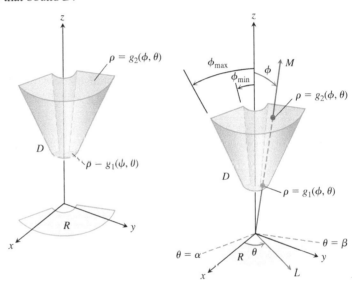

2. *Find the ρ-limits of integration.* Draw a ray M from the origin through D, making an angle ϕ with the positive z-axis. Also draw the projection of M on the xy-plane (call the projection L). The ray L makes an angle θ with the positive x-axis. As ρ increases, M enters D at $\rho = g_1(\phi, \theta)$ and leaves at $\rho = g_2(\phi, \theta)$. These are the ρ-limits of integration shown in the above figure.

3. *Find the ϕ-limits of integration.* For any given θ, the angle ϕ that M makes with the z-axis runs from $\phi = \phi_{min}$ to $\phi = \phi_{max}$. These are the ϕ-limits of integration.

4. *Find the θ-limits of integration.* The ray L sweeps over R as θ runs from α to β. These are the θ-limits of integration. The integral is

$$\iiint_D f(\rho, \phi, \theta) \, dV = \int_{\theta = \alpha}^{\theta = \beta} \int_{\phi = \phi_{min}}^{\phi = \phi_{max}} \int_{\rho = g_1(\phi, \theta)}^{\rho = g_2(\phi, \theta)} f(\rho, \phi, \theta) \, \rho^2 \sin \phi \, d\rho \, d\phi \, d\theta.$$

EXAMPLE 5   Find the volume of the "ice cream cone" D cut from the solid sphere $\rho \le 1$ by the cone $\phi = \pi/3$.

Solution   The volume is $V = \iiint_D \rho^2 \sin \phi \, d\rho \, d\phi \, d\theta$, the integral of $f(\rho, \phi, \theta) = 1$ over D.

To find the limits of integration for evaluating the integral, we begin by sketching D and its projection R on the xy-plane (Figure 15.56).

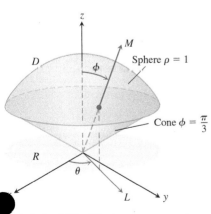

FIGURE 15.56   The ice cream cone in Example 5.

The ρ-limits of integration. We draw a ray M from the origin through D, making an angle ϕ with the positive z-axis. We also draw L, the projection of M on the xy-plane, along with the angle θ that L makes with the positive x-axis. Ray M enters D at $\rho = 0$ and leaves at $\rho = 1$.

The φ-limits of integration. The cone $\phi = \pi/3$ makes an angle of $\pi/3$ with the positive z-axis. For any given θ, the angle ϕ can run from $\phi = 0$ to $\phi = \pi/3$.

The θ-limits of integration. The ray L sweeps over R as θ runs from 0 to 2π. The volume is

$$V = \iiint_D \rho^2 \sin\phi \, d\rho \, d\phi \, d\theta = \int_0^{2\pi} \int_0^{\pi/3} \int_0^1 \rho^2 \sin\phi \, d\rho \, d\phi \, d\theta$$

$$= \int_0^{2\pi} \int_0^{\pi/3} \left[\frac{\rho^3}{3}\right]_0^1 \sin\phi \, d\phi \, d\theta = \int_0^{2\pi} \int_0^{\pi/3} \frac{1}{3} \sin\phi \, d\phi \, d\theta$$

$$= \int_0^{2\pi} \left[-\frac{1}{3}\cos\phi\right]_0^{\pi/3} d\theta = \int_0^{2\pi} \left(-\frac{1}{6} + \frac{1}{3}\right) d\theta = \frac{1}{6}(2\pi) = \frac{\pi}{3}. \quad \blacksquare$$

EXAMPLE 6   A solid of constant density $\delta = 1$ occupies the region D in Example 5. Find the solid's moment of inertia about the z-axis.

Solution   In rectangular coordinates, the moment is

$$I_z = \iiint_D (x^2 + y^2) \, dV.$$

In spherical coordinates, $x^2 + y^2 = (\rho \sin\phi \cos\theta)^2 + (\rho \sin\phi \sin\theta)^2 = \rho^2 \sin^2\phi$. Hence,

$$I_z = \iiint_D (\rho^2 \sin^2\phi)\, \rho^2 \sin\phi \, d\rho \, d\phi \, d\theta = \iiint_D \rho^4 \sin^3\phi \, d\rho \, d\phi \, d\theta.$$

For the region D in Example 5, this becomes

$$I_z = \int_0^{2\pi} \int_0^{\pi/3} \int_0^1 \rho^4 \sin^3\phi \, d\rho \, d\phi \, d\theta = \int_0^{2\pi} \int_0^{\pi/3} \left[\frac{\rho^5}{5}\right]_0^1 \sin^3\phi \, d\phi \, d\theta$$

$$= \frac{1}{5} \int_0^{2\pi} \int_0^{\pi/3} (1 - \cos^2\phi) \sin\phi \, d\phi \, d\theta = \frac{1}{5} \int_0^{2\pi} \left[-\cos\phi + \frac{\cos^3\phi}{3}\right]_0^{\pi/3} d\theta$$

$$= \frac{1}{5} \int_0^{2\pi} \left(-\frac{1}{2} + 1 + \frac{1}{24} - \frac{1}{3}\right) d\theta = \frac{1}{5} \int_0^{2\pi} \frac{5}{24} d\theta = \frac{1}{24}(2\pi) = \frac{\pi}{12}. \quad \blacksquare$$

Coordinate Conversion Formulas

CYLINDRICAL TO RECTANGULAR	SPHERICAL TO RECTANGULAR	SPHERICAL TO CYLINDRICAL
$x = r\cos\theta$	$x = \rho\sin\phi\cos\theta$	$r = \rho\sin\phi$
$y = r\sin\theta$	$y = \rho\sin\phi\sin\theta$	$z = \rho\cos\phi$
$z = z$	$z = \rho\cos\phi$	$\theta = \theta$

Corresponding formulas for dV in triple integrals:

$$dV = dx\, dy\, dz$$
$$= dz\, r\, dr\, d\theta$$
$$= \rho^2 \sin\phi \, d\rho \, d\phi \, d\theta$$

In the next section we offer a more general procedure for determining dV in cylindrical and spherical coordinates. The results, of course, will be the same.

EXERCISES 15.7

In Exercises 1–12, sketch the graph described by the following cylindrical coordinates in three-dimensional space.

1. $r = 2$

2. $\theta = \dfrac{\pi}{4}$

3. $z = -1$

4. $z = r$

5. $r = \theta$

6. $z = r \sin \theta$

7. $r^2 + z^2 = 4$

8. $1 \le r \le 2, \quad 0 \le \theta \le \dfrac{\pi}{3}$

9. $r \le z \le \sqrt{9 - r^2}$

10. $0 \le r \le 2 \sin \theta, \quad 1 \le z \le 3$

11. $0 \le r \le 4 \cos \theta, \quad 0 \le \theta \le \dfrac{\pi}{2}, \quad 0 \le z \le 5$

12. $0 \le r \le 3, \quad \dfrac{-\pi}{2} \le \theta \le \dfrac{\pi}{2}, \quad 0 \le z \le r \cos \theta$

In Exercises 13–22, sketch the graph described by the following spherical coordinates in three-dimensional space.

13. $\rho = 3$

14. $\phi = \dfrac{\pi}{6}$

15. $\theta = \dfrac{2}{3} \pi$

16. $\rho = \csc \phi$

17. $\rho \cos \phi = 4$

18. $1 \le \rho \le 2 \sec \phi, \quad 0 \le \phi \le \dfrac{\pi}{4}$

19. $0 \le \rho \le 3 \csc \phi$

20. $0 \le \rho \le 1, \quad \dfrac{\pi}{2} \le \phi \le \pi, \quad 0 \le \theta \le \pi$

21. $0 \le \rho \cos \theta \sin \phi \le 2, \quad 0 \le \rho \sin \theta \sin \phi \le 3,$
$\quad 0 \le \rho \cos \phi \le 4$

22. $4 \sec \phi \le \rho \le 5$

Evaluating Integrals in Cylindrical Coordinates

Evaluate the cylindrical coordinate integrals in Exercises 23–28.

23. $\displaystyle\int_0^{2\pi}\int_0^1\int_r^{\sqrt{2-r^2}} dz \, r \, dr \, d\theta$

24. $\displaystyle\int_0^{2\pi}\int_0^3\int_{r^2/3}^{\sqrt{18-r^2}} dz \, r \, dr \, d\theta$

25. $\displaystyle\int_0^{2\pi}\int_0^{\theta/2\pi}\int_0^{3+24r^2} dz \, r \, dr \, d\theta$

26. $\displaystyle\int_0^{\pi}\int_0^{\theta/\pi}\int_{-\sqrt{4-r^2}}^{3\sqrt{4-r^2}} z \, dz \, r \, dr \, d\theta$

27. $\displaystyle\int_0^{2\pi}\int_0^1\int_r^{1/\sqrt{2-r^2}} 3 \, dz \, r \, dr \, d\theta$

28. $\displaystyle\int_0^{2\pi}\int_0^1\int_{-1/2}^{1/2} (r^2 \sin^2 \theta + z^2) \, dz \, r \, dr \, d\theta$

Changing the Order of Integration in Cylindrical Coordinates

The integrals we have seen so far suggest that there are preferred orders of integration for cylindrical coordinates, but other orders usually work well and are occasionally easier to evaluate. Evaluate the integrals in Exercises 29–32.

29. $\displaystyle\int_0^{2\pi}\int_0^3\int_0^{z/3} r^3 \, dr \, dz \, d\theta$

30. $\displaystyle\int_{-1}^1\int_0^{2\pi}\int_0^{1+\cos\theta} 4r \, dr \, d\theta \, dz$

31. $\displaystyle\int_0^1\int_0^{\sqrt{z}}\int_0^{2\pi} (r^2 \cos^2 \theta + z^2) r \, d\theta \, dr \, dz$

32. $\displaystyle\int_0^2\int_{r-2}^{\sqrt{4-r^2}}\int_0^{2\pi} (r \sin \theta + 1) r \, d\theta \, dz \, dr$

33. Let D be the region bounded below by the plane $z = 0$, above by the sphere $x^2 + y^2 + z^2 = 4$, and on the sides by the cylinder $x^2 + y^2 = 1$. Set up the triple integrals in cylindrical coordinates that give the volume of D using the following orders of integration.

   a. $dz \, dr \, d\theta$ **b.** $dr \, dz \, d\theta$ **c.** $d\theta \, dz \, dr$

34. Let D be the region bounded below by the cone $z = \sqrt{x^2 + y^2}$ and above by the paraboloid $z = 2 - x^2 - y^2$. Set up the triple integrals in cylindrical coordinates that give the volume of D using the following orders of integration.

   a. $dz \, dr \, d\theta$ **b.** $dr \, dz \, d\theta$ **c.** $d\theta \, dz \, dr$

Finding Iterated Integrals in Cylindrical Coordinates

35. Give the limits of integration for evaluating the integral

$$\iiint f(r, \theta, z) \, dz \, r \, dr \, d\theta$$

as an iterated integral over the region that is bounded below by the plane $z = 0$, on the side by the cylinder $r = \cos \theta$, and on top by the paraboloid $z = 3r^2$.

36. Convert the integral

$$\int_{-1}^1\int_0^{\sqrt{1-y^2}}\int_0^x (x^2 + y^2) \, dz \, dx \, dy$$

to an equivalent integral in cylindrical coordinates and evaluate the result.

In Exercises 37–42, set up the iterated integral for evaluating $\iiint_D f(r, \theta, z) \, dz \, r \, dr \, d\theta$ over the given region D.

37. D is the right circular cylinder whose base is the circle $r = 2 \sin \theta$ in the xy-plane and whose top lies in the plane $z = 4 - y$.

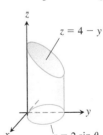

38. D is the right circular cylinder whose base is the circle $r = 3 \cos \theta$ and whose top lies in the plane $z = 5 - x$.

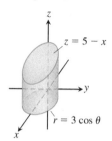

39. D is the solid right cylinder whose base is the region in the xy-plane that lies inside the cardioid $r = 1 + \cos \theta$ and outside the circle $r = 1$ and whose top lies in the plane $z = 4$.

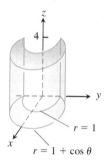

40. D is the solid right cylinder whose base is the region between the circles $r = \cos \theta$ and $r = 2 \cos \theta$ and whose top lies in the plane $z = 3 - y$.

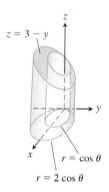

41. D is the prism whose base is the triangle in the xy-plane bounded by the x-axis and the lines $y = x$ and $x = 1$ and whose top lies in the plane $z = 2 - y$.

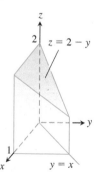

42. D is the prism whose base is the triangle in the xy-plane bounded by the y-axis and the lines $y = x$ and $y = 1$ and whose top lies in the plane $z = 2 - x$.

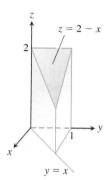

Evaluating Integrals in Spherical Coordinates

Evaluate the spherical coordinate integrals in Exercises 43–48.

43. $\displaystyle\int_0^\pi \int_0^\pi \int_0^{2 \sin \phi} \rho^2 \sin \phi \, d\rho \, d\phi \, d\theta$

44. $\displaystyle\int_0^{2\pi} \int_0^{\pi/4} \int_0^2 (\rho \cos \phi) \, \rho^2 \sin \phi \, d\rho \, d\phi \, d\theta$

45. $\displaystyle\int_0^{2\pi} \int_0^\pi \int_0^{(1-\cos \phi)/2} \rho^2 \sin \phi \, d\rho \, d\phi \, d\theta$

46. $\displaystyle\int_0^{3\pi/2} \int_0^\pi \int_0^1 5\rho^3 \sin^3 \phi \, d\rho \, d\phi \, d\theta$

47. $\displaystyle\int_0^{2\pi} \int_0^{\pi/3} \int_{\sec \phi}^2 3\rho^2 \sin \phi \, d\rho \, d\phi \, d\theta$

48. $\displaystyle\int_0^{2\pi} \int_0^{\pi/4} \int_0^{\sec \phi} (\rho \cos \phi) \, \rho^2 \sin \phi \, d\rho \, d\phi \, d\theta$

Changing the Order of Integration in Spherical Coordinates

The previous integrals suggest there are preferred orders of integration for spherical coordinates, but other orders give the same value and are occasionally easier to evaluate. Evaluate the integrals in Exercises 49–52.

49. $\displaystyle\int_0^2 \int_{-\pi}^0 \int_{\pi/4}^{\pi/2} \rho^3 \sin 2\phi \, d\phi \, d\theta \, d\rho$

50. $\displaystyle\int_{\pi/6}^{\pi/3} \int_{\csc \phi}^{2 \csc \phi} \int_0^{2\pi} \rho^2 \sin \phi \, d\theta \, d\rho \, d\phi$

51. $\displaystyle\int_0^1 \int_0^\pi \int_0^{\pi/4} 12\rho \sin^3 \phi \, d\phi \, d\theta \, d\rho$

52. $\displaystyle\int_{\pi/6}^{\pi/2} \int_{-\pi/2}^{\pi/2} \int_{\csc \phi}^2 5\rho^4 \sin^3 \phi \, d\rho \, d\theta \, d\phi$

53. Let D be the region in Exercise 33. Set up the triple integrals in spherical coordinates that give the volume of D using the following orders of integration.

    a. $d\rho \, d\phi \, d\theta$ **b.** $d\phi \, d\rho \, d\theta$

54. Let D be the region bounded below by the cone $z = \sqrt{x^2 + y^2}$ and above by the plane $z = 1$. Set up the triple integrals in spherical coordinates that give the volume of D using the following orders of integration.

   a. $d\rho\, d\phi\, d\theta$
   b. $d\phi\, d\rho\, d\theta$

Finding Iterated Integrals in Spherical Coordinates

In Exercises 55–60, **(a)** find the spherical coordinate limits for the integral that calculates the volume of the given solid and then **(b)** evaluate the integral.

55. The solid between the sphere $\rho = \cos\phi$ and the hemisphere $\rho = 2, z \geq 0$

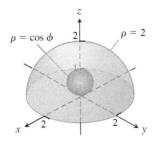

56. The solid bounded below by the hemisphere $\rho = 1, z \geq 0$, and above by the cardioid of revolution $\rho = 1 + \cos\phi$

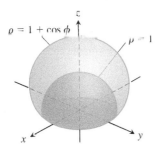

57. The solid enclosed by the cardioid of revolution $\rho = 1 - \cos\phi$

58. The upper portion cut from the solid in Exercise 57 by the xy-plane

59. The solid bounded below by the sphere $\rho = 2\cos\phi$ and above by the cone $z = \sqrt{x^2 + y^2}$

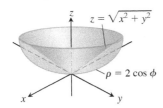

60. The solid bounded below by the xy-plane, on the sides by the sphere $\rho = 2$, and above by the cone $\phi = \pi/3$

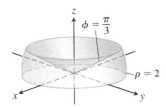

Finding Triple Integrals

61. Set up triple integrals for the volume of the sphere $\rho = 2$ in **(a)** spherical, **(b)** cylindrical, and **(c)** rectangular coordinates.

62. Let D be the region in the first octant that is bounded below by the cone $\phi = \pi/4$ and above by the sphere $\rho = 3$. Express the volume of D as an iterated triple integral in **(a)** cylindrical and **(b)** spherical coordinates. Then **(c)** find V.

63. Let D be the smaller cap cut from a solid ball of radius 2 units by a plane 1 unit from the center of the sphere. Express the volume of D as an iterated triple integral in **(a)** spherical, **(b)** cylindrical, and **(c)** rectangular coordinates. Then **(d)** find the volume by evaluating one of the three triple integrals.

64. Express the moment of inertia I_z of the solid hemisphere $x^2 + y^2 + z^2 \leq 1, z \geq 0$, as an iterated integral in **(a)** cylindrical and **(b)** spherical coordinates. Then **(c)** find I_z.

Volumes

Find the volumes of the solids in Exercises 65–70.

65.
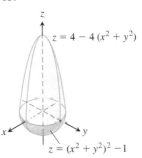

66.
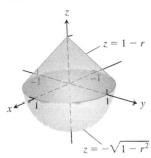

67.
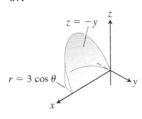

68.
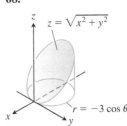

69.
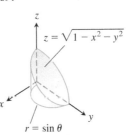

70.
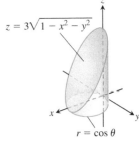

71. Sphere and cones Find the volume of the portion of the solid sphere $\rho \leq a$ that lies between the cones $\phi = \pi/3$ and $\phi = 2\pi/3$.

72. Sphere and half-planes Find the volume of the region cut from the solid sphere $\rho \leq a$ by the half-planes $\theta = 0$ and $\theta = \pi/6$ in the first octant.

73. Sphere and plane Find the volume of the smaller region cut from the solid sphere $\rho \le 2$ by the plane $z = 1$.

74. Cone and planes Find the volume of the solid enclosed by the cone $z = \sqrt{x^2 + y^2}$ between the planes $z = 1$ and $z = 2$.

75. Cylinder and paraboloid Find the volume of the region bounded below by the plane $z = 0$, laterally by the cylinder $x^2 + y^2 = 1$, and above by the paraboloid $z = x^2 + y^2$.

76. Cylinder and paraboloids Find the volume of the region bounded below by the paraboloid $z = x^2 + y^2$, laterally by the cylinder $x^2 + y^2 = 1$, and above by the paraboloid $z = x^2 + y^2 + 1$.

77. Cylinder and cones Find the volume of the solid cut from the thick-walled cylinder $1 \le x^2 + y^2 \le 2$ by the cones $z = \pm \sqrt{x^2 + y^2}$.

78. Sphere and cylinder Find the volume of the region that lies inside the sphere $x^2 + y^2 + z^2 = 2$ and outside the cylinder $x^2 + y^2 = 1$.

79. Cylinder and planes Find the volume of the region enclosed by the cylinder $x^2 + y^2 = 4$ and the planes $z = 0$ and $y + z = 4$.

80. Cylinder and planes Find the volume of the region enclosed by the cylinder $x^2 + y^2 = 4$ and the planes $z = 0$ and $x + y + z = 4$.

81. Region trapped by paraboloids Find the volume of the region bounded above by the paraboloid $z = 5 - x^2 - y^2$ and below by the paraboloid $z = 4x^2 + 4y^2$.

82. Paraboloid and cylinder Find the volume of the region bounded above by the paraboloid $z = 9 - x^2 - y^2$, below by the xy-plane, and lying *outside* the cylinder $x^2 + y^2 = 1$.

83. Cylinder and sphere Find the volume of the region cut from the solid cylinder $x^2 + y^2 \le 1$ by the sphere $x^2 + y^2 + z^2 = 4$.

84. Sphere and paraboloid Find the volume of the region bounded above by the sphere $x^2 + y^2 + z^2 = 2$ and below by the paraboloid $z = x^2 + y^2$.

Average Values

85. Find the average value of the function $f(r, \theta, z) = r$ over the region bounded by the cylinder $r = 1$ between the planes $z = -1$ and $z = 1$.

86. Find the average value of the function $f(r, \theta, z) = r$ over the solid ball bounded by the sphere $r^2 + z^2 = 1$. (This is the sphere $x^2 + y^2 + z^2 = 1$.)

87. Find the average value of the function $f(\rho, \phi, \theta) = \rho$ over the solid ball $\rho \le 1$.

88. Find the average value of the function $f(\rho, \phi, \theta) = \rho \cos \phi$ over the solid upper ball $\rho \le 1, 0 \le \phi \le \pi/2$.

Masses, Moments, and Centroids

89. Center of mass A solid of constant density is bounded below by the plane $z = 0$, above by the cone $z = r, r \ge 0$, and on the sides by the cylinder $r = 1$. Find the center of mass.

90. Centroid Find the centroid of the region in the first octant that is bounded above by the cone $z = \sqrt{x^2 + y^2}$, below by the plane $z = 0$, and on the sides by the cylinder $x^2 + y^2 = 4$ and the planes $x = 0$ and $y = 0$.

91. Centroid Find the centroid of the solid in Exercise 60.

92. Centroid Find the centroid of the solid bounded above by the sphere $\rho = a$ and below by the cone $\phi = \pi/4$.

93. Centroid Find the centroid of the region that is bounded above by the surface $z = \sqrt{r}$, on the sides by the cylinder $r = 4$, and below by the xy-plane.

94. Centroid Find the centroid of the region cut from the solid ball $r^2 + z^2 \le 1$ by the half-planes $\theta = -\pi/3, r \ge 0$, and $\theta = \pi/3, r \ge 0$.

95. Moment of inertia of solid cone Find the moment of inertia of a right circular cone of base radius 1 and height 1 about an axis through the vertex parallel to the base. (Take $\delta = 1$.)

96. Moment of inertia of solid sphere Find the moment of inertia of a solid sphere of radius a about a diameter. (Take $\delta = 1$.)

97. Moment of inertia of solid cone Find the moment of inertia of a right circular cone of base radius a and height h about its axis. (*Hint:* Place the cone with its vertex at the origin and its axis along the z-axis.)

98. Variable density A solid is bounded on the top by the paraboloid $z = r^2$, on the bottom by the plane $z = 0$, and on the sides by the cylinder $r = 1$. Find the center of mass and the moment of inertia about the z-axis if the density is

   a. $\delta(r, \theta, z) = z$

   b. $\delta(r, \theta, z) = r$.

99. Variable density A solid is bounded below by the cone $z = \sqrt{x^2 + y^2}$ and above by the plane $z = 1$. Find the center of mass and the moment of inertia about the z-axis if the density is

   a. $\delta(r, \theta, z) = z$

   b. $\delta(r, \theta, z) = z^2$.

100. Variable density A solid ball is bounded by the sphere $\rho = a$. Find the moment of inertia about the z-axis if the density is

   a. $\delta(\rho, \phi, \theta) = \rho^2$

   b. $\delta(\rho, \phi, \theta) = r = \rho \sin \phi$.

101. Centroid of solid semiellipsoid Show that the centroid of the solid semiellipsoid of revolution $(r^2/a^2) + (z^2/h^2) \le 1, z \ge 0$, lies on the z-axis three-eighths of the way from the base to the top. The special case $h = a$ gives a solid hemisphere. Thus, the centroid of a solid hemisphere lies on the axis of symmetry three-eighths of the way from the base to the top.

102. Centroid of solid cone Show that the centroid of a solid right circular cone is one-fourth of the way from the base to the vertex. (In general, the centroid of a solid cone or pyramid is one-fourth of the way from the centroid of the base to the vertex.)

103. Density of center of a planet A planet is in the shape of a sphere of radius R and total mass M with spherically symmetric density distribution that increases linearly as one approaches its center. What is the density at the center of this planet if the density at its edge (surface) is taken to be zero?

104. Mass of planet's atmosphere A spherical planet of radius R has an atmosphere whose density is $\mu = \mu_0 e^{-ch}$, where h is the altitude above the surface of the planet, μ_0 is the density at sea level, and c is a positive constant. Find the mass of the planet's atmosphere.

Theory and Examples

105. Vertical planes in cylindrical coordinates

    a. Show that planes perpendicular to the x-axis have equations of the form $r = a \sec \theta$ in cylindrical coordinates.

    b. Show that planes perpendicular to the y-axis have equations of the form $r = b \csc \theta$.

106. (*Continuation of Exercise 105.*) Find an equation of the form $r = f(\theta)$ in cylindrical coordinates for the plane $ax + by = c$, $c \neq 0$.

107. Symmetry What symmetry will you find in a surface that has an equation of the form $r = f(z)$ in cylindrical coordinates? Give reasons for your answer.

108. Symmetry What symmetry will you find in a surface that has an equation of the form $\rho = f(\phi)$ in spherical coordinates? Give reasons for your answer.

15.8 Substitutions in Multiple Integrals

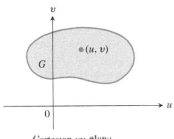

Cartesian uv-plane

$$x = g(u, v)$$
$$y = h(u, v)$$

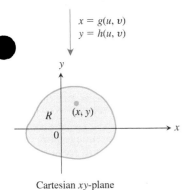

Cartesian xy-plane

FIGURE 15.57 The equations $x = g(u, v)$ and $y = h(u, v)$ allow us to change an integral over a region R in the xy-plane into an integral over a region G in the uv-plane.

This section introduces the ideas involved in coordinate transformations to evaluate multiple integrals by substitution. The method replaces complicated integrals by ones that are easier to evaluate. Substitutions accomplish this by simplifying the integrand, the limits of integration, or both. A thorough discussion of multivariable transformations and substitutions is best left to a more advanced course, but our introduction here shows how the substitutions just studied reflect the general idea derived for single integral calculus.

Substitutions in Double Integrals

The polar coordinate substitution of Section 15.4 is a special case of a more general substitution method for double integrals, a method that pictures changes in variables as transformations of regions.

    Suppose that a region G in the uv-plane is transformed into the region R in the xy-plane by equations of the form

$$x = g(u, v), \qquad y = h(u, v),$$

as suggested in Figure 15.57. We assume the transformation is one-to-one on the interior of G. We call R the **image** of G under the transformation, and G the **preimage** of R. Any function $f(x, y)$ defined on R can be thought of as a function $f(g(u, v), h(u, v))$ defined on G as well. How is the integral of $f(x, y)$ over R related to the integral of $f(g(u, v), h(u, v))$ over G?

    To gain some insight into the question, we look again at the single variable case. To be consistent with how we are using them now, we interchange the variables x and u used in the substitution method for single integrals in Chapter 5, so the equation is

$$\int_{g(a)}^{g(b)} f(x)\, dx = \int_a^b f(g(u))\, g'(u)\, du. \qquad {\scriptstyle x = g(u),\ dx = g'(u)\, du}$$

To propose an analogue for substitution in a double integral $\iint_R f(x, y)\, dx\, dy$, we need a derivative factor like $g'(u)$ as a multiplier that transforms the area element $du\, dv$ in the region G to its corresponding area element $dx\, dy$ in the region R. We denote this factor by J. In continuing with our analogy, it is reasonable to assume that J is a function of both variables u and v, just as g' is a function of the single variable u. Moreover, J should register instantaneous change, so partial derivatives are going to be involved in its expression. Since four partial derivatives are associated with the transforming equations $x = g(u, v)$ and $y = h(u, v)$, it is also reasonable to assume that the factor $J(u, v)$ we seek includes them all. These features are captured in the following definition constructed from the partial derivatives, and named after the German mathematician Carl Jacobi.

DEFINITION   The **Jacobian determinant** or **Jacobian** of the coordinate transformation $x = g(u, v), y = h(u, v)$ is

$$J(u, v) = \begin{vmatrix} \dfrac{\partial x}{\partial u} & \dfrac{\partial x}{\partial v} \\[2mm] \dfrac{\partial y}{\partial u} & \dfrac{\partial y}{\partial v} \end{vmatrix} = \frac{\partial x}{\partial u}\frac{\partial y}{\partial v} - \frac{\partial y}{\partial u}\frac{\partial x}{\partial v}. \tag{1}$$

The Jacobian can also be denoted by

$$J(u, v) = \frac{\partial(x, y)}{\partial(u, v)}$$

Differential Area Change Substituting
$x = g(u, v), y = h(u, v)$

$$dx\, dy = \left| \frac{\partial(x, y)}{\partial(u, v)} \right| du\, dv$$

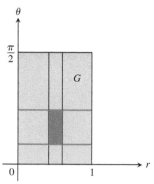

Cartesian $r\theta$-plane

$$x = r \cos \theta$$
$$y = r \sin \theta$$

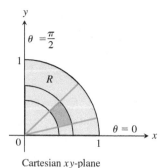

Cartesian xy-plane

FIGURE 15.58 The equations $x = r\cos\theta, y = r\sin\theta$ transform G into R. The Jacobian factor r, calculated in Example 1, scales the differential rectangle $dr\, d\theta$ in G to match with the differential area element $dx\, dy$ in R.

to help us remember how the determinant in Equation (1) is constructed from the partial derivatives of x and y. The array of partial derivatives in Equation (1) behaves just like the derivative g' in the single variable situation. The Jacobian measures how much the transformation is expanding or contracting the area around the point (u, v). Effectively, the factor $|J|$ converts the area of the differential rectangle $du\, dv$ in G to match with its corresponding differential area $dx\, dy$ in R. We note that, in general, the value of the scaling factor $|J|$ depends on the point (u, v) in G; that is, the scaling changes as the point (u, v) varies through the region G. Our examples to follow will show how it scales the differential area $du\, dv$ for specific transformations.

Now we can answer our original question concerning the relationship of the integral of $f(x, y)$ over the region R to the integral of $f(g(u, v), h(u, v))$ over G.

THEOREM 3—Substitution for Double Integrals
Suppose that $f(x, y)$ is continuous over the region R. Let G be the preimage of R under the transformation $x = g(u, v), y = h(u, v)$, which is assumed to be one-to-one on the interior of G. If the functions g and h have continuous first partial derivatives within the interior of G, then

$$\iint\limits_{R} f(x, y)\, dx\, dy = \iint\limits_{G} f(g(u, v), h(u, v)) \left| \frac{\partial(x, y)}{\partial(u, v)} \right| du\, dv. \tag{2}$$

The derivation of Equation (2) is intricate and properly belongs to a course in advanced calculus, so we do not derive it here. We now present examples illustrating the substitution method defined by the equation.

EXAMPLE 1   Find the Jacobian for the polar coordinate transformation $x = r\cos\theta$, $y = r\sin\theta$, and use Equation (2) to write the Cartesian integral $\iint_R f(x, y)\, dx\, dy$ as a polar integral.

Solution   Figure 15.58 shows how the equations $x = r\cos\theta, y = r\sin\theta$ transform the rectangle $G: 0 \le r \le 1, 0 \le \theta \le \pi/2$, into the quarter circle R bounded by $x^2 + y^2 = 1$ in the first quadrant of the xy-plane.

For polar coordinates, we have r and θ in place of u and v. With $x = r \cos \theta$ and $y = r \sin \theta$, the Jacobian is

$$J(r, \theta) = \begin{vmatrix} \dfrac{\partial x}{\partial r} & \dfrac{\partial x}{\partial \theta} \\[2mm] \dfrac{\partial y}{\partial r} & \dfrac{\partial y}{\partial \theta} \end{vmatrix} = \begin{vmatrix} \cos \theta & -r \sin \theta \\ \sin \theta & r \cos \theta \end{vmatrix} = r(\cos^2 \theta + \sin^2 \theta) = r.$$

Since we assume $r \geq 0$ when integrating in polar coordinates, $|J(r, \theta)| = |r| = r$, so that Equation (2) gives

$$\iint\limits_{R} f(x, y) \, dx \, dy = \iint\limits_{G} f(r \cos \theta, r \sin \theta) \, r \, dr \, d\theta. \tag{3}$$

This is the same formula we derived independently using a geometric argument for polar area in Section 15.4.   ∎

Here is an example of a substitution in which the image of a rectangle under the coordinate transformation is a trapezoid. Transformations like this one are called **linear transformations** and their Jacobians are constant throughout G.

EXAMPLE 2   Evaluate

$$\int_{0}^{4} \int_{x=y/2}^{x=(y/2)+1} \frac{2x - y}{2} \, dx \, dy$$

by applying the transformation

$$u = \frac{2x - y}{2}, \qquad v = \frac{y}{2} \tag{4}$$

and integrating over an appropriate region in the uv-plane.

Solution   We sketch the region R of integration in the xy-plane and identify its boundaries (Figure 15.59).

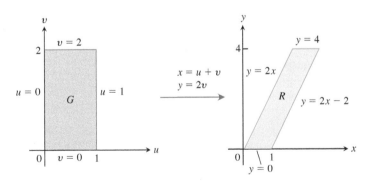

FIGURE 15.59   The equations $x = u + v$ and $y = 2v$ transform G into R. Reversing the transformation by the equations $u = (2x - y)/2$ and $v = y/2$ transforms R into G (Example 2).

To apply Equation (2), we need to find the corresponding uv-region G and the Jacobian of the transformation. To find them, we first solve Equations (4) for x and y in terms of u and v. From those equations it is easy to find algebraically that

$$x = u + v, \qquad y = 2v. \tag{5}$$

We then find the boundaries of G by substituting these expressions into the equations for the boundaries of R (Figure 15.59)

xy-equations for the boundary of R	Corresponding uv-equations for the boundary of G	Simplified uv-equations
$x = y/2$	$u + v = 2v/2 = v$	$u = 0$
$x = (y/2) + 1$	$u + v = (2v/2) + 1 = v + 1$	$u = 1$
$y = 0$	$2v = 0$	$v = 0$
$y = 4$	$2v = 4$	$v = 2$

From Equations (5) the Jacobian of the transformation is

$$J(u, v) = \begin{vmatrix} \dfrac{\partial x}{\partial u} & \dfrac{\partial x}{\partial v} \\[2mm] \dfrac{\partial y}{\partial u} & \dfrac{\partial y}{\partial v} \end{vmatrix} = \begin{vmatrix} \dfrac{\partial}{\partial u}(u + v) & \dfrac{\partial}{\partial v}(u + v) \\[2mm] \dfrac{\partial}{\partial u}(2v) & \dfrac{\partial}{\partial v}(2v) \end{vmatrix} = \begin{vmatrix} 1 & 1 \\ 0 & 2 \end{vmatrix} = 2.$$

We now have everything we need to apply Equation (2):

$$\int_0^4 \int_{x=y/2}^{x=(y/2)+1} \frac{2x - y}{2} \, dx \, dy = \int_{v=0}^{v=2} \int_{u=0}^{u=1} u \, |J(u, v)| \, du \, dv$$

$$= \int_0^2 \int_0^1 (u)(2) \, du \, dv = \int_0^2 \left[u^2 \right]_0^1 dv = \int_0^2 dv = 2. \quad \blacksquare$$

EXAMPLE 3   Evaluate

$$\int_0^1 \int_0^{1-x} \sqrt{x + y}\,(y - 2x)^2 \, dy \, dx.$$

Solution   We sketch the region R of integration in the xy-plane and identify its boundaries (Figure 15.60). The integrand suggests the transformation $u = x + y$ and $v = y - 2x$. Routine algebra produces x and y as functions of u and v:

$$x = \frac{u}{3} - \frac{v}{3}, \qquad y = \frac{2u}{3} + \frac{v}{3}. \tag{6}$$

From Equations (6), we can find the boundaries of the uv-region G (Figure 15.60).

xy-equations for the boundary of R	Corresponding uv-equations for the boundary of G	Simplified uv-equations
$x + y = 1$	$\left(\dfrac{u}{3} - \dfrac{v}{3}\right) + \left(\dfrac{2u}{3} + \dfrac{v}{3}\right) = 1$	$u = 1$
$x = 0$	$\dfrac{u}{3} - \dfrac{v}{3} = 0$	$v = u$
$y = 0$	$\dfrac{2u}{3} + \dfrac{v}{3} = 0$	$v = -2u$

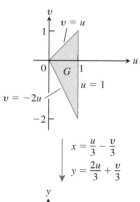

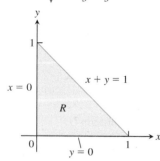

FIGURE 15.60   The equations $x = (u/3) - (v/3)$ and $y = (2u/3) + (v/3)$ transform G into R. Reversing the transformation by the equations $u = x + y$ and $v = y - 2x$ transforms R into G (Example 3).

The Jacobian of the transformation in Equations (6) is

$$J(u, v) = \begin{vmatrix} \dfrac{\partial x}{\partial u} & \dfrac{\partial x}{\partial v} \\[2ex] \dfrac{\partial y}{\partial u} & \dfrac{\partial y}{\partial v} \end{vmatrix} = \begin{vmatrix} \dfrac{1}{3} & -\dfrac{1}{3} \\[2ex] \dfrac{2}{3} & \dfrac{1}{3} \end{vmatrix} = \dfrac{1}{3}.$$

Applying Equation (2), we evaluate the integral:

$$\int_0^1 \int_0^{1-x} \sqrt{x+y}\,(y-2x)^2\,dy\,dx = \int_{u=0}^{u=1} \int_{v=-2u}^{v=u} u^{1/2} v^2 \left| J(u,v) \right| dv\,du$$

$$= \int_0^1 \int_{-2u}^u u^{1/2} v^2 \left(\dfrac{1}{3}\right) dv\,du = \dfrac{1}{3} \int_0^1 u^{1/2} \left[\dfrac{1}{3} v^3\right]_{v=-2u}^{v=u} du$$

$$= \dfrac{1}{9} \int_0^1 u^{1/2}(u^3 + 8u^3)\,du = \int_0^1 u^{7/2}\,du = \dfrac{2}{9} u^{9/2} \Big]_0^1 = \dfrac{2}{9}.$$ ∎

In the next example we illustrate a nonlinear transformation of coordinates resulting from simplifying the form of the integrand. Like the polar coordinates' transformation, nonlinear transformations can map a straight-line boundary of a region into a curved boundary (or vice versa with the inverse transformation). In general, nonlinear transformations are more complex to analyze than linear ones, and a complete treatment is left to a more advanced course.

EXAMPLE 4   Evaluate the integral

$$\int_1^2 \int_{1/y}^y \sqrt{\dfrac{y}{x}}\, e^{\sqrt{xy}}\,dx\,dy.$$

Solution   The square root terms in the integrand suggest that we might simplify the integration by substituting $u = \sqrt{xy}$ and $v = \sqrt{y/x}$. Squaring these equations gives $u^2 = xy$ and $v^2 = y/x$, which imply that $u^2 v^2 = y^2$ and $u^2/v^2 = x^2$. So we obtain the transformation (in the same ordering of the variables as discussed before)

$$x = \dfrac{u}{v} \qquad \text{and} \qquad y = uv,$$

with $u > 0$ and $v > 0$. Let's first see what happens to the integrand itself under this transformation. The Jacobian of the transformation is not constant:

$$J(u, v) = \begin{vmatrix} \dfrac{\partial x}{\partial u} & \dfrac{\partial x}{\partial v} \\[2ex] \dfrac{\partial y}{\partial u} & \dfrac{\partial y}{\partial v} \end{vmatrix} = \begin{vmatrix} \dfrac{1}{v} & \dfrac{-u}{v^2} \\[2ex] v & u \end{vmatrix} = \dfrac{2u}{v}.$$

If G is the region of integration in the uv-plane, then by Equation (2) the transformed double integral under the substitution is

$$\iint_R \sqrt{\dfrac{y}{x}}\, e^{\sqrt{xy}}\,dx\,dy = \iint_G v e^u \dfrac{2u}{v}\,du\,dv = \iint_G 2u e^u\,du\,dv.$$

The transformed integrand function is easier to integrate than the original one, so we proceed to determine the limits of integration for the transformed integral.

The region of integration R of the original integral in the xy-plane is shown in Figure 15.61. From the substitution equations $u = \sqrt{xy}$ and $v = \sqrt{y/x}$, we see that the image of the left-hand boundary $xy = 1$ for R is the vertical line segment $u = 1, 2 \geq v \geq 1$, in G (see Figure 15.62). Likewise, the right-hand boundary $y = x$ of R maps to the horizontal line segment $v = 1, 1 \leq u \leq 2$, in G. Finally, the horizontal top boundary $y = 2$ of R

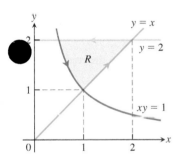

FIGURE 15.61   The region of integration R in Example 4.

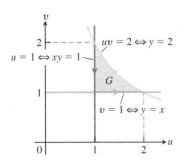

FIGURE 15.62   The boundaries of the region G correspond to those of region R in Figure 15.61. Notice as we move counterclockwise around the region R, we also move counterclockwise around the region G. The inverse transformation equations $u = \sqrt{xy}, v = \sqrt{y/x}$ produce the region G from the region R.

maps to $uv = 2, 1 \le v \le 2$, in G. As we move counterclockwise around the boundary of the region R, we also move counterclockwise around the boundary of G, as shown i Figure 15.62. Knowing the region of integration G in the uv-plane, we can now write equivalent iterated integrals:

$$\int_1^2 \int_{1/y}^y \sqrt{\frac{y}{x}} \, e^{\sqrt{xy}} \, dx \, dy = \int_1^2 \int_1^{2/u} 2ue^u \, dv \, du. \qquad \text{Note the order of integration.}$$

We now evaluate the transformed integral on the right-hand side,

$$\int_1^2 \int_1^{2/u} 2ue^u \, dv \, du = 2\int_1^2 \left[vue^u \right]_{v=1}^{v=2/u} du$$

$$= 2\int_1^2 (2e^u - ue^u) \, du$$

$$= 2\int_1^2 (2 - u)e^u \, du$$

$$= 2\left[(2 - u)e^u + e^u \right]_{u=1}^{u=2} \qquad \text{Integrate by parts.}$$

$$= 2(e^2 - (e + e)) = 2e(e - 2). \qquad \blacksquare$$

Substitutions in Triple Integrals

The cylindrical and spherical coordinate substitutions in Section 15.7 are special cases of a substitution method that pictures changes of variables in triple integrals as transformations of three-dimensional regions. The method is like the method for double integrals given by Equation (2) except that now we work in three dimensions instead of two.

Suppose that a region G in uvw-space is transformed one-to-one into the region D i xyz-space by differentiable equations of the form

$$x = g(u, v, w), \qquad y = h(u, v, w), \qquad z = k(u, v, w),$$

as suggested in Figure 15.63. Then any function $F(x, y, z)$ defined on D can be thought of as a function

$$F(g(u, v, w), h(u, v, w), k(u, v, w)) = H(u, v, w)$$

defined on G. If g, h, and k have continuous first partial derivatives, then the integral of $F(x, y, z)$ over D is related to the integral of $H(u, v, w)$ over G by the equation

$$\iiint_D F(x, y, z) \, dx \, dy \, dz = \iiint_G H(u, v, w) \, |J(u, v, w)| \, du \, dv \, dw. \qquad (7)$$

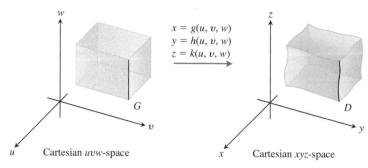

FIGURE 15.63 The equations $x = g(u, v, w)$, $y = h(u, v, w)$, and $z = k(u, v, w)$ allow us to change an integral over a region D in Cartesian xyz-space into an integral over a region G in Cartesian uvw-space using Equation (7).

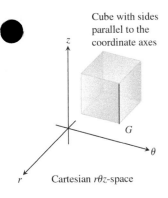

Cube with sides parallel to the coordinate axes

Cartesian $r\theta z$-space

$$x = r \cos \theta$$
$$y = r \sin \theta$$
$$z = z$$

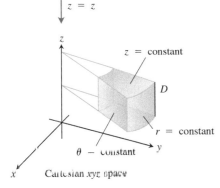

Cartesian xyz-space

FIGURE 15.64 The equations $x = r \cos \theta$, $y = r \sin \theta$, and $z = z$ transform the cube G into a cylindrical wedge D.

The factor $J(u, v, w)$, whose absolute value appears in this equation, is the **Jacobian determinant**

$$J(u, v, w) = \begin{vmatrix} \dfrac{\partial x}{\partial u} & \dfrac{\partial x}{\partial v} & \dfrac{\partial x}{\partial w} \\[2mm] \dfrac{\partial y}{\partial u} & \dfrac{\partial y}{\partial v} & \dfrac{\partial y}{\partial w} \\[2mm] \dfrac{\partial z}{\partial u} & \dfrac{\partial z}{\partial v} & \dfrac{\partial z}{\partial w} \end{vmatrix} = \frac{\partial(x, y, z)}{\partial(u, v, w)}.$$

This determinant measures how much the volume near a point in G is being expanded or contracted by the transformation from (u, v, w) to (x, y, z) coordinates. As in the two-dimensional case, the derivation of the change-of-variable formula in Equation (7) is omitted.

For cylindrical coordinates, r, θ, and z take the place of u, v, and w. The transformation from Cartesian $r\theta z$-space to Cartesian xyz-space is given by the equations

$$x = r \cos \theta, \qquad y = r \sin \theta, \qquad z = z$$

(Figure 15.64). The Jacobian of the transformation is

$$J(r, \theta, z) = \begin{vmatrix} \dfrac{\partial x}{\partial r} & \dfrac{\partial x}{\partial \theta} & \dfrac{\partial x}{\partial z} \\[2mm] \dfrac{\partial y}{\partial r} & \dfrac{\partial y}{\partial \theta} & \dfrac{\partial y}{\partial z} \\[2mm] \dfrac{\partial z}{\partial r} & \dfrac{\partial z}{\partial \theta} & \dfrac{\partial z}{\partial z} \end{vmatrix} = \begin{vmatrix} \cos \theta & -r \sin \theta & 0 \\ \sin \theta & r \cos \theta & 0 \\ 0 & 0 & 1 \end{vmatrix}$$

$$= r \cos^2 \theta + r \sin^2 \theta = r.$$

The corresponding version of Equation (7) is

$$\iiint\limits_{D} F(x, y, z) \, dx \, dy \, dz = \iiint\limits_{G} H(r, \theta, z) |r| \, dr \, d\theta \, dz.$$

We can drop the absolute value signs because $r \geq 0$.

For spherical coordinates, ρ, ϕ, and θ take the place of u, v, and w. The transformation from Cartesian $\rho\phi\theta$-space to Cartesian xyz-space is given by

$$x = \rho \sin \phi \cos \theta, \qquad y = \rho \sin \phi \sin \theta, \qquad z = \rho \cos \phi$$

(Figure 15.65). The Jacobian of the transformation (see Exercise 23) is

$$J(\rho, \phi, \theta) = \begin{vmatrix} \dfrac{\partial x}{\partial \rho} & \dfrac{\partial x}{\partial \phi} & \dfrac{\partial x}{\partial \theta} \\[2mm] \dfrac{\partial y}{\partial \rho} & \dfrac{\partial y}{\partial \phi} & \dfrac{\partial y}{\partial \theta} \\[2mm] \dfrac{\partial z}{\partial \rho} & \dfrac{\partial z}{\partial \phi} & \dfrac{\partial z}{\partial \theta} \end{vmatrix} = \rho^2 \sin \phi.$$

The corresponding version of Equation (7) is

$$\iiint\limits_{D} F(x, y, z) \, dx \, dy \, dz = \iiint\limits_{G} H(\rho, \phi, \theta) |\rho^2 \sin \phi| \, d\rho \, d\phi \, d\theta.$$

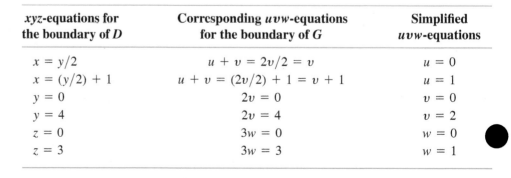

FIGURE 15.65 The equations $x = \rho \sin\phi \cos\theta$, $y = \rho \sin\phi \sin\theta$, and $z = \rho \cos\phi$ transform the cube G into the spherical wedge D.

We can drop the absolute value signs because $\sin\phi$ is never negative for $0 \le \phi \le \pi$. Note that this is the same result we obtained in Section 15.7.

Here is an example of another substitution. Although we could evaluate the integral in this example directly, we have chosen it to illustrate the substitution method in a simple (and fairly intuitive) setting.

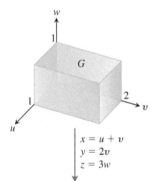

EXAMPLE 5   Evaluate

$$\int_0^3 \int_0^4 \int_{x=y/2}^{x=(y/2)+1} \left(\frac{2x-y}{2} + \frac{z}{3} \right) dx\, dy\, dz$$

by applying the transformation

$$u = (2x - y)/2, \qquad v = y/2, \qquad w = z/3 \tag{8}$$

and integrating over an appropriate region in uvw-space.

Solution   We sketch the region D of integration in xyz-space and identify its boundaries (Figure 15.66). In this case, the bounding surfaces are planes.

To apply Equation (7), we need to find the corresponding uvw-region G and the Jacobian of the transformation. To find them, we first solve Equations (8) for x, y, and z in terms of u, v, and w. Routine algebra gives

$$x = u + v, \qquad y = 2v, \qquad z = 3w. \tag{9}$$

We then find the boundaries of G by substituting these expressions into the equations for the boundaries of D:

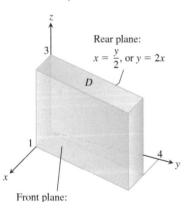

FIGURE 15.66 The equations $x = u + v$, $y = 2v$, and $z = 3w$ transform G into D. Reversing the transformation by the equations $u = (2x - y)/2$, $v = y/2$, and $w = z/3$ transforms D into G (Example 5).

xyz-equations for the boundary of D	Corresponding uvw-equations for the boundary of G	Simplified uvw-equations
$x = y/2$	$u + v = 2v/2 = v$	$u = 0$
$x = (y/2) + 1$	$u + v = (2v/2) + 1 = v + 1$	$u = 1$
$y = 0$	$2v = 0$	$v = 0$
$y = 4$	$2v = 4$	$v = 2$
$z = 0$	$3w = 0$	$w = 0$
$z = 3$	$3w = 3$	$w = 1$

The Jacobian of the transformation, again from Equations (9), is

$$J(u, v, w) = \begin{vmatrix} \dfrac{\partial x}{\partial u} & \dfrac{\partial x}{\partial v} & \dfrac{\partial x}{\partial w} \\[2mm] \dfrac{\partial y}{\partial u} & \dfrac{\partial y}{\partial v} & \dfrac{\partial y}{\partial w} \\[2mm] \dfrac{\partial z}{\partial u} & \dfrac{\partial z}{\partial v} & \dfrac{\partial z}{\partial w} \end{vmatrix} = \begin{vmatrix} 1 & 1 & 0 \\ 0 & 2 & 0 \\ 0 & 0 & 3 \end{vmatrix} = 6.$$

We now have everything we need to apply Equation (7):

$$\int_0^3 \int_0^4 \int_{x=y/2}^{x=(y/2)+1} \left(\frac{2x - y}{2} + \frac{z}{3} \right) dx\, dy\, dz$$

$$= \int_0^1 \int_0^2 \int_0^1 (u + w)\, |J(u, v, w)|\, du\, dv\, dw$$

$$= \int_0^1 \int_0^2 \int_0^1 (u + w)(6)\, du\, dv\, dw = 6 \int_0^1 \int_0^2 \left[\frac{u^2}{2} + uw \right]_0^1 dv\, dw$$

$$= 6 \int_0^1 \int_0^2 \left(\frac{1}{2} + w \right) dv\, dw = 6 \int_0^1 \left[\frac{v}{2} + vw \right]_0^2 dw = 6 \int_0^1 (1 + 2w)\, dw$$

$$= 6 \left[w + w^2 \right]_0^1 = 6(2) = 12.$$

EXERCISES 15.8

Jacobians and Transformed Regions in the Plane

1. a. Solve the system

$$u = x - y, \qquad v = 2x + y$$

for x and y in terms of u and v. Then find the value of the Jacobian $\partial(x, y)/\partial(u, v)$.

b. Find the image under the transformation $u = x - y$, $v = 2x + y$ of the triangular region with vertices $(0, 0)$, $(1, 1)$, and $(1, -2)$ in the xy-plane. Sketch the transformed region in the uv-plane.

2. a. Solve the system

$$u = x + 2y, \qquad v = x - y$$

for x and y in terms of u and v. Then find the value of the Jacobian $\partial(x, y)/\partial(u, v)$.

b. Find the image under the transformation $u = x + 2y$, $v = x - y$ of the triangular region in the xy-plane bounded by the lines $y = 0$, $y = x$, and $x + 2y = 2$. Sketch the transformed region in the uv-plane.

3. a. Solve the system

$$u = 3x + 2y, \qquad v = x + 4y$$

for x and y in terms of u and v. Then find the value of the Jacobian $\partial(x, y)/\partial(u, v)$.

b. Find the image under the transformation $u = 3x + 2y$, $v = x + 4y$ of the triangular region in the xy-plane bounded

by the x-axis, the y-axis, and the line $x + y = 1$. Sketch the transformed region in the uv-plane.

4. a. Solve the system

$$u = 2x - 3y, \qquad v = -x + y$$

for x and y in terms of u and v. Then find the value of the Jacobian $\partial(x, y)/\partial(u, v)$.

b. Find the image under the transformation $u = 2x - 3y$, $v = -x + y$ of the parallelogram R in the xy-plane with boundaries $x = -3$, $x = 0$, $y = x$, and $y = x + 1$. Sketch the transformed region in the uv-plane.

Substitutions in Double Integrals

5. Evaluate the integral

$$\int_0^4 \int_{x-y/2}^{x=(y/2)+1} \frac{2x - y}{2}\, dx\, dy$$

from Example 1 directly by integration with respect to x and y to confirm that its value is 2.

6. Use the transformation in Exercise 1 to evaluate the integral

$$\iint_R (2x^2 - xy - y^2)\, dx\, dy$$

for the region R in the first quadrant bounded by the lines $y = -2x + 4$, $y = -2x + 7$, $y = x - 2$, and $y = x + 1$.

7. Use the transformation in Exercise 3 to evaluate the integral

$$\iint_R (3x^2 + 14xy + 8y^2)\, dx\, dy$$

for the region R in the first quadrant bounded by the lines $y = -(3/2)x + 1$, $y = -(3/2)x + 3$, $y = -(1/4)x$, and $y = -(1/4)x + 1$.

8. Use the transformation and parallelogram R in Exercise 4 to evaluate the integral

$$\iint_R 2(x - y)\, dx\, dy.$$

9. Let R be the region in the first quadrant of the xy-plane bounded by the hyperbolas $xy = 1$, $xy = 9$ and the lines $y = x$, $y = 4x$. Use the transformation $x = u/v$, $y = uv$ with $u > 0$ and $v > 0$ to rewrite

$$\iint_R \left(\sqrt{\frac{y}{x}} + \sqrt{xy} \right) dx\, dy$$

as an integral over an appropriate region G in the uv-plane. Then evaluate the uv-integral over G.

10. a. Find the Jacobian of the transformation $x = u$, $y = uv$ and sketch the region G: $1 \le u \le 2$, $1 \le uv \le 2$, in the uv-plane.

b. Then use Equation (2) to transform the integral

$$\int_1^2 \int_1^2 \frac{y}{x}\, dy\, dx$$

into an integral over G, and evaluate both integrals.

11. Polar moment of inertia of an elliptical plate A thin plate of constant density covers the region bounded by the ellipse $x^2/a^2 + y^2/b^2 = 1$, $a > 0$, $b > 0$, in the xy-plane. Find the second moment of the plate about the origin. (*Hint:* Use the transformation $x = ar\cos\theta$, $y = br\sin\theta$.)

12. The area of an ellipse The area πab of the ellipse $x^2/a^2 + y^2/b^2 = 1$ can be found by integrating the function $f(x, y) = 1$ over the region bounded by the ellipse in the xy-plane. Evaluating the integral directly requires a trigonometric substitution. An easier way to evaluate the integral is to use the transformation $x = au$, $y = bv$ and evaluate the transformed integral over the disk G: $u^2 + v^2 \le 1$ in the uv-plane. Find the area this way.

13. Use the transformation in Exercise 2 to evaluate the integral

$$\int_0^{2/3} \int_y^{2-2y} (x + 2y)e^{(y-x)}\, dx\, dy$$

by first writing it as an integral over a region G in the uv-plane.

14. Use the transformation $x = u + (1/2)v$, $y = v$ to evaluate the integral

$$\int_0^2 \int_{y/2}^{(y+4)/2} y^3(2x - y)e^{(2x-y)^2}\, dx\, dy$$

by first writing it as an integral over a region G in the uv-plane.

15. Use the transformation $x = u/v$, $y = uv$ to evaluate the integral sum

$$\int_1^2 \int_{1/y}^y (x^2 + y^2)\, dx\, dy + \int_2^4 \int_{y/4}^{4/y} (x^2 + y^2)\, dx\, dy.$$

16. Use the transformation $x = u^2 - v^2$, $y = 2uv$ to evaluate the integral

$$\int_0^1 \int_0^{2\sqrt{1-x}} \sqrt{x^2 + y^2}\, dy\, dx.$$

(*Hint:* Show that the image of the triangular region G with vertices $(0, 0)$, $(1, 0)$, $(1, 1)$ in the uv-plane is the region of integration R in the xy-plane defined by the limits of integration.)

Substitutions in Triple Integrals

17. Evaluate the integral in Example 5 by integrating with respect to x, y, and z.

18. Volume of an ellipsoid Find the volume of the ellipsoid

$$\frac{x^2}{a^2} + \frac{y^2}{b^2} + \frac{z^2}{c^2} = 1.$$

(*Hint:* Let $x = au$, $y = bv$, and $z = cw$. Then find the volume of an appropriate region in uvw-space.)

19. Evaluate

$$\iiint |xyz|\, dx\, dy\, dz$$

over the solid ellipsoid

$$\frac{x^2}{a^2} + \frac{y^2}{b^2} + \frac{z^2}{c^2} \le 1.$$

(*Hint:* Let $x = au$, $y = bv$, and $z = cw$. Then integrate over an appropriate region in uvw-space.)

20. Let D be the region in xyz-space defined by the inequalities

$$1 \le x \le 2, \quad 0 \le xy \le 2, \quad 0 \le z \le 1.$$

Evaluate

$$\iiint_D (x^2y + 3xyz)\, dx\, dy\, dz$$

by applying the transformation

$$u = x, \quad v = xy, \quad w = 3z$$

and integrating over an appropriate region G in uvw-space.

Theory and Examples

21. Find the Jacobian $\partial(x, y)/\partial(u, v)$ of the transformation
a. $x = u\cos v$, $y = u\sin v$.
b. $x = u\sin v$, $y = u\cos v$.

22. Find the Jacobian $\partial(x, y, z)/\partial(u, v, w)$ of the transformation
a. $x = u\cos v$, $y = u\sin v$, $z = w$.
b. $x = 2u - 1$, $y = 3v - 4$, $z = (1/2)(w - 4)$.

23. Evaluate the appropriate determinant to show that the Jacobian of the transformation from Cartesian $\rho\phi\theta$-space to Cartesian xyz-space is $\rho^2 \sin\phi$.

24. Substitutions in single integrals How can substitutions in single definite integrals be viewed as transformations of regions? What is the Jacobian in such a case? Illustrate with an example.

25. Centroid of a solid semiellipsoid Assuming the result that the centroid of a solid hemisphere lies on the axis of symmetry three-eighths of the way from the base toward the top, show, by transforming the appropriate integrals, that the center of mass of a solid semiellipsoid $(x^2/a^2) + (y^2/b^2) + (z^2/c^2) \le 1$, $z \ge 0$, lies on the z-axis three-eighths of the way from the base toward the top. (You can do this without evaluating any of the integrals.)

26. Cylindrical shells In Section 6.2, we learned how to find the volume of a solid of revolution using the shell method; namely, if the region between the curve $y = f(x)$ and the x-axis from a to b $(0 < a < b)$ is revolved about the y-axis, the volume of the resulting solid is $\int_a^b 2\pi x f(x)\, dx$. Prove that finding volumes by using triple integrals gives the same result. (*Hint:* Use cylindrical coordinates with the roles of y and z changed.)

27. Inverse transform The equations $x = g(u, v), y = h(u, v)$ in Figure 15.57 transform the region G in the uv-plane into the region R in the xy-plane. Since the substitution transformation is one-to-one with continuous first partial derivatives, it has an inverse transformation and there are equations $u = \alpha(x, y)$, $v = \beta(x, y)$ with continuous first partial derivatives transforming R back into G. Moreover, the Jacobian determinants of the transformations are related reciprocally by

$$\frac{\partial(x, y)}{\partial(u, v)} = \left(\frac{\partial(u, v)}{\partial(x, y)}\right)^{-1}. \tag{10}$$

Equation (10) is proved in advanced calculus. Use it to find the area of the region R in the first quadrant of the xy-plane bounded by the lines $y = 2x, 2y = x$, and the curves $xy = 2, 2xy = 1$ for $u = xy$ and $v = y/x$.

28. (*Continuation of Exercise 27.*) For the region R described in Exercise 27, evaluate the integral $\iint_R y^2\, dA$.

Questions to Guide Your Review

1. Define the double integral of a function of two variables over a bounded region in the coordinate plane.

2. How are double integrals evaluated as iterated integrals? Does the order of integration matter? How are the limits of integration determined? Give examples.

3. How are double integrals used to calculate areas and average values. Give examples.

4. How can you change a double integral in rectangular coordinates into a double integral in polar coordinates? Why might it be worthwhile to do so? Give an example.

5. Define the triple integral of a function $f(x, y, z)$ over a bounded region in space.

6. How are triple integrals in rectangular coordinates evaluated? How are the limits of integration determined? Give an example.

7. How are double and triple integrals in rectangular coordinates used to calculate volumes, average values, masses, moments, and centers of mass? Give examples.

8. How are triple integrals defined in cylindrical and spherical coordinates? Why might one prefer working in one of these coordinate systems to working in rectangular coordinates?

9. How are triple integrals in cylindrical and spherical coordinates evaluated? How are the limits of integration found? Give examples.

10. How are substitutions in double integrals pictured as transformations of two-dimensional regions? Give a sample calculation.

11. How are substitutions in triple integrals pictured as transformations of three-dimensional regions? Give a sample calculation.

Practice Exercises

Evaluating Double Iterated Integrals

In Exercises 1–4, sketch the region of integration and evaluate the double integral.

1. $\displaystyle\int_1^{10}\int_0^{1/y} y e^{xy}\, dx\, dy$

2. $\displaystyle\int_0^1\int_0^{x^3} e^{y/x}\, dy\, dx$

3. $\displaystyle\int_0^{3/2}\int_{-\sqrt{9-4t^2}}^{\sqrt{9-4t^2}} t\, ds\, dt$

4. $\displaystyle\int_0^1\int_{\sqrt{y}}^{2-\sqrt{y}} xy\, dx\, dy$

In Exercises 5–8, sketch the region of integration and write an equivalent integral with the order of integration reversed. Then evaluate both integrals.

5. $\displaystyle\int_0^4\int_{-\sqrt{4-y}}^{(y-4)/2} dx\, dy$

6. $\displaystyle\int_0^1\int_{x^2}^x \sqrt{x}\, dy\, dx$

7. $\displaystyle\int_0^{3/2}\int_{-\sqrt{9-4y^2}}^{\sqrt{9-4y^2}} y\, dx\, dy$

8. $\displaystyle\int_0^2\int_0^{4-x^2} 2x\, dy\, dx$

Evaluate the integrals in Exercises 9–12.

9. $\displaystyle\int_0^1 \int_{2y}^2 4 \cos(x^2) \, dx \, dy$

10. $\displaystyle\int_0^2 \int_{y/2}^1 e^{x^2} \, dx \, dy$

11. $\displaystyle\int_0^8 \int_{\sqrt[3]{x}}^2 \frac{dy \, dx}{y^4 + 1}$

12. $\displaystyle\int_0^1 \int_{\sqrt[3]{y}}^1 \frac{2\pi \sin \pi x^2}{x^2} dx \, dy$

Areas and Volumes Using Double Integrals

13. Area between line and parabola Find the area of the region enclosed by the line $y = 2x + 4$ and the parabola $y = 4 - x^2$ in the xy-plane.

14. Area bounded by lines and parabola Find the area of the "triangular" region in the xy-plane that is bounded on the right by the parabola $y = x^2$, on the left by the line $x + y = 2$, and above by the line $y = 4$.

15. Volume of the region under a paraboloid Find the volume under the paraboloid $z = x^2 + y^2$ above the triangle enclosed by the lines $y = x, x = 0$, and $x + y = 2$ in the xy-plane.

16. Volume of the region under a parabolic cylinder Find the volume under the parabolic cylinder $z = x^2$ above the region enclosed by the parabola $y = 6 - x^2$ and the line $y = x$ in the xy-plane.

Average Values

Find the average value of $f(x, y) = xy$ over the regions in Exercises 17 and 18.

17. The square bounded by the lines $x = 1, y = 1$ in the first quadrant

18. The quarter circle $x^2 + y^2 \le 1$ in the first quadrant

Polar Coordinates

Evaluate the integrals in Exercises 19 and 20 by changing to polar coordinates.

19. $\displaystyle\int_{-1}^1 \int_{-\sqrt{1-x^2}}^{\sqrt{1-x^2}} \frac{2 \, dy \, dx}{(1 + x^2 + y^2)^2}$

20. $\displaystyle\int_{-1}^1 \int_{-\sqrt{1-y^2}}^{\sqrt{1-y^2}} \ln(x^2 + y^2 + 1) \, dx \, dy$

21. Integrating over a lemniscate Integrate the function $f(x, y) = 1/(1 + x^2 + y^2)^2$ over the region enclosed by one loop of the lemniscate $(x^2 + y^2)^2 - (x^2 - y^2) = 0$.

22. Integrate $f(x, y) = 1/(1 + x^2 + y^2)^2$ over

    a. Triangular region The triangle with vertices $(0, 0), (1, 0)$, and $(1, \sqrt{3})$.

    b. First quadrant The first quadrant of the xy-plane.

Evaluating Triple Iterated Integrals

Evaluate the integrals in Exercises 23–26.

23. $\displaystyle\int_0^\pi \int_0^\pi \int_0^\pi \cos(x + y + z) \, dx \, dy \, dz$

24. $\displaystyle\int_{\ln 6}^{\ln 7} \int_0^{\ln 2} \int_{\ln 4}^{\ln 5} e^{(x+y+z)} \, dz \, dy \, dx$

25. $\displaystyle\int_0^1 \int_0^{x^2} \int_0^{x+y} (2x - y - z) \, dz \, dy \, dx$

26. $\displaystyle\int_1^e \int_1^x \int_0^z \frac{2y}{z^3} \, dy \, dz \, dx$

Volumes and Average Values Using Triple Integrals

27. Volume Find the volume of the wedge-shaped region enclosed on the side by the cylinder $x = -\cos y, -\pi/2 \le y \le \pi/2$, on the top by the plane $z = -2x$, and below by the xy-plane.

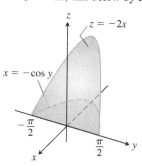

28. Volume Find the volume of the solid that is bounded above by the cylinder $z = 4 - x^2$, on the sides by the cylinder $x^2 + y^2 = 4$, and below by the xy-plane.

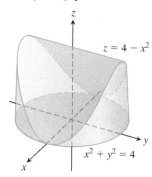

29. Average value Find the average value of $f(x, y, z) = 30xz\sqrt{x^2 + y}$ over the rectangular solid in the first octant bounded by the coordinate planes and the planes $x = 1, y = 3, z = 1$.

30. Average value Find the average value of ρ over the solid sphere $\rho \le a$ (spherical coordinates).

Cylindrical and Spherical Coordinates

31. Cylindrical to rectangular coordinates Convert

$$\int_0^{2\pi} \int_0^{\sqrt 2} \int_r^{\sqrt{4-r^2}} 3 \, dz \, r \, dr \, d\theta, \qquad r \ge 0$$

to **(a)** rectangular coordinates with the order of integration $dz \, dx \, dy$ and **(b)** spherical coordinates. Then **(c)** evaluate one of the integrals.

32. Rectangular to cylindrical coordinates **(a)** Convert to cylindrical coordinates. Then **(b)** evaluate the new integral.

$$\int_0^1 \int_{-\sqrt{1-x^2}}^{\sqrt{1-x^2}} \int_{-(x^2+y^2)}^{(x^2+y^2)} 21xy^2 \, dz \, dy \, dx$$

33. Rectangular to spherical coordinates **(a)** Convert to spherical coordinates. Then **(b)** evaluate the new integral.

$$\int_{-1}^1 \int_{-\sqrt{1-x^2}}^{\sqrt{1-x^2}} \int_{\sqrt{x^2+y^2}}^1 dz \, dy \, dx$$

34. Rectangular, cylindrical, and spherical coordinates Write an iterated triple integral for the integral of $f(x, y, z) = 6 + 4y$ over the region in the first octant bounded by the cone $z = \sqrt{x^2 + y^2}$, the cylinder $x^2 + y^2 = 1$, and the coordinate planes in **(a)** rect-angular coordinates, **(b)** cylindrical coordinates, and **(c)** spherical coordinates. Then **(d)** find the integral of f by evaluating one of the triple integrals.

35. Cylindrical to rectangular coordinates Set up an integral in rectangular coordinates equivalent to the integral

$$\int_0^{\pi/2} \int_1^{\sqrt{3}} \int_1^{\sqrt{4-r^2}} r^3(\sin\theta \cos\theta)z^2 \, dz \, dr \, d\theta.$$

Arrange the order of integration to be z first, then y, then x.

36. Rectangular to cylindrical coordinates The volume of a solid is

$$\int_0^2 \int_0^{\sqrt{2x-x^2}} \int_{-\sqrt{4-x^2-y^2}}^{\sqrt{4-x^2-y^2}} dz \, dy \, dx.$$

a. Describe the solid by giving equations for the surfaces that form its boundary.

b. Convert the integral to cylindrical coordinates but do not evaluate the integral.

37. Spherical versus cylindrical coordinates Triple integrals involving spherical shapes do not always require spherical coor-dinates for convenient evaluation. Some calculations may be ac-complished more easily with cylindrical coordinates. As a case in point, find the volume of the region bounded above by the sphere $x^2 + y^2 + z^2 = 8$ and below by the plane $z = 2$ by using **(a)** cy-lindrical coordinates and **(b)** spherical coordinates.

Masses and Moments

38. Finding I_z in spherical coordinates Find the moment of inertia about the z-axis of a solid of constant density $\delta = 1$ that is bound-ed above by the sphere $\rho = 2$ and below by the cone $\phi = \pi/3$ (spherical coordinates).

39. Moment of inertia of a "thick" sphere Find the moment of inertia of a solid of constant density δ bounded by two concentric spheres of radii a and b $(a < b)$ about a diameter.

40. Moment of inertia of an apple Find the moment of inertia about the z-axis of a solid of density $\delta = 1$ enclosed by the spheri-cal coordinate surface $\rho = 1 - \cos\phi$. The solid is the red curve rotated about the z-axis in the accompanying figure.

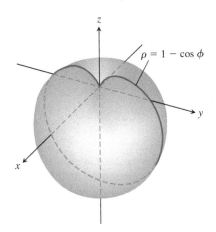

$\rho = 1 - \cos\phi$

41. Centroid Find the centroid of the "triangular" region bound-ed by the lines $x = 2$, $y = 2$ and the hyperbola $xy = 2$ in the xy-plane.

42. Centroid Find the centroid of the region between the parabola $x + y^2 - 2y = 0$ and the line $x + 2y = 0$ in the xy-plane.

43. Polar moment Find the polar moment of inertia about the origin of a thin triangular plate of constant density $\delta = 3$ bounded by the y-axis and the lines $y = 2x$ and $y = 4$ in the xy-plane.

44. Polar moment Find the polar moment of inertia about the cen-ter of a thin rectangular sheet of constant density $\delta = 1$ bounded by the lines

a. $x = \pm 2$, $y = \pm 1$ in the xy-plane

b. $x = \pm a$, $y = \pm b$ in the xy-plane.

(*Hint:* Find I_x. Then use the formula for I_x to find I_y, and add the two to find I_0.)

45. Inertial moment Find the moment of inertia about the x-axis of a thin plate of constant density δ covering the triangle with verti-ces $(0, 0)$, $(3, 0)$, and $(3, 2)$ in the xy-plane.

46. Plate with variable density Find the center of mass and the mo-ments of inertia about the coordinate axes of a thin plate bounded by the line $y = x$ and the parabola $y = x^2$ in the xy-plane if the density is $\delta(x, y) = x + 1$.

47. Plate with variable density Find the mass and first moments about the coordinate axes of a thin square plate bounded by the lines $x = \pm 1$, $y = \pm 1$ in the xy-plane if the density is $\delta(x, y) = x^2 + y^2 + 1/3$.

48. Triangles with same inertial moment Find the moment of in-ertia about the x-axis of a thin triangular plate of constant den-sity δ whose base lies along the interval $[0, b]$ on the x-axis and whose vertex lies on the line $y = h$ above the x-axis. As you will see, it does not matter where on the line this vertex lies. All such triangles have the same moment of inertia about the x-axis.

49. Centroid Find the centroid of the region in the polar coordinate plane defined by the inequalities $0 \le r \le 3$, $-\pi/3 \le \theta \le \pi/3$.

50. Centroid Find the centroid of the region in the first quadrant bounded by the rays $\theta = 0$ and $\theta = \pi/2$ and the circles $r = 1$ and $r = 3$.

51. a. Centroid Find the centroid of the region in the polar coor-dinate plane that lies inside the cardioid $r = 1 + \cos\theta$ and outside the circle $r = 1$.

b. Sketch the region and show the centroid in your sketch.

52. a. Centroid Find the centroid of the plane region defined by the polar coordinate inequalities $0 \le r \le a$, $-\alpha \le \theta \le \alpha$ $(0 < \alpha \le \pi)$. How does the centroid move as $\alpha \to \pi^-$?

b. Sketch the region for $\alpha = 5\pi/6$ and show the centroid in your sketch.

Substitutions

53. Show that if $u = x - y$ and $v = y$, then for any continuous f

$$\int_0^\infty \int_0^x e^{-sx} f(x - y, y) \, dy \, dx = \int_0^\infty \int_0^\infty e^{-s(u+v)} f(u, v) \, du \, dv.$$

54. What relationship must hold between the constants a, b, and c to make

$$\int_{-\infty}^\infty \int_{-\infty}^\infty e^{-(ax^2+2bxy+cy^2)} \, dx \, dy = 1?$$

(*Hint:* Let $s = \alpha x + \beta y$ and $t = \gamma x + \delta y$, where $(\alpha\delta - \beta\gamma)^2 = ac - b^2$. Then $ax^2 + 2bxy + cy^2 = s^2 + t^2$.)

CHAPTER 15   Additional and Advanced Exercises

Volumes

1. Sand pile: double and triple integrals   The base of a sand pile covers the region in the xy-plane that is bounded by the parabola $x^2 + y = 6$ and the line $y = x$. The height of the sand above the point (x, y) is x^2. Express the volume of sand as **(a)** a double integral, **(b)** a triple integral. Then **(c)** find the volume.

2. Water in a hemispherical bowl   A hemispherical bowl of radius 5 cm is filled with water to within 3 cm of the top. Find the volume of water in the bowl.

3. Solid cylindrical region between two planes   Find the volume of the portion of the solid cylinder $x^2 + y^2 \le 1$ that lies between the planes $z = 0$ and $x + y + z = 2$.

4. Sphere and paraboloid   Find the volume of the region bounded above by the sphere $x^2 + y^2 + z^2 = 2$ and below by the paraboloid $z = x^2 + y^2$.

5. Two paraboloids   Find the volume of the region bounded above by the paraboloid $z = 3 - x^2 - y^2$ and below by the paraboloid $z = 2x^2 + 2y^2$.

6. Spherical coordinates   Find the volume of the region enclosed by the spherical coordinate surface $\rho = 2 \sin \phi$ (see accompanying figure).

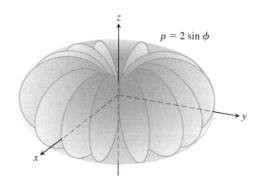

$\rho = 2 \sin \phi$

7. Hole in sphere   A circular cylindrical hole is bored through a solid sphere, the axis of the hole being a diameter of the sphere. The volume of the remaining solid is

$$V = 2 \int_0^{2\pi} \int_0^{\sqrt{3}} \int_1^{\sqrt{4-z^2}} r \, dr \, dz \, d\theta.$$

a. Find the radius of the hole and the radius of the sphere.

b. Evaluate the integral.

8. Sphere and cylinder   Find the volume of material cut from the solid sphere $r^2 + z^2 \le 9$ by the cylinder $r = 3 \sin \theta$.

9. Two paraboloids   Find the volume of the region enclosed by the surfaces $z = x^2 + y^2$ and $z = (x^2 + y^2 + 1)/2$.

10. Cylinder and surface $z = xy$ Find the volume of the region in the first octant that lies between the cylinders $r = 1$ and $r = 2$ and that is bounded below by the xy-plane and above by the surface $z = xy$.

Changing the Order of Integration

11. Evaluate the integral

$$\int_0^\infty \frac{e^{-ax} - e^{-bx}}{x} \, dx.$$

(*Hint:* Use the relation

$$\frac{e^{-ax} - e^{-bx}}{x} = \int_a^b e^{-xy} \, dy$$

to form a double integral and evaluate the integral by changing the order of integration.)

12. a. Polar coordinates   Show, by changing to polar coordinates, that

$$\int_0^{a \sin \beta} \int_{y \cot \beta}^{\sqrt{a^2-y^2}} \ln (x^2 + y^2) \, dx \, dy = a^2 \beta \left(\ln a - \frac{1}{2} \right),$$

where $a > 0$ and $0 < \beta < \pi/2$.

b. Rewrite the Cartesian integral with the order of integration reversed.

13. Reducing a double to a single integral   By changing the order of integration, show that the following double integral can be reduced to a single integral:

$$\int_0^x \int_0^u e^{m(x-t)} f(t) \, dt \, du = \int_0^x (x - t) e^{m(x-t)} f(t) \, dt.$$

Similarly, it can be shown that

$$\int_0^x \int_0^v \int_0^u e^{m(x-t)} f(t) \, dt \, du \, dv = \int_0^x \frac{(x - t)^2}{2} e^{m(x-t)} f(t) \, dt.$$

14. Transforming a double integral to obtain constant limit   Sometimes a multiple integral with variable limits can be changed into one with constant limits. By changing the order of integration, show that

$$\int_0^1 f(x) \left(\int_0^x g(x - y) f(y) \, dy \right) dx$$

$$= \int_0^1 f(y) \left(\int_y^1 g(x - y) f(x) \, dx \right) dy$$

$$= \frac{1}{2} \int_0^1 \int_0^1 g(|x - y|) f(x) f(y) \, dx \, dy.$$

Masses and Moments

15. Minimizing polar inertia   A thin plate of constant density is to occupy the triangular region in the first quadrant of the xy-plane having vertices $(0, 0)$, $(a, 0)$, and $(a, 1/a)$. What value of a will minimize the plate's polar moment of inertia about the origin?

16. Polar inertia of triangular plate   Find the polar moment of inertia about the origin of a thin triangular plate of constant density $\delta = 3$ bounded by the y-axis and the lines $y = 2x$ and $y = 4$ in the xy-plane.

17. **Mass and polar inertia of a counterweight** The counterweight of a flywheel of constant density 1 has the form of the smaller segment cut from a circle of radius a by a chord at a distance b from the center $(b < a)$. Find the mass of the counterweight and its polar moment of inertia about the center of the wheel.

18. **Centroid of a boomerang** Find the centroid of the boomerang-shaped region between the parabolas $y^2 = -4(x - 1)$ and $y^2 = -2(x - 2)$ in the xy-plane.

Theory and Examples

19. Evaluate

$$\int_0^a \int_0^b e^{\max(b^2x^2,\, a^2y^2)}\, dy\, dx,$$

where a and b are positive numbers and

$$\max(b^2x^2, a^2y^2) = \begin{cases} b^2x^2 & \text{if } b^2x^2 \geq a^2y^2 \\ a^2y^2 & \text{if } b^2x^2 < a^2y^2. \end{cases}$$

20. Show that

$$\iint \frac{\partial^2 F(x, y)}{\partial x\, \partial y}\, dx\, dy$$

over the rectangle $x_0 \leq x \leq x_1$, $y_0 \leq y \leq y_1$, is

$$F(x_1, y_1) - F(x_0, y_1) - F(x_1, y_0) + F(x_0, y_0).$$

21. Suppose that $f(x, y)$ can be written as a product $f(x, y) = F(x)G(y)$ of a function of x and a function of y. Then the integral of f over the rectangle $R\colon a \leq x \leq b, c \leq y \leq d$ can be evaluated as a product as well, by the formula

$$\iint_R f(x, y)\, dA = \left(\int_a^b F(x)\, dx\right)\left(\int_c^d G(y)\, dy\right). \tag{1}$$

The argument is that

$$\iint_R f(x, y)\, dA = \int_c^d \left(\int_a^b F(x)G(y)\, dx\right) dy \tag{i}$$

$$= \int_c^d \left(G(y)\int_a^b F(x)\, dx\right) dy \tag{ii}$$

$$= \int_c^d \left(\int_a^b F(x)\, dx\right)G(y)\, dy \tag{iii}$$

$$= \left(\int_a^b F(x)\, dx\right)\int_c^d G(y)\, dy. \tag{iv}$$

a. Give reasons for steps (i) through (iv).

When it applies, Equation (1) can be a time-saver. Use it to evaluate the following integrals.

b. $\displaystyle\int_0^{\ln 2} \int_0^{\pi/2} e^x \cos y\, dy\, dx$ c. $\displaystyle\int_1^2 \int_{-1}^1 \frac{x}{y^2}\, dx\, dy$

22. Let $D_{\mathbf{u}}f$ denote the derivative of $f(x, y) = (x^2 + y^2)/2$ in the direction of the unit vector $\mathbf{u} = u_1\mathbf{i} + u_2\mathbf{j}$.

a. **Finding average value** Find the average value of $D_{\mathbf{u}}f$ over the triangular region cut from the first quadrant by the line $x + y = 1$.

b. **Average value and centroid** Show in general that the average value of $D_{\mathbf{u}}f$ over a region in the xy-plane is the value of $D_{\mathbf{u}}f$ at the centroid of the region.

23. **The value of $\Gamma(1/2)$** The gamma function,

$$\Gamma(x) = \int_0^\infty t^{x-1} e^{-t}\, dt,$$

extends the factorial function from the nonnegative integers to other real values. Of particular interest in the theory of differential equations is the number

$$\Gamma\left(\frac{1}{2}\right) = \int_0^\infty t^{(1/2)-1} e^{-t}\, dt = \int_0^\infty \frac{e^{-t}}{\sqrt{t}}\, dt. \tag{2}$$

a. If you have not yet done Exercise 41 in Section 15.4, do it now to show that

$$I = \int_0^\infty e^{-y^2}\, dy = \frac{\sqrt{\pi}}{2}.$$

b. Substitute $y = \sqrt{t}$ in Equation (2) to show that $\Gamma(1/2) = 2I = \sqrt{\pi}$.

24. **Total electrical charge over circular plate** The electrical charge distribution on a circular plate of radius R meters is $\sigma(r, \theta) = kr(1 - \sin \theta)$ coulomb/m^2 (k a constant). Integrate σ over the plate to find the total charge Q.

25. **A parabolic rain gauge** A bowl is in the shape of the graph of $z = x^2 + y^2$ from $z = 0$ to $z = 10$ in. You plan to calibrate the bowl to make it into a rain gauge. What height in the bowl would correspond to 1 in. of rain? 3 in. of rain?

26. **Water in a satellite dish** A parabolic satellite dish is 2 m wide and $1/2$ m deep. Its axis of symmetry is tilted 30 degrees from the vertical.

a. Set up, but do not evaluate, a triple integral in rectangular coordinates that gives the amount of water the satellite dish will hold. (*Hint:* Put your coordinate system so that the satellite dish is in "standard position" and the plane of the water level is slanted.) (*Caution:* The limits of integration are not "nice.")

b. What would be the smallest tilt of the satellite dish so that it holds no water?

27. **An infinite half-cylinder** Let D be the interior of the infinite right circular half-cylinder of radius 1 with its single-end face suspended 1 unit above the origin and its axis the ray from $(0, 0, 1)$ to ∞. Use cylindrical coordinates to evaluate

$$\iiint_D z(r^2 + z^2)^{-5/2}\, dV.$$

28. **Hypervolume** We have learned that $\int_a^b 1\, dx$ is the length of the interval $[a, b]$ on the number line (one-dimensional space), $\iint_R 1\, dA$ is the area of region R in the xy-plane (two-dimensional space), and $\iiint_D 1\, dV$ is the volume of the region D in three-dimensional space (xyz-space). We could continue: If Q is a region in 4-space ($xyzw$-space), then $\iiiint_Q 1\, dV$ is the "hyper-volume" of Q. Use your generalizing abilities and a Cartesian coordinate system of 4-space to find the hypervolume inside the unit 4-dimensional sphere $x^2 + y^2 + z^2 + w^2 = 1$.

CHAPTER 15   Technology Application Projects

Mathematica/Maple Projects

Projects can be found within MyMathLab.

- *Take Your Chances: Try the Monte Carlo Technique for Numerical Integration in Three Dimensions*
  Use the Monte Carlo technique to integrate numerically in three dimensions.

- *Means and Moments and Exploring New Plotting Techniques, Part II*
  Use the method of moments in a form that makes use of geometric symmetry as well as multiple integration.

16

Integrals and Vector Fields

OVERVIEW In this chapter we extend the theory of integration to general curves and surfaces in space. The resulting line and surface integrals give powerful mathematical tools for science and engineering. Line integrals are used to find the work done by a force in moving an object along a path, and to find the mass of a curved wire with variable density. Surface integrals are used to find the rate of flow of a fluid across a surface and to describe the interactions of electric and magnetic forces. We present the fundamental theorems of vector integral calculus, and discuss their mathematical consequences and physical applications. The theorems of vector calculus are then shown to be generalized versions of the Fundamental Theorem of Calculus.

16.1 Line Integrals of Scalar Functions

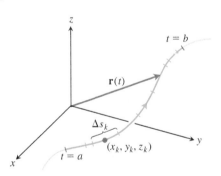

FIGURE 16.1 The curve $\mathbf{r}(t)$ partitioned into small arcs from $t = a$ to $t = b$. The length of a typical subarc is Δs_k.

To calculate the total mass of a wire lying along a curve in space, or to find the work done by a variable force acting along such a curve, we need a more general notion of integral than was defined in Chapter 5. We need to integrate over a curve C rather than over an interval $[a, b]$. These more general integrals are called *line integrals* (although *path* integrals might be more descriptive). We make our definitions for space curves, with curves in the xy-plane being the special case with z-coordinate identically zero.

Suppose that $f(x, y, z)$ is a real-valued function we wish to integrate over the curve C lying within the domain of f and parametrized by $\mathbf{r}(t) = g(t)\mathbf{i} + h(t)\mathbf{j} + k(t)\mathbf{k}$, $a \leq t \leq b$. The values of f along the curve are given by the composite function $f(g(t), h(t), k(t))$. We are going to integrate this composition with respect to arc length from $t = a$ to $t = b$. To begin, we first partition the curve C into a finite number n of subarcs (Figure 16.1). The typical subarc has length Δs_k. In each subarc we choose a point (x_k, y_k, z_k) and form the sum

$$S_n = \sum_{k=1}^{n} \underbrace{f(x_k, y_k, z_k)}_{\substack{\text{value of } f \text{ at a point} \\ \text{on the subarc}}} \underbrace{\Delta s_k}_{\substack{\text{length of a small} \\ \text{subarc of the curve}}},$$

which is similar to a Riemann sum. Depending on how we partition the curve C and pick (x_k, y_k, z_k) in the kth subarc, we may get different values for S_n. If f is continuous and the functions g, h, and k have continuous first derivatives, then these sums approach a limit as n increases and the lengths Δs_k approach zero. This leads to the following definition, which is similar to that for a single integral. In the definition, we assume that the norm of the partition approaches zero as $n \to \infty$, so that the length of the longest subarc approaches zero.

> **DEFINITION** If f is defined on a curve C given parametrically by $\mathbf{r}(t) = g(t)\mathbf{i} + h(t)\mathbf{j} + k(t)\mathbf{k}$, $a \le t \le b$, then the **line integral of f over C** is
>
> $$\int_C f(x, y, z)\, ds = \lim_{n \to \infty} \sum_{k=1}^{n} f(x_k, y_k, z_k)\, \Delta s_k, \qquad (1)$$
>
> provided this limit exists.

If the curve C is smooth for $a \le t \le b$ (so $\mathbf{v} = d\mathbf{r}/dt$ is continuous and never $\mathbf{0}$) and the function f is continuous on C, then the limit in Equation (1) can be shown to exist. We can then apply the Fundamental Theorem of Calculus to differentiate the arc length equation,

$$s(t) = \int_a^t |\mathbf{v}(\tau)|\, d\tau, \qquad \text{Eq. (3) of Section 13.3 with } t_0 = a$$

to express ds in Equation (1) as $ds = |\mathbf{v}(t)|\, dt$ and evaluate the integral of f over C as

$$\frac{ds}{dt} = |\mathbf{v}| = \sqrt{\left(\frac{dx}{dt}\right)^2 + \left(\frac{dy}{dt}\right)^2 + \left(\frac{dz}{dt}\right)^2}$$

$$\int_C f(x, y, z)\, ds = \int_a^b f(g(t), h(t), k(t)) |\mathbf{v}(t)|\, dt. \qquad (2)$$

The integral on the right side of Equation (2) is just an ordinary definite integral, as defined in Chapter 5, where we are integrating with respect to the parameter t. The formula evaluates the line integral on the left side correctly no matter what smooth parametrization is used. Note that the parameter t defines a direction along the path. The starting point on C is the position $\mathbf{r}(a)$, and movement along the path is in the direction of increasing t (see Figure 16.1).

> **How to Evaluate a Line Integral**
> To integrate a continuous function $f(x, y, z)$ over a curve C:
>
> **1.** Find a smooth parametrization of C,
>
> $$\mathbf{r}(t) = g(t)\mathbf{i} + h(t)\mathbf{j} + k(t)\mathbf{k}, \qquad a \le t \le b.$$
>
> **2.** Evaluate the integral as
>
> $$\int_C f(x, y, z)\, ds = \int_a^b f(g(t), h(t), k(t)) |\mathbf{v}(t)|\, dt.$$

$$f(\mathbf{r}(t)) = f(g(t), h(t), k(t))$$

If f has the constant value 1, then the integral of f over C gives the length of C from $t = a$ to $t = b$. We also write $f(\mathbf{r}(t))$ for the evaluation $f(g(t), h(t), k(t))$ along the curve $\mathbf{r}$.

EXAMPLE 1 Integrate $f(x, y, z) = x - 3y^2 + z$ over the line segment C joining the origin to the point $(1, 1, 1)$ (Figure 16.2).

Solution Since any choice of parametrization will give the same answer, we choose the simplest parametrization we can think of:

$$\mathbf{r}(t) = t\mathbf{i} + t\mathbf{j} + t\mathbf{k}, \qquad 0 \le t \le 1.$$

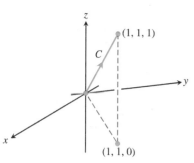

FIGURE 16.2 The integration path in Example 1.

The components have continuous first derivatives and $|\mathbf{v}(t)| = |\mathbf{i} + \mathbf{j} + \mathbf{k}| = \sqrt{1^2 + 1^2 + 1^2} = \sqrt{3}$ is never 0, so the parametrization is smooth. The integral of f over C is

$$\int_C f(x, y, z)\, ds = \int_0^1 f(t, t, t)\sqrt{3}\, dt \qquad \text{Eq. (2), } ds = |\mathbf{v}(t)|\, dt = \sqrt{3}\, dt$$

$$= \int_0^1 (t - 3t^2 + t)\sqrt{3}\, dt$$

$$= \sqrt{3}\int_0^1 (2t - 3t^2)\, dt = \sqrt{3}\left[t^2 - t^3\right]_0^1 = 0. \qquad \blacksquare$$

Additivity

Line integrals have the useful property that if a piecewise smooth curve C is made by joining a finite number of smooth curves $C_1, C_2, \ldots, C_n$ end to end (Section 13.1), then the integral of a function over C is the sum of the integrals over the curves that make it up:

$$\int_C f\, ds = \int_{C_1} f\, ds + \int_{C_2} f\, ds + \cdots + \int_{C_n} f\, ds. \qquad (3)$$

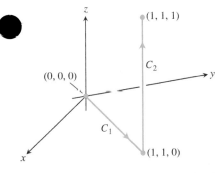

FIGURE 16.3 The path of integration in Example 2.

EXAMPLE 2 Figure 16.3 shows another path from the origin to $(1, 1, 1)$, formed from two line segments C_1 and C_2. Integrate $f(x, y, z) = x - 3y^2 + z$ over $C_1 \cup C_2$.

Solution We choose the simplest parametrizations for C_1 and C_2 we can find, calculating the lengths of the velocity vectors as we go along:

$$C_1: \quad \mathbf{r}(t) = t\mathbf{i} + t\mathbf{j}, \quad 0 \le t \le 1; \quad |\mathbf{v}| = \sqrt{1^2 + 1^2} = \sqrt{2}$$
$$C_2: \quad \mathbf{r}(t) = \mathbf{i} + \mathbf{j} + t\mathbf{k}, \quad 0 \le t \le 1; \quad |\mathbf{v}| = \sqrt{0^2 + 0^2 + 1^2} = 1.$$

With these parametrizations we find that

$$\int_{C_1 \cup C_2} f(x, y, z)\, ds = \int_{C_1} f(x, y, z)\, ds + \int_{C_2} f(x, y, z)\, ds \qquad \text{Eq. (3)}$$

$$= \int_0^1 f(t, t, 0)\sqrt{2}\, dt + \int_0^1 f(1, 1, t)(1)\, dt \qquad \text{Eq. (2)}$$

$$= \int_0^1 (t - 3t^2 + 0)\sqrt{2}\, dt + \int_0^1 (1 - 3 + t)(1)\, dt$$

$$= \sqrt{2}\left[\frac{t^2}{2} - t^3\right]_0^1 + \left[\frac{t^2}{2} - 2t\right]_0^1 = -\frac{\sqrt{2}}{2} - \frac{3}{2}. \qquad \blacksquare$$

Notice three things about the integrations in Examples 1 and 2. First, as soon as the components of the appropriate curve were substituted into the formula for f, the integration became a standard integration with respect to t. Second, the integral of f over $C_1 \cup C_2$ was obtained by integrating f over each section of the path and adding the results. Third, the integrals of f over C and $C_1 \cup C_2$ had different values. We investigate this third observation in Section 16.3.

The value of the line integral along a path joining two points can change if you change the path between them.

FIGURE 16.4 A line integral is taken over a curve such as this helix from Example 3.

EXAMPLE 3 Find the line integral of $f(x, y, z) = 2xy + \sqrt{z}$ over the helix $\mathbf{r}(t) = \cos t\mathbf{i} + \sin t\mathbf{j} + t\mathbf{k}, 0 \le t \le \pi$.

Solution For the helix (Figure 16.4) we find $\mathbf{v}(t) = \mathbf{r}'(t) = -\sin t\mathbf{i} + \cos t\mathbf{j} + \mathbf{k}$ and $|\mathbf{v}(t)| = \sqrt{(-\sin t)^2 + (\cos t)^2 + 1} = \sqrt{2}$. Evaluating the function f at the point $\mathbf{r}(t)$, we obtain

$$f(\mathbf{r}(t)) = f(\cos t, \sin t, t) = 2\cos t \sin t + \sqrt{t} = \sin 2t + \sqrt{t}.$$

The line integral is given by

$$\int_C f(x, y, z)\, ds = \int_0^\pi \left(\sin 2t + \sqrt{t}\right)\sqrt{2}\, dt$$

$$= \sqrt{2}\left[-\frac{1}{2}\cos 2t + \frac{2}{3}t^{3/2}\right]_0^\pi$$

$$= \frac{2\sqrt{2}}{3}\pi^{3/2} \approx 5.25. \qquad \blacksquare$$

Mass and Moment Calculations

We treat coil springs and wires as masses distributed along smooth curves in space. The distribution is described by a continuous density function $\delta(x, y, z)$ representing mass per unit length. When a curve C is parametrized by $\mathbf{r}(t) = x(t)\mathbf{i} + y(t)\mathbf{j} + z(t)\mathbf{k}, a \le t \le b$, then x, y, and z are functions of the parameter t, the density is the function $\delta(x(t), y(t), z(t))$, and the arc length differential is given by

$$ds = \sqrt{\left(\frac{dx}{dt}\right)^2 + \left(\frac{dy}{dt}\right)^2 + \left(\frac{dz}{dt}\right)^2}\, dt.$$

(See Section 13.3.) The spring's or wire's mass, center of mass, and moments are then calculated using the formulas in Table 16.1, with the integrations in terms of the parameter t over the interval $[a, b]$. For example, the formula for mass becomes

$$M = \int_a^b \delta(x(t), y(t), z(t))\sqrt{\left(\frac{dx}{dt}\right)^2 + \left(\frac{dy}{dt}\right)^2 + \left(\frac{dz}{dt}\right)^2}\, dt.$$

These formulas also apply to thin rods, and their derivations are similar to those in Section 6.6. Notice how alike the formulas are to those in Tables 15.1 and 15.2 for double and triple integrals. The double integrals for planar regions, and the triple integrals for solids, become line integrals for coil springs, wires, and thin rods.

Notice that the element of mass dm is equal to $\delta\, ds$ in the table rather than $\delta\, dV$ as in Table 15.1, and that the integrals are taken over the curve C.

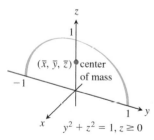

FIGURE 16.5 Example 4 shows how to find the center of mass of a circular arch of variable density.

EXAMPLE 4 A slender metal arch, denser at the bottom than top, lies along the semicircle $y^2 + z^2 = 1, z \ge 0$, in the yz-plane (Figure 16.5). Find the center of the arch's mass if the density at the point (x, y, z) on the arch is $\delta(x, y, z) = 2 - z$.

Solution We know that $\bar{x} = 0$ and $\bar{y} = 0$ because the arch lies in the yz-plane with its mass distributed symmetrically about the z-axis. To find $\bar{z}$, we parametrize the circle as

$$\mathbf{r}(t) = (\cos t)\mathbf{j} + (\sin t)\mathbf{k}, \qquad 0 \le t \le \pi.$$

TABLE 16.1 Mass and moment formulas for coil springs, wires, and thin rods lying along a smooth curve C in space

Mass: $\quad M = \displaystyle\int_C \delta\, ds \qquad \delta = \delta(x, y, z) \text{ is the density at } (x, y, z)$

First moments about the coordinate planes:

$$M_{yz} = \int_C x\, \delta\, ds, \qquad M_{xz} = \int_C y\, \delta\, ds, \qquad M_{xy} = \int_C z\, \delta\, ds$$

Coordinates of the center of mass:

$$\bar{x} = M_{yz}/M, \qquad \bar{y} = M_{xz}/M, \qquad \bar{z} = M_{xy}/M$$

Moments of inertia about axes and other lines:

$$I_x = \int_C (y^2 + z^2)\delta\, ds, \qquad I_y = \int_C (x^2 + z^2)\delta\, ds, \qquad I_z = \int_C (x^2 + y^2)\delta\, ds,$$

$$I_L = \int_C r^2 \delta\, ds \qquad r(x, y, z) = \text{distance from the point } (x, y, z) \text{ to line } L$$

For this parametrization,

$$|\mathbf{v}(t)| = \sqrt{\left(\frac{dx}{dt}\right)^2 + \left(\frac{dy}{dt}\right)^2 + \left(\frac{dz}{dt}\right)^2} = \sqrt{(0)^2 + (-\sin t)^2 + (\cos t)^2} = 1,$$

so $ds = |\mathbf{v}|\, dt = dt$.

The formulas in Table 16.1 then give

$$M = \int_C \delta\, ds = \int_C (2 - z)\, ds = \int_0^\pi (2 - \sin t)\, dt = 2\pi - 2$$

$$M_{xy} = \int_C z\, \delta\, ds = \int_C z(2 - z)\, ds = \int_0^\pi (\sin t)(2 - \sin t)\, dt$$

$$= \int_0^\pi (2\sin t - \sin^2 t)\, dt = \frac{8 - \pi}{2} \qquad \text{Routine integration}$$

$$\bar{z} = \frac{M_{xy}}{M} = \frac{8 - \pi}{2} \cdot \frac{1}{2\pi - 2} = \frac{8 - \pi}{4\pi - 4} \approx 0.57.$$

With $\bar{z}$ to the nearest hundredth, the center of mass is $(0, 0, 0.57)$. $\quad\blacksquare$

Line Integrals in the Plane

Line integrals for curves in the plane have a natural geometric interpretation. If C is a smooth curve in the xy-plane parametrized by $\mathbf{r}(t) = x(t)\mathbf{i} + y(t)\mathbf{j}$, $a \le t \le b$, we generate a cylindrical surface by moving a straight line along C orthogonal to the plane, holding the line parallel to the z-axis, as in Figure 16.6. If $z = f(x, y)$ is a nonnegative continuous function over a region in the plane containing the curve C, then the graph of f is a surface that lies above the plane. The cylinder cuts through this surface, forming a curve on it that lies above the curve C and follows its winding nature. The part of the cylindrical surface that lies beneath the surface curve and above the xy-plane forms a "curved wall" or "fence"

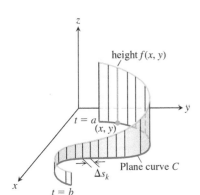

FIGURE 16.6 The line integral $\displaystyle\int_C f\, ds$ gives the area of the portion of the cylindrical surface or "wall" beneath $z = f(x, y) \ge 0$.

standing on the curve C and orthogonal to the plane. At any point (x, y) along the curve, the height of the wall is $f(x, y)$. From the definition

$$\int_C f\,ds = \lim_{n\to\infty} \sum_{k=1}^{n} f(x_k, y_k)\,\Delta s_k,$$

where $\Delta s_k \to 0$ as $n \to \infty$, we see that the line integral $\int_C f\,ds$ is the area of the wall shown in the figure.

EXERCISES 16.1

Graphs of Vector Equations

Match the vector equations in Exercises 1–8 with the graphs (a)–(h) given here.

a.
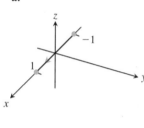

b.
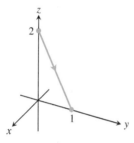

c.

d.
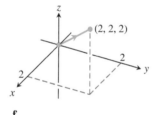

e.
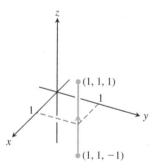

f.

g.
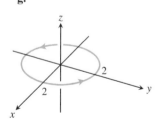

h.

1. $\mathbf{r}(t) = t\mathbf{i} + (1 - t)\mathbf{j}, \quad 0 \le t \le 1$
2. $\mathbf{r}(t) = \mathbf{i} + \mathbf{j} + t\mathbf{k}, \quad -1 \le t \le 1$
3. $\mathbf{r}(t) = (2\cos t)\mathbf{i} + (2\sin t)\mathbf{j}, \quad 0 \le t \le 2\pi$
4. $\mathbf{r}(t) = t\mathbf{i}, \quad -1 \le t \le 1$
5. $\mathbf{r}(t) = t\mathbf{i} + t\mathbf{j} + t\mathbf{k}, \quad 0 \le t \le 2$
6. $\mathbf{r}(t) = t\mathbf{j} + (2 - 2t)\mathbf{k}, \quad 0 \le t \le 1$
7. $\mathbf{r}(t) = (t^2 - 1)\mathbf{j} + 2t\mathbf{k}, \quad -1 \le t \le 1$
8. $\mathbf{r}(t) = (2\cos t)\mathbf{i} + (2\sin t)\mathbf{k}, \quad 0 \le t \le \pi$

Evaluating Line Integrals over Space Curves

9. Evaluate $\int_C (x + y)\,ds$ where C is the straight-line segment $x = t, y = (1 - t), z = 0$, from $(0, 1, 0)$ to $(1, 0, 0)$.

10. Evaluate $\int_C (x - y + z - 2)\,ds$ where C is the straight-line segment $x = t, y = (1 - t), z = 1$, from $(0, 1, 1)$ to $(1, 0, 1)$.

11. Evaluate $\int_C (xy + y + z)\,ds$ along the curve $\mathbf{r}(t) = 2t\mathbf{i} + t\mathbf{j} + (2 - 2t)\mathbf{k}, 0 \le t \le 1$.

12. Evaluate $\int_C \sqrt{x^2 + y^2}\,ds$ along the curve $\mathbf{r}(t) = (4\cos t)\mathbf{i} + (4\sin t)\mathbf{j} + 3t\mathbf{k}, -2\pi \le t \le 2\pi$.

13. Find the line integral of $f(x, y, z) = x + y + z$ over the straight-line segment from $(1, 2, 3)$ to $(0, -1, 1)$.

14. Find the line integral of $f(x, y, z) = \sqrt{3}/(x^2 + y^2 + z^2)$ over the curve $\mathbf{r}(t) = t\mathbf{i} + t\mathbf{j} + t\mathbf{k}, 1 \le t \le \infty$.

15. Integrate $f(x, y, z) = x + \sqrt{y} - z^2$ over the path from $(0, 0, 0)$ to $(1, 1, 1)$ (see accompanying figure) given by

$$C_1: \quad \mathbf{r}(t) = t\mathbf{i} + t^2\mathbf{j}, \quad 0 \le t \le 1$$
$$C_2: \quad \mathbf{r}(t) = \mathbf{i} + \mathbf{j} + t\mathbf{k}, \quad 0 \le t \le 1$$

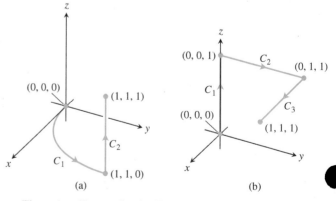

(a)

(b)

The paths of integration for Exercises 15 and 16.

16. Integrate $f(x, y, z) = x + \sqrt{y} - z^2$ over the path from $(0, 0, 0)$ to $(1, 1, 1)$ (see accompanying figure) given by

$$C_1: \quad \mathbf{r}(t) = t\mathbf{k}, \quad 0 \le t \le 1$$
$$C_2: \quad \mathbf{r}(t) = t\mathbf{j} + \mathbf{k}, \quad 0 \le t \le 1$$
$$C_3: \quad \mathbf{r}(t) = t\mathbf{i} + \mathbf{j} + \mathbf{k}, \quad 0 \le t \le 1$$

17. Integrate $f(x, y, z) = (x + y + z)/(x^2 + y^2 + z^2)$ over the path $\mathbf{r}(t) = t\mathbf{i} + t\mathbf{j} + t\mathbf{k}, 0 < a \le t \le b$.

18. Integrate $f(x, y, z) = -\sqrt{x^2 + z^2}$ over the circle

$$\mathbf{r}(t) = (a \cos t)\mathbf{j} + (a \sin t)\mathbf{k}, \quad 0 \le t \le 2\pi.$$

Line Integrals over Plane Curves

19. Evaluate $\int_C x \, ds$, where C is

a. the straight-line segment $x = t, y = t/2$, from $(0, 0)$ to $(4, 2)$.

b. the parabolic curve $x = t, y = t^2$, from $(0, 0)$ to $(2, 4)$.

20. Evaluate $\int_C \sqrt{x + 2y} \, ds$, where C is

a. the straight-line segment $x = t, y = 4t$, from $(0, 0)$ to $(1, 4)$.

b. $C_1 \cup C_2$; C_1 is the line segment from $(0, 0)$ to $(1, 0)$ and C_2 is the line segment from $(1, 0)$ to $(1, 2)$.

21. Find the line integral of $f(x, y) = ye^{x^2}$ along the curve $\mathbf{r}(t) = 4t\mathbf{i} - 3t\mathbf{j}, -1 \le t \le 2$.

22. Find the line integral of $f(x, y) = x - y + 3$ along the curve $\mathbf{r}(t) = (\cos t)\mathbf{i} + (\sin t)\mathbf{j}, 0 \le t \le 2\pi$.

23. Evaluate $\int_C \dfrac{x^2}{y^{4/3}} \, ds$, where C is the curve $x = t^2, y = t^3$, for $1 \le t \le 2$.

24. Find the line integral of $f(x, y) = \sqrt{y}/x$ along the curve $\mathbf{r}(t) = t^3\mathbf{i} + t^4\mathbf{j}, 1/2 \le t \le 1$.

25. Evaluate $\int_C (x + \sqrt{y}) \, ds$ where C is given in the accompanying figure.

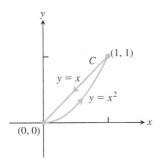

26. Evaluate $\int_C \dfrac{1}{x^2 + y^2 + 1} \, ds$ where C is given in the accompanying figure.

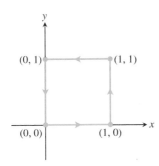

In Exercises 27–30, integrate f over the given curve.

27. $f(x, y) = x^3/y, \quad C: \quad y = x^2/2, \quad 0 \le x \le 2$

28. $f(x, y) = (x + y^2)/\sqrt{1 + x^2}, \quad C: \quad y = x^2/2$ from $(1, 1/2)$ to $(0, 0)$

29. $f(x, y) = x + y, \quad C: \quad x^2 + y^2 = 4$ in the first quadrant from $(2, 0)$ to $(0, 2)$

30. $f(x, y) = x^2 - y, \quad C: \quad x^2 + y^2 = 4$ in the first quadrant from $(0, 2)$ to $(\sqrt{2}, \sqrt{2})$

31. Find the area of one side of the "winding wall" standing orthogonally on the curve $y = x^2, 0 \le x \le 2$, and beneath the curve on the surface $f(x, y) = x + \sqrt{y}$.

32. Find the area of one side of the "wall" standing orthogonally on the curve $2x + 3y = 6, 0 \le x \le 6$, and beneath the curve on the surface $f(x, y) = 4 + 3x + 2y$.

Masses and Moments

33. Mass of a wire Find the mass of a wire that lies along the curve $\mathbf{r}(t) = (t^2 - 1)\mathbf{j} + 2t\mathbf{k}, 0 \le t \le 1$, if the density is $\delta = (3/2)t$.

34. Center of mass of a curved wire A wire of density $\delta(x, y, z) = 15\sqrt{y + 2}$ lies along the curve $\mathbf{r}(t) = (t^2 - 1)\mathbf{j} + 2t\mathbf{k}, -1 \le t \le 1$. Find its center of mass. Then sketch the curve and center of mass together.

35. Mass of wire with variable density Find the mass of a thin wire lying along the curve $\mathbf{r}(t) = \sqrt{2}t\mathbf{i} + \sqrt{2}t\mathbf{j} + (4 - t^2)\mathbf{k}$, $0 \le t \le 1$, if the density is **(a)** $\delta = 3t$ and **(b)** $\delta = 1$.

36. Center of mass of wire with variable density Find the center of mass of a thin wire lying along the curve $\mathbf{r}(t) = t\mathbf{i} + 2t\mathbf{j} + (2/3)t^{3/2}\mathbf{k}, 0 \le t \le 2$, if the density is $\delta = 3\sqrt{5 + t}$.

37. Moment of inertia of wire hoop A circular wire hoop of constant density δ lies along the circle $x^2 + y^2 = a^2$ in the xy-plane. Find the hoop's moment of inertia about the z-axis.

38. Inertia of a slender rod A slender rod of constant density lies along the line segment $\mathbf{r}(t) = t\mathbf{j} + (2 - 2t)\mathbf{k}, 0 \le t \le 1$, in the yz-plane. Find the moments of inertia of the rod about the three coordinate axes.

39. Two springs of constant density A spring of constant density δ lies along the helix

$$\mathbf{r}(t) = (\cos t)\mathbf{i} + (\sin t)\mathbf{j} + t\mathbf{k}, \quad 0 \le t \le 2\pi.$$

a. Find I_z.

b. Suppose that you have another spring of constant density δ that is twice as long as the spring in part (a) and lies along the helix for $0 \le t \le 4\pi$. Do you expect I_z for the longer spring to be the same as that for the shorter one, or should it be different? Check your prediction by calculating I_z for the longer spring.

40. Wire of constant density A wire of constant density $\delta = 1$ lies along the curve

$$\mathbf{r}(t) = (t \cos t)\mathbf{i} + (t \sin t)\mathbf{j} + (2\sqrt{2}/3)t^{3/2}\mathbf{k}, \quad 0 \le t \le 1.$$

Find $\bar{z}$ and I_z.

41. The arch in Example 4 Find I_x for the arch in Example 4.

42. Center of mass and moments of inertia for wire with variable density Find the center of mass and the moments of inertia about the coordinate axes of a thin wire lying along the curve

$$\mathbf{r}(t) = t\mathbf{i} + \frac{2\sqrt{2}}{3}t^{3/2}\mathbf{j} + \frac{t^2}{2}\mathbf{k}, \qquad 0 \le t \le 2,$$

if the density is $\delta = 1/(t + 1)$.

COMPUTER EXPLORATIONS

In Exercises 43–46, use a CAS to perform the following steps to evaluate the line integrals.

a. Find $ds = |\mathbf{v}(t)|\, dt$ for the path $\mathbf{r}(t) = g(t)\mathbf{i} + h(t)\mathbf{j} + k(t)\mathbf{k}$.

b. Express the integrand $f(g(t), h(t), k(t))|\mathbf{v}(t)|$ as a function of the parameter t.

c. Evaluate $\int_C f\, ds$ using Equation (2) in the text.

43. $f(x, y, z) = \sqrt{1 + 30x^2 + 10y};\quad \mathbf{r}(t) = t\mathbf{i} + t^2\mathbf{j} + 3t^2\mathbf{k},$
$0 \le t \le 2$

44. $f(x, y, z) = \sqrt{1 + x^3 + 5y^3};\quad \mathbf{r}(t) = t\mathbf{i} + \frac{1}{3}t^2\mathbf{j} + \sqrt{t}\mathbf{k},$
$0 \le t \le 2$

45. $f(x, y, z) = x\sqrt{y} - 3z^2;\quad \mathbf{r}(t) = (\cos 2t)\mathbf{i} + (\sin 2t)\mathbf{j} + 5t\mathbf{k},$
$0 \le t \le 2\pi$

46. $f(x, y, z) = \left(1 + \frac{9}{4}z^{1/3}\right)^{1/4};\quad \mathbf{r}(t) = (\cos 2t)\mathbf{i} + (\sin 2t)\mathbf{j} +$
$t^{5/2}\mathbf{k}, \quad 0 \le t \le 2\pi$

16.2 Vector Fields and Line Integrals: Work, Circulation, and Flux

Gravitational and electric forces have both a direction and a magnitude. They are represented by a vector at each point in their domain, producing a *vector field*. In this section we show how to compute the work done in moving an object through such a field by using a line integral involving the vector field. We also discuss velocity fields, such as the vector field representing the velocity of a flowing fluid in its domain. A line integral can be used to find the rate at which the fluid flows along or across a curve within the domain.

Vector Fields

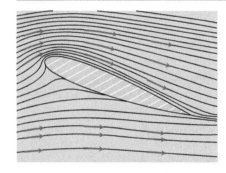

FIGURE 16.7 Velocity vectors of a flow around an airfoil in a wind tunnel.

FIGURE 16.8 Streamlines in a contracting channel. The water speeds up as the channel narrows and the velocity vectors increase in length.

Suppose a region in the plane or in space is occupied by a moving fluid, such as air or water. The fluid is made up of a large number of particles, and at any instant of time, a particle has a velocity $\mathbf{v}$. At different points of the region at a given (same) time, these velocities can vary. We can think of a velocity vector being attached to each point of the fluid representing the velocity of a particle at that point. Such a fluid flow is an example of a *vector field*. Figure 16.7 shows a velocity vector field obtained from air flowing around an airfoil in a wind tunnel. Figure 16.8 shows a vector field of velocity vectors along the streamlines of water moving through a contracting channel. Vector fields are also associated with forces such as gravitational attraction (Figure 16.9), and with magnetic fields, electric fields, and there are also purely mathematical fields.

Generally, a **vector field** is a function that assigns a vector to each point in its domain. A vector field on a three-dimensional domain in space might have a formula like

$$\mathbf{F}(x, y, z) = M(x, y, z)\mathbf{i} + N(x, y, z)\mathbf{j} + P(x, y, z)\mathbf{k}.$$

The vector field is **continuous** if the **component functions** M, N, and P are continuous; it is **differentiable** if each of the component functions is differentiable. The formula for a field of two-dimensional vectors could look like

$$\mathbf{F}(x, y) = M(x, y)\mathbf{i} + N(x, y)\mathbf{j}.$$

We encountered another type of vector field in Chapter 13. The tangent vectors $\mathbf{T}$ and normal vectors $\mathbf{N}$ for a curve in space both form vector fields along the curve. Along a curve $\mathbf{r}(t)$ they might have a component formula similar to the velocity field expression

$$\mathbf{v}(t) = f(t)\mathbf{i} + g(t)\mathbf{j} + h(t)\mathbf{k}.$$

If we attach the gradient vector ∇f of a scalar function $f(x, y, z)$ to each point of a level surface of the function, we obtain a three-dimensional field on the surface. If we attach the velocity vector to each point of a flowing fluid, we have a three-dimensional field defined on a region in space. These and other fields are illustrated in Figures 16.7–16.16. To sketch the fields, we picked a representative selection of domain points and drew the vectors attached to them. The arrows are drawn with their tails, not their heads, attached to the points where the vector functions are evaluated.

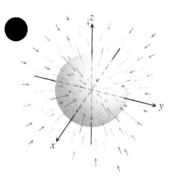

FIGURE 16.9 Vectors in a gravitational field point toward the center of mass that gives the source of the field.

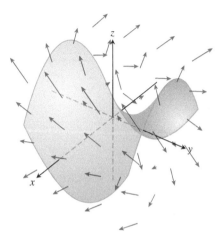

FIGURE 16.10 A surface might represent a filter, or a net, or a parachute, in a vector field representing water or wind flow velocity vectors. The arrows show the direction of fluid flow, and their lengths indicate speed.

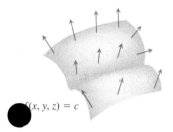

$f(x, y, z) = c$

FIGURE 16.11 The field of gradient vectors ∇f on a level surface $f(x, y, z) = c$. The function f is constant on the surface, and each vector points in the direction where f is increasing fastest.

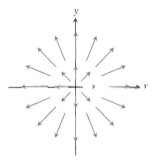

FIGURE 16.12 The radial field $\mathbf{F} = x\mathbf{i} + y\mathbf{j}$ formed by the position vectors of points in the plane. Notice the convention that an arrow is drawn with its tail, not its head, at the point where $\mathbf{F}$ is evaluated.

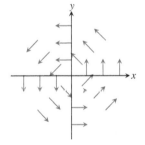

FIGURE 16.13 A "spin" field of rotating unit vectors

$$\mathbf{F} = (-y\mathbf{i} + x\mathbf{j})/(x^2 + y^2)^{1/2}$$

in the plane. The field is not defined at the origin.

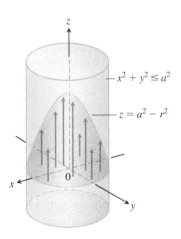

FIGURE 16.14 The flow of fluid in a long cylindrical pipe. The vectors $\mathbf{v} = (a^2 - r^2)\mathbf{k}$ inside the cylinder that have their bases in the xy-plane have their tips on the paraboloid $z = a^2 - r^2$.

Gradient Fields

The gradient vector of a differentiable scalar-valued function at a point gives the direction of greatest increase of the function. An important type of vector field is formed by all the gradient vectors of the function (see Section 14.5). We define the **gradient field** of a differentiable function $f(x, y, z)$ to be the field of gradient vectors

$$\nabla f = \frac{\partial f}{\partial x}\mathbf{i} + \frac{\partial f}{\partial y}\mathbf{j} + \frac{\partial f}{\partial z}\mathbf{k}.$$

At each point (x, y, z), the gradient field gives a vector pointing in the direction of greatest increase of f, with magnitude being the value of the directional derivative in that direction. The gradient field might represent a force field, or a velocity field that gives the motion of

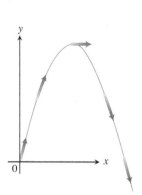

FIGURE 16.15 The velocity vectors $\mathbf{v}(t)$ of a projectile's motion make a vector field along the trajectory.

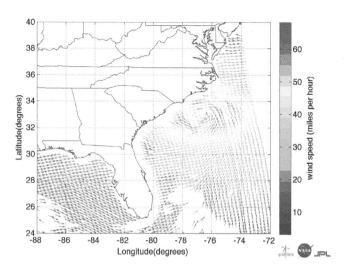

FIGURE 16.16 Data from NASA's QuikSCAT satellite were used to create this representation of windspeed and wind direction in Hurricane Irene approximately six hours before it made landfall in North Carolina on August 27, 2011. The arrows show wind direction, while speed is indicated by color (rather than length). The maximum wind speeds (over 130 km/hour) occurred over a region too small to be resolved in this illustration.

a fluid, or the flow of heat through a medium, depending on the application being considered. In many physical applications, f represents a potential energy, and the gradient vector field indicates the corresponding force. In such situations, f is often taken to be negative, so that the force gives the direction of decreasing potential energy.

EXAMPLE 1   Suppose that a material is heated, that the resulting temperature T each point (x, y, z) in a region of space is given by

$$T = 100 - x^2 - y^2 - z^2,$$

and that $\mathbf{F}(x, y, z)$ is defined to be the gradient of T. Find the vector field $\mathbf{F}$.

Solution   The gradient field $\mathbf{F}$ is the field $\mathbf{F} = \nabla T = -2x\mathbf{i} - 2y\mathbf{j} - 2z\mathbf{k}$. At each point in the region, the vector field $\mathbf{F}$ gives the direction for which the increase in temperature is greatest. The vectors point toward the origin, where the temperature is greatest. See Figure 16.17. ∎

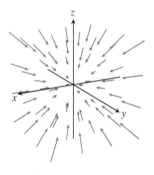

FIGURE 16.17 The vectors in a temperature gradient field point in the direction of greatest increase in temperature. In this case they are pointing toward the origin.

Line Integrals of Vector Fields

In Section 16.1 we defined the line integral of a scalar function $f(x, y, z)$ over a path C. We turn our attention now to the idea of a line integral of a vector field $\mathbf{F}$ along the curve C. Such line integrals have important applications in studying fluid flows, work and energy, and electrical or gravitational fields.

Assume that the vector field $\mathbf{F} = M(x, y, z)\mathbf{i} + N(x, y, z)\mathbf{j} + P(x, y, z)\mathbf{k}$ has continuous components, and that the curve C has a smooth parametrization $\mathbf{r}(t) = g(t)\mathbf{i} + h(t)\mathbf{j} + k(t)\mathbf{k}$, $a \leq t \leq b$. As discussed in Section 16.1, the parametrization $\mathbf{r}(t)$ defines a direction (or orientation) along C which we call the **forward direction**. At each point along the path C, the tangent vector $\mathbf{T} = d\mathbf{r}/ds = \mathbf{v}/|\mathbf{v}|$ is a unit vector tangent to the path and pointing in this forward direction. (The vector $\mathbf{v} = d\mathbf{r}/dt$ is the velocity vector tangent to C at the point, as discussed in Sections 13.1 and 13.3.) The line integral of the vector field is the line integral of the scalar tangential component of $\mathbf{F}$ along C. This tangential component is given by the dot product

$$\mathbf{F} \cdot \mathbf{T} = \mathbf{F} \cdot \frac{d\mathbf{r}}{ds},$$

so we are led to the following definition.

DEFINITION Let **F** be a vector field with continuous components defined along a smooth curve C parametrized by $\mathbf{r}(t)$, $a \leq t \leq b$. Then the **line integral of F along C** is

$$\int_C \mathbf{F} \cdot \mathbf{T} \, ds = \int_C \left(\mathbf{F} \cdot \frac{d\mathbf{r}}{ds} \right) ds = \int_C \mathbf{F} \cdot d\mathbf{r}. \qquad (1)$$

We evaluate line integrals of vector fields in a way similar to how we evaluate line integrals of scalar functions (Section 16.1).

Evaluating the Line Integral of $\mathbf{F} = M\mathbf{i} + N\mathbf{j} + P\mathbf{k}$ Along
$C: \mathbf{r}(t) = g(t)\mathbf{i} + h(t)\mathbf{j} + k(t)\mathbf{k}$

1. Express the vector field **F** along the parametrized curve C as $\mathbf{F}(\mathbf{r}(t))$ by substituting the components $x = g(t)$, $y = h(t)$, $z = k(t)$ of $\mathbf{r}$ into the scalar components $M(x, y, z)$, $N(x, y, z)$, $P(x, y, z)$ of **F**.
2. Find the derivative (velocity) vector $d\mathbf{r}/dt$.
3. Evaluate the line integral with respect to the parameter t, $a \leq t \leq b$, to obtain

$$\int_C \mathbf{F} \cdot d\mathbf{r} = \int_a^b \mathbf{F}(\mathbf{r}(t)) \cdot \frac{d\mathbf{r}}{dt} \, dt. \qquad (2)$$

EXAMPLE 2   Evaluate $\int_C \mathbf{F} \cdot d\mathbf{r}$, where $\mathbf{F}(x, y, z) = z\mathbf{i} + xy\mathbf{j} - y^2\mathbf{k}$ along the curve C given by $\mathbf{r}(t) = t^2\mathbf{i} + t\mathbf{j} + \sqrt{t}\,\mathbf{k}$, $0 \leq t \leq 1$ and shown in Figure 16.18.

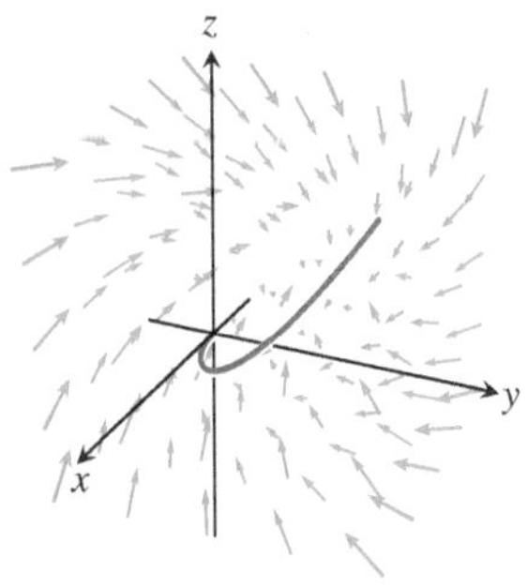

Solution We have

$$\mathbf{F}(\mathbf{r}(t)) = \sqrt{t}\,\mathbf{i} + t^3\mathbf{j} - t^2\mathbf{k} \qquad z = \sqrt{t}, \; xy = t^3, \; -y^2 = -t^2$$

and

$$\frac{d\mathbf{r}}{dt} = 2t\mathbf{i} + \mathbf{j} + \frac{1}{2\sqrt{t}}\mathbf{k}.$$

FIGURE 16.18 The curve (in red) winds through the vector field in Example 2. The line integral is determined by the vectors that lie along the curve.

Thus,

$$\int_C \mathbf{F} \cdot d\mathbf{r} = \int_0^1 \mathbf{F}(\mathbf{r}(t)) \cdot \frac{d\mathbf{r}}{dt} \, dt \qquad \text{Eq. (2)}$$

$$= \int_0^1 \left(2t^{3/2} + t^3 - \frac{1}{2}t^{3/2} \right) dt$$

$$= \left[\left(\frac{3}{2}\right)\left(\frac{2}{5}t^{5/2}\right) + \frac{1}{4}t^4 \right]_0^1 = \frac{17}{20}. \qquad \blacksquare$$

Line Integrals with Respect to dx, dy, or dz

When analyzing forces or flows, it is often useful to consider each component direction separately. For example, when analyzing the effect of a gravitational force, we might want to consider motion and forces in the vertical direction, while ignoring horizontal motions. Or we might be interested only in the force exerted horizontally by water pushing against the face of a dam or in wind affecting the course of a plane. In such situations we want to evaluate a line integral of a scalar function with respect to only one of the coordinates, such as $\int_C M\,dx$. This type of integral is not the same as the arc length line integral $\int_C M\,ds$ we defined in Section 16.1, since it picks out displacement in the direction of only one coordinate. To define the integral $\int_C M\,dx$ for the scalar function $M(x, y, z)$, we specify a vector field $\mathbf{F} = M(x, y, z)\mathbf{i}$ having a component only in the x-direction, and none in the y- or z-direction. Then, over the curve C parametrized by $\mathbf{r}(t) = g(t)\mathbf{i} + h(t)\mathbf{j} + k(t)\mathbf{k}$ for $a \le t \le b$, we have $x = g(t)$, $dx = g'(t)\,dt$, and

$$\mathbf{F}\cdot d\mathbf{r} = \mathbf{F}\cdot\frac{d\mathbf{r}}{dt}\,dt = M(x, y, z)\mathbf{i}\cdot\left(g'(t)\mathbf{i} + h'(t)\mathbf{j} + k'(t)\mathbf{k}\right)dt$$

$$= M(x, y, z)\,g'(t)\,dt = M(x, y, z)\,dx.$$

As in the definition of the line integral of $\mathbf{F}$ along C, we define

$$\int_C M(x, y, z)\,dx = \int_C \mathbf{F}\cdot d\mathbf{r}, \quad \text{where} \quad \mathbf{F} = M(x, y, z)\mathbf{i}.$$

In the same way, by defining $\mathbf{F} = N(x, y, z)\mathbf{j}$ with a component only in the y-direction, or $\mathbf{F} = P(x, y, z)\mathbf{k}$ with a component only in the z-direction, we obtain the line integrals $\int_C N\,dy$ and $\int_C P\,dz$. Expressing everything in terms of the parameter t along the curve C, we have the following formulas for these three integrals:

$$\int_C M(x, y, z)\,dx = \int_a^b M(g(t), h(t), k(t))\,g'(t)\,dt \qquad (3)$$

$$\int_C N(x, y, z)\,dy = \int_a^b N(g(t), h(t), k(t))\,h'(t)\,dt \qquad (4)$$

$$\int_C P(x, y, z)\,dz = \int_a^b P(g(t), h(t), k(t))\,k'(t)\,dt \qquad (5)$$

Line Integral Notation

The commonly occurring expression

$$\int_C M\,dx + N\,dy + P\,dz$$

is a short way of expressing the sum of three line integrals, one for each coordinate direction:

$$\int_C M(x, y, z)\,dx + \int_C N(x, y, z)\,dy$$

$$+ \int_C P(x, y, z)\,dz.$$

To evaluate these integrals, we parametrize C as $g(t)\mathbf{i} + h(t)\mathbf{j} + k(t)\mathbf{k}$ and use Equations (3), (4), and (5).

It often happens that these line integrals occur in combination, and we abbreviate the notation by writing

$$\int_C M(x, y, z)\,dx + \int_C N(x, y, z)\,dy + \int_C P(x, y, z)\,dz = \int_C M\,dx + N\,dy + P\,dz.$$

EXAMPLE 3   Evaluate the line integral $\int_C -y\,dx + z\,dy + 2x\,dz$, where C is the helix $\mathbf{r}(t) = (\cos t)\mathbf{i} + (\sin t)\mathbf{j} + t\mathbf{k}$, $0 \le t \le 2\pi$.

Solution   We express everything in terms of the parameter t, so $x = \cos t$, $y = \sin t$, $z = t$, and $dx = -\sin t\,dt$, $dy = \cos t\,dt$, $dz = dt$. Then,

$$\int_C -y\,dx + z\,dy + 2x\,dz = \int_0^{2\pi}\left[(-\sin t)(-\sin t) + t\cos t + 2\cos t\right]dt$$

$$= \int_0^{2\pi}\left[2\cos t + t\cos t + \sin^2 t\right]dt$$

$$= \left[2\sin t + (t\sin t + \cos t) + \left(\frac{t}{2} - \frac{\sin 2t}{4}\right)\right]_0^{2\pi}$$

$$= \left[0 + (0 + 1) + (\pi - 0)\right] - \left[0 + (0 + 1) + (0 - 0)\right]$$

$$= \pi.$$

Work Done by a Force over a Curve in Space

Suppose that the vector field $\mathbf{F} = M(x, y, z)\mathbf{i} + N(x, y, z)\mathbf{j} + P(x, y, z)\mathbf{k}$ represents a force throughout a region in space (it might be the force of gravity or an electromagnetic force) and that

$$\mathbf{r}(t) = g(t)\mathbf{i} + h(t)\mathbf{j} + k(t)\mathbf{k}, \qquad a \le t \le b,$$

is a smooth curve in the region. The formula for the work done by the force in moving an object along the curve is motivated by the same kind of reasoning we used in Chapter 6 to derive the ordinary single integral for the work done by a continuous force of magnitude $F(x)$ directed along an interval of the x-axis. For a curve C in space, we define the work done by a continuous force field $\mathbf{F}$ to move an object along C from a point A to another point B as follows.

We divide C into n subarcs $P_{k-1}P_k$ with lengths Δs_k, starting at A and ending at B. We choose any point (x_k, y_k, z_k) in the subarc $P_{k-1}P_k$ and let $\mathbf{T}(x_k, y_k, z_k)$ be the unit tangent vector at the chosen point. The work W_k done to move the object along the subarc $P_{k-1}P_k$ is approximated by the tangential component of the force $\mathbf{F}(x_k, y_k, z_k)$ times the arclength Δs_k approximating the distance the object moves along the subarc (see Figure 16.19). The total work done in moving the object from point A to point B is then approximated by summing the work done along each of the subarcs, so

$$W \approx \sum_{k=1}^{n} W_k \approx \sum_{k=1}^{n} \mathbf{F}(x_k, y_k, z_k) \cdot \mathbf{T}(x_k, y_k, z_k) \, \Delta s_k.$$

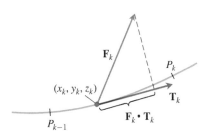

FIGURE 16.19 The work done along the subarc shown here is approximately $\mathbf{F}_k \cdot \mathbf{T}_k \, \Delta s_k$, where $\mathbf{F}_k = \mathbf{F}(x_k, y_k, z_k)$ and $\mathbf{T}_k = \mathbf{T}(x_k, y_k, z_k)$.

For any subdivision of C into n subarcs, and for any choice of the points (x_k, y_k, z_k) within each subarc, as $n \to \infty$ and $\Delta s_k \to 0$, these sums approach the line integral

$$\int_C \mathbf{F} \cdot \mathbf{T} \, ds.$$

This is the line integral of $\mathbf{F}$ along C, which now defines the total work done.

DEFINITION   Let C be a smooth curve parametrized by $\mathbf{r}(t)$, $a \le t \le b$, and let $\mathbf{F}$ be a continuous force field over a region containing C. Then the **work** done in moving an object from the point $A = \mathbf{r}(a)$ to the point $B = \mathbf{r}(b)$ along C is

$$W = \int_C \mathbf{F} \cdot \mathbf{T} \, ds = \int_a^b \mathbf{F}(\mathbf{r}(t)) \cdot \frac{d\mathbf{r}}{dt} \, dt. \tag{6}$$

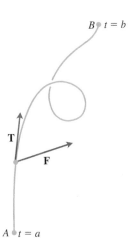

FIGURE 16.20 The work done by a force $\mathbf{F}$ is the line integral of the scalar component $\mathbf{F} \cdot \mathbf{T}$ over the smooth curve from A to B.

The sign of the number we calculate with this integral depends on the direction in which the curve is traversed. If we reverse the direction of motion, then we reverse the direction of $\mathbf{T}$ in Figure 16.20 and change the sign of $\mathbf{F} \cdot \mathbf{T}$ and its integral.

Using the notations we have presented, we can express the work integral in a variety of ways, depending upon what seems most suitable or convenient for a particular discussion. Table 16.2 shows five ways we can write the work integral in Equation (6). In the table, the field components M, N, and P are functions of the intermediate variables x, y, and z, which in turn are functions of the independent variable t along the curve C in the vector field. So along the curve, $x = g(t)$, $y = h(t)$, and $z = k(t)$ with $dx = g'(t) \, dt$, $dy = h'(t) \, dt$, and $dz = k'(t) \, dt$.

TABLE 16.2 Different ways to write the work integral for $\mathbf{F} = M\mathbf{i} + N\mathbf{j} + P\mathbf{k}$ over the curve $C\!:\mathbf{r}(t) = g(t)\mathbf{i} + h(t)\mathbf{j} + k(t)\mathbf{k},\, a \leq t \leq b$

$$W = \int_C \mathbf{F} \cdot \mathbf{T}\, ds \qquad\qquad \text{The definition}$$

$$= \int_C \mathbf{F} \cdot d\mathbf{r} \qquad\qquad \text{Vector differential form}$$

$$= \int_a^b \mathbf{F} \cdot \frac{d\mathbf{r}}{dt}\, dt \qquad\qquad \text{Parametric vector evaluation}$$

$$= \int_a^b \big(Mg'(t) + Nh'(t) + Pk'(t)\big)\, dt \qquad \text{Parametric scalar evaluation}$$

$$= \int_C M\, dx + N\, dy + P\, dz \qquad \text{Scalar differential form}$$

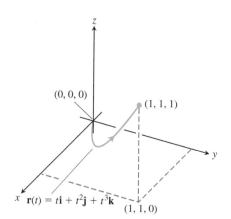

$\mathbf{r}(t) = t\mathbf{i} + t^2\mathbf{j} + t^3\mathbf{k}$

FIGURE 16.21   The curve in Example 4.

EXAMPLE 4   Find the work done by the force field $\mathbf{F} = (y - x^2)\mathbf{i} + (z - y^2)\mathbf{j} + (x - z^2)\mathbf{k}$ in moving an object along the curve $\mathbf{r}(t) = t\mathbf{i} + t^2\mathbf{j} + t^3\mathbf{k}, 0 \leq t \leq 1$, from $(0, 0, 0)$ to $(1, 1, 1)$ (Figure 16.21).

Solution   First we evaluate $\mathbf{F}$ on the curve $\mathbf{r}(t)$:

$$\mathbf{F} = (y - x^2)\mathbf{i} + (z - y^2)\mathbf{j} + (x - z^2)\mathbf{k}$$
$$= \underbrace{(t^2 - t^2)}_{0}\mathbf{i} + (t^3 - t^4)\mathbf{j} + (t - t^6)\mathbf{k}. \qquad \text{Substitute } x = t, y = t^2, z = t^3.$$

Then we find $d\mathbf{r}/dt$,

$$\frac{d\mathbf{r}}{dt} = \frac{d}{dt}(t\mathbf{i} + t^2\mathbf{j} + t^3\mathbf{k}) = \mathbf{i} + 2t\mathbf{j} + 3t^2\mathbf{k}.$$

Finally, we find $\mathbf{F} \cdot d\mathbf{r}/dt$ and integrate from $t = 0$ to $t = 1$:

$$\mathbf{F} \cdot \frac{d\mathbf{r}}{dt} = \big[(t^3 - t^4)\mathbf{j} + (t - t^6)\mathbf{k}\big] \cdot (\mathbf{i} + 2t\mathbf{j} + 3t^2\mathbf{k})$$
$$= (t^3 - t^4)(2t) + (t - t^6)(3t^2) = 2t^4 - 2t^5 + 3t^3 - 3t^8. \qquad \text{Evaluate dot product.}$$

So,

$$\text{Work} = \int_a^b \mathbf{F} \cdot \frac{d\mathbf{r}}{dt}\, dt = \int_0^1 (2t^4 - 2t^5 + 3t^3 - 3t^8)\, dt$$
$$= \left[\frac{2}{5}t^5 - \frac{2}{6}t^6 + \frac{3}{4}t^4 - \frac{3}{9}t^9\right]_0^1 = \frac{29}{60}.$$

■

EXAMPLE 5   Find the work done by the force field $\mathbf{F} = x\mathbf{i} + y\mathbf{j} + z\mathbf{k}$ in moving an object along the curve C parametrized by $\mathbf{r}(t) = \cos(\pi t)\mathbf{i} + t^2\mathbf{j} + \sin(\pi t)\mathbf{k}, 0 \leq t \leq 1$.

Solution   We begin by writing $\mathbf{F}$ along C as a function of t,

$$\mathbf{F}(\mathbf{r}(t)) = \cos(\pi t)\mathbf{i} + t^2\mathbf{j} + \sin(\pi t)\mathbf{k}.$$

Next we compute $d\mathbf{r}/dt$,

$$\frac{d\mathbf{r}}{dt} = -\pi \sin(\pi t)\mathbf{i} + 2t\mathbf{j} + \pi \cos(\pi t)\mathbf{k}.$$

We then calculate the dot product,

$$\mathbf{F}(\mathbf{r}(t)) \cdot \frac{d\mathbf{r}}{dt} = -\pi \sin(\pi t) \cos(\pi t) + 2t^3 + \pi \sin(\pi t) \cos(\pi t) = 2t^3.$$

The work done is the line integral

$$\int_a^b \mathbf{F}(\mathbf{r}(t)) \cdot \frac{d\mathbf{r}}{dt}\, dt = \int_0^1 2t^3\, dt = \frac{t^4}{2}\Big]_0^1 = \frac{1}{2}.$$ ∎

Flow Integrals and Circulation for Velocity Fields

Suppose that $\mathbf{F}$ represents the velocity field of a fluid flowing through a region in space (a tidal basin or the turbine chamber of a hydroelectric generator, for example). Under these circumstances, the integral of $\mathbf{F} \cdot \mathbf{T}$ along a curve in the region gives the fluid's flow along, or *circulation* around, the curve. For instance, the vector field in Figure 16.12 gives zero circulation around the unit circle in the plane. By contrast, the vector field in Figure 16.13 gives a nonzero circulation around the unit circle.

DEFINITION   If $\mathbf{r}(t)$ parametrizes a smooth curve C in the domain of a continuous velocity field $\mathbf{F}$, the **flow** along the curve from $A = \mathbf{r}(a)$ to $B = \mathbf{r}(b)$ is

$$\text{Flow} = \int_C \mathbf{F} \cdot \mathbf{T}\, ds. \qquad (7)$$

The integral is called a **flow integral**. If the curve starts and ends at the same point, so that $A = B$, the flow is called the **circulation** around the curve.

The direction we travel along C matters. If we reverse the direction, then $\mathbf{T}$ is replaced by $-\mathbf{T}$ and the sign of the integral changes. We evaluate flow integrals the same way we evaluate work integrals.

EXAMPLE 6   A fluid's velocity field is $\mathbf{F} = x\mathbf{i} + z\mathbf{j} + y\mathbf{k}$. Find the flow along the helix $\mathbf{r}(t) = (\cos t)\mathbf{i} + (\sin t)\mathbf{j} + t\mathbf{k}$, $0 \le t \le \pi/2$.

Solution   We evaluate $\mathbf{F}$ on the curve $\mathbf{r}(t)$,

$$\mathbf{F} = x\mathbf{i} + z\mathbf{j} + y\mathbf{k} = (\cos t)\mathbf{i} + t\mathbf{j} + (\sin t)\mathbf{k} \qquad \text{Substitute } x = \cos t, z = t, y = \sin t.$$

and then find $d\mathbf{r}/dt$:

$$\frac{d\mathbf{r}}{dt} = (-\sin t)\mathbf{i} + (\cos t)\mathbf{j} + \mathbf{k}.$$

The dot product of $\mathbf{F}$ with $d\mathbf{r}/dt$ is

$$\mathbf{F} \cdot \frac{d\mathbf{r}}{dt} = (\cos t)(-\sin t) + (t)(\cos t) + (\sin t)(1)$$

$$= -\sin t \cos t + t \cos t + \sin t.$$

Finally, we integrate $\mathbf{F} \cdot (d\mathbf{r}/dt)$ from $t = 0$ to $t = \dfrac{\pi}{2}$:

$$\text{Flow} = \int_{t=a}^{t=b} \mathbf{F} \cdot \frac{d\mathbf{r}}{dt}\, dt = \int_0^{\pi/2} (-\sin t \cos t + t \cos t + \sin t)\, dt$$

$$= \left[\frac{\cos^2 t}{2} + t \sin t \right]_0^{\pi/2} = \left(0 + \frac{\pi}{2} \right) - \left(\frac{1}{2} + 0 \right) = \frac{\pi}{2} - \frac{1}{2}.$$ ∎

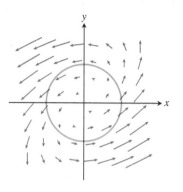

FIGURE 16.22 The vector field **F** and curve $\mathbf{r}(t)$ in Example 7.

EXAMPLE 7   Find the circulation of the field $\mathbf{F} = (x - y)\mathbf{i} + x\mathbf{j}$ around the circle $\mathbf{r}(t) = (\cos t)\mathbf{i} + (\sin t)\mathbf{j}, 0 \le t \le 2\pi$ (Figure 16.22).

Solution   On the circle, $\mathbf{F} = (x - y)\mathbf{i} + x\mathbf{j} = (\cos t - \sin t)\mathbf{i} + (\cos t)\mathbf{j}$, and

$$\frac{d\mathbf{r}}{dt} = (-\sin t)\mathbf{i} + (\cos t)\mathbf{j}.$$

Then

$$\mathbf{F} \cdot \frac{d\mathbf{r}}{dt} = -\sin t \cos t + \sin^2 t + \cos^2 t$$

gives

$$\text{Circulation} = \int_0^{2\pi} \mathbf{F} \cdot \frac{d\mathbf{r}}{dt} dt = \int_0^{2\pi} (1 - \sin t \cos t)\, dt$$

$$= \left[t - \frac{\sin^2 t}{2} \right]_0^{2\pi} = 2\pi.$$

As Figure 16.22 suggests, a fluid with this velocity field is circulating *counterclockwise* around the circle, so the circulation is positive. ∎

Flux Across a Simple Closed Plane Curve

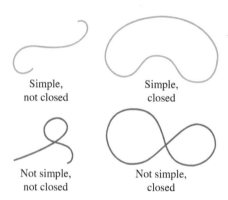

Simple, not closed

Simple, closed

Not simple, not closed

Not simple, closed

FIGURE 16.23 Distinguishing curves that are simple or closed. Closed curves are also called loops.

A curve in the xy-plane is **simple** if it does not cross itself (Figure 16.23). When a curve starts and ends at the same point, it is a **closed curve** or **loop**. To find the rate at which fluid is entering or leaving a region enclosed by a smooth simple closed curve C in the xy-plane, we calculate the line integral over C of $\mathbf{F} \cdot \mathbf{n}$, the scalar component of the fluid's velocity field in the direction of the curve's outward-pointing normal vector. We use only the normal component of $\mathbf{F}$, while ignoring the tangential component, because the normal component leads to the flow across C. The value of this integral is the *flux* of $\mathbf{F}$ across C. *Flux* is Latin for *flow*, but many flux calculations involve no motion at all. When $\mathbf{F}$ is an electric or magnetic field, for instance, the integral of $\mathbf{F} \cdot \mathbf{n}$ is still called the flux of the field across C.

> DEFINITION   If C is a smooth simple closed curve in the domain of a continuous vector field $\mathbf{F} = M(x, y)\mathbf{i} + N(x, y)\mathbf{j}$ in the plane, and if $\mathbf{n}$ is the outward-pointing unit normal vector on C, the **flux** of $\mathbf{F}$ across C is
>
> $$\text{Flux of } \mathbf{F} \text{ across } C = \int_C \mathbf{F} \cdot \mathbf{n}\, ds. \qquad (8)$$

Notice the difference between flux and circulation. The flux of $\mathbf{F}$ across C is the line integral with respect to arc length of $\mathbf{F} \cdot \mathbf{n}$, the scalar component of $\mathbf{F}$ in the direction of the outward normal. The circulation of $\mathbf{F}$ around C is the line integral with respect to arc length of $\mathbf{F} \cdot \mathbf{T}$, the scalar component of $\mathbf{F}$ in the direction of the unit tangent vector. Flux is the integral of the normal component of $\mathbf{F}$; circulation is the integral of the tangential component of $\mathbf{F}$. In Section 16.6 we define flux across a surface.

To evaluate the integral for flux in Equation (8), we begin with a smooth parametrization

$$x = g(t), \qquad y = h(t), \qquad a \le t \le b,$$

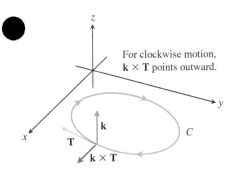

For clockwise motion, $\mathbf{k} \times \mathbf{T}$ points outward.

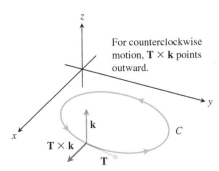

For counterclockwise motion, $\mathbf{T} \times \mathbf{k}$ points outward.

FIGURE 16.24 To find an outward unit normal vector for a smooth simple curve C in the xy-plane that is traversed counterclockwise as t increases, we take $\mathbf{n} = \mathbf{T} \times \mathbf{k}$. For clockwise motion, we take $\mathbf{n} = \mathbf{k} \times \mathbf{T}$.

that traces the curve C exactly once as t increases from a to b. We can find the outward unit normal vector $\mathbf{n}$ by crossing the curve's unit tangent vector $\mathbf{T}$ with the vector $\mathbf{k}$. But which order do we choose, $\mathbf{T} \times \mathbf{k}$ or $\mathbf{k} \times \mathbf{T}$? Which one points outward? It depends on which way C is traversed as t increases. If the motion is clockwise, $\mathbf{k} \times \mathbf{T}$ points outward; if the motion is counterclockwise, $\mathbf{T} \times \mathbf{k}$ points outward (Figure 16.24). The usual choice is $\mathbf{n} = \mathbf{T} \times \mathbf{k}$, the choice that assumes counterclockwise motion. Thus, although the value of the integral in Equation (8) does not depend on which way C is traversed, the formulas we are about to derive for computing $\mathbf{n}$ and evaluating the integral assume counterclockwise motion.

In terms of components,

$$\mathbf{n} = \mathbf{T} \times \mathbf{k} = \left(\frac{dx}{ds}\mathbf{i} + \frac{dy}{ds}\mathbf{j}\right) \times \mathbf{k} = \frac{dy}{ds}\mathbf{i} - \frac{dx}{ds}\mathbf{j}. \qquad \begin{vmatrix} \mathbf{i} & \mathbf{j} & \mathbf{k} \\ \frac{dx}{ds} & \frac{dy}{ds} & 0 \\ 0 & 0 & 1 \end{vmatrix}$$

If $\mathbf{F} = M(x, y)\mathbf{i} + N(x, y)\mathbf{j}$, then

$$\mathbf{F} \cdot \mathbf{n} = M(x, y)\frac{dy}{ds} - N(x, y)\frac{dx}{ds}.$$

Hence,

$$\int_C \mathbf{F} \cdot \mathbf{n} \, ds = \int_C \left(M\frac{dy}{ds} - N\frac{dx}{ds}\right) ds = \oint_C M \, dy - N \, dx.$$

We put a directed circle $\circlearrowleft$ on the last integral as a reminder that the integration around the closed curve C is to be in the counterclockwise direction. To evaluate this integral, we express M, dy, N, and dx in terms of the parameter t and integrate from $t = a$ to $t = b$. We do not need to know $\mathbf{n}$ or ds explicitly to find the flux.

Calculating Flux Across a Smooth Closed Plane Curve

$$\text{Flux of } \mathbf{F} = M\mathbf{i} + N\mathbf{j} \text{ across } C = \oint_C M \, dy - N \, dx \qquad (9)$$

The integral can be evaluated from any smooth parametrization $x = g(t)$, $y = h(t)$, $a \le t \le b$, that traces C counterclockwise exactly once.

EXAMPLE 8   Find the flux of $\mathbf{F} = (x - y)\mathbf{i} + x\mathbf{j}$ across the circle $x^2 + y^2 = 1$ in the xy-plane. (The vector field and curve were shown previously in Figure 16.22.)

Solution   The parametrization $\mathbf{r}(t) = (\cos t)\mathbf{i} + (\sin t)\mathbf{j}, 0 \le t \le 2\pi$, traces the circle counterclockwise exactly once. We can therefore use this parametrization in Equation (9). With

$$M = x - y = \cos t - \sin t, \qquad dy = d(\sin t) = \cos t \, dt,$$
$$N = x = \cos t, \qquad dx = d(\cos t) = -\sin t \, dt,$$

we find

$$\text{Flux} = \oint_C M \, dy - N \, dx = \int_0^{2\pi} (\cos^2 t - \sin t \cos t + \cos t \sin t) \, dt \qquad \text{Eq. (9)}$$

$$= \int_0^{2\pi} \cos^2 t \, dt = \int_0^{2\pi} \frac{1 + \cos 2t}{2} \, dt = \left[\frac{t}{2} + \frac{\sin 2t}{4}\right]_0^{2\pi} = \pi.$$

The flux of $\mathbf{F}$ across the circle is π. Since the answer is positive, the net flow across the curve is outward. A net inward flow would have given a negative flux. ∎

EXERCISES 16.2

Vector Fields

Find the gradient fields of the functions in Exercises 1–4.

1. $f(x, y, z) = (x^2 + y^2 + z^2)^{-1/2}$

2. $f(x, y, z) = \ln\sqrt{x^2 + y^2 + z^2}$

3. $g(x, y, z) = e^z - \ln(x^2 + y^2)$

4. $g(x, y, z) = xy + yz + xz$

5. Give a formula $\mathbf{F} = M(x, y)\mathbf{i} + N(x, y)\mathbf{j}$ for the vector field in the plane that has the property that $\mathbf{F}$ points toward the origin with magnitude inversely proportional to the square of the distance from (x, y) to the origin. (The field is not defined at $(0, 0)$.)

6. Give a formula $\mathbf{F} = M(x, y)\mathbf{i} + N(x, y)\mathbf{j}$ for the vector field in the plane that has the properties that $\mathbf{F} = \mathbf{0}$ at $(0, 0)$ and that at any other point (a, b), $\mathbf{F}$ is tangent to the circle $x^2 + y^2 = a^2 + b^2$ and points in the clockwise direction with magnitude $|\mathbf{F}| = \sqrt{a^2 + b^2}$.

Line Integrals of Vector Fields

In Exercises 7–12, find the line integrals of $\mathbf{F}$ from $(0, 0, 0)$ to $(1, 1, 1)$ over each of the following paths in the accompanying figure.

a. The straight-line path C_1: $\mathbf{r}(t) = t\mathbf{i} + t\mathbf{j} + t\mathbf{k}$, $\quad 0 \leq t \leq 1$

b. The curved path C_2: $\mathbf{r}(t) = t\mathbf{i} + t^2\mathbf{j} + t^4\mathbf{k}$, $\quad 0 \leq t \leq 1$

c. The path $C_3 \cup C_4$ consisting of the line segment from $(0, 0, 0)$ to $(1, 1, 0)$ followed by the segment from $(1, 1, 0)$ to $(1, 1, 1)$

7. $\mathbf{F} = 3y\mathbf{i} + 2x\mathbf{j} + 4z\mathbf{k}$ 8. $\mathbf{F} = [1/(x^2 + 1)]\mathbf{j}$

9. $\mathbf{F} = \sqrt{z}\mathbf{i} - 2x\mathbf{j} + \sqrt{y}\mathbf{k}$ 10. $\mathbf{F} = xy\mathbf{i} + yz\mathbf{j} + xz\mathbf{k}$

11. $\mathbf{F} = (3x^2 - 3x)\mathbf{i} + 3z\mathbf{j} + \mathbf{k}$

12. $\mathbf{F} = (y + z)\mathbf{i} + (z + x)\mathbf{j} + (x + y)\mathbf{k}$

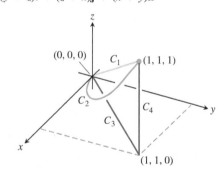

Line Integrals with Respect to x, y, and z

In Exercises 13–16, find the line integrals along the given path C.

13. $\displaystyle\int_C (x - y)\, dx$, where C: $x = t$, $y = 2t + 1$, for $0 \leq t \leq 3$

14. $\displaystyle\int_C \frac{x}{y}\, dy$, where C: $x = t$, $y = t^2$, for $1 \leq t \leq 2$

15. $\displaystyle\int_C (x^2 + y^2)\, dy$, where C is given in the accompanying figure

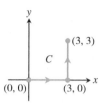

16. $\displaystyle\int_C \sqrt{x + y}\, dx$, where C is given in the accompanying figure

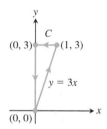

17. Along the curve $\mathbf{r}(t) = t\mathbf{i} - \mathbf{j} + t^2\mathbf{k}$, $0 \leq t \leq 1$, evaluate each of the following integrals.

a. $\displaystyle\int_C (x + y - z)\, dx$ b. $\displaystyle\int_C (x + y - z)\, dy$

c. $\displaystyle\int_C (x + y - z)\, dz$

18. Along the curve $\mathbf{r}(t) = (\cos t)\mathbf{i} + (\sin t)\mathbf{j} - (\cos t)\mathbf{k}$, $0 \leq t \leq \pi$, evaluate each of the following integrals.

a. $\displaystyle\int_C xz\, dx$ b. $\displaystyle\int_C xz\, dy$ c. $\displaystyle\int_C xyz\, dz$

Work

In Exercises 19–22, find the work done by $\mathbf{F}$ over the curve in the direction of increasing t.

19. $\mathbf{F} = xy\mathbf{i} + y\mathbf{j} - yz\mathbf{k}$

$\mathbf{r}(t) = t\mathbf{i} + t^2\mathbf{j} + t\mathbf{k}$, $\quad 0 \leq t \leq 1$

20. $\mathbf{F} = 2y\mathbf{i} + 3x\mathbf{j} + (x + y)\mathbf{k}$

$\mathbf{r}(t) = (\cos t)\mathbf{i} + (\sin t)\mathbf{j} + (t/6)\mathbf{k}$, $\quad 0 \leq t \leq 2\pi$

21. $\mathbf{F} = z\mathbf{i} + x\mathbf{j} + y\mathbf{k}$

$\mathbf{r}(t) = (\sin t)\mathbf{i} + (\cos t)\mathbf{j} + t\mathbf{k}$, $\quad 0 \leq t \leq 2\pi$

22. $\mathbf{F} = 6z\mathbf{i} + y^2\mathbf{j} + 12x\mathbf{k}$

$\mathbf{r}(t) = (\sin t)\mathbf{i} + (\cos t)\mathbf{j} + (t/6)\mathbf{k}$, $\quad 0 \leq t \leq 2\pi$

Line Integrals in the Plane

23. Evaluate $\int_C xy\,dx + (x + y)\,dy$ along the curve $y = x^2$ from $(-1, 1)$ to $(2, 4)$.

24. Evaluate $\int_C (x - y)\,dx + (x + y)\,dy$ counterclockwise around the triangle with vertices $(0, 0)$, $(1, 0)$, and $(0, 1)$.

25. Evaluate $\int_C \mathbf{F} \cdot \mathbf{T}\,ds$ for the vector field $\mathbf{F} = x^2\mathbf{i} - y\mathbf{j}$ along the curve $x = y^2$ from $(4, 2)$ to $(1, -1)$.

26. Evaluate $\int_C \mathbf{F} \cdot d\mathbf{r}$ for the vector field $\mathbf{F} = y\mathbf{i} - x\mathbf{j}$ counterclockwise along the unit circle $x^2 + y^2 = 1$ from $(1, 0)$ to $(0, 1)$.

Work, Circulation, and Flux in the Plane

27. Work Find the work done by the force $\mathbf{F} = xy\mathbf{i} + (y - x)\mathbf{j}$ over the straight line from $(1, 1)$ to $(2, 3)$.

28. Work Find the work done by the gradient of $f(x, y) = (x + y)^2$ counterclockwise around the circle $x^2 + y^2 = 4$ from $(2, 0)$ to itself.

29. Circulation and flux Find the circulation and flux of the fields

$$\mathbf{F}_1 = x\mathbf{i} + y\mathbf{j} \qquad \text{and} \qquad \mathbf{F}_2 = -y\mathbf{i} + x\mathbf{j}$$

around and across each of the following curves.

a. The circle $\mathbf{r}(t) = (\cos t)\mathbf{i} + (\sin t)\mathbf{j}, \quad 0 \le t \le 2\pi$

b. The ellipse $\mathbf{r}(t) = (\cos t)\mathbf{i} + (4 \sin t)\mathbf{j}, \quad 0 \le t \le 2\pi$

30. Flux across a circle Find the flux of the fields

$$\mathbf{F}_1 = 2x\mathbf{i} - 3y\mathbf{j} \qquad \text{and} \qquad \mathbf{F}_2 = 2x\mathbf{i} + (x - y)\mathbf{j}$$

across the circle

$$\mathbf{r}(t) = (a \cos t)\mathbf{i} + (a \sin t)\mathbf{j}, \qquad 0 \le t \le 2\pi.$$

In Exercises 31–34, find the circulation and flux of the field $\mathbf{F}$ around and across the closed semicircular path that consists of the semicircular arch $\mathbf{r}_1(t) = (a \cos t)\mathbf{i} + (a \sin t)\mathbf{j}, 0 \le t \le \pi$, followed by the line segment $\mathbf{r}_2(t) = t\mathbf{i}, -a \le t \le a$.

31. $\mathbf{F} = x\mathbf{i} + y\mathbf{j}$ **32.** $\mathbf{F} = x^2\mathbf{i} + y^2\mathbf{j}$

33. $\mathbf{F} = -y\mathbf{i} + x\mathbf{j}$ **34.** $\mathbf{F} = -y^2\mathbf{i} + x^2\mathbf{j}$

35. Flow integrals Find the flow of the velocity field $\mathbf{F} = (x + y)\mathbf{i} - (x^2 + y^2)\mathbf{j}$ along each of the following paths from $(1, 0)$ to $(-1, 0)$ in the xy-plane.

a. The upper half of the circle $x^2 + y^2 = 1$

b. The line segment from $(1, 0)$ to $(-1, 0)$

c. The line segment from $(1, 0)$ to $(0, -1)$ followed by the line segment from $(0, -1)$ to $(-1, 0)$

36. Flux across a triangle Find the flux of the field $\mathbf{F}$ in Exercise 35 outward across the triangle with vertices $(1, 0)$, $(0, 1)$, $(-1, 0)$.

37. The flow of a gas with a density of $\delta = 0.001\,\text{kg/m}^2$ over the closed curve $\mathbf{r}(t) = (-\sin t)\mathbf{i} + (\cos t)\mathbf{j}, \ 0 \le t \le 2\pi$, is given by the vector field $\mathbf{F} = \delta\mathbf{v}$, where $\mathbf{v} = x\mathbf{i} + y^2\mathbf{j}$ is a velocity field measured in meters per second. Find the flux of $\mathbf{F}$ across the curve $\mathbf{r}(t)$.

38. The flow of a gas with a density of $\delta = 0.3\,\text{kg/m}^2$ over the closed curve $\mathbf{r}(t) = (\cos t)\mathbf{i} + (\sin t)\mathbf{j}, \ 0 \le t \le 2\pi$, is given by the vector field $\mathbf{F} = \delta\mathbf{v}$, where $\mathbf{v} = x^2\mathbf{i} - y\mathbf{j}$ is a velocity field measured in meters per second. Find the flux of $\mathbf{F}$ across the curve $\mathbf{r}(t)$.

39. Find the flow of the velocity field $\mathbf{F} = y^2\mathbf{i} + 2xy\mathbf{j}$ along each of the following paths from $(0, 0)$ to $(2, 4)$.

a. 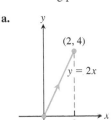   **b.**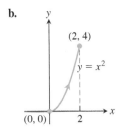

c. Use any path from $(0, 0)$ to $(2, 4)$ different from parts (a) and (b).

40. Find the circulation of the field $\mathbf{F} = y\mathbf{i} + (x + 2y)\mathbf{j}$ around each of the following closed paths.

a.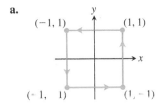

b.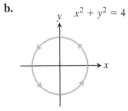

c. Use any closed path different from parts (a) and (b).

41. Find the work done by the force $\mathbf{F} = y^2\mathbf{i} + x^3\mathbf{j}$, where force is measured in newtons, in moving an object over the curve $\mathbf{r}(t) = 2t\mathbf{i} + t^2\mathbf{j}, \ 0 \le t \le 2$, where distance is measured in meters.

42. Find the work done by the force $\mathbf{F} = e^y\mathbf{i} + (\ln x)\mathbf{j} + 3z\mathbf{k}$, where force is measured in newtons, in moving an object over the curve $\mathbf{r}(t) = e^t\mathbf{i} + (\ln t)\mathbf{j} + t^2\mathbf{k}, \ 1 \le t \le e$, where distance is measured in meters.

43. Find the flow of the velocity field $\mathbf{F} = \dfrac{x}{y + 1}\mathbf{i} + \dfrac{y}{x + 1}\mathbf{j}$, where velocity is measured in meters per second, over the curve $\mathbf{r}(t) = t^2\mathbf{i} + t\mathbf{j}, 0 \le t \le 1$.

44. Find the flow of the velocity field $\mathbf{F} = (y + z)\mathbf{i} + x\mathbf{j} - y\mathbf{k}$, where velocity is measured in meters per second, over the curve $\mathbf{r}(t) = e^t\mathbf{i} - e^{2t}\mathbf{j} + e^{-t}\mathbf{k}$, $0 \le t \le \ln 2$.

45. Salt water with a density of $\delta = 0.25\,\text{g/cm}^2$ flows over the curve $\mathbf{r}(t) = \sqrt{t}\mathbf{i} + t\mathbf{j}$, $0 \le t \le 4$, according to the vector field $\mathbf{F} = \delta\mathbf{v}$, where $\mathbf{v} = xy\mathbf{i} + (y - x)\mathbf{j}$ is a velocity field measured in centimeters per second. Find the flow of $\mathbf{F}$ over the curve $\mathbf{r}(t)$.

46. Propyl alcohol with a density of $\delta = 0.2\,\text{g/cm}^2$ flows over the closed curve $\mathbf{r}(t) = (\sin t)\mathbf{i} - (\cos t)\mathbf{j}$, $0 \le t \le 2\pi$, according to the vector field $\mathbf{F} = \delta\mathbf{v}$, where $\mathbf{v} = (x - y)\mathbf{i} + x^2\mathbf{j}$ is a velocity field measured in centimeters per second. Find the circulation of $\mathbf{F}$ around the curve $\mathbf{r}(t)$.

Vector Fields in the Plane

47. Spin field   Draw the spin field

$$\mathbf{F} = -\frac{y}{\sqrt{x^2 + y^2}}\mathbf{i} + \frac{x}{\sqrt{x^2 + y^2}}\mathbf{j}$$

(see Figure 16.13) along with its horizontal and vertical components at a representative assortment of points on the circle $x^2 + y^2 = 4$.

48. Radial field   Draw the radial field

$$\mathbf{F} = x\mathbf{i} + y\mathbf{j}$$

(see Figure 16.12) along with its horizontal and vertical components at a representative assortment of points on the circle $x^2 + y^2 = 1$.

49. A field of tangent vectors

a. Find a field $\mathbf{G} = P(x, y)\mathbf{i} + Q(x, y)\mathbf{j}$ in the xy-plane with the property that at any point $(a, b) \ne (0, 0)$, $\mathbf{G}$ is a vector of magnitude $\sqrt{a^2 + b^2}$ tangent to the circle $x^2 + y^2 = a^2 + b^2$ and pointing in the counterclockwise direction. (The field is undefined at $(0, 0)$.)

b. How is $\mathbf{G}$ related to the spin field $\mathbf{F}$ in Figure 16.13?

50. A field of tangent vectors

a. Find a field $\mathbf{G} = P(x, y)\mathbf{i} + Q(x, y)\mathbf{j}$ in the xy-plane with the property that at any point $(a, b) \ne (0, 0)$, $\mathbf{G}$ is a unit vector tangent to the circle $x^2 + y^2 = a^2 + b^2$ and pointing in the clockwise direction.

b. How is $\mathbf{G}$ related to the spin field $\mathbf{F}$ in Figure 16.13?

51. Unit vectors pointing toward the origin   Find a field $\mathbf{F} = M(x, y)\mathbf{i} + N(x, y)\mathbf{j}$ in the xy-plane with the property that at each point $(x, y) \ne (0, 0)$, $\mathbf{F}$ is a unit vector pointing toward the origin. (The field is undefined at $(0, 0)$.)

52. Two "central" fields   Find a field $\mathbf{F} = M(x, y)\mathbf{i} + N(x, y)\mathbf{j}$ in the xy-plane with the property that at each point $(x, y) \ne (0, 0)$, $\mathbf{F}$ points toward the origin and $|\mathbf{F}|$ is (a) the distance from (x, y) to the origin, (b) inversely proportional to the distance from (x, y) to the origin. (The field is undefined at $(0, 0)$.)

53. Work and area   Suppose that $f(t)$ is differentiable and positive for $a \le t \le b$. Let C be the path $\mathbf{r}(t) = t\mathbf{i} + f(t)\mathbf{j}$, $a \le t \le b$ and $\mathbf{F} = y\mathbf{i}$. Is there any relation between the value of the work integral

$$\int_C \mathbf{F} \cdot d\mathbf{r}$$

and the area of the region bounded by the t-axis, the graph of f, and the lines $t = a$ and $t = b$? Give reasons for your answer.

54. Work done by a radial force with constant magnitude   A particle moves along the smooth curve $y = f(x)$ from $(a, f(a))$ to $(b, f(b))$. The force moving the particle has constant magnitude k and always points away from the origin. Show that the work done by the force is

$$\int_C \mathbf{F} \cdot \mathbf{T}\,ds = k\left[(b^2 + (f(b))^2)^{1/2} - (a^2 + (f(a))^2)^{1/2} \right].$$

Flow Integrals in Space

In Exercises 55–58, $\mathbf{F}$ is the velocity field of a fluid flowing through a region in space. Find the flow along the given curve in the direction of increasing t.

55. $\mathbf{F} = -4xy\mathbf{i} + 8y\mathbf{j} + 2\mathbf{k}$

   $\mathbf{r}(t) = t\mathbf{i} + t^2\mathbf{j} + \mathbf{k}, \quad 0 \le t \le 2$

56. $\mathbf{F} = x^2\mathbf{i} + yz\mathbf{j} + y^2\mathbf{k}$

   $\mathbf{r}(t) = 3t\mathbf{j} + 4t\mathbf{k}, \quad 0 \le t \le 1$

57. $\mathbf{F} = (x - z)\mathbf{i} + x\mathbf{k}$

   $\mathbf{r}(t) = (\cos t)\mathbf{i} + (\sin t)\mathbf{k}, \quad 0 \le t \le \pi$

58. $\mathbf{F} = -y\mathbf{i} + x\mathbf{j} + 2\mathbf{k}$

   $\mathbf{r}(t) = (-2\cos t)\mathbf{i} + (2\sin t)\mathbf{j} + 2t\mathbf{k}, \quad 0 \le t \le 2\pi$

59. Circulation   Find the circulation of $\mathbf{F} = 2x\mathbf{i} + 2z\mathbf{j} + 2y\mathbf{k}$ around the closed path consisting of the following three curves traversed in the direction of increasing t.

   $C_1:\quad \mathbf{r}(t) = (\cos t)\mathbf{i} + (\sin t)\mathbf{j} + t\mathbf{k}, \quad 0 \le t \le \pi/2$

   $C_2:\quad \mathbf{r}(t) = \mathbf{j} + (\pi/2)(1 - t)\mathbf{k}, \quad 0 \le t \le 1$

   $C_3:\quad \mathbf{r}(t) = t\mathbf{i} + (1 - t)\mathbf{j}, \quad 0 \le t \le 1$

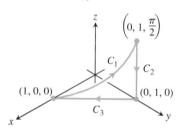

60. Zero circulation   Let C be the ellipse in which the plane $2x + 3y - z = 0$ meets the cylinder $x^2 + y^2 = 12$. Show, without evaluating either line integral directly, that the circulation of the field $\mathbf{F} = x\mathbf{i} + y\mathbf{j} + z\mathbf{k}$ around C in either direction is zero.

61. Flow along a curve The field $\mathbf{F} = xy\mathbf{i} + y\mathbf{j} - yz\mathbf{k}$ is the velocity field of a flow in space. Find the flow from $(0, 0, 0)$ to $(1, 1, 1)$ along the curve of intersection of the cylinder $y = x^2$ and the plane $z = x$. (*Hint:* Use $t = x$ as the parameter.)

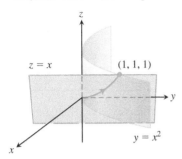

62. Flow of a gradient field Find the flow of the field $\mathbf{F} = \nabla(xy^2z^3)$:

a. Once around the curve C in Exercise 58, clockwise as viewed from above

b. Along the line segment from $(1, 1, 1)$ to $(2, 1, -1)$.

COMPUTER EXPLORATIONS

In Exercises 63–68, use a CAS to perform the following steps for finding the work done by force $\mathbf{F}$ over the given path:

a. Find $d\mathbf{r}$ for the path $\mathbf{r}(t) = g(t)\mathbf{i} + h(t)\mathbf{j} + k(t)\mathbf{k}$.

b. Evaluate the force $\mathbf{F}$ along the path.

c. Evaluate $\displaystyle\int_C \mathbf{F} \cdot d\mathbf{r}$.

63. $\mathbf{F} = xy^6\mathbf{i} + 3x(xy^5 + 2)\mathbf{j}; \quad \mathbf{r}(t) = (2\cos t)\mathbf{i} + (\sin t)\mathbf{j}$,
$0 \le t \le 2\pi$

64. $\mathbf{F} = \dfrac{3}{1 + x^2}\mathbf{i} + \dfrac{2}{1 + y^2}\mathbf{j}; \quad \mathbf{r}(t) = (\cos t)\mathbf{i} + (\sin t)\mathbf{j}$,
$0 \le t \le \pi$

65. $\mathbf{F} = (y + yz\cos xyz)\mathbf{i} + (x^2 + xz\cos xyz)\mathbf{j} + (z + xy\cos xyz)\mathbf{k}; \quad \mathbf{r}(t) = (2\cos t)\mathbf{i} + (3\sin t)\mathbf{j} + \mathbf{k}$,
$0 \le t \le 2\pi$

66. $\mathbf{F} = 2xy\mathbf{i} - y^2\mathbf{j} + ze^x\mathbf{k}; \quad \mathbf{r}(t) = -t\mathbf{i} + \sqrt{t}\mathbf{j} + 3t\mathbf{k}$,
$1 \le t \le 4$

67. $\mathbf{F} = (2y + \sin x)\mathbf{i} + (z^2 + (1/3)\cos y)\mathbf{j} + x^4\mathbf{k}; \quad \mathbf{r}(t) = (\sin t)\mathbf{i} + (\cos t)\mathbf{j} + (\sin 2t)\mathbf{k}, \quad -\pi/2 \le t \le \pi/2$

68. $\mathbf{F} = (x^2y)\mathbf{i} + \dfrac{1}{3}x^3\mathbf{j} + xy\mathbf{k}; \quad \mathbf{r}(t) = (\cos t)\mathbf{i} + (\sin t)\mathbf{j} + (2\sin^2 t - 1)\mathbf{k}, \quad 0 \le t \le 2\pi$

16.3 Path Independence, Conservative Fields, and Potential Functions

A **gravitational field G** is a vector field that represents the effect of gravity at a point in space due to the presence of a massive object. The gravitational force on a body of mass m placed in the field is given by $\mathbf{F} = m\mathbf{G}$. Similarly, an **electric field E** is a vector field in space that represents the effect of electric forces on a charged particle placed within it. The force on a body of charge q placed in the field is given by $\mathbf{F} = q\mathbf{E}$. In gravitational and electric fields, the amount of work it takes to move a mass or charge from one point to another depends on the initial and final positions of the object—not on which path is taken between these positions. In this section we study vector fields with this independence-of-path property and the calculation of work integrals associated with them.

Path Independence

If A and B are two points in an open region D in space, the line integral of $\mathbf{F}$ along C from A to B for a field $\mathbf{F}$ defined on D usually depends on the path C taken, as we saw in Section 16.1. For some special fields, however, the integral's value is the same for all paths from A to B.

> DEFINITIONS Let $\mathbf{F}$ be a vector field defined on an open region D in space, and suppose that for any two points A and B in D the line integral $\int_C \mathbf{F} \cdot d\mathbf{r}$ along a path C from A to B in D is the same over all paths from A to B. Then the integral $\int_C \mathbf{F} \cdot d\mathbf{r}$ is **path independent in D** and the field $\mathbf{F}$ is **conservative on D**.

The word *conservative* comes from physics, where it refers to fields in which the principle of conservation of energy holds. When a line integral is independent of the path C from point A to point B, we sometimes represent the integral by the symbol $\int_A^B$ rather than the usual line integral symbol $\int_C$. This substitution helps us remember the path-independence property by indicating that the integral depends only on the initial and final points, and not on the path connecting them.

Under reasonable differentiability conditions that we will specify, we will show that a field **F** is conservative if and only if it is the gradient field of a scalar function f—that is, if and only if **F** = ∇f for some f. The function f then has a special name.

> **DEFINITION** If **F** is a vector field defined on D and **F** = ∇f for some scalar function f on D, then f is called a **potential function for F**.

A gravitational potential is a scalar function whose gradient field is a gravitational field, an electric potential is a scalar function whose gradient field is an electric field, and so on. As we will see, once we have found a potential function f for a field **F**, we can evaluate all the line integrals in the domain of **F** over any path between A and B by

$$\int_A^B \mathbf{F} \cdot d\mathbf{r} = \int_A^B \nabla f \cdot d\mathbf{r} = f(B) - f(A). \tag{1}$$

If you think of ∇f for functions of several variables as analogous to the derivative f' for functions of a single variable, then you see that Equation (1) is the vector calculus rendition of the Fundamental Theorem of Calculus formula

$$\int_a^b f'(x)\, dx = f(b) - f(a).$$

Conservative fields have other important properties. For example, saying that **F** is conservative on D is equivalent to saying that the integral of **F** around every closed path in D is zero. Certain conditions on the curves, fields, and domains must be satisfied for Equation (1) to be valid. We discuss these conditions next.

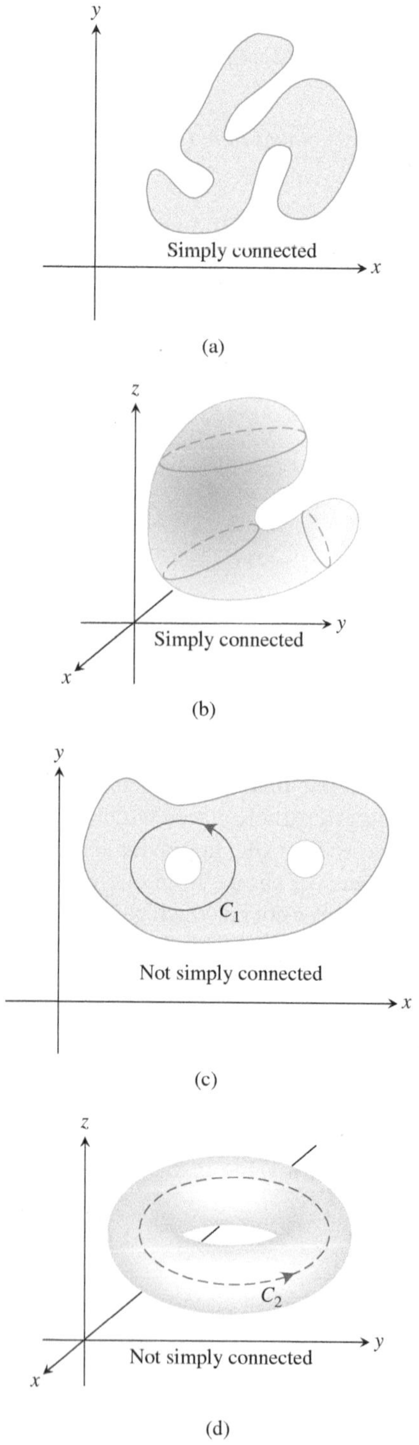

Assumptions on Curves, Vector Fields, and Domains

In order for the computations and results we derive below to be valid, we must assume certain properties for the curves, surfaces, domains, and vector fields we consider. We give these assumptions in the statements of theorems, and they also apply to the examples and exercises unless otherwise stated.

The curves we consider are **piecewise smooth**. Such curves are made up of finitely many smooth pieces connected end to end, as discussed in Section 13.1. For such curves we can compute lengths and, except at finitely many points where the smooth pieces connect, tangent vectors. We consider vector fields **F** whose components have continuous first partial derivatives.

The domains D we consider are **connected**. For an open region, this means that any two points in D can be joined by a smooth curve that lies in the region. Some results require D to be **simply connected**, which means that every loop in D can be contracted to a point in D without ever leaving D. The plane with a disk removed is a two-dimensional region that is *not* simply connected; a loop in the plane that goes around the disk cannot be contracted to a point without going into the "hole" left by the removed disk (see Figure 16.25c). Similarly, if we remove a line from space, the remaining region D is *not* simply connected. A curve encircling the line cannot be shrunk to a point while remaining inside D.

Connectivity and simple connectivity are not the same, and neither property implies the other. Think of connected regions as being in "one piece" and simply connected regions as not having any "loop-catching holes." All of space itself is both connected and simply connected. Figure 16.25 illustrates some of these properties.

Caution Some of the results in this chapter can fail to hold if applied to situations where the conditions we've imposed do not hold. In particular, the component test for conservative fields, given later in this section, is not valid on domains that are not simply connected (see Example 5). The condition will be stated when needed.

FIGURE 16.25 Four connected regions. In (a) and (b), the regions are simply connected. In (c) and (d), the regions are not simply connected because the curves C_1 and C_2 cannot be contracted to a point inside the regions containing them.

Line Integrals in Conservative Fields

Gradient fields $\mathbf{F}$ are obtained by differentiating a scalar function f. A theorem analogous to the Fundamental Theorem of Calculus gives a way to evaluate the line integrals of gradient fields.

Like the Fundamental Theorem of Calculus, Theorem 1 gives a direct way to evaluate line integrals, without having to take limits of Riemann sums, and without needing to compute a line integral by the procedure used in Section 16.2. Before proving Theorem 1, we give an example.

THEOREM 1—Fundamental Theorem of Line Integrals

Let C be a smooth curve joining the point A to the point B in the plane or in space and parametrized by $\mathbf{r}(t)$. Let f be a differentiable function with a continuous gradient vector $\mathbf{F} = \nabla f$ on a domain D containing C. Then

$$\int_C \mathbf{F} \cdot d\mathbf{r} = f(B) - f(A).$$

EXAMPLE 1   Suppose the force field $\mathbf{F} = \nabla f$ is the gradient of the function

$$f(x, y, z) = -\frac{1}{x^2 + y^2 + z^2}.$$

Find the work done by $\mathbf{F}$ in moving an object along a smooth curve C joining $(1, 0, 0)$ to $(0, 0, 2)$ that does not pass through the origin.

Solution   An application of Theorem 1 shows that the work done by $\mathbf{F}$ along any smooth curve C joining the two points and not passing through the origin is

$$\int_C \mathbf{F} \cdot d\mathbf{r} = f(0, 0, 2) - f(1, 0, 0) = -\frac{1}{4} - (-1) = \frac{3}{4}. \qquad \blacksquare$$

The gravitational force due to a planet, and the electric force associated with a charged particle, can both be modeled by the field $\mathbf{F}$ given in Example 1 up to a constant that depends on the units of measurement. When used to model gravity, the function f in Example 1 represents gravitational potential energy. The sign of f is negative, and f approaches $-\infty$ near the origin. This choice ensures that the gravitational force $\mathbf{F}$, the gradient of f, points toward the origin, so that objects fall down rather than up.

Proof of Theorem 1   Suppose that A and B are two points in the region D and that $C: \mathbf{r}(t) = g(t)\mathbf{i} + h(t)\mathbf{j} + k(t)\mathbf{k}$, $a \le t \le b$, is a smooth curve in D joining A to B. In Section 14.5 we found that the derivative of a scalar function f along a path C is the dot product $\nabla f(\mathbf{r}(t)) \cdot \mathbf{r}'(t)$, so we have

$$\int_C \mathbf{F} \cdot d\mathbf{r} = \int_A^B \nabla f \cdot d\mathbf{r} \qquad \text{F} = \nabla f$$

$$= \int_{t=a}^{t=b} \nabla f(\mathbf{r}(t)) \cdot \mathbf{r}'(t)\, dt \qquad \text{Eq. (2) of Section 16.2 for computing } d\mathbf{r}$$

$$= \int_a^b \frac{d}{dt} f(\mathbf{r}(t))\, dt \qquad \text{Eq. (7) of Section 14.5 giving derivative along a path}$$

$$= f(\mathbf{r}(b)) - f(\mathbf{r}(a)) \qquad \text{Fundamental Theorem of Calculus}$$

$$= f(B) - f(A). \qquad \mathbf{r}(a) = A, \mathbf{r}(b) = B \qquad \blacksquare$$

We see from Theorem 1 that the line integral of a gradient field $\mathbf{F} = \nabla f$ is straightforward to compute once we know the function f. Many important vector fields arising in applications are indeed gradient fields. The next result, which follows from Theorem 1, shows that any conservative field is of this type.

THEOREM 2—Conservative Fields are Gradient Fields

Let $\mathbf{F} = M\mathbf{i} + N\mathbf{j} + P\mathbf{k}$ be a vector field whose components are continuous throughout an open connected region D in space. Then $\mathbf{F}$ is conservative if and only if $\mathbf{F}$ is a gradient field ∇f for a differentiable function f.

Theorem 2 says that $\mathbf{F} = \nabla f$ if and only if for any two points A and B in the region D, the value of the line integral $\int_C \mathbf{F} \cdot d\mathbf{r}$ is independent of the path C joining A to B in D.

Proof of Theorem 2   If $\mathbf{F}$ is a gradient field, then $\mathbf{F} = \nabla f$ for a differentiable function f, and Theorem 1 shows that $\int_C \mathbf{F} \cdot d\mathbf{r} = f(B) - f(A)$. The value of the line integral does not depend on C, but only on its endpoints A and B. So the line integral is path independent and $\mathbf{F}$ satisfies the definition of a conservative field.

On the other hand, suppose that $\mathbf{F}$ is a conservative vector field. We want to find a function f on D satisfying $\nabla f = \mathbf{F}$. First, pick a point A in D and set $f(A) = 0$. For any other point B in D define $f(B)$ to equal $\int_C \mathbf{F} \cdot d\mathbf{r}$, where C is *any* smooth path in D from A to B. The value of $f(B)$ does not depend on the choice of C, since $\mathbf{F}$ is conservative. To show that $\nabla f = \mathbf{F}$ we need to demonstrate that $\partial f/\partial x = M$, $\partial f/\partial y = N$, and $\partial f/\partial z = P$.

Suppose that B has coordinates (x, y, z). By the definition of f, the value of the function f at a nearby point B_0 located at (x_0, y, z) is $\int_{C_0} \mathbf{F} \cdot d\mathbf{r}$, where C_0 is any path from A to B_0. We take a path $C = C_0 \cup L$ from A to B formed by first traveling along C_0 to arrive at B_0 and then traveling along the line segment L from B_0 to B (Figure 16.26). When B_0 is close to B, the segment L lies in D and, since the value $f(B)$ is independent of the path from A to B,

$$f(x, y, z) = \int_{C_0} \mathbf{F} \cdot d\mathbf{r} + \int_L \mathbf{F} \cdot d\mathbf{r}.$$

Differentiating, we have

$$\frac{\partial}{\partial x} f(x, y, z) = \frac{\partial}{\partial x}\left(\int_{C_0} \mathbf{F} \cdot d\mathbf{r} + \int_L \mathbf{F} \cdot d\mathbf{r} \right).$$

Only the last term on the right depends on x, so

$$\frac{\partial}{\partial x} f(x, y, z) = \frac{\partial}{\partial x} \int_L \mathbf{F} \cdot d\mathbf{r}.$$

Now parametrize L as $\mathbf{r}(t) = t\mathbf{i} + y\mathbf{j} + z\mathbf{k}$, $x_0 \leq t \leq x$. Then $d\mathbf{r}/dt = \mathbf{i}$, and since $\mathbf{F} = M\mathbf{i} + N\mathbf{j} + P\mathbf{k}$, it follows that $\mathbf{F} \cdot d\mathbf{r}/dt = M$ and $\int_L \mathbf{F} \cdot d\mathbf{r} = \int_{x_0}^{x} M(t, y, z)\, dt$. Differentiating then gives

$$\frac{\partial}{\partial x} f(x, y, z) = \frac{\partial}{\partial x} \int_{x_0}^{x} M(t, y, z)\, dt = M(x, y, z)$$

by the Fundamental Theorem of Calculus. The partial derivatives $\partial f/\partial y = N$ and $\partial f/\partial z = P$ follow similarly, showing that $\mathbf{F} = \nabla f$.

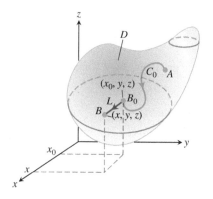

FIGURE 16.26 The function $f(x, y, z)$ in the proof of Theorem 2 is computed by a line integral $\int_{C_0} \mathbf{F} \cdot d\mathbf{r} = f(B_0)$ from A to B_0, plus a line integral $\int_L \mathbf{F} \cdot d\mathbf{r}$ along a line segment L parallel to the x-axis and joining B_0 to B located at (x, y, z). The value of f at A is $f(A) = 0$.

EXAMPLE 2   Find the work done by the conservative field

$$\mathbf{F} = yz\mathbf{i} + xz\mathbf{j} + xy\mathbf{k} = \nabla f, \quad \text{where} \quad f(x, y, z) = xyz,$$

in moving an object along any smooth curve C joining the point $A(-1, 3, 9)$ to $B(1, 6, -4)$.

Solution   With $f(x, y, z) = xyz$, we have

$$\int_C \mathbf{F} \cdot d\mathbf{r} = \int_A^B \nabla f \cdot d\mathbf{r} \qquad \text{\small F = ∇f and path independence}$$

$$= f(B) - f(A) \qquad \text{\small Theorem 1}$$

$$= xyz\big|_{(1,6,-4)} - xyz\big|_{(-1,3,9)}$$

$$= (1)(6)(-4) - (-1)(3)(9)$$

$$= -24 + 27 = 3.$$

A very useful property of line integrals in conservative fields comes into play when the path of integration is a closed curve, or loop. We often use the notation $\oint_C$ for integration around a closed path (discussed with more detail in the next section).

> **THEOREM 3—Loop Property of Conservative Fields**
> The following statements are equivalent.
> 1. $\oint_C \mathbf{F} \cdot d\mathbf{r} = 0$ around every loop (that is, closed curve C) in D.
> 2. The field $\mathbf{F}$ is conservative on D.

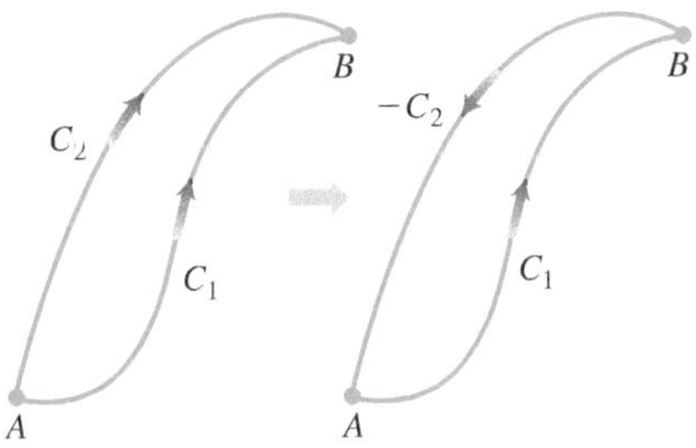

FIGURE 16.27   If we have two paths from A to B, one of them can be reversed to make a loop.

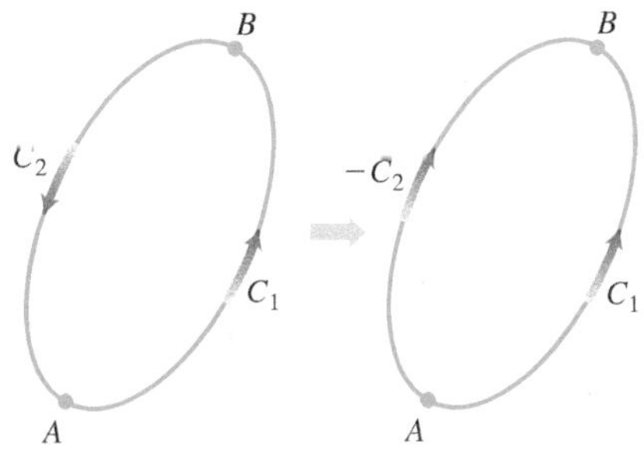

FIGURE 16.28   If A and B lie on a loop, we can reverse part of the loop to make two paths from A to B.

Proof that Part 1 ⇒ Part 2   We want to show that for any two points A and B in D, the integral of $\mathbf{F} \cdot d\mathbf{r}$ has the same value over any two paths C_1 and C_2 from A to B. We reverse the direction on C_2 to make a path $-C_2$ from B to A (Figure 16.27). Together, C_1 and $-C_2$ make a closed loop C, and by assumption,

$$\int_{C_1} \mathbf{F} \cdot d\mathbf{r} - \int_{C_2} \mathbf{F} \cdot d\mathbf{r} = \int_{C_1} \mathbf{F} \cdot d\mathbf{r} + \int_{-C_2} \mathbf{F} \cdot d\mathbf{r} = \int_C \mathbf{F} \cdot d\mathbf{r} = 0.$$

Thus, the integrals over C_1 and C_2 give the same value. Note that the definition of $\mathbf{F} \cdot d\mathbf{r}$ shows that changing the direction along a curve reverses the sign of the line integral.

Proof that Part 2 ⇒ Part 1   We want to show that the integral of $\mathbf{F} \cdot d\mathbf{r}$ is zero over any closed loop C. We pick two points A and B on C and use them to break C into two pieces: C_1 from A to B followed by C_2 from B back to A (Figure 16.28). Then

$$\oint_C \mathbf{F} \cdot d\mathbf{r} = \int_{C_1} \mathbf{F} \cdot d\mathbf{r} + \int_{C_2} \mathbf{F} \cdot d\mathbf{r} = \int_A^B \mathbf{F} \cdot d\mathbf{r} - \int_A^B \mathbf{F} \cdot d\mathbf{r} = 0$$

The following diagram summarizes the results of Theorems 2 and 3.

$$\mathbf{F} = \nabla f \text{ on } D \quad \Longleftrightarrow \quad \begin{matrix}\mathbf{F} \text{ conservative} \\ \text{on } D\end{matrix} \quad \Longleftrightarrow \quad \oint_C \mathbf{F} \cdot d\mathbf{r} = 0 \\ \text{over any loop in } D$$

Two questions arise:

1. How do we know whether a given vector field $\mathbf{F}$ is conservative?

2. If $\mathbf{F}$ is in fact conservative, how do we find a potential function f (so that $\mathbf{F} = \nabla f$)?

Finding Potentials for Conservative Fields

The test for a vector field being conservative involves the equivalence of certain first partial derivatives of the field components.

Component Test for Conservative Fields

Let $\mathbf{F} = M(x, y, z)\mathbf{i} + N(x, y, z)\mathbf{j} + P(x, y, z)\mathbf{k}$ be a field on an open simply connected domain whose component functions have continuous first partial derivatives. Then, $\mathbf{F}$ is conservative if and only if

$$\frac{\partial P}{\partial y} = \frac{\partial N}{\partial z}, \qquad \frac{\partial M}{\partial z} = \frac{\partial P}{\partial x}, \qquad \text{and} \qquad \frac{\partial N}{\partial x} = \frac{\partial M}{\partial y}. \qquad (2)$$

We can view the component test as saying that on a simply connected region, the vector

$$\left(\frac{\partial P}{\partial y} - \frac{\partial N}{\partial z} \right)\mathbf{i} + \left(\frac{\partial M}{\partial z} - \frac{\partial P}{\partial x} \right)\mathbf{j} + \left(\frac{\partial N}{\partial x} - \frac{\partial M}{\partial y} \right)\mathbf{k}$$

is zero if and only if $\mathbf{F}$ is conservative. This interesting vector is called the curl of $\mathbf{F}$, and we study it in Section 16.7.

Proof that Equations (2) hold if F is conservative   If $\mathbf{F}$ is conservative, then there is a potential function f such that

$$\mathbf{F} = M\mathbf{i} + N\mathbf{j} + P\mathbf{k} = \nabla f = \frac{\partial f}{\partial x}\mathbf{i} + \frac{\partial f}{\partial y}\mathbf{j} + \frac{\partial f}{\partial z}\mathbf{k}.$$

Hence,

$$\frac{\partial P}{\partial y} = \frac{\partial}{\partial y}\left(\frac{\partial f}{\partial z} \right) = \frac{\partial^2 f}{\partial y\, \partial z}$$

$$= \frac{\partial^2 f}{\partial z\, \partial y} \qquad \text{Mixed Derivative Theorem, Section 14.3}$$

$$= \frac{\partial}{\partial z}\left(\frac{\partial f}{\partial y} \right) = \frac{\partial N}{\partial z}.$$

The others in Equations (2) are proved similarly.   ∎

The second half of the proof, that Equations (2) imply that $\mathbf{F}$ is conservative, is a consequence of Stokes' Theorem, taken up in Section 16.7, and requires our assumption that the domain of $\mathbf{F}$ be simply connected.

Once we know that $\mathbf{F}$ is conservative, we often want to find a potential function for $\mathbf{F}$. This requires solving the equation $\nabla f = \mathbf{F}$ or

$$\frac{\partial f}{\partial x}\mathbf{i} + \frac{\partial f}{\partial y}\mathbf{j} + \frac{\partial f}{\partial z}\mathbf{k} = M\mathbf{i} + N\mathbf{j} + P\mathbf{k}$$

for f. We accomplish this by integrating the three equations

$$\frac{\partial f}{\partial x} = M, \qquad \frac{\partial f}{\partial y} = N, \qquad \frac{\partial f}{\partial z} = P,$$

as illustrated in the next example.

EXAMPLE 3   Show that $\mathbf{F} = (e^x \cos y + yz)\mathbf{i} + (xz - e^x \sin y)\mathbf{j} + (xy + z)\mathbf{k}$ is conservative over its natural domain and find a potential function for it.

Solution   The natural domain of $\mathbf{F}$ is all of space, which is open and simply connected. We apply the test in Equations (2) to

$$M = e^x \cos y + yz, \qquad N = xz - e^x \sin y, \qquad P = xy + z$$

and calculate

$$\frac{\partial P}{\partial y} = x = \frac{\partial N}{\partial z}, \qquad \frac{\partial M}{\partial z} = y = \frac{\partial P}{\partial x}, \qquad \frac{\partial N}{\partial x} = -e^x \sin y + z = \frac{\partial M}{\partial y}.$$

The partial derivatives are continuous, so these equalities tell us that $\mathbf{F}$ is conservative, so there is a function f with $\nabla f = \mathbf{F}$ (Theorem 2).
     We find f by integrating the equations

$$\frac{\partial f}{\partial x} = e^x \cos y + yz, \qquad \frac{\partial f}{\partial y} = xz - e^x \sin y, \qquad \frac{\partial f}{\partial z} = xy + z. \qquad (3)$$

We integrate the first equation with respect to x, holding y and z fixed, to get

$$f(x, y, z) = e^x \cos y + xyz + g(y, z).$$

We write the constant of integration as a function of y and z because its value may depend on y and z, though not on x. We then calculate $\partial f / \partial y$ from this equation and match it with the expression for $\partial f / \partial y$ in Equations (3). This gives

$$-e^x \sin y + xz + \frac{\partial g}{\partial y} = xz - e^x \sin y,$$

so $\partial g / \partial y = 0$. Therefore, g is a function of z alone, and

$$f(x, y, z) = e^x \cos y + xyz + h(z).$$

We now calculate $\partial f / \partial z$ from this equation and match it to the formula for $\partial f / \partial z$ in Equations (3). This gives

$$xy + \frac{dh}{dz} = xy + z, \qquad \text{or} \qquad \frac{dh}{dz} = z,$$

so

$$h(z) = \frac{z^2}{2} + C.$$

Hence,

$$f(x, y, z) = e^x \cos y + xyz + \frac{z^2}{2} + C.$$

We found infinitely many potential functions of $\mathbf{F}$, one for each value of C.

EXAMPLE 4 Show that $\mathbf{F} = (2x - 3)\mathbf{i} - z\mathbf{j} + (\cos z)\mathbf{k}$ is not conservative.

Solution We apply the Component Test in Equations (2) and find immediately that

$$\frac{\partial P}{\partial y} = \frac{\partial}{\partial y}(\cos z) = 0, \qquad \frac{\partial N}{\partial z} = \frac{\partial}{\partial z}(-z) = -1.$$

The two are unequal, so $\mathbf{F}$ is not conservative. No further testing is required. ∎

EXAMPLE 5 Show that the vector field

$$\mathbf{F} = \frac{-y}{x^2 + y^2}\mathbf{i} + \frac{x}{x^2 + y^2}\mathbf{j} + 0\mathbf{k}$$

satisfies the equations in the Component Test, but is not conservative over its natural domain. Explain why this is possible.

Solution We have $M = -y/(x^2 + y^2)$, $N = x/(x^2 + y^2)$, and $P = 0$. If we apply the Component Test, we find

$$\frac{\partial P}{\partial y} = 0 = \frac{\partial N}{\partial z}, \qquad \frac{\partial P}{\partial x} = 0 = \frac{\partial M}{\partial z}, \qquad \text{and} \qquad \frac{\partial M}{\partial y} = \frac{y^2 - x^2}{(x^2 + y^2)^2} = \frac{\partial N}{\partial x}.$$

So it may appear that the field $\mathbf{F}$ passes the Component Test. However, the test assumes that the domain of $\mathbf{F}$ is simply connected, which is not the case here. Since $x^2 + y^2$ cannot equal zero, the natural domain is the complement of the z-axis and contains loops that cannot be contracted to a point. One such loop is the unit circle C in the xy-plane. The circle is parametrized by $\mathbf{r}(t) = (\cos t)\mathbf{i} + (\sin t)\mathbf{j}$, $0 \le t \le 2\pi$. This loop wraps around the z-axis and cannot be contracted to a point while staying within the complement of the z-axis.

To show that $\mathbf{F}$ is not conservative, we compute the line integral $\oint_C \mathbf{F} \cdot d\mathbf{r}$ around the loop C. First we write the field in terms of the parameter t:

$$\mathbf{F} = \frac{-y}{x^2 + y^2}\mathbf{i} + \frac{x}{x^2 + y^2}\mathbf{j} = \frac{-\sin t}{\sin^2 t + \cos^2 t}\mathbf{i} + \frac{\cos t}{\sin^2 t + \cos^2 t}\mathbf{j} = (-\sin t)\mathbf{i} + (\cos t)\mathbf{j}.$$

Next we find $d\mathbf{r}/dt = (-\sin t)\mathbf{i} + (\cos t)\mathbf{j}$, and then calculate the line integral as

$$\oint_C \mathbf{F} \cdot d\mathbf{r} = \oint_C \mathbf{F} \cdot \frac{d\mathbf{r}}{dt}\, dt = \int_0^{2\pi} \left(\sin^2 t + \cos^2 t\right) dt = 2\pi.$$

Since the line integral of $\mathbf{F}$ around the loop C is not zero, the field $\mathbf{F}$ is not conservative, by Theorem 3. The field $\mathbf{F}$ is displayed in Figure 16.31d in the next section. ∎

Example 5 shows that the Component Test does not apply when the domain of the field is not simply connected. However, if we change the domain in the example so that it is restricted to the ball of radius 1 centered at the point $(2, 2, 2)$, or to any similar ball-shaped region which does not contain a piece of the z-axis, then this new domain D *is* simply connected. Now the partial derivative Equations (2), as well as all the assumptions of the Component Test, are satisfied. In this new situation, the field $\mathbf{F}$ in Example 5 is conservative on D. Just as we must be careful with a function when determining if it satisfies a property throughout its domain (like continuity or the Intermediate Value Property), so must we also be careful with a vector field in determining the properties it may or may not have over its assigned domain.

Exact Differential Forms

It is often convenient to express work and circulation integrals in the differential form

$$\int_C M\,dx + N\,dy + P\,dz$$

discussed in Section 16.2. Such line integrals are relatively easy to evaluate if $M\,dx + N\,dy + P\,dz$ is the total differential of a function f and C is any path joining the two points from A to B. For then

$$\int_C M\,dx + N\,dy + P\,dz = \int_C \frac{\partial f}{\partial x}dx + \frac{\partial f}{\partial y}dy + \frac{\partial f}{\partial z}dz$$

$$= \int_A^B \nabla f \cdot d\mathbf{r} \qquad \nabla f \text{ is conservative.}$$

$$= f(B) - f(A). \qquad \text{Theorem 1}$$

Thus,

$$\int_A^B df = f(B) - f(A),$$

just as with differentiable functions of a single variable.

DEFINITIONS Any expression $M(x, y, z)\,dx + N(x, y, z)\,dy + P(x, y, z)\,dz$ is a **differential form**. A differential form is **exact** on a domain D in space if

$$M\,dx + N\,dy + P\,dz = \frac{\partial f}{\partial x}dx + \frac{\partial f}{\partial y}dy + \frac{\partial f}{\partial z}dz = df$$

for some scalar function f throughout D.

Notice that if $M\,dx + N\,dy + P\,dz = df$ on D, then $\mathbf{F} = M\mathbf{i} + N\mathbf{j} + P\mathbf{k}$ is the gradient field of f on D. Conversely, if $\mathbf{F} = \nabla f$, then the form $M\,dx + N\,dy + P\,dz$ is exact. The test for the form being exact is therefore the same as the test for $\mathbf{F}$ being conservative.

Component Test for Exactness of $M\,dx + N\,dy + P\,dz$

The differential form $M\,dx + N\,dy + P\,dz$ is exact on an open simply connected domain if and only if

$$\frac{\partial P}{\partial y} = \frac{\partial N}{\partial z}, \qquad \frac{\partial M}{\partial z} = \frac{\partial P}{\partial x}, \qquad \text{and} \qquad \frac{\partial N}{\partial x} = \frac{\partial M}{\partial y}.$$

This is equivalent to saying that the field $\mathbf{F} = M\mathbf{i} + N\mathbf{j} + P\mathbf{k}$ is conservative.

EXAMPLE 6   Show that $y\,dx + x\,dy + 4\,dz$ is exact and evaluate the integral

$$\int_{(1,1,1)}^{(2,3,-1)} y\,dx + x\,dy + 4\,dz$$

over any path from $(1, 1, 1)$ to $(2, 3, -1)$.

Solution We let $M = y$, $N = x$, $P = 4$ and apply the Test for Exactness:

$$\frac{\partial P}{\partial y} = 0 = \frac{\partial N}{\partial z}, \qquad \frac{\partial M}{\partial z} = 0 = \frac{\partial P}{\partial x}, \qquad \frac{\partial N}{\partial x} = 1 = \frac{\partial M}{\partial y}.$$

These equalities tell us that $y\,dx + x\,dy + 4\,dz$ is exact, so

$$y\,dx + x\,dy + 4\,dz = df$$

for some function f, and the integral's value is $f(2, 3, -1) - f(1, 1, 1)$.
We find f up to a constant by integrating the equations

$$\frac{\partial f}{\partial x} = y, \qquad \frac{\partial f}{\partial y} = x, \qquad \frac{\partial f}{\partial z} = 4. \tag{4}$$

From the first equation we get

$$f(x, y, z) = xy + g(y, z).$$

The second equation tells us that

$$\frac{\partial f}{\partial y} = x + \frac{\partial g}{\partial y} = x, \qquad \text{or} \qquad \frac{\partial g}{\partial y} = 0.$$

Hence, g is a function of z alone, and

$$f(x, y, z) = xy + h(z).$$

The third of Equations (4) tells us that

$$\frac{\partial f}{\partial z} = 0 + \frac{dh}{dz} = 4, \qquad \text{or} \qquad h(z) = 4z + C.$$

Therefore,

$$f(x, y, z) = xy + 4z + C.$$

The value of the line integral is independent of the path taken from $(1, 1, 1)$ to $(2, 3, -1)$, and equals

$$f(2, 3, -1) - f(1, 1, 1) = 2 + C - (5 + C) = -3.$$

EXERCISES 16.3

Testing for Conservative Fields
Which fields in Exercises 1–6 are conservative, and which are not?

1. $\mathbf{F} = yz\mathbf{i} + xz\mathbf{j} + xy\mathbf{k}$

2. $\mathbf{F} = (y\sin z)\mathbf{i} + (x\sin z)\mathbf{j} + (xy\cos z)\mathbf{k}$

3. $\mathbf{F} = y\mathbf{i} + (x + z)\mathbf{j} - y\mathbf{k}$

4. $\mathbf{F} = -y\mathbf{i} + x\mathbf{j}$

5. $\mathbf{F} = (z + y)\mathbf{i} + z\mathbf{j} + (y + x)\mathbf{k}$

6. $\mathbf{F} = (e^x\cos y)\mathbf{i} - (e^x\sin y)\mathbf{j} + z\mathbf{k}$

Finding Potential Functions
In Exercises 7–12, find a potential function f for the field $\mathbf{F}$.

7. $\mathbf{F} = 2x\mathbf{i} + 3y\mathbf{j} + 4z\mathbf{k}$

8. $\mathbf{F} = (y + z)\mathbf{i} + (x + z)\mathbf{j} + (x + y)\mathbf{k}$

9. $\mathbf{F} = e^{y+2z}(\mathbf{i} + x\mathbf{j} + 2x\mathbf{k})$

10. $\mathbf{F} = (y\sin z)\mathbf{i} + (x\sin z)\mathbf{j} + (xy\cos z)\mathbf{k}$

11. $\mathbf{F} = (\ln x + \sec^2(x + y))\mathbf{i} +$
$$\left(\sec^2(x + y) + \frac{y}{y^2 + z^2}\right)\mathbf{j} + \frac{z}{y^2 + z^2}\mathbf{k}$$

12. $\mathbf{F} = \dfrac{y}{1 + x^2y^2}\mathbf{i} + \left(\dfrac{x}{1 + x^2y^2} + \dfrac{z}{\sqrt{1 - y^2z^2}}\right)\mathbf{j} +$
$$\left(\frac{y}{\sqrt{1 - y^2z^2}} + \frac{1}{z}\right)\mathbf{k}$$

Exact Differential Forms
In Exercises 13–17, show that the differential forms in the integrals are exact. Then evaluate the integrals.

13. $\displaystyle\int_{(0,0,0)}^{(2,3,-6)} 2x\,dx + 2y\,dy + 2z\,dz$

14. $\displaystyle\int_{(1,1,2)}^{(3,5,0)} yz\,dx + xz\,dy + xy\,dz$

15. $\int_{(0,0,0)}^{(1,2,3)} 2xy\,dx + (x^2 - z^2)\,dy - 2yz\,dz$

16. $\int_{(0,0,0)}^{(3,3,1)} 2x\,dx - y^2\,dy - \dfrac{4}{1+z^2}\,dz$

17. $\int_{(1,0,0)}^{(0,1,1)} \sin y \cos x\,dx + \cos y \sin x\,dy + dz$

Finding Potential Functions to Evaluate Line Integrals
Although they are not defined on all of space R^3, the fields associated with Exercises 18–22 are conservative. Find a potential function for each field and evaluate the integrals as in Example 6.

18. $\int_{(0,2,1)}^{(1,\pi/2,2)} 2\cos y\,dx + \left(\dfrac{1}{y} - 2x\sin y\right)dy + \dfrac{1}{z}dz$

19. $\int_{(1,1,1)}^{(1,2,3)} 3x^2\,dx + \dfrac{z^2}{y}dy + 2z\ln y\,dz$

20. $\int_{(1,2,1)}^{(2,1,1)} (2x\ln y - yz)\,dx + \left(\dfrac{x^2}{y} - xz\right)dy - xy\,dz$

21. $\int_{(1,1,1)}^{(2,2,2)} \dfrac{1}{y}dx + \left(\dfrac{1}{z} - \dfrac{x}{y^2}\right)dy - \dfrac{y}{z^2}dz$

22. $\int_{(-1,-1,-1)}^{(2,2,2)} \dfrac{2x\,dx + 2y\,dy + 2z\,dz}{x^2 + y^2 + z^2}$

Applications and Examples
23. Revisiting Example 6 Evaluate the integral

$$\int_{(1,1,1)}^{(2,3,-1)} y\,dx + x\,dy + 4\,dz$$

from Example 6 by finding parametric equations for the line segment from $(1, 1, 1)$ to $(2, 3, -1)$ and evaluating the line integral of $\mathbf{F} = y\mathbf{i} + x\mathbf{j} + 4\mathbf{k}$ along the segment. Since $\mathbf{F}$ is conservative, the integral is independent of the path.

24. Evaluate

$$\int_C x^2\,dx + yz\,dy + (y^2/2)\,dz$$

along the line segment C joining $(0, 0, 0)$ to $(0, 3, 4)$.

Independence of path Show that the values of the integrals in Exercises 25 and 26 do not depend on the path taken from A to B.

25. $\int_A^B z^2\,dx + 2y\,dy + 2xz\,dz$ **26.** $\int_A^B \dfrac{x\,dx + y\,dy + z\,dz}{\sqrt{x^2 + y^2 + z^2}}$

In Exercises 27 and 28, find a potential function for $\mathbf{F}$.

27. $\mathbf{F} = \dfrac{2x}{y}\mathbf{i} + \left(\dfrac{1-x^2}{y^2}\right)\mathbf{j}$, $\{(x, y): y > 0\}$

28. $\mathbf{F} = (e^x \ln y)\mathbf{i} + \left(\dfrac{e^x}{y} + \sin z\right)\mathbf{j} + (y\cos z)\mathbf{k}$

29. Work along different paths Find the work done by $\mathbf{F} = (x^2 + y)\mathbf{i} + (y^2 + x)\mathbf{j} + ze^z\mathbf{k}$ over the following paths from $(1, 0, 0)$ to $(1, 0, 1)$.

  a. The line segment $x = 1, y = 0, 0 \le z \le 1$

  b. The helix $\mathbf{r}(t) = (\cos t)\mathbf{i} + (\sin t)\mathbf{j} + (t/2\pi)\mathbf{k}, 0 \le t \le 2\pi$

  c. The x-axis from $(1, 0, 0)$ to $(0, 0, 0)$ followed by the parabola $z = x^2, y = 0$ from $(0, 0, 0)$ to $(1, 0, 1)$

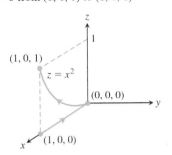

30. Work along different paths Find the work done by $\mathbf{F} = e^{yz}\mathbf{i} + (xze^{yz} + z\cos y)\mathbf{j} + (xye^{yz} + \sin y)\mathbf{k}$ over the following paths from $(1, 0, 1)$ to $(1, \pi/2, 0)$.

  a. The line segment $x = 1, y = \pi t/2, z = 1 - t, 0 \le t \le 1$

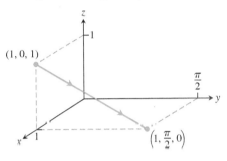

  b. The line segment from $(1, 0, 1)$ to the origin followed by the line segment from the origin to $(1, \pi/2, 0)$

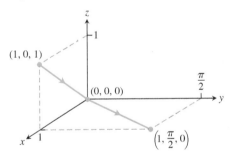

  c. The line segment from $(1, 0, 1)$ to $(1, 0, 0)$, followed by the x-axis from $(1, 0, 0)$ to the origin, followed by the parabola $y = \pi x^2/2, z = 0$ from there to $(1, \pi/2, 0)$

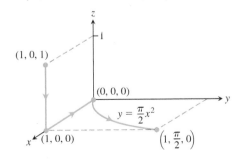

31. Evaluating a work integral two ways Let $\mathbf{F} = \nabla(x^3 y^2)$ and let C be the path in the xy-plane from $(-1, 1)$ to $(1, 1)$ that consists of the line segment from $(-1, 1)$ to $(0, 0)$ followed by the line segment from $(0, 0)$ to $(1, 1)$. Evaluate $\int_C \mathbf{F} \cdot d\mathbf{r}$ in two ways.

a. Find parametrizations for the segments that make up C and evaluate the integral.

b. Use $f(x, y) = x^3 y^2$ as a potential function for $\mathbf{F}$.

32. Integral along different paths Evaluate the line integral $\int_C 2x \cos y \, dx - x^2 \sin y \, dy$ along the following paths C in the xy-plane.

a. The parabola $y = (x - 1)^2$ from $(1, 0)$ to $(0, 1)$

b. The line segment from $(-1, \pi)$ to $(1, 0)$

c. The x-axis from $(-1, 0)$ to $(1, 0)$

d. The astroid $\mathbf{r}(t) = (\cos^3 t)\mathbf{i} + (\sin^3 t)\mathbf{j}$, $0 \le t \le 2\pi$, counterclockwise from $(1, 0)$ back to $(1, 0)$

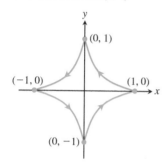

33. a. Exact differential form How are the constants a, b, and c related if the following differential form is exact?

$$(ay^2 + 2czx) \, dx + y(bx + cz) \, dy + (ay^2 + cx^2) \, dz$$

b. Gradient field For what values of b and c will

$$\mathbf{F} = (y^2 + 2czx)\mathbf{i} + y(bx + cz)\mathbf{j} + (y^2 + cx^2)\mathbf{k}$$

be a gradient field?

34. Gradient of a line integral Suppose that $\mathbf{F} = \nabla f$ is a conservative vector field and

$$g(x, y, z) = \int_{(0,0,0)}^{(x,y,z)} \mathbf{F} \cdot d\mathbf{r}.$$

Show that $\nabla g = \mathbf{F}$.

35. Path of least work You have been asked to find the path along which a force field $\mathbf{F}$ will perform the least work in moving a particle between two locations. A quick calculation on your part shows $\mathbf{F}$ to be conservative. How should you respond? Give reasons for your answer.

36. A revealing experiment By experiment, you find that a force field $\mathbf{F}$ performs only half as much work in moving an object along path C_1 from A to B as it does in moving the object along path C_2 from A to B. What can you conclude about $\mathbf{F}$? Give reasons for your answer.

37. Work by a constant force Show that the work done by a constant force field $\mathbf{F} = a\mathbf{i} + b\mathbf{j} + c\mathbf{k}$ in moving a particle along any path from A to B is $W = \mathbf{F} \cdot \overrightarrow{AB}$.

38. Gravitational field

a. Find a potential function for the gravitational field

$$\mathbf{F} = -GmM \frac{x\mathbf{i} + y\mathbf{j} + z\mathbf{k}}{(x^2 + y^2 + z^2)^{3/2}}$$

$(G, m, \text{ and } M \text{ are constants}).$

b. Let P_1 and P_2 be points at distance s_1 and s_2 from the origin. Show that the work done by the gravitational field in part (a) in moving a particle from P_1 to P_2 is

$$GmM\left(\frac{1}{s_2} - \frac{1}{s_1}\right).$$

16.4 Green's Theorem in the Plane

If $\mathbf{F}$ is a conservative field, then we know $\mathbf{F} = \nabla f$ for a differentiable function f, and we can calculate the line integral of $\mathbf{F}$ over any path C joining point A to point B as $\int_C \mathbf{F} \cdot d\mathbf{r} = f(B) - f(A)$. In this section we derive a method for computing a work or flux integral over a *closed* curve C in the plane when the field $\mathbf{F}$ is *not* conservative. This method comes from Green's Theorem, which allows us to convert the line integral into a double integral over the region enclosed by C.

The discussion is given in terms of velocity fields of fluid flows (a fluid is a liquid or a gas) because they are easy to visualize. However, Green's Theorem applies to any vector field, independent of any particular interpretation of the field, provided the assumptions of the theorem are satisfied. We introduce two new ideas for Green's Theorem: *circulation density* around an axis perpendicular to the plane and *divergence* (or *flux density*).

Spin Around an Axis: The k-Component of Curl

Suppose that $\mathbf{F}(x, y) = M(x, y)\mathbf{i} + N(x, y)\mathbf{j}$ is the velocity field of a fluid flowing in the plane and that the first partial derivatives of M and N are continuous at each point of a region R. Let (x, y) be a point in R and let A be a small rectangle with one corner at (x, y)

that, along with its interior, lies entirely in R. The sides of the rectangle, parallel to the coordinate axes, have lengths of Δx and Δy. Assume that the components M and N do not change sign throughout a small region containing the rectangle A. The first idea we use to convey Green's Theorem quantifies the rate at which a floating paddle wheel, with axis perpendicular to the plane, spins at a point in a fluid flowing in a plane region. This idea gives some sense of how the fluid is circulating around axes located at different points and perpendicular to the plane. Physicists sometimes refer to this as the *circulation density* of a vector field $\mathbf{F}$ at a point. To obtain it, we consider the velocity field

$$\mathbf{F}(x, y) = M(x, y)\mathbf{i} + N(x, y)\mathbf{j}$$

and the rectangle A in Figure 16.29 (where we assume both components of $\mathbf{F}$ are positive).

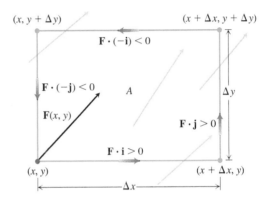

FIGURE 16.29 The rate at which a fluid flows along the bottom edge of a rectangular region A in the direction $\mathbf{i}$ is approximately $\mathbf{F}(x, y) \cdot \mathbf{i} \, \Delta x$, which is positive for the vector field $\mathbf{F}$ shown here. To approximate the rate of circulation at the point (x, y), we calculate the (approximate) flow rates along each edge in the directions of the red arrows, sum these rates, and then divide the sum by the area of A. Taking the limit as $\Delta x \to 0$ and $\Delta y \to 0$ gives the rate of the circulation per unit area.

The circulation rate of $\mathbf{F}$ around the boundary of A is the sum of flow rates along the sides in the tangential direction. For the bottom edge, the flow rate is approximately

$$\mathbf{F}(x, y) \cdot \mathbf{i} \, \Delta x = M(x, y)\Delta x.$$

This is the scalar component of the velocity $\mathbf{F}(x, y)$ in the tangent direction $\mathbf{i}$ times the length of the segment. The flow rates may be positive or negative depending on the components of $\mathbf{F}$. We approximate the net circulation rate around the rectangular boundary of A by summing the flow rates along the four edges as defined by the following dot products.

Top: $\mathbf{F}(x, y + \Delta y) \cdot (-\mathbf{i})\Delta x = -M(x, y + \Delta y)\,\Delta x$

Bottom: $\mathbf{F}(x, y) \cdot \mathbf{i} \, \Delta x = M(x, y)\Delta x$

Right: $\mathbf{F}(x + \Delta x, y) \cdot \mathbf{j} \, \Delta y = N(x + \Delta x, y)\Delta y$

Left: $\mathbf{F}(x, y) \cdot (-\mathbf{j})\Delta y = -N(x, y)\Delta y$

We sum opposite pairs to get

Top and bottom: $-(M(x, y + \Delta y) - M(x, y))\, \Delta x \approx -\left(\dfrac{\partial M}{\partial y}\Delta y\right)\Delta x$

Right and left: $(N(x + \Delta x, y) - N(x, y))\, \Delta y \approx \left(\dfrac{\partial N}{\partial x}\Delta x\right)\Delta y.$

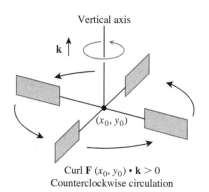

Curl **F** $(x_0, y_0) \cdot \mathbf{k} > 0$
Counterclockwise circulation

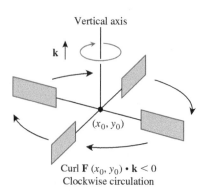

Curl **F** $(x_0, y_0) \cdot \mathbf{k} < 0$
Clockwise circulation

FIGURE 16.30 In the flow of an incompressible fluid over a plane region, the **k**-component of the curl measures the rate of the fluid's rotation at a point. The **k**-component of the curl is positive at points where the rotation is counterclockwise and negative where the rotation is clockwise.

Adding these last two equations gives the net circulation rate relative to the counterclockwise orientation,

$$\text{Circulation rate around rectangle} \approx \left(\frac{\partial N}{\partial x} - \frac{\partial M}{\partial y} \right) \Delta x \, \Delta y.$$

We now divide by $\Delta x \, \Delta y$ to estimate the circulation rate per unit area or *circulation density* for the rectangle:

$$\frac{\text{Circulation around rectangle}}{\text{rectangle area}} \approx \frac{\partial N}{\partial x} - \frac{\partial M}{\partial y}.$$

We let Δx and Δy approach zero to define the *circulation density* of **F** at the point (x, y).

If we see a counterclockwise rotation looking downward onto the xy-plane from the tip of the unit **k** vector, then the circulation density is positive (Figure 16.30). The value of the circulation density is the **k**-component of a more general circulation vector field we define in Section 16.7, called the *curl* of the vector field **F**. For Green's Theorem, we need only this **k**-component, obtained by taking the dot product of curl **F** with **k**.

DEFINITION   The **circulation density** of a vector field $\mathbf{F} = M\mathbf{i} + N\mathbf{j}$ at the point (x, y) is the scalar expression

$$\frac{\partial N}{\partial x} - \frac{\partial M}{\partial y}. \tag{1}$$

This expression is also called **the k-component of the curl**, denoted by (curl **F**) $\cdot$ **k**.

If water is moving about a region in the xy-plane in a thin layer, then the **k**-component of the curl at a point (x_0, y_0) gives a way to measure how fast and in what direction a small paddle wheel spins if it is put into the water at (x_0, y_0) with its axis perpendicular to the plane, parallel to **k** (Figure 16.30). Looking downward onto the xy-plane, it spins counterclockwise when (curl **F**) $\cdot$ **k** is positive and clockwise when the **k**-component is negative.

EXAMPLE 1   The following vector fields represent the velocity of a gas flowing in the xy-plane. Find the circulation density of each vector field and interpret its physical meaning. Figure 16.31 displays the vector fields.

(a) *Uniform expansion or compression:* $\mathbf{F}(x, y) = cx\mathbf{i} + cy\mathbf{j}$ c a constant

(b) *Uniform rotation:* $\mathbf{F}(x, y) = -cy\mathbf{i} + cx\mathbf{j}$

(c) *Shearing flow:* $\mathbf{F}(x, y) = y\mathbf{i}$

(d) *Whirlpool effect:* $\mathbf{F}(x, y) = \dfrac{-y}{x^2 + y^2}\mathbf{i} + \dfrac{x}{x^2 + y^2}\mathbf{j}$

Solution

(a) *Uniform expansion:* (curl **F**) $\cdot$ **k** $= \dfrac{\partial}{\partial x}(cy) - \dfrac{\partial}{\partial y}(cx) = 0$. The gas is not circulating at very small scales.

(b) *Rotation:* (curl **F**) $\cdot$ **k** $= \dfrac{\partial}{\partial x}(cx) - \dfrac{\partial}{\partial y}(-cy) = 2c$. The constant circulation density indicates rotation around every point. If $c > 0$, the rotation is counterclockwise; if $c < 0$, the rotation is clockwise.

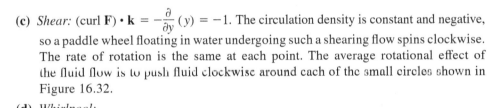

FIGURE 16.31 Velocity fields of a gas flowing in the plane (Example 1).

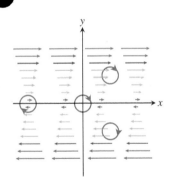

FIGURE 16.32 A shearing flow pushes the fluid clockwise around each point (Example 1c).

(c) *Shear:* $(\text{curl } \mathbf{F}) \cdot \mathbf{k} = -\dfrac{\partial}{\partial y}(y) = -1$. The circulation density is constant and negative, so a paddle wheel floating in water undergoing such a shearing flow spins clockwise. The rate of rotation is the same at each point. The average rotational effect of the fluid flow is to push fluid clockwise around each of the small circles shown in Figure 16.32.

(d) *Whirlpool:*

$$(\text{curl } \mathbf{F}) \cdot \mathbf{k} = \frac{\partial}{\partial x}\left(\frac{x}{x^2 + y^2}\right) - \frac{\partial}{\partial y}\left(\frac{-y}{x^2 + y^2}\right) = \frac{y^2 - x^2}{(x^2 + y^2)^2} - \frac{y^2 - x^2}{(x^2 + y^2)^2} = 0.$$

The circulation density is 0 at every point away from the origin (where the vector field is undefined and the whirlpool effect is taking place), and the gas is not circulating at any point for which the vector field is defined. ∎

One form of Green's Theorem tells us how circulation density can be used to calculate the line integral for flow in the *xy*-plane. (The flow integral was defined in Section 16.2.) A second form of the theorem tells us how we can calculate the flux integral, which gives the flow across the boundary, from *flux density*. We define this idea next and then present both versions of the theorem.

Divergence

Consider again the velocity field $\mathbf{F}(x, y) = M(x, y)\mathbf{i} + N(x, y)\mathbf{j}$ in a domain containing the rectangle *A*, as shown in Figure 16.33. As before, we assume the field components do not change sign throughout a small region containing the rectangle *A*. Our interest now is to determine the rate at which the fluid leaves *A* by flowing across its boundary.

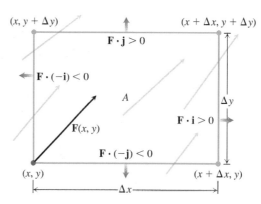

FIGURE 16.33 The rate at which the fluid leaves the rectangular region A across the bottom edge in the direction of the outward normal $-\mathbf{j}$ is approximately $\mathbf{F}(x, y) \cdot (-\mathbf{j}) \, \Delta x$, which is negative for the vector field $\mathbf{F}$ shown here. To approximate the flow rate at the point (x, y), we calculate the (approximate) flow rates across each edge in the directions of the red arrows, sum these rates, and then divide the sum by the area of A. Taking the limit as $\Delta x \to 0$ and $\Delta y \to 0$ gives the flow rate per unit area.

The rate at which fluid leaves the rectangle across the bottom edge is approximately (Figure 16.33)

$$\mathbf{F}(x, y) \cdot (-\mathbf{j}) \, \Delta x = -N(x, y)\Delta x.$$

This is the scalar component of the velocity at (x, y) in the direction of the outward normal times the length of the segment. If the velocity is in meters per second, for example, the flow rate will be in meters per second times meters or square meters per second. The rate at which the fluid crosses the other three sides in the directions of their outward normals can be estimated in a similar way. The flow rates may be positive or negative depending on the signs of the components of $\mathbf{F}$. We approximate the net flow rate across the rectangular boundary of A by summing the flow rates across the four edges as defined by the following dot products.

Fluid Flow Rates: Top: $\mathbf{F}(x, y + \Delta y) \cdot \mathbf{j} \, \Delta x = N(x, y + \Delta y) \, \Delta x$

Bottom: $\mathbf{F}(x, y) \cdot (-\mathbf{j})\Delta x = -N(x, y) \, \Delta x$

Right: $\mathbf{F}(x + \Delta x, y) \cdot \mathbf{i} \, \Delta y = M(x + \Delta x, y) \, \Delta y$

Left: $\mathbf{F}(x, y) \cdot (-\mathbf{i})\Delta y = -M(x, y) \, \Delta y$

Summing opposite pairs gives

Top and bottom: $(N(x, y + \Delta y) - N(x, y)) \, \Delta x \approx \left(\frac{\partial N}{\partial y}\Delta y\right)\Delta x$

Right and left: $(M(x + \Delta x, y) - M(x, y)) \, \Delta y \approx \left(\frac{\partial M}{\partial x}\Delta x\right)\Delta y.$

Adding these last two equations gives the net effect of the flow rates, or the

Flux across rectangle boundary $\approx \left(\frac{\partial M}{\partial x} + \frac{\partial N}{\partial y}\right)\Delta x \, \Delta y.$

We now divide by $\Delta x\Delta y$ to estimate the total flux per unit area or *flux density* for the rectangle:

$$\frac{\text{Flux across rectangle boundary}}{\text{rectangle area}} \approx \left(\frac{\partial M}{\partial x} + \frac{\partial N}{\partial y}\right).$$

div **F** is the symbol for divergence.

Finally, we let Δx and Δy approach zero to define the flux density of **F** at the point (x, y). The mathematical term for the flux density is the *divergence* of **F**. The symbol for it is div **F**, pronounced "divergence of **F**" or "div **F**."

DEFINITION   The **divergence (flux density)** of a vector field $\mathbf{F} = M\mathbf{i} + N\mathbf{j}$ at the point (x, y) is

$$\text{div } \mathbf{F} = \frac{\partial M}{\partial x} + \frac{\partial N}{\partial y}. \tag{2}$$

A gas is compressible, unlike a liquid, and the divergence of its velocity field measures to what extent it is expanding or compressing at each point. Intuitively, if a gas is expanding at the point (x_0, y_0), the lines of flow would diverge there (hence the name) and, since the gas would be flowing out of a small rectangle about (x_0, y_0), the divergence of **F** at (x_0, y_0) would be positive. If the gas were compressing instead of expanding, the divergence would be negative (Figure 16.34).

EXAMPLE 2   Find the divergence, and interpret what it means, for each vector field in Example 1 representing the velocity of a gas flowing in the xy-plane.

Solution

(a) div $\mathbf{F} = \frac{\partial}{\partial x}(cx) + \frac{\partial}{\partial y}(cy) = 2c$: If $c > 0$, the gas is undergoing uniform expansion; if $c < 0$, it is undergoing uniform compression.

(b) div $\mathbf{F} = \frac{\partial}{\partial x}(-cy) + \frac{\partial}{\partial y}(cx) = 0$: The gas is neither expanding nor compressing.

(c) div $\mathbf{F} = \frac{\partial}{\partial x}(y) = 0$: The gas is neither expanding nor compressing.

(d) div $\mathbf{F} = \frac{\partial}{\partial x}\left(\frac{-y}{x^2 + y^2}\right) + \frac{\partial}{\partial y}\left(\frac{x}{x^2 + y^2}\right) = \frac{2xy}{(x^2 + y^2)^2} - \frac{2xy}{(x^2 + y^2)^2} = 0$: Again, the divergence is zero at all points in the domain of the velocity field.   ■

Cases (b), (c), and (d) of Figure 16.31 are plausible models for the two-dimensional flow of a liquid. In fluid dynamics, when the velocity field of a flowing liquid always has divergence equal to zero, as in those cases, the liquid is said to be **incompressible**.

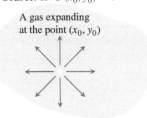

Source: div $\mathbf{F}(x_0, y_0) > 0$

A gas expanding at the point (x_0, y_0)

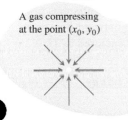

Sink: div $\mathbf{F}(x_0, y_0) < 0$

A gas compressing at the point (x_0, y_0)

FIGURE 16.34   If a gas is expanding at a point (x_0, y_0), the lines of flow have positive divergence; if the gas is compressing, the divergence is negative.

Two Forms for Green's Theorem

A simple closed curve C can be traversed in two possible directions. (Recall that a curve is simple if it does not cross itself.) The curve is traversed counterclockwise, and said to be *positively oriented*, if the region it encloses is always to the left when moving along the curve. If the curve is traversed clockwise then the enclosed region is on the right when moving along the curve and the curve is said to be *negatively oriented*. The line integral of a vector field **F** along C reverses sign if we change the orientation. We use the notation

$$\oint_C \mathbf{F}(x, y) \cdot d\mathbf{r}$$

for the line integral when the simple closed curve C is traversed counterclockwise, with its positive orientation.

In one form, Green's Theorem says that the counterclockwise circulation of a vector field around a simple closed curve is the double integral of the **k**-component of the curl of the field over the region enclosed by the curve. Recall the defining Equation (5) for circulation in Section 16.2.

Circulation and Curl

Circulation around $C = \oint_C \mathbf{F} \cdot \mathbf{T}\, ds$

$(\text{curl } \mathbf{F}) \cdot \mathbf{k} = \dfrac{\partial N}{\partial x} - \dfrac{\partial M}{\partial y}$

THEOREM 4—Green's Theorem (Circulation-Curl or Tangential Form)

Let C be a piecewise smooth, simple closed curve enclosing a region R in the plane. Let $\mathbf{F} = M\mathbf{i} + N\mathbf{j}$ be a vector field with M and N having continuous first partial derivatives in an open region containing R. Then the counterclockwise circulation of $\mathbf{F}$ around C equals the double integral of $(\text{curl } \mathbf{F}) \cdot \mathbf{k}$ over R.

$$\underbrace{\oint_C \mathbf{F} \cdot \mathbf{T}\, ds = \oint_C M\, dx + N\, dy}_{\text{Counterclockwise circulation}} = \underbrace{\iint_R \left(\frac{\partial N}{\partial x} - \frac{\partial M}{\partial y} \right) dx\, dy}_{\text{Curl integral}} \qquad (3)$$

A second form of Green's Theorem says that the outward flux of a vector field across a simple closed curve in the plane equals the double integral of the divergence of the field over the region enclosed by the curve. Recall the formulas for flux in Equations (8) and (9) in Section 16.2.

Flux and Divergence

Flux of $\mathbf{F}$ across $C = \oint_C \mathbf{F} \cdot \mathbf{n}\, ds$

$\text{div } \mathbf{F} = \dfrac{\partial M}{\partial x} + \dfrac{\partial N}{\partial y}$

THEOREM 5—Green's Theorem (Flux-Divergence or Normal Form)

Let C be a piecewise smooth, simple closed curve enclosing a region R in the plane. Let $\mathbf{F} = M\mathbf{i} + N\mathbf{j}$ be a vector field with M and N having continuous first partial derivatives in an open region containing R. Then the outward flux of $\mathbf{F}$ across C equals the double integral of div $\mathbf{F}$ over the region R enclosed by C.

$$\underbrace{\oint_C \mathbf{F} \cdot \mathbf{n}\, ds = \oint_C M\, dy - N\, dx}_{\text{Outward flux}} = \underbrace{\iint_R \left(\frac{\partial M}{\partial x} + \frac{\partial N}{\partial y} \right) dx\, dy}_{\text{Divergence integral}} \qquad (4)$$

The two forms of Green's Theorem are equivalent. Applying Equation (3) to the field $\mathbf{G}_1 = -N\mathbf{i} + M\mathbf{j}$ gives Equation (4), and applying Equation (4) to $\mathbf{G}_2 = N\mathbf{i} - M\mathbf{j}$ gives Equation (3).

Both forms of Green's Theorem can be viewed as two-dimensional generalizations of the Fundamental Theorem of Calculus from Section 5.4. The counterclockwise circulation of $\mathbf{F}$ around C, defined by the line integral on the left-hand side of Equation (3), is the integral of its rate of change (circulation density) over the region R enclosed by C, which is the double integral on the right-hand side of Equation (3). Likewise, the outward flux of $\mathbf{F}$ across C, defined by the line integral on the left-hand side of Equation (4), is the integral of its rate of change (flux density) over the region R enclosed by C, which is the double integral on the right-hand side of Equation (4).

EXAMPLE 3 Verify both forms of Green's Theorem for the vector field

$$\mathbf{F}(x, y) = (x - y)\mathbf{i} + x\mathbf{j}$$

and the region R bounded by the unit circle

$$C: \quad \mathbf{r}(t) = (\cos t)\mathbf{i} + (\sin t)\mathbf{j}, \quad 0 \le t \le 2\pi.$$

Solution First we evaluate the counterclockwise circulation of $\mathbf{F} = M\mathbf{i} + N\mathbf{j}$ around C. On the curve C we have $x = \cos t$ and $y = \sin t$. Evaluating $\mathbf{F}(\mathbf{r}(t))$ and computing the partial derivatives of the components of $\mathbf{F}$, we have

$$M = x - y = \cos t - \sin t, \qquad dx = d(\cos t) = -\sin t\, dt,$$

$$N = x = \cos t, \qquad\qquad dy = d(\sin t) = \cos t\, dt.$$

Therefore,

$$\oint_C \mathbf{F} \cdot \mathbf{T} \, ds = \oint_C M \, dx + N \, dy$$

$$= \int_{t=0}^{t=2\pi} (\cos t - \sin t)(-\sin t) \, dt + (\cos t)(\cos t) \, dt$$

$$= \int_0^{2\pi} (-\sin t \cos t + 1) \, dt = 2\pi.$$

This gives the left side of Equation (3). Next we find the curl integral, the right side of Equation (3). Since $M = x - y$ and $N = x$, we have

$$\frac{\partial M}{\partial x} = 1, \qquad \frac{\partial M}{\partial y} = -1, \qquad \frac{\partial N}{\partial x} = 1, \qquad \frac{\partial N}{\partial y} = 0.$$

Therefore,

$$\iint_R \left(\frac{\partial N}{\partial x} - \frac{\partial M}{dy} \right) dx \, dy = \iint_R (1 - (-1)) \, dx \, dy$$

$$= 2 \iint_R dx \, dy = 2(\text{area inside the unit circle}) = 2\pi.$$

Thus, the right and left sides of Equation (3) both equal 2π, as asserted by the circulation-flux version of Green's Theorem.

Figure 16.35 displays the vector field and circulation around C.

Now we compute the two sides of Equation (4) in the flux-divergence form of Green's Theorem, starting with the outward flux:

$$\oint_C M \, dy - N \, dx = \int_{t=0}^{t=2\pi} (\cos t - \sin t)(\cos t \, dt) - (\cos t)(-\sin t \, dt)$$

$$= \int_0^{2\pi} \cos^2 t \, dt = \pi.$$

Next we compute the divergence integral:

$$\iint_R \left(\frac{\partial M}{\partial x} + \frac{\partial N}{\partial y} \right) dx \, dy = \iint_R (1 + 0) \, dx \, dy = \iint_R dx \, dy = \pi.$$

Hence the right and left sides of Equation (4) both equal π, as asserted by the flux-divergence version of Green's Theorem.

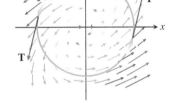

FIGURE 16.35 The vector field in Example 3 has a counterclockwise circulation of 2π around the unit circle.

Using Green's Theorem to Evaluate Line Integrals

If we construct a closed curve C by piecing together a number of different curves end to end, the process of evaluating a line integral over C can be lengthy because there are so many different integrals to evaluate. If C bounds a region R to which Green's Theorem applies, however, we can use Green's Theorem to change the line integral around C into one double integral over R.

EXAMPLE 4   Evaluate the line integral

$$\oint_C xy \, dy - y^2 \, dx,$$

where C is the square cut from the first quadrant by the lines $x = 1$ and $y = 1$.

Solution   We can use either form of Green's Theorem to change the line integral into a double integral over the square, where C is the square's boundary and R is its interior.

1. *With the Tangential Form* Equation (3): Taking $M = -y^2$ and $N = xy$ gives the result:

$$\oint_C -y^2\,dx + xy\,dy = \iint_R \left(\frac{\partial N}{\partial x} - \frac{\partial M}{\partial y}\right) dx\,dy = \iint_R (y - (-2y))\,dx\,dy$$

$$= \int_0^1 \int_0^1 3y\,dx\,dy = \int_0^1 \left[3xy\right]_{x=0}^{x=1} dy = \int_0^1 3y\,dy = \frac{3}{2}y^2\Big]_0^1 = \frac{3}{2}.$$

2. *With the Normal Form* Equation (4): Taking $M = xy$, $N = y^2$, gives the same result:

$$\oint_C xy\,dy - y^2\,dx = \iint_R \left(\frac{\partial M}{\partial x} + \frac{\partial N}{\partial y}\right) dx\,dy = \iint_R (y + 2y)\,dx\,dy = \frac{3}{2}.$$ ∎

EXAMPLE 5   Calculate the outward flux of the vector field $\mathbf{F}(x, y) = 2e^{xy}\mathbf{i} + y^3\mathbf{j}$ across the square bounded by the lines $x = \pm 1$ and $y = \pm 1$.

Solution   Calculating the flux with a line integral would take four integrations, one for each side of the square. With Green's Theorem, we can change the line integral to one double integral. With $M = 2e^{xy}$, $N = y^3$, C the square, and R the square's interior, we have

$$\text{Flux} = \oint_C \mathbf{F} \cdot \mathbf{n}\,ds = \oint_C M\,dy - N\,dx$$

$$= \iint_R \left(\frac{\partial M}{\partial x} + \frac{\partial N}{\partial y}\right) dx\,dy \qquad \text{Green's Theorem, Eq. (4)}$$

$$= \int_{-1}^1 \int_{-1}^1 (2ye^{xy} + 3y^2)\,dx\,dy = \int_{-1}^1 \left[2e^{xy} + 3xy^2\right]_{x=-1}^{x=1} dy$$

$$= \int_{-1}^1 (2e^y + 6y^2 - 2e^{-y})\,dy = \left[2e^y + 2y^3 + 2e^{-y}\right]_{-1}^1 = 4.$$ ∎

Proof of Green's Theorem for Special Regions

Let C be a smooth simple closed curve in the xy-plane with the property that lines parallel to the axes cut it at no more than two points. Let R be the region enclosed by C and suppose that M, N, and their first partial derivatives are continuous at every point of some open region containing C and R. We want to prove the circulation-curl form of Green's Theorem,

$$\oint_C M\,dx + N\,dy = \iint_R \left(\frac{\partial N}{\partial x} - \frac{\partial M}{\partial y}\right) dx\,dy. \tag{5}$$

Figure 16.36 shows C made up of two directed parts:

$$C_1: \quad y = f_1(x), \quad a \leq x \leq b, \qquad C_2: \quad y = f_2(x), \quad b \geq x \geq a.$$

For any x between a and b, we can integrate $\partial M/\partial y$ with respect to y from $y = f_1(x)$ to $y = f_2(x)$ and obtain

$$\int_{f_1(x)}^{f_2(x)} \frac{\partial M}{\partial y}\,dy = M(x, y)\bigg]_{y=f_1(x)}^{y=f_2(x)} = M(x, f_2(x)) - M(x, f_1(x)).$$

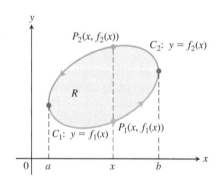

FIGURE 16.36   The boundary curve C is made up of C_1, the graph of $y = f_1(x)$, and C_2, the graph of $y = f_2(x)$.

We can then integrate this with respect to x from a to b:

$$\int_a^b \int_{f_1(x)}^{f_2(x)} \frac{\partial M}{\partial y} dy\, dx = \int_a^b \left[M(x, f_2(x)) - M(x, f_1(x)) \right] dx$$

$$= -\int_b^a M(x, f_2(x))\, dx - \int_a^b M(x, f_1(x))\, dx$$

$$= -\int_{C_2} M\, dx - \int_{C_1} M\, dx$$

$$= -\oint_C M\, dx.$$

Therefore, reversing the order of the equations, we have

$$\oint_C M\, dx = \iint_R \left(-\frac{\partial M}{\partial y} \right) dx\, dy. \qquad (6)$$

Equation (6) is half the result we need for Equation (5). We derive the other half by integrating $\partial N/\partial x$ first with respect to x and then with respect to y, as suggested by Figure 16.37. This shows the curve C of Figure 16.36 decomposed into the two directed parts C_1': $x = g_1(y), d \geq y \geq c$ and C_2': $x = g_2(y), c \leq y \leq d$. The result of this double integration is

$$\oint_C N\, dy = \iint_R \frac{\partial N}{\partial x} dx\, dy. \qquad (7)$$

Summing Equations (6) and (7) gives Equation (5). This concludes the proof.   ■

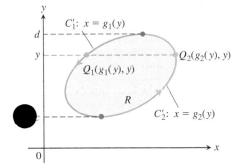

FIGURE 16.37 The boundary curve C is made up of C_1', the graph of $x = g_1(y)$, and C_2', the graph of $x = g_2(y)$.

Green's Theorem also holds for more general regions, such as those shown in Figure 16.38. Notice that the region in Figure 16.38c is not simply connected. The curves C_1 and C_h on its boundary are oriented so that the region R is always on the left-hand side as the curves are traversed in the directions shown, and cancelation occurs over common boundary arcs traversed in opposite directions. With this convention, Green's Theorem is valid for regions that are not simply connected. The proof proceeds by summing the contributions to the Integral of a collection of special regions, which overlap along their boundaries. Cancelation occurs along arcs that are traversed twice, one in each direction, as in Figure 16.38c. We do not give the full proof here.

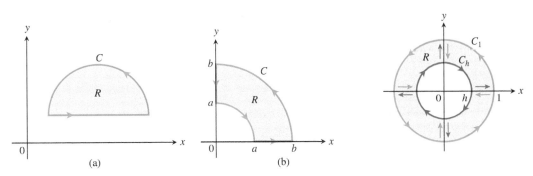

FIGURE 16.38 Other regions to which Green's Theorem applies. In (c) the axes convert the region into four simply connected regions, and we sum the line integrals along the oriented boundaries.

EXERCISES 16.4

Computing the k-Component of Curl(F)

In Exercises 1–6, find the **k**-component of curl(**F**) for the following vector fields on the plane.

1. $\mathbf{F} = (x + y)\mathbf{i} + (2xy)\mathbf{j}$
2. $\mathbf{F} = (x^2 - y)\mathbf{i} + (y^2)\mathbf{j}$
3. $\mathbf{F} = (xe^y)\mathbf{i} + (ye^x)\mathbf{j}$
4. $\mathbf{F} = (x^2y)\mathbf{i} + (xy^2)\mathbf{j}$
5. $\mathbf{F} = (y\sin x)\mathbf{i} + (x\sin y)\mathbf{j}$
6. $\mathbf{F} = (x/y)\mathbf{i} - (y/x)\mathbf{j}$

Verifying Green's Theorem

In Exercises 7–10, verify the conclusion of Green's Theorem by evaluating both sides of Equations (3) and (4) for the field $\mathbf{F} = M\mathbf{i} + N\mathbf{j}$. Take the domains of integration in each case to be the disk $R: x^2 + y^2 \leq a^2$ and its bounding circle $C: \mathbf{r} = (a\cos t)\mathbf{i} + (a\sin t)\mathbf{j}, 0 \leq t \leq 2\pi$.

7. $\mathbf{F} = -y\mathbf{i} + x\mathbf{j}$
8. $\mathbf{F} = y\mathbf{i}$
9. $\mathbf{F} = 2x\mathbf{i} - 3y\mathbf{j}$
10. $\mathbf{F} = -x^2y\mathbf{i} + xy^2\mathbf{j}$

Circulation and Flux

In Exercises 11–20, use Green's Theorem to find the counterclockwise circulation and outward flux for the field **F** and curve C.

11. $\mathbf{F} = (x - y)\mathbf{i} + (y - x)\mathbf{j}$

    C: The square bounded by $x = 0, x = 1, y = 0, y = 1$
12. $\mathbf{F} = (x^2 + 4y)\mathbf{i} + (x + y^2)\mathbf{j}$

    C: The square bounded by $x = 0, x = 1, y = 0, y = 1$
13. $\mathbf{F} = (y^2 - x^2)\mathbf{i} + (x^2 + y^2)\mathbf{j}$

    C: The triangle bounded by $y = 0, x = 3$, and $y = x$
14. $\mathbf{F} = (x + y)\mathbf{i} - (x^2 + y^2)\mathbf{j}$

    C: The triangle bounded by $y = 0, x = 1$, and $y = x$
15. $\mathbf{F} = (xy + y^2)\mathbf{i} + (x - y)\mathbf{j}$ 16. $\mathbf{F} = (x + 3y)\mathbf{i} + (2x - y)\mathbf{j}$

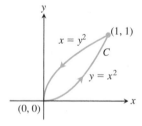

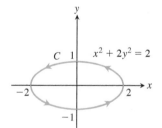

17. $\mathbf{F} = x^3y^2\mathbf{i} + \frac{1}{2}x^4y\,\mathbf{j}$ 18. $\mathbf{F} = \frac{x}{1 + y^2}\mathbf{i} + (\tan^{-1}y)\mathbf{j}$

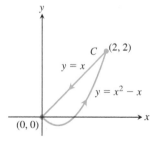

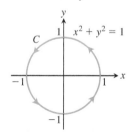

19. $\mathbf{F} = (x + e^x \sin y)\mathbf{i} + (x + e^x \cos y)\mathbf{j}$

    C: The right-hand loop of the lemniscate $r^2 = \cos 2\theta$
20. $\mathbf{F} = \left(\tan^{-1}\dfrac{y}{x}\right)\mathbf{i} + \ln(x^2 + y^2)\mathbf{j}$

    C: The boundary of the region defined by the polar coordinate inequalities $1 \leq r \leq 2, 0 \leq \theta \leq \pi$
21. Find the counterclockwise circulation and outward flux of the field $\mathbf{F} = xy\mathbf{i} + y^2\mathbf{j}$ around and over the boundary of the region enclosed by the curves $y = x^2$ and $y = x$ in the first quadrant.
22. Find the counterclockwise circulation and the outward flux of the field $\mathbf{F} = (-\sin y)\mathbf{i} + (x\cos y)\mathbf{j}$ around and over the square cut from the first quadrant by the lines $x = \pi/2$ and $y = \pi/2$.
23. Find the outward flux of the field

$$\mathbf{F} = \left(3xy - \frac{x}{1 + y^2}\right)\mathbf{i} + (e^x + \tan^{-1}y)\mathbf{j}$$

    across the cardioid $r = a(1 + \cos\theta), a > 0$.
24. Find the counterclockwise circulation of $\mathbf{F} = (y + e^x \ln y)\mathbf{i} + (e^x/y)\mathbf{j}$ around the boundary of the region that is bounded above by the curve $y = 3 - x^2$ and below by the curve $y = x^4 + 1$.

Work

In Exercises 25 and 26, find the work done by **F** in moving a particle once counterclockwise around the given curve.

25. $\mathbf{F} = 2xy^3\mathbf{i} + 4x^2y^2\mathbf{j}$

    C: The boundary of the "triangular" region in the first quadrant enclosed by the x-axis, the line $x = 1$, and the curve $y = x^3$
26. $\mathbf{F} = (4x - 2y)\mathbf{i} + (2x - 4y)\mathbf{j}$

    C: The circle $(x - 2)^2 + (y - 2)^2 = 4$

Using Green's Theorem

Apply Green's Theorem to evaluate the integrals in Exercises 27–30.

27. $\displaystyle\oint_C (y^2\,dx + x^2\,dy)$

    C: The triangle bounded by $x = 0, x + y = 1, y = 0$
28. $\displaystyle\oint_C (3y\,dx + 2x\,dy)$

    C: The boundary of $0 \leq x \leq \pi, 0 \leq y \leq \sin x$
29. $\displaystyle\oint_C (6y + x)\,dx + (y + 2x)\,dy$

    C: The circle $(x - 2)^2 + (y - 3)^2 = 4$
30. $\displaystyle\oint_C (2x + y^2)\,dx + (2xy + 3y)\,dy$

    C: Any simple closed curve in the plane for which Green's Theorem holds

Calculating Area with Green's Theorem   If a simple closed curve C in the plane and the region R it encloses satisfy the hypotheses of Green's Theorem, the area of R is given by

---

Green's Theorem Area Formula

$$\text{Area of } R = \frac{1}{2} \oint_C x \, dy - y \, dx$$

The reason is that by Equation (4), run backward,

$$\text{Area of } R = \iint_R dy \, dx = \iint_R \left(\frac{1}{2} + \frac{1}{2}\right) dy \, dx$$

$$= \oint_C \frac{1}{2} x \, dy - \frac{1}{2} y \, dx.$$

Use the Green's Theorem area formula given above to find the areas of the regions enclosed by the curves in Exercises 31–34.

31. The circle $\mathbf{r}(t) = (a \cos t)\mathbf{i} + (a \sin t)\mathbf{j}, \quad 0 \le t \le 2\pi$

32. The ellipse $\mathbf{r}(t) = (a \cos t)\mathbf{i} + (b \sin t)\mathbf{j}, \quad 0 \le t \le 2\pi$

33. The astroid $\mathbf{r}(t) = (\cos^3 t)\mathbf{i} + (\sin^3 t)\mathbf{j}, \quad 0 \le t \le 2\pi$

34. One arch of the cycloid $x = t - \sin t, \quad y = 1 - \cos t$

35. Let C be the boundary of a region on which Green's Theorem holds. Use Green's Theorem to calculate

a. $\oint_C f(x) \, dx + g(y) \, dy$

b. $\oint_C ky \, dx + hx \, dy \quad$ (k and h constants).

36. Integral dependent only on area   Show that the value of

$$\oint_C xy^2 \, dx + (x^2 y + 2x) \, dy$$

around any square depends only on the area of the square and not on its location in the plane.

37. Evaluate the integral

$$\oint_C 4x^3 y \, dx + x^4 \, dy$$

for any closed path C.

38. Evaluate the integral

$$\oint_C -y^3 \, dy + x^3 \, dx$$

for any closed path C.

39. Area as a line integral   Show that if R is a region in the plane bounded by a piecewise smooth, simple closed curve C, then

$$\text{Area of } R = \oint_C x \, dy = -\oint_C y \, dx.$$

40. Definite integral as a line integral   Suppose that a nonnegative function $y = f(x)$ has a continuous first derivative on $[a, b]$. Let C be the boundary of the region in the xy-plane that is bounded below by the x-axis, above by the graph of f, and on the sides by the lines $x = a$ and $x = b$. Show that

$$\int_a^b f(x) \, dx = -\oint_C y \, dx.$$

41. Area and the centroid   Let A be the area and $\bar{x}$ the x-coordinate of the centroid of a region R that is bounded by a piecewise smooth, simple closed curve C in the xy-plane. Show that

$$\frac{1}{2}\oint_C x^2 \, dy = -\oint_C xy \, dx = \frac{1}{3}\oint_C x^2 \, dy - xy \, dx = A\bar{x}.$$

42. Moment of inertia   Let I_y be the moment of inertia about the y-axis of the region in Exercise 41. Show that

$$\frac{1}{3}\oint_C x^3 \, dy = -\oint_C x^2 y \, dx = \frac{1}{4}\oint_C x^3 \, dy - x^2 y \, dx = I_y.$$

43. Green's Theorem and Laplace's equation   Assuming that all the necessary derivatives exist and are continuous, show that if $f(x, y)$ satisfies the Laplace equation

$$\frac{\partial^2 f}{\partial x^2} + \frac{\partial^2 f}{\partial y^2} = 0,$$

then

$$\oint_C \frac{\partial f}{\partial y} dx - \frac{\partial f}{\partial x} dy = 0$$

for all closed curves C to which Green's Theorem applies. (The converse is also true: If the line integral is always zero, then f satisfies the Laplace equation.)

44. Maximizing work   Among all smooth, simple closed curves in the plane, oriented counterclockwise, find the one along which the work done by

$$\mathbf{F} = \left(\frac{1}{4}x^2 y + \frac{1}{3}y^3\right)\mathbf{i} + x\mathbf{j}$$

is greatest. (*Hint:* Where is (curl $\mathbf{F}$) $\cdot$ $\mathbf{k}$ positive?)

45. Regions with many holes   Green's Theorem holds for a region R with any finite number of holes as long as the bounding curves are smooth, simple, and closed and we integrate over each component of the boundary in the direction that keeps R on our immediate left as we go along (see accompanying figure).

a. Let $f(x, y) = \ln (x^2 + y^2)$ and let C be the circle $x^2 + y^2 = a^2$. Evaluate the flux integral

$$\oint_C \nabla f \cdot \mathbf{n} \, ds.$$

b. Let K be an arbitrary smooth, simple closed curve in the plane that does not pass through $(0, 0)$. Use Green's Theorem to show that

$$\oint_K \nabla f \cdot \mathbf{n}\, ds$$

has two possible values, depending on whether $(0, 0)$ lies inside K or outside K.

46. Bendixson's criterion The *streamlines* of a planar fluid flow are the smooth curves traced by the fluid's individual particles. The vectors $\mathbf{F} = M(x, y)\mathbf{i} + N(x, y)\mathbf{j}$ of the flow's velocity field are the tangent vectors of the streamlines. Show that if the flow takes place over a simply connected region R (no holes or missing points) and that if $M_x + N_y \ne 0$ throughout R, then none of the streamlines in R is closed. In other words, no particle of fluid ever has a closed trajectory in R. The criterion $M_x + N_y \ne 0$ is called **Bendixson's criterion** for the nonexistence of closed trajectories.

47. Establish Equation (7) to finish the proof of the special case of Green's Theorem.

48. Curl component of conservative fields Can anything be said about the curl component of a conservative two-dimensional vector field? Give reasons for your answer.

COMPUTER EXPLORATIONS

In Exercises 49–52, use a CAS and Green's Theorem to find the counterclockwise circulation of the field $\mathbf{F}$ around the simple closed curve C. Perform the following CAS steps.

   a. Plot C in the xy-plane.

   b. Determine the integrand $(\partial N/\partial x) - (\partial M/\partial y)$ for the tangential form of Green's Theorem.

   c. Determine the (double integral) limits of integration from your plot in part (a) and evaluate the curl integral for the circulation.

49. $\mathbf{F} = (2x - y)\mathbf{i} + (x + 3y)\mathbf{j}$, C: The ellipse $x^2 + 4y^2 = 4$

50. $\mathbf{F} = (2x^3 - y^3)\mathbf{i} + (x^3 + y^3)\mathbf{j}$, C: The ellipse $\dfrac{x^2}{4} + \dfrac{y^2}{9} = 1$

51. $\mathbf{F} = x^{-1}e^y\mathbf{i} + (e^y \ln x + 2x)\mathbf{j}$,

C: The boundary of the region defined by $y = 1 + x^4$ (below) and $y = 2$ (above)

52. $\mathbf{F} = xe^y\mathbf{i} + (4x^2 \ln y)\mathbf{j}$,

C: The triangle with vertices $(0, 0)$, $(2, 0)$, and $(0, 4)$

16.5 Surfaces and Area

We have defined curves in the plane in three different ways:

Explicit form:	$y = f(x)$
Implicit form:	$F(x, y) = 0$
Parametric vector form:	$\mathbf{r}(t) = f(t)\mathbf{i} + g(t)\mathbf{j}, \quad a \le t \le b.$

We have analogous definitions of surfaces in space:

Explicit form:	$z = f(x, y)$
Implicit form:	$F(x, y, z) = 0.$

There is also a parametric form for surfaces that gives the position of a point on the surface as a vector function of two variables. We discuss this new form in this section and apply the form to obtain the area of a surface as a double integral. Double integral formulas for areas of surfaces given in implicit and explicit forms are then obtained as special cases of the more general parametric formula.

Parametrizations of Surfaces

Suppose

$$\mathbf{r}(u, v) = f(u, v)\mathbf{i} + g(u, v)\mathbf{j} + h(u, v)\mathbf{k} \tag{1}$$

is a continuous vector function that is defined on a region R in the uv-plane and one-to-one on the interior of R (Figure 16.39). We call the range of $\mathbf{r}$ the **surface** S defined or traced by $\mathbf{r}$. Equation (1) together with the domain R constitutes a **parametrization** of the surface. The variables u and v are the **parameters**, and R is the **parameter domain**. To simplify our discussion, we take R to be a rectangle defined by inequalities of the form $a \le u \le b$, $c \le v \le d$. The requirement that $\mathbf{r}$ be one-to-one on the interior of R ensures that S does not cross itself. Notice that Equation (1) is the vector equivalent of *three* parametric equations:

$$x = f(u, v), \qquad y = g(u, v), \qquad z = h(u, v).$$

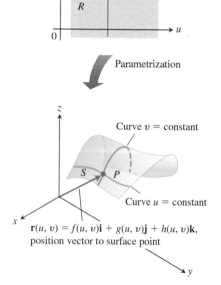

FIGURE 16.39 A parametrized surface S expressed as a vector function of two variables defined on a region R.

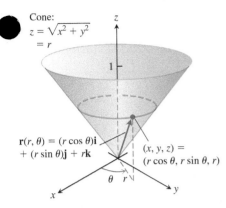

FIGURE 16.40 The cone in Example 1 can be parametrized using cylindrical coordinates.

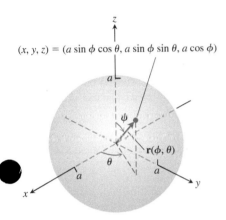

FIGURE 16.41 The sphere in Example 2 can be parametrized using spherical coordinates.

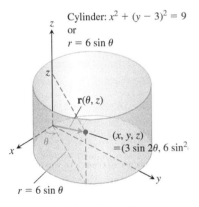

FIGURE 16.42 The cylinder in Example 3 can be parametrized using cylindrical coordinates.

EXAMPLE 1 Find a parametrization of the cone

$$z = \sqrt{x^2 + y^2}, \quad 0 \le z \le 1.$$

Solution Here, cylindrical coordinates provide a parametrization. A typical point (x, y, z) on the cone (Figure 16.40) has $x = r \cos \theta$, $y = r \sin \theta$, and $z = \sqrt{x^2 + y^2} = r$, with $0 \le r \le 1$ and $0 \le \theta \le 2\pi$. Taking $u = r$ and $v = \theta$ in Equation (1) gives the parametrization

$$\mathbf{r}(r, \theta) = (r \cos \theta)\mathbf{i} + (r \sin \theta)\mathbf{j} + r\mathbf{k}, \quad 0 \le r \le 1, \quad 0 \le \theta \le 2\pi.$$

The parametrization is one-to-one on the interior of the domain R, though not on the boundary tip of its cone where $r = 0$. ∎

EXAMPLE 2 Find a parametrization of the sphere $x^2 + y^2 + z^2 = a^2$.

Solution Spherical coordinates provide what we need. A typical point (x, y, z) on the sphere (Figure 16.41) has $x = a \sin \phi \cos \theta$, $y = a \sin \phi \sin \theta$, and $z = a \cos \phi$, $0 \le \phi \le \pi$, $0 \le \theta \le 2\pi$. Taking $u = \phi$ and $v = \theta$ in Equation (1) gives the parametrization

$$\mathbf{r}(\phi, \theta) = (a \sin \phi \cos \theta)\mathbf{i} + (a \sin \phi \sin \theta)\mathbf{j} + (a \cos \phi)\mathbf{k},$$
$$0 \le \phi \le \pi, \quad 0 \le \pi \le 2\pi,$$

Again, the parametrization is one-to-one on the interior of the domain R, though not on its boundary "poles" where $\phi = 0$ or $\phi = \pi$. ∎

EXAMPLE 3 Find a parametrization of the cylinder

$$x^2 + (y - 3)^2 = 9, \quad 0 \le z \le 5.$$

Solution In cylindrical coordinates, a point (x, y, z) has $x = r \cos \theta$, $y = r \sin \theta$, and $z = z$. For points on the cylinder $x^2 + (y - 3)^2 = 9$ (Figure 16.42), the equation is the same as the polar equation for the cylinder's base in the xy-plane:

$$x^2 + (y^2 - 6y + 9) = 9$$
$$r^2 - 6r \sin \theta = 0 \qquad {\scriptstyle x^2 + y^2 = r^2,\; y = r \sin \theta}$$

or

$$r = 6 \sin \theta, \quad 0 \le \theta \le \pi.$$

A typical point on the cylinder therefore has

$$x = r \cos \theta = 6 \sin \theta \cos \theta = 3 \sin 2\theta$$
$$y = r \sin \theta = 6 \sin^2 \theta$$
$$z = z.$$

Taking $u = \theta$ and $v = z$ in Equation (1) gives the one-to-one parametrization

$$\mathbf{r}(\theta, z) = (3 \sin 2\theta)\mathbf{i} + (6 \sin^2 \theta)\mathbf{j} + z\mathbf{k}, \quad 0 \le \theta \le \pi, \quad 0 \le z \le 5. \quad ∎$$

Surface Area

Our goal is to find a double integral for calculating the area of a curved surface S based on the parametrization

$$\mathbf{r}(u, v) = f(u, v)\mathbf{i} + g(u, v)\mathbf{j} + h(u, v)\mathbf{k}, \quad a \le u \le b, \quad c \le v \le d.$$

We need S to be smooth for the construction we are about to carry out. The definition of smoothness involves the partial derivatives of $\mathbf{r}$ with respect to u and v:

$$\mathbf{r}_u = \frac{\partial \mathbf{r}}{\partial u} = \frac{\partial f}{\partial u}\mathbf{i} + \frac{\partial g}{\partial u}\mathbf{j} + \frac{\partial h}{\partial u}\mathbf{k}$$

$$\mathbf{r}_v = \frac{\partial \mathbf{r}}{\partial v} = \frac{\partial f}{\partial v}\mathbf{i} + \frac{\partial g}{\partial v}\mathbf{j} + \frac{\partial h}{\partial v}\mathbf{k}.$$

> **DEFINITION** A parametrized surface $\mathbf{r}(u, v) = f(u, v)\mathbf{i} + g(u, v)\mathbf{j} + h(u, v)\mathbf{k}$ is **smooth** if $\mathbf{r}_u$ and $\mathbf{r}_v$ are continuous and $\mathbf{r}_u \times \mathbf{r}_v$ is never zero on the interior of the parameter domain.

The condition that $\mathbf{r}_u \times \mathbf{r}_v$ is never the zero vector in the definition of smoothness means that the two vectors $\mathbf{r}_u$ and $\mathbf{r}_v$ are nonzero and never lie along the same line, so they always determine a plane tangent to the surface. We relax this condition on the boundary of the domain, but this does not affect the area computations.

Now consider a small rectangle ΔA_{uv} in R with sides on the lines $u = u_0$, $u = u_0 + \Delta u$, $v = v_0$, and $v = v_0 + \Delta v$ (Figure 16.43). Each side of ΔA_{uv} maps to a curve on the surface S, and together these four curves bound a "curved patch element" $\Delta \sigma_{uv}$. In the notation of the figure, the side $v = v_0$ maps to curve C_1, the side $u = u_0$ maps to C_2, and their common vertex (u_0, v_0) maps to P_0.

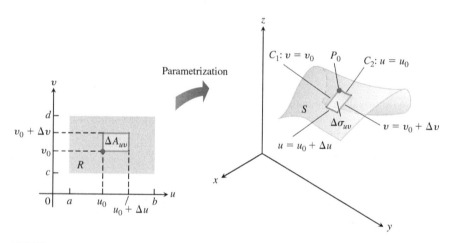

FIGURE 16.43 A rectangular area element ΔA_{uv} in the uv-plane maps onto a curved patch element $\Delta \sigma_{uv}$ on S.

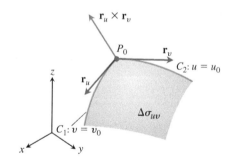

FIGURE 16.44 A magnified view of a surface patch element $\Delta \sigma_{uv}$.

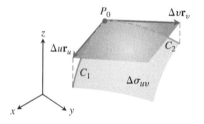

FIGURE 16.45 The area of the parallelogram determined by the vectors $\Delta u\, \mathbf{r}_u$ and $\Delta v\, \mathbf{r}_v$ approximates the area of the surface patch element $\Delta \sigma_{uv}$.

Figure 16.44 shows an enlarged view of $\Delta \sigma_{uv}$. The partial derivative vector $\mathbf{r}_u(u_0, v_0)$ is tangent to C_1 at P_0. Likewise, $\mathbf{r}_v(u_0, v_0)$ is tangent to C_2 at P_0. The cross product $\mathbf{r}_u \times \mathbf{r}_v$ is normal to the surface at P_0. (Here is where we begin to use the assumption that S is smooth. We want to be sure that $\mathbf{r}_u \times \mathbf{r}_v \neq \mathbf{0}$.)

We next approximate the surface patch element $\Delta \sigma_{uv}$ by the parallelogram on the tangent plane whose sides are determined by the vectors $\Delta u\, \mathbf{r}_u$ and $\Delta v\, \mathbf{r}_v$ (Figure 16.45). The area of this parallelogram is

$$\left| \Delta u\, \mathbf{r}_u \times \Delta v\, \mathbf{r}_v \right| = \left| \mathbf{r}_u \times \mathbf{r}_v \right| \Delta u\, \Delta v. \qquad (2)$$

A partition of the region R in the uv-plane by rectangular regions ΔA_{uv} induces a partition of the surface S into surface patch elements $\Delta \sigma_{uv}$. We approximate the area of each surface patch element $\Delta\sigma_{uv}$ by the parallelogram area in Equation (2) and sum these areas together to obtain an approximation of the surface area of S:

$$\sum_{n} \left| \mathbf{r}_u \times \mathbf{r}_v \right| \Delta u \, \Delta v. \tag{3}$$

As Δu and Δv approach zero independently, the number of area elements n tends to ∞ and the continuity of $\mathbf{r}_u$ and $\mathbf{r}_v$ guarantees that the sum in Equation (3) approaches the double integral $\int_c^d \int_a^b \left| \mathbf{r}_u \times \mathbf{r}_v \right| du \, dv$. This double integral over the region R defines the area of the surface S.

DEFINITION   The **area** of the smooth surface

$$\mathbf{r}(u, v) = f(u, v)\mathbf{i} + g(u, v)\mathbf{j} + h(u, v)\mathbf{k}, \qquad a \le u \le b, \quad c \le v \le d$$

is

$$A = \iint\limits_R \left| \mathbf{r}_u \times \mathbf{r}_v \right| dA = \int_c^d \int_a^b \left| \mathbf{r}_u \times \mathbf{r}_v \right| du \, dv. \tag{4}$$

We can abbreviate the integral in Equation (4) by writing $d\sigma$ for $\left| \mathbf{r}_u \times \mathbf{r}_v \right| du \, dv$. The surface area differential $d\sigma$ is analogous to the arc length differential ds in Section 13.3.

Surface Area Differential for a Parametrized Surface

$$d\sigma = \left| \mathbf{r}_u \times \mathbf{r}_v \right| du \, dv \qquad\qquad \iint\limits_S d\sigma \tag{5}$$

Surface area differential, also   Differential formula
called surface area element   for surface area

EXAMPLE 4   Find the surface area of the cone in Example 1 (Figure 16.40).

Solution   In Example 1, we found the parametrization

$$\mathbf{r}(r, \theta) = (r \cos \theta)\mathbf{i} + (r \sin \theta)\mathbf{j} + r\mathbf{k}, \qquad 0 \le r \le 1, \quad 0 \le \theta \le 2\pi.$$

To apply Equation (4), we first find $\mathbf{r}_r \times \mathbf{r}_\theta$:

$$\mathbf{r}_r \times \mathbf{r}_\theta = \begin{vmatrix} \mathbf{i} & \mathbf{j} & \mathbf{k} \\ \cos \theta & \sin \theta & 1 \\ -r \sin \theta & r \cos \theta & 0 \end{vmatrix}$$

$$= -(r \cos \theta)\mathbf{i} - (r \sin \theta)\mathbf{j} + \underbrace{(r \cos^2 \theta + r \sin^2 \theta)}_{r}\mathbf{k}.$$

Thus, $\left| \mathbf{r}_r \times \mathbf{r}_\theta \right| = \sqrt{r^2 \cos^2 \theta + r^2 \sin^2 \theta + r^2} = \sqrt{2r^2} = \sqrt{2}\,r$. The area of the cone is

$$A = \int_0^{2\pi} \int_0^1 \left| \mathbf{r}_r \times \mathbf{r}_\theta \right| dr \, d\theta \qquad \text{Eq. (4) with } u = r, v = \theta$$

$$= \int_0^{2\pi} \int_0^1 \sqrt{2}\, r \, dr \, d\theta = \int_0^{2\pi} \frac{\sqrt{2}}{2} d\theta = \frac{\sqrt{2}}{2}(2\pi) = \pi\sqrt{2} \text{ square units.} \qquad \blacksquare$$

EXAMPLE 5   Find the surface area of a sphere of radius a.

Solution   We use the parametrization from Example 2:

$$\mathbf{r}(\phi, \theta) = (a \sin \phi \cos \theta)\mathbf{i} + (a \sin \phi \sin \theta)\mathbf{j} + (a \cos \phi)\mathbf{k},$$
$$0 \leq \phi \leq \pi, \quad 0 \leq \theta \leq 2\pi.$$

For $\mathbf{r}_\phi \times \mathbf{r}_\theta$, we get

$$\mathbf{r}_\phi \times \mathbf{r}_\theta = \begin{vmatrix} \mathbf{i} & \mathbf{j} & \mathbf{k} \\ a \cos \phi \cos \theta & a \cos \phi \sin \theta & -a \sin \phi \\ -a \sin \phi \sin \theta & a \sin \phi \cos \theta & 0 \end{vmatrix}$$

$$= (a^2 \sin^2 \phi \cos \theta)\mathbf{i} + (a^2 \sin^2 \phi \sin \theta)\mathbf{j} + (a^2 \sin \phi \cos \phi)\mathbf{k}.$$

Thus,

$$|\mathbf{r}_\phi \times \mathbf{r}_\theta| = \sqrt{a^4 \sin^4 \phi \cos^2 \theta + a^4 \sin^4 \phi \sin^2 \theta + a^4 \sin^2 \phi \cos^2 \phi}$$

$$= \sqrt{a^4 \sin^4 \phi + a^4 \sin^2 \phi \cos^2 \phi} = \sqrt{a^4 \sin^2 \phi (\sin^2 \phi + \cos^2 \phi)}$$

$$= a^2 \sqrt{\sin^2 \phi} = a^2 \sin \phi,$$

since $\sin \phi \geq 0$ for $0 \leq \phi \leq \pi$. Therefore, the area of the sphere is

$$A = \int_0^{2\pi} \int_0^{\pi} a^2 \sin \phi \, d\phi \, d\theta$$

$$= \int_0^{2\pi} \left[-a^2 \cos \phi \right]_0^{\pi} d\theta = \int_0^{2\pi} 2a^2 \, d\theta = 4\pi a^2 \quad \text{square units.}$$

This gives the well-known formula for the surface area of a sphere.

EXAMPLE 6   Let S be the "football" surface formed by rotating the curve $x = \cos z$, $y = 0$, $-\pi/2 \leq z \leq \pi/2$ around the z-axis (see Figure 16.46). Find a parametrization for S and compute its surface area.

Solution   Example 2 suggests finding a parametrization of S based on its rotation around the z-axis. If we rotate a point $(x, 0, z)$ on the curve $x = \cos z$, $y = 0$ about the z-axis, we obtain a circle at height z above the xy-plane that is centered on the z-axis and has radius $r = \cos z$ (see Figure 16.46). The point sweeps out the circle through an angle of rotation θ, $0 \leq \theta \leq 2\pi$. We let (x, y, z) be an arbitrary point on this circle, and define the parameters $u = z$ and $v = \theta$. Then we have $x = r \cos \theta = \cos u \cos v$, $y = r \sin \theta = \cos u \sin v$, and $z = u$ giving a parametrization for S as

$$\mathbf{r}(u, v) = \cos u \cos v \, \mathbf{i} + \cos u \sin v \, \mathbf{j} + u \mathbf{k}, \quad -\frac{\pi}{2} \leq u \leq \frac{\pi}{2}, \quad 0 \leq v \leq 2\pi.$$

Next we use Equation (5) to find the surface area of S. Differentiation of the parametrization gives

$$\mathbf{r}_u = -\sin u \cos v \, \mathbf{i} - \sin u \sin v \, \mathbf{j} + \mathbf{k}$$

and

$$\mathbf{r}_v = -\cos u \sin v \, \mathbf{i} + \cos u \cos v \, \mathbf{j}.$$

Computing the cross product we have

$$\mathbf{r}_u \times \mathbf{r}_v = \begin{vmatrix} \mathbf{i} & \mathbf{j} & \mathbf{k} \\ -\sin u \cos v & -\sin u \sin v & 1 \\ -\cos u \sin v & \cos u \cos v & 0 \end{vmatrix}$$

$$= -\cos u \cos v \, \mathbf{i} - \cos u \sin v \, \mathbf{j} - (\sin u \cos u \cos^2 v + \cos u \sin u \sin^2 v)\mathbf{k}.$$

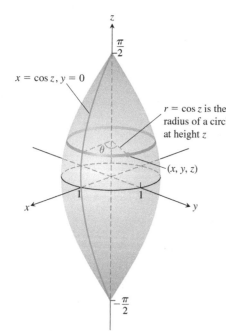

$x = \cos z, y = 0$

$r = \cos z$ is the radius of a circle at height z

(x, y, z)

FIGURE 16.46 The "football" surface in Example 6 obtained by rotating the curve $x = \cos z$ about the z-axis.

Taking the magnitude of the cross product gives

$$|\mathbf{r}_u \times \mathbf{r}_v| = \sqrt{\cos^2 u \, (\cos^2 v + \sin^2 v) + \sin^2 u \cos^2 u}$$

$$= \sqrt{\cos^2 u \, (1 + \sin^2 u)}$$

$$= \cos u \, \sqrt{1 + \sin^2 u}. \qquad \cos u \geq 0 \text{ for } -\frac{\pi}{2} \leq u \leq \frac{\pi}{2}$$

From Equation (4) the surface area is given by the integral

$$A = \int_0^{2\pi} \int_{-\pi/2}^{\pi/2} \cos u \, \sqrt{1 + \sin^2 u} \, du \, dv.$$

To evaluate the integral, we substitute $w = \sin u$ and $dw = \cos u \, du$, $-1 \leq w \leq 1$. Since the surface S is symmetric across the xy-plane, we need only integrate with respect to w from 0 to 1, and multiply the result by 2. In summary, we have

$$A = 2 \int_0^{2\pi} \int_0^1 \sqrt{1 + w^2} \, dw \, dv$$

$$= 2 \int_0^{2\pi} \left[\frac{w}{2} \sqrt{1 + w^2} + \frac{1}{2} \ln \left(w + \sqrt{1 + w^2} \right) \right]_0^1 dv \qquad \text{Integral Table Formula 35}$$

$$= \int_0^{2\pi} 2 \left[\frac{1}{2} \sqrt{2} + \frac{1}{2} \ln \left(1 + \sqrt{2} \right) \right] dv$$

$$= 2\pi \left[\sqrt{2} + \ln \left(1 + \sqrt{2} \right) \right].$$

Implicit Surfaces

Surfaces are often presented as level sets of a function, described by an equation such as

$$F(x, y, z) = c,$$

for some constant c. Such a level surface does not come with an explicit parametrization, and is called an *implicitly defined surface*. Implicit surfaces arise, for example, as equipotential surfaces in electric or gravitational fields. Figure 16.47 shows a piece of such a surface. It may be difficult to find explicit formulas for the functions f, g, and h that describe the surface in the form $\mathbf{r}(u, v) = f(u, v)\mathbf{i} + g(u, v)\mathbf{j} + h(u, v)\mathbf{k}$. We now show how to compute the surface area differential $d\sigma$ for implicit surfaces.

Figure 16.47 shows a piece of an implicit surface S that lies above its "shadow" region R in the plane beneath it. The surface is defined by the equation $F(x, y, z) = c$ and we choose $\mathbf{p}$ to be a unit vector normal to the plane region R. We assume that the surface is **smooth** (F is differentiable and ∇F is nonzero and continuous on S) and that $\nabla F \cdot \mathbf{p} \neq 0$, so the surface never folds back over itself.

Assume that the normal vector $\mathbf{p}$ is the unit vector $\mathbf{k}$, so the region R in Figure 16.47 lies in the xy-plane. By assumption, we then have $\nabla F \cdot \mathbf{p} = \nabla F \cdot \mathbf{k} = F_z \neq 0$ on S. The Implicit Function Theorem (see Section 14.4) implies that S is then the graph of a differentiable function $z = h(x, y)$, although the function $h(x, y)$ is not explicitly known. Define the parameters u and v by $u = x$ and $v = y$. Then $z = h(u, v)$ and

$$\mathbf{r}(u, v) = u\mathbf{i} + v\mathbf{j} + h(u, v)\mathbf{k} \qquad (6)$$

gives a parametrization of the surface S. We use Equation (4) to find the area of S.

Calculating the partial derivatives of $\mathbf{r}$, we find

$$\mathbf{r}_u = \mathbf{i} + \frac{\partial h}{\partial u} \mathbf{k} \qquad \text{and} \qquad \mathbf{r}_v = \mathbf{j} + \frac{\partial h}{\partial v} \mathbf{k}.$$

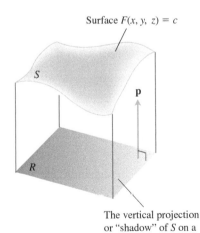

Surface $F(x, y, z) = c$

The vertical projection or "shadow" of S on a coordinate plane

FIGURE 16.47 As we soon see, the area of a surface S in space can be calculated by evaluating a related double integral over the vertical projection or "shadow" of S on a coordinate plane. The unit vector $\mathbf{p}$ is normal to the plane.

Applying the Chain Rule for implicit differentiation (see Equation (2) in Section 14.4) to $F(x, y, z) = c$, where $x = u, y = v$, and $z = h(u, v)$, we obtain the partial derivatives

$$\frac{\partial h}{\partial u} = -\frac{F_x}{F_z} \quad \text{and} \quad \frac{\partial h}{\partial v} = -\frac{F_y}{F_z}. \qquad F_z \neq 0$$

Substitution of these derivatives into the derivatives of **r** gives

$$\mathbf{r}_u = \mathbf{i} - \frac{F_x}{F_z}\mathbf{k} \quad \text{and} \quad \mathbf{r}_v = \mathbf{j} - \frac{F_y}{F_z}\mathbf{k}.$$

From a routine calculation of the cross product we find

$$\mathbf{r}_u \times \mathbf{r}_v = \frac{F_x}{F_z}\mathbf{i} + \frac{F_y}{F_z}\mathbf{j} + \mathbf{k} \qquad \begin{vmatrix} \mathbf{i} & \mathbf{j} & \mathbf{k} \\ 1 & 0 & -F_x/F_z \\ 0 & 1 & -F_y/F_z \end{vmatrix} \begin{array}{l} \text{cross product of} \\ \mathbf{r}_u \\ \mathbf{r}_v \end{array}$$

$$= \frac{1}{F_z}(F_x\mathbf{i} + F_y\mathbf{j} + F_z\mathbf{k})$$

$$= \frac{\nabla F}{F_z} = \frac{\nabla F}{\nabla F \cdot \mathbf{k}}$$

$$= \frac{\nabla F}{\nabla F \cdot \mathbf{p}}. \qquad \mathbf{p} = \mathbf{k}$$

Therefore, the surface area differential is given by

$$d\sigma = |\mathbf{r}_u \times \mathbf{r}_v|\, du\, dv = \frac{|\nabla F|}{|\nabla F \cdot \mathbf{p}|}\, dx\, dy. \qquad u = x \text{ and } v = y$$

We obtain similar calculations if instead the vector $\mathbf{p} = \mathbf{j}$ is normal to the xz-plane when $F_y \neq 0$ on S, or if $\mathbf{p} = \mathbf{i}$ is normal to the yz-plane when $F_x \neq 0$ on S. Combining these results with Equation (4) then gives the following general formula.

Formula for the Surface Area of an Implicit Surface
The area of the surface $F(x, y, z) = c$ over a closed and bounded plane region R is

$$\text{Surface area} = \iint\limits_R \frac{|\nabla F|}{|\nabla F \cdot \mathbf{p}|}\, dA, \qquad (7)$$

where $\mathbf{p} = \mathbf{i}, \mathbf{j},$ or $\mathbf{k}$ is normal to R and $\nabla F \cdot \mathbf{p} \neq 0$.

Thus, the area is the double integral over R of the magnitude of ∇F divided by the magnitude of the scalar component of ∇F normal to R.

We reached Equation (7) under the assumption that $\nabla F \cdot \mathbf{p} \neq 0$ throughout R and that ∇F is continuous. Whenever the integral exists, however, we define its value to be the area of the portion of the surface $F(x, y, z) = c$ that lies over R. (Recall that the projection is assumed to be one-to-one.)

EXAMPLE 7 Find the area of the surface cut from the bottom of the paraboloid $x^2 + y^2 - z = 0$ by the plane $z = 4$.

Solution We sketch the surface S and the region R below it in the xy-plane (Figure 16.48). The surface S is part of the level surface $F(x, y, z) = x^2 + y^2 - z = 0$, and R is the disk $x^2 + y^2 \leq 4$ in the xy-plane. To get a unit vector normal to the plane of R, we can take $\mathbf{p} = \mathbf{k}$.

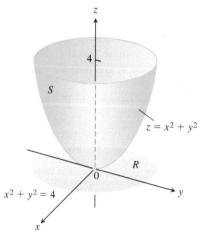

FIGURE 16.48 The area of this parabolic surface is calculated in Example 7.

At any point (x, y, z) on the surface, we have

$$F(x, y, z) = x^2 + y^2 - z$$

$$\nabla F = 2x\mathbf{i} + 2y\mathbf{j} - \mathbf{k}$$

$$|\nabla F| = \sqrt{(2x)^2 + (2y)^2 + (-1)^2}$$

$$= \sqrt{4x^2 + 4y^2 + 1}$$

$$|\nabla F \cdot \mathbf{p}| = |\nabla F \cdot \mathbf{k}| = |-1| = 1.$$

In the region R, $dA = dx \, dy$. Therefore,

$$\text{Surface area} = \iint\limits_{R} \frac{|\nabla F|}{|\nabla F \cdot \mathbf{p}|} dA \qquad\qquad \text{Eq. (7)}$$

$$= \iint\limits_{x^2+y^2\le 4} \sqrt{4x^2 + 4y^2 + 1} \, dx \, dy$$

$$= \int_0^{2\pi} \int_0^2 \sqrt{4r^2 + 1} \; r \, dr \, d\theta \qquad\qquad \text{Polar coordinates}$$

$$= \int_0^{2\pi} \left[\frac{1}{12}(4r^2 + 1)^{3/2} \right]_0^2 d\theta$$

$$= \int_0^{2\pi} \frac{1}{12}(17^{3/2} - 1) \, d\theta = \frac{\pi}{6}\left(17\sqrt{17} - 1\right). \qquad\blacksquare$$

Example 7 illustrates how to find the surface area for a function $z = f(x, y)$ over a region R in the xy-plane. Actually, the surface area differential can be obtained in two ways, and we show this in the next example.

EXAMPLE 8   Derive the surface area differential $d\sigma$ of the surface $z = f(x, y)$ over a region R in the xy-plane **(a)** parametrically using Equation (5), and **(b)** implicitly, as in Equation (7).

Solution

(a) We parametrize the surface by taking $x = u$, $y = v$, and $z = f(x, y)$ over R. This gives the parametrization

$$\mathbf{r}(u, v) = u\mathbf{i} + v\mathbf{j} + f(u, v)\mathbf{k}.$$

Computing the partial derivatives gives $\mathbf{r}_u = \mathbf{i} + f_u\mathbf{k}$, $\mathbf{r}_v = \mathbf{j} + f_v\mathbf{k}$ and

$$\mathbf{r}_u \times \mathbf{r}_v = -f_u\mathbf{i} - f_v\mathbf{j} + \mathbf{k}. \qquad \begin{vmatrix} \mathbf{i} & \mathbf{j} & \mathbf{k} \\ 1 & 0 & f_u \\ 0 & 1 & f_v \end{vmatrix}$$

Then $|\mathbf{r}_u \times \mathbf{r}_v| \, du \, dv = \sqrt{f_u^2 + f_v^2 + 1} \, du \, dv$. Substituting for u and v then gives the surface area differential

$$d\sigma = \sqrt{f_x^2 + f_y^2 + 1} \, dx \, dy.$$

(b) We define the implicit function $F(x, y, z) = f(x, y) - z$. Since (x, y) belongs to the region R, the unit normal to the plane of R is $\mathbf{p} = \mathbf{k}$. Then $\nabla F = f_x\mathbf{i} + f_y\mathbf{j} - \mathbf{k}$ so that $|\nabla F \cdot \mathbf{p}| = |-1| = 1$, $|\nabla F| = \sqrt{f_x{}^2 + f_y{}^2 + 1}$, and $|\nabla F| / |\nabla F \cdot \mathbf{p}| = |\nabla F|$. The surface area differential is again given by

$$d\sigma = \sqrt{f_x{}^2 + f_y{}^2 + 1}\, dx\, dy.$$

The surface area differential derived in Example 8 gives the following formula for calculating the surface area of the graph of a function defined explicitly as $z = f(x, y)$.

Formula for the Surface Area of a Graph $z = f(x, y)$

For a graph $z = f(x, y)$ over a region R in the xy-plane, the surface area formula is

$$A = \iint_R \sqrt{f_x{}^2 + f_y{}^2 + 1}\, dx\, dy. \tag{8}$$

EXERCISES 16.5

Finding Parametrizations

In Exercises 1–16, find a parametrization of the surface. (There are many correct ways to do these, so your answers may not be the same as those in the back of the book.)

1. The paraboloid $z = x^2 + y^2$, $z \le 4$

2. The paraboloid $z = 9 - x^2 - y^2$, $z \ge 0$

3. **Cone frustum** The first-octant portion of the cone $z = \sqrt{x^2 + y^2}/2$ between the planes $z = 0$ and $z = 3$

4. **Cone frustum** The portion of the cone $z = 2\sqrt{x^2 + y^2}$ between the planes $z = 2$ and $z = 4$

5. **Spherical cap** The cap cut from the sphere $x^2 + y^2 + z^2 = 9$ by the cone $z = \sqrt{x^2 + y^2}$

6. **Spherical cap** The portion of the sphere $x^2 + y^2 + z^2 = 4$ in the first octant between the xy-plane and the cone $z = \sqrt{x^2 + y^2}$

7. **Spherical band** The portion of the sphere $x^2 + y^2 + z^2 = 3$ between the planes $z = \sqrt{3}/2$ and $z = -\sqrt{3}/2$

8. **Spherical cap** The upper portion cut from the sphere $x^2 + y^2 + z^2 = 8$ by the plane $z = -2$

9. **Parabolic cylinder between planes** The surface cut from the parabolic cylinder $z = 4 - y^2$ by the planes $x = 0$, $x = 2$, and $z = 0$

10. **Parabolic cylinder between planes** The surface cut from the parabolic cylinder $y = x^2$ by the planes $z = 0$, $z = 3$, and $y = 2$

11. **Circular cylinder band** The portion of the cylinder $y^2 + z^2 = 9$ between the planes $x = 0$ and $x = 3$

12. **Circular cylinder band** The portion of the cylinder $x^2 + z^2 = 4$ above the xy-plane between the planes $y = -2$ and $y = 2$

13. **Tilted plane inside cylinder** The portion of the plane $x + y + z = 1$
   a. Inside the cylinder $x^2 + y^2 = 9$
   b. Inside the cylinder $y^2 + z^2 = 9$

14. **Tilted plane inside cylinder** The portion of the plane $x - y + 2z = 2$
   a. Inside the cylinder $x^2 + z^2 = 3$
   b. Inside the cylinder $y^2 + z^2 = 2$

15. **Circular cylinder band** The portion of the cylinder $(x - 2)^2 + z^2 = 4$ between the planes $y = 0$ and $y = 3$

16. **Circular cylinder band** The portion of the cylinder $y^2 + (z - 5)^2 = 25$ between the planes $x = 0$ and $x = 10$

Surface Area of Parametrized Surfaces

In Exercises 17–26, use a parametrization to express the area of the surface as a double integral. Then evaluate the integral. (There are many correct ways to set up the integrals, so your integrals may not be the same as those in the back of the book. They should have the same values, however.)

17. **Tilted plane inside cylinder** The portion of the plane $y + 2z = 2$ inside the cylinder $x^2 + y^2 = 1$

18. **Plane inside cylinder** The portion of the plane $z = -x$ inside the cylinder $x^2 + y^2 = 4$

19. **Cone frustum** The portion of the cone $z = 2\sqrt{x^2 + y^2}$ between the planes $z = 2$ and $z = 6$

20. **Cone frustum** The portion of the cone $z = \sqrt{x^2 + y^2}/3$ between the planes $z = 1$ and $z = 4/3$

21. **Circular cylinder band** The portion of the cylinder $x^2 + y^2 = 1$ between the planes $z = 1$ and $z = 4$

22. **Circular cylinder band** The portion of the cylinder $x^2 + z^2 = 10$ between the planes $y = -1$ and $y = 1$

23. **Parabolic cap** The cap cut from the paraboloid $z = 2 - x^2 - y^2$ by the cone $z = \sqrt{x^2 + y^2}$

24. **Parabolic band** The portion of the paraboloid $z = x^2 + y^2$ between the planes $z = 1$ and $z = 4$

25. **Sawed-off sphere** The lower portion cut from the sphere $x^2 + y^2 + z^2 = 2$ by the cone $z = \sqrt{x^2 + y^2}$

26. **Spherical band** The portion of the sphere $x^2 + y^2 + z^2 = 4$ between the planes $z = -1$ and $z = \sqrt{3}$

Planes Tangent to Parametrized Surfaces

The tangent plane at a point $P_0(f(u_0, v_0), g(u_0, v_0), h(u_0, v_0))$ on a parametrized surface $\mathbf{r}(u, v) = f(u, v)\mathbf{i} + g(u, v)\mathbf{j} + h(u, v)\mathbf{k}$ is the plane through P_0 normal to the vector $\mathbf{r}_u(u_0, v_0) \times \mathbf{r}_v(u_0, v_0)$, the cross product of the tangent vectors $\mathbf{r}_u(u_0, v_0)$ and $\mathbf{r}_v(u_0, v_0)$ at P_0. In Exercises 27–30, find an equation for the plane tangent to the surface at P_0. Then find a Cartesian equation for the surface and sketch the surface and tangent plane together.

27. Cone The cone $\mathbf{r}(r, \theta) = (r \cos \theta)\mathbf{i} + (r \sin \theta)\mathbf{j} + r\mathbf{k}, r \geq 0,$ $0 \leq \theta \leq 2\pi$ at the point $P_0(\sqrt{2}, \sqrt{2}, 2)$ corresponding to $(r, \theta) = (2, \pi/4)$

28. Hemisphere The hemisphere surface $\mathbf{r}(\phi, \theta) = (4 \sin \phi \cos \theta)\mathbf{i} + (4 \sin \phi \sin \theta)\mathbf{j} + (4 \cos \phi)\mathbf{k}, 0 \leq \phi \leq \pi/2, 0 \leq \theta \leq 2\pi,$ at the point $P_0(\sqrt{2}, \sqrt{2}, 2\sqrt{3})$ corresponding to $(\phi, \theta) = (\pi/6, \pi/4)$

29. Circular cylinder The circular cylinder $\mathbf{r}(\theta, z) = (3 \sin 2\theta)\mathbf{i} + (6 \sin^2 \theta)\mathbf{j} + z\mathbf{k}, 0 \leq \theta \leq \pi,$ at the point $P_0(3\sqrt{3}/2, 9/2, 0)$ corresponding to $(\theta, z) = (\pi/3, 0)$ (See Example 3.)

30. Parabolic cylinder The parabolic cylinder surface $\mathbf{r}(x, y) = x\mathbf{i} + y\mathbf{j} - x^2\mathbf{k}, -\infty < x < \infty, -\infty < y < \infty,$ at the point $P_0(1, 2, -1)$ corresponding to $(x, y) = (1, 2)$

More Parametrizations of Surfaces

31. a. A *torus of revolution* (doughnut) is obtained by rotating a circle C in the xz-plane about the z-axis in space. (See the accompanying figure.) If C has radius $r > 0$ and center $(R, 0, 0)$, show that a parametrization of the torus is

$$\mathbf{r}(u, v) = ((R + r \cos u)\cos v)\mathbf{i}$$
$$+ ((R + r \cos u)\sin v)\mathbf{j} + (r \sin u)\mathbf{k},$$

where $0 \leq u \leq 2\pi$ and $0 \leq v \leq 2\pi$ are the angles in the figure.

b. Show that the surface area of the torus is $A = 4\pi^2 Rr$.

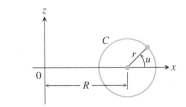

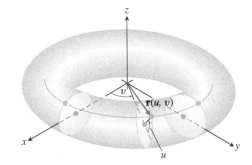

32. Parametrization of a surface of revolution Suppose that the parametrized curve C: $(f(u), g(u))$ is revolved about the x-axis, where $g(u) > 0$ for $a \leq u \leq b$.

a. Show that

$$\mathbf{r}(u, v) = f(u)\mathbf{i} + (g(u)\cos v)\mathbf{j} + (g(u)\sin v)\mathbf{k}$$

is a parametrization of the resulting surface of revolution, where $0 \leq v \leq 2\pi$ is the angle from the xy-plane to the point $\mathbf{r}(u, v)$ on the surface. (See the accompanying figure.) Notice that $f(u)$ measures distance *along* the axis of revolution and $g(u)$ measures distance *from* the axis of revolution.

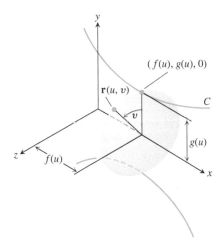

b. Find a parametrization for the surface obtained by revolving the curve $x = y^2, y \geq 0$, about the x-axis.

33. a. Parametrization of an ellipsoid The parametrization $x = a \cos \theta, y = b \sin \theta, 0 \leq \theta \leq 2\pi$ gives the ellipse $(x^2/a^2) + (y^2/b^2) = 1$. Using the angles θ and ϕ in spherical coordinates, show that

$$\mathbf{r}(\theta, \phi) = (a \cos \theta \sin \phi)\mathbf{i} + (b \sin \theta \sin \phi)\mathbf{j} + (c \cos \phi)\mathbf{k}$$

is a parametrization of the ellipsoid $(x^2/a^2) + (y^2/b^2) + (z^2/c^2) = 1$.

b. Write an integral for the surface area of the ellipsoid, but do not evaluate the integral.

34. Hyperboloid of one sheet

a. Find a parametrization for the hyperboloid of one sheet $x^2 + y^2 - z^2 = 1$ in terms of the angle θ associated with the circle $x^2 + y^2 = r^2$ and the hyperbolic parameter u associated with the hyperbolic function $r^2 - z^2 = 1$. (*Hint*: $\cosh^2 u - \sinh^2 u = 1$.)

b. Generalize the result in part (a) to the hyperboloid $(x^2/a^2) + (y^2/b^2) - (z^2/c^2) = 1$.

35. (*Continuation of Exercise 34.*) Find a Cartesian equation for the plane tangent to the hyperboloid $x^2 + y^2 - z^2 = 25$ at the point $(x_0, y_0, 0)$, where $x_0^2 + y_0^2 = 25$.

36. Hyperboloid of two sheets Find a parametrization of the hyperboloid of two sheets $(z^2/c^2) - (x^2/a^2) - (y^2/b^2) = 1$.

Surface Area for Implicit and Explicit Forms

37. Find the area of the surface cut from the paraboloid $x^2 + y^2 - z = 0$ by the plane $z = 2$.

38. Find the area of the band cut from the paraboloid $x^2 + y^2 - z = 0$ by the planes $z = 2$ and $z = 6$.

39. Find the area of the region cut from the plane $x + 2y + 2z = 5$ by the cylinder whose walls are $x = y^2$ and $x = 2 - y^2$.

40. Find the area of the portion of the surface $x^2 - 2z = 0$ that lies above the triangle bounded by the lines $x = \sqrt{3}, y = 0$, and $y = x$ in the xy-plane.

41. Find the area of the surface $x^2 - 2y - 2z = 0$ that lies above the triangle bounded by the lines $x = 2, y = 0$, and $y = 3x$ in the xy-plane.

42. Find the area of the cap cut from the sphere $x^2 + y^2 + z^2 = 2$ by the cone $z = \sqrt{x^2 + y^2}$.

43. Find the area of the ellipse cut from the plane $z = cx$ (c a constant) by the cylinder $x^2 + y^2 = 1$.

44. Find the area of the upper portion of the cylinder $x^2 + z^2 = 1$ that lies between the planes $x = \pm 1/2$ and $y = \pm 1/2$.

45. Find the area of the portion of the paraboloid $x = 4 - y^2 - z^2$ that lies above the ring $1 \le y^2 + z^2 \le 4$ in the yz-plane.

46. Find the area of the surface cut from the paraboloid $x^2 + y + z^2 = 2$ by the plane $y = 0$.

47. Find the area of the surface $x^2 - 2 \ln x + \sqrt{15}y - z = 0$ above the square $R: 1 \le x \le 2, 0 \le y \le 1$, in the xy-plane.

48. Find the area of the surface $2x^{3/2} + 2y^{3/2} - 3z = 0$ above the square $R: 0 \le x \le 1, 0 \le y \le 1$, in the xy-plane.

Find the area of the surfaces in Exercises 49–54.

49. The surface cut from the bottom of the paraboloid $z = x^2 + y^2$ by the plane $z = 3$

50. The surface cut from the "nose" of the paraboloid $x = 1 - y^2 - z^2$ by the yz-plane

51. The portion of the cone $z = \sqrt{x^2 + y^2}$ that lies over the region between the circle $x^2 + y^2 = 1$ and the ellipse $9x^2 + 4y^2 = 36$ in the xy-plane. (*Hint:* Use formulas from geometry to find the area of the region.)

52. The triangle cut from the plane $2x + 6y + 3z = 6$ by the bounding planes of the first octant. Calculate the area three ways, using different explicit forms.

53. The surface in the first octant cut from the cylinder $y = (2/3)z^{3/2}$ by the planes $x = 1$ and $y = 16/3$

54. The portion of the plane $y + z = 4$ that lies above the region cut from the first quadrant of the xz-plane by the parabola $x = 4 - z^2$

55. Use the parametrization

$$\mathbf{r}(x, z) = x\mathbf{i} + f(x, z)\mathbf{j} + z\mathbf{k}$$

and Equation (5) to derive a formula for $d\sigma$ associated with the explicit form $y = f(x, z)$.

56. Let S be the surface obtained by rotating the smooth curve $y = f(x), a \le x \le b$, about the x-axis, where $f(x) \ge 0$.

a. Show that the vector function

$$\mathbf{r}(x, \theta) = x\mathbf{i} + f(x) \cos \theta\mathbf{j} + f(x) \sin \theta\mathbf{k}$$

is a parametrization of S, where θ is the angle of rotation around the x-axis (see the accompanying figure).

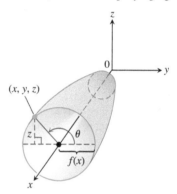

b. Use Equation (4) to show that the surface area of this surface of revolution is given by

$$A = \int_a^b 2\pi f(x)\sqrt{1 + [f'(x)]^2}\,dx.$$

16.6 Surface Integrals

To compute the mass of a surface, the flow of a liquid across a curved membrane, or the total electrical charge on a surface, we need to integrate a function over a curved surface in space. Such a *surface integral* is the two-dimensional extension of the line integral concept used to integrate over a one-dimensional curve. Like line integrals, surface integrals arise in two forms. The first occurs when we integrate a scalar function over a surface, such as integrating a mass density function defined on a surface to find its total mass. This form corresponds to line integrals of scalar functions defined in Section 16.1, and can be used to find the mass of a thin wire. The second form involves surface integrals of vector fields, analogous to the line integrals for vector fields defined in Section 16.2. An example occurs when we want to measure the net flow of a fluid across a surface submerged in the fluid (just as we previously defined the flux of **F** across a curve). In this section we investigate these ideas and their applications.

Surface Integrals

Suppose that the function $G(x, y, z)$ gives the *mass density* (mass per unit area) at each point on a surface S. Then we can calculate the total mass of S as an integral in the following way.

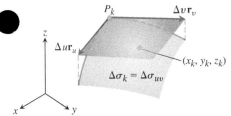

FIGURE 16.49 The area of the patch $\Delta\sigma_k$ is approximated by the area of the tangent parallelogram determined by the vectors $\Delta u\, \mathbf{r}_u$ and $\Delta v\, \mathbf{r}_v$. The point (x_k, y_k, z_k) lies on the surface patch, beneath the parallelogram shown here.

Assume, as in Section 16.5, that the surface S is defined parametrically on a region R in the uv-plane,

$$\mathbf{r}(u, v) = f(u, v)\mathbf{i} + g(u, v)\mathbf{j} + h(u, v)\mathbf{k}, \qquad (u, v) \in R.$$

In Figure 16.49, we see how a subdivision of R (considered as a rectangle for simplicity) divides the surface S into corresponding curved surface elements, or patches, of area

$$\Delta\sigma_{uv} \approx |\mathbf{r}_u \times \mathbf{r}_v|\, du\, dv.$$

As we did for the subdivisions when defining double integrals in Section 15.2, we number the surface element patches in some order with their areas given by $\Delta\sigma_1, \Delta\sigma_2, \ldots, \Delta\sigma_n$. To form a Riemann sum over S, we choose a point (x_k, y_k, z_k) in the kth patch, multiply the value of the function G at that point by the area $\Delta\sigma_k$, and add together the products:

$$\sum_{k=1}^{n} G(x_k, y_k, z_k)\, \Delta\sigma_k.$$

Depending on how we pick (x_k, y_k, z_k) in the kth patch, we may get different values for this Riemann sum. Then we take the limit as the number of surface patches increases, their areas shrink to zero, and both $\Delta u \to 0$ and $\Delta v \to 0$. This limit, whenever it exists independent of all choices made, defines the **surface integral of G over the surface S** as

$$\iint_S G(x, y, z)\, d\sigma = \lim_{n \to \infty} \sum_{k=1}^{n} G(x_k, y_k, z_k)\, \Delta\sigma_k. \qquad (1)$$

Notice the analogy with the definition of the double integral (Section 15.2) and with the line integral (Section 16.1). If S is a piecewise smooth surface, and G is continuous over S, then the surface integral defined by Equation (1) can be shown to exist.

The formula for evaluating the surface integral depends on the manner in which S is described, parametrically, implicitly or explicitly, as discussed in Section 16.5.

Formulas for a Surface Integral of a Scalar Function

1. For a smooth surface S defined **parametrically** as $\mathbf{r}(u, v) = f(u, v)\mathbf{i} + g(u, v)\mathbf{j} + h(u, v)\mathbf{k}$, $(u, v) \in R$, and a continuous function $G(x, y, z)$ defined on S, the surface integral of G over S is given by the double integral over R,

$$\iint_S G(x, y, z)\, d\sigma = \iint_R G(f(u, v), g(u, v), h(u, v))\, |\mathbf{r}_u \times \mathbf{r}_v|\, du\, dv. \qquad (2)$$

2. For a surface S given **implicitly** by $F(x, y, z) = c$, where F is a continuously differentiable function, with S lying above its closed and bounded shadow region R in the coordinate plane beneath it, the surface integral of the continuous function G over S is given by the double integral over R,

$$\iint_S G(x, y, z)\, d\sigma = \iint_R G(x, y, z)\, \frac{|\nabla F|}{|\nabla F \cdot \mathbf{p}|}\, dA, \qquad (3)$$

where $\mathbf{p}$ is a unit vector normal to R and $\nabla F \cdot \mathbf{p} \neq 0$.

3. For a surface S given **explicitly** as the graph of $z = f(x, y)$, where f is a continuously differentiable function over a region R in the xy-plane, the surface integral of the continuous function G over S is given by the double integral over R,

$$\iint_S G(x, y, z)\, d\sigma = \iint_R G(x, y, f(x, y))\, \sqrt{f_x^2 + f_y^2 + 1}\, dx\, dy. \qquad (4)$$

The surface integral in Equation (1) takes on different meanings in different applications. If G has the constant value 1, the integral gives the area of S. If G gives the mass density of a thin shell of material modeled by S, the integral gives the mass of the shell. If G gives the charge density of a thin shell, then the integral gives the total charge.

EXAMPLE 1   Integrate $G(x, y, z) = x^2$ over the cone $z = \sqrt{x^2 + y^2}, 0 \le z \le 1$.

Solution   Using Equation (2) and the calculations from Example 4 in Section 16.5, we have $|\mathbf{r}_r \times \mathbf{r}_\theta| = \sqrt{2}r$ and

$$\iint_S x^2 \, d\sigma = \int_0^{2\pi} \int_0^1 (r^2 \cos^2\theta)(\sqrt{2}r) \, dr \, d\theta \qquad x = r\cos\theta$$

$$= \sqrt{2} \int_0^{2\pi} \int_0^1 r^3 \cos^2\theta \, dr \, d\theta$$

$$= \frac{\sqrt{2}}{4} \int_0^{2\pi} \cos^2\theta \, d\theta = \frac{\sqrt{2}}{4}\left[\frac{\theta}{2} + \frac{1}{4}\sin 2\theta\right]_0^{2\pi} = \frac{\pi\sqrt{2}}{4}. \qquad \blacksquare$$

Surface integrals behave like other double integrals, the integral of the sum of two functions being the sum of their integrals and so on. The domain Additivity Property takes the form

$$\iint_S G \, d\sigma = \iint_{S_1} G \, d\sigma + \iint_{S_2} G \, d\sigma + \cdots + \iint_{S_n} G \, d\sigma.$$

When S is partitioned by smooth curves into a finite number of smooth patches with nonoverlapping interiors (i.e., if S is piecewise smooth), then the integral over S is the sum of the integrals over the patches. Thus, the integral of a function over the surface of a cube is the sum of the integrals over the faces of the cube. We integrate over a turtle shell of welded plates by integrating over one plate at a time and adding the results.

EXAMPLE 2   Integrate $G(x, y, z) = xyz$ over the surface of the cube cut from the first octant by the planes $x = 1, y = 1$, and $z = 1$ (Figure 16.50).

Solution   We integrate xyz over each of the six sides and add the results. Since $xyz = 0$ on the sides that lie in the coordinate planes, the integral over the surface of the cube reduces to

$$\iint_{\substack{\text{Cube} \\ \text{surface}}} xyz \, d\sigma = \iint_{\text{Side } A} xyz \, d\sigma + \iint_{\text{Side } B} xyz \, d\sigma + \iint_{\text{Side } C} xyz \, d\sigma.$$

Side A is the surface $f(x, y, z) = z = 1$ over the square region $R_{xy}: 0 \le x \le 1$, $0 \le y \le 1$, in the xy-plane. For this surface and region,

$$\mathbf{p} = \mathbf{k}, \qquad \nabla f = \mathbf{k}, \qquad |\nabla f| = 1, \qquad |\nabla f \cdot \mathbf{p}| = |\mathbf{k} \cdot \mathbf{k}| = 1$$

$$d\sigma = \frac{|\nabla f|}{|\nabla f \cdot \mathbf{p}|} dA = \frac{1}{1} dx \, dy = dx \, dy \qquad \text{Eq. (3)}$$

$$xyz = xy(1) = xy$$

and

$$\iint_{\text{Side } A} xyz \, d\sigma = \iint_{R_{xy}} xy \, dx \, dy = \int_0^1 \int_0^1 xy \, dx \, dy = \int_0^1 \frac{y}{2} dy = \frac{1}{4}.$$

Symmetry tells us that the integrals of xyz over sides B and C are also $1/4$. Hence,

$$\iint_{\substack{\text{Cube} \\ \text{surface}}} xyz \, d\sigma = \frac{1}{4} + \frac{1}{4} + \frac{1}{4} = \frac{3}{4}.$$

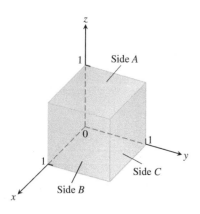

FIGURE 16.50   The cube in Example 2.

EXAMPLE 3 Integrate $G(x, y, z) = \sqrt{1 - x^2 - y^2}$ over the "football" surface S formed by rotating the curve $x = \cos z$, $y = 0, -\pi/2 \le z \le \pi/2$, around the z-axis.

Solution The surface is displayed in Figure 16.46, and in Example 6 of Section 16.5 we found the parametrization

$$x = \cos u \cos v, \quad y = \cos u \sin v, \quad z = u, \quad -\frac{\pi}{2} \le u \le \frac{\pi}{2} \quad \text{and} \quad 0 \le v \le 2\pi,$$

where v represents the angle of rotation from the xz-plane about the z-axis. Substituting this parametrization into the expression for G gives

$$\sqrt{1 - x^2 - y^2} = \sqrt{1 - (\cos^2 u)(\cos^2 v + \sin^2 v)} = \sqrt{1 - \cos^2 u} = |\sin u|.$$

The surface area differential for the parametrization was found to be (Example 6, Section 16.5)

$$d\sigma = \cos u \sqrt{1 + \sin^2 u} \, du \, dv.$$

These calculations give the surface integral

$$\iint_S \sqrt{1 - x^2 - y^2} \, d\sigma = \int_0^{2\pi} \int_{-\pi/2}^{\pi/2} |\sin u| \cos u \sqrt{1 + \sin^2 u} \, du \, dv$$

$$= 2 \int_0^{2\pi} \int_0^{\pi/2} \sin u \cos u \sqrt{1 + \sin^2 u} \, du \, dv$$

$$= \int_0^{2\pi} \int_1^2 \sqrt{w} \, dw \, dv \qquad \begin{array}{l} w = 1 + \sin^2 u, \\ dw = 2 \sin u \cos u \, du \\ \text{When } u = 0, w = 1. \\ \text{When } u = \pi/2, w = 2. \end{array}$$

$$= 2\pi \cdot \frac{2}{3} w^{3/2} \Big]_1^2 = \frac{4\pi}{3}\left(2\sqrt{2} - 1\right). \qquad \blacksquare$$

EXAMPLE 4 Evaluate $\iint_S \sqrt{x(1 + 2z)} \, d\sigma$ on the portion of the cylinder $z = y^2/2$ over the triangular region $R: x \ge 0, y \ge 0, x + y \le 1$ in the xy-plane (Figure 16.51).

Solution The function G on the surface S is given by

$$G(x, y, z) = \sqrt{x(1 + 2z)} = \sqrt{x}\sqrt{1 + y^2}.$$

With $z = f(x, y) = y^2/2$, we use Equation (4) to evaluate the surface integral:

$$d\sigma = \sqrt{f_x^2 + f_y^2 + 1} \, dx \, dy = \sqrt{0 + y^2 + 1} \, dx \, dy$$

and

$$\iint_S G(x, y, z) \, d\sigma = \iint_R \left(\sqrt{x}\sqrt{1 + y^2}\right)\sqrt{1 + y^2} \, dx \, dy$$

$$= \int_0^1 \int_0^{1-x} \sqrt{x}\left(1 + y^2\right) dy \, dx$$

$$= \int_0^1 \sqrt{x}\left[(1 - x) + \frac{1}{3}(1 - x)^3\right] dx \qquad \text{Integrate and evaluate.}$$

$$= \int_0^1 \left(\frac{4}{3}x^{1/2} - 2x^{3/2} + x^{5/2} - \frac{1}{3}x^{7/2}\right) dx \qquad \text{Routine algebra}$$

$$= \left[\frac{8}{9}x^{3/2} - \frac{4}{5}x^{5/2} + \frac{2}{7}x^{7/2} - \frac{2}{27}x^{9/2}\right]_0^1$$

$$= \frac{8}{9} - \frac{4}{5} + \frac{2}{7} - \frac{2}{27} = \frac{284}{945} \approx 0.30. \qquad \blacksquare$$

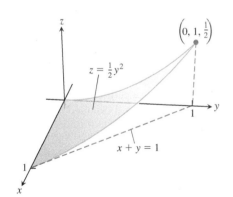

$\left(0, 1, \frac{1}{2}\right)$

$z = \frac{1}{2}y^2$

$x + y = 1$

FIGURE 16.51 The surface S in Example 4.

Orientation of a Surface

A curve C with a parametrization $\mathbf{r}(t)$ has a natural orientation, or direction, that comes from the direction of increasing t. The unit tangent vector $\mathbf{T}$ along C points in this forward direction at each point on the curve. There are two possible orientations for a curve, corresponding to whether we follow the direction of the tangent vector $\mathbf{T}$ at each point, or the direction of $-\mathbf{T}$.

To specify an orientation on a surface in space S, we do something similar, but this time we specify a normal vector at each point on the surface. A parametrization of a surface $\mathbf{r}(u, v)$ gives a vector $\mathbf{r}_u \times \mathbf{r}_v$ that is normal to the surface, and so gives an orientation wherever the parametrization applies. A second choice of orientation is found by taking $-(\mathbf{r}_u \times \mathbf{r}_v)$, giving a vector that points to the opposite side of the surface at each point. In essence, an orientation is a way of consistently choosing one of the two sides of a surface. Not all surfaces have orientations, but a surface that does have one also has a second, opposite orientation.

Each point on the sphere in Figure 16.52 has one normal vector pointing inward, toward the center of the sphere, and another opposite normal vector pointing outward. We specify one of two possible orientations for the sphere by choosing either the inward vector at each point, or alternatively the outward vector at each point.

When we can choose a continuous field of unit normal vectors $\mathbf{n}$ on a smooth surface S then we say that S is *orientable* (or *two-sided*). Spheres and other smooth surfaces that are the boundaries of regions in space are orientable, since we can choose an outward-pointing unit vector $\mathbf{n}$ at each point to specify an orientation.

A surface together with its normal field $\mathbf{n}$, or, equivalently, a surface with a consistent choice of sides, is called an oriented surface. The vector $\mathbf{n}$ at any point gives the positive direction or positively oriented side at that point (Figure 16.52). Not all surfaces can be oriented. The Mobius band in Figure 16.53 is an example of a surface that is not orientable. No matter how you try to construct a continuous unit normal vector field (shown as the shafts of thumbtacks in the figure), starting at one point and moving the vector continuously around the surface in the manner shown will return it to the starting point, but pointing in the opposite direction. No choice of a vectors can give a continuous normal vector field on the Mobius band, so the Mobius band is not orientable.

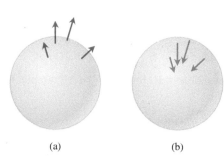

FIGURE 16.52   An outward-pointing vector field (a) and an inward-pointing vector field (b) give the two possible orientations of a sphere.

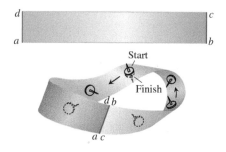

FIGURE 16.53   To make a Möbius band, take a rectangular strip of paper *abcd*, give the end *bc* a single twist, and paste the ends of the strip together to match *a* with *c* and *b* with *d*. The Möbius band is a nonorientable or one-sided surface.

Surface Integrals of Vector Fields

In Section 16.2 we defined the line integral of a vector field along a path C as $\int_C \mathbf{F} \cdot \mathbf{T} \, ds$, where $\mathbf{T}$ is the unit tangent vector to the path pointing in the forward oriented direction. We have a similar definition for surface integrals.

> **DEFINITION**   Let $\mathbf{F}$ be a vector field in three-dimensional space with continuous components defined over a smooth surface S having a chosen field of normal unit vectors $\mathbf{n}$ orienting S. Then the **surface integral of F over S** is
> $$\iint_S \mathbf{F} \cdot \mathbf{n} \, d\sigma. \tag{5}$$
> This integral is also called the **flux** of the vector field $\mathbf{F}$ across S.

If $\mathbf{F}$ is the velocity field of a three-dimensional fluid flow, then the flux of $\mathbf{F}$ across S is the net rate at which fluid is crossing S per unit time in the chosen positive direction $\mathbf{n}$ defined by the orientation of S. Fluid flows are discussed in more detail in Section 16.7.

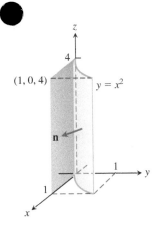

FIGURE 16.54 Finding the flux through the surface of a parabolic cylinder (Example 5).

Computing a Surface Integral for a Parametrized Surface

EXAMPLE 5   Find the flux of $\mathbf{F} = yz\mathbf{i} + x\mathbf{j} - z^2\mathbf{k}$ through the parabolic cylinder $y = x^2$, $0 \le x \le 1$, $0 \le z \le 4$, in the direction $\mathbf{n}$ indicated in Figure 16.54.

Solution   On the surface we have $x = x$, $y = x^2$, and $z = z$, so we automatically have the parametrization $\mathbf{r}(x, z) = x\mathbf{i} + x^2\mathbf{j} + z\mathbf{k}$, $0 \le x \le 1$, $0 \le z \le 4$. The cross product of tangent vectors is

$$\mathbf{r}_x \times \mathbf{r}_z = \begin{vmatrix} \mathbf{i} & \mathbf{j} & \mathbf{k} \\ 1 & 2x & 0 \\ 0 & 0 & 1 \end{vmatrix} = 2x\mathbf{i} - \mathbf{j}.$$

The unit normal vectors pointing outward from the surface as indicated in Figure 16.54 are

$$\mathbf{n} = \frac{\mathbf{r}_x \times \mathbf{r}_z}{|\mathbf{r}_x \times \mathbf{r}_z|} = \frac{2x\mathbf{i} - \mathbf{j}}{\sqrt{4x^2 + 1}}.$$

On the surface, $y = x^2$, so the vector field there is

$$\mathbf{F} = yz\mathbf{i} + x\mathbf{j} - z^2\mathbf{k} = x^2z\mathbf{i} + x\mathbf{j} - z^2\mathbf{k}.$$

Thus,

$$\mathbf{F} \cdot \mathbf{n} = \frac{1}{\sqrt{4x^2 + 1}}((x^2z)(2x) + (x)(-1) + (-z^2)(0)) = \frac{2x^3z - x}{\sqrt{4x^2 + 1}}.$$

The flux of $\mathbf{F}$ outward through the surface is

$$\iint_S \mathbf{F} \cdot \mathbf{n} \, d\sigma = \int_0^4 \int_0^1 \frac{2x^3z - x}{\sqrt{4x^2 + 1}} |\mathbf{r}_x \times \mathbf{r}_z| \, dx \, dz \qquad d\sigma = |\mathbf{r}_x \times \mathbf{r}_z| \, dx \, dz$$

$$= \int_0^4 \int_0^1 \frac{2x^3z - x}{\sqrt{4x^2 + 1}} \sqrt{4x^2 + 1} \, dx \, dz$$

$$= \int_0^4 \int_0^1 (2x^3z - x) \, dx \, dz = \int_0^4 \left[\frac{1}{2}x^4z - \frac{1}{2}x^2 \right]_{x=0}^{x=1} dz$$

$$= \int_0^4 \frac{1}{2}(z - 1) \, dz = \frac{1}{4}(z - 1)^2 \Big|_0^4$$

$$= \frac{1}{4}(9) - \frac{1}{4}(1) = 2. \qquad \blacksquare$$

There is a simple formula for the flux of $\mathbf{F}$ across a parametrized surface $\mathbf{r}(u, v)$. Since

$$d\sigma = |\mathbf{r}_u \times \mathbf{r}_v| \, du \, dv,$$

with the orientation

$$\mathbf{n} = \frac{\mathbf{r}_u \times \mathbf{r}_v}{|\mathbf{r}_u \times \mathbf{r}_v|}$$

it follows that

$$\iint_S \mathbf{F} \cdot \mathbf{n} \, d\sigma = \iint_R \mathbf{F} \cdot \frac{\mathbf{r}_u \times \mathbf{r}_v}{|\mathbf{r}_u \times \mathbf{r}_v|} |\mathbf{r}_u \times \mathbf{r}_v| \, du \, dv = \iint_R \mathbf{F} \cdot (\mathbf{r}_u \times \mathbf{r}_v) \, du \, dv.$$

Flux Across a Parametrized Surface

$$\text{Flux} = \iint_R \mathbf{F} \cdot (\mathbf{r}_u \times \mathbf{r}_v) \, du \, dv$$

This integral for flux simplifies the computation in Example 5 by eliminating the need to compute the canceled term $|\mathbf{r}_u \times \mathbf{r}_v|$. Since

$$\mathbf{F} \cdot (\mathbf{r}_x \times \mathbf{r}_z) = (x^2z)(2x) + (x)(-1) = 2x^3z - x,$$

we obtain directly

$$\text{Flux} = \iint_S \mathbf{F} \cdot \mathbf{n} \, d\sigma = \int_0^4 \int_0^1 (2x^3z - x) \, dx \, dz = 2$$

in Example 5.

Computing a Surface Integral for a Level Surface

If S is part of a level surface $g(x, y, z) = c$, then $\mathbf{n}$ may be taken to be one of the two fields

$$\mathbf{n} = \pm \frac{\nabla g}{|\nabla g|}, \qquad (6)$$

depending on which one gives the preferred direction. The corresponding flux is

$$\begin{aligned}
\text{Flux} &= \iint_S \mathbf{F} \cdot \mathbf{n} \, d\sigma \\[2mm]
&= \iint_R \left(\mathbf{F} \cdot \frac{\pm \nabla g}{|\nabla g|} \right) \frac{|\nabla g|}{|\nabla g \cdot \mathbf{p}|} \, dA \qquad \text{Eqs. (6) and (3)} \\[2mm]
&= \iint_R \mathbf{F} \cdot \frac{\pm \nabla g}{|\nabla g \cdot \mathbf{p}|} \, dA. \qquad\qquad (7)
\end{aligned}$$

EXAMPLE 6   Find the flux of $\mathbf{F} = yz\mathbf{j} + z^2\mathbf{k}$ outward through the surface S cut from the cylinder $y^2 + z^2 = 1$, $z \geq 0$, by the planes $x = 0$ and $x = 1$.

Solution   The outward normal field on S (Figure 16.55) may be calculated from the gradient of $g(x, y, z) = y^2 + z^2$ to be

$$\mathbf{n} = +\frac{\nabla g}{|\nabla g|} = \frac{2y\mathbf{j} + 2z\mathbf{k}}{\sqrt{4y^2 + 4z^2}} = \frac{2y\mathbf{j} + 2z\mathbf{k}}{2\sqrt{1}} = y\mathbf{j} + z\mathbf{k}.$$

With $\mathbf{p} = \mathbf{k}$, we also have

$$d\sigma = \frac{|\nabla g|}{|\nabla g \cdot \mathbf{k}|} \, dA = \frac{2}{|2z|} \, dA = \frac{1}{z} \, dA. \qquad \text{Eq. (3)}$$

We can drop the absolute value bars because $z \geq 0$ on S.
The value of $\mathbf{F} \cdot \mathbf{n}$ on the surface is

$$\begin{aligned}
\mathbf{F} \cdot \mathbf{n} &= (yz\mathbf{j} + z^2\mathbf{k}) \cdot (y\mathbf{j} + z\mathbf{k}) \\
&= y^2z + z^3 = z(y^2 + z^2) \qquad \\
&= z. \qquad\qquad\qquad\qquad {\scriptstyle y^2 + z^2 = 1 \text{ on } S}
\end{aligned}$$

The surface projects onto the shadow region R_{xy}, which is the rectangle in the xy-plane shown in Figure 16.55. Therefore, the flux of $\mathbf{F}$ outward through S is

$$\iint_S \mathbf{F} \cdot \mathbf{n} \, d\sigma = \iint_{R_{xy}} (z) \left(\frac{1}{z} \, dA \right) = \iint_{R_{xy}} dA = \text{area}(R_{xy}) = 2. \qquad \blacksquare$$

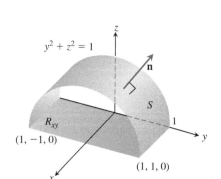

FIGURE 16.55 Calculating the flux of a vector field outward through the surface S. The area of the shadow region R_{xy} is 2 (Example 6).

Moments and Masses of Thin Shells

Thin shells of material like bowls, metal drums, and domes are modeled with surfaces. Their moments and masses are calculated with the formulas in Table 16.3. The derivations are similar to those in Section 6.6. The formulas are like those for line integrals in Table 16.1, Section 16.1.

TABLE 16.3 **Mass and moment formulas for very thin shells**

Mass: $\quad M = \iint\limits_S \delta \, d\sigma$ $\quad$ $\delta = \delta(x, y, z) =$ density at (x, y, z) is mass per unit area

First moments about the coordinate planes:

$$M_{yz} = \iint\limits_S x \, \delta \, d\sigma, \qquad M_{xz} = \iint\limits_S y \, \delta \, d\sigma, \qquad M_{xy} = \iint\limits_S z \, \delta \, d\sigma$$

Coordinates of center of mass:

$$\bar{x} = M_{yz}/M, \qquad \bar{y} = M_{xz}/M, \qquad \bar{z} = M_{xy}/M$$

Moments of inertia about coordinate axes:

$$I_x = \iint\limits_S (y^2 + z^2) \, \delta \, d\sigma, \quad I_y = \iint\limits_S (x^2 + z^2) \, \delta \, d\sigma, \quad I_z = \iint\limits_S (x^2 + y^2) \, \delta \, d\sigma,$$

$$I_L = \iint\limits_S r^2 \delta \, d\sigma \qquad r(x, y, z) = \text{distance from point } (x, y, z) \text{ to line } L$$

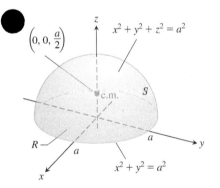

FIGURE 16.56 The center of mass of a thin hemispherical shell of constant density lies on the axis of symmetry halfway from the base to the top (Example 7).

EXAMPLE 7 Find the center of mass of a thin hemispherical shell of radius a and constant density δ.

Solution We model the shell with the hemisphere

$$f(x, y, z) = x^2 + y^2 + z^2 = a^2, \qquad z \geq 0$$

(Figure 16.56). The symmetry of the surface about the z-axis tells us that $\bar{x} = \bar{y} = 0$. It remains only to find $\bar{z}$ from the formula $\bar{z} = M_{xy}/M$.

The mass of the shell is

$$M = \iint\limits_S \delta \, d\sigma = \delta \iint\limits_S d\sigma = (\delta)(\text{area of } S) = 2\pi a^2 \delta. \qquad \delta = \text{constant}$$

To evaluate the integral for M_{xy}, we take $\mathbf{p} = \mathbf{k}$ and calculate

$$|\nabla f| = |2x\mathbf{i} + 2y\mathbf{j} + 2z\mathbf{k}| = 2\sqrt{x^2 + y^2 + z^2} = 2a$$

$$|\nabla f \cdot \mathbf{p}| = |\nabla f \cdot \mathbf{k}| = |2z| = 2z$$

$$d\sigma = \frac{|\nabla f|}{|\nabla f \cdot \mathbf{p}|} dA = \frac{a}{z} dA. \qquad \text{Eq. (3)}$$

Then

$$M_{xy} = \iint\limits_S z\delta \, d\sigma = \delta \iint\limits_R z\frac{a}{z} dA = \delta a \iint\limits_R dA = \delta a(\pi a^2) = \delta\pi a^3$$

$$\bar{z} = \frac{M_{xy}}{M} = \frac{\pi a^3 \delta}{2\pi a^2 \delta} = \frac{a}{2}.$$

The shell's center of mass is the point $(0, 0, a/2)$.

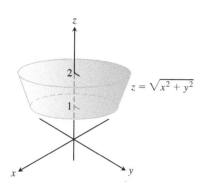

FIGURE 16.57 The cone frustum formed when the cone $z = \sqrt{x^2 + y^2}$ is cut by the planes $z = 1$ and $z = 2$ (Example 8).

EXAMPLE 8 Find the center of mass of a thin shell of density $\delta = 1/z^2$ cut from the cone $z = \sqrt{x^2 + y^2}$ by the planes $z = 1$ and $z = 2$ (Figure 16.57).

Solution The symmetry of the surface about the z-axis tells us that $\bar{x} = \bar{y} = 0$. We find $\bar{z} = M_{xy}/M$. Working as in Example 4 of Section 16.5, we have

$$\mathbf{r}(r, \theta) = (r \cos \theta)\mathbf{i} + (r \sin \theta)\mathbf{j} + r\mathbf{k}, \qquad 1 \le r \le 2, \quad 0 \le \theta \le 2\pi,$$

and

$$|\mathbf{r}_r \times \mathbf{r}_\theta| = \sqrt{2}\,r.$$

Therefore,

$$M = \iint_S \delta\, d\sigma = \int_0^{2\pi} \int_1^2 \frac{1}{r^2}\sqrt{2}\,r\, dr\, d\theta$$

$$= \sqrt{2} \int_0^{2\pi} \Big[\ln r\Big]_1^2 d\theta = \sqrt{2}\int_0^{2\pi} \ln 2\, d\theta$$

$$= 2\pi\sqrt{2}\ln 2,$$

$$M_{xy} = \iint_S \delta z\, d\sigma = \int_0^{2\pi} \int_1^2 \frac{1}{r^2}\,r\sqrt{2}\,r\, dr\, d\theta$$

$$= \sqrt{2}\int_0^{2\pi}\int_1^2 dr\, d\theta$$

$$= \sqrt{2}\int_0^{2\pi} d\theta = 2\pi\sqrt{2},$$

$$\bar{z} = \frac{M_{xy}}{M} = \frac{2\pi\sqrt{2}}{2\pi\sqrt{2}\ln 2} = \frac{1}{\ln 2}.$$

The shell's center of mass is the point $(0, 0, 1/\ln 2)$.

EXERCISES 16.6

Surface Integrals of Scalar Functions

In Exercises 1–8, integrate the given function over the given surface.

1. **Parabolic cylinder** $G(x, y, z) = x$, over the parabolic cylinder $y = x^2, 0 \le x \le 2, 0 \le z \le 3$

2. **Circular cylinder** $G(x, y, z) = z$, over the cylindrical surface $y^2 + z^2 = 4, z \ge 0, 1 \le x \le 4$

3. **Sphere** $G(x, y, z) = x^2$, over the unit sphere $x^2 + y^2 + z^2 = 1$

4. **Hemisphere** $G(x, y, z) = z^2$, over the hemisphere $x^2 + y^2 + z^2 = a^2, z \ge 0$

5. **Portion of plane** $F(x, y, z) = z$, over the portion of the plane $x + y + z = 4$ that lies above the square $0 \le x \le 1$, $0 \le y \le 1$, in the xy-plane

6. **Cone** $F(x, y, z) = z - x$, over the cone $z = \sqrt{x^2 + y^2}$, $0 \le z \le 1$

7. **Parabolic dome** $H(x, y, z) = x^2\sqrt{5 - 4z}$, over the parabolic dome $z = 1 - x^2 - y^2, z \ge 0$

8. **Spherical cap** $H(x, y, z) = yz$, over the part of the sphere $x^2 + y^2 + z^2 = 4$ that lies above the cone $z = \sqrt{x^2 + y^2}$

9. Integrate $G(x, y, z) = x + y + z$ over the surface of the cube cut from the first octant by the planes $x = a, y = a, z = a$.

10. Integrate $G(x, y, z) = y + z$ over the surface of the wedge in the first octant bounded by the coordinate planes and the planes $x = 2$ and $y + z = 1$.

11. Integrate $G(x, y, z) = xyz$ over the surface of the rectangular solid cut from the first octant by the planes $x = a, y = b$, and $z = c$.

12. Integrate $G(x, y, z) = xyz$ over the surface of the rectangular solid bounded by the planes $x = \pm a, y = \pm b$, and $z = \pm c$.

13. Integrate $G(x, y, z) = x + y + z$ over the portion of the plane $2x + 2y + z = 2$ that lies in the first octant.

14. Integrate $G(x, y, z) = x\sqrt{y^2 + 4}$ over the surface cut from the parabolic cylinder $y^2 + 4z = 16$ by the planes $x = 0$, $x = 1$, and $z = 0$.

15. Integrate $G(x, y, z) = z - x$ over the portion of the graph of $z = x + y^2$ above the triangle in the xy-plane having vertices $(0, 0, 0)$, $(1, 1, 0)$, and $(0, 1, 0)$. (See accompanying figure.)

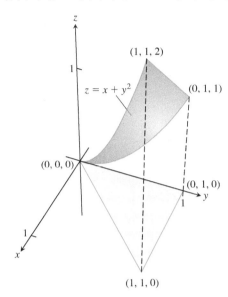

16. Integrate $G(x, y, z) = x$ over the surface given by

$$z = x^2 + y \quad \text{for} \quad 0 \le x \le 1, \quad -1 \le y \le 1.$$

17. Integrate $G(x, y, z) = xyz$ over the triangular surface with vertices $(1, 0, 0)$, $(0, 2, 0)$, and $(0, 1, 1)$.

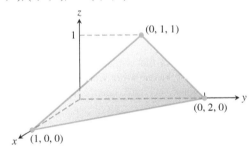

18. Integrate $G(x, y, z) = x - y - z$ over the portion of the plane $x + y = 1$ in the first octant between $z = 0$ and $z = 1$ (see the accompanying figure below).

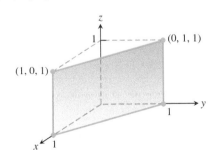

Finding Flux or Surface Integrals of Vector Fields
In Exercises 19–28, use a parametrization to find the flux $\iint_S \mathbf{F} \cdot \mathbf{n} \, d\sigma$ across the surface in the specified direction.

19. Parabolic cylinder $\mathbf{F} = z^2\mathbf{i} + x\mathbf{j} - 3z\mathbf{k}$ outward (normal away from the x-axis) through the surface cut from the parabolic cylinder $z = 4 - y^2$ by the planes $x = 0$, $x = 1$, and $z = 0$

20. Parabolic cylinder $\mathbf{F} = x^2\mathbf{j} - xz\mathbf{k}$ outward (normal away from the yz-plane) through the surface cut from the parabolic cylinder $y = x^2$, $-1 \le x \le 1$, by the planes $z = 0$ and $z = 2$

21. Sphere $\mathbf{F} = z\mathbf{k}$ across the portion of the sphere $x^2 + y^2 + z^2 = a^2$ in the first octant in the direction away from the origin

22. Sphere $\mathbf{F} = x\mathbf{i} + y\mathbf{j} + z\mathbf{k}$ across the sphere $x^2 + y^2 + z^2 = a^2$ in the direction away from the origin

23. Plane $\mathbf{F} = 2xy\mathbf{i} + 2yz\mathbf{j} + 2xz\mathbf{k}$ upward across the portion of the plane $x + y + z = 2a$ that lies above the square $0 \le x \le a$, $0 \le y \le a$, in the xy-plane

24. Cylinder $\mathbf{F} = x\mathbf{i} + y\mathbf{j} + z\mathbf{k}$ outward through the portion of the cylinder $x^2 + y^2 = 1$ cut by the planes $z = 0$ and $z = a$

25. Cone $\mathbf{F} = xy\mathbf{i} - z\mathbf{k}$ outward (normal away from the z-axis) through the cone $z = \sqrt{x^2 + y^2}$, $0 \le z \le 1$

26. Cone $\mathbf{F} = y^2\mathbf{i} + xz\mathbf{j} - \mathbf{k}$ outward (normal away from the z-axis) through the cone $z = 2\sqrt{x^2 + y^2}$, $0 \le z \le 2$

27. Cone frustum $\mathbf{F} = -x\mathbf{i} - y\mathbf{j} + z^2\mathbf{k}$ outward (normal away from the z-axis) through the portion of the cone $z = \sqrt{x^2 + y^2}$ between the planes $z = 1$ and $z = 2$

28. Paraboloid $\mathbf{F} = 4x\mathbf{i} + 4y\mathbf{j} + 2\mathbf{k}$ outward (normal away from the z-axis) through the surface cut from the bottom of the paraboloid $z = x^2 + y^2$ by the plane $z = 1$

In Exercises 29 and 30, find the surface integral of the field $\mathbf{F}$ over the portion of the given surface in the specified direction.

29. $\mathbf{F}(x, y, z) = -\mathbf{i} + 2\mathbf{j} + 3\mathbf{k}$

S: rectangular surface $z = 0$, $\quad 0 \le x \le 2$, $\quad 0 \le y \le 3$, direction $\mathbf{k}$

30. $\mathbf{F}(x, y, z) = yx^2\mathbf{i} - 2\mathbf{j} + xz\mathbf{k}$

S: rectangular surface $y = 0$, $\quad -1 \le x \le 2$, $\quad 2 \le z \le 7$, direction $-\mathbf{j}$

In Exercises 31–36, use Equation (7) to find the surface integral of the field $\mathbf{F}$ over the portion of the sphere $x^2 + y^2 + z^2 = a^2$ in the first octant in the direction away from the origin.

31. $\mathbf{F}(x, y, z) = z\mathbf{k}$

32. $\mathbf{F}(x, y, z) = -y\mathbf{i} + x\mathbf{j}$

33. $\mathbf{F}(x, y, z) = y\mathbf{i} - x\mathbf{j} + \mathbf{k}$

34. $\mathbf{F}(x, y, z) = zx\mathbf{i} + zy\mathbf{j} + z^2\mathbf{k}$

35. $\mathbf{F}(x, y, z) = x\mathbf{i} + y\mathbf{j} + z\mathbf{k}$

36. $\mathbf{F}(x, y, z) = \dfrac{x\mathbf{i} + y\mathbf{j} + z\mathbf{k}}{\sqrt{x^2 + y^2 + z^2}}$

37. Find the flux of the field $\mathbf{F}(x, y, z) = z^2\mathbf{i} + x\mathbf{j} - 3z\mathbf{k}$ outward through the surface cut from the parabolic cylinder $z = 4 - y^2$ by the planes $x = 0$, $x = 1$, and $z = 0$.

38. Find the flux of the field $\mathbf{F}(x, y, z) = 4x\mathbf{i} + 4y\mathbf{j} + 2\mathbf{k}$ outward (away from the z-axis) through the surface cut from the bottom of the paraboloid $z = x^2 + y^2$ by the plane $z = 1$.

39. Let S be the portion of the cylinder $y = e^x$ in the first octant that projects parallel to the x-axis onto the rectangle R_{yz}: $1 \leq y \leq 2$, $0 \leq z \leq 1$ in the yz-plane (see the accompanying figure). Let $\mathbf{n}$ be the unit vector normal to S that points away from the yz-plane. Find the flux of the field $\mathbf{F}(x, y, z) = -2\mathbf{i} + 2y\mathbf{j} + z\mathbf{k}$ across S in the direction of $\mathbf{n}$.

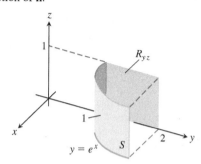

40. Let S be the portion of the cylinder $y = \ln x$ in the first octant whose projection parallel to the y-axis onto the xz-plane is the rectangle R_{xz}: $1 \leq x \leq e, 0 \leq z \leq 1$. Let $\mathbf{n}$ be the unit vector normal to S that points away from the xz-plane. Find the flux of $\mathbf{F} = 2y\mathbf{j} + z\mathbf{k}$ through S in the direction of $\mathbf{n}$.

41. Find the outward flux of the field $\mathbf{F} = 2xy\mathbf{i} + 2yz\mathbf{j} + 2xz\mathbf{k}$ across the surface of the cube cut from the first octant by the planes $x = a, y = a, z = a$.

42. Find the outward flux of the field $\mathbf{F} = xz\mathbf{i} + yz\mathbf{j} + \mathbf{k}$ across the surface of the upper cap cut from the solid sphere $x^2 + y^2 + z^2 \leq 25$ by the plane $z = 3$.

Moments and Masses

43. Centroid Find the centroid of the portion of the sphere $x^2 + y^2 + z^2 = a^2$ that lies in the first octant.

44. Centroid Find the centroid of the surface cut from the cylinder $y^2 + z^2 = 9, z \geq 0$, by the planes $x = 0$ and $x = 3$ (resembles the surface in Example 6).

45. Thin shell of constant density Find the center of mass and the moment of inertia about the z-axis of a thin shell of constant density δ cut from the cone $x^2 + y^2 - z^2 = 0$ by the planes $z = 1$ and $z = 2$.

46. Conical surface of constant density Find the moment of inertia about the z-axis of a thin shell of constant density δ cut from

the cone $4x^2 + 4y^2 - z^2 = 0, z \geq 0$, by the circular cylinder $x^2 + y^2 = 2x$ (see the accompanying figure).

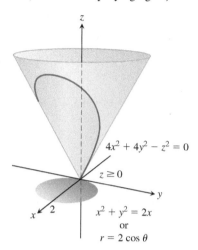

47. Spherical shells

a. Find the moment of inertia about a diameter of a thin spherical shell of radius a and constant density δ. (Work with a hemispherical shell and double the result.)

b. Use the Parallel Axis Theorem (Exercises 15.6) and the result in part (a) to find the moment of inertia about a line tangent to the shell.

48. Conical Surface Find the centroid of the lateral surface of a solid cone of base radius a and height h (cone surface minus the base).

49. A surface S lies on the plane $2x + 3y + 6z = 12$ directly above the rectangle in the xy-plane with vertices $(0, 0)$, $(1, 0)$, $(0, 2)$, and $(1, 2)$. If the density at a point (x, y, z) on S is given by $\delta(x, y, z) = 4xy + 6z \, \text{mg/cm}^2$, find the total mass of S.

50. A surface S lies on the paraboloid $z = \frac{1}{2}x^2 + \frac{1}{2}y^2$ directly above the triangle in the xy-plane with vertices $(0, 0)$, $(2, 0)$, and $(2, 4)$. If the density at a point (x, y, z) on S is given by $\delta(x, y, z) = 9xy \, \text{g/cm}^2$, find the total mass of S.

16.7 Stokes' Theorem

To calculate the counterclockwise circulation of a two-dimensional vector field $\mathbf{F} = M\mathbf{i} + N\mathbf{j}$ around a simple closed curve in the plane, Green's Theorem says we can compute the double integral over the region enclosed by the curve of the scalar quantity $(\partial N/\partial x - \partial M/\partial y)$. This expression is the $\mathbf{k}$-component of a *curl vector* field, which we define in this section, and it measures the rate of rotation of $\mathbf{F}$ at each point in the region around an axis parallel to $\mathbf{k}$. For a vector field on three-dimensional space, the rotation at each point is around an axis that is parallel to the curl vector at that point. When a closed curve C in space is the boundary of an oriented surface, we will see that the circulation of $\mathbf{F}$ around C is equal to the surface integral of the curl vector field. This result extends Green's Theorem from regions in the plane to general surfaces in space having a smooth boundary curve.

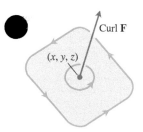

FIGURE 16.58 The circulation vector at a point (x, y, z) in a plane in a three-dimensional fluid flow. Notice its right-hand relation to the rotating particles in the fluid.

∇ is the symbol "del."

The Curl Vector Field

Suppose that **F** is the velocity field of a fluid flowing in space. Particles near the point (x, y, z) in the fluid tend to rotate around an axis through (x, y, z) that is parallel to a certain vector we are about to define. This vector points in the direction for which the rotation is counterclockwise when viewed looking down onto the plane of the circulation from the tip of the arrow representing the vector. This is the direction your right-hand thumb points when your fingers curl around the axis of rotation in the way consistent with the rotating motion of the particles in the fluid (see Figure 16.58). The length of the vector measures the rate of rotation. The vector is called the **curl vector**, and for the vector field $\mathbf{F} = M\mathbf{i} + N\mathbf{j} + P\mathbf{k}$ it is defined to be

$$\text{curl } \mathbf{F} = \left(\frac{\partial P}{\partial y} - \frac{\partial N}{\partial z}\right)\mathbf{i} + \left(\frac{\partial M}{\partial z} - \frac{\partial P}{\partial x}\right)\mathbf{j} + \left(\frac{\partial N}{\partial x} - \frac{\partial M}{\partial y}\right)\mathbf{k}. \tag{1}$$

This information is a consequence of Stokes' Theorem, the generalization to space of the circulation-curl form of Green's Theorem and is the subject of this section.

Notice that $(\text{curl } \mathbf{F}) \cdot \mathbf{k} = (\partial N/\partial x - \partial M/\partial y)$, which is consistent with our definition in Section 16.4 when $\mathbf{F} = M(x, y)\mathbf{i} + N(x, y)\mathbf{j}$. The formula for curl **F** in Equation (1) is often expressed using the symbol

$$\nabla = \mathbf{i}\frac{\partial}{\partial x} + \mathbf{j}\frac{\partial}{\partial y} + \mathbf{k}\frac{\partial}{\partial z}. \tag{2}$$

The symbol ∇ is pronounced "del," and we can use this symbol to compute the curl of **F** with the formula

$$\nabla \times \mathbf{F} = \begin{vmatrix} \mathbf{i} & \mathbf{j} & \mathbf{k} \\ \frac{\partial}{\partial x} & \frac{\partial}{\partial y} & \frac{\partial}{\partial z} \\ M & N & P \end{vmatrix}$$

$$= \left(\frac{\partial P}{\partial y} - \frac{\partial N}{\partial z}\right)\mathbf{i} + \left(\frac{\partial M}{\partial z} - \frac{\partial P}{\partial x}\right)\mathbf{j} + \left(\frac{\partial N}{\partial x} - \frac{\partial M}{\partial y}\right)\mathbf{k}.$$

We often use this cross product notation to write the curl symbolically as "del cross **F**."

$$\text{curl } \mathbf{F} = \nabla \times \mathbf{F} \tag{3}$$

EXAMPLE 1   Find the curl of $\mathbf{F} = (x^2 - z)\mathbf{i} + xe^z\mathbf{j} + xy\mathbf{k}$.

Solution   We use Equation (3) and the determinant form for the cross product, which gives,

$$\text{curl } \mathbf{F} = \nabla \times \mathbf{F}$$

$$= \begin{vmatrix} \mathbf{i} & \mathbf{j} & \mathbf{k} \\ \frac{\partial}{\partial x} & \frac{\partial}{\partial y} & \frac{\partial}{\partial z} \\ x^2 - z & xe^z & xy \end{vmatrix}$$

$$= \left(\frac{\partial}{\partial y}(xy) - \frac{\partial}{\partial z}(xe^z)\right)\mathbf{i} - \left(\frac{\partial}{\partial x}(xy) - \frac{\partial}{\partial z}(x^2 - z)\right)\mathbf{j}$$

$$+ \left(\frac{\partial}{\partial x}(xe^z) - \frac{\partial}{\partial y}(x^2 - z)\right)\mathbf{k} \qquad \text{Notice that curl } \mathbf{F} \text{ is a vector, not a scalar.}$$

$$= (x - xe^z)\mathbf{i} - (y + 1)\mathbf{j} + (e^z - 0)\mathbf{k}$$

$$= x(1 - e^z)\mathbf{i} - (y + 1)\mathbf{j} + e^z\mathbf{k}. \qquad \blacksquare$$

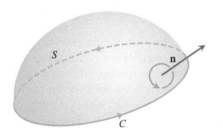

FIGURE 16.59 The orientation of the bounding curve C gives it a right-handed relation to the normal field $\mathbf{n}$. If the thumb of a right hand points along $\mathbf{n}$, the fingers curl in the direction of C.

As we will see, the operator ∇ has a number of other applications. For instance, when applied to a scalar function $f(x, y, z)$, it gives the gradient of f:

$$\nabla f = \frac{\partial f}{\partial x}\mathbf{i} + \frac{\partial f}{\partial y}\mathbf{j} + \frac{\partial f}{\partial z}\mathbf{k}.$$

In this setting it is sometimes read as "del f" and sometimes as "grad f."

Stokes' Theorem

Stokes' Theorem generalizes Green's Theorem to three dimensions. The circulation-curl form of Green's Theorem relates the counterclockwise circulation of a vector field around a simple closed curve C in the xy-plane to a double integral over the plane region R enclosed by C. Stokes' Theorem relates the circulation of a vector field around the boundary C of an oriented surface S in space (Figure 16.59) to a surface integral over the surface S. We require that the surface be **piecewise smooth**, which means that it is a finite union of smooth surfaces joining along smooth curves.

THEOREM 6—Stokes' Theorem

Let S be a piecewise smooth oriented surface having a piecewise smooth boundary curve C. Let $\mathbf{F} = M\mathbf{i} + N\mathbf{j} + P\mathbf{k}$ be a vector field whose components have continuous first partial derivatives on an open region containing S. Then the circulation of $\mathbf{F}$ around C in the direction counterclockwise with respect to the surface's unit normal vector $\mathbf{n}$ equals the integral of the curl vector field $\nabla \times \mathbf{F}$ over S:

$$\underbrace{\oint_C \mathbf{F} \cdot d\mathbf{r}}_{\substack{\text{Counterclockwise} \\ \text{circulation}}} = \underbrace{\iint_S (\nabla \times \mathbf{F}) \cdot \mathbf{n}\, d\sigma}_{\text{Curl integral}} \qquad (4)$$

Notice from Equation (4) that if two different oriented surfaces S_1 and S_2 have the same boundary C, their curl integrals are equal:

$$\iint_{S_1} (\nabla \times \mathbf{F}) \cdot \mathbf{n}_1\, d\sigma = \iint_{S_2} (\nabla \times \mathbf{F}) \cdot \mathbf{n}_2\, d\sigma.$$

Both curl integrals equal the counterclockwise circulation integral on the left side of Equation (4) as long as the unit normal vectors $\mathbf{n}_1$ and $\mathbf{n}_2$ correctly orient the surfaces. So the curl integral is independent of the surface and depends only on circulation along the boundary curve. This independence of surface resembles the path independence for the flow integral of a conservative velocity field along a curve, where the value of the flow integral depends only on the endpoints (that is, the boundary points) of the path. The curl field $\nabla \times \mathbf{F}$ is analogous to the gradient field ∇f of a scalar function f.

If C is a curve in the xy-plane, oriented counterclockwise, and R is the region in the xy-plane bounded by C, then $d\sigma = dx\,dy$ and

$$(\nabla \times \mathbf{F}) \cdot \mathbf{n} = (\nabla \times \mathbf{F}) \cdot \mathbf{k} = \left(\frac{\partial N}{\partial x} - \frac{\partial M}{\partial y}\right).$$

Under these conditions, Stokes' equation becomes

$$\oint_C \mathbf{F} \cdot d\mathbf{r} = \iint_R \left(\frac{\partial N}{\partial x} - \frac{\partial M}{\partial y}\right) dx\,dy,$$

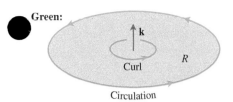

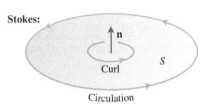

FIGURE 16.60 When applied to curves and surfaces in the plane, Stokes' Theorem gives the circulation-curl version of Green's Theorem. But Stokes' Theorem also applies more generally, to curves and surfaces not lying in the plane.

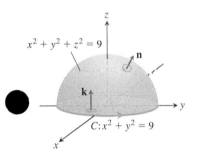

FIGURE 16.61 A hemisphere and a disk, each with boundary C (Examples 2 and 3).

which is the circulation-curl form of the equation in Green's Theorem. Conversely, by reversing these steps we can rewrite the circulation-curl form of Green's Theorem for two-dimensional fields in del notation as

$$\oint_C \mathbf{F} \cdot d\mathbf{r} = \iint_R (\nabla \times \mathbf{F}) \cdot \mathbf{k} \, dA. \qquad (5)$$

See Figure 16.60.

EXAMPLE 2 Evaluate Equation (4) for the hemisphere $S: x^2 + y^2 + z^2 = 9, z \geq 0$, its bounding circle $C: x^2 + y^2 = 9, z = 0$, and the field $\mathbf{F} = y\mathbf{i} - x\mathbf{j}$.

Solution The hemisphere looks much like the surface in Figure 16.59 with the bounding circle C in the xy-plane (see Figure 16.61). We calculate the counterclockwise circulation around C (as viewed from above) using the parametrization $\mathbf{r}(\theta) = (3 \cos \theta)\mathbf{i} + (3 \sin \theta)\mathbf{j}, 0 \leq \theta \leq 2\pi$:

$$d\mathbf{r} = (-3 \sin \theta \, d\theta)\mathbf{i} + (3 \cos \theta \, d\theta)\mathbf{j}$$

$$\mathbf{F} = y\mathbf{i} - x\mathbf{j} = (3 \sin \theta)\mathbf{i} - (3 \cos \theta)\mathbf{j}$$

$$\mathbf{F} \cdot d\mathbf{r} = -9 \sin^2 \theta \, d\theta - 9 \cos^2 \theta \, d\theta = -9 \, d\theta$$

$$\oint_C \mathbf{F} \cdot d\mathbf{r} = \int_0^{2\pi} -9 \, d\theta = -18\pi.$$

For the curl integral of $\mathbf{F}$, we have

$$\nabla \times \mathbf{F} = \left(\frac{\partial P}{\partial y} - \frac{\partial N}{\partial z} \right)\mathbf{i} + \left(\frac{\partial M}{\partial z} - \frac{\partial P}{\partial x} \right)\mathbf{j} + \left(\frac{\partial N}{\partial x} - \frac{\partial M}{\partial y} \right)\mathbf{k}$$

$$= (0 - 0)\mathbf{i} + (0 - 0)\mathbf{j} + (-1 - 1)\mathbf{k} = -2\mathbf{k}$$

$$\mathbf{n} = \frac{x\mathbf{i} + y\mathbf{j} + z\mathbf{k}}{\sqrt{x^2 + y^2 + z^2}} = \frac{x\mathbf{i} + y\mathbf{j} + z\mathbf{k}}{3} \qquad \text{Outer unit normal}$$

$$d\sigma = \frac{3}{z} \, dA \qquad \text{Section 16.6, Example 7, with } a = 3$$

$$(\nabla \times \mathbf{F}) \cdot \mathbf{n} \, d\sigma = (-2\mathbf{k}) \times \left(\frac{x\mathbf{i} + y\mathbf{j} + z\mathbf{k}}{3} \right) d\sigma = -\frac{2z}{3} \frac{3}{z} \, dA = -2 \, dA$$

and

$$\iint_S (\nabla \times \mathbf{F}) \cdot \mathbf{n} \, d\sigma = \iint_{x^2 + y^2 \leq 9} -2 \, dA = -18\pi.$$

The circulation around the circle equals the integral of the curl over the hemisphere, as it should from Stokes' Theorem. ∎

The surface integral in Stokes' Theorem can be computed using any surface having boundary curve C, provided the surface is properly oriented and lies within the domain of the field $\mathbf{F}$. The next example illustrates this fact for the circulation around the curve C in Example 2.

EXAMPLE 3 Calculate the circulation around the bounding circle C in Example 2 using the disk of radius 3 centered at the origin in the xy-plane as the surface S (instead of the hemisphere). See Figure 16.61.

Solution As in Example 2, $\nabla \times \mathbf{F} = -2\mathbf{k}$. For the surface being the described disk in the xy-plane, we have the normal vector $\mathbf{n} = \mathbf{k}$ so that

$$(\nabla \times \mathbf{F}) \cdot \mathbf{n} \, d\sigma = -2\mathbf{k} \cdot \mathbf{k} \, dA = -2 \, dA$$

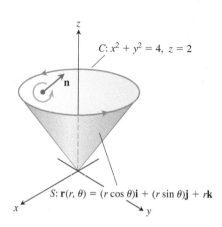

FIGURE 16.62 The curve C and cone S in Example 4.

and

$$\iint_S (\nabla \times \mathbf{F}) \cdot \mathbf{n} \, d\sigma = \iint_{x^2+y^2 \le 9} -2 \, dA = -18\pi,$$

a simpler calculation than before.

■

EXAMPLE 4 Find the circulation of the field $\mathbf{F} = (x^2 - y)\mathbf{i} + 4z\mathbf{j} + x^2\mathbf{k}$ around the curve C in which the plane $z = 2$ meets the cone $z = \sqrt{x^2 + y^2}$, counterclockwise as viewed from above (Figure 16.62).

Solution Stokes' Theorem enables us to find the circulation by integrating over the surface of the cone. Traversing C in the counterclockwise direction viewed from above corresponds to taking the *inner* normal $\mathbf{n}$ to the cone, the normal with a positive $\mathbf{k}$-component.

We parametrize the cone as

$$\mathbf{r}(r, \theta) = (r \cos \theta)\mathbf{i} + (r \sin \theta)\mathbf{j} + r\mathbf{k}, \qquad 0 \le r \le 2, \quad 0 \le \theta \le 2\pi.$$

We then have

$$\mathbf{n} = \frac{\mathbf{r}_r \times \mathbf{r}_\theta}{|\mathbf{r}_r \times \mathbf{r}_\theta|} = \frac{-(r \cos \theta)\mathbf{i} - (r \sin \theta)\mathbf{j} + r\mathbf{k}}{r\sqrt{2}} \qquad \text{Section 16.5, Example 4}$$

$$= \frac{1}{\sqrt{2}}\left(-(\cos \theta)\mathbf{i} - (\sin \theta)\mathbf{j} + \mathbf{k}\right)$$

$$d\sigma = r\sqrt{2} \, dr \, d\theta \qquad \text{Section 16.5, Example 4}$$

$$\nabla \times \mathbf{F} = -4\mathbf{i} - 2x\mathbf{j} + \mathbf{k} \qquad \text{Routine calculation}$$

$$= -4\mathbf{i} - 2r \cos \theta \mathbf{j} + \mathbf{k}. \qquad x = r \cos \theta$$

Accordingly,

$$(\nabla \times \mathbf{F}) \cdot \mathbf{n} = \frac{1}{\sqrt{2}}\left(4 \cos \theta + 2r \cos \theta \sin \theta + 1\right)$$

$$= \frac{1}{\sqrt{2}}\left(4 \cos \theta + r \sin 2\theta + 1\right)$$

and the circulation is

$$\oint_C \mathbf{F} \cdot d\mathbf{r} = \iint_S (\nabla \times \mathbf{F}) \cdot \mathbf{n} \, d\sigma \qquad \text{Stokes' Theorem, Eq. (4)}$$

$$= \int_0^{2\pi} \int_0^2 \frac{1}{\sqrt{2}}\left(4 \cos \theta + r \sin 2\theta + 1\right)\left(r\sqrt{2} \, dr \, d\theta\right) = 4\pi. \qquad ■$$

EXAMPLE 5 The cone used in Example 4 is not the easiest surface to use for calculating the circulation around the bounding circle C lying in the plane $z = 2$. If instead we use the flat disk of radius 2 centered on the z-axis and lying in the plane $z = 2$, then the normal vector to the surface S is $\mathbf{n} = \mathbf{k}$. Just as in the computation for Example 4, we still have $\nabla \times \mathbf{F} = -4\mathbf{i} - 2x\mathbf{j} + \mathbf{k}$. However, now we get $(\nabla \times \mathbf{F}) \cdot \mathbf{n} = 1$, so that

$$\iint_S (\nabla \times \mathbf{F}) \cdot \mathbf{n} \, d\sigma = \iint_{x^2+y^2 \le 4} 1 \, dA = 4\pi. \qquad \text{The shadow is the disk of radius 2 in the } xy\text{-plane.}$$

This result agrees with the circulation value found in Example 4.

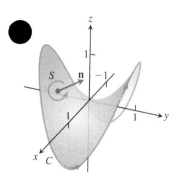

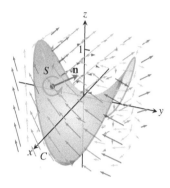

FIGURE 16.63 The surface and vector field for Example 6.

EXAMPLE 6 Find a parametrization for the surface S formed by the part of the hyperbolic paraboloid $z = y^2 - x^2$ lying inside the cylinder of radius one around the z-axis and for the boundary curve C of S. (See Figure 16.63.) Then verify Stokes' Theorem for S using the normal having positive $\mathbf{k}$-component and the vector field $\mathbf{F} = y\mathbf{i} - x\mathbf{j} + x^2\mathbf{k}$.

Solution As the unit circle is traversed counterclockwise in the xy-plane, the z-coordinate of the surface with the curve C as boundary is given by $y^2 - x^2$. A parametrization of C is given by

$$\mathbf{r}(t) = (\cos t)\mathbf{i} + (\sin t)\mathbf{j} + (\sin^2 t - \cos^2 t)\mathbf{k}, \quad 0 \le t \le 2\pi$$

with

$$\frac{d\mathbf{r}}{dt} = (-\sin t)\mathbf{i} + (\cos t)\mathbf{j} + (4 \sin t \cos t)\mathbf{k}, \; 0 \le t \le 2\pi.$$

Along the curve $\mathbf{r}(t)$ the formula for the vector field $\mathbf{F}$ is

$$\mathbf{F} = (\sin t)\mathbf{i} - (\cos t)\mathbf{j} + (\cos^2 t)\mathbf{k}.$$

The counterclockwise circulation along C is the value of the line integral

$$\int_0^{2\pi} \mathbf{F} \cdot \frac{d\mathbf{r}}{dt}\, dt = \int_0^{2\pi} \left(-\sin^2 t - \cos^2 t + 4 \sin t \cos^3 t\right) dt$$

$$= \int_0^{2\pi} \left(4 \sin t \cos^3 t - 1\right) dt$$

$$= \left[-\cos^4 t - t\right]_0^{2\pi} = -2\pi.$$

We now compute the same quantity by integrating $(\nabla \times \mathbf{F}) \cdot \mathbf{n}$ over the surface S. We use polar coordinates and parametrize S by noting that above the point (r, θ) in the plane, the z–coordinate of S is $y^2 - x^2 = r^2 \sin^2 \theta - r^2 \cos^2 \theta$. A parametrization of S is

$$\mathbf{r}(r, \theta) = (r \cos \theta)\mathbf{i} + (r \sin \theta)\mathbf{j} + r^2(\sin^2 \theta - \cos^2 \theta)\mathbf{k}, \quad 0 \le r \le 1, \quad 0 \le \theta \le 2\pi.$$

We next compute $(\nabla \times \mathbf{F}) \cdot \mathbf{n}\, d\sigma$. We have

$$\nabla \times \mathbf{F} = \begin{vmatrix} \mathbf{i} & \mathbf{j} & \mathbf{k} \\ \dfrac{\partial}{\partial x} & \dfrac{\partial}{\partial y} & \dfrac{\partial}{\partial z} \\ y & -x & x^2 \end{vmatrix} = -2x\mathbf{j} - 2\mathbf{k} = -(2r \cos \theta)\mathbf{j} - 2\mathbf{k}$$

and

$$\mathbf{r}_r = (\cos \theta)\mathbf{i} + (\sin \theta)\mathbf{j} + 2r(\sin^2 \theta - \cos^2 \theta)\mathbf{k}$$

$$\mathbf{r}_\theta = (-r \sin \theta)\mathbf{i} + (r \cos \theta)\mathbf{j} + 4r^2(\sin \theta \cos \theta)\mathbf{k}$$

$$\mathbf{r}_r \times \mathbf{r}_\theta = \begin{vmatrix} \mathbf{i} & \mathbf{j} & \mathbf{k} \\ \cos \theta & \sin \theta & 2r(\sin^2 \theta - \cos^2 \theta) \\ -r \sin \theta & r \cos \theta & 4r^2(\sin \theta \cos \theta) \end{vmatrix}$$

$$= 2r^2(2 \sin^2 \theta \cos \theta - \sin^2 \theta \cos \theta + \cos^3 \theta)\mathbf{i}$$
$$- 2r^2(2 \sin \theta \cos^2 \theta + \sin^3 \theta + \sin \theta \cos^2 \theta)\mathbf{j} + r\mathbf{k}.$$

We now obtain

$$\iint_S (\nabla \times \mathbf{F}) \cdot \mathbf{n} \, d\sigma = \int_0^{2\pi} \int_0^1 (\nabla \times \mathbf{F}) \cdot \frac{\mathbf{r}_r \times \mathbf{r}_\theta}{|\mathbf{r}_r \times \mathbf{r}_\theta|} |\mathbf{r}_r \times \mathbf{r}_\theta| \, dr \, d\theta$$

$$= \int_0^{2\pi} \int_0^1 (\nabla \times \mathbf{F}) \cdot (\mathbf{r}_r \times \mathbf{r}_\theta) \, dr \, d\theta$$

$$= \int_0^{2\pi} \int_0^1 \left[4r^3 \left(2 \sin\theta \cos^3\theta + \sin^3\theta \cos\theta + \sin\theta \cos^3\theta \right) - 2r \right] dr \, d\theta$$

$$= \int_0^{2\pi} \left[r^4 (3 \sin\theta \cos^3\theta + \sin^3\theta \cos\theta) - r^2 \right]_{r=0}^{r=1} d\theta. \qquad \text{Integrate}$$

$$= \int_0^{2\pi} (3 \sin\theta \cos^3\theta + \sin^3\theta \cos\theta - 1) \, d\theta \qquad \text{Evaluate.}$$

$$= \left[-\frac{3}{4} \cos^4\theta + \frac{1}{4} \sin^4\theta - \theta \right]_0^{2\pi}$$

$$= \left(-\frac{3}{4} + 0 - 2\pi + \frac{3}{4} - 0 + 0 \right) = -2\pi.$$

So the surface integral of $(\nabla \times \mathbf{F}) \cdot \mathbf{n}$ over S equals the counterclockwise circulation of $\mathbf{F}$ along C, as asserted by Stokes' Theorem. ∎

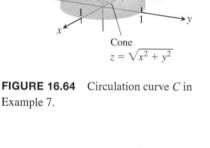

Sphere
$x^2 + y^2 + z^2 = 1$

Circle C in the plane $z = \frac{1}{\sqrt{2}}$

Cone
$z = \sqrt{x^2 + y^2}$

FIGURE 16.64 Circulation curve C in Example 7.

EXAMPLE 7 Calculate the circulation of the vector field

$$\mathbf{F} = (x^2 + z)\mathbf{i} + (y^2 + 2x)\mathbf{j} + (z^2 - y)\mathbf{k}$$

along the curve of intersection of the sphere $x^2 + y^2 + z^2 = 1$ with the cone $z = \sqrt{x^2 + y^2}$ traversed in the counterclockwise direction around the z-axis when viewed from above.

Solution The sphere and cone intersect when $1 = (x^2 + y^2) + z^2 = z^2 + z^2 = 2z^2$, or $z = 1/\sqrt{2}$ (see Figure 16.64). We apply Stokes' Theorem to the curve of intersection $x^2 + y^2 = 1/2$ considered as the boundary of the enclosed disk in the plane $z = 1/\sqrt{2}$. The normal vector to the surface is then $\mathbf{n} = \mathbf{k}$. We calculate the curl vector as

$$\nabla \times \mathbf{F} = \begin{vmatrix} \mathbf{i} & \mathbf{j} & \mathbf{k} \\ \dfrac{\partial}{\partial x} & \dfrac{\partial}{\partial y} & \dfrac{\partial}{\partial z} \\ x^2 + z & y^2 + 2x & z^2 - y \end{vmatrix} = -\mathbf{i} + \mathbf{j} + 2\mathbf{k}, \qquad \text{Routine calculation}$$

so that $(\nabla \times \mathbf{F}) \cdot \mathbf{k} = 2$. The circulation around the disk is

$$\oint_C \mathbf{F} \cdot d\mathbf{r} = \iint_S (\nabla \times \mathbf{F}) \cdot \mathbf{k} \, d\sigma$$

$$= \iint_S 2 \, d\sigma = 2 \cdot \text{area of disk} = 2 \cdot \pi \left(\frac{1}{\sqrt{2}} \right)^2 = \pi. \qquad ∎$$

Paddle Wheel Interpretation of $\nabla \times \mathbf{F}$

Suppose that $\mathbf{F}$ is the velocity field of a fluid moving in a region R in space containing the closed curve C. Then

$$\oint_C \mathbf{F} \cdot d\mathbf{r}$$

is the circulation of the fluid around C. By Stokes' Theorem, the circulation is equal to the flux of $\nabla \times \mathbf{F}$ through any suitably oriented surface S with boundary C:

$$\oint_C \mathbf{F} \cdot d\mathbf{r} = \iint_S (\nabla \times \mathbf{F}) \cdot \mathbf{n} \, d\sigma.$$

Suppose we fix a point Q in the region R and a direction $\mathbf{u}$ at Q. Take C to be a circle of radius ρ, with center at Q, whose plane is normal to $\mathbf{u}$. If $\nabla \times \mathbf{F}$ is continuous at Q, the average value of the $\mathbf{u}$-component of $\nabla \times \mathbf{F}$ over the circular disk S bounded by C approaches the $\mathbf{u}$-component of $\nabla \times \mathbf{F}$ at Q as the radius $\rho \to 0$:

$$((\nabla \times \mathbf{F}) \cdot \mathbf{u})(Q) = \lim_{\rho \to 0} \frac{1}{\pi \rho^2} \iint_S (\nabla \times \mathbf{F}) \cdot \mathbf{u} \, d\sigma.$$

If we apply Stokes' Theorem and replace the surface integral by a line integral over C, we get

$$((\nabla \times \mathbf{F}) \cdot \mathbf{u})(Q) = \lim_{\rho \to 0} \frac{1}{\pi \rho^2} \oint_C \mathbf{F} \cdot d\mathbf{r}. \tag{6}$$

The left-hand side of Equation (6) has its maximum value when $\mathbf{u}$ is the direction of $\nabla \times \mathbf{F}$. When ρ is small, the limit on the right-hand side of Equation (6) is approximately

$$\frac{1}{\pi \rho^2} \oint_C \mathbf{F} \cdot d\mathbf{r},$$

which is the circulation around C divided by the area of the disk (circulation density). Suppose that a small paddle wheel of radius ρ is introduced into the fluid at Q, with its axle directed along $\mathbf{u}$ (Figure 16.65). The circulation of the fluid around C affects the rate of spin of the paddle wheel. The wheel spins fastest when the circulation integral is maximized; therefore it spins fastest when the axle of the paddle wheel points in the direction of $\nabla \times \mathbf{F}$.

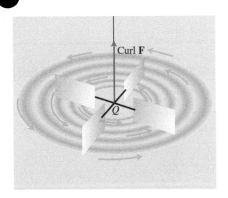

FIGURE 16.65 A small paddle wheel in a fluid spins fastest at point Q when its axle points in the direction of curl $\mathbf{F}$.

EXAMPLE 8   A fluid of constant density rotates around the z-axis with velocity $\mathbf{F} = \omega(-y\mathbf{i} + x\mathbf{j})$, where ω is a positive constant called the *angular velocity* of the rotation (Figure 16.66). Find $\nabla \times \mathbf{F}$ and relate it to the circulation density.

Solution   With $\mathbf{F} = -\omega y\mathbf{i} + \omega x\mathbf{j}$, we find the curl

$$\nabla \times \mathbf{F} = \left(\frac{\partial P}{\partial y} - \frac{\partial N}{\partial z} \right)\mathbf{i} + \left(\frac{\partial M}{\partial z} - \frac{\partial P}{\partial x} \right)\mathbf{j} + \left(\frac{\partial N}{\partial x} - \frac{\partial M}{\partial y} \right)\mathbf{k}$$

$$= (0 - 0)\mathbf{i} + (0 - 0)\mathbf{j} + (\omega - (-\omega))\mathbf{k} = 2\omega\mathbf{k},$$

and therefore $(\nabla \times \mathbf{F}) \cdot \mathbf{k} = 2\omega$. By Stokes' Theorem, the circulation of $\mathbf{F}$ around a circle C of radius ρ bounding a disk S in a plane normal to $\nabla \times \mathbf{F}$, say the xy-plane, is

$$\oint_C \mathbf{F} \cdot d\mathbf{r} = \iint_S (\nabla \times \mathbf{F}) \cdot \mathbf{n} \, d\sigma = \iint_S 2\omega\mathbf{k} \cdot \mathbf{k} \, dx \, dy = (2\omega)(\pi \rho^2).$$

Solving this last equation for 2ω, we see that

$$(\nabla \times \mathbf{F}) \cdot \mathbf{k} = 2\omega = \frac{1}{\pi \rho^2} \oint_C \mathbf{F} \cdot d\mathbf{r},$$

which is consistent with Equation (6) when $\mathbf{u} = \mathbf{k}$.   ∎

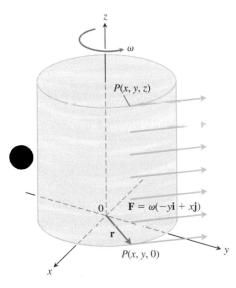

FIGURE 16.66 A steady rotational flow parallel to the xy-plane, with constant angular velocity ω in the positive (counterclockwise) direction (Example 8).

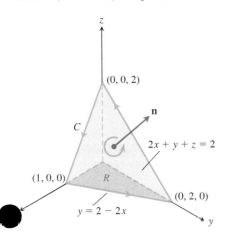

FIGURE 16.67 The planar surface in Example 9.

EXAMPLE 9   Use Stokes' Theorem to evaluate $\int_C \mathbf{F} \cdot d\mathbf{r}$, if $\mathbf{F} = xz\mathbf{i} + xy\mathbf{j} + 3xz\mathbf{k}$ and C is the boundary of the portion of the plane $2x + y + z = 2$ in the first octant, traversed counterclockwise as viewed from above (Figure 16.67).

Solution   The plane is the level surface $f(x, y, z) = 2$ of the function $f(x, y, z) = 2x + y + z$. The unit normal vector

$$\mathbf{n} = \frac{\nabla f}{|\nabla f|} = \frac{(2\mathbf{i} + \mathbf{j} + \mathbf{k})}{|2\mathbf{i} + \mathbf{j} + \mathbf{k}|} = \frac{1}{\sqrt{6}}\left(2\mathbf{i} + \mathbf{j} + \mathbf{k}\right)$$

is consistent with the counterclockwise motion around C. To apply Stokes' Theorem, we find

$$\text{curl } \mathbf{F} = \nabla \times \mathbf{F} = \begin{vmatrix} \mathbf{i} & \mathbf{j} & \mathbf{k} \\ \frac{\partial}{\partial x} & \frac{\partial}{\partial y} & \frac{\partial}{\partial z} \\ xz & xy & 3xz \end{vmatrix} = (x - 3z)\mathbf{j} + y\mathbf{k}.$$

On the plane, z equals $2 - 2x - y$, so

$$\nabla \times \mathbf{F} = (x - 3(2 - 2x - y))\mathbf{j} + y\mathbf{k} = (7x + 3y - 6)\mathbf{j} + y\mathbf{k}$$

and

$$(\nabla \times \mathbf{F}) \cdot \mathbf{n} = \frac{1}{\sqrt{6}}\left(7x + 3y - 6 + y\right) = \frac{1}{\sqrt{6}}\left(7x + 4y - 6\right).$$

The surface area differential is

$$d\sigma = \frac{|\nabla f|}{|\nabla f \cdot \mathbf{k}|} dA = \frac{\sqrt{6}}{1} dx\, dy. \qquad \text{Formula (7) in Section 16.5}$$

The circulation is

$$\oint_C \mathbf{F} \cdot d\mathbf{r} = \iint_S (\nabla \times \mathbf{F}) \cdot \mathbf{n}\, d\sigma \qquad \text{Stokes' Theorem, Eq. (4)}$$

$$= \int_0^1 \int_0^{2-2x} \frac{1}{\sqrt{6}}\left(7x + 4y - 6\right)\sqrt{6}\, dy\, dx$$

$$= \int_0^1 \int_0^{2-2x} (7x + 4y - 6)\, dy\, dx = -1. \qquad \blacksquare$$

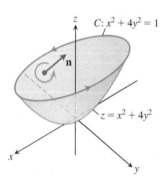

FIGURE 16.68 The portion of the ellipitical paraboloid in Example 10, showing its curve of intersection C with the plane $z = 1$ and its inner normal orientation by $\mathbf{n}$.

EXAMPLE 10   Let the surface S be the ellipitical paraboloid $z = x^2 + 4y^2$ lying beneath the plane $z = 1$ (Figure 16.68). We define the orientation of S by taking the *inner* normal vector $\mathbf{n}$ to the surface, which is the normal having a positive $\mathbf{k}$-component. Find the flux of $\nabla \times \mathbf{F}$ across S in the direction $\mathbf{n}$ for the vector field $\mathbf{F} = y\mathbf{i} - xz\mathbf{j} + xz^2\mathbf{k}$.

Solution   We use Stokes' Theorem to calculate the curl integral by finding the equivalent counterclockwise circulation of $\mathbf{F}$ around the curve of intersection C of the paraboloid $z = x^2 + 4y^2$ and the plane $z = 1$, as shown in Figure 16.68. Note that the orientation of S is consistent with traversing C in a counterclockwise direction around the z-axis. The curve C is the ellipse $x^2 + 4y^2 = 1$ in the plane $z = 1$. We can parametrize the ellipse by $x = \cos t$, $y = \frac{1}{2}\sin t$, $z = 1$ for $0 \leq t \leq 2\pi$, so C is given by

$$\mathbf{r}(t) = (\cos t)\mathbf{i} + \frac{1}{2}(\sin t)\mathbf{j} + \mathbf{k}, \qquad 0 \leq t \leq 2\pi.$$

To compute the circulation integral $\oint_C \mathbf{F} \cdot d\mathbf{r}$, we evaluate $\mathbf{F}$ along C and find the velocity vector $d\mathbf{r}/dt$:

$$\mathbf{F}(\mathbf{r}(t)) = \frac{1}{2}(\sin t)\mathbf{i} - (\cos t)\mathbf{j} + (\cos t)\mathbf{k}$$

and

$$\frac{d\mathbf{r}}{dt} = -(\sin t)\mathbf{i} + \frac{1}{2}(\cos t)\mathbf{j}.$$

Then,

$$\oint_C \mathbf{F} \cdot d\mathbf{r} = \int_0^{2\pi} \mathbf{F}(\mathbf{r}(t)) \cdot \frac{d\mathbf{r}}{dt}\, dt$$

$$= \int_0^{2\pi} \left(-\frac{1}{2}\sin^2 t - \frac{1}{2}\cos^2 t\right) dt$$

$$= -\frac{1}{2}\int_0^{2\pi} dt = -\pi.$$

Therefore, by Stokes' Theorem the flux of the curl across S in the direction $\mathbf{n}$ for the field $\mathbf{F}$ is

$$\iint_S (\nabla \times \mathbf{F}) \cdot \mathbf{n}\, d\sigma = -\pi. \qquad \blacksquare$$

Proof Outline of Stokes' Theorem for Polyhedral Surfaces

Let S be a polyhedral surface consisting of a finite number of plane regions or faces. (See Figure 16.69 for examples.) We apply Green's Theorem to each separate face of S. There are two types of faces:

1. Those that are surrounded on all sides by other faces.

2. Those that have one or more edges that are not adjacent to other faces.

The boundary of S consists of those edges of the type 2 faces that are not adjacent to other faces. In Figure 16.69a, the triangles EAB, BCE, and CDE represent a part of S, with $ABCD$ part of the boundary of the surface, boundary(S). Although Green's Theorem was stated for curves in the xy-plane, a generalized form applies to curves that lie in a plane in space. In the generalized form, the theorem asserts that the line integral of $\mathbf{F}$ around the curve enclosing the plane region R normal to $\mathbf{n}$ equals the double integral of (curl $\mathbf{F}$) $\cdot$ $\mathbf{n}$ over R. Applying this generalized form to the three triangles of Figure 16.69a in turn, and adding the results, gives

$$\left(\oint_{EAB} + \oint_{BCE} + \oint_{CDE}\right) \mathbf{F} \cdot d\mathbf{r} = \left(\iint_{EAB} + \iint_{BCE} + \iint_{CDE}\right)(\nabla \times \mathbf{F}) \cdot \mathbf{n}\, d\sigma. \qquad (7)$$

The three line integrals on the left-hand side of Equation (7) combine into a single line integral taken around the periphery $ABCDE$ because the integrals along interior segments cancel in pairs. For example, the integral along segment BE in triangle ABE is opposite in sign to the integral along the same segment in triangle EBC. The same holds for segment CE. Hence, Equation (7) reduces to

$$\oint_{ABCDE} \mathbf{F} \cdot d\mathbf{r} = \iint_{ABCDE} (\nabla \times \mathbf{F}) \cdot \mathbf{n}\, d\sigma.$$

When we apply Green's Theorem to all the faces and add the results, we get

$$\oint_{\text{boundary}(S)} \mathbf{F} \cdot d\mathbf{r} = \iint_S (\nabla \times \mathbf{F}) \cdot \mathbf{n}\, d\sigma.$$

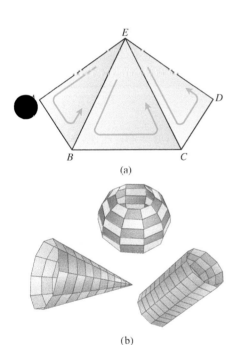

FIGURE 16.69 (a) Part of a polyhedral surface. (b) Other polyhedral surfaces.

This is Stokes' Theorem for the polyhedral surface S in Figure 16.69a. More general polyhedral surfaces are shown in Figure 16.69b and the proof can be extended to them. General smooth surfaces can be obtained as limits of polyhedral surfaces.

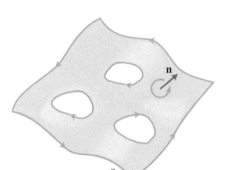

Stokes' Theorem for Surfaces with Holes

Stokes' Theorem holds for an oriented surface S that has one or more holes (Figure 16.70). The surface integral over S of the normal component of $\nabla \times \mathbf{F}$ equals the sum of the line integrals around all the boundary curves of the tangential component of $\mathbf{F}$, where the curves are to be traced in the direction induced by the orientation of S. For such surfaces the theorem is unchanged, but C is considered as a union of simple closed curves.

An Important Identity

The following identity arises frequently in mathematics and the physical sciences.

$$\text{curl grad } f = \mathbf{0} \qquad \text{or} \qquad \nabla \times \nabla f = \mathbf{0} \tag{8}$$

FIGURE 16.70 Stokes' Theorem also holds for oriented surfaces with holes. Consistent with the orientation of S, the outer curve is traversed counterclockwise around $\mathbf{n}$ and the inner curves surrounding the holes are traversed clockwise.

Forces arising in the study of electromagnetism and gravity are often associated with a potential function f. The identity (8) says that these forces have curl equal to zero. The identity (8) holds for any function $f(x, y, z)$ whose second partial derivatives are continuous. The proof goes like this:

$$\nabla \times \nabla f = \begin{vmatrix} \mathbf{i} & \mathbf{j} & \mathbf{k} \\ \dfrac{\partial}{\partial x} & \dfrac{\partial}{\partial y} & \dfrac{\partial}{\partial z} \\ \dfrac{\partial f}{\partial x} & \dfrac{\partial f}{\partial y} & \dfrac{\partial f}{\partial z} \end{vmatrix} = (f_{zy} - f_{yz})\mathbf{i} - (f_{zx} - f_{xz})\mathbf{j} + (f_{yx} - f_{xy})\mathbf{k}.$$

If the second partial derivatives are continuous, the mixed second derivatives in parentheses are equal (Theorem 2, Section 14.3) and the vector is zero.

Conservative Fields and Stokes' Theorem

In Section 16.3, we found that a field $\mathbf{F}$ being conservative in an open region D in space is equivalent to the integral of $\mathbf{F}$ around every closed loop in D being zero. This, in turn, is equivalent in *simply connected* open regions to saying that $\nabla \times \mathbf{F} = \mathbf{0}$ (which gives a test for determining if $\mathbf{F}$ is conservative for such regions).

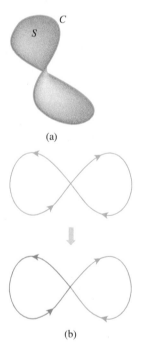

THEOREM 7—Curl F = 0 Related to the Closed-Loop Property
If $\nabla \times \mathbf{F} = \mathbf{0}$ at every point of a simply connected open region D in space, then on any piecewise-smooth closed path C in D,

$$\oint_C \mathbf{F} \cdot d\mathbf{r} = 0.$$

FIGURE 16.71 (a) In a simply connected open region in space, a simple closed curve C is the boundary of a smooth surface S. (b) Smooth curves that cross themselves can be divided into loops to which Stokes' Theorem applies.

Sketch of a Proof Theorem 7 can be proved in two steps. The first step is for simple closed curves (loops that do not cross themselves), like the one in Figure 16.71a. A theorem from topology, a branch of advanced mathematics, states that every smooth simple closed curve C in a simply connected open region D is the boundary of a smooth two-sided surface S that also lies in D. Hence, by Stokes' Theorem,

$$\oint_C \mathbf{F} \cdot d\mathbf{r} = \iint_S (\nabla \times \mathbf{F}) \cdot \mathbf{n} \, d\sigma = 0.$$

The second step is for curves that cross themselves, like the one in Figure 16.71b. The idea is to break these into simple loops spanned by orientable surfaces, apply Stokes' Theorem one loop at a time, and add the results.

The following diagram summarizes the results for conservative fields defined on connected, simply connected open regions. For such regions, the four statements are equivalent to each other.

Theorem 2,
Section 16.3

F conservative on D $\Longleftrightarrow$ $\mathbf{F} = \nabla f$ on D

Theorem 3,
Section 16.3 $\updownarrow$

Vector identity (Eq. 8)
(continuous second
partial derivatives)

$\oint_C \mathbf{F} \cdot d\mathbf{r} = 0$
over any closed
path in D

$\Longleftarrow$

Theorem 7
Domain's simple
connectivity and
Stokes' Theorem

$\nabla \times \mathbf{F} = \mathbf{0}$ throughout D

In Exercises 1–6, find the curl of each vector field **F**.

1. $\mathbf{F} = (x + y - z)\mathbf{i} + (2x - y + 3z)\mathbf{j} + (3x + 2y + z)\mathbf{k}$

2. $\mathbf{F} = (x^2 - y)\mathbf{i} + (y^2 - z)\mathbf{j} + (z^2 - x)\mathbf{k}$

3. $\mathbf{F} = (xy + z)\mathbf{i} + (yz + x)\mathbf{j} + (xz + y)\mathbf{k}$

4. $\mathbf{F} = ye^z\mathbf{i} + ze^x\mathbf{j} - xe^y\mathbf{k}$

5. $\mathbf{F} = x^2yz\mathbf{i} + xy^2z\mathbf{j} + xyz^2\mathbf{k}$

6. $\mathbf{F} = \dfrac{x}{yz}\mathbf{i} - \dfrac{y}{xz}\mathbf{j} + \dfrac{z}{xy}\mathbf{k}$

Using Stokes' Theorem to Find Line Integrals
In Exercises 7–12, use the surface integral in Stokes' Theorem to calculate the circulation of the field **F** around the curve C in the indicated direction.

7. $\mathbf{F} = x^2\mathbf{i} + 2x\mathbf{j} + z^2\mathbf{k}$

C: The ellipse $4x^2 + y^2 = 4$ in the xy-plane, counterclockwise when viewed from above

8. $\mathbf{F} = 2y\mathbf{i} + 3x\mathbf{j} - z^2\mathbf{k}$

C: The circle $x^2 + y^2 = 9$ in the xy-plane, counterclockwise when viewed from above

9. $\mathbf{F} = y\mathbf{i} + xz\mathbf{j} + x^2\mathbf{k}$

C: The boundary of the triangle cut from the plane $x + y + z = 1$ by the first octant, counterclockwise when viewed from above

10. $\mathbf{F} = (y^2 + z^2)\mathbf{i} + (x^2 + z^2)\mathbf{j} + (x^2 + y^2)\mathbf{k}$

C: The boundary of the triangle cut from the plane $x + y + z = 1$ by the first octant, counterclockwise when viewed from above

11. $\mathbf{F} = (y^2 + z^2)\mathbf{i} + (x^2 + y^2)\mathbf{j} + (x^2 + y^2)\mathbf{k}$

C: The square bounded by the lines $x = \pm 1$ and $y = \pm 1$ in the xy-plane, counterclockwise when viewed from above

12. $\mathbf{F} = x^2y^3\mathbf{i} + \mathbf{j} + z\mathbf{k}$

C: The intersection of the cylinder $x^2 + y^2 = 4$ and the hemisphere $x^2 + y^2 + z^2 = 16$, $z \geq 0$, counterclockwise when viewed from above

Integral of the Curl Vector Field
13. Let **n** be the outer unit normal of the elliptical shell

$$S: \quad 4x^2 + 9y^2 + 36z^2 = 36, \qquad z \geq 0,$$

and let

$$\mathbf{F} = y\mathbf{i} + x^2\mathbf{j} + (x^2 + y^4)^{3/2} \sin e^{\sqrt{xyz}} \, \mathbf{k}.$$

Find the value of

$$\iint_S (\nabla \times \mathbf{F}) \cdot \mathbf{n} \, d\sigma.$$

(*Hint:* One parametrization of the ellipse at the base of the shell is $x = 3 \cos t, y = 2 \sin t, 0 \leq t \leq 2\pi$.)

14. Let **n** be the outer unit normal (normal away from the origin) of the parabolic shell

$$S: \quad 4x^2 + y + z^2 = 4, \qquad y \geq 0,$$

and let

$$\mathbf{F} = \left(-z + \frac{1}{2+x}\right)\mathbf{i} + (\tan^{-1}y)\mathbf{j} + \left(x + \frac{1}{4+z}\right)\mathbf{k}.$$

Find the value of

$$\iint_S (\nabla \times \mathbf{F}) \cdot \mathbf{n}\, d\sigma.$$

15. Let S be the cylinder $x^2 + y^2 = a^2$, $0 \le z \le h$, together with its top, $x^2 + y^2 \le a^2$, $z = h$. Let $\mathbf{F} = -y\mathbf{i} + x\mathbf{j} + x^2\mathbf{k}$. Use Stokes' Theorem to find the flux of $\nabla \times \mathbf{F}$ outward through S.

16. Evaluate

$$\iint_S (\nabla \times (y\mathbf{i})) \cdot \mathbf{n}\, d\sigma,$$

where S is the hemisphere $x^2 + y^2 + z^2 = 1$, $z \ge 0$.

17. Suppose $\mathbf{F} = \nabla \times \mathbf{A}$, where

$$\mathbf{A} = (y + \sqrt{z})\mathbf{i} + e^{xyz}\mathbf{j} + \cos(xz)\mathbf{k}.$$

Determine the flux of $\mathbf{F}$ outward through the hemisphere $x^2 + y^2 + z^2 = 1$, $z \ge 0$.

18. Repeat Exercise 17 for the flux of $\mathbf{F}$ across the entire unit sphere.

Stokes' Theorem for Parametrized Surfaces

In Exercises 19–24, use the surface integral in Stokes' Theorem to calculate the flux of the curl of the field $\mathbf{F}$ across the surface S in the direction of the outward unit normal $\mathbf{n}$.

19. $\mathbf{F} = 2z\mathbf{i} + 3x\mathbf{j} + 5y\mathbf{k}$

S: $\mathbf{r}(r, \theta) = (r \cos \theta)\mathbf{i} + (r \sin \theta)\mathbf{j} + (4 - r^2)\mathbf{k}$,
$0 \le r \le 2$, $0 \le \theta \le 2\pi$

20. $\mathbf{F} = (y - z)\mathbf{i} + (z - x)\mathbf{j} + (x + z)\mathbf{k}$

S: $\mathbf{r}(r, \theta) = (r \cos \theta)\mathbf{i} + (r \sin \theta)\mathbf{j} + (9 - r^2)\mathbf{k}$,
$0 \le r \le 3$, $0 \le \theta \le 2\pi$

21. $\mathbf{F} = x^2 y\mathbf{i} + 2y^3 z\mathbf{j} + 3z\mathbf{k}$

S: $\mathbf{r}(r, \theta) = (r \cos \theta)\mathbf{i} + (r \sin \theta)\mathbf{j} + r\mathbf{k}$,
$0 \le r \le 1$, $0 \le \theta \le 2\pi$

22. $\mathbf{F} = (x - y)\mathbf{i} + (y - z)\mathbf{j} + (z - x)\mathbf{k}$

S: $\mathbf{r}(r, \theta) = (r \cos \theta)\mathbf{i} + (r \sin \theta)\mathbf{j} + (5 - r)\mathbf{k}$,
$0 \le r \le 5$, $0 \le \theta \le 2\pi$

23. $\mathbf{F} = 3y\mathbf{i} + (5 - 2x)\mathbf{j} + (z^2 - 2)\mathbf{k}$

S: $\mathbf{r}(\phi, \theta) = (\sqrt{3} \sin \phi \cos \theta)\mathbf{i} + (\sqrt{3} \sin \phi \sin \theta)\mathbf{j} + (\sqrt{3} \cos \phi)\mathbf{k}$, $0 \le \phi \le \pi/2$, $0 \le \theta \le 2\pi$

24. $\mathbf{F} = y^2\mathbf{i} + z^2\mathbf{j} + x\mathbf{k}$

S: $\mathbf{r}(\phi, \theta) = (2 \sin \phi \cos \theta)\mathbf{i} + (2 \sin \phi \sin \theta)\mathbf{j} + (2 \cos \phi)\mathbf{k}$, $0 \le \phi \le \pi/2$, $0 \le \theta \le 2\pi$

Theory and Examples

25. Let C be the smooth curve $\mathbf{r}(t) = (2 \cos t)\mathbf{i} + (2 \sin t)\mathbf{j} + (3 - 2 \cos^3 t)\mathbf{k}$, oriented to be traversed counterclockwise around the z-axis when viewed from above. Let S be the piecewise smooth cylindrical surface $x^2 + y^2 = 4$, below the curve for $z \ge 0$, together with the base disk in the xy-plane. Note that C lies

on the cylinder S and above the xy-plane (see the accompanying figure). Verify Equation (4) in Stokes' Theorem for the vector field $\mathbf{F} = y\mathbf{i} - x\mathbf{j} + x^2\mathbf{k}$.

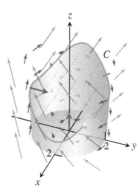

26. Verify Stokes' Theorem for the vector field $\mathbf{F} = 2xy\mathbf{i} + x\mathbf{j} + (y + z)\mathbf{k}$ and surface $z = 4 - x^2 - y^2$, $z \ge 0$, oriented with unit normal $\mathbf{n}$ pointing upward.

27. **Zero circulation** Use Equation (8) and Stokes' Theorem to show that the circulations of the following fields around the boundary of any smooth orientable surface in space are zero.

a. $\mathbf{F} = 2x\mathbf{i} + 2y\mathbf{j} + 2z\mathbf{k}$ b. $\mathbf{F} = \nabla(xy^2 z^3)$
c. $\mathbf{F} = \nabla \times (x\mathbf{i} + y\mathbf{j} + z\mathbf{k})$ d. $\mathbf{F} = \nabla f$

28. **Zero circulation** Let $f(x, y, z) = (x^2 + y^2 + z^2)^{-1/2}$. Show that the clockwise circulation of the field $\mathbf{F} = \nabla f$ around the circle $x^2 + y^2 = a^2$ in the xy-plane is zero.

a. by taking $\mathbf{r} = (a \cos t)\mathbf{i} + (a \sin t)\mathbf{j}$, $0 \le t \le 2\pi$, and integrating $\mathbf{F} \cdot d\mathbf{r}$ over the circle.

b. by applying Stokes' Theorem.

29. Let C be a simple closed smooth curve in the plane $2x + 2y + z = 2$, oriented as shown here. Show that

$$\oint_C 2y\, dx + 3z\, dy - x\, dz$$

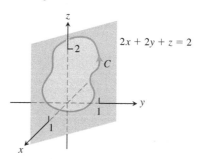

depends only on the area of the region enclosed by C and not on the position or shape of C.

30. Show that if $\mathbf{F} = x\mathbf{i} + y\mathbf{j} + z\mathbf{k}$, then $\nabla \times \mathbf{F} = \mathbf{0}$.

31. Find a vector field with twice-differentiable components whose curl is $x\mathbf{i} + y\mathbf{j} + z\mathbf{k}$ or prove that no such field exists.

32. Does Stokes' Theorem say anything special about circulation in a field whose curl is zero? Give reasons for your answer.

33. Let R be a region in the xy-plane that is bounded by a piecewise smooth simple closed curve C and suppose that the moments of inertia of R about the x- and y-axes are known to be I_x and I_y.

Evaluate the integral

$$\oint_C \nabla(r^4) \cdot \mathbf{n}\, ds,$$

where $r = \sqrt{x^2 + y^2}$, in terms of I_x and I_y.

34. Zero curl, yet the field is not conservative Show that the curl of

$$\mathbf{F} = \frac{-y}{x^2 + y^2}\mathbf{i} + \frac{x}{x^2 + y^2}\mathbf{j} + z\mathbf{k}$$

is zero but that

$$\oint_C \mathbf{F} \cdot d\mathbf{r}$$

is not zero if C is the circle $x^2 + y^2 = 1$ in the xy-plane. (Theorem 7 does not apply here because the domain of $\mathbf{F}$ is not simply connected. The field $\mathbf{F}$ is not defined along the z-axis so there is no way to contract C to a point without leaving the domain of $\mathbf{F}$.)

16.8 The Divergence Theorem and a Unified Theory

The divergence form of Green's Theorem in the plane states that the net outward flux of a vector field across a simple closed curve can be calculated by integrating the divergence of the field over the region enclosed by the curve. The corresponding theorem in three dimensions, called the *Divergence Theorem*, states that the net outward flux of a vector field across a closed surface in space can be calculated by integrating the divergence of the field over the region enclosed by the surface. In this section we prove the Divergence Theorem and show how it simplifies the calculation of flux, which is the integral of the field over the closed oriented surface. We also derive Gauss's law for flux in an electric field and the continuity equation of hydrodynamics. Finally, we summarize the chapter's vector integral theorems in a single unifying principle generalizing the Fundamental Theorem of Calculus.

Divergence in Three Dimensions

The **divergence** of a vector field $\mathbf{F} = M(x, y, z)\mathbf{i} + N(x, y, z)\mathbf{j} + P(x, y, z)\mathbf{k}$ is the scalar function

$$\text{div } \mathbf{F} = \nabla \cdot \mathbf{F} = \frac{\partial M}{\partial x} + \frac{\partial N}{\partial y} + \frac{\partial P}{\partial z}. \tag{1}$$

The symbol "div $\mathbf{F}$" is read as "divergence of $\mathbf{F}$" or "div $\mathbf{F}$." The notation $\nabla \cdot \mathbf{F}$ is read "del dot $\mathbf{F}$."

Div $\mathbf{F}$ has the same physical interpretation in three dimensions that it does in two. If $\mathbf{F}$ is the velocity field of a flowing gas, the value of div $\mathbf{F}$ at a point (x, y, z) is the rate at which the gas is compressing or expanding at (x, y, z). The divergence is the flux per unit volume or *flux density* at the point.

EXAMPLE 1 The following vector fields represent the velocity of a gas flowing in space. Find the divergence of each vector field and interpret its physical meaning. Figure 16.72 displays the vector fields.

(a) Expansion: $\mathbf{F}(x, y, z) = x\mathbf{i} + y\mathbf{j} + z\mathbf{k}$

(b) Compression: $\mathbf{F}(x, y, z) = -x\mathbf{i} - y\mathbf{j} - z\mathbf{k}$

(c) Rotation about the z-axis: $\mathbf{F}(x, y, z) = -y\mathbf{i} + x\mathbf{j}$

(d) Shearing along parallel horizontal planes: $\mathbf{F}(x, y, z) = z\mathbf{j}$

Solution

(a) $\text{div } \mathbf{F} = \frac{\partial}{\partial x}(x) + \frac{\partial}{\partial y}(y) + \frac{\partial}{\partial z}(z) = 3$: The gas is undergoing constant uniform expansion at all points.

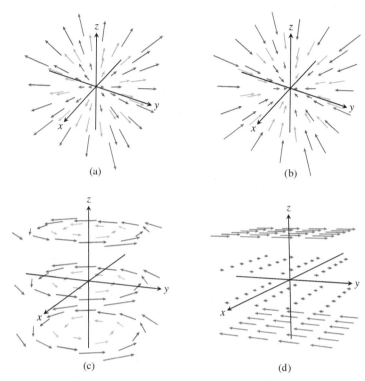

FIGURE 16.72 Velocity fields of a gas flowing in space (Example 1).

(b) div $\mathbf{F} = \dfrac{\partial}{\partial x}(-x) + \dfrac{\partial}{\partial y}(-y) + \dfrac{\partial}{\partial z}(-z) = -3$: The gas is undergoing constant uniform compression at all points.

(c) div $\mathbf{F} = \dfrac{\partial}{\partial x}(-y) + \dfrac{\partial}{\partial y}(x) = 0$: The gas is neither expanding nor compressing at any point.

(d) div $\mathbf{F} = \dfrac{\partial}{\partial y}(z) = 0$: Again, the divergence is zero at all points in the domain of the velocity field, so the gas is neither expanding nor compressing at any point. ∎

Divergence Theorem

The Divergence Theorem says that under suitable conditions, the outward flux of a vector field across a closed surface equals the triple integral of the divergence of the field over the three-dimensional region enclosed by the surface.

THEOREM 8—Divergence Theorem

Let $\mathbf{F}$ be a vector field whose components have continuous first partial derivatives, and let S be a piecewise smooth oriented closed surface. The flux of $\mathbf{F}$ across S in the direction of the surface's outward unit normal field $\mathbf{n}$ equals the triple integral of the divergence $\nabla \cdot \mathbf{F}$ over the region D enclosed by the surface:

$$\iint_S \mathbf{F} \cdot \mathbf{n}\, d\sigma = \iiint_D \nabla \cdot \mathbf{F}\, dV. \tag{2}$$

$\underset{\text{Outward flux}}{} \qquad \underset{\text{Divergence integral}}{}$

EXAMPLE 2   Evaluate both sides of Equation (2) for the expanding vector field $\mathbf{F} = x\mathbf{i} + y\mathbf{j} + z\mathbf{k}$ over the sphere $x^2 + y^2 + z^2 = a^2$ (Figure 16.73).

Solution   The outer unit normal to S, calculated from the gradient of $f(x, y, z) = x^2 + y^2 + z^2 - a^2$, is

$$\mathbf{n} = \frac{2(x\mathbf{i} + y\mathbf{j} + z\mathbf{k})}{\sqrt{4(x^2 + y^2 + z^2)}} = \frac{x\mathbf{i} + y\mathbf{j} + z\mathbf{k}}{a}. \qquad x^2 + y^2 + z^2 = a^2 \text{ on } S$$

It follows that

$$\mathbf{F} \cdot \mathbf{n}\, d\sigma = \frac{x^2 + y^2 + z^2}{a}\, d\sigma = \frac{a^2}{a}\, d\sigma = a\, d\sigma.$$

Therefore, the outward flux is

$$\iint_S \mathbf{F} \cdot \mathbf{n}\, d\sigma = \iint_S a\, d\sigma = a \iint_S d\sigma = a(4\pi a^2) = 4\pi a^3. \qquad \text{Area of } S \text{ is } 4\pi a^2.$$

For the right-hand side of Equation (2), the divergence of $\mathbf{F}$ is

$$\nabla \cdot \mathbf{F} = \frac{\partial}{\partial x}(x) + \frac{\partial}{\partial y}(y) + \frac{\partial}{\partial z}(z) = 3,$$

so we obtain the divergence integral,

$$\iiint_D \nabla \cdot \mathbf{F}\, dV = \iiint_D 3\, dV = 3\left(\frac{4}{3}\pi a^3\right) = 4\pi a^3. \qquad \blacksquare$$

Many vector fields of interest in applied science have zero divergence at each point. A common example is the velocity field of a circulating incompressible liquid, since it is neither expanding nor contracting. Other examples include constant vector fields $\mathbf{F} = a\mathbf{i} + b\mathbf{j} + c\mathbf{k}$, and velocity fields for shearing action along a fixed plane (see Example 1d). If $\mathbf{F}$ is a vector field whose divergence is zero at each point in the region D, then the integral on the right-hand side of Equation (2) equals 0. So if S is any closed surface for which the Divergence Theorem applies, then the outward flux of $\mathbf{F}$ across S is zero. We state this important application of the Divergence Theorem.

> **COROLLARY**   The outward flux across a piecewise smooth oriented closed surface S is zero for any vector field $\mathbf{F}$ having zero divergence at every point of the region enclosed by the surface.

EXAMPLE 3   Find the flux of $\mathbf{F} = xy\mathbf{i} + yz\mathbf{j} + xz\mathbf{k}$ outward through the surface of the cube cut from the first octant by the planes $x = 1$, $y = 1$, and $z = 1$.

Solution   Instead of calculating the flux as a sum of six separate integrals, one for each face of the cube, we can calculate the flux by integrating the divergence

$$\nabla \cdot \mathbf{F} = \frac{\partial}{\partial x}(xy) + \frac{\partial}{\partial y}(yz) + \frac{\partial}{\partial z}(xz) = y + z + x$$

over the cube's interior:

$$\text{Flux} = \iint_{\substack{\text{Cube} \\ \text{surface}}} \mathbf{F} \cdot \mathbf{n}\, d\sigma = \iiint_{\substack{\text{Cube} \\ \text{interior}}} \nabla \cdot \mathbf{F}\, dV \qquad \text{The Divergence Theorem}$$

$$= \int_0^1 \int_0^1 \int_0^1 (x + y + z)\, dx\, dy\, dz = \frac{3}{2}. \qquad \text{Routine integration} \qquad \blacksquare$$

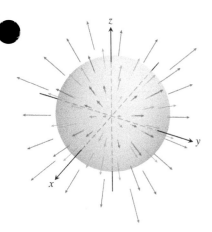

FIGURE 16.73   A uniformly expanding vector field and a sphere (Example 2).

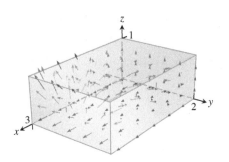

FIGURE 16.74   The integral of div **F** over this region equals the total flux across the six sides (Example 4).

EXAMPLE 4

(a) Calculate the flux of the vector field

$$\mathbf{F} = x^2\mathbf{i} + 4xyz\mathbf{j} + ze^x\mathbf{k}$$

out of the box-shaped region D: $0 \le x \le 3, 0 \le y \le 2, 0 \le z \le 1$. (See Figure 16.74.)

(b) Integrate div **F** over this region and show that the result is the same value as in part (a), as asserted by the Divergence Theorem.

Solution

(a) The region D has six sides. We calculate the flux across each side in turn. Consider the top side in the plane $z = 1$, having outward normal $\mathbf{n} = \mathbf{k}$. The flux across this side is given by $\mathbf{F} \cdot \mathbf{n} = ze^x$. Since $z = 1$ on this side, the flux at a point (x, y, z) on the top is e^x. The total outward flux across this side is given by the surface integral

$$\int_0^2 \int_0^3 e^x \, dx \, dy = 2e^3 - 2. \qquad \text{\small Routine integration}$$

The outward flux across the other sides is computed similarly, and the results are summarized in the following table.

Side	Unit normal n	F · n	Flux across side
$x = 0$	$-\mathbf{i}$	$-x^2 = 0$	0
$x = 3$	$\mathbf{i}$	$x^2 = 9$	18
$y = 0$	$-\mathbf{j}$	$-4xyz = 0$	0
$y = 2$	$\mathbf{j}$	$4xyz = 8xz$	18
$z = 0$	$-\mathbf{k}$	$-ze^x = 0$	0
$z = 1$	$\mathbf{k}$	$ze^x = e^x$	$2e^3 - 2$

The total outward flux is obtained by adding the terms for each of the six sides, giving

$$18 + 18 + 2e^3 - 2 = 34 + 2e^3.$$

(b) We first compute the divergence of **F**, obtaining

$$\text{div } \mathbf{F} = \nabla \cdot \mathbf{F} = 2x + 4xz + e^x.$$

The integral of the divergence of **F** over D is

$$\iiint_D \text{div } \mathbf{F} \, dV = \int_0^1 \int_0^2 \int_0^3 (2x + 4xz + e^x) \, dx \, dy \, dz$$

$$= \int_0^1 \int_0^2 (8 + 18z + e^3) \, dy \, dz$$

$$= \int_0^1 (16 + 36z + 2e^3) \, dz$$

$$= 34 + 2e^3.$$

As asserted by the Divergence Theorem, the integral of the divergence over D equals the outward flux across the boundary surface of D. ∎

Divergence and the Curl

If $\mathbf{F}$ is a vector field on three-dimensional space, then the curl $\nabla \times \mathbf{F}$ is also a vector field on three-dimensional space. So we can calculate the divergence of $\nabla \times \mathbf{F}$ using Equation (1). The result of this calculation is always 0.

THEOREM 9   If $\mathbf{F} = M\mathbf{i} + N\mathbf{j} + P\mathbf{k}$ is a vector field with continuous second partial derivatives, then
$$\text{div (curl } \mathbf{F}) = \nabla \cdot (\nabla \times \mathbf{F}) = 0.$$

Proof   From the definitions of the divergence and curl, we have

$$\text{div (curl } \mathbf{F}) = \nabla \cdot (\nabla \times \mathbf{F})$$

$$= \frac{\partial}{\partial x}\left(\frac{\partial P}{\partial y} - \frac{\partial N}{\partial z}\right) + \frac{\partial}{\partial y}\left(\frac{\partial M}{\partial z} - \frac{\partial P}{\partial x}\right) + \frac{\partial}{\partial z}\left(\frac{\partial N}{\partial x} - \frac{\partial M}{\partial y}\right)$$

$$= \frac{\partial^2 P}{\partial x\,\partial y} - \frac{\partial^2 N}{\partial x\,\partial z} + \frac{\partial^2 M}{\partial y\,\partial z} - \frac{\partial^2 P}{\partial y\,\partial x} + \frac{\partial^2 N}{\partial z\,\partial x} - \frac{\partial^2 M}{\partial z\,\partial y}$$

$$= 0,$$

because the mixed second partial derivatives cancel by the Mixed Derivative Theorem in Section 14.3. ∎

Theorem 9 has some interesting applications. If a vector field $\mathbf{G} = \text{curl } \mathbf{F}$, then the field $\mathbf{G}$ must have divergence 0. Saying this another way, if div $\mathbf{G} \neq 0$, then $\mathbf{G}$ cannot be the curl of any vector field $\mathbf{F}$ having continuous second partial derivatives. Moreover, if $\mathbf{G} = \text{curl } \mathbf{F}$, then the outward flux of $\mathbf{G}$ across any closed surface S is zero by the corollary to the Divergence Theorem, provided the conditions of the theorem are satisfied. So if there is a closed surface for which the surface integral of the vector field $\mathbf{G}$ is nonzero, we can conclude that $\mathbf{G}$ is *not* the curl of some vector field $\mathbf{F}$.

Proof of the Divergence Theorem for Special Regions

To prove the Divergence Theorem, we take the components of $\mathbf{F}$ to have continuous first partial derivatives. We first assume that D is a convex region with no holes or bubbles, such as a solid ball, cube, or ellipsoid, and that S is a piecewise smooth surface. In addition, we assume that any line perpendicular to the xy-plane at an interior point of the region R_{xy} that is the projection of D on the xy-plane intersects the surface S in exactly two points, producing surfaces

$$S_1: \quad z = f_1(x, y), \qquad (x, y) \text{ in } R_{xy}$$
$$S_2: \quad z = f_2(x, y), \qquad (x, y) \text{ in } R_{xy},$$

with $f_1 \leq f_2$. We make similar assumptions about the projection of D onto the other coordinate planes. See Figure 16.75, which illustrates these assumptions.

The components of the unit normal vector $\mathbf{n} = n_1\mathbf{i} + n_2\mathbf{j} + n_3\mathbf{k}$ are the cosines of the angles α, β, and γ that $\mathbf{n}$ makes with $\mathbf{i}$, $\mathbf{j}$, and $\mathbf{k}$ (Figure 16.76). This is true because all the vectors involved are unit vectors, giving the *direction cosines*

$$n_1 = \mathbf{n} \cdot \mathbf{i} = |\mathbf{n}||\mathbf{i}| \cos \alpha = \cos \alpha$$
$$n_2 = \mathbf{n} \cdot \mathbf{j} = |\mathbf{n}||\mathbf{j}| \cos \beta = \cos \beta$$
$$n_3 = \mathbf{n} \cdot \mathbf{k} = |\mathbf{n}||\mathbf{k}| \cos \gamma = \cos \gamma.$$

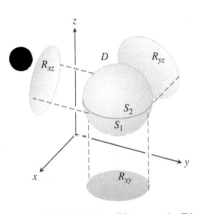

FIGURE 16.75   We prove the Divergence Theorem for the kind of three-dimensional region shown here.

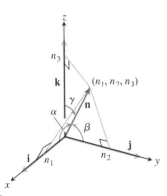

FIGURE 16.76   The components of $\mathbf{n}$ are the cosines of the angles α, β, and γ that it makes with $\mathbf{i}$, $\mathbf{j}$, and $\mathbf{k}$.

Thus, the unit normal vector is given by

$$\mathbf{n} = (\cos \alpha)\mathbf{i} + (\cos \beta)\mathbf{j} + (\cos \gamma)\mathbf{k}$$

and

$$\mathbf{F} \cdot \mathbf{n} = M \cos \alpha + N \cos \beta + P \cos \gamma.$$

In component form, the Divergence Theorem states that

$$\iint_S \underbrace{(M \cos \alpha + N \cos \beta + P \cos \gamma)}_{\mathbf{F} \cdot \mathbf{n}} d\sigma = \iiint_D \underbrace{\left(\frac{\partial M}{\partial x} + \frac{\partial N}{\partial y} + \frac{\partial P}{\partial z} \right)}_{\text{div } \mathbf{F}} dx\, dy\, dz.$$

We prove the theorem by establishing the following three equations:

$$\iint_S M \cos \alpha \, d\sigma = \iiint_D \frac{\partial M}{\partial x} dx\, dy\, dz \tag{3}$$

$$\iint_S N \cos \beta \, d\sigma = \iiint_D \frac{\partial N}{\partial y} dx\, dy\, dz \tag{4}$$

$$\iint_S P \cos \gamma \, d\sigma = \iiint_D \frac{\partial P}{\partial z} dx\, dy\, dz \tag{5}$$

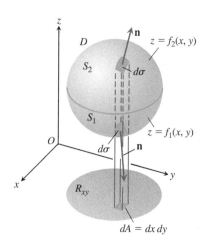

FIGURE 16.77 The region D enclosed by the surfaces S_1 and S_2 projects vertically onto R_{xy} in the xy-plane.

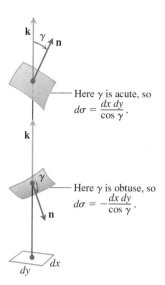

FIGURE 16.78 An enlarged view of the area patches in Figure 16.77. The relations $d\sigma = \pm dx\, dy/\cos \gamma$ come from Eq. (7) in Section 16.5 with $F = \mathbf{F} \cdot \mathbf{n}$.

Proof of Equation (5) We prove Equation (5) by converting the surface integral on the left to a double integral over the projection R_{xy} of D on the xy-plane (Figure 16.77). The surface S consists of an upper part S_2 whose equation is $z = f_2(x, y)$ and a lower part S_1 whose equation is $z = f_1(x, y)$. On S_2, the outer normal $\mathbf{n}$ has a positive $\mathbf{k}$-component and

$$\cos \gamma \, d\sigma = dx\, dy \quad \text{because} \quad d\sigma = \frac{dA}{|\cos \gamma|} = \frac{dx\, dy}{\cos \gamma}.$$

See Figure 16.78. On S_1, the outer normal $\mathbf{n}$ has a negative $\mathbf{k}$-component and

$$\cos \gamma \, d\sigma = -dx\, dy.$$

Therefore,

$$\iint_S P \cos \gamma \, d\sigma = \iint_{S_2} P \cos \gamma \, d\sigma + \iint_{S_1} P \cos \gamma \, d\sigma$$

$$= \iint_{R_{xy}} P(x, y, f_2(x, y)) \, dx\, dy - \iint_{R_{xy}} P(x, y, f_1(x, y)) \, dx\, dy$$

$$= \iint_{R_{xy}} \left[P(x, y, f_2(x, y)) - P(x, y, f_1(x, y)) \right] dx\, dy$$

$$= \iint_{R_{xy}} \left[\int_{f_1(x, y)}^{f_2(x, y)} \frac{\partial P}{\partial z} dz \right] dx\, dy = \iiint_D \frac{\partial P}{\partial z} dz\, dx\, dy.$$

This proves Equation (5). The proofs for Equations (3) and (4) follow the same pattern; or just permute x, y, z; M, N, P; α, β, γ, in order, and get those results from Equation (5). This proves the Divergence Theorem for these special regions. ∎

Divergence Theorem for Other Regions

The Divergence Theorem can be extended to regions that can be partitioned into a finite number of simple regions of the type just discussed and to regions that can be defined as limits of simpler regions in certain ways. For an example of one step in such a splitting process, suppose that D is the region between two concentric spheres and that $\mathbf{F}$ has continuously differentiable components throughout D and on the bounding surfaces. Split D by an equatorial plane and apply the Divergence Theorem to each half separately. The bottom half, D_1, is shown in Figure 16.79. The surface S_1 that bounds D_1 consists of an outer hemisphere, a plane washer-shaped base, and an inner hemisphere. The Divergence Theorem says that

$$\iint_{S_1} \mathbf{F} \cdot \mathbf{n}_1 \, d\sigma_1 = \iiint_{D_1} \nabla \cdot \mathbf{F} \, dV_1. \tag{6}$$

The unit normal $\mathbf{n}_1$ that points outward from D_1 points away from the origin along the outer surface, equals $\mathbf{k}$ along the flat base, and points toward the origin along the inner surface. Next apply the Divergence Theorem to D_2, and its surface S_2 (Figure 16.80):

$$\iint_{S_2} \mathbf{F} \cdot \mathbf{n}_2 \, d\sigma_2 = \iiint_{D_2} \nabla \cdot \mathbf{F} \, dV_2. \tag{7}$$

As we follow $\mathbf{n}_2$ over S_2, pointing outward from D_2, we see that $\mathbf{n}_2$ equals $-\mathbf{k}$ along the washer-shaped base in the xy-plane, points away from the origin on the outer sphere, and points toward the origin on the inner sphere. When we add Equations (6) and (7), the integrals over the flat base cancel because of the opposite signs of $\mathbf{n}_1$ and $\mathbf{n}_2$. We thus arrive at the result

$$\iint_S \mathbf{F} \cdot \mathbf{n} \, d\sigma = \iiint_D \nabla \cdot \mathbf{F} \, dV,$$

with D the region between the spheres, S the boundary of D consisting of two spheres, and $\mathbf{n}$ the unit normal to S directed outward from D.

EXAMPLE 5   Find the net outward flux of the field

$$\mathbf{F} = \frac{x\mathbf{i} + y\mathbf{j} + z\mathbf{k}}{\rho^3}, \qquad \rho = \sqrt{x^2 + y^2 + z^2} \tag{8}$$

across the boundary of the region $D: 0 < b^2 \le x^2 + y^2 + z^2 \le a^2$ (Figure 16.81).

Solution   The flux can be calculated by integrating $\nabla \cdot \mathbf{F}$ over D. Note that $\rho \ne 0$ in D. We have

$$\frac{\partial \rho}{\partial x} = \frac{1}{2}(x^2 + y^2 + z^2)^{-1/2}(2x) = \frac{x}{\rho}$$

and

$$\frac{\partial M}{\partial x} = \frac{\partial}{\partial x}(x\rho^{-3}) = \rho^{-3} - 3x\rho^{-4}\frac{\partial \rho}{\partial x} = \frac{1}{\rho^3} - \frac{3x^2}{\rho^5}.$$

Similarly,

$$\frac{\partial N}{\partial y} = \frac{1}{\rho^3} - \frac{3y^2}{\rho^5} \quad \text{and} \quad \frac{\partial P}{\partial z} = \frac{1}{\rho^3} - \frac{3z^2}{\rho^5}.$$

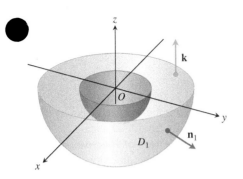

FIGURE 16.79   The lower half of the solid region between two concentric spheres.

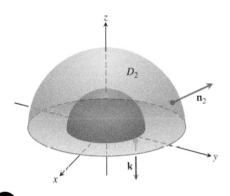

FIGURE 16.80   The upper half of the solid region between two concentric spheres.

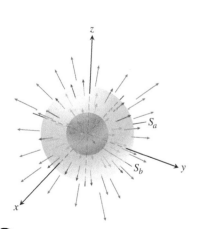

FIGURE 16.81   Two concentric spheres in an expanding vector field. The outer sphere S_a surrounds the inner sphere S_b.

Hence,

$$\text{div}\, \mathbf{F} = \frac{\partial M}{\partial x} + \frac{\partial N}{\partial y} + \frac{\partial P}{\partial z} = \frac{3}{\rho^3} - \frac{3}{\rho^5}(x^2 + y^2 + z^2) = \frac{3}{\rho^3} - \frac{3\rho^2}{\rho^5} = 0.$$

So the net outward flux of $\mathbf{F}$ across the boundary of D is zero by the corollary to the Divergence Theorem. There is more to learn about this vector field $\mathbf{F}$, though. The flux leaving D across the inner sphere S_b is the negative of the flux leaving D across the outer sphere S_a (because the sum of these fluxes is zero). Hence, the flux of $\mathbf{F}$ across S_b in the direction away from the origin equals the flux of $\mathbf{F}$ across S_a in the direction away from the origin. Thus, the flux of $\mathbf{F}$ across a sphere centered at the origin is independent of the radius of the sphere. What is this flux?

To find it, we evaluate the flux integral directly for an arbitrary sphere S_a. The outward unit normal on the sphere of radius a is

$$\mathbf{n} = \frac{x\mathbf{i} + y\mathbf{j} + z\mathbf{k}}{\sqrt{x^2 + y^2 + z^2}} = \frac{x\mathbf{i} + y\mathbf{j} + z\mathbf{k}}{a}.$$

Hence, on the sphere,

$$\mathbf{F} \cdot \mathbf{n} = \frac{x\mathbf{i} + y\mathbf{j} + z\mathbf{k}}{a^3} \cdot \frac{x\mathbf{i} + y\mathbf{j} + z\mathbf{k}}{a} = \frac{x^2 + y^2 + z^2}{a^4} = \frac{a^2}{a^4} = \frac{1}{a^2}$$

and

$$\iint_{S_a} \mathbf{F} \cdot \mathbf{n}\, d\sigma = \frac{1}{a^2} \iint_{S_a} d\sigma = \frac{1}{a^2}(4\pi a^2) = 4\pi.$$

The outward flux of $\mathbf{F}$ in Equation (8) across any sphere centered at the origin is 4π. This result does not contradict the Divergence Theorem because $\mathbf{F}$ is not continuous at the origin. ∎

Gauss's Law: One of the Four Great Laws of Electromagnetic Theory

In electromagnetic theory, the electric field created by a point charge q located at the origin is

$$\mathbf{E}(x, y, z) = \frac{1}{4\pi\varepsilon_0} \frac{q}{|\mathbf{r}|^2}\left(\frac{\mathbf{r}}{|\mathbf{r}|}\right) = \frac{q}{4\pi\varepsilon_0} \frac{\mathbf{r}}{|\mathbf{r}|^3} = \frac{q}{4\pi\varepsilon_0} \frac{x\mathbf{i} + y\mathbf{j} + z\mathbf{k}}{\rho^3},$$

where ε_0 is a physical constant, $\mathbf{r}$ is the position vector of the point (x, y, z), and $\rho = |\mathbf{r}| = \sqrt{x^2 + y^2 + z^2}$. From Equation (8),

$$\mathbf{E} = \frac{q}{4\pi\varepsilon_0}\mathbf{F}.$$

The calculations in Example 5 show that the outward flux of $\mathbf{E}$ across any sphere centered at the origin is q/ε_0, but this result is not confined to spheres. The outward flux of $\mathbf{E}$ across any closed surface S that encloses the origin (and to which the Divergence Theorem applies) is also q/ε_0. To see why, we have only to imagine a large sphere S_a centered at the origin and enclosing the surface S (see Figure 16.82). Since

$$\nabla \cdot \mathbf{E} = \nabla \cdot \frac{q}{4\pi\varepsilon_0}\mathbf{F} = \frac{q}{4\pi\varepsilon_0}\nabla \cdot \mathbf{F} = 0$$

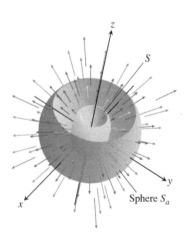

FIGURE 16.82 A sphere S_a surrounding another surface S. The tops of the surfaces are removed for visualization.

when $\rho > 0$, the triple integral of $\nabla \cdot \mathbf{E}$ over the region D between S and S_a is zero. Hence, by the Divergence Theorem,

$$\iint_{\substack{\text{Boundary} \\ \text{of } D}} \mathbf{E} \cdot \mathbf{n} \, d\sigma = 0.$$

So the flux of $\mathbf{E}$ across S in the direction away from the origin must be the same as the flux of $\mathbf{E}$ across S_a in the direction away from the origin, which is q/ε_0. This statement, called *Gauss's law*, also applies to charge distributions that are more general than the one assumed here, as shown in most physics texts. For any closed surface that encloses the origin, we have

Gauss's law: $$\iint_S \mathbf{E} \cdot \mathbf{n} \, d\sigma = \frac{q}{\varepsilon_0}.$$

Continuity Equation of Hydrodynamics

Let D be a region in space bounded by a closed oriented surface S. If $\mathbf{v}(x, y, z)$ is the velocity field of a fluid flowing smoothly through D, $\delta = \delta(t, x, y, z)$ is the fluid's density at (x, y, z) at time t, and $\mathbf{F} = \delta \mathbf{v}$, then the **continuity equation** of hydrodynamics states that

$$\nabla \cdot \mathbf{F} + \frac{\partial \delta}{\partial t} = 0.$$

If the functions involved have continuous first partial derivatives, the equation evolves naturally from the Divergence Theorem, as we now demonstrate.

First, the integral

$$\iint_S \mathbf{F} \cdot \mathbf{n} \, d\sigma$$

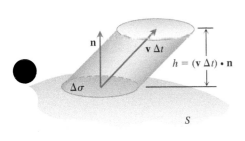

FIGURE 16.83 The fluid that flows upward through the patch $\Delta\sigma$ in a short time Δt fills a "cylinder" whose volume is approximately base $\times$ height $=$ $\mathbf{v} \cdot \mathbf{n} \, \Delta\sigma \, \Delta t$.

is the rate at which mass leaves D across S (leaves because $\mathbf{n}$ is the outer normal). To see why, consider a patch of area $\Delta\sigma$ on the surface (Figure 16.83). In a short time interval Δt, the volume ΔV of fluid that flows across the patch is approximately equal to the volume of a cylinder with base area $\Delta\sigma$ and height $(\mathbf{v}\Delta t) \cdot \mathbf{n}$, where $\mathbf{v}$ is a velocity vector rooted at a point of the patch:

$$\Delta V \approx \mathbf{v} \cdot \mathbf{n} \, \Delta\sigma \, \Delta t.$$

The mass of this volume of fluid is about

$$\Delta m \approx \delta \mathbf{v} \cdot \mathbf{n} \, \Delta\sigma \, \Delta t,$$

so the rate at which mass is flowing out of D across the patch is about

$$\frac{\Delta m}{\Delta t} \approx \delta \mathbf{v} \cdot \mathbf{n} \, \Delta\sigma.$$

This leads to the approximation

$$\frac{\sum \Delta m}{\Delta t} \approx \sum \delta \mathbf{v} \cdot \mathbf{n} \, \Delta\sigma$$

as an estimate of the average rate at which mass flows across S. Finally, letting $\Delta\sigma \to 0$ and $\Delta t \to 0$ gives the instantaneous rate at which mass leaves D across S as

$$\frac{dm}{dt} = \iint_S \delta \mathbf{v} \cdot \mathbf{n} \, d\sigma,$$

which for our particular flow is

$$\frac{dm}{dt} = \iint_S \mathbf{F} \cdot \mathbf{n} \, d\sigma.$$

Now let B be a solid sphere centered at a point Q in the flow. The average value of $\nabla \cdot \mathbf{F}$ over B is

$$\frac{1}{\text{volume of } B} \iiint_B \nabla \cdot \mathbf{F} \, dV.$$

It is a consequence of the continuity of the divergence that $\nabla \cdot \mathbf{F}$ actually takes on this value at some point P in B. Thus, by the Divergence Theorem Equation (2),

$$(\nabla \cdot \mathbf{F})(P) = \frac{1}{\text{volume of } B} \iiint_B \nabla \cdot \mathbf{F} \, dV = \frac{\displaystyle\iint_S \mathbf{F} \cdot \mathbf{n} \, d\sigma}{\text{volume of } B}$$

$$= \frac{\text{rate at which mass leaves } B \text{ across its surface } S}{\text{volume of } B}. \tag{9}$$

The last term of the equation describes decrease in mass per unit volume.

Now let the radius of B approach zero while the center Q stays fixed. The left side of Equation (9) converges to $(\nabla \cdot \mathbf{F})_Q$, and the right side converges to $(-\partial\delta/\partial t)_Q$, since $\delta = m/V$. The equality of these two limits is the continuity equation

$$\nabla \cdot \mathbf{F} = -\frac{\partial\delta}{\partial t}.$$

The continuity equation "explains" $\nabla \cdot \mathbf{F}$: The divergence of $\mathbf{F}$ at a point is the rate at which the density of the fluid is decreasing there. The Divergence Theorem

$$\iint_S \mathbf{F} \cdot \mathbf{n} \, d\sigma = \iiint_D \nabla \cdot \mathbf{F} \, dV$$

now says that the net decrease in density of the fluid in region D (divergence integral) is accounted for by the mass transported across the surface S (outward flux integral). So, the theorem is a statement about conservation of mass (Exercise 35).

Unifying the Integral Theorems

If we think of a two-dimensional field $\mathbf{F} = M(x, y)\mathbf{i} + N(x, y)\mathbf{j}$ as a three-dimensional field whose $\mathbf{k}$-component is zero, then $\nabla \cdot \mathbf{F} = (\partial M/\partial x) + (\partial N/\partial y)$ and the normal form of Green's Theorem can be written as

$$\oint_C \mathbf{F} \cdot \mathbf{n} \, ds = \iint_R \left(\frac{\partial M}{\partial x} + \frac{\partial N}{\partial y}\right) dx \, dy = \iint_R \nabla \cdot \mathbf{F} \, dA.$$

Similarly, $\nabla \times \mathbf{F} \cdot \mathbf{k} = (\partial N/\partial x) - (\partial M/\partial y)$, so the tangential form of Green's Theorem can be written as

$$\oint_C \mathbf{F} \cdot \mathbf{T} \, ds = \iint_R \left(\frac{\partial N}{\partial x} - \frac{\partial M}{\partial y}\right) dx \, dy = \iint_R (\nabla \times \mathbf{F}) \cdot \mathbf{k} \, dA.$$

With the equations of Green's Theorem now in del notation, we can see their relationships to the equations in Stokes' Theorem and the Divergence Theorem, all summarized here.

Green's Theorem and Its Generalization to Three Dimensions

Tangential form of Green's Theorem:
$$\oint_C \mathbf{F} \cdot \mathbf{T} \, ds = \iint_R (\nabla \times \mathbf{F}) \cdot \mathbf{k} \, dA$$

Stokes' Theorem:
$$\oint_C \mathbf{F} \cdot \mathbf{T} \, ds = \iint_S (\nabla \times \mathbf{F}) \cdot \mathbf{n} \, d\sigma$$

Normal form of Green's Theorem:
$$\oint_C \mathbf{F} \cdot \mathbf{n} \, ds = \iint_R \nabla \cdot \mathbf{F} \, dA$$

Divergence Theorem:
$$\iint_S \mathbf{F} \cdot \mathbf{n} \, d\sigma = \iiint_D \nabla \cdot \mathbf{F} \, dV$$

Notice how Stokes' Theorem generalizes the tangential (curl) form of Green's Theorem from a flat surface in the plane to a surface in three-dimensional space. In each case, the surface integral of curl $\mathbf{F}$ over the interior of the oriented surface equals the circulation of $\mathbf{F}$ around the boundary.

Likewise, the Divergence Theorem generalizes the normal (flux) form of Green's Theorem from a two-dimensional region in the plane to a three-dimensional region in space. In each case, the integral of $\nabla \cdot \mathbf{F}$ over the interior of the region equals the total flux of the field across the boundary enclosing the region.

All these results can be thought of as forms of a *single fundamental theorem*. The Fundamental Theorem of Calculus in Section 5.4 says that if $f(x)$ is differentiable on (a, b) and continuous on $[a, b]$, then

$$\int_a^b \frac{df}{dx} \, dx = f(b) - f(a).$$

FIGURE 16.84 The outward unit normals at the boundary of $[a, b]$ in one-dimensional space.

If we let $\mathbf{F} = f(x)\mathbf{i}$ throughout $[a, b]$, then $df/dx = \nabla \cdot \mathbf{F}$. If we define the unit vector field $\mathbf{n}$ normal to the boundary of $[a, b]$ to be $\mathbf{i}$ at b and $-\mathbf{i}$ at a (Figure 16.84), then

$$f(b) - f(a) = f(b)\mathbf{i} \cdot (\mathbf{i}) + f(a)\mathbf{i} \cdot (-\mathbf{i})$$
$$= \mathbf{F}(b) \cdot \mathbf{n} + \mathbf{F}(a) \cdot \mathbf{n}$$
$$= \text{total outward flux of } \mathbf{F} \text{ across the boundary of } [a, b].$$

The Fundamental Theorem now says that

$$\mathbf{F}(b) \cdot \mathbf{n} + \mathbf{F}(a) \cdot \mathbf{n} = \int_{[a, b]} \nabla \cdot \mathbf{F} \, dx.$$

The Fundamental Theorem of Calculus, the normal form of Green's Theorem, and the Divergence Theorem all say that the integral of the differential operator $\nabla \cdot$ operating on a field $\mathbf{F}$ over a region equals the sum of the normal field components over the boundary enclosing the region. (Here we are interpreting the line integral in Green's Theorem and the surface integral in the Divergence Theorem as "sums" over the boundary.)

Stokes' Theorem and the tangential form of Green's Theorem say that, when things are properly oriented, the surface integral of the differential operator $\nabla \times$ operating on a field equals the sum of the tangential field components over the boundary of the surface.

The beauty of these interpretations is the observance of a single unifying principle, which we can state as follows.

A Unifying Fundamental Theorem of Vector Integral Calculus

The integral of a differential operator acting on a field over a region equals the sum of the field components appropriate to the operator over the boundary of the region.

EXERCISES 16.8

Calculating Divergence

In Exercises 1–8, find the divergence of the field.

1. $\mathbf{F} = (x - y + z)\mathbf{i} + (2x + y - z)\mathbf{j} + (3x + 2y - 2z)\mathbf{k}$

2. $\mathbf{F} = (x \ln y)\mathbf{i} + (y \ln z)\mathbf{j} + (z \ln x)\mathbf{k}$

3. $\mathbf{F} = ye^{xyz}\mathbf{i} + ze^{xyz}\mathbf{j} + xe^{xyz}\mathbf{k}$

4. $\mathbf{F} = \sin(xy)\mathbf{i} + \cos(yz)\mathbf{j} + \tan(xz)\mathbf{k}$

5. The spin field in Figure 16.13

6. The radial field in Figure 16.12

7. The gravitational field in Figure 16.9 and Exercise 38a in Section 16.3

8. The velocity field in Figure 16.14

Calculating Flux Using the Divergence Theorem

In Exercises 9–20, use the Divergence Theorem to find the outward flux of **F** across the boundary of the region D.

9. **Cube** $\mathbf{F} = (y - x)\mathbf{i} + (z - y)\mathbf{j} + (y - x)\mathbf{k}$

   D: The cube bounded by the planes $x = \pm 1, y = \pm 1,$ and $z = \pm 1$

10. $\mathbf{F} = x^2\mathbf{i} + y^2\mathbf{j} + z^2\mathbf{k}$

    a. **Cube** D: The cube cut from the first octant by the planes $x = 1, y = 1,$ and $z = 1$

    b. **Cube** D: The cube bounded by the planes $x = \pm 1, y = \pm 1,$ and $z = \pm 1$

    c. **Cylindrical can** D: The region cut from the solid cylinder $x^2 + y^2 \leq 4$ by the planes $z = 0$ and $z = 1$

11. **Cylinder and paraboloid** $\mathbf{F} = y\mathbf{i} + xy\mathbf{j} - z\mathbf{k}$

    D: The region inside the solid cylinder $x^2 + y^2 \leq 4$ between the plane $z = 0$ and the paraboloid $z = x^2 + y^2$

12. **Sphere** $\mathbf{F} = x^2\mathbf{i} + xz\mathbf{j} + 3z\mathbf{k}$

    D: The solid sphere $x^2 + y^2 + z^2 \leq 4$

13. **Portion of sphere** $\mathbf{F} = x^2\mathbf{i} - 2xy\mathbf{j} + 3xz\mathbf{k}$

    D: The region cut from the first octant by the sphere $x^2 + y^2 + z^2 = 4$

14. **Cylindrical can** $\mathbf{F} = (6x^2 + 2xy)\mathbf{i} + (2y + x^2z)\mathbf{j} + 4x^2y^3\mathbf{k}$

    D: The region cut from the first octant by the cylinder $x^2 + y^2 = 4$ and the plane $z = 3$

15. **Wedge** $\mathbf{F} = 2xz\mathbf{i} - xy\mathbf{j} - z^2\mathbf{k}$

    D: The wedge cut from the first octant by the plane $y + z = 4$ and the elliptical cylinder $4x^2 + y^2 = 16$

16. **Sphere** $\mathbf{F} = x^3\mathbf{i} + y^3\mathbf{j} + z^3\mathbf{k}$

    D: The solid sphere $x^2 + y^2 + z^2 \leq a^2$

17. **Thick sphere** $\mathbf{F} = \sqrt{x^2 + y^2 + z^2}\,(x\mathbf{i} + y\mathbf{j} + z\mathbf{k})$

    D: The region $1 \leq x^2 + y^2 + z^2 \leq 2$

18. **Thick sphere** $\mathbf{F} = (x\mathbf{i} + y\mathbf{j} + z\mathbf{k})/\sqrt{x^2 + y^2 + z^2}$

    D: The region $1 \leq x^2 + y^2 + z^2 \leq 4$

19. **Thick sphere** $\mathbf{F} = (5x^3 + 12xy^2)\mathbf{i} + (y^3 + e^y \sin z)\mathbf{j} + (5z^3 + e^y \cos z)\mathbf{k}$

    D: The solid region between the spheres $x^2 + y^2 + z^2 = 1$ and $x^2 + y^2 + z^2 = 2$

20. **Thick cylinder** $\mathbf{F} = \ln(x^2 + y^2)\mathbf{i} - \left(\dfrac{2z}{x}\tan^{-1}\dfrac{y}{x}\right)\mathbf{j} + z\sqrt{x^2 + y^2}\,\mathbf{k}$

    D: The thick-walled cylinder $1 \leq x^2 + y^2 \leq 2, -1 \leq z \leq 2$

Theory and Examples

21. **a.** Show that the outward flux of the position vector field $\mathbf{F} = x\mathbf{i} + y\mathbf{j} + z\mathbf{k}$ through a smooth closed surface S is three times the volume of the region enclosed by the surface.

    b. Let **n** be the outward unit normal vector field on S. Show that it is not possible for **F** to be orthogonal to **n** at every point of S.

22. The base of the closed cubelike surface shown here is the unit square in the xy-plane. The four sides lie in the planes $x = 0$, $x = 1, y = 0,$ and $y = 1$. The top is an arbitrary smooth surface whose identity is unknown. Let $\mathbf{F} = x\mathbf{i} - 2y\mathbf{j} + (z + 3)\mathbf{k}$ and suppose the outward flux of **F** through Side A is 1 and through Side B is -3. Can you conclude anything about the outward flux through the top? Give reasons for your answer.

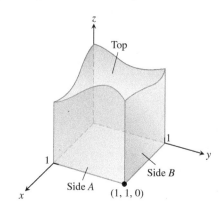

23. Let $\mathbf{F} = (y \cos 2x)\mathbf{i} + (y^2 \sin 2x)\mathbf{j} + (x^2y + z)\mathbf{k}$. Is there a vector field **A** such that $\mathbf{F} = \nabla \times \mathbf{A}$? Explain your answer.

24. **Outward flux of a gradient field** Let S be the surface of the portion of the solid sphere $x^2 + y^2 + z^2 \leq a^2$ that lies in the first octant and let $f(x, y, z) = \ln\sqrt{x^2 + y^2 + z^2}$. Calculate

$$\iint_S \nabla f \cdot \mathbf{n}\, d\sigma.$$

    ($\nabla f \cdot \mathbf{n}$ is the derivative of f in the direction of outward normal **n**.)

25. Let **F** be a field whose components have continuous first partial derivatives throughout a portion of space containing a region D bounded by a smooth closed surface S. If $|\mathbf{F}| \leq 1$, can any bound be placed on the size of

$$\iiint_D \nabla \cdot \mathbf{F}\, dV?$$

    Give reasons for your answer.

26. **Maximum flux** Among all rectangular solids defined by the inequalities $0 \leq x \leq a, 0 \leq y \leq b, 0 \leq z \leq 1$, find the one for which the total flux of $\mathbf{F} = (-x^2 - 4xy)\mathbf{i} - 6yz\mathbf{j} + 12z\mathbf{k}$ outward through the six sides is greatest. What *is* the greatest flux?

27. Calculate the net outward flux of the vector field

$$\mathbf{F} = xy\mathbf{i} + (\sin xz + y^2)\mathbf{j} + (e^{xy^2} + x)\mathbf{k}$$

over the surface S surrounding the region D bounded by the planes $y = 0, z = 0, z = 2 - y$ and the parabolic cylinder $z = 1 - x^2$.

28. Compute the net outward flux of the vector field $\mathbf{F} = (x\mathbf{i} + y\mathbf{j} + z\mathbf{k})/(x^2 + y^2 + z^2)^{3/2}$ across the ellipsoid $9x^2 + 4y^2 + 6z^2 = 36$.

29. Let $\mathbf{F}$ be a differentiable vector field and let $g(x, y, z)$ be a differentiable scalar function. Verify the following identities.

a. $\nabla \cdot (g\mathbf{F}) = g\nabla \cdot \mathbf{F} + \nabla g \cdot \mathbf{F}$

b. $\nabla \times (g\mathbf{F}) = g\nabla \times \mathbf{F} + \nabla g \times \mathbf{F}$

30. Let $\mathbf{F}_1$ and $\mathbf{F}_2$ be differentiable vector fields and let a and b be arbitrary real constants. Verify the following identities.

a. $\nabla \cdot (a\mathbf{F}_1 + b\mathbf{F}_2) = a\nabla \cdot \mathbf{F}_1 + b\nabla \cdot \mathbf{F}_2$

b. $\nabla \times (a\mathbf{F}_1 + b\mathbf{F}_2) = a\nabla \times \mathbf{F}_1 + b\nabla \times \mathbf{F}_2$

c. $\nabla \cdot (\mathbf{F}_1 \times \mathbf{F}_2) = \mathbf{F}_2 \cdot \nabla \times \mathbf{F}_1 - \mathbf{F}_1 \cdot \nabla \times \mathbf{F}_2$

31. If $\mathbf{F} = M\mathbf{i} + N\mathbf{j} + P\mathbf{k}$ is a differentiable vector field, we define the notation $\mathbf{F} \cdot \nabla$ to mean

$$M\frac{\partial}{\partial x} + N\frac{\partial}{\partial y} + P\frac{\partial}{\partial z}.$$

For differentiable vector fields $\mathbf{F}_1$ and $\mathbf{F}_2$, verify the following identities.

a. $\nabla \times (\mathbf{F}_1 \times \mathbf{F}_2) = (\mathbf{F}_2 \cdot \nabla)\mathbf{F}_1 - (\mathbf{F}_1 \cdot \nabla)\mathbf{F}_2 + (\nabla \cdot \mathbf{F}_2)\mathbf{F}_1 - (\nabla \cdot \mathbf{F}_1)\mathbf{F}_2$

b. $\nabla(\mathbf{F}_1 \cdot \mathbf{F}_2) = (\mathbf{F}_1 \cdot \nabla)\mathbf{F}_2 + (\mathbf{F}_2 \cdot \nabla)\mathbf{F}_1 + \mathbf{F}_1 \times (\nabla \times \mathbf{F}_2) + \mathbf{F}_2 \times (\nabla \times \mathbf{F}_1)$

32. Harmonic functions A function $f(x, y, z)$ is said to be *harmonic* in a region D in space if it satisfies the Laplace equation

$$\nabla^2 f = \nabla \cdot \nabla f = \frac{\partial^2 f}{\partial x^2} + \frac{\partial^2 f}{\partial y^2} + \frac{\partial^2 f}{\partial z^2} = 0$$

throughout D.

a. Suppose that f is harmonic throughout a bounded region D enclosed by a smooth surface S and that $\mathbf{n}$ is the chosen unit normal vector on S. Show that the integral over S of $\nabla f \cdot \mathbf{n}$, the derivative of f in the direction of $\mathbf{n}$, is zero.

b. Show that if f is harmonic on D, then

$$\iint_S f\nabla f \cdot \mathbf{n}\, d\sigma = \iiint_D |\nabla f|^2\, dV.$$

33. Green's first formula Suppose that f and g are scalar functions with continuous first- and second-order partial derivatives throughout a region D that is bounded by a closed piecewise smooth surface S. Show that

$$\iint_S f\nabla g \cdot \mathbf{n}\, d\sigma = \iiint_D (f\nabla^2 g + \nabla f \cdot \nabla g)\, dV. \qquad (10)$$

Equation (10) is **Green's first formula**. (*Hint:* Apply the Divergence Theorem to the field $\mathbf{F} = f\nabla g$.)

34. Green's second formula (*Continuation of Exercise 33.*) Interchange f and g in Equation (10) to obtain a similar formula. Then subtract this formula from Equation (10) to show that

$$\iint_S (f\nabla g - g\nabla f) \cdot \mathbf{n}\, d\sigma = \iiint_D (f\nabla^2 g - g\nabla^2 f)\, dV. \qquad (11)$$

This equation is **Green's second formula**.

35. Conservation of mass Let $\mathbf{v}(t, x, y, z)$ be a continuously differentiable vector field over the region D in space and let $p(t, x, y, z)$ be a continuously differentiable scalar function. The variable t represents the time domain. The Law of Conservation of Mass asserts that

$$\frac{d}{dt}\iiint_D p(t, x, y, z)\, dV = -\iint_S p\mathbf{v} \cdot \mathbf{n}\, d\sigma,$$

where S is the surface enclosing D.

a. Give a physical interpretation of the conservation of mass law if $\mathbf{v}$ is a velocity flow field and p represents the density of the fluid at point (x, y, z) at time t.

b. Use the Divergence Theorem and Leibniz's Rule,

$$\frac{d}{dt}\iiint_D p(t, x, y, z)\, dV = \iiint_D \frac{\partial p}{\partial t}\, dV,$$

to show that the Law of Conservation of Mass is equivalent to the continuity equation,

$$\nabla \cdot p\mathbf{v} + \frac{\partial p}{\partial t} = 0.$$

(In the first term $\nabla \cdot p\mathbf{v}$, the variable t is held fixed, and in the second term $\partial p/\partial t$, it is assumed that the point (x, y, z) in D is held fixed.)

36. The heat diffusion equation Let $T(t, x, y, z)$ be a function with continuous second derivatives giving the temperature at time t at the point (x, y, z) of a solid occupying a region D in space. If the solid's heat capacity and mass density are denoted by the constants c and ρ, respectively, the quantity $c\rho T$ is called the solid's **heat energy per unit volume**.

a. Explain why $-\nabla T$ points in the direction of heat flow.

b. Let $-k\nabla T$ denote the **energy flux vector**. (Here the constant k is called the **conductivity**.) Assuming the Law of Conservation of Mass with $-k\nabla T = \mathbf{v}$ and $c\rho T = p$ in Exercise 35, derive the diffusion (heat) equation

$$\frac{\partial T}{\partial t} = K\nabla^2 T,$$

where $K = k/(c\rho) > 0$ is the *diffusivity* constant. (Notice that if $T(t, x)$ represents the temperature at time t at position x in a uniform conducting rod with perfectly insulated sides, then $\nabla^2 T = \partial^2 T/\partial x^2$ and the diffusion equation reduces to the one-dimensional heat equation in Chapter 14's Additional Exercises.)

CHAPTER 16   Questions to Guide Your Review

1. What are line integrals of scalar functions? How are they evaluated? Give examples.

2. How can you use line integrals to find the centers of mass of springs or wires? Explain.

3. What is a vector field? What is the line integral of a vector field? What is a gradient field? Give examples.

4. What is the flow of a vector field along a curve? What is the work done by vector field moving an object along a curve? How do you calculate the work done? Give examples.

5. What is the Fundamental Theorem of line integrals? Explain how it relates to the Fundamental Theorem of Calculus.

6. Specify three properties that are special about conservative fields. How can you tell when a field is conservative?

7. What is special about path independent fields?

8. What is a potential function? Show by example how to find a potential function for a conservative field.

9. What is a differential form? What does it mean for such a form to be exact? How do you test for exactness? Give examples.

10. What is Green's Theorem? Discuss how the two forms of Green's Theorem extend the Net Change Theorem in Chapter 5.

11. How do you calculate the area of a parametrized surface in space? Of an implicitly defined surface $F(x, y, z) = 0$? Of the surface which is the graph of $z = f(x, y)$? Give examples.

12. How do you integrate a scalar function over a parametrized surface? Of surfaces that are defined implicitly or in explicit form? Give examples.

13. What is an oriented surface? What is the surface integral of a vector field in three-dimensional space over an oriented surface? How is it related to the net outward flux of the field? Give examples.

14. What is the curl of a vector field? How can you interpret it?

15. What is Stokes' Theorem? Explain how it generalizes Green's Theorem to three dimensions.

16. What is the divergence of a vector field? How can you interpret it?

17. What is the Divergence Theorem? Explain how it generalizes Green's Theorem to three dimensions.

18. How do Green's Theorem, Stokes' Theorem, and the Divergence Theorem relate to the Fundamental Theorem of Calculus for ordinary single integrals?

CHAPTER 16   Practice Exercises

Evaluating Line Integrals

1. The accompanying figure shows two polygonal paths in space joining the origin to the point (1, 1, 1). Integrate $f(x, y, z) = 2x - 3y^2 - 2z + 3$ over each path.

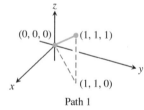 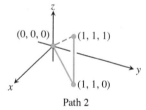

Path 1   Path 2

2. The accompanying figure shows three polygonal paths joining the origin to the point (1, 1, 1). Integrate $f(x, y, z) = x^2 + y - z$ over each path.

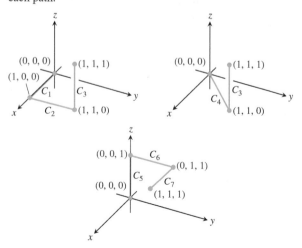

3. Integrate $f(x, y, z) = \sqrt{x^2 + z^2}$ over the circle
$$\mathbf{r}(t) = (a \cos t)\mathbf{j} + (a \sin t)\mathbf{k}, \qquad 0 \le t \le 2\pi.$$

4. Integrate $f(x, y, z) = \sqrt{x^2 + y^2}$ over the involute curve
$$\mathbf{r}(t) = (\cos t + t \sin t)\mathbf{i} + (\sin t - t \cos t)\mathbf{j}, \qquad 0 \le t \le \sqrt{3}.$$

Evaluate the integrals in Exercises 5 and 6.

5. $\displaystyle\int_{(-1,1,1)}^{(4,-3,0)} \frac{dx + dy + dz}{\sqrt{x + y + z}}$ 6. $\displaystyle\int_{(1,1,1)}^{(10,3,3)} dx - \sqrt{\frac{z}{y}}dy - \sqrt{\frac{y}{z}}dz$

7. Integrate $\mathbf{F} = -(y \sin z)\mathbf{i} + (x \sin z)\mathbf{j} + (xy \cos z)\mathbf{k}$ around the circle cut from the sphere $x^2 + y^2 + z^2 = 5$ by the plane $z = -1$, clockwise as viewed from above.

8. Integrate $\mathbf{F} = 3x^2y\mathbf{i} + (x^3 + 1)\mathbf{j} + 9z^2\mathbf{k}$ around the circle cut from the sphere $x^2 + y^2 + z^2 = 9$ by the plane $x = 2$.

Evaluate the integrals in Exercises 9 and 10.

9. $\displaystyle\int_C 8x \sin y \, dx - 8y \cos x \, dy$

   C is the square cut from the first quadrant by the lines $x = \pi/2$ and $y = \pi/2$.

10. $\displaystyle\int_C y^2 \, dx + x^2 \, dy$

    C is the circle $x^2 + y^2 = 4$.

Finding and Evaluating Surface Integrals

11. **Area of an elliptical region**   Find the area of the elliptical region cut from the plane $x + y + z = 1$ by the cylinder $x^2 + y^2 = 1$.

12. **Area of a parabolic cap**   Find the area of the cap cut from the paraboloid $y^2 + z^2 = 3x$ by the plane $x = 1$.

13. Area of a spherical cap Find the area of the cap cut from the top of the sphere $x^2 + y^2 + z^2 = 1$ by the plane $z = \sqrt{2}/2$.

14. a. Hemisphere cut by cylinder Find the area of the surface cut from the hemisphere $x^2 + y^2 + z^2 = 4, z \geq 0$, by the cylinder $x^2 + y^2 = 2x$.

  b. Find the area of the portion of the cylinder that lies inside the hemisphere. (*Hint:* Project onto the xz-plane. Or evaluate the integral $\int h\, ds$, where h is the altitude of the cylinder and ds is the element of arc length on the circle $x^2 + y^2 = 2x$ in the xy-plane.)

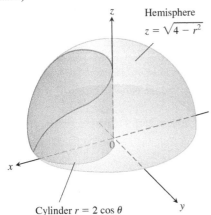

Hemisphere $z = \sqrt{4 - r^2}$

Cylinder $r = 2\cos\theta$

15. Area of a triangle Find the area of the triangle in which the plane $(x/a) + (y/b) + (z/c) = 1 \ (a, b, c > 0)$ intersects the first octant. Check your answer with an appropriate vector calculation.

16. Parabolic cylinder cut by planes Integrate

  a. $g(x, y, z) = \dfrac{yz}{\sqrt{4y^2 + 1}}$ **b.** $g(x, y, z) = \dfrac{z}{\sqrt{4y^2 + 1}}$

  over the surface cut from the parabolic cylinder $y^2 - z = 1$ by the planes $x = 0, x = 3$, and $z = 0$.

17. Circular cylinder cut by planes Integrate $g(x, y, z) = x^4 y(y^2 + z^2)$ over the portion of the cylinder $y^2 + z^2 = 25$ that lies in the first octant between the planes $x = 0$ and $x = 1$ and above the plane $z = 3$.

18. Area of Wyoming The state of Wyoming is bounded by the meridians $111°3'$ and $104°3'$ west longitude and by the circles $41°$ and $45°$ north latitude. Assuming that Earth is a sphere of radius $R = 3959$ mi, find the area of Wyoming.

Parametrized Surfaces

Find parametrizations for the surfaces in Exercises 19–24. (There are many ways to do these, so your answers may not be the same as those in the back of the book.)

19. Spherical band The portion of the sphere $x^2 + y^2 + z^2 = 36$ between the planes $z = -3$ and $z = 3\sqrt{3}$

20. Parabolic cap The portion of the paraboloid $z = -(x^2 + y^2)/2$ above the plane $z = -2$

21. Cone The cone $z = 1 + \sqrt{x^2 + y^2}, z \leq 3$

22. Plane above square The portion of the plane $4x + 2y + 4z = 12$ that lies above the square $0 \leq x \leq 2, 0 \leq y \leq 2$ in the first quadrant

23. Portion of paraboloid The portion of the paraboloid $y = 2(x^2 + z^2), \ y \leq 2$, that lies above the xy-plane

24. Portion of hemisphere The portion of the hemisphere $x^2 + y^2 + z^2 = 10, y \geq 0$, in the first octant

25. Surface area Find the area of the surface
$$\mathbf{r}(u, v) = (u + v)\mathbf{i} + (u - v)\mathbf{j} + v\mathbf{k},$$
$$0 \leq u \leq 1, \ 0 \leq v \leq 1.$$

26. Surface integral Integrate $f(x, y, z) = xy - z^2$ over the surface in Exercise 25.

27. Area of a helicoid Find the surface area of the helicoid $\mathbf{r}(r, \theta) = (r\cos\theta)\mathbf{i} + (r\sin\theta)\mathbf{j} + \theta\mathbf{k}, \ 0 \leq \theta \leq 2\pi, \ 0 \leq r \leq 1$, in the accompanying figure.

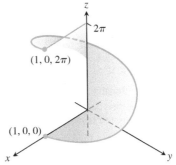

2π

$(1, 0, 2\pi)$

$(1, 0, 0)$

28. Surface integral Evaluate the integral $\iint_S \sqrt{x^2 + y^2 + 1}\, d\sigma$, where S is the helicoid in Exercise 27.

Conservative Fields

Which of the fields in Exercises 29–32 are conservative, and which are not?

29. $\mathbf{F} = x\mathbf{i} + y\mathbf{j} + z\mathbf{k}$

30. $\mathbf{F} = (x\mathbf{i} + y\mathbf{j} + z\mathbf{k})/(x^2 + y^2 + z^2)^{3/2}$

31. $\mathbf{F} = xe^y\mathbf{i} + ye^z\mathbf{j} + ze^x\mathbf{k}$

32. $\mathbf{F} = (\mathbf{i} + z\mathbf{j} + y\mathbf{k})/(x + yz)$

Find potential functions for the fields in Exercises 33 and 34.

33. $\mathbf{F} = 2\mathbf{i} + (2y + z)\mathbf{j} + (y + 1)\mathbf{k}$

34. $\mathbf{F} = (z\cos xz)\mathbf{i} + e^y\mathbf{j} + (x\cos xz)\mathbf{k}$

Work and Circulation

In Exercises 35 and 36, find the work done by each field along the paths from $(0, 0, 0)$ to $(1, 1, 1)$ in Exercise 1.

35. $\mathbf{F} = 2xy\mathbf{i} + \mathbf{j} + x^2\mathbf{k}$ **36.** $\mathbf{F} = 2xy\mathbf{i} + x^2\mathbf{j} + \mathbf{k}$

37. Finding work in two ways Find the work done by
$$\mathbf{F} = \frac{x\mathbf{i} + y\mathbf{j}}{(x^2 + y^2)^{3/2}}$$
over the plane curve $\mathbf{r}(t) = (e^t\cos t)\mathbf{i} + (e^t\sin t)\mathbf{j}$ from the point $(1, 0)$ to the point $(e^{2\pi}, 0)$ in two ways:

  a. By using the parametrization of the curve to evaluate the work integral.

  b. By evaluating a potential function for $\mathbf{F}$.

38. Flow along different paths Find the flow of the field $\mathbf{F} = \nabla(x^2 z e^y)$

  a. once around the ellipse C in which the plane $x + y + z = 1$ intersects the cylinder $x^2 + z^2 = 25$, clockwise as viewed from the positive y-axis.

b. along the curved boundary of the helicoid in Exercise 27 from $(1, 0, 0)$ to $(1, 0, 2\pi)$.

In Exercises 39 and 40, use the curl integral in Stokes' Theorem to find the circulation of the field **F** around the curve C in the indicated direction.

39. Circulation around an ellipse $\quad \mathbf{F} = y^2\mathbf{i} - y\mathbf{j} + 3z^2\mathbf{k}$

C: The ellipse in which the plane $2x + 6y - 3z = 6$ meets the cylinder $x^2 + y^2 = 1$, counterclockwise as viewed from above

40. Circulation around a circle $\quad \mathbf{F} = (x^2 + y)\mathbf{i} + (x + y)\mathbf{j} + (4y^2 - z)\mathbf{k}$

C: The circle in which the plane $z = -y$ meets the sphere $x^2 + y^2 + z^2 = 4$, counterclockwise as viewed from above

Masses and Moments

41. Wire with different densities $\quad$ Find the mass of a thin wire lying along the curve $\mathbf{r}(t) = \sqrt{2}t\mathbf{i} + \sqrt{2}t\mathbf{j} + (4 - t^2)\mathbf{k}$, $0 \le t \le 1$, if the density at t is (**a**) $\delta = 3t$ and (**b**) $\delta = 1$.

42. Wire with variable density $\quad$ Find the center of mass of a thin wire lying along the curve $\mathbf{r}(t) = t\mathbf{i} + 2t\mathbf{j} + (2/3)t^{3/2}\mathbf{k}$, $0 \le t \le 2$, if the density at t is $\delta = 3\sqrt{5 + t}$.

43. Wire with variable density $\quad$ Find the center of mass and the moments of inertia about the coordinate axes of a thin wire lying along the curve

$$\mathbf{r}(t) = t\mathbf{i} + \frac{2\sqrt{2}}{3}t^{3/2}\mathbf{j} + \frac{t^2}{2}\mathbf{k}, \qquad 0 \le t \le 2,$$

if the density at t is $\delta = 1/(t + 1)$.

44. Center of mass of an arch $\quad$ A slender metal arch lies along the semicircle $y = \sqrt{a^2 - x^2}$ in the xy-plane. The density at the point (x, y) on the arch is $\delta(x, y) = 2a - y$. Find the center of mass.

45. Wire with constant density $\quad$ A wire of constant density $\delta = 1$ lies along the curve $\mathbf{r}(t) = (e^t \cos t)\mathbf{i} + (e^t \sin t)\mathbf{j} + e^t\mathbf{k}$, $0 \le t \le \ln 2$. Find $\bar{z}$ and I_z.

46. Helical wire with constant density $\quad$ Find the mass and center of mass of a wire of constant density δ that lies along the helix $\mathbf{r}(t) = (2 \sin t)\mathbf{i} + (2 \cos t)\mathbf{j} + 3t\mathbf{k}$, $0 \le t \le 2\pi$.

47. Inertia and center of mass of a shell $\quad$ Find I_z and the center of mass of a thin shell of density $\delta(x, y, z) = z$ cut from the upper portion of the sphere $x^2 + y^2 + z^2 = 25$ by the plane $z = 3$.

48. Moment of inertia of a cube $\quad$ Find the moment of inertia about the z-axis of the surface of the cube cut from the first octant by the planes $x = 1$, $y = 1$, and $z = 1$ if the density is $\delta = 1$.

Flux Across a Plane Curve or Surface

Use Green's Theorem to find the counterclockwise circulation and outward flux for the fields and curves in Exercises 49 and 50.

49. Square $\quad \mathbf{F} = (2xy + x)\mathbf{i} + (xy - y)\mathbf{j}$

C: The square bounded by $x = 0$, $x = 1$, $y = 0$, $y = 1$

50. Triangle $\quad \mathbf{F} = (y - 6x^2)\mathbf{i} + (x + y^2)\mathbf{j}$

C: The triangle made by the lines $y = 0$, $y = x$, and $x = 1$

51. Zero line integral $\quad$ Show that

$$\oint_C \ln x \sin y \, dy - \frac{\cos y}{x}dx = 0$$

for any closed curve C to which Green's Theorem applies.

52. a. Outward flux and area $\quad$ Show that the outward flux of the position vector field $\mathbf{F} = x\mathbf{i} + y\mathbf{j}$ across any closed curve to which Green's Theorem applies is twice the area of the region enclosed by the curve.

b. Let **n** be the outward unit normal vector to a closed curve to which Green's Theorem applies. Show that it is not possible for $\mathbf{F} = x\mathbf{i} + y\mathbf{j}$ to be orthogonal to **n** at every point of C.

In Exercises 53–56, find the outward flux of **F** across the boundary of D.

53. Cube $\quad \mathbf{F} = 2xy\mathbf{i} + 2yz\mathbf{j} + 2xz\mathbf{k}$

D: The cube cut from the first octant by the planes $x = 1$, $y = 1$, $z = 1$

54. Spherical cap $\quad \mathbf{F} = xz\mathbf{i} + yz\mathbf{j} + \mathbf{k}$

D: The entire surface of the upper cap cut from the solid sphere $x^2 + y^2 + z^2 \le 25$ by the plane $z = 3$

55. Spherical cap $\quad \mathbf{F} = -2x\mathbf{i} - 3y\mathbf{j} + z\mathbf{k}$

D: The upper region cut from the solid sphere $x^2 + y^2 + z^2 \le 2$ by the paraboloid $z = x^2 + y^2$

56. Cone and cylinder $\quad \mathbf{F} = (6x + y)\mathbf{i} - (x + z)\mathbf{j} + 4yz\mathbf{k}$

D: The region in the first octant bounded by the cone $z = \sqrt{x^2 + y^2}$, the cylinder $x^2 + y^2 = 1$, and the coordinate planes

57. Hemisphere, cylinder, and plane $\quad$ Let S be the surface that is bounded on the left by the hemisphere $x^2 + y^2 + z^2 = a^2$, $y \le 0$, in the middle by the cylinder $x^2 + z^2 = a^2$, $0 \le y \le a$, and on the right by the plane $y = a$. Find the flux of $\mathbf{F} = y\mathbf{i} + z\mathbf{j} + x\mathbf{k}$ outward across S.

58. Cylinder and planes $\quad$ Find the outward flux of the field $\mathbf{F} = 3xz^2\mathbf{i} + y\mathbf{j} - z^3\mathbf{k}$ across the surface of the solid in the first octant that is bounded by the cylinder $x^2 + 4y^2 = 16$ and the planes $y = 2z$, $x = 0$, and $z = 0$.

59. Cylindrical can $\quad$ Use the Divergence Theorem to find the flux of $\mathbf{F} = xy^2\mathbf{i} + x^2y\mathbf{j} + y\mathbf{k}$ outward through the surface of the region enclosed by the cylinder $x^2 + y^2 = 1$ and the planes $z = 1$ and $z = -1$.

60. Hemisphere $\quad$ Find the flux of $\mathbf{F} = (3z + 1)\mathbf{k}$ upward across the hemisphere $x^2 + y^2 + z^2 = a^2$, $z \ge 0$ (**a**) with the Divergence Theorem and (**b**) by evaluating the flux integral directly.

CHAPTER 16 Additional and Advanced Exercises

Finding Areas with Green's Theorem

Use the Green's Theorem area formula in Exercises 16.4 to find the areas of the regions enclosed by the curves in Exercises 1–4.

1. The limaçon $x = 2 \cos t - \cos 2t$, $y = 2 \sin t$, $0 \le t \le 2\pi$

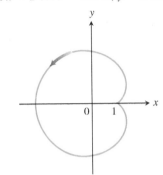

2. The deltoid $x = 2 \cos t + \cos 2t$, $y = 2 \sin t - \sin 2t$, $0 \le t \le 2\pi$

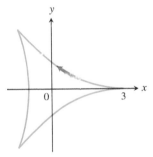

3. The eight curve $x = (1/2) \sin 2t$, $y = \sin t$, $0 \le t \le \pi$ (one loop)

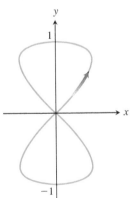

4. The teardrop $x = 2a \cos t - a \sin 2t$, $y = b \sin t$, $0 \le t \le 2\pi$

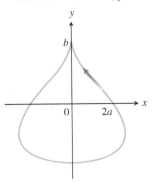

Theory and Applications

5. a. Give an example of a vector field $\mathbf{F}\ (x, y, z)$ that has value $\mathbf{0}$ at only one point and such that curl $\mathbf{F}$ is nonzero everywhere. Be sure to identify the point and compute the curl.

  b. Give an example of a vector field $\mathbf{F}\ (x, y, z)$ that has value $\mathbf{0}$ on precisely one line and such that curl $\mathbf{F}$ is nonzero everywhere. Be sure to identify the line and compute the curl.

  c. Give an example of a vector field $\mathbf{F}\ (x, y, z)$ that has value $\mathbf{0}$ on a surface and such that curl $\mathbf{F}$ is nonzero everywhere. Be sure to identify the surface and compute the curl.

6. Find all points (a, b, c) on the sphere $x^2 + y^2 + z^2 = R^2$ where the vector field $\mathbf{F} = yz^2\mathbf{i} + xz^2\mathbf{j} + 2xyz\mathbf{k}$ is normal to the surface and $\mathbf{F}(a, b, c) \ne \mathbf{0}$.

7. Find the mass of a spherical shell of radius R such that at each point (x, y, z) on the surface the mass density $\delta(x, y, z)$ is its distance to some fixed point (a, b, c) of the surface.

8. Find the mass of a helicoid

$$\mathbf{r}(r, \theta) = (r \cos \theta)\mathbf{i} + (r \sin \theta)\mathbf{j} + \theta\mathbf{k},$$

$0 \le r \le 1, 0 \le \theta \le 2\pi$, if the density function is $\delta(x, y, z) = 2\sqrt{x^2 + y^2}$. See Practice Exercise 27 for a figure.

9. Among all rectangular regions $0 \le x \le a, 0 \le y \le b$, find the one for which the total outward flux of $\mathbf{F} = (x^2 + 4xy)\mathbf{i} - 6y\mathbf{j}$ across the four sides is least. What *is* the least flux?

10. Find an equation for the plane through the origin such that the circulation of the flow field $\mathbf{F} = z\mathbf{i} + x\mathbf{j} + y\mathbf{k}$ around the circle of intersection of the plane with the sphere $x^2 + y^2 + z^2 = 4$ is a maximum.

11. A string lies along the circle $x^2 + y^2 = 4$ from $(2, 0)$ to $(0, 2)$ in the first quadrant. The density of the string is $\rho\ (x, y) = xy$.

  a. Partition the string into a finite number of subarcs to show that the work done by gravity to move the string straight down to the x-axis is given by

$$\text{Work} = \lim_{n\to\infty} \sum_{k=1}^{n} g\, x_k y_k^2 \Delta s_k = \int_C g\, xy^2\, ds,$$

  where g is the gravitational constant.

  b. Find the total work done by evaluating the line integral in part (a).

  c. Show that the total work done equals the work required to move the string's center of mass $(\bar{x}, \bar{y})$ straight down to the x-axis.

12. A thin sheet lies along the portion of the plane $x + y + z = 1$ in the first octant. The density of the sheet is $\delta\ (x, y, z) = xy$.

  a. Partition the sheet into a finite number of subpieces to show that the work done by gravity to move the sheet straight down to the xy-plane is given by

$$\text{Work} = \lim_{n\to\infty} \sum_{k=1}^{n} g\, x_k y_k z_k\, \Delta\sigma_k = \iint_S g\, xyz\, d\sigma,$$

  where g is the gravitational constant.

  b. Find the total work done by evaluating the surface integral in part (a).

c. Show that the total work done equals the work required to move the sheet's center of mass $(\bar{x}, \bar{y}, \bar{z})$ straight down to the xy-plane.

13. Archimedes' principle If an object such as a ball is placed in a liquid, it will either sink to the bottom, float, or sink a certain distance and remain suspended in the liquid. Suppose a fluid has constant weight density w and that the fluid's surface coincides with the plane $z = 4$. A spherical ball remains suspended in the fluid and occupies the region $x^2 + y^2 + (z - 2)^2 \leq 1$.

a. Show that the surface integral giving the magnitude of the total force on the ball due to the fluid's pressure is

$$\text{Force} = \lim_{n \to \infty} \sum_{k=1}^{n} w(4 - z_k)\,\Delta\sigma_k = \iint_S w(4 - z)\,d\sigma.$$

b. Since the ball is not moving, it is being held up by the buoyant force of the liquid. Show that the magnitude of the buoyant force on the sphere is

$$\text{Buoyant force} = \iint_S w(z - 4)\mathbf{k} \cdot \mathbf{n}\,d\sigma,$$

where $\mathbf{n}$ is the outer unit normal at (x, y, z). This illustrates Archimedes' principle that the magnitude of the buoyant force on a submerged solid equals the weight of the displaced fluid.

c. Use the Divergence Theorem to find the magnitude of the buoyant force in part (b).

14. Fluid force on a curved surface A cone in the shape of the surface $z = \sqrt{x^2 + y^2}$, $0 \leq z \leq 2$ is filled with a liquid of constant weight density w. Assuming the xy-plane is "ground level," show that the total force on the portion of the cone from $z = 1$ to $z = 2$ due to liquid pressure is the surface integral

$$F = \iint_S w(2 - z)\,d\sigma.$$

Evaluate the integral.

15. Faraday's law If $\mathbf{E}(t, x, y, z)$ and $\mathbf{B}(t, x, y, z)$ represent the electric and magnetic fields at point (x, y, z) at time t, a basic principle of electromagnetic theory says that $\nabla \times \mathbf{E} = -\partial\mathbf{B}/\partial t$. In this expression $\nabla \times \mathbf{E}$ is computed with t held fixed and $\partial\mathbf{B}/\partial t$ is calculated with (x, y, z) fixed. Use Stokes' Theorem to derive Faraday's law,

$$\oint_C \mathbf{E} \cdot d\mathbf{r} = -\frac{\partial}{\partial t} \iint_S \mathbf{B} \cdot \mathbf{n}\,d\sigma,$$

where C represents a wire loop through which current flows counterclockwise with respect to the surface's unit normal $\mathbf{n}$, giving rise to the voltage

$$\oint_C \mathbf{E} \cdot d\mathbf{r}$$

around C. The surface integral on the right side of the equation is called the *magnetic flux*, and S is any oriented surface with boundary C.

16. Let

$$\mathbf{F} = -\frac{GmM}{|\mathbf{r}|^3}\mathbf{r}$$

be the gravitational force field defined for $\mathbf{r} \neq \mathbf{0}$. Use Gauss's law in Section 16.8 to show that there is no continuously differentiable vector field $\mathbf{H}$ satisfying $\mathbf{F} = \nabla \times \mathbf{H}$.

17. If $f(x, y, z)$ and $g(x, y, z)$ are continuously differentiable scalar functions defined over the oriented surface S with boundary curve C, prove that

$$\iint_S (\nabla f \times \nabla g) \cdot \mathbf{n}\,d\sigma = \oint_C f\,\nabla g \cdot d\mathbf{r}.$$

18. Suppose that $\nabla \cdot \mathbf{F}_1 = \nabla \cdot \mathbf{F}_2$ and $\nabla \times \mathbf{F}_1 = \nabla \times \mathbf{F}_2$ over a region D enclosed by the oriented surface S with outward unit normal $\mathbf{n}$ and that $\mathbf{F}_1 \cdot \mathbf{n} = \mathbf{F}_2 \cdot \mathbf{n}$ on S. Prove that $\mathbf{F}_1 = \mathbf{F}_2$ throughout D.

19. Prove or disprove that if $\nabla \cdot \mathbf{F} = 0$ and $\nabla \times \mathbf{F} = \mathbf{0}$, then $\mathbf{F} = \mathbf{0}$.

20. Let S be an oriented surface parametrized by $\mathbf{r}(u, v)$. Define the notation $d\boldsymbol{\sigma} = \mathbf{r}_u\,du \times \mathbf{r}_v\,dv$ so that $d\boldsymbol{\sigma}$ is a vector normal to the surface. Also, the magnitude $d\sigma = |d\boldsymbol{\sigma}|$ is the element of surface area (by Equation 5 in Section 16.5). Derive the identity

$$d\sigma = (EG - F^2)^{1/2}\,du\,dv$$

where

$$E = |\mathbf{r}_u|^2, \quad F = \mathbf{r}_u \cdot \mathbf{r}_v, \quad \text{and} \quad G = |\mathbf{r}_v|^2.$$

21. Show that the volume V of a region D in space enclosed by the oriented surface S with outward normal $\mathbf{n}$ satisfies the identity

$$V = \frac{1}{3} \iint_S \mathbf{r} \cdot \mathbf{n}\,d\sigma,$$

where $\mathbf{r}$ is the position vector of the point (x, y, z) in D.

CHAPTER 16   Technology Application Projects

Mathematica/Maple Projects

Projects can be found within MyMathLab.

- **Work in Conservative and Nonconservative Force Fields**
  Explore integration over vector fields and experiment with conservative and nonconservative force functions along different paths in the field.

- **How Can You Visualize Green's Theorem?**
  Explore integration over vector fields and use parametrizations to compute line integrals. Both forms of Green's Theorem are explored.

- **Visualizing and Interpreting the Divergence Theorem**
  Verify the Divergence Theorem by formulating and evaluating certain divergence and surface integrals.

Appendices

This section reviews real numbers, inequalities, intervals, and absolute values.

Real Numbers

Much of calculus is based on properties of the real number system. **Real numbers** are numbers that can be expressed as decimals, such as

$$-\frac{3}{4} = -0.75000 \ldots$$

$$\frac{1}{3} = 0.33333 \ldots$$

$$\sqrt{2} = 1.4142 \ldots$$

The dots ... in each case indicate that the sequence of decimal digits goes on forever. Every conceivable decimal expansion represents a real number, although some numbers have two representations. For instance, the infinite decimals .999 ... and 1.000 ... represent the same real number 1. A similar statement holds for any number with an infinite tail of 9's.

The real numbers can be represented geometrically as points on a number line called the **real line**.

$$\begin{array}{ccccccccc} \hline & & & & & & & & \\ -2 & & -1 \ -\tfrac{3}{4} & & 0 \ \tfrac{1}{3} & & 1 \ \sqrt{2} & 2 & 3 \ \pi & 4 \end{array}$$

The symbol $\mathbb{R}$ denotes either the real number system or, equivalently, the real line.

The properties of the real number system fall into three categories: algebraic properties, order properties, and completeness. The **algebraic properties** say that the real numbers can be added, subtracted, multiplied, and divided (except by 0) to produce more real numbers under the usual rules of arithmetic. *You can never divide by* 0.

The **order properties** of real numbers are given in Appendix 6. The useful rules at the left can be derived from them, where the symbol $\Rightarrow$ means "implies."

Notice the rules for multiplying an inequality by a number. Multiplying by a positive number preserves the inequality; multiplying by a negative number reverses the inequality. Also, reciprocation reverses the inequality for numbers of the same sign. For example, $2 < 5$ but $-2 > -5$ and $1/2 > 1/5$.

The **completeness property** of the real number system is deeper and harder to define precisely. However, the property is essential to the idea of a limit (Chapter 2). Roughly speaking, it says that there are enough real numbers to "complete" the real number line, in

RULES FOR INEQUALITIES

If a, b, and c are real numbers, then:
1. $a < b \Rightarrow a + c < b + c$
2. $a < b \Rightarrow a - c < b - c$
3. $a < b$ and $c > 0 \Rightarrow ac < bc$
4. $a < b$ and $c < 0 \Rightarrow bc < ac$
   Special case. $a < b \Rightarrow -b < -a$
5. $a > 0 \Rightarrow \dfrac{1}{a} > 0$
6. If a and b are both positive or both negative, then $a < b \Rightarrow \dfrac{1}{b} < \dfrac{1}{a}$.

the sense that there are no "holes" or "gaps" in it. Many theorems of calculus would fail if the real number system were not complete. Appendix 6 introduces the ideas involved and discusses how the real numbers are constructed.

We distinguish three special subsets of real numbers.

1. The **natural numbers**, namely 1, 2, 3, 4, . . .

2. The **integers**, namely 0, ±1, ±2, ±3, . . .

3. The **rational numbers**, namely the numbers that can be expressed in the form of a fraction m/n, where m and n are integers and $n \neq 0$. Examples are

$$\frac{1}{3}, \quad -\frac{4}{9} = \frac{-4}{9} = \frac{4}{-9}, \quad \frac{200}{13}, \quad \text{and} \quad 57 = \frac{57}{1}.$$

The rational numbers are precisely the real numbers with decimal expansions that are either

a. terminating (ending in an infinite string of zeros), for example,

$$\frac{3}{4} = 0.75000 \ldots = 0.75 \quad \text{or}$$

b. eventually repeating (ending with a block of digits that repeats over and over), for example,

$$\frac{23}{11} = 2.090909 \ldots = 2.\overline{09} \qquad \text{The bar indicates the block of repeating digits.}$$

A terminating decimal expansion is a special type of repeating decimal, since the ending zeros repeat.

The set of rational numbers has all the algebraic and order properties of the real numbers but lacks the completeness property. For example, there is no rational number whose square is 2; there is a "hole" in the rational line where $\sqrt{2}$ should be.

Real numbers that are not rational are called **irrational numbers**. They are characterized by having nonterminating and nonrepeating decimal expansions. Examples are π, $\sqrt{2}$, $\sqrt[3]{5}$, and $\log_{10} 3$. Since every decimal expansion represents a real number, there are infinitely many irrational numbers. Both rational and irrational numbers are found arbitrarily close to any given point on the real line.

Set notation is very useful for specifying sets of real numbers. A **set** is a collection of objects, and these objects are the **elements** of the set. If S is a set, the notation $a \in S$ means that a is an element of S, and $a \notin S$ means that a is not an element of S. If S and T are sets, then $S \cup T$ is their **union** and consists of all elements belonging to either S or T (or to both S and T). The **intersection** $S \cap T$ consists of all elements belonging to both S and T. The **empty set** $\varnothing$ is the set that contains no elements. For example, the intersection of the rational numbers and the irrational numbers is the empty set.

Some sets can be described by *listing* their elements in braces. For instance, the set A consisting of the natural numbers (or positive integers) less than 6 can be expressed as

$$A = \{1, 2, 3, 4, 5\}.$$

The entire set of integers is written as

$$\{0, \pm 1, \pm 2, \pm 3, \ldots\}.$$

Another way to describe a set is to enclose in braces a rule that generates all the elements of the set. For instance, the set

$$A = \{x \mid x \text{ is an integer and } 0 < x < 6\}$$

is the set of positive integers less than 6.

Intervals

A subset of the real line is called an **interval** if it contains at least two numbers and contains all the real numbers lying between any two of its elements. For example, the set of all real numbers x such that $x > 6$ is an interval, as is the set of all x such that $-2 \le x \le 5$. The set of all nonzero real numbers is not an interval; since 0 is absent, the set fails to contain every real number between -1 and 1 (for example).

Geometrically, intervals correspond to rays and line segments on the real line, along with the real line itself. Intervals of numbers corresponding to line segments are **finite intervals**; intervals corresponding to rays and the real line are **infinite intervals**.

A finite interval is said to be **closed** if it contains both of its endpoints, **half-open** if it contains one endpoint but not the other, and **open** if it contains neither endpoint. The endpoints are also called **boundary points**; they make up the interval's **boundary**. The remaining points of the interval are **interior points** and together compose the interval's **interior**. Infinite intervals are closed if they contain a finite endpoint, and open otherwise. The entire real line $\mathbb{R}$ is an infinite interval that is both open and closed. Table A.1 summarizes the various types of intervals.

TABLE A.1 Types of intervals

Notation	Set description	Type	Picture
(a, b)	$\{x \mid a < x < b\}$	Open	
$[a, b]$	$\{x \mid a \le x \le b\}$	Closed	
$[a, b)$	$\{x \mid a \le x < b\}$	Half-open	
$(a, b]$	$\{x \mid a < x \le b\}$	Half-open	
(a, ∞)	$\{x \mid x > a\}$	Open	
$[a, \infty)$	$\{x \mid x \ge a\}$	Closed	
$(-\infty, b)$	$\{x \mid x < b\}$	Open	
$(-\infty, b]$	$\{x \mid x \le b\}$	Closed	
$(-\infty, \infty)$	$\mathbb{R}$ (set of all real numbers)	Both open and closed	

Solving Inequalities

The process of finding the interval or intervals of numbers that satisfy an inequality in x is called **solving** the inequality.

EXAMPLE 1   Solve the following inequalities and show their solution sets on the real line.

(a) $2x - 1 < x + 3$ **(b)** $\dfrac{6}{x-1} \ge 5$

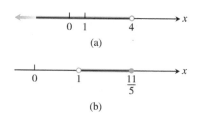

FIGURE A.1 Solution sets for the inequalities in Example 1. Hollow circles indicate endpoints that are not included in the interval, and solid dots indicate included endpoints.

Solution

(a)
$$2x - 1 < x + 3$$
$$2x < x + 4 \qquad \text{Add 1 to both sides.}$$
$$x < 4 \qquad \text{Subtract } x \text{ from both sides.}$$

The solution set is the open interval $(-\infty, 4)$ (Figure A.1a).

(b) The inequality $6/(x - 1) \geq 5$ can hold only if $x > 1$, because otherwise $6/(x - 1)$ is undefined or negative. Therefore, $(x - 1)$ is positive and the inequality will be preserved if we multiply both sides by $(x - 1)$:

$$\frac{6}{x - 1} \geq 5$$
$$6 \geq 5x - 5 \qquad \text{Multiply both sides by } (x - 1).$$
$$11 \geq 5x \qquad \text{Add 5 to both sides.}$$
$$\frac{11}{5} \geq x. \qquad \text{Or } x \leq \frac{11}{5}.$$

The solution set is the half-open interval $(1, 11/5]$ (Figure A.1b). ∎

Absolute Value

The **absolute value** of a number x, denoted by $|x|$, is defined by the formula

$$|x| = \begin{cases} x, & x \geq 0 \\ -x, & x < 0. \end{cases}$$

EXAMPLE 2 $|3| = 3, \quad |0| = 0, \quad |-5| = -(-5) = 5, \quad |-|a|| = |a|$ ∎

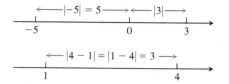

FIGURE A.2 Absolute values give distances between points on the number line.

Geometrically, the absolute value of x is the distance from x to 0 on the real number line. Since distances are always positive or 0, we see that $|x| \geq 0$ for every real number x, and $|x| = 0$ if and only if $x = 0$. Also,

$$|x - y| = \text{the distance between } x \text{ and } y$$

on the real line (Figure A.2).

Since the symbol $\sqrt{a}$ always denotes the *nonnegative* square root of a, an alternate definition of $|x|$ is

$$|x| = \sqrt{x^2}.$$

It is important to remember that $\sqrt{a^2} = |a|$. Do not write $\sqrt{a^2} = a$ unless you already know that $a \geq 0$.

The absolute value function has the following properties. (You are asked to prove these properties in the exercises.)

Absolute Value Properties

1. $|-a| = |a|$

A number and its negative have the same absolute value.

2. $|ab| = |a||b|$

The absolute value of a product is the product of the absolute values.

3. $\left|\dfrac{a}{b}\right| = \dfrac{|a|}{|b|}$

The absolute value of a quotient is the quotient of the absolute values.

4. $|a + b| \leq |a| + |b|$

The **triangle inequality**. The absolute value of the sum of two numbers is less than or equal to the sum of their absolute values.

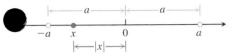

FIGURE A.3 $|x| < a$ means x lies between $-a$ and a.

Note that $|-a| \neq -|a|$. For example, $|-3| = 3$, whereas $-|3| = -3$. If a and b differ in sign, then $|a + b|$ is less than $|a| + |b|$. In all other cases, $|a + b|$ equals $|a| + |b|$. Absolute value bars in expressions like $|-3 + 5|$ work like parentheses: We do the arithmetic inside *before* taking the absolute value.

EXAMPLE 3

$$|-3 + 5| = |2| = 2 < |-3| + |5| = 8$$
$$|3 + 5| = |8| = |3| + |5|$$
$$|-3 - 5| = |-8| = 8 = |-3| + |-5|$$ ∎

ABSOLUTE VALUES AND INTERVALS

If a is any positive number, then

5. $|x| = a \iff x = \pm a$
6. $|x| < a \iff -a < x < a$
7. $|x| > a \iff x > a \text{ or } x < -a$
8. $|x| \leq a \iff -a \leq x \leq a$
9. $|x| \geq a \iff x \geq a \text{ or } x \leq -a$

The inequality $|x| < a$ says that the distance from x to 0 is less than the positive number a. This means that x must lie between $-a$ and a, as we can see from Figure A.3.

Statements 5–9 in the table at left are all consequences of the definition of absolute value and are often helpful when solving equations or inequalities involving absolute values. The symbol $\iff$ that appears in the table is often used by mathematicians to denote the "if and only if" logical relationship. It also means "implies and is implied by."

EXAMPLE 4   Solve the equation $|2x - 3| = 7$.

Solution   By Property 5, $2x - 3 = \pm 7$, so there are two possibilities:

$$2x - 3 = 7 \qquad 2x - 3 = -7 \qquad \text{Equivalent equations without absolute values}$$
$$2x = 10 \qquad 2x = -4 \qquad \text{Solve as usual.}$$
$$x = 5 \qquad x = -2$$

The solutions of $|2x - 3| = 7$ are $x = 5$ and $x = -2$. ∎

EXAMPLE 5   Solve the inequality $\left| 5 - \dfrac{2}{x} \right| < 1$.

Solution   We have

$$\left| 5 - \frac{2}{x} \right| < 1 \iff -1 < 5 - \frac{2}{x} < 1 \qquad \text{Property 6}$$

$$\iff -6 < -\frac{2}{x} < -4 \qquad \text{Subtract 5.}$$

$$\iff 3 > \frac{1}{x} > 2 \qquad \text{Multiply by } -\frac{1}{2}.$$

$$\iff \frac{1}{3} < x < \frac{1}{2}. \qquad \text{Take reciprocals.}$$

Notice how the various rules for inequalities were used here. Multiplying by a negative number reverses the inequality. So does taking reciprocals in an inequality in which both sides are positive. The original inequality holds if and only if $(1/3) < x < (1/2)$. The solution set is the open interval $(1/3, 1/2)$. ∎

1. Express $1/9$ as a repeating decimal, using a bar to indicate the repeating digits. What are the decimal representations of $2/9$? $3/9$? $8/9$? $9/9$?

2. If $2 < x < 6$, which of the following statements about x are necessarily true, and which are not necessarily true?

   a. $0 < x < 4$ b. $0 < x - 2 < 4$

   c. $1 < \dfrac{x}{2} < 3$ d. $\dfrac{1}{6} < \dfrac{1}{x} < \dfrac{1}{2}$

   e. $1 < \dfrac{6}{x} < 3$ f. $|x - 4| < 2$

   g. $-6 < -x < 2$ h. $-6 < -x < -2$

In Exercises 3–6, solve the inequalities and show the solution sets on the real line.

3. $-2x > 4$ 4. $5x - 3 \le 7 - 3x$

5. $2x - \dfrac{1}{2} \ge 7x + \dfrac{7}{6}$ 6. $\dfrac{4}{5}(x - 2) < \dfrac{1}{3}(x - 6)$

Solve the equations in Exercises 7–9.

7. $|y| = 3$ 8. $|2t + 5| = 4$ 9. $|8 - 3s| = \dfrac{9}{2}$

Solve the inequalities in Exercises 10–17, expressing the solution sets as intervals or unions of intervals. Also, show each solution set on the real line.

10. $|x| < 2$ 11. $|t - 1| \le 3$ 12. $|3y - 7| < 4$

13. $\left|\dfrac{z}{5} - 1\right| \le 1$ 14. $\left|3 - \dfrac{1}{x}\right| < \dfrac{1}{2}$ 15. $|2s| \ge 4$

16. $|1 - x| > 1$ 17. $\left|\dfrac{r + 1}{2}\right| \ge 1$

Solve the inequalities in Exercises 18–21. Express the solution sets as intervals or unions of intervals and show them on the real line. Use the result $\sqrt{a^2} = |a|$ as appropriate.

18. $x^2 < 2$ 19. $4 < x^2 < 9$

20. $(x - 1)^2 < 4$ 21. $x^2 - x < 0$

22. Do not fall into the trap of thinking $|-a| = a$. For what real numbers a is this equation true? For what real numbers is it false?

23. Solve the equation $|x - 1| = 1 - x$.

24. **A proof of the triangle inequality**   Give the reason justifying each of the numbered steps in the following proof of the triangle inequality.

$$|a + b|^2 = (a + b)^2 \tag{1}$$
$$= a^2 + 2ab + b^2$$
$$\le a^2 + 2|a||b| + b^2 \tag{2}$$
$$= |a|^2 + 2|a||b| + |b|^2 \tag{3}$$
$$= (|a| + |b|)^2$$
$$|a + b| \le |a| + |b| \tag{4}$$

25. Prove that $|ab| = |a||b|$ for any numbers a and b.

26. If $|x| \le 3$ and $x > -1/2$, what can you say about x?

27. Graph the inequality $|x| + |y| \le 1$.

28. For any number a, prove that $|-a| = |a|$.

29. Let a be any positive number. Prove that $|x| > a$ if and only if $x > a$ or $x < -a$.

29. a. If b is any nonzero real number, prove that $|1/b| = 1/|b|$.

    b. Prove that $\left|\dfrac{a}{b}\right| = \dfrac{|a|}{|b|}$ for any numbers a and $b \ne 0$.

A.2   Mathematical Induction

Many formulas, like

$$1 + 2 + \cdots + n = \frac{n(n + 1)}{2},$$

can be shown to hold for every positive integer n by applying an axiom called the *mathematical induction principle*. A proof that uses this axiom is called a *proof by mathematical induction* or a *proof by induction*.

The steps in proving a formula by induction are the following:

1. Check that the formula holds for $n = 1$.

2. Prove that *if* the formula holds for any positive integer $n = k$, *then* it also holds for the next integer, $n = k + 1$.

The induction axiom says that once these steps are completed, the formula holds for all positive integers n. By Step 1 it holds for $n = 1$. By Step 2 it holds for $n = 2$, and therefore by Step 2 also for $n = 3$, and by Step 2 again for $n = 4$, and so on. If the first domino falls, and if the kth domino always knocks over the $(k + 1)$st when it falls, then all the dominoes fall.

From another point of view, suppose we have a sequence of statements $S_1, S_2, \ldots, S_n, \ldots$, one for each positive integer. Suppose we can show that assuming any one of the statements to be true implies that the next statement in line is true. Suppose that we can also show that S_1 is true. Then we may conclude that the statements are true from S_1 on.

EXAMPLE 1   Use mathematical induction to prove that for every positive integer n,

$$1 + 2 + \cdots + n = \frac{n(n + 1)}{2}.$$

Solution   We accomplish the proof by carrying out the two steps above.

1. The formula holds for $n = 1$ because

$$1 = \frac{1(1 + 1)}{2}.$$

2. If the formula holds for $n = k$, does it also hold for $n = k + 1$? The answer is yes, as we now show. If it is the case that

$$1 + 2 + \cdots + k = \frac{k(k + 1)}{2},$$

then it follows that

$$1 + 2 + \cdots + k + (k + 1) = \frac{k(k + 1)}{2} + (k + 1) = \frac{k^2 + k + 2k + 2}{2}$$
$$= \frac{(k + 1)(k + 2)}{2} = \frac{(k + 1)((k + 1) + 1)}{2}.$$

The last expression in this string of equalities is the expression $n(n + 1)/2$ for $n = (k + 1)$.

The mathematical induction principle now guarantees the original formula for all positive integers n.   ■

In Example 4 of Section 5.2 we gave another proof for the formula giving the sum of the first n integers. However, proofs by mathematical induction can also be used to find the sums of the squares and cubes of the first n integers (Exercises 9 and 10). Here is another example of a proof by induction.

EXAMPLE 2   Show by mathematical induction that for all positive integers n,

$$\frac{1}{2^1} + \frac{1}{2^2} + \cdots + \frac{1}{2^n} = 1 - \frac{1}{2^n}.$$

Solution   We accomplish the proof by carrying out the two steps of mathematical induction.

1. The formula holds for $n = 1$ because

$$\frac{1}{2^1} = 1 - \frac{1}{2^1}.$$

2. If it is the case that

$$\frac{1}{2^1} + \frac{1}{2^2} + \cdots + \frac{1}{2^k} = 1 - \frac{1}{2^k},$$

then it follows that

$$\frac{1}{2^1} + \frac{1}{2^2} + \cdots + \frac{1}{2^k} + \frac{1}{2^{k+1}} = 1 - \frac{1}{2^k} + \frac{1}{2^{k+1}} = 1 - \frac{1 \cdot 2}{2^k \cdot 2} + \frac{1}{2^{k+1}}$$

$$= 1 - \frac{2}{2^{k+1}} + \frac{1}{2^{k+1}} = 1 - \frac{1}{2^{k+1}}.$$

Thus, the original formula holds for $n = (k + 1)$ whenever it holds for $n = k$.

With these steps verified, the mathematical induction principle now guarantees the formula for every positive integer n. ∎

Other Starting Integers

Instead of starting at $n = 1$ some induction arguments start at another integer. The steps for such an argument are as follows.

1. Check that the formula holds for $n = n_1$ (the first appropriate integer).
2. Prove that if the formula holds for any integer $n = k \geq n_1$, then it also holds for $n = (k + 1)$.

Once these steps are completed, the mathematical induction principle guarantees the formula for all $n \geq n_1$.

EXAMPLE 3   Show that $n! > 3^n$ if n is large enough.

Solution   How large is large enough? We experiment:

n	1	2	3	4	5	6	7
$n!$	1	2	6	24	120	720	5040
3^n	3	9	27	81	243	729	2187

It looks as if $n! > 3^n$ for $n \geq 7$. To be sure, we apply mathematical induction. We take $n_1 = 7$ in Step 1 and complete Step 2.

Suppose $k! > 3^k$ for some $k \geq 7$. Then

$$(k + 1)! = (k + 1)(k!) > (k + 1)3^k > 7 \cdot 3^k > 3^{k+1}.$$

Thus, for $k \geq 7$,

$$k! > 3^k \quad \text{implies} \quad (k + 1)! > 3^{k+1}.$$

The mathematical induction principle now guarantees $n! \geq 3^n$ for all $n \geq 7$. ∎

Proof of the Derivative Sum Rule for Sums of Finitely Many Functions

We prove the statement

$$\frac{d}{dx}(u_1 + u_2 + \cdots + u_n) = \frac{du_1}{dx} + \frac{du_2}{dx} + \cdots + \frac{du_n}{dx}$$

by mathematical induction. The statement is true for $n = 2$, as was proved in Section 3.3. This is Step 1 of the induction proof.

Step 2 is to show that if the statement is true for any positive integer $n = k$, where $k \geq n_0 = 2$, then it is also true for $n = k + 1$. So suppose that

$$\frac{d}{dx}(u_1 + u_2 + \cdots + u_k) = \frac{du_1}{dx} + \frac{du_2}{dx} + \cdots + \frac{du_k}{dx}. \tag{1}$$

Then

$$\frac{d}{dx}\underbrace{(u_1 + u_2 + \cdots + u_k}_{\substack{\text{Call the function} \\ \text{defined by this sum } u}} + \underbrace{u_{k+1})}_{\substack{\text{Call this} \\ \text{function } v.}}$$

$$= \frac{d}{dx}(u_1 + u_2 + \cdots + u_k) + \frac{du_{k+1}}{dx} \qquad \text{Sum Rule for } \frac{d}{dx}(u + v)$$

$$= \frac{du_1}{dx} + \frac{du_2}{dx} + \cdots + \frac{du_k}{dx} + \frac{du_{k+1}}{dx}. \qquad \text{Eq. (1)}$$

With these steps verified, the mathematical induction principle now guarantees the Sum Rule for every integer $n \geq 2$.

EXERCISES A.2

1. Assuming that the triangle inequality $|a + b| \leq |a| + |b|$ holds for any two numbers a and b, show that

$$|x_1 + x_2 + \cdots + x_n| \leq |x_1| + |x_2| + \cdots + |x_n|$$

for any n numbers.

2. Show that if $r \neq 1$, then

$$1 + r + r^2 + \cdots + r^n = \frac{1 - r^{n+1}}{1 - r}$$

for every positive integer n.

3. Use the Product Rule, $\frac{d}{dx}(uv) = u\frac{dv}{dx} + v\frac{du}{dx}$, and the fact that $\frac{d}{dx}(x) = 1$ to show that $\frac{d}{dx}(x^n) = nx^{n-1}$ for every positive integer n.

4. Suppose that a function $f(x)$ has the property that $f(x_1x_2) = f(x_1) + f(x_2)$ for any two positive numbers x_1 and x_2. Show that

$$f(x_1x_2 \cdots x_n) = f(x_1) + f(x_2) + \cdots + f(x_n)$$

for the product of any n positive numbers $x_1, x_2, \ldots, x_n$.

5. Show that

$$\frac{2}{3^1} + \frac{2}{3^2} + \cdots + \frac{2}{3^n} = 1 - \frac{1}{3^n}$$

for all positive integers n.

6. Show that $n! > n^3$ if n is large enough.

7. Show that $2^n > n^2$ if n is large enough.

8. Show that $2^n \geq 1/8$ for $n \geq -3$.

9. **Sums of squares** Show that the sum of the squares of the first n positive integers is

$$\frac{n\left(n + \dfrac{1}{2}\right)(n + 1)}{3}.$$

10. **Sums of cubes** Show that the sum of the cubes of the first n positive integers is $(n(n + 1)/2)^2$.

11. **Rules for finite sums** Show that the following finite sum rules hold for every positive integer n. (See Section 5.2.)

    a. $\displaystyle\sum_{k=1}^{n}(a_k + b_k) = \sum_{k=1}^{n} a_k + \sum_{k=1}^{n} b_k$

    b. $\displaystyle\sum_{k=1}^{n}(a_k - b_k) = \sum_{k=1}^{n} a_k - \sum_{k=1}^{n} b_k$

    c. $\displaystyle\sum_{k=1}^{n} ca_k = c \cdot \sum_{k=1}^{n} a_k$ (any number c)

    d. $\displaystyle\sum_{k=1}^{n} a_k = n \cdot c$ (if a_k has the constant value c)

12. Show that $|x^n| = |x|^n$ for every positive integer n and every real number x.

A.3 Lines, Circles, and Parabolas

This section reviews coordinates, lines, distance, circles, and parabolas in the plane. The notion of increment is also discussed.

Cartesian Coordinates in the Plane

In Appendix 1 we identified the points on the line with real numbers by assigning them coordinates. Points in the plane can be identified with ordered pairs of real numbers. To begin, we draw two perpendicular coordinate lines that intersect at the 0-point of each line.

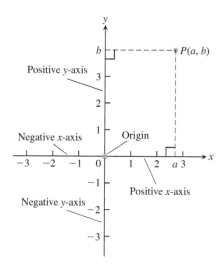

FIGURE A.4 Cartesian coordinates in the plane are based on two perpendicular axes intersecting at the origin.

HISTORICAL BIOGRAPHY
René Descartes
(1596–1650)
www.goo.gl/XSzlEA

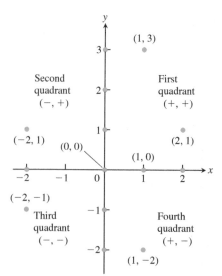

FIGURE A.5 Points labeled in the xy-coordinate or Cartesian plane. The points on the axes all have coordinate pairs but are usually labeled with single real numbers, (so (1, 0) on the x-axis is labeled as 1). Notice the coordinate sign patterns of the quadrants.

These lines are called **coordinate axes**. On the horizontal x-axis, numbers are denoted by x and increase to the right. On the vertical y-axis, numbers are denoted by y and increase upward (Figure A.4). Thus "upward" and "to the right" are positive directions, whereas "downward" and "to the left" are considered negative. The **origin** O, also labeled 0, of the coordinate system is the point in the plane where x and y are both zero.

If P is any point in the plane, it can be located by exactly one ordered pair of real numbers in the following way. Draw lines through P perpendicular to the two coordinate axes. These lines intersect the axes at points with coordinates a and b (Figure A.4). The ordered pair (a, b) is assigned to the point P and is called its **coordinate pair**. The first number a is the **x-coordinate** (or **abscissa**) of P; the second number b is the **y-coordinate** (or **ordinate**) of P. The x-coordinate of every point on the y-axis is 0. The y-coordinate of every point on the x-axis is 0. The origin is the point (0, 0).

Starting with an ordered pair (a, b), we can reverse the process and arrive at a corresponding point P in the plane. Often we identify P with the ordered pair and write P(a, b). We sometimes also refer to "the point (a, b)" and it will be clear from the context when (a, b) refers to a point in the plane and not to an open interval on the real line. Several points labeled by their coordinates are shown in Figure A.5.

This coordinate system is called the **rectangular coordinate system** or **Cartesian coordinate system** (after the sixteenth-century French mathematician René Descartes). The coordinate axes of this coordinate or Cartesian plane divide the plane into four regions called **quadrants**, numbered counterclockwise as shown in Figure A.5.

The **graph** of an equation or inequality in the variables x and y is the set of all points P(x, y) in the plane whose coordinates satisfy the equation or inequality. When we plot data in the coordinate plane or graph formulas whose variables have different units of measure, we do not need to use the same scale on the two axes. If we plot time vs. thrust for a rocket motor, for example, there is no reason to place the mark that shows 1 sec on the time axis the same distance from the origin as the mark that shows 1 lb on the thrust axis.

Usually when we graph functions whose variables do not represent physical measurements and when we draw figures in the coordinate plane to study their geometry and trigonometry, we make the scales on the axes identical. A vertical unit of distance then looks the same as a horizontal unit. As on a surveyor's map or a scale drawing, line segments that are supposed to have the same length will look as if they do and angles that are supposed to be congruent will look congruent.

Computer displays and calculator displays are another matter. The vertical and horizontal scales on machine-generated graphs usually differ, and there are corresponding distortions in distances, slopes, and angles. Circles may look like ellipses, rectangles may look like squares, right angles may appear to be acute or obtuse, and so on. We discuss these displays and distortions in greater detail in Section 1.4.

Increments and Straight Lines

When a particle moves from one point in the plane to another, the net changes in its coordinates are called *increments*. They are calculated by subtracting the coordinates of the starting point from the coordinates of the ending point. If x changes from x_1 to x_2, the **increment** in x is

$$\Delta x = x_2 - x_1.$$

EXAMPLE 1   As shown in Figure A.6, in going from the point A(4, −3) to the point B(2, 5) the increments in the x- and y-coordinates are

$$\Delta x = 2 - 4 = -2, \qquad \Delta y = 5 - (-3) = 8.$$

From C(5, 6) to D(5, 1) the coordinate increments are

$$\Delta x = 5 - 5 = 0, \qquad \Delta y = 1 - 6 = -5.$$

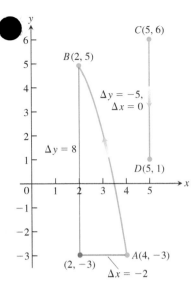

FIGURE A.6 Coordinate increments may be positive, negative, or zero (Example 1).

Given two points $P_1(x_1, y_1)$ and $P_2(x_2, y_2)$ in the plane, we call the increments $\Delta x = x_2 - x_1$ and $\Delta y = y_2 - y_1$ the **run** and the **rise**, respectively, between P_1 and P_2. Two such points always determine a unique straight line (usually called simply a line) passing through them both. We call the line P_1P_2.

Any nonvertical line in the plane has the property that the ratio

$$m = \frac{\text{rise}}{\text{run}} = \frac{\Delta y}{\Delta x} = \frac{y_2 - y_1}{x_2 - x_1}$$

has the same value for every choice of the two points $P_1(x_1, y_1)$ and $P_2(x_2, y_2)$ on the line (Figure A.7). This is because the ratios of corresponding sides for similar triangles are equal.

DEFINITION The constant ratio

$$m = \frac{\text{rise}}{\text{run}} = \frac{\Delta y}{\Delta x} = \frac{y_2 - y_1}{x_2 - x_1}$$

is the **slope** of the nonvertical line P_1P_2.

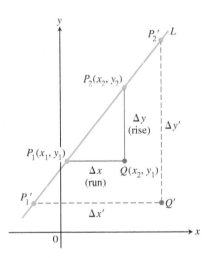

FIGURE A.7 Triangles P_1QP_2 and $P_1'Q'P_2'$ are similar, so the ratio of their sides has the same value for any two points on the line. This common value is the line's slope.

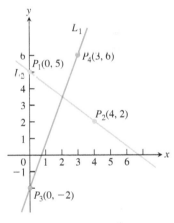

FIGURE A.8 The slope of L_1 is

$$m = \frac{\Delta y}{\Delta x} = \frac{6 - (-2)}{3 - 0} = \frac{8}{3}.$$

That is, y increases 8 units every time x increases 3 units. The slope of L_2 is

$$m = \frac{\Delta y}{\Delta x} = \frac{2 - 5}{4 - 0} = \frac{-3}{4}.$$

That is, y decreases 3 units every time x increases 4 units.

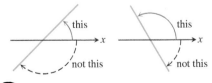

FIGURE A.9 Angles of inclination are measured counterclockwise from the x-axis.

The slope tells us the direction (uphill, downhill) and steepness of a line. A line with positive slope rises uphill to the right; one with negative slope falls downhill to the right (Figure A.8). The greater the absolute value of the slope, the more rapid the rise or fall. The slope of a vertical line is *undefined*. Since the run Δx is zero for a vertical line, we cannot form the slope ratio m.

The direction and steepness of a line can also be measured with an angle. The **angle of inclination** of a line that crosses the x-axis is the smallest counterclockwise angle from the x-axis to the line (Figure A.9). The inclination of a horizontal line is 0°. The inclination of a vertical line is 90°. If ϕ (the Greek letter phi) is the inclination of a line, then $0 \le \phi < 180°$.

The relationship between the slope m of a nonvertical line and the line's angle of inclination ϕ is shown in Figure A.10:

$$m = \tan \phi.$$

Straight lines have relatively simple equations. All points on the *vertical line* through the point a on the x-axis have x-coordinates equal to a. Thus, $x = a$ is an equation for the vertical line. Similarly, $y = b$ is an equation for the *horizontal line* meeting the y-axis at b. (See Figure A.11.)

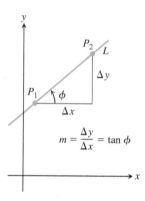

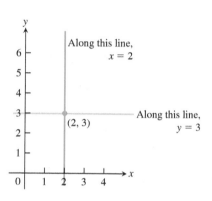

FIGURE A.10   The slope of a nonvertical line is the tangent of its angle of inclination.

FIGURE A.11   The standard equations for the vertical and horizontal lines through $(2, 3)$ are $x = 2$ and $y = 3$.

We can write an equation for a nonvertical straight line L if we know its slope m and the coordinates of one point $P_1(x_1, y_1)$ on it. If $P(x, y)$ is *any* other point on L, then we can use the two points P_1 and P to compute the slope,

$$m = \frac{y - y_1}{x - x_1}$$

so that

$$y - y_1 = m(x - x_1), \qquad \text{or} \qquad y = y_1 + m(x - x_1).$$

The equation

$$y = y_1 + m(x - x_1)$$

is the **point-slope equation** of the line that passes through the point (x_1, y_1) and has slope m.

EXAMPLE 2   Write an equation for the line through the point $(2, 3)$ with slope $-3/2$.

Solution   We substitute $x_1 = 2$, $y_1 = 3$, and $m = -3/2$ into the point-slope equation and obtain

$$y = 3 - \frac{3}{2}(x - 2), \qquad \text{or} \qquad y = -\frac{3}{2}x + 6.$$

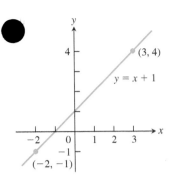

FIGURE A.12 The line in Example 3.

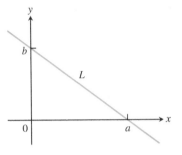

FIGURE A.13 Line L has x-intercept a and y-intercept b.

EXAMPLE 3   Write an equation for the line through $(-2, -1)$ and $(3, 4)$.

Solution   The line's slope is

$$m = \frac{-1 - 4}{-2 - 3} = \frac{-5}{-5} = 1.$$

We can use this slope with either of the two given points in the point-slope equation:

With $(x_1, y_1) = (-2, -1)$	With $(x_1, y_1) = (3, 4)$
$y = -1 + 1 \cdot (x - (-2))$	$y = 4 + 1 \cdot (x - 3)$
$y = -1 + x + 2$	$y = 4 + x - 3$
$y = x + 1$	$y = x + 1$

Same result

Either way, we see that $y = x + 1$ is an equation for the line (Figure A.12). ■

The y-coordinate of the point where a nonvertical line intersects the y-axis is called the **y-intercept** of the line. Similarly, the **x-intercept** of a nonhorizontal line is the x-coordinate of the point where it crosses the x-axis (Figure A.13). A line with slope m and y-intercept b passes through the point $(0, b)$, so it has equation

$$y = b + m(x - 0), \qquad \text{or, more simply,} \qquad y = mx + b.$$

> The equation
>
> $$y = mx + b$$
>
> is called the **slope-intercept equation** of the line with slope m and y-intercept b.

Lines with equations of the form $y = mx$ have y-intercept 0 and so pass through the origin. Equations of lines are called **linear** equations.
  The equation

$$Ax + By = C \qquad (A \text{ and } B \text{ not both } 0)$$

is called the **general linear equation** in x and y because its graph always represents a line and every line has an equation in this form (including lines with undefined slope).

Parallel and Perpendicular Lines

Lines that are parallel have equal angles of inclination, so they have the same slope (if they are not vertical). Conversely, lines with equal slopes have equal angles of inclination and so are parallel.
  If two nonvertical lines L_1 and L_2 are perpendicular, their slopes m_1 and m_2 satisfy $m_1 m_2 = -1$, so each slope is the *negative reciprocal* of the other:

$$m_1 = -\frac{1}{m_2}, \qquad m_2 = -\frac{1}{m_1}.$$

To see this, notice by inspecting similar triangles in Figure A.14 that $m_1 = a/h$, and $m_2 = -h/a$. Hence, $m_1 m_2 = (a/h)(-h/a) = -1$.

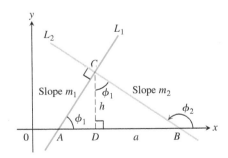

FIGURE A.14 $\triangle ADC$ is similar to $\triangle CDB$. Hence ϕ_1 is also the upper angle in $\triangle CDB$. From the sides of $\triangle CDB$, we read $\tan \phi_1 = a/h$.

Distance and Circles in the Plane

The distance between points in the plane is calculated with a formula that comes from the Pythagorean theorem (Figure A.15).

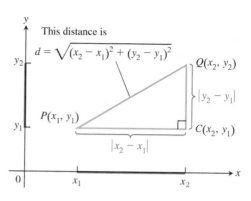

FIGURE A.15   To calculate the distance between $P(x_1, y_1)$ and $Q(x_2, y_2)$, apply the Pythagorean theorem to triangle PCQ.

Distance Formula for Points in the Plane

The distance between $P(x_1, y_1)$ and $Q(x_2, y_2)$ is

$$d = \sqrt{(\Delta x)^2 + (\Delta y)^2} = \sqrt{(x_2 - x_1)^2 + (y_2 - y_1)^2}.$$

By definition, a **circle** of radius a is the set of all points $P(x, y)$ whose distance from some center $C(h, k)$ equals a (Figure A.16). From the distance formula, P lies on the circle if and only if

$$\sqrt{(x - h)^2 + (y - k)^2} = a,$$

so

$$(x - h)^2 + (y - k)^2 = a^2. \qquad (1)$$

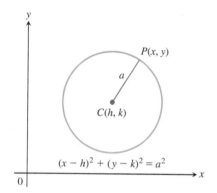

FIGURE A.16   A circle of radius a in the xy-plane, with center at (h, k).

Equation (1) is the **standard equation** of a circle with center (h, k) and radius a. The circle of radius $a = 1$ and centered at the origin is the **unit circle** with equation

$$x^2 + y^2 = 1.$$

EXAMPLE 4

(a) The standard equation for the circle of radius 2 centered at $(3, 4)$ is

$$(x - 3)^2 + (y - 4)^2 = 2^2 = 4.$$

(b) The circle

$$(x - 1)^2 + (y + 5)^2 = 3$$

has $h = 1$, $k = -5$, and $a = \sqrt{3}$. The center is the point $(h, k) = (1, -5)$ and the radius is $a = \sqrt{3}$.   ∎

If an equation for a circle is not in standard form, we can find the circle's center and radius by first converting the equation to standard form. The algebraic technique for doing so is *completing the square*.

EXAMPLE 5   Find the center and radius of the circle

$$x^2 + y^2 + 4x - 6y - 3 = 0.$$

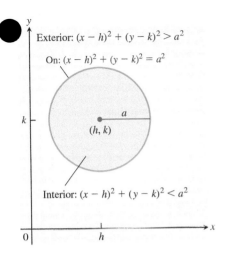

FIGURE A.17 The interior and exterior of the circle $(x - h)^2 + (y - k)^2 = a^2$.

Solution We convert the equation to standard form by completing the squares in x and y:

$$x^2 + y^2 + 4x - 6y - 3 = 0$$

Start with the given equation.

$$(x^2 + 4x) + (y^2 - 6y) = 3$$

Gather terms. Move the constant to the right-hand side.

$$\left(x^2 + 4x + \left(\frac{4}{2}\right)^2\right) + \left(y^2 - 6y + \left(\frac{-6}{2}\right)^2\right) =$$

$$3 + \left(\frac{4}{2}\right)^2 + \left(\frac{-6}{2}\right)^2$$

Add the square of half the coefficient of x to each side of the equation. Do the same for y. The parenthetical expressions on the left-hand side are now perfect squares.

$$(x^2 + 4x + 4) + (y^2 - 6y + 9) = 3 + 4 + 9$$

$$(x + 2)^2 + (y - 3)^2 = 16$$

Write each quadratic as a squared linear expression.

The center is $(-2, 3)$ and the radius is $a = 4$. ∎

The points (x, y) satisfying the inequality

$$(x - h)^2 + (y - k)^2 < a^2$$

make up the **interior** region of the circle with center (h, k) and radius a (Figure A.17). The circle's **exterior** consists of the points (x, y) satisfying

$$(x - h)^2 + (y - k)^2 > a^2.$$

Parabolas

The geometric definition and properties of general parabolas are reviewed in Chapter 11. Here we look at parabolas arising as the graphs of equations of the form $y = ax^2 + bx + c$.

EXAMPLE 6 Consider the equation $y = x^2$. Some points whose coordinates satisfy this equation are $(0, 0)$, $(1, 1)$, $\left(\frac{3}{2}, \frac{9}{4}\right)$, $(-1, 1)$, $(2, 4)$, and $(-2, 4)$. These points (and all others satisfying the equation) make up a smooth curve called a parabola (Figure A.18). ∎

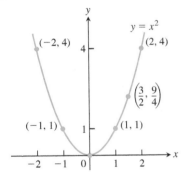

FIGURE A.18 The parabola $y = x^2$ (Example 6).

The graph of an equation of the form

$$y = ax^2$$

is a **parabola** whose **axis** (axis of symmetry) is the y-axis. The parabola's **vertex** (point where the parabola and axis cross) lies at the origin. The parabola opens upward if $a > 0$ and downward if $a < 0$. The larger the value of $|a|$, the narrower the parabola (Figure A.19).

Generally, the graph of $y = ax^2 + bx + c$ is a shifted and scaled version of the parabola $y = x^2$. We discuss shifting and scaling of graphs in more detail in Section 1.2.

The Graph of $y = ax^2 + bx + c$, $a \neq 0$

The graph of the equation $y = ax^2 + bx + c, a \neq 0$, is a parabola. The parabola opens upward if $a > 0$ and downward if $a < 0$. The **axis** is the line

$$x = -\frac{b}{2a}. \tag{2}$$

The **vertex** of the parabola is the point where the axis and parabola intersect. Its x-coordinate is $x = -b/2a$; its y-coordinate is found by substituting $x = -b/2a$ in the parabola's equation.

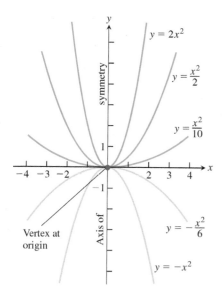

FIGURE A.19   Besides determining the direction in which the parabola $y = ax^2$ opens, the number a is a scaling factor. The parabola widens as a approaches zero and narrows as $|a|$ becomes large.

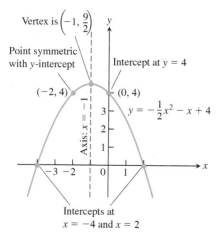

FIGURE A.20   The parabola in Example 7.

Notice that if $a = 0$, then we have $y = bx + c$, which is an equation for a line. The axis, given by Equation (2), can be found by completing the square.

EXAMPLE 7   Graph the equation $y = -\dfrac{1}{2}x^2 - x + 4$.

Solution   Comparing the equation with $y = ax^2 + bx + c$ we see that

$$a = -\frac{1}{2}, \qquad b = -1, \qquad c = 4.$$

Since $a < 0$, the parabola opens downward. From Equation (2) the axis is the vertical line

$$x = -\frac{b}{2a} = -\frac{(-1)}{2(-1/2)} = -1.$$

When $x = -1$, we have

$$y = -\frac{1}{2}(-1)^2 - (-1) + 4 = \frac{9}{2}.$$

The vertex is $(-1, 9/2)$.
   The x-intercepts are where $y = 0$:

$$-\frac{1}{2}x^2 - x + 4 = 0$$
$$x^2 + 2x - 8 = 0$$
$$(x - 2)(x + 4) = 0$$
$$x = 2, \qquad x = -4$$

We plot some points, sketch the axis, and use the direction of opening to complete the graph in Figure A.20.

Ellipses

The geometric definition and properties of general ellipses are reviewed in Chapter 11. Here we relate them to circles. Although they are not the graphs of functions, circles can be stretched horizontally or vertically in the same way as the graphs of functions. The standard equation for a circle of radius r centered at the origin is

$$x^2 + y^2 = r^2.$$

Substituting cx for x in the standard equation for a circle (Figure A.21) gives

$$c^2x^2 + y^2 = r^2. \tag{3}$$

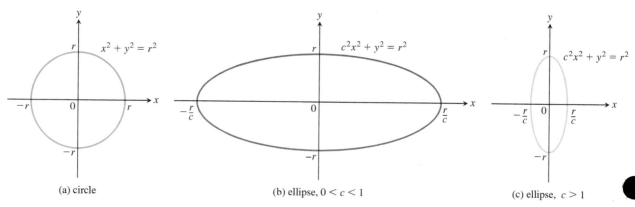

FIGURE A.21   Horizontal stretching or compression of a circle produces graphs of ellipses.

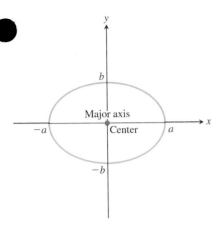

FIGURE A.22 Graph of the ellipse $\dfrac{x^2}{a^2} + \dfrac{y^2}{b^2} = 1$, $a > b$, where the major axis is horizontal.

If $0 < c < 1$, the graph of Equation (3) horizontally stretches the circle; if $c > 1$ the circle is compressed horizontally. In either case, the graph of Equation (3) is an ellipse (Figure A.21). Notice in Figure A.21 that the y-intercepts of all three graphs are always $-r$ and r. In Figure A.21b, the line segment joining the points $(\pm r/c, 0)$ is called the **major axis** of the ellipse; the **minor axis** is the line segment joining $(0, \pm r)$. The axes of the ellipse are reversed in Figure A.21c: The major axis is the line segment joining the points $(0, \pm r)$, and the minor axis is the line segment joining the points $(\pm r/c, 0)$. In both cases, the major axis is the longer line segment.

If we divide both sides of Equation (3) by r^2, we obtain

$$\frac{x^2}{a^2} + \frac{y^2}{b^2} = 1 \qquad (4)$$

where $a = r/c$ and $b = r$. If $a > b$, the major axis is horizontal; if $a < b$, the major axis is vertical. The **center** of the ellipse given by Equation (4) is the origin (Figure A.22).

Substituting $x - h$ for x, and $y - k$ for y, in Equation (4) results in

$$\frac{(x - h)^2}{a^2} + \frac{(y - k)^2}{b^2} = 1. \qquad (5)$$

Equation (5) is the **standard equation of an ellipse** with center at (h, k).

EXERCISES A.3

Distance, Slopes, and Lines

In Exercises 1 and 2, a particle moves from A to B in the coordinate plane. Find the increments Δx and Δy in the particle's coordinates. Also find the distance from A to B.

1. $A(-3, 2)$, $B(-1, -2)$ **2.** $A(-3.2, -2)$, $B(-8.1, -2)$

Describe the graphs of the equations in Exercises 3 and 4.

3. $x^2 + y^2 = 1$ **4.** $x^2 + y^2 \le 3$

Plot the points in Exercises 5 and 6 and find the slope (if any) of the line they determine. Also find the common slope (if any) of the lines perpendicular to line AB.

5. $A(-1, 2)$, $B(-2, -1)$ **6.** $A(2, 3)$, $B(-1, 3)$

In Exercises 7 and 8, find an equation for **(a)** the vertical line and **(b)** the horizontal line through the given point.

7. $(-1, 4/3)$ **8.** $\left(0, -\sqrt{2}\right)$

In Exercises 9–15, write an equation for each line described.

9. Passes through $(-1, 1)$ with slope -1

10. Passes through $(3, 4)$ and $(-2, 5)$

11. Has slope $-5/4$ and y-intercept 6

12. Passes through $(-12, -9)$ and has slope 0

13. Has y-intercept 4 and x-intercept -1

14. Passes through $(5, -1)$ and is parallel to the line $2x + 5y = 15$

15. Passes through $(4, 10)$ and is perpendicular to the line $6x - 3y = 5$

In Exercises 16 and 17, find the line's x- and y-intercepts and use this information to graph the line.

16. $3x + 4y = 12$ **17.** $\sqrt{2}x - \sqrt{3}y = \sqrt{6}$

18. Is there anything special about the relationship between the lines $Ax + By = C_1$ and $Bx - Ay = C_2$ ($A \ne 0, B \ne 0$)? Give reasons for your answer.

19. A particle starts at $A(-2, 3)$ and its coordinates change by increments $\Delta x = 5$, $\Delta y = -6$. Find its new position.

20. The coordinates of a particle change by $\Delta x = 5$ and $\Delta y = 6$ as it moves from $A(x, y)$ to $B(3, -3)$. Find x and y.

Circles

In Exercises 21–23, find an equation for the circle with the given center $C(h, k)$ and radius a. Then sketch the circle in the xy-plane. Include the circle's center in your sketch. Also, label the circle's x- and y-intercepts, if any, with their coordinate pairs.

21. $C(0, 2)$, $a = 2$ **22.** $C(-1, 5)$, $a = \sqrt{10}$

23. $C\left(-\sqrt{3}, -2\right)$, $a = 2$

Graph the circles whose equations are given in Exercises 24–26. Label each circle's center and intercepts (if any) with their coordinate pairs.

24. $x^2 + y^2 + 4x - 4y + 4 = 0$

25. $x^2 + y^2 - 3y - 4 = 0$ **26.** $x^2 + y^2 - 4x + 4y = 0$

Parabolas

Graph the parabolas in Exercises 27–30. Label the vertex, axis, and intercepts in each case.

27. $y = x^2 - 2x - 3$ **28.** $y = -x^2 + 4x$

29. $y = -x^2 - 6x - 5$ **30.** $y = \dfrac{1}{2}x^2 + x + 4$

Inequalities

Describe the regions defined by the inequalities and pairs of inequalities in Exercises 31–34.

31. $x^2 + y^2 > 7$ **32.** $(x - 1)^2 + y^2 \leq 4$

33. $x^2 + y^2 > 1, \quad x^2 + y^2 < 4$

34. $x^2 + y^2 + 6y < 0, \quad y > -3$

35. Write an inequality that describes the points that lie inside the circle with center $(-2, 1)$ and radius $\sqrt{6}$.

36. Write a pair of inequalities that describe the points that lie inside or on the circle with center $(0, 0)$ and radius $\sqrt{2}$, and on or to the right of the vertical line through $(1, 0)$.

Theory and Examples

In Exercises 37–40, graph the two equations and find the points at which the graphs intersect.

37. $y = 2x, \quad x^2 + y^2 = 1$ **38.** $y - x = 1, \quad y = x^2$

39. $y = -x^2, \quad y = 2x^2 - 1$

40. $x^2 + y^2 = 1, \quad (x - 1)^2 + y^2 = 1$

41. Insulation By measuring slopes in the figure, estimate the temperature change in degrees per inch for **(a)** the gypsum wallboard; **(b)** the fiberglass insulation; **(c)** the wood sheathing.

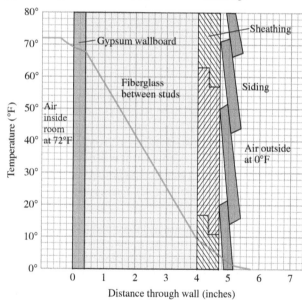

The temperature changes in the wall in Exercises 41 and 42.

42. Insulation According to the figure in Exercise 41, which of the materials is the best insulator? The poorest? Explain.

43. Pressure under water The pressure p experienced by a diver under water is related to the diver's depth d by an equation of the form $p = kd + 1$ (k a constant). At the surface, the pressure is 1 atmosphere. The pressure at 100 meters is about 10.94 atmospheres. Find the pressure at 50 meters.

44. Reflected light A ray of light comes in along the line $x + y = 1$ from the second quadrant and reflects off the x-axis (see the accompanying figure). The angle of incidence is equal to the angle

of reflection. Write an equation for the line along which the departing light travels.

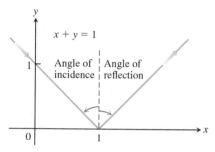

The path of the light ray in Exercise 44. Angles of incidence and reflection are measured from the perpendicular.

45. Fahrenheit vs. Celsius In the FC-plane, sketch the graph of the equation

$$C = \frac{5}{9}(F - 32)$$

linking Fahrenheit and Celsius temperatures. On the same graph sketch the line $C = F$. Is there a temperature at which a Celsius thermometer gives the same numerical reading as a Fahrenheit thermometer? If so, find it.

46. The Mt. Washington Cog Railway Civil engineers calculate the slope of roadbed as the ratio of the distance it rises or falls to the distance it runs horizontally. They call this ratio the **grade** of the roadbed, usually written as a percentage. Along the coast, commercial railroad grades are usually less than 2%. In the mountains, they may go as high as 4%. Highway grades are usually less than 5%.

The steepest part of the Mt. Washington Cog Railway in New Hampshire has an exceptional 37.1% grade. Along this part of the track, the seats in the front of the car are 14 ft above those in the rear. About how far apart are the front and rear rows of seats?

47. By calculating the lengths of its sides, show that the triangle with vertices at the points $A(1, 2)$, $B(5, 5)$, and $C(4, -2)$ is isosceles but not equilateral.

48. Show that the triangle with vertices $A(0, 0)$, $B(1, \sqrt{3})$, and $C(2, 0)$ is equilateral.

49. Show that the points $A(2, -1)$, $B(1, 3)$, and $C(-3, 2)$ are vertices of a square, and find the fourth vertex.

50. Three different parallelograms have vertices at $(-1, 1)$, $(2, 0)$, and $(2, 3)$. Sketch them and find the coordinates of the fourth vertex of each.

51. For what value of k is the line $2x + ky = 3$ perpendicular to the line $4x + y = 1$? For what value of k are the lines parallel?

52. Midpoint of a line segment Show that the point with coordinates

$$\left(\frac{x_1 + x_2}{2}, \frac{y_1 + y_2}{2} \right)$$

is the midpoint of the line segment joining $P(x_1, y_1)$ to $Q(x_2, y_2)$.

A.4 Proofs of Limit Theorems

This appendix proves Theorem 1, Parts 2–5, and Theorem 4 from Section 2.2.

THEOREM 1—Limit Laws

If L, M, c, and k are real numbers and

$$\lim_{x \to c} f(x) = L \quad \text{and} \quad \lim_{x \to c} g(x) = M, \quad \text{then}$$

1. *Sum Rule*: $\quad \lim_{x \to c} (f(x) + g(x)) = L + M$

2. *Difference Rule*: $\quad \lim_{x \to c} (f(x) - g(x)) = L - M$

3. *Constant Multiple Rule*: $\quad \lim_{x \to c} (k\,f(x)) = kL$

4. *Product Rule*: $\quad \lim_{x \to c} (f(x)\,g(x)) = LM$

5. *Quotient Rule*: $\quad \lim_{x \to c} \dfrac{f(x)}{g(x)} = \dfrac{L}{M}, \quad M \neq 0$

6. *Power Rule*: $\quad \lim_{x \to c} [\,f(x)\,]^n = L^n$, n a positive integer

7. *Root Rule*: $\quad \lim_{x \to c} \sqrt[n]{f(x)} = \sqrt[n]{L} = L^{1/n}$, n a positive integer

(If n is even, we assume that $\lim_{x \to c} f(x) = L > 0$.)

We proved the Sum Rule in Section 2.3, and the Power and Root Rules are proved in more advanced texts. We obtain the Difference Rule by replacing $g(x)$ by $-g(x)$ and M by $-M$ in the Sum Rule. The Constant Multiple Rule is the special case $g(x) = k$ of the Product Rule. This leaves only the Product and Quotient Rules.

Proof of the Limit Product Rule We show that for any $\varepsilon > 0$ there exists a $\delta > 0$ such that for all x in the intersection D of the domains of f and g,

$$|f(x)\,g(x) - LM| < \varepsilon \quad \text{whenever} \quad 0 < |x - c| < \delta.$$

Suppose then that ε is a positive number, and write $f(x)$ and $g(x)$ as

$$f(x) = L + (f(x) - L), \quad g(x) = M + (g(x) - M).$$

Multiply these expressions together and subtract LM:

$$f(x)\,g(x) - LM = (L + (f(x) - L))(M + (g(x) - M)) - LM$$

$$= LM + L(g(x) - M) + M(f(x) - L)$$

$$\quad + (f(x) - L)(g(x) - M) - LM$$

$$= L(g(x) - M) + M(f(x) - L) + (f(x) - L)(g(x) - M). \quad (1)$$

Since f and g have limits L and M as $x \to c$, there exist positive numbers $\delta_1, \delta_2, \delta_3,$ and δ_4 such that

$$\begin{aligned}
|f(x) - L| &< \sqrt{\varepsilon/3} & \text{whenever} && 0 < |x - c| < \delta_1 \\
|g(x) - M| &< \sqrt{\varepsilon/3} & \text{whenever} && 0 < |x - c| < \delta_2 \\
|f(x) - L| &< \varepsilon/(3(1 + |M|)) & \text{whenever} && 0 < |x - c| < \delta_3 \\
|g(x) - M| &< \varepsilon/(3(1 + |L|)) & \text{whenever} && 0 < |x - c| < \delta_4.
\end{aligned} \quad (2)$$

If we take δ to be the smallest of the numbers δ_1 through δ_4, the inequalities on the right-hand side of the Implications (2) will hold simultaneously for $0 < |x - c| < \delta$. Therefore, for all x in D, if $0 < |x - c| < \delta$ then

$$|f(x)g(x) - LM| \qquad \text{\small Triangle inequality applied to Eq. (1)}$$
$$\leq |L||g(x) - M| + |M||f(x) - L| + |f(x) - L||g(x) - M|$$
$$\leq (1 + |L|)|g(x) - M| + (1 + |M|)|f(x) - L| + |f(x) - L||g(x) - M|$$
$$< \frac{\varepsilon}{3} + \frac{\varepsilon}{3} + \sqrt{\frac{\varepsilon}{3}}\sqrt{\frac{\varepsilon}{3}} = \varepsilon. \qquad \text{\small Values from (2)}$$

This completes the proof of the Limit Product Rule. ∎

Proof of the Limit Quotient Rule   We show that $\lim_{x \to c}(1/g(x)) = 1/M$. We can then conclude that

$$\lim_{x \to c}\frac{f(x)}{g(x)} = \lim_{x \to c}\left(f(x) \cdot \frac{1}{g(x)}\right) = \lim_{x \to c} f(x) \cdot \lim_{x \to c}\frac{1}{g(x)} = L \cdot \frac{1}{M} = \frac{L}{M}$$

by the Limit Product Rule.

Let $\varepsilon > 0$ be given. To show that $\lim_{x \to c}(1/g(x)) = 1/M$, we need to show that there exists a $\delta > 0$ such that

$$\left|\frac{1}{g(x)} - \frac{1}{M}\right| < \varepsilon \qquad \text{whenever} \qquad 0 < |x - c| < \delta.$$

Since g has the limit M as $x \to c$ and since $|M| > 0$, there exists a positive number δ_1 such that

$$|g(x) - M| < \frac{M}{2} \qquad \text{whenever} \qquad 0 < |x - c| < \delta_1. \qquad (3)$$

For any numbers A and B, the triangle inequality implies that $|A| - |B| \leq |A - B|$ and $|B| - |A| \leq |A - B|$, from which it follows that $||A| - |B|| \leq |A - B|$. With $A = g(x)$ and $B = M$, this becomes

$$\big||g(x)| - |M|\big| \leq |g(x) - M|,$$

which can be combined with the inequality on the right in Implication (3) to get, in turn,

$$\big||g(x)| - |M|\big| < \frac{|M|}{2}$$
$$-\frac{|M|}{2} < |g(x)| - |M| < \frac{|M|}{2}$$
$$\frac{|M|}{2} < |g(x)| < \frac{3|M|}{2}$$
$$|M| < 2|g(x)| < 3|M|$$
$$\frac{1}{|g(x)|} < \frac{2}{|M|} < \frac{3}{|g(x)|}. \qquad (4)$$

Therefore, $0 < |x - c| < \delta_1$ implies that

$$\left|\frac{1}{g(x)} - \frac{1}{M}\right| = \left|\frac{M - g(x)}{Mg(x)}\right| \leq \frac{1}{|M|} \cdot \frac{1}{|g(x)|} \cdot |M - g(x)|$$

$$< \frac{1}{|M|} \cdot \frac{2}{|M|} \cdot |M - g(x)|. \qquad \text{\small Inequality (4)} \qquad (5)$$

Since $(1/2)|M|^2\varepsilon > 0$, there exists a number $\delta_2 > 0$ such that

$$|M - g(x)| < \frac{\varepsilon}{2}|M|^2 \quad \text{whenever} \quad 0 < |x - c| < \delta_2. \tag{6}$$

If we take δ to be the smaller of δ_1 and δ_2, the conclusions in (5) and (6) both hold whenever $0 < |x - c| < \delta$. Combining these conclusions gives

$$\left|\frac{1}{g(x)} - \frac{1}{M}\right| < \varepsilon \quad \text{whenever} \quad 0 < |x - c| < \delta.$$

This concludes the proof of the Limit Quotient Rule.   ∎

THEOREM 4—The Sandwich Theorem

Suppose that $g(x) \le f(x) \le h(x)$ for all x in some open interval I containing c, except possibly at $x = c$ itself. Suppose also that $\lim_{x\to c} g(x) = \lim_{x\to c} h(x) = L$. Then $\lim_{x\to c} f(x) = L$.

Proof for Right-Hand Limits   Suppose $\lim_{x\to c^+} g(x) = \lim_{x\to c^+} h(x) = L$. Then for any $\varepsilon > 0$ there exists a $\delta > 0$ such that the interval $(c, c + \delta)$ is contained in I and

$$L - \varepsilon < g(x) < L + \varepsilon \quad \text{and} \quad L - \varepsilon < h(x) < L + \varepsilon$$

whenever $c < x < c + \delta$. Since we always have $g(x) \le f(x) \le h(x)$ it follows that if $c < x < c + \delta$, then

$$L - \varepsilon < g(x) \le f(x) \le h(x) < L + \varepsilon,$$
$$L - \varepsilon < f(x) < L + \varepsilon,$$
$$-\varepsilon < f(x) - L < \varepsilon.$$

Therefore $|f(x) - L| < \varepsilon$ whenever $c < x < c + \delta$.

Proof for Left-Hand Limits   Suppose $\lim_{x\to c^-} g(x) = \lim_{x\to c^-} h(x) = L$. Then for any $\varepsilon > 0$ there exists a $\delta > 0$ such that the interval $(c - \delta, c)$ is contained in I and

$$L - \varepsilon < g(x) < L + \varepsilon \quad \text{and} \quad L - \varepsilon < h(x) < L + \varepsilon$$

whenever $c - \delta < x < c$. We conclude as before that $|f(x) - L| < \varepsilon$ whenever $c - \delta < x < c$.

Proof for Two-Sided Limits   If $\lim_{x\to c} g(x) = \lim_{x\to c} h(x) = L$, then $g(x)$ and $h(x)$ both approach L as $x \to c^+$ and as $x \to c^-$; so $\lim_{x\to c^+} f(x) = L$ and $\lim_{x\to c^-} f(x) = L$. Hence $\lim_{x\to c} f(x)$ exists and equals L.   ∎

EXERCISES A.4

1. Suppose that functions $f_1(x)$, $f_2(x)$, and $f_3(x)$ have limits L_1, L_2, and L_3, respectively, as $x \to c$. Show that their sum has limit $L_1 + L_2 + L_3$. Use mathematical induction (Appendix 2) to generalize this result to the sum of any finite number of functions.

2. Use mathematical induction and the Limit Product Rule in Theorem 1 to show that if functions $f_1(x), f_2(x), \ldots, f_n(x)$ have limits $L_1, L_2, \ldots, L_n$ as $x \to c$, then

$$\lim_{x\to c} f_1(x) \cdot f_2(x) \cdot \cdots \cdot f_n(x) = L_1 \cdot L_2 \cdot \cdots \cdot L_n.$$

3. Use the fact that $\lim_{x\to c} x = c$ and the result of Exercise 2 to show that $\lim_{x\to c} x^n = c^n$ for any integer $n > 1$.

4. **Limits of polynomials**   Use the fact that $\lim_{x\to c}(k) = k$ for any number k together with the results of Exercises 1 and 3 to show that $\lim_{x\to c} f(x) = f(c)$ for any polynomial function

$$f(x) = a_n x^n + a_{n-1} x^{n-1} + \cdots + a_1 x + a_0.$$

5. **Limits of rational functions**   Use Theorem 1 and the result of Exercise 4 to show that if $f(x)$ and $g(x)$ are polynomial functions and $g(c) \neq 0$, then

$$\lim_{x \to c} \frac{f(x)}{g(x)} = \frac{f(c)}{g(c)}.$$

6. **Composites of continuous functions**   Figure A.23 gives the diagram for a proof that the composite of two continuous func-tions is continuous. Reconstruct the proof from the diagram. The statement to be proved is this: If f is continuous at $x = c$ and g is continuous at $f(c)$, then $g \circ f$ is continuous at c.

Assume that c is an interior point of the domain of f and that $f(c)$ is an interior point of the domain of g. This will make the limits involved two-sided. (The arguments for the cases that involve one-sided limits are similar.)

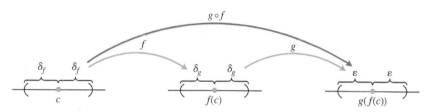

FIGURE A.23   The diagram for a proof that the composite of two continuous func-tions is continuous.

A.5 Commonly Occurring Limits

This appendix verifies limits (4)–(6) in Theorem 5 of Section 10.1.

Limit 4: If $|x| < 1$, $\lim\limits_{n \to \infty} x^n = 0$ We need to show that to each $\varepsilon > 0$ there corresponds an integer N so large that $|x^n| < \varepsilon$ for all n greater than N. Since $\varepsilon^{1/n} \to 1$, while $|x| < 1$, there exists an integer N for which $\varepsilon^{1/N} > |x|$. In other words,

$$|x^N| = |x|^N < \varepsilon. \tag{1}$$

This is the integer we seek because, if $|x| < 1$, then

$$|x^n| < |x^N| \quad \text{for all } n > N. \tag{2}$$

Combining (1) and (2) produces $|x^n| < \varepsilon$ for all $n > N$, concluding the proof.   ∎

Limit 5: For any number x, $\lim\limits_{n \to \infty} \left(1 + \dfrac{x}{n}\right)^n = e^x$ Let

$$a_n = \left(1 + \frac{x}{n}\right)^n.$$

Then

$$\ln a_n = \ln\left(1 + \frac{x}{n}\right)^n = n \ln\left(1 + \frac{x}{n}\right) \to x,$$

as we can see by the following application of L'Hôpital's Rule, in which we differentiate with respect to n:

$$\lim_{n \to \infty} n \ln\left(1 + \frac{x}{n}\right) = \lim_{n \to \infty} \frac{\ln(1 + x/n)}{1/n}$$

$$= \lim_{n \to \infty} \frac{\left(\dfrac{1}{1 + x/n}\right) \cdot \left(-\dfrac{x}{n^2}\right)}{-1/n^2} = \lim_{n \to \infty} \frac{x}{1 + x/n} = x.$$

Apply Theorem 3, Section 10.1, with $f(x) = e^x$ to conclude that

$$\left(1 + \frac{x}{n}\right)^n = a_n = e^{\ln a_n} \to e^x.$$

Limit 6: For any number x, $\displaystyle\lim_{n\to\infty} \frac{x^n}{n!} = 0$ Since

$$-\frac{|x|^n}{n!} \le \frac{x^n}{n!} \le \frac{|x|^n}{n!},$$

all we need to show is that $|x|^n/n! \to 0$. We can then apply the Sandwich Theorem for Sequences (Section 10.1, Theorem 2) to conclude that $x^n/n! \to 0$.

The first step in showing that $|x|^n/n! \to 0$ is to choose an integer $M > |x|$, so that $(|x|/M) < 1$. By Limit 4, just proved, we then have $(|x|/M)^n \to 0$. We then restrict our attention to values of $n > M$. For these values of n, we can write

$$\frac{|x|^n}{n!} = \frac{|x|^n}{1 \cdot 2 \cdots \cdot \underbrace{M \cdot (M+1) \cdot (M+2) \cdots \cdot n}_{(n-M) \text{ factors}}}$$

$$\le \frac{|x|^n}{M! M^{n-M}} = \frac{|x|^n M^M}{M! M^n} = \frac{M^M}{M!}\left(\frac{|x|}{M}\right)^n.$$

Thus,

$$0 \le \frac{|x|^n}{n!} \le \frac{M^M}{M!}\left(\frac{|x|}{M}\right)^n.$$

Now, the constant $M^M/M!$ does not change as n increases. Thus the Sandwich Theorem tells us that $|x|^n/n! \to 0$ because $(|x|/M)^n \to 0$.   ∎

A.6 Theory of the Real Numbers

A rigorous development of calculus is based on properties of the real numbers. Many results about functions, derivatives, and integrals would be false if stated for functions defined only on the rational numbers. In this appendix we briefly examine some basic concepts of the theory of the reals that hint at what might be learned in a deeper, more theoretical study of calculus.

Three types of properties make the real numbers what they are. These are the **algebraic**, **order**, and **completeness** properties. The algebraic properties involve addition and multiplication, subtraction and division. They apply to rational or complex numbers (discussed in Appendix A.7) as well as to the reals.

The structure of numbers is built around a set with addition and multiplication operations. The following properties are required of addition and multiplication.

A1 $a + (b + c) = (a + b) + c$ for all a, b, c.

A2 $a + b = b + a$ for all a, b.

A3   There is a number called "0" such that $a + 0 = a$ for all a.

A4   For each number a, there is a number b such that $a + b = 0$.

M1 $a(bc) = (ab)c$ for all a, b, c.

M2 $ab = ba$ for all a, b.

M3   There is a number called "1" such that $a \cdot 1 = a$ for all a.

M4   For each nonzero number a, there is a number b such that $ab = 1$.

D $a(b + c) = ab + ac$ for all a, b, c.

A1 and M1 are *associative laws*, A2 and M2 are *commutativity laws*, A3 and M3 are *identity laws*, and D is the *distributive law*. Sets that have these algebraic properties are examples of **fields**, and are studied in depth in the area of theoretical mathematics called abstract algebra.

The **order** properties allow us to compare the size of any two numbers. The order properties are

O1   For any a and b, either $a \leq b$ or $b \leq a$ or both.

O2   If $a \leq b$ and $b \leq a$ then $a = b$.

O3   If $a \leq b$ and $b \leq c$ then $a \leq c$.

O4   If $a \leq b$ then $a + c \leq b + c$.

O5   If $a \leq b$ and $0 \leq c$ then $ac \leq bc$.

O3 is the *transitivity law*, and O4 and O5 relate ordering to addition and multiplication.

We can order the reals, the integers, and the rational numbers, but we cannot order the complex numbers (there is no reasonable way to decide whether a number like $i = \sqrt{-1}$ is bigger or smaller than zero). A field in which the size of any two elements can be compared as above is called an **ordered field**. Both the rational numbers and the real numbers are ordered fields, and there are many others.

We can think of real numbers geometrically, lining them up as points on a line. The **completeness property** says that the real numbers correspond to all points on the line, with no "holes" or "gaps." The rationals, in contrast, omit points such as $\sqrt{2}$ and π, and the integers even leave out fractions like $1/2$. The reals, having the completeness property, omit no points.

What exactly do we mean by this vague idea of missing holes? To answer this we must give a more precise description of completeness. A number M is an **upper bound** for a set of numbers if all numbers in the set are smaller than or equal to M. M is a **least upper bound** if it is the smallest upper bound. For example, $M = 2$ is an upper bound for the negative numbers. So is $M = 1$, showing that 2 is not a least upper bound. The least upper bound for the set of negative numbers is $M = 0$. We define a **complete** ordered field to be one in which every nonempty set bounded above has a least upper bound.

If we work with just the rational numbers, the set of numbers less than $\sqrt{2}$ is bounded, but it does not have a rational least upper bound, since any rational upper bound M can be replaced by a slightly smaller rational number that is still larger than $\sqrt{2}$. So the rationals are not complete. In the real numbers, a set that is bounded above always has a least upper bound. The reals are a complete ordered field.

The completeness property is at the heart of many results in calculus. One example occurs when searching for a maximum value for a function on a closed interval $[a, b]$, as in Section 4.1. The function $y = x - x^3$ has a maximum value on $[0, 1]$ at the point x satisfying $1 - 3x^2 = 0$, or $x = \sqrt{1/3}$. If we limited our consideration to functions defined only on rational numbers, we would have to conclude that the function has no maximum, since $\sqrt{1/3}$ is irrational (Figure A.24). The Extreme Value Theorem (Section 4.1), which implies that continuous functions on closed intervals $[a, b]$ have a maximum value, is not true for functions defined only on the rationals.

The Intermediate Value Theorem implies that a continuous function f on an interval $[a, b]$ with $f(a) < 0$ and $f(b) > 0$ must be zero somewhere in $[a, b]$. The function values cannot jump from negative to positive without there being some point x in $[a, b]$ where $f(x) = 0$. The Intermediate Value Theorem also relies on the completeness of the real numbers and is false for continuous functions defined only on the rationals. The function $f(x) = 3x^2 - 1$ has $f(0) = -1$ and $f(1) = 2$, but if we consider f only on the rational numbers, it never equals zero. The only value of x for which $f(x) = 0$ is $x = \sqrt{1/3}$, an irrational number.

We have captured the desired properties of the reals by saying that the real numbers are a complete ordered field. But we're not quite finished. Greek mathematicians in the

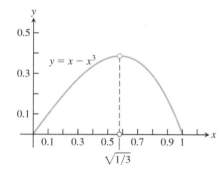

FIGURE A.24 The maximum value of $y = x - x^3$ on $[0, 1]$ occurs at the irrational number $x = \sqrt{1/3}$.

school of Pythagoras tried to impose another property on the numbers of the real line, the condition that all numbers are ratios of integers. They learned that their effort was doomed when they discovered irrational numbers such as $\sqrt{2}$. How do we know that our efforts to specify the real numbers are not also flawed, for some unseen reason? The artist Escher drew optical illusions of spiral staircases that went up and up until they rejoined themselves at the bottom. An engineer trying to build such a staircase would find that no structure realized the plans the architect had drawn. Could it be that our design for the reals contains some subtle contradiction, and that no construction of such a number system can be made?

We resolve this issue by giving a specific description of the real numbers and verifying that the algebraic, order, and completeness properties are satisfied in this model. This is called a **construction** of the reals, and just as stairs can be built with wood, stone, or steel, there are several approaches to constructing the reals. One construction treats the reals as all the infinite decimals,

$$a.d_1d_2d_3d_4\ldots$$

In this approach a real number is an integer a followed by a sequence of decimal digits $d_1, d_2, d_3, \ldots$, each between 0 and 9. This sequence may stop, or repeat in a periodic pattern, or keep going forever with no pattern. In this form, $2.00, 0.3333333\ldots$ and $3.1415926535898\ldots$ represent three familiar real numbers. The real meaning of the dots "$\ldots$" following these digits requires development of the theory of sequences and series, as in Chapter 10. Each real number is constructed as the limit of a sequence of rational numbers given by its finite decimal approximations. An infinite decimal is then the same as a series

$$a + \frac{d_1}{10} + \frac{d_2}{100} + \cdots.$$

This decimal construction of the real numbers is not entirely straightforward. It's easy enough to check that it gives numbers that satisfy the completeness and order properties, but verifying the algebraic properties is rather involved. Even adding or multiplying two numbers requires an infinite number of operations. Making sense of division requires a careful argument involving limits of rational approximations to infinite decimals.

A different approach was taken by Richard Dedekind (1831–1916), a German mathematician, who gave the first rigorous construction of the real numbers in 1872. Given any real number x, we can divide the rational numbers into two sets: those less than or equal to x and those greater. Dedekind cleverly reversed this reasoning and defined a real number to be a division of the rational numbers into two such sets. This seems like a strange approach, but such indirect methods of constructing new structures from old are powerful tools in theoretical mathematics.

These and other approaches can be used to construct a system of numbers having the desired algebraic, order, and completeness properties. A final issue that arises is whether all the constructions give the same thing. Is it possible that different constructions result in different number systems satisfying all the required properties? If yes, which of these is the real numbers? Fortunately, the answer turns out to be no. The reals are the only number system satisfying the algebraic, order, and completeness properties.

Confusion about the nature of the numbers and about limits caused considerable controversy in the early development of calculus. Calculus pioneers such as Newton, Leibniz, and their successors, when looking at what happens to the difference quotient

$$\frac{\Delta y}{\Delta x} = \frac{f(x + \Delta x) - f(x)}{\Delta x}$$

as each of Δy and Δx approach zero, talked about the resulting derivative being a quotient of two infinitely small quantities. These "infinitesimals," written dx and dy, were thought to be some new kind of number, smaller than any fixed number but not zero. Similarly, a definite integral was thought of as a sum of an infinite number of infinitesimals

$$f(x) \cdot dx$$

as x varied over a closed interval. While the approximating difference quotients $\Delta y / \Delta x$ were understood much as today, it was the quotient of infinitesimal quantities, rather than a limit, that was thought to encapsulate the meaning of the derivative. This way of thinking led to logical difficulties, as attempted definitions and manipulations of infinitesimals ran into contradictions and inconsistencies. The more concrete and computable difference quotients did not cause such trouble, but they were thought of merely as useful calculation tools. Difference quotients were used to work out the numerical value of the derivative and to derive general formulas for calculation, but were not considered to be at the heart of the question of what the derivative actually was. Today we realize that the logical problems associated with infinitesimals can be avoided by *defining* the derivative to be the limit of its approximating difference quotients. The ambiguities of the old approach are no longer present, and in the standard theory of calculus, infinitesimals are neither needed nor used.

A.7 Complex Numbers

Complex numbers are expressed in the form $a + ib$, or $a + bi$, where a and b are real numbers and i is a symbol for $\sqrt{-1}$. Unfortunately, the words "real" and "imaginary" have connotations that somehow place $\sqrt{-1}$ in a less favorable position in our minds than $\sqrt{2}$. As a matter of fact, a good deal of imagination, in the sense of *inventiveness*, has been required to construct the *real* number system, which forms the basis of calculus (see Appendix 6). In this appendix we review the various stages of these inventions.

The Hierarchy of Numbers

The first stage of number development was the recognition of the counting numbers 1, 2, 3, . . . , which we now call the **natural numbers** or the **positive integers**. Certain arithmetical operations on the positive integers, such as addition and multiplication, keep us entirely within this system. That is, if m and n are any positive integers, then their sum $m + n$ and product mn are also positive integers.

Some equations can be solved entirely within the system of positive integers. For example, we can solve $3 + x = 7$ using only positive integers. But other simple equations, such as $7 + x = 3$, cannot be solved if positive integers are the only numbers at our disposal. The number zero and the negative numbers were invented to solve equations such as $7 + x = 3$. Using the **integers**

$$\ldots, -3, -2, -1, 0, 1, 2, 3, \ldots,$$

we can always find the missing integer x that solves the equation $m + x = p$ when we are given the other two integers m and p in the equation.

Addition and multiplication of integers always keep us within the system of integers. However, division does not, and so fractions m/n, where m and n are integers with n nonzero, were invented. This system, which is called the **rational numbers**, is rich enough to perform all of the **rational operations** of arithmetic, including addition, subtraction, multiplication, and division (although division by zero is excluded since it is meaningless).

Yet there are still simple polynomial equations that cannot be solved within the system of rational numbers. The ancient Greeks realized that there is no *rational* number that solves the equation $x^2 = 2$, even though the Pythagorean Theorem implies that the length x of the diagonal of the unit square satisfies this equation! (See Figure A.25.) To see why $x^2 = 2$ has no rational solution, consider the following argument.

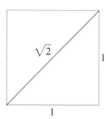

FIGURE A.25 The diagonal of the unit square has irrational length.

Suppose that there did exist some integers p and q with no common factor other than 1 such that the fraction $x = p/q$ satisfied $x^2 = 2$. Writing this out, we see that $p^2/q^2 = 2$, and therefore

$$p^2 = 2q^2.$$

Thus p^2 is an even integer. Since the square of an odd number is odd, we conclude that p itself must be an even number (for if p were odd, then p^2 would also be odd). Hence p is divisible by 2, and therefore $p = 2k$ for some integer k. Hence $p^2 = 4k^2$. Since we already saw that $p^2 = 2q^2$, it follows that $2q^2 = p^2 = 4k^2$, and therefore

$$q^2 = 2k^2.$$

Hence q^2 is an even number. This requires that q itself be even. Therefore, both p and q are divisible by 2, which is contrary to our assumption that they contain no common factors other than 1. Since we have arrived at a contradiction, there cannot exist any such integers p and q, and therefore there is no rational number that solves the equation $x^2 = 2$.

The invention of real numbers addressed this issue (and others). Using real numbers, we can represent every possible physical length. As we saw in Appendix A.6, each real number can be represented as an infinite decimal $a.d_1d_2d_3d_4\ldots$, where a is an integer followed by a sequence of decimal digits each between 0 and 9. If the sequence stops or repeats in a periodic pattern, then the decimal represents a rational number. An irrational number is represented by a nonterminating and nonrepeating decimal. The rational and irrational numbers together make up the real number system. Unlike the rational numbers, the real numbers have the **completeness property**, meaning that there are no "holes" or "gaps" in the real line. Yet for all of its utility, there are still simple equations that cannot be solved within the real number system alone. For example, the polynomial equation $x^2 + 1 = 0$ has no real solutions.

The Complex Numbers

We have discussed three invented systems of numbers that form a hierarchy in which each system contains the previous system. Each system is richer than its predecessor in that it permits additional operations to be performed without going outside the system.

1. Using the integer system we can solve all equations of the form

$$x + a = 0, \tag{1}$$

where a is an integer.

2. Using the rational numbers we can solve all equations of the form

$$ax + b = 0, \tag{2}$$

provided that a and b are rational numbers and $a \neq 0$.

3. Using the real numbers, we can solve all of Equations (1) and (2) and, in addition, all quadratic equations

$$ax^2 + bx + c = 0 \quad \text{provided that} \quad a \neq 0 \quad \text{and} \quad b^2 - 4ac \geq 0. \tag{3}$$

The **quadratic formula**

$$x = \frac{-b \pm \sqrt{b^2 - 4ac}}{2a} \tag{4}$$

gives the solutions to Equation (3). When $b^2 - 4ac$ is negative there are no real number solutions to the equation $ax^2 + bx + c = 0$. In particular, the simple quadratic equation $x^2 + 1 = 0$ cannot be solved using any of the three invented systems of numbers that we have discussed.

Thus we come to the fourth invented system, which is the set of **complex numbers** $a + ib$. The symbol i represents a new number whose square equals -1. We call a the **real part** and b the **imaginary part** of the complex number $a + ib$. Sometimes it is convenient to write $a + bi$ instead of $a + ib$; both notations describe the same complex number.

We define equality and addition for complex numbers in the following way.

Equality $a + ib = c + id$
                    if and only if
                    $a = c$ and $b = d$

Two complex numbers $a + ib$ and $c + id$ are equal if and only if their real parts are equal and their imaginary parts are equal.

Addition $(a + ib) + (c + id)$
                    $= (a + c) + i(b + d)$

We sum the real parts and separately sum the imaginary parts.

To multiply two complex numbers, we multiply using the distributive rule and then simplify using $i^2 = -1$:

Multiplication $(a + ib)(c + id)$
                      $= ac + iad + ibc + i^2bd$
                      $= (ac - bd) + i(ad + bc)$ $i^2 = -1.$

The set of all complex numbers $a + i0$, where the second number b is zero, has all of the properties of the set of real numbers. For example, addition and multiplication as complex numbers give

$$(a + i0) + (c + i0) = (a + c) + i0, \qquad (a + i0)(c + i0) = ac + i0,$$

which are numbers of the same type with imaginary part zero. We usually just write a instead of $a + i0$, and in this sense the real number system is "embedded" into the complex number system.

If we multiply a "real number" $a = a + i0$ by a complex number $c + id$, we get $a(c + id) = (a + i0)(c + id) = ac + iad$. In particular, the number $0 = 0 + i0$ plays the role of zero in the complex number system, and the complex number $1 = 1 + i0$ plays the role of unity, or one, in the complex number system.

The complex number $i = 0 + i1$, which has real part zero and imaginary part one, has the property that its square is

$$i^2 = (0 + i1)^2 = (0 + i1)(0 + i1) = (-1) + i0 = -1.$$

Thus $x = i$ is a solution to the quadratic equation $x^2 + 1 = 0$. Using the complex number system, there are exactly two solutions to this equation, the other solution being $x = -i = 0 + i(-1)$.

We can divide any two complex numbers as long as we do not divide by the number $0 = 0 + i0$. As long as $a + ib \neq 0$ (meaning that *either $a \neq 0$ or $b \neq 0$ or both $a \neq 0$ and $b \neq 0$*), we carry out division as follows:

$$\frac{c + id}{a + ib} = \frac{(c + id)(a - ib)}{(a + ib)(a - ib)} = \frac{(ac + bd) + i(ad - bc)}{a^2 + b^2} = \frac{ac + bd}{a^2 + b^2} + i\frac{ad - bc}{a^2 + b^2}.$$

Note that $a^2 + b^2 \neq 0$ since we stipulated that a and b cannot both be zero.

The number $a - ib$ that is used as the multiplier to clear the i from the denominator is called the **complex conjugate** of $a + ib$. If we denote the original complex number by $z = a + ib$, then it is customary to write $\bar{z}$ (read "z bar") to denote its complex conjugate:

$$z = a + ib, \qquad \bar{z} = a - ib.$$

Multiplying the numerator and denominator of a fraction $(c + id)/(a + ib)$ by the complex conjugate of the denominator will always replace the denominator by a real number.

EXAMPLE 1   We give some illustrations of the arithmetic operations with complex numbers.

(a) $(2 + 3i) + (6 - 2i) = (2 + 6) + (3 - 2)i = 8 + i$

(b) $(2 + 3i) - (6 - 2i) = (2 - 6) + (3 - (-2))i = -4 + 5i$

(c) $(2 + 3i)(6 - 2i) = (2)(6) + (2)(-2i) + (3i)(6) + (3i)(-2i)$
$$= 12 - 4i + 18i - 6i^2 = 12 + 14i + 6 = 18 + 14i$$

(d) $\dfrac{2 + 3i}{6 - 2i} = \dfrac{2 + 3i}{6 - 2i}\dfrac{6 + 2i}{6 + 2i} = \dfrac{12 + 4i + 18i + 6i^2}{36 + 12i - 12i - 4i^2} = \dfrac{6 + 22i}{40} = \dfrac{3}{20} + \dfrac{11}{20}i$ ∎

Argand Diagrams

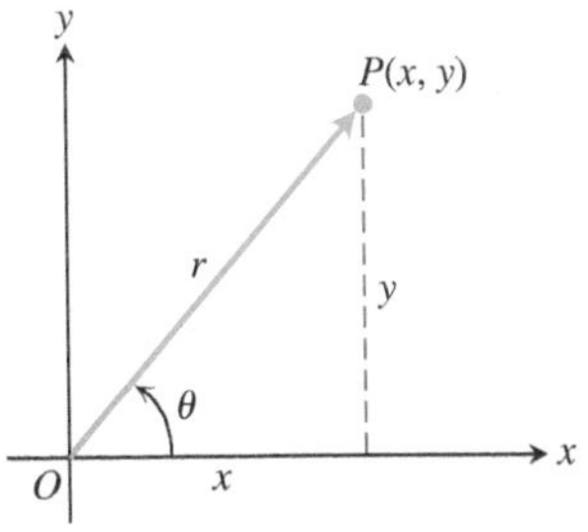

There are two geometric representations of the complex number $z = x + iy$:

1. as the point $P(x, y)$ in the xy-plane
2. as the vector $\overrightarrow{OP}$ from the origin to P.

In each representation, the x-axis is called the **real axis** and the y-axis is the **imaginary axis**. Both representations are **Argand diagrams** for $x + iy$ (Figure A.26).

In terms of the polar coordinates of x and y, we have

$$x = r \cos\theta, \qquad y = r \sin\theta,$$

and

$$z = x + iy = r(\cos\theta + i \sin\theta). \qquad (5)$$

We define the **absolute value** of a complex number $x + iy$ to be the length r of a vector $\overrightarrow{OP}$ from the origin to $P(x, y)$. We denote the absolute value by vertical bars; thus,

$$|x + iy| = \sqrt{x^2 + y^2}.$$

If we always choose the polar coordinates r and θ so that r is nonnegative, then

$$r = |x + iy|.$$

The polar angle θ is called the **argument** of z and is written $\theta = \arg z$. Of course, any integer multiple of 2π may be added to θ to produce another appropriate angle.

The following equation gives a useful formula connecting a complex number z, its conjugate $\bar{z}$, and its absolute value $|z|$:

$$z \cdot \bar{z} = |z|^2.$$

Euler's Formula

The identity

$$e^{i\theta} = \cos\theta + i \sin\theta, \qquad (6)$$

is called **Euler's formula**. We show an Argand diagram for $e^{i\theta}$ in Figure A.27. Using Equation (6), we can write Equation (5) as

$$z = re^{i\theta}.$$

This formula, in turn, leads to the following rules for calculating products, quotients, powers, and roots of complex numbers.

The notation exp (A) is also used for e^A.

FIGURE A.26 This Argand diagram represents $z = x + iy$ both as a point $P(x, y)$ and as a vector $\overrightarrow{OP}$.

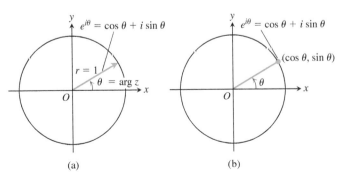

(a)　　　　　(b)

FIGURE A.27 Argand diagrams for $e^{i\theta} = \cos\theta + i \sin\theta$ (a) as a vector and (b) as a point.

Products

To multiply two complex numbers, we multiply their absolute values and add their angles. To see why, let

$$z_1 = r_1 e^{i\theta_1}, \qquad z_2 = r_2 e^{i\theta_2}, \tag{7}$$

so that

$$|z_1| = r_1, \qquad \arg z_1 = \theta_1; \qquad |z_2| = r_2, \qquad \arg z_2 = \theta_2.$$

Then

$$z_1 z_2 = r_1 e^{i\theta_1} \cdot r_2 e^{i\theta_2} = r_1 r_2 e^{i(\theta_1 + \theta_2)}$$

and hence

$$|z_1 z_2| = r_1 r_2 = |z_1| \cdot |z_2|$$
$$\arg (z_1 z_2) = \theta_1 + \theta_2 = \arg z_1 + \arg z_2. \tag{8}$$

Thus, the product of two complex numbers is represented by a vector whose length is the product of the lengths of the two factors and whose argument is the sum of their arguments (Figure A.28). In particular, from Equation (8) a vector may be rotated counterclockwise through an angle θ by multiplying it by $e^{i\theta}$. Multiplication by i rotates 90°, by -1 rotates 180°, by $-i$ rotates 270°, and so on.

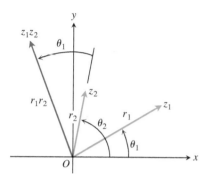

FIGURE A.28 When z_1 and z_2 are multiplied, $|z_1 z_2| = r_1 \cdot r_2$ and $\arg (z_1 z_2) = \theta_1 + \theta_2$.

EXAMPLE 2   Let $z_1 = 1 + i, z_2 = \sqrt{3} - i$. We plot these complex numbers in an Argand diagram (Figure A.29) from which we read off the polar representations

$$z_1 = \sqrt{2}e^{i\pi/4}, \qquad z_2 = 2e^{-i\pi/6}.$$

Then

$$z_1 z_2 = 2\sqrt{2} \exp\left(\frac{i\pi}{4} - \frac{i\pi}{6}\right) = 2\sqrt{2} \exp\left(\frac{i\pi}{12}\right)$$

$$= 2\sqrt{2}\left(\cos \frac{\pi}{12} + i \sin \frac{\pi}{12}\right) \approx 2.73 + 0.73i.$$

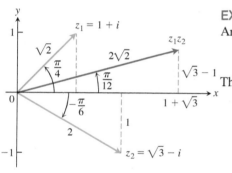

FIGURE A.29   To multiply two complex numbers, multiply their absolute values and add their arguments.

Quotients

Suppose $r_2 \neq 0$ in Equation (7). Then

$$\frac{z_1}{z_2} = \frac{r_1 e^{i\theta_1}}{r_2 e^{i\theta_2}} = \frac{r_1}{r_2} e^{i(\theta_1 - \theta_2)}.$$

Hence

$$\left|\frac{z_1}{z_2}\right| = \frac{r_1}{r_2} = \frac{|z_1|}{|z_2|} \quad \text{and} \quad \arg\left(\frac{z_1}{z_2}\right) = \theta_1 - \theta_2 = \arg z_1 - \arg z_2.$$

That is, we divide lengths and subtract angles for the quotient of complex numbers.

EXAMPLE 3   Let $z_1 = 1 + i$ and $z_2 = \sqrt{3} - i$, as in Example 2. Then

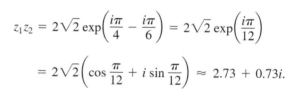

$$\approx 0.183 + 0.683i.$$

Powers

If n is a positive integer, we may apply the product formulas in Equation (8) to find

$$z^n = z \cdot z \cdot \cdots \cdot z. \qquad n \text{ factors}$$

With $z = re^{i\theta}$, we obtain

$$z^n = (re^{i\theta})^n = r^n e^{i(\theta + \theta + \cdots + \theta)} \qquad n \text{ summands}$$

$$= r^n e^{in\theta}. \qquad (9)$$

The length $r = |z|$ is raised to the nth power and the angle $\theta = \arg z$ is multiplied by n.

If we take $r = 1$ in Equation (9), we obtain De Moivre's Theorem.

De Moivre's Theorem

$$(\cos \theta + i \sin \theta)^n = \cos n\theta + i \sin n\theta. \qquad (10)$$

If we expand the left side of De Moivre's equation above by the Binomial Theorem and reduce it to the form $a + ib$, we obtain formulas for $\cos n\theta$ and $\sin n\theta$ as polynomials of degree n in $\cos \theta$ and $\sin \theta$.

EXAMPLE 4 If $n = 3$ in Equation (10), we have

$$(\cos \theta + i \sin \theta)^3 = \cos 3\theta + i \sin 3\theta.$$

The left side of this equation expands to

$$\cos^3 \theta + 3i \cos^2 \theta \sin \theta - 3 \cos \theta \sin^2 \theta - i \sin^3 \theta.$$

The real part of this must equal $\cos 3\theta$ and the imaginary part must equal $\sin 3\theta$. Therefore,

$$\cos 3\theta = \cos^3 \theta - 3 \cos \theta \sin^2 \theta,$$
$$\sin 3\theta = 3 \cos^2 \theta \sin \theta - \sin^3 \theta. \qquad \blacksquare$$

Roots

If $z = re^{i\theta}$ is a complex number different from zero and n is a positive integer, then there are precisely n different complex numbers $w_0, w_1, \ldots, w_{n-1}$, that are nth roots of z. To see why, let $w = \rho e^{i\alpha}$ be an nth root of $z = re^{i\theta}$. Then

$$w^n = z$$

or

$$\rho^n e^{in\alpha} = re^{i\theta}.$$

Since both r and ρ^n are positive, this implies that $\rho^n = r$, and so

$$\rho = \sqrt[n]{r}$$

is the real, positive nth root of r. For the argument, although we cannot say that $n\alpha$ and θ must be equal, we can say that they may differ only by an integer multiple of 2π. That is,

$$n\alpha = \theta + 2k\pi, \qquad k = 0, \pm 1, \pm 2, \ldots.$$

Therefore,

$$\alpha = \frac{\theta}{n} + k\frac{2\pi}{n}.$$

Hence, all the nth roots of $z = re^{i\theta}$ are given by

$$\sqrt[n]{re^{i\theta}} = \sqrt[n]{r} \exp i\left(\frac{\theta}{n} + k\frac{2\pi}{n}\right), \qquad k = 0, \pm 1, \pm 2, \ldots. \qquad (11)$$

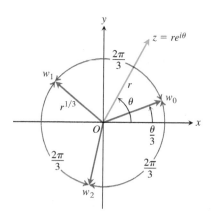

FIGURE A.30   The three cube roots of $z = re^{i\theta}$.

There might appear to be infinitely many different answers corresponding to the infinitely many possible values of k, but $k = n + m$ gives the same answer as $k = m$ in Equation (11). Thus, we need only take n consecutive values for k to obtain all the different nth roots of z. For convenience, we take

$$k = 0, 1, 2, \ldots, n - 1.$$

All the nth roots of $re^{i\theta}$ lie on a circle centered at the origin and having radius equal to the real, positive nth root of r. One of them has argument $\alpha = \theta/n$. The others are uniformly spaced around the circle, each being separated from its neighbors by an angle equal to $2\pi/n$. Figure A.30 illustrates the placement of the three cube roots, w_0, w_1, w_2, of the complex number $z = re^{i\theta}$.

EXAMPLE 5   Find the four fourth roots of -16.

Solution   As our first step, we plot the number -16 in an Argand diagram (Figure A.31) and determine its polar representation $re^{i\theta}$. Here, $z = -16$, $r = +16$, and $\theta = \pi$. One of the fourth roots of $16e^{i\pi}$ is $2e^{i\pi/4}$. We obtain others by successive additions of $2\pi/4 = \pi/2$ to the argument of this first one. Hence,

$$\sqrt[4]{16 \exp i\pi} = 2 \exp i\left(\frac{\pi}{4}, \frac{3\pi}{4}, \frac{5\pi}{4}, \frac{7\pi}{4}\right),$$

and the four roots are

$$w_0 = 2\left[\cos\frac{\pi}{4} + i\sin\frac{\pi}{4}\right] = \sqrt{2}(1 + i)$$

$$w_1 = 2\left[\cos\frac{3\pi}{4} + i\sin\frac{3\pi}{4}\right] = \sqrt{2}(-1 + i)$$

$$w_2 = 2\left[\cos\frac{5\pi}{4} + i\sin\frac{5\pi}{4}\right] = \sqrt{2}(-1 - i)$$

$$w_3 = 2\left[\cos\frac{7\pi}{4} + i\sin\frac{7\pi}{4}\right] = \sqrt{2}(1 - i).$$

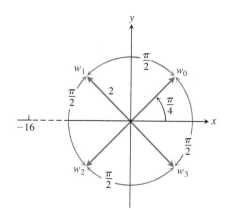

FIGURE A.31   The four fourth roots of -16.

The Fundamental Theorem of Algebra

One might say that the invention of $\sqrt{-1}$ is all well and good and leads to a number system that is richer than the real number system alone; but where will this process end? Are we also going to invent still more systems so as to obtain $\sqrt[4]{-1}$, $\sqrt[6]{-1}$, and so on? But it turns out this is not necessary. These numbers are already expressible in terms of the complex number system $a + ib$. In fact, the Fundamental Theorem of Algebra says that with the introduction of the complex numbers we now have enough numbers to factor every polynomial into a product of linear factors and so enough numbers to solve every possible polynomial equation.

> **The Fundamental Theorem of Algebra**
> Every polynomial equation of the form
>
> $$a_n z^n + a_{n-1} z^{n-1} + \cdots + a_1 z + a_0 = 0,$$
>
> in which the coefficients $a_0, a_1, \ldots, a_n$ are any complex numbers, whose degree n is greater than or equal to one, and whose leading coefficient a_n is not zero, has exactly n roots in the complex number system, provided each multiple root of multiplicity m is counted as m roots.

A proof of this theorem can be found in most texts on the theory of functions of a complex variable.

EXERCISES A.7

Operations with Complex Numbers

1. Find the following products of complex numbers

   a. $(2 + 3i)(4 - 2i)$ b. $(2 - i)(-2 - 3i)$

   c. $(-1 - 2i)(2 + i)$

2. Solve the following equations for the real numbers, x and y.

   a. $(3 + 4i)^2 - 2(x - iy) = x + iy$

   b. $\left(\dfrac{1 + i}{1 - i}\right)^2 + \dfrac{1}{x + iy} = 1 + i$

   c. $(3 - 2i)(x + iy) = 2(x - 2iy) + 2i - 1$

Graphing and Geometry

3. How may the following complex numbers be obtained from $z = x + iy$ geometrically? Sketch.

   a. $\bar{z}$ b. $\overline{(-z)}$

   c. $-z$ d. $1/z$

4. Show that the distance between the two points z_1 and z_2 in an Argand diagram is $|z_1 - z_2|$.

In Exercises 5–10, graph the points $z = x + iy$ that satisfy the given conditions.

5. a. $|z| = 2$ b. $|z| < 2$ c. $|z| > 2$

6. $|z - 1| = 2$ 7. $|z + 1| = 1$

8. $|z + 1| = |z - 1|$ 9. $|z + i| = |z - 1|$

10. $|z + 1| \geq |z|$

Express the complex numbers in Exercises 11–14 in the form $re^{i\theta}$, with $r \geq 0$ and $-\pi < \theta \leq \pi$. Draw an Argand diagram for each calculation.

11. $\left(1 + \sqrt{-3}\right)^2$ 12. $\dfrac{1 + i}{1 - i}$

13. $\dfrac{1 + i\sqrt{3}}{1 - i\sqrt{3}}$ 14. $(2 + 3i)(1 - 2i)$

Powers and Roots

Use De Moivre's Theorem to express the trigonometric functions in Exercises 15 and 16 in terms of $\cos \theta$ and $\sin \theta$.

15. $\cos 4\theta$ 16. $\sin 4\theta$

17. Find the three cube roots of 1.

18. Find the two square roots of i.

19. Find the three cube roots of $-8i$.

20. Find the six sixth roots of 64.

21. Find the four solutions of the equation $z^4 - 2z^2 + 4 = 0$.

22. Find the six solutions of the equation $z^6 + 2z^3 + 2 = 0$.

23. Find all solutions of the equation $x^4 + 4x^2 + 16 = 0$.

24. Solve the equation $x^4 + 1 = 0$.

Theory and Examples

25. **Complex numbers and vectors in the plane** Show with an Argand diagram that the law for adding complex numbers is the same as the parallelogram law for adding vectors.

26. **Complex arithmetic with conjugates** Show that the conjugate of the sum (product, or quotient) of two complex numbers, z_1 and z_2, is the same as the sum (product, or quotient) of their conjugates.

27. **Complex roots of polynomials with real coefficients come in complex-conjugate pairs**

   a. Extend the results of Exercise 26 to show that $f(\bar{z}) = \overline{f(z)}$ when

   $$f(z) = a_n z^n + a_{n-1} z^{n-1} + \cdots + a_1 z + a_0$$

   is a polynomial with real coefficients $a_0, \ldots, a_n$.

   b. If z is a root of the equation $f(z) = 0$, where $f(z)$ is a polynomial with real coefficients as in part (a), show that the conjugate $\bar{z}$ is also a root of the equation. (*Hint:* Let $f(z) = u + iv = 0$; then both u and v are zero. Use the fact that $f(\bar{z}) = \overline{f(z)} = u - iv$.)

28. Absolute value of a conjugate   Show that $|\bar{z}| = |z|$.

29. When $z = \bar{z}$ If z and $\bar{z}$ are equal, what can you say about the location of the point z in the complex plane?

30. Real and imaginary parts   Let Re(z) denote the real part of z and Im(z) the imaginary part. Show that the following relations hold for any complex numbers z, z_1, and z_2.

a. $z + \bar{z} = 2\text{Re}(z)$

b. $z - \bar{z} = 2i\text{Im}(z)$

c. $|\text{Re}(z)| \leq |z|$

d. $|z_1 + z_2|^2 = |z_1|^2 + |z_2|^2 + 2\text{Re}(z_1\bar{z}_2)$

e. $|z_1 + z_2| \leq |z_1| + |z_2|$

A.8   The Distributive Law for Vector Cross Products

In this appendix we prove the Distributive Law

$$\mathbf{u} \times (\mathbf{v} + \mathbf{w}) = \mathbf{u} \times \mathbf{v} + \mathbf{u} \times \mathbf{w},$$

which is Property 2 in Section 12.4.

Proof   To derive the Distributive Law, we construct $\mathbf{u} \times \mathbf{v}$ a new way. We draw $\mathbf{u}$ and $\mathbf{v}$ from the common point O and construct a plane M perpendicular to $\mathbf{u}$ at O (Figure A.32). We then project $\mathbf{v}$ orthogonally onto M, yielding a vector $\mathbf{v}'$ with length $|\mathbf{v}|\sin\theta$. We rotate $\mathbf{v}'$ 90° about $\mathbf{u}$ in the positive sense to produce a vector $\mathbf{v}''$. Finally, we multiply $\mathbf{v}''$ by the length of $\mathbf{u}$. The resulting vector $|\mathbf{u}|\mathbf{v}''$ is equal to $\mathbf{u} \times \mathbf{v}$ since $\mathbf{v}''$ has the same direction as $\mathbf{u} \times \mathbf{v}$ by its construction (Figure A.32) and

$$|\mathbf{u}||\mathbf{v}''| = |\mathbf{u}||\mathbf{v}'| = |\mathbf{u}||\mathbf{v}|\sin\theta = |\mathbf{u} \times \mathbf{v}|.$$

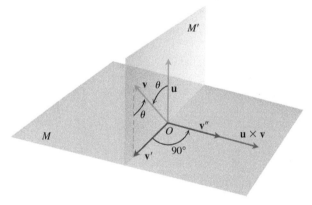

FIGURE A.32   As explained in the text, $\mathbf{u} \times \mathbf{v} = |\mathbf{u}|\mathbf{v}''$. (The primes used here are purely notational and do not denote derivatives.)

Now each of these three operations, namely,

1.   projection onto M

2.   rotation about $\mathbf{u}$ through 90°

3.   multiplication by the scalar $|\mathbf{u}|$

when applied to a triangle whose plane is not parallel to $\mathbf{u}$, will produce another triangle. If we start with the triangle whose sides are $\mathbf{v}$, $\mathbf{w}$, and $\mathbf{v} + \mathbf{w}$ (Figure A.33) and apply these three steps, we successively obtain the following:

1.   A triangle whose sides are $\mathbf{v}'$, $\mathbf{w}'$, and $(\mathbf{v} + \mathbf{w})'$ satisfying the vector equation

$$\mathbf{v}' + \mathbf{w}' = (\mathbf{v} + \mathbf{w})'$$

2.   A triangle whose sides are $\mathbf{v}''$, $\mathbf{w}''$, and $(\mathbf{v} + \mathbf{w})''$ satisfying the vector equation

$$\mathbf{v}'' + \mathbf{w}'' = (\mathbf{v} + \mathbf{w})''$$

(The double prime on each vector has the same meaning as in Figure A.32.)

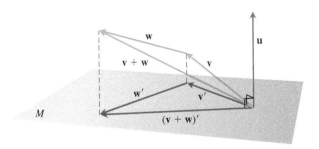

FIGURE A.33   The vectors, **v**, **w**, **v** + **w**, and their projections onto a plane perpendicular to **u**.

3.   A triangle whose sides are $|\mathbf{u}|\mathbf{v}''$, $|\mathbf{u}|\mathbf{w}''$, and $|\mathbf{u}|(\mathbf{v} + \mathbf{w})''$ satisfying the vector equation

$$|\mathbf{u}|\mathbf{v}'' + |\mathbf{u}|\mathbf{w}'' = |\mathbf{u}|(\mathbf{v} + \mathbf{w})''.$$

Substituting $|\mathbf{u}|\mathbf{v}'' = \mathbf{u} \times \mathbf{v}$, $|\mathbf{u}|\mathbf{w}'' = \mathbf{u} \times \mathbf{w}$, and $|\mathbf{u}|(\mathbf{v} + \mathbf{w})'' = \mathbf{u} \times (\mathbf{v} + \mathbf{w})$ from our discussion above into this last equation gives

$$\mathbf{u} \times \mathbf{v} + \mathbf{u} \times \mathbf{w} = \mathbf{u} \times (\mathbf{v} + \mathbf{w}),$$

which is the law we wanted to establish.   ∎

A.9   The Mixed Derivative Theorem and the Increment Theorem

This appendix derives the Mixed Derivative Theorem (Theorem 2, Section 14.3) and the Increment Theorem for Functions of Two Variables (Theorem 3, Section 14.3). Euler first published the Mixed Derivative Theorem in 1734, in a series of papers he wrote on hydrodynamics.

THEOREM 2—The Mixed Derivative Theorem
If $f(x, y)$ and its partial derivatives f_x, f_y, f_{xy}, and f_{yx} are defined throughout an open region containing a point (a, b) and are all continuous at (a, b), then

$$f_{xy}(a, b) = f_{yx}(a, b).$$

Proof   The equality of $f_{xy}(a, b)$ and $f_{yx}(a, b)$ can be established by four applications of the Mean Value Theorem (Theorem 4, Section 4.2). By hypothesis, the point (a, b) lies in the interior of a rectangle R in the xy-plane on which f, f_x, f_y, f_{xy}, and f_{yx} are all defined. We let h and k be the numbers such that the point $(a + h, b + k)$ also lies in R, and we consider the difference

$$\Delta = F(a + h) - F(a), \tag{1}$$

where

$$F(x) = f(x, b + k) - f(x, b). \tag{2}$$

We apply the Mean Value Theorem to F, which is continuous because it is differentiable. Then Equation (1) becomes

$$\Delta = hF'(c_1), \tag{3}$$

where c_1 lies between a and $a + h$. From Equation (2),

$$F'(x) = f_x(x, b + k) - f_x(x, b),$$

so Equation (3) becomes

$$\Delta = h[f_x(c_1, b + k) - f_x(c_1, b)]. \tag{4}$$

Now we apply the Mean Value Theorem to the function $g(y) = f_x(c_1, y)$ and have

$$g(b + k) - g(b) = kg'(d_1),$$

or

$$f_x(c_1, b + k) - f_x(c_1, b) = kf_{xy}(c_1, d_1)$$

for some d_1 between b and $b + k$. By substituting this into Equation (4), we get

$$\Delta = hkf_{xy}(c_1, d_1) \tag{5}$$

for some point (c_1, d_1) in the rectangle R' whose vertices are the four points (a, b), $(a + h, b)$, $(a + h, b + k)$, and $(a, b + k)$. (See Figure A.34.)

By substituting from Equation (2) into Equation (1), we may also write

$$\begin{aligned}
\Delta &= f(a + h, b + k) - f(a + h, b) - f(a, b + k) + f(a, b) \\
&= [f(a + h, b + k) - f(a, b + k)] - [f(a + h, b) - f(a, b)] \\
&= \phi(b + k) - \phi(b),
\end{aligned} \tag{6}$$

where

$$\phi(y) = f(a + h, y) - f(a, y). \tag{7}$$

The Mean Value Theorem applied to Equation (6) now gives

$$\Delta = k\phi'(d_2) \tag{8}$$

for some d_2 between b and $b + k$. By Equation (7),

$$\phi'(y) = f_y(a + h, y) - f_y(a, y). \tag{9}$$

Substituting from Equation (9) into Equation (8) gives

$$\Delta = k[f_y(a + h, d_2) - f_y(a, d_2)].$$

Finally, we apply the Mean Value Theorem to the expression in brackets and get

$$\Delta = khf_{yx}(c_2, d_2) \tag{10}$$

for some c_2 between a and $a + h$.

Together, Equations (5) and (10) show that

$$f_{xy}(c_1, d_1) = f_{yx}(c_2, d_2), \tag{11}$$

where (c_1, d_1) and (c_2, d_2) both lie in the rectangle R' (Figure A.34). Equation (11) is not quite the result we want, since it says only that f_{xy} has the same value at (c_1, d_1) that f_{yx} has at (c_2, d_2). The numbers h and k in our discussion, however, may be made as small as we wish. The hypothesis that f_{xy} and f_{yx} are both continuous at (a, b) means that $f_{xy}(c_1, d_1) = f_{xy}(a, b) + \varepsilon_1$ and $f_{yx}(c_2, d_2) = f_{yx}(a, b) + \varepsilon_2$, where each of $\varepsilon_1, \varepsilon_2 \to 0$ as both $h, k \to 0$. Hence, if we let h and $k \to 0$, we have $f_{xy}(a, b) = f_{yx}(a, b)$. ∎

The equality of $f_{xy}(a, b)$ and $f_{yx}(a, b)$ can be proved with hypotheses weaker than the ones we assumed. For example, it is enough for f, f_x, and f_y to exist in R and for f_{xy} to be continuous at (a, b). Then f_{yx} will exist at (a, b) and equal f_{xy} at that point.

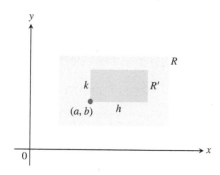

FIGURE A.34 The key to proving $f_{xy}(a, b) = f_{yx}(a, b)$ is that no matter how small R' is, f_{xy} and f_{yx} take on equal values somewhere inside R' (although not necessarily at the same point).

> **THEOREM 3—The Increment Theorem for Functions of Two Variables**
> Suppose that the first partial derivatives of $f(x, y)$ are defined throughout an open region R containing the point (x_0, y_0) and that f_x and f_y are continuous at (x_0, y_0). Then the change
>
> $$\Delta z = f(x_0 + \Delta x, y_0 + \Delta y) - f(x_0, y_0)$$
>
> in the value of f that results from moving from (x_0, y_0) to another point $(x_0 + \Delta x, y_0 + \Delta y)$ in R satisfies an equation of the form
>
> $$\Delta z = f_x(x_0, y_0)\, \Delta x + f_y(x_0, y_0)\, \Delta y + \varepsilon_1 \Delta x + \varepsilon_2 \Delta y$$
>
> in which each of $\varepsilon_1, \varepsilon_2 \to 0$ as both $\Delta x, \Delta y \to 0$.

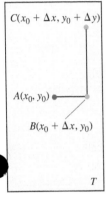

FIGURE A.35 The rectangular region T in the proof of the Increment Theorem. The figure is drawn for Δx and Δy positive, but either increment might be zero or negative.

Proof   We work within a rectangle T centered at $A(x_0, y_0)$ and lying within R, and we assume that Δx and Δy are already so small that the line segment joining A to $B(x_0 + \Delta x, y_0)$ and the line segment joining B to $C(x_0 + \Delta x, y_0 + \Delta y)$ lie in the interior of T (Figure A.35).

We may think of Δz as the sum $\Delta z = \Delta z_1 + \Delta z_2$ of two increments, where

$$\Delta z_1 = f(x_0 + \Delta x, y_0) - f(x_0, y_0)$$

is the change in the value of f from A to B and

$$\Delta z_2 = f(x_0 + \Delta x, y_0 + \Delta y) - f(x_0 + \Delta x, y_0)$$

is the change in the value of f from B to C (Figure A.36).

On the closed interval of x-values joining x_0 to $x_0 + \Delta x$, the function $F(x) = f(x, y_0)$ is a differentiable (and hence continuous) function of x, with derivative

$$F'(x) = f_x(x, y_0).$$

By the Mean Value Theorem (Theorem 4, Section 4.2), there is an x-value c between x_0 and $x_0 + \Delta x$ at which

$$F(x_0 + \Delta x) - F(x_0) = F'(c)\, \Delta x$$

or

$$f(x_0 + \Delta x, y_0) - f(x_0, y_0) = f_x(c, y_0)\, \Delta x$$

or

$$\Delta z_1 = f_x(c, y_0)\, \Delta x. \tag{12}$$

Similarly, $G(y) = f(x_0 + \Delta x, y)$ is a differentiable (and hence continuous) function of y on the closed y-interval joining y_0 and $y_0 + \Delta y$, with derivative

$$G'(y) = f_y(x_0 + \Delta x, y).$$

Hence, there is a y-value d between y_0 and $y_0 + \Delta y$ at which

$$G(y_0 + \Delta y) - G(y_0) = G'(d)\, \Delta y$$

or

$$f(x_0 + \Delta x, y_0 + \Delta y) - f(x_0 + \Delta x, y) = f_y(x_0 + \Delta x, d)\, \Delta y$$

or

$$\Delta z_2 = f_y(x_0 + \Delta x, d)\, \Delta y. \tag{13}$$

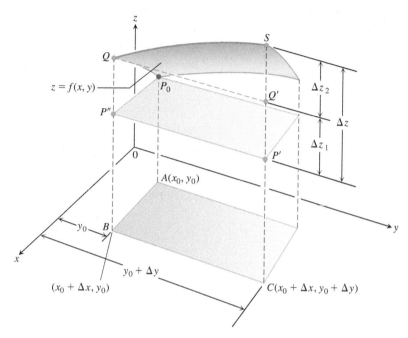

FIGURE A.36   Part of the surface $z = f(x, y)$ near $P_0(x_0, y_0, f(x_0, y_0))$. The points P_0, P', and P'' have the same height $z_0 = f(x_0, y_0)$ above the xy-plane. The change in z is $\Delta z = P'S$. The change

$$\Delta z_1 = f(x_0 + \Delta x, y_0) - f(x_0, y_0),$$

shown as $P''Q = P'Q'$, is caused by changing x from x_0 to $x_0 + \Delta x$ while holding y equal to y_0. Then, with x held equal to $x_0 + \Delta x$,

$$\Delta z_2 = f(x_0 + \Delta x, y_0 + \Delta y) - f(x_0 + \Delta x, y_0)$$

is the change in z caused by changing y_0 from $y_0 + \Delta y$, which is represented by $Q'S$. The total change in z is the sum of Δz_1 and Δz_2.

Now, as both Δx and $\Delta y \to 0$, we know that $c \to x_0$ and $d \to y_0$. Therefore, since f_x and f_y are continuous at (x_0, y_0), the quantities

$$\varepsilon_1 = f_x(c, y_0) - f_x(x_0, y_0),$$
$$\varepsilon_2 = f_y(x_0 + \Delta x, d) - f_y(x_0, y_0) \tag{14}$$

both approach zero as both Δx and $\Delta y \to 0$.

Finally,

$$\begin{aligned}
\Delta z &= \Delta z_1 + \Delta z_2 \\
&= f_x(c, y_0)\Delta x + f_y(x_0 + \Delta x, d)\Delta y && \text{From Eqs. (12) and (13)} \\
&= \left[f_x(x_0, y_0) + \varepsilon_1 \right]\Delta x + \left[f_y(x_0, y_0) + \varepsilon_2 \right]\Delta y && \text{From Eq. (14)} \\
&= f_x(x_0, y_0)\Delta x + f_y(x_0, y_0)\Delta y + \varepsilon_1\Delta x + \varepsilon_2\Delta y,
\end{aligned}$$

where both ε_1 and $\varepsilon_2 \to 0$ as both Δx and $\Delta y \to 0$, which is what we set out to prove. ∎

Analogous results hold for functions of any finite number of independent variables. Suppose that the first partial derivatives of $w = f(x, y, z)$ are defined throughout an open region containing the point (x_0, y_0, z_0) and that f_x, f_y, and f_z are continuous at (x_0, y_0, z_0). Then

$$\begin{aligned}
\Delta w &= f(x_0 + \Delta x, y_0 + \Delta y, z_0 + \Delta z) - f(x_0, y_0, z_0) \\
&= f_x\Delta x + f_y\Delta y + f_z\Delta z + \varepsilon_1\Delta x + \varepsilon_2\Delta y + \varepsilon_3\Delta z, \tag{15}
\end{aligned}$$

where $\varepsilon_1, \varepsilon_2, \varepsilon_3 \to 0$ as Δx, Δy, and $\Delta z \to 0$.

The partial derivatives f_x, f_y, f_z in Equation (15) are to be evaluated at the point (x_0, y_0, z_0).

Equation (15) can be proved by treating Δw as the sum of three increments,

$$\Delta w_1 = f(x_0 + \Delta x, y_0, z_0) - f(x_0, y_0, z_0) \tag{16}$$

$$\Delta w_2 = f(x_0 + \Delta x, y_0 + \Delta y, z_0) - f(x_0 + \Delta x, y_0, z_0) \tag{17}$$

$$\Delta w_3 = f(x_0 + \Delta x, y_0 + \Delta y, z_0 + \Delta z) - f(x_0 + \Delta x, y_0 + \Delta y, z_0), \tag{18}$$

and applying the Mean Value Theorem to each of these separately. Two coordinates remain constant and only one varies in each of these partial increments $\Delta w_1, \Delta w_2, \Delta w_3$. In Equation (17), for example, only y varies, since x is held equal to $x_0 + \Delta x$ and z is held equal to z_0. Since $f(x_0 + \Delta x, y, z_0)$ is a continuous function of y with a derivative f_y, it is subject to the Mean Value Theorem, and we have

$$\Delta w_2 = f_y(x_0 + \Delta x, y_1, z_0)\,\Delta y$$

for some y_1 between y_0 and $y_0 + \Delta y$.

ANSWERS TO ODD-NUMBERED EXERCISES

Chapter 10

SECTION 10.1, pp. 586–590

1. $a_1 = 0, a_2 = -1/4, a_3 = -2/9, a_4 = -3/16$

3. $a_1 = 1, a_2 = -1/3, a_3 = 1/5, a_4 = -1/7$

5. $a_1 = 1/2, a_2 = 1/2, a_3 = 1/2, a_4 = 1/2$

7. $1, \dfrac{3}{2}, \dfrac{7}{4}, \dfrac{15}{8}, \dfrac{31}{16}, \dfrac{63}{32}, \dfrac{127}{64}, \dfrac{255}{128}, \dfrac{511}{256}, \dfrac{1023}{512}$

9. $2, 1, -\dfrac{1}{2}, -\dfrac{1}{4}, \dfrac{1}{8}, \dfrac{1}{16}, -\dfrac{1}{32}, -\dfrac{1}{64}, \dfrac{1}{128}, \dfrac{1}{256}$

11. $1, 1, 2, 3, 5, 8, 13, 21, 34, 55$

13. $a_n = (-1)^{n+1}, n \geq 1$

15. $a_n = (-1)^{n+1}(n)^2, n \geq 1$ **17.** $a_n = \dfrac{2^{n-1}}{3(n+2)}, n \geq 1$

19. $a_n = n^2 - 1, n \geq 1$ **21.** $a_n = 4n - 3, n \geq 1$

23. $a_n = \dfrac{3n+2}{n!}, n \geq 1$ **25.** $a_n = \dfrac{1 + (-1)^{n+1}}{2}, n \geq 1$

27. $a_n = \dfrac{1}{(n+1)(n+2)}$ **29.** $a_n = \sin\left(\dfrac{\sqrt{n+1}}{1 + (n+1)^2}\right)$

31. Converges, 2   **33.** Converges, -1 **35.** Converges, -5
37. Diverges   **39.** Diverges   **41.** Converges, $1/2$
43. Converges, 0   **45.** Converges, $\sqrt{2}$ **47.** Converges, 1
49. Converges, 0   **51.** Converges, 0   **53.** Converges, 0
55. Converges, 1   **57.** Converges, e^7 **59.** Converges, 1
61. Converges, 1   **63.** Diverges   **65.** Converges, 4
67. Converges, 0   **69.** Diverges   **71.** Converges, e^{-1}
73. Diverges   **75.** Converges, 0   **77.** Diverges
79. Converges, $e^{2/3}$ **81.** Converges, $x \, (x > 0)$
83. Converges, 0   **85.** Converges, 1   **87.** Converges, $1/2$
89. Converges, 1   **91.** Converges, $\pi/2$ **93.** Converges, 0
95. Converges, 0   **97.** Converges, $1/2$ **99.** Converges, 0

101. 8   **103.** 4   **105.** 5   **107.** $1 + \sqrt{2}$ **109.** $x_n = 2^{n-2}$
111. (a) $f(x) = x^2 - 2$, $1.414213562 \approx \sqrt{2}$
    (b) $f(x) = \tan(x) - 1$, $0.7853981635 \approx \pi/4$
    (c) $f(x) = e^x$, diverges
113. 1
121. Nondecreasing, bounded
123. Not nondecreasing, bounded
125. Converges, nondecreasing sequence theorem
127. Converges, nondecreasing sequence theorem
129. Diverges, definition of divergence
131. Converges
133. Converges
145. (b) $\sqrt{3}$

SECTION 10.2, pp. 597–599

1. $s_n = \dfrac{2(1 - (1/3)^n)}{1 - (1/3)}, 3$ **3.** $s_n = \dfrac{1 - (-1/2)^n}{1 - (-1/2)}, 2/3$

5. $s_n = \dfrac{1}{2} - \dfrac{1}{n+2}, \dfrac{1}{2}$ **7.** $1 - \dfrac{1}{4} + \dfrac{1}{16} - \dfrac{1}{64} + \cdots, \dfrac{4}{5}$

9. $-\dfrac{3}{4} + \dfrac{9}{16} + \dfrac{57}{64} + \dfrac{249}{256} + \cdots$, diverges.

11. $(5 + 1) + \left(\dfrac{5}{2} + \dfrac{1}{3}\right) + \left(\dfrac{5}{4} + \dfrac{1}{9}\right) + \left(\dfrac{5}{8} + \dfrac{1}{27}\right) + \cdots, \dfrac{23}{2}$

13. $(1 + 1) + \left(\dfrac{1}{2} - \dfrac{1}{5}\right) + \left(\dfrac{1}{4} + \dfrac{1}{25}\right) + \left(\dfrac{1}{8} - \dfrac{1}{125}\right) + \cdots, \dfrac{17}{6}$

15. Converges, $5/3$ **17.** Converges, $1/7$

19. Converges, $\dfrac{e}{e + 2}$ **21.** Diverges   **23.** $23/99$

25. $7/9$ **27.** $1/15$ **29.** $41333/33300$ **31.** Diverges
33. Inconclusive   **35.** Diverges   **37.** Diverges

39. $s_n = 1 - \dfrac{1}{n+1}$; converges, 1

41. $s_n = \ln\sqrt{n+1}$; diverges

43. $s_n = \dfrac{\pi}{3} - \cos^{-1}\left(\dfrac{1}{n+2}\right)$; converges, $-\dfrac{\pi}{6}$ **45.** 1   **47.** 5

49. 1   **51.** $-\dfrac{1}{\ln 2}$ **53.** Converges, $2 + \sqrt{2}$
55. Converges, 1   **57.** Diverges
59. Converges, $\dfrac{e^2}{e^2 - 1}$
61. Converges, $2/9$ **63.** Converges, $3/2$ **65.** Diverges
67. Converges, 4   **69.** Diverges   **71.** Converges, $\dfrac{\pi}{\pi - e}$
73. Converges, $-5/6$ **75.** Diverges
77. $a = 1, r = -x$; converges to $1/(1 + x)$ for $|x| < 1$
79. $a = 3, r = (x - 1)/2$; converges to $6/(3 - x)$ for x in $(-1, 3)$
81. $|x| < \dfrac{1}{2}, \dfrac{1}{1 - 2x}$ **83.** $-2 < x < 0, \dfrac{1}{2 + x}$
85. $x \neq (2k + 1)\dfrac{\pi}{2}, k$ an integer; $\dfrac{1}{1 - \sin x}$

87. (a) $\displaystyle\sum_{n=-2}^{\infty} \dfrac{1}{(n+4)(n+5)}$ (b) $\displaystyle\sum_{n=0}^{\infty} \dfrac{1}{(n+2)(n+3)}$
    (c) $\displaystyle\sum_{n=5}^{\infty} \dfrac{1}{(n-3)(n-2)}$

97. (a) $r = 3/5$ (b) $r = -3/10$ **99.** $|r| < 1, \dfrac{1 + 2r}{1 - r^2}$
101. (a) 16.84 mg, 17.79 mg (b) 17.84 mg
103. (a) $0, \dfrac{1}{27}, \dfrac{2}{27}, \dfrac{1}{9}, \dfrac{2}{9}, \dfrac{7}{27}, \dfrac{8}{27}, \dfrac{1}{3}, \dfrac{2}{3}, \dfrac{7}{9}, \dfrac{8}{9}, 1$
    (b) $\displaystyle\sum_{n=1}^{\infty} \dfrac{1}{2}\left(\dfrac{2}{3}\right)^{n-1} = 1$ **105.** $(1/3)\pi$

SECTION 10.3, pp. 604–606

1. Converges   **3.** Converges   **5.** Converges   **7.** Diverges
9. Converges   **11.** Diverges
13. Converges; geometric series, $r = \dfrac{1}{10} < 1$
15. Diverges; $\lim_{n\to\infty} \dfrac{n}{n+1} = 1 \neq 0$
17. Diverges; p-series, $p < 1$
19. Converges; geometric series, $r = \dfrac{1}{8} < 1$
21. Diverges; Integral Test
23. Converges; geometric series, $r = 2/3 < 1$
25. Diverges; Integral Test
27. Diverges; $\lim_{n\to\infty} \dfrac{2^n}{n+1} \neq 0$
29. Diverges; $\lim_{n\to\infty} (\sqrt{n}/\ln n) \neq 0$
31. Diverges; geometric series, $r = \dfrac{1}{\ln 2} > 1$
33. Converges; Integral Test
35. Diverges; nth-Term Test
37. Converges; Integral Test
39. Diverges; nth-Term Test
41. Converges; by taking limit of partial sums
43. Converges; Integral Test
45. Converges; Integral Test   **47.** $a = 1$

49. (a)

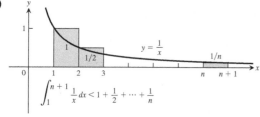

$$\int_1^{n+1} \frac{1}{x}\,dx < 1 + \frac{1}{2} + \cdots + \frac{1}{n}$$

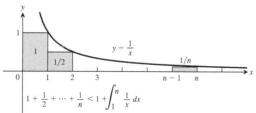

$$1 + \frac{1}{2} + \cdots + \frac{1}{n} < 1 + \int_1^n \frac{1}{x}\,dx$$

(b) ≈ 41.55

51. True   **53.** $n \geq 251{,}415$

55. $s_8 = \sum_{n=1}^{8} \frac{1}{n^3} \approx 1.195$ **57.** 10^{60}

65. (a) $1.20166 \leq S \leq 1.20253$
    (b) $S \approx 1.2021$, error < 0.0005

67. $\left(\dfrac{\pi^2}{6} - 1\right) \approx 0.64493$

SECTION 10.4, pp. 610–611

1. Converges; compare with $\sum(1/n^2)$
3. Diverges; compare with $\sum\left(1/\sqrt{n}\right)$
5. Converges; compare with $\sum(1/n^{3/2})$
7. Converges; compare with $\sum\sqrt{\dfrac{n+4n}{n^4+0}} = \sqrt{5}\,\sum\dfrac{1}{n^{3/2}}$
9. Converges
11. Diverges; limit comparison with $\sum(1/n)$
13. Diverges; limit comparison with $\sum\left(1/\sqrt{n}\right)$
15. Diverges
17. Diverges; limit comparison with $\sum\left(1/\sqrt{n}\right)$
19. Converges; compare with $\sum(1/2^n)$
21. Diverges; nth-Term Test
23. Converges; compare with $\sum(1/n^2)$
25. Converges; $\left(\dfrac{n}{3n+1}\right)^n < \left(\dfrac{n}{3n}\right)^n = \left(\dfrac{1}{3}\right)^n$
27. Diverges; direct comparison with $\sum(1/n)$
29. Diverges; limit comparison with $\sum(1/n)$
31. Diverges; limit comparison with $\sum(1/n)$
33. Converges; compare with $\sum(1/n^{3/2})$
35. Converges; $\dfrac{1}{n2^n} \leq \dfrac{1}{2^n}$ **37.** Converges; $\dfrac{1}{3^{n-1}+1} < \dfrac{1}{3^{n-1}}$
39. Converges; comparison with $\sum(1/5n^2)$
41. Diverges; comparison with $\sum(1/n)$
43. Converges; comparison with $\sum\dfrac{1}{n(n-1)}$ or limit comparison with $\sum(1/n^2)$
45. Diverges; limit comparison with $\sum(1/n)$
47. Converges; $\dfrac{\tan^{-1}n}{n^{1.1}} < \dfrac{\pi/2}{n^{1.1}}$
49. Converges; compare with $\sum(1/n^2)$
51. Diverges; limit comparison with $\sum(1/n)$
53. Converges; limit comparison with $\sum(1/n^2)$
55. Diverges nth-Term Test
67. Converges   **69.** Answers   **71.** Converges

SECTION 10.5, pp. 616–617

1. Converges   **3.** Diverges   **5.** Converges
7. Converges   **9.** Converges   **11.** Diverges
13. Converges   **15.** Converges
17. Converges; Ratio Test   **19.** Diverges; Ratio Test
21. Converges; Ratio Test
23. Converges; compare with $\sum(3/(1.25)^n)$
25. Diverges; $\lim\limits_{n\to\infty}\left(1-\dfrac{3}{n}\right)^n = e^{-3} \neq 0$
27. Converges; compare with $\sum(1/n^2)$
29. Diverges; compare with $\sum(1/(2n))$ **31.** Diverges; $a_n \nrightarrow 0$
33. Converges; Ratio Test   **35.** Converges; Ratio Test
37. Converges; Ratio Test   **39.** Converges; Root Test
41. Converges; compare with $\sum(1/n^2)$
43. Converges; Ratio Test   **45.** Diverges; Ratio Test
47. Converges; Ratio Test   **49.** Diverges; Ratio Test
51. Converges; Ratio Test   **53.** Converges; Ratio Test
55. Diverges; $a_n = \left(\dfrac{1}{3}\right)^{(1/n!)} \to 1$ **57.** Converges; Ratio Test
59. Diverges; Root Test   **61.** Converges; Root Test
63. Converges; Ratio Test
65. (a) Diverges; nth-Term Test
    (b) Diverges; Root Test
    (c) Converges; Root Test
    (d) Converges; Ratio Test
69. Yes

SECTION 10.6, pp. 622–624

1. Converges by Alternating Series Test
3. Converges; Alternating Series Test
5. Converges; Alternating Series Test
7. Diverges; $a_n \nrightarrow 0$
9. Diverges; $a_n \nrightarrow 0$
11. Converges; Alternating Series Test
13. Converges by Alternating Series Test
15. Converges absolutely. Series of absolute s is a convergent geometric series.
17. Converges conditionally; $1/\sqrt{n} \to 0$ but $\sum_{n=1}^{\infty}\dfrac{1}{\sqrt{n}}$ diverges.
19. Converges absolutely; compare with $\sum_{n=1}^{\infty}(1/n^2)$.
21. Converges conditionally; $1/(n+3) \to 0$ but $\sum_{n=1}^{\infty}\dfrac{1}{n+3}$ diverges (compare with $\sum_{n=1}^{\infty}(1/n)$).
23. Diverges; $\dfrac{3+n}{5+n} \to 1$
25. Converges conditionally; $\left(\dfrac{1}{n^2} + \dfrac{1}{n}\right) \to 0$ but $(1+n)/n^2 > 1/n$
27. Converges absolutely; Ratio Test
29. Converges absolutely by Integral Test
31. Diverges; $a_n \nrightarrow 0$
33. Converges absolutely by Ratio Test
35. Converges absolutely, since $\left|\dfrac{\cos n\pi}{n\sqrt{n}}\right| = \left|\dfrac{(-1)^{n+1}}{n^{3/2}}\right| = \dfrac{1}{n^{3/2}}$ (convergent p-series)
37. Converges absolutely by Root Test
39. Diverges; $a_n \to \infty$
41. Converges conditionally; $\sqrt{n+1} - \sqrt{n} = 1/(\sqrt{n} + \sqrt{n+1}) \to 0$, but series of absolute values diverges $\left(\text{compare with } \sum\left(1/\sqrt{n}\right)\right)$.

43. Diverges, $a_n \to 1/2 \neq 0$

45. Converges absolutely; sech $n = \dfrac{2}{e^n + e^{-n}} = \dfrac{2e^n}{e^{2n} + 1} <$
$\dfrac{2e^n}{e^{2n}} = \dfrac{2}{e^n}$, a term from a convergent geometric series.

47. Converges conditionally; $\Sigma(-1)^{n+1}\dfrac{1}{2(n+1)}$ converges by
Alternating Series Test; $\Sigma\dfrac{1}{2(n+1)}$ diverges by limit comparison
with $\Sigma(1/n)$.

49. $|\text{Error}| < 0.2$ **51.** $|\text{Error}| < 2 \times 10^{-11}$
53. $n \geq 31$ **55.** $n \geq 4$ **57.** Converges; Root Test
59. Converges; Limit of Partial Sums
61. Converges; Ratio Test   **63.** Diverges; p-series Test
65. Converges; Root Test   **67.** Converges; Limit Comparison Test
69. Diverges; Limit of Partial Sums
71. Diverges; Limit Comparison Test
73. Diverges; nth-Term Test   **75.** Diverges; Limit of Partial Sums
77. Converges; Limit Comparison Test
79. Converges; Limit Comparison Test
81. Converges; Ratio Test
83. 0.54030   **85. (a)** $a_n \geq a_{n+1}$ **(b)** $-1/2$

SECTION 10.7, pp. 633–636
1. (a) $1, -1 < x < 1$ **(b)** $-1 < x < 1$ **(c)** none
3. (a) $1/4, -1/2 < x < 0$ **(b)** $-1/2 < x < 0$ **(c)** none
5. (a) $10, -8 < x < 12$ **(b)** $-8 < x < 12$ **(c)** none
7. (a) $1, -1 < x < 1$ **(b)** $-1 < x < 1$ **(c)** none
9. (a) $3, -3 \leq x \leq 3$ **(b)** $-3 \leq x \leq 3$ **(c)** none
11. (a) ∞, for all x **(b)** for all x **(c)** none
13. (a) $1/2, -1/2 < x < 1/2$ **(b)** $-1/2 < x < 1/2$ **(c)** none
15. (a) $1, -1 \leq x < 1$ **(b)** $-1 < x < 1$ **(c)** $x = -1$
17. (a) $5, -8 < x < 2$ **(b)** $-8 < x < 2$ **(c)** none
19. (a) $3, -3 < x < 3$ **(b)** $-3 < x < 3$ **(c)** none
21. (a) $1, -2 < x < 0$ **(b)** $-2 < x < 0$ **(c)** none
23. (a) $1, -1 < x < 1$ **(b)** $-1 < x < 1$ **(c)** none
25. (a) $0, x = 0$ **(b)** $x = 0$ **(c)** none
27. (a) $2, -4 < x \leq 0$ **(b)** $-4 < x < 0$ **(c)** $x = 0$
29. (a) $1, -1 \leq x \leq 1$ **(b)** $-1 \leq x \leq 1$ **(c)** none
31. (a) $1/4, 1 \leq x \leq 3/2$ **(b)** $1 \leq x \leq 3/2$ **(c)** none
33. (a) ∞, for all x **(b)** for all x **(c)** none
35. (a) $1, -1 \leq x < 1$ **(b)** $-1 < x < 1$ **(c)** -1
37. 3   **39.** 8   **41.** $-1/3 < x < 1/3, 1/(1 - 3x)$
43. $-1 < x < 3, 4/(3 + 2x - x^2)$
45. $0 < x < 16, 2/(4 - \sqrt{x})$
47. $-\sqrt{2} < x < \sqrt{2}, 3/(2 - x^2)$

49. $\dfrac{2}{x} = \displaystyle\sum_{n=0}^{\infty} 2(-1)^n(x-1)^n, \ 0 < x < 2$

51. $\displaystyle\sum_{n=0}^{\infty}(-\tfrac{1}{3})^n(x-5)^n, \ 2 < x < 8$

53. $1 < x < 5, 2/(x-1), \displaystyle\sum_{n=1}^{\infty}(-\tfrac{1}{2})^n n(x-3)^{n-1}$,
$1 < x < 5, -2/(x-1)^2$

55. (a) $\cos x = 1 - \dfrac{x^2}{2!} + \dfrac{x^4}{4!} - \dfrac{x^6}{6!} + \dfrac{x^8}{8!} - \dfrac{x^{10}}{10!} + \cdots$; converges
for all x
(b) Same answer as part (c)
(c) $2x - \dfrac{2^3 x^3}{3!} + \dfrac{2^5 x^5}{5!} - \dfrac{2^7 x^7}{7!} + \dfrac{2^9 x^9}{9!} - \dfrac{2^{11} x^{11}}{11!} + \cdots$

57. (a) $\dfrac{x^2}{2} + \dfrac{x^4}{12} + \dfrac{x^6}{45} + \dfrac{17x^8}{2520} + \dfrac{31x^{10}}{14175}, -\dfrac{\pi}{2} < x < \dfrac{\pi}{2}$
(b) $1 + x^2 + \dfrac{2x^4}{3} + \dfrac{17x^6}{45} + \dfrac{62x^8}{315} + \cdots, -\dfrac{\pi}{2} < x < \dfrac{\pi}{2}$
63. (a) T **(b)** T **(c)** F **(d)** T **(e)** N **(f)** F **(g)** N **(h)** T

SECTION 10.8, pp. 640–641
1. $P_0(x) = 1, P_1(x) = 1 + 2x, P_2(x) = 1 + 2x + 2x^2$,
$P_3(x) = 1 + 2x + 2x^2 + \dfrac{4}{3}x^3$

3. $P_0(x) = 0, P_1(x) = x - 1, P_2(x) = (x-1) - \dfrac{1}{2}(x-1)^2$,
$P_3(x) = (x-1) - \dfrac{1}{2}(x-1)^2 + \dfrac{1}{3}(x-1)^3$

5. $P_0(x) = \dfrac{1}{2}, P_1(x) = \dfrac{1}{2} - \dfrac{1}{4}(x-2)$,
$P_2(x) = \dfrac{1}{2} - \dfrac{1}{4}(x-2) + \dfrac{1}{8}(x-2)^2$,
$P_3(x) = \dfrac{1}{2} - \dfrac{1}{4}(x-2) + \dfrac{1}{8}(x-2)^2 - \dfrac{1}{16}(x-2)^3$

7. $P_0(x) = \dfrac{\sqrt{2}}{2}, P_1(x) = \dfrac{\sqrt{2}}{2} + \dfrac{\sqrt{2}}{2}\left(x - \dfrac{\pi}{4}\right)$,
$P_2(x) = \dfrac{\sqrt{2}}{2} + \dfrac{\sqrt{2}}{2}\left(x - \dfrac{\pi}{4}\right) - \dfrac{\sqrt{2}}{4}\left(x - \dfrac{\pi}{4}\right)^2$,
$P_3(x) = \dfrac{\sqrt{2}}{2} + \dfrac{\sqrt{2}}{2}\left(x - \dfrac{\pi}{4}\right) - \dfrac{\sqrt{2}}{4}\left(x - \dfrac{\pi}{4}\right)^2$
$- \dfrac{\sqrt{2}}{12}\left(x - \dfrac{\pi}{4}\right)^3$

9. $P_0(x) = 2, P_1(x) = 2 + \dfrac{1}{4}(x-4)$,
$P_2(x) = 2 + \dfrac{1}{4}(x-4) - \dfrac{1}{64}(x-4)^2$,
$P_3(x) = 2 + \dfrac{1}{4}(x-4) - \dfrac{1}{64}(x-4)^2 + \dfrac{1}{512}(x-4)^3$

11. $\displaystyle\sum_{n=0}^{\infty}\dfrac{(-x)^n}{n!} = 1 - x + \dfrac{x^2}{2!} - \dfrac{x^3}{3!} + \dfrac{x^4}{4!} \cdots$

13. $\displaystyle\sum_{n=0}^{\infty}(-1)^n x^n = 1 - x + x^2 - x^3 + \cdots$

15. $\displaystyle\sum_{n=0}^{\infty}\dfrac{(-1)^n 3^{2n+1} x^{2n+1}}{(2n+1)!}$ **17.** $7\displaystyle\sum_{n=0}^{\infty}\dfrac{(-1)^n x^{2n}}{(2n)!}$ **19.** $\displaystyle\sum_{n=0}^{\infty}\dfrac{x^{2n}}{(2n)!}$

21. $x^4 - 2x^3 - 5x + 4$ **23.** $\displaystyle\sum_{n=1}^{\infty}(-1)^{n+1}\dfrac{x^{2n}}{(2n-1)!}$

25. $8 + 10(x-2) + 6(x-2)^2 + (x-2)^3$
27. $21 - 36(x+2) + 25(x+2)^2 - 8(x+2)^3 + (x+2)^4$

29. $\displaystyle\sum_{n=0}^{\infty}(-1)^n(n+1)(x-1)^n$ **31.** $\displaystyle\sum_{n=0}^{\infty}\dfrac{e^2}{n!}(x-2)^n$

33. $\displaystyle\sum_{n=0}^{\infty}(-1)^{n+1}\dfrac{2^{2n}}{(2n)!}\left(x - \dfrac{\pi}{4}\right)^{2n}$

35. $-1 - 2x - \dfrac{5}{2}x^2 - \cdots, -1 < x < 1$

37. $x^2 - \dfrac{1}{2}x^3 + \dfrac{1}{6}x^4 + \cdots, -1 < x < 1$

39. $x^4 + x^6 + \dfrac{x^8}{2} + \cdots, (-\infty, \infty)$

45. $L(x) = 0, Q(x) = -x^2/2$ **47.** $L(x) = 1, Q(x) = 1 + x^2/2$
49. $L(x) = x, Q(x) = x$

SECTION 10.9, pp. 647–648

1. $\displaystyle\sum_{n=0}^{\infty} \frac{(-5x)^n}{n!} = 1 - 5x + \frac{5^2 x^2}{2!} - \frac{5^3 x^3}{3!} + \cdots$

3. $\displaystyle\sum_{n=0}^{\infty} \frac{5(-1)^n(-x)^{2n+1}}{(2n+1)!} = \sum_{n=0}^{\infty} \frac{5(-1)^{n+1}x^{2n+1}}{(2n+1)!}$

$\quad = -5x + \frac{5x^3}{3!} - \frac{5x^5}{5!} + \frac{5x^7}{7!} + \cdots$

5. $\displaystyle\sum_{n=0}^{\infty} \frac{(-1)^n(5x^2)^{2n}}{(2n)!} = 1 - \frac{25x^4}{2!} + \frac{625x^8}{4!} - \cdots$

7. $\displaystyle\sum_{n=1}^{\infty} (-1)^{n+1}\frac{x^{2n}}{n} = x^2 - \frac{x^4}{2} + \frac{x^6}{3} - \frac{x^8}{4} + \cdots$

9. $\displaystyle\sum_{n=0}^{\infty} (-1)^n \left(\frac{3}{4}\right)^n x^{3n} = 1 - \frac{3}{4}x^3 + \frac{3^2}{4^2}x^6 - \frac{3^3}{4^3}x^9 + \cdots$

11. $\displaystyle\ln 3 + \sum_{n=1}^{\infty} (-1)^{n+1}\frac{2^n x^n}{n} = \ln 3 + 2x - 2x^2 + \frac{8}{3}x^3 - \cdots$

13. $\displaystyle\sum_{n=0}^{\infty} \frac{x^{n+1}}{n!} = x + x^2 + \frac{x^3}{2!} + \frac{x^4}{3!} + \frac{x^5}{4!} + \cdots$

15. $\displaystyle\sum_{n=2}^{\infty} \frac{(-1)^n x^{2n}}{(2n)!} = \frac{x^4}{4!} - \frac{x^6}{6!} + \frac{x^8}{8!} - \frac{x^{10}}{10!} + \cdots$

17. $\displaystyle x - \frac{\pi^2 x^3}{2!} + \frac{\pi^4 x^5}{4!} - \frac{\pi^6 x^7}{6!} + \cdots = \sum_{n=0}^{\infty} \frac{(-1)^n \pi^{2n} x^{2n+1}}{(2n)!}$

19. $\displaystyle 1 + \sum_{n=1}^{\infty} \frac{(-1)^n(2x)^{2n}}{2 \cdot (2n)!} =$

$\quad 1 - \frac{(2x)^2}{2 \cdot 2!} + \frac{(2x)^4}{2 \cdot 4!} - \frac{(2x)^6}{2 \cdot 6!} + \frac{(2x)^8}{2 \cdot 8!} - \cdots$

21. $\displaystyle x^2 \sum_{n=0}^{\infty} (2x)^n = x^2 + 2x^3 + 4x^4 + \cdots$

23. $\displaystyle\sum_{n=1}^{\infty} nx^{n-1} = 1 + 2x + 3x^2 + 4x^3 + \cdots$

25. $\displaystyle\sum_{n=1}^{\infty} (-1)^{n+1}\frac{x^{4n-1}}{2n-1} = x^3 - \frac{x^7}{3} + \frac{x^{11}}{5} - \frac{x^{15}}{7} + \cdots$

27. $\displaystyle\sum_{n=0}^{\infty} \left(\frac{1}{n!} + (-1)^n\right)x^n = 2 + \frac{3}{2}x^2 - \frac{5}{6}x^3 + \frac{25}{24}x^4 - \cdots$

29. $\displaystyle\sum_{n=1}^{\infty} \frac{(-1)^{n-1}x^{2n+1}}{3n} = \frac{x^3}{3} - \frac{x^5}{6} + \frac{x^7}{9} - \cdots$

31. $\displaystyle x + x^2 + \frac{x^3}{3} - \frac{x^5}{30} + \cdots$

33. $\displaystyle x^2 - \frac{2}{3}x^4 + \frac{23}{45}x^6 - \frac{44}{105}x^8 + \cdots$

35. $\displaystyle 1 + x + \frac{1}{2}x^2 - \frac{1}{8}x^4 + \cdots$ **37.** $\displaystyle 1 - \frac{x^2}{2} - \frac{x^3}{2} - \frac{x^4}{4} - \cdots$

39. $\displaystyle |\text{Error}| \le \frac{1}{10^4 \cdot 4!} < 4.2 \times 10^{-6}$

41. $\displaystyle |x| < (0.06)^{1/5} < 0.56968$

43. $\displaystyle |\text{Error}| < (10^{-3})^3/6 < 1.67 \times 10^{-10}, \quad -10^{-3} < x < 0$

45. $\displaystyle |\text{Error}| < (3^{0.1})(0.1)^3/6 < 1.87 \times 10^{-4}$

53. (a) $Q(x) = 1 + kx + \dfrac{k(k-1)}{2}x^2$ (b) $0 \le x < 100^{-1/3}$

SECTION 10.10, pp. 655–657

1. $\displaystyle 1 + \frac{x}{2} - \frac{x^2}{8} + \frac{x^3}{16}$ **3.** $1 + 3x + 6x^2 + 10x^3$

5. $\displaystyle 1 - x + \frac{3x^2}{4} - \frac{x^3}{2}$ **7.** $\displaystyle 1 - \frac{x^3}{2} + \frac{3x^6}{8} - \frac{5x^9}{16}$

9. $\displaystyle 1 + \frac{1}{2x} - \frac{1}{8x^2} + \frac{1}{16x^3}$

11. $(1 + x)^4 = 1 + 4x + 6x^2 + 4x^3 + x^4$

13. $(1 - 2x)^3 = 1 - 6x + 12x^2 - 8x^3$

15. 0.0713362   **17.** 0.4969536   **19.** 0.0999445   **21.** 0.10000

23. $\displaystyle\frac{1}{13 \cdot 6!} \approx 0.00011$ **25.** $\displaystyle\frac{x^3}{3} - \frac{x^7}{7 \cdot 3!} + \frac{x^{11}}{11 \cdot 5!}$

27. (a) $\displaystyle\frac{x^2}{2} - \frac{x^4}{12}$

$\quad$ (b) $\displaystyle\frac{x^2}{2} - \frac{x^4}{3 \cdot 4} + \frac{x^6}{5 \cdot 6} - \frac{x^8}{7 \cdot 8} + \cdots + (-1)^{15}\frac{x^{32}}{31 \cdot 32}$

29. $1/2$ **31.** $-1/24$ **33.** $1/3$ **35.** -1 **37.** 2

39. $3/2$ **41.** e **43.** $\cos\dfrac{3}{4}$ **45.** $\dfrac{\sqrt{3}}{2}$ **47.** $\dfrac{x^3}{1-x}$

49. $\displaystyle\frac{x^3}{1+x^2}$ **51.** $\displaystyle\frac{-1}{(1+x)^2}$ **55.** 500 terms   **57.** 4 terms

59. (a) $\displaystyle x + \frac{x^3}{6} + \frac{3x^5}{40} + \frac{5x^7}{112}$, radius of convergence $= 1$

$\quad$ (b) $\displaystyle\frac{\pi}{2} - x - \frac{x^3}{6} - \frac{3x^5}{40} - \frac{5x^7}{112}$

61. $1 - 2x + 3x^2 - 4x^3 + \cdots$

67. (a) -1 (b) $(1/\sqrt{2})(1 + i)$ (c) $-i$

71. $\displaystyle x + x^2 + \frac{1}{3}x^3 - \frac{1}{30}x^5 + \cdots$, for all x

PRACTICE EXERCISES, pp. 658–660

1. Converges to 1   **3.** Converges to -1 **5.** Diverges
7. Converges to 0   **9.** Converges to 1   **11.** Converges o e^{-5}
13. Converges to 3   **15.** Converges to $\ln 2$ **17.** Diverges
19. $1/6$ **21.** $3/2$ **23.** $e/(e-1)$ **25.** Diverges
27. Converges conditionally   **29.** Converges conditionally
31. Converges absolutely   **33.** Converges absolutely
35. Converges absolutely   **37.** Converges absolutely
39. Converges absolutely   **41.** Converges absolutely
43. Diverges
45. (a) $3, -7 \le x < -1$ (b) $-7 < x < -1$ (c) $x = -7$
47. (a) $1/3, 0 \le x \le 2/3$ (b) $0 \le x \le 2/3$ (c) None
49. (a) ∞, for all x (b) For all x (c) None
51. (a) $\sqrt{3}, -\sqrt{3} < x < \sqrt{3}$ (b) $-\sqrt{3} < x < \sqrt{3}$ (c) None
53. (a) $e, -e < x < e$ (b) $-e < x < e$ (c) Empty set

55. $\displaystyle\frac{1}{1+x}, \frac{1}{4}, \frac{4}{5}$ **57.** $\sin x, \pi, 0$ **59.** $e^x, \ln 2, 2$ **61.** $\displaystyle\sum_{n=0}^{\infty} 2^n x^n$

63. $\displaystyle\sum_{n=0}^{\infty} \frac{(-1)^n \pi^{2n+1} x^{2n+1}}{(2n+1)!}$ **65.** $\displaystyle\sum_{n=0}^{\infty} \frac{(-1)^n x^{10n/3}}{(2n)!}$ **67.** $\displaystyle\sum_{n=0}^{\infty} \frac{((\pi x)/2)^n}{n!}$

69. $\displaystyle 2 - \frac{(x+1)}{2 \cdot 1!} + \frac{3(x+1)^2}{2^3 \cdot 2!} - \frac{9(x+1)^3}{2^5 \cdot 3!} + \cdots$

71. $\displaystyle\frac{1}{4} - \frac{1}{4^2}(x-3) + \frac{1}{4^3}(x-3)^2 - \frac{1}{4^4}(x-3)^3$

73. 0.4849171431   **75.** 0.4872223583   **77.** $7/2$ **79.** $1/12$
81. -2 **83.** $r = -3, s = 9/2$ **85.** $2/3$

87. $\displaystyle\ln\left(\frac{n+1}{2n}\right)$; the series converges to $\ln\left(\dfrac{1}{2}\right)$.

89. (a) ∞ (b) $a = 1, b = 0$ **91.** It converges.
99. (a) Converges; Limit Comparison Test
$\quad$ (b) Converges; Direct Comparison Test
$\quad$ (c) Diverges; nth-Term Test
101. 2

ADDITIONAL AND ADVANCED EXERCISES, pp. 660–662

1. Converges; Comparison Test
3. Diverges; nth-Term Test
5. Converges; Comparison Test
7. Diverges; nth-Term Test

9. With $a = \pi/3$, $\cos x = \dfrac{1}{2} - \dfrac{\sqrt{3}}{2}(x - \pi/3) - \dfrac{1}{4}(x - \pi/3)^2$
$+ \dfrac{\sqrt{3}}{12}(x - \pi/3)^3 + \cdots$

11. With $a = 0$, $e^x = 1 + x + \dfrac{x^2}{2!} + \dfrac{x^3}{3!} + \cdots$

13. With $a = 22\pi$, $\cos x = 1 - \dfrac{1}{2}(x - 22\pi)^2 + \dfrac{1}{4!}(x - 22\pi)^4$
$- \dfrac{1}{6!}(x - 22\pi)^6 + \cdots$

15. Converges, limit $= b$ **17.** $\pi/2$ **21.** $b = \pm\dfrac{1}{5}$

23. $a = 2, L = -7/6$ **27. (b)** Yes

31. (a) $\displaystyle\sum_{n=1}^{\infty} nx^{n-1}$ **(b)** 6 **(c)** $\dfrac{1}{q}$

33. (a) $R_n = C_0 e^{-kt_0}\big(1 - e^{-nkt_0}\big)/\big(1 - e^{-kt_0}\big)$,
$R = C_0\big(e^{-kt_0}\big)/\big(1 - e^{-kt_0}\big) = C_0/\big(e^{kt_0} - 1\big)$
(b) $R_1 = 1/e \approx 0.368$,
$R_{10} = R(1 - e^{-10}) \approx R(0.9999546) \approx 0.58195$;
$R \approx 0.58198$; $0 < (R - R_{10})/R < 0.0001$
(c) 7

Chapter 11

SECTION 11.1, pp. 669–672

1.

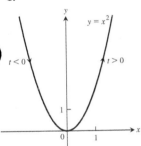

3.

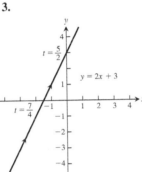

5.

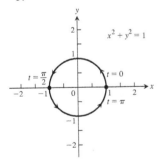

7.

9.

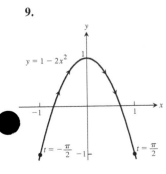

11.

13.

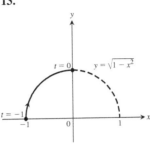

15.

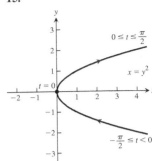

17.

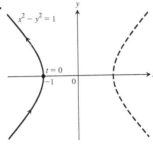

19. D **21.** E **23.** C

25.

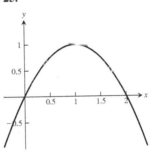

27.

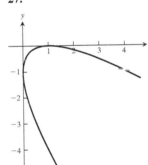

29. (a) $x = a \cos t$, $y = -a \sin t$, $0 \le t < 2\pi$
(b) $x = a \cos t$, $y = a \sin t$, $0 \le t \le 2\pi$
(c) $x = a \cos t$, $y = -a \sin t$, $0 \le t \le 4\pi$
(d) $x = a \cos t$, $y = a \sin t$, $0 \le t \le 4\pi$

31. Possible answer: $x = -1 + 5t$, $y = -3 + 4t$, $0 \le t \le 1$

33. Possible answer: $x = t^2 + 1$, $y = t$, $t \le 0$

35. Possible answer: $x = 2 - 3t$, $y = 3 - 4t$, $t \ge 0$

37. Possible answer: $x = 2 \cos t$, $y = 2|\sin t|$, $0 \le t \le 4\pi$

39. Possible answer: $x = \dfrac{-at}{\sqrt{1 + t^2}}$, $y = \dfrac{a}{\sqrt{1 + t^2}}$, $-\infty < t < \infty$

41. Possible answer: $x = \dfrac{4}{1 + 2\tan\theta}$, $y = \dfrac{4\tan\theta}{1 + 2\tan\theta}$,
$0 \le \theta < \pi/2$ and $x = 0$, $y = 2$ if $\theta = \pi/2$

43. Possible answer: $x = 2 - \cos t$, $y = \sin t$, $0 \le t \le 2\pi$

45. $x = 2 \cot t$, $y = 2 \sin^2 t$, $0 < t < \pi$

47. $x = a \sin^2 t \tan t$, $y = a \sin^2 t$, $0 \le t < \pi/2$ **49.** (1, 1)

SECTION 11.2, pp. 680–681

1. $y = -x + 2\sqrt{2}$, $\dfrac{d^2y}{dx^2} = -\sqrt{2}$

3. $y = -\dfrac{1}{2}x + 2\sqrt{2}$, $\dfrac{d^2y}{dx^2} = -\dfrac{\sqrt{2}}{4}$

5. $y = x + \dfrac{1}{4}$, $\dfrac{d^2y}{dx^2} = -2$ **7.** $y = 2x - \sqrt{3}$, $\dfrac{d^2y}{dx^2} = -3\sqrt{3}$

9. $y = x - 4$, $\dfrac{d^2y}{dx^2} = \dfrac{1}{2}$

11. $y = \sqrt{3}x - \dfrac{\pi\sqrt{3}}{3} + 2$, $\dfrac{d^2y}{dx^2} = -4$

13. $y = 9x - 1$, $\dfrac{d^2y}{dx^2} = 108$ **15.** $-\dfrac{3}{16}$ **17.** -6

19. 1   **21.** $3a^2\pi$ **23.** $|ab|\pi$ **25.** 4   **27.** 12

29. π^2 **31.** $8\pi^2$ **33.** $\dfrac{52\pi}{3}$ **35.** $3\pi\sqrt{5}$

37. $(\bar{x}, \bar{y}) = \left(\dfrac{12}{\pi} - \dfrac{24}{\pi^2}, \dfrac{24}{\pi^2} - 2\right)$

39. $(\bar{x}, \bar{y}) = \left(\dfrac{1}{3}, \pi - \dfrac{4}{3}\right)$ **41. (a)** π **(b)** π

43. (a) $x = 1$, $y = 0$, $\dfrac{dy}{dx} = \dfrac{1}{2}$ **(b)** $x = 0$, $y = 3$, $\dfrac{dy}{dx} = 0$

     (c) $x = \dfrac{\sqrt{3} - 1}{2}$, $y = \dfrac{3 - \sqrt{3}}{2}$, $\dfrac{dy}{dx} = \dfrac{2\sqrt{3} - 1}{\sqrt{3} - 2}$

45. $\left(\dfrac{\sqrt{2}}{2}, 1\right)$, $y = 2x$ at $t = 0$, $y = -2x$ at $t = \pi$

47. (a) 8a   **(b)** $\dfrac{64\pi}{3}$ **49.** $32\pi/15$

SECTION 11.3, pp. 684–685

1. a, e; b, g; c, h; d, f   **3.**

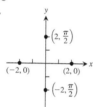

   (a) $\left(2, \dfrac{\pi}{2} + 2n\pi\right)$ and $\left(-2, \dfrac{\pi}{2} + (2n + 1)\pi\right)$, n an integer

   (b) $(2, 2n\pi)$ and $(-2, (2n + 1)\pi)$, n an integer

   (c) $\left(2, \dfrac{3\pi}{2} + 2n\pi\right)$ and $\left(-2, \dfrac{3\pi}{2} + (2n + 1)\pi\right)$, n an integer

   (d) $(2, (2n + 1)\pi)$ and $(-2, 2n\pi)$, n an integer

5. (a) $(3, 0)$ **(b)** $(-3, 0)$ **(c)** $\left(-1, \sqrt{3}\right)$ **(d)** $\left(1, \sqrt{3}\right)$
   (e) $(3, 0)$ **(f)** $\left(1, \sqrt{3}\right)$ **(g)** $(-3, 0)$ **(h)** $\left(-1, \sqrt{3}\right)$

7. (a) $\left(\sqrt{2}, \dfrac{\pi}{4}\right)$ **(b)** $(3, \pi)$ **(c)** $\left(2, \dfrac{11\pi}{6}\right)$

   (d) $\left(5, \pi - \tan^{-1}\dfrac{4}{3}\right)$

9. (a) $\left(-3\sqrt{2}, \dfrac{5\pi}{4}\right)$ **(b)** $(-1, 0)$ **(c)** $\left(-2, \dfrac{5\pi}{3}\right)$

   (d) $\left(-5, \pi - \tan^{-1}\dfrac{3}{4}\right)$

11.

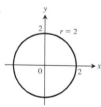

13.

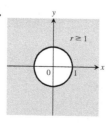

15.

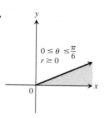

17.

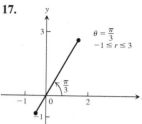

19.

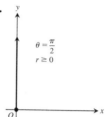

21.

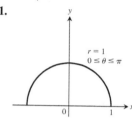

23.

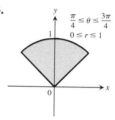

25.

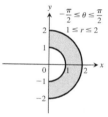

27. $x = 2$, vertical line through $(2, 0)$ **29.** $y = 0$, the x-axis

31. $y = 4$, horizontal line through $(0, 4)$

33. $x + y = 1$, line, $m = -1$, $b = 1$

35. $x^2 + y^2 = 1$, circle, $C(0, 0)$, radius 1

37. $y - 2x = 5$, line, $m = 2$, $b = 5$

39. $y^2 = x$, parabola, vertex $(0, 0)$, opens right

41. $y = e^x$, graph of natural exponential function

43. $x + y = \pm 1$, two straight lines of slope -1, y-intercepts $b = \pm 1$

45. $(x + 2)^2 + y^2 = 4$, circle, $C(-2, 0)$, radius 2

47. $x^2 + (y - 4)^2 = 16$, circle, $C(0, 4)$, radius 4

49. $(x - 1)^2 + (y - 1)^2 = 2$, circle, $C(1, 1)$, radius $\sqrt{2}$

51. $\sqrt{3}y + x = 4$ **53.** $r\cos\theta = 7$ **55.** $\theta = \pi/4$

57. $r = 2$ or $r = -2$ **59.** $4r^2\cos^2\theta + 9r^2\sin^2\theta = 36$

61. $r\sin^2\theta = 4\cos\theta$ **63.** $r = 4\sin\theta$

65. $r^2 = 6r\cos\theta - 2r\sin\theta - 6$

67. $(0, \theta)$, where θ is any angle

SECTION 11.4, pp. 688–689

1. x-axis

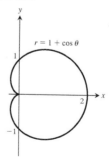

3. y-axis

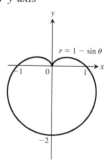

5. y-axis

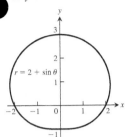

7. x-axis, y-axis, origin

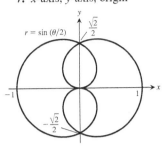

9. x-axis, y-axis, origin

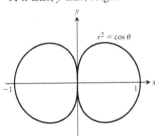

11. y-axis, x-axis, origin

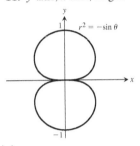

13. x-axis, y-axis, origin   **15.** Origin

17. The slope at $(-1, \pi/2)$ is -1, at $(-1, -\pi/2)$ is 1.

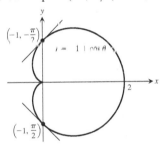

19. The slope at $(1, \pi/4)$ is -1, at $(-1, -\pi/4)$ is 1, at $(-1, 3\pi/4)$ is 1, at $(1, -3\pi/4)$ is -1.

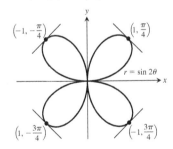

21. At $\pi/6$: slope $\sqrt{3}$, concavity 16 (concave up); at $\pi/3$: slope $-\sqrt{3}$, concavity -16 (concave down).

23. At 0: slope 0, concavity 2 (concave up); at $\pi/2$: slope $-2/\pi$, concavity $-2(8 + \pi^2)/\pi^3$ (concave down).

25. (a)

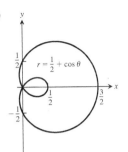

(b)

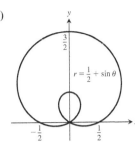

27. (a)

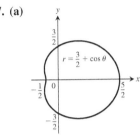

(b)

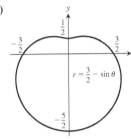

29.

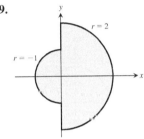

31.

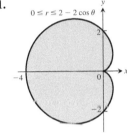

33. Equation (a)

SECTION 11.5, pp. 693–694

1. $\dfrac{1}{6}\pi^3$ **3.** 18π **5.** $\dfrac{\pi}{8}$ **7.** 2   **9.** $\dfrac{\pi}{2} - 1$

11. $5\pi - 8$ **13.** $3\sqrt{3} - \pi$ **15.** $\dfrac{\pi}{3} + \dfrac{\sqrt{3}}{2}$

17. $\dfrac{8\pi}{3} + \sqrt{3}$ **19.** (a) $\dfrac{3}{2} - \dfrac{\pi}{4}$ **21.** $19/3$ **23.** 8

25. $3\left(\sqrt{2} + \ln\left(1 + \sqrt{2}\right)\right)$ **27.** $\dfrac{\pi}{8} + \dfrac{3}{8}$

31. (a) a (b) a (c) $2a/\pi$

SECTION 11.6, pp. 700–702

1. $y^2 = 8x$, $F(2, 0)$, directrix: $x = -2$

3. $x^2 = -6y$, $F(0, -3/2)$, directrix: $y = 3/2$

5. $\dfrac{x^2}{4} - \dfrac{y^2}{9} = 1$, $F\left(\pm\sqrt{13}, 0\right)$, $V(\pm2, 0)$,

asymptotes: $y = \pm\dfrac{3}{2}x$

7. $\dfrac{x^2}{2} + y^2 = 1$, $F(\pm1, 0)$, $V\left(\pm\sqrt{2}, 0\right)$

9.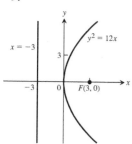

11.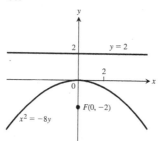

13.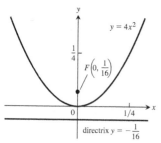

15.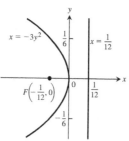

17.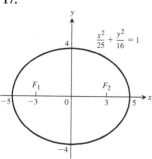

19.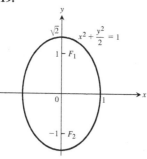

21.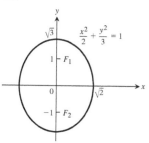

23.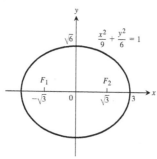

25. $\dfrac{x^2}{4} + \dfrac{y^2}{2} = 1$

27. Asymptotes: $y = \pm x$

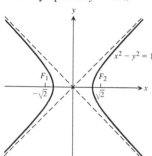

29. Asymptotes: $y = \pm x$

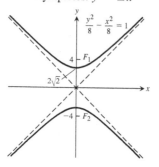

31. Asymptotes: $y = \pm 2x$

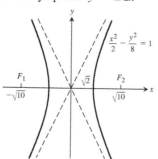

33. Asymptotes: $y = \pm x/2$

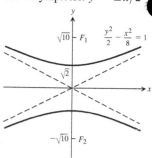

35. $y^2 - x^2 = 1$

37. $\dfrac{x^2}{9} - \dfrac{y^2}{16} = 1$

39. **(a)** Vertex: $(1, -2)$; focus: $(3, -2)$; directrix: $x = -1$

(b)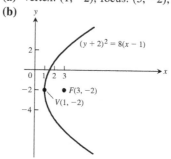

41. **(a)** Foci: $\left(4 \pm \sqrt{7}, 3\right)$; vertices: $(8, 3)$ and $(0, 3)$; center: $(4, 3)$

(b)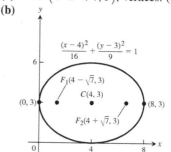

43. **(a)** Center: $(2, 0)$; foci: $(7, 0)$ and $(-3, 0)$; vertices: $(6, 0)$ and $(-2, 0)$; asymptotes: $y = \pm \dfrac{3}{4}(x - 2)$

(b)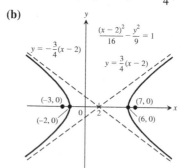

45. $(y + 3)^2 = 4(x + 2)$, $V(-2, -3)$, $F(-1, -3)$, directrix: $x = -3$

47. $(x - 1)^2 = 8(y + 7)$, $V(1, -7)$, $F(1, -5)$, directrix: $y = -9$

49. $\dfrac{(x + 2)^2}{6} + \dfrac{(y + 1)^2}{9} = 1$, $F\left(-2, \pm\sqrt{3} - 1\right)$, $V(-2, \pm 3 - 1)$, $C(-2, -1)$

51. $\dfrac{(x-2)^2}{3} + \dfrac{(y-3)^2}{2} = 1$, $F(3,3)$ and $F(1,3)$,
$V(\pm\sqrt{3}+2,3)$, $C(2,3)$

53. $\dfrac{(x-2)^2}{4} - \dfrac{(y-2)^2}{5} = 1$, $C(2,2)$, $F(5,2)$ and $F(-1,2)$,
$V(4,2)$ and $V(0,2)$; asymptotes: $(y-2) = \pm\dfrac{\sqrt{5}}{2}(x-2)$

55. $(y+1)^2 - (x+1)^2 = 1$, $C(-1,-1)$, $F(-1,\sqrt{2}-1)$
and $F(-1,-\sqrt{2}-1)$, $V(-1,0)$ and $V(-1,-2)$; asymptotes
$(y+1) = \pm(x+1)$

57. $C(-2,0)$, $a=4$ **59.** $V(-1,1)$, $F(-1,0)$

61. Ellipse: $\dfrac{(x+2)^2}{5} + y^2 = 1$, $C(-2,0)$, $F(0,0)$ and
$F(-4,0)$, $V(\sqrt{5}-2,0)$ and $V(-\sqrt{5}-2,0)$

63. Ellipse: $\dfrac{(x-1)^2}{2} + (y-1)^2 = 1$, $C(1,1)$, $F(2,1)$ and
$F(0,1)$, $V(\sqrt{2}+1,1)$ and $V(-\sqrt{2}+1,1)$

65. Hyperbola: $(x-1)^2 - (y-2)^2 = 1$, $C(1,2)$,
$F(1+\sqrt{2},2)$ and $F(1-\sqrt{2},2)$, $V(2,2)$ and
$V(0,2)$;   asymptotes: $(y-2) = \pm(x-1)$

67. Hyperbola: $\dfrac{(y-3)^2}{6} - \dfrac{x^2}{3} = 1$, $C(0,3)$, $F(0,6)$
and $F(0,0)$, $V(0,\sqrt{6}+3)$ and $V(0,-\sqrt{6}+3)$;
asymptotes: $y = \sqrt{2}x + 3$ or $y = -\sqrt{2}x + 3$

69. (b) 1:1   **73.** Length $= 2\sqrt{2}$, width $= \sqrt{2}$, area $= 4$

75. 24π

77. $x=0, y=0: y=-2x; x=0, y=2: y=2x+2;$
$x=4, y=0: y=2x-8$

79. $\bar{x}=0$, $\bar{y}=\dfrac{16}{3\pi}$

SECTION 11.7, pp. 707–708

1. $e=\dfrac{3}{5}$, $F(\pm3,0)$;
directrices are $x = \pm\dfrac{25}{3}$.

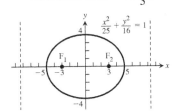

5. $e=\dfrac{1}{\sqrt{3}}$; $F(0,\pm1)$;
directrices are $y=\pm3$.

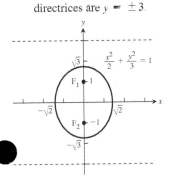

3. $e=\dfrac{1}{\sqrt{2}}$; $F(0,\pm1)$;
directrices are $y=\pm2$.

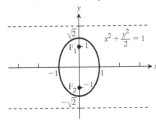

7. $e=\dfrac{\sqrt{3}}{3}$; $F(\pm\sqrt{3},0)$;
directrices are
$x=\pm3\sqrt{3}$.

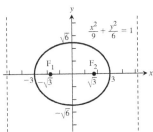

9. $\dfrac{x^2}{27} + \dfrac{y^2}{36} = 1$ **11.** $\dfrac{x^2}{4851} + \dfrac{y^2}{4900} = 1$

13. $\dfrac{x^2}{9} + \dfrac{y^2}{4} = 1$ **15.** $\dfrac{x^2}{64} + \dfrac{y^2}{48} = 1$

17. $e=\sqrt{2}$; $F(\pm\sqrt{2},0)$;
directrices are $x = \pm\dfrac{1}{\sqrt{2}}$.

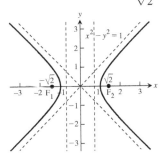

19. $e=\sqrt{2}$; $F(0,\pm4)$;
directrices are $y=\pm2$.

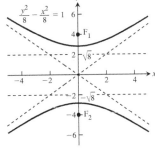

21. $e=\sqrt{5}$; $F(\pm\sqrt{10},0)$;
directrices are $x = \pm\dfrac{2}{\sqrt{10}}$.

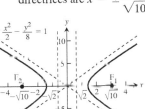

23. $e=\sqrt{5}$; $F(0,\pm\sqrt{10})$;
directrices are $y = \pm\dfrac{2}{\sqrt{10}}$.

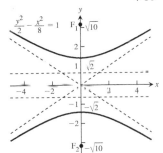

25. $y^2 - \dfrac{x^2}{8} = 1$ **27.** $x^2 - \dfrac{y^2}{8} = 1$ **29.** $r = \dfrac{2}{1+\cos\theta}$

31. $r = \dfrac{30}{1-5\sin\theta}$ **33.** $r = \dfrac{1}{2+\cos\theta}$ **35.** $r = \dfrac{10}{5-\sin\theta}$

37.

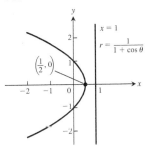

39.

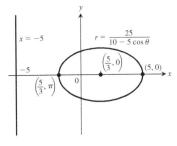

41.

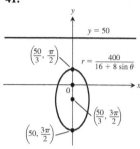

43.

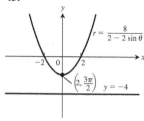

45. $y = 2 - x$

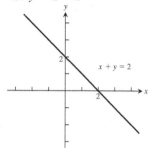

47. $y = \dfrac{\sqrt{3}}{3}x + 2\sqrt{3}$

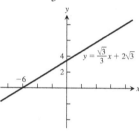

49. $r\cos\left(\theta - \dfrac{\pi}{4}\right) = 3$ **51.** $r\cos\left(\theta + \dfrac{\pi}{2}\right) = 5$

53.

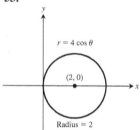

55.

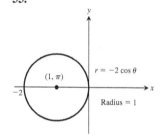

57. $r = 12\cos\theta$

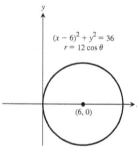

59. $r = 10\sin\theta$

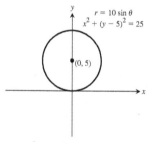

61. $r = -2\cos\theta$

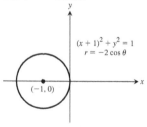

63. $r = -\sin\theta$

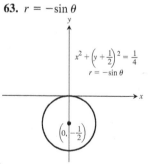

65.

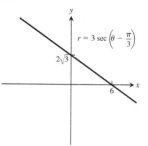

67.

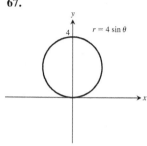

69.

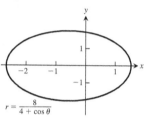

71.

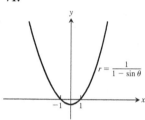

73.

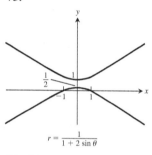

75. (b)

Planet	Perihelion	Aphelion
Mercury	0.3075 AU	0.4667 AU
Venus	0.7184 AU	0.7282 AU
Earth	0.9833 AU	1.0167 AU
Mars	1.3817 AU	1.6663 AU
Jupiter	4.9512 AU	5.4548 AU
Saturn	9.0210 AU	10.0570 AU
Uranus	18.2977 AU	20.0623 AU
Neptune	29.8135 AU	30.3065 AU

PRACTICE EXERCISES, pp. 709–711

1.

3.

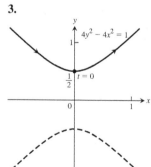

5.

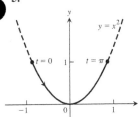

7. $x = 3 \cos t, \quad y = 4 \sin t, \quad 0 \le t \le 2\pi$

9. $y = \dfrac{\sqrt{3}}{2}x + \dfrac{1}{4}, \dfrac{1}{4}$

11. **(a)** $y = \dfrac{\pm |x|^{3/2}}{8} - 1$ **(b)** $y = \dfrac{\pm\sqrt{1-x^2}}{x}$

13. $\dfrac{10}{3}$ **15.** $\dfrac{285}{8}$ **17.** 10 **19.** $\dfrac{9\pi}{2}$ **21.** $\dfrac{76\pi}{3}$

23. $y = \dfrac{\sqrt{3}}{3}x - 4$

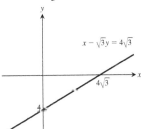

25. $x = 2$

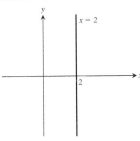

27. $y = -\dfrac{3}{2}$

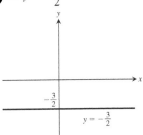

29. $x^2 + (y+2)^2 = 4$

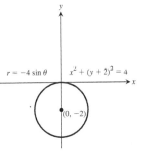

31. $\left(x - \sqrt{2}\right)^2 + y^2 = 2$

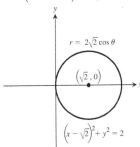

33. $r = -5 \sin \theta$

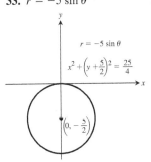

35. $r = 3 \cos \theta$

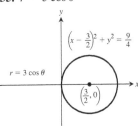

37.

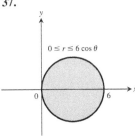

39. d   **41.** 1   **43.** k   **45.** i   **47.** $\dfrac{9}{2}\pi$ **49.** $2 + \dfrac{\pi}{4}$

51. 8   **53.** $\pi - 3$

55. Focus is $(0, -1)$, directrix is $y = 1$.

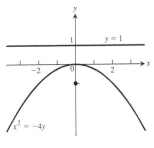

57. Focus is $\left(\dfrac{3}{4}, 0\right)$, directrix is $x = -\dfrac{3}{4}$.

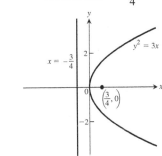

59. $e = \dfrac{3}{4}$

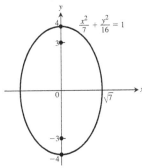

61. $c = 2$; the asymptotes are $y = \pm\sqrt{3}\,x$.

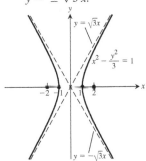

63. $(x - 2)^2 = -12(y - 3)$, $V(2, 3)$, $F(2, 0)$, directrix is $y = 6$.

65. $\dfrac{(x+3)^2}{9} + \dfrac{(y+5)^2}{25} = 1$, $C(-3, -5)$, $F(-3, -1)$ and $F(-3, -9)$, $V(-3, -10)$ and $V(-3, 0)$.

67. $\dfrac{\left(y - 2\sqrt{2}\right)^2}{8} - \dfrac{(x - 2)^2}{2} = 1$, $C\left(2, 2\sqrt{2}\right)$, $F\left(2, 2\sqrt{2} \pm \sqrt{10}\right)$, $V\left(2, 4\sqrt{2}\right)$ and $V(2, 0)$, the asymptotes are $y = 2x - 4 + 2\sqrt{2}$ and $y = -2x + 4 + 2\sqrt{2}$.

69. Hyperbola: $C(2, 0)$, $V(0, 0)$ and $V(4, 0)$, the foci are $F\left(2 \pm \sqrt{5}, 0\right)$, and the asymptotes are $y = \pm\dfrac{x - 2}{2}$.

71. Parabola: $V(-3, 1)$, $F(-7, 1)$, and the directrix is $x = 1$.

73. Ellipse: $C(-3, 2)$, $F\left(-3 \pm \sqrt{7}, 2\right)$, $V(1, 2)$ and $V(-7, 2)$

75. Circle: $C(1, 1)$ and radius $= \sqrt{2}$

77. $V(1, 0)$

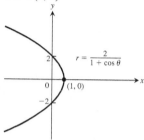

79. $V(2, \pi)$ and $V(6, \pi)$

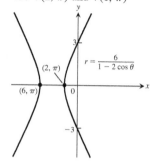

81. $r = \dfrac{4}{1 + 2\cos\theta}$ **83.** $r = \dfrac{2}{2 + \sin\theta}$

85. (a) 24π **(b)** 16π

ADDITIONAL AND ADVANCED EXERCISES, pp. 711–713

1. $x - \dfrac{7}{2} = \dfrac{y^2}{2}$

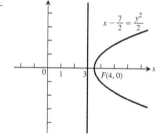

3. $3x^2 + 3y^2 - 8y + 4 = 0$ **5.** $F(0, \pm 1)$

7. (a) $\dfrac{(y - 1)^2}{16} - \dfrac{x^2}{48} = 1$ **(b)** $\dfrac{\left(y + \dfrac{3}{4}\right)^2}{\left(\dfrac{25}{16}\right)} - \dfrac{x^2}{\left(\dfrac{75}{2}\right)} = 1$

11.

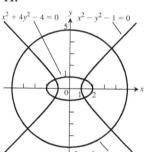

13.

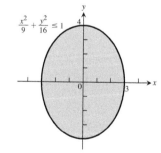

15.

17. (a) $r = e^{2\theta}$ **(b)** $\dfrac{\sqrt{5}}{2}\left(e^{4\pi} - 1\right)$

19. $r = \dfrac{4}{1 + 2\cos\theta}$ **21.** $r = \dfrac{2}{2 + \sin\theta}$

23. $x = (a + b)\cos\theta - b\cos\left(\dfrac{a + b}{b}\theta\right),$

$y = (a + b)\sin\theta - b\sin\left(\dfrac{a + b}{b}\theta\right)$

27. $\dfrac{\pi}{2}$

Chapter 12

SECTION 12.1, pp. 717–719

1. The line through the point $(2, 3, 0)$ parallel to the z-axis

3. The x-axis

5. The circle $x^2 + y^2 = 4$ in the xy-plane

7. The circle $x^2 + z^2 = 4$ in the xz-plane

9. The circle $y^2 + z^2 = 1$ in the yz-plane

11. The circle $x^2 + y^2 = 16$ in the xy-plane

13. The ellipse formed by the intersection of the cylinder $x^2 + y^2 = 4$ and the plane $z = y$

15. The parabola $y = x^2$ in the xy-plane

17. (a) The first quadrant of the xy-plane

   (b) The fourth quadrant of the xy-plane

19. (a) The ball of radius 1 centered at the origin

   (b) All points more than 1 unit from the origin

21. (a) The ball of radius 2 centered at the origin with the interior of the ball of radius 1 centered at the origin removed

   (b) The solid upper hemisphere of radius 1 centered at the origin

23. (a) The region on or inside the parabola $y = x^2$ in the xy-plane and all points above this region

   (b) The region on or to the left of the parabola $x = y^2$ in the xy-plane and all points above it that are 2 units or less away from the xy-plane

25. 3   **27.** 7   **29.** $2\sqrt{3}$ **31. (a)** 2   **(b)** 3   **(c)** 4

33. (a) 3   **(b)** 4   **(c)** 5

35. (a) $x = 3$ **(b)** $y = -1$ **(c)** $z = -2$

37. (a) $z = 1$ **(b)** $x = 3$ **(c)** $y = -1$

A-15 Chapter 12: Answers to Odd-Numbered Exercises

39. (a) $x^2 + (y - 2)^2 = 4, z = 0$
   (b) $(y - 2)^2 + z^2 = 4, x = 0$ (c) $x^2 + z^2 = 4, y = 2$
41. (a) $y = 3, z = -1$ (b) $x = 1, z = -1$ (c) $x = 1, y = 3$
43. $x^2 + y^2 + z^2 = 25, z = 3$ **45.** $0 \le z \le 1$ **47.** $z \le 0$
49. (a) $(x - 1)^2 + (y - 1)^2 + (z - 1)^2 < 1$
   (b) $(x - 1)^2 + (y - 1)^2 + (z - 1)^2 > 1$
51. $C(-2, 0, 2), a = 2\sqrt{2}$ **53.** $C(\sqrt{2}, \sqrt{2}, -\sqrt{2}), a = \sqrt{2}$
55. $C(-2, 0, 2), a = \sqrt{8}$ **57.** $C\left(-\dfrac{1}{4}, -\dfrac{1}{4}, -\dfrac{1}{4}\right), a = \dfrac{5\sqrt{3}}{4}$
59. $C(2, -3, 5), a = 7$
61. $(x - 1)^2 + (y - 2)^2 + (z - 3)^2 = 14$
63. $(x + 1)^2 + \left(y - \dfrac{1}{2}\right)^2 + \left(z + \dfrac{2}{3}\right)^2 = \dfrac{16}{81}$
65. (a) $\sqrt{y^2 + z^2}$ (b) $\sqrt{x^2 + z^2}$ (c) $\sqrt{x^2 + y^2}$
67. $\sqrt{17} + \sqrt{33} + 6$ **69.** $y = 1$
71. (a) $(0, 3, -3)$ (b) $(0, 5, -5)$
73. $z = x^2/4 + 1$ **75.** (a) $z^2 = x^2$ (b) $y^2 = x^2$

SECTION 12.2, pp. 726–728
  1. (a) $\langle 9, -6 \rangle$ (b) $3\sqrt{13}$ **3.** (a) $\langle 1, 3 \rangle$ (b) $\sqrt{10}$
  5. (a) $\langle 12, -19 \rangle$ (b) $\sqrt{505}$
  7. (a) $\left\langle \dfrac{1}{5}, \dfrac{14}{5} \right\rangle$ (b) $\dfrac{\sqrt{197}}{5}$ **9.** $\langle 1, -4 \rangle$
  11. $\langle -2, -3 \rangle$ **13.** $\left\langle -\dfrac{1}{2}, \dfrac{\sqrt{3}}{2} \right\rangle$ **15.** $\left\langle -\dfrac{\sqrt{3}}{2}, -\dfrac{1}{2} \right\rangle$
  17. $-3\mathbf{i} + 2\mathbf{j} - \mathbf{k}$ **19.** $-3\mathbf{i} + 16\mathbf{j}$
  21. $3\mathbf{i} + 5\mathbf{j} - 8\mathbf{k}$
  23. The vector v is horizontal and 1 in. long. The vectors **u** and **w**
   are $\dfrac{11}{16}$ in. long. **w** is vertical and **u** makes a 45° angle with the
   horizontal. All vectors must be drawn to scale.
   (a) 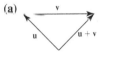   (b)
   (c) 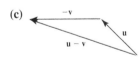   (d)
  25. $3\left(\dfrac{2}{3}\mathbf{i} + \dfrac{1}{3}\mathbf{j} - \dfrac{2}{3}\mathbf{k}\right)$ **27.** $5(\mathbf{k})$
  29. $\sqrt{\dfrac{1}{2}}\left(\dfrac{1}{\sqrt{3}}\mathbf{i} - \dfrac{1}{\sqrt{3}}\mathbf{j} - \dfrac{1}{\sqrt{3}}\mathbf{k}\right)$
  31. (a) $2\mathbf{i}$ (b) $-\sqrt{3}\mathbf{k}$ (c) $\dfrac{3}{10}\mathbf{j} + \dfrac{2}{5}\mathbf{k}$ (d) $6\mathbf{i} - 2\mathbf{j} + 3\mathbf{k}$
  33. $\dfrac{7}{13}(12\mathbf{i} - 5\mathbf{k})$

35. (a) $\dfrac{3}{5\sqrt{2}}\mathbf{i} + \dfrac{4}{5\sqrt{2}}\mathbf{j} - \dfrac{1}{\sqrt{2}}\mathbf{k}$ (b) $(1/2, 3, 5/2)$
37. (a) $-\dfrac{1}{\sqrt{3}}\mathbf{i} - \dfrac{1}{\sqrt{3}}\mathbf{j} - \dfrac{1}{\sqrt{3}}\mathbf{k}$ (b) $\left(\dfrac{5}{2}, \dfrac{7}{2}, \dfrac{9}{2}\right)$
39. $A(4, -3, 5)$ **41.** $a = \dfrac{3}{2}, b = \dfrac{1}{2}$
43. $a = -1, b = 2, c = 1$ **45.** $\approx \langle -338.095, 725.046 \rangle$
47. $|\mathbf{F}_1| = \dfrac{100 \cos 45°}{\sin 75°} \approx 73.205$ N,
   $|\mathbf{F}_2| = \dfrac{100 \cos 30°}{\sin 75°} \approx 89.658$ N,
   $\mathbf{F}_1 = \langle -|\mathbf{F}_1| \cos 30°, |\mathbf{F}_1| \sin 30° \rangle \approx \langle -63.397, 36.603 \rangle$,
   $\mathbf{F}_2 = \langle |\mathbf{F}_2| \cos 45°, |\mathbf{F}_2| \sin 45° \rangle \approx \langle 63.397, 63.397 \rangle$
49. $w = \dfrac{100 \sin 75°}{\cos 40°} \approx 126.093$ N,
   $|\mathbf{F}_1| = \dfrac{w \cos 35°}{\sin 75°} \approx 106.933$ N
51. (a) $(5 \cos 60°, 5 \sin 60°) = \left(\dfrac{5}{2}, \dfrac{5\sqrt{3}}{2}\right)$
   (b) $(5 \cos 60° + 10 \cos 315°, 5 \sin 60° + 10 \sin 315°) =$
   $\left(\dfrac{5 + 10\sqrt{2}}{2}, \dfrac{5\sqrt{3} - 10\sqrt{2}}{2}\right)$
53. (a) $\dfrac{3}{2}\mathbf{i} + \dfrac{3}{2}\mathbf{j} - 3\mathbf{k}$ (b) $\mathbf{i} + \mathbf{j} - 2\mathbf{k}$ (c) $(2, 2, 1)$
59. (a) $\langle 0, 0, 0 \rangle$ (b) $\langle 0, 0, 0 \rangle$

SECTION 12.3, pp. 734–736
  1. (a) $-25, 5, 5$ (b) -1 (c) -5 (d) $2\mathbf{i} + 4\mathbf{j} - \sqrt{5}\mathbf{k}$
  3. (a) $25, 15, 5$ (b) $\dfrac{1}{3}$ (c) $\dfrac{5}{3}$ (d) $\dfrac{1}{9}(10\mathbf{i} + 11\mathbf{j} - 2\mathbf{k})$
  5. (a) $2, \sqrt{34}, \sqrt{3}$ (b) $\dfrac{2}{\sqrt{3}\sqrt{34}}$ (c) $\dfrac{2}{\sqrt{34}}$
   (d) $\dfrac{1}{17}(5\mathbf{j} - 3\mathbf{k})$
  7. (a) $10 + \sqrt{17}, \sqrt{26}, \sqrt{21}$ (b) $\dfrac{10 + \sqrt{17}}{\sqrt{546}}$ (c) $\dfrac{10 + \sqrt{17}}{\sqrt{26}}$
   (d) $\dfrac{10 + \sqrt{17}}{26}(5\mathbf{i} + \mathbf{j})$
  9. 0.75 rad   **11.** 1.77 rad
  13. Angle at $A = \cos^{-1}\left(\dfrac{1}{\sqrt{5}}\right) \approx 63.435$ degrees, angle at
   $B = \cos^{-1}\left(\dfrac{3}{5}\right) \approx 53.130$ degrees, angle at
   $C = \cos^{-1}\left(\dfrac{1}{\sqrt{5}}\right) \approx 63.435$ degrees.
  17. $\cos^{-1}\left(\dfrac{3}{\sqrt{10}}\right) \approx 0.322$ radian or 18.43 degrees

25. Horizontal component: ≈ 1188 ft/sec, vertical component: ≈ 167 ft/sec

27. (a) Since $|\cos \theta| \le 1$, we have $|\mathbf{u} \cdot \mathbf{v}| = |\mathbf{u}| |\mathbf{v}| |\cos \theta| \le |\mathbf{u}| |\mathbf{v}| (1) = |\mathbf{u}| |\mathbf{v}|$.

(b) We have equality precisely when $|\cos \theta| = 1$ or when one or both of $\mathbf{u}$ and v are $\mathbf{0}$. In the case of nonzero vectors, we have equality when $\theta = 0$ or π, that is, when the vectors are parallel.

29. a

35. $x + 2y = 4$

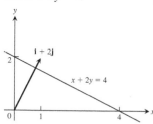

37. $-2x + y = -3$

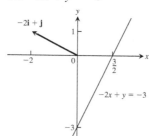

39. $x + y = -1$

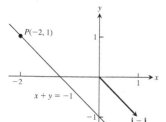

41. $2x - y = 0$

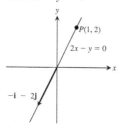

43. 5 J   **45.** 3464 J   **47.** $\dfrac{\pi}{4}$ **49.** $\dfrac{\pi}{6}$ **51.** 0.14

SECTION 12.4, pp. 741–742

1. $|\mathbf{u} \times \mathbf{v}| = 3$, direction is $\dfrac{2}{3}\mathbf{i} + \dfrac{1}{3}\mathbf{j} + \dfrac{2}{3}\mathbf{k}$; $|\mathbf{v} \times \mathbf{u}| = 3$, direction is $-\dfrac{2}{3}\mathbf{i} - \dfrac{1}{3}\mathbf{j} - \dfrac{2}{3}\mathbf{k}$

3. $|\mathbf{u} \times \mathbf{v}| = 0$, no direction; $|\mathbf{v} \times \mathbf{u}| = 0$, no direction

5. $|\mathbf{u} \times \mathbf{v}| = 6$, direction is $-\mathbf{k}$; $|\mathbf{v} \times \mathbf{u}| = 6$, direction is $\mathbf{k}$

7. $|\mathbf{u} \times \mathbf{v}| = 6\sqrt{5}$, direction is $\dfrac{1}{\sqrt{5}}\mathbf{i} - \dfrac{2}{\sqrt{5}}\mathbf{k}$; $|\mathbf{v} \times \mathbf{u}| = 6\sqrt{5}$, direction is $-\dfrac{1}{\sqrt{5}}\mathbf{i} + \dfrac{2}{\sqrt{5}}\mathbf{k}$

9.

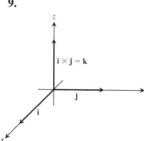

11.

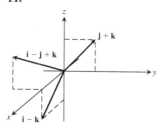

13.

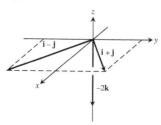

15. (a) $2\sqrt{6}$ **(b)** $\pm \dfrac{1}{\sqrt{6}} (2\mathbf{i} + \mathbf{j} + \mathbf{k})$

17. (a) $\dfrac{\sqrt{2}}{2}$ **(b)** $\pm \dfrac{1}{\sqrt{2}} (\mathbf{i} - \mathbf{j})$

19. 8   **21.** 7   **23. (a)** None   **(b)** $\mathbf{u}$ and $\mathbf{w}$

25. $10\sqrt{3}$ ft-lb

27. (a) True   **(b)** Not always true   **(c)** True   **(d)** True
(e) Not always true   **(f)** True   **(g)** True   **(h)** True

29. (a) $\text{proj}_{\mathbf{v}} \mathbf{u} = \dfrac{\mathbf{u} \cdot \mathbf{v}}{\mathbf{v} \cdot \mathbf{v}} \mathbf{v}$ **(b)** $\pm \mathbf{u} \times \mathbf{v}$ **(c)** $\pm (\mathbf{u} \times \mathbf{v}) \times \mathbf{w}$
(d) $|(\mathbf{u} \times \mathbf{v}) \cdot \mathbf{w}|$ **(e)** $(\mathbf{u} \times \mathbf{v}) \times (\mathbf{u} \times \mathbf{w})$ **(f)** $|\mathbf{u}| \dfrac{\mathbf{v}}{|\mathbf{v}|}$

31. (a) Yes   **(b)** No   **(c)** Yes   **(d)** No

33. No, v need not equal $\mathbf{w}$. For example, $\mathbf{i} + \mathbf{j} \ne -\mathbf{i} + \mathbf{j}$, but $\mathbf{i} \times (\mathbf{i} + \mathbf{j}) = \mathbf{i} \times \mathbf{i} + \mathbf{i} \times \mathbf{j} = \mathbf{0} + \mathbf{k} = \mathbf{k}$ and $\mathbf{i} \times (-\mathbf{i} + \mathbf{j}) = -\mathbf{i} \times \mathbf{i} + \mathbf{i} \times \mathbf{j} = \mathbf{0} + \mathbf{k} = \mathbf{k}$.

35. 2   **37.** 13   **39.** $\sqrt{129}$ **41.** $\dfrac{11}{2}$ **43.** $\dfrac{25}{2}$

45. $\dfrac{3}{2}$ **47.** $\dfrac{\sqrt{21}}{2}$

49. If $\mathbf{A} = a_1\mathbf{i} + a_2\mathbf{j}$ and $\mathbf{B} = b_1\mathbf{i} + b_2\mathbf{j}$, then

$$\mathbf{A} \times \mathbf{B} = \begin{vmatrix} \mathbf{i} & \mathbf{j} & \mathbf{k} \\ a_1 & a_2 & 0 \\ b_1 & b_2 & 0 \end{vmatrix} = \begin{vmatrix} a_1 & a_2 \\ b_1 & b_2 \end{vmatrix} \mathbf{k}$$

and the triangle's area is

$$\frac{1}{2} |\mathbf{A} \times \mathbf{B}| = \pm \frac{1}{2} \begin{vmatrix} a_1 & a_2 \\ b_1 & b_2 \end{vmatrix}.$$

The applicable sign is $(+)$ if the acute angle from $\mathbf{A}$ to $\mathbf{B}$ runs counterclockwise in the xy-plane, and $(-)$ if it runs clockwise.

51. 4   **53.** 44/3   **55.** Coplanar   **57.** Not coplanar

SECTION 12.5, pp. 749–751

1. $x = 3 + t$, $y = -4 + t$, $z = -1 + t$

3. $x = -2 + 5t$, $y = 5t$, $z = 3 - 5t$

5. $x = 0$, $y = 2t$, $z = t$

7. $x = 1$, $y = 1$, $z = 1 + t$

9. $x = t$, $y = -7 + 2t$, $z = 2t$

11. $x = t$, $y = 0$, $z = 0$

13. $x = t,\quad y = t,\quad z = \dfrac{3}{2}t,$
$0 \le t \le 1$

15. $x = 1,\quad y = 1 + t,$
$z = 0,\quad -1 \le t \le 0$

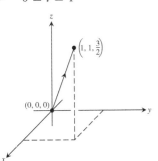

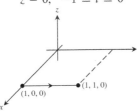

17. $x = 0,\quad y = 1 - 2t,$
$z = 1,\quad 0 \le t \le 1$

19. $x = 2 - 2t,\quad y = 2t,$
$z = 2 - 2t,\quad 0 \le t \le 1$

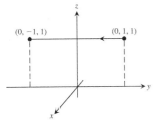

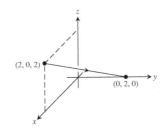

21. $3x - 2y - z = -3$ **23.** $7x - 5y - 4z = 6$
25. $x + 3y + 4z = 34$ **27.** $(1, 2, 3), -20x + 12y + z = 7$
29. $y + z = 3$ **31.** $x - y + z = 0$ **33.** $2\sqrt{30}$ **35.** 0
37. $\dfrac{9\sqrt{42}}{7}$ **39.** 3   **41.** 19/5   **43.** 5/3   **45.** $9/\sqrt{41}$
47. $\pi/4$ **49.** $\arccos(-1/6) \approx 1.738$ radians
51. $\arcsin(2/\sqrt{154}) \approx 0.161$ radians   **53.** 1.38 rad
55. 0.82 rad   **57.** $\left(\dfrac{3}{2}, -\dfrac{3}{2}, \dfrac{1}{2}\right)$ **59.** $(1, 1, 0)$
61. $x = 1 - t,\quad y = 1 + t,\quad z = -1$
63. $x = 4,\quad y = 3 + 6t,\quad z = 1 + 3t$
65. $L1$ intersects $L2$; $L2$ is parallel to $L3$, $\sqrt{5}/3$; $L1$ and $L3$ are skew, $10\sqrt{2}/3$
67. $x = 2 + 2t,\quad y = -4 - t,\quad z = 7 + 3t;\quad x = -2 - t,$
$y = -2 + (1/2)t,\quad z = 1 - (3/2)t$
69. $\left(0, -\dfrac{1}{2}, -\dfrac{3}{2}\right), (-1, 0, -3), (1, -1, 0)$
73. Many possible answers. One possibility: $x + y = 3$ and $2y + z = 7$.
75. $(x/a) + (y/b) + (z/c) = 1$ describes all planes *except* those through the origin or parallel to a coordinate axis.

SECTION 12.6, pp. 755–757
1. (d), ellipsoid   **3.** (a), cylinder   **5.** (l), hyperbolic paraboloid
7. (b), cylinder   **9.** (k), hyperbolic paraboloid   **11.** (h), cone

13.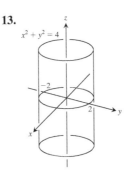
15.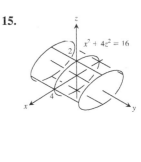
17.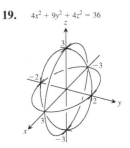
19.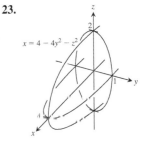
21.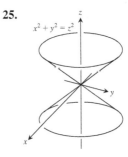
23.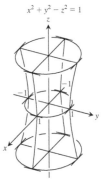
25.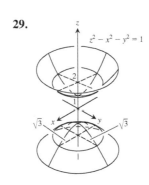
27.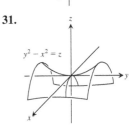
29.
31.

33. $z^2 = 1 + y^2 - x^2$
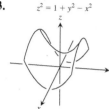

35. $y = -(x^2 + z^2)$
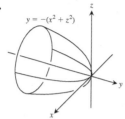

37. $x^2 + y^2 - z^2 = 4$
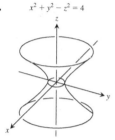

39. $x^2 + z^2 = 1$
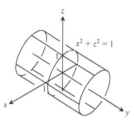

41. $z = -(x^2 + y^2)$
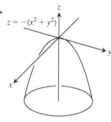

43. $4y^2 + z^2 - 4x^2 = 4$
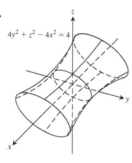

45. (a) $\dfrac{2\pi(9 - c^2)}{9}$ **(b)** 8π **(c)** $\dfrac{4\pi abc}{3}$

PRACTICE EXERCISES, pp. 757–759

1. (a) $\langle -17, 32 \rangle$ **(b)** $\sqrt{1313}$

3. (a) $\langle 6, -8 \rangle$ **(b)** 10

5. $\left\langle -\dfrac{\sqrt{3}}{2}, -\dfrac{1}{2} \right\rangle$ [assuming counterclockwise]

7. $\left\langle \dfrac{8}{\sqrt{17}}, -\dfrac{2}{\sqrt{17}} \right\rangle$

9. Length = 2, direction is $\dfrac{1}{\sqrt{2}}\mathbf{i} + \dfrac{1}{\sqrt{2}}\mathbf{j}$.

11. $\mathbf{v}\,(\pi/2) = 2(-\mathbf{i})$

13. Length = 7, direction is $\dfrac{2}{7}\mathbf{i} - \dfrac{3}{7}\mathbf{j} + \dfrac{6}{7}\mathbf{k}$.

15. $\dfrac{8}{\sqrt{33}}\mathbf{i} - \dfrac{2}{\sqrt{33}}\mathbf{j} + \dfrac{8}{\sqrt{33}}\mathbf{k}$

17. $|\mathbf{v}| = \sqrt{2},\ |\mathbf{u}| = 3,\ \mathbf{v}\cdot\mathbf{u} = \mathbf{u}\cdot\mathbf{v} = 3,\ \mathbf{v}\times\mathbf{u} = -2\mathbf{i} + 2\mathbf{j} - \mathbf{k},$
$\mathbf{u}\times\mathbf{v} = 2\mathbf{i} - 2\mathbf{j} + \mathbf{k},\ |\mathbf{v}\times\mathbf{u}| = 3,\ \theta = \cos^{-1}\left(\dfrac{1}{\sqrt{2}}\right) = \dfrac{\pi}{4},$
$|\mathbf{u}|\cos\theta = \dfrac{3}{\sqrt{2}},\ \text{proj}_{\mathbf{v}}\,\mathbf{u} = \dfrac{3}{2}\,(\mathbf{i} + \mathbf{j})$

19. $\dfrac{4}{3}\,(2\mathbf{i} + \mathbf{j} - \mathbf{k})$

21. $\mathbf{u}\times\mathbf{v} = \mathbf{k}$
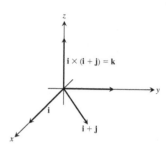

23. $2\sqrt{7}$ **25. (a)** $\sqrt{14}$ **(b)** 1   **29.** $\sqrt{78}/3$

31. $x = 1 - 3t,\ y = 2,\ z = 3 + 7t$ **33.** $\sqrt{2}$

35. $2x + y + z = 5$ **37.** $-9x + y + 7z = 4$

39. $\left(0, -\dfrac{1}{2}, -\dfrac{3}{2}\right), (-1, 0, -3), (1, -1, 0)$ **41.** $\pi/3$

43. $x = -5 + 5t,\ y = 3 - t,\ z = -3t$

45. (b) $x = -12t,\ y = 19/12 + 15t,\ z = 1/6 + 6t$

47. Yes; v is parallel to the plane.

49. 3   **51.** $-3\mathbf{j} + 3\mathbf{k}$

53. $\dfrac{2}{\sqrt{35}}\,(5\mathbf{i} - \mathbf{j} - 3\mathbf{k})$ **55.** $\left(\dfrac{11}{9}, \dfrac{26}{9}, -\dfrac{7}{9}\right)$

57. $(1, -2, -1);\ x = 1 - 5t,\ y = -2 + 3t,\ z = -1 + 4t$

59. $2x + 7y + 2z + 10 = 0$

61. (a) No **(b)** No **(c)** No **(d)** No **(e)** Yes

63. $11/\sqrt{107}$

65. $x^2 + y^2 + z^2 = 4$
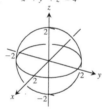

67. $4x^2 + 4y^2 + z^2 = 4$

69. $z = -(x^2 + y^2)$

71. $x^2 + y^2 = z^2$
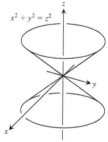

73. $x^2 + y^2 - z^2 = 4$
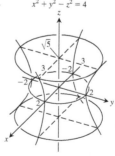

75. $y^2 - x^2 - z^2 = 1$
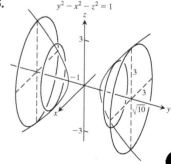

ADDITIONAL AND ADVANCED EXERCISES, pp. 759–762

1. $(26, 23, -1/3)$ **3.** $|\mathbf{F}| = 20$ lb

5. (a) $|\mathbf{F}_1| = 80$ lb, $|\mathbf{F}_2| = 60$ lb, $\mathbf{F}_1 = \langle -48, 64 \rangle$,

$\mathbf{F}_2 = \langle 48, 36 \rangle$, $\alpha = \tan^{-1}\dfrac{4}{3}$, $\beta = \tan^{-1}\dfrac{3}{4}$

(b) $|\mathbf{F}_1| = \dfrac{2400}{13} \approx 184.615$ lb, $|\mathbf{F}_2| = \dfrac{1000}{13} \approx 76.923$ lb,

$\mathbf{F}_1 = \left\langle \dfrac{-12{,}000}{169}, \dfrac{28{,}800}{169} \right\rangle \approx \langle -71.006, 170.414 \rangle$,

$\mathbf{F}_2 = \left\langle \dfrac{12{,}000}{169}, \dfrac{5000}{169} \right\rangle \approx \langle 71.006, 29.586 \rangle$,

$\alpha = \tan^{-1}\dfrac{12}{5}$, $\beta = \tan^{-1}\dfrac{5}{12}$

9. (a) $\theta = \tan^{-1}\sqrt{2} \approx 54.74°$ (b) $\theta = \tan^{-1} 2\sqrt{2} \approx 70.53°$

13. (a) $\dfrac{6}{\sqrt{14}}$ (b) $2x - y + 2z = 8$

(c) $x - 2y + z = 3 + 5\sqrt{6}$ and $x - 2y + z = 3 - 5\sqrt{6}$

15. $\dfrac{32}{41}\mathbf{i} + \dfrac{23}{41}\mathbf{j} - \dfrac{13}{41}\mathbf{k}$

17. (a) $0, 0$ (b) $-10\mathbf{i} - 2\mathbf{j} + 6\mathbf{k}, -9\mathbf{i} - 2\mathbf{j} + 7\mathbf{k}$

(c) $-4\mathbf{i} - 6\mathbf{j} + 2\mathbf{k}, \mathbf{i} - 2\mathbf{j} - 4\mathbf{k}$

(d) $-10\mathbf{i} - 10\mathbf{k}, -12\mathbf{i} - 4\mathbf{j} - 8\mathbf{k}$

19. The formula is always true.

Chapter 13

SECTION 13.1, pp. 770–772

1. $\mathbf{i} - \dfrac{1}{2}\mathbf{j} + \mathbf{k}$ **3.** $2\mathbf{i} + \dfrac{1}{2}\mathbf{j} + \dfrac{\pi}{4}\mathbf{k}$

5. $y = x^2 - 2x$, $\mathbf{v} = \mathbf{i} + 2\mathbf{j}$, $\mathbf{a} = 2\mathbf{j}$

7. $y = \dfrac{2}{9}x^2$, $\mathbf{v} = 3\mathbf{i} + 4\mathbf{j}$, $\mathbf{a} = 3\mathbf{i} + 8\mathbf{j}$

9. $t = \dfrac{\pi}{4}: \mathbf{v} = \dfrac{\sqrt{2}}{2}\mathbf{i} - \dfrac{\sqrt{2}}{2}\mathbf{j}$, $\mathbf{a} = \dfrac{-\sqrt{2}}{2}\mathbf{i} - \dfrac{\sqrt{2}}{2}\mathbf{j}$;

$t = \pi/2: \mathbf{v} = -\mathbf{j}$, $\mathbf{a} = -\mathbf{i}$

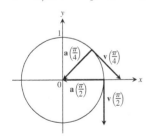

11. $t = \pi: \mathbf{v} = 2\mathbf{i}$, $\mathbf{a} = -\mathbf{j}$; $t = \dfrac{3\pi}{2}: \mathbf{v} = \mathbf{i} - \mathbf{j}$, $\mathbf{a} = -\mathbf{i}$

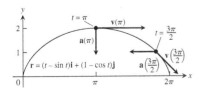

13. $\mathbf{v} = \mathbf{i} + 2t\mathbf{j} + 2\mathbf{k}; \mathbf{a} = 2\mathbf{j}$; speed: 3; direction: $\dfrac{1}{3}\mathbf{i} + \dfrac{2}{3}\mathbf{j} + \dfrac{2}{3}\mathbf{k}$;

$\mathbf{v}(1) = 3\left(\dfrac{1}{3}\mathbf{i} + \dfrac{2}{3}\mathbf{j} + \dfrac{2}{3}\mathbf{k}\right)$

15. $\mathbf{v} = (-2 \sin t)\mathbf{i} + (3 \cos t)\mathbf{j} + 4\mathbf{k}$;

$\mathbf{a} = (-2 \cos t)\mathbf{i} - (3 \sin t)\mathbf{j}$; speed: $2\sqrt{5}$;

direction: $\left(-1/\sqrt{5}\right)\mathbf{i} + \left(2/\sqrt{5}\right)\mathbf{k}$;

$\mathbf{v}(\pi/2) = 2\sqrt{5}\left[\left(-1/\sqrt{5}\right)\mathbf{i} + \left(2/\sqrt{5}\right)\mathbf{k}\right]$

17. $\mathbf{v} = \left(\dfrac{2}{t+1}\right)\mathbf{i} + 2t\mathbf{j} + t\mathbf{k}; \mathbf{a} = \left(\dfrac{-2}{(t+1)^2}\right)\mathbf{i} + 2\mathbf{j} + \mathbf{k}$;

speed: $\sqrt{6}$; direction: $\dfrac{1}{\sqrt{6}}\mathbf{i} + \dfrac{2}{\sqrt{6}}\mathbf{j} + \dfrac{1}{\sqrt{6}}\mathbf{k}$;

$\mathbf{v}(1) = \sqrt{6}\left(\dfrac{1}{\sqrt{6}}\mathbf{i} + \dfrac{2}{\sqrt{6}}\mathbf{j} + \dfrac{1}{\sqrt{6}}\mathbf{k}\right)$

19. $\pi/2$ **21.** $\pi/2$ **23.** $x = t$, $y = -1$, $z = 1 + t$

25. $x = t$, $y = \dfrac{1}{3}t$, $z = t$ **27.** $4, -2$ **29.** $2, -2$

31. E   **33.** D   **35.** C

37. (a) (i): It has constant speed 1.   (ii): Yes

(iii): Counterclockwise   (iv): Yes

(b) (i): It has constant speed 2.   (ii): Yes

(iii): Counterclockwise   (iv): Yes

(c) (i): It has constant speed 1.   (ii): Yes

(iii): Counterclockwise

(iv): It starts at $(0, -1)$ instead of $(1, 0)$.

(d) (i): It has constant speed 1.   (ii): Yes

(iii): Clockwise   (iv): Yes

(i): It has variable speed.   (ii): No

(iii): Counterclockwise   (iv): Yes

39. $\mathbf{v} = 2\sqrt{5}\mathbf{i} + \sqrt{5}\mathbf{j}$

SECTION 13.2, pp. 777–781

1. $(1/4)\mathbf{i} + 7\mathbf{j} + (3/2)\mathbf{k}$ **3.** $\left(\dfrac{\pi + 2\sqrt{2}}{2}\right)\mathbf{j} + 2\mathbf{k}$

5. $(\ln 4)\mathbf{i} + (\ln 4)\mathbf{j} + (\ln 2)\mathbf{k}$

7. $\dfrac{e-1}{2}\mathbf{i} + \dfrac{e-1}{e}\mathbf{j} + \mathbf{k}$ **9.** $\mathbf{i} - \mathbf{j} + \dfrac{\pi}{4}\mathbf{k}$

11. $\mathbf{r}(t) = \left(\dfrac{-t^2}{2} + 1\right)\mathbf{i} + \left(\dfrac{-t^2}{2} + 2\right)\mathbf{j} + \left(\dfrac{-t^2}{2} + 3\right)\mathbf{k}$

13. $\mathbf{r}(t) = ((t+1)^{3/2} - 1)\mathbf{i} + (-e^{-t} + 1)\mathbf{j} + (\ln(t+1) + 1)\mathbf{k}$

15. $\mathbf{r}(t) = (3 + \ln|\sec t|)\mathbf{i} + (-2 + 2\sin(t/2))\mathbf{j}$

$+ (1 - (1/2) \ln|\sec 2t| + \tan 2t|)\mathbf{k}$

17. $\mathbf{r}(t) = 8t\mathbf{i} + 8t\mathbf{j} + (-16t^2 + 100)\mathbf{k}$

19. $\mathbf{r}(t) = (e^t - 2t + 2)\mathbf{i} + (-e^{-t} + 3t + 2)\mathbf{j} + (e^{2t} - 2t + 1)\mathbf{k}$

21. $\mathbf{r}(t) = \left(\dfrac{3}{2}t^2 + \dfrac{6}{\sqrt{11}}t + 1\right)\mathbf{i} - \left(\dfrac{1}{2}t^2 + \dfrac{2}{\sqrt{11}}t - 2\right)\mathbf{j}$

$+ \left(\dfrac{1}{2}t^2 + \dfrac{2}{\sqrt{11}}t + 3\right)\mathbf{k} = \left(\dfrac{1}{2}t^2 + \dfrac{2t}{\sqrt{11}}\right)(3\mathbf{i} - \mathbf{j} + \mathbf{k})$

$+ (\mathbf{i} + 2\mathbf{j} + 3\mathbf{k})$

23. 50 sec

25. (a) 72.2 sec; 25,510 m   (b) 4020 m   (c) 6378 m

27. (a) $v_0 \approx 9.9$ m/sec   (b) $\alpha \approx 18.4°$ or $71.6°$

29. $39.3°$ or $50.7°$ **35.** (b) $\mathbf{v}_0$ would bisect $\angle AOR$.

37. (a) (Assuming that "x" is zero at the point of impact)
$\mathbf{r}(t) = (x(t))\mathbf{i} + (y(t))\mathbf{j}$, where $x(t) = (35 \cos 27°)t$ and
$y(t) = 4 + (35 \sin 27°)t - 16t^2$.

(b) At $t \approx 0.497$ sec, it reaches its maximum height of about
7.945 ft.

(c) Range $\approx$ 37.45 ft; flight time $\approx$ 1.201 sec

(d) At $t \approx 0.254$ and $t \approx 0.740$ sec, when it is ≈ 29.532 and
≈ 14.376 ft from where it will land

(e) Yes. It changes things because the ball won't clear the net.

39. 4.00 ft, 7.80 ft / sec

47. (a) $\mathbf{r}(t) = (x(t))\mathbf{i} + (y(t))\mathbf{j}$; where

$$x(t) = \left(\frac{1}{0.08}\right)(1 - e^{-0.08t})(152 \cos 20° - 17.6) \text{ and}$$

$$y(t) = 3 + \left(\frac{152}{0.08}\right)(1 - e^{-0.08t})(\sin 20°)$$

$$+\left(\frac{32}{0.08^2}\right)(1 - 0.08t - e^{-0.08t})$$

(b) At $t \approx 1.527$ sec it reaches a maximum height of about
41.893 feet.

(c) Range $\approx$ 351.734 ft; flight time $\approx$ 3.181 sec

(d) At $t \approx 0.877$ and 2.190 sec, when it is about 106.028 and
251.530 ft from home plate

(e) No

SECTION 13.3, pp. 784–785

1. $\mathbf{T} = \left(-\frac{2}{3}\sin t\right)\mathbf{i} + \left(\frac{2}{3}\cos t\right)\mathbf{j} + \frac{\sqrt{5}}{3}\mathbf{k}, \ 3\pi$

3. $\mathbf{T} = \frac{1}{\sqrt{1+t}}\mathbf{i} + \frac{\sqrt{t}}{\sqrt{1+t}}\mathbf{k}, \ \frac{52}{3}$

5. $\mathbf{T} = -\cos t\,\mathbf{j} + \sin t\,\mathbf{k}, \ \frac{3}{2}$

7. $\mathbf{T} = \left(\frac{\cos t - t\sin t}{t+1}\right)\mathbf{i} + \left(\frac{\sin t + t\cos t}{t+1}\right)\mathbf{j}$

$+\left(\frac{\sqrt{2}t^{1/2}}{t+1}\right)\mathbf{k}, \ \frac{\pi^2}{2} + \pi$

9. $(0, 5, 24\pi)$

11. $s(t) = 5t, \quad L = \frac{5\pi}{2}$

13. $s(t) = \sqrt{3}e^t - \sqrt{3}, \quad L = \frac{3\sqrt{3}}{4}$

15. $\sqrt{2} + \ln(1 + \sqrt{2})$

17. (a) Cylinder is $x^2 + y^2 = 1$; plane is $x + z = 1$.

(b) and **(c)**

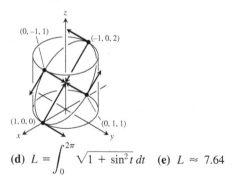

(d) $L = \displaystyle\int_0^{2\pi} \sqrt{1 + \sin^2 t}\, dt$ **(e)** $L \approx 7.64$

SECTION 13.4, pp. 790–791

1. $\mathbf{T} = (\cos t)\mathbf{i} - (\sin t)\mathbf{j}, \quad \mathbf{N} = (-\sin t)\mathbf{i} - (\cos t)\mathbf{j}, \quad \kappa = \cos t$

3. (a) $\mathbf{T} = \dfrac{1}{\sqrt{1+t^2}}\mathbf{i} - \dfrac{t}{\sqrt{1+t^2}}\mathbf{j}, \ \mathbf{N} = \dfrac{-t}{\sqrt{1+t^2}}\mathbf{i} - $

$\dfrac{1}{\sqrt{1+t^2}}\mathbf{j}, \ \kappa = \dfrac{1}{2(\sqrt{1+t^2})^3}$

5. (b) $\cos x$

7. (b) $\mathbf{N} = \dfrac{-2e^{2t}}{\sqrt{1+4e^{4t}}}\mathbf{i} + \dfrac{1}{\sqrt{1+4e^{4t}}}\mathbf{j}$

(c) $\mathbf{N} = -\dfrac{1}{2}\left(\sqrt{4 - t^2}\,\mathbf{i} + t\mathbf{j}\right)$

9. $\mathbf{T} = \dfrac{3\cos t}{5}\mathbf{i} - \dfrac{3\sin t}{5}\mathbf{j} + \dfrac{4}{5}\mathbf{k}$,

$\mathbf{N} = (-\sin t)\mathbf{i} - (\cos t)\mathbf{j}, \ \kappa = \dfrac{3}{25}$

11. $\mathbf{T} = \left(\dfrac{\cos t - \sin t}{\sqrt{2}}\right)\mathbf{i} + \left(\dfrac{\cos t + \sin t}{\sqrt{2}}\right)\mathbf{j}$,

$\mathbf{N} = \left(\dfrac{-\cos t - \sin t}{\sqrt{2}}\right)\mathbf{i} + \left(\dfrac{-\sin t + \cos t}{\sqrt{2}}\right)\mathbf{j}, \ \kappa = \dfrac{1}{e^t\sqrt{2}}$

13. $\mathbf{T} = \dfrac{t}{\sqrt{t^2+1}}\mathbf{i} + \dfrac{1}{\sqrt{t^2+1}}\mathbf{j}$,

$\mathbf{N} = \dfrac{\mathbf{i}}{\sqrt{t^2+1}} - \dfrac{t\mathbf{j}}{\sqrt{t^2+1}}, \ \kappa = \dfrac{1}{t(t^2+1)^{3/2}}$

15. $\mathbf{T} = \left(\text{sech}\,\dfrac{t}{a}\right)\mathbf{i} + \left(\tanh\dfrac{t}{a}\right)\mathbf{j}$,

$\mathbf{N} = \left(-\tanh\dfrac{t}{a}\right)\mathbf{i} + \left(\text{sech}\,\dfrac{t}{a}\right)\mathbf{j}$,

$\kappa = \dfrac{1}{a}\text{sech}^2\dfrac{t}{a}$

19. $1/(2b)$

21. $\left(x - \dfrac{\pi}{2}\right)^2 + y^2 = 1$

23. $\kappa(x) = 2/(1 + 4x^2)^{3/2}$

25. $\kappa(x) = |\sin x|/(1 + \cos^2 x)^{3/2}$

27. maximum curvature $2/(3\sqrt{3})$ at $x = 1/\sqrt{2}$

SECTION 13.5, p. 797

1. $\mathbf{a} = |a|\mathbf{N}$ **3.** $\mathbf{a}(1) = \dfrac{4}{3}\mathbf{T} + \dfrac{2\sqrt{5}}{3}\mathbf{N}$ **5.** $\mathbf{a}(0) = 2\mathbf{N}$

7. $\mathbf{r}\left(\dfrac{\pi}{4}\right) = \dfrac{\sqrt{2}}{2}\mathbf{i} + \dfrac{\sqrt{2}}{2}\mathbf{j} - \mathbf{k}, \ \mathbf{T}\left(\dfrac{\pi}{4}\right) = -\dfrac{\sqrt{2}}{2}\mathbf{i} + \dfrac{\sqrt{2}}{2}\mathbf{j}$,

$\mathbf{N}\left(\dfrac{\pi}{4}\right) = -\dfrac{\sqrt{2}}{2}\mathbf{i} - \dfrac{\sqrt{2}}{2}\mathbf{j}, \ \mathbf{B}\left(\dfrac{\pi}{4}\right) = \mathbf{k}$; osculating plane:

$z = -1$; normal plane: $-x + y = 0$; rectifying plane:
$x + y = \sqrt{2}$

9. $\mathbf{B} = \left(\dfrac{4}{5}\cos t\right)\mathbf{i} - \left(\dfrac{4}{5}\sin t\right)\mathbf{j} - \dfrac{3}{5}\mathbf{k}, \ \tau = -\dfrac{4}{25}$

11. $\mathbf{B} = \mathbf{k}, \tau = 0$ **13.** $\mathbf{B} = -\mathbf{k}, \tau = 0$ **15.** $\mathbf{B} = \mathbf{k}, \tau = 0$

17. Yes. If the car is moving on a curved path ($\kappa \neq 0$), then
$a_N = \kappa|\mathbf{v}|^2 \neq 0$ and $\mathbf{a} \neq \mathbf{0}$.

23. $\kappa = \dfrac{1}{t}, \rho = t$

27. Components of **v**: $-1.8701, 0.7089, 1.0000$
Components of **a**: $-1.6960, -2.0307, 0$
Speed: 2.2361; Components of **T**: $-0.8364, 0.3170, 0.4472$
Components of **N**: $-0.4143, -0.8998, -0.1369$
Components of **B**: $0.3590, -0.2998, 0.8839$; Curvature: 0.5060
Torsion: 0.2813; Tangential component of acceleration: 0.7746
Normal component of acceleration: 2.5298

29. Components of **v**: $2.0000, 0, -0.1629$
Components of **a**: $0, -1.0000, -0.0086$; Speed: 2.0066
Components of **T**: $0.9967, 0, -0.0812$
Components of **N**: $-0.0007, -1.0000, -0.0086$
Components of **B**: $-0.0812, 0.0086, 0.9967$;
Curvature: 0.2484
Torsion: 0.0411; Tangential component of acceleration: 0.0007
Normal component of acceleration: 1.0000

SECTION 13.6, p. 801

1. $\mathbf{v} = 2\mathbf{u}_r + 2\theta\mathbf{u}_\theta$
$\mathbf{a} = -4\theta\mathbf{u}_r + 8\mathbf{u}_\theta$
3. $\mathbf{v} = (3a\sin\theta)\mathbf{u}_r + 3a(1 - \cos\theta)\mathbf{u}_\theta$
$\mathbf{a} = 9a(2\cos\theta - 1)\mathbf{u}_r + (18a\sin\theta)\mathbf{u}_\theta$
5. $\mathbf{v} = 2ae^{a\theta}\mathbf{u}_r + 2e^{a\theta}\mathbf{u}_\theta$
$\mathbf{a} = 4e^{a\theta}(a^2 - 1)\mathbf{u}_r + 8ae^{a\theta}\mathbf{u}_\theta$
7. $\mathbf{v} = (-8\sin 4t)\mathbf{u}_r + (4\cos 4t)\mathbf{u}_\theta$
$\mathbf{a} = (-40\cos 4t)\mathbf{u}_r - (32\sin 4t)\mathbf{u}_\theta$
13. $\approx 29.93 \times 10^{10}$ m   **15.** $\sim 2.25 \times 10^9$ km²/sec
17. $\approx 1.876 \times 10^{27}$ kg

PRACTICE EXERCISES, pp. 802–803

1. $\dfrac{x^2}{16} + \dfrac{y^2}{2} = 1$

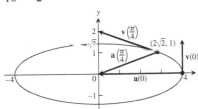

At $t = 0$: $a_T = 0$, $a_N = 4$, $\kappa = 2$;
At $t = \dfrac{\pi}{4}$: $a_T = \dfrac{7}{3}$, $a_N = \dfrac{4\sqrt{2}}{3}$, $\kappa = \dfrac{4\sqrt{2}}{27}$

3. $|\mathbf{v}|_{\max} = 1$ **5.** $\kappa = 1/5$ **7.** $dy/dt = -x$; clockwise
11. Shot put is on the ground, about 66 ft 3 in. from the stopboard.

15. Length $= \dfrac{\pi}{4}\sqrt{1 + \dfrac{\pi^2}{16}} + \ln\left(\dfrac{\pi}{4} + \sqrt{1 + \dfrac{\pi^2}{16}}\right)$

17. $\mathbf{T}(0) = \dfrac{2}{3}\mathbf{i} - \dfrac{2}{3}\mathbf{j} + \dfrac{1}{3}\mathbf{k}$; $\mathbf{N}(0) = \dfrac{1}{\sqrt{2}}\mathbf{i} + \dfrac{1}{\sqrt{2}}\mathbf{j}$;
$\mathbf{B}(0) = -\dfrac{1}{3\sqrt{2}}\mathbf{i} + \dfrac{1}{3\sqrt{2}}\mathbf{j} + \dfrac{4}{3\sqrt{2}}\mathbf{k}$; $\kappa = \dfrac{\sqrt{2}}{3}$; $\tau = \dfrac{1}{6}$

19. $\mathbf{T}(\ln 2) = \dfrac{1}{\sqrt{17}}\mathbf{i} + \dfrac{4}{\sqrt{17}}\mathbf{j}$; $\mathbf{N}(\ln 2) = -\dfrac{4}{\sqrt{17}}\mathbf{i} + \dfrac{1}{\sqrt{17}}\mathbf{j}$;
$\mathbf{B}(\ln 2) = \mathbf{k}$; $\kappa = \dfrac{8}{17\sqrt{17}}$; $\tau = 0$

21. $\mathbf{a}(0) = 10\mathbf{T} + 6\mathbf{N}$

23. $\mathbf{T} = \left(\dfrac{1}{\sqrt{2}}\cos t\right)\mathbf{i} - (\sin t)\mathbf{j} + \left(\dfrac{1}{\sqrt{2}}\cos t\right)\mathbf{k}$;

$\mathbf{N} = \left(-\dfrac{1}{\sqrt{2}}\sin t\right)\mathbf{i} - (\cos t)\mathbf{j} - \left(\dfrac{1}{\sqrt{2}}\sin t\right)\mathbf{k}$;

$\mathbf{B} = \dfrac{1}{\sqrt{2}}\mathbf{i} - \dfrac{1}{\sqrt{2}}\mathbf{k}$; $\kappa = \dfrac{1}{\sqrt{2}}$; $\tau = 0$

25. $\dfrac{\pi}{3}$ **27.** $x = 1 + t$, $y = t$, $z = -t$ **31.** $\kappa = \dfrac{1}{a}$

ADDITIONAL AND ADVANCED EXERCISES, pp. 804–805

1. (a) $\dfrac{d\theta}{dt}\bigg|_{\theta=2\pi} = 2\sqrt{\dfrac{\pi gb}{a^2 + b^2}}$

(b) $\theta = \dfrac{gbt^2}{2(a^2 + b^2)}$, $z = \dfrac{gb^2t^2}{2(a^2 + b^2)}$

(c) $\mathbf{v}(t) = \dfrac{gbt}{\sqrt{a^2 + b^2}}\mathbf{T}$;

$\dfrac{d^2\mathbf{r}}{dt^2} = \dfrac{bg}{\sqrt{a^2 + b^2}}\mathbf{T} + a\left(\dfrac{bgt}{a^2 + b^2}\right)^2\mathbf{N}$

There is no component in the direction of **B**.

5. (a) $\dfrac{dx}{dt} = \dot{r}\cos\theta - r\dot{\theta}\sin\theta, \dfrac{dy}{dt} = \dot{r}\sin\theta + r\dot{\theta}\cos\theta$

(b) $\dfrac{dr}{dt} = \dot{x}\cos\theta + \dot{y}\sin\theta, r\dfrac{d\theta}{dt} = -\dot{x}\sin\theta + \dot{y}\cos\theta$

7. (a) $\mathbf{a}(1) = -9\mathbf{u}_r - 6\mathbf{u}_\theta, \mathbf{v}(1) = -\mathbf{u}_r + 3\mathbf{u}_\theta$

(b) 6.5 in.

9. (c) $\mathbf{v} = \dot{r}\mathbf{u}_r + r\dot{\theta}\mathbf{u}_\theta + \dot{z}\mathbf{k}, \mathbf{a} = (\ddot{r} - r\dot{\theta}^2)\mathbf{u}_r + (r\ddot{\theta} + 2\dot{r}\dot{\theta})\mathbf{u}_\theta + \ddot{z}\mathbf{k}$

Chapter 14

SECTION 14.1, pp. 812–814

1. (a) 0   **(b)** 0   **(c)** 58   **(d)** 33
3. (a) $4/5$ **(b)** $8/5$ **(c)** 3   **(d)** 0
5. Domain: all points (x, y) on or above line $y = x + 2$
7. Domain: all points (x, y) not lying on the graph of $y = x$ or $y = x^3$

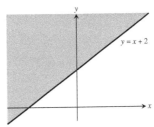

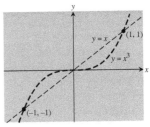

9. Domain: all points (x, y) satisfying $x^2 - 1 \le y \le x^2 + 1$

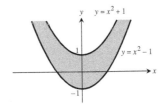

11. Domain: all points (x, y) for which
$(x - 2)(x + 2)(y - 3)(y + 3) \geq 0$

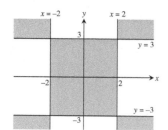

13. 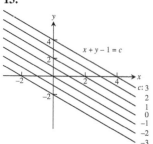 **15.**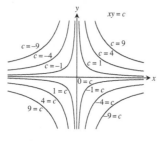

17. (a) All points in the xy-plane (b) All reals
  (c) The lines $y - x = c$ (d) No boundary points
  (e) Both open and closed (f) Unbounded
19. (a) All points in the xy-plane (b) $z \geq 0$
  (c) For $f(x, y) = 0$, the origin; for $f(x, y) \neq 0$, ellipses with
    the center $(0, 0)$, and major and minor axes along the x- and
    y-axes, respectively
  (d) No boundary points (e) Both open and closed
  (f) Unbounded
21. (a) All points in the xy-plane (b) All reals
  (c) For $f(x, y) = 0$, the x- and y-axes; for $f(x, y) \neq 0$, hyperbo-
    las with the x- and y-axes as asymptotes
  (d) No boundary points (e) Both open and closed
  (f) Unbounded
23. (a) All (x, y) satisfying $x^2 + y^2 < 16$ (b) $z \geq 1/4$
  (c) Circles centered at the origin with radii $r < 4$
  (d) Boundary is the circle $x^2 + y^2 = 16$
  (e) Open (f) Bounded
25. (a) $(x, y) \neq (0, 0)$ (b) All reals
  (c) The circles with center $(0, 0)$ and radii $r > 0$
  (d) Boundary is the single point $(0, 0)$
  (e) Open (f) Unbounded
27. (a) All (x, y) satisfying $-1 \leq y - x \leq 1$
  (b) $-\pi/2 \leq z \leq \pi/2$
  (c) Straight lines of the form $y - x = c$ where $-1 \leq c \leq 1$
  (d) Boundary is two straight lines $y = 1 + x$ and $y = -1 + x$
  (e) Closed (f) Unbounded
29. (a) Domain: all points (x, y) outside the circle $x^2 + y^2 = 1$
  (b) Range: all reals
  (c) Circles centered at the origin with radii $r > 1$
  (d) Boundary: $x^2 + y^2 = 1$
  (e) Open (f) Unbounded
31. (f), (h) **33.** (a), (i) **35.** (d), (j)

37. (a) 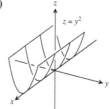 (b)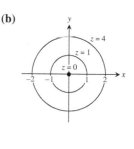

39. (a)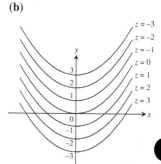

41. (a) 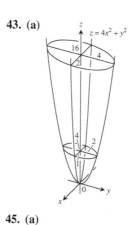 (b)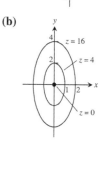

43. (a)

45. (a) 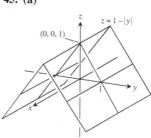 (b)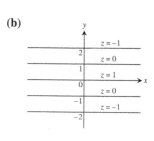

47. (a)   **(b)**

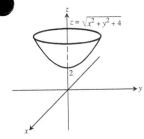
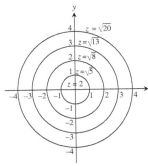

49. $x^2 + y^2 = 10$ **51.** $x + y^2 = 4$

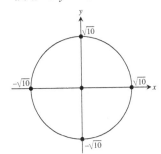
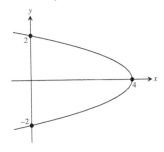

53.   **55.**

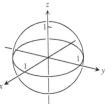
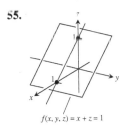

57.   **59.**

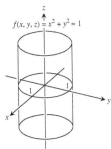
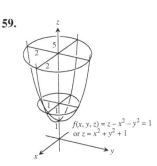

61. $\sqrt{x - y} - \ln z = 2$ **63.** $x^2 + y^2 + z^2 = 4$

65. Domain: all points (x, y) satisfying $|x| < |y|$

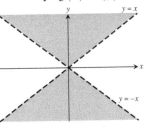

level curve: $y = 2x$

67. Domain: all points (x, y) satisfying $-1 \le x \le 1$ and $-1 \le y \le 1$

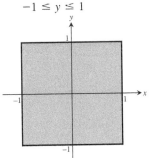

level curve:
$$\sin^{-1} y - \sin^{-1} x = \frac{\pi}{2}$$

SECTION 14.2, pp. 820–823

1. $5/2$ **3.** $2\sqrt{6}$ **5.** 1 **7.** $1/2$ **9.** 1
11. $1/4$ **13.** 0 **15.** -1 **17.** 2 **19.** $1/4$
21. 1 **23.** 3 **25.** $19/12$ **27.** 2 **29.** 3
31. (a) All (x, y) **(b)** All (x, y) except $(0, 0)$
33. (a) All (x, y) except where $x = 0$ or $y = 0$ **(b)** All (x, y)
35. (a) All (x, y, z)
     (b) All (x, y, z) except the interior of the cylinder $x^2 + y^2 = 1$
37. (a) All (x, y, z) with $z \ne 0$ **(b)** All (x, y, z) with $x^2 + z^2 \ne 1$
39. (a) All points (x, y, z) satisfying $z > x^2 + y^2 + 1$
     (b) All points (x, y, z) satisfying $z \ne \sqrt{x^2 + y^2}$
41. Consider paths along $y = x, x > 0$, and along $y = x, x < 0$.
43. Consider the paths $y = kx^2$, k a constant.
45. Consider the paths $y = mx$, m a constant, $m \ne -1$.
47. Consider the paths $y = kx^2$, k a constant, $k \ne 0$.
49. Consider the paths $x = 1$ and $y = x$.
51. Along $y = 1$ the limit is 0; along $y = e^x$ the limit is $1/2$.
53. Along $y = 0$ the limit is 1; along $y = -\sin x$ the limit is 0.
55. (a) 1   **(b)** 0   **(c)** Does not exist
59. The limit is 1.   **61.** The limit is 0.
63. (a) $f(x, y)\big|_{y=mx} = \sin 2\theta$ where $\tan\theta = m$ **65.** 0
67. Does not exist   **69.** $\pi/2$ **71.** $f(0, 0) = \ln 3$
73. $\delta = 0.1$ **75.** $\delta = 0.005$ **77.** $\delta = 0.04$
79. $\delta = \sqrt{0.015}$ **81.** $\delta = 0.005$

SECTION 14.3, pp. 833–835

1. $\dfrac{\partial f}{\partial x} = 4x, \dfrac{\partial f}{\partial y} = -3$ **3.** $\dfrac{\partial f}{\partial x} = 2x(y + 2), \dfrac{\partial f}{\partial y} = x^2 - 1$

5. $\dfrac{\partial f}{\partial x} = 2y(xy - 1), \dfrac{\partial f}{\partial y} = 2x(xy - 1)$

7. $\dfrac{\partial f}{\partial x} = \dfrac{x}{\sqrt{x^2 + y^2}}, \dfrac{\partial f}{\partial y} = \dfrac{y}{\sqrt{x^2 + y^2}}$

9. $\dfrac{\partial f}{\partial x} = \dfrac{-1}{(x + y)^2}, \dfrac{\partial f}{\partial y} = \dfrac{-1}{(x + y)^2}$

11. $\dfrac{\partial f}{\partial x} = \dfrac{-y^2 - 1}{(xy - 1)^2}, \dfrac{\partial f}{\partial y} = \dfrac{-x^2 - 1}{(xy - 1)^2}$

13. $\dfrac{\partial f}{\partial x} = e^{x+y+1}, \dfrac{\partial f}{\partial y} = e^{x+y+1}$ **15.** $\dfrac{\partial f}{\partial x} = \dfrac{1}{x + y}, \dfrac{\partial f}{\partial y} = \dfrac{1}{x + y}$

17. $\dfrac{\partial f}{\partial x} = 2\sin(x - 3y)\cos(x - 3y),$

$\dfrac{\partial f}{\partial y} = -6\sin(x - 3y)\cos(x - 3y)$

19. $\dfrac{\partial f}{\partial x} = yx^{y-1}, \dfrac{\partial f}{\partial y} = x^y \ln x$ **21.** $\dfrac{\partial f}{\partial x} = -g(x), \dfrac{\partial f}{\partial y} = g(y)$

23. $f_x = y^2, f_y = 2xy, f_z = -4z$

25. $f_x = 1, f_y = -y(y^2 + z^2)^{-1/2}, f_z = -z(y^2 + z^2)^{-1/2}$

27. $f_x = \dfrac{yz}{\sqrt{1 - x^2 y^2 z^2}}, f_y = \dfrac{xz}{\sqrt{1 - x^2 y^2 z^2}}, f_z = \dfrac{xy}{\sqrt{1 - x^2 y^2 z^2}}$

29. $f_x = \dfrac{1}{x + 2y + 3z}, f_y = \dfrac{2}{x + 2y + 3z}, f_z = \dfrac{3}{x + 2y + 3z}$

31. $f_x = -2xe^{-(x^2+y^2+z^2)}, f_y = -2ye^{-(x^2+y^2+z^2)}, f_z = -2ze^{-(x^2+y^2+z^2)}$

33. $f_x = \text{sech}^2(x + 2y + 3z), f_y = 2\,\text{sech}^2(x + 2y + 3z),$
$f_z = 3\,\text{sech}^2(x + 2y + 3z)$

35. $\dfrac{\partial f}{\partial t} = -2\pi \sin(2\pi t - \alpha), \dfrac{\partial f}{\partial \alpha} = \sin(2\pi t - \alpha)$

37. $\dfrac{\partial h}{\partial \rho} = \sin\phi\cos\theta, \dfrac{\partial h}{\partial \phi} = \rho\cos\phi\cos\theta, \dfrac{\partial h}{\partial \theta} = -\rho\sin\phi\sin\theta$

39. $W_P(P, V, \delta, v, g) = V, W_V(P, V, \delta, v, g) = P + \dfrac{\delta v^2}{2g},$

$W_\delta(P, V, \delta, v, g) = \dfrac{Vv^2}{2g}, W_v(P, V, \delta, v, g) = \dfrac{V\delta v}{g},$

$W_g(P, V, \delta, v, g) = -\dfrac{V\delta v^2}{2g^2}$

41. $\dfrac{\partial f}{\partial x} = 1 + y, \dfrac{\partial f}{\partial y} = 1 + x, \dfrac{\partial^2 f}{\partial x^2} = 0, \dfrac{\partial^2 f}{\partial y^2} = 0, \dfrac{\partial^2 f}{\partial y \partial x} = \dfrac{\partial^2 f}{\partial x \partial y} = 1$

43. $\dfrac{\partial g}{\partial x} = 2xy + y\cos x, \dfrac{\partial g}{\partial y} = x^2 - \sin y + \sin x,$

$\dfrac{\partial^2 g}{\partial x^2} = 2y - y\sin x, \dfrac{\partial^2 g}{\partial y^2} = -\cos y,$

$\dfrac{\partial^2 g}{\partial y \partial x} = \dfrac{\partial^2 g}{\partial x \partial y} = 2x + \cos x$

45. $\dfrac{\partial r}{\partial x} = \dfrac{1}{x + y}, \dfrac{\partial r}{\partial y} = \dfrac{1}{x + y}, \dfrac{\partial^2 r}{\partial x^2} = \dfrac{-1}{(x + y)^2}, \dfrac{\partial^2 r}{\partial y^2} = \dfrac{-1}{(x + y)^2},$

$\dfrac{\partial^2 r}{\partial y \partial x} = \dfrac{\partial^2 r}{\partial x \partial y} = \dfrac{-1}{(x + y)^2}$

47. $\dfrac{\partial w}{\partial x} = x^2 y \sec^2(xy) + 2x\tan(xy), \dfrac{\partial w}{\partial y} = x^3 \sec^2(xy),$

$\dfrac{\partial^2 w}{\partial y \partial x} = \dfrac{\partial^2 w}{\partial x \partial y} = 2x^3 y \sec^2(xy)\tan(xy) + 3x^2 \sec^2(xy)$

$\dfrac{\partial^2 w}{\partial x^2} = 4xy\sec^2(xy) + 2x^2 y^2 \sec^2(xy)\tan(xy) + 2\tan(xy)$

$\dfrac{\partial^2 w}{\partial y^2} = 2x^4 \sec^2(xy)\tan(xy)$

49. $\dfrac{\partial w}{\partial x} = \sin(x^2 y) + 2x^2 y\cos(x^2 y), \dfrac{\partial w}{\partial y} = x^3 \cos(x^2 y),$

$\dfrac{\partial^2 w}{\partial y \partial x} = \dfrac{\partial^2 w}{\partial x \partial y} = 3x^2 \cos(x^2 y) - 2x^4 y\sin(x^2 y)$

$\dfrac{\partial^2 w}{\partial x^2} = 6xy\cos(x^2 y) - 4x^3 y^2 \sin(x^2 y)$

$\dfrac{\partial^2 w}{\partial y^2} = -x^5 \sin(x^2 y)$

51. $\dfrac{\partial f}{\partial x} = 2xy^3 - 4x^3, \dfrac{\partial f}{\partial y} = 3x^2 y^2 + 5y^4,$

$\dfrac{\partial^2 f}{\partial x^2} = 2y^3 - 12x^2, \dfrac{\partial^2 f}{\partial y^2} = 6x^2 y + 20y^3,$

$\dfrac{\partial^2 f}{\partial y \partial x} = \dfrac{\partial^2 f}{\partial x \partial y} = 6xy^2$

53. $\dfrac{\partial z}{\partial x} = 2x\cos(2x - y^2) + \sin(2x - y^2),$

$\dfrac{\partial z}{\partial y} = -2xy\cos(2x - y^2),$

$\dfrac{\partial^2 z}{\partial x^2} = 4\cos(2x - y^2) - 4x\sin(2x - y^2),$

$\dfrac{\partial^2 z}{\partial y^2} = -4xy^2 \sin(2x - y^2) - 2x\cos(2x - y^2),$

$\dfrac{\partial^2 z}{\partial x \partial y} = \dfrac{\partial^2 z}{\partial y \partial x} = 4xy\sin(2x - y^2) - 2y\cos(2x - y^2)$

55. $\dfrac{\partial w}{\partial x} = \dfrac{2}{2x + 3y}, \dfrac{\partial w}{\partial y} = \dfrac{3}{2x + 3y}, \dfrac{\partial^2 w}{\partial y \partial x} = \dfrac{\partial^2 w}{\partial x \partial y} = \dfrac{-6}{(2x + 3y)^2}$

57. $\dfrac{\partial w}{\partial x} = y^2 + 2xy^3 + 3x^2 y^4, \dfrac{\partial w}{\partial y} = 2xy + 3x^2 y^2 + 4x^3 y^3,$

$\dfrac{\partial^2 w}{\partial y \partial x} = \dfrac{\partial^2 w}{\partial x \partial y} = 2y + 6xy^2 + 12x^2 y^3$

59. $\dfrac{\partial \omega}{\partial x} = \dfrac{2x}{y^3}, \dfrac{\partial \omega}{\partial y} = \dfrac{-3x^2}{y^4}$

$\dfrac{\partial^2 \omega}{\partial y \partial x} = \dfrac{-6x}{y^4}, \dfrac{\partial^2 \omega}{\partial x \partial y} = \dfrac{-6x}{y^4}$

61. (a) x first   **(b)** y first   **(c)** x first
   (d) x first   **(e)** y first   **(f)** y first

63. $f_x(1, 2) = -13, f_y(1, 2) = -2$

65. $f_x(-2, 3) = 1/2, f_y(-2, 3) = 3/4$

67. (a) 3   **(b)** 2   **69.** 12

71. $\dfrac{\partial f}{\partial x} = 3x^2y^2 - 2x \Rightarrow$

$f(x, y) = x^3y^2 - x^2 + g(y) \Rightarrow$

$\dfrac{\partial f}{\partial y} = 2x^3y + g'(y) = 2x^3y + 64 \Rightarrow$

$g'(y) = 6y \Rightarrow g(y) = 3y^2$ works $\Rightarrow$

$f(x, y) = x^3y^2 - x^2 + 3y^2$ works

73. $\dfrac{\partial^2 f}{\partial y\,\partial x} = \dfrac{2x - 2y}{(x + y)^3} \neq \dfrac{\partial^2 f}{\partial x\,\partial y} = \dfrac{2y - 2x}{(x + y)^3}$ so impossible   **75.** -2

77. $\dfrac{\partial A}{\partial a} = \dfrac{a}{bc \sin A}, \dfrac{\partial A}{\partial b} = \dfrac{c \cos A - b}{bc \sin A}$

79. $v_x = \dfrac{\ln v}{(\ln u)(\ln v) - 1}$

81. $f_x(x, y) = 0$ for all points (x, y),

$f_y(x, y) = \begin{cases} 3y^2, & y \geq 0 \\ -2y, & y < 0 \end{cases}$,

$f_{xy}(x, y) = f_{yx}(x, y) = 0$ for all points (x, y)

99. Yes

SECTION 14.4, pp. 842–844

1. (a) $\dfrac{dw}{dt} = 0$,   (b) $\dfrac{dw}{dt}(\pi) = 0$

3. (a) $\dfrac{dw}{dt} = 1$,   (b) $\dfrac{dw}{dt}(3) = 1$

5. (a) $\dfrac{dw}{dt} = 4t \tan^{-1} t + 1$,   (b) $\dfrac{dw}{dt}(1) = \pi + 1$

7. (a) $\dfrac{\partial z}{\partial u} = 4 \cos v \ln(u \sin v) + 4 \cos v$,

$\dfrac{\partial z}{\partial v} = -4u \sin v \ln(u \sin v) + \dfrac{4u \cos^2 v}{\sin v}$

(b) $\dfrac{\partial z}{\partial u} = \sqrt{2}(\ln 2 + 2), \dfrac{\partial z}{\partial v} = -2\sqrt{2}(\ln 2 - 2)$

9. (a) $\dfrac{\partial w}{\partial u} = 2u + 4uv, \dfrac{\partial w}{\partial v} = -2v + 2u^2$

(b) $\dfrac{\partial w}{\partial u} = 3, \dfrac{\partial w}{\partial v} = -\dfrac{3}{2}$

11. (a) $\dfrac{\partial u}{\partial x} = 0, \dfrac{\partial u}{\partial y} = \dfrac{z}{(z - y)^2}, \dfrac{\partial u}{\partial z} = \dfrac{-y}{(z - y)^2}$

(b) $\dfrac{\partial u}{\partial x} = 0, \dfrac{\partial u}{\partial y} = 1, \dfrac{\partial u}{\partial z} = -2$

13. $\dfrac{dz}{dt} = \dfrac{\partial z}{\partial x}\dfrac{dx}{dt} + \dfrac{\partial z}{\partial y}\dfrac{dy}{dt}$

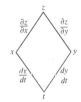

15. $\dfrac{\partial w}{\partial u} = \dfrac{\partial w}{\partial x}\dfrac{\partial x}{\partial u} + \dfrac{\partial w}{\partial y}\dfrac{\partial y}{\partial u} + \dfrac{\partial w}{\partial z}\dfrac{\partial z}{\partial u}$,

$\dfrac{\partial w}{\partial v} = \dfrac{\partial w}{\partial x}\dfrac{\partial x}{\partial v} + \dfrac{\partial w}{\partial y}\dfrac{\partial y}{\partial v} + \dfrac{\partial w}{\partial z}\dfrac{\partial z}{\partial v}$

 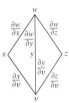

17. $\dfrac{\partial w}{\partial u} = \dfrac{\partial w}{\partial x}\dfrac{\partial x}{\partial u} + \dfrac{\partial w}{\partial y}\dfrac{\partial y}{\partial u}, \dfrac{\partial w}{\partial v} = \dfrac{\partial w}{\partial x}\dfrac{\partial x}{\partial v} + \dfrac{\partial w}{\partial y}\dfrac{\partial y}{\partial v}$.

 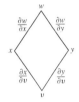

19. $\dfrac{\partial z}{\partial t} = \dfrac{\partial z}{\partial x}\dfrac{\partial x}{\partial t} + \dfrac{\partial z}{\partial y}\dfrac{\partial y}{\partial t}, \dfrac{\partial z}{\partial s} = \dfrac{\partial z}{\partial x}\dfrac{\partial x}{\partial s} + \dfrac{\partial z}{\partial y}\dfrac{\partial y}{\partial s}$

 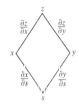

21. $\dfrac{\partial w}{\partial s} = \dfrac{dw}{du}\dfrac{\partial u}{\partial s}, \dfrac{\partial w}{\partial t} = \dfrac{dw}{du}\dfrac{\partial u}{\partial t}$

 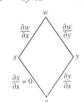

23. $\dfrac{\partial w}{\partial r} = \dfrac{\partial w}{\partial x}\dfrac{\partial x}{\partial r} + \dfrac{\partial w}{\partial y}\dfrac{\partial y}{\partial r} = \dfrac{\partial w}{\partial x}\dfrac{\partial x}{\partial r}$ since $\dfrac{\partial y}{\partial r} = 0$,

$\dfrac{\partial w}{\partial s} = \dfrac{\partial w}{\partial x}\dfrac{\partial x}{\partial s} + \dfrac{\partial w}{\partial y}\dfrac{\partial y}{\partial s} = \dfrac{\partial w}{\partial y}\dfrac{\partial y}{\partial s}$ since $\dfrac{\partial x}{\partial s} = 0$

25. $4/3$ **27.** $-4/5$ **29.** 20 **31.** $\dfrac{\partial z}{\partial x} = \dfrac{1}{4}, \dfrac{\partial z}{\partial y} = -\dfrac{3}{4}$

33. $\dfrac{\partial z}{\partial x} = -1, \dfrac{\partial z}{\partial y} = -1$ **35.** 12 **37.** -7

39. $\dfrac{\partial z}{\partial u} = 2, \dfrac{\partial z}{\partial v} = 1$ **41.** $\dfrac{\partial w}{\partial t} = 2t\, e^{s^3 + t^2}, \dfrac{\partial w}{\partial s} = 3s^2\, e^{s^3 + t^2}$

43. 23 **45.** $-16, 2$ **47.** -0.00005 amp/sec

53. $(\cos 1, \sin 1, 1)$ and $(\cos(-2), \sin(-2), -2)$

55. (a) Maximum at $\left(-\dfrac{\sqrt{2}}{2}, \dfrac{\sqrt{2}}{2}\right)$ and $\left(\dfrac{\sqrt{2}}{2}, -\dfrac{\sqrt{2}}{2}\right)$; minimum

at $\left(\dfrac{\sqrt{2}}{2}, \dfrac{\sqrt{2}}{2}\right)$ and $\left(-\dfrac{\sqrt{2}}{2}, -\dfrac{\sqrt{2}}{2}\right)$

(b) Max $= 6$, min $= 2$

57. $5°$C/sec   **59.** $2x\sqrt{x^8 + x^3} + \displaystyle\int_0^{x^2} \dfrac{3x^2}{2\sqrt{t^4 + x^3}}\, dt$

SECTION 14.5, pp. 852–853

1.

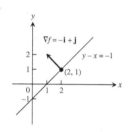

3.

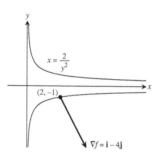

5.

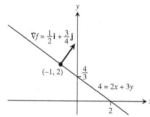

7. $\nabla f = 3\mathbf{i} + 2\mathbf{j} - 4\mathbf{k}$ **9.** $\nabla f = -\dfrac{26}{27}\mathbf{i} + \dfrac{23}{54}\mathbf{j} - \dfrac{23}{54}\mathbf{k}$

11. -4 **13.** $21/13$ **15.** 3 **17.** 2

19. $\mathbf{u} = -\dfrac{1}{\sqrt{2}}\mathbf{i} + \dfrac{1}{\sqrt{2}}\mathbf{j}, (D_{\mathbf{u}}f)_{P_0} = \sqrt{2}; -\mathbf{u} = \dfrac{1}{\sqrt{2}}\mathbf{i} - \dfrac{1}{\sqrt{2}}\mathbf{j},$

$(D_{-\mathbf{u}}f)_{P_0} = -\sqrt{2}$

21. $\mathbf{u} = \dfrac{1}{3\sqrt{3}}\mathbf{i} - \dfrac{5}{3\sqrt{3}}\mathbf{j} - \dfrac{1}{3\sqrt{3}}\mathbf{k}, (D_{\mathbf{u}}f)_{P_0} = 3\sqrt{3};$

$-\mathbf{u} = -\dfrac{1}{3\sqrt{3}}\mathbf{i} + \dfrac{5}{3\sqrt{3}}\mathbf{j} + \dfrac{1}{3\sqrt{3}}\mathbf{k}, (D_{-\mathbf{u}}f)_{P_0} = -3\sqrt{3}$

23. $\mathbf{u} = \dfrac{1}{\sqrt{3}}(\mathbf{i} + \mathbf{j} + \mathbf{k}), (D_{\mathbf{u}}f)_{P_0} = 2\sqrt{3};$

$-\mathbf{u} = -\dfrac{1}{\sqrt{3}}(\mathbf{i} + \mathbf{j} + \mathbf{k}), (D_{-\mathbf{u}}f)_{P_0} = -2\sqrt{3}$

25.

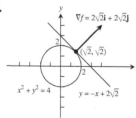

27.

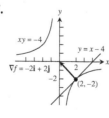

29. (a) $\mathbf{u} = \dfrac{3}{5}\mathbf{i} - \dfrac{4}{5}\mathbf{j}, D_{\mathbf{u}}\, f(1, -1) = 5$

(b) $\mathbf{u} = -\dfrac{3}{5}\mathbf{i} + \dfrac{4}{5}\mathbf{j}, D_{\mathbf{u}}\, f(1, -1) = -5$

(c) $\mathbf{u} = \dfrac{4}{5}\mathbf{i} + \dfrac{3}{5}\mathbf{j}, \mathbf{u} = -\dfrac{4}{5}\mathbf{i} - \dfrac{3}{5}\mathbf{j}$

(d) $\mathbf{u} = -\mathbf{j}, \mathbf{u} = \dfrac{24}{25}\mathbf{i} - \dfrac{7}{25}\mathbf{j}$

(e) $\mathbf{u} = -\mathbf{i}, \mathbf{u} = \dfrac{7}{25}\mathbf{i} + \dfrac{24}{25}\mathbf{j}$

31. $\mathbf{u} = \dfrac{7}{\sqrt{53}}\mathbf{i} - \dfrac{2}{\sqrt{53}}\mathbf{j}, -\mathbf{u} = -\dfrac{7}{\sqrt{53}}\mathbf{i} + \dfrac{2}{\sqrt{53}}\mathbf{j}$

33. No, the maximum rate of change is $\sqrt{185} < 14$.

35. $-7/\sqrt{5}$ **41.** $r(t) = (-3 - 6t)\mathbf{i} + (4 + 8t)\mathbf{j}, -\infty < t < \infty$

43. $r(t) = (3 + 6t)\mathbf{i} + (-2 - 4t)\mathbf{j} + (1 + 2t)\mathbf{k}, -\infty < t < \infty$

SECTION 14.6, pp. 860–863

1. (a) $x + y + z = 3$

(b) $x = 1 + 2t, y = 1 + 2t, z = 1 + 2t$

3. (a) $2x - z - 2 = 0$

(b) $x = 2 - 4t, y = 0, z = 2 + 2t$

5. (a) $2x + 2y + z - 4 = 0$

(b) $x = 2t, y = 1 + 2t, z = 2 + t$

7. (a) $x + y + z - 1 = 0$

(b) $x = t, y = 1 + t, z = t$

9. (a) $-x + 3y + z/e = 2$

(b) $x = 2 - t, y = 1 + 3t, z = e + (1/e)t$

11. $2x - z - 2 = 0$

13. $x - y + 2z - 1 = 0$

15. $x = 1, y = 1 + 2t, z = 1 - 2t$

17. $x = 1 - 2t, y = 1, z = \dfrac{1}{2} + 2t$

19. $x = 1 + 90t, y = 1 - 90t, z = 3$

21. $df = \dfrac{9}{11,830} \approx 0.0008$ **23.** $dg = 0$

25. (a) $\dfrac{\sqrt{3}}{2}\sin\sqrt{3} - \dfrac{1}{2}\cos\sqrt{3} \approx 0.935°$C/ft

(b) $\sqrt{3}\sin\sqrt{3} - \cos\sqrt{3} \approx 1.87°$C/sec

27. (a) $L(x, y) = 1$ **(b)** $L(x, y) = 2x + 2y - 1$

29. (a) $L(x, y) = 3x - 4y + 5$ **(b)** $L(x, y) = 3x - 4y + 5$

31. (a) $L(x, y) = 1 + x$ **(b)** $L(x, y) = -y + \dfrac{\pi}{2}$

33. (a) $W(20, 25) = 11°$F, $W(30, -10) = -39°$F, $W(15, 15) = 0°$F

(b) $W(10, -40) \approx -65.5°$F, $W(50, -40) \approx -88°$F,
$W(60, 30) \approx 10.2°$F

(c) $L(v, T) \approx -0.36\,(v - 25) + 1.337(T - 5) - 17.4088$

(d)   **i)** $L(24, 6) \approx -15.7°$F
    ii) $L(27, 2) \approx -22.1°$F
    iii) $L(5, -10) \approx -30.2°$F

35. $L(x, y) = 7 + x - 6y; 0.06$ **37.** $L(x, y) = x + y + 1; 0.08$

39. $L(x, y) = 1 + x; 0.0222$

41. (a) $L(x, y, z) = 2x + 2y + 2z - 3$ **(b)** $L(x, y, z) = y + z$

(c) $L(x, y, z) = 0$

43. (a) $L(x, y, z) = x$

(b) $L(x, y, z) = \dfrac{1}{\sqrt{2}}x + \dfrac{1}{\sqrt{2}}y$

(c) $L(x, y, z) = \dfrac{1}{3}x + \dfrac{2}{3}y + \dfrac{2}{3}z$

45. (a) $L(x, y, z) = 2 + x$

   (b) $L(x, y, z) = x - y - z + \dfrac{\pi}{2} + 1$

   (c) $L(x, y, z) = x - y - z + \dfrac{\pi}{2} + 1$

47. $L(x, y, z) = 2x - 6y - 2z + 6,\ 0.0024$

49. $L(x, y, z) = x + y - z - 1,\ 0.00135$

51. Maximum error (estimate) ≤ 0.31 in magnitude

53. Pay more attention to the smaller of the two dimensions. It will generate the larger partial derivative.

55. f is most sensitive to a change in d.

61. (a) 1.75%   **(b)** 1.75%

SECTION 14.7, pp. 870–872

1. $f(-3, 3) = -5$, local minimum   **3.** $f(-2, 1)$, saddle point

5. $f\left(3, \dfrac{3}{2}\right) = \dfrac{17}{2}$, local maximum

7. $f(2, -1) = -6$, local minimum   **9.** $f(1, 2)$, saddle point

11. $f\left(\dfrac{16}{7}, 0\right) = -\dfrac{16}{7}$, local maximum

13. $f(0, 0)$, saddle point; $f\left(-\dfrac{2}{3}, \dfrac{2}{3}\right) = \dfrac{170}{27}$, local maximum

15. $f(0, 0) = 0$, local minimum; $f(1, -1)$, saddle point

17. $f(0, \pm\sqrt{5})$, saddle points; $f(-2, -1) = 30$, local maximum; $f(2, 1) = -30$, local minimum

19. $f(0, 0)$, saddle point; $f(1, 1) = 2,\ f(-1, -1) = 2$, local maxima

21. $f(0, 0) = -1$, local maximum

23. $f(n\pi, 0)$, saddle points, for every integer n

25. $f(2, 0) = e^{-4}$, local minimum

27. $f(0, 0) = 0$, local minimum; $f(0, 2)$, saddle point

29. $f\left(\dfrac{1}{2}, 1\right) = \ln\left(\dfrac{1}{4}\right) - 3$, local maximum

31. Absolute maximum: 1 at $(0, 0)$; absolute minimum: -5 at $(1, 2)$

33. Absolute maximum: 4 at $(0, 2)$; absolute minimum: 0 at $(0, 0)$

35. Absolute maximum: 11 at $(0, -3)$; absolute minimum: 10 at $(4, -2)$

37. Absolute maximum: 4 at $(2, 0)$; absolute minimum: $\dfrac{3\sqrt{2}}{2}$ at $\left(3, -\dfrac{\pi}{4}\right), \left(3, \dfrac{\pi}{4}\right), \left(1, -\dfrac{\pi}{4}\right),$ and $\left(1, \dfrac{\pi}{4}\right)$

39. $a = -3,\ b = 2$

41. Hottest is $2\dfrac{1}{4}°$ at $\left(-\dfrac{1}{2}, \dfrac{\sqrt{3}}{2}\right)$ and $\left(-\dfrac{1}{2}, -\dfrac{\sqrt{3}}{2}\right)$; coldest is $-\dfrac{1}{4}°$ at $\left(\dfrac{1}{2}, 0\right)$.

43. (a) $f(0, 0)$, saddle point   **(b)** $f(1, 2)$, local minimum

   (c) $f(1, -2)$, local minimum; $f(-1, -2)$, saddle point

49. $\left(\dfrac{1}{6}, \dfrac{1}{3}, \dfrac{355}{36}\right)$ **51.** $\left(\dfrac{9}{7}, \dfrac{6}{7}, \dfrac{3}{7}\right)$ **53.** 3, 3, 3   **55.** 12

57. $\dfrac{4}{\sqrt{3}} \times \dfrac{4}{\sqrt{3}} \times \dfrac{4}{\sqrt{3}}$ **59.** 2 ft $\times$ 2 ft $\times$ 1 ft

61. Points $(0, 2, 0)$ and $(0, -2, 0)$ have distance 2 from the origin.

63. (a) On the semicircle, max $f = 2\sqrt{2}$ at $t = \pi/4$, min $f = -2$ at $t = \pi$. On the quarter circle, max $f = 2\sqrt{2}$ at $t = \pi/4$, min $f = 2$ at $t = 0, \pi/2$.

   (b) On the semicircle, max $g = 2$ at $t = \pi/4$, min $g = -2$ at $t = 3\pi/4$. On the quarter circle, max $g = 2$ at $t = \pi/4$, min $g = 0$ at $t = 0, \pi/2$.

   (c) On the semicircle, max $h = 8$ at $t = 0, \pi$; min $h = 4$ at $t = \pi/2$. On the quarter circle, max $h = 8$ at $t = 0$, min $h = 4$ at $t = \pi/2$.

65. i) min $f = -1/2$ at $t = -1/2$; no max

  ii) max $f = 0$ at $t = -1, 0$; min $f = -1/2$ at $t = -1/2$

  iii) max $f = 4$ at $t = 1$; min $f = 0$ at $t = 0$

69. $y = -\dfrac{20}{13}x + \dfrac{9}{13},\ \left. y \right|_{x=4} = -\dfrac{71}{13}$

SECTION 14.8, pp. 879–882

1. $\left(\pm\dfrac{1}{\sqrt{2}}, \dfrac{1}{2}\right), \left(\pm\dfrac{1}{\sqrt{2}}, -\dfrac{1}{2}\right)$ **3.** 39   **5.** $\left(3, \pm 3\sqrt{2}\right)$

7. (a) 8   **(b)** 64

9. $r = 2$ cm, $h = 4$ cm

11. Length $= 4\sqrt{2}$, width $= 3\sqrt{2}$

13. $f(0, 0) = 0$ is minimum; $f(2, 4) = 20$ is maximum.

15. Lowest $= 0°$, highest $= 125°$

17. $\left(\dfrac{3}{2}, 2, \dfrac{5}{2}\right)$ **19.** 1   **21.** $(0, 0, 2), (0, 0, -2)$

23. $f(1, -2, 5) = 30$ is maximum; $f(-1, 2, -5) = -30$ is minimum.

25. 3, 3, 3   **27.** $\dfrac{2}{\sqrt{3}}$ by $\dfrac{2}{\sqrt{3}}$ by $\dfrac{2}{\sqrt{3}}$ units

29. $(\pm 4/3, -4/3, -4/3)$ **31.** $\approx 24{,}322$ units

33. $U(8, 14) = \$128$ **37.** $f(2/3, 4/3, -4/3) = \dfrac{4}{3}$

39. $(2, 4, 4)$ **41.** Maximum is $1 + 6\sqrt{3}$ at $\left(\pm\sqrt{6}, \sqrt{3}, 1\right)$; minimum is $1 - 6\sqrt{3}$ at $\left(\pm\sqrt{6}, -\sqrt{3}, 1\right)$.

43. Maximum is 4 at $(0, 0, \pm 2)$; minimum is 2 at $\left(\pm\sqrt{2}, \pm\sqrt{2}, 0\right)$.

SECTION 14.9, p. 886

1. Quadratic: $x + xy$; cubic: $x + xy + \dfrac{1}{2}xy^2$

3. Quadratic: xy; cubic: xy

5. Quadratic: $y + \dfrac{1}{2}(2xy - y^2)$;

   cubic: $y + \dfrac{1}{2}(2xy - y^2) + \dfrac{1}{6}(3x^2y - 3xy^2 + 2y^3)$

7. Quadratic: $\dfrac{1}{2}(2x^2 + 2y^2) = x^2 + y^2$; cubic: $x^2 + y^2$

9. Quadratic: $1 + (x + y) + (x + y)^2$;

   cubic: $1 + (x + y) + (x + y)^2 + (x + y)^3$

11. Quadratic: $1 - \dfrac{1}{2}x^2 - \dfrac{1}{2}y^2$; $E(x, y) \leq 0.00134$

SECTION 14.10, p. 890

1. (a) 0   **(b)** $1 + 2z$ **(c)** $1 + 2z$

3. (a) $\dfrac{\partial U}{\partial P} + \dfrac{\partial U}{\partial T}\left(\dfrac{V}{nR}\right)$ **(b)** $\dfrac{\partial U}{\partial P}\left(\dfrac{nR}{V}\right) + \dfrac{\partial U}{\partial T}$

5. (a) 5   **(b)** 5   **7.** $\left(\dfrac{\partial x}{\partial r}\right)_\theta = \cos\theta$ $\left(\dfrac{\partial r}{\partial x}\right)_y = \dfrac{x}{\sqrt{x^2 + y^2}}$

PRACTICE EXERCISES, pp. 891–894

1. Domain: all points in the xy-plane; range: $z \geq 0$. Level curves are ellipses with major axis along the y-axis and minor axis along the x-axis.

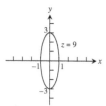

3. Domain: all (x, y) such that $x \neq 0$ and $y \neq 0$; range: $z \neq 0$. Level curves are hyperbolas with the x- and y-axes as asymptotes.

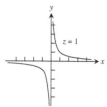

5. Domain: all points in xyz-space; range: all real numbers. Level surfaces are paraboloids of revolution with the z-axis as axis.

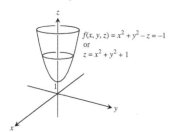

7. Domain: all (x, y, z) such that $(x, y, z) \neq (0, 0, 0)$; range: positive real numbers. Level surfaces are spheres with center $(0, 0, 0)$ and radius $r > 0$.

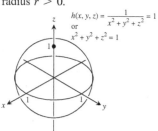

9. -2 **11.** $1/2$ **13.** 1

15. Let $y = kx^2$, $k \neq 1$

17. No; $\lim_{(x,y)\to(0,0)} f(x, y)$ does not exist.

19. $\dfrac{\partial g}{\partial r} = \cos\theta + \sin\theta$, $\dfrac{\partial g}{\partial \theta} = -r\sin\theta + r\cos\theta$

21. $\dfrac{\partial f}{\partial R_1} = -\dfrac{1}{R_1^2}$, $\dfrac{\partial f}{\partial R_2} = -\dfrac{1}{R_2^2}$, $\dfrac{\partial f}{\partial R_3} = -\dfrac{1}{R_3^2}$

23. $\dfrac{\partial P}{\partial n} = \dfrac{RT}{V}$, $\dfrac{\partial P}{\partial R} = \dfrac{nT}{V}$, $\dfrac{\partial P}{\partial T} = \dfrac{nR}{V}$, $\dfrac{\partial P}{\partial V} = -\dfrac{nRT}{V^2}$

25. $\dfrac{\partial^2 g}{\partial x^2} = 0$, $\dfrac{\partial^2 g}{\partial y^2} = \dfrac{2x}{y^3}$, $\dfrac{\partial^2 g}{\partial y \, \partial x} = \dfrac{\partial^2 g}{\partial x \, \partial y} = -\dfrac{1}{y^2}$

27. $\dfrac{\partial^2 f}{\partial x^2} = -30x + \dfrac{2 - 2x^2}{(x^2 + 1)^2}$, $\dfrac{\partial^2 f}{\partial y^2} = 0$, $\dfrac{\partial^2 f}{\partial y \, \partial x} = \dfrac{\partial^2 f}{\partial x \, \partial y} = 1$

29. $\dfrac{dw}{dt}\bigg|_{t=0} = -1$

31. $\dfrac{\partial w}{\partial r}\bigg|_{(r, s)=(\pi, 0)} = 2$, $\dfrac{\partial w}{\partial s}\bigg|_{(r, s)=(\pi, 0)} = 2 - \pi$

33. $\dfrac{df}{dt}\bigg|_{t=1} = -(\sin 1 + \cos 2)(\sin 1) + (\cos 1 + \cos 2)(\cos 1)$
$\qquad\qquad -2(\sin 1 + \cos 1)(\sin 2)$

35. $\dfrac{dy}{dx}\bigg|_{(x, y)=(0,1)} = -1$

37. Increases most rapidly in the direction $\mathbf{u} = -\dfrac{\sqrt{2}}{2}\mathbf{i} - \dfrac{\sqrt{2}}{2}\mathbf{j}$;

decreases most rapidly in the direction $-\mathbf{u} = \dfrac{\sqrt{2}}{2}\mathbf{i} + \dfrac{\sqrt{2}}{2}\mathbf{j}$;

$D_{\mathbf{u}}f = \dfrac{\sqrt{2}}{2}$; $D_{-\mathbf{u}}f = -\dfrac{\sqrt{2}}{2}$; $D_{\mathbf{u}_1}f = -\dfrac{7}{10}$ where $\mathbf{u}_1 = \dfrac{\mathbf{v}}{|\mathbf{v}|}$

39. Increases most rapidly in the direction $\mathbf{u} = \dfrac{2}{7}\mathbf{i} + \dfrac{3}{7}\mathbf{j} + \dfrac{6}{7}\mathbf{k}$;

decreases most rapidly in the direction $-\mathbf{u} = -\dfrac{2}{7}\mathbf{i} - \dfrac{3}{7}\mathbf{j} - \dfrac{6}{7}\mathbf{k}$;

$D_{\mathbf{u}}f = 7$; $D_{-\mathbf{u}}f = -7$; $D_{\mathbf{u}_1}f = 7$ where $\mathbf{u}_1 = \dfrac{\mathbf{v}}{|\mathbf{v}|}$

41. $\pi/\sqrt{2}$ **43. (a)** $f_x(1, 2) = f_y(1, 2) = 2$ **(b)** $14/5$

45.

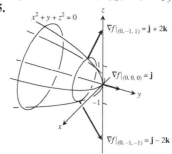

47. Tangent: $4x - y - 5z = 4$; normal line:
$x = 2 + 4t$, $y = -1 - t$, $z = 1 - 5t$

49. $2y - z - 2 = 0$

51. Tangent: $x + y = \pi + 1$; normal line: $y = x - \pi + 1$

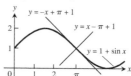

53. $x = 1 - 2t$, $y = 1$, $z = 1/2 + 2t$

55. Answers will depend on the upper bound used for $|f_{xx}|$, $|f_{xy}|$, $|f_{yy}|$. With $M = \sqrt{2}/2$, $|E| \leq 0.0142$. With $M = 1$, $|E| \leq 0.02$.

57. $L(x, y, z) = y - 3z$, $L(x, y, z) = x + y - z - 1$

59. Be more careful with the diameter.

61. $dI = 0.038$, % change in $I = 15.83\%$, more sensitive to voltage change

63. (a) 5% **65.** Local minimum of -8 at $(-2, -2)$

67. Saddle point at $(0, 0)$, $f(0, 0) = 0$; local maximum of $1/4$ at $(-1/2, -1/2)$

69. Saddle point at $(0, 0)$, $f(0, 0) = 0$; local minimum of -4 at $(0, 2)$; local maximum of 4 at $(-2, 0)$; saddle point at $(-2, 2)$, $f(-2, 2) = 0$

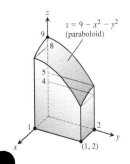

71. Absolute maximum: 28 at $(0, 4)$; absolute minimum: $-9/4$ at $(3/2, 0)$

73. Absolute maximum: 18 at $(2, -2)$; absolute minimum: $-17/4$ at $(-2, 1/2)$

75. Absolute maximum: 8 at $(-2, 0)$; absolute minimum: -1 at $(1, 0)$

77. Absolute maximum: 4 at $(1, 0)$; absolute minimum: -4 at $(0, -1)$

79. Absolute maximum: 1 at $(0, \pm 1)$ and $(1, 0)$; absolute minimum: -1 at $(-1, 0)$

81. Maximum: 5 at $(0, 1)$; minimum: $-1/3$ at $(0, -1/3)$

83. Maximum: $\sqrt{3}$ at $\left(\dfrac{1}{\sqrt{3}}, -\dfrac{1}{\sqrt{3}}, \dfrac{1}{\sqrt{3}} \right)$; minimum: $-\sqrt{3}$ at $\left(-\dfrac{1}{\sqrt{3}}, \dfrac{1}{\sqrt{3}}, -\dfrac{1}{\sqrt{3}} \right)$

85. Width $= \left(\dfrac{c^2 V}{ab} \right)^{1/3}$, depth $= \left(\dfrac{b^2 V}{ac} \right)^{1/3}$, height $= \left(\dfrac{a^2 V}{bc} \right)^{1/3}$

87. Maximum: $\dfrac{3}{2}$ at $\left(\dfrac{1}{\sqrt{2}}, \dfrac{1}{\sqrt{2}}, \sqrt{2} \right)$ and $\left(-\dfrac{1}{\sqrt{2}}, -\dfrac{1}{\sqrt{2}}, -\sqrt{2} \right)$; minimum: $\dfrac{1}{2}$ at $\left(-\dfrac{1}{\sqrt{2}}, \dfrac{1}{\sqrt{2}}, -\sqrt{2} \right)$ and $\left(\dfrac{1}{\sqrt{2}}, -\dfrac{1}{\sqrt{2}}, \sqrt{2} \right)$

89. $\dfrac{\partial w}{\partial x} = \cos\theta \dfrac{\partial w}{\partial r} - \dfrac{\sin\theta}{r} \dfrac{\partial w}{\partial \theta}, \dfrac{\partial w}{\partial y} = \sin\theta \dfrac{\partial w}{\partial r} + \dfrac{\cos\theta}{r} \dfrac{\partial w}{\partial \theta}$

95. $(t, -t \pm 4, t)$, t a real number

101. (a) $(2y + x^2 z)e^{yz}$ (b) $x^2 e^{yz}\left(y - \dfrac{z}{2y} \right)$ (c) $(1 + x^2 y)e^{yz}$

ADDITIONAL AND ADVANCED EXERCISES, pp. 894–896

1. $f_{xy}(0, 0) = -1, f_{yx}(0, 0) = 1$

7. (c) $\dfrac{r^2}{2} = \dfrac{1}{2}(x^2 + y^2 + z^2)$ **13.** $V = \dfrac{\sqrt{3}abc}{2}$

17. $f(x, y) = \dfrac{y}{2} + 4, g(x, y) = \dfrac{x}{2} + \dfrac{9}{2}$

19. $y = 2\ln|\sin x| + \ln 2$

21. (a) $\dfrac{1}{\sqrt{53}}(2\mathbf{i} + 7\mathbf{j})$ (b) $\dfrac{-1}{\sqrt{29{,}097}}(98\mathbf{i} - 127\mathbf{j} + 58\mathbf{k})$

23. $w = e^{-c^2 \pi^2 t} \sin \pi x$

Chapter 15

SECTION 15.1, pp. 901–902

1. 24   **3.** 1   **5.** 16   **7.** $2\ln 2 - 1$ **9.** $(3/2)(5 - e)$

11. $3/2$ **13.** $\ln 2$ **15.** $3/2, -2$ **17.** 14   **19.** 0

21. $1/2$ **23.** $2\ln 2$ **25.** $(\ln 2)^2$

27.

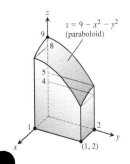

29. $8/3$ **31.** 1   **33.** $\sqrt{2}$ **35.** $2/27$

37. $\dfrac{3}{2}\ln 3 - 1$ **39.** (a) $1/3$ (b) $2/3$

SECTION 15.2, pp. 909–911

1.

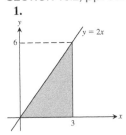

3.

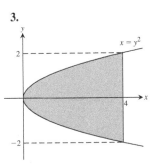

5.

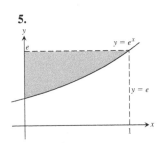

7.

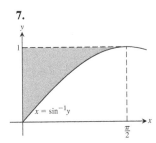

9. (a) $0 \le x \le 2, x^3 \le y \le 8$
(b) $0 \le y \le 8, 0 \le x \le y^{1/3}$

11. (a) $0 \le x \le 3, x^2 \le y \le 3x$
(b) $0 \le y \le 9, \dfrac{y}{3} \le x \le \sqrt{y}$

13. (a) $0 \le x \le 9, 0 \le y \le \sqrt{x}$
(b) $0 \le y \le 3, y^2 \le x \le 9$

15. (a) $0 \le x \le \ln 3, e^{-x} \le y \le 1$
(b) $\dfrac{1}{3} \le y \le 1, -\ln y \le x \le \ln 3$

17. (a) $0 \le x \le 1, x \le y \le 3 - 2x$
(b) $0 \le y \le 1, 0 \le x \le y \cup 1 \le y \le 3, 0 \le x \le \dfrac{3 - y}{2}$

19. $\dfrac{\pi^2}{2} + 2$

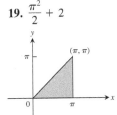

21. $8\ln 8 - 16 + e$

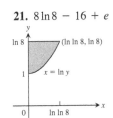

23. $e - 2$

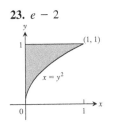

25. $\dfrac{3}{2}\ln 2$ **27.** $-1/10$

29. 8

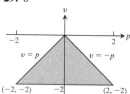

31. 2π

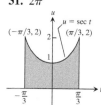

33. $\int_{2}^{4}\int_{0}^{(4-y)/2} dx\, dy$

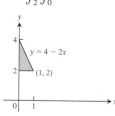

35. $\int_{0}^{1}\int_{x^2}^{x} dy\, dx$

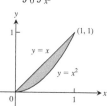

37. $\int_{1}^{e}\int_{\ln y}^{1} dx\, dy$

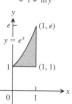

39. $\int_{0}^{9}\int_{0}^{(\sqrt{9-y})/2} 16x\, dx\, dy$

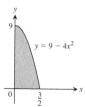

41. $\int_{-1}^{1}\int_{0}^{\sqrt{1-x^2}} 3y\, dy\, dx$

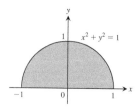

43. $\int_{0}^{1}\int_{e^y}^{e} xy\, dx\, dy$

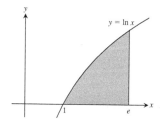

45. $\int_{1}^{e^3}\int_{\ln x}^{3} (x+y)\, dy\, dx$

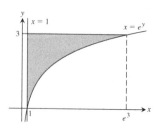

47. 2

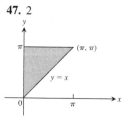

49. $\dfrac{e-2}{2}$

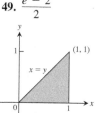

51. 2

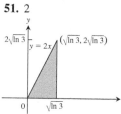

53. $1/(80\pi)$

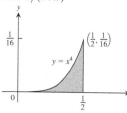

55. $-2/3$

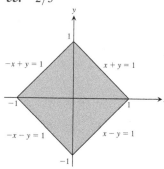

57. $4/3$ **59.** $625/12$ **61.** 16   **63.** 20   **65.** $2(1 + \ln 2)$

67.

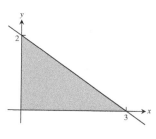

69. 1   **71.** π^2 **73.** $-\dfrac{3}{32}$ **75.** $\dfrac{20\sqrt{3}}{9}$

77. $\int_{0}^{1}\int_{x}^{2-x} (x^2 + y^2)\, dy\, dx = \dfrac{4}{3}$

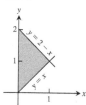

79. R is the set of points (x, y) such that $x^2 + 2y^2 < 4$.

81. No, by Fubini's Theorem, the two orders of integration must give the same result.

85. 0.603   **87.** 0.233

SECTION 15.3, p. 914

1. $\int_0^2 \int_0^{2-x} dy\, dx = 2$ or **3.** $\int_{-2}^1 \int_{y-2}^{-y^2} dx\, dy = \frac{9}{2}$

$\int_0^2 \int_0^{2-y} dx\, dy = 2$

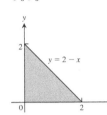

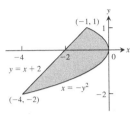

5. $\int_0^{\ln 2} \int_0^{e^x} dy\, dx = 1$ **7.** $\int_0^1 \int_{y^2}^{2y-y^2} dx\, dy = \frac{1}{3}$

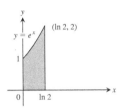

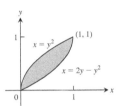

9. $\int_0^2 \int_y^{3y} 1\, dx\, dy = 4$ or

$\int_0^2 \int_{x/3}^x 1\, dy\, dx + \int_2^6 \int_{x/3}^2 1\, dy\, dx = 4$

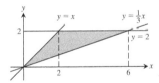

11. $\int_0^1 \int_{x/2}^{2x} 1\, dy\, dx + \int_1^2 \int_{x/2}^{3-x} 1\, dy\, dx = \frac{3}{2}$ or

$\int_0^1 \int_{y/2}^{2y} 1\, dx\, dy + \int_1^2 \int_{y/2}^{3-y} 1\, dx\, dy = \frac{3}{2}$

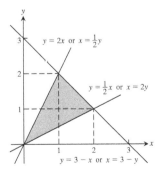

13. 12

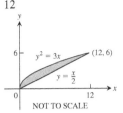

NOT TO SCALE

15. $\sqrt{2} - 1$

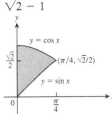

17. $\frac{3}{2}$

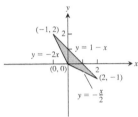

19. (a) 0   **(b)** $4/\pi^2$ **21.** $8/3$ **23.** $\pi - 2$
25. $40{,}000(1 - e^{-2})\ln(7/2) \approx 43{,}329$

SECTION 15.4, pp. 919–921

1. $\frac{\pi}{2} \le \theta \le 2\pi, 0 \le r \le 9$ **3.** $\frac{\pi}{4} \le \theta \le \frac{3\pi}{4}, 0 \le r \le \csc\theta$

5. $0 \le \theta \le \frac{\pi}{6}, 1 \le r \le 2\sqrt{3}\sec\theta$;

$\frac{\pi}{6} \le \theta \le \frac{\pi}{2}, 1 \le r \le 2\csc\theta$

7. $-\frac{\pi}{2} \le \theta \le \frac{\pi}{2}, 0 \le r \le 2\cos\theta$ **9.** $\frac{\pi}{2}$

11. 2π **13.** 36   **15.** $2 - \sqrt{3}$ **17.** $(1 - \ln 2)\pi$

19. $(2\ln 2 - 1)(\pi/2)$ **21.** $\dfrac{2(1 + \sqrt{2})}{3}$

23.

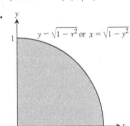

$\int_0^1 \int_0^{\sqrt{1-x^2}} xy\, dy\, dx$ or $\int_0^1 \int_0^{\sqrt{1-y^2}} xy\, dx\, dy$

25.

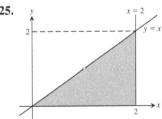

$\int_0^2 \int_0^x y^2(x^2 + y^2)\, dy\, dx$ or $\int_0^2 \int_y^2 y^2(x^2 + y^2)\, dx\, dy$

27. $2(\pi - 2)$ **29.** 12π **31.** $(3\pi/8) + 1$ **33.** $\dfrac{2a}{3}$

35. $\dfrac{2a}{3}$ **37.** $2\pi\left(2 - \sqrt{e}\right)$ **39.** $\dfrac{4}{3} + \dfrac{5\pi}{8}$

41. (a) $\dfrac{\sqrt{\pi}}{2}$ **(b)** 1   **43.** $\pi \ln 4$, no   **45.** $\dfrac{1}{2}\left(a^2 + 2h^2\right)$

47. $\dfrac{8}{9}(3\pi - 4)$

SECTION 15.5, 929–931

1. $1/6$

3. $\displaystyle\int_0^1\!\!\int_0^{2-2x}\!\!\int_0^{3-3x-3y/2} dz\,dy\,dx,\quad \int_0^2\!\!\int_0^{1-y/2}\!\!\int_0^{3-3x-3y/2} dz\,dx\,dy,$

$\displaystyle\int_0^1\!\!\int_0^{3-3x}\!\!\int_0^{2-2x-2z/3} dy\,dz\,dx,\quad \int_0^3\!\!\int_0^{1-z/3}\!\!\int_0^{2-2x-2z/3} dy\,dx\,dz,$

$\displaystyle\int_0^2\!\!\int_0^{3-3y/2}\!\!\int_0^{1-y/2-z/3} dx\,dz\,dy,\quad \int_0^3\!\!\int_0^{2-2z/3}\!\!\int_0^{1-y/2-z/3} dx\,dy\,dz.$

The value of all six integrals is 1.

5. $\displaystyle\int_{-2}^2\!\!\int_{-\sqrt{4-x^2}}^{\sqrt{4-x^2}}\!\!\int_{x^2+y^2}^{8-x^2-y^2} 1\,dz\,dx\,dy,\quad \int_{-2}^2\!\!\int_{-\sqrt{4-y^2}}^{\sqrt{4-y^2}}\!\!\int_{x^2+y^2}^{8-x^2-y^2} 1\,dz\,dx\,dy,$

$\displaystyle\int_{-2}^2\!\!\int_4^{8-y^2}\!\!\int_{-\sqrt{8-z-y^2}}^{\sqrt{8-z-y^2}} 1\,dx\,dz\,dy + \int_{-2}^2\!\!\int_{y^2}^4\!\!\int_{-\sqrt{z-y^2}}^{\sqrt{z-y^2}} 1\,dx\,dz\,dy,$

$\displaystyle\int_4^8\!\!\int_{-\sqrt{8-z}}^{\sqrt{8-z}}\!\!\int_{-\sqrt{8-z-y^2}}^{\sqrt{8-z-y^2}} 1\,dx\,dy\,dz + \int_0^4\!\!\int_{-\sqrt{z}}^{\sqrt{z}}\!\!\int_{-\sqrt{z-y^2}}^{\sqrt{z-y^2}} 1\,dx\,dy\,dz,$

$\displaystyle\int_{-2}^2\!\!\int_4^{8-x^2}\!\!\int_{-\sqrt{8-z-x^2}}^{\sqrt{8-z-x^2}} 1\,dy\,dz\,dx + \int_{-2}^2\!\!\int_{x^2}^4\!\!\int_{-\sqrt{z-x^2}}^{\sqrt{z-x^2}} 1\,dy\,dz\,dx,$

$\displaystyle\int_4^8\!\!\int_{-\sqrt{8-z}}^{\sqrt{8-z}}\!\!\int_{-\sqrt{8-z-x^2}}^{\sqrt{8-z-x^2}} 1\,dy\,dx\,dz + \int_0^4\!\!\int_{-\sqrt{z}}^{\sqrt{z}}\!\!\int_{-\sqrt{z-x^2}}^{\sqrt{z-x^2}} 1\,dy\,dx\,dz.$

The value of all six integrals is 16π.

7. 1   **9.** 6   **11.** $\dfrac{5\left(2 - \sqrt{3}\right)}{4}$ **13.** 18

15. $7/6$ **17.** 0   **19.** $\dfrac{1}{2} - \dfrac{\pi}{8}$

21. (a) $\displaystyle\int_{-1}^1\!\!\int_0^{1-x^2}\!\!\int_{x^2}^{1-z} dy\,dz\,dx$ **(b)** $\displaystyle\int_0^1\!\!\int_{-\sqrt{1-z}}^{\sqrt{1-z}}\!\!\int_{x^2}^{1-z} dy\,dx\,dz$

(c) $\displaystyle\int_0^1\!\!\int_0^{1-z}\!\!\int_{-\sqrt{y}}^{\sqrt{y}} dx\,dy\,dz$ **(d)** $\displaystyle\int_0^1\!\!\int_0^{1-y}\!\!\int_{-\sqrt{y}}^{\sqrt{y}} dx\,dz\,dy$

(e) $\displaystyle\int_0^1\!\!\int_{-\sqrt{y}}^{\sqrt{y}}\!\!\int_0^{1-y} dz\,dx\,dy$

23. $2/3$ **25.** $20/3$ **27.** 1   **29.** $16/3$ **31.** $8\pi - \dfrac{32}{3}$
33. 2   **35.** 4π **37.** $31/3$ **39.** 1   **41.** $2\sin 4$
43. 4   **45.** $a = 3$ or $a = 13/3$
47. The domain is the set of all points (x, y, z) such that
$4x^2 + 4y^2 + z^2 \le 4$.

SECTION 15.6, pp. 938–941

1. $\bar{x} = 5/14, \bar{y} = 38/35$ **3.** $\bar{x} = 64/35, \bar{y} = 5/7$
5. $\bar{x} = \bar{y} = 4a/(3\pi)$
7. $I_x = I_y = 4\pi$ gm/cm², $I_0 = 8\pi$ gm/cm²
9. $\bar{x} = -1, \bar{y} = 1/4$ **11.** $I_x = 64/105$
13. $\bar{x} = 3/8, \bar{y} = 17/16$ **15.** $\bar{x} = 11/3, \bar{y} = 14/27, I_y = 432$
17. $\bar{x} = 0, \bar{y} = 13/31, I_y = 7/5$
19. $\bar{x} = 0, \bar{y} = 7/10; I_x = 9/10$ kg/m², $I_y = 3/10$ kg/m²,
   $I_0 = 6/5$ kg/m²

21. $I_x = \dfrac{M}{3}\left(b^2 + c^2\right), I_y = \dfrac{M}{3}\left(a^2 + c^2\right), I_z = \dfrac{M}{3}\left(a^2 + b^2\right)$

23. $\bar{x} = \bar{y} = 0, \bar{z} = 12/5, I_x = 7904/105 \approx 75.28,$
   $I_y = 4832/63 \approx 76.70, I_z = 256/45 \approx 5.69$
25. (a) $\bar{x} = \bar{y} = 0, \bar{z} = 8/3$ **(b)** $c = 2\sqrt{2}$
27. $I_L = 1386$
29. (a) $4/3$ gm   **(b)** $\bar{x} = 4/5$ cm, $\bar{y} = \bar{z} = 2/5$ cm
31. (a) $5/2$ **(b)** $\bar{x} = \bar{y} = \bar{z} = 8/15$ **(c)** $I_x = I_y = I_z = 11/6$
33. 3 kg

37. (a) $I_{\text{c.m.}} = \dfrac{abc(a^2 + b^2)}{12}, R_{\text{c.m.}} = \sqrt{\dfrac{a^2 + b^2}{12}}$

   (b) $I_L = \dfrac{abc(a^2 + 7b^2)}{3}, R_L = \sqrt{\dfrac{a^2 + 7b^2}{3}}$

39. $\mu_x = \mu_y = 7/12$
41. $\mu_x = 3/4, \mu_y = 2/3$
43. $f(x, y) = 1/6, P(X < Y) = 2/3$

SECTION 15.7, pp. 949–953

1.

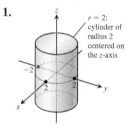

$r = 2$:
cylinder of
radius 2
centered on
the z-axis

3. $z = -1$:
plane parallel
to the xy-plane

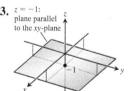

5.

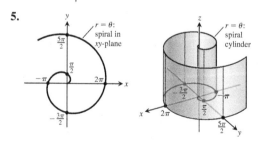

$r = \theta$:
spiral in
xy-plane

$r = \theta$:
spiral
cylinder

7.

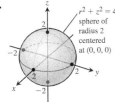

$r^2 + z^2 = 4$
sphere of
radius 2
centered
at (0, 0, 0)

9. $r \le z \le \sqrt{9 - r^2}$: cone with vertex angle $\frac{\pi}{2}$ below a sphere of radius 3 centered at (0, 0, 0), and its interior

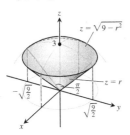

$z = \sqrt{9 - r^2}$

$z = r$

11. $0 \le r \le 4 \cos \theta, 0 \le \theta \le \frac{\pi}{2}, 0 \le z \le 5$: half-cylinder of height 5, radius 2, and tangent to the z-axis, and its interior

13. $\rho = 3$: sphere of radius 3 centered at (0, 0, 0)

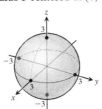

15. $\theta = \frac{2}{3}\pi$: closed half-plane along the z-axis

$\theta = \frac{2}{3}\pi$

17. $\rho \cos \phi = 4$: plane with z-intercept 4 and parallel to the xy-plane

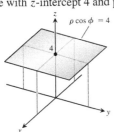

$\rho \cos \phi = 4$

19. $0 \le \rho \le 3 \csc \phi \Rightarrow 0 \le \rho \sin \phi \le 3$: cylinder of radius 3 centered on the z-axis, and its interior

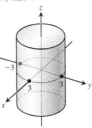

21. $0 \le \rho \cos \theta \sin \phi \le 2, 0 \le \rho \sin \theta \sin \phi \le 3,$
$0 \le \rho \cos \phi \le 4$: rectangular box $2 \times 3 \times 4$, and its interior

23. $\dfrac{4\pi(\sqrt{2} - 1)}{3}$ **25.** $\dfrac{17\pi}{5}$ **27.** $\pi(6\sqrt{2} - 8)$ **29.** $\dfrac{3\pi}{10}$

31. $\pi/3$

33. (a) $\displaystyle\int_0^{2\pi}\int_0^1\int_0^{\sqrt{4-r^2}} r \, dz \, dr \, d\theta$

(b) $\displaystyle\int_0^{2\pi}\int_0^{\sqrt{3}}\int_0^1 r \, dr \, dz \, d\theta + \int_0^{2\pi}\int_{\sqrt{3}}^2\int_0^{\sqrt{4-z^2}} r \, dr \, dz \, d\theta$

(c) $\displaystyle\int_0^1\int_0^{\sqrt{4-r^2}}\int_0^{2\pi} r \, d\theta \, dz \, dr$

35. $\displaystyle\int_{-\pi/2}^{\pi/2}\int_0^{\cos\theta}\int_0^{3r^2} f(r, \theta, z) \, dz \, r \, dr \, d\theta$

37. $\displaystyle\int_0^{\pi}\int_0^{2\sin\theta}\int_0^{4-r\sin\theta} f(r, \theta, z) \, dz \, r \, dr \, d\theta$

39. $\displaystyle\int_{-\pi/2}^{\pi/2}\int_1^{1+\cos\theta}\int_0^4 f(r, \theta, z) \, dz \, r \, dr \, d\theta$

41. $\displaystyle\int_0^{\pi/4}\int_0^{\sec\theta}\int_0^{2-r\sin\theta} f(r, \theta, z) \, dz \, r \, dr \, d\theta$ **43.** π^2

45. $\pi/3$ **47.** 5π **49.** 2π **51.** $\left(\dfrac{8 - 5\sqrt{2}}{2}\right)\pi$

53. (a) $\displaystyle\int_0^{2\pi}\int_0^{\pi/6}\int_0^2 \rho^2 \sin\phi \, d\rho \, d\phi \, d\theta +$
$\displaystyle\int_0^{2\pi}\int_{\pi/6}^{\pi/2}\int_0^{\csc\phi} \rho^2 \sin\phi \, d\rho \, d\phi \, d\theta$

(b) $\displaystyle\int_0^{2\pi}\int_1^2\int_{\pi/6}^{\sin^{-1}(1/\rho)} \rho^2 \sin\phi \, d\phi \, d\rho \, d\theta +$
$\displaystyle\int_0^{2\pi}\int_0^2\int_0^{\pi/6} \rho^2 \sin\phi \, d\phi \, d\rho \, d\theta +$
$\displaystyle\int_0^{2\pi}\int_0^1\int_{\pi/6}^{\pi/2} \rho^2 \sin\phi \, d\phi \, d\rho \, d\theta$

55. $\displaystyle\int_0^{2\pi}\int_0^{\pi/2}\int_{\cos\phi}^2 \rho^2 \sin\phi\, d\rho\, d\phi\, d\theta = \frac{31\pi}{6}$

57. $\displaystyle\int_0^{2\pi}\int_0^{\pi}\int_0^{1-\cos\phi} \rho^2 \sin\phi\, d\rho\, d\phi\, d\theta = \frac{8\pi}{3}$

59. $\displaystyle\int_0^{2\pi}\int_{\pi/4}^{\pi/2}\int_0^{2\cos\phi} \rho^2 \sin\phi\, d\rho\, d\phi\, d\theta = \frac{\pi}{3}$

61. (a) $\displaystyle 8\int_0^{\pi/2}\int_0^{\pi/2}\int_0^2 \rho^2 \sin\phi\, d\rho\, d\phi\, d\theta$

(b) $\displaystyle 8\int_0^{\pi/2}\int_0^2\int_0^{\sqrt{4-r^2}} r\, dz\, dr\, d\theta$

(c) $\displaystyle 8\int_0^2\int_0^{\sqrt{4-x^2}}\int_0^{\sqrt{4-x^2-y^2}} dz\, dy\, dx$

63. (a) $\displaystyle \int_0^{2\pi}\int_0^{\pi/3}\int_{\sec\phi}^2 \rho^2 \sin\phi\, d\rho\, d\phi\, d\theta$

(b) $\displaystyle \int_0^{2\pi}\int_0^{\sqrt3}\int_1^{\sqrt{4-r^2}} r\, dz\, dr\, d\theta$

(c) $\displaystyle \int_{-\sqrt3}^{\sqrt3}\int_{-\sqrt{3-x^2}}^{\sqrt{3-x^2}}\int_1^{\sqrt{4-x^2-y^2}} dz\, dy\, dx$ **(d)** $5\pi/3$

65. $8\pi/3$ **67.** $9/4$ **69.** $\dfrac{3\pi-4}{18}$ **71.** $\dfrac{2\pi a^3}{3}$

73. $5\pi/3$ **75.** $\pi/2$ **77.** $\dfrac{4(2\sqrt2-1)\pi}{3}$ **79.** 16π

81. $5\pi/2$ **83.** $\dfrac{4\pi(8-3\sqrt3)}{3}$ **85.** $2/3$ **87.** $3/4$

89. $\bar x = \bar y = 0, \bar z = 3/8$ **91.** $(\bar x, \bar y, \bar z) = (0,0,3/8)$

93. $\bar x = \bar y = 0, \bar z = 5/6$ **95.** $I_x = \pi/4$ **97.** $\dfrac{a^4 h\pi}{10}$

99. (a) $(\bar x, \bar y, \bar z) = \left(0,0,\dfrac{4}{5}\right), I_z = \dfrac{\pi}{12}$

(b) $(\bar x, \bar y, \bar z) = \left(0,0,\dfrac{5}{6}\right), I_z = \dfrac{\pi}{14}$

101. $\dfrac{3M}{\pi R^3}$

103. The surface's equation $r = f(z)$ tells us that the point $(r, \theta, z) = (f(z), \theta, z)$ will lie on the surface for all θ. In particular, $(f(z), \theta + \pi, z)$ lies on the surface whenever $(f(z), \theta, z)$ lies on the surface, so the surface is symmetric with respect to the z-axis.

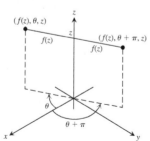

SECTION 15.8, pp. 961–963

1. (a) $x = \dfrac{u+v}{3}, y = \dfrac{v-2u}{3}; \dfrac{1}{3}$

(b) Triangular region with boundaries $u = 0, v = 0$, and $u + v = 3$

3. (a) $x = \dfrac{1}{5}(2u-v), y = \dfrac{1}{10}(3v-u); \dfrac{1}{10}$

(b) Triangular region with boundaries $3v = u, v = 2u$, and $3u + v = 10$

7. $64/5$ **9.** $\displaystyle\int_1^2\int_1^3 (u+v)\dfrac{2u}{v}\, du\, dv = 8 + \dfrac{52}{3}\ln 2$

11. $\dfrac{\pi ab(a^2 + b^2)}{4}$ **13.** $\dfrac{1}{3}\left(1 + \dfrac{3}{e^2}\right) \approx 0.4687$

15. $\dfrac{225}{16}$ **17.** 12 **19.** $\dfrac{a^2 b^2 c^2}{6}$

21. (a) $\begin{vmatrix} \cos v & -u\sin v \\ \sin v & u\cos v \end{vmatrix} = u\cos^2 v + u\sin^2 v = u$

(b) $\begin{vmatrix} \sin v & u\cos v \\ \cos v & -u\sin v \end{vmatrix} = -u\sin^2 v - u\cos^2 v = -u$

27. $\dfrac{3}{2}\ln 2$

PRACTICE EXERCISES, pp. 963–965

1. $9e - 9$ **3.** $9/2$

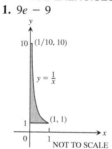

NOT TO SCALE

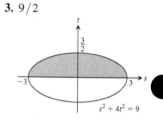

5. $\displaystyle\int_{-2}^0\int_{2x+4}^{4-x^2} dy\, dx = \dfrac{4}{3}$ **7.** $\displaystyle\int_{-3}^3\int_0^{(1/2)\sqrt{9-x^2}} y\, dy\, dx = \dfrac{9}{2}$

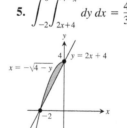

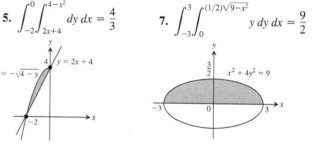

9. $\sin 4$ **11.** $\dfrac{\ln 17}{4}$ **13.** $4/3$ **15.** $4/3$ **17.** $1/4$

19. π **21.** $\dfrac{\pi-2}{4}$ **23.** 0 **25.** $8/35$ **27.** $\pi/2$

29. $\dfrac{2(31-3^{5/2})}{3}$

31. (a) $\displaystyle\int_{-\sqrt2}^{\sqrt2}\int_{-\sqrt{2-y^2}}^{\sqrt{2-y^2}}\int_{\sqrt{x^2+y^2}}^{\sqrt{4-x^2-y^2}} 3\, dz\, dx\, dy$

(b) $\displaystyle\int_0^{2\pi}\int_0^{\pi/4}\int_0^2 3\rho^2 \sin\phi\, d\rho\, d\phi\, d\theta$ **(c)** $2\pi(8 - 4\sqrt2)$

33. $\int_0^{2\pi} \int_0^{\pi/4} \int_0^{\sec\phi} \rho^2 \sin\phi \, d\rho \, d\phi \, d\theta = \dfrac{\pi}{3}$

35. $\int_0^1 \int_{\sqrt{1-x^2}}^{\sqrt{3-x^2}} \int_1^{\sqrt{4-x^2-y^2}} z^2 xy \, dz \, dy \, dx$

$+ \int_1^{\sqrt{3}} \int_0^{\sqrt{3-x^2}} \int_1^{\sqrt{4-x^2-y^2}} z^2 xy \, dz \, dy \, dx$

37. (a) $\dfrac{8\pi(4\sqrt{2}-5)}{3}$ **(b)** $\dfrac{8\pi(4\sqrt{2}-5)}{3}$

39. $I_z = \dfrac{8\delta(b^5-a^5)}{15}$

41. $\bar{x} = \bar{y} = \dfrac{1}{2-\ln 4}$ **43.** $I_0 = 104$ **45.** $I_x = 2\delta$

47. $M = 4, M_x = 0, M_y = 0$

49. $\bar{x} = \dfrac{3\sqrt{3}}{\pi}, \bar{y} = 0$

51. (a) $\bar{x} = \dfrac{15\pi + 32}{6\pi + 48}, \bar{y} = 0$

(b)

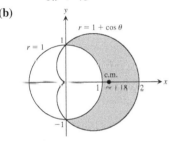

$r = 1 + \cos\theta$
$r = 1$
c.m.

ADDITIONAL AND ADVANCED EXERCISES, pp. 966–967

1. (a) $\int_{-3}^2 \int_x^{6-x^2} x^2 \, dy \, dx$ **(b)** $\int_{-3}^2 \int_x^{6-x^2} \int_0^{x^2} dz \, dy \, dx$

(c) $125/4$

3. 2π **5.** $3\pi/2$

7. (a) Hole radius = 1, sphere radius = 2   **(b)** $4\sqrt{3}\pi$

9. $\pi/4$ **11.** $\ln\left(\dfrac{b}{a}\right)$ **15.** $1/\sqrt[4]{3}$

17. Mass $= a^2 \cos^{-1}\left(\dfrac{b}{a}\right) - b\sqrt{a^2-b^2}$,

$I_0 = \dfrac{a^4}{2}\cos^{-1}\left(\dfrac{b}{a}\right) - \dfrac{b^3}{2}\sqrt{a^2-b^2} - \dfrac{b^3}{6}(a^2-b^2)^{3/2}$

19. $\dfrac{1}{ab}(e^{a^2b^2}-1)$ **21. (b)** 1   **(c)** 0

25. $h = \sqrt{20}$ in., $h = \sqrt{60}$ in.   **27.** $2\pi\left[\dfrac{1}{3} - \left(\dfrac{1}{3}\right)\dfrac{\sqrt{2}}{2}\right]$

Chapter 16

SECTION 16.1, pp. 974–976

1. Graph (c)   **3.** Graph (g)   **5.** Graph (d)   **7.** Graph (f)

9. $\sqrt{2}$ **11.** $\dfrac{13}{2}$ **13.** $3\sqrt{14}$ **15.** $\dfrac{1}{6}(5\sqrt{5}+9)$

17. $\sqrt{3}\ln\left(\dfrac{b}{a}\right)$ **19. (a)** $4\sqrt{5}$ **(b)** $\dfrac{1}{12}(17^{3/2}-1)$

21. $\dfrac{15}{32}(e^{16}-e^{64})$ **23.** $\dfrac{1}{27}(40^{3/2}-13^{3/2})$

25. $\dfrac{1}{6}(5^{3/2}+7\sqrt{2}-1)$ **27.** $\dfrac{10\sqrt{5}-2}{3}$ **29.** 8

31. $\dfrac{1}{6}(17^{3/2}-1)$ **33.** $2\sqrt{2}-1$

35. (a) $4\sqrt{2}-2$ **(b)** $\sqrt{2}+\ln(1+\sqrt{2})$ **37.** $I_z = 2\pi\delta a^3$

39. (a) $I_z = 2\pi\sqrt{2}\delta$ **(b)** $I_z = 4\pi\sqrt{2}\delta$ **41.** $I_x = 2\pi - 2$

SECTION 16.2, pp. 986–989

1. $\nabla f = -(x\mathbf{i} + y\mathbf{j} + z\mathbf{k})(x^2+y^2+z^2)^{-3/2}$

3. $\nabla g = -\left(\dfrac{2x}{x^2+y^2}\right)\mathbf{i} - \left(\dfrac{2y}{x^2+y^2}\right)\mathbf{j} + e^z\mathbf{k}$

5. $\mathbf{F} = -\dfrac{kx}{(x^2+y^2)^{3/2}}\mathbf{i} - \dfrac{ky}{(x^2+y^2)^{3/2}}\mathbf{j}$, any $k > 0$

7. (a) $9/2$ **(b)** $13/3$ **(c)** $9/2$

9. (a) $1/3$ **(b)** $-1/5$ **(c)** 0

11. (a) 2   **(b)** $3/2$ **(c)** $1/2$

13. $-15/2$ **15.** 36   **17. (a)** $-5/6$ **(b)** 0   **(c)** $-7/12$

19. $1/2$ **21.** $-\pi$ **23.** $69/4$ **25.** $-39/2$ **27.** $25/6$

29. (a) $\text{Circ}_1 = 0, \text{circ}_2 = 2\pi, \text{flux}_1 = 2\pi, \text{flux}_2 = 0$
   (b) $\text{Circ}_1 = 0, \text{circ}_2 = 8\pi, \text{flux}_1 = 8\pi, \text{flux}_2 = 0$

31. $\text{Circ} = 0, \text{flux} = a^2\pi$ **33.** $\text{Circ} = a^2\pi, \text{flux} = 0$

35. (a) $-\dfrac{\pi}{2}$ **(b)** 0   **(c)** 1   **37.** $(.0001)\pi\,\text{kg/s}$

39. (a) 32   **(b)** 32   **(c)** 32   **41.** $115.2\,\text{J}$

43. $5/3 - (3/2)\ln 2\ \text{m}^2/\text{s}$ **45.** $5/3\,\text{g/s}$

47.

$x^2 + y^2 = 4$

49. (a) $\mathbf{G} = -y\mathbf{i} + x\mathbf{j}$ **(b)** $\mathbf{G} = \sqrt{x^2+y^2}\,\mathbf{F}$

51. $\mathbf{F} = -\dfrac{x\mathbf{i}+y\mathbf{j}}{\sqrt{x^2+y^2}}$ **55.** 48   **57.** π **59.** 0   **61.** $\dfrac{1}{2}$

SECTION 16.3, pp. 998–1000

1. Conservative   **3.** Not conservative   **5.** Not conservative

7. $f(x,y,z) = x^2 + \dfrac{3y^2}{2} + 2z^2 + C$ **9.** $f(x,y,z) = xe^{y+2z} + C$

11. $f(x,y,z) = x\ln x - x + \tan(x+y) + \dfrac{1}{2}\ln(y^2+z^2) + C$

13. 49   **15.** -16 **17.** 1   **19.** $9\ln 2$ **21.** 0   **23.** -3

27. $\mathbf{F} = \nabla\left(\dfrac{x^2-1}{y}\right)$ **29. (a)** 1   **(b)** 1   **(c)** 1

31. (a) 2   **(b)** 2   **33. (a)** $c = b = 2a$ **(b)** $c = b = 2$

35. It does not matter what path you use. The work will be the same on any path because the field is conservative.

37. The force $\mathbf{F}$ is conservative because all partial derivatives of M, N, and P are zero. $f(x, y, z) = ax + by + cz + C$; $A = (xa, ya, za)$ and $B = (xb, yb, zb)$. Therefore, $\int \mathbf{F} \cdot d\mathbf{r} = f(B) - f(A) = a(xb - xa) + b(yb - ya) + c(zb - za) = \mathbf{F} \cdot \vec{AB}$.

SECTION 16.4, pp. 1010–1012

1. $2y - 1$ **3.** $ye^x - xe^y$ **5.** $\sin y - \sin x$
7. Flux $= 0$, circ $= 2\pi a^2$ **9.** Flux $= -\pi a^2$, circ $= 0$
11. Flux $= 2$, circ $= 0$ **13.** Flux $= -9$, circ $= 9$
15. Flux $= -11/60$, circ $= -7/60$
17. Flux $= 64/9$, circ $= 0$ **19.** Flux $= 1/2$, circ $= 1/2$
21. Flux $= 1/5$, circ $= -1/12$ **23.** 0   **25.** $2/33$ **27.** 0
29. -16π **31.** πa^2 **33.** $3\pi/8$
35. (a) 0 if C is traversed counterclockwise
   (b) $(h - k)$(area of the region)   **45. (a)** 0

SECTION 16.5, pp. 1020–1022

1. $\mathbf{r}(r, \theta) = (r \cos \theta)\mathbf{i} + (r \sin \theta)\mathbf{j} + r^2\mathbf{k}, 0 \le r \le 2,$
   $0 \le \theta \le 2\pi$
3. $\mathbf{r}(r, \theta) = (r \cos \theta)\mathbf{i} + (r \sin \theta)\mathbf{j} + (r/2)\mathbf{k}, 0 \le r \le 6,$
   $0 \le \theta \le \pi/2$
5. $\mathbf{r}(r, \theta) = (r \cos \theta)\mathbf{i} + (r \sin \theta)\mathbf{j} + \sqrt{9 - r^2}\,\mathbf{k},$
   $0 \le r \le 3\sqrt{2}/2, 0 \le \theta \le 2\pi$; Also:
   $\mathbf{r}(\phi, \theta) = (3 \sin \phi \cos \theta)\mathbf{i} + (3 \sin \phi \sin \theta)\mathbf{j} +$
   $(3 \cos \phi)\mathbf{k}, 0 \le \phi \le \pi/4, 0 \le \theta \le 2\pi$
7. $\mathbf{r}(\phi, \theta) = \left(\sqrt{3} \sin \phi \cos \theta\right)\mathbf{i} + \left(\sqrt{3} \sin \phi \sin \theta\right)\mathbf{j} +$
   $\left(\sqrt{3} \cos \phi\right)\mathbf{k}, \pi/3 \le \phi \le 2\pi/3, 0 \le \theta \le 2\pi$
9. $\mathbf{r}(x, y) = x\mathbf{i} + y\mathbf{j} + (4 - y^2)\mathbf{k}, 0 \le x \le 2, -2 \le y \le 2$
11. $\mathbf{r}(u, v) = u\mathbf{i} + (3 \cos v)\mathbf{j} + (3 \sin v)\mathbf{k}, 0 \le u \le 3,$
   $0 \le v \le 2\pi$
13. (a) $\mathbf{r}(r, \theta) = (r \cos \theta)\mathbf{i} + (r \sin \theta)\mathbf{j} + (1 - r \cos \theta - r \sin \theta)\mathbf{k},$
   $0 \le r \le 3, 0 \le \theta \le 2\pi$
   (b) $\mathbf{r}(u, v) = (1 - u \cos v - u \sin v)\mathbf{i} + (u \cos v)\mathbf{j} +$
   $(u \sin v)\mathbf{k}, 0 \le u \le 3, 0 \le v \le 2\pi$
15. $\mathbf{r}(u, v) = (4 \cos^2 v)\mathbf{i} + u\mathbf{j} + (4 \cos v \sin v)\mathbf{k}, 0 \le u \le 3,$
   $-(\pi/2) \le v \le (\pi/2)$; Another way: $\mathbf{r}(u, v) = (2 + 2 \cos v)\mathbf{i}$
   $+ u\mathbf{j} + (2 \sin v)\mathbf{k}, 0 \le u \le 3, 0 \le v \le 2\pi$
17. $\displaystyle\int_0^{2\pi} \int_0^1 \frac{\sqrt{5}}{2} r \, dr \, d\theta = \frac{\pi\sqrt{5}}{2}$
19. $\displaystyle\int_0^{2\pi} \int_1^3 r\sqrt{5} \, dr \, d\theta = 8\pi\sqrt{5}$ **21.** $\displaystyle\int_0^{2\pi} \int_1^4 1 \, du \, dv = 6\pi$
23. $\displaystyle\int_0^{2\pi} \int_0^1 u\sqrt{4u^2 + 1} \, du \, dv = \frac{(5\sqrt{5} - 1)}{6}\pi$
25. $\displaystyle\int_0^{2\pi} \int_{\pi/4}^{\pi} 2 \sin \phi \, d\phi \, d\theta = \left(4 + 2\sqrt{2}\right)\pi$

27.

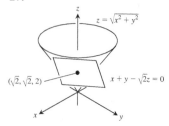

29.

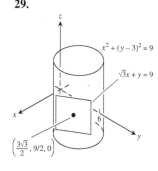

33. (b) $A = \displaystyle\int_0^{2\pi} \int_0^{\pi} \left[a^2b^2 \sin^2 \phi \cos^2 \phi + b^2c^2 \cos^4 \phi \cos^2 \theta + a^2c^2 \cos^4 \phi \sin^2 \theta\right]^{1/2} d\phi \, d\theta$

35. $x_0 x + y_0 y = 25$ **37.** $13\pi/3$ **39.** 4
41. $6\sqrt{6} - 2\sqrt{2}$ **43.** $\pi\sqrt{c^2 + 1}$
45. $\frac{\pi}{6}\left(17\sqrt{17} - 5\sqrt{5}\right)$ **47.** $3 + 2\ln 2$
49. $\frac{\pi}{6}\left(13\sqrt{13} - 1\right)$ **51.** $5\pi\sqrt{2}$ **53.** $\frac{2}{3}\left(5\sqrt{5} - 1\right)$

SECTION 16.6, pp. 1030–1032

1. $\displaystyle\iint\limits_S x \, d\sigma = \int_0^3 \int_0^2 u\sqrt{4u^2 + 1} \, du \, dv = \frac{17\sqrt{17} - 1}{4}$

3. $\displaystyle\iint\limits_S x^2 \, d\sigma = \int_0^{2\pi} \int_0^{\pi} \sin^3 \phi \cos^2 \theta \, d\phi \, d\theta = \frac{4\pi}{3}$

5. $\displaystyle\iint\limits_S z \, d\sigma = \int_0^1 \int_0^1 (4 - u - v)\sqrt{3} \, dv \, du = 3\sqrt{3}$
   (for $x = u, y = v$)

7. $\displaystyle\iint\limits_S x^2\sqrt{5 - 4z} \, d\sigma = \int_0^1 \int_0^{2\pi} u^2 \cos^2 v \cdot \sqrt{4u^2 + 1} \cdot$
   $u\sqrt{4u^2 + 1} \, dv \, du = \int_0^1 \int_0^{2\pi} u^3(4u^2 + 1) \cos^2 v \, dv \, du = \frac{11\pi}{12}$

9. $9a^3$ **11.** $\frac{abc}{4}(ab + ac + bc)$ **13.** 2
15. $\frac{1}{30}\left(\sqrt{2} + 6\sqrt{6}\right)$ **17.** $\sqrt{6}/30$ **19.** -32 **21.** $\frac{\pi a^3}{6}$
23. $13a^4/6$ **25.** $2\pi/3$ **27.** $-73\pi/6$ **29.** 18
31. $\frac{\pi a^3}{6}$ **33.** $\frac{\pi a^2}{4}$ **35.** $\frac{\pi a^3}{2}$ **37.** -32 **39.** -4
41. $3a^4$ **43.** $\left(\frac{a}{2}, \frac{a}{2}, \frac{a}{2}\right)$
45. $(\bar{x}, \bar{y}, \bar{z}) = \left(0, 0, \frac{14}{9}\right), I_z = \frac{15\pi\sqrt{2}}{2}\delta$
47. (a) $\frac{8\pi}{3}a^4\delta$ **(b)** $\frac{20\pi}{3}a^4\delta$ **49.** $70/3\,\text{mg}$

SECTION 16.7, pp. 1043–1045

1. $-\mathbf{i} - 4\mathbf{j} + \mathbf{k}$ **3.** $(1 - y)\mathbf{i} + (1 - z)\mathbf{j} + (1 - x)\mathbf{k}$
5. $x(z^2 - y^2)\mathbf{i} + y(x^2 - z^2)\mathbf{j} + z(y^2 - x^2)\mathbf{k}$ **7.** 4π
9. $-5/6$ **11.** 0   **13.** -6π **15.** $2\pi a^2$ **17.** $-\pi$
19. 12π **21.** $-\pi/4$ **23.** -15π **25.** -8π
33. $16I_y + 16I_x$

SECTION 16.8, pp. 1056–1057

1. 0   **3.** $(y^2z + xz^2 + x^2y)e^{xyz}$ **5.** 0   **7.** 0
9. -16 **11.** -8π **13.** 3π **15.** $-40/3$ **17.** 12π
19. $12\pi(4\sqrt{2} - 1)$ **23.** No
25. The integral's value never exceeds the surface area of S.
27. $184/35$

PRACTICE EXERCISES, pp. 1058–1060

1. Path 1: $2\sqrt{3}$; path 2: $1 + 3\sqrt{2}$ **3.** $4a^2$ **5.** 0
7. $8\pi \sin(1)$ **9.** 0   **11.** $\pi\sqrt{3}$
13. $2\pi\left(1 - \frac{1}{\sqrt{2}}\right)$ **15.** $\frac{abc}{2}\sqrt{\frac{1}{a^2} + \frac{1}{b^2} + \frac{1}{c^2}}$ **17.** 50

19. $\mathbf{r}(\phi, \theta) = (6 \sin \phi \cos \theta)\mathbf{i} + (6 \sin \phi \sin \theta)\mathbf{j} + (6 \cos \phi)\mathbf{k}$,
$\dfrac{\pi}{6} \le \phi \le \dfrac{2\pi}{3}, 0 \le \theta \le 2\pi$

21. $\mathbf{r}(r, \theta) = (r \cos \theta)\mathbf{i} + (r \sin \theta)\mathbf{j} + (1 + r)\mathbf{k}, 0 \le r \le 2$,
$0 \le \theta \le 2\pi$

23. $\mathbf{r}(u, v) = (u \cos v)\mathbf{i} + 2u^2\mathbf{j} + (u \sin v)\mathbf{k}, 0 \le u \le 1$,
$0 \le v \le \pi$

25. $\sqrt{6}$ **27.** $\pi\left[\sqrt{2} + \ln\left(1 + \sqrt{2}\right)\right]$ **29.** Conservative

31. Not conservative   **33.** $f(x, y, z) = y^2 + yz + 2x + z$

35. Path 1: 2; path 2: 8/3   **37. (a)** $1 - e^{-2\pi}$ **(b)** $1 - e^{-2\pi}$

39. 0   **41. (a)** $4\sqrt{2} - 2$ **(b)** $\sqrt{2} + \ln\left(1 + \sqrt{2}\right)$

43. $(\bar{x}, \bar{y}, \bar{z}) = \left(1, \dfrac{16}{15}, \dfrac{2}{3}\right); I_x = \dfrac{232}{45}, I_y = \dfrac{64}{15}, I_z = \dfrac{56}{9}$

45. $\bar{z} = \dfrac{3}{2}, I_z = \dfrac{7\sqrt{3}}{3}$ **47.** $(\bar{x}, \bar{y}, \bar{z}) = (0, 0, 49/12), I_z = 640\pi$

49. Flux: 3/2; circ: $-1/2$ **53.** 3

55. $\dfrac{2\pi}{3}\left(7 - 8\sqrt{2}\right)$ **57.** 0   **59.** π

ADDITIONAL AND ADVANCED EXERCISES, pp. 1061–1062

1. 6π **3.** 2/3

5. (a) $\mathbf{F}(x, y, z) = z\mathbf{i} + x\mathbf{j} + y\mathbf{k}$ **(b)** $\mathbf{F}(x, y, z) = z\mathbf{i} + y\mathbf{k}$
   (c) $\mathbf{F}(x, y, z) = z\mathbf{i}$

7. $\dfrac{16\pi R^3}{3}$ **9.** $a = 2, b = 1$. The minimum flux is -4.

11. (b) $\dfrac{16}{3}g$ **(c)** Work $= \left(\displaystyle\int_C gxy \, ds\right) \bar{y} = g \displaystyle\int_C xy^2 \, ds = \dfrac{16}{3}g$

13. (c) $\dfrac{4}{3}\pi w$ **19.** False if $\mathbf{F} = y\mathbf{i} + x\mathbf{j}$

Appendices

APPENDIX 1, p. AP-6

1. $0.\overline{1}, 0.\overline{2}, 0.\overline{3}, 0.\overline{8}, 0.\overline{9}$ or 1

3. $x < -2$ **5.** $x \le -\dfrac{1}{3}$

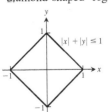

7. $3, -3$ **9.** $7/6, 25/6$

11. $-2 \le t \le 4$ **13.** $0 \le z \le 10$

15. $(-\infty, -2] \cup [2, \infty)$ **17.** $(-\infty, -3] \cup [1, \infty)$

19. $(-3, -2) \cup (2, 3)$ **21.** $(0, 1)$ **23.** $(-\infty, 1]$

27. The graph of $|x| + |y| \le 1$ is the interior and boundary of the "diamond-shaped" region.

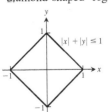

APPENDIX 3, pp. AP-17–AP-18

1. $2, -4; 2\sqrt{5}$ **3.** Unit circle

5. $m_\perp = -\dfrac{1}{3}$

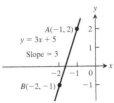

7. (a) $x = -1$ **(b)** $y = 4/3$ **9.** $y = -x$

11. $y = -\dfrac{5}{4}x + 6$ **13.** $y = 4x + 4$ **15.** $y = -\dfrac{x}{2} + 12$

17. x-intercept $= \sqrt{3}$, y-intercept $= -\sqrt{2}$

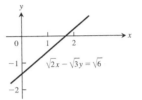

19. $(3, -3)$

21. $x^2 + (y - 2)^2 = 4$ **23.** $\left(x + \sqrt{3}\right)^2 + (y + 2)^2 = 4$

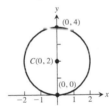
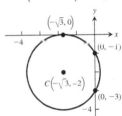

25. $x^2 + (y - 3/2)^2 = 25/4$ **27.**

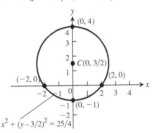
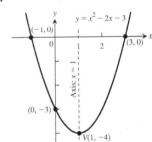

29.

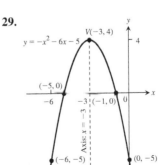

31. Exterior points of a circle of radius $\sqrt{7}$, centered at the origin

33. The washer between the circles $x^2 + y^2 = 1$ and $x^2 + y^2 = 4$
(points with distance from the origin between 1 and 2)

35. $(x + 2)^2 + (y - 1)^2 < 6$

37. $\left(\dfrac{1}{\sqrt{5}}, \dfrac{2}{\sqrt{5}}\right)$, $\left(-\dfrac{1}{\sqrt{5}}, -\dfrac{2}{\sqrt{5}}\right)$

39. $\left(-\dfrac{1}{\sqrt{3}}, -\dfrac{1}{3}\right)$, $\left(\dfrac{1}{\sqrt{3}}, -\dfrac{1}{3}\right)$

41. (a) ≈ -2.5 degrees/inch **(b)** ≈ -16.1 degrees/inch
(c) ≈ -8.3 degrees/inch

43. 5.97 atm

45. Yes: $C = F = -40°$

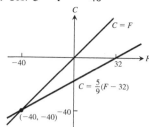

51. $k = -8$, $k = 1/2$

1. (a) $14 + 8i$ **(b)** $-7 - 4i$ **(c)** $-5i$

3. (a) By reflecting z across the real axis
(b) By reflecting z across the imaginary axis
(c) By reflecting z across the real axis and then multiplying the length of the vector by $1/|z|^2$

5. (a) Points on the circle $x^2 + y^2 = 4$
(b) Points inside the circle $x^2 + y^2 = 4$
(c) Points outside the circle $x^2 + y^2 = 4$

7. Points on a circle of radius 1, center $(-1, 0)$

9. Points on the line $y = -x$ **11.** $4e^{2\pi i/3}$ **13.** $1e^{2\pi i/3}$

15. $\cos^4\theta - 6\cos^2\theta\sin^2\theta + \sin^4\theta$

17. $1, -\dfrac{1}{2} \pm \dfrac{\sqrt{3}}{2}i$ **19.** $2i, -\sqrt{3} - i, \sqrt{3} - i$

21. $\dfrac{\sqrt{6}}{2} \pm \dfrac{\sqrt{2}}{2}i, -\dfrac{\sqrt{6}}{2} \pm \dfrac{\sqrt{2}}{2}i$ **23.** $1 \pm \sqrt{3}i, -1 \pm \sqrt{3}i$

Applications Index

Subject Index

Credits

A Brief Table of Integrals

Basic Forms

1. $\int k \, dx = kx + C, \quad k \text{ any number}$

2. $\int x^n \, dx = \dfrac{x^{n+1}}{n+1} + C, \quad n \neq -1$

3. $\int \dfrac{dx}{x} = \ln|x| + C$

4. $\int e^x \, dx = e^x + C$

5. $\int a^x \, dx = \dfrac{a^x}{\ln a} + C \quad (a > 0, a \neq 1)$

6. $\int \sin x \, dx = -\cos x + C$

7. $\int \cos x \, dx = \sin x + C$

8. $\int \sec^2 x \, dx = \tan x + C$

9. $\int \csc^2 x \, dx = -\cot x + C$

10. $\int \sec x \tan x \, dx = \sec x + C$

11. $\int \csc x \cot x \, dx = -\csc x + C$

12. $\int \tan x \, dx = \ln|\sec x| + C$

13. $\int \cot x \, dx = \ln|\sin x| + C$

14. $\int \sinh x \, dx = \cosh x + C$

15. $\int \cosh x \, dx = \sinh x + C$

16. $\int \dfrac{dx}{\sqrt{a^2 - x^2}} = \sin^{-1}\dfrac{x}{a} + C$

17. $\int \dfrac{dx}{a^2 + x^2} = \dfrac{1}{a}\tan^{-1}\dfrac{x}{a} + C$

18. $\int \dfrac{dx}{x\sqrt{x^2 - a^2}} = \dfrac{1}{a}\sec^{-1}\left|\dfrac{x}{a}\right| + C$

19. $\int \dfrac{dx}{\sqrt{a^2 + x^2}} = \sinh^{-1}\dfrac{x}{a} + C \quad (a > 0)$

20. $\int \dfrac{dx}{\sqrt{x^2 - a^2}} = \cosh^{-1}\dfrac{x}{a} + C \quad (x > a > 0)$

Forms Involving $ax + b$

21. $\int (ax + b)^n \, dx = \dfrac{(ax + b)^{n+1}}{a(n + 1)} + C, \quad n \neq -1$

22. $\int x(ax + b)^n \, dx = \dfrac{(ax + b)^{n+1}}{a^2}\left[\dfrac{ax + b}{n + 2} - \dfrac{b}{n + 1}\right] + C, \quad n \neq -1, -2$

23. $\int (ax + b)^{-1} \, dx = \dfrac{1}{a}\ln|ax + b| + C$

24. $\int x(ax + b)^{-1} \, dx = \dfrac{x}{a} - \dfrac{b}{a^2}\ln|ax + b| + C$

25. $\int x(ax + b)^{-2} \, dx = \dfrac{1}{a^2}\left[\ln|ax + b| + \dfrac{b}{ax + b}\right] + C$

26. $\int \dfrac{dx}{x(ax + b)} = \dfrac{1}{b}\ln\left|\dfrac{x}{ax + b}\right| + C$

27. $\int \left(\sqrt{ax + b}\right)^n \, dx = \dfrac{2}{a}\dfrac{\left(\sqrt{ax + b}\right)^{n+2}}{n + 2} + C, \quad n \neq -2$

28. $\int \dfrac{\sqrt{ax + b}}{x} \, dx = 2\sqrt{ax + b} + b\int \dfrac{dx}{x\sqrt{ax + b}}$

29. (a) $\displaystyle\int \frac{dx}{x\sqrt{ax+b}} = \frac{1}{\sqrt{b}} \ln \left| \frac{\sqrt{ax+b} - \sqrt{b}}{\sqrt{ax+b} + \sqrt{b}} \right| + C$ **(b)** $\displaystyle\int \frac{dx}{x\sqrt{ax-b}} = \frac{2}{\sqrt{b}} \tan^{-1} \sqrt{\frac{ax-b}{b}} + C$

30. $\displaystyle\int \frac{\sqrt{ax+b}}{x^2}\, dx = -\frac{\sqrt{ax+b}}{x} + \frac{a}{2} \int \frac{dx}{x\sqrt{ax+b}} + C$ **31.** $\displaystyle\int \frac{dx}{x^2\sqrt{ax+b}} = -\frac{\sqrt{ax+b}}{bx} - \frac{a}{2b} \int \frac{dx}{x\sqrt{ax+b}} + C$

Forms Involving $a^2 + x^2$

32. $\displaystyle\int \frac{dx}{a^2+x^2} = \frac{1}{a} \tan^{-1} \frac{x}{a} + C$ **33.** $\displaystyle\int \frac{dx}{(a^2+x^2)^2} = \frac{x}{2a^2(a^2+x^2)} + \frac{1}{2a^3} \tan^{-1} \frac{x}{a} + C$

34. $\displaystyle\int \frac{dx}{\sqrt{a^2+x^2}} = \sinh^{-1} \frac{x}{a} + C = \ln\left(x + \sqrt{a^2+x^2}\right) + C$

35. $\displaystyle\int \sqrt{a^2+x^2}\, dx = \frac{x}{2}\sqrt{a^2+x^2} + \frac{a^2}{2} \ln\left(x + \sqrt{a^2+x^2}\right) + C$

36. $\displaystyle\int x^2\sqrt{a^2+x^2}\, dx = \frac{x}{8}(a^2+2x^2)\sqrt{a^2+x^2} - \frac{a^4}{8} \ln\left(x + \sqrt{a^2+x^2}\right) + C$

37. $\displaystyle\int \frac{\sqrt{a^2+x^2}}{x}\, dx = \sqrt{a^2+x^2} - a \ln \left| \frac{a + \sqrt{a^2+x^2}}{x} \right| + C$

38. $\displaystyle\int \frac{\sqrt{a^2+x^2}}{x^2}\, dx = \ln\left(x + \sqrt{a^2+x^2}\right) - \frac{\sqrt{a^2+x^2}}{x} + C$

39. $\displaystyle\int \frac{x^2}{\sqrt{a^2+x^2}}\, dx = -\frac{a^2}{2} \ln\left(x + \sqrt{a^2+x^2}\right) + \frac{x\sqrt{a^2+x^2}}{2} + C$

40. $\displaystyle\int \frac{dx}{x\sqrt{a^2+x^2}} = -\frac{1}{a} \ln \left| \frac{a + \sqrt{a^2+x^2}}{x} \right| + C$ **41.** $\displaystyle\int \frac{dx}{x^2\sqrt{a^2+x^2}} = -\frac{\sqrt{a^2+x^2}}{a^2 x} + C$

Forms Involving $a^2 - x^2$

42. $\displaystyle\int \frac{dx}{a^2-x^2} = \frac{1}{2a} \ln \left| \frac{x+a}{x-a} \right| + C$ **43.** $\displaystyle\int \frac{dx}{(a^2-x^2)^2} = \frac{x}{2a^2(a^2-x^2)} + \frac{1}{4a^3} \ln \left| \frac{x+a}{x-a} \right| + C$

44. $\displaystyle\int \frac{dx}{\sqrt{a^2-x^2}} = \sin^{-1} \frac{x}{a} + C$ **45.** $\displaystyle\int \sqrt{a^2-x^2}\, dx = \frac{x}{2}\sqrt{a^2-x^2} + \frac{a^2}{2} \sin^{-1} \frac{x}{a} + C$

46. $\displaystyle\int x^2\sqrt{a^2-x^2}\, dx = \frac{a^4}{8} \sin^{-1} \frac{x}{a} - \frac{1}{8} x\sqrt{a^2-x^2}\,(a^2 - 2x^2) + C$

47. $\displaystyle\int \frac{\sqrt{a^2-x^2}}{x}\, dx = \sqrt{a^2-x^2} - a \ln \left| \frac{a + \sqrt{a^2-x^2}}{x} \right| + C$ **48.** $\displaystyle\int \frac{\sqrt{a^2-x^2}}{x^2}\, dx = -\sin^{-1} \frac{x}{a} - \frac{\sqrt{a^2-x^2}}{x} + C$

49. $\displaystyle\int \frac{x^2}{\sqrt{a^2-x^2}}\, dx = \frac{a^2}{2} \sin^{-1} \frac{x}{a} - \frac{1}{2} x\sqrt{a^2-x^2} + C$ **50.** $\displaystyle\int \frac{dx}{x\sqrt{a^2-x^2}} = -\frac{1}{a} \ln \left| \frac{a + \sqrt{a^2-x^2}}{x} \right| + C$

51. $\displaystyle\int \frac{dx}{x^2\sqrt{a^2-x^2}} = -\frac{\sqrt{a^2-x^2}}{a^2 x} + C$

Forms Involving $x^2 - a^2$

52. $\displaystyle\int \frac{dx}{\sqrt{x^2-a^2}} = \ln \left| x + \sqrt{x^2-a^2} \right| + C$

53. $\displaystyle\int \sqrt{x^2-a^2}\, dx = \frac{x}{2}\sqrt{x^2-a^2} - \frac{a^2}{2} \ln \left| x + \sqrt{x^2-a^2} \right| + C$

54. $\displaystyle \int \left(\sqrt{x^2 - a^2}\right)^n dx = \frac{x\left(\sqrt{x^2 - a^2}\right)^n}{n + 1} - \frac{na^2}{n + 1}\int \left(\sqrt{x^2 - a^2}\right)^{n-2} dx, \quad n \neq -1$

55. $\displaystyle \int \frac{dx}{\left(\sqrt{x^2 - a^2}\right)^n} = \frac{x\left(\sqrt{x^2 - a^2}\right)^{2-n}}{(2 - n)a^2} - \frac{n - 3}{(n - 2)a^2}\int \frac{dx}{\left(\sqrt{x^2 - a^2}\right)^{n-2}}, \quad n \neq 2$

56. $\displaystyle \int x\left(\sqrt{x^2 - a^2}\right)^n dx = \frac{\left(\sqrt{x^2 - a^2}\right)^{n+2}}{n + 2} + C, \quad n \neq -2$

57. $\displaystyle \int x^2\sqrt{x^2 - a^2}\, dx = \frac{x}{8}(2x^2 - a^2)\sqrt{x^2 - a^2} - \frac{a^4}{8}\ln\left|x + \sqrt{x^2 - a^2}\right| + C$

58. $\displaystyle \int \frac{\sqrt{x^2 - a^2}}{x}\, dx = \sqrt{x^2 - a^2} - a\sec^{-1}\left|\frac{x}{a}\right| + C$

59. $\displaystyle \int \frac{\sqrt{x^2 - a^2}}{x^2}\, dx = \ln\left|x + \sqrt{x^2 - a^2}\right| - \frac{\sqrt{x^2 - a^2}}{x} + C$

60. $\displaystyle \int \frac{x^2}{\sqrt{x^2 - a^2}}\, dx = \frac{a^2}{2}\ln\left|x + \sqrt{x^2 - a^2}\right| + \frac{x}{2}\sqrt{x^2 - a^2} + C$

61. $\displaystyle \int \frac{dx}{x\sqrt{x^2 - a^2}} = \frac{1}{a}\sec^{-1}\left|\frac{x}{a}\right| + C = \frac{1}{a}\cos^{-1}\left|\frac{a}{x}\right| + C$ **62.** $\displaystyle \int \frac{dx}{x^2\sqrt{x^2 - a^2}} = \frac{\sqrt{x^2 - a^2}}{a^2 x} + C$

Trigonometric Forms

63. $\displaystyle \int \sin ax\, dx = -\frac{1}{a}\cos ax + C$ **64.** $\displaystyle \int \cos ax\, dx = \frac{1}{a}\sin ax + C$

65. $\displaystyle \int \sin^2 ax\, dx = \frac{x}{2} - \frac{\sin 2ax}{4a} + C$ **66.** $\displaystyle \int \cos^2 ax\, dx = \frac{x}{2} + \frac{\sin 2ax}{4a} + C$

67. $\displaystyle \int \sin^n ax\, dx = -\frac{\sin^{n-1} ax \cos ax}{na} + \frac{n - 1}{n}\int \sin^{n-2} ax\, dx$

68. $\displaystyle \int \cos^n ax\, dx = \frac{\cos^{n-1} ax \sin ax}{na} + \frac{n - 1}{n}\int \cos^{n-2} ax\, dx$

69. (a) $\displaystyle \int \sin ax \cos bx\, dx = -\frac{\cos(a + b)x}{2(a + b)} - \frac{\cos(a - b)x}{2(a - b)} + C, \quad a^2 \neq b^2$

    (b) $\displaystyle \int \sin ax \sin bx\, dx = \frac{\sin(a - b)x}{2(a - b)} - \frac{\sin(a + b)x}{2(a + b)} + C, \quad a^2 \neq b^2$

    (c) $\displaystyle \int \cos ax \cos bx\, dx = \frac{\sin(a - b)x}{2(a - b)} + \frac{\sin(a + b)x}{2(a + b)} + C, \quad a^2 \neq b^2$

70. $\displaystyle \int \sin ax \cos ax\, dx = -\frac{\cos 2ax}{4a} + C$ **71.** $\displaystyle \int \sin^n ax \cos ax\, dx = \frac{\sin^{n+1} ax}{(n + 1)a} + C, \quad n \neq -1$

72. $\displaystyle \int \frac{\cos ax}{\sin ax}\, dx = \frac{1}{a}\ln\left|\sin ax\right| + C$ **73.** $\displaystyle \int \cos^n ax \sin ax\, dx = -\frac{\cos^{n+1} ax}{(n + 1)a} + C, \quad n \neq -1$

74. $\displaystyle \int \frac{\sin ax}{\cos ax}\, dx = -\frac{1}{a}\ln\left|\cos ax\right| + C$

75. $\displaystyle \int \sin^n ax \cos^m ax\, dx = -\frac{\sin^{n-1} ax \cos^{m+1} ax}{a(m + n)} + \frac{n - 1}{m + n}\int \sin^{n-2} ax \cos^m ax\, dx, \quad n \neq -m \quad \text{(reduces } \sin^n ax)$

76. $\displaystyle \int \sin^n ax \cos^m ax\, dx = \frac{\sin^{n+1} ax \cos^{m-1} ax}{a(m + n)} + \frac{m - 1}{m + n}\int \sin^n ax \cos^{m-2} ax\, dx, \quad m \neq -n \quad \text{(reduces } \cos^m ax)$

77. $\displaystyle\int \frac{dx}{b + c \sin ax} = \frac{-2}{a\sqrt{b^2 - c^2}} \tan^{-1}\left[\sqrt{\frac{b - c}{b + c}} \tan\left(\frac{\pi}{4} - \frac{ax}{2}\right)\right] + C, \quad b^2 > c^2$

78. $\displaystyle\int \frac{dx}{b + c \sin ax} = \frac{-1}{a\sqrt{c^2 - b^2}} \ln\left|\frac{c + b \sin ax + \sqrt{c^2 - b^2}\cos ax}{b + c \sin ax}\right| + C, \quad b^2 < c^2$

79. $\displaystyle\int \frac{dx}{1 + \sin ax} = -\frac{1}{a}\tan\left(\frac{\pi}{4} - \frac{ax}{2}\right) + C$

80. $\displaystyle\int \frac{dx}{1 - \sin ax} = \frac{1}{a}\tan\left(\frac{\pi}{4} + \frac{ax}{2}\right) + C$

81. $\displaystyle\int \frac{dx}{b + c \cos ax} = \frac{2}{a\sqrt{b^2 - c^2}} \tan^{-1}\left[\sqrt{\frac{b - c}{b + c}} \tan\frac{ax}{2}\right] + C, \quad b^2 > c^2$

82. $\displaystyle\int \frac{dx}{b + c \cos ax} = \frac{1}{a\sqrt{c^2 - b^2}} \ln\left|\frac{c + b \cos ax + \sqrt{c^2 - b^2}\sin ax}{b + c \cos ax}\right| + C, \quad b^2 < c^2$

83. $\displaystyle\int \frac{dx}{1 + \cos ax} = \frac{1}{a}\tan\frac{ax}{2} + C$

84. $\displaystyle\int \frac{dx}{1 - \cos ax} = -\frac{1}{a}\cot\frac{ax}{2} + C$

85. $\displaystyle\int x \sin ax\, dx = \frac{1}{a^2}\sin ax - \frac{x}{a}\cos ax + C$

86. $\displaystyle\int x \cos ax\, dx = \frac{1}{a^2}\cos ax + \frac{x}{a}\sin ax + C$

87. $\displaystyle\int x^n \sin ax\, dx = -\frac{x^n}{a}\cos ax + \frac{n}{a}\int x^{n-1}\cos ax\, dx$

88. $\displaystyle\int x^n \cos ax\, dx = \frac{x^n}{a}\sin ax - \frac{n}{a}\int x^{n-1}\sin ax\, dx$

89. $\displaystyle\int \tan ax\, dx = \frac{1}{a}\ln|\sec ax| + C$

90. $\displaystyle\int \cot ax\, dx = \frac{1}{a}\ln|\sin ax| + C$

91. $\displaystyle\int \tan^2 ax\, dx = \frac{1}{a}\tan ax - x + C$

92. $\displaystyle\int \cot^2 ax\, dx = -\frac{1}{a}\cot ax - x + C$

93. $\displaystyle\int \tan^n ax\, dx = \frac{\tan^{n-1} ax}{a(n - 1)} - \int \tan^{n-2} ax\, dx, \quad n \neq 1$

94. $\displaystyle\int \cot^n ax\, dx = -\frac{\cot^{n-1} ax}{a(n - 1)} - \int \cot^{n-2} ax\, dx, \quad n \neq 1$

95. $\displaystyle\int \sec ax\, dx = \frac{1}{a}\ln|\sec ax + \tan ax| + C$

96. $\displaystyle\int \csc ax\, dx = -\frac{1}{a}\ln|\csc ax + \cot ax| + C$

97. $\displaystyle\int \sec^2 ax\, dx = \frac{1}{a}\tan ax + C$

98. $\displaystyle\int \csc^2 ax\, dx = -\frac{1}{a}\cot ax + C$

99. $\displaystyle\int \sec^n ax\, dx = \frac{\sec^{n-2} ax \tan ax}{a(n - 1)} + \frac{n - 2}{n - 1}\int \sec^{n-2} ax\, dx, \quad n \neq 1$

100. $\displaystyle\int \csc^n ax\, dx = -\frac{\csc^{n-2} ax \cot ax}{a(n - 1)} + \frac{n - 2}{n - 1}\int \csc^{n-2} ax\, dx, \quad n \neq 1$

101. $\displaystyle\int \sec^n ax \tan ax\, dx = \frac{\sec^n ax}{na} + C, \quad n \neq 0$

102. $\displaystyle\int \csc^n ax \cot ax\, dx = -\frac{\csc^n ax}{na} + C, \quad n \neq 0$

Inverse Trigonometric Forms

103. $\displaystyle\int \sin^{-1} ax\, dx = x \sin^{-1} ax + \frac{1}{a}\sqrt{1 - a^2 x^2} + C$

104. $\displaystyle\int \cos^{-1} ax\, dx = x \cos^{-1} ax - \frac{1}{a}\sqrt{1 - a^2 x^2} + C$

105. $\displaystyle\int \tan^{-1} ax\, dx = x \tan^{-1} ax - \frac{1}{2a}\ln(1 + a^2 x^2) + C$

106. $\displaystyle\int x^n \sin^{-1} ax\, dx = \frac{x^{n+1}}{n + 1}\sin^{-1} ax - \frac{a}{n + 1}\int \frac{x^{n+1}\, dx}{\sqrt{1 - a^2 x^2}}, \quad n \neq -1$

107. $\displaystyle\int x^n \cos^{-1} ax \, dx = \frac{x^{n+1}}{n+1} \cos^{-1} ax + \frac{a}{n+1} \int \frac{x^{n+1} \, dx}{\sqrt{1-a^2x^2}}, \quad n \neq -1$

108. $\displaystyle\int x^n \tan^{-1} ax \, dx = \frac{x^{n+1}}{n+1} \tan^{-1} ax - \frac{a}{n+1} \int \frac{x^{n+1} \, dx}{1+a^2x^2}, \quad n \neq -1$

Exponential and Logarithmic Forms

109. $\displaystyle\int e^{ax} \, dx = \frac{1}{a} e^{ax} + C$

110. $\displaystyle\int b^{ax} \, dx = \frac{1}{a} \frac{b^{ax}}{\ln b} + C, \quad b > 0, b \neq 1$

111. $\displaystyle\int xe^{ax} \, dx = \frac{e^{ax}}{a^2} (ax - 1) + C$

112. $\displaystyle\int x^n e^{ax} \, dx = \frac{1}{a} x^n e^{ax} - \frac{n}{a} \int x^{n-1} e^{ax} \, dx$

113. $\displaystyle\int x^n b^{ax} \, dx = \frac{x^n b^{ax}}{a \ln b} - \frac{n}{a \ln b} \int x^{n-1} b^{ax} \, dx, \quad b > 0, b \neq 1$

114. $\displaystyle\int e^{ax} \sin bx \, dx = \frac{e^{ax}}{a^2 + b^2} (a \sin bx - b \cos bx) + C$

115. $\displaystyle\int e^{ax} \cos bx \, dx = \frac{e^{ax}}{a^2 + b^2} (a \cos bx + b \sin bx) + C$

116. $\displaystyle\int \ln ax \, dx = x \ln ax - x + C$

117. $\displaystyle\int x^n (\ln ax)^m \, dx = \frac{x^{n+1}(\ln ax)^m}{n+1} - \frac{m}{n+1} \int x^n (\ln ax)^{m-1} \, dx, \quad n \neq -1$

118. $\displaystyle\int x^{-1}(\ln ax)^m \, dx = \frac{(\ln ax)^{m+1}}{m+1} + C, \quad m \neq -1$

119. $\displaystyle\int \frac{dx}{x \ln ax} = \ln |\ln ax| + C$

Forms Involving $\sqrt{2ax - x^2}, \, a > 0$

120. $\displaystyle\int \frac{dx}{\sqrt{2ax - x^2}} = \sin^{-1}\left(\frac{x-a}{a}\right) + C$

121. $\displaystyle\int \sqrt{2ax - x^2} \, dx = \frac{x-a}{2}\sqrt{2ax - x^2} + \frac{a^2}{2} \sin^{-1}\left(\frac{x-a}{a}\right) + C$

122. $\displaystyle\int \left(\sqrt{2ax - x^2}\right)^n dx = \frac{(x-a)\left(\sqrt{2ax - x^2}\right)^n}{n+1} + \frac{na^2}{n+1} \int \left(\sqrt{2ax - x^2}\right)^{n-2} dx$

123. $\displaystyle\int \frac{dx}{\left(\sqrt{2ax - x^2}\right)^n} = \frac{(x-a)\left(\sqrt{2ax - x^2}\right)^{2-n}}{(n-2)a^2} + \frac{n-3}{(n-2)a^2} \int \frac{dx}{\left(\sqrt{2ax - x^2}\right)^{n-2}}$

124. $\displaystyle\int x\sqrt{2ax - x^2} \, dx = \frac{(x+a)(2x - 3a)\sqrt{2ax - x^2}}{6} + \frac{a^3}{2} \sin^{-1}\left(\frac{x-a}{a}\right) + C$

125. $\displaystyle\int \frac{\sqrt{2ax - x^2}}{x} \, dx = \sqrt{2ax - x^2} + a \sin^{-1}\left(\frac{x-a}{a}\right) + C$

126. $\displaystyle\int \frac{\sqrt{2ax - x^2}}{x^2} \, dx = -2\sqrt{\frac{2a-x}{x}} - \sin^{-1}\left(\frac{x-a}{a}\right) + C$

127. $\displaystyle\int \frac{x \, dx}{\sqrt{2ax - x^2}} = a \sin^{-1}\left(\frac{x-a}{a}\right) - \sqrt{2ax - x^2} + C$

128. $\displaystyle\int \frac{dx}{x\sqrt{2ax - x^2}} = -\frac{1}{a}\sqrt{\frac{2a-x}{x}} + C$

Hyperbolic Forms

129. $\displaystyle\int \sinh ax \, dx = \frac{1}{a} \cosh ax + C$

130. $\displaystyle\int \cosh ax \, dx = \frac{1}{a} \sinh ax + C$

131. $\displaystyle\int \sinh^2 ax \, dx = \frac{\sinh 2ax}{4a} - \frac{x}{2} + C$

132. $\displaystyle\int \cosh^2 ax \, dx = \frac{\sinh 2ax}{4a} + \frac{x}{2} + C$

133. $\displaystyle\int \sinh^n ax \, dx = \frac{\sinh^{n-1} ax \cosh ax}{na} - \frac{n-1}{n} \int \sinh^{n-2} ax \, dx, \quad n \neq 0$

134. $\displaystyle\int \cosh^n ax \, dx = \frac{\cosh^{n-1} ax \sinh ax}{na} + \frac{n-1}{n} \int \cosh^{n-2} ax \, dx, \quad n \neq 0$

135. $\displaystyle\int x \sinh ax \, dx = \frac{x}{a} \cosh ax - \frac{1}{a^2} \sinh ax + C$ **136.** $\displaystyle\int x \cosh ax \, dx = \frac{x}{a} \sinh ax - \frac{1}{a^2} \cosh ax + C$

137. $\displaystyle\int x^n \sinh ax \, dx = \frac{x^n}{a} \cosh ax - \frac{n}{a} \int x^{n-1} \cosh ax \, dx$ **138.** $\displaystyle\int x^n \cosh ax \, dx = \frac{x^n}{a} \sinh ax - \frac{n}{a} \int x^{n-1} \sinh ax \, dx$

139. $\displaystyle\int \tanh ax \, dx = \frac{1}{a} \ln (\cosh ax) + C$ **140.** $\displaystyle\int \coth ax \, dx = \frac{1}{a} \ln |\sinh ax| + C$

141. $\displaystyle\int \tanh^2 ax \, dx = x - \frac{1}{a} \tanh ax + C$ **142.** $\displaystyle\int \coth^2 ax \, dx = x - \frac{1}{a} \coth ax + C$

143. $\displaystyle\int \tanh^n ax \, dx = -\frac{\tanh^{n-1} ax}{(n-1)a} + \int \tanh^{n-2} ax \, dx, \quad n \neq 1$

144. $\displaystyle\int \coth^n ax \, dx = -\frac{\coth^{n-1} ax}{(n-1)a} + \int \coth^{n-2} ax \, dx, \quad n \neq 1$

145. $\displaystyle\int \operatorname{sech} ax \, dx = \frac{1}{a} \sin^{-1} (\tanh ax) + C$ **146.** $\displaystyle\int \operatorname{csch} ax \, dx = \frac{1}{a} \ln \left| \tanh \frac{ax}{2} \right| + C$

147. $\displaystyle\int \operatorname{sech}^2 ax \, dx = \frac{1}{a} \tanh ax + C$ **148.** $\displaystyle\int \operatorname{csch}^2 ax \, dx = -\frac{1}{a} \coth ax + C$

149. $\displaystyle\int \operatorname{sech}^n ax \, dx = \frac{\operatorname{sech}^{n-2} ax \tanh ax}{(n-1)a} + \frac{n-2}{n-1} \int \operatorname{sech}^{n-2} ax \, dx, \quad n \neq 1$

150. $\displaystyle\int \operatorname{csch}^n ax \, dx = -\frac{\operatorname{csch}^{n-2} ax \coth ax}{(n-1)a} - \frac{n-2}{n-1} \int \operatorname{csch}^{n-2} ax \, dx, \quad n \neq 1$

151. $\displaystyle\int \operatorname{sech}^n ax \tanh ax \, dx = -\frac{\operatorname{sech}^n ax}{na} + C, \quad n \neq 0$ **152.** $\displaystyle\int \operatorname{csch}^n ax \coth ax \, dx = -\frac{\operatorname{csch}^n ax}{na} + C, \quad n \neq 0$

153. $\displaystyle\int e^{ax} \sinh bx \, dx = \frac{e^{ax}}{2} \left[\frac{e^{bx}}{a+b} - \frac{e^{-bx}}{a-b} \right] + C, \quad a^2 \neq b^2$

154. $\displaystyle\int e^{ax} \cosh bx \, dx = \frac{e^{ax}}{2} \left[\frac{e^{bx}}{a+b} + \frac{e^{-bx}}{a-b} \right] + C, \quad a^2 \neq b^2$

Some Definite Integrals

155. $\displaystyle\int_0^\infty x^{n-1} e^{-x} \, dx = \Gamma(n) = (n-1)!, \quad n > 0$ **156.** $\displaystyle\int_0^\infty e^{-ax^2} \, dx = \frac{1}{2} \sqrt{\frac{\pi}{a}}, \quad a > 0$

157. $\displaystyle\int_0^{\pi/2} \sin^n x \, dx = \int_0^{\pi/2} \cos^n x \, dx = \begin{cases} \dfrac{1 \cdot 3 \cdot 5 \cdot \,\cdots\, \cdot (n-1)}{2 \cdot 4 \cdot 6 \cdot \,\cdots\, \cdot n} \cdot \dfrac{\pi}{2}, & \text{if } n \text{ is an even integer} \geq 2 \\[3mm] \dfrac{2 \cdot 4 \cdot 6 \cdot \,\cdots\, \cdot (n-1)}{3 \cdot 5 \cdot 7 \cdot \,\cdots\, \cdot n}, & \text{if } n \text{ is an odd integer} \geq 3 \end{cases}$

Basic Algebra Formulas

Arithmetic Operations

$$a(b + c) = ab + ac, \qquad \frac{a}{b} \cdot \frac{c}{d} = \frac{ac}{bd}$$

$$\frac{a}{b} + \frac{c}{d} = \frac{ad + bc}{bd}, \qquad \frac{a/b}{c/d} = \frac{a}{b} \cdot \frac{d}{c}$$

Laws of Signs

$$-(-a) = a, \qquad \frac{-a}{b} = -\frac{a}{b} = \frac{a}{-b}$$

Zero   Division by zero is not defined.

$$\text{If } a \neq 0: \quad \frac{0}{a} = 0, \quad a^0 = 1, \quad 0^a = 0$$

$$\text{For any number } a: a \cdot 0 = 0 \cdot a = 0$$

Laws of Exponents

$$a^m a^n = a^{m+n}, \qquad (ab)^m = a^m b^m, \qquad (a^m)^n = a^{mn}, \qquad a^{m/n} = \sqrt[n]{a^m} = \left(\sqrt[n]{a}\right)^m$$

If $a \neq 0$, then

$$\frac{a^m}{a^n} = a^{m-n}, \qquad a^0 = 1, \qquad a^{-m} = \frac{1}{a^m}.$$

The Binomial Theorem   For any positive integer n,

$$(a + b)^n = a^n + na^{n-1}b + \frac{n(n-1)}{1 \cdot 2}a^{n-2}b^2$$

$$+ \frac{n(n-1)(n-2)}{1 \cdot 2 \cdot 3}a^{n-3}b^3 + \cdots + nab^{n-1} + b^n.$$

For instance,

$$(a + b)^2 = a^2 + 2ab + b^2, \qquad\qquad (a - b)^2 = a^2 - 2ab + b^2$$
$$(a + b)^3 = a^3 + 3a^2b + 3ab^2 + b^3, \qquad (a - b)^3 = a^3 - 3a^2b + 3ab^2 - b^3.$$

Factoring the Difference of Like Integer Powers, $n > 1$

$$a^n - b^n = (a - b)(a^{n-1} + a^{n-2}b + a^{n-3}b^2 + \cdots + ab^{n-2} + b^{n-1})$$

For instance,

$$a^2 - b^2 = (a - b)(a + b),$$
$$a^3 - b^3 = (a - b)(a^2 + ab + b^2),$$
$$a^4 - b^4 = (a - b)(a^3 + a^2b + ab^2 + b^3).$$

Completing the Square   If $a \neq 0$, then

$$ax^2 + bx + c = au^2 + C \qquad \left(u = x + (b/2a), C = c - \frac{b^2}{4a}\right)$$

The Quadratic Formula
If $a \neq 0$ and $ax^2 + bx + c = 0$, then

$$x = \frac{-b \pm \sqrt{b^2 - 4ac}}{2a}.$$

Geometry Formulas

A = area, B = area of base, C = circumference, S = surface area, V = volume

Triangle

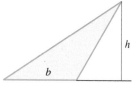

$$A = \frac{1}{2}bh$$

Similar Triangles

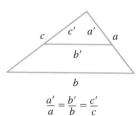

$$\frac{a'}{a} = \frac{b'}{b} = \frac{c'}{c}$$

Pythagorean Theorem

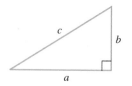

$$a^2 + b^2 = c^2$$

Parallelogram

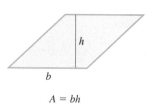

$$A = bh$$

Trapezoid

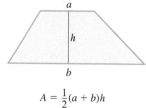

$$A = \frac{1}{2}(a + b)h$$

Circle

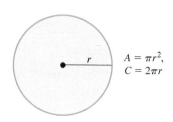

$$A = \pi r^2,$$
$$C = 2\pi r$$

Any Cylinder or Prism with Parallel Bases

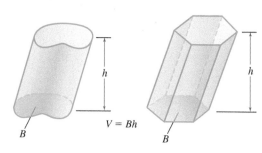

$$V = Bh$$

Right Circular Cylinder

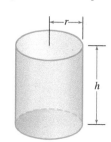

$$V = \pi r^2 h$$
$$S = 2\pi rh = \text{Area of side}$$

Any Cone or Pyramid

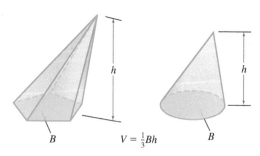

$$V = \frac{1}{3}Bh$$

Right Circular Cone

$$V = \frac{1}{3}\pi r^2 h$$
$$S = \pi rs = \text{Area of side}$$

Sphere

$$V = \frac{4}{3}\pi r^3, \; S = 4\pi r^2$$

Trigonometry Formulas

Definitions and Fundamental Identities

Sine:	$\sin\theta = \dfrac{y}{r} = \dfrac{1}{\csc\theta}$
Cosine:	$\cos\theta = \dfrac{x}{r} = \dfrac{1}{\sec\theta}$
Tangent:	$\tan\theta = \dfrac{y}{x} = \dfrac{1}{\cot\theta}$

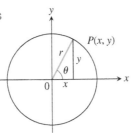

Identities

$\sin(-\theta) = -\sin\theta, \quad \cos(-\theta) = \cos\theta$

$\sin^2\theta + \cos^2\theta = 1, \quad \sec^2\theta = 1 + \tan^2\theta, \quad \csc^2\theta = 1 + \cot^2\theta$

$\sin 2\theta = 2\sin\theta\cos\theta, \quad \cos 2\theta = \cos^2\theta - \sin^2\theta$

$\cos^2\theta = \dfrac{1 + \cos 2\theta}{2}, \quad \sin^2\theta = \dfrac{1 - \cos 2\theta}{2}$

$\sin(A + B) = \sin A\cos B + \cos A\sin B$

$\sin(A - B) = \sin A\cos B - \cos A\sin B$

$\cos(A + B) = \cos A\cos B - \sin A\sin B$

$\cos(A - B) = \cos A\cos B + \sin A\sin B$

$\tan(A + B) = \dfrac{\tan A + \tan B}{1 - \tan A\tan B}$

$\tan(A - B) = \dfrac{\tan A - \tan B}{1 + \tan A\tan B}$

$\sin\left(A - \dfrac{\pi}{2}\right) = -\cos A, \qquad \cos\left(A - \dfrac{\pi}{2}\right) = \sin A$

$\sin\left(A + \dfrac{\pi}{2}\right) = \cos A, \qquad \cos\left(A + \dfrac{\pi}{2}\right) = -\sin A$

$\sin A\sin B = \dfrac{1}{2}\cos(A - B) - \dfrac{1}{2}\cos(A + B)$

$\cos A\cos B = \dfrac{1}{2}\cos(A - B) + \dfrac{1}{2}\cos(A + B)$

$\sin A\cos B = \dfrac{1}{2}\sin(A - B) + \dfrac{1}{2}\sin(A + B)$

$\sin A + \sin B = 2\sin\dfrac{1}{2}(A + B)\cos\dfrac{1}{2}(A - B)$

$\sin A - \sin B = 2\cos\dfrac{1}{2}(A + B)\sin\dfrac{1}{2}(A - B)$

$\cos A + \cos B = 2\cos\dfrac{1}{2}(A + B)\cos\dfrac{1}{2}(A - B)$

$\cos A - \cos B = -2\sin\dfrac{1}{2}(A + B)\sin\dfrac{1}{2}(A - B)$

Trigonometric Functions

Radian Measure

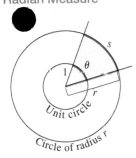

Unit circle

Circle of radius r

$\dfrac{s}{r} = \dfrac{\theta}{1} = \theta \quad \text{or} \quad \theta = \dfrac{s}{r},$

$180° = \pi$ radians.

Degrees	Radians
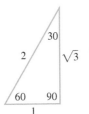	

The angles of two common triangles, in degrees and radians.

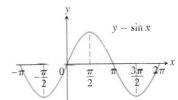

Domain: $(-\infty, \infty)$
Range: $[-1, 1]$

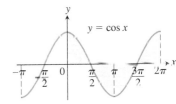

Domain: $(-\infty, \infty)$
Range: $[-1, 1]$

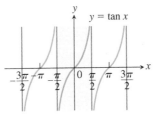

Domain: All real numbers except odd integer multiples of $\pi/2$
Range: $(-\infty, \infty)$

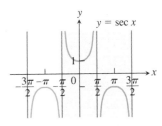

Domain: All real numbers except odd integer multiples of $\pi/2$
Range: $(-\infty, -1] \cup [1, \infty)$

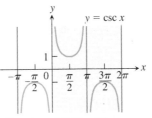

Domain: $x \neq 0, \pm\pi, \pm 2\pi, \ldots$
Range: $(-\infty, -1] \cup [1, \infty)$

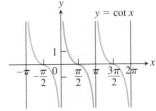

Domain: $x \neq 0, \pm\pi, \pm 2\pi, \ldots$
Range: $(-\infty, \infty)$

Series

Tests for Convergence of Infinite Series

1. **The nth-Term Test:** Unless $a_n \to 0$, the series diverges.
2. **Geometric series:** $\sum ar^n$ converges if $|r| < 1$; otherwise it diverges.
3. **p-series:** $\sum 1/n^p$ converges if $p > 1$; otherwise it diverges.
4. **Series with nonnegative terms:** Try the Integral Test, Ratio Test, or Root Test. Try comparing to a known series with the Comparison Test or the Limit Comparison Test.

5. **Series with some negative terms:** Does $\sum |a_n|$ converge? If yes, so does $\sum a_n$ because absolute convergence implies convergence.
6. **Alternating series:** $\sum a_n$ converges if the series satisfies the conditions of the Alternating Series Test.

Taylor Series

$$\frac{1}{1 - x} = 1 + x + x^2 + \cdots + x^n + \cdots = \sum_{n=0}^{\infty} x^n, \qquad |x| < 1$$

$$\frac{1}{1 + x} = 1 - x + x^2 - \cdots + (-x)^n + \cdots = \sum_{n=0}^{\infty} (-1)^n x^n, \qquad |x| < 1$$

$$e^x = 1 + x + \frac{x^2}{2!} + \cdots + \frac{x^n}{n!} + \cdots = \sum_{n=0}^{\infty} \frac{x^n}{n!}, \qquad |x| < \infty$$

$$\sin x = x - \frac{x^3}{3!} + \frac{x^5}{5!} - \cdots + (-1)^n \frac{x^{2n+1}}{(2n + 1)!} + \cdots = \sum_{n=0}^{\infty} \frac{(-1)^n x^{2n+1}}{(2n + 1)!}, \qquad |x| < \infty$$

$$\cos x = 1 - \frac{x^2}{2!} + \frac{x^4}{4!} - \cdots + (-1)^n \frac{x^{2n}}{(2n)!} + \cdots = \sum_{n=0}^{\infty} \frac{(-1)^n x^{2n}}{(2n)!}, \qquad |x| < \infty$$

$$\ln (1 + x) = x - \frac{x^2}{2} + \frac{x^3}{3} - \cdots + (-1)^{n-1} \frac{x^n}{n} + \cdots = \sum_{n=1}^{\infty} \frac{(-1)^{n-1} x^n}{n}, \qquad -1 < x \leq 1$$

$$\ln \frac{1 + x}{1 - x} = 2 \tanh^{-1} x = 2 \left(x + \frac{x^3}{3} + \frac{x^5}{5} + \cdots + \frac{x^{2n+1}}{2n + 1} + \cdots \right) = 2 \sum_{n=0}^{\infty} \frac{x^{2n+1}}{2n + 1}, \qquad |x| < 1$$

$$\tan^{-1} x = x - \frac{x^3}{3} + \frac{x^5}{5} - \cdots + (-1)^n \frac{x^{2n+1}}{2n + 1} + \cdots = \sum_{n=0}^{\infty} \frac{(-1)^n x^{2n+1}}{2n + 1}, \qquad |x| \leq 1$$

Binomial Series

$$(1 + x)^m = 1 + mx + \frac{m(m - 1)x^2}{2!} + \frac{m(m - 1)(m - 2)x^3}{3!} + \cdots + \frac{m(m - 1)(m - 2) \cdots (m - k + 1)x^k}{k!} + \cdots$$

$$= 1 + \sum_{k=1}^{\infty} \binom{m}{k} x^k, \qquad |x| < 1,$$

where

$$\binom{m}{1} = m, \qquad \binom{m}{2} = \frac{m(m - 1)}{2!}, \qquad \binom{m}{k} = \frac{m(m - 1) \cdots (m - k + 1)}{k!} \qquad \text{for } k \geq 3.$$

Vector Operator Formulas (Cartesian Form)

Formulas for Grad, Div, Curl, and the Laplacian

Cartesian (x, y, z) $\mathbf{i}$, $\mathbf{j}$, and $\mathbf{k}$ are unit vectors in the directions of increasing x, y, and z. M, N, and P are the scalar components of $\mathbf{F}(x, y, z)$ in these directions.

Gradient $\qquad \nabla f = \dfrac{\partial f}{\partial x}\mathbf{i} + \dfrac{\partial f}{\partial y}\mathbf{j} + \dfrac{\partial f}{\partial z}\mathbf{k}$

Divergence $\quad \nabla \cdot \mathbf{F} = \dfrac{\partial M}{\partial x} + \dfrac{\partial N}{\partial y} + \dfrac{\partial P}{\partial z}$

Curl $\qquad \nabla \times \mathbf{F} = \begin{vmatrix} \mathbf{i} & \mathbf{j} & \mathbf{k} \\ \dfrac{\partial}{\partial x} & \dfrac{\partial}{\partial y} & \dfrac{\partial}{\partial z} \\ M & N & P \end{vmatrix}$

Laplacian $\quad \nabla^2 f = \dfrac{\partial^2 f}{\partial x^2} + \dfrac{\partial^2 f}{\partial y^2} + \dfrac{\partial^2 f}{\partial z^2}$

Vector Triple Products

$$(\mathbf{u} \times \mathbf{v}) \cdot \mathbf{w} = (\mathbf{v} \times \mathbf{w}) \cdot \mathbf{u} = (\mathbf{w} \times \mathbf{u}) \cdot \mathbf{v}$$

$$\mathbf{u} \times (\mathbf{v} \times \mathbf{w}) = (\mathbf{u} \cdot \mathbf{w})\mathbf{v} - (\mathbf{u} \cdot \mathbf{v})\mathbf{w}$$

The Fundamental Theorem of Line Integrals

Part 1 Let $\mathbf{F} = M\mathbf{i} + N\mathbf{j} + P\mathbf{k}$ be a vector field whose components are continuous throughout an open connected region D in space. Then there exists a differentiable function f such that

$$\mathbf{F} = \nabla f = \frac{\partial f}{\partial x}\mathbf{i} + \frac{\partial f}{\partial y}\mathbf{j} + \frac{\partial f}{\partial z}\mathbf{k}$$

if and only if for all points A and B in D, the value of $\int_A^B \mathbf{F} \cdot d\mathbf{r}$ is independent of the path joining A to B in D.

Part 2 If the integral is independent of the path from A to B, its value is

$$\int_A^B \mathbf{F} \cdot d\mathbf{r} = f(B) - f(A).$$

Green's Theorem and Its Generalization to Three Dimensions

Tangential form of Green's Theorem: $\qquad \displaystyle\oint_C \mathbf{F} \cdot \mathbf{T}\, ds = \iint_R (\nabla \times \mathbf{F}) \cdot \mathbf{k}\, dA$

Stokes' Theorem: $\qquad \displaystyle\oint_C \mathbf{F} \cdot \mathbf{T}\, ds = \iint_S (\nabla \times \mathbf{F}) \cdot \mathbf{n}\, d\sigma$

Normal form of Green's Theorem: $\qquad \displaystyle\oint_C \mathbf{F} \cdot \mathbf{n}\, ds = \iint_R (\nabla \cdot \mathbf{F})\, dA$

Divergence Theorem: $\qquad \displaystyle\iint_S \mathbf{F} \cdot \mathbf{n}\, d\sigma = \iiint_D \nabla \cdot \mathbf{F}\, dV$

Vector Identities

In the identities here, f and g are differentiable scalar functions; $\mathbf{F}$, $\mathbf{F}_1$, and $\mathbf{F}_2$ are differentiable vector fields; and a and b are real constants.

$$\nabla \times (\nabla f) = \mathbf{0}$$

$$\nabla(fg) = f\nabla g + g\nabla f$$

$$\nabla \cdot (g\mathbf{F}) = g\nabla \cdot \mathbf{F} + \nabla g \cdot \mathbf{F}$$

$$\nabla \times (g\mathbf{F}) = g\nabla \times \mathbf{F} + \nabla g \times \mathbf{F}$$

$$\nabla \cdot (a\mathbf{F}_1 + b\mathbf{F}_2) = a\nabla \cdot \mathbf{F}_1 + b\nabla \cdot \mathbf{F}_2$$

$$\nabla \times (a\mathbf{F}_1 + b\mathbf{F}_2) = a\nabla \times \mathbf{F}_1 + b\nabla \times \mathbf{F}_2$$

$$\nabla(\mathbf{F}_1 \cdot \mathbf{F}_2) = (\mathbf{F}_1 \cdot \nabla)\mathbf{F}_2 + (\mathbf{F}_2 \cdot \nabla)\mathbf{F}_1 + \mathbf{F}_1 \times (\nabla \times \mathbf{F}_2) + \mathbf{F}_2 \times (\nabla \times \mathbf{F}_1)$$

$$\nabla \cdot (\mathbf{F}_1 \times \mathbf{F}_2) = \mathbf{F}_2 \cdot (\nabla \times \mathbf{F}_1) - \mathbf{F}_1 \cdot (\nabla \times \mathbf{F}_2)$$

$$\nabla \times (\mathbf{F}_1 \times \mathbf{F}_2) = (\mathbf{F}_2 \cdot \nabla)\mathbf{F}_1 - (\mathbf{F}_1 \cdot \nabla)\mathbf{F}_2 + (\nabla \cdot \mathbf{F}_2)\mathbf{F}_1 - (\nabla \cdot \mathbf{F}_1)\mathbf{F}_2$$

$$\nabla \times (\nabla \times \mathbf{F}) = \nabla(\nabla \cdot \mathbf{F}) - (\nabla \cdot \nabla)\mathbf{F} = \nabla(\nabla \cdot \mathbf{F}) - \nabla^2\mathbf{F}$$

$$(\nabla \times \mathbf{F}) \times \mathbf{F} = (\mathbf{F} \cdot \nabla)\mathbf{F} - \frac{1}{2}\nabla(\mathbf{F} \cdot \mathbf{F})$$

Limits

General Laws

If L, M, c, and k are real numbers and

$$\lim_{x \to c} f(x) = L \quad \text{and} \quad \lim_{x \to c} g(x) = M, \quad \text{then}$$

Sum Rule: $\qquad\qquad\qquad \lim_{x \to c} (f(x) + g(x)) = L + M$

Difference Rule: $\qquad\qquad \lim_{x \to c} (f(x) - g(x)) = L - M$

Product Rule: $\qquad\qquad\; \lim_{x \to c} (f(x) \cdot g(x)) = L \cdot M$

Constant Multiple Rule: $\quad\; \lim_{x \to c} (k \cdot f(x)) = k \cdot L$

Quotient Rule: $\qquad\qquad\; \lim_{x \to c} \dfrac{f(x)}{g(x)} = \dfrac{L}{M}, \quad M \neq 0$

The Sandwich Theorem

If $g(x) \leq f(x) \leq h(x)$ in an open interval containing c, except possibly at $x = c$, and if

$$\lim_{x \to c} g(x) = \lim_{x \to c} h(x) = L,$$

then $\lim_{x \to c} f(x) = L$.

Inequalities

If $f(x) \leq g(x)$ in an open interval containing c, except possibly at $x = c$, and both limits exist, then

$$\lim_{x \to c} f(x) \leq \lim_{x \to c} g(x).$$

Continuity

If g is continuous at L and $\lim_{x \to c} f(x) = L$, then

$$\lim_{x \to c} g(f(x)) = g(L).$$

Specific Formulas

If $P(x) = a_n x^n + a_{n-1} x^{n-1} + \cdots + a_0$, then

$$\lim_{x \to c} P(x) = P(c) = a_n c^n + a_{n-1} c^{n-1} + \cdots + a_0.$$

If $P(x)$ and $Q(x)$ are polynomials and $Q(c) \neq 0$, then

$$\lim_{x \to c} \frac{P(x)}{Q(x)} = \frac{P(c)}{Q(c)}.$$

If $f(x)$ is continuous at $x = c$, then

$$\lim_{x \to c} f(x) = f(c).$$

$$\lim_{x \to 0} \frac{\sin x}{x} = 1 \quad \text{and} \quad \lim_{x \to 0} \frac{1 - \cos x}{x} = 0$$

L'Hôpital's Rule

If $f(a) = g(a) = 0$, both f' and g' exist in an open interval I containing a, and $g'(x) \neq 0$ on I if $x \neq a$, then

$$\lim_{x \to a} \frac{f(x)}{g(x)} = \lim_{x \to a} \frac{f'(x)}{g'(x)},$$

assuming the limit on the right side exists.

Differentiation Rules

General Formulas

Assume u and v are differentiable functions of x.

Constant: $\dfrac{d}{dx}(c) = 0$

Sum: $\dfrac{d}{dx}(u + v) = \dfrac{du}{dx} + \dfrac{dv}{dx}$

Difference: $\dfrac{d}{dx}(u - v) = \dfrac{du}{dx} - \dfrac{dv}{dx}$

Constant Multiple: $\dfrac{d}{dx}(cu) = c\dfrac{du}{dx}$

Product: $\dfrac{d}{dx}(uv) = u\dfrac{dv}{dx} + \dfrac{du}{dx}v$

Quotient: $\dfrac{d}{dx}\left(\dfrac{u}{v}\right) = \dfrac{v\dfrac{du}{dx} - u\dfrac{dv}{dx}}{v^2}$

Power: $\dfrac{d}{dx}x^n = nx^{n-1}$

Chain Rule: $\dfrac{d}{dx}(f(g(x)) = f'(g(x)) \cdot g'(x)$

Trigonometric Functions

$\dfrac{d}{dx}(\sin x) = \cos x \qquad \dfrac{d}{dx}(\cos x) = -\sin x$

$\dfrac{d}{dx}(\tan x) = \sec^2 x \qquad \dfrac{d}{dx}(\sec x) = \sec x \tan x$

$\dfrac{d}{dx}(\cot x) = -\csc^2 x \qquad \dfrac{d}{dx}(\csc x) = -\csc x \cot x$

Exponential and Logarithmic Functions

$\dfrac{d}{dx}e^x = e^x \qquad \dfrac{d}{dx}\ln x = \dfrac{1}{x}$

$\dfrac{d}{dx}a^x = a^x \ln a \qquad \dfrac{d}{dx}(\log_a x) = \dfrac{1}{x \ln a}$

Inverse Trigonometric Functions

$\dfrac{d}{dx}(\sin^{-1} x) = \dfrac{1}{\sqrt{1 - x^2}} \qquad \dfrac{d}{dx}(\cos^{-1} x) = -\dfrac{1}{\sqrt{1 - x^2}}$

$\dfrac{d}{dx}(\tan^{-1} x) = \dfrac{1}{1 + x^2} \qquad \dfrac{d}{dx}(\sec^{-1} x) = \dfrac{1}{|x|\sqrt{x^2 - 1}}$

$\dfrac{d}{dx}(\cot^{-1} x) = -\dfrac{1}{1 + x^2} \qquad \dfrac{d}{dx}(\csc^{-1} x) = -\dfrac{1}{|x|\sqrt{x^2 - 1}}$

Hyperbolic Functions

$\dfrac{d}{dx}(\sinh x) = \cosh x \qquad \dfrac{d}{dx}(\cosh x) = \sinh x$

$\dfrac{d}{dx}(\tanh x) = \operatorname{sech}^2 x \qquad \dfrac{d}{dx}(\operatorname{sech} x) = -\operatorname{sech} x \tanh x$

$\dfrac{d}{dx}(\coth x) = -\operatorname{csch}^2 x \qquad \dfrac{d}{dx}(\operatorname{csch} x) = -\operatorname{csch} x \coth x$

Inverse Hyperbolic Functions

$\dfrac{d}{dx}(\sinh^{-1} x) = \dfrac{1}{\sqrt{1 + x^2}} \qquad \dfrac{d}{dx}(\cosh^{-1} x) = \dfrac{1}{\sqrt{x^2 - 1}}$

$\dfrac{d}{dx}(\tanh^{-1} x) = \dfrac{1}{1 - x^2} \qquad \dfrac{d}{dx}(\operatorname{sech}^{-1} x) = -\dfrac{1}{x\sqrt{1 - x^2}}$

$\dfrac{d}{dx}(\coth^{-1} x) = \dfrac{1}{1 - x^2} \qquad \dfrac{d}{dx}(\operatorname{csch}^{-1} x) = -\dfrac{1}{|x|\sqrt{1 + x^2}}$

Parametric Equations

If $x = f(t)$ and $y = g(t)$ are differentiable, then

$$y' = \dfrac{dy}{dx} = \dfrac{dy/dt}{dx/dt} \qquad \text{and} \qquad \dfrac{d^2y}{dx^2} = \dfrac{dy'/dt}{dx/dt}.$$

Integration Rules

General Formulas

Zero:
$$\int_a^a f(x)\, dx = 0$$

Order of Integration:
$$\int_b^a f(x)\, dx = -\int_a^b f(x)\, dx$$

Constant Multiples:
$$\int_a^b k f(x)\, dx = k\int_a^b f(x)\, dx, \qquad k \text{ any number}$$

$$\int_a^b -f(x)\, dx = -\int_a^b f(x)\, dx, \qquad k = -1$$

Sums and Differences:
$$\int_a^b (f(x) \pm g(x))\, dx = \int_a^b f(x)\, dx \pm \int_a^b g(x)\, dx$$

Additivity:
$$\int_a^b f(x)\, dx + \int_b^c f(x)\, dx = \int_a^c f(x)\, dx$$

Max-Min Inequality: If max f and min f are the maximum and minimum values of f on $[a, b]$, then

$$\min f \cdot (b - a) \le \int_a^b f(x)\, dx \le \max f \cdot (b - a).$$

Domination:
$$f(x) \ge g(x) \quad \text{on} \quad [a, b] \quad \text{implies} \quad \int_a^b f(x)\, dx \ge \int_a^b g(x)\, dx$$

$$f(x) \ge 0 \quad \text{on} \quad [a, b] \quad \text{implies} \quad \int_a^b f(x)\, dx \ge 0$$

The Fundamental Theorem of Calculus

Part 1 If f is continuous on $[a, b]$, then $F(x) = \int_a^x f(t)\, dt$ is continuous on $[a, b]$ and differentiable on (a, b) and its derivative is $f(x)$:

$$F'(x) = \frac{d}{dx}\int_a^x f(t)\, dt = f(x).$$

Part 2 If f is continuous at every point of $[a, b]$ and F is any antiderivative of f on $[a, b]$, then

$$\int_a^b f(x)\, dx = F(b) - F(a).$$

Substitution in Definite Integrals

$$\int_a^b f(g(x)) \cdot g'(x)\, dx = \int_{g(a)}^{g(b)} f(u)\, du$$

Integration by Parts

$$\int_a^b u(x)\, v'(x)\, dx = u(x)\, v(x) \Big]_a^b - \int_a^b v(x)\, u'(x)\, dx$$

A Brief Table of Integrals follows the Index at the back of the text.